Electric Cables Handbook

Third Edition

BICCCables

Edited by

G. F. Moore BSc, PhD, FInstP, FIEE
*Formerly Director of Technology,
BICC Cables Limited
Royal Academy of Engineering Visiting Professor
University of Liverpool*

b

Blackwell
Science

© 1982, 1990, 1997 by
Blackwell Science Ltd
Editorial Offices:
Osney Mead, Oxford OX2 0EL
25 John Street, London WC1N 2BL
23 Ainslie Place, Edinburgh EH3 6AL
350 Main Street, Malden
 MA 02148 5018, USA
54 University Street, Carlton
 Victoria 3053, Australia
10, rue Casimir Delavigne
 75006 Paris, France

Other Editorial Offices:

Blackwell Wissenschafts-Verlag GmbH
Kurfürstendamm 57
10707 Berlin, Germany

Blackwell Science KK
MG Kodenmacho Building
7–10 Kodenmacho Nihombashi
Chuo-ku, Tokyo 104, Japan

The right of the Author to be identified
as the Author of this Work has been
asserted in accordance with the Copyright,
Designs and Patents Act 1988.

All rights reserved. No part of
this publication may be reproduced,
stored in a retrieval system, or
transmitted, in any form or by any
means, electronic, mechanical,
photocopying, recording or otherwise,
except as permitted by the UK
Copyright, Designs and Patents Act
1988, without prior permission
of the publisher.

First published by
 Granada Publishing 1992
Reprinted 1983, 1984, 1987
Second edition published by
 Blackwell Science 1990
Reprinted 1992, 1994, 1995, 1996
Third edition 1997
Reprinted 1998, 1999

Set in 10½/12 pt Times by Aarontype Ltd,
Bristol, England
Printed and bound in Great Britain by
MPG Books Ltd, Bodmin, Cornwall

The Blackwell Science logo is a trade mark
of Blackwell Science Ltd, registered at the
United Kingdom Trade Marks Registry

DISTRIBUTORS

Marston Book Services Ltd
PO Box 269
Abingdon
Oxon OX14 4YN
(*Orders:* Tel: 01235 465500
           Fax: 01235 465555)

USA
Blackwell Science, Inc.
Commerce Place
350 Main Street
Malden, MA 02148 5018
(*Orders:* Tel: 800 759 6102
               781 388 8250
           Fax: 781 388 8255)

Canada
Login Brothers Book Company
324 Saulteaux Crescent
Winnipeg, Manitoba R3J 3T2
(*Orders:* Tel: 204 837 2987
           Fax: 204 837 3116)

Australia
Blackwell Science Pty Ltd
54 University Street
Carlton, Victoria 3053
(*Orders:* Tel: 03 9347 0300
           Fax: 03 9347 5001)

A catalogue record for this title
is available from the British Library
ISBN 0–632–04075–0

Library of Congress
Cataloging-in-Publication Data

Electric cables handbook/BICC Cables;
 edited by G. F. Moore. – 3rd ed.
  p.   cm.
 Includes bibliographical references
 (p.   ) and index.
 ISBN 0-632-04075-0
 1. Electric cables - Handbooks,
  manuals, etc. II. Moore, G. F.
 (George F.) II. BICC Cables Ltd.
 TK3351.E43 1997
 621.319'34–dc21   97-24044
                                        CIP

For further information on Blackwell Science,
visit our website:
www.blackwell-science.com

Contents

Preface to Third Edition — xi

Contributors — xiii

1 Introduction — 1

PART 1: THEORY, DESIGN AND PRINCIPLES COMMON TO ALL CABLE TYPES

2 Basic Electrical Theory Applicable to Cable Design — 9
3 Materials Used in Cables — 33
4 Conductors — 69
5 Armour and Protective Finishes — 81
6 Cables in Fires – Material and Design Considerations — 92
7 Cable Standards and Quality Assurance — 108
8 Current Carrying Capacity — 121
9 Short-circuit Ratings — 151
10 Technical Data Applicable to Cable Planning and Usage — 170

PART 2: WIRING CABLES, FLEXIBLE CABLES AND CABLES FOR GENERAL INDUSTRIAL USE

11 Cables for Fixed Installations — 205
12 Flexible Cables and Cords — 230
13 Auxiliary Cables (Pilot and Telephone) — 251
14 Manufacture of General Wiring Cables — 255
15 Installation of General Wiring Cables — 262
16 Mineral Insulated Cables — 271

PART 3: SUPPLY DISTRIBUTION SYSTEMS AND CABLES

17	Supply Distribution Systems	283
18	Distribution Cable Types, Design and Applications	301
19	Paper Insulated Distribution Cables	316
20	PVC Insulated Cables	326
21	Thermoset Insulated Cables up to 3.3 kV	335
22	600/1000 V Cables with Combined Neutral and Earth for Public Supply	339
23	Service Distribution Cables	346
24	Polymeric Insulated Distribution Cables for 6–30 kV	351
25	Manufacture of Distribution Cables	365
26	Installation of Distribution Cables	381
27	Joints and Terminations for Distribution Cables	401
28	Testing of Distribution Cables	418

PART 4: TRANSMISSION SYSTEMS AND CABLES

29	Basic Cable Types for A.C. Transmission	431
30	Self-contained Fluid-filled Cables	440
31	Gas Pressure Cables	456
32	High Pressure Fluid-filled Pipe Cables	461
33	Polymeric Insulated Cables for Transmission Voltages	468
34	Techniques for Increasing Current Carrying Capacity	479
35	Transmission Cable Accessories and Jointing for Pressure-assisted and Polymeric Cables	503
36	Installation of Transmission Cables	549
37	Thermomechanical Design	562
38	D.C. Cables	574
39	Testing of Transmission Cable Systems	581
40	Fault Location for Transmission Cables	597
41	Recent Improvements and Development of Transmission Cables	603

PART 5: SUBMARINE DISTRIBUTION AND TRANSMISSION

| 42 | Submarine Cables and Systems | 621 |
| 43 | Submarine Cable Installation | 646 |

PART 6: HIGH TEMPERATURE SUPERCONDUCTIVITY

44	Introduction to Superconductivity	659
45	High Temperature Superconductors	666
46	High Temperature Superconducting Power Cables	676

PART 7: OPTICAL FIBRES IN POWER TRANSMISSION SYSTEMS

47	Introduction to Part 7	685
48	Principles of Optical Fibre Transmission and Manufacture	688
49	Optical Fibre Cable Construction	700
50	Composite Overhead Conductors	709
51	All-dielectric Self-supporting Cables	730
52	Wrap Cable	745

PART 8: CABLES FOR COMMUNICATION APPLICATIONS

53	Communication Systems	755
54	Datacommunication Cables	779
55	Twisted Pair Telecommunication Cables	788

APPENDICES

A1	Abbreviations		795
A2	Symbols Used		800
A3	Conversion Factors and Multiple Metric Units		803
A4	Conductor Data		804
	A4.1	Metric conductor sizes and resistances for fixed wiring	804
	A4.2	Metric sizes and resistances for flexible conductors	805
	A4.3	USA stranded conductors sizes and resistances for fixed wiring	806
	A4.4	Temperature correction factors for conductor resistance	808
	A4.5	Maximum diameters of circular copper conductors	809
	A4.6	Minimum and maximum diameters of circular aluminium conductors	810
A5	Industrial Cables for Fixed Supply		811
	PVC Armoured Power or Control Cables		
	A5.1	Current ratings and volt drop	811
	A5.2	Dimensions and weights	812

viii   *Contents*

    PVC Armoured Auxiliary Multicore Cables
    A5.3   Dimensions and weights   813
    XLPE Armoured Power or Control Cables
    A5.4   Current ratings and volt drop   815
    A5.5–A5.6   Dimensions and weights   816

A6 Cables for Fixed Installation in Buildings   818
    PVC Wiring Cables
    A6.1   Current ratings for single-core non-sheathed conduit cables and single-core sheathed cables   819
    A6.2   Voltage drop for single-core non-sheathed conduit cables and single-core sheathed cables   820
    A6.3   Current ratings and volt drop for flat cables   821
    A6.4–A6.6   Dimensions and weights   822
    Elastomeric Insulated Cables
    A6.7   Current ratings and volt drop   825
    A6.8   Dimensions and weights   826
    XLPE Insulated Floorwarming Cables
    A6.9   General data   827
    XL-LSF Cables
    A6.10   Current ratings for single-core non-sheathed conduit cables and single-core sheathed cables   829
    A6.11   Voltage drop for single-core non-sheathed conduit cables and single-core sheathed cables   830
    A6.12   Current ratings and volt drop for flat cables   831
    A6.13–A6.15   Dimensions and weights   832

A7 Cables for Fixed Installation (Shipwiring and Offshore)   835
    Elastomeric Shipwiring Power and Instrumentation Cables
    A7.1   Maximum a.c. continuous current ratings in ships and offshore   835
    A7.2   Maximum a.c. continuous current ratings for specific installation conditions   836
    A7.3–A7.5   Dimensions and weights   840
    Elastomeric High Voltage Cables
    A7.6   Dimensions and weights   850

A8 Flexible Cords and Cables   852
    PVC Insulated Flexible Cords
    A8.1   Current ratings   852
    A8.2–A8.4   Dimensions and weights   854
    Elastomeric Insulated Flexible Cords
    A8.5–A8.10   Dimensions and weights   858
    PVC Insulated Flexible Cords
    A8.11   Conductors and ratings   863
    A8.12   Dimensions and weights   864
    Elastomeric Insulated Flexible Cables
    A8.13–A8.14   Current ratings and volt drop   865
    A8.15–A8.17   Dimensions and weights   868

Contents

	Heat Resisting Flexible Cables for 250°C Operation		
	A8.18	Range and dimensions	873
	A8.19	Current ratings	874
	EPR Insulated Linaflex Flat Flexible Cords and Cables		
	A8.20	Current ratings	875
	A8.21	Range and dimensions	876
A9	**Industrial Cables for Special Applications**		877
	Welding Cables		
	A9.1	Range and dimensions	877
	A9.2	Resistance and volt drop	878
	A9.3	Current ratings	879
A10	**Mining Cables**		881
	Unarmoured Flexible Trailing Cables		
	A10.1	Construction	881
	A10.2	Current ratings	882
	A10.3	Range and dimensions	883
	Pliable Armoured Flexible Trailing Cables		
	A10.4	Construction	886
	A10.5	Current ratings	886
	A10.6	Range and dimensions	887
	Pliable Armoured Flexible Multicore Auxiliary Cables		
	A10.7	Range and dimensions	889
	Unarmoured Flexible Cables		
	A10.8	Range and dimensions	890
	Pliable Armoured Flexible Cables		
	A10.9	Current ratings	891
	A10.10	Range and dimensions	892
	PVC Insulated Armoured Mining Power Cables		
	A10.11	Current ratings	895
	A10.12–A10.13	Range and dimensions	896
	EPR Insulated Armoured Mining Power Cables		
	A10.14–A10.15	Range and dimensions	900
A11	**Mineral Insulated Wiring Cables**		902
	500 V Light Duty M.I. Cable		
	A11.1–A11.2	Current ratings	903
	A11.3	Dimensions and weights	905
	750 V Heavy Duty M.I. Cable		
	A11.4–A11.5	Current ratings	906
	A11.6–A11.7	Volt drops	910
	A11.8	Dimensions and weights	912
A12	**Paper Insulated Distribution Cables**		913
	A12.1–A12.8	Dimensions and weights	916
	A12.9–A12.14	Sustained ratings	934
	A12.15	Voltage drop	944
	A12.16–A12.21	Electrical characteristics	945

A13	PVC Insulated Distribution Cables		955
	A13.1–A13.3	Dimensions and weights	958
	A13.4–A13.6	Sustained ratings	964
	A13.7	Voltage drop	969
	A13.8–A13.9	Electrical characteristics	971
A14	XLPE Insulated Distribution Cables		973
	A14.1–A14.8	Dimensions and weights	978
	A14.9–A14.13	Sustained ratings	988
	A14.14	Voltage drop	993
	A14.15–A14.19	Electrical characteristics	994
A15	PVC Insulated House Service Cables		1001
	A15.1	Dimensions and weights	1002
	A15.2	Ratings and characteristics	1003
A16	Self-contained Fluid-filled Paper Insulated Cables		1004
	A16.1	Conditions applicable	1004
	Figs A16.1–A16.8	Power ratings	1005
	A16.2–A16.8	Technical data	1013
A17	PPL Insulated Self-contained Fluid-filled Transmission Cables		1020
	A17.1	Conditions applicable	1020
	Figs A17.1–A17.4	Power ratings	1021
	A17.2–A17.3	Technical data	1025
A18	XLPE Insulated Transmission Cables		1027
	A18.1	Conditions applicable	1027
	Figs A18.1–A18.4	Power ratings	1028
	A18.2–A18.5	Technical data	1032
A19	Minimum Installation Bending Radii		1036
	General Wiring Cables		
	A19.1	Cables for fixed wiring	1036
	A19.2	EPR insulated cables for installation in ships	1037
	A19.3	PVC and EPR insulated wire armoured mining cables	1037
	Distribution Cables		
	A19.4	Paper insulated cables	1037
	A19.5	PVC and XLPE insulated cables rated at 600/1000 V and 1.9/3.3 kV	1038
	A19.6	XLPE insulated cables for 6.6–22 kV	1038
	Transmission Cables		
	A19.7	Fluid-filled cables	1039
	A19.8	XLPE insulated cables	1039
A20	Weibull distribution		1040
A21	Bibliography		1048
	Index		1083

Preface to Third Edition

The first and second editions of the *Electric Cables Handbook* were well received and there were many requests for a revised edition. The third edition provides a substantial, comprehensive and up-to-date review, reflecting the changes which have occurred in the cable industry during the last seven years. The Handbook therefore continues to be a major reference book for professional engineers and electrical contractors involved in cables.

The Handbook covers all types of energy cables, from wiring and flexible cables for general use to distribution, transmission and submarine cables. It includes information on materials, design principles, installation and standards, and the many appendices contain extensive tables of data on commonly used cable types.

Once again it has been difficult to decide which older types of cable should be excluded and which new types should be included. While the Handbook has been comprehensively revised, the main aim has been to describe adequately those new energy cables which reflect the present dynamic state of the industry. The description of cables in fires has been completely rewritten; the significant developments in high performance materials, both polymeric and paper laminate, are described in detail; the application of new materials to high voltage systems is thoroughly covered; for the first time there is an extensive description of the recent developments in high temperature superconductivity, with the emphasis on potential applications in power engineering.

The Handbook has also been significantly extended to give a broad view of communication cables for the power engineer. Recognising the increasing impact of telecommunications in the power industry, a new and substantive section describes the application of optical fibres in power transmission systems. There is also an additional section on communication cables; this includes a revision of the original chapter on electronic cables.

Reflecting the extensive revision of the text, extra material has also been added to the appendices, references and bibliography. Much of the original text has been retained in the interest of providing the necessary background to the continuing technical developments in the industry.

As in previous editions, this Handbook draws on the extensive expertise and experience of the staff of BICC Cables Ltd and the editor would like to thank the many contributors for their text, their collaboration, and many helpful suggestions.

A particular tribute is due to the editorial committee of Dr D. G. Dalgoutte, Dr A. W. Field, and Dr K. Julian who organised the work as well as providing their own detailed technical knowledge, and to D. McAllister for his work on the first two editions and the late E. W. G. Bungay who co-edited the second edition. Finally, thanks are also due to the directors and management of BICC Cables Ltd for the provision of time and facilities for preparing the manuscript and illustrations.

G. F. Moore

Contributors

C. A. Arkell, BSc, CEng, MIEE
Formerly Chief Engineer, BICC Cables Limited, Power Division, Supertension Cables

J. R. Attwood, BSc, MSc,
Engineering Manager, Submarine Cables, BICC Cables Limited, Erith Technology Centre, Supertension Cable Systems

H. Baker, BA, FInstP
Formerly Chief Engineer, BICC Pyrotenax Limited

V. A. A. Banks, BSc
Senior Projects Engineer, BICC Cables Limited, Power Division, Mains Cables

A. L. Barclay, BSc, CPhys, MInstP
High Voltage Engineer, BICC Cables Limited, Supertension Cable Systems

T. P. Beales, BSc, PhD
Manager, Superconductivity, BICC Cables Limited, Wrexham Technology Centre

Z. Bonikowski, BSc, CEng, MIMechE, FIEE
Formerly Technology and Planning Manager, BICC Cables Limited

R. T. Brown, BSc
Formerly Cable Assessment Manager, BICC Elastomeric Cables Limited

E. W. G. Bungay, BSc, CEng, MIEE
Formerly Chief Engineer, BICC Power Cables Limited
(E. W. G. Bungay was co-editor of the second edition of the *Electric Cables Handbook*.)

P. Calton, IEng, MIEIE
Cable Design Manager, BICC Cables Limited, Distribution Cable Systems

R. G. Cottignies, CEng, BEng, MIEE
Chief Engineer, BICC Cables Limited, Construction & Industrial Cable Systems

D. G. Dalgoutte, BSc, PhD, MIEE
General Manager, BICC Cables Limited, Helsby Technology Centre

A. C. Dawson, CEng, MIEE
Technical Manager, BICC Cables Limited, BICC Flexible Cables

G. R. M. Dench, TEng(CEI), MITE
Formerly Specifications Engineer, BICC Supertension Cables Limited

H. Downey, BSc, AMIEE
Senior Research Engineer, BICC Cables Limited, Helsby Technology Centre

T. M. Endersby, MEng, MIEE, CEng
Projects Design Manager, BICC Cables Limited, Erith Technology Centre, Supertension Cable Systems

J. E. Evans, BSc
Technical Manager, BICC Cables Limited, BICC PALCO Ltd

A. W. Field, PhD, CEng, FIM
Technology Manager, Energy Cables and Strategic Materials, BICC Cables Limited

P. H. Fraser, BSc, MIEE
Technical Standards Manager, BICC Cables Limited, Construction & Industrial Cable Systems

A. Friday, BSc, MSc, PhD
Principal Research Engineer, BICC Cables Limited, Wrexham Technology Centre

P. F. Gale, BTech, PhD, CEng, MIEE
Formerly Chief Engineer, Biccotest Limited

S. G. Galloway, BSc
Systems Installation Engineer, BICC Cables Limited, Erith Technology Centre, Supertension Cable Systems

B. Gregory, BSc, CEng, MIEE
Chief Engineer, BICC Cables Limited, Erith Technology Centre, Supertension Cable Systems

D. R. Groombridge
Technology Manager, Analytical Services, BICC Cables Limited, Wrexham Technology Centre

R. N. Hampton, BSc, PhD, CPhys
Manager, Research, BICC Cables Limited, Erith Technology Centre, Supertension Cable Systems

J. E. Hawkes, TEng(CEI), MIEE
Formerly High Voltage Laboratory Engineer, BICC Cables Limited, Power Division, Supertension Cables

J. T. Henderson, BSc, CEng, FIE
Formerly Marketing Manager (Far East and Pacific), BICC Supertension Cables Limited

P. M. Henstock,
Technical Officer, Wiring Cables, Construction and Cabling Division

N. M. Hodgkinson, BSc, PhD
Senior Research Engineer, BICC Cables Limited, Wrexham Technology Centre

S. H. Jagger, BSc, CEng, FIEE
Formerly Contracts Manager, Balfour Beatty Power Limited, Traction and General Division

F. E. Keen, BSc, PhD, CChem
Principal Research Engineer, BICC Cables Limited, Wrexham Technology Centre

I. G. Knight, MSc
Technical Manager, Cable Systems, BICC Cables Limited, Helsby Technology Centre

C. D. Knipe
Technical Manager (Export), BICC Cables Limited, Telecommunication Cable Systems

R. S. Linton, BSc, PhD
Senior Research Engineer, BICC Cables Limited, Helsby Technology Centre

D. McAllister
Formerly Assistant Chief Engineer, BICC Power Cables Limited
(D. McAllister was co-editor of the second edition of the *Electric Cables Handbook*.)

P. L. Mayhew, BSc, CEng, MIEE, MIIM
Formerly Site Director and General Manager, Elastomeric Cables Unit, BICC Cables Limited

A. P. Morris, BSc
Manager, Product Applications and Development, Optical Fibres

D. H. M. Munro
Divisional Technical Manager, Cabling Division, Balfour Kilpatrick Limited

S. Phillips, MEng, AMIEE
Section Leader, Installation Engineering, BICC Cables Limited, Erith Technology Centre, Supertension Cable Systems

D Pollard, CEng, FIEE
Formerly Technical Manager, Leigh Works, BICC General Cables Limited

I. B. Riley, BSc
Projects Manager, BICC Cables Limited, Construction and Wiring Division

B. E. Roberts, BEng, CEng, FIEE
Formerly Director and General Manager, BICC Cables Limited, Power Division, Mains Cables (25, 36)

D. G. Roberts, CChem, MRSC
Director of Research, BICC Cables Corporation, USA

S. M. Rowland, BSc, PhD, MIEE, CEng
Programmes Manager, BICC Cables Limited, Helsby Technology Centre

G. J. Smee
Director and General Manager, BICC Cables Limited, Erith Technology Centre, Supertension Cable Systems

S. T. Spedding, BSc, MSc, ARCS
Technical Manager, Facilities Planning, BICC Cables Limited

L. R. Trim, BEng
Senior Development Engineer, BICC Cables Limited, Erith Technology Centre, Supertension Cable Systems

J. Ward, BA
Technical Standards Manager, BICC Cables Limited, Construction & Industrial Cable Systems

N. H. Waterhouse, BSc
Formerly Engineering Manager, Submarine Cables Unit, Balfour Kilpatrick Limited

S. Wigginton
Marine Superintendent, Undersea Cabling Services, Transmission and Cabling Division, Balfour Kilpatrick Limited

G. R. Williams
Formerly Contracts Manager, BICC Cables Limited, Power Division, Supertension Cables

A. J. Willis, CEng, MIEE
Chief Engineer, BICC Cables Limited, BICC Brand Rex Ltd

W. J. Willis, BSc
Formerly Sales Engineer, Biccotest Limited

H. S. Wood, CEng, MIEE
Formerly Development Engineer, Merseyside and North Wales Electricity Board

D. E. E. Wolmer, CEng, MIEE
Formerly Chief Cabling Engineer, Balfour Kilpatrick Limited

CHAPTER 1
Introduction

SCOPE OF THE THIRD EDITION

The first edition of the book covered all types of insulated cable for the supply of electrical energy for voltages from about 100 V to 525 kV. The second edition covered a similar range of cables but also reflected the significant changes in materials, the application of polymeric cables at higher voltages and the increasing technology content of high voltage (HV) transmission systems. Owing to the subsequent extensive growth of cables for electronic equipment, and their similarity to some existing energy cables, the second edition included a chapter on such applications at lower voltages.

The third edition describes the further elimination of the traditional distinctions between materials, the wider use of fire retardant cables and other environmental issues, the significant advances in dielectric materials at high voltages, and the major technology developments in supertension systems. High temperature superconductivity is discussed with the emphasis on practical applications.

We have resisted calls for a comprehensive coverage of communication cables. Nevertheless the scope of the edition has been widened substantially to include some communication cables, in particular the use of optical fibres within energy cables.

FORMAT

The division into chapters has followed the principles established in the earlier editions and the specialist contributors have reflected the current patterns for manufacturing and marketing cables. For the benefit of the reader, and for ease of reference, each chapter is almost self-contained; there is therefore some repetition but it is hoped that this is not excessive.

Organisation of the book and the cable industry

There are no sharp distinctions between cable types and applications and in practice there is considerable overlap; this presents problems in chapter sequence. Operating voltage provides a rough guide but does not represent a clear division between cable types; nevertheless progress through the book broadly follows increasing voltage.

The cable making industry, together with its relationships with users and standardising authorities, was built mainly around specific factories for established groupings of cable types. Historically these groupings arose because of the materials used in the cables and the types of manufacturing plant adopted; size and weight of the cables can be allied with the same pattern.

The division of the book is set out below.

Part 1
Many aspects of cable design are common to all types; new developments and trends in usage continue to eliminate the traditional distinctions between materials. Part 1 deals extensively with materials and design features which are reasonably applicable to most cables.

Part 2
Historically a group of cables generally known as 'wiring and general' grew around cables mainly with rubber insulation; these contrasted with power distribution cables with impregnated paper insulation. Whereas paper cables were usually bought directly by the end-user, the wiring cables were commonly marketed through distributors and wholesalers. Although the main product types still remain, the insulants used in the two fields are often similar, i.e. thermoplastics and thermosets (rubbers and crosslinked thermoplastics).

These cables are often further subdivided by technology or factory, e.g. cables having thermoset insulation and sheaths, cables produced in large quantity for specific applications (such as PVC insulated cables for fixed wiring), and flexible cables.

Part 3
This part describes cables required for public supply and heavy industrial distribution. However, the latter are designed for a wide range of power requirements and do overlap part 2. For example the British Standard for PVC armoured cables for industrial use is common to cables covered by parts 2 and 3 of the book. It is common for cables with conductor sizes of 25 mm^2 and above to be classed as power distribution cables.

Part 4
This covers cables for public supply transmission systems, but below 132 kV there is some overlap with part 3. Historically, transmission cables have been of the pressure-assisted paper insulated type and major developments in paper laminated dielectrics are included. With the further development of polymeric materials, polyethylene (PE), crosslinked polyethylene (XLPE), and EPR are now well established as insulants for voltages of 132 kV and above; they are described in some detail and extend the use covered earlier in the book.

Part 5
This covers the very specialist, and growing area, of submarine cables; it deals with system design, manufacture and installation.

Part 6
This is a new addition to the handbook and reflects the worldwide interest in high temperature superconductivity (HTS). The background to development at 4 K is

covered followed by a description of materials, conductor fabrication, superconducting cables and their economics; the emphasis is on practical applications in power engineering.

Part 7
This is a new addition and describes the application of optical fibres in power transmission systems. It begins with the principles of optical fibre transmission and then describes the various methods of incorporating fibres into long span power lines.

Part 8
This part brings up to date the chapter on data communication cables from the previous handbook and includes a chapter on metallic telephone cables.

Appendices
As this edition continues to serve as both a reference book and a handbook the substantial list of appendices has been retained. Much of the tabular data presented provides information on the range of cables (and their properties) available in the most widely used fields. Those engineers dealing with cables on a regular basis will have manufacturers' catalogues available which give other more detailed information.

HISTORICAL SURVEY

This edition gives an up-to-date account of the present position on the types of cable used and their applications; an outline is also given of the stages which have led to current practices. Some of the more important dates and periods which have been significant in past developments are indicated in table 1.1. An extensive bibliography and references are also included.

DIFFERENCES IN CABLE DESIGN THROUGHOUT THE WORLD

Transmission cable practices are similar throughout the world. For wiring type cables many countries have preferences for particular designs or materials but the differences are not fundamental. Similar preferences apply to distribution cables but there are some major differences allied with the systems; these may be divided into two broad categories: those countries following British and European practice, and those which have adopted the USA system designs.

There is some overlap between distribution practices but the differences are such that it was not possible to produce a book which would adequately deal with both. Hence the coverage, which is mainly for British practice, also reflects the philosophies in Europe and the majority of countries throughout the world.

In the British and European systems the distribution cables have been installed underground in urban areas – hence the avoidance of unsightly poles, wires and overhead transformers. The USA practices are followed in countries and areas that have been more closely allied to the American economy, such as South America and the Philippines. The most notable difference in these systems is that in urban areas, apart from the concept of underground residential distribution (URD), the vast majority of

Table 1.1 Significant dates in cable developments

1880s	First gutta percha electric cable followed by rubber and vulcanised bitumen insulation
1890	Ferranti 10 kV tubular cable and the introduction of paper insulation
1914	Hochstadter development of screening which enabled distribution voltage to be increased to 33 kV
1926	Emanueli provided the principle of pressurisation with fluid-filled paper cables for voltages of 66 kV upwards
1930s	PVC insulation first tried out in Germany
1943	First 3-core 132 kV pressure cable in service
1949	Introduction of the mass-impregnated non-draining cable in the UK to overcome the problems of drainage of oil–rosin impregnant with cables installed on slopes
1950s	Full commercial introduction of PVC and later thermoset insulation for wiring cables PVC for power cables followed at the end of the decade Successful development of aluminium sheaths, initially for pressure-assisted cables, and gradual adoption of aluminium conductors for power cables First 275 kV FF cable (1954), operational use in 1959
1960s	Significant distribution economies obtained by the use of combined neutral and earth cables England/France ± 100 kV submarine d.c. link inaugurated in 1961 First 400 kV FF cable, operational in 1969
1970	Metrication of British Standards
1970s	Gradual extension of the use of thermoset insulation, mainly XLPE, as an alternative to paper insulation. Large commercial applications up to 15 kV but also experimental installations at higher voltages including transmission up to 132 kV
1980s	Introduction of optical fibre into overhead lines Very widespread use of XLPE in the 11–33 kV range with significant quantities installed for transmission voltages of 66–240 kV Discovery of high temperature superconducting materials Development and growing use of cables designed to alleviate the effects when cables are involved in fires; properties include reduced flame propagation, low smoke emission, reduced emission of noxious fumes and corrosive gases and combinations of these characteristics
1990s	Widespread use of optical fibres in overhead lines Extension of polymerics to EHV and the commercialisation of PPL Practical demonstrations of superconducting cables

distribution is via overhead cables; undergrounding is only adopted in limited areas in the innermost parts of cities and large towns. Even there the concept often differs from undergrounding in Europe, as it often uses conductors similar to those used overhead but having insulation, i.e. single-core cables, which are installed in ducts. The transformers often consist of small single-phase types, in contrast to the larger 3-phase transformers used elsewhere. The use of single-core cables in ducts has an effect on cable designs, particularly concerning the neutral and/or protective conductor and armour.

In the early stages of development, the American systems favoured rubber and so paper insulation was never developed to the same extent as in Europe. The introduction of thermoplastic and thermoset insulation, coupled with the large output from the American chemical industry, also favoured the development of single-core polymeric insulated cables and this has led the way to changes which are steadily being adopted throughout the world.

PART 1
Theory, Design and Principles Common to all Cable Types

CHAPTER 2
Basic Electrical Theory Applicable to Cable Design

In all engineering undertakings, economical, technical and practical aspects are taken into consideration to establish the optimum solution or design. For the transmission, distribution and utilisation of electrical power, the choice normally lies between the use of overhead lines and underground cables.

For economic reasons, overhead lines are used extensively for the transmission and distribution of electricity in rural areas where environmental or practical considerations do not dictate otherwise. However, in urban areas it is more usual to install insulated cables which, in the main, are buried underground. The utilisation of electricity in factories, domestic premises and other locations is also mainly by cables as they present the most practical means of conveying electrical power to equipment, tools and appliances of all types. Cable designs vary enormously to meet the diverse requirements but there are certain components which are common to all.

All types of electric cable consist essentially of a low resistance conductor to carry the current except in special cases, such as heating cables, and insulation to isolate the conductors from each other and from their surroundings. In several types, such as single-core wiring cables, the two components form the finished cable, but generally as the voltage increases the construction becomes much more complex.

Other main components may include screening to obtain a radial electrostatic field, a metal sheath to keep out moisture or to contain a pressurising medium, armouring for mechanical protection, corrosion protection for the metallic components and a variety of additions extending, for example, to internal and external pipes to remove the heat generated in the cable.

This chapter contains some of the electrical theory applicable to all cable types. Further details of individual cable designs and components are given in later chapters.

VOLTAGE DESIGNATION

In the early days of electric power utilisation, direct current was widely used, but little now remains except for special applications and for a few interconnections in transmission networks. Alternating current has many advantages and 3-phase alternating current is used almost exclusively throughout the world.

So that suitable insulation and cable construction can be specified for the required 3-phase a.c. service performance, the design voltages for cables are expressed in the form U_0/U (formerly E_0/E). U_0 is the power frequency voltage between conductor and earth and U is the power frequency voltage between phase conductors for which the cable is designed, U_0 and U both being r.m.s. values.

Power cables in British Standards are thus designated 600/1000 V, 1900/3300 V, 3800/6600 V, 6350/11000 V, 8700/15000 V, 12700/22000 V and 19000/33000 V. For transmission voltages above this it is normal to quote only the value of U and thus the higher standard voltages in the UK are 66, 132, 275 and 400 kV. The maximum voltage can be 10% greater than the above values for voltages up to and including 275 kV and 5% greater for 400 kV.

Although the local distribution voltage in the UK is 240/415 V, the cables are designed for 600/1000 V, largely because during manufacture and installation this grade of cable requires an insulation designed on mechanical rather than electrical parameters.

Standardisation of system voltages has not been achieved worldwide although there is some move towards this. IEC has published voltage designations which are approaching universal acceptance.

D.C. system voltages, by which is meant d.c. voltages with not more than 3% ripple, are designated by the positive and negative value of the voltage above and below earth potential. The symbol U_0 is used for the rated d.c. voltage between conductor and the earthed core screen.

Many references will be found to cables described as low voltage (LV), medium voltage (MV), high voltage (HV) and even EHV or UHV. Apart from low voltage, which is defined internationally, these terms do not have generally accepted precise meanings and can be misleading. In some countries MV has in the past applied to 600/1000 V cables (these now fall clearly within the LV designation), whereas others have taken it to mean 6/10 kV or 8.7/15 kV and misunderstanding could arise. For precision it is best to use the actual voltage rating of the cable.

CONDUCTOR RESISTANCE

D.C. resistance

Factors affecting d.c. conductor resistance in terms of material resistivity and purity are discussed in chapter 3 and those relating to conductor design in chapter 4. The latter are associated with the fact that the prime path of the current is a helical one following the individual wires in the conductor. Hence if an attempt is made to calculate the resistance of a length of stranded conductor a factor must be applied to cater for the linear length of wire in the conductor to allow for extra length caused by the stranding effect. In a multicore cable an additional factor must be applied to allow for the additional length due to the lay of the cores.

The d.c. resistance is also dependent on temperature as given by

$$R_t = R_{20}[1 + \alpha_{20}(t - 20)] \tag{2.1}$$

where R_t = conductor resistance at $t°C$ (Ω)
R_{20} = conductor resistance at 20°C (Ω)
α_{20} = temperature coefficient of resistance of the conductor material at 20°C
t = conductor temperature (°C)

A.C. resistance

If a conductor is carrying high alternating currents, the distribution of current is not evenly disposed throughout the cross-section of the conductor. This is due to two independent effects known as the 'skin effect' and the 'proximity effect'.

If the conductor is considered to be composed of a large number of concentric circular elements, those at the centre of the conductor will be enveloped by a greater magnetic flux than those on the outside. Consequently the self-induced back e.m.f. will be greater towards the centre of the conductor, thus causing the current density to be less at the centre than at the conductor surface. This extra concentration at the surface is the skin effect and it results in an increase in the effective resistance of the conductor. The magnitude of the skin effect is influenced by the frequency, the size of the conductor, the amount of current flowing and the diameter of the conductor.

The proximity effect also increases the effective resistance and is associated with the magnetic fields of two conductors which are close together. If each carries a current in the same direction, the halves of the conductors in close proximity are cut by more magnetic flux than the remote halves. Consequently, the current distribution is not even throughout the cross-section, a greater proportion being carried by the remote halves. If the currents are in opposite directions the halves in closer proximity carry the greater density of current. In both cases the overall effect results in an increase in the effective resistance of the conductor. The proximity effect decreases with increase in spacing between cables.

Mathematical treatment of these effects is complicated because of the large number of possible variations but Arnold[1] has produced a comprehensive report (see also chapter 8).

Skin and proximity effects may be ignored with small conductors carrying low currents. They become increasingly significant with larger conductors and it is often desirable for technical and economic reasons to design the conductors to minimise them. The Milliken conductor, which reduces skin and proximity effects, is described in chapter 4.

A.C. resistances are important for calculation of current carrying capacity. Values for standard designs of distribution and transmission cables are included in the tables in appendices 12–16.

INDUCTANCE

The inductance L per core of a 3-core cable or of three single-core cables comprises two parts, the self-inductance of the conductor and the mutual inductance with other cores. It is given by

$$L = K + 0.2 \log_e \frac{2S}{d} \quad \text{(mH/km)} \tag{2.2}$$

where K = a constant relating to the conductor formation (table 2.1)
     S = axial spacing between conductors within the cable (mm), or
       = axial spacing between conductors of a trefoil group of single-core cables (mm), or
       = 1.26 × phase spacing for a flat formation of three single-core cables (mm)
     d = conductor diameter or for shaped designs the diameter of an equivalent circular conductor (mm)

For 2-core, 3-core and 4-core cables, the inductance obtained from the formula should be multiplied by 1.02 if the conductors are circular or sector-shaped, and by 0.97 for 3-core oval conductors.

Table 2.1 Typical values for constant K for different stranded conductors (at 50 Hz)

Number of wires in conductor	K
3	0.0778
7	0.0642
19	0.0554
37	0.0528
61 and over	0.0514
1 (solid)	0.0500
Hollow-core conductor, 12 mm duct	0.0383

REACTANCE

The inductive reactance per phase of a cable, or of the single-core cables comprising the circuit, may be obtained from the formula

$$X = 2\pi f L \times 10^{-3} \qquad (\Omega/\text{km}) \qquad (2.3)$$

where f = frequency (Hz)
     L = inductance (mH/km)

IMPEDANCE

The phase impedance Z of a cable, or of the single-core cables comprising the circuit, is given by

$$Z = (R^2 + X^2)^{1/2} \qquad (\Omega/\text{km}) \qquad (2.4)$$

where R = a.c. resistance at operating temperature (Ω/km)
     X = reactance (Ω/km)

INSULATION RESISTANCE

Insulation resistance is the resistance to the passage of direct current through the dielectric between two electrodes. In the case of an electric cable it is the value of the resistance between the conductor and the earthed core screen, metallic sheath, armour or adjacent conductors.

Consider a unit length of single-core cable with conductor radius r and radius over insulation R (fig. 2.1). The surface area of a ring of insulation of radial thickness δx at radius x is $2\pi x$ times the unit length. The insulation resistance δR of this ring is given by

$$\delta R = \frac{\rho \delta x}{2\pi x} \quad (\Omega) \tag{2.5}$$

where ρ is the specific resistivity ($\Omega\,\text{m}$).

Thus the insulation resistance D_R of radial thickness $R - r$ for 1 m cable length is given by

$$D_R = \frac{\rho}{2\pi} \int_r^R \frac{dx}{x} \quad (\Omega) \tag{2.6}$$

which evolves to

$$D_R = \frac{\rho}{2\pi} \log_e \frac{R}{r} \quad (\Omega)$$

or for cable length l

$$\frac{\rho}{2\pi l} \log_e \frac{R}{r} \quad (\Omega) \tag{2.7}$$

Correction for temperature may be made according to

$$\rho_t = \rho_{20} \exp(-\alpha t)$$

where ρ_{20} = specific resistivity at 20°C
α = temperature coefficient of resistivity per degree Celsius at 20°C
t = temperature (°C)

Specific resistivity is also dependent on electric stress and hence the above derivation is a simplification. However, the effect is much less than that of temperature and in most cases it can be neglected. It can be of importance in the design of high voltage d.c.

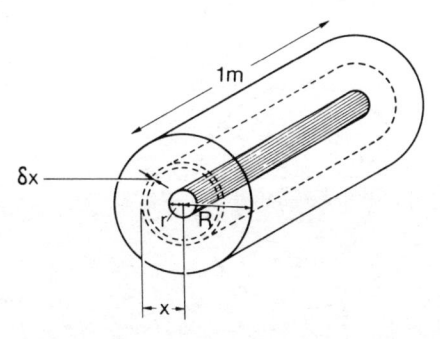

Fig. 2.1 Calculation of insulation resistance of an insulated conductor

paper cables and this is dealt with in chapter 38. In making measurements of insulation resistance it is necessary to maintain the d.c. test voltage for sufficient time to ensure that any transient currents associated with charging of the cable are of negligible value.

In a.c. cables there will be additional currents charging the insulation due to the capacitance of the insulation and for high voltage d.c. paper cables an extra factor is introduced to take account of the variation of resistivity with applied voltage (see chapter 38).

CAPACITANCE

Single-core cables or circular screened cores

Medium voltage and high voltage cable may be single-core cable with an earthed metallic layer around the core, or 3-core cable with an earthed metallic screen around each core. In both cases, the electrostatic field is contained within the earthed screen and is substantially radial.

Consider a single-core cable with a smooth conductor of radius r (m) and an internal screen radius of R (m) (fig. 2.2). Assume that the conductor carries a charge of q (coulomb/m), then the electric flux emanates from the conductor radially, giving a flux density at radius x (m) from the centre of the conductor of

$$D_x = \frac{q}{2\pi x} \quad \text{(coulomb/m}^2\text{)}$$

The electric field intensity at radius x is

$$E_x = \frac{D_x}{\epsilon_0 \epsilon_r} = \frac{q}{2\pi x \epsilon_0 \epsilon_r} \tag{2.8}$$

where ϵ_0 = permittivity of free space, $10^{-9}/36\pi$
  ϵ_r = relative permittivity of the insulation

The work done in moving a unit positive charge the distance dx in an electric field of intensity E is given by

$$dW = -E_x \, dx$$

i.e. the change in potential along dx is

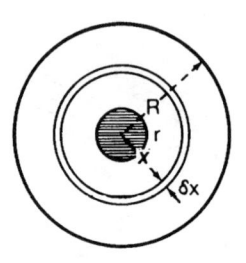

$$dV = -E_x \, dx$$

Fig. 2.2 Cross-section of cable core for derivation of capacitance

Therefore the work done in moving a unit charge from the conductor surface to the outer surface of the insulation is governed by

$$V = \int_R^r -E_x \, dx$$

$$= -\frac{q}{2\pi\epsilon_0\epsilon_r} \int_R^r \frac{dx}{x}$$

$$= \frac{q}{2\pi\epsilon_0\epsilon_r} \log_e\left(\frac{R}{r}\right) \quad \text{(V)} \tag{2.9}$$

The capacitance of the cable per metre length is given by

$$C = \frac{q}{V}$$

$$= \frac{2\pi\epsilon_0\epsilon_r}{\log_e(R/r)} \quad \text{(F/m)}$$

$$= \frac{2\pi\epsilon_r \times 10^{-9}}{36\pi \log_e(R/r)} \quad \text{(F/m)}$$

or

$$C = \frac{\epsilon_r}{18 \log_e(D/d)} \quad (\mu\text{F/km}) \tag{2.10}$$

where D = diameter over the insulation (m)
 d = diameter over the conductor (m)
 ϵ_r = relative permittivity

The relative permittivity is a characteristic of the insulation material and is dependent on temperature and frequency. For power frequencies and normal operating temperatures the effect is small and can be ignored for most engineering calculations.

The capacitance calculated using equation (2.10) is referred to as either the nominal capacitance per core or the nominal star capacitance. This equation is also used to calculate the capacitance of low voltage single-core cables without any metallic outer layer. This is justified on the basis that it is impossible to know what the additional capacitance is in a particular installation between the outer surface of the insulated core and any surrounding earth reference. In any case, such an additional capacitance in series with C would reduce the total star capacitance.

Three-core belted type cables

The equation for calculating the capacitance of a belted type cable is not readily formulated (see later for field theory for paper insulated cables), but an approximation of the capacitance between one conductor and the other conductors connected to the metallic sheath, screen or armour can be obtained from equation (2.10) if D is taken as

 D = diameter of one conductor plus the thickness of insulation between conductors plus the thickness of insulation between any conductor and the metal sheath, screen or armour

The various other capacitances of a belted type cable may be obtained, to a close approximation, by calculating C by equation (2.10) and using the following factors:

$C_1 = 0.83 \times C =$ capacitance of one conductor to all other conductors and outer metallic layer

$C_2 = 0.50 \times C =$ capacitance of one conductor to one other conductor, with remaining conductors and outer metallic layer floating

$C_3 = 1.50 \times C =$ capacitance of all conductors (bunched) to outer metallic layer

DIELECTRIC POWER FACTOR (DIELECTRIC LOSS ANGLE)

It is of great importance that the power factor of the dielectric of cables for voltages of 33 kV and above is of a very low value. The power factor of a dielectric is the ratio

$$\frac{\text{loss in dielectric (watt)}}{\text{volts} \times \text{amps}}$$

Referring to fig. 2.3, when a voltage is applied to a cable with a 'perfect' dielectric, a charging current I_C flows which is in leading quadrature with the voltage. In such a 'perfect' dielectric there would be no component of this current in phase with U. However, perfection in dielectrics has not been achieved and there is a small current I_R which is in phase with U. (See later section on dielectric losses.) This current causes losses $I_R U$ in the dielectric which generate heat. The losses in the dielectric are proportional to the cosine of the angle between the resultant current I_t and applied voltage U.

Now

$$I_t = (I_R{}^2 + I_C{}^2)^{1/2}$$

$$I_R U = I_t U \cos \phi$$

and

$$\cos \phi = \frac{I_R}{(I_R{}^2 + I_C{}^2)^{1/2}} \tag{2.11}$$

As ϕ is close to 90°, $\cos \phi$ equates approximately to $\tan(90 - \phi)$, i.e. equates (approximately) to $\tan \delta$, and the dielectric power factor of a cable is frequently referred to as $\tan \delta$, where δ is known as the dielectric loss angle (DLA).

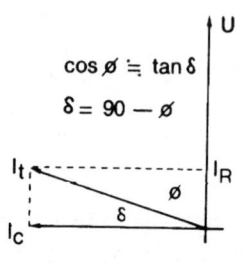

Fig. 2.3 Vector diagram to represent dielectric loss angle

The dielectric loss in watts per kilometre per phase is given by

$$D = 2\pi f C U_0^2 \tan \delta \, 10^{-6} \qquad \text{(watt/km per phase)} \qquad (2.12)$$

It will be seen from this equation that, for a specified design of cable in which values of f, C and U_0 are fixed, the dielectric loss angle must be kept to an absolute minimum to achieve low dielectric losses.

In addition to establishing the dielectric loss angle of the cable to determine the dielectric losses for cable rating purposes, valuable information can be obtained by testing the power factor of the cable in discrete voltage steps within the voltage range $0.5U_0$–$2U_0$. Such a test gives information on the ionisation which takes place in the insulation because ionisation increases with increase in applied voltage (see later under 'Mechanism of insulation breakdown in paper insulated cables').

In the case of paper insulated cables the DLA is a function of the density of the paper and the contamination in the fluid and paper. The fluid can readily be cleaned and the contaminants arise mostly from the paper as ionisable salts. To obtain the lowest DLA in transmission cables at high temperature, deionised water is used in paper manufacture.

ELECTRICAL STRESS DISTRIBUTION AND CALCULATION

The flux distribution in a.c. belted cable insulation is complex and is shown diagrammatically in fig. 2.4. The stress is a maximum at the conductor surface and varies throughout the insulation, decreasing with distance from the conductor surface, but not in a clearly defined manner because of the differing permittivities of the components and the distribution of the flux at various times during the voltage phase rotation. The screened cable used for alternating voltages has a clearly defined stress pattern, while that of cables used on d.c. transmission has a changing pattern depending on temperature due to cable loading.

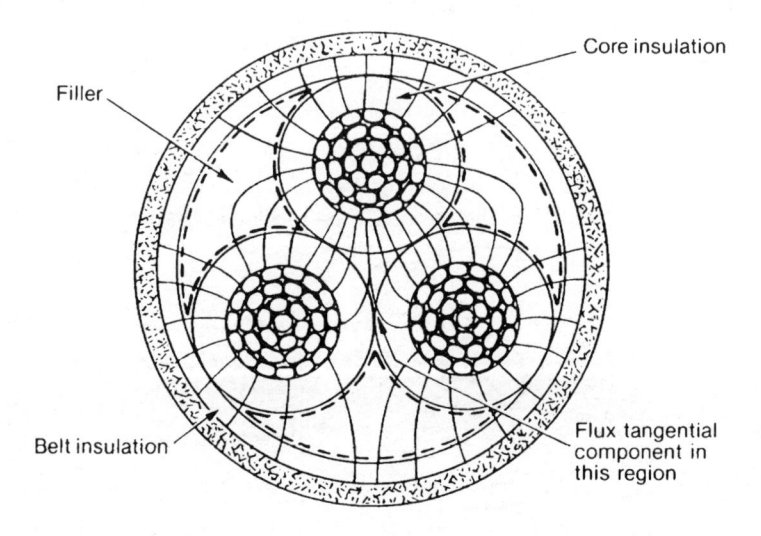

Fig. 2.4 Paper insulated belted cable with top conductor at peak potential

A.C. stress distribution in single-core and screened multicore cables

The stranding effect of unscreened conductors gives a slight unevenness of stress distribution around the periphery of the conductor because of the small radius of the individual wires. Neglecting this effect, a screened core consists of a cylindrical capacitor with the conductor as the inner electrode and the core screen as the outer electrode. Assuming a uniform permittivity, the radial stress distribution curve of a circular core is derived as below. It will be seen to be maximum at the conductor surface, reducing in a hyperbolic curve (fig. 2.5). As the permittivity of the dielectric is substantially constant throughout the operating temperature range of the cable, the stress distribution remains constant at all operating conditions.

Using the notation given previously, i.e.

r = radius of conductor (m)
R = internal radius of sheath (m)
ϵ_0 = permittivity of free space
ϵ_r = relative permittivity of the dielectric
V = potential of conductor relative to the sheath (V)
q = charge (coulomb/m of axial length)

from equation (2.9)

$$V = \frac{q}{2\pi\epsilon_0\epsilon_r} \log_e\left(\frac{R}{r}\right) \quad \text{(V)}$$

Therefore

$$\frac{q}{2\pi\epsilon_0\epsilon_r} = \frac{V}{\log_e(R/r)}$$

Stress may be calculated from equation (2.8) which showed the electric field intensity E_x at radius x to be

$$E_x = \frac{q}{2\pi x \epsilon_0 \epsilon_r}$$

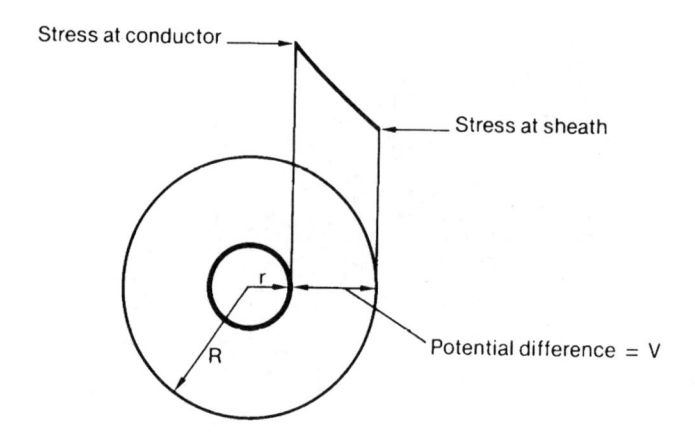

Fig. 2.5 Electrostatic stress in single-core cable

By substitution, the stress at radius x is

$$E_x = \frac{V}{x \log_e(R/r)} \qquad \text{(MV/m)} \qquad (2.13)$$

It will be noted that the range is

$$\text{maximum} = \frac{V}{r \log_e(R/r)} \qquad \text{at conductor surface}$$

$$\text{minimum} = \frac{V}{R \log_e(R/r)} \qquad \text{at sheath inner surface}$$

Several features arising from this are as follows.

Stress at conductor surface

In the above derivation of the stress distribution it was assumed that the conductor surface was smooth. However, the effects of the radius of the wires in a stranded, uncompacted conductor may increase the stress substantially. In high and medium voltage cables, therefore, it is usual to apply a semiconducting screen to a stranded conductor to obtain a smooth surface. This screen would be a carbon paper tape on paper cables or an extruded semiconducting polymer on polymeric cables. When using the equation (2.13), the value of r would be the radius over this conductor screen.

Conductor diameter

The equation (2.13) derived for the stress within the cable dielectric showed the stress to be a maximum at the conductor surface. The equation may be developed further to obtain the ratio of the diameter over the conductor to that over the insulation and hence the conductor diameter which gives the minimum stress at the conductor surface for a specified voltage and diameter over the insulation:

$$E_r = \frac{V}{r \log_e(R/r)}$$

For minimum stress at the conductor for constant values of V and R

$$\frac{dE_r}{dr} = 0$$

i.e.

$$\frac{d}{dr}\left[\frac{V}{r \log_e(R/r)}\right] = 0$$

$$\frac{d}{dr}\left[\frac{V}{r \log_e R - r \log_e r}\right] = 0$$

$$\frac{-V(\log_e R - \log_e r - 1)}{(r \log_e R - r \log_e r)^2} = 0$$

Therefore

$$\log_e R - \log_e r - 1 = 0$$

$$\log_e(R/r) = 1$$

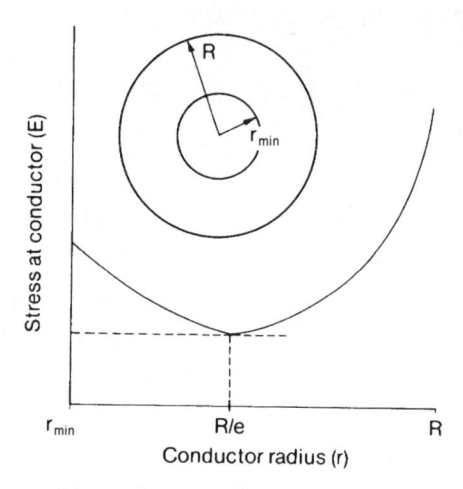

[Figure 2.6: plot of Stress at conductor (E) vs Conductor radius (r), with minimum at R/e]

Fig. 2.6 Variation of stress with conductor radius

and

$$R/r = e \tag{2.14}$$

The stress at the conductor is at a minimum when the ratio R/r equals e. By substituting this in the stress equation it will be seen that the value of the stress is V/r. This is illustrated in fig. 2.6.

The conductor cross-sectional area is determined by the current which has to be carried. When designing cables, particularly for the higher transmission voltages, it is possible that the radius of the conductor size needed to give the current carrying capacity called for will be smaller than that to give optimum ratio of diameter over conductor to diameter over insulation. This necessitates the application of thicker insulation to maintain an equal maximum stress. Frequently the smaller conductors within a specific voltage range have a greater insulation thickness than the larger sizes. An alternative design procedure is to increase the conductor diameter to attain the optimum ratio. For example, in the case of fluid-filled cables, the diameter of the central fluid duct may be increased.

Grading of insulation
In high voltage impregnated paper insulated cables it is an advantage to have material of high electric strength near the conductor to reduce the total insulation thickness. This is normally achieved by having thinner high quality, higher density papers in this region. The use of graded insulation also gives improved bending properties (chapter 19).

D.C. stress distribution

The stress distribution within a d.c. cable is determined, *inter alia*, by insulation resistivity, which, as indicated previously, is influenced by the temperature of the insulation and also the stress. The stress pattern within a d.c. cable thus alters with the load. This is discussed in chapter 38.

Basic Electrical Theory Applicable to Cable Design

FIELD CONTROL IN PAPER INSULATED CABLES

In the early days of electrical power transmission the belted type cable was used extensively but, with the increase in system voltage to 33 kV in the early 1920s, the shortcomings of the belted construction became apparent and the screened type (sometimes called H or Hochstadter type) cable was introduced and has been used successfully for this voltage to the present day. As transmission voltages increased, 'pressure-assisted' type cables using the screened construction were designed and are employed at voltages of 33 kV and above.

Belted type cables

The construction of the 3-core belted type cable is shown in fig. 2.7 (left). The insulation between conductors is twice the radial thickness of the insulation around each conductor and the insulation between any conductor and the earthed sheath is the radial thickness of the insulation around a conductor plus the thickness of the belt. As the phase to phase voltage is $\sqrt{3}$ times the phase to earth voltage, the total thickness of insulation between conductors and that between conductors and sheath are designed in approximately the same ratio.

For voltages up to 22 kV the belted construction has been used as the electrical stresses are acceptably low. For cables used at higher voltages it is necessary to raise the operating stresses for economical and practical reasons and this brings to light defects in the belted cable design which are described in the later section, 'Mechanism of insulation breakdown in paper insulated cables'. Nowadays screened cables are generally used at 15 kV and above.

Screened or H type cable

To eliminate the weaknesses of the belted type cable a design of cable in which each core is individually surrounded by an earthed metallic layer was introduced and is shown in fig. 2.7 (right). This design of cable, first patented by Hochstadter in 1914,

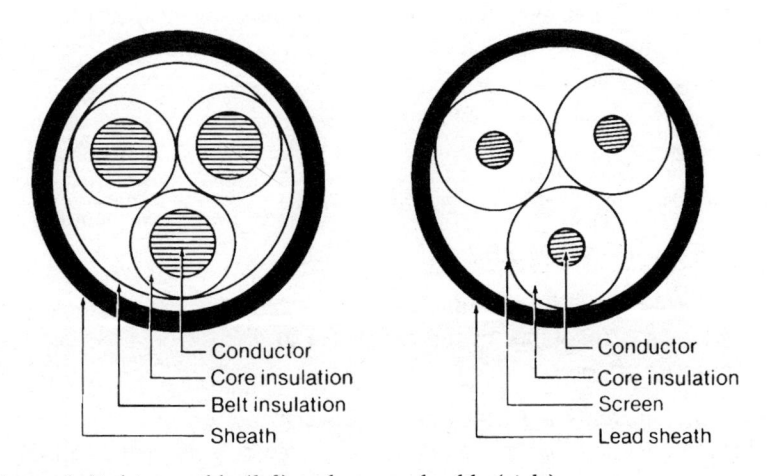

Fig. 2.7 3-core belted type cable (*left*) and screened cable (*right*)

ensures that the stress is substantially radial and hence normal to the paper surface and is also contained within the machine lapped insulation, which is electrically strong. The possibility of separation of the cores due to thermal excursions and mechanical handling, while not eliminated, does not present a hazard as all electrical flux is contained within the earthed screens.

FIELD CONTROL IN POLYMERIC INSULATED CABLES

During the development of medium voltage XLPE cables, it was soon realised that belted designs above 1.9/3.3 kV rating exhibited similar weaknesses to those of belted paper cables above 12.7/22 kV rating. That is, the electrical stress in the fillers around the cores produced partial discharge activity which impinged on the surface of the XLPE insulation, steadily eroding the dielectric and resulting in premature cable failure.

For most normal applications, therefore, polymeric cables rated at 3.8/6.6 kV and above have screened cores. Furthermore, these cables nearly always have circular cores so the cable section is similar to that in Fig. 2.7 (right).

The main functional element of the core screen is a semiconducting layer in intimate contact with the insulation. This can comprise a semiconducting varnish in combination with a semiconducting tape but, nowadays, is invariably an extruded semiconducting layer. The combination of conductor screen, insulation and core screen are produced in one production process to ensure a composite extrusion without voids at the insulation surfaces.

As a result, the electric field in these cables is kept within a homogeneous dielectric and has a radial pattern. In order to provide a continuous earth reference for the semiconducting core screen and to supply the charging currents for all parts of the cable length, a layer of metallic tape or wires is applied over the core screen.

SOURCES OF ELECTRICAL LOSSES

An electric power cable consists of three basic components, namely the conductor(s), the dielectric and the outer metallic layer(s). When the cable is energised and carrying load, heat, which must be dissipated to the surrounding medium, is generated by each of these components. The effects on current carrying capacity are discussed in chapter 8.

Conductor losses

The conductor losses are ohmic losses and are equal to

$$nI^2 R_\theta \quad \text{(watt)}$$

where I = current carried by the conductor (A)
R_θ = ohmic a.c. resistance of the conductor at $\theta°C$ (Ω)
n = number of cores

When high a.c. currents are transmitted, the distribution of current is not evenly disposed throughout the cross-section of the conductor. This is due to the skin effect and the proximity effect as discussed earlier in relation to a.c. resistance.

Dielectric losses

The dielectric losses of an a.c. cable are proportional to the capacitance, the frequency, the phase voltage and the power factor. They are given by

$$D = n\omega C U_0^2 \tan\delta \, 10^{-6} \quad \text{(watt/km)} \quad \text{(see equation (2.12))}$$

where $\omega = 2\pi$ multiplied by frequency
$C =$ capacitance to neutral (μF/km)
$U_0 =$ phase to neutral voltage (V)
$\tan\delta =$ dielectric power factor

The loss component of the power factor (i.e. the current in phase with the applied voltage) is made up of

(a) leakage current flowing through the dielectric which is independent of frequency and consequently occurs with both a.c. and d.c. voltage applications;
(b) dielectric hysteresis, by far the largest effect, which is caused by the interaction of the alternating field with the molecules of the constituents of the insulation and is only present with a.c. voltage application;
(c) ionisation, i.e. partial discharge in the dielectric.

The power factor of the cable insulation is dependent on frequency, temperature and applied voltage. It is of a very low order and consequently for cables of up to 50 kV operating voltage the dielectric losses are small in comparison with conductor losses. However, for cables for operation above this level the losses rise rapidly with voltage and this must be taken into consideration when calculating the current carrying capacity of the cables.

Sheath losses

When single-core cables carry alternating current the magnetic field associated with the current induces e.m.f.s in the sheath of the cable and also in the sheaths of surrounding cables. Two types of sheath losses are possible resulting from such e.m.f.s: sheath eddy current loss and sheath circuit loss.

Sheath eddy current loss
Eddy currents are induced by the current or currents in the conductors of the cables in close proximity to the sheath. Consider a three-phase circuit with cables R, Y and B. The flux of cables Y and B cuts the sheath of cable R. More lines of flux cut the sections of the sheath of R closer to Y and B than that section remote from Y and B. Thus a resultant e.m.f. is induced which causes current (eddy current) to flow along one side of the sheath and return along the other.

The integral of such currents over the sheath cross-section is zero. These eddy currents are independent of the type of sheath bonding which is applied and decrease with the distance between the cables.

The sheath eddy current losses are given by

$$S_e = I^2 \left[\frac{3\omega^2}{R_s} \left(\frac{d_m}{2S}\right)^2 \times 10^{-8} \right] \quad \text{(watt/km per phase)} \quad (2.15)$$

where S_e = sheath eddy current losses
 I = current (A)
 $\omega = 2\pi$ multiplied by frequency
 d_m = mean diameter of the sheath (m)
 S = distance between cable centres (m)
 R_s = sheath resistance (Ω/km)

For single-core lead sheathed cables these losses are normally small compared with conductor losses, but are considerably higher with aluminium sheathed cables when they are in close proximity.

Sheath circuit loss
When the sheath of a single-core cable is bonded to earth or to other sheaths at more than one point, a current flows in the sheath due to the e.m.f. induced by the a.c. conductor current by 'transformer' action. This is because the sheath and return path, to which each end of the sheath is bonded, form a closed loop which is cut by the flux associated with the current in the conductor. The magnitude of the flux which cuts the sheath is dependent on the size of the loop which, in turn, is dependent on the spacing between the cables or between the sheath and the mean return path of the current through the earth or other medium.

The voltage induced in the sheath is given by

$$E_s = IX_m$$

where I = conductor current (A)
 $X_m = 2\pi f M \times 10^{-3}$ (Ω/km)

The mutual inductance M between conductor and sheath is given by

$$M = 0.2 \log_e \left(\frac{2S}{d_m}\right) \quad \text{(mH/km)}$$

The impedance of the sheath Z_s (Ω/km) is given by

$$Z_s = (R_s^2 + X_m^2)^{1/2}$$

where R_s is the sheath resistance (Ω/km). Therefore the sheath current I_s is equal to

$$I_s = \frac{E_s}{(R_s^2 + X_m^2)^{1/2}}$$

$$= \frac{IX_m}{(R_s^2 + X_m^2)^{1/2}} \quad \text{(A)}$$

The sheath current losses per phase are given by

$$I_s^2 R_s = \frac{I^2 X_m^2 R_s}{R_s^2 + X_m^2} \quad \text{(watt/km)} \tag{2.16}$$

Therefore total sheath losses, i.e. sheath circuit losses plus sheath eddy current losses, are given by

$$I^2 R_s \left\{ \frac{X_m^2}{R_s^2 + X_m^2} + \left[\frac{3\omega^2}{R_s^2}\left(\frac{d_m}{2S}\right)^2 \times 10^{-8}\right] \right\} \tag{2.17}$$

The heat generated by losses in the conductor, the dielectric, the sheath and armour has to pass to the surrounding medium, which may be the ground, air, water or some other material. As the current carrying capacity of an electric cable is normally dictated by the maximum temperature of the conductor, the components of the cable, in addition to meeting the electrical requirements, must also have as low a thermal resistivity as possible to ensure that the heat can be dissipated efficiently. The subjects of heat dissipation and operating temperatures of cable components are discussed in chapter 8.

MECHANISM OF INSULATION BREAKDOWN IN PAPER INSULATED CABLES

In addition to the obvious and by far the most usual reasons for failure, such as mechanical damage to the insulation or ingress of moisture, there are three basic reasons for failure:

(a) breakdown due to ionisation;
(b) thermal breakdown;
(c) breakdown under transient voltage conditions.

Breakdown due to ionisation

The 'perfect' belted solid type cable would be so manufactured that the impregnant would completely fill the interstices between the wires of the conductor, the fibres of the paper, the gaps between papers and the filler material. In short, the whole volume contained within the lead or aluminium sheath would be completely void free. During installation, however, the 'perfect' belted cable undergoes mechanical manipulation, and movement of the cores relative to each other takes place. Also, when the cable is loaded electrically the conductors, insulation, free impregnant and lead sheath expand, but the lead sheath does not expand with temperature to the same extent as the interior components of the cable. The sheath is thus extended physically by the pressure exerted by the inner components. On cooling, the lead sheath does not return to its original dimensions and consequently the interior of the sheath is no longer completely occupied, there being voids formed within the cable.

Installation, repeated load cycling and migration of compound of inclined routes can therefore all cause voids within the cable. Such voids are particularly hazardous when they occur within the highly stressed zones of the insulation. As transmission voltages increased, the insulation of the belted cables was increased to meet the higher stresses involved. On the introduction of 33 kV belted cables in the early 1920s, however, it was found that merely increasing the insulation thickness did not give satisfactory performance as there was a high failure rate. On studying the problem, certain weaknesses were found in the belted construction of cables for this voltage. Because of the design of the cable, part of the flux due to the 3-phase voltage, at a certain instant of time during the voltage phase rotation, passes radially through electrically strong core insulation; at other parts of the insulation there is a radial and a tangential component of the flux, and at still other parts the flux passes through the sound core insulation and into the inferior insulation (fig. 2.4).

The tangential strength, i.e. the strength along the surface of the papers, of lapped dielectric is only about one-fifteenth of the radial strength and also the filler insulation is much weaker than the normal core insulation; consequently these weaknesses were highlighted by the higher stresses involved in 33 kV cables.

A more serious weakness in the belted construction occurs, however, when the cores which were originally in close contact with each other move or are forced apart, either by mechanical manipulation or by cable loading. When this happens, the flux passes through sound core insulation, through the space between the two cores and then through the insulation of the second core. The space between the cores which is likely to be devoid of compound is therefore highly stressed, and any gas which may be present will ionise. The adjacent core insulation is thereby weakened by ionic bombardment and failure can occur.

The screened cable was introduced to eliminate the weakness of the belted type cable caused by the weak filler insulation, the spaces between cores and the tangential flux. The screen round the individual cores confines the stress to sound core insulation and also ensures that the flux is substantially radial.

With the introduction of the screened or H type cable, many of the weaknesses of the belted construction were eliminated and 33 kV solid H type cables have given extremely good service performance for many years. On increasing the transmission voltage above 33 kV, however, the one weakness remaining, though not manifestly harmful at 33 kV, had to be eliminated. This final weakness was the gaseous ionisation in voids formed within the insulation by compound migration resulting from cable loading and steep inclines.

The breakdown of paper insulation due to ionisation occurs through the formation of carbonaceous 'fronds' on the insulation papers (fig. 2.8). This is generally known as

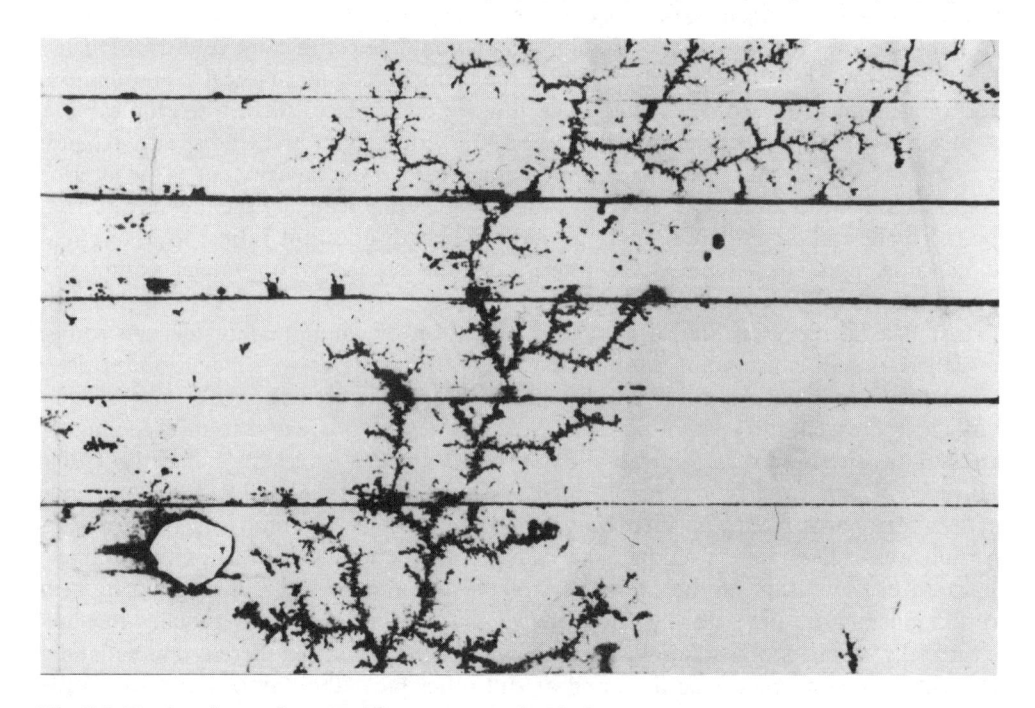

Fig. 2.8 Treeing, i.e. carbon tracking, on paper insulating tapes

'treeing'. The carbonaceous paths start at an almost imperceptible carbon core, generally at the conductor surface, and gradually spread outwards through the insulation, increasing in width and complexity as progression takes place.

Figure 2.9 depicts the development of tree paths. The steps which lead to breakdown comprise the following.

(a) Ionisation takes place within a gaseous void in the gap between the paper next to the conductor and the conductor.
(b) Owing to the ionic bombardment of the second paper, the impregnant is partly pushed out and partly condensed into cable wax with the formation of more gas. Eventually, if the ionic bombardment is sufficiently severe, a carbonaceous path will penetrate between the fibres of the second paper.
(c) There is now a conducting path extending from the conductor, through the butt gap of the first paper and reaching the surface of the third paper nearest to the conductor. Neglecting the voltage drop along this path, the potential of the path front is substantially the same as that of the conductor; thus there is a point on the conductor side of the third paper which is at conductor potential, i.e. V_1 in fig. 2.9. There is a potential gradient throughout the dielectric and thus the third paper is at a potential lower than that of the conductor, i.e. V_2 in fig. 2.9. Therefore there exists a tangential stress $V_1 - V_2$ across the surface of the third paper.
(d) Ionisation due to this tangential stress sweeps away some of the compound and condenses some of the remainder, forming wax.

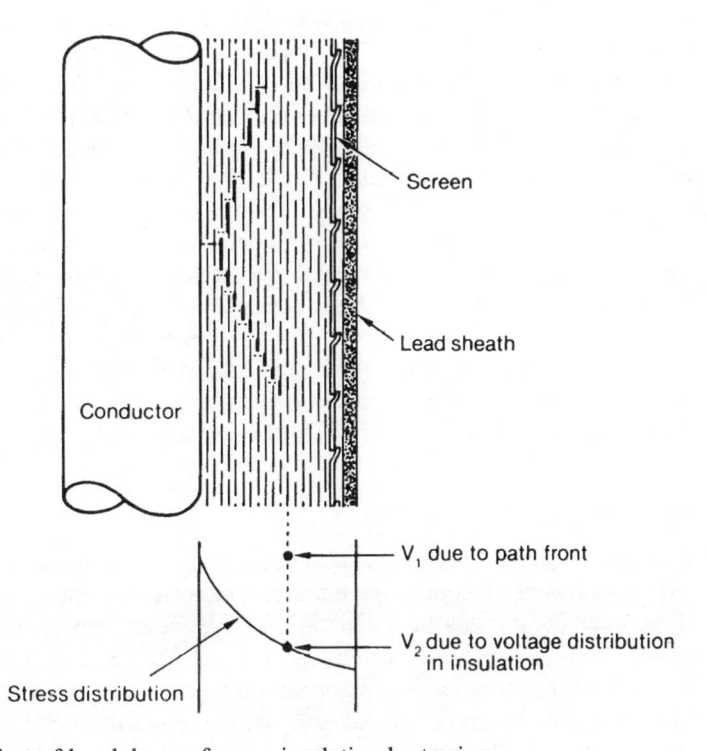

Fig. 2.9 Mechanism of breakdown of paper insulation by treeing

(e) Eventually carbonisation of the compound takes place. 'Fronds' of carbon spread out until the gap in the third paper is reached and the path proceeds, through the gap where there is no fibrous barrier, to the fourth paper.
(f) The treeing mechanism thus progresses, increasing in severity as the distance from the conductor increases owing to the greater tangential stress which exists.
(g) The carbon fronds at each paper continue in length along the surface of the paper until the voltage drop due to the current within the main track and the frond lowers the tangential stress to a value below ionising level.

Several methods of controlling ionisation may be used. One method is to prevent the formation of voids throughout the life of the cable by impregnating the cable with a low viscosity fluid and maintaining a positive fluid pressure within the cable over the complete operating temperature range of the cable. This is attained by fitting pressurised reservoir tanks at strategic positions throughout the cable route. At periods when the cable is loaded, the components within the aluminium or reinforced lead sheath expand and the displaced fluid passes along the cable via the ducts into the pressure tanks which accept the fluid and retain it until the cable cools. On cooling, the pressure tanks force the fluid back into the cable, thus retaining void-free insulation. Such a cable is called a self-contained fluid-filled cable.

Other methods of controlling ionisation are (a) to fill the voids with pressurised gas and (b) to apply sufficient pressure externally to the cable to close up the voids. The high pressure gas-filled cable controls ionisation by having the conductors and gap spaces between the pre-impregnated paper tapes filled with an inert gas under high pressure. Ionisation does not occur in the gaps because of the high pressure involved. During cable heating, the gas expands more in the gaps close to the conductor than in those at the outside of the insulation owing to the temperature differential, and the gas moves towards the aluminium or reinforced lead sheath. On cooling, the gas is forced by the differential pressure back from the outside of the insulation towards the conductor. The pressure within the cable is kept at all times at a value sufficient to prevent ionisation.

Ionisation within voids is prevented in the pipe type compression cable by the application of external pressure to the cable. In this case, the mass-impregnated single-core cables are shaped so that the sheath can be deformed under pressure. The 3-core or three single-core cables are all contained within a steel pipe and the intervening space is filled with a gas under high pressure. When the cable expands on load, the gas absorbs the expansion and on cooling the gas pressurises the cable and prevents void formation.

The pipe type high pressure fluid-filled cable is constructed with three unsheathed cores within a steel pipe. The intervening space within the pipe is filled with a low viscosity fluid under pressure which prevents the formation of voids.

Thermal breakdown in paper insulation

The dielectric loss angle for fluid/paper insulation has a minimum value in the region of 50–60°C. Thus, at around the operating temperature, a rise in temperature increases the dielectric loss, so giving a larger heat generation. The rise in temperature also increases the temperature gradient to the surroundings and raises the rate of heat dissipation. If the rate of rise of heat generation is greater than the rate of rise of heat dissipation, the cable temperature will continue to increase, with the result that the dielectric will overheat and fail electrically.

Figure 2.10 illustrates the dielectric loss angle versus temperature characteristics of two types of impregnated paper. Paper A is typical of paper in use some years ago, while paper B represents paper of the type now used for very high voltage cables. Paper B exhibits comparatively little variation of DLA with temperature over the important temperature range, whereas the DLA of paper A increases considerably above 60°C.

Figure 2.11 indicates the effect of using these papers on the thermal stability of 400 kV 2000 mm^2 self-contained fluid-filled cable buried at standard depth in ground of normal thermal resistivity. The straight line shows the relationship between the cable sheath temperature and the amount of heat which can be dissipated through the ground. The other curves show the relationship between the total losses generated with the cable and the sheath temperature. These losses comprise two main components: the I^2R loss in the conductor and the dielectric loss. The I^2R loss is approximately linearly dependent on the cable temperature, because of the increase in conductor resistance with temperature. The dielectric loss is proportional to the DLA in accordance with

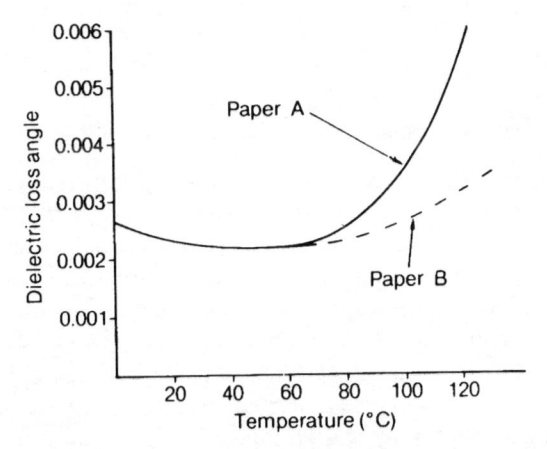

Fig. 2.10 Dielectric loss angle versus temperature characteristics of two types of impregnated paper for high voltage cables

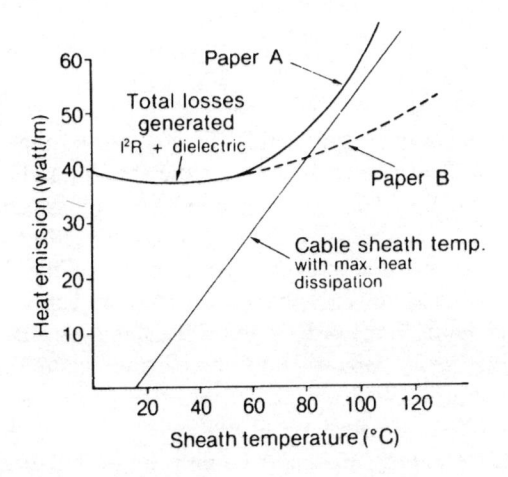

Fig. 2.11 Thermal stability of a buried 2000 mm^2, 400 kV cable with a current of 1400 A

equation (2.12). If the current is taken to be 1400 A, applied at ambient temperature, the cable with the insulation comprising paper B would at first rise rapidly in temperature because the total losses generated would greatly exceed the amount of heat which could be dissipated from the sheath. The general cable temperature would continue to rise until the heat generation and losses were in equilibrium, i.e. at the intersection of the two lines. The temperature of the cable would then remain steady (at a temperature of 80°C on the graph).

If the cable had insulation corresponding to paper A, the initial temperature rise would follow a similar pattern. However, when the temperature exceeded 60°C, the rate of rise of the dielectric losses would exceed the increase in the ability of the sheath to dissipate the losses and the cable temperature would rise. The losses generated would continue to exceed the losses dissipated, with the result that the progressive temperature rise would continue until thermal breakdown of the cable occurred. In this particular example the current loading condition is somewhat exaggerated but the principle illustrates the importance of selecting paper having suitable characteristics. It has also been assumed that the sheath bonding and spacing between cables is such that sheath losses can be ignored.

Breakdown under transient conditions

Cables are designed to withstand transient conditions appropriate to their operating voltage, and this is an important aspect in the case of pressure assisted transmission cables (chapter 29). However, should the cable insulation be subjected to transient voltages such as lightning or switching surges, which are higher than the impulse voltage for which the cable is designed, a failure may occur. Such a breakdown takes the form of a puncture of the insulation and is usually very localised in nature.

BREAKDOWN OF PLASTIC INSULATION

This subject is covered in chapters 3 and 24.

ELECTROMAGNETIC FIELDS

Insulated distribution and transmission cables have an advantage over overhead lines; the external electrostatic field is zero because of the shielding effect of the conducting insulation screen within the cable. The magnetic field external to a three-core distribution cable carrying balanced load currents rapidly reduces to zero because the vector sum of the spatial and time resolved components of the field is zero. A useful degree of ferromagnetic shielding is achieved for three-core cables by the application of steel wire armour which helps to contain the flux. The shielding effect can be significantly increased by eliminating the air gaps with steel tape armour (suitable for small diameter cables) or by the installation of the cable within a steel pipe (as employed with high pressure fluid-filled and high pressure gas-filled cables).

The magnetic field external to single-core cables laid in flat formation does not sum to zero close to the cables because of the geometric asymmetry. The distribution of flux density can be calculated analytically by the application of the Biot–Savart law[2] to each individual current carrying conductor and metallic sheath and by making a vectorially and temporally resolved summation (equations (2.18) and (2.19)). The simple analytical method can be used in those applications which have one value of permeability and which do not use eddy current shielding. For more complex application, such as those employing ferrous materials, specialised computer programs are required which usually employ a finite element algorithm with the ability to model a non-linear B–H hysteresis curve. It is usual in calculations to use the peak value of current. The waveform of the resultant flux density is complex, comprising both sinusoidal and bias components and with a polarised vector rotating about an axis and pulsating in magnitude. In consequence it is usual to quote either the r.m.s. value or the mean value of flux density, the preferred unit being μT (1 μT = 10 mG).

$$\bar{B} = \bar{B}_1 + \bar{B}_2 + \cdots + \bar{B}_n \tag{2.18}$$

$$\bar{B}_r = \frac{\mu_0 \mu_r I_1 \sin(\omega t + \phi_1)}{2\pi r_1} \tag{2.19}$$

where $\bar{B}_r$ = resultant flux density at r due to $I_1, I_2, \ldots, I_n$, f(t, x, y) (T)
 $\bar{B}_1$ = flux density at r due to I_1, f(t, x, y) (T)
 I_1 = conductor sheath current (A$_{peak}$)
 r_1 = distance from centre-line of conductor 1, f(x, y) (m)
 x, y = co-ordinates of horizontal and vertical distance (m)
 μ_0 = magnetic permeability of free space (T/A)
 μ_r = relative magnetic permeability
 ω = angular frequency of supply (radians/s)
 ϕ_1 = phase displacement of I_1 (radians)

For comparative purposes it has become practice to calculate the magnetic flux density above a buried cable circuit at a height of 1 m above ground level. At this height the ratio of distance to cable centre-line spacing is comparatively large such that the maximum magnitude of the flux density is low and, compared with an overhead line,

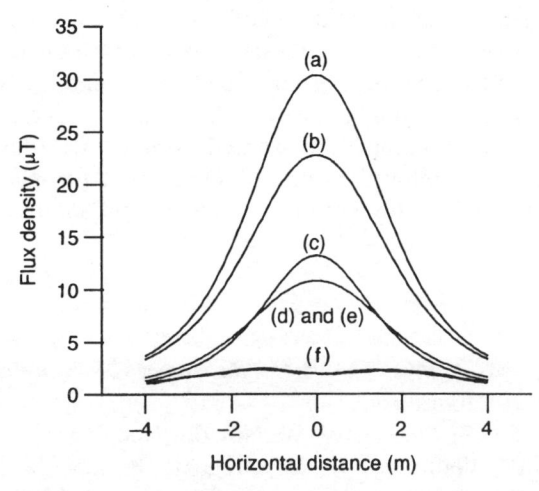

Fig. 2.12 Horizontal flux density distribution 1 m above ground level for 400 kV underground double cable circuits carrying 1000 A. (The legend is given in table 2.2.)

Table 2.2 Effect of installation configuration on magnetic flux density and cable rating (see also fig. 2.12)

Configuration (Double circuit)	Cable spacing (mm)	Circuit spacing (mm)	Depth (mm)	Flux density at 1000 A (µT) Max.	Fig.	Rating per circuit (A)
Flat – XB[a]	300	1500	900	30	2.12a	2432
Flat – XB[b]	300	1500	900	23	2.12b	2432
Flat – XB	200	800	900	13	2.12c	1934
Flat – XB	300	1500	1800	11	2.12d	1886
Trefoil – XB	300	1500	900	11	2.12e	2248
Flat – SB	300	1500	900	2	2.12f	1370

[a] Reference case
[b] With steel plates
XB Cross bonded sheath
SB Solidly bonded sheath

rapidly reduces in magnitude on both sides of the cable circuit, i.e. within the width of a roadway (fig. 2.12). Should it be required, significant further reduction in flux density can be achieved, however this is in varying degrees detrimental to the cable thermal rating and to the cost of the circuit.[3] Examples of flux density distributions are shown in fig. 2.12 for the 400 kV double circuit configuration given in table 2.2. The simplest methods are to lay the cables closer together and at greater depth, the most effective compromise being to lay the cables in an open trefoil formation.

A degree of cancellation of the magnetic field is obtained by solidly bonding the metallic sheaths of single core cables thereby permitting the induced voltages to drive a current which, in ideal theoretical circumstances, would be of equal magnitude and in antiphase to the conductor current. In practice the finite resistances of the metallic sheath and earth return wires, if present, reduce the magnitude and alter the phase of the sheath current thereby achieving only partial magnetic screening and with the disadvantage of generating sheath heat loss of comparatively high magnitude. Ferrous screening can be employed by positioning steel plates above the cables, a practice also used to protect cables from vertical ground loading at road crossings. The screening effect can be increased by positioning inverted, U-shaped, steel troughs over the cables but to the detriment of trapping an unwanted layer of air above the cables. Eddy current screening can be employed by positioning non-ferrous plates of high electrical conductivity over the cables, however this is an inefficient solution of high cost.

REFERENCES

(1) Arnold, A. H. M. (1946) *The A.C. Resistance of Non-magnetic Conductors*. National Physical Laboratory.
(2) Ohanian, H. C. (1989) *Physics*. W. W. Norton, New York.
(3) Endersby, T. M., Galloway, S. J., Gregory, B. and Mohan, N. C. (1993) 'Environmental compatibility of supertension cables'. *3rd IEE Int. Conf. on Power Cables and Accessories 10 kV–500 kV*.

CHAPTER 3
Materials Used in Cables

METALS

Electrical properties

Table 3.1 indicates the electrical properties of the common metals used in cables. Taking price into consideration, copper and aluminium are clearly the best choice for conductors but there has been some experience with sodium. Reference to this is made in chapter 4 which also contains information on the variation of resistance with temperature.

Physical properties

The physical properties of metals used for conductors and sheaths are given in table 3.2. Except for the conductors of self-supporting overhead cables, copper is invariably used in the annealed condition. Solid aluminium conductors are also mainly in a soft condition but stranded aluminium conductors are $\frac{3}{4}$H (hard) to H. Aluminium sheaths

Table 3.1 Electrical properties of metals

Metal	Relative conductivity (copper = 100)	Electrical resistivity at 20°C (Ω m, 10^{-8})	Temperature coefficient of resistance (per °C)
Silver	106	1.626	0.0041
Copper (HC, annealed)	100	1.724	0.0039
Copper (HC, hard drawn)	97	1.777	0.0039
Tinned copper	95–99	1.741–1.814	0.0039
Aluminium (EC grade, soft)	61	2.803	0.0040
Aluminium (EC grade, $\frac{1}{2}$H–H)	61	2.826	0.0040
Sodium	35	4.926	0.0054
Mild steel	12	13.80	0.0045
Lead	8	21.4	0.0040

Table 3.2 Physical properties of metals used in cables

Property	Unit	Copper	Aluminium	Lead
Density at 20°C	kg/m^3	8890	2703	11370
Coefficient of thermal expansion per °C	×10^{-6}	17	23	29
Melting point	°C	1083	659	327
Thermal conductivity	W/cm °C	3.8	2.4	0.34
Ultimate tensile stress				
soft temper	MN/m^2	225	70–90	–
$\frac{3}{4}$H to H	MN/m^2	385	125–205	–
Elastic modulus	MN/m^2	26	14	–
Hardness				
soft	DPHN	50	20–25	5
$\frac{3}{4}$H to H	DPHN	115	30–40	–
Stress fatigue endurance limit				
(approximate)	MN/m^2	±65	±40	±2.8

are now extruded directly onto cables and hence of soft temper but a small amount of work hardening occurs during corrugation.

Copper conductors

Because of the way it can readily be rolled into rod and then drawn to wire, together with its excellent electrical conductivity, copper was virtually unchallenged as a conductor for all types of insulated cable for well over 50 years. Indeed, in the electrical world, the International Electrotechnical Commission established an International Annealed Copper Standard (IACS) with copper of resistivity of 1.724 μΩ cm at 20°C assigned as 100%. National specifications for copper can be found in the publication *High Conductivity Coppers, Technical Data TN27* which is produced and issued by the Copper Development Association. This publication also forms part of the contents of a CD-ROM, *Megabytes on Copper*, available from the same Association.

The grade and quality of copper is very important and the high conductivity copper used for electrical purposes comfortably exceeds the 100% IACS value. Conductivity is greatly influenced by impurities and by mechanical working. Consequently, the purity is of the order of 99.99%, which nowadays is obtained by final electrolytic refining.

Fortunately, the mechanical strength of annealed wire is adequate for nearly all types of insulated cable. If any minor working of the material occurs during conductor manufacture, e.g. in compacting to reduce the overall dimensions, allowance has to be made for work hardening by increasing the copper volume to compensate for the reduction in conductance. In an extreme case, such as the use of hard drawn copper for self-supporting overhead lines, this may amount to as much as 3%.

Almost the only unsatisfactory feature of copper is the way that the price fluctuates widely with the world supply and demand. An example of the scale of fluctuations that can be experienced is shown in fig. 3.1, where the monthly low and high cash trends for 1995 are presented. The extreme values for the year are £1680 and £2030 per tonne. If its use could always be justified economically it would not have competitors. However, over the last two decades aluminium has become a replacement in the power distribution field solely on the basis of cost.

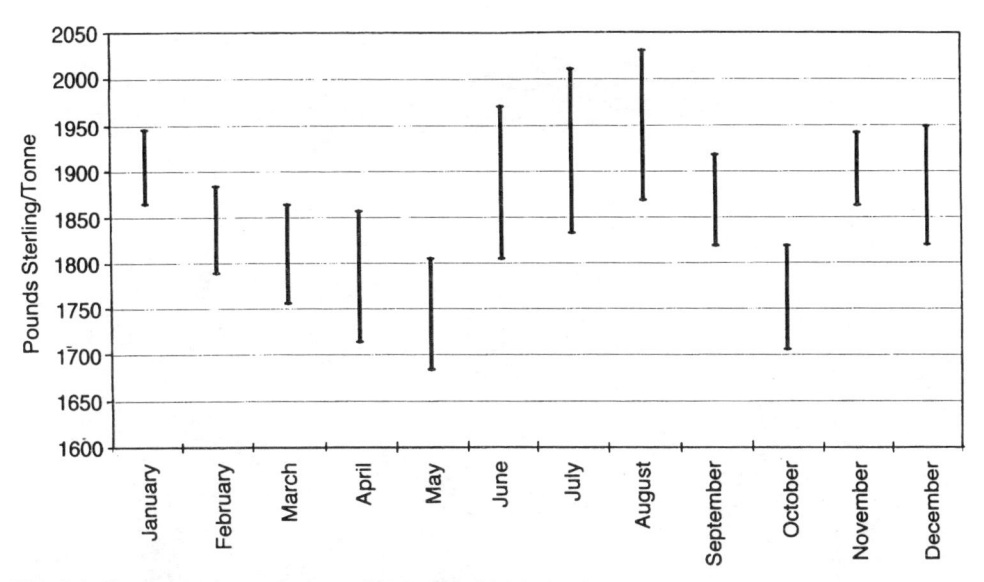

Fig. 3.1 Copper cash trends (monthly low and high figures, 1995)

Aluminium conductors

Although aluminium did not make much serious impact until the price of copper soared in the late 1950s, it is surprising to find that a relatively substantial amount was used between 1909 and 1912. Figure 3.2 shows a sophisticated low voltage paper insulated d.c. cable made in quantity by British Insulated Cables at Prescot during this period. This particular cable was taken out of service in 1967 and was just as perfect as when installed. Even the jointing was of interest as each shaped wire had an individual sleeve secured by pinch-screws.

Unlike copper, the mechanical strength of annealed aluminium is rather low for soft single wire or stranded conductors but fortunately conductivity is little changed by working. Consequently the temper of wire used is commonly known as a broad $\frac{3}{4}$-hard grade. Tensile strength is relatively unimportant and this grade covers ranges which in the aluminium industry spread from $\frac{1}{2}$-hard to fully hard, the main requirement being to specify a maximum of 205 MN/m^2, which avoids the use of wire having too little elongation before fracture on bending. Aluminium, however, can also be used in solid as distinct from stranded form and in this case a soft temper is desirable to keep the stiffness of the conductor to a minimum. Such conductors are mostly produced by hot extrusion, which provides an almost annealed temper, and the upper limit of tensile strength is fixed at 80 MN/m^2 for sizes above 35 mm^2. For small sizes this limit is rather low to exclude some stretching during cable processing and also there are advantages in producing the conductors by drawing. The maximum is therefore in the range 125–165 MN/m^2 according to size.

Compared with copper, aluminium has a number of technical disadvantages, all of which can be satisfactorily overcome to benefit from its economic attraction. The advantage of a low density of one-third that of copper is partly offset by its low conductivity of 61% that of copper, but for equal conductance the weight of the expensive conductor metal required is almost exactly halved. However, the cross-sectional conductor area has to be increased by a factor of 1.6 and this means extra

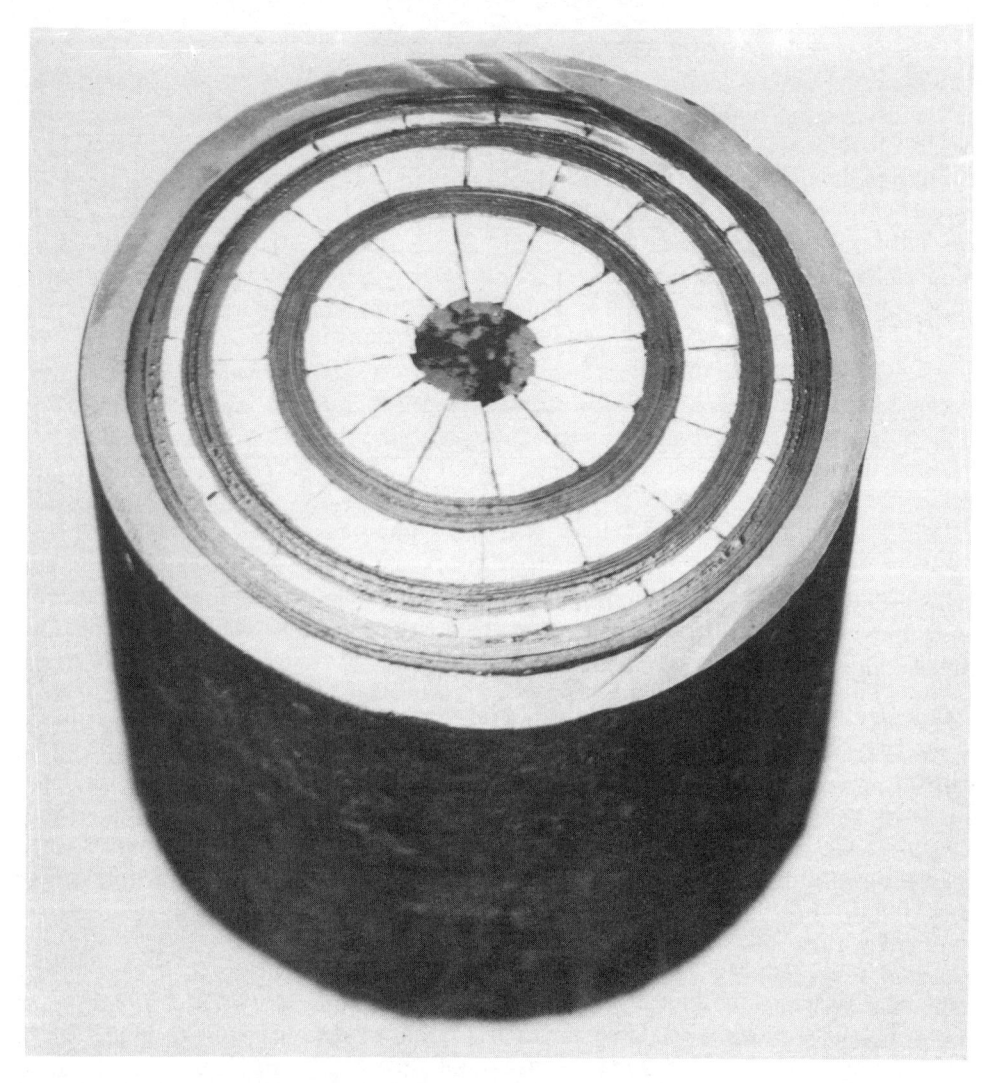

Fig. 3.2 ±230 V d.c. cable installed in London in 1911 and removed from service in 1967

usage of insulation, sheathing and armouring material. With conventional paper insulated, lead sheathed and armoured constructions, there is no overall saving in cable weight but the situation is quite different if aluminium is also used for sheathing. The above is a generalisation based on cables having equal conductance. In practice there are other factors to be considered such as the cable voltage, current carrying capacity and voltage drop. In general, current ratings of aluminium cables are about 78%–80% of those of copper cables of the same conductor size.

As with copper the addition of alloying metals to provide high tensile strength has no application for insulated cables. There is a conductivity penalty which is only justified for all-aluminium overhead lines. Incidentally, steel-cored aluminium (non-alloyed) has been used for such lines for over 50 years because the economic aspects are very different from those with insulated cables and the use of massive compression jointing sleeves presented no problems.

Impurities, as distinct from alloying additions, do not greatly affect the conductivity of aluminium but a grade known as EC (electrical conductivity) is commonly accepted in the aluminium industry. The basic British Standards for wrought aluminium for electrical purposes, i.e. BS 2627, 'Wire', and BS 3988, 'Solid conductors for insulated cables', define the purity as 99.5% minimum, with limits for individual impurities. More recently, in 1991, the British Standard BS 6360 'Specification for Conductors in Insulated Cables and Cords' was issued. This includes the requirements for both aluminium and copper conductors.

By controlling the amounts of copper, silicon and iron and/or small amounts of other metals, it is possible to create a range of 'dilute alloys', without much sacrifice of conductivity but with particular strength properties to cater for special requirements in manufacture and jointing, e.g. for single-wire conductors for telephone cables. No applications have been found for such compositions for power or wiring cables.

A particular disadvantage of aluminium for electrical purposes is the thin, hard protective oxide skin which is so valuable in giving corrosion protection to aluminium installed above ground. Satisfactory techniques for removing it for making soldered joints or to break it up by appropriate designs of compression or mechanical joint have long been developed, but it is always necessary to emphasise that jointers must exercise care and attention when following prescribed instructions in detail. This emphasis applied more particularly to soldering and plumbing, and modern techniques for power cable conductor jointing have been centred on mechanical methods capable of easy adoption by less expert operatives.

Apart from the oxide film, there is another factor which has contributed to lack of success in one particular field of use, namely wiring cables. The low yield or proof stress of aluminium means that if the conductor is only held, for example, by a single pinch-screw, as is common in many wiring cable accessories, there can be relaxation and development of a high resistance joint. Overheating and failure may then follow. It is difficult to ensure that all accessories are specially designed for aluminium, and although large-scale installations have been undertaken in North America the problems have not been overcome and the use of aluminium wiring cable has been halted.

Corrosion protection is covered elsewhere and all that need be mentioned here is that, whilst claims for aluminium having excellent corrosion resistance are true if made in relation to dry indoor situations, or even outdoors if drying off is rapid after rain, cable engineers must ensure that bare aluminium is not left exposed anywhere. Cable situations are rarely dry; for example in the base of outdoor pillars corrosion can be rapid. Protection in a form such as heat-shrink sleeving should always be used.

Copper-clad aluminium conductors

Copper-clad aluminium comprises an aluminium core with a heavy layer of copper which is metallurgically bonded to the aluminium. The thickness of copper may be varied but for cable purposes it is defined by BS 4990 as 10% by volume, i.e. 27% by weight. Conductor wire may be produced in various ways but the techniques most commonly used have been to start from a composite billet with either subsequent rolling to rod and then drawing to wire or extrusion.

The whole purpose of the cladding with copper is to overcome the problems in making mechanical connections to aluminium, particularly in wiring type cable accessories, which are predominantly of the pinch-screw type. Copper-clad may be treated in the same way as copper conductors.

Lead sheaths

Lead has served the cable industry well since the end of the last century, but, because of its weight and softness and the need for good plumbing skill, few users will regret that its use has diminished significantly. Although the overall incidence of faults associated with the lead sheath itself has been quite small, large users are likely to have met them at some time. They can be divided into four categories:

(a) fatigue cracking
(b) extrusion defects
(c) fractures associated with internal pressure
(d) corrosion

Fatigue

Lead has comparatively low resistance to fatigue and fig. 3.3 shows a typical case of sheath fracture associated with cracks in the crystal boundaries. Figure 3.4 indicates the effect after etching and illustrates why such failures used to be erroneously attributed to crystallisation. Most sheath fatigue fractures are caused by vibration, either during transport over a long distance to site, or by installation on bridges, in ships, alongside railways or in overhead catenary systems etc. In such cases the vibration frequency is usually high and the amplitude low. Other situations arise where the frequency is low and the amplitude high, so that the effect is more akin to repeated bending, e.g. due to thermal expansion and contraction with bodily longitudinal movement. This can occur in jointing manholes for cables in ducts or with unsecured cables on racks or trays exposed to solar heating and/or having pronounced load cycles. In all such installations it is important to prevent the expansion being accommodated in a local loop or offset.

Much can be done to prevent such failures. If long intercontinental transport is involved it is important for the cable ends to be fixed tightly so that there is no loose cable which can vibrate separately from the whole drum. Installations should also be planned to accommodate expansion and contraction uniformly along the whole length. However, the main factor is that lead alloys are available to provide increased fatigue resistance according to the requirement.

British practice is based on the use of unalloyed lead, alloy E (0.4% tin, 0.2% antimony) and alloy B (0.85% antimony), which have respective fatigue endurance limits (10^6 cycles) of 2.8, 6.3 and 9.3 MN/m^2. An exception is that $\frac{1}{2}$C (0.2% tin, 0.1% cadmium) has better properties for internal pressure cables. Unalloyed lead is quite satisfactory for most cables, including armoured cables to be shipped overseas. Alloy E should be used for unarmoured cables and when vibration is suspected. Alloy B is only required for severe cases, e.g. cables on bridges and for catenary suspension. If there is any doubt, advice should be sought from the cablemaker. The use of high purity lead is undesirable (see later).

Different alloys are often favoured in other countries. For the duct jointing manhole requirements in the USA, various combinations of lead/arsenic/tellurium/bismuth compositions have shown particularly good properties, the contents of each alloying

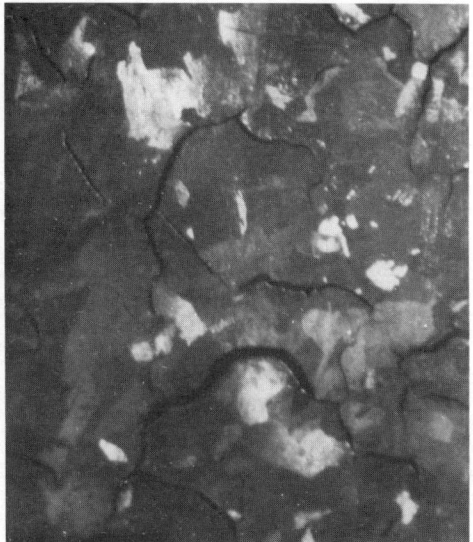

Fig. 3.3 Surface of lead sheath showing typical cracking due to fatigue

Fig. 3.4 Etched surface of lead sheath to show the fatigue cracks at crystal boundaries

metal varying between 0.05% and 0.15%. In continental Europe there has been some preference for alloys containing copper (0.03%–0.05%), often with an addition of tellurium (0.04%). Alloys containing antimony in proportions from 0.35% to 1.0% are also used.

Extrusion defects
When lead sheathing was carried out on hydraulic vertical ram presses, defects were sometimes caused by the entrapment of impurities as the press cylinder was refilled with molten lead. This can be overcome by the Glover tray principle in which the charging is automatic from a bowl of liquid lead on the top of the extrusion cylinder. Such presses have now largely been replaced by continuous screw type extruders which also overcame the occasional faulty stop mark that occurred when ram presses were stopped for recharging with lead.

Effects of internal pressure
Lead has no elastic limit and will deform slowly (creep) at very low stresses. The complex stress–strain relationship has an important bearing on the requirements for sheath composition of cables in which hoop stresses arise as a result of internal pressure. Even though sheaths for gas- and fluid-filled cables are reinforced to withstand the pressure, the sheath itself may be subject to a low stress for a long time and therefore will need to expand by a small amount. Under high stresses lead sheaths will show considerable extensibility and burst with a knife-edge fracture, but at low stresses the fracture with some alloys may be of a blunt intercrystalline type with little expansion. Ageing effects such as recrystallisation, precipitation of impurities and alloying metals in grain boundaries, and change of creep rate in circumferential zones due to bending also have an important influence and affect the choice of alloy. In the

case of fluid-filled cables it is important to select an alloy which is not subject to grain growth during the heating period that occurs with one method of impregnation.

Correct choice of alloy for all internal pressure cables is therefore vital. Much testing is required with conditions relating not only to cable design and processing but also to the particular type of extruder to be used for sheathing. Extrusion temperature, grain size and segregation effects all have a bearing on subsequent stability. Nowadays most cables in this category are of the fluid-filled type, lead sheathed on a screw press, and $\frac{1}{2}$C alloy has given outstanding service.

Corrosion

Some lead water pipes have survived since Roman times but lead may corrode in peaty or highly alkaline soils. Protection in the form of bitumen or bituminised hessian tape plus bitumen is normally adequate. Where corrosion does occur the analysis of corrosion products will normally show the cause and over the years the main source of trouble has been stray direct currents, e.g. from tramways, leaving the sheath.

Purity

Some cable users had a preference for specifying high purity virgin lead. Much commercial lead is now refined to 99.99% purity and such a material is really best avoided as it can lead to very large grain size with a single crystal across the sheath. Such a structure is particularly weak in fatigue and creep properties. BS 801 allows a maximum impurity content of 0.1% and some alloying metals such as tin or antimony are often added with advantage up to this limit. While being convenient for the utilisation of manufacturing scrap it also provides a better sheath than very high purity lead.

Aluminium sheaths

A move to reduce the use of lead in UK cables was made in the early 1950s. This involved marketing 1 kV paper cables which utilised a novel and successful sheathing procedure involving pulling the insulated cores into a length of oversize aluminium tubing and then drawing either through a die to reduce the diameter or through a rotating nut to form a cable sheath with a corrugated rib.

Unfortunately initially it was not appreciated that the apparently good corrosion resistance of aluminium in free air did not apply to buried conditions. Moreover it had not then been established that relatively short time tests on small buried samples were not representative of the differential concentration cell type of corrosion mechanism which applied with long lengths of aluminium. Consequently the bituminous type finishes, satisfactory for lead sheaths, were used with somewhat disastrous consequences.

Nevertheless, it was immediately apparent that aluminium sheaths were quite practicable for handling, up to at least 100 mm diameter, and that they had an economic attraction for internal pressure cables because of the lack of need for sheath reinforcement. The corrosion problem was first overcome by the use of plastic and rubber tapes (chapter 5). Extruded PVC was then coming in as an insulating material and was an obvious choice for providing a reasonably tough protective jacket for aluminium sheaths. The combination became an accepted alternative to lead sheaths in the late 1950s and within a decade had become the preferred choice for pressurised cables.

For solid type cables there was little further activity until the later 1960s when, following experience in Germany, interest developed in the use of combined neutral/

earth distribution cables. The Consac cable (chapter 22) used the aluminium sheath as a conductor and created an economic design which was soon used in large quantity (BS 5593 1978 (1991)). By this time sheathing by direct extrusion was replacing the tube sinking process, with the advantages of reduced cost and a lower permissible sheath thickness. The latter arose because thickness is determined by the ability of the sheath to bend without any serious buckling and the softer extruded material is more favourable in this respect.

In the early 1970s the economic advantage of aluminium was extended to 11 kV paper insulated cables, as used by the UK Electricity Companies. A price saving of the order of 20% was obtained at the time.

Sheath composition and properties
The extrusion pressures and temperatures (approximately 500°C) are quite critical because of the high stress on the extrusion tools. To keep stresses to a minimum, aluminium of 99.7% minimum purity was formerly used but the standard grade of 99.5% minimum has since been proved to be acceptable. At the temperature used the resultant sheath mechanical properties are equivalent to those of material of annealed temper.

Compared with lead sheaths, the most notable difference in properties of aluminium is the greater force that can be developed owing to thermal expansion. As a result of load cycles this can give rise to increased thrust and mechanical stresses in joints. If the sheath is of the corrugated design no problems arise but in the case of solid type cables with smooth sheaths some strengthening may be necessary if operation to full rating is required. Use of a cast resin filling material suffices.

Corrosion
Much may be written and indeed has been published about the vulnerability of aluminium to corrosion and the mechanism by which it may be quickly penetrated by pinholes. Various aspects are the formation of corrosion cells by differential concentration and aeration conditions, crevice attack and avoidance of contact with more electropositive metals. For cables, however, discussions can be avoided by stipulating that aluminium must never be left exposed and sheaths must always have appropriate protective finishes. Extruded PVC, polyethylene and high density polyethylene (chapter 5) have been proved by experience.

Armour for distribution cables

Power cables are usually armoured to carry earth fault currents and to give some protection against mechanical damage both during installation and in service. However, it is difficult to define an optimum requirement or how long it should last in service before the armour material is destroyed by corrosion. National practices vary widely but the materials used consist mainly of steel tape or strip and galvanised steel wire.

Steel tape
This is mild steel of thickness 0.5 mm to 0.8 mm according to cable diameter. The smaller sizes are cold rolled and the larger hot rolled. They are coated with bitumen by the supplier and again subsequently during the armouring process. On the basis that

most external damage to cables occurs in the first few years of life it probably does not matter that corrosion may be severe after a few more years. On the other hand it can be argued, as it was with CNE cables (chapter 22), that at 0.5 m depth of burial most damage is due to heavy mechanical plant and the cable would suffer severely whether armoured or not, i.e. steel tape armour fulfils little purpose.

Continental European practice tends to favour narrow flat steel strip instead of wire armour but this is probably due to history rather than to any technical or economic difference. If there is a good case for ensuring that either strip or tape armour has long life then galvanising is possible and, although much more expensive, such material is often preferred in the Middle and Far East.

Galvanised steel wire
Wire to BS 1442: 1969 (1986) is used for cables having no metallic sheath and is usually preferred to steel tape for solid type lead sheathed cables for 11 kV upwards. If higher tensile strength is required, e.g. for submarine cables, the requirements for carbon steel wire are covered by BS 1441: 1969 (1988).

Non-magnetic armour
For single-core cables in a.c. circuits, it is usually preferable to use non-magnetic material. Wire rather than tape is generally adopted to secure adequate mechanical protection. Stainless steel is difficult to justify on cost grounds and aluminium is the normal choice.

BS 6346: 1989 caters for PVC cables with aluminium strip armour. Whilst useful if it is necessary to carry high earth fault currents, the degree of mechanical protection obtained is not as good as with steel wire armour. Such strip armour now also tends to be more expensive and is little used.

Mechanical protection for wiring cables

Metal tapes and even helically applied wires are not suitable for very small cables, especially when good flexibility is required. Braided constructions using plain or tinned copper wire are usually adopted.

IMPREGNATED PAPER INSULATION

Paper

Paper for cable making consists of a felted mat of long cellulose fibres derived by chemical treatment of wood pulp, which mostly comes from North America and Scandinavia. Digestion with sodium sulphide and caustic soda at high temperature and pressure removes impurities such as lignin and resins. The paper construction is generally of 2-ply form, but some 3-ply is used at the highest voltages. The important physical properties required are controlled by the amount of beating of the pulp, together with the quantity of thin pulp (93%–95% water) fed to a rotating wire mesh on which the paper is formed and the amount of subsequent calendering. These factors determine the thickness, apparent density and impermeability, all of which have to be adjusted according to the cable type and voltage.

The thicknesses of paper normally used are from 65 to 190 µm. The density varies from 650 to 1000 kg/m^3, and as the density of the actual fibres is around 1500 kg/m^3 a considerable volume of the space within the fibres is available for filling with impregnating compound. The impermeability, i.e. the porosity, of the paper can be adjusted independently of density and can have important effects on final mechanical and electrical properties. Tensile strength and, more particularly, tearing resistance also have to be controlled according to application and bending requirements and may have to take account of the width of the tape which is used. The tensile strength in the longitudinal and transverse directions is in a ratio of about 2:1.

Vast quantities of water are used in paper making, up to 200 or even 300 tonnes per tonne of paper, and the quality of the water is important for obtaining the required electrical properties of the paper. For very high voltage transmission cables, the water has to be of deionised quality to achieve the highest degree of purity.

The actual properties of the paper selected by the cable manufacturer have also to be related to the paper–impregnant combination and this is discussed later. Among many important factors is the ability to obtain a good cable bending performance because individual layers must slide over each other. Insulation thicknesses vary from 0.6 to over 30 mm according to voltage and electrical stress and the bending problems increase progressively with thickness. In addition, to achieve strict control over the individual paper tape widths and application tensions, the surface finish of the paper is highly important, as is the tensile strength and thickness, these being increased towards the outside of the insulation from mechanical considerations. Adjacent to the conductor, where the electrical stress is highest, thinner papers may be used.

Polypropylene paper laminate

Polypropylene paper laminate – variously termed PPL, PPLP or BICCLAM – is a low loss dielectric material in tape form. It comprises of a layer of extruded polypropylene to which two thin layers of insulating paper are bonded. Externally the tape has the appearance of a paper tape and can be applied on to a cable using the same paper lapping and impregnating technology.

The PP (polypropylene) component of the laminate has the properties of:

(a) low dielectric loss angle
(b) low permittivity
(c) high operating temperature
(d) high mechanical strength.

The paper component has the properties of:

(a) low elasticity and high tensile strength, these acting to accommodate the PP thermal expansion
(b) a fibrous composition to provide a passage for the cable fluid during service operation
(c) high resistance to partial discharge in the oil-filled butt gaps, such that a high impulse design stress is achieved.

The resulting dielectric permittivity of the PPL fluid-filled insulation is dependent upon the precise ratio of the thickness of the tape. For a 50% ratio tape the measured relative permittivity is 2.6, compared with 3.1–3.4 for typical paper insulation. The reduction in the permittivity of PPL cable compared to paper cable gives the benefit of a reduction in the stress in the oil butt gaps. Refer to table 30.2 for typical properties of PPL insulation.

Impregnating fluids and compounds

Paper has good electrical properties only when dry and the impregnation with suitable fluids and compounds, which is necessary for electrical reasons, also helps to reduce moisture absorption. Impregnation, which is covered in chapter 25, is preceded by heating to 120°C and evacuation to a pressure of the order of $10-20 \text{N/m}^2$ so as to remove both air and moisture from the paper web and ensure that the whole of the matrix is filled with impregnant. The initial moisture content of the paper of around 2–7% is reduced to of the order of 0.01%–0.5% according to voltage. For voltages of 6.6 kV and above, where partial discharge in any vacuous voids could be important, the impregnation treatment must also ensure that the butt-gap spaces between tapes are full of impregnant. When the compounds exhibit a change of state, with highly increased viscosity as temperature is reduced, very slow cooling is necessary and preferably the compound should be circulated through a heat exchanger to achieve uniform cooling.

The impregnant for solid type distribution cables is based on refined mineral oil derived from petroleum crudes. According to compound formulation these may be of basically paraffinic or naphthenic type. After final distillation the oil, and subsequently the mixed compound, are subjected to clay treatment (Fullers' earth) to remove remaining impurities which affect electrical quality.

In the conventional oil–rosin impregnant, the mineral oil is thickened by the addition of refined gum rosin to increase the viscosity in the cable working temperature range. Gum rosin is a material which exudes from pine trees and after initial distillation is selected according to shade of colour before further refining to high quality. In addition to its function for viscosity control, the addition of rosin increases electric strength and it gives substantial improvements in resistance to oxidation. This is of importance in the manufacturing process, where large quantities of impregnants are held in tanks for long periods with frequent cycles of heating.

The increase in viscosity which can be obtained by addition of rosin is still insufficient to prevent drainage of impregnant from cables installed vertically or on steep slopes and the development of non-draining compounds for this purpose is discussed in chapter 19.

The fundamental characteristic of mass-impregnated non-draining (MIND) compounds is that while they have the same fluidity as oil–rosin compounds, at the impregnating temperature of 120°C, and therefore good impregnating properties, they set to a waxy solid at maximum cable operating temperature. The formulation is controlled to produce a plastic consistency which provides good cable bending behaviour. The somewhat greater stiffness of MIND compounds at very low temperature is no handicap to normal installation but excessively severe bending at such temperatures would cause more damage than with oil–rosin impregnant. Satisfactory bending test performance at −10°C can be achieved but demands a higher standard of manufacture in which the quality of paper and paper lapping application play an important part.

The gelling of the mineral oil to a consistency akin to that of petroleum jelly is achieved by the addition of such materials as microcrystalline waxes, polyethylene, polyisobutylene and a small amount of rosin. These constituents possess high electrical

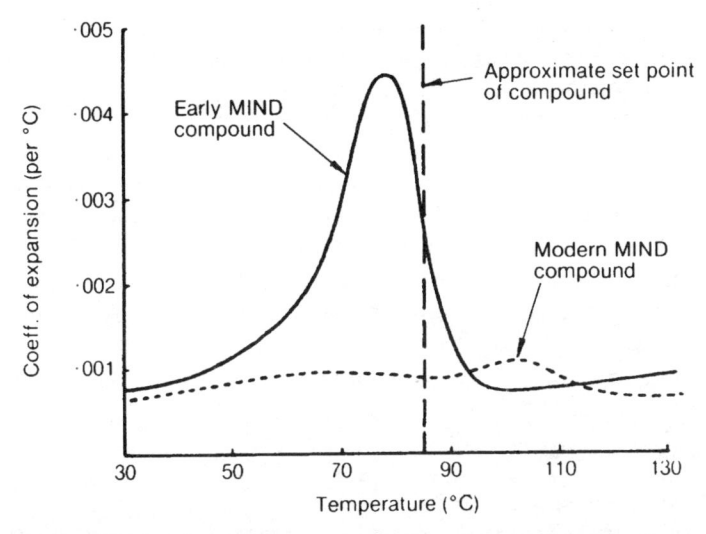

Fig. 3.5 Typical coefficient of expansion versus temperature relationships for mass-impregnated non-draining compound

strength and the properties are chosen to give an elevated softening temperature with suitable plasticity over the operational temperature range.

Some forms of microcrystalline waxes are obtained from the residues resulting from the distillation of crude petroleum oils after a solvent extraction process. Others are derived when the Sasol technique is used in the production of petrol from coal, principally at present in South Africa. The important feature of microwaxes is their ability to hold or occlude oil, i.e. to stop the oil from separating from the mixture. This property is not possessed by macrowaxes such as paraffin wax or the common mineral waxes, which have a much more coarse crystal structure. Because non-draining compounds are of complex composition, there may nevertheless be some slight movement of some of the constituents at elevated temperatures just below the set point of the compound and consequently a slight loss of relatively fluid impregnant when a vertical sample of cable is heated in an oven for a drainage test. The total amount is quite small and together with the expulsion due to thermal expansion of the compound an amount of 1%–3% is permitted by cable specifications.

Progressive improvements to MIND compounds were made during the 1950s and 1960s. The early compounds contained a high proportion of microcrystalline waxes having a high coefficient of expansion. This meant that during the impregnation cooling process, after the compound had reached its set point, no further compound could feed into the insulation and in the final cooling to room temperature some very small contraction voids were formed. Figure 3.5 shows that the problem was overcome by later developments using different waxes. The coefficient of expansion has been much reduced and the shape of the expansion coefficient–temperature relationship has been changed so that more expansion occurs above the set point temperature of the compound. As the compounds were patented, full details have not been published and this is probably the reason that statements have been made in other countries that non-draining cables suffer from lower breakdown strength and short-circuit performance. UK experience has shown that, although the earlier compounds also had somewhat lower impulse and a.c. breakdown strength than oil–rosin compounds, the long-term performance has been much better because the insulation remained fully impregnated with no loss of impregnant.

Different circumstances apply to high voltage pressure type cables because of the better electrical quality necessary and the fact that the drainage problem does not arise. The pre-impregnated paper construction for gas-filled cables uses compounds with petroleum jelly and polyisobutylene for viscosity control. For fluid-filled cables, the impregnant must be very fluid so that it readily flows from the cable into pressure tanks during heating and back on cooling. Resistance to gas evolution under electrical stress and to oxidation are other key properties necessary. Although highly refined mineral oils were used for many decades, the increasing stress requirement for modern cables has led to a change to the use of synthetic alkylates of dodecylbenzene type.

Impregnated paper dielectrics

Although various important characteristics have been mentioned it is ultimately the electrical properties of the complete dielectric which are paramount. In this respect the effects in the butt-gap spaces between the layers of paper tapes are a critical feature as voltage increases. Such spaces are necessary so that individual papers may slide as the cable is bent but, except for low voltage cables, should be well filled with impregnating compound by careful control of impregnating conditions, particularly to ensure slow cooling. They are inevitably a source of electrical weakness and hence the registration of paper tapes compared with underlying layers should be chosen so that a minimum of gaps is superimposed in a radial direction. Figure 3.6 compares a registration of 35:65 (35% of the tape lying in front of and 65% behind the trailing edge of the tape beneath it) with a registration of 50:50, in which the butt gaps are superimposed every alternate layer. Theoretically, 20:80 registration would be better than 35:65 but as a tolerance is necessary there is a danger of some drift towards 0/100, with a butt gap of double depth.

The electric strength of impregnated paper in sheet form is very high. Impulse strength increases with density, and with impermeability up to an optimum value, and also increases as paper thickness is reduced. A level of 200 MV/m can readily be obtained. Similarly, a.c. strength also increases with density and a level of 50 MV/m is possible. However, other factors have to be considered for high voltage cables, particularly the effects in butt-gap spaces, the relationship between density and dielectric losses and the moisture content.

The relevant requirements also have to be aligned with the operating voltage of the cable. A moisture content up to 1% can be tolerated at 1 kV and 0.2% is satisfactory

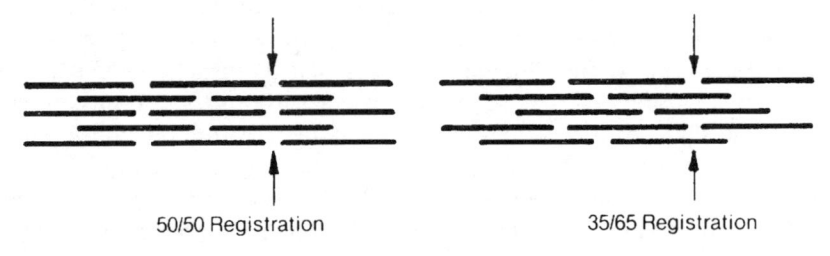

Fig. 3.6 Effect of paper layer registrations on butt-gap alignment: one paper between aligning gaps with 50/50 registration and two papers with 35/65

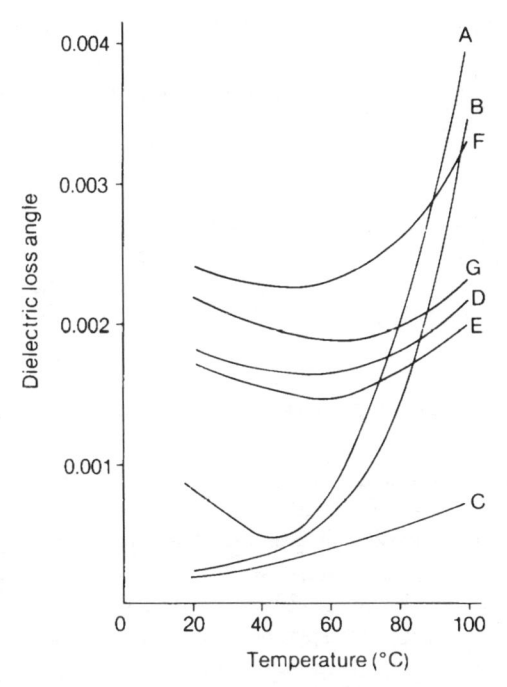

Fig. 3.7 Typical dielectric loss angle versus temperature characteristics of oil–rosin compound (curve A), MIND compound (curve B), FF cable oil (curve C), paper for solid type cables (curve D), paper for FF cables (curve E), impregnated paper for solid cables (curve F), impregnated paper for FF cables (curve G)

for the 10–30 kV range but for transmission cables it needs to be kept below 0.1%. Increase in paper density increases the dielectric losses and also the differential between the permittivity of the paper and the impregnant, so causing greater stresses in the latter. Effects of partial discharge in butt gaps and dielectric losses are of particular importance for transmission cables and are considered in greater detail in chapter 33.

Figure 3.7 illustrates typical examples of the dielectric loss angle versus temperature relationship for dry paper, impregnating compounds and impregnated paper for various types of cables. The actual values may be somewhat displaced according to

Table 3.3 Permittivities of cable components

Material	Temperature	
	20°C	100°C
Oil–rosin compound	2.5	2.4
MIND compound	2.4	2.2
FF cable oil	2.2	2.0
Impregnated paper		
low density	3.3	3.3
high density	3.8	3.8

moisture content. The loss angle and permittivity of the composite dielectric are related to the density of the paper. As the density of the paper increases, and the space occupied by the impregnant decreases, the higher loss angle and permittivity of the paper predominate and hence they increase in the composite material. However, other factors are also involved such as the purity of the paper pulp, to which reference has been made.

Table 3.3 gives similar data for permittivity. The relative permittivity of dry paper is related to the ratio of paper fibres and air, as the permittivity of the fibres is of the order of 5.5 compared with 1.0 for air. With increasing density of paper the fibre ratio increases, and the permittivity of a paper tape is within the range 1.7–3.2 according to density.

POLYMERIC INSULATION AND SHEATHING MATERIALS

In tonnage terms, synthetic polymers have replaced natural materials such as paper, mineral oil and natural rubber for the insulation of distribution and wiring type cables and for the oversheathing of cables in general. The range of polymers available is extensive and variations in chemical composition enable specific mechanical, electrical and thermal properties to be obtained. Where appropriate, these properties may be further modified by the addition of specific fillers, plasticisers, softness extenders, colourants, antioxidants and many other ingredients.

As the variety of polymeric materials has grown, it has become increasingly difficult to place them all within easy general definitions. A polymer is a molecule, or a substance consisting of molecules, composed of one or a few structural units repeated many times. However, this definition embraces not only synthetic materials but also materials such as paper (cellulose).

In the cable industry the term polymeric material is taken to signify polymers which are plastics or rubbers. British Standards define plastics as materials based on synthetic or modified natural polymers which at some stage of manufacture can be formed to shape by flow, aided in many cases by heat and pressure. Rubbers are considered to be solid materials, with elastic properties, which are made from latex derived from living plants or synthetically and used in the manufacture of rubber products.

An elastomer is a material which returns rapidly to approximately its initial shape after substantial deformation at room temperature by a weak stress and release of that stress. In cable technology the terms 'rubber' and 'elastomer' are used synonymously and interchangeably, although 'rubber' to many implies 'natural rubber'.

Plastics may be further divided into thermoplastics and thermosets. A thermoplastic is a material in which the molecules are held together by physical rather than chemical bonds. This means that once the material is above its melting point it can flow. The process is reversible and upon cooling the material hardens. A typical example is polyvinyl chloride (PVC). The molecules in a thermoset are held together by chemical bonds which are not easily broken. This means that on heating the polymer does not soften sufficiently to be reshaped. Typical examples are crosslinked polyethylene (XLPE) and elastomers. Unlike thermoplastics, thermosets are insoluble and infusible, i.e. they will not fuse together. Many thermoplastics may be converted to thermosets by appropriate treatment to induce 'crosslinking', e.g. by the addition of a suitable chemical crosslinking agent or by irradiation. Rubbers for cable insulation and sheath, whether natural or synthetic, are normally crosslinked.

The definition of the terms used in the rubber industry is given in BS 3558: 1980. Common names and abbreviations for plastics and rubbers are given in BS 3502: Parts 1 and 2: 1991.

PHYSICAL PROPERTIES OF THERMOPLASTIC AND THERMOSETTING INSULATING AND SHEATHING MATERIALS

Table 3.4 shows the typical properties of some cable insulating and sheathing materials.

Table 3.4 Physical properties of polymeric materials

Material	Use	Type	Tensile strength (min) (N/mm^2)	Elongation at break (min) (%)	Limiting temperature[a] Rating (°C)	Installation (°C)
Thermoplastic[b]						
Polyvinyl chloride	insulation	TI 1	12.5	125	70	5
Polyvinyl chloride	sheath	TM 1	12.5	125	70	5
Polyvinyl chloride	insulation	TI 2	10	150	70	5
Polyvinyl chloride	sheath	TM 2	10	150	70	5
Polyvinyl chloride	insulation	TI 3	15	150	90	5
Polyvinyl chloride	insulation	TI 4	12.5	125	70	−25
Polyvinyl chloride	sheath	TM 3	10	150	85	5
Polyvinyl chloride	sheath	5	12.5	125	85	5
Polyethylene[c]	insulation	PE 03	7	300	70	−60
Polyethylene[c]	insulation	PE 2	7	300	70	−60
Polyethylene	sheath	TS 2	12.5	300	90	
Ordinary duty elastomeric[b]						
Rubber	insulation	EI 4	5	200	60	−25
Rubber	sheath	EM 1	7	300	60	−25
Heat resistant	insulation	GP 4	6.5	200	90	−25
Hard	insulation	GP 6	8.5	200	90	−25
Low temperature	insulation	EI 6	5	200	90	−40
Flame retardant	insulation	OR 1	7	200	85	−30
Oil resistant, flame retardant	sheath	EM 2	10	300	60	−25
Silicone rubber	insulation	EI 2	5	150	180	5
Ethylene vinyl acetate	insulation	EI 3	6.5	200	110	−25
Crosslinked polyethylene	insulation	GP 8	12.5	200	90	−40

[a] Maximum temperature for sustained operation and minimum temperature for installation of cables from BS 7540: 1994
[b] BS 7655: 1994 and amendments where appropriate
[c] BS 6234 for insulation types of polythylene

Additionally, for sheathing materials, properties such as abrasion resistance, fluid resistance, or flame retardancy are often required.

ELECTRICAL PROPERTIES OF THERMOPLASTIC AND THERMOSETTING INSULATING MATERIALS

Table 3.5 shows the electrical properties of typical insulating materials used in cables.

Compounds with improved electrical properties are required for power distribution cables operating above 3 kV. IEC 502 requires the product of permittivity and dielectric loss angle to be below 0.75 within the temperature range from ambient to 85°C (chapter 20). In addition the DLA at 80°C must not exceed the value at 60°C.

The major reason for these electrical requirements is that when a length of open circuited PVC cable is energised, provided that the voltage is sufficiently high, the insulation will become warm even though no current is passing through the conductor. The effect is caused by dielectric heating and the energy loss is given by the equation

$$\text{power loss per phase} = 2\pi f C U_0^2 \tan \delta \qquad (3.1)$$

where f = supply frequency (Hz)
C = capacitance per core (F/m)
U_0 = voltage to earth (V)
$\tan \delta$ = dielectric loss angle (DLA)

There are significant differences between PVC, XLPE and EPR in this respect.

Table 3.5 Electrical properties of polymeric materials

Material	Type	Volume resistivity (min) at 20°C (Ω m)	Permittivity at 50 Hz	Tan δ at 50 Hz
Thermoplastic				
Polyvinyl chloride	TI 1	2×10^{11}	6–7	0.1
Polyvinyl chloride	2	1×10^{12}	4–6	0.08–0.1
Polyvinyl chloride	TI 2	2×10^{11}	6–7	0.09–0.1
Polyvinyl chloride	4	1×10^{9}	5–6	0.07–0.13
Polyvinyl chloride	5	5×10^{11}	6	0.9
Polyethylene LD	PE 03	1×10^{16}	2.35	0.0003
Polyethylene LD	PE 2	1×10^{16}	2.35	0.0003
Polyethylene HD		1×10^{16}	2.35	0.0006
Polypropylene		1×10^{16}	2.25	0.0005
Elastomeric				
General purpose GP rubber	EI 1	2×10^{12}	4–4.5	0.01–0.03
Heat-resisting MEPR rubber	GP 4	7×10^{12}	3–4	0.01–0.02
Flame-retardant rubber	FR 1	5×10^{12}	4.5–5	0.02–0.04
Flame-retardant rubber	FR 2	1×10^{13}	4–5	0.015–0.035
OFR rubber	OR 1	1×10^{10}	8–11	0.05–0.10
Silicone rubber	EI 2	2×10^{12}	2.9–3.5	0.002–0.02
Ethylene–vinyl acetate	EI 3	2×10^{12}	2.5–3.5	0.002–0.02
Hard ethylene–propylene rubber	GP 6	2×10^{13}	3.2	0.01
Crosslinked polyethylene	GP 8	1×10^{14}	2.3–5.2	0.0004–0.005

BS 7655 Parts 3 and 4 for PVC types; Parts 1 and 2 for elastomeric types; and BS 6234 for polyethylene types

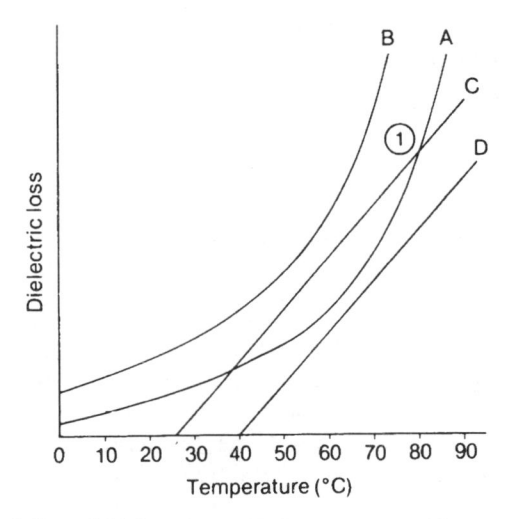

Fig. 3.8 Thermal stability of high voltage PVC compounds illustrated by dependence on the dielectric loss–temperature relationship

Typical values of relative permittivity and tan δ are as follows:

	PVC	XLPE	EPR
Permittivity (at 50 Hz)	6–8	2.3	3.5
Tan δ	0.08	0.0003	0.004

The heat generated in this way is dissipated through the cable into the surroundings and a condition of equilibrium exists between the two. However, if the voltage is increased or if excessive current is allowed to flow through the conductor, or part of the cable is lagged, it may not then be possible to maintain the balance with the result that the temperature of the dielectric continues to rise until the dielectric fails.

Figure 3.8 depicts the relationship between the dielectric loss and temperature. Curves A and B are idealised DLA–temperature curves. As the rate of dissipation of heat to the surroundings is more or less proportional to the temperature rise, lines C and D relate to this condition, starting from two different ambient temperatures. Considering cable A and the rate of heat dissipation shown in C, the rate of heat dissipation will exceed the rate of heat generation until the two curves intersect at (1). Above this temperature heat is generated faster than it can be dissipated and, as the losses further increase, a runaway condition develops. Ultimately, the dielectric will fail. In these circumstances the characteristics of cable B are totally unsatisfactory. Similarly, cable A can only operate at an ambient temperature of 40°C if the cable is derated. Hence, materials for use at high voltage require careful selection.

FURTHER PROPERTIES OF THERMOPLASTIC AND THERMOSETTING INSULATING AND SHEATHING MATERIALS

Thermoplastic materials

Many of the major thermoplastics in use today were developed in the 1930s and although some were used for specialised applications during the 1940s it was not until the early 1950s that polyvinyl chloride (PVC) and polyethylene (PE) came into widespread use for electric cables.

Polyvinyl chloride (PVC)
The basic unit which is repeated in the PVC chain is

$$-CH_2-CH- \atop |\atop Cl$$

PVC polymer cannot be processed by extrusion without the addition of materials which, *inter alia*, act as processing aids, e.g. plasticisers and lubricants. PVC is also strongly polar, because of the C—Cl dipole moment, and the addition of plasticisers also shifts the electrical loss peaks to a lower temperature at constant frequency. The properties of PVC are such that some grades are suitable as insulation materials up to 6 kV.

Apart from the plasticisers, other additions to the PVC resin to produce compounds for electrical applications include fillers and stabilisers. The resin itself is characterised by its molecular weight and freedom from impurities. The most common plasticiser for general purpose compounds is dioctyl phthalate and it is often used in conjunction with a secondary plasticiser such as a chlorinated extender (e.g. Cereclor).

A common filler is calcium carbonate, usually whiting, and it may be coated with a lubricant such as calcium stearate to aid extrudability. It is not just a diluent but increases the resistance of the compound to hot deformation. The most common general purpose stabilisers are lead carbonate and dibasic lead sulphate.

The whole formulation is optimised for mechanical and electrical properties, ease of processing and cost. The requirements for PVC insulation and sheathing compounds are given in BS 7655. In the IEC format they are included in the standards for the particular cable types, e.g. IEC 502.

In addition to the general purpose compounds, other formulations are available for particular applications. Where higher temperature resistance is required, e.g. for the oversheathing of cables with thermosetting insulation, a less volatile plasticiser such as didecyl phthalate is beneficial. For high temperature insulation, di-tridecyl phthalate may be used. For low temperature applications, PVC can be compounded to be flexible at −40°C by omitting fillers and substituting sebacate or adipate ester plasticisers for part of the normal phthalates. It is now more usual to use rubber or other thermosetting materials for such situations. Where PVC is used in power distribution cables operating above 3 kV the filler level is reduced to reduce water absorption under service conditions, and the secondary plasticiser is omitted to reduce the dielectric loss angle. The DLA–temperature curves of PVC compounds suitable for low voltage and high voltage use are illustrated in fig. 3.9.

Coloured oversheaths are sometimes used in accordance with an identification code, and when buried they may darken or change to a black colour. This is caused by a reaction between sulphur compounds in the ground and lead stabilisers, with the formation of lead sulphide. The problem was originally overcome by the use of the more expensive calcium/barium/cadmium stabiliser systems but the elimination of cadmium on health and safety grounds has resulted in modern coloured oversheaths being stabilised by calcium/zinc type systems. While PVC is not as susceptible to

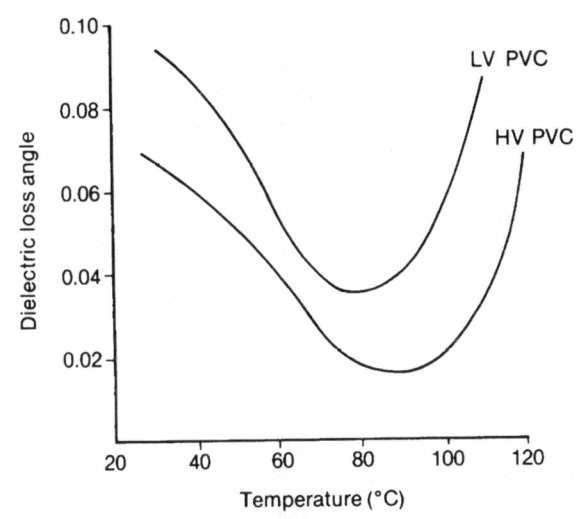

Fig. 3.9 Relationship between DLA and temperature for PVC compounds suitable for low voltage and high voltage uses

degradation by ultraviolet light as the polyolefins it is recommended that black compound should be used wherever possible for service involving exposure to sunlight.

An important field of development has been the formulation of PVC compounds with modified burning characteristics (chapter 6). Resistance to flame propagation can be improved by the use of fillers such as aluminium trihydrate and antimony trioxide, and phosphate and halogenated plasticisers. The concentration of hydrogen chloride released during burning can be reduced by the use of calcium carbonate filler of very fine particle size. These are known as acid-binding PVC compounds. Reduction of smoke generation can be achieved by the use of certain metal salts, but unfortunately a compromise of properties is always necessary.

Polyethylene
Polyethylene is a semicrystalline polymer with the repeat unit —CH$_2$—CH$_2$—. It is available in a variety of grades, differing in chemical structure, molecular weight and density, which are obtained by the use of different methods of polymerisation. High density polyethylenes (HDPE), with densities of 945–960 kg/m^3, have fewer and shorter chain branches than low density polyethylenes (LDPE) having densities of 916–930 kg/m^3. Very low density polyethylenes (VLDPE) have densities of 854–910 kg/m^3. A higher density is a direct consequence of higher crystallinity resulting from reduced chain branching. Among other properties which increase with crystallinity are stiffness and melting point. The crystalline melting point increases from 110 to 130°C as the density increases from 916 to 960 kg/m^3. The crystallinity of LDPE is typically 45%–55% while that of HDPE is 70%–80%.

Two newer classes of polyethylenes which are now being used for cable sheathing are linear low density polyethylene (LLDPE), made by a low pressure process similar to that used for HDPE, and medium density polyethylene (MDPE).

Although the properties of polyethylene are often suitable for sheathing applications, it is more often used as an insulation as it is a non-polar material with excellent electrical characteristics. Much of the small electrical loss (DLA approximately 10^{-4}) is due to impurities such as oxidation products generated during processing or residues from the polymerisation process. LDPE is used for both insulation and oversheathing. HDPE finds fewer applications and is mainly used for oversheathing (chapter 5). When polyethylene was first introduced for oversheathing some failures occurred due to environmental stress cracking but this problem was overcome by the incorporation of 5% butyl rubber. More recently the use of copolymers and higher molecular weight polymers having melt flow indices (MFI) of 0.3 or less (2.16 kg load at 190°C) has also been successful. For insulation grades attention is given to the molecular weight distribution and the purity of the polymers. For many applications the only compounding ingredient is a small amount of antioxidant to prevent oxidation during processing and subsequent use. UK grades are covered by BS 6234.

Normal grades of LDPE have melting points in the range 110–115°C but begin to soften in the 80–90°C region. For this reason the maximum continuous operating temperature of LDPE is limited to 70°C unless it is crosslinked (XLPE), when the operating temperature can be increased to 90°C. HDPE softens above 110°C.

In direct contact with copper conductors or screening tapes, some embrittlement may occur and this can be rapid at elevated temperatures. This is caused by oxidation, which is catalysed by copper, and is inhibited by suitable choice of antioxidant. Polymerised 1,2-dihydro-2,2,4-trimethylquinoline type is used for general purpose applications but for more severe conditions a copper inhibitor of bishydrazine type is also used.

Polyethylenes are also susceptible to oxidative degradation induced by ultraviolet radiation and for this reason the addition is recommended to oversheaths of 2.5% of a fine particle size (20 µm) carbon black.

A wide range of compositions can be obtained by the incorporation of other materials. The addition of fillers is limited because they lead to brittleness but small amounts of butyl or ethylene–propylene rubber may be added to maintain flexibility. An alternative approach is to use an ethylene copolymer, in which case the second component increases the flexibility. Vinyl acetate and alkyl acrylates are widely used as the comonomers and the resulting polymers, when filled with a suitable carbon black, provide a semiconducting grade with reasonable flexibility. Copolymerisation may also give considerable improvement in environmental stress cracking resistance for oversheathing applications. Like all hydrocarbons polyethylene burns readily, which limits its use in applications where flame retardancy is important.

Polypropylene
The repeated basic unit of polypropylene is

$$-CH_2-\underset{\underset{CH_3}{|}}{CH}-$$

Polypropylene is generally considerably harder and stiffer than polyethylene despite recent advances in polymerisation technology which have produced much more flexible polypropylene and interest arises because the commercial grades of the homopolymer do not melt until about 160°C, in comparison with approximately 110–115°C for LDPE and 135°C for HDPE. Polypropylenes are not widely used at present.

Thermoplastic elastomers
Thermoplastic elastomers are processible on thermoplastic and modified rubber equipment and have the feel and recovery of vulcanised rubber. A unique property

balance includes outstanding oil resistance, heat ageing and weather resistance coupled with good general rubber properties. Several types are available, block copolymers and physical blends of thermoplastics or elastomers dispersed in a thermoplastic matrix.

Nylon
The group of nylon materials contain units which are similar to

$$-R'-\overset{O}{\underset{\|}{C}}-NH-R-NH-\overset{O}{\underset{\|}{C}}-R'-\overset{O}{\underset{\|}{C}}-NH-$$

where R and R' are usually alkylene groups. These condensation polymers may be formed from a dibasic acid and a diamine to produce nylon 6, nylon 66 or nylon 11, where the digits represent the number of carbon atoms in the polymer chain. These materials have differing softening points, abrasion resistance, and water absorption and are used in applications where tough, abrasion resistant materials are required (e.g. oversheathing). Some have good resistance to softening up to about 200°C.

Polyurethanes
Polyurethanes are characterised by the urethane group

$$-NH-\overset{O}{\underset{\|}{C}}-O-$$

and are manufactured by the reaction of hydroxyl terminated polyester or polyether prepolymers with isocyanate. Thermoplastic polyurethanes are used for special oversheathing applications because of their excellent abrasion resistance.

Polyester block copolymers (PEE)
Polyester block copolymers consist of a hard (crystalline) segment of polybutylene terephthalate and a soft (amorphous) segment based on long-chain polyether glycols. The properties are determined by the ratio of hard to soft segments and the make-up of the segments. They offer ease of processing, chemical resistance and wide temperature resistance whilst retaining mechanical strength and durability. The commercial name is Hytrel (DuPont).

Fluorinated polymers
Polytetrafluoroethylene (PTFE) is by far the most important member of this group and has a repeat unit of

$$-CF_2-CF_2-$$

The polymer is non-polar and has outstanding electrical properties. It cannot really be considered to be a true thermoplastic because the melt viscosity is so high that extrusion is not possible. The insulation has to be applied to conductors by a cold shaping operation followed by a sintering process to cause the polymer particles to fuse and coalesce. The good electrical properties are combined with very high resistance to chemicals and temperature: service operation up to 260°C is possible.

Because of the high cost and difficulty in processing PTFE, much attention has been given to other fluorinated monomers which are melt processable. Commercial names of materials include FEP (DuPont, TFE copolymer with hexafluoropropylene), Tefzel (DuPont, TFE copolymer with ethylene), Halar (Allied Chemical, ethylene copolymer with chlorotrifluoroethylene), PFA (DuPont, perfluoroalkoxy branched polymers) and Dyflor (Dynamit Nobel, polyvinylidene fluoride).

Thermoplastics in tape form
Some materials are used for insulation, bedding and sheathing as tapes or strings, polypropylene being an example of the latter. Polyethylene terephthalate (PET), under the trade name Melinex (ICI) or Mylar (DuPont), is widely used as a binder tape for multicore cables and as a barrier layer to retard the migration of components such as oils and plasticisers from beddings and sheaths into insulation and to separate insulation from stranded conductors.

Polyimide, e.g. Kapton (DuPont), in tape form is being used increasingly for high temperature applications where light weight is important. It has the structure

Thermoset materials

In the cable industry thermoset polymers have two distinct fields of application. Firstly rubber type materials having intrinsic flexibility or other required property (e.g. oil resistance) need to be crosslinked to cater for the thermal conditions arising when some types of sheath are applied, as well as to provide the required service performance. Secondly crosslinking provides greater resistance to thermal deformation at higher cable operating temperature. The need for crosslinking tends to increase cost and so when a thermoplastic material such as PVC can be used satisfactorily, e.g. for domestic and industrial wiring cables, it is generally preferred.

In general, specifications are prepared for a particular polymer type but some, such as BS 7655, 'Specification of Insulating and Sheathing Materials for Cables', define performance characteristics.

Rubbers and elastomers
Natural rubber (NR)
Natural rubber is a dried or coagulated solid obtained from the latex exuded from certain trees (usually *Hevea brasiliensis*). In the past, crude methods of collection and preparation caused variability in compounding and quality but natural rubber is now produced to well defined specifications, e.g. to the Standard Malaysian Rubber (SMR) scheme. This defines maximum levels for dirt, ash, nitrogen and volatiles, and controls the plasticity.

In chemical terms natural rubber is *cis*-1,4-polyisoprene with the structural unit

$$-CH_2-\underset{\underset{CH_3}{|}}{C}=CH-CH_2-$$

Raw rubber is difficult to extrude satisfactorily. It needs to be masticated to the required viscosity and compounded with fillers, extenders and other additives. Cross-linking of natural rubber is generally achieved by sulphur or sulphur-bearing chemicals and as these materials may react with copper such conductors are usually tinned.

Natural rubber tends to compare unfavourably with synthetic polymers in terms of maximum operating temperature (approximately 60°C) and ageing performance. It is also susceptible to cracking in the presence of ozone and so is rarely used in modern insulation or sheathing materials. Thermal stability in air depends on the method of crosslinking and the antioxidant used. Much testing is devoted to ageing and, by experience, correlations can be obtained between natural and accelerated ageing conditions. In general silicone rubber is the least susceptible to ageing followed in order by EVA, XLPE, EPR, butyl rubber and, finally, natural rubber, which is the most susceptible.

Synthetic versions of *cis*-1,4-polyisoprene polymerised by transition metal catalysis in solution are available but although they are free from dirt and nitrogenous material they do not match natural rubber in all respects. They are usually used as a part replacement.

Styrene–butadiene rubber (SBR)
SBR is an amorphous rubber manufactured usually by the emulsion copolymerisation of styrene and butadiene. The structural units are

$$-CH_2-CH(C_6H_5)- \quad \text{and/or} \quad -CH_2-CH=CH-CH_2- \quad \text{or} \quad -CH_2-CH(CH=CH_2)-$$

The properties of SBR are affected by the styrene-to-butadiene ratio and also by the butadiene configurations (1,2, *cis*-1,4, *trans*-1,4). SBR was used for the same applications as natural rubber and the two were often used in blends according to the economic conditions. Nowadays, SBR tends to be used as a modifier in rubber formulations.

Butyl rubber (IIR)
Butyl rubber is a copolymer of isobutylene (97%) and isoprene (3%), with a structural unit of

$$-CH_2-\underset{\underset{CH_3}{|}}{\overset{\overset{CH_3}{|}}{C}}-$$

It is manufactured in solution with an aluminium chloride catalyst. The isoprene is present to provide suitable sites for crosslinking and this may be effected either by the normal sulphur donor/accelerator systems, or preferably with the dibenzoyl quinone dioxime (dibenzo GMF)/red lead system. The latter gives improved resistance to ageing.

Because the cured butyl rubber is substantially free from unsaturation (carbon–carbon double bonds), it has excellent resistance to ozone and weathering. IIR has low resistance to radiation.

Being suitable for operation to 85°C, compared with 60°C for natural rubber, it was used for higher voltage cables and cables in ships but now has been superseded by EPR because of the latter's ease of processing and improved performance.

Ethylene–propylene rubber (EPR)

The term EPR encompasses a range of polymers which fall into two classes. Firstly the saturated copolymers of ethylene and propylene (EPM) form a minor class in which organic peroxides have to be used for crosslinking. The structural units may be arranged randomly or in blocks:

$$-CH_2-CH_2- \quad \text{and} \quad -CH_2-CH(CH_3)-$$

The second class is formed by the incorporation of a third monomer, a non-conjugated diene, to provide unsaturation for crosslinking by sulphur compounds as well as by peroxides. These materials are termed EPDM, and are available with dicyclopentadiene (DCPD), cyclooctadiene (COD), ethylidene norbornene (ENB) and 1,4-hexadiene (HD) as comonomer. The latter two are now normally preferred because of increased rate of cure.

The use of peroxides for crosslinking provides better long-term resistance to ageing and electrical properties. They decompose under the action of heat to give radicals which react with the polymer and result in stable carbon–carbon crosslinks. Improved properties are obtained through the use of coagents with the peroxides, such as ethylene glycol dimethylacrylate or triallylcyanurate. Dicumyl peroxide is a common crosslinking agent but others are available.

Calcium carbonate incorporated in larger quantities reduces the cost. Better physical properties and reduced moisture absorption are obtained by replacing calcium carbonate with calcined clay. An unsaturated silane, which binds to the filler, also improves mechanical properties. Other additives include red lead to give better water resistance and oils to help processing. These have to be selected with care to obviate reaction with peroxides.

Compounds have also been designed which use other polymers such as polyethylene and polypropylene to provide reinforcement, and these blends when crosslinked by peroxide or the newer silane systems meet all the physical requirements of mineral filled systems but have superior electrical properties.

Because of its superior performance, with suitability for continuous operation at 90°C, EPR has gradually displaced butyl rubber for insulation and is now being considered for oversheaths.

Crosslinked polyethylene (XLPE)
Despite its excellent electrical properties the use of polyethylene has been limited by an upper operational temperature of about 70°C, due to its thermoplasticity. By crosslinking, this constraint is removed and the working temperature is increased to 90°C.

For general purpose low voltage cables it is possible to incorporate up to 30% calcium carbonate into XLPE to reduce the cost. However, to maintain the best electrical properties, especially when immersed in water, the filled compound should not be used.

In the USA, compounds incorporating approximately 30% thermal carbon black are used. These have the advantage of improved resistance to hot deformation and cut-through resistance.

Currently XLPE is primarily used for insulation but in some countries, notably Sweden, it has some application for oversheathing as an alternative to PVC and LDPE.

Crosslinking in pressurised tubes
The most common method of crosslinking is by the incorporation of peroxides into the polymer followed, after extrusion, by heating under pressure to activate the peroxide. If steam is used for heating, the pressure is correspondingly high (18–20 bar) so as to achieve a temperature of the order of 210°C. If an electrically heated tube with inert gas atmosphere is used, the gas pressure needs to be around 5–10 bar to prevent the formation of voids due to peroxide decomposition products.

A suitable peroxide has to be selected to give fast crosslinking without precuring in the extruder, and dicumyl peroxide is widely used. For compounding in-line during the extrusion process, there are advantages in using a liquid instead of a powder. Di-*tert*-butyl peroxide, which decomposes more slowly, is frequently adopted.

The crosslinking mechanism of polyethylene with dicumyl peroxide is illustrated in fig. 3.10. Similar schemes apply to rubbers such as EPR and EVA. The diagram indicates how volatile by-products such as acetophenone, cumyl alcohol, methyl styrene and water vapour are formed. It is also important with peroxide systems to choose the antioxidant system carefully because they may react with the free radicals (equation 6 in fig. 3.10) to reduce the number of crosslinks. Phenolic antioxidants are particularly reactive and only two are in common use: Santonox R (Flexsys) and Irganox 1010 (Ciba).

A very suitable antioxidant is 1,2-dihydro-2,2,4-trimethylquinoline, e.g. Flectol pastilles (Flexsys), but it can increase the dielectric loss angle of XLPE and the phenolic materials are preferred for high voltage cables.

Chemical crosslinking using silanes
A newer process, using the well established technology of silicones, is now replacing the conventional crosslinking procedure for low voltage cables. The basic system (Sioplas) was developed by Dow Corning in the UK in the early 1970s. This is a two-component system for which two materials are first prepared: a crosslinkable graft polymer and a catalyst master batch. These are blended together at the fabricating machine and the product is subsequently crosslinked by immersion in water or low pressure steam.

Typical reaction sequences are shown in fig. 3.11. Because the polyethylene radicals are constantly being regenerated in the (3) and (4) reaction cycles, only small amounts of dicumyl peroxide are required in comparison with peroxide linking under heat and

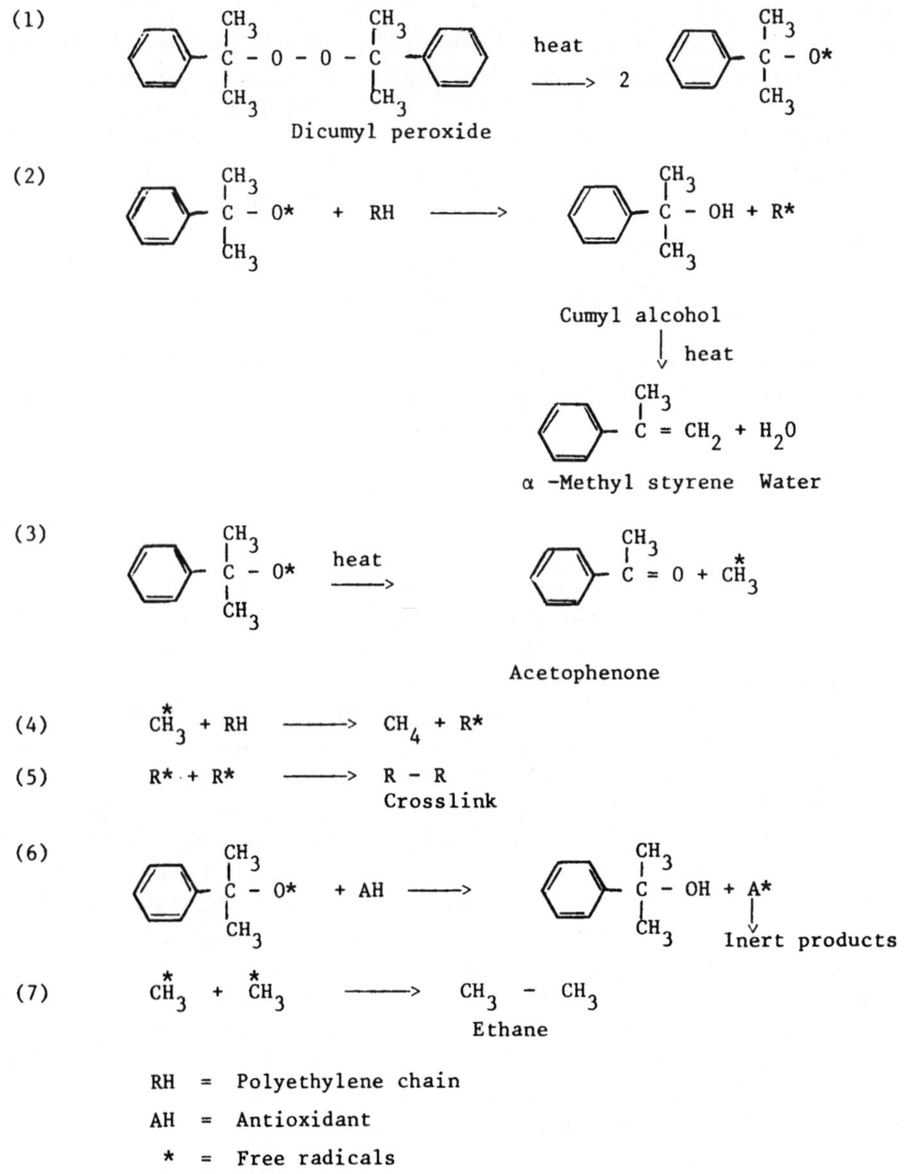

(1) Dicumyl peroxide $\xrightarrow{heat}$ 2 cumyloxy radicals

(2) Cumyloxy radical + RH → Cumyl alcohol + R*

Cumyl alcohol $\xrightarrow{heat}$ α-Methyl styrene + Water

(3) Cumyloxy radical $\xrightarrow{heat}$ Acetophenone + $\overset{*}{C}H_3$

(4) $\overset{*}{C}H_3$ + RH → CH_4 + R*

(5) R* + R* → R–R Crosslink

(6) Cumyloxy radical + AH → Cumyl alcohol + A* → Inert products

(7) $\overset{*}{C}H_3$ + $\overset{*}{C}H_3$ → CH_3–CH_3 Ethane

RH = Polyethylene chain
AH = Antioxidant
* = Free radicals

Fig. 3.10 Schematic crosslinking mechanism for polyethylene with peroxides

pressure (typically 0.1% instead of 2.5%). Constraints concerning antioxidant choice apply equally. Water is the actual crosslinking agent and a relatively large amount is required (3000 ppm) but as polyethylene can only absorb about 100 ppm of water at 80°C, and this is constantly being converted into methanol, the finished XLPE always has a low water content.

The crosslinks formed are of the Si—O—Si type common to silicone rubbers and are thus thermally and hydrolytically stable.

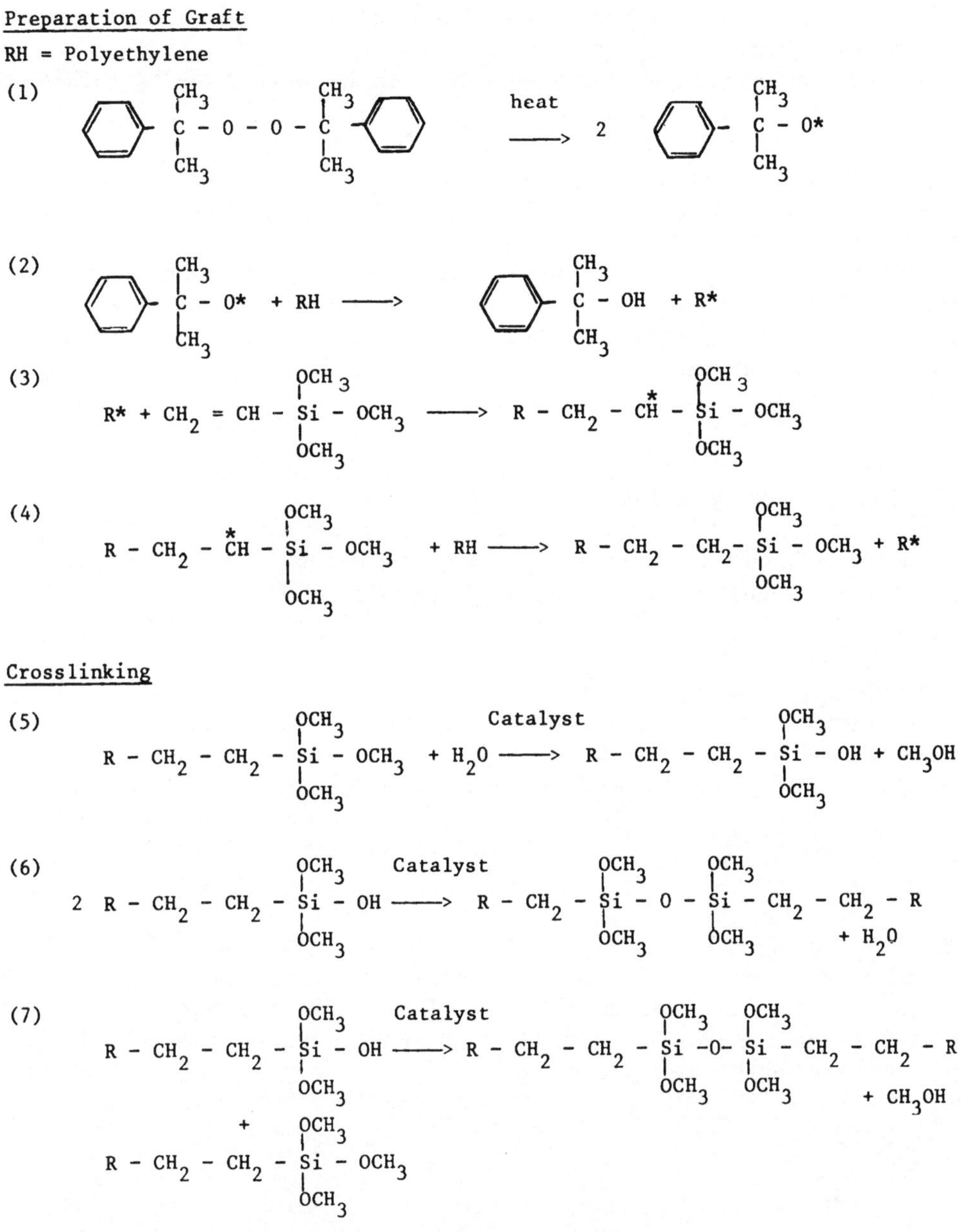

Preparation of Graft

RH = Polyethylene

(1) $(C_6H_5)-C(CH_3)_2-O-O-C(CH_3)_2-(C_6H_5) \xrightarrow{heat} 2 \; (C_6H_5)-C(CH_3)_2-O^*$

(2) $(C_6H_5)-C(CH_3)_2-O^* + RH \longrightarrow (C_6H_5)-C(CH_3)_2-OH + R^*$

(3) $R^* + CH_2=CH-Si(OCH_3)_3 \longrightarrow R-CH_2-\overset{*}{C}H-Si(OCH_3)_3$

(4) $R-CH_2-\overset{*}{C}H-Si(OCH_3)_3 + RH \longrightarrow R-CH_2-CH_2-Si(OCH_3)_3 + R^*$

Crosslinking

(5) $R-CH_2-CH_2-Si(OCH_3)_3 + H_2O \xrightarrow{Catalyst} R-CH_2-CH_2-Si(OCH_3)_2-OH + CH_3OH$

(6) $2\; R-CH_2-CH_2-Si(OCH_3)_2-OH \xrightarrow{Catalyst} R-CH_2-Si(OCH_3)_2-O-Si(OCH_3)_2-CH_2-CH_2-R + H_2O$

(7) $R-CH_2-CH_2-Si(OCH_3)_2-OH + R-CH_2-CH_2-Si(OCH_3)_3 \xrightarrow{Catalyst} R-CH_2-CH_2-Si(OCH_3)_2-O-Si(OCH_3)_2-CH_2-CH_2-R + CH_3OH$

Fig. 3.11 Schematic crosslinking mechanism for polyethylene with silane

It is important to carry out tests to determine the degree of crosslinking achieved because this governs the physical properties of this material. At one time it was usual to use a solvent extraction technique to determine the proportion of uncrosslinked material. However, it became apparent that, because of the different crosslink structures, the alignment between extraction proportion and properties such as high temperature modulus was different for silane and peroxide crosslinked materials. For this reason the extraction method has now been superseded by an elongation test under a fixed load at 200°C.

A further development of the Sioplas materials has been made by BICC and Maillefer. This process (Monosil) introduces all the ingredients together by metering them into the extruder so that the separate grafting stage is eliminated. This ensures that the material is free from contamination and is thus suitable for cables in the 10–30 kV range. It also removes problems due to the limited storage life of the graft polymer and reduces processing costs.

Because water remains as the crosslinking agent and has to diffuse into the material, the cure time required is proportional to the square of the thickness. This places a limitation on extrusion thicknesses which can be handled within a reasonable time scale. Currently, cure times are of the order of 4 hours at 90°C in water for insulation of 2.5 mm thickness.

Crosslinking of silane copolymers
Recently, copolymers of ethylene and vinyl trimethoxysilane have been developed. With the addition of a suitable crosslinking catalyst these materials are cured by immersion in hot water in a similar manner to the grafted silane (fig. 3.11).

Crosslinking by irradiation
Irradiation with fast electrons or γ-rays produces crosslinking by a mechanism which is chemically similar to the use of peroxides. It works by a radical process with hydrogen being evolved. The material to be irradiated comprises only polyethylene with an antioxidant and so there is no problem of pre-cure in the extruder. However, the process is only economically suitable for relatively small insulation thicknesses.

Ethyl vinyl acetate (EVA)
Ethylene copolymers containing small amounts of vinyl acetate have been known for many years. More flexible grades of EVA suitable for cables became available around 1970, including amorphous polymers with vinyl acetate contents in excess of 40%. Such materials have vinyl acetate units in the form

$$-CH_2-CH-\atop{\displaystyle |\atop\displaystyle O\atop\displaystyle |\atop\displaystyle CO\atop\displaystyle |\atop\displaystyle CH_3}$$

The polymers are crosslinked by peroxides or by radiation and the formulation is generally similar to that for EPR. Resistance to ageing in air is slightly superior to that of EPR, service temperatures up to 110°C being possible, but the electrical properties such as permittivity are inferior. EVA is also used as the base for flame-retardant sheaths and extruded semiconducting dielectric screening materials, especially those designed to be strippable from XLPE and EPR insulations.

Silicone rubber

Silicone rubbers have the general chemical structure

$$-\underset{R}{\overset{R}{\underset{|}{\overset{|}{Si}}}}-O-$$

where R is usually methyl or phenyl. Some vinyl groups are also introduced to facilitate peroxide crosslinking, which with silicones, unlike other rubbers, can be carried out in hot air.

Silicones have outstanding properties, especially in resistance to high temperatures, 180°C being the normal operating limit. At ambient temperatures the mechanical properties are somewhat inferior to those of the more commonly used materials, particularly for sheathing applications. Silica fillers are generally used to provide reinforcement.

Chloroprene rubber (CR or PCP)

Polychloroprene, otherwise known as neoprene, was the first commercial synthetic rubber. It has the structure

$$-CH_2-\underset{Cl}{\overset{|}{C}}=CH-CH_2-$$

It has rarely been used by itself for insulation but is often used blended with natural rubber. Its major use is as a very tough flexible sheathing material. Polychloroprene compounds have good abrasion and tear resistance together with good resistance to swelling and to chemical attack by a wide range of natural oils and aliphatic hydrocarbons. They do not normally support combustion.

Chlorosulphonated polyethylene rubber (CSP, CSM)

CSP is obtained when polyethylene is reacted in a solvent with chlorine and sulphur dioxide. The polymer has the following units arranged randomly in the approximate ratio of A:B:C of 45:15:1:

$$-CH_2-CH_2- \qquad -CH_2-\underset{Cl}{\overset{|}{CH}}- \qquad -CH_2-\underset{\underset{Cl}{\overset{|}{SO_2}}}{\overset{|}{CH}}-$$

(A)   (B)   (C)

CSP compounds have superior electrical properties to compounds based on PCP and are particularly advantageous for insulation and sheathing which is required to be oil-resisting. CSP also has good resistance to ozone and weathering.

When blended with EVA or EPR and filled with a suitable carbon black, CSP compounds provide a strippable dielectric screening material for XLPE and EPR cables in the 10–30 kV range.

Acrylonitrile–butadiene rubber (NBR/PVC blends)

The copolymerisation of acrylonitrile with butadiene produces a range of polymers characterised by good oil resistance. They have the structural units

$$-CH_2-CH=CH-CH_2-$$

and

$$-CH_2-\underset{\underset{CN}{|}}{CH}-$$

and

$$-CH_2-\underset{\underset{\underset{CH_2}{\|}}{CH}}{\overset{}{CH}}-$$

The addition of PVC improves resistance to ozone, weathering and abrasion. By suitable choice of plasticisers improved processability and flame retardance are also obtained. These materials are used solely for sheathing.

Fluorocarbon rubbers

The fluorocarbon rubbers find application for sheathing where very good resistance to oils is required at high temperatures. The best known material is a copolymer of vinylidene fluoride and hexafluoropropylene (Viton).

Ethylene–acrylic elastomers (EMA)

Ethylene–acrylic elastomers are heat- and oil-resistant non-halogen synthetic rubbers which can be compounded to resist ignition in the presence of flame and have low smoke generation when burned. They are suitable for service temperatures of 40–170°C. The commercial name is Vamac (DuPont).

SUMMARY

Tables 3.6 and 3.7 summarise the main uses of thermoplastic and thermoset materials in cables.

ENVIRONMENTAL DEGRADATION OF POLYMERS

There are many factors which determine the maximum operating temperature of a material. For some thermoplastics, such as polyethylene, the main determining factor is resistance to deformation. For most thermosets and indeed some thermoplastics, service life is determined by the susceptibility of the material to thermal degradation at elevated temperatures. The most common form of degradation is that due to heat and oxygen. Hydrocarbon polymers oxidise thermally and, because the process is autocatalytic, large changes in mechanical and electrical properties usually accompany the onset of oxidation. The chemical process is illustrated in fig. 3.12. The number of

Materials Used in Cables

Table 3.6 Summary of use of thermoplastics in cable applications

Material	Conductor temperature rating (°C)	Additives	Electrical properties	Comments
Polyvinyl chloride	70–90	Plasticisers, fillers	Plasticisers and polar nature of PVC increase DLA and limits use as insulation to 6 kV. Used as sheathing for higher voltage cables	Most insulation grades rated at 70°C. Burns with dense smoke. Used as insulation and sheathing
Polyethylene	70	Antioxidant, carbon black	Good but limited by low melting point	Susceptible to UV degradation and burns readily unless filled which restricts electrical properties and flexibility. Mainly used as a carbon black filled sheathing material
Polypropylene	80	Antioxidant	Good, but use limited by other factors	Less flexible than most polyethylenes, very susceptible to UV degradation, so only used as a bedding material
Nylon	Not applicable		Polar nature of nylon increases DLA	Used as a sheathing material where abrasion resistance or resistance to high temperatures, e.g. 150°C is required

Table 3.7 Summary of use of thermosets in cable applications

Material	Conductor temperature rating (°C)	Typical additives	Electrical properties	Typical properties when crosslinked	Crosslinking system
Ethylene–propylene elastomers	90–105	Processing aids, fillers, antioxidant	Non-polar polymer, suitable for LV and MV insulations	Used as insulation and sheathing, flexible, low and high temperature grades available	Peroxide, often enhanced by coagents
Polyethylene	90	Antioxidant	Non-polar polymer, suitable for LV, MV and EHV applications	Less flexible than ethylene–propylene	Peroxide, silane or irradiation
Ethylene vinyl acetate	110	Processing aids, fillers	Polar polymer, higher DLA than ethylene–propylene or polyethylene, suitable for LV or sheathing materials	Compound properties depend on grade, may be used in low smoke and fume applications, or as sheathing materials with enhanced properties	Peroxide, silane or irradiation
Silicone rubber	180	Silica fillers	Good, but use limited to LV	Mechanical properties at ambient temperatures are inferior to other thermosets, but material used as insulation in high-temperature environments	Peroxide

A. Initiation

$$\text{Polymer} \longrightarrow \text{R}^\bullet$$
$$\text{Polymer radical}$$

B. Propagation

$$\text{R}^\bullet + \text{O}_2 \longrightarrow \text{ROO}^\bullet$$

$$\text{ROO}^\bullet + \text{RH} \longrightarrow \text{ROOH} + \text{R}^\bullet$$
$$\text{Polymer}$$

$$\text{ROOH} \longrightarrow \text{RO}^\bullet + {}^\bullet\text{OH}$$

C. Termination

$$\text{ROO}^\bullet \longrightarrow \text{unreactive products}$$

$$\text{ROO}^\bullet + \text{HA} \longrightarrow \text{ROOH} + \text{A}^\bullet$$
$$\text{Antioxidant} \qquad\qquad \text{Inactive product}$$

Fig. 3.12 Auto-oxidation mechanism of hydrocarbon polymers

radicals generated multiplies through the decomposition of the hydroperoxide (ROOH). The most efficient method of breaking the multiplication process is through the use of suitable antioxidants. Termination reactions can result in both polymer crosslinking and chain scission. Chain scission does not necessarily result in the destruction of radicals:

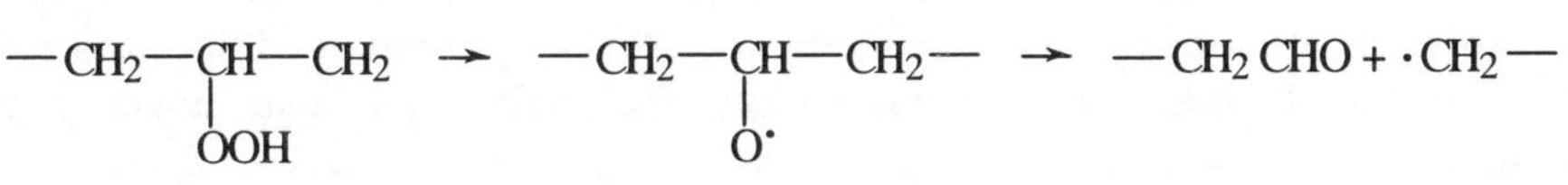

$$-\text{CH}_2-\underset{\underset{\text{OOH}}{|}}{\text{CH}}-\text{CH}_2- \;\longrightarrow\; -\text{CH}_2-\underset{\underset{\text{O}^\bullet}{|}}{\text{CH}}-\text{CH}_2- \;\longrightarrow\; -\text{CH}_2\text{CHO} + {}^\bullet\text{CH}_2-$$

The oxidative stability of a polymer depends upon the availability of hydrogen atoms which can be abstracted by the peroxy (ROO$^\bullet$) radicals. This depends upon two factors: the carbon–hydrogen bond dissociation energies and the steric position of the labile hydrogens. Some typical bond dissociation energies of hydrocarbons are given in table 3.8. Thus polyisoprene (natural rubber) oxidises more easily than polyethylene:

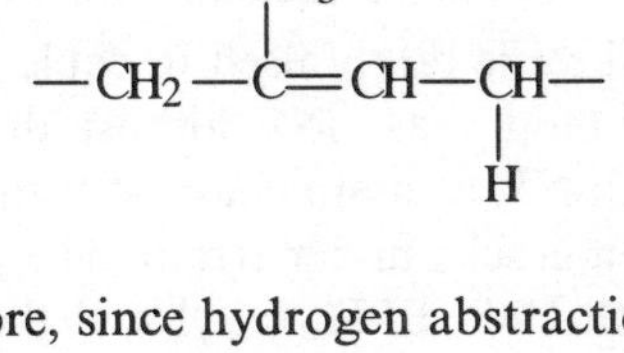

$$-\text{CH}_2-\underset{\underset{\text{H}}{|}}{\overset{\overset{\text{CH}_3}{|}}{\text{C}}}=\text{CH}-\text{CH}- \quad \text{less stable than} \quad -\text{CH}_2-\underset{\underset{\text{H}}{|}}{\text{CH}}-$$

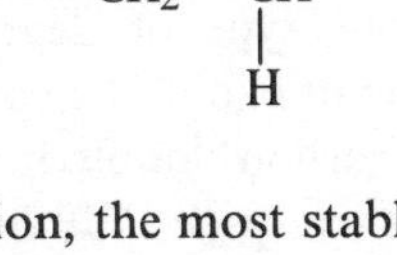

Furthermore, since hydrogen abstraction is the key reaction, the most stable polymers will be those which either have no hydrogen atoms or other labile groups

Table 3.8 Bond dissociation energies

R—H	Dissociation (kJ/mole)
CH$_3$—H	435
C$_6$H$_5$—H	435
CH$_3$CH$_2$CH$_2$—H	410
(CH$_3$)$_2$CH—H	393
(CH$_3$)$_3$C—H	381
CH$_2$=CH—CH$_2$—H	356

(e.g. PTFE) or have hydrogen atoms which are relatively inert in methyl or phenyl groups (e.g. silicones). Antioxidants have removable hydrogens, usually in amine or phenol moieties. However, the resultant antioxidant radicals are relatively stable and propagation ceases.

Accelerated ageing tests are often performed to assess the thermal endurance properties of the material. Guidance on these tests for electrical insulation materials is given in IEC 214. In these tests it is often found in practice that a plot of the logarithm of material life against the reciprocal of the absolute temperature is linear. However, it is unlikely that extrapolations can be readily justified, as the physical manifestation of degradation is often the result of many reactions of differing importance at high and low temperatures. For example, at low temperatures chain scission predominates in the degradation of polyethylene in the early stages with crosslinking predominating after long time periods. At higher temperatures crosslinking occurs from the start. Also, as oxygen is mainly taken up by the amorphous regions, the life of semicrystalline polymers such as polyethylene may be considerably increased below their melting points (about 100°C). The diffusion of air through cables is relatively slow and for this reason ageing studies on materials should always be accompanied by studies on complete cables. Contact with incompatible materials may also affect ageing performance.

In certain polymers other chemical reactions occur alongside those of chain scission and crosslinking outlined above. For example, PVC compounds evolve hydrogen chloride at high temperatures and, using time to first evolution as the failure criterion, Arrhenius plots can be constructed. In this type of polymer metallic compounds are often used as the chain terminators.

Polymers differ in their susceptibility to ultraviolet radiation. The problem is usually confined to a thin surface craze but cracks can propagate through the harder materials such as polypropylene. The addition of suitable carbon black provides adequate screening. The organic chromophore type of UV stabiliser is rarely used in cable sheathing.

Finally, one type of degradation, once common, has now almost disappeared. Cracking due to ozone is a common problem with NR as main chain scission occurs at the carbon–carbon double bonds which results in cracks under strain. The problem is much less severe with PCP and non-existent with CSP, EPR, butyl and the common thermoplastics.

CHAPTER 4
Conductors

Conductors are designed to conform to a range of nominal areas in graduated steps. Outside North and South America, and countries following their techniques, the rest of the world now adopts essentially common practices and standards based on IEC 228, i.e. metric (mm^2) sizes.

American usage continues to be based on American Wire Gauge (AWG) up to 4/0, i.e. 107 mm^2, and larger sizes are in thousand circular mils (kcmil). A 'mil' is 1/1000 inch and circular mils represent the area of an equivalent solid rod with a diameter expressed in mils.

It is important to bear in mind the connotation of 'nominal' as applied to the standard sizes because it is impracticable for manufacturers to produce conductors to precise areas. One factor is that in a stranded conductor the current flows essentially along the wires and as the outer wires are longer than the conductor, due to helical application of the wires, they have increased resistance. As a very large number of sizes of conductor have to be made it is uneconomical to produce special wire sizes for each conductor area and it is necessary to limit the total number of wire sizes handled. This always has to be taken into account in revision of conductor standards. Manufacturers, therefore, adjust their wire sizes and manufacturing processes to meet a specified maximum resistance rather than an area. For this purpose it is also necessary in specifications to allow manufacturers to choose the number of wires, within limits, for shaped and compacted conductors and for flexible conductors. The effective electrical areas are thus based on the maximum d.c. resistance and are slightly different from the nominal areas. The divergences are upwards and downwards on individual sizes and there is no regular pattern.

The overall system has evolved over many years with arbitrary rules. Until 1978 the IEC practice provided different maximum resistance values for single-core and multicore cables, thus allowing for the increased conductor length due to laying-up the cores. This has now changed, however, and a single maximum d.c. resistance is specified for each size of conductor of a given material. American specifications still adhere to the pattern of a 'nominal' resistance, together with a tolerance to provide a maximum resistance for single-core cable and a further tolerance for multicore cable. These tolerances also vary with conductor classification.

Table A4.1 in appendix A4 shows the standard conductor sizes for fixed wiring cables included in IEC 228 and the maximum resistances for three conductor materials. These resistances also apply to the size range which is applicable for solid conductors, i.e. essentially 0.5–16 mm² for copper and 1.5–300 mm² for aluminium. Table A4.2 provides similar information for flexible conductors and applies to conductors having two classifications of flexibility, covered by the standard.

Table A4.3 indicates corresponding information for fixed wiring cables to a typical American specification. In trying to select cables which correspond between metric and American practice much care is necessary, however, and table A4.3 merely extracts data from one of three tables in the specification to which reference is made in the footnote. A column has been included with the heading 'equivalent metric area' and this is merely an arithmetic conversion from a 'nominal area'. As explained above the nominal area has no exactitude and in choosing corresponding sizes it is necessary to work from the specified maximum resistances.

Other points in the metric–American comparison are as follows.

(a) Table A4.3 caters only for stranded conductors and different resistances apply to solid conductors.
(b) There are many classifications of conductors in American practice, each with different standard resistances. Classes B, C and D cover power cables, C having more flexibility than B, and D still further flexibility. All three classes have 'concentric-lay' conductors (see later). Classes G, H, I, K, M cater for 'rope-lay' or 'bunch-stranded' flexible cords and cables with varying flexibility for particular applications. Table A4.3 is based on classes B to D and so illustrates only a small proportion of the total range.
(c) The tolerances between 'nominal' and 'maximum' resistances also vary with some types of conductors and cables.

For sizes up to 16 mm² and for higher voltages, where electrical stress at the conductor surface is important, the construction is circular. For the majority of multicore power cables up to 11 kV a sector shape is used (fig. 4.1) to keep the cable dimensions to a minimum. The corner radii are adjusted according to the cable voltage for reasons of electrical stress and influence on the dielectric during bending. A 'D' shape is frequently used for 2-core cables and an oval construction for 33 kV paper insulated cables, the latter largely to provide better uniformity of frictional forces between layers of paper.

In this chapter the references to resistance are mainly concerned with d.c. resistance. The a.c. resistance is also very important in many applications and is covered in chapter 2.

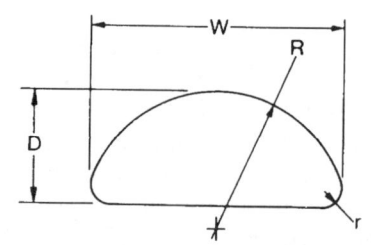

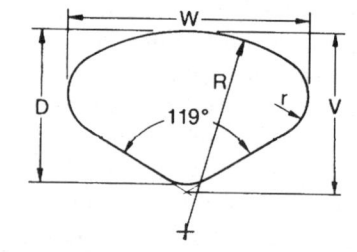

Fig. 4.1 Shapes for multicore power cables: W, width; D, depth; R, back radius; r corner radius; V, V gauge depth

INTERNATIONAL STANDARDISATION

British practice is based on BS 6360: 1991, 'Conductors in insulated cables', which implements CENELEC HD 383 S2, and which is in line with IEC 228 and IEC 228A. As IEC 228 reflects practice throughout the world (outside the USA), it can be taken as a basis for comments.

IEC 228 covers conductors from 0.5 to 2000 mm^2 and specifies the maximum d.c. resistance values for insulated conductors and flexible cords generally but excluding pressurised transmission cables. Class 1 for solid conductors and Class 2 for stranded conductors both deal with single-core and multicore cables for fixed installations and define identical maximum resistance values for each size and type in copper, metal-coated copper and aluminium.

Class 3 and 4 of the previous editions have been deleted. Classes 5 and 6 cater for flexible copper and metal-coated copper conductors of two ranges of flexibility. The maximum resistances in the two classes are the same for each, but differ from those of conductors for fixed installations.

The wires used in stranded conductors are specified by designation of a minimum number of wires in the conductors for fixed installations. This minimum number is varied according to whether the conductor is circular (non-compacted), circular (compacted) or shaped. For circular non-compacted conductors all wires must be of the same diameter but a ratio of 2:1 is permitted for the other two types. For flexible conductors the control is by specification of the maximum diameter of the wires.

A new feature of the 1978 edition of IEC 228 was the inclusion of metal-coated aluminium, with the object of catering for copper-clad aluminium. However, IEC 228 Amendment 1: 1993 again restricted aluminium conductors to plain aluminium or aluminium alloy.

From the fact that conductor resistances for single-core and multicore cables are the same, it may be considered that the resistances are now more arbitrary than previously. Originally the resistances were derived from the assumption that, for example in the case of stranded copper conductors, the wire had 100% International Annealed Copper Standard (IACS) conductivity and various factors were applied, e.g.

$$R = \frac{4A}{n\pi d^2} K_1 K_2 K_3 \qquad (\Omega/\text{km}) \qquad (4.1)$$

where R = maximum conductor resistance at 20°C (Ω/km)
A = volume resistivity at 20°C ($\Omega\,\text{mm}^2$/km)
n = number of wires in conductor
d = nominal diameter of wires in conductor (mm)

K_1, K_2 and K_3 are constants of specified values. K_1 varies with wire diameter and allows for tolerances on the nominal diameter and the effect of metal coating when appropriate. K_2 allows for the fact that in stranding the length of individual wires is longer than the length of the finished conductor. K_3 allows similarly for laying-up the cores in multicore cable.

When effects of actual wire conductivity and compacting were superimposed, the cablemaker still had to derive his own requirements for input wire diameter and number of wires and the allowances for K_1, K_2 and K_3 had limited practical significance in the published form.

The resistances of copper and aluminium vary significantly with temperature and measured values on cables need to be corrected to the specified values at 20°C. At this temperature, the temperature coefficient of resistance per degree Celsius for copper is 0.00393 and for aluminium is 0.00403. For most purposes the value for each metal can be taken as 0.004 and table A4.4 (appendix A4) provides factors to convert to or from 20°C over the range 5–85°C. More exact formulae for conversion of a measured conductor resistance to a basis of 20°C and a length of 1000 m are as follows.

$$\text{Copper:} \quad R_{20} = R_t \frac{254.5}{234.5 + t} \frac{1000}{L} \quad (\Omega/\text{km})$$

$$\text{Aluminium:} \quad R_{20} = R_t \frac{248}{228 + t} \frac{1000}{L} \quad (\Omega/\text{km})$$

where t = conductor temperature (°C)
 R_t = measured resistance (Ω/km)
 L = cable length (m)
 R_{20} = conductor resistance at 20°C (Ω/km)

BRITISH STANDARDS

Whilst IEC 228 prescribes the basic essentials in terms of the finished resistance and control of the wire diameters used, British practice has always provided some additional requirements:

(a) properties of the wire or material from which the conductor is made;
(b) restrictions applicable to joints in individual wires and solid conductors;
(c) control of the dimensions of shaped solid aluminium conductors to ensure compatibility with sleeves fitted for compression jointing.

Largely for the purpose of providing guidance to engineers involved in the preparation of standards for conductors, some further information on the principles of standardisation and the use of formula (4.1) was provided in BS 6360: 1991. This refers to a 'conceptual' construction of conductors and provides more data on the sizes used in standards. These data comprise for non-compacted stranded conductors a number and nominal diameter of wires, nominal conductor diameters and weights. Stranded copper conductors for fixed installation and for flexible cables are included, together with plain aluminium stranded conductors for fixed installation. Further reference is also made to formula (4.1) and the values for the constants K_1, K_2 and K_3.

CONDUCTOR DIMENSIONS

When conventional practice for joining conductors and terminating them on appliances was by soldering for the larger conductors and use of pinch-screw type fitting for small conductors, no problems arose in relation to the tolerance between conductor and ferrule or tunnel diameter. However, with the introduction of compression and crimping techniques for straight ferrules and termination lugs, the clearance between conductor and fitting became important. Cases arose when conductors and fittings from one manufacturer were not compatible with those from another, e.g. the conductor was too large to enter the bore of the ferrule.

The problem was tackled nationally and internationally and the outcome was the issue in 1982 of a supplement (IEC 228A) to IEC 228 which provided what is termed a 'Guide to the dimensional limits of circular conductors'.

The tables included in BS 6360: 1991 are reproduced in appendix A4, tables A4.5 and A4.6, and some further comments, largely relating to circular conductors, are printed under the tables. BS 6360 has a cross-reference to BS 3988 which gives dimensional data for shaped solid aluminium conductors used in cables covered by British Standards. The data in BS 3988 apply to conductors for use in making the cables and not directly to conductors in the manufactured cables. However, the dimensions of the conductor envelope will not increase during cable manufacture and hence are a good guide for connectors to fit the conductors.

As shaped stranded copper and aluminium conductors are normally circularised before connection it is considered unnecessary to standardise dimensional limits for these conductors.

STRANDED CONDUCTORS

Conventional practice is illustrated in fig. 4.2(a) in which there is a centre wire and then concentric layers of nominally 6, 12, 18, 24, 30, 36 and 42 wires – hence the term 'concentric-lay' conductor often used in American terminology. In practice fewer wires are used in the outer layers, e.g. 17 rather than 18 in fig. 4.2(a), due to compression/compaction. Usually each layer is applied with alternate direction of lay as this provides the most stable construction with resistance to 'birdcaging'. For some special applications unidirectional lay may be used. Instead of one wire there may be three or four wires in the centre.

However, it is more usual nowadays to pass such conductors through shaping rolls or pull them through a die to obtain some compacting (fig. 4.2(b)). This has the advantage of providing a smoother surface and reducing the overall size of the cable. Sector conductors are basically stranded in the same way and the shape is obtained by passage through shaping rolls (fig. 4.3) preferably after application of each layer of wires. With copper conductors work hardening in compacting has to be kept to a minimum. This factor also limits the degree of compaction to a normal range of 85–90% (expressed against the volume of the circumscribing envelope). At higher degrees of compaction the extra copper required to meet the specified resistance may be greater in cost than the saving in material usage outside the insulation obtained by the reduced conductor diameter.

Compacted conductors normally have fewer wires than uncompacted conductors and this is one reason why manufacturers deprecate the former practice of denoting conductor sizes by their wire formation, e.g. 37/1.78 mm. Nevertheless, as already mentioned, it is not practicable for manufacturers to use an ideal wire diameter for every single conductor and it is very important to keep the total number of wire sizes to a minimum. This applies particularly when there is a storage system for wire on bobbins. Figure 4.4 illustrates the automatic loading of a bobbin of wire on to the strander carriage.

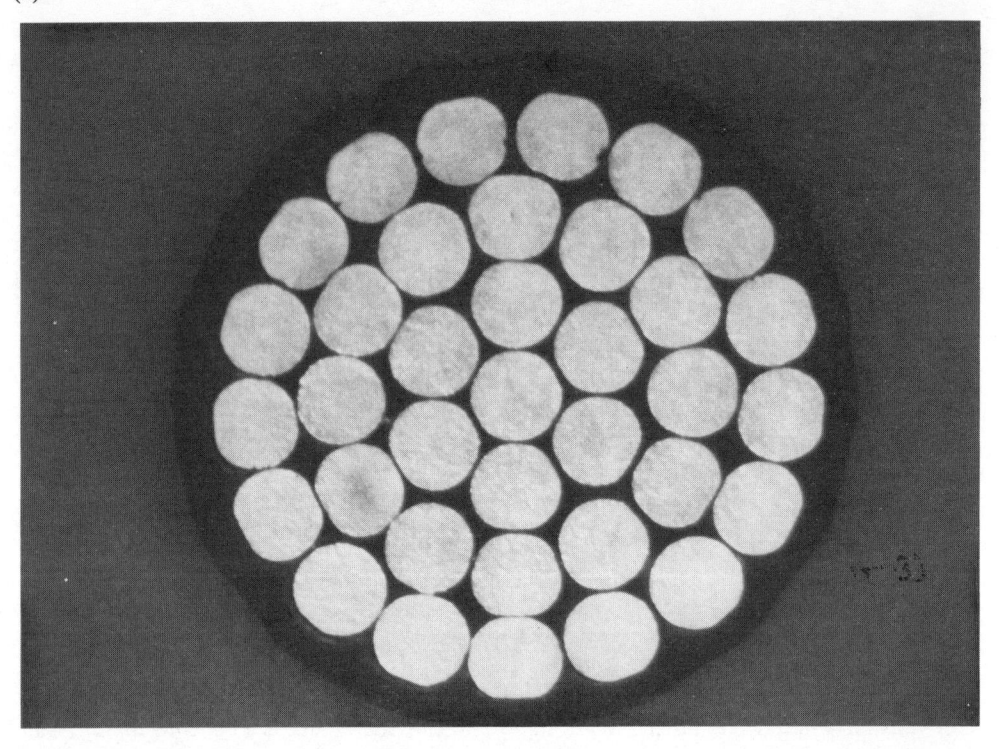

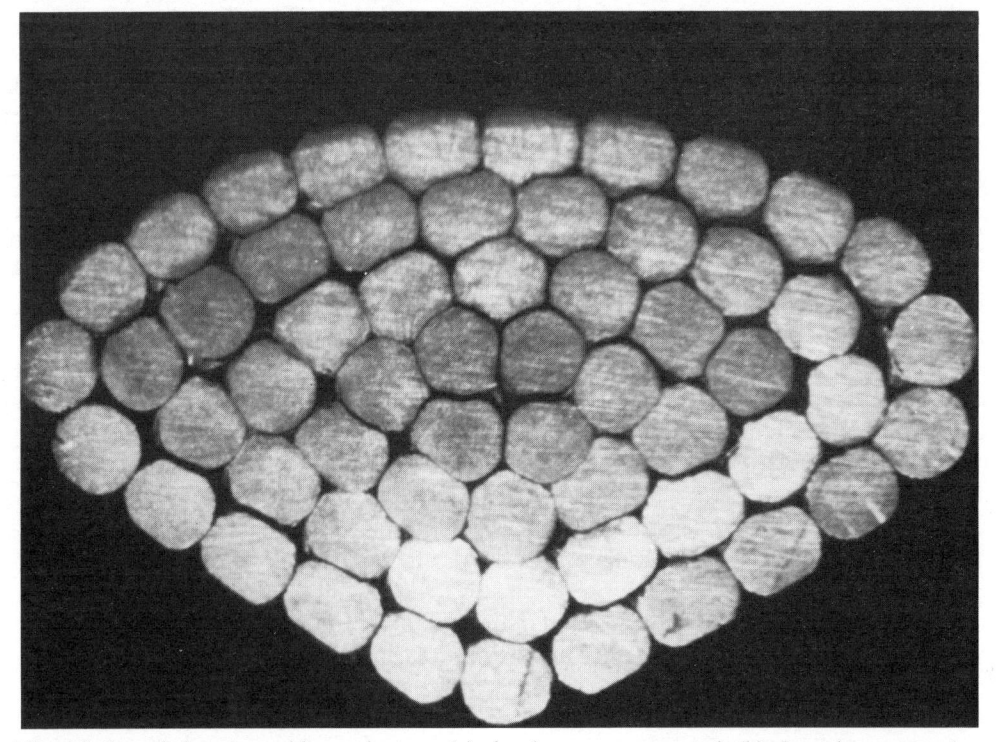

Fig. 4.2 Stranded power cable conductors: (a) circular non-compacted; (b) shaped compacted

Fig. 4.3 Rotating shaping and compacting rolls for power cable conductors

The rolls used for shaping and compacting normally rotate around the conductor to impart a pre-spiral lay (fig. 4.5). This avoids the need to twist the cores when laying-up into the multicore form, thus providing a more stable construction with less possibility of damage to the insulation, both in manufacture and in subsequent bending.

Milliken conductors

With alternating current there is a tendency for more of the current to be carried on the outside of the conductor than in the centre (skin effect), and to overcome this problem the larger sizes of conductor are frequently of Milliken construction (fig. 4.6). Such conductors are formed from several individual sector shapes, usually four for power distribution cables and six for hollow-core fluid filled cables. A thin layer of paper or other suitable insulation is applied over alternate sectors. There is insufficient economic advantage to use this construction below 900 mm^2 but the Milliken design may also be used to obtain increased conductor flexibility. In the USA such conductors are often termed to be of 'segmental' construction.

FLEXIBLE CONDUCTORS

Flexibility may be achieved by stranding the conductors as described above and merely using a larger number of smaller diameter wires. The smaller the wire diameter for a

Fig. 4.4 Automatic loading of a bobbin of wire on to the head of a stranding machine

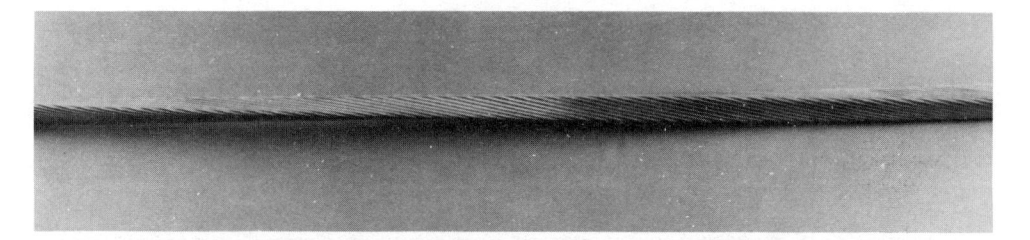

Fig. 4.5 Shaped conductor with pre-spiralled lay

Fig. 4.6 400 kV fluid filled cable with Milliken conductor

conductor of given area, the greater is the flexibility, and hence the main specification requirement for flexibility is the maximum diameter of wire used. IEC 228 does not go beyond this but some specifications differ between concentric-lay and rope-lay or bunch-stranding.

Concentric-lay refers to the conventional stranding practice in which there is a single wire or a group of wires in the centre and then successive complete layers of wires are applied on each other. Normally the direction of application is reversed between layers but in some special cases it may be the same, i.e. unidirectional stranding. In bunched conductors the total number of individual wires is merely twisted together, all in the same direction and with the same length of lay. The position of any one wire with respect to any other is largely fortuitous. The construction may be used for both flexible cords and cables, typically with up to 56 wires.

There are limitations to the total number of wires which can be twisted together to form a single satisfactory bunch and so for the larger conductors small individual bunches are stranded together in layers, i.e. 'multiple bunch stranded'. It is common in cable conductors for the direction of lay in the bunch and in the layer to be the same and for unidirectional stranding to be used. American practice, however, is to reverse the direction of lay in alternate layers.

Fig. 4.7 600/1000 V, 4-core unarmoured cable with solid aluminium conductors and XLPE insulation

SOLID CONDUCTORS

Solid copper conductors are permitted up to 16 mm^2 in British Standards, but with the exception of mineral insulated cables are rarely used above 6 mm^2 because of reduced flexibility. IEC 228 caters for solid conductors up to 150 mm^2 for special applications.

The position is quite different with aluminium as used for power distribution cables. There is an economic advantage for using solid conductors up to the maximum size which provides a cable which can be readily handled and also a technical advantage in that both soldering and compression jointing are easier and more reliable with the solid form. Solid conductors cause some increase in stiffness but this is not usually significant up to 185 mm^2 with paper insulation or 240 mm^2 with plastic insulation. In the case of paper insulated cables the use of shaped solid conductors is limited to 600/1000 V because no satisfactory way of pre-spiralling them has been found. Solid conductors may be circular or sector shaped up to 300 mm^2. For larger sizes of circular conductor the stiffness problem is overcome by the solid sectoral construction comprising four 90° sector shaped sections laid up together. There are six sizes between 380 and 1200 mm^2.

The use of solid conductors, mostly of sector shape, has been extensive in the UK for 600/1000 V cables with paper and polymeric insulation (fig. 4.7), but less so throughout the rest of the world. For economic reasons the corner radii are relatively sharp. The German specification requires larger radii and this design is more suitable for 11 kV cables with XLPE insulation. Conductor shape is important because ferrules for compression fitting need to fit closely on the conductor.

CONDUCTORS FOR TRANSMISSION CABLES

Most of this chapter has been concerned with conductors for solid type cables. The special requirements for pressure cables are covered in part 4.

USE OF ALUMINIUM IN RELATION TO COPPER FOR DISTRIBUTION CABLES

Usage of aluminium has varied and still varies from country to country. UK utilities having rationalised on aluminium in the early 1960s have remained with it for almost all their cables, other than small sizes of the house-service type. The relative prices of copper and aluminium frequently vary but over many years copper cables have rarely been cheaper than aluminium. Much, of course, depends on cable type and size and the nature of the complete installation.

For industrial applications the position is similar in that large users have been better able to deal with the different circumstances arising and have adopted solid aluminium conductor cable with PVC or, increasingly, XLPE insulation. But as the cost of joints and terminations tend to be higher than for copper cables and there are more of them in relation to cable length, small industrial users have tended to stay with copper. Space availability in terminal boxes or motors and distribution boards has also been a factor: more space is required to deal with solid aluminium conductors and it is necessary to stipulate the use of aluminium when ordering the equipment. BS 5372: 1976, 'Cable terminations for electrical equipment', was issued in an attempt to provide guidance in this respect. Many other detailed points arise such as making alternative or appropriate arrangements for dealing with pinch-screw fittings and, whilst all can be dealt with satisfactorily, it is only economic to devote the extra effort when large-scale installations are involved.

Following the successful experience in public supply systems in the UK, similar usage of aluminium became the general practice throughout Europe and today aluminium is predominant in the USA and throughout much of the world. A notable contribution to this has been the development of compression jointing instead of soldering. For industrial applications, however, comparatively little aluminium is now used.

COPPER-CLAD ALUMINIUM CONDUCTORS

Large-scale attempts were made in the USA, and also more particularly in India, to use aluminium conductors in general house wiring type systems. Owing to overheating in accessories having pinch-screw type connections, the results were unsatisfactory and much of the cable had to be replaced to avoid excessive maintenance cost.

Copper-clad aluminium conductors were developed mainly to overcome these problems but they have had only very limited service. Operational experience was satisfactory, the main disadvantage being that the increase in size over a copper cable of equal rating caused conduit occupancy to be reduced. They are now no longer included in IEC 228.

For power cables there is no justification for increase in conductor cost to obtain more satisfactory jointing. Another factor with stranded conductors of the larger sizes is that with the conventional plant used for stranding there is a possible danger of local damage to the copper surface and any exposure of aluminium could lead to corrosion due to the bimetallic effect.

SODIUM CONDUCTORS

Because of the low density and price of sodium, coupled with its reasonable conductivity of 35% IACS, interest developed in the early years of cable making and patents were first filed in 1905. It was not until the 1960s, however, that experimental cables were made in the UK and the USA. The significant aspect was that by then

Table 4.1 Comparative data on conductor metals (sodium taken as 100)

	Sodium	Aluminium	Copper
Density	100	280	910
Weight/unit resistance	100	160	300
Diameter/unit resistance	100	74	57

polyethylene had been developed for insulation and a convenient way to use sodium as a conductor was to inject it into a polyethylene tube at the same time as the tube was being extruded, i.e. to make a complete cable core in one operation from basic raw materials. In the USA a considerable amount of cable was put into service to obtain operating experience.

Table 4.1 indicates some of the relevant data.

Whilst the cost of conductor metal was comparatively low, there was a much larger penalty than with aluminium in terms of conductor diameter. Nevertheless, for fairly simple cables, especially single-core cables, popular in the USA, there appeared to be a strong economic case. Even for 15 kV cables it was only necessary to apply some copper wires outside the insulation to obtain URD cable comparable to the design with copper conductors.

No great difficulty was found in producing the cables. Joints and terminations could also readily be made by sealing a cap on to the cables and fitting a corkscrew arrangement into the conductor to achieve good electrical contact. The performance under service conditions was quite satisfactory provided that suitable protection was used to cater for the low melting point of sodium (98°C). Tests showed that short circuits to higher cable temperature could be sustained without dangerous effect.

However, sodium is an extremely reactive metal and when exposed to water it reacts vigorously with the formation of hydrogen and sodium hydroxide. Whilst such action was not catastrophic, even if a cable was cut when immersed in water, there were obvious handling problems for cable installation and jointing personnel. Disposal of cut ends required special care and attention and difficulties arose in making repairs to damaged cable. Largely for these reasons it became apparent that in comparison with aluminium the handling difficulties exceeded what could be justified by the saving in the initial cost of cable. The use of sodium subsequently lapsed.

CHAPTER 5
Armour and Protective Finishes

ARMOUR

This chapter applies mainly to power distribution cables, certainly in relation to armour, as special considerations apply to wiring type cables and to transmission cables. Features relating to the design of both these groups of cables are given with the design aspects in the relevant chapters.

For lead sheathed paper insulated cables, the two universal types of armour are steel tape (STA) and galvanised steel wire (GSW), usually referred to as single wire armour (SWA). Steel tapes are applied over a cushion of bituminised textile materials which also contribute to corrosion protection. Two tapes are applied helically, each tape having a gap between turns of up to half the width of the tape and the second tape covering the gap and overlapping the edges of the first tape. By applying the two tapes from the same taping head of the armouring machine the lay length of each tape is identical and the tapes register correctly with each other. Although the tapes and the underlying bedding are flooded with bitumen during application the tapes are pre-coated by the supplier with a bitumen varnish to prevent rusting during delivery and storage and to ensure that the underside of the tape is always coated.

Although bitumen provides reasonably good protection of steel in many soils, it has to be remembered that a cable life of 40–50 years is not unusual and in some conditions severe corrosion of the steel may occur after long periods of burial. Before the introduction of extruded plastic layers for protective finishes, some cable users called for the much more expensive galvanised steel tape. This material is seldom adopted now but may still be valuable in special cases, e.g. where the metallic protection is of particular importance on polymeric insulated cables and it is not desirable to use bitumen under an extruded thermoplastic oversheath (see later).

Other users have taken the view that mechanical protection of buried cables is primarily of value during the early years in the life of the cable, e.g. in the development of the site for housing or industrial purposes. After a few years the ground is more rarely disturbed and this has been shown by fault statistics.

Wire armour consists of a layer of galvanised steel wires applied with a fairly long lay. It is more expensive than STA but has several advantages such as (a) better corrosion protection and hence longer armour life, (b) greatly increased longitudinal reinforcement of the cable, (c) avoidance of problems due to armour displacement which can occur under difficult laying conditions and (d) better compatibility with extruded thermoplastic oversheathing layers, as discussed later. It may be argued that SWA is not so resistant to sharp spikes which may pierce between armour wires but in practice it appears that such damage seldom occurs.

The longitudinal reinforcement aspect covers such advantages as the ability to withstand higher pulling loads, e.g. for cables drawn into ducts, and better support of cable in cleats or hangers, particularly if vertical runs are involved. SWA is essential for cables to be laid in ground liable to subsidence.

A further, often important, feature of armour design is to provide effective conductance of earth fault currents. If the cable has a lead sheath, the sheath alone is adequate in most cases; steel tape armour does nothing to extend the sheath current carrying capacity. In the absence of a lead sheath, as with PVC and XLPE insulated cables, it is generally desirable or necessary to use wire armour to deal with fault currents.

Before the 1970s in the UK, when PILS cables were invariably used for power distribution, it was common practice for 600/1000 V cables to have STA and the more important higher voltage cables to have SWA. Central London was an exception, because with a preponderance of installations of cables in ducts, SWA was preferred. The armour was usually left bare. Outside London the two different types of armour provided a good guide during digging operations to indicate whether any cables exposed were low or high voltage. When the PILS design disappeared it was necessary to take other steps for easy cable identification and the supply authorities standardised on a red PVC oversheath for 11–33 kV cables.

Wire armour normally comprises a single layer of wires but there are a few special applications where a double layer (DWA) is preferred to provide increased robustness. The main field is for cables in coal mines, not only for the in-coming supply cables which need to be self-supporting in the primary vertical pit shafts but also for cables which are laid along roadways. At one time all the roadway cables had DWA but this caused increased rigidity, which was a problem in coiling the cables for transport down the lift shafts. It is now often considered that SWA is adequate for fixed cables in tunnels. DWA comprises two layers of wires, with opposite directions of lay and a separator layer such as bituminised hessian tape applied between the wire layers.

Single-core cables

A problem arises with single-core cables because in their installation formation the armour is situated between the conductors and if a magnetic material is used it causes high induced currents in the armour. These result in high electrical losses. Consequently if armour is essential it is necessary to use a non-magnetic material such as aluminium. However, few single-core cables are buried directly in the ground and as the main application is for short interconnectors armour is not usually essential.

Armour conductance

Technical details of the importance of armour conductance with particular reference to wiring cables are discussed in chapter 10, and armour conductance requirements for colliery cables are detailed in chapter 18.

Armouring for polymeric insulated cables

The lower voltage (i.e. up to 3.3 kV) PVC or XLPE cables which have replaced PILS cables require mechanical protection of a similar standard. For various reasons, such as ease of cable handling and armour conductance, SWA is normally used. Such cables usually have an extruded PVC oversheath (other extruded materials are used for cables to meet certain performance requirements in fires – see chapter 6), and particularly from overseas there are users who express a preference for steel tape armour. Whilst PVC can certainly be applied over STA, and much such cable is supplied, it is more satisfactory to apply the PVC over the smooth surface provided by SWA. In extruding over the stepped surface of STA there is a tendency for thinning of the PVC to occur at the edges of the steel tape.

Nowadays a significant proportion of cables for 11 kV and upwards has XLPE insulation instead of paper and metallic sheath and it is mandatory for such cables to have an earthed screen over each core and/or collectively over the laid-up cores. SWA can serve quite well as the metallic component of the earthed screen and it has the advantage of providing good earth fault current carrying capacity. As discussed in chapter 24 a common cable design is to have an extruded PVC bedding followed by SWA and PVC oversheath. Some users overseas adopt STA and deal with earth fault currents by a thin metallic tape on each core. Tape thickness is restricted by the need to maintain intimate contact with the core surface, and some caution is necessary with such design as fault current carrying capacity is necessarily limited.

Nevertheless, with polymeric insulated HV cables, there is a greater tendency to use single-core cables, rather than multicore, and for these it is not practicable to use SWA. Furthermore, as there is no lead sheath, protection against mechanical damage and penetration of water may be of lesser importance than with PILS cables and so there is less need for increased robustness, especially as the whole cable design is more resilient. A frequently used cable design is therefore to replace armour by a layer of copper wires spaced apart round the cable (chapter 24). Nevertheless, with such cables, and for HV polymeric insulated cables generally, the design of protective finish has to take account not just of corrosion protection but also of the effects of water penetration on the electrical properties of the insulation ('water treeing' – see chapter 24).

PROTECTIVE FINISHES

Protective finishes are of particular interest for the protection of metal sheaths, reinforcement and armour in buried distribution and transmission cables. Whether the installation is below or above ground, very few cables are supplied without a protective finish. For the newer designs of the above categories of cable, an oversheath of extruded PVC or MDPE has become nearly universal.

When wiring type cables need to be of robust construction, the cable design normally comprises an extruded oversheath, but one important difference is that this oversheath can be of a tougher thermoset material such as PCP, CSP or NBR/PVC. These materials are also used where special properties are required, e.g. for oil resisting and flame retardant (OFR) finishes and heat and oil resisting and flame retardant (HOFR) finishes as discussed in chapter 3. Such materials could have advantages for distribution and transmission cables but they can rarely be adopted because the high temperature required by most methods of curing would cause damage to the cable.

A special situation arises when the protective finish both has to be flame retardant and has to emit a minimum of toxic fumes and smoke in a fire. This is covered in chapter 6.

The UK attitude for distribution cables has always been that after burial the user expects to forget about them for the rest of their life, and with paper insulated cables life expectancy has steadily been extended to over 40 years. In spite of the generally good corrosion resistance of metals like lead and copper they can fail when buried in some types of ground, and in the UK such metals have rarely been left bare. Much attention has been paid to corrosion protection, and cable failures due to this cause have been rare. In the USA the practice has been somewhat different and in the case of cables having a copper concentric neutral conductor, it is not unusual to install cables without protection. The literature has reported a substantial level of cable failures and, although it is relatively easy with a duct system to replace the cable, the economics of savings against subsequent expenditure would seem to be finely balanced.

The protection of aluminium against corrosion is of particular importance because aluminium may corrode very quickly when buried, or even when near the surface and exposed to damp conditions, e.g. where cables descend from cabinets into the ground. At these positions, it is most important to apply protection over exposed metal right up to and over the termination.

The corrosion of aluminium usually takes the form of local pits which may quickly penetrate a sheath, although general surface attack may be quite small. The mechanism of pitting is associated with the local breakdown of the protective oxide film, in conditions which do not allow its repair, followed by cell action due to differential conditions of electrolyte concentrations or of aeration. The presence of other underground services containing metals anodic to aluminium, such as lead, steel or copper, may accelerate the attack. While some soils, such as in made-up ground, are worse than others, it is always essential to consider that any ground is aggressive and to ensure that good protection exists. Protection is also necessary for above ground installation because in damp conditions corrosion can occur in crevices, owing to differential aeration conditions, and it is difficult to avoid crevice conditions at points of support or in contact with walls.

It is convenient to review protective finishes in two categories: first, the bitumen–textile type used traditionally with lead sheathed paper cables, and second the extruded thermoplastic oversheaths which were first developed to provide satisfactory corrosion protection for aluminium sheathed cables and were then adopted for all newer cable designs as well as for paper/lead cables when improved protection was desirable. For armoured cables, the protective finish comprises a bedding under the armour and a layer over the armour, usually termed a serving when textile and an oversheath when it is extruded.

BITUMINOUS FINISHES

Constructions

The earliest paper/lead cables at the end of the last century were protected with coal tar pitch and fibrous textile materials such as cotton and jute in yarn or woven form, i.e. hessian. The basic design has changed little over the years but the once popular jute yarn roving has largely given way to woven hessian. Replacement of coal tars by bitumens provided better uniformity and enabled particular grades to be used according to their function within the construction. The introduction of two impregnated paper tapes adjacent to the lead sheath was also beneficial. For many years stray direct currents from tramway systems caused corrosion problems with underground cables due to electrolytic effects. The combination of impregnated paper and bitumen improved the consistency of the protective layer.

For armoured cables, the most common construction today is for the bedding to consist of two bituminised paper tapes plus two bituminised cotton or hessian tapes with a serving of two bituminised hessian tapes. A suitable grade of bitumen is applied over each layer and a coating of limewash is given overall to stop rings sticking together on the drum.

Bitumen

A thick layer of bitumen would give excellent corrosion protection, as it does on rigid steel underground pipes, but no method has been found of producing such a layer to withstand the bending and other conditions necessary with cable. Many grades of bitumen are available, with a wide range of viscosities and pliability. Careful selection has to be made to suit the requirement for impregnating the textiles of the particular layer in the cable and to suit the subsequent service conditions, bearing in mind that some cables have to operate in the tropics whilst others may be in polar regions. On the whole the bitumen should be as hard and rubbery as possible, but must not crack when bent at low temperatures. If the viscosity is too low, troublesome 'bleeding' may ensue at high ambient temperatures.

Microbiological degradation of textiles

The jute or other textiles in the finish are present largely to support the bitumen and even if they are partially rotted the bitumen layer continues to give a reasonable measure of protection. In tropical countries such rotting can develop very quickly, even to a serious extent on drums held in a stockyard for a year or so, caused by the acidic degradation products of jute and other cellulosic constituents, namely acetic and butyric acids.

The immediate solution was to substitute PVC tapes for the paper tapes adjacent to the lead sheath, so as to create a barrier. Rot proofing of hessian, by the incorporation of a material such as zinc naphthenate, has also been popular for cables in tropical countries but a much better solution is to use extruded PVC instead of bitumen–textile finishes.

EXTRUDED THERMOPLASTIC FINISHES

PVC oversheaths

Following many years' successful use on wiring cables, PVC was established as an oversheath for power distribution and transmission cables in the mid-1950s and since then PVC sheaths have become the most commonly used type of protective finish for power cables.

Grade of material

For wiring cables, flexibility and easy sheath removal are important and the grade of PVC used for sheathing is softer than that used for insulation. However, for the heavier power cables different criteria apply. Toughness and resistance to deformation both during and after installation are of greater significance and it is usual to use a similar grade of material to that used for insulation. For most purposes oversheaths are black, in order to give good resistance to sunlight, but where this is not important colours may be adopted as a means of cable identification. When red sheaths were first used for buried cables some darkening of colour occurred due to sulphides in the ground and special compositions were found to be necessary.

Whilst the amount of plasticiser in the PVC may theoretically be varied to suit a wide range of high and low ambient temperature conditions, the practical advantages are limited because of other aspects such as the resistance to deformation previously mentioned. This is of particular significance when low temperatures are encountered because at around 0°C PVC becomes somewhat brittle and may crack either by a sudden blow or by rapid bending during handling operations. For this reason it is important not to install cables at low temperatures.

Construction

When applied over any type of cable, extruded PVC provides a clean finish which is attractive and contributes to ease of handling. Whilst the external profile is smooth if the application is over a normal metal sheath or wire armour, this is not the case with corrugated aluminium sheaths or steel tape armour. Extrusion is by the 'tubing-on technique' (chapter 24) and the external profile follows the contour of the underlying surface, thus creating small steps over steel tape armour. In the case of corrugated sheaths the oversheath tends to be slightly thicker in the troughs.

Initially, when PVC oversheaths were first applied over steel wire armour, no change was made in the use of bitumen, i.e. the oversheath was extruded directly over armour which was flooded with bitumen. Such cables are frequently installed vertically in factories. When bitumen is in contact with PVC it absorbs some plasticiser from the PVC. Whilst the amount of loss is insufficient to have much effect on the properties of the PVC, it is enough to cause considerable reduction in the viscosity of the bitumen. In fact, the bitumen may become so fluid that it can run out of the cable at terminations situated below the cable run. Subsequent practice has therefore been to omit bitumen between wire armour and PVC unless a preference for inclusion is expressed by the user. A particular example is the use of bitumen in the single or double wire armoured PVC insulated cables used underground in coal mines. The cables are usually installed horizontally and as the oversheaths may be damaged it is useful to have the benefit of bitumen as secondary protection. Similarly, in the UK, bitumen is normally applied over steel tape armoured cables, as these are invariably buried. It is not so necessary and normally not used over galvanised steel tape.

When PVC is applied directly over aluminium sheaths it is particularly desirable to have a uniform coating of bitumen on the aluminium to provide an interface seal and mitigate against spread of corrosion resulting from damage to the PVC. At one time it was a practice to include a small amount of a sparingly soluble chromate in the bitumen to serve as a corrosion inhibitor but, whilst this was undoubtedly beneficial, it is no longer used because of possible toxic effects. With corrugated sheaths the tendency is for an excess of bitumen to remain in the troughs and the corrugation effect is less visually apparent on the finished cable. Some German manufacturers aim at complete filling of the troughs on such cables and also apply a layer of a thin plastic tape to assist in maintaining a thick bitumen layer.

Characteristics of PVC

In combination with a good measure of flame retardance, and resistance to oils and chemicals, PVC contributes much to robustness and ease of handling. Very few chemicals normally found in the ground on cable routes have significant effect on PVC. Investigations have been carried out to determine whether any substances found in chemical factories and oil refineries are particularly aggressive to PVC. It was found that the only chemicals which caused attack were high concentrations of chlorinated solvents, esters, ketones, phenols, nitro compounds, cyclic ethers, aromatic amino compounds, pyridine, acetic anhydride and acetic acid. Except for perhaps occasional splashes, it is unlikely that such materials will be in contact with cables. Prolonged immersion in oils and creosote may cause swelling and softening but there is usually considerable recovery when the source is removed.

Loss of plasticiser from PVC to other materials in contact with cables may cause some problems with wiring type cables. One example of this relates to cables in roof spaces where polystyrene granules have been used for thermal insulation over ceilings. The granules in contact with the oversheath may become soft and tacky due to leaching of plasticiser from the PVC, which itself will lose some flexibility. Direct contact between the granules and cables should be avoided. Similar remarks apply to some fittings which are in contact with cables. In addition to polystyrene and expanded polystyrene, acrylonitrile–butadiene–styrene (ABS), polyphenol oxide and polycarbonate are also affected. Nylon, polyester, polyethylene, polypropylene, rigid PVC and most thermosetting plastics are little affected.

With the greater thickness of oversheath on distribution and transmission cables the rate of loss of plasticiser from the body of the oversheath is much slower and in any case subsequent disturbance by bending is unlikely. Although the PVC becomes somewhat harder as plasticiser is removed, the properties of permeability to moisture and protection against corrosion are not impaired.

A special situation arises when PVC oversheathed cables are in ground which may contain hydrocarbons such as petrol. This can happen in oil refineries as a result of spillage. Although only a small amount of hydrocarbon is absorbed by the PVC, such materials can diffuse through the sheath and be taken up by materials inside, or condense in any air spaces available. The PVC itself is unaffected, except for some loss of plasticiser. However, the petrol may then flow along the cable and run out of joints and terminations. For such conditions it is general practice, therefore, for cables to have a lead sheath, even if the insulation is of PVC. By sealing the lead sheath to the equipment, no hydrocarbon will flow into the equipment. Flow will still take place along the armour, and if it is important that no leakage should occur at the armour termination this may also need to be specially sealed.

Polyethylene oversheaths

Although from the outset PVC became the established material in the UK for power cables, polyethylene has tended to be preferred in the USA and also throughout the world for telephone cables. Most of what has been stated for PVC applies equally to polyethylene but, whereas PVC has to be compounded with plasticisers and fillers etc., the only additions normally made to polyethylene are antioxidant and carbon black. Problems due to such aspects as loss of plasticiser do not arise. Furthermore, polyethylene does not suffer from effects of brittleness at low temperatures and so has a particular advantage for installation in countries where sub-zero temperatures exist for long periods. Polyethylene sheaths have to contain carbon black and this prevents their use when coloured sheaths are required.

On balance the overall advantages and disadvantages of the two materials are marginal and it is more economic for manufacturers to standardise on one material. Polyethylene is flammable and this is one of the main reasons for the preference for PVC. Another is that resistance to thermal deformation is an important characteristic for distribution cables. Whilst polyethylene may be marginally better for temperatures within the normal continuous operating range, it softens more than PVC above 80°C, and at 130°C it flows readily under mechanical load.

Many grades of polyethylene are available and for oversheaths it is necessary to select one which has good resistance to environmental stress cracking.

High density and medium density polyethylene oversheaths

Frequent references have been made in this chapter to the importance of damage to extruded oversheaths due to mechanical deformation. Because of the susceptibility of aluminium to corrosion, this is a matter of particular importance with heavy and expensive aluminium sheathed transmission cables. Damage to the oversheath may occur, for example, through the presence of sharp stones in the cable trench, especially when cables are being installed at high ambient temperatures in tropical conditions. To ensure freedom from damage an electrical test is carried out on the oversheath after laying and in bad conditions much time may be spent in locating and repairing faults. In general a material which is tougher than PVC or polyethylene would be desirable.

High density polyethylene (HDPE) and medium density polyethylene (MDPE) are materials which are used frequently for transmission cables. They are extremely hard and difficult to damage even with a sharp spade thrown directly at the cable. They are particularly advantageous for conditions involving high ambient temperatures, e.g. in tropical and subtropical countries where, at the time of laying, PVC is much softer than it is in the UK. However, they are not suitable for universal adoption as an alternative to PVC because their very rigidity causes a penalty due to increased cable stiffness. This is not a serious disadvantage with transmission cables because the proportionate increase in stiffness is only modest and bending to small radii is not required.

HDPE and MDPE oversheaths cost a little more than PVC oversheaths and this is also a factor which has restricted their use on the less expensive types of cable. Even more care has to be taken in grade selection than with PE to ensure good resistance to cracking in some environments.

A further feature which may add marginally to overall cost is that, because of the enhanced coefficient of expansion and rigidity, it is important for the oversheath to be secured rigidly at joints and terminations; otherwise retraction may occur on cooling and the metal sheath could be left exposed. No difficulties in this respect arise with FF cables because it is fairly standard practice for sheath plumbs to be reinforced by resin and glass fibre and the oversheath is readily included in the arrangement.

Although not important with the majority of oversheathing applications, another feature of HDPE and MDPE is that they are notch sensitive. Cracking may occur if there is a sudden change of section due, for example, to steps in the surface on which the oversheath is applied.

FINISHES FOR PROTECTION AGAINST INSECTS

Although instances have arisen of damage caused to cables and cable drums by various insects and by gnawing from rodents, the most common form of attack is from ants, termites and teredos.

Protection against ants and termites

Termites are essentially subterranean in habit and exist only in tropical and subtropical regions up to 40° on each side of the equator. Most reports of damage to cables have come from Malaysia and Australia. Termites feed on cellulose, generally from wood, and attack on cables is not for food, but because their path is obstructed. They have hard saw-tooth jaws capable of tearing through soft materials including some metals and plastics.

Ants are much more common but again problems with cables are confined to warmer climates. They exist on the ground surface but may extend underground to a depth of about 120 mm in loose sandy soils.

Measures required for protective finishes are the same for both ants and termites:[1,2]

(a) Use of a metal protection layer.
(b) Use of a damage resistant polymeric oversheath.
(c) Treatment of the cable backfill material.
(d) Inclusion of insecticide in the polymeric oversheath.

The use of metal barriers has been shown historically to be effective provided no gaps exist. However additional metallic layers add expense, complexity, increased heat losses, and the need to electrically bond metallic layers to the cable. It may not be possible to prevent corrosion of the metallic layer because of termite attack to the polymeric oversheath.

High- or medium-density polyethylenes with a Shore D Hardness of 60 provide adequate resistance to attack by some termites,[3] however, the surface of the oversheath must be smooth to prevent termites from initiating their attack. Softer materials such as PVC provide little protection against the more aggressive species of termites. Harder materials such as nylon, which have hardness in excess of 65 show absence of termite attack.[2] Due to its high cost nylon is usually applied as an additional thin layer over the normal cable oversheath, then to prevent abrasion damage or wrinkling of this layer during installation, a further protective oversheath is necessary.[4] This solution has tended to be used for small diameter cables.

The chemical treatment of trench backfills to deter termites is, at first sight, a simple solution and has been widely used to provide protection to timber buildings. However, treating the quantity of backfill required to protect a cable route is expensive and the introduction of quantities of chemicals into the ground is undesirable. Three insecticides which have been successfully used in the past, i.e. aldrin, dieldrin and lindane, have become increasingly unacceptable because of their accumulation in the environment resulting from their overuse in agriculture and timber protection.

The compounding of insecticides into the cable oversheath is a more efficient method that has been used for many years with additives such as aldrin, dieldrin, lindane and naphthenates. These materials are becoming untenable for environmental reasons. New more environmentally acceptable materials (e.g. synthetic pyrethoids, metal naphthenates and organophosphates) are used to achieve termite resistance in other industries and these have also been investigated for incorporation into cables.

A new anti-termite oversheathing material has been developed specifically for cables.[5,6,7] The technology is based on a novel masterbatch formulation containing an environmentally acceptable anti-termite agent; the masterbatch is mixed into the sheathing polymer immediately prior to extrusion. Extensive laboratory and field testing has demonstrated that some termite protection is given to the soft PVC oversheaths, but that full protection is given to polyethylene oversheaths.

Protection against teredos

Teredos are a form of marine life existing in relatively shallow sea-water near to land. They were often called ship-worms because of the attack on the hulks of wooden sailing ships. The shore ends of submarine cables usually have to be protected and the standard method is to include a layer of tin–bronze tapes in the protective finish. Adequate protection under and over this layer has to be considered according to the other circumstances arising.

BRAIDED FINISHES

Braided finishes used to be popular before extruded finishes came into general use. Because of the slow manufacturing process, however, they were expensive and could only be justified when there was a need for improved abrasion resistance. For power cables, hessian, cotton and asbestos braids were superseded by extruded PVC and the special formulations for flame retardance discussed in chapter 6. Health and safety problems also arose with asbestos.

For flexible cables, braided finishes have continued to find fairly standard application for domestic appliances where there is a possibility of the cable coming into contact with hot metal and an extruded finish might be damaged, e.g. for electric irons and toasters etc. Continuous filament of the regenerated cellulose type is now the most common material for such braids. When there is justification for a more robust finish, as for aircraft cables, nylon is used.

TESTS ON PROTECTIVE FINISHES

Apart from tests on the thickness and properties of the materials used, the only tests of importance are those given in IEC 229. These cover requirements when the quality of the finish is especially significant, such as for the protection of aluminium sheaths or the sheath and reinforcement of transmission cables.

D.C. voltage test

When the requirements are onerous it is usual to check that no damage has occurred during laying and often subsequently during the life of the cable. This is done by a voltage test and for the purpose a conducting layer of graphite type varnish is applied on the outer surface of the oversheath at the time of cable manufacture. The routine test in the factory is at a voltage $(2.5t + 5)$ kV, applied for 1 min, where t is the minimum average thickness of the oversheath.

Test for protection of aluminium

In the event of local damage to the oversheath, it is important that soil water should not penetrate beyond the exposed area and reference has been made to the application of bitumen to provide such secondary protection. The test for this purpose is carried out on a length of cable previously submitted to a bend test. Four circular pieces of oversheath 10 mm in diameter are cut out with a cork borer and the exposed aluminium is cleaned. The sample is placed in 1% sodium sulphate solution with the ends protruding and 100 V d.c. is applied between the aluminium sheath and the bath, a resistor being used so that the current is 100 mA. After 100 hours the oversheath is removed and no signs of corrosion should be visible beyond 10 mm from the initial holes.

Abrasion and penetration tests

Tests are included in IEC 229 to demonstrate that the finish has adequate resistance to abrasion and to penetration by sharp objects. They are described in chapter 39. These tests were originally developed to create a standard at the time when finishes included layers of unvulcanised or vulcanised rubber together with outer layers of hessian. They are of less significance with modern extruded oversheaths.

References

(1) Field, A. W. and Philbrick, S. E. (1991) 'Protecting Cables from Insect Attack'. *JiCable Conference Proceedings*, 447–452.
(2) Report of CIGRE WG21-07, 'Prevention of Termite Attack on HV Power Cables'. *Electra*, No. 157, December 1994.
(3) Bell, A. (1985) 'Plastics that Termites Can't Chew'. *Plastics News*, July 1985.
(4) Nakamura, T. et al. (1934) 'The Practical Application of Nylon Jacketed Termite Resistant Power Cables'. *Sumitomo Electrical Technical Review*, No. 124, 1984.
(5) Blackwell, N., Gregory, B., Head, J. G., Mohan, N. C., Porro, R. and Rosevear, R. D. (1996) *Termite Protection of Underground Distribution and Transmission Cables*. Electropic 96 Conference, 1996.
(6) Process for Making Termite Resistant Cable Sheaths. UK Patent No. 2276171B.
(7) PANGOLIN® registered product of BICC Cables Limited.

CHAPTER 6
Cables in Fires – Material and Design Considerations

Each year fires result in thousands of fatalities worldwide with many more injured. The financial cost is large. Although cables seldom cause fire, they are an integral part of property and equipment and are often subjected to fire resulting from other causes. In the majority of cases the continued electrical operation of the cables during a fire is not required. However, certain circuits, such as those for emergency lighting or for the safe shutdown of equipment, will have to remain in service during the fire and for these requirements special cables with fire survival properties are available.

The behaviour of cables in a fire varies and depends on a number of factors. The component materials and the construction of the cable are very important as is the particular nature of the fire. Depending on their location and means of installation, cables can contribute to a fire in a number of ways. For example, burning cables can propagate flames from one area to another, can add to the fuel available for combustion and can liberate smoke and toxic and corrosive gases. The hazard from burning cables should be put in the context of the surroundings. Sometimes, cables form a very small proportion of the combustible material, while in other situations they can form the majority. Considerable effort has been made in recent years to develop cables and cable materials which constitute a reduced hazard in a fire and the performance of some of the latest materials is far superior to what was achievable only a few years ago. Mineral insulated cables represent a negligible fire hazard, as they have a copper sheath and contain no combustible materials; such cables can withstand very high temperatures and are used as fire survival cables (chapter 16). Polymeric insulated cables cannot be regarded as hazard free, as all organic materials will burn under most fire conditions and will liberate heat and toxic gases, such as carbon monoxide.

The difficulty in trying to predict how a cable will perform in a fire is now fully appreciated. Nevertheless, there are numerous small and larger scale standard tests to which cables are subjected to determine their fire behaviour, although laboratory test results often do not correlate to real life situations. The present emphasis in assessing fire performance is to develop more realistic tests in which cables are evaluated in the context of a simulated installation environment.

LEVELS OF CABLE FIRE PERFORMANCE

A wide spectrum of fire performance is available from the many types of cables on the market. This can range from cables at one extreme which have no enhanced fire properties, which are readily ignitable and burn with ease, to, at the other extreme,

Table 6.1 Levels of fire performance for different cable types

Cable type	Fire characteristics	Application
Mineral insulated (copper sheathed)	• Fire survival and circuit integrity up to the melting point of copper • Negligible fire hazard	For maintaining essential circuits such as emergency lighting and fire alarms, circuits for the safe shutdown of critical processes, etc.
Limited circuit integrity, low fire hazard, zero halogen	• Limited fire survival • Flame retardant • Low smoke and acid gas emission	As above but circuit integrity maintained for shorter time periods. Reduced hazard from cable combustion.
Limited circuit integrity, reduced hazard (halogen containing)	• Limited fire survival • Flame retardant • Reduced acid gas emission • Reduced smoke emission	As above, but increased hazard from smoke and acid gas emission.
Low fire hazard, low smoke, zero halogen	• Flame retardant, low smoke and acid gas emission	For installation in areas where smoke and acid gas evolution could pose a hazard to personnel or sensitive equipment, but where circuit integrity is not needed.
Low emission PVC based (or chlorinated polymer)	• Flame retardant grades possible • Reduced smoke and/or acid gas • Reduced flame propagation possible	In situations where reduced levels of smoke and corrosive gases are needed, compared to ordinary PVC or chlorinated polymer based cables.
PVC or chlorinated polymer	• Flame retardant	Where flame retardance is desirable, but smoke and acid gas evolution is not considered to pose a serious hazard.
Fluoropolymer based	• Inherently flame retardant	Where cables are exposed to high temperatures or aggressive environments in normal use.
Non-flame retarded (e.g. polyethylene or EPR based)	• Readily combustible	In situations when fire performance requirements are low and where cable combustion poses little hazard.

fire survival mineral insulated cables which contain no combustible materials and which present no hazard in a fire. The choice of cable for a given application depends on the degree of hazard which can be tolerated and the level of performance required. The level of fire performance and the potential hazard resulting from the combustion of a given cable depend on the materials from which the cable is made and the cable construction. Table 6.1 summarises the different levels of performance that can be achieved by different categories of cables, along with typical areas of application.

The hazard from cables in a fire not only depends on their constituent materials and construction, but also on their location and installation geometry. Where cables form a very small part of the potential fuel load for the fire, or where they are buried underground for example, it is clear the potential hazard is minimal. On the other hand, the combustion of cables could pose a more serious threat in situations where they run through ducting or roof spaces connecting different parts of a building. In these situations the cables could propagate and spread the fire more rapidly. In public buildings and mass transit installations the evolution of smoke from burning cables could cause a serious hazard in terms of preventing the rapid escape of people from the fire. Evolution of corrosive acid gases from burning cables is becoming of increasing concern because of the damage that can be caused to sensitive electronics and computer systems, not to mention people. In such situations, since the halogen content can generate corrosive acid gases, halogen free cables are usually specified.

There are, therefore, two aspects worth considering with respect to the behaviour of cables in a fire. The first is the level of fire resistance or fire survivability required, and the second is the hazard posed by the combustion of cables in the fire. Both considerations are important in determining the potential cost in human and material terms resulting from a fire in which cables are involved.

MATERIAL CONSIDERATIONS

A wide range of materials is used in the manufacture of cables, which, in the context of fire performance, can be broadly separated into those which are combustible and those which are not. For the majority of cable types, polymers form the bulk of the combustible material, although some more traditional cable constructions utilise paper insulation, for example. Polymers offer a number of advantages in terms of cable manufacture and performance, generally exhibiting a very high level of electrical resistance and ease of processing. Unfortunately, being organic materials, polymers are all to a greater or lesser extent flammable. The range of flammability is wide however and many polymeric cable components are formulated so as to reduce their tendency to burn. It should be noted that polymeric materials overall are no more hazardous in their combustion behaviour than other flammable materials such as wood, paper, cotton or wool.

There are several factors that describe a material's flammability and combustion behaviour, although how these are measured and their relevance is often a cause of debate. The major factors are:

(1) Ease of ignition (flammability)
(2) Resistance to propagation (flame spread)
(3) Heat of combustion (heat release)

(4) Smoke emission
(5) Toxic gas evolution
(6) Corrosive gas evolution.

All of the above will determine the nature and extent of the hazard presented by the exposure to fire of a given material. The ease of ignition of a wide range of materials has been determined, and is often indicated by the flash ignition temperature and the self-ignition temperature. The flash ignition temperature is the temperature at which gases evolved from the material can be ignited by a spark, and the self-ignition temperature is the temperature at which the material will spontaneously ignite. Other measures of flammability are the limiting oxygen index and temperature index; the former is the proportion of oxygen in the atmosphere for which the combustion of a material can be sustained at room temperature and the latter indicates the temperature at which the combustion of a material can be sustained in a standard atmosphere. The relevance of such test methods with respect to real fire behaviour will be discussed in a later section. Table 6.2 shows the flash ignition and self-ignition temperatures of a number of polymers and other natural materials, along with oxygen index values and measured heats of combustion.[1,2,3]

It is interesting to observe in table 6.2 that wood, although readily ignitable, generates much less heat during combustion than polystyrene which has a higher ignition temperature. Polyvinyl chloride offers the advantage of resistance to ignition and a low heat of combustion but does, however, tend to liberate large amounts of black smoke, along with corrosive hydrogen chloride gas and other toxic gases. The important point clearly, is to consider all aspects of a material's fire behaviour and put

Table 6.2 Flash ignition and self-ignition temperatures of a number of polymers

Material	Flash ignition temperature (°C)	Self-ignition temperature (°C)	Limiting oxygen index (%O$_2$)	Heat of combustion (MJ/kg)
Polyethylene	341–357	349	18	46
Polypropylene	320	350	18	46
Ethylene–propylene rubber (EPR)	—	—	18	—
Chlorosulphonated polyethylene (CSP)	—	—	27	28
Polychloroprene (PCP)	—	—	26–40	24
Silicone rubber	490	550	26–39	—
Polyvinyl chloride (PVC) plasticised	391	454	47	19
Polystyrene	350	490	18	40
Nylon 66	421	424	20	33
Polytetrafluoroethylene (PTFE)	560	580	95	endothermic
Polyurethane	310	416	16.5	28
Wood	220–264	260–416	22–25	18.5
Wool	200	590	24–25	20
Cotton	230–266	254–400	18–27	16.7

this in the context of the application and installation of the cable. For a given material it is important to consider the contribution that each of the factors listed above make to the total fire hazard. For example, it has been estimated that 80% of fire deaths in the United States are a result of smoke and toxic fume inhalation.[4] Importantly, smoke obscuration also serves to prevent escape from burning buildings.

All combustible materials liberate toxic gases when burned. The amount and type of gases vary from one material to another, but at the very least all organic materials will release carbon dioxide and carbon monoxide. Carbon dioxide, although not poisonous, acts as an asphyxiant in large concentrations; carbon monoxide, liberated in situations where the level of oxygen has been depleted (as usually happens in a fire), is very toxic and lethal in relatively small concentrations: more than 1.3% of carbon monoxide in the atmosphere can kill within a few minutes. (Oxygen depletion in itself is a serious problem, but this is common to all fires in confined spaces and is not particularly dependent on the materials involved.) A wide range of other toxic substances can be generated, many of which cause irritation as well as poisoning. The effect of toxic fire gases is sometimes overestimated, however, fatalities caused by the inhalation of hot air and other gases forming a significant proportion of fire deaths. It has been shown that, in many cases, the minimum oxygen content for survival (or the maximum breathing temperature level) is reached before any toxic gases attain a lethal concentration.[1]

All polymeric cable components will burn under fire conditions and very few, if any, will survive temperatures of more than 500°C; often the temperature in a fire can reach 1000°C or more. Many polymers will ignite and burn readily while others will resist ignition, either inherently (as in PTFE) or by design, but sooner or later they will decompose and burn. Considerable effort has been expended in recent years to modify and improve the fire behaviour of polymeric cable materials and often the first step is to reduce the flammability of a material by the incorporation of flame retardant additives, of which there is a wide range. There are several means by which this can be achieved, but only a few of the more commonly used methods will be discussed here. Further details of flame retardant additives and mechanisms can be found in abundance in the technical literature.[1,2,3]

It has been known for many years that halogen-containing materials exhibit a high degree of flame retardance, and polymers such as PVC, CSP and PTFE fall into this category. It is not surprising, therefore, that many flame retardant additives are also chlorine- or bromine-containing compounds. Mechanistically, it is believed that the halogen atoms interfere with the decomposition of the polymer and reduce the amount of combustible gases liberated but, as an often undesirable consequence of the reaction, corrosive hydrogen halides are produced. Halogenated compounds can be formulated to reduce the emission of acid gas but this can compromise the flame retardant performance. Antimony trioxide is often incorporated into halogen containing polymers as it behaves synergistically with chlorine and bromine to further reduce flammability and much work has been carried out in order to understand the mechanism of this flame retardant system.[2]

In recent years there has been increasing concern about the emission of smoke and corrosive gases from burning cables. In order to address this problem, a wide range of cables has been developed which are often termed low fire hazard (LFH) or low smoke and fume (LSF). Such cables have been introduced to comply with specifications which set limits on the level of halogen acid gas evolution permissible, the levels of smoke generation, and other requirements for flame retardance and flame propagation. Many

LFH cable materials are halogen-free and this is usually implicit with LSF products, as these compounds are based on polyolefins such as EVA, polyethylene and other ethylene copolymers. Unfilled, these polymers are readily flammable, and to achieve an appropriate level of fire performance, it is necessary to incorporate high loadings of mineral fillers with alumina trihydrate (ATH) and magnesium hydroxide being the most commonly used, both of which liberate water on heating to temperatures in excess of 200°C and 300°C respectively. The water that is produced as steam from the decomposition of ATH and magnesium hydroxide dilutes the combustible gases generated from the base polymer and cools the surrounding area. The residue from the filler decomposition reaction in both cases is the oxide, which is thermally stable and incombustible, and which helps to form a fire resistant char; typically addition levels of over 60% by weight of these minerals are required to give acceptable flame retardance. Early compounds of this type suffered from a deterioration in physical properties and processibility because of the high level of filler, in comparison to halogen based systems, for example. However, recent advances in polymer technology and processing mean that compounds containing 65% by weight of filler are possible while still retaining good physical properties and a melt viscosity which allows fast extrusion speeds. These materials exhibit resistance to ignition and combustion, reduced flame propagation, low levels of smoke emission and lower level of toxic fumes. Importantly, LSF materials do not evolve corrosive halogen acid gases. LSF cables were used in the Channel Tunnel, and although all cables were destroyed at the hottest point of the November 1996 fire, the LSF cables did not propagate the fire along the tunnel.

There are a number of other fire retardant systems available, but these are not as widely used in cable applications. The mode of operation of these is often the formation of a physical barrier to the fire in the form of a thermally stable char. The mechanism of char formation is termed intumescence. The precise ingredients in the formulation which give rise to the char formation vary, but usually three basic components are required which act in combination: a carbon source, a dehydrating agent and a blowing agent. When added to a polymer base in the correct proportions it is possible to form an expanded char on exposure to flame which can act as a barrier to prevent damage to underlying layers. Camino and Costa have undertaken a considerable amount of work in this area of fire retardant technology.[5,6] A number of factors have perhaps combined to limit the use of such additives in cable compounds. Often the char formed can be quite fragile and so does not form a very robust barrier to the fire. Although work has been undertaken to address this issue,[7] there is some doubt as to the ability of intumescent materials to resist flame propagation, since the resistance to ignition is limited in the early stages of combustion until a char has formed. The structure of the char formed can also depend on the temperatures to which the materials are exposed. In terms of overall fire hazard, there are further concerns with carbonific intumescent additives because many are based on phosphorous- and nitrogen-containing chemicals, which can give rise to toxic gases on combustion, such as nitrogen oxides, ammonia and phosphoric acid. Avoiding some of the disadvantages just mentioned, inorganic char forming materials are also available. On exposure to high temperatures these glass-like materials can ceramify and yield a hard thermally stable substance. As with the organic intumescent additives, these ceramifiable materials do not impart resistance to ignition as such, but form a barrier to fire at high temperatures.[8,9]

A number of polymers are inherently flame retardant and require no further additives. PTFE is an example of such a material, the combustion of which is an endothermic reaction, that is, it requires an input of heating to maintain burning. Despite being highly flame retardant, at very high temperatures the polymer will decompose into the monomer and other toxic gaseous fluorine compounds. Silicone rubber is another polymer that exhibits distinctive combustion behaviour. Silicones are inorganic materials which degrade to form primarily silicon dioxide (silica) at elevated temperatures. This is an inert, thermally stable, electrically insulating material, which is produced in the form of an ash on burning and for this reason silicone elastomers are often used as cable components to provide a limited fire survival capability.[10,11] Inorganic tapes are applied over the silicone layer in order to keep the rather fragile ash in position following combustion of the polymer. Other methods have also been proposed for improving the structural stability of the char.[12,13,14]

It is important to mention that, in addition to fire performance considerations, the choice of polymeric components for the majority of cables is governed by the use for which they are intended, and takes into account also the environment in which they must operate. Cables can be required to perform in a very wide range of climatic conditions and may have to resist attack by aggressive oils or chemicals. Often it is necessary to compromise fire performance in order to achieve the most effective balance of properties for a given application.

While the electrical insulation, bedding and sheathing materials are usually polymeric, clearly a large proportion of a cable can be made up of metallic components, in particular the conductors and armour layers where present. Most conductors are copper although aluminium is also quite widely used, and steel is often used as a protective armour in the form of a braid or helically wound single wire. Metallic components do not in themselves pose a fire hazard but they can still be damaged, especially at high temperatures. The electrical integrity of a cable will be affected if the temperature within the cable exceeds the melting point of the metallic conductors. Aluminium melts at 660°C and copper at 1085°C. Steel has a higher melting point, in the region of 1500°C. Other inorganic materials may be used as part of a cable construction, often in the form of tapes or braids and some, which have especially good thermal stability, are used to enhance the fire survival characteristics in conjunction with low fire hazard polymeric components. A tape composed of mica impregnated glass fibre is often wound around conductors as a means of imparting limited circuit integrity characteristics.[15,16] The glass melts in fire conditions to encapsulate the mica and form a hard thermally stable barrier to prevent contact between conductors and hence prevent short circuits.

THE EFFECT OF CABLE DESIGN ON FIRE PERFORMANCE

The previous section described some material considerations but the construction or design of a cable also has a major bearing on how that cable will perform in a fire. The construction is particularly important for those cables designed to exhibit fire survival or circuit integrity characteristics. Mineral insulated cables are a good example of cables designed to operate under fire conditions, as they use a mineral powder insulant, often magnesium oxide, separating the copper conductors inside a tubular copper sheath. The only combustible material content is a polymeric oversheath sometimes used for identification purposes or for when the cable is to be used in an environment hostile to copper. Cables without any polymeric layer will not liberate smoke or toxic fumes in a fire. These cables are robust, can withstand a high level of mechanical impact or crushing, and their fire performance is limited only by the melting point of copper. Provided the temperature does not approach 1085°C, and there is no major structural damage to the installation, these cables can continue operating during and after the fire, so maintaining essential electrical circuits.

Aside from mineral insulated cables, there are several other means by which the design of a cable can be engineered to offer limited fire survival performance. One effective way is to employ mica glass tapes (as briefly mentioned earlier[15,16]) which are applied around individual conductors, underneath polymeric insulation. The insulation material varies, but is commonly EPR or crosslinked polyethylene (XLPE), both of which are not fire retardant. The principle of operation is that the fused mica–glass layer remains when all polymeric components have been consumed by the fire. The fused inorganic layer is thermally stable and very robust, but is brittle and so could easily be damaged by mechanical impact or bending. The other components of mica–glass based cables can influence the fire performance also and often a combination of fire resistant tapes and reduced fire hazard polymeric materials are employed in conjunction with the mica tapes.

Armoured cables can also give fire performance benefits. First, metallic armour acts to conduct heat away from the immediate area affected by the fire and reduces the rate of heat build up; second, armour provides a degree of protection against mechanical shock and also helps to maintain the structure of the cable.

In a similar vein to mica–glass tapes is the principle of inorganic refractory coatings applied to the conductors using techniques such as sputter deposition. The circuit integrity performance is achieved in a similar way to mica tapes and similar comments apply; a number of patents have been issued describing this technology.[17,18,19]

Another common way of providing a limited fire survival capability is to use silicone elastomers in combination with fire resistant tapes.[10,11] The combustion behaviour of silicones is mentioned in the previous section, and it is the formation of the silica ash which is utilised to maintain circuit integrity. The construction of these cable types is especially important if enhanced fire performance is to be achieved. The use of woven glass-fibre tapes or other fire resistant tapes is necessary in order to enclose the powdery ash formed on combustion around the conductors. Clearly, the use of metallic armour components in these cables will help to maintain the structural integrity of the cable during a fire.

The phenomenon of intumescence mentioned in the previous section is also utilised to improve fire resistance, and this is often achieved by the use of intumescent tapes,[20] applied over the outer polymeric sheath. There are a number of other means by which the behaviour in a fire can be influenced, but it is often the case that cables employ several construction and material elements combined in order to achieve a better overall performance.

FIRE TEST METHODS AND STANDARDS

It is clear that the hazard from cables involved in a fire can take many forms, from the ease of ignition and flame propagation to the evolution of smoke and toxic gases. Many tests have been developed over the years in an attempt to evaluate the potential performance of both individual materials and complete cables. It has been an ongoing problem to find test methods that can not only differentiate between materials, but which are capable of yielding useful information with respect to real fire behaviour. Many of the small laboratory-scale methods, while being useful for material comparison, often cannot be directly correlated to real fires. Part of the problem is that no two fires are the same, making accurate simulation very difficult. Efforts in recent years have concentrated on tests that try to simulate the installation environment of cables and this approach is leading to ever more realistic methods of fire performance evaluation. In recognition of the fact that many small-scale tests are not greatly relevant to real fire behaviour, the aim is now to identify ways of assessing the total fire hazard.

Fire test methods relating to cables can be split into two categories, those which test the whole cable and those which evaluate individual component materials. The tests on materials are not specific to cable standards but are often specified therein. There is a wide range of tests to assess many of the aspects of fire behaviour mentioned above. A detailed review of existing cable test methods was compiled in 1991 by the National Institute of Standards and Technology (NIST),[21] whilst the *International Plastics Flammability Handbook*[22] gives a good overview on test methods for materials.

Material test methods

Flammability/Ignitability
Oxygen index is perhaps the most widely used indicator of a material's flammability. It is the minimum percentage of oxygen in an oxygen/nitrogen mixture required to support combustion of a given material at room temperature. Whilst this measure allows comparison of different materials, it does not provide an indication of how that material will behave in a real fire. Flame retardant materials require a level of oxygen higher than that normally present in the atmosphere (21%) for burning to be maintained and a material having an oxygen index of 26 or above is considered to be self-extinguishing. There are a number of reasons why this test fails to characterise fully the burning behaviour of materials. The test uses a small sample which is held vertically and ignited at the top and the direction of flame propagation is therefore downwards. In most fires, vertical propagation of fire predominates and material ahead of the flame front is heated, leading to an increased rate of decomposition. Another problem is that the oxygen index test is carried out at room temperature but the ease of ignition of materials is usually increased with increasing temperature.

To address the effect of temperature, another test was developed using a similar apparatus. In this case a sample is again held in the same configuration and ignited at the top, but the test is carried out at elevated temperatures in a standard 21% oxygen atmosphere; the temperature index value is the minimum temperature which is required to support combustion. This test is now specified by London Underground Limited and is becoming more widely used. Additionally, a requirement for a temperature index of 280°C is often stipulated. Despite being more widely used, this test still cannot provide any reliable prediction of how a material will behave in a fire.

The flash and self-ignition temperatures of materials are often quoted (see table 6.2), but such values are rarely specified in cable standards. Most small-scale laboratory tests such as these have limited use in relation to understanding real fire performance.

Smoke evolution

Smoke evolution is another critical performance indicator which needs to be evaluated on a laboratory scale and there are a number of methods used, based either on gravimetric or optical techniques. The Arapahoe smoke chamber is the best known gravimetric method and involves the collection of smoke particles on a filter at the top of a cylinder attached to a combustion chamber. The specimen is exposed to a small propane gas flame at the base of the chamber. However, because of the problems arising from smoke particle flocculation and the variable nature of smoke generated, optical methods are preferred. The NBS smoke chamber test, according to ASTM E-662-79, measures the density of smoke produced from a 76 mm square sample exposed to a fixed heat flux using a vertical photometer. Flaming and non-flaming modes can be used.

Corrosivity

The corrosivity of gases produced during burning can be measured in a number of ways.[23] The concern regarding the evolution of halogen acids was addressed by a technique specified in IEC 754-1 and BS 6425: Part 1. This test measures the levels of gases liberated from the pyrolysis of 1 g of material at 800°C and a material must liberate less than 0.5% HCl to be classed as zero halogen. Hydrogen bromide and chloride cannot, however, be analysed separately. In addition, non-halogenated acid gases cannot be analysed so other methods have been developed to overcome this deficiency. Tests are available to measure the pH and ionic conductivity of aqueous solutions of combustion products thus giving an indication of levels of corrosive gases. The corrosion of metals can also be evaluated, by measuring the yield of soluble metal ions. As a result of the concern relating to the corrosion of sensitive electronic equipment during fires, a test (DEC-0611/C) has been developed by the French National Centre for Telecommunications Studies (CNET), whereby the change in resistance of a circuit exposed to combustion products is measured.

Toxicity

The toxicity of combustion products is perhaps one of the most difficult things to measure, and has been the subject of considerable research.[3,24,25] There is a very wide range of different toxic gases liberated from all burning materials; the particular chemical cocktail depends on the type of polymer and the nature of additives present, and on the conditions under which the combustion occurs. Oxygen levels and temperatures can have a large effect on the type and level of the gases evolved. The identification and measurement of decomposition products is often very complicated and much work has been carried out in this area. The *Flammability Handbook for Plastics*[1] lists, for each of a range of polymers, the gases that have been identified as being produced during combustion. For many of these chemicals, the physiological effect on humans has been identified. However, historically it has been common to look at the effect of toxic combustion fumes on laboratory animals. The toxic hazard from the combustion of a given material is often indicated by LC_{50}, which is the concentration of material which causes a mortality rate of 50% in a controlled test. Clearly, the effect of toxic gases is likely to be different in humans than in rats or mice.

Little is known though, about the toxicity of mixtures of gases, but attempts have been made to analyse the combustion products for a range of toxic gases and then to sum the toxicities to obtain a 'hazard index'. Such methods may be of some use, but it is very difficult to simulate the conditions experienced in a real fire and measurement of toxic gas emissions in a controlled environment will be dependent upon the conditions used. Another approach to investigating combustion toxicology was adopted by Professor Einhorn[25] of the Flammability Research Centre of the University of Utah. In a study of fires involving PVC wiring, fire victims were examined on admission to hospital and breath and blood samples analysed. This method could help to provide very useful information relating to real fire situations in terms of toxic gas evolution.

Clearly, it could be many years before the full relationship between the chemical nature of cable materials and their toxicity can be determined. For further reference more detailed studies have been carried out by Punderson[26,27] and other task forces.[28] As mentioned in the section on materials, inhalation of hot and/or toxic gases has been attributed to a large proportion of fire deaths, but it is the problem of smoke obscuration that is a major factor responsible for preventing the safe escape from a fire situation.

Heat release

Another technique that is being increasingly used to evaluate the burning behaviour of materials on a small scale is the cone calorimeter. The apparatus consists of a chamber in which a 100 mm square sample can be exposed to a range of radiant heat levels; the pyrolysis gases are ignited by a spark and combustion gases are extracted through an exhaust system. The calorimeter measures the amount of oxygen consumed, based on the principle that the net heat of combustion is proportional to oxygen consumption. The sample is mounted either horizontally or vertically on to a load cell to allow weight loss measurements with time. It is possible to obtain a number of useful parameters, such as the peak rate of heat release, the time to peak rate of heat release, total heat release, smoke release data, and weight loss due to combustion. The technique therefore allows a large amount of information to be obtained for a given material under a range of conditions. Work has been undertaken in an attempt to correlate data from the laboratory cone calorimeter to that obtained from large-scale tests and reasonable agreement has been obtained.[29,30]

Tests on cables

Table 6.3 summarises the principal standards used to assess the performance of cables and their constituent materials.

There are both small- and large-scale cable tests to evaluate fire performance, but all fall into two categories: reaction to fire and resistance to fire. The tests which determine how a cable reacts to fire generally involve observing the flame propagation or smoke evolution behaviour in either horizontal, inclined, or a vertical configuration. Methods investigating fire resistance or fire survivability usually entail the application of an applied voltage in order to test for the maintenance of circuit integrity.

Reaction to fire – propagation

Some of the most widely applied tests are those as specified in the IEC 332 series (based on BS 4066), which evaluates the small-scale vertical fire propagation behaviour of cables. IEC 332: Part 1 is a test on a single vertical wire or cable. A 600 mm long cable sample is suspended vertically in a draught-free enclosure and the lower end is exposed to a gas burner angled at 45° to the horizontal. The requirement is that after a specified

Table 6.3 Major standards for assessing fire performance of cables

Standard	Title	Test methods referred to
BS 7835 (1996)	Cables with extruded crosslinked polyethylene or ethylene propylene rubber insulation having low emission of smoke and corrosive gases when affected by fire for rated voltages from 3.8/6 kV up to 19/33 kV.	BS 6425: Part 1 (corrosive and acid gas) BS 7622 (smoke emission – 3 m^3)
BS 6724 (1990)	Specification for armoured cables having thermosetting insulation with low emission of smoke and corrosive gases when affected by fire for electricity supply.	BS 6425: Part 1 BS 7622
BS 7211 (1994)	Specification for thermosetting insulated cables (non-armoured) with low emission of smoke and corrosive gases when affected by fire for electric power and lighting.	BS 6425: Part 1 BS 4066 (vertical wire) BS 7622
BS 7629 (1993)	Specification for thermosetting cables with limited circuit integrity when affected by fire.	—
BS 6387 (1994)	Specification for performance requirements for cables required to maintain circuit integrity under fire conditions.	BS 4066 (Part 1) IEC 331 BS 6207 (MI copper sheathed)
BS 6883 (1991)	Specification for elastomeric insulated cables for fixed wiring on ships and on mobile and fixed offshore units.	—
BS 7622 Parts 1, 2 (1993)	Measurement of smoke density of electric cables burning under defined conditions. (Technically equivalent to IEC 1034, Parts 1, 2.)	—
BS 6425 Parts 1, 2 (1990/93)	Tests on gases evaluated during the combustion of materials from cables. Method for evaluating level of HCl gas emission (Part 1), and determination of degree of corrosivity by pH and conductivity measurements (Part 2). (Similar to IEC 754 (Parts 1, 2).)	—
BS 7655 (1993/4) (several Parts)	Insulating and sheathing materials for cables. Series of standards specifying cable materials having, low smoke, and low corrosive gas emission	BS 6425

(cont.)

Table 6.3 continued

Standard	Title	Test methods referred to
IEC 331 (1070)	Fire resisting characteristics of electric cables.	—
IEC 332 Parts 1–3 (1989–93)	See BS 4066: Parts 1–3.	—
IEC 754 Part 1 (1994)	Tests on gases evolved during combustion of electric cables – determination of the halogen acid gas evolved during the combustion of polymeric materials taken from cables. (Similar to BS 6425: Part 1 (1990).)	—
NES 518 (1993)	Requirements for limited fire hazard (LFH) sheathing for electric cables.	NES 525 NES 711 (Determination of smoke index) NES 713 (Determination of toxicity index) NES 715 (Temperature Index Determination)
NES 527	Requirements for cables, electric, fire survival, high temperature zones and LFH sheathed.	—

time the cable shall not have charred to a distance of less than 50 mm from the upper clamp. The burner and the fuel mixture are specified to ensure the reproducibility of the test. IEC 332: Part 2 is basically the same test, but is used for smaller diameter wires such that the conductors could melt in the Part 1 test, and specifies a lower energy burner. Part 3 differs in that bunched cables are tested.

During the 1960s and 1970s there were several fires involving cables at power stations, where the fire was propagated along ducts and tunnels at speeds of up to 10 m/minute. In response to the incidents, new tests were developed to measure fire propagation, and new cables designed to exhibit reduced propagation behaviour. These tests have evolved in a number of stages over the years, culminating in the international standard IEC 332: Part 3. This test attempts to simulate a real installation environment and uses bunched cables, 3.5 m long, fixed in a vertical arrangement. The ignition source is a ribbon type propane/air burner with a fuel input of 73.7 MJ/hour. The burner is arranged horizontally at the foot of the ladder to which the cables are fastened and is applied for 20 or 40 minutes, depending on the category. In order to pass the test the cables should not propagate fire vertically more than 2.5 m. Different configurations of cables on the ladder are used depending on the size of cables and the ratio of combustible material they contain. Work is ongoing to further develop the test so that cable loading configurations can simulate different installation environments such as railways and ships.

Although the IEC 332 test is widely recognised internationally, there are a number of national variants based on the theme of measuring cable propagation behaviour. One example in which a horizontal configuration is used is the American Steiner Tunnel test, UL910. The French and Belgian authorities employ a test using a radiant heat furnace in which the bundled cables are effectively enclosed. This test is particularly applicable for testing cables specified for use in the railway industry. A variation of this test has been developed in Sweden using a tray of alcohol as a fire source and this test is employed by the UK Ministry of Defence for evaluating shipboard cables (NES 641).

Reaction to fire – smoke evolution
The methods mentioned earlier for quantifying smoke evolution are small-scale laboratory tests and use artificial conditions, which are difficult to apply to complete cable samples. However, a large-scale test has been developed by London Underground Limited which allows the evaluation of smoke emission behaviour for cables intended for use in mass transit systems. This test is now internationally recognised under the designation of IEC 1034. The test is carried out in a chamber which forms a cube having 3 m sides and walls and for this reason the test is often referred to as the 3m^3 test. A 1 m length of cable or bunch of cables is suspended horizontally above a metal bath containing 1 litre of alcohol which is ignited to form the fire source. The smoke generated is monitored throughout the test by the use of a light beam and photocell arrange horizontally across the chamber at a height of 2.15 m. A fan is used to distribute the smoke uniformly but is screened from the burning cable to prevent fanning of the flames. The 1 litre of alcohol provides an energy input of 23.5 MJ and burns for a period of approximately 30 minutes. The smoke density is measured during this time and for 10 minutes after the fuel has been consumed. For all cable sizes the specification sets a limit of 60% light transmittance.

Fire resistance
IEC 331, which is based on BS 6387, is the most widely used test to investigate the fire resistance or survival characteristics of cables. The test is intended to evaluate cables which are designed to be used for fire alarm circuits, emergency lighting, and cables for other emergency services. The test establishes whether a cable can maintain electrical circuit integrity for up to 3 hours at temperatures ranging from 750°C up to in excess of 950°C. The energised cable (at the rated voltage) is mounted horizontally in a test chamber and is exposed to a gas flame from a ribbon type burner, adjusted to give the appropriate temperature (fig. 6.1). No electrical failure must occur during the test, and up to 12 hours after the test when the cable is re-energised. There are many variations of this test using different conditions and a cable is rated depending on how the cable performs in the various categories. At its most stringent, the test requires a cable to maintain circuit integrity for 3 hours at temperatures of 950°C or above and also to withstand mechanical impact and a water spray while still maintaining electrical integrity.

In response to the requirements of the construction industry, and to some perceived deficiencies in this small-scale test, the use of large-scale furnace testing for fire resistance is being developed, especially in Germany and Belgium, in which the complete cable assembly, including ceiling, ladder and tray support systems are evaluated. The whole assembly is subjected to the standard time/temperature curve as specified in ISO 834 Part 1. The trend to develop ever more realistic tests for cables as well as other products is well established and the subject of considerable worldwide effort.

Fig. 6.1 Cable undergoing fire resistance tests to IEC Recommendation 331

SUMMARY

Cables have been developed which have a significantly reduced tendency to propagate fire, even under onerous conditions, and major reductions have been achieved in the emission of smoke and toxic and corrosive gases from burning cables. Cables and systems have been developed which will provide rapid detection and location of fires.

Considerable efforts have been made to understand better the combustion behaviour of cables and their constituent materials, and much work has concentrated on developing new test methods which better simulate real fire situations. The application of mathematical modelling of fire behaviour is another area which is also making a contribution to our ability to understand and predict how materials will behave in reality. However, despite the advances in both cables and cable test methods, care must always be taken when interpreting the result of tests, especially small-scale tests, when selecting cables for a particular application. Good performance in a laboratory test can never give a guarantee of how a cable will perform in a real fire, it can only give an indication. Research and development will continue to allow improvements to be made in cable fire performance and help to reduce further the risks from cables involved in fire.

REFERENCES

(1) Hilado, C. J. (1990) *Flammability Handbook for Plastics*, 4th Edn. Technomic Publishing Co. Inc.

(2) Cullis, C. F. and Hirschler, M. M. (1981) *The Combustion of Organic Polymers*, International Series of Monographs on Chemistry. Clarendon Press.
(3) Landrock, A. H. (1983) *Handbook of Plastics Flammability and Combustion Toxicology*. Noyes Publications.
(4) Lewis, C. F. (1989) *Materials Engineering*, **106**, 31.
(5) Camino, G. and Costa, L. (1986) *Reviews in Inorganic Chemistry*, **8**, 69.
(6) Camino, G. et al. (1989) *Die Ang. Makromol. Chemie*, **172**, 153.
(7) Halpern, Y. et al. (1984) *Ind. Eng. Chem. Prod. Res. Dev.*, **23**, 233.
(8) GB Patent 2226022A (Vactite Ltd 1990).
(9) European Patent 0559382 A1 (AT&T 1993).
(10) European Patent 0628972 A2 (BICC Cables Ltd 1994).
(11) Belgian Patent 880326 (Soc. Belge. Fab. Cable 1980).
(12) GB Patent 2046771A (Toray Silicones Ltd 1980).
(13) US Patent 1559454 (General Electric Co. 1980).
(14) GB Patent 2238547A (VSEL 1991).
(15) GB Patent 2050041 (Pirelli Cables 1979).
(16) European Patent 2050041 A2 (Int. Standard Electric Corp. 1979).
(17) European Patent 0249252 A1 (Raychem 1984).
(18) WO Patent 89/00545 (Raychem 1989).
(19) US 4985313 (Raychem 1995).
(20) European Patent A2 (3M 1984).
(21) Babrauskas, V. et al. (1991) *Fire Performance of Wire and Cable: Reaction to Fire Tests – A Critical Review of the Existing Methods and of New Concepts*, NIST Technical Note 1291. US Department of Commerce.
(22) Troitzsch, J. (1990) *International Plastics Flammability Handbook, Principles-Regulations-Testing and Approval*, 2nd Edn. Hanser Publications.
(23) Patton, J. S. (1979) *J. Fire Sciences*, **9**, 149.
(24) Sasse, H. R. (1985) *J. Fire Sciences*, **6**, 42.
(25) Einhorn, I. N. and Grunnet, M. L. (1979) *The physiological and toxicological aspects of degradation products produced during the combustion of polyvinyl chloride polymers: flammability of solid polymer cable dielectrics*, EPRI Report No. EL-1263, TPS 77-738.
(26) Punderson, J. O. (1977) 'Toxicity and Fire Safety of Wire Insulation: A State-of-the-Art Review', *Wire Tech.*, **64** (Sep/Oct).
(27) Punderson, J. O. (1981) 'Toxic Fumes – Forget Standards, Moderate the Fire', *Electr. Rev.*, **208**, 24.
(28) National Materials Advisory Board (1978) *Flammability, Smoke, Toxicity and Corrosive Gases of Electric Cable Materials*. Washington DC: National Academy of Sciences, Publication No. NMAB-342.
(29) Hirschler, M. M. (1996) *Fire and Materials*, **20**, 1.
(30) Hirschler, M. M. (1994) *Fire and Materials*, **18**, 16.

CHAPTER 7
Cable Standards and Quality Assurance

STANDARDS

The most universal standardising authority for cables is the International Electrotechnical Commission (IEC), although comparatively little commercial business is placed directly against the IEC standards. This arises because IEC standards cater for a large variety of permissible options and serve mainly as a basis for the preparation of national standards, which are usually prepared in accordance with the IEC requirements. Furthermore the IEC standards represent a consensus of national opinions and hence take several years both to prepare initially and for agreement to be reached on amendments. If all minor points were to be included, the time period for resolution would be extremely lengthy, especially in dealing with new developments. Countries such as the UK which have always been well to the fore in cable development are therefore able to issue much more comprehensive and up-to-date standards.

This can be seen from the number of items in the lists below of the relevant IEC and British Standards. Whilst it might have been of interest to include standards from other countries, these tend to be even more numerous, e.g. in the USA where in addition to national standards for materials and components there is widespread use by industry at large of cable standards issued by four bodies: Underwriter's Laboratories (UL), Association of Edison Illuminating Companies (AEIC) and jointly by the Insulated Cables Engineers' Association and the National Electrical Manufacturers' Association (ICEA/NEMA). In the UK some large organisations have separate specifications for their own use but many adopt the available British Standards and only add any requirements necessary for their particular purposes.

British and IEC Standards

Cables and flexible cords

BS 638: Part 4: 1979	Welding cables
BS 4553: 1991	600/1000 V PVC-insulated single-phase split concentric cables with copper conductors for electricity supply
BS 5055: 1991	Elastomer-insulated cables for electric signs and high-voltage luminous-discharge-tube installations

BS 5308	Instrumentation cables: Part 1: 1986 (1993) Polyethylene insulated cables Part 2: 1986 (1993) PVC insulated cables
BS 5467: 1989	Cables with thermosetting insulation for electricity supply for rated voltages of up to and including 600/1000 V and up to and including 1900/3300 V
BS 5593: 1978 (1991)	Impregnated paper-insulated cables with aluminium sheath/neutral conductor and three shaped solid aluminium phase conductors (CONSAC), 600/1000 V, for electricity supply
BS 6004: 1995	PVC-insulated cables (non-armoured) for electric power and lighting
BS 6007: 1993	Rubber-insulated cables for electric power and lighting
BS 6141: 1991	Insulated cables and flexible cords for use in high temperature zones
BS 6195: 1993	Insulated flexible cables and cords for coil leads
BS 6207	Mineral insulated cables with a rated voltage not exceeding 750 V: Part 1: 1995 Cables Part 2: 1995 Terminations
BS 6231: 1990	PVC-insulated cables for switchgear and controlgear wiring
BS 6346: 1989	PVC-insulated cables for electricity supply
BS 6480: 1988	Impregnated paper-insulated lead or lead-alloy sheathed electric cables of rated voltages up to and including 33000 V
BS 6500: 1994	Insulated flexible cords and cables
BS 6622: 1991	Cables with extruded crosslinked polyethylene or ethylene propylene rubber insulation for rated voltages from 3800/6600 V up to 19000/33000 V
BS 6708: 1991	Flexible cables for use in mines and quarries
BS 6724: 1990	Armoured cables for electricity supply having thermosetting insulation with low emission of smoke and corrosive gases when affected by fire
BS 6726: 1991	Festoon and temporary lighting cables and cords
BS 6862	Cables for vehicles Part 1: 1971 (1990) cables with copper conductors
BS 6883: 1991	Elastomer insulated cables for fixed wiring in ships and on mobile and fixed offshore units
BS 6977: 1991	Insulated flexible cables for lifts and for other flexible connections
BS 7211: 1994	Thermosetting insulated cables (non-armoured) for electric power and lighting with low emission of smoke and corrosive gases when affected by fire
BS 7629: 1993	Thermosetting insulated cables with limited circuit integrity when affected by fire
IEC 55	Paper-insulated metal-sheathed cables for rated voltages up to 18/30 kV (with copper or aluminium conductors and excluding gas pressure and oil-filled cables)

	55-1 Part 1 – Tests
	55-2 Part 2 – Construction
IEC 92	Electrical installations in ships:
	There are many parts of this standard, of which those relevant are 92-3, 92-351, 92-352, 92-359, 92-373, 92-375 and 92-376
IEC 227	Poly(vinyl chloride)-insulated cables of rated voltages up to and including 450/750 V:
	227-1 Part 1 – General requirements
	227-2 Part 2 – Test methods
	227-3 Part 3 – Non-sheathed cables for fixed wiring
	227-4 Part 4 – Sheathed cables for fixed wiring
	227-5 Part 5 – Flexible cables (cords)
	227-6 Part 6 – Lift cables and cables for flexible connections
IEC 245	Rubber-insulated cables of rated voltages up to and including 450/750 V
	245-1 Part 1 – General requirements
	245-2 Part 2 – Test methods
	245-3 Part 3 – Heat-resisting silicone-insulated cables
	245-4 Part 4 – Cords and flexible cables
	245-5 Part 5 – Lift cables
	245-6 Part 6 – Arc welding electrode cables
IEC 502	Extruded solid dielectric insulated power cables for rated voltages from 1 kV to 30 kV
IEC 541	Comparative information on IEC and North American flexible cord types
IEC 702	Mineral insulated cables and their terminations with a rated voltage not exceeding 750 V
	702-1 Part 1 – Cables
	702-2 Part 2 – Terminations
IEC 800	Heating cables with a rated voltage of 300/500 V for comfort heating and prevention of ice formation
IEC 1138	Cables for portable earthing and short-circuiting equipment

Conductors

BS 2627: 1970 (1985)	Wrought aluminium for electrical purposes. Wire
BS 3988: 1970	Wrought aluminium for electrical purposes. Solid conductors for insulated cables
BS 4109: 1970 (1991)	Copper for electrical purposes. Wire for general electrical purposes and for insulated cables and flexible cords
BS 5714: 1979 (1987)	Method of measurement of resistivity of metallic materials
BS 6360: 1991	Conductors in insulated cables and cords
IEC 228	Conductors of insulated cables
	228A First Supplement: Guide to the dimensional limits of circular conductors

Cable Standards and Quality Assurance

Insulation and sheathing (non-metallic)

BS 6234: 1987	Polyethylene insulation and sheath of electric cables
BS 6746: 1990	PVC insulation and sheath of electric cables
BS 6746C: 1993	Colour chart for insulation and sheath of electric cables
BS 6899: 1991	Rubber insulation and sheath of electric cables
BS 7655	Insulating and sheathing materials for cables

BS 7655 comprises a number of parts and sections and will replace both BS 6746 and BS 6899 in cable specifications.

Part 0: 1993 General introduction
Part 1: Elastomeric insulating compounds
   Section 1.1: 1993 Harmonised types
   Section 1.2: 1993 General 90°C application
   Section 1.3: 1993 XLPE
   Section 1.4: 1993 Oil resisting types
   Section 1.5: 1993 Flame retardant composites
   Section 1.6: 1993 Coil end lead types
Part 2: Elastomeric sheathing compounds
   Section 2.1: 1993 Harmonised types
   Section 2.2: 1993 Heat resisting types
   Section 2.3: 1993 General application
   Section 2.4: 1993 Welding cable covering
   Section 2.5: 1993 Sheathing compounds having low smoke and acid gas emission for general applications
   Section 2.6: 1993 Sheathing compounds for ships' wiring and offshore applications
Part 3: PVC insulating compounds
   Section 3.1: 1993 Harmonised types
   Section 3.2: 1993 Hard grade types
Part 4: PVC sheathing compounds
   Section 4.1: 1993 Harmonised types
   Section 4.2: 1993 General application
   Section 4.3: 1993 Special applications – RF cables
Part 5: Crosslinked insulating compounds having low emission of corrosive gases, and suitable for use in cables having low emission of smoke when affected by fire
   Section 5.1: 1993 Harmonised crosslinked types
Part 6: Thermoplastic sheathing compounds having low emission of corrosive gases, and suitable for use in cables having low emission of smoke when affected by fire
   Section 6.1: 1994 General application thermoplastic types

IEC 173	Colours of the cores of flexible cables and cords
IEC 304	Standard colours for PVC insulation for low frequency cables and wires
IEC 391	Marking of insulated conductors
IEC 446	Identification of insulated and bare conductors by colour

Tests on cables and materials

BS EN 10002	Tensile testing of metallic materials
BS 903	Physical testing of rubber
BS 2782	Methods of testing plastics

BS EN 10002, BS 903 and BS 2782 contain many parts and sections

BS 4066   Tests on electric cables under fire conditions
- Part 1: 1980 (1995) Method of test on a single vertical insulated wire or cable
- Part 2: 1989 Method of test on a single small vertical insulated wire or cable
- Part 3: 1994 Tests on bunched wires or cables

BS 5099: 1992   Spark testing of electric cables

BS 6387: 1994   Performance requirements for cables required to maintain circuit integrity under fire conditions

BS 6425   Tests on gases evolved during the combustion of materials from cables
- Part 1: 1990 Method for determination of amount of halogen acid gas evolved during combustion of polymeric materials taken from cables
- Part 2: 1993 Determination of degree of acidity (corrosivity) of gases by measuring pH and conductivity

BS 6469   Insulating and sheathing materials of electric cables

BS 6469 comprises several parts and sections, it is being replaced by BS EN 60811, with the exception of:
- Part 5: Methods of test specific to filling compounds
  - Section 5.1: 1992 Drop point. Separation of oil. Lower temperature brittleness. Total acid number. Absence of corrosive components. Permittivity at 23°C. D.C. resistivity at 23°C and 100°C
- Part 99: Test methods used in the United Kingdom but not specified in Parts 1 to 5
  - Section 99.1: 1992 Non-electrical tests
  - Section 99.2: 1992 Electrical tests

BS 6470: 1984 (1991)   Method for determination of water in insulating oils, and in oil-impregnated paper and pressboard

BS 7622   Measurement of smoke density of electric cables burning under defined conditions
- Part 1: 1993 Test apparatus
- Part 2: 1993 Test procedure and requirements

BS EN 60811   Insulating and sheathing materials of electric cables – Common test methods
- BS EN 60811-1: General application
  - BS EN 60811-1-1: 1995 Measurement of thickness and overall dimensions – Tests for determining the mechanical properties

Cable Standards and Quality Assurance

	BS EN 60811-1-2: 1995	Thermal ageing methods
	BS EN 60811-1-3: 1995	Methods for determining the density – Water absorption tests – Shrinkage test
	BS EN 60811-1-4: 1995	Tests at low temperatures
	BS EN 60811-2:	Methods specific to elastomeric compounds
	BS EN 60811-2-1: 1995	Ozone resistance test – Hot set test – Mineral oil immersion test
	BS EN 60811-3:	Methods specific to PVC compounds
	BS EN 60811-3-1: 1995	Pressure test at high temperature – Tests for resistance to cracking
	BS EN 60811-3-2: 1995	Loss of mass test – Thermal stability test
	BS EN 60811-4:	Methods specific to polyethylene and polypropylene compounds
	BS EN 60811-4-1: 1995	Resistance to environmental stress cracking Wrapping test after thermal ageing in air – Measurement of the melt flow index – Carbon black and/or mineral content measurement in PE
IEC 55	See above	
IEC 60	High voltage test techniques:	
	60-1 Part 1 – General definitions and test requirements	
	60-2 Part 2 – Test procedures	
	60-3 Part 3 – Measuring devices	
	60-4 Part 4 – Application guide for measuring devices	
IEC 141	Tests on oil-filled and gas pressure cable and their accessories:	
	141-1 Part 1 – Oil-filled, paper-insulated, metal-sheathed cables for alternating voltages up to and including 400 kV	
	141-2 Part 2 – Internal gas pressure cables and accessories for alternating voltages up to and including 275 kV	
	141-3 Part 3 – External gas pressure (gas compression) cables and accessories for alternating voltages up to 275 kV	
	141-4 Part 4 – Oil-impregnated paper-insulated high pressure oil-filled pipe-type cables and accessories for alternating voltages up to and including 400 kV	
IEC 229	Tests on cable oversheaths which have a special protective function and are applied by extrusion	

IEC 230	Impulse tests on cables and their accessories
IEC 332	Tests on electric cables under fire conditions:
	332-1 Part 1 – Test on a single vertical insulated wire or cable
	332-2 Part 2 – Test on a single small vertical insulated copper wire or cable
	332-3 Part 3 – Tests on bunched wires or cables
IEC 538	Electric cables, wires and cords: methods of test for polyethylene insulation and sheath
IEC 540	Test methods for insulation and sheaths of electric cables and cords (elastomeric and thermoplastic compounds)
IEC 754	Test on gases evolved during combustion of electric cables:
	754-1 Part 1 – Determination of the amount of halogen acid gas evolved during the combustion of polymeric materials taken from cables
	754-2 Part 2 – Determination of degree of acidity of gases evolved during the combustion of materials taken from electric cables by measuring pH and conductivity
IEC 811	Common test methods for insulating and sheathing materials of electric cables
	811-1 Part 1 – Methods for general application
	811-2 Part 2 – Methods specific to elastomeric compounds
	811-4 Part 4 – Methods specific to polyethylene and polypropylene compounds
	811-5 Part 5 – Methods specific to filling compounds
IEC 840	Tests for power cables with extruded insulation for rated voltages above 30 kV (U = 36 kV) up to 150 kV (U = 170 kV)
IEC 885	Electrical test methods for electric cables
	885-1 Part 1 – Electrical tests for cables, cords and wires for voltages up to and including 450/750 V
	885-2 Part 2 – Partial discharge tests
	885-3 Part 3 – Test methods for partial discharge measurements on lengths of extruded power cables
IEC 1034	Measurement of smoke density of electric cables burning under defined conditions
	1034-1 Part 1 – Test apparatus
	1034-2 Part 2 – Test procedures and requirements

Jointing and accessories

BS 4579	Performance of mechanical and compression joints in electric cable and wire connectors
	Part 1: 1970 (1988) Compression joints in copper conductors
	Part 2: 1973 (1990) Compression joints in nickel, iron and plated copper conductors
	Part 3: 1976 (1988) Mechanical and compression joints in aluminium conductors

Cable Standards and Quality Assurance 115

BS 5372: 1989	Dimensions of cable terminations for 3-core and 4-core polymeric insulated cables of rated voltages 600/1000 V and 1900/3300 V having aluminium conductors
BS 6121	Mechanical cable glands
	Part 1: 1989 Metallic glands
	Part 2: 1989 Polymeric glands
	Part 3: 1990 Special corrosion resistant glands
	Part 5: 1992 Code of practice for selection, installation and inspection of cable glands used in electrical installations
IEC 702	Part 2 See above

Miscellaneous

BS 801: 1984 (1991)	Composition of lead and lead alloy sheaths of electric cables
BS 1441: 1969 (1988)	Galvanised steel wire for armouring submarine cables
BS 1442: 1969 (1986)	Galvanised mild steel wire for armouring cables
BS 2897: 1970 (1985)	Wrought aluminium for electrical purposes. Strip with drawn or rolled edges
BS 7450: 1991	Method for determination of economic optimisation of power cable size
BS 7454: 1991	Method for calculation of thermally permissible short-circuit currents, taking into account non-adiabatic heating effects
BS 7540: 1994	Guide to use of cables with a rated voltage not exceeding 450/750
BS 7769	Electric cables. Calculation of the current rating
IEC 38	IEC standard voltages
IEC 71	Insulation co-ordination
	71-1 Part 1 – Terms, definitions, principles and rules
	71-2 Part 2 – Application guide
	71-3 Part 3 – Phase-to-phase insulation co-ordination: principles, rules and application guide
IEC 183	Guide to the selection of high voltage cables
IEC 287	Calculation of the continuous current rating of cables (100% load factor)
IEC 331	Fire-resisting characteristics of electric cables
IEC 364	Electrical installations of buildings. This has a number of parts, which are subdivided into chapters, and sections, of which some have a bearing on cables; the following, which supersedes IEC 448, is particularly relevant:
	364-5-523 Part 5 – Selection and erection of electrical equipment
	Chapter 52 Wiring systems
	Section 523 Current carrying capacities
IEC 724	Guide to the short-circuit temperature limits of electric cables with a rated voltage not exceeding 0.6/1.0 kV

IEC 949   Calculation of thermally permissible short-circuit currents, taking into account non-adiabatic heating effects
IEC 1059   Economic optimisation of power cable size

Influence of CENELEC

CENELEC, the European Committee for Electrotechnical Standardisation, has an important effect on the preparation and issue of new cable standards in Europe. Membership consists of the electrotechnical standards organisations of the countries of the European Union together with those of the European Free Trade Association (EFTA) outside the EU. One of the main aims of CENELEC is to harmonise national standards in order to remove technical barriers to trading. For cables the basic work is done by a technical committee, TC20, on which all countries are represented. Working groups are established for individual subjects. The activities of CENELEC embrace a wide field of electrical equipment and regulations concerning its use, including the harmonisation of rules for electrical installations, at present up to 1000 V. The latter is covered by Technical Committee TC64, and the outcome of this committee's work has an important bearing on the IEE Wiring Regulations, BS 7671: 1992. Generally, as for TC20 and TC64, the numbering of the CENELEC technical committees is the same as for the IEC technical committees dealing with the same subjects and, because of the correlation between the work of the two organisations, it is convenient for representation for European countries to be the same, at least in part, in the two bodies.

To achieve the aims, Harmonisation Documents or European Standards are prepared taking account of IEC requirements and they are published after approval by the technical committee and other overall committees. Subsequently, and within a limited time scale, all member countries have to bring their national specifications into line, without deviations (other than any which may be justified by special national conditions, which should be only temporary, if possible). To date, in the cable field, flexible cables and cords, mineral insulated cables and some types of wiring cables have been harmonised. Harmonisation Documents have also been produced for certain components or aspects which relate to all or several types of cable, so that reference can be made to them in the documents for the cables. These include Harmonisation Documents for conductors, test methods for thermoplastic and elastomeric insulations and sheaths, the method of numbering of small cores in cables with more than five cores and a standardised system of coded designations of cables. In some fields IEC standards are adopted verbatim as Harmonisation Documents or European Standards. In the cable field this applies to the Harmonisation Documents for conductors and test methods mentioned above, which constitute endorsements of IEC 228 and IEC 811 respectively, but the Harmonisation Documents for the cables themselves, while conforming in most respects with the corresponding IEC standards, include some differences and additions agreed between the CENELEC countries.

One effect of the CENELEC procedure can be a delay in the up-dating of national standards. This arises because when work is announced on a new subject a stand-still arrangement is imposed and no changes may be made in national standards until after harmonisation has been agreed, unless special permission has been obtained from CENELEC.

Harmonised types of cable may be marketed without restriction in any of the EU and EFTA countries and attempts are made to keep the number of types to a reasonable minimum. Provided that agreement has been obtained within CENELEC it is still possible to retain non-harmonised designs as recognised national types if they are not of interest to members of other CENELEC countries. These tend to be for a wider range of conductor sizes or for particular wiring practices which are specific to national standards. Permission would not be given to types which would inhibit the use of harmonised designs.

Individual customers may still obtain cables manufactured to their own specification but it is one of the aims to keep these to a minimum and to regard them as specials for small-scale local use.

QUALITY ASSURANCE

In general the importance of quality has always been fully recognised in the cable industry and quality assurance, which, as the term implies, comprises the planned and systematic actions designed to give confidence that a product or service will satisfy the requirements for quality, is an intrinsic feature of a cablemaker's activities, as it is of reputable suppliers of most other products or services. However, quality has been the subject of increasing publicity, with governmental bodies lending their support to, and taking initiatives in, promoting its importance. There has been a growing emphasis on the formalising and documentation of quality assurance procedures, not only to enable suppliers to satisfy themselves that their system of quality assurance is effective, but also to enable them to demonstrate this to their customers and others.

Certification

A means of providing evidence that a manufacturer's quality management system and/or their product conforms to recognised standards is through certification to that effect by a body recognised to be competent to apply examinations and tests to verify it. Similarly, for a product for which no recognised standard exists, evidence of its suitability for the function claimed for it and for its safety may be provided by an approval certificate from a body recognised to be competent to make a judgement to that effect.

There are, then, three main categories of certification related to quality.

(a) Certification of the quality management system: this is a verification that the supplier's organisation, planning and system of quality control and its operation provide confidence that the supplier will satisfy requirements for quality.
(b) Certification of product conformity: this is a verification that, in so far as is reasonably ascertainable, the supplier's product conforms with the standard with which it is intended to comply, and is based upon the examination and testing of actual samples of the product taken from the supplier's production or purchased in the market.
(c) Product approval certification: for products outside the scope of existing standards, this is a certification that a product can confidently be expected to perform safely and reliably as required of it. To provide certification in this category the approvals

organisation needs to go beyond satisfying itself that the product meets criteria embodied in a standard: it needs to determine the criteria to be used to assess the likely operational performance and satisfy itself that these are met.

Over a long period in the UK cable industry, in addition to elements of self-certification on the part of manufacturers, 'second party' certification has been practised by some purchasers. Certification of a supplier by a purchaser of his products is often referred to as 'second party' certification. The Ministry of Defence and British Telecom are prominent examples of organisations that have long operated a practice of auditing the quality assurance systems and their operation of their suppliers. They, and several other major cable users who operate similar schemes, are concerned with the manufacturer's quality assurance as it affects the products supplied to them. However, the likelihood is that a manufacturer who applies a system to products for some customers will apply the same system to his production as a whole and those customers who have not the resources and/or the inclination to audit a supplier's quality assurance themselves might reasonably be influenced by the knowledge that the supplier is approved by one or more other purchasers who do operate an auditing scheme and dispense certificates. Indeed there has been a register of approvals, Register of Quality Assessed United Kingdom Manufacturers issued by the Department of Trade (at that time) and published by HM Stationery Office, which can be consulted to ascertain who has approved whom. In so far as purchasers are influenced by second party certification by another purchaser, the second party certification is used, in effect, as 'third party' certification.

Strictly, 'third party' certification is certification by an independent approvals organisation and not an individual purchaser. The certification may cover the supplier's quality management system and/or a product or range of products and it applies to these irrespective of who the purchasers may be. Any customer can therefore accept third party certification of a manufacturer as applying to his purchases, provided, of course, that the product bought is within the range covered by the certification.

In the UK cable industry third party certification was little used until the 1970s and 1980s. For example, not much use has been made for cables of the Kite Mark scheme administered by the British Standards Institution (BSI). This is in contrast to many other countries. In some, such as Canada, Denmark and Sweden, there have been national approval organisations backed by government. In some instances approval has been required by law. In some other countries third party certification has been almost essential for marketing; for example, in Germany VDE (Verband Deutscher Elektrotechniker) approval is required by many cable users.

The approvals organisation which has become most widely used in the field of cables in the UK is the British Approvals Service for Electric Cables (BASEC). This organisation is an independent approvals organisation which utilises the BSI's laboratories at Hemel Hempstead. BASEC was the approvals organisation appointed for administration in the UK of the ◁HAR▷ mark, devised by CENELEC as a reciprocal product conformity certification scheme for harmonised cables, which is explained more fully later. As well as this, BASEC's activities now cover certification of a manufacturer's quality management systems, certification of product conformity for non-harmonised cables to British Standards and product approval.

The British Standards which specify requirements for effective quality management systems are contained in the BS EN ISO 9000 series, which has now superseded BS 5750.

BS EN ISO 9000-1 gives guidance on the selection and use of the other standards in the series. Of these, the most appropriate for cable manufacture and supply are BS EN ISO 9001 and 9002. ISO 9001 is the most demanding in that it requires total quality assurance in design, development, production, installation and servicing, where appropriate. ISO 9002 does not include the requirements for design and development. It is therefore used where products are manufactured to a recognised standard and it is assumed that compliance with the standard gives an assurance that the design of the product is suitable for the intended use.

Certification of a manufacturer's quality management system by BASEC will normally be to the effect that it conforms to BS EN ISO 9001 or 9002 as the case may be. The first such BASEC certificates, denoting at that time conformity with BS 5750: Part 1 or Part 2, were issued to a number of cablemakers in 1986.

Product conformity certification has been provided by BASEC for cables to several British Standards over a period from 1973, mainly for cables of rated voltages up to 1000 V. The schemes for appraisal include type approval and regular surveillances; they began with flexibles and wiring cables and have extended to PVC and XLPE insulated 600/1000 V mains cables.

In the field of product approval BASEC is nominated in the IEE Wiring Regulations as the body to assess whether a new type of cable, not included in a British Standard, can be regarded as providing equivalent safety to types covered by the regulations, which are all required to be to British Standards. At one time this provision in the IEE Wiring Regulations, to allow for the use of newly developed equipment not strictly in accordance with the regulations, was through the Assessment of New Techniques (ANT) scheme, but this has been abandoned and, where cables are concerned, BASEC provides the equivalent. In carrying out this function BASEC may recruit experts from outside its regular staff.

Accreditation

One facet of government interest in quality assurance and certification was the setting up of a body to accredit certification bodies. This body, which has merged the activities of the NACCB and NAMAS, is now known as the UK Accreditation Service (UKAS). A body which itself issues certificates to suppliers is able to apply for a certificate of its own to signify that it is accredited by the UKAS. It is appraised in a way not unlike its own appraisal of its clients, the basic criteria being impartiality and competence, and is certified as accredited, if appropriate. It may be accredited to operate in one or more of the areas of certification of quality management systems, certification of product conformity and certification of product approval, but accreditation for product conformity or product approval certification is generally not given unless the certification body requires of its clients that their quality management systems be approved by an accredited certification body. BASEC was accredited for certification of quality management systems of suppliers of electric cables in 1986.

A certification body may operate without accreditation, relying on its established prestige within the part of industry where it operates, but obviously accreditation by an official body is seen as conferring greater status.

Governmental papers on the subject of accreditation and certification envisage the possibility of more than one certification body being accredited for operation in a given product area. Indeed there is implied encouragement of this in the interests of competition. For suppliers, however, an advantage of third party certification by a body recognised in the industry and acceptable to customers generally is that this can lead to a single assessment based upon standardised criteria in place of a number of second party appraisals which may differ from each other in detail and anyway cause duplication of time spent and expense incurred in external studies. This advantage would be lost if there were too many bodies providing third party certification in the same product area and individual customers required certification of their suppliers by different third parties.

As with certification, accreditation is on-going. After the first accreditation, appraisals are carried out periodically for maintenance of the status.

The ◁HAR▷ mark

A particular and significant form of product conformity certification is a licence to use the ◁HAR▷ mark. This is a certification system devised in CENELEC to apply to harmonised cables. While harmonisation of standards is a major step in removing technical barriers to trade between the member countries, it was recognised that differences in certification procedures, with some customers in some countries insisting on certification by their own national body, could inhibit marketing by one country in others. For this reason a number of bodies entered into mutual recognition agreements under the auspices of CENELEC and one of these, the HAR Group, relates to cables. The ◁HAR▷ mark scheme for cables, drawn up in collaboration between CENELEC TC20 and the HAR Group, is in effect a harmonised product conformity certification scheme.

Licences to use the ◁HAR▷ mark on harmonised types of cable are granted by the nominated National Approval Organisation (NAO) in each participating country. In the UK this is BASEC. The mark consists of ◁HAR▷ preceded by the mark of the NAO printed or otherwise displayed on the outside of the cable, or it may be signified by a coloured thread within the cable. The colours of the thread are yellow, red and black and the lengths of the three colours indicate the country of the NAO.

Requirements for the NAO to issue approval are common for all member countries and comprise initial inspection of manufacturing and testing facilities, the testing of samples for initial approval and subsequent surveillance by the periodic testing of samples. The numbers of samples, related to production volume, the tests to be carried out and their frequency and the bases of assessment are the same for all countries and there is reciprocal acceptance of the mark between the countries. For example, the BASEC◁HAR▷ mark would be accepted in Germany as equivalent to the VDE◁HAR▷ mark, in The Netherlands as equivalent to the KEMA◁HAR▷ mark and so on.

To achieve common safety requirements, a low voltage directive was issued at an early stage by the European Commission to establish safety standards between 50 V and 1000 V and the requirements have to be incorporated in the national laws of the EU countries. The low voltage directive was amended in 1993 to include a requirement for a mark of conformity with the essential requirements of the directive, as a pre-requisite to marketing of a product. The CE marking is not intended as a mark of quality but merely to indicate that the product is conforming with relevant directives. For the foreseeable future quality will continue to be indicated by the well known quality marks such as the ◁HAR▷ mark.

CHAPTER 8
Current Carrying Capacity

To achieve maximum economy in first cost and subsequent operation of cables, an important aspect is the selection of the optimum size of conductor. Several factors are involved in this and whilst the continuous current carrying capacity is paramount, other factors such as voltage drop, cost of losses and ability to carry short-circuit currents must not be neglected. In this chapter on current rating aspects, particular emphasis is placed on data concerning supply distribution cables but the principles are equally applicable to general wiring and transmission cables. For the latter, however, other more specialised features arise and further information is given in chapter 34.

For reasons which are discussed later, the most convenient way to establish a rating for a particular cable design is to calculate an amperage which can be carried continuously (often called a sustained rating) under prescribed standard conditions. Appropriate factors may then be applied to cater for the actual installation conditions and mode of operation.

AVAILABILITY OF PUBLISHED RATINGS

As is usual with most cable matters, the basic source of reference is an IEC specification and IEC 287, 'Electric Cables – Calculation of the current rating', provides in great detail the theory and mathematical treatment for most situations. IEC 287 has been divided into a number of parts so that more complex calculations, for particular conditions, can be included as separate sections. The three basic parts are: IEC 287-1-1 which deals with the calculation of losses and contains the current rating equations; IEC 287-2-1 which covers thermal resistance; and IEC 287-3-1 which gives reference conditions for various countries. IEC 287-1-2 covers eddy current losses for two circuits and a section dealing with current sharing and circulating current losses in parallel cables will follow. IEC 287-2-2 and IEC 287-3-2 are re-issues of IEC 1042 and IEC 1059 which covered factors for groups of cables in air and the selection of cable size from economic considerations respectively. IEC 364, 'Electrical installations of buildings', Part 5: section 523, gives tabulated ratings for standard cable designs up to 1000 V (unarmoured only) under standard conditions. For other types of cable and installation it is necessary to look elsewhere but all recognised publications provide figures which are deduced substantially in accordance with IEC 287.

Two of the important parameters in establishing ratings for standard operating conditions for particular installations are the ambient temperature and the permissible temperature rise. Therefore in selecting or comparing figures from published sources it is important to take account of the temperature in assessing the information provided. The most commonly used sources of tabulated ratings are as follows.

(a) The Institution of Electrical Engineers in the UK publishes BS 7671: 1992, 'Requirements for electrical installations' (IEE Wiring Regulations) and this contains tables for most standard cable types up to 1000 V, including mineral insulated. The tabulated ratings are for cables 'in air', i.e. not buried, and are calculated for a base ambient temperature of 30°C. In the case of general supply distribution cables most other published ratings are based on an ambient temperature of 25°C, and hence a greater permissible temperature rise. When corrected to any specific ambient temperature there is alignment with IEC 287.

(b) Since the very early years of cable utilisation, the Electrical Research Association in the UK (now ERA Technology Ltd) has specialised in methods of calculation and practical work for verification. It has become a recognised authority and many reports of its work have been published (see Bibliography). Report ERA 69-30 provides ratings in Part 1 for paper insulated cables up to 33 kV, in Part 2 for 600/1000 V Consac cables, in Part 3 for PVC insulated cables up to 3.3 kV, in Part 5 for armoured cables with thermosetting insulation to BS 5467: 1989. Ratings for multilayer groups of cables to BS 6346: 1989 and BS 5467: 1989 on trays are covered in Part 6 and Part 7 respectively. Parts 8 and 9 set out a method for determining ratings for mixed groups of cables, carrying mixed loads, in steel trunking. Part 7 covers PVC insulated cables to BS 6004: 1995 and Part 8 covers cables with thermosetting insulation to BS 7211: 1994. ERA Report 69-30 Part 4 sets out a method of calculating cyclic ratings for buried cables up to 19/33 kV.

(c) Most cable manufacturers issue catalogues which contain ratings for the cables which they supply.

(d) Other sources exist for more specialised installations. Cables for ships are based on an ambient temperature of 45°C with somewhat lower maximum temperatures for continuous operation than permitted elsewhere. Ratings are provided in the IEE 'Regulations for the electrical and electronic equipment of ships' and IEC 92-352 'Electrical installations in ships'.

(e) While all the above references apply generally throughout the world, the types of cable and systems involved with USA practice are slightly different and reference should be made to NEMA/ICEA publications such as ICEA P53-426/NEMA WC 50-1976, which contains details for XLPE/EPR insulated cables from 15 to 69 kV, or the National Electrical Code published by the National Fire Protection Association. Another different feature of American practice is that the published data allow for limited periods of emergency overload for a specific number of hours per year to a higher cable temperature. Whilst it is recognised that such operation could have an effect on the life of a cable, the conditions are chosen to ensure that only limited ageing is likely to occur. International practice may well move in this direction in the future.

The above sources generally contain ratings for individual cable types and sizes installed under specified conditions in air, in buried ducts and buried directly in the ground. The use of multiplying factors for variations in the conditions is discussed in a later section.

TYPICAL VALUES OF SUSTAINED RATINGS UNDER STANDARD CONDITIONS

Tables in the appendices contain sustained ratings and other data such as a.c. resistance at maximum operating temperature as required for rating calculations. The ratings conform to the principles of IEC 287 and are for cables in air and cables directly buried in the ground. Relevant installation and operating conditions for each table are given. The types of cable covered are essentially those to British Standards, or otherwise of recognised designs as used in the UK.

GENERAL BASIS OF RATING DETERMINATION

During service operation, cables suffer electrical losses which appear as heat in the conductor, insulation and metallic components. The current rating is dependent on the way this heat is transmitted to the cable surface and then dissipated to the surroundings. Temperature is clearly an important factor and is expressed as a conductor temperature to establish a datum for the cable itself. A maximum temperature is fixed which is commonly the limit for the insulation material, without undue ageing, for a reasonable maximum life. Then, by choosing a base ambient temperature for the surroundings, a permissible temperature rise is available from which a maximum cable rating can be calculated for a particular environment.

Under steady state conditions the difference between the conductor temperature and the external ground or ambient temperature is related to the total heat losses and the law of heat flow is very similar to Ohm's law. Heat flow corresponds to current, temperature

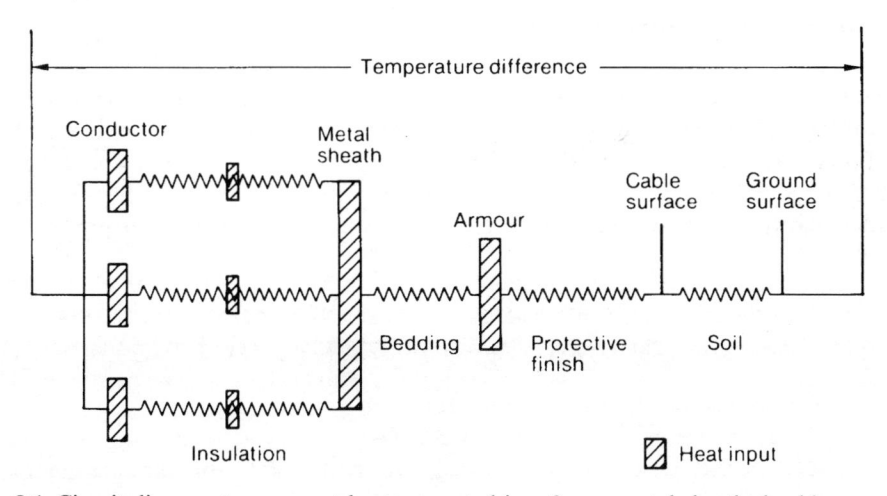

Fig. 8.1 Circuit diagram to represent heat generated in a 3-core metal sheathed cable

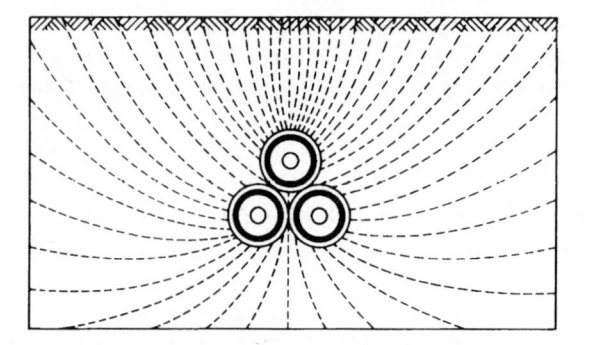

Fig. 8.2 Heat flow from a circuit of single-core cables installed in trefoil

difference to voltage and the total thermal resistance in the cable and surroundings to electrical resistance. Using this analogy it is possible to construct a circuit diagram as illustrated in fig. 8.1. This shows how the heat generated at several positions has to flow through a number of layers of different thermal resistances. By measuring values for the materials, rating calculations can then be made. Thermal resistivity is defined as the difference in temperature in kelvins between opposite faces of a metre cube of material caused by the transference of 1 watt of heat – hence the units K m/W.

The heat flow within a cable is radial but externally it is not so and allowance must be made for the method of installation. Figure 8.2, which shows the pattern of heat flow for three buried single-core cables, illustrates the importance of making allowance for the depth of burial and could be extended to show the effects of other cables in close proximity.

Mathematical treatment is most conveniently expressed for steady state conditions, i.e. for continuous (sustained) ratings. A small cable in air will heat up very quickly to a steady state condition but a large buried power cable may take very many hours. Hence for most types of operation for supply distribution cables laid direct, the continuous ratings may be conservative and allowance can be made for cyclic operation as discussed later.

MATHEMATICAL TREATMENT

The temperature rise in the cable is due to the heat generated in the conductors (I^2R), in the insulation (W) and in the sheath and armour ($\lambda I^2 R$), with allowance being made by multiplying each of these by the thermal resistance of the layers through which the heat flows (T). More detailed derivation of these components is discussed in the next section but the following formula shows how they can be used for calculation purposes for a.c. cables:

$$\Delta\theta = (I^2 R + \tfrac{1}{2} W_d) T_1 + [I^2 R(1 + \lambda_1) + W_d] n T_2 \\ + [I^2 R(1 + \lambda_1 + \lambda_2) + W_d] n (T_3 + T_4) \tag{8.1}$$

where $\Delta\theta$ = conductor temperature rise (K)
I = current flowing in one conductor (A)
R = alternating current resistance per unit length of the conductor at maximum operating temperature (Ω/m)

W_d = dielectric loss per unit length for the insulation surrounding the conductor (W/m)

T_1 = thermal resistance per unit length between one conductor and the sheath (K m/W)

T_2 = thermal resistance per unit length of the bedding between sheath and armour (K m/W)

T_3 = thermal resistance per unit length of the external serving of the cable (K m/W)

T_4 = thermal resistance per unit length between the cable surface and the surrounding medium (K m/W)

n = number of load-carrying conductors in the cable (conductors of equal size and carrying the same load)

λ_1 = ratio of losses in the metal sheath to total losses in all conductors in that cable

λ_2 = ratio of losses in the armouring to total losses in all conductors in that cable

This formula may be rewritten as follows to obtain the permissible current rating:

$$I = \left\{ \frac{\Delta\theta - W_d[\frac{1}{2}T_1 + n(T_2 + T_3 + T_4)]}{RT_1 + nR(1 + \lambda_1)T_2 + nR(1 + \lambda_1 + \lambda_2)(T_3 + T_4)} \right\}^{1/2} \quad (8.2)$$

In using this formula account needs to be taken of the fact that it only provides ratings for the prescribed representative conditions. It does not allow for heat generation from any other source, such as other cables in close proximity, or from exposure to direct solar radiation. More detailed treatment for the latter is given in IEC 287.

In the case of 1 kV 4-core cables, n may be assumed to be 3 if the fourth conductor is neutral or is a protective conductor. This assumes that the neutral conductor is not carrying currents which are due to the presence of harmonics. Where triple harmonic currents, particularly the third harmonic, are present in a system they do not cancel in the neutral. This means that all four conductors will be loaded and measurements have shown that the current in the neutral conductor may be higher than the 50 Hz current in the phase conductors.

For d.c. cables some of the losses are not applicable and for up to 5 kV formula (8.2) may be simplified to

$$I = \left[\frac{\Delta\theta}{R'T_1 + nR'T_2 + nR'(T_3 + T_4)} \right]^{1/2} \quad (8.3)$$

where R' = d.c. resistance per unit length of the conductor at maximum operating temperature (Ω/m)

CALCULATION OF LOSSES

Conductor resistance

It must be noted that R in the formula is the resistance at the maximum operating temperature and for a.c. operation allowance must be made for skin and proximity effects.

The d.c. resistance (Ω/km) at temperature θ is

$$R' = R_{20}[1 + \alpha_{20}(\theta - 20)] \tag{8.4}$$

Values for R_{20} are given in appendix A4. The temperature coefficient per degree Celsius at 20°C (α_{20}) for copper is 0.00393 and for aluminium is 0.00403. Reference to values for θ is made in the next section.

The a.c. resistance at temperature θ is

$$R = R'(1 + y_s + y_p) \quad (\Omega/\text{km}) \tag{8.5}$$

where y_s = the skin factor
y_p = the proximity effect factor

At power frequencies of 50–60 Hz the skin effect factor is small for conductors smaller than about 150 mm². Above this size it may be taken as

$$y_s = \frac{x_s^4}{192 + 0.8 x_s^4} \tag{8.6}$$

$$x_s^2 = \frac{8\pi f}{R'} \times 10^{-7} k_s \tag{8.7}$$

where f = supply frequency (Hz)
k_s = a constant for cable type (see IEC 287)

These formulae are accurate provided that x_s does not exceed 2.8.

Proximity effects are due to mutual effects between the main cable conductors themselves plus inductive currents in any metallic sheath and eddy currents in both metallic sheaths and armour. They can be neglected for small conductor sizes at power frequencies.

If detailed calculation is necessary reference should be made to IEC 287, but for standard cables figures for the total effective a.c. resistance at maximum operating temperature are included in the tables in the appendices.

Dielectric losses in a.c. cables

The dielectric loss in each phase is

$$W_d = \omega C U_0^2 \tan \delta \quad (\text{W/m}) \tag{8.8}$$

where $\omega = 2\pi f$ (1/s) in which f is frequency (Hz) (s = second)
C = capacitance (F/m)

Values for $\tan \delta$ are given in table 8.1.

The capacitance for cables with circular conductors (F/m) is given by the formula below and this may also be applied for oval conductors if the geometric mean diameter is used:

$$C = \frac{\epsilon}{18 \log_e(D_i d_c)} \times 10^{-9} \tag{8.9}$$

Table 8.1 Nominal values for relative permittivity and loss factor

Type of cable		Permittivity	$\tan \delta$
Solid type paper insulated		4	0.01
Fluid-filled paper	up to $U_0 = 36$ kV	3.6	0.0035
	up to $U_0 = 87$ kV	3.6	0.0033
	up to $U_0 = 160$ kV	3.5	0.003
	up to $U_0 = 220$ kV	3.5	0.0028
Fluid-pressure, pipe type/paper		3.7	0.0045
External gas pressure/paper		3.6	0.004
Internal gas pressure/paper		3.4	0.0045
Butyl rubber		4	0.05
EPR	up to 18/30 (36) kV	3	0.02
	above 18/30 (36) kV	3	0.005
PVC		8	0.1
PE (HD and LD)		2.3	0.001
XLPE	up to and including 18/30 (36) kV (unfilled)	2.5	0.004
	above 18/30 (36) kV (unfilled)	2.5	0.001
	above 18/30 (36) kV (filled)	3	0.005

where ϵ = relative permittivity of insulation (table 8.1)
D_i = external diameter of insulation excluding screen (mm)
d_c = diameter of conductor including screen (mm)

It is not usually necessary to calculate the capacitance for cables with shaped conductors because they are only used in cables for which dielectric losses may be neglected, i.e. for values of U_0 below 38 kV for paper cables, 18 kV for butyl rubber, 63.5 kV for EPR, 6 kV for PVC, 127 kV for PE, 63.5 kV for filled XLPE and 127 kV for unfilled XLPE.

Losses in metal sheaths and armour (a.c. cables)

In multicore cables, sheath and armour losses may make some contribution to total losses but the effects are not of very great significance. However, with single-core cables the situation is very different and substantial losses may result from circulating currents and eddy currents in the sheaths and non-magnetic armour. Eddy current losses may be ignored when cables are bonded at both ends.

Sheath circulating currents are of particular importance but losses can be reduced to zero by single-point bonding or by carrying out cross-bonding of the sheaths as described in chapter 34. Allowance still has to be made for eddy current losses. Equations for calculations of all the losses relating to sheaths and armour are available but because of the many different possible combinations of circumstances they are difficult to summarise concisely and reference should be made to IEC 287 which contains full details.

Similar remarks apply to magnetic armour on single-core cables and in this case the effect on rating is so great that it is seldom possible to use such armour. Non-magnetic aluminium or bronze is normally adopted and the losses are then much lower, but allowance still has to be made for circulating and eddy currents.

Table 8.2 Thermal resistivities of materials

Material	Thermal resistivity (K m/W)
Insulation	
Paper (varies with cable type)	5.0–6.0
PE and XLPE	3.5
PVC – up to and including 3 kV	5.0
– over 3 kV	6.0
EPR – up to and including 3 kV	3.5
– over 3 kV	5.0
Butyl rubber and natural rubber	5.0
Protective coverings	
Compounded jute and fibrous materials	6.0
PCP	5.5
PVC – up to and including 35 kV cables	5.0
– over 35 kV cables	6.0
PE	3.5
Materials for ducts	
Concrete	1.0
Fibre	4.8
Asbestos	2.0
Earthenware	1.2
PVC	6.0
PE	3.5

CALCULATION OF THERMAL RESISTANCES

In order to use equation (8.2) it is necessary to calculate the thermal resistances of the different parts of the cable (T_1, T_2 and T_3). Representative values for the resistivity of the individual materials used in cables are included in IEC 287, of which table 8.2 is a summary.

In making calculations, the thermal resistance of metallic layers such as screens or sheaths is ignored but the semiconducting screens are considered to be part of the insulation. With cables which have corrugated metallic sheaths, the thickness of insulation is based on the mean internal diameter of the sheath.

Thermal resistance between one conductor and sheath (T_1)

The equations for various cable constructions are outlined below, IEC 287 contains further information, including data for the geometric factor G and an additional screening factor which is necessary for screened cables.

Single-core cables

$$T_1 = \frac{\rho_T}{2\pi} \log_e \left(1 + \frac{2t_1}{d_c}\right) \quad (8.10)$$

where ρ_τ = thermal resistivity of insulation (K m/W)
d_c = diameter of conductor (mm)
t_1 = thickness of insulation, conductor to sheath (mm)

Multicore belted cables

$$T_1 = \frac{\rho_\tau}{2\pi} G \tag{8.11}$$

where G is a geometric factor from IEC 287.

Multicore screened cables

$$T_1 = \frac{\rho_\tau}{2\pi} G \times \text{screening factor} \tag{8.12}$$

SL and SA type cables
These are treated as single-core cables.

Thermal resistance between sheath and armour (T_2)

Single-core and multicore cables

$$T_2 = \frac{\rho_\tau}{2\pi} \log_e\left(1 + \frac{2t_2}{D_s}\right) \tag{8.13}$$

where t_2 = thickness of bedding (mm)
D_s = external diameter of sheath (mm)

SL and SA type cables

$$T_2 = \frac{\rho_\tau}{2\pi} G' \tag{8.14}$$

where G' is a geometric factor from IEC 287.

Thermal resistance of outer coverings (T_3)

$$T_3 = \frac{\rho_\tau}{2\pi} \log_e\left(1 + \frac{2t_3}{D'_a}\right) \tag{8.15}$$

where t_3 = thickness of outer covering (mm)
D'_a = external diameter of armour (mm)

For corrugated sheaths reference should be made to IEC 287.

External thermal resistance in free air (T_4)

Assuming protection from solar radiation,

$$T_4 = \frac{1}{\pi D_e h (\Delta\theta_s)^{1/4}} \tag{8.16}$$

where D_e = external diameter of cable (mm)
h = heat dissipation coefficient from IEC 287
$\Delta\theta_s$ = excess of surface temperature above ambient (K)

Because $\Delta\theta_s$ is a function of the heat generated by the cable, which is proportional to the current rating, the calculation of T_4 for cables in air has to be an iterative process. Allowance can be made for exposure to solar radiation and details are given in IEC 287.

External thermal resistance for buried cables (T_4)

Single isolated buried cables

$$T_4 = \frac{\rho_T}{2\pi} \log_e[\mu + (\mu^2 - 1)^{1/2}] \qquad (8.17)$$

where ρ_T = the thermal resistivity of the soil (K m/W)
$\mu = 2L/D_e$
L = distance from ground surface to cable axis (mm)
D_e = external diameter of the cable (mm)

IEC 287 contains further information for groups of cables, touching and non-touching, with equal and unequal loading, together with a reference to unfilled troughs at surface level.

The range of values for soil thermal resistivity is given later, and the subject is of considerable importance, particularly for transmission cables. For cables in buried troughs careful attention must be given to the filling medium and the value of resistivity selected.

Cables in buried ducts (T_4)

In this case T_4 is the sum of the thermal resistance of the air space between cable and duct (T'_4), the thermal resistance of the duct itself (T''_4) and the external thermal resistance (T'''_4).

Thermal resistance of air space (T'_4)
For cable diameters of 25 to 100 mm

$$T'_4 = \frac{U}{1 + 0.1(V + Y\theta_m)D_e} \qquad (8.18)$$

where U, V and Y are constants for various types of duct (IEC 187)
D_e = external diameter of cable (mm)
θ_m = mean temperature of the air space (°C)

Thermal resistance of the duct (T''_4)

$$T''_4 = \frac{\rho_T}{2\pi} \log_e\left(\frac{D_o}{D_d}\right) \qquad (8.19)$$

where D_o = outside diameter of duct (mm)
      D_d = inside diameter of duct (mm)
      ρ_τ = thermal resistivity of duct material as given in table 8.2 (K m/W)

Thermal resistance of the external medium (T_4''')

In general this is treated as though the duct represents a cable in equation (8.17). If the ducts are surrounded by concrete, allowance has to be made for the composite surrounding of concrete and soil.

IMPORTANT PARAMETERS WHICH AFFECT RATINGS

The main factors which have effects on ratings may be split into a number of groups as follows.

Temperature

Primarily it is the temperature rise which is important but this is governed by the base ambient temperature for the given cable location and the maximum temperature applicable to the insulation and cable construction. This is discussed in the next section.

Cable design

Apart from the temperature limit, the other effect of cable design is the ability to transfer heat from the conductors to the outer surface. This varies with the materials used and the number of layers in the construction. Details have been given in the sections on mathematical treatment.

Conditions of installation

On the whole, a cable in air can dissipate heat better than a cable in the ground but in this respect the cable diameter, or more particularly the surface area, is important. Up to a certain size cables in air have a lower rating than buried cables and when cables are buried the rating decreases with depth of burial. Further details are given in a later section.

Effects of neighbouring cables

Any other heat input from hot pipes or other cables in the vicinity has to be taken into account. In the case of other cables, allowances can usually be made by the use of correction factors, as discussed later, and reference may be made to IEC 287 for the basis of calculation.

Correction factors for deviation from standard conditions

In addition to the above there are many other conditions such as ambient temperature, etc. for which rating correction factors can be applied and these are given later.

Table 8.3 Ambient air and ground temperature (°C)

Climate	Air temperature		Ground temperature (at 1 m depth)	
	Minimum	Maximum	Minimum	Maximum
Tropical	25	55	25	40
Subtropical	10	40	15	30
Temperate	0	25	10	20

AMBIENT AND CABLE OPERATING TEMPERATURES

Ambient temperature

Representative average ambient temperatures may vary within any individual country, e.g. according to whether the cables are buried or in air outdoors or within a building, and between countries according to the geographical climate. For convenience, the normal tabulated ratings in the UK are based on 15°C for cables in the ground, 25°C outdoors in air, 30°C in air within buildings and 45°C for conditions in ships. For guidance purposes IEC 287 has made an attempt to provide representative average values for other countries. These are included in table 8.3 which illustrates the general overall range throughout the world and table 8.4 which aims to provide data for individual countries. Table 8.4 is only very approximate, however, and subject to many variations.

In using information from these tables several points must be kept in mind. Cable ratings must be applicable for the worst conditions throughout the year and hence for the highest temperatures. Minimum temperatures are only of interest if specific winter ratings are being considered. Also many countries have important differences in climate across the country. Although different countries quote different values of soil resistivity for calculation purposes it is important to investigate the conditions for the individual cable circuit.

It will be apparent from table 8.3 that the rating penalty, when referred to permissible temperature rise, is very high for cables at high ambient temperatures. There is clearly a considerable advantage in using a type of insulation which can sustain a high operating temperature, e.g. XLPE at 90°C instead of solid type paper cable at 33 kV with a temperature of 65°C.

Maximum cable operating temperature

Maximum cable operating temperatures according to insulation material, cable design and voltage have been agreed in IEC and the standard values are almost universally accepted throughout the world for continuous operation (table 8.5).

In using these values an important proviso is that attention must be given to soil resistivity. Continuous operation at cable surface temperatures above 50°C will cause movement of moisture away from the cables and, with many types of cable, drying out of the backfill may occur and the cable could exceed the permissible temperature.

Taken at face value, the figures in table 8.5 indicate a major advantage for the materials which can be operated at high temperature but it must be remembered that

Table 8.4 Representative conditions for various countries

Country	Ambient air temperature (°C)	Ambient ground temperature (°C)	Soil thermal resistivity (K m/W)	Depth of burial (mm)
Australia	40 summer 30 winter	25 summer 18 winter	1.2	
below 11 kV				500–750
11 kV				800
33 kV and above				1000
Austria	20 average	20 maximum	0.7	700
1 kV				
3–6 kV	20 average	20 maximum	0.7	800
10 kV	20 average	20 maximum	0.7	1000
pressure	20 average	20 maximum	1.0	1200
Canada	40 maximum −40 minimum	20 maximum −5 minimum	1.2 ref[a]	
paper to 69 kV				1100
polymeric to 46 kV				900
pressure				1100
Finland	25 ref[a] (−20 to 35)	15 maximum 0 minimum	1.0 ref[a]	
below 36 kV				700
36–52 kV				1000
52–123 kV				1300
123–245 kV				1500
France	30 summer 20 winter	20 summer 10 winter	1.2 summer 0.85 winter	
Germany	30 maximum (−20 to +20 average)	20 maximum 0 minimum	1.0	
below 20 kV				700 ref[a]
above 60 kV				1200
Italy	30 ref[a] (0–30)	20 ref[a] 5 minimum	1.0 maximum	
below 12 kV				800
12–17.5 kV				1000
17.5–24 kV				1200
24–36 kV				1500
36–72 kV				1800
72–220 kV				2200
Japan	40 summer 30 winter	25 summer 15 winter	1.0 average	
up to 33 kV				1200
pressure				1500

(*cont.*)

Table 8.4 continued

Country	Ambient air temperature (°C)	Ambient ground temperature (°C)	Soil thermal resistivity (K m/W)	Depth of burial (mm)
The Netherlands	20 average (−5 to 30)	15 average (5–20)	0.5–0.8	
up to 10 kV				700
over 10 kV				1000
Poland	25 ref[a]	20 ref[a]	1.0	
up to 1 kV				700
1–15 kV				800
over 15 kV				1000
Sweden		15 maximum 0 minimum	1.0 ref[a]	
up to 52 kV				700
pressure				1000–1500
Switzerland	25 ref[a]	20 ref[a]	1.0 ref[a]	1000
UK	25 outdoors 30 buildings	15	1.2 ref[a]	
1 kV				500
3.3–33 kV				800
pressure				900
USA	40 ref[a]	20	0.9 ref[a]	900

[a] ref signifies reference value for rating purposes

Table 8.5 Conductor temperature limits for standard cable types

Insulation	Cable design	Maximum conductor temperature (°C)
Impregnated paper (U_0/U)		
0.6/1, 1.8/3, 3.6/6	Belted	80
6/10	Belted	65
6/10, 8.7/15	Screened	70
12/20, 18/30 MIND	Screened	65
Polyvinyl chloride	All	70
Polyethylene	All	70
Butyl rubber	All	85
Ethylene–propylene rubber	All	90
Crosslinked polyethylene	All	90
Natural rubber	All	60

there are other factors involved in the choice of cable size. For cables in buildings in the UK (base ambient temperature of 30°C) the permissible temperature rise for XLPE is 60°C, compared with 40°C for PVC. However, in such installations it is often voltage drop which is the determining factor. Furthermore, the extra cost of electrical losses due to increase in conductor resistance may not be insignificant with relatively continuous operation.

EFFECT OF INSTALLATION CONDITIONS ON RATINGS

Reference has already been made to a conductor size effect in rating differences between installation in air and in the ground, this being associated with the cable surface area. Some other aspects arising are covered below.

Depth of burial

The depth of laying is governed primarily by what is considered to be the most advisable to minimise effects of damage and generally increases with cable voltage. Values adopted in various countries have been given in table 8.4. An equation covering the effect on rating has already been quoted and for most purposes the thermal resistance of the soil may be simplified to

$$T_4 = \frac{\rho_T}{2\pi} \log_e \left(\frac{4L}{D_e}\right) \qquad (8.20)$$

where ρ_T = soil thermal resistivity (K m/W)
L = depth of burial to cable axis (mm)
D_e = cable external diameter (mm)

In this formula variations of ρ_T may be extremely important but variations of laying depth have less effect. As the depth increases the maximum ambient ground temperature decreases and also the moisture content increases, so that this improves the soil resistivity. Therefore for conditions where the temperature can be taken as 15°C and ρ_T is around 1.2 K m/W, it is only with transmission cables that much account need be taken of depth of laying, and with these cables the subject of external thermal resistance has to be thoroughly evaluated anyway because of the need to select appropriate backfill.

Thermal resistivity of the soil

Provided that it is possible to arrive at a reasonably representative value, it is not normally necessary with supply distribution cables to devote much attention to soil thermal resistivity, unless because of fully continuous operation there is a danger of the soil drying out.

The presence of moisture has a predominant effect on the resistivity of any type of soil and so it is necessary to take the weather conditions into account. IEC 287 gives guidance as indicated in table 8.6 and ignores the make-up of particular ground types. However, the steps quoted are rather broad and certainly for transmission cables greater precision is necessary.

Table 8.6 Soil thermal resistivities

Thermal resistivity (K m/W)	Soil conditions	Weather conditions
0.7	Very moist	Continuously moist
1.0	Moist	Regular rainfall
2.0	Dry	Seldom rains
3.0	Very dry	Little or no rain

At one time it was common to take direct measurements using a needle-probe technique but such results are very difficult to interpret because judgement is necessary on whether the moisture content at the time of measurement is representative and because the resistivity is strongly dependent on the amount of compaction; this can be changed when the probe is inserted and the soil put back round the cable will be in a different condition. It is better, therefore, to adopt a more empirical approach according to whether the cable operating conditions will or will not cause drying out of the soil. IEC 287 gives a method for calculating the current rating by which drying out of the soil can be avoided and a method for calculating the effect of partial drying out on the cable rating.

If drying out is not a problem, as with nearly all distribution cable installations, the question is the likely moisture content. In some cases it will be known that the ground will remain wet or fairly moist and in these situations it is reasonable to adopt 0.8–1.0 K m/W.

In the more general case where soils are not always moist but the texture is of an average clay or loam type, a good representative figure is 1.2 K m/W. In the UK this is usually taken as a standard value for the preparation of tabulated ratings. The situation is more difficult if the soil consists of sand, shingle or made-up ground, i.e. with a large air-space content after water has drained away. If such drainage can occur during some months of the year the value used should be between 2.0 and 3.0 according to circumstances.

Guidance is given in ERA Report 69–30: Part 1 on the following lines.

Type A – cables carrying constant load throughout the year

Whether the load is sustained or cyclic, allowance needs to be made for maximum values of soil resistivity which may occur in some years only and for relatively short periods in summer and autumn. Values recommended are as follows:

All soils except those below	1.5 K m/W
Chalk soil with crushed chalk backfill	1.2 K m/W
Peat	1.2 K m/W
Very stony soil or ballast	1.5 K m/W
Well-drained sand	2.5 K m/W
Made-up soils	1.8 K m/W

The value for the 'all soils' category may be reduced to 1.2 if the soil is under impermeable cover such as asphalt or concrete.

Type B – cables with varying load and maximum in summer

If the load is mixed, advantage may be taken of the fact that in all probability the maximum load in summer will not coincide with the dry periods. During the summer periods, recommended values are as follows:

All soils except those below	1.2 K m/W
Stony soils or ballast	1.3 K m/W
Well-drained sand	2.0 K m/W
Made-up soils	1.6 K m/W

The value for the 'all soils' category may be reduced to 1.0 if the soil is under impermeable cover and assumes that in chalky ground the backfill is crushed.

Type C – cables with varying load and maximum in winter

For the winter period it is safe to use rather lower values:

All soils except those below	1.0 K m/W
Clay	0.9 K m/W
Chalk soil with crushed chalk backfill	1.2 K m/W
Well-drained sand	1.5 K m/W
Made-up soil	1.2 K m/W

The value for clay soils may be reduced to 0.8 if the soil is under impermeable cover.

The drying out of soil is quite a complex subject because, in addition to the cable temperature and natural drainage, drying is also caused by tree roots and natural vegetation. A solid cover on the surface, e.g. asphalt or concrete, restricts drying out.

As already mentioned, a cable surface temperature of 50°C is sufficient to lead to progressive drying out, and if the soil is well drained, e.g. sandy, drying out can occur at an even lower temperature. The most favourable dried out natural soil and sands are unlikely to have a thermal resistivity lower than 2.0 K m/W and values of 2.5–3.0 K m/W are probable. To obtain optimum cable rating in situations where drying out is possible, it is now general practice to surround the cable with imported material of known properties. Such material is usually known as thermal or stabilised backfill and the key to successful formulation is minimum air space together with good compaction. These materials are always designed to have a good thermal resistivity when dry and measurement of the dry density at a defined compaction provides a reasonable means of evaluation. Nowadays they fall into two groups: one in which there is control of particle sizes and another using a weak mix of cement. The dry mixture is composed of blends of shingle and sand each having a wide range of particle sizes with the object of obtaining very good packing. The alternative material, now more widely used, is a 20:1 ratio of suitable mixed particle size sand and cement with optimum water content.

Further details of the importance of the use of controlled backfill material are given in chapter 34.

STANDARD OPERATING CONDITIONS AND RATING FACTORS FOR SUPPLY DISTRIBUTION CABLES

The standard conditions for the ratings given in the tables in the appendices are included with the tables and in general are as given below.

The standard conditions, and particularly the rating factors, vary with the cable design and the operating parameters. The data given apply primarily to supply distribution cables and are averages for such cables. They may be taken as a general guide for wiring type cables but some special comments on wiring type cables are given in a later section.

In the case of transmission cables it is not possible to obtain sufficient accuracy by the use of rating factors and the rating for each installation needs to be calculated directly.

Cables installed in air

Standard conditions

(a) Ambient air temperature is taken to be 25°C for paper insulated cables and for XLPE insulated cables above 1.9/3.3 kV. 30°C is chosen for PVC insulated cables and for XLPE cables of 1.9/3.3 kV and below in order to be in conformity with the IEE Wiring Regulations.
(b) Air circulation is not restricted significantly, e.g. if cables are fastened to a wall they should be spaced at least 20 mm from it.
(c) Adjacent circuits are spaced at least 150 mm apart and suitable disposed to prevent mutual heating.
(d) Cables are shielded from direct sunshine.

Rating factors for other ambient air temperatures

Factors to correct from a base of 25 or 30°C to other temperatures are given in table 8.7. For the types and voltages of paper insulated cables to which the various operating temperatures apply reference should be made to table 8.5.

Group rating factors for cables in air

When groups of multicore power cables are installed in air it is necessary to have a sufficient air space for dissipation of heat. No reduction in rating is necessary provided that:

(a) the horizontal clearance between circuits is not less than twice the overall diameter of an individual cable;

Table 8.7 Rating factors for ambient temperature

Cable insulation	Maximum conductor operating temperature (°C)	Ambient air temperature (°C)					
		25	30	35	40	45	50	55
Paper	65	1.0	0.93	0.85	0.77	0.68	0.58	0.47
Paper	70	1.0	0.93	0.87	0.80	0.72	0.64	0.55
Paper	80	1.0	0.94	0.89	0.84	0.77	0.72	0.65
PVC	70	1.06	1.0	0.94	0.87	0.79	0.71	0.61
XLPE[a]	90	1.0	0.95	0.91	0.86	0.80	0.75	0.69
XLPE[b]	90	1.04	1.0	0.96	0.91	0.87	0.82	0.76

[a] Above 1.9/3.3 kV
[b] 1.9/3.3 kV and below

(b) the vertical clearance between circuits is not less than four times the diameter of an individual cable;
(c) if the number of circuits exceeds three, they are installed in a horizontal plane.

For smaller clearances reference should be made to ERA Report 74-27, 'Heat dissipation for cables in air', and ERA Report 74-28, 'Heat dissipation for cables on perforated steel trays'.

Further information relative to the smaller sizes of PVC and XLPE cables, as used in buildings, is given later in this chapter.

Cables laid direct in ground

Standard conditions
(a) Ground temperature   15°C
(b) Soil thermal resistivity   1.2 K m/W
(c) Adjacent circuits   at least 1.8 m distance
(d) Depth of laying   0.5 m for 1 kV cables
    0.8 m for cables above 1 kV and up to 33 kV

Rating factors
Factors for ground temperature, soil thermal resistivity, grouped cables and depth of laying are given in tables 8.8–8.12.

Cables installed in ducts

Standard conditions
(a) Ground temperature, 15°C
(b) Thermal resistivity of ground and ducts, 1.2 K m/W
(c) Adjacent circuits, at least 1.8 m distance
(d) Depth of laying, 0.5 m, except for paper insulated cables above 1 kV and up to 33 kV, for which the value is 0.8 m

Rating factors
Factors for variation of ground temperatures are the same as in table 8.8 for cables laid directly in the ground. Factors for soil thermal resistivity, groups of cables and depth of laying are given in tables 8.13–8.16.

Table 8.8 Rating factor for ground temperature

Cable insulation	Maximum conductor operating temperature (°C)	Ground temperature (°C)							
		10	15	20	25	30	35	40	45
Paper	65	1.05	1.0	0.95	0.89	0.84	0.77	0.71	0.63
Paper	70	1.04	1.0	0.95	0.90	0.85	0.80	0.74	0.67
Paper	75	1.04	1.0	0.96	0.92	0.88	0.83	0.78	0.73
PVC	70	1.04	1.0	0.95	0.90	0.85	0.80	0.74	0.67
XLPE	90	1.03	1.0	0.97	0.93	0.89	0.85	0.81	0.77

Table 8.9 Rating factors for thermal resistivity of soil (average values)

Conductor size (mm^2)	Soil thermal resistivity (K m/W)						
	0.8	0.9	1.0	1.5	2.0	2.5	3.0
Single-core cables							
Up to 150	1.16	1.11	1.07	0.91	0.81	0.73	0.67
From 185 to 400	1.17	1.12	1.07	0.90	0.80	0.72	0.66
From 500 to 1200	1.18	1.13	1.08	0.90	0.79	0.71	0.65
Multicore cables							
Up to 16	1.09	1.06	1.04	0.95	0.86	0.79	0.74
From 25 to 150	1.14	1.10	1.07	0.93	0.84	0.76	0.70
From 185 to 400	1.16	1.11	1.07	0.92	0.82	0.74	0.68

Table 8.10 Group rating factors for circuits of three single-core cables, in trefoil and laid flat touching, horizontal formation

Cable voltage (kV)	Number of circuits	Spacing of circuits (between centres of cable groups)					
		Touching		0.15 m[a]	0.3 m	0.45 m	0.6 m
		Trefoil	Laid flat				
0.6/1	2	0.77	0.80	0.82	0.88	0.90	0.93
	3	0.65	0.68	0.72	0.79	0.83	0.87
	4	0.59	0.63	0.67	0.75	0.81	0.85
	5	0.55	0.58	0.63	0.72	0.78	0.83
	6	0.52	0.56	0.60	0.70	0.77	0.82
1.9/3.3 to 12.7/22	2	0.78	0.80	0.81	0.85	0.88	0.90
	3	0.66	0.69	0.71	0.76	0.80	0.83
	4	0.60	0.63	0.65	0.72	0.76	0.80
	5	0.55	0.58	0.61	0.68	0.73	0.77
	6	0.52	0.55	0.58	0.66	0.72	0.76
19/33	2	0.79	0.81	0.81	0.85	0.88	0.90
	3	0.67	0.70	0.71	0.76	0.80	0.83
	4	0.62	0.65	0.65	0.72	0.76	0.80
	5	0.57	0.60	0.60	0.68	0.73	0.77
	6	0.54	0.57	0.57	0.66	0.72	0.76

[a] This spacing will not be possible for some of the larger diameter cables

SUSTAINED RATINGS FOR WIRING TYPE CABLES

Ratings for the most commonly used types of cables in the UK are given in the tables in appendices A5–A11. Standard conditions vary according to cable type and application and the various relevant conditions applicable are included with the tables. Some of the more important aspects are given below.

Table 8.11 Group rating factors for multicore cables in horizontal formation

Cable voltage (kV)	Number of cables in group	Spacing (between cable centres)				
		Touching	0.15 m	0.3 m	0.45 m	0.6 m
0.6/1	2	0.81	0.87	0.91	0.93	0.94
	3	0.70	0.78	0.84	0.87	0.90
	4	0.63	0.74	0.81	0.86	0.89
	5	0.59	0.70	0.78	0.83	0.87
	6	0.55	0.67	0.76	0.82	0.86
1.9/3.3 to 12.7/22	2	0.80	0.85	0.89	0.90	0.92
	3	0.69	0.75	0.80	0.84	0.86
	4	0.63	0.70	0.77	0.80	0.84
	5	0.57	0.66	0.73	0.78	0.81
	6	0.55	0.63	0.71	0.76	0.80
19/33	2	0.80	0.83	0.87	0.89	0.91
	3	0.70	0.73	0.78	0.82	0.85
	4	0.64	0.68	0.74	0.78	0.82
	5	0.59	0.63	0.70	0.75	0.79
	6	0.56	0.60	0.68	0.74	0.78

Table 8.12 Rating factors for depth of laying (to centre of cable or trefoil group of cables)

Depth of laying (m)	0.6/1 kV cables			1.9/3.3 kV to 19/33 kV cables	
	Up to 50 mm^2	70 mm to 300 mm^2	Above 300 mm^2	Up to 300 mm^2	Above 300 mm^2
0.50	1.00	1.00	1.00	–	–
0.60	0.99	0.98	0.97	–	–
0.80	0.97	0.96	0.94	1.00	1.00
1.00	0.95	0.94	0.92	0.98	0.97
1.25	0.94	0.92	0.90	0.96	0.95
1.50	0.93	0.91	0.89	0.95	0.94
1.75	0.92	0.89	0.87	0.94	0.92
2.00	0.91	0.88	0.86	0.92	0.90
2.50	0.90	0.87	0.85	0.91	0.89
3.0 or more	0.89	0.86	0.83	0.90	0.88

Ambient temperature

The tabulated ratings are based on an ambient temperature, and hence temperature rise, according to the primary application of the cable, e.g. 25°C where no regulations apply, 30°C for cables in buildings subject to the IEE Wiring Regulations and 45°C for cables in ships subject to IEE Regulations. When applying temperature correction factors it is important to choose the correct factor which takes into account the base ambient temperature and the maximum permissible conductor temperature.

Table 8.13 Rating factor for soil thermal resistivity (average values)

Conductor size (mm^2)	\multicolumn{7}{c}{Soil thermal resistivity (K m/W)}
	0.8	0.9	1.0	1.5	2.0	2.5	3.0
Single-core cables							
Up to 150	1.10	1.07	1.04	0.94	0.87	0.81	0.75
185 to 400	1.11	1.08	1.05	0.94	0.86	0.79	0.73
500 to 1200	1.13	1.09	1.06	0.93	0.84	0.77	0.70
Multicore cables							
Up to 16	1.05	1.04	1.03	0.97	0.92	0.87	0.83
25 to 150	1.07	1.05	1.03	0.96	0.90	0.85	0.78
185 to 400	1.09	1.06	1.04	0.95	0.87	0.82	0.76

Table 8.14 Group rating factors for single-core cables in trefoil single-way ducts, horizontal formation

Cable voltage (kV)	Number of circuits	\multicolumn{3}{c}{Spacing (between duct centres)}
		Touching	0.45 m	0.60 m
0.6/1	2	0.86	0.90	0.93
	3	0.77	0.83	0.87
	4	0.73	0.81	0.85
	5	0.70	0.78	0.83
	6	0.68	0.77	0.82
1.9/3.3 to 12.7/22	2	0.85	0.88	0.90
	3	0.75	0.80	0.83
	4	0.70	0.76	0.80
	5	0.67	0.73	0.77
	6	0.64	0.71	0.76
19/33	2	0.85	0.88	0.90
	3	0.76	0.80	0.83
	4	0.71	0.76	0.80
	5	0.67	0.73	0.77
	6	0.65	0.71	0.76

Excess current protection

The IEE Wiring Regulations require that the overload protective device should operate in conventional time at 1.45 times the current rating of the cable. As discussed in chapter 10, if the protection is by semi-enclosed rewirable fuse, which may require a current up to twice its own rating to operate it in conventional time, a cable with a higher rating, in comparison with the fuse rating, is required than for other standard protective devices.

Table 8.15 Group rating factors for multicore cables in single-way ducts, horizontal formation

Cable voltage (kV)	Number of ducts in groups	Spacing (between duct centres)			
		Touching	0.30 m	0.45 m	0.60 m
0.6/1	2	0.90	0.93	0.95	0.96
	3	0.82	0.87	0.90	0.93
	4	0.78	0.85	0.89	0.91
	5	0.75	0.82	0.87	0.90
	6	0.72	0.81	0.86	0.90
1.9/3.3 to 12.7/22	2	0.88	0.91	0.93	0.94
	3	0.80	0.84	0.87	0.89
	4	0.75	0.81	0.84	0.87
	5	0.71	0.77	0.82	0.85
	6	0.69	0.75	0.80	0.84
19/33	2	0.87	0.89	0.92	0.93
	3	0.78	0.82	0.85	0.87
	4	0.73	0.78	0.82	0.85
	5	0.69	0.75	0.79	0.83
	6	0.67	0.73	0.78	0.82

Table 8.16 Rating factors for depth of laying (to centre of duct or trefoil group of ducts)

Depth of laying (m)	0.6/1 kV cables		1.9/3.3 to 19/33 kV cables	
	Single-core	Multicore	Single-core	Multicore
0.50	1.00	1.00		
0.60	0.98	0.99		
0.80	0.95	0.97	1.00	1.00
1.00	0.93	0.96	0.98	0.99
1.25	0.90	0.95	0.95	0.97
1.50	0.89	0.94	0.93	0.96
1.75	0.88	0.94	0.92	0.95
2.00	0.87	0.93	0.90	0.94
2.50	0.86	0.93	0.89	0.93
3.00 or more	0.85	0.92	0.88	0.92

Selection of cable size

The rating factors for ambient temperature and installation conditions are factors by which the tabulated ratings, as given in the appendices, should be multiplied in order to determine the rating under the applicable conditions. For distribution cables the procedure for determining the size of cable is often to make a first judgement of the size required, multiply its tabulated rating by any applicable factors and check the result against the circuit requirement. If necessary, adjacent sizes are considered.

For installations to the IEE Wiring Regulations, the Regulations recommend an alternative procedure. The rating of the overload protective device to be used, decided according to the circuit current, is divided by the applicable factors and the result is compared with the ratings listed in the tables to determine the cable size, i.e. the principles for wiring cables are as outlined below.

Protective device being a fuse to BS 88 or BS 1361 or a circuit breaker to BS EN 60898: 1991 or BS EN 60947-2: 1992
When the protective device is a fuse or circuit breaker to the above specifications, it is necessary to ascertain the cable rating required by one of the following methods.

(a) Select a protective device rating I_n adequate for the design current I_b which the circuit is to carry.
(b) Divide the nominal current of the protective device (I_n) by any applicable correction factor for ambient temperature (C_a), if this is other than 30°C.
(c) Further divide by any applicable correction factor for thermal insulation (C_i).
(d) Divide by any applicable correction factor (C_g) for cable grouping, when such groups are liable to simultaneous overload. (See the alternative method below for groups which are *not* liable to simultaneous overload.)
(e) The size of cable required is such that its tabulated current carrying capacity I_t is not less than the value of the nominal current of the protective device (I_n) adjusted as above, i.e. $I_t \geq I_n / C_a C_i C_g$.

Alternative method for cable grouping when groups are not liable to simultaneous overload
Provided that the circuits of the groups are *not* liable to simultaneous overload, the tabulated current rating I_t may be calculated by the following formulae:

$$I_t \geq \frac{I_b}{C_g} \tag{8.21}$$

and

$$I_t \geq \left(I_n^2 + 0.48 I_b^2 \frac{1 - C_g^2}{C_g^2} \right)^{1/2} \tag{8.22}$$

The size of the cable required is such that its tabulated single-circuit current carrying capacity is not less than the larger of the two values of I_t given by equations (8.21) and (8.22).

Where any further correction factor is applicable, such as for ambient temperature or thermal insulation, this must be applied as a divisor to the value of I_t derived by the methods indicated above in respect of the group rating factor.

Protective device being a semi-enclosed fuse to BS 3036: 1952 (1992)
When the protective device is a semi-enclosed fuse to BS 3036: 1952 (1992) (i.e. formerly coarse excess current protection) it is necessary to ascertain the cable rating required by one of the following methods.

(a) Select a fuse I_n required from BS 3036: 1952 (1992) adequate for the design current I_b which the circuit is to carry.
(b) Divide the nominal current of the protective device (I_n) by any applicable correction factor for ambient temperature, if this is other than 30°C (C_a).

(c) Further divide by any applicable correction factor for thermal insulation (C_i).
(d) Further divide by 0.725.
(e) Divide by any applicable correction factor C_g for cable grouping when groups are liable to simultaneous overload. (See the alternative method for groups which are *not* liable to simultaneous overload.)
(f) The size of cable required is such that its tabulated current carrying capacity I_t is not less than the value of nominal current of the protective device (I_n) adjusted as above, i.e. $I_t \geq I_n/0.725\, C_a C_i C_g$.

Alternative method for cable grouping when groups are not liable to simultaneous overload
Provided that the circuits of the groups are *not* liable to simultaneous overload, the tabulated current rating I_t may be calculated by the following formulae:

$$I_t \geq \frac{I_b}{C_g} \tag{8.23}$$

and

$$I_t \geq \left(1.9 I_n^2 + 0.48 I_b^2 \, \frac{1 - C_g^2}{C_g^2}\right)^{1/2} \tag{8.24}$$

The size of the cable required is such that its tabulated single-circuit current carrying capacity is not less than the larger of the two values of I_t given by equations (8.23) and (8.24).

Group rating factors
Correction factors C_g for groups of cables are given in tables 8.17 and 8.18. With reference to the tables the following should be noted.

(a) The factors are applicable to uniform groups of cables, equally loaded.
(b) If, with known operating conditions, a cable is expected to carry a current of not more than 30% of its grouped rating, it may be ignored for the purpose of obtaining the rating factor for the rest of the group. For example, a group of N loaded cables would normally require a group reduction factor of C_g applied to the tabulated I_t. However, if M cables in the group carry loads which are not greater than $0.3 C_g I_t$ amperes, the other cables can be sized by using the group rating factor corresponding to $N - M$ cables.
(c) The factors have been calculated on the basis of prolonged steady state operation at 100% load factor for all live conductors.
(d) When cables with different maximum conductor operating temperatures are grouped together the current ratings for the cables with the lowest operating temperature should be used for all the cables. For example if a group contains both XLPE and PVC insulated cables the ratings for PVC cables should be applied to the XLPE cables.

Thermal insulation
For cable installed in a thermally insulating wall or installed above a thermally insulated ceiling and in contact with a thermally conductive surface on one side, in the absence of more precise information the rating may be taken as 0.75 times the current

Table 8.17 Group correction factors for cables having extruded insulation

Arrangement of cables	\multicolumn{9}{c}{Number of circuits or multicore cables}
	2	3	4	5	6	8	10	12	14	16
Enclosed in conduit or trunking, or bunched and clipped direct	0.80	0.70	0.65	0.60	0.57	0.52	0.48	0.45	0.43	0.41
Single layer clipped direct to or lying on a non-metallic surface:										
Touching	0.85	0.79	0.75	0.73	0.72	0.71	–	–	–	–
Spaced[a]	0.94	0.90	0.90	0.90	0.90	0.90	0.90	0.90	0.90	0.90
Single layer on a perforated metal cable tray, vertical or horizontal:										
Touching	0.86	0.81	0.77	0.75	0.74	0.73	0.71	0.70	–	–
Spaced[a]	0.91	0.89	0.88	0.87	0.87	–	–	–	–	–
Single layer touching on ladder supports	0.86	0.82	0.80	0.79	0.78	–	–	–	–	–

[a] Spaced means a clearance between adjacent surfaces of at least one cable diameter D_e. Where the horizontal clearance between adjacent cables exceeds twice their overall diameter, no correction factor need be applied.

Table 8.18 Group correction factors for mineral insulated cables on perforated tray

Tray orientation	Arrangement of cables	Number of trays	\multicolumn{6}{c}{Number of multicore cables or circuits}
			1	2	3	4	6	9
Horizontal	Multiconductor cables touching	1	1	0.9	0.8	0.8	0.75	0.75
Horizontal	Multiconductor cables spaced	1	1	1	1	0.95	0.9	–
Vertical	Multiconductor cables touching	1	1	0.9	0.8	0.75	0.75	0.7
Vertical	Multiconductor cables spaced	1	1	1	0.95	–	–	–
Horizontal	Single conductor cables trefoil separated	1	1	0.9	0.9	–	–	–
Vertical	Single conductor cables trefoil separated	1	1					

Table 8.19 Facxtors for ambient temperature where semi-enclosed fuses to BS 3036 are used

Insulation	Ambient temperature (°C)								
	25	35	40	45	50	55	65	75	85
60°C rubber	1.04	0.96	0.91	0.87	0.79	0.56			
70°C PVC	1.03	0.97	0.94	0.91	0.87	0.84	0.48		
Paper	1.02	0.97	0.95	0.92	0.90	0.87	0.76	0.43	
85°C rubber	1.02	0.97	0.95	0.93	0.91	0.88	0.83	0.58	
90°C PVC	1.03	0.97	0.94	0.91	0.87	0.84	0.76	0.68	0.49
Thermoset	1.02	0.98	0.95	0.93	0.91	0.89	0.85	0.69	0.39
Mineral:									
70°C sheath	1.03	0.96	0.93	0.89	0.86	0.79	0.42		
105°C sheath	1.02	0.98	0.96	0.93	0.91	0.89	0.84	0.79	0.64

Table 8.20 Factors for ambient temperature where the protection is a fuse to BS 88 or BS 1361 or a circuit breaker to BS 3871 or BS 4752

Insulation	Ambient temperature (°C)								
	25	35	40	45	50	55	65	75	85
60°C rubber	1.04	0.91	0.82	0.71	0.58	0.41			
70°C PVC	1.03	0.94	0.87	0.79	0.71	0.61	0.35		
Paper	1.02	0.95	0.89	0.84	0.72	0.71	0.55	0.32	
85°C rubber	1.02	0.95	0.90	0.85	0.80	0.74	0.60	0.43	
90°C PVC	1.03	0.97	0.94	0.91	0.87	0.84	0.76	0.61	0.35
Thermoset	1.02	0.96	0.91	0.87	0.82	0.76	0.65	0.50	0.29
Mineral:									
70°C sheath	1.03	0.93	0.85	0.77	0.67	0.57			
105°C sheath	1.02	0.96	0.92	0.88	0.84	0.80	0.70	0.60	0.47

carrying capacity for that cable when clipped direct to a surface and unenclosed. For a cable likely to be totally surrounded by thermally insulating material over an appreciable length, the applicable rating factor may be as low as 0.5. For a cable surrounded by thermal insulation over a short length, e.g. passing through an insulated wall, advice is given in the IEE Wiring Regulations.

Rating factors C_a for ambient temperature
For ambient temperatures other than 30°C the tabulated rating may be adjusted by the temperature rating factors in tables 8.19 or 8.20 according to the type of protective device.

CABLES FOR SHIPWIRING AND OFFSHORE INSTALLATIONS

Wiring in ships usually has to comply with regulations such as the IEE Regulations for the Electrical and Electronic Equipment of Ships. As well as basing tabulated ratings on an ambient temperature of 45°C, to which reference has been made, there is also a specified maximum cable conductor temperature of 5°C or 10°C below the values given earlier in table 8.5, e.g. 85°C for EPR insulated cables. cables of this type are now quite capable of operation to 90°C and offshore applications in most countries are at ambient temperatures well below 45°C. Much higher ratings are therefore possible for the many offshore installations for which it is considered that shipwiring regulations do not apply. Tables of ratings for a number of cable types for shipwiring and offshore installation are included in appendix A7 and details of the conditions applicable to offshore installations follow table A7.2.

SHORT TIME AND CYCLIC RATINGS

It very often happens that loads are cyclic rather than sustained. many cables, particularly when buried, may take up to 24 hours or even longer for the temperature to build up to the equilibrium conditions on which sustained ratings are based. Allowance may be made for this, therefore, together with the cooling period between loads, to derive a cyclic rating which will be higher than the value for sustained operation.

Figure 8.3 indicates the temperature rise of the conductor of a typical buried cable. It will be noted that the heating is exponential and hence during sustained loading the temperature change when nearing equilibrium is slow.

If the loading is only at maximum for a few hours, or is at a level below maximum for a longer period, it is possible to calculate a rating to suit the circumstances. Public supply cables having daily cycles with morning and afternoon peaks represent one application where such treatment is beneficial. Another, of a different type, relates to requirements for machines where the loading may be for minutes rather than hours, e.g. arc welding.

Many factors have to be taken into account in calculations for such ratings. Cable diameter in relation to the environment has a major effect, because surface area increases with diameter and if the cable is in air heat may be dissipated quickly. This is the reason, for example, that small size cables have a lower rating in air than in the ground, whereas the reverse applies for large cables. Cables in air heat up very quickly compared with buried cables. Cables so installed may therefore have an allowance for short time currents but not for cyclic loads over a 24 hour period.

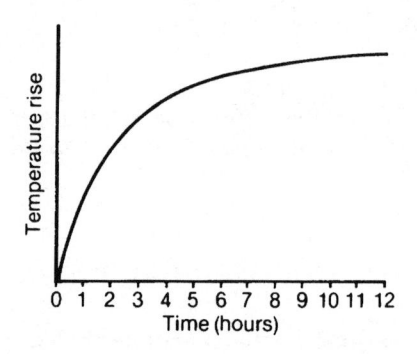

Fig. 8.3 Rate of conductor temperature rise for typical buried power cable

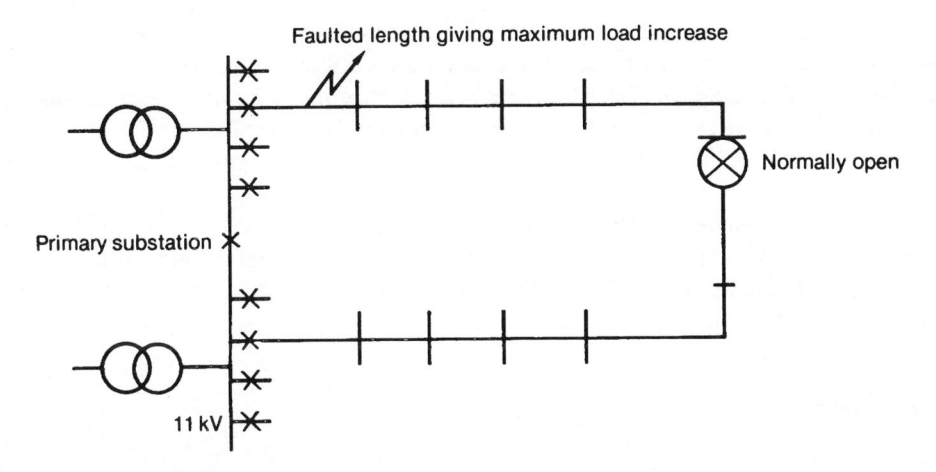

Fig. 8.4 Typical open ring system in 11 kV UK public supply distribution (Courtesy of Institution of Electrical Engineers)

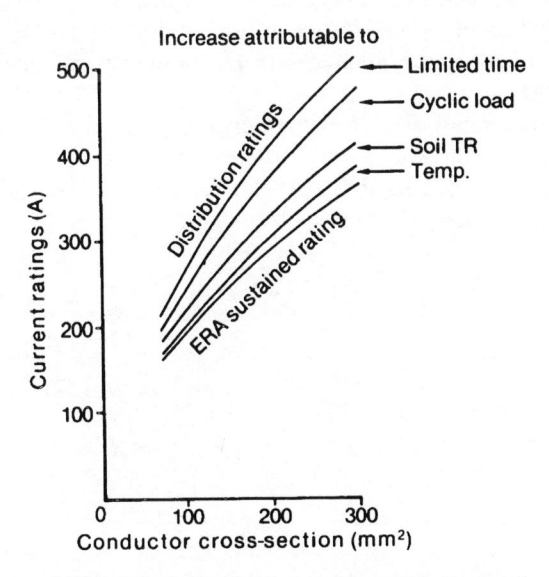

Fig. 8.5 Comparison of UK 11 kV distribution ratings and standard ratings for PILS belted cables with aluminium conductors (Courtesy of Institution of Electrical Engineers)

Other factors are the type and reproducibility of the cycle, the effect of any other cables in the vicinity and the thermal resistivity of the soil. The mathematics is rather voluminous and complicated. Until the mid-1980s a standard work of reference was a report by Goldenberg[1] which uses the concept of a 'loss load factor' representative of the loading cycle. The principles advanced by Goldenberg are now set out in two IEC publications, the first[2] for cables up to 30 kV and the second[3] for cables over 30 kV.

A report by Gosden and Kendall[4] deals specifically with ratings for 11 kV public supply distribution cables in the UK and explains the background to the utilisation of considerably higher ratings[5] than are published for sustained operation. Account is also taken of the fact that common practice is to install 11 kV cables in normally open rings (fig. 8.4). In such an operation any point in the ring is fed by two cables and each carries half the load of the ring, i.e. the cables have only 50% utilisation. In the rare occurrence of a cable fault in the ring the link is closed and the load on the cable beyond the fault is back-fed from the unfaulted cable, which thus carries an increased load until repair is completed.

Compared with sustained ratings, fig. 8.5 indicates the magnitude of the improvement possible for public supply operation. This figure also includes contributions from other variations from standard conditions, i.e. a soil temperature and soil thermal resistivity (TR) that are lower than standard.

REFERENCES

(1) Goldenberg, H. (1958) *Methods for the calculation of cyclic rating factors and emergency loading for cables direct in ground or in ducts.* ERA Report No. F/T 186.
(2) IEC 853-1 (1985) 'Calculation of the cyclic and emergency current rating of cables', Part 1, 'Cyclic rating factor for cables up to and including 30 kV'.
(3) IEC 853-2 (1989) 'Calculation of the cyclic and emergency current rating of cables', Part 2, 'Cyclic rating of cables greater than 18/30(36) kV and emergency ratings for cables of all voltages'.
(4) Gosden, J. H. and Kendall, P. G. (1976) 'Current rating of 11 kV cables'. *IEE Conf. on Distribution Cables and Jointing Techniques for Systems up to 11 kV.*
(5) Engineering Recommendation P17 (1977) *Current Rating Guide for Distribution Cables.* London: Electricity Council.

CHAPTER 9
Short-circuit Ratings

It happens frequently that the conductor size necessary for an installation is dictated by its ability to carry short-circuit rather than sustained current. During a short-circuit there is a sudden inrush of current for a few cycles followed by a steadier flow for a short period until the protection operates, normally between 0.2 and 3 seconds. During this period the current falls off slightly due to the increase in conductor resistance with temperature but for calculation purposes it is assumed to remain steady.

At the commencement of the short circuit the cable may be operating at its maximum permissible continuous temperature and the increase in temperature caused by the short circuit is a main factor in deriving acceptable ratings. However, the current may be twenty or more times greater than the sustained current and it produces thermo-mechanical and electromagnetic forces proportional to the square of the current. The stresses induced may themselves impose an operating limit unless they can be contained adequately by the whole installation. This requires checks on cable design, joints, terminations and installation conditions.

TEMPERATURE

As the time involved is short and cooling follows rapidly, the cable insulation can withstand much higher temperatures than are allowed for sustained operation. Table 9.1 shows the values, related to a conductor temperature, used for transmission and distribution cables in the UK. These temperatures are in accordance with IEC Publication 724, 'Guide to the short-circuit temperature limits of electric cables with a rated voltage not exceeding 0.6/1.0 kV', first issued in 1982 and revised in 1984.

For convenience, table 9.1 includes temperatures for materials and components other than insulation so as to indicate other constraints. Figures for oversheaths are included to allow for the fact that the material is in contact with armour wires and a higher temperature value may be assigned. In the absence of armour the oversheath should be treated as insulation. It is important to appreciate that the temperatures in table 9.1 for components cannot be adopted if the insulation dictates a lower temperature.

The difference between the maximum conductor temperatures for sustained rating, given in chapter 8, and the above temperatures, provides a maximum temperature rise which can be used in short-circuit rating calculations.

Table 9.1 Short-circuit temperature limits

Material or component	Temperature (°C)
Paper insulation	250
PVC – insulation up to 300 mm^2	160[a]
PVC – insulation above 300 mm^2	140[a]
PVC – insulation 6.6 kV and above	160[a]
PVC – oversheath	200
Natural rubber	200
Butyl rubber	220
Polyethylene – oversheath	150
XLPE and EPR	250
Silicone rubber	350
CSP – oversheath	220
Soldered conductor joints	160
Compression joints	250
Lead sheaths – unalloyed	170
Lead sheaths – alloyed	200

[a]For grades TI 1 and TI 2: not applicable to non-standard soft grades

RATINGS DERIVED ON A TEMPERATURE BASIS

Short-circuit ratings can be calculated using either the adiabatic method, which assumes that all of the heat generated remains trapped within the current carrying component, or non-adiabatic methods, which allow for heat absorption by adjacent materials.

The adiabatic method may be used when the ratio of short-circuit duration to conductor cross-sectional area is less than 0.1 s/mm^2. On smaller conductors such as screen wires, as the short-circuit duration increases the loss of heat from the conductor becomes more significant. In such cases the non-adiabatic method can be used to provide a significant increase in permissible short-circuit current.

Adiabatic method

By ignoring heat loss an equation can be derived which equates heat input (I^2RT) to heat absorbed into the current carrying component (product of mass, specific heat and temperature rise). The adiabatic temperature rise formula given in IEC 724 is:

$$I^2 = \frac{K^2 S^2}{T} \log_e\left(\frac{\theta_1 + \beta}{\theta_0 + \beta}\right) \quad (9.1)$$

where I = short-circuit current (r.m.s. over duration) (A)
T = duration of short circuit (second)
K = constant for the material of the conductor
S = area of conductor (mm^2)
θ_1 = final temperature (°C)
θ_0 = initial temperature (°C)
β = reciprocal of the temperature coefficient of resistance (α) of the conductor (per degree Celsius at 0°C)

In the above, 'conductor' refers to the current carrying component. The constants for the usual metals are given in table 9.2 in which

$$K^2 = \frac{Q_c(\beta + 20)}{\rho_{20}}$$

where Q_c = volumetric specific heat of the conductor at 20°C (J/°C mm^3)
ρ_{20} = resistivity of conductor metal at 20°C (Ω mm)

Non-adiabatic method

IEC 949 gives a non-adiabatic method of calculating the thermally permissible short-circuit current allowing for heat transfer from the current carrying component to adjacent materials. The non-adiabatic method is valid for all short-circuit durations and provides a significant increase in permissible short-circuit current for screens, metallic sheaths and some small conductors.

The approach adopted is to calculate the adiabatic short-circuit current, using equation (9.1), and a modifying factor which takes into account the heat lost to adjacent materials. The adiabatic short-circuit current is multiplied by the modifying factor to obtain the permissible non-adiabatic short-circuit current.

The equations used to calculate the non-adiabatic factor are given in IEC 949. For conductors and spaced screen wires fully surrounded by non-metallic materials the equation for the non-adiabatic factor (ϵ) is:

$$\epsilon = [1 + X(T/S)^{1/2} + Y(T/S)]^{1/2} \qquad (9.2)$$

where X and Y are given in table 9.3.

Table 9.2 Constants for short-circuit calculation

Material	K	β	Q_c	ρ_{20}
Copper	226	234.5	3.45×10^{-3}	17.241×10^{-6}
Aluminium	148	228	2.5×10^{-3}	28.264×10^{-6}
Lead	42	230	1.45×10^{-3}	214×10^{-6}
Steel	78	202	3.8×10^{-3}	138×10^{-6}

Table 9.3 Constants for use in equation for conductors and spaced screen wires

Insulation	Constants for copper		Constants for aluminium	
	X (mm^2/s)$^{1/2}$	Y mm^2/s	X (mm^2/s)$^{1/2}$	Y mm^2/s
PVC – under 3 kV	0.29	0.06	0.4	0.08
PVC – above 3 kV	0.27	0.05	0.37	0.07
XLPE	0.41	0.12	0.57	0.16
EPR – under 3 kV	0.38	0.1	0.52	0.14
EPR – above 3 kV	0.32	0.07	0.44	0.1
Paper – fluid-filled	0.45	0.14	0.62	0.2
Paper – others	0.29	0.06	0.4	0.08

Table 9.4 Thermal constants for use in equation (9.3)

Material	Thermal resistivity, ρ (K m/W)	Volumetric specific heat, σ (J/K m^3)
Impregnated paper in solid type cable	6	2.0×10^6
Impregnated paper in fluid-filled cable	5	2.0×10^6
Insulating fluid	7	1.7×10^6
PE and XLPE insulation	3.5	2.4×10^6
PVC insulation in cables up to 3 kV	5	1.7×10^6
PVC insulation in cables above 3 kV	6	1.7×10^6
EPR insulation in cables up to 3 kV	3.5	2.0×10^6
EPR insulation in cables above 3 kV	5	2.0×10^6
Butyl and natural rubber insulation	5	2.0×10^6
Compounded jute coverings	6	2.0×10^6
Polychloroprene sheaths	5.5	2.0×10^6
PVC sheath on cables up to 35 kV	5	1.7×10^6
PVC sheath on cable above 35 kV	6	1.7×10^6
PE sheath	3.5	2.4×10^6
Semi-conducting PE and XLPE	2.5	2.4×10^6
Semi-conducting EPR	3.5	2.1×10^6

For sheaths, screens and armour the equation for the non-adiabatic factor is:

$$\epsilon = 1 + 0.61 M\sqrt{T} - 0.069(M\sqrt{T})^2 + 0.0043(M\sqrt{T})^3 \qquad (9.3)$$

where

$$M = \frac{(\sqrt{\sigma_2/\rho_2} + \sqrt{\sigma_3/\rho_3})}{2\sigma_1 \delta x 10^{-3}} F \qquad (s^{1/2}) \qquad (9.4)$$

σ_1 = volumetric specific heat of screen, sheath or amour (J/K m^3)
σ_2, σ_3 = volumetric specific heat of materials each side of screen, sheath or armour (J/K m^3)
δ = thickness of screen, sheath or armour (mm)
ρ_2, ρ_3 = thermal resistivity of materials each side of screen, sheath or armour (K m/W)
F = factor to allow for imperfect thermal contact with adjacent materials

The contact factor F is normally 0.7, however there are some exceptions. For example, for a current carrying component such as a metallic foil sheath, completely bonded on one side to the outer non-metallic sheath, a contact factor of 0.9 is used.

Thermal constants for common materials are shown in table 9.4.

Power distribution cables

For specific conditions of temperature rise in accordance with table 9.1 the adiabatic formula may be further adapted as indicated in table 9.5. In this table, as is usual for short-circuit calculations, it is assumed that the cable is operating at maximum permissible continuous temperature when the short circuit occurs. If the system design dictates a lower temperature, a factor may be applied.

Short-circuit Ratings

Table 9.5 Short-circuit currents for various cables

Cable insulation/Type	Conductor metal	Temperature rise (°C)	Short-circuit current (A)
Paper			
1–6 kV: belted	Copper	80–160	$108ST^{-1/2}$
	Aluminium	80–160	$70ST^{-1/2}$
10–15 kV: belted	Copper	65–160	$119ST^{-1/2}$
	Aluminium	65–160	$77ST^{-1/2}$
10–15 kV: screened	Copper	70–160	$116ST^{-1/2}$
	Aluminium	70–160	$74ST^{-1/2}$
20–30 kV: screened	Copper	65–160	$119ST^{-1/2}$
	Aluminium	65–160	$77ST^{-1/2}$
PVC: 1 and 3 kV			
Up to 300 mm²	Copper	70–160	$115ST^{-1/2}$
	Aluminium	70–160	$76ST^{-1/2}$
Over 300 mm²	Copper	70–140	$103ST^{-1/2}$
	Aluminium	70–140	$68ST^{-1/2}$
XLPE and EPR	Copper	90–250	$114ST^{-1/2}$
	Aluminium	90–250	$92ST^{-1/2}$

Care must be exercised in using the figures derived on this basis, e.g. for paper cables it is assumed that the conductor joints are soldered, and hence subject to a 160°C limitation, whereas mechanical joints suitable for 250°C are adopted with XLPE and EPR. Other considerations are given in the next section.

An alternative way of expressing the data available from the last column of table 9.5 is the graphical presentation given in figs 9.1 and 9.2 for paper insulated cables, figs 9.3 and 9.4 for PVC insulation cables and figs 9.5 and 9.6 for XLPE insulated cables.

Fluid-filled cables

Short-circuit ratings for fluid-filled cables are calculated from a temperature rise of 90–160°C and are shown in figs 9.7–9.10 for single- and 3-core cables with copper and aluminium conductors.

Wiring cables

For wiring cables the basic method of calculation given earlier applies but the actual method of implementation is slightly different. Table 9.6, which is in the same form as table 9.5, includes data as used in the IEE Wiring Regulations. An average is taken for 65°C rubber and 85°C rubber (which includes EPR). Although the short-circuit temperatures for the two materials are different, so also are their operating temperatures under sustained conditions; hence the permissible temperature rise is similar in each case (see also chapter 18).

A typical graphical presentation applicable to PVC insulated conduit wires is shown in fig. 9.11.

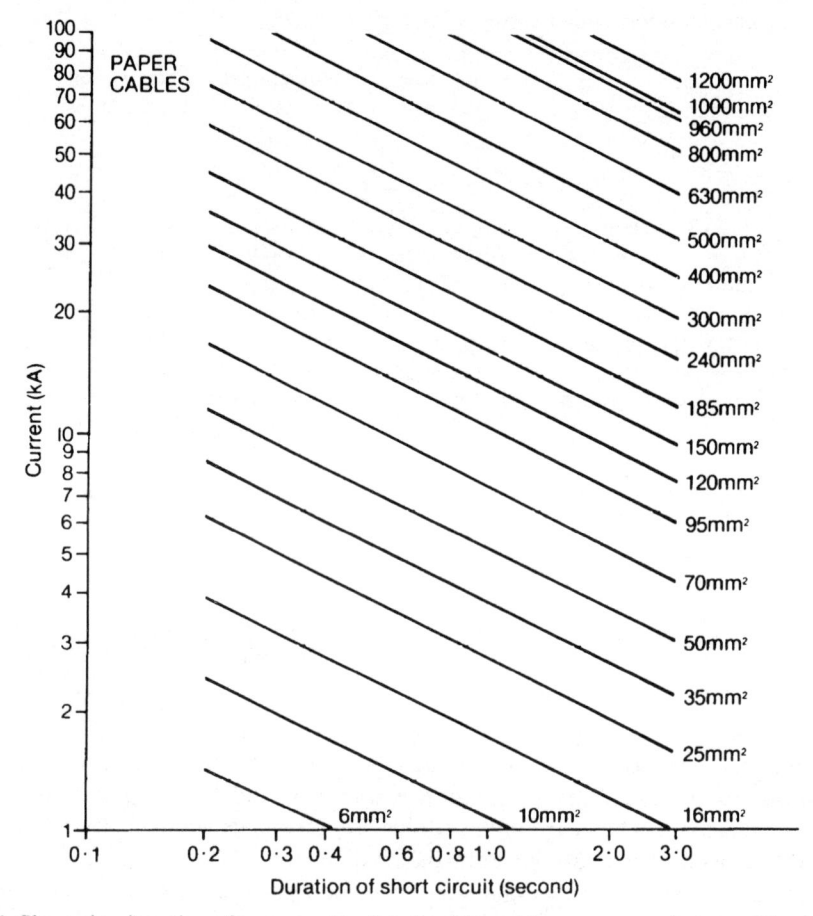

Fig. 9.1 Short-circuit ratings for paper insulated cables with copper conductors. These ratings apply to cables for voltages up to and including 3.8/6.6 kV. For higher voltages they may be increased by the factors shown below:

Cable type	Voltage	Factor
3-core (belted)	6.35/11 kV	1.10
Single- and 3-core (screened)	6.35/11 and 8.7/15 kV	1.07
Single- and 3-core	12.7/22 and 19/33 kV	1.10

ASYMMETRICAL FAULTS

The previous section was concerned with symmetrical 3-phase faults, with a short circuit between the phase conductors. In the case of asymmetrical, i.e. earth, faults other factors have to be taken into account because the current is carried by the lead sheath and/or armour. In general, for small conductor sizes, the conductor temperature rise is still the limiting factor, but with the larger sizes a lower limit is imposed by lead

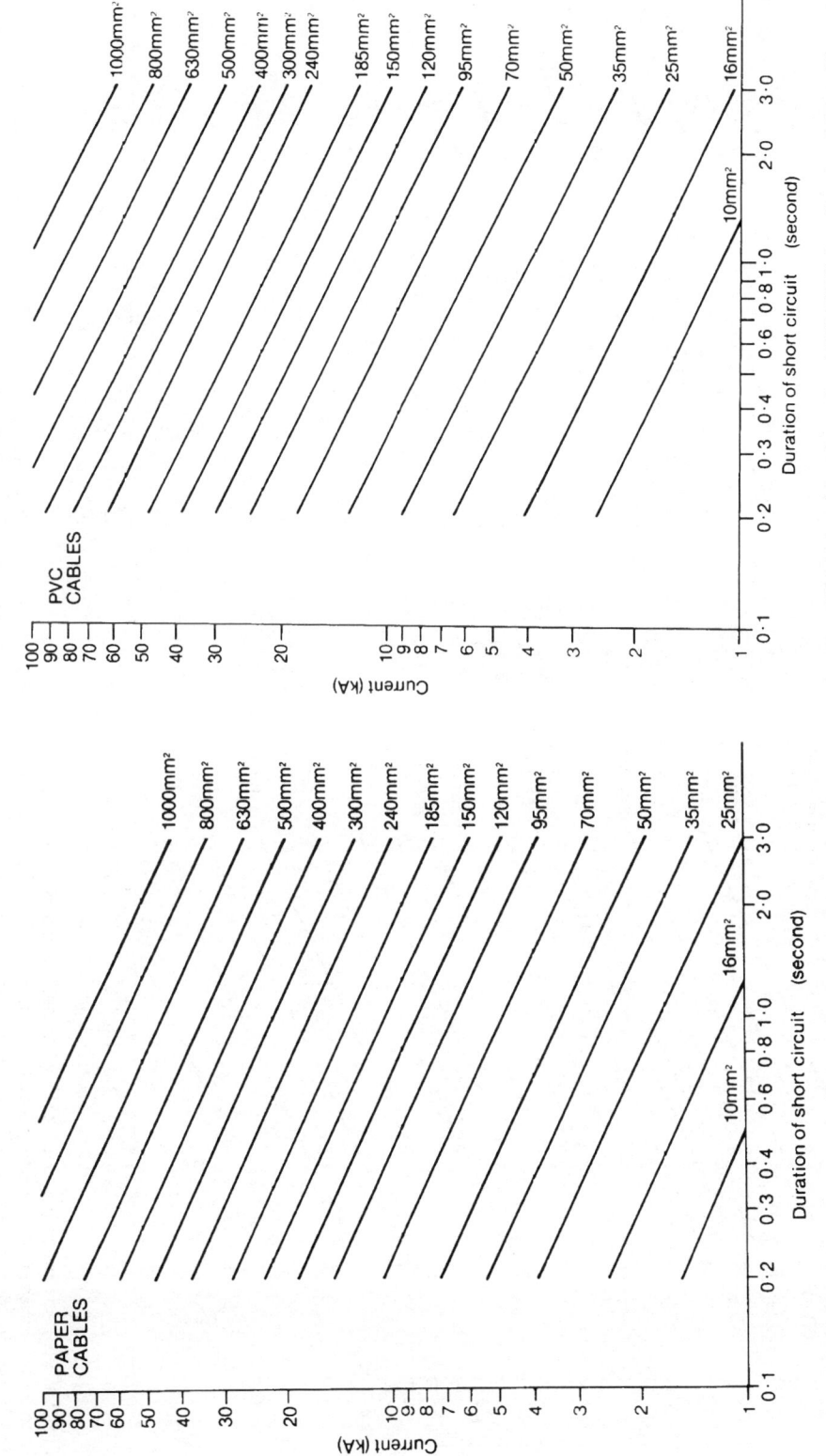

Fig. 9.2 Short-circuit ratings for paper insulated cables with aluminium conductors. The factors given under fig. 9.1 for higher voltages are applicable

Fig. 9.3 Short-circuit ratings for 1 kV PVC insulated cables with copper conductors (based on a final conductor temperature of 160°C for sizes up to and including 300 mm² and 140°C for conductors above 300 mm²)

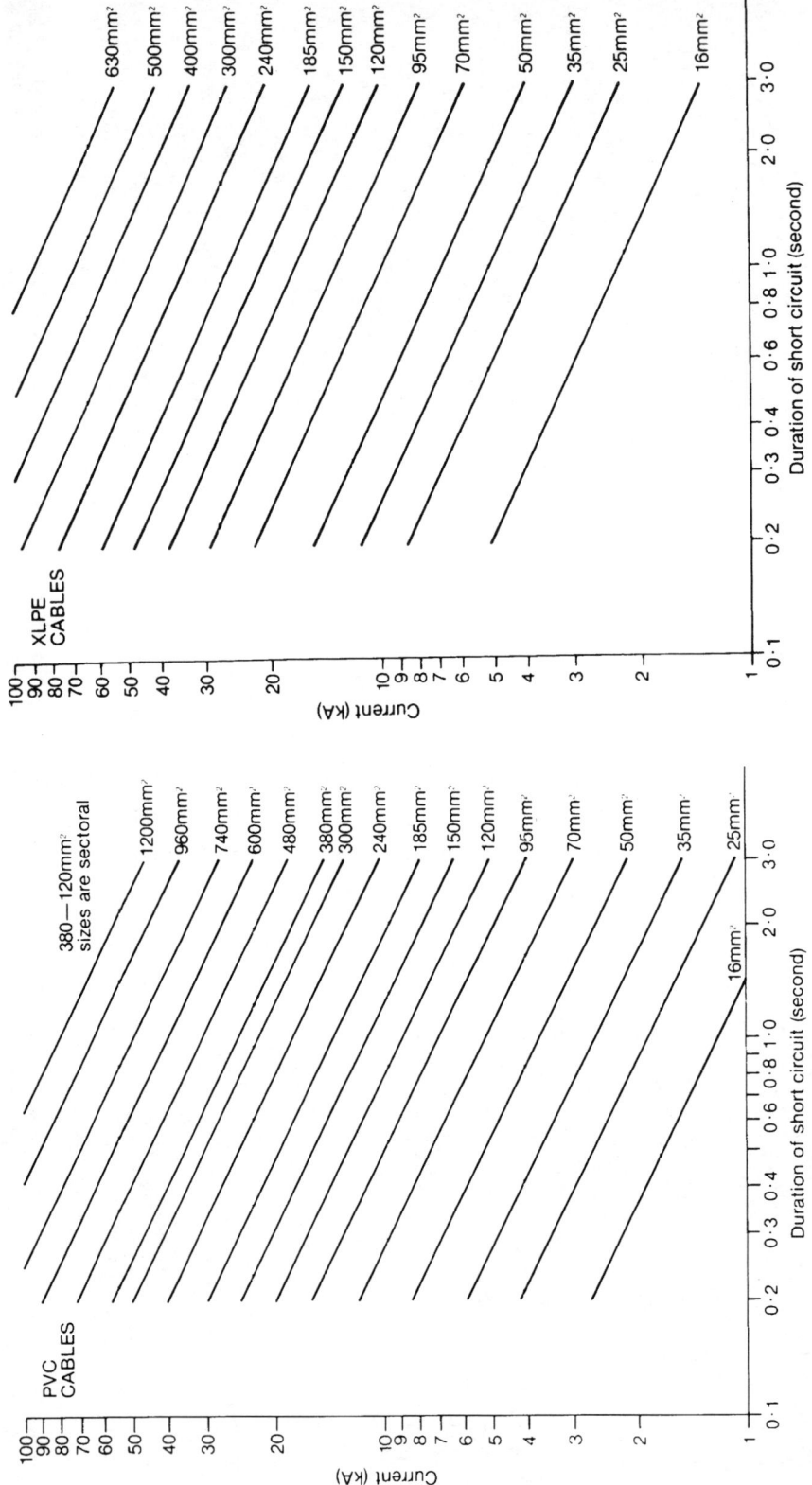

Fig. 9.4 Short-circuit ratings for 1 kV PVC insulated cables with aluminium conductors (based on a final conductor temperature of 160°C for sizes up to and including 300 mm² and 140°C for conductors above 300 mm²)

Fig. 9.5 Short-circuit ratings for XLPE insulated cables with copper conductors (based on a temperature rise from 90 to 250°C, i.e. 160°C)

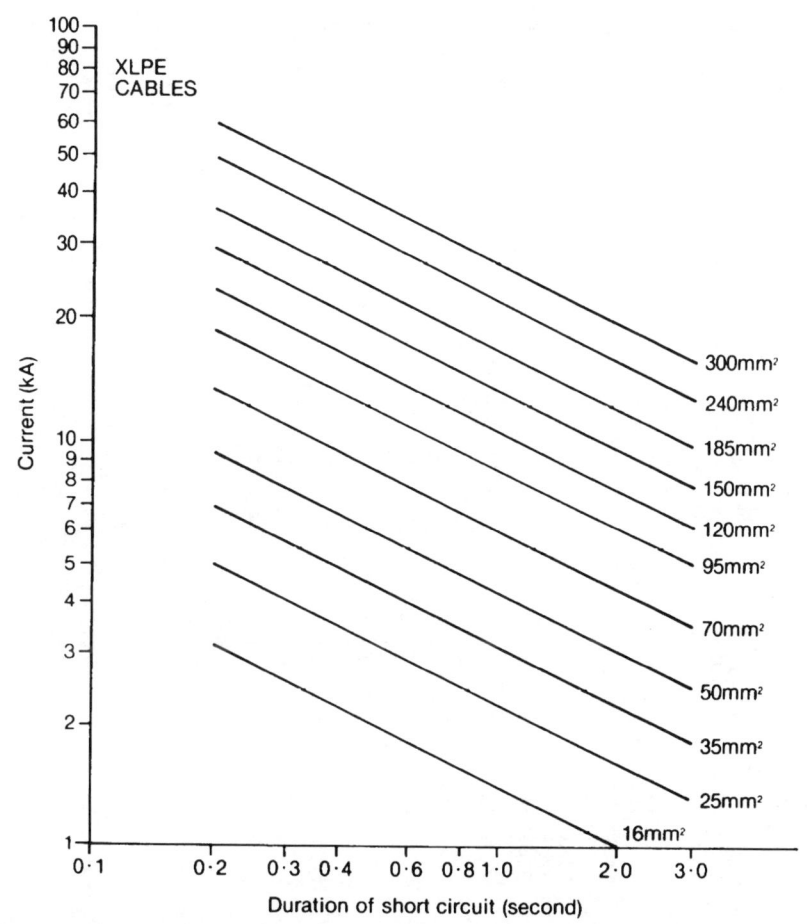

Fig. 9.6 Short-circuit ratings for XLPE insulated cables with solid aluminium conductors (based on a temperature rise of 160°C)

sheath and armour considerations, as indicated in table 9.1. Lead sheaths are liable to be damaged if they are suddenly heated to temperatures above those shown in the table and the armour temperature may be controlled by the performance of the extruded PVC in contact with it.

Maximum allowable asymmetrical fault currents for the more common distribution cable types are given in tables 9.7–9.12. Unless otherwise stated, the ratings apply to cables with stranded conductors. They are also for a fault duration of 1 s. For other periods the values should be divided by the square root of the time in seconds. In the case of lead sheathed wire armoured cable it is assumed that the current is shared between the lead sheath and the armour.

ELECTROMAGNETIC FORCES AND CABLE BURSTING

In multicore cables a short circuit produces electromagnetic forces which repel the cores from each other, and if they are not adequately bound together the cable will tend to

Fig. 9.7 Short-circuit ratings for single-core fluid-filled cables with copper conductors (based on conductor temperature of 90°C at start of short circuit and final conductor temperature of 160°C)

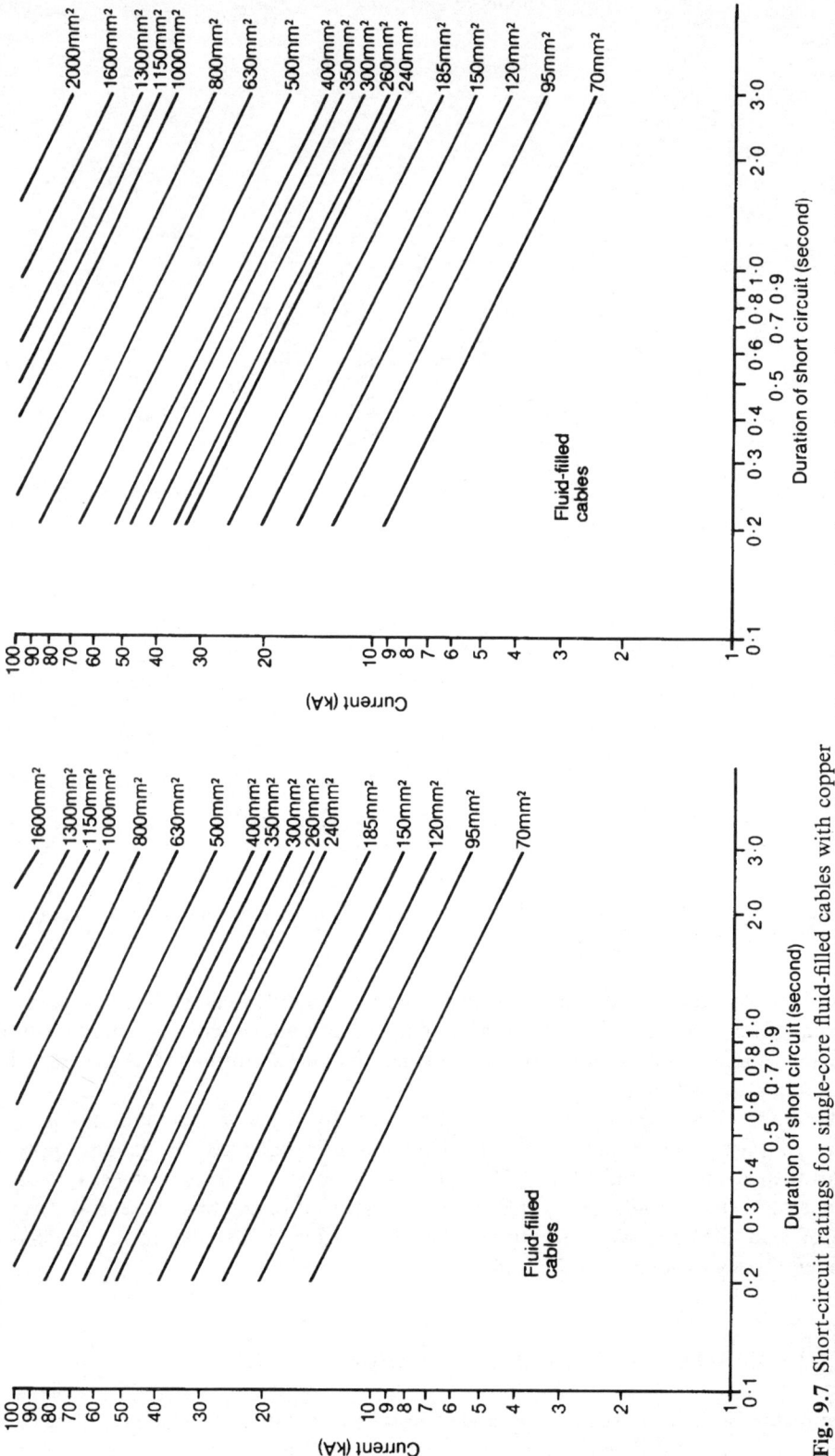

Fig. 9.8 Short-circuit ratings for single-core fluid-filled cables with aluminium conductors (conditions as given under fig. 9.7)

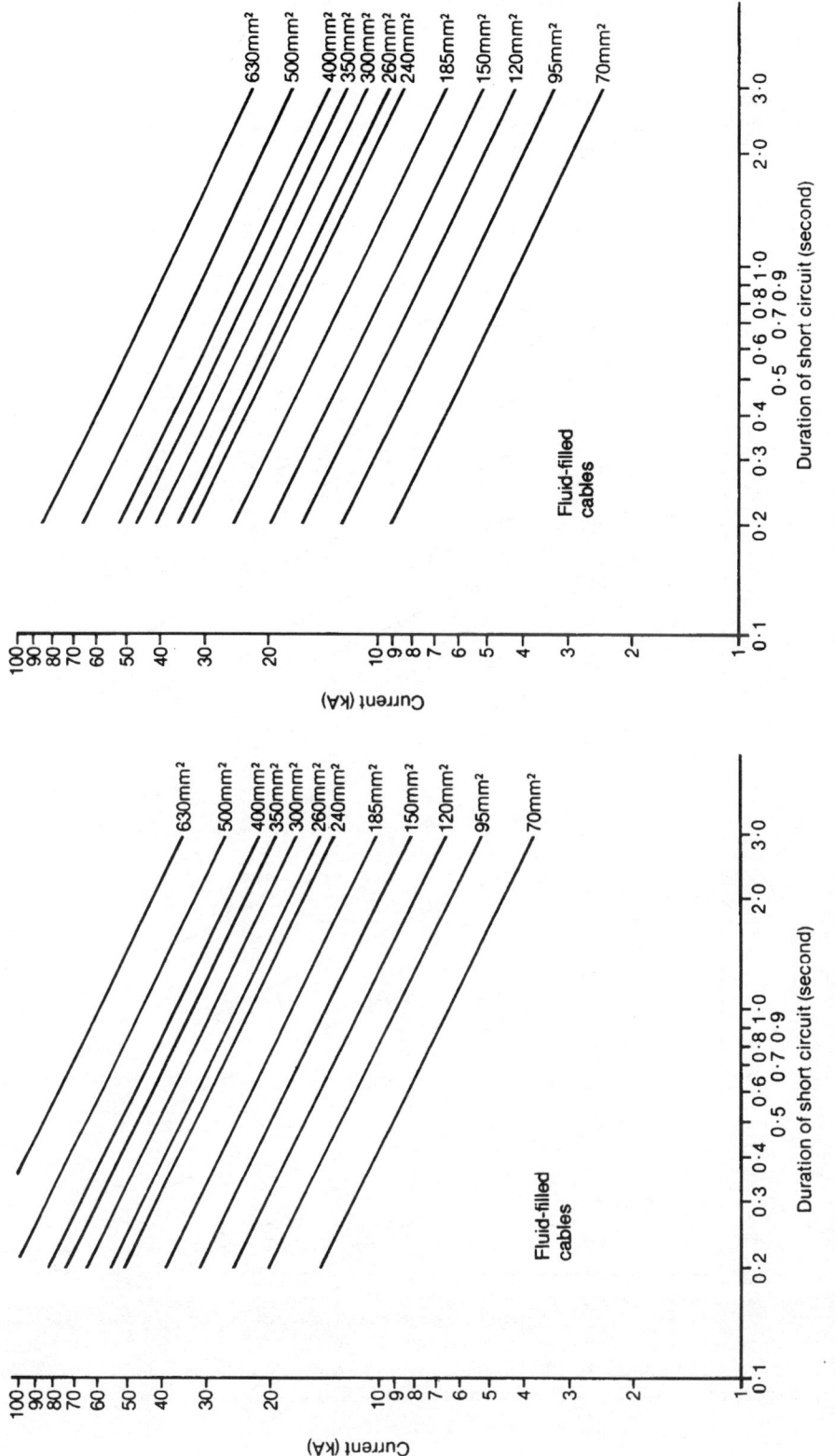

Fig. 9.9 Short-circuit ratings for 3-core fluid-filled cables with copper conductors (conditions as given under fig. 9.7)

Fig. 9.10 Short-circuit ratings for 3-core fluid-filled cables with aluminium conductors (conditions as given under fig. 9.7)

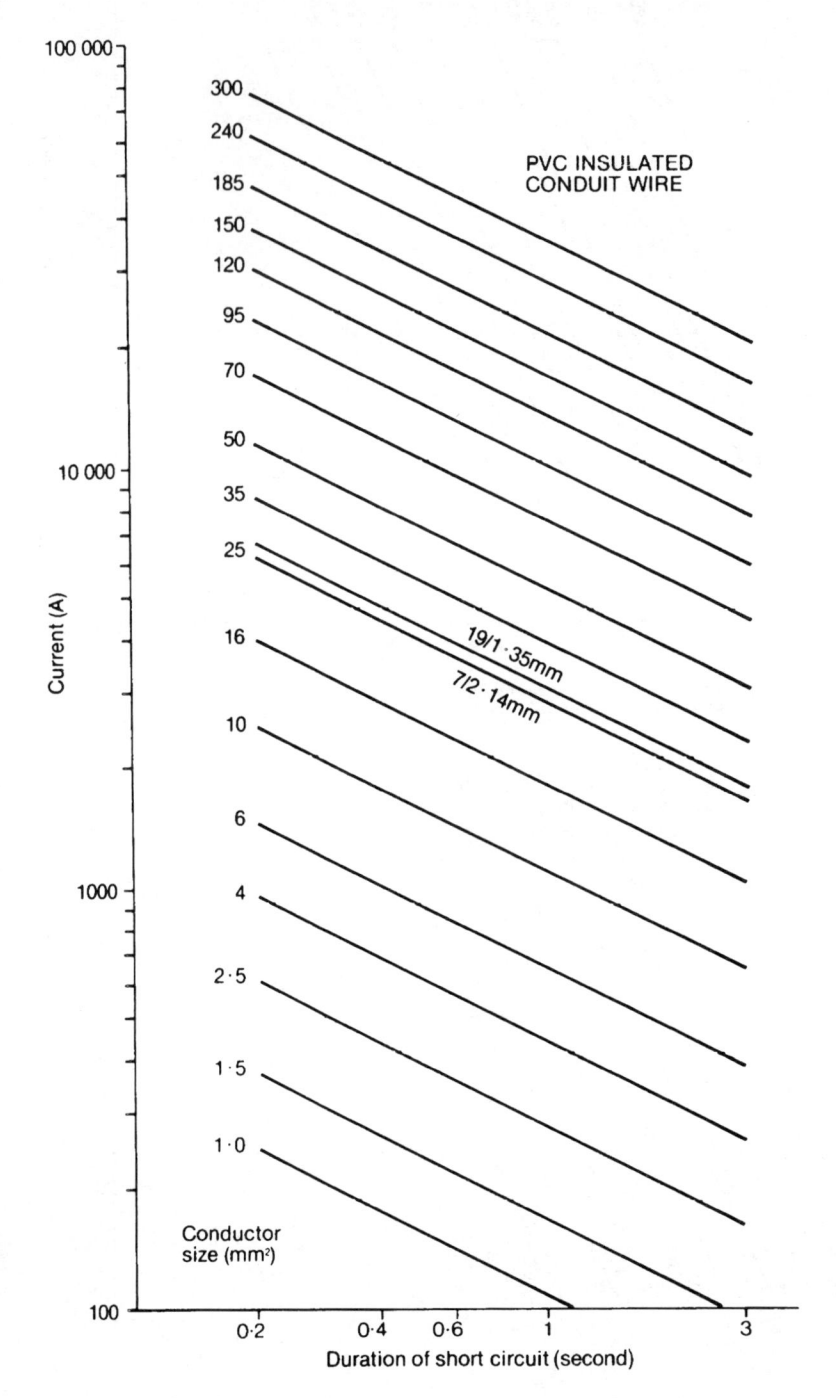

Fig. 9.11 Short-circuit ratings for PVC insulated conduit wires with copper conductors (70°C at start of short circuit and final conductor temperature of 160°C)

Table 9.6 Short-circuit ratings for wiring cables

Cable insulation	Conductor metal	Short-circuit current (A)
PVC	Copper	$115/103 ST^{-1/2}$ [a]
	Aluminium	$76/68 ST^{-1/2}$ [a]
65°C rubber	Copper	$134 ST^{-1/2}$
85°C rubber	Aluminium	$89 ST^{-1/2}$
Mineral	Copper	$135 ST^{-1/2}$
	Aluminium	$87 ST^{-1/2}$

[a] The lower value applies to conductors over 300 mm^2.

burst. The effect is of importance with unarmoured paper insulated cables, as the insulation may be damaged in the process. For conductor sizes of 185 mm^2 and above and currents above about 30 kA, the damage which can be caused by bursting forces imposes limitations on the ratings deduced by the thermal basis used in table 9.3 and figs 9.1–9.6, i.e. there is a cut-off point irrespective of fault duration. For unarmoured 3-core belted paper cables with stranded conductors these limits are given in table 9.13.

Armouring provides sufficient reinforcement to prevent damage due to these bursting forces. Screened paper insulated cables are similarly affected but it is seldom that such cables are used in the unarmoured condition.

Table 9.7 Maximum allowable asymmetrical current to earth[a] (single-core PILS cables)

Conductor size (mm^2)	0.6/1 kV (kA)	1.9/3.3 kV (kA)	3.3/6.6 kV (kA)	6.35/11 kV (kA)	8.7/15 kV (kA)	12.7/22 kV (kA)	19/33 kV (kA)
50	1.4	1.6	1.7	1.9	2.3	2.9	3.8
70	1.6	1.8	1.9	2.3	2.5	3.1	4.0
95	1.9	2.0	2.3	2.5	2.9	3.6	4.6
120	2.2	2.4	2.5	2.9	3.2	3.8	4.8
150	2.5	2.6	3.0	3.1	3.6	4.3	5.3
185	2.9	3.0	3.2	3.6	3.9	4.6	5.8
240	3.3	3.4	3.8	4.0	4.6	5.3	6.3
300	3.9	4.0	4.2	4.7	4.9	5.7	6.8
400	4.7	4.7	4.9	5.5	5.7	6.6	7.7
500	5.6	5.6	5.7	6.1	6.8	7.6	8.8
630	6.6	6.6	6.8	7.2	7.9	8.8	10.1
800	7.7	7.7	7.9	8.3	9.1	10.1	11.4
1000	9.0	9.0	9.2	10.2	10.5	11.6	13.0
960[b]				9.8	10.6	11.7	13.2
1200[b]				11.8	12.2	13.4	14.9

[a] For 1 s rating
[b] Milliken type conductors (copper only)

Table 9.8 Maximum allowable asymmetrical fault currents to earth[a] (multicore PILS/SWA cable)

Conductor size (mm²)	0.6/1 kV 3-core (kA)	0.6/1 kV 4-core (kA)	1.9/3.3 kV 3-core (kA)	3.8/6.6 kV 3-core (kA)	6.35/11 kV 3-core (belted) (kA)	6.35/11 kV 3-core (screened) (kA)	8.7/15 kV 3-core (screened) (kA)	12.7/22 kV 3-core (screened) (kA)	19/33 kV 3-core (screened) (kA)
4	3.1	3.4							
6	3.4	3.6							
10	3.8	4.9	5.0	6.4					
16	4.4	5.0	5.1	7.3	10.1	10.1			
25	5.2	6.1	6.0	8.7	11.3	11.3	12.8		
35	6.0	6.7	6.6	9.6	11.3	10.9	16.2		
50	6.8	8.9	8.5	10.3	14.3	12.0	16.4	19.5	
70	8.7	10.3	9.9	11.6	15.7	15.4	17.8	20.6	
95	10.1	12.0	11.3	15.0	17.5	17.1	20.1	21.3	26.9
120	11.4	15.4	14.4	16.5	18.6	18.7	21.3	23.2	27.5
150	15.3	17.9	16.0	17.9	20.2	20.2	22.9	24.9	33.5
185	17.1	20.1	17.5	19.8	22.1	22.1	24.9	26.7	35.7
240	19.4	23.6	19.8	22.6	25.0	24.6	32.2	33.2	37.0
300	21.8	26.3	22.6	25.0	31.6	31.6	35.1	36.8	39.5
400	25.0	34.5	29.8	32.6	35.6	35.7	39.3	39.9	43.0
								44.4	46.3
									51.0

[a] For 1 s rating

Short-circuit Ratings

Table 9.9 Maximum allowable asymmetrical fault currents to earth[a] (PVC insulated wire armoured cables with solid aluminium conductors)

Conductor size (mm^2)	Aluminium armour 0.6/1 kV single-core (kA)	Aluminium armour 1.9/3.3 kV single-core (kA)	Steel armour 0.6/1 kV 2-core (kA)	Steel armour 0.6/1 kV 3-core (kA)	Steel armour 0.6/1 kV 4-core (kA)	Steel armour 1.9/3.3 kV 3-core (kA)
16			1.6	1.8	2.7	3.3
25			2.4	2.7	3.2	3.5
35			2.6	3.1	3.5	3.8
50	2.8	3.2	4.0	3.5	5.0	5.1
70	3.2	3.6	4.4	5.0	5.5	5.7
95	3.6	5.2	4.8	5.7	6.5	6.2
120	5.2	5.6		6.1	8.9	8.4
150	5.7	5.9		8.4	9.7	9.1
185	6.2	6.4		9.5	10.8	9.7
240	7.0	7.0		10.6	12.1	10.6
300	7.6	7.6		11.7	13.4	11.7
380	10.9	10.9				
480	12.2	12.2				
600	12.9	12.9				
740	17.8	17.8				
960	20.2	20.2				
1200	22.1	22.1				

[a] For 1 s rating

Cables with polymeric insulation, either thermoplastic or thermoset, are more resistant to damage than paper cables but would have a short-circuit rating limitation due to bursting forces if they were unarmoured.

When installing single-core cables it is necessary to hold the cores together with adequate binding straps, or to use suitable cleats, to withstand bursting forces.[1] This is an aspect of particular importance for cables in generating stations.

THERMOMECHANICAL EFFECTS

The high temperature rise resulting from a short circuit produces expansion of the conductors which may cause problems due to longitudinal thrust of the conductors and bodily movement of the cable if it is not adequately supported. The possibility of excessive thrust is also important if the cable has a solid conductor.

Design of joints and terminations

The effect on joints is more critical with buried cables, i.e. because of the ground restraint on the cable surface the cores may tend to move longitudinally within the cable and into the accessory. The magnitude of this thrust can be very high, e.g. 50 N/mm^2 of conductor area, and is particularly important with large conductors.

Table 9.10 Maximum allowable asymmetrical fault currents to earth[a] (PVC insulated wire armoured cables with copper conductors)

Conductor size (mm²)	Aluminium armour		Steel wire					
	0.6/1 kV single-core (kA)	1.9/3.3 kV single-core (kA)	0.6/1 kV				1.9/3.3 kV 3-core (kA)
			2-core (kA)	3-core (kA)	4-core (kA)	4-core (reduced neutral) (kA)	
1.5			0.7	0.7	0.7		
2.5			0.8	0.8	0.9		
4			0.9	1.0	1.5		
6			1.0	1.5	1.7		
10			1.8	1.9	2.1		
16			1.7	1.9	2.7		3.3
25			2.7	2.9	3.4	3.4	3.6
35			2.9	3.3	3.7	3.6	4.0
50	3.1	3.5	3.3	3.7	5.4	4.2	5.4
70	3.5	3.9	3.7	5.3	6.1	5.9	6.1
95	4.0	5.7	5.4	6.1	7.0	6.9	6.6
120	5.7	6.2	5.8	6.6	9.7	9.5	9.1
150	6.4	6.5	6.4	9.3	10.8	10.4	9.7
185	7.0	7.0	8.9	10.2	11.7	11.4	10.4
240	7.8	7.8	9.9	11.4	13.2	12.7	11.4
300	8.6	8.6	11.0	12.7	14.7	14.3[b] 14.7[c]	12.7
400	12.2	12.2	12.3	14.0	20.6	19.9	14.0
500	13.4	13.4					
630	14.6	14.6					
800	20.6	20.6					
1000	22.9	22.9					

[a] For 1 s rating
[b] 300/150 mm²
[c] 300/185 mm²

If the filling material in the accessory is sufficiently soft to permit movement of the cores, the force may cause buckling and collapse within the accessory. Once this has occurred, a tension will develop as the conductors cool down and this can create further problems such as a stress on the ferrules which is sufficiently high for the conductors to pull out. This is the reason for the 160°C temperature limit on soldered conductor connections.

Other factors arising are that (a) the design of fittings applied by mechanical and compression techniques must be adequate to cater for stability of electrical resistance at the temperature generated, and (b) in the case of paper insulated cables, the outer

Table 9.11 Maximum allowable asymmetrical fault currents to earth[a] (XLPE insulated wire armoured cables with solid aluminium conductors)

Conductor size (mm^2)	Aluminium armour 0.6/1 kV single-core (kA)	1.9/3.3 kV single-core (kA)	Steel armour 0.6/1 kV 2-core (kA)	3-core (kA)	4-core (kA)	1.9/3.3 kV 3-core (kA)
16			1.2	1.4	1.6	2.7
25			1.5	2.3	2.6	3.0
35			2.2	2.6	2.9	3.2
50	1.6	2.8	2.4	2.9	3.3	4.5
70	2.6	3.1	2.8	3.3	4.9	5.0
95	3.0	3.1	4.1	4.8	5.4	5.5
120	3.2	4.3		5.2	7.6	7.4
150	4.8	4.6		7.4	8.4	8.0
185	5.2	5.2		8.2	9.4	8.6
240	5.7	5.7		9.2	10.5	9.6
300	6.3	6.3		10.1	11.7	10.3

[a] For 1 s rating

casing must be able to deal with the high fluid pressure caused by expansion of the cable impregnant and its possible flow into the accessory, with consequent softening of the joint filling compound.

Differences between copper and aluminium conductors

Although the coefficient of expansion of aluminium is somewhat higher than that of copper, the effect as far as stress is concerned is largely balanced by the lower elastic modulus of aluminium. Forces causing buckling are therefore of a similar order. On the other hand, in the temper used for solid conductors, aluminium is a softer metal and the yield under compressive forces during heating could result in higher tensile forces on cooling. When such conductors were first introduced it was considered that a lower permissible short-circuit temperature might be necessary but subsequent experience has not indicated need for special treatment.

When limitations are imposed by lead sheaths or electromagnetic forces, the type of conductor metal is irrelevant in theory but, in respect of bursting forces, aluminium is at some disadvantage compared with copper because of the physical size effect for equal rating.

Installation conditions

As mentioned above, the effects of longitudinal thrust are most important for cables which are buried in the ground. However, as discussed in chapter 26, it is also important that cables installed in air should have adequate support spacings and/or

Table 9.12 Maximum allowable asymmetrical fault currents to earth[a] (XLPE insulated wire armoured cables with copper conductors)

Conductor size (mm²)	Aluminium armour		Steel armour				
	0.6/1 kV single-core (kA)	1.9/3.3 kV single-core (kA)	0.6/1 kV				1.9/3.3 kV 3-core (kA)
			2-core (kA)	3-core (kA)	4-core (equal) (kA)	4-core (reduced neutral) (kA)	
16			1.7	1.7	1.9		3.1
25			1.7	2.4	2.7	2.7	3.1
35			2.4	2.7	3.1	3.0	3.3
50	1.8	2.7	2.6	3.0	3.5	3.3	4.6
70	2.7	3.1	3.1	3.5	5.1	5.0	5.1
95	3.1	3.3	4.4	5.0	5.7	5.6	5.7
120	3.3	4.8	4.9	5.5	8.0	6.3	7.8
150	4.8	5.1	5.4	7.8	9.0	8.6	8.4
185	5.4	5.7	7.4	8.6	9.9	9.7	9.0
240	6.0	6.0	8.4	9.7	11.3	10.9	9.9
300	6.4	6.8	9.2	10.5	12.4	11.8[b]	10.9
300	6.4	6.8	9.2	10.5	12.4	12.4[c]	10.9
400	9.1	9.1					
500	10.5	10.5					
630	11.8	11.8					

[a] For 1 s rating
[b] 300/150 mm²
[c] 300/185 mm²

rigid cleating to prevent excessive generation of expansion and contraction in local spans, particularly associated with accessories.

In the case of cables with thermoplastic insulation and oversheaths, it is also important to prevent excessive local pressure on the cable which could cause deformation of the material. This could occur due to small bending radii, by unsuitable fixing arrangements at bends or by unsatisfactory clamping devices. Similar remarks apply to thermosetting insulation with large conductor sizes because of the high temperature of 250°C which is quoted for XLPE and EPR.

GENERAL COMMENTS

It is usual to quote short-circuit ratings, as is done in the tables and figures, relative to the basic cable design and type of insulation. It must be stressed that, as outlined in the latter part of this chapter, such values are only applicable if the method of installation and the accessories are appropriate. These aspects may represent the weakest links and are often overlooked.

Table 9.13 Limitation due to bursting of unarmoured belted multicore paper-insulated cables

Voltage (kV)	Aluminium conductor Size (mm^2)	Short-circuit rating (kA)	Copper conductor Size (mm^2)	Short-circuit rating (kA)
0.6/1	240	33	120	25
	300	35	150	27
	400	37	185	29
			240	33
			300	36
			400	38
1.9/3.3 and 3.8/6.6	240	33	185	33
	300	35	240	35
	400	38	300	37
			400	38
6.35/11	240	39	185	36
	300	41	240	39
	400	43	300	41
			400	43

However, there may be some mitigating circumstances. In order to have a standardised basis for the publication of short-circuit ratings, it is assumed that the short circuit occurs whilst the cable is already at the maximum temperature for continuous operation. This is seldom the case. When it is important to design or choose a cable to secure a very high short-circuit rating and all other circumstances can be predicted fairly accurately, the tabulated figures may even be increased by an appropriate factor.

REFERENCE

(1) Foulsham, N., Metcalf, J. C. and Philbrick, S. E. (1974) 'Proposals for installation practice of single-core cables'. *Proc. IEE* **121** (10).

CHAPTER 10
Technical Data Applicable to Cable Planning and Usage

Many of the cable characteristics which are mentioned and for which methods of calculation are given in chapter 2 also provide data required in the planning of installations and give information to cable users for controlling operation. The d.c. and a.c. resistance of the conductors, the inductance and the inductive reactance calculated from it and the impedance, derived from the resistance and reactance, are data required by the designer of the installation and likely to be of interest to the user. This also applies to the capacitance, from which charging current can be calculated, and the power factor, from which the dielectric losses can be derived. The other losses, in conductor and metallic coverings, in addition to their effect on current ratings, are again of interest to designers and users on their own account.

Similarly the sustained current carrying capacity and short-circuit rating, dealt with in detail in chapters 8 and 9 respectively, constitute technical data applicable to installation planning and usage.

There is no need to dwell further on aspects fully covered in the chapters mentioned. This chapter is confined to a few additional aspects and to the application of the parameters of cables rated up to 600/1000 V for compliance with BS 7671: 1992, 'Requirements for Electrical Installations, including Amendment No. 1' (1994), also known as the IEE Wiring Regulations Sixteenth Edition.

CHARGING CURRENT

The charging current is the capacitive current which flows when an a.c. voltage is applied to the cable as a result of the capacitance between the conductor(s) and earth and, for a multicore cable in which the individual cores are not screened, between conductors. The value can be derived from the equation

$$I_c = \omega C V \times 10^{-6} \quad \text{(A/km)} \quad (10.1)$$

where I_c = charging current (A/km)
$\omega = 2\pi$ times the frequency of the applied voltage
C = capacitance between the electrodes between which the voltage is applied (μF/km)
V = applied voltage (V)

In normal operation, C is the capacitance to neutral (known as the star capacitance), V is the voltage to neutral, i.e. U_0, and the charging current is the current in each phase. The expression $\omega C U_0 \times 10^{-6}$ is then recognisable as part of equation (2.12) in chapter 2, where it is multiplied by U_0 and the power factor to give the dielectric loss.

In addition to its relevance to operating conditions, a knowledge of the prospective charging current is required in deciding the currents that have to be provided by test transformers. For cables which do not have individually screened cores, e.g. belted cables, the test voltages may be applied in sequence as single-phase voltages between each conductor and the others connected to the metal covering, and between the bunched conductors and the metal covering. The values of capacitance for estimating charging current are then the values given in chapter 2 appropriate to these forms of connection, instead of the equivalent star capacitance (capacitance to neutral) given the symbol C in chapter 2, applicable under 3-phase voltage conditions.

VOLTAGE DROP

When current flows in a cable conductor there is a voltage drop between the ends of the conductor which is the product of the current and the impedance. If the voltage drop were excessive, it could result in the voltage at the equipment being supplied being too low for proper operation. The voltage drop is of more consequence at the low end of the voltage range of supply voltages than it is at higher voltages, and generally it is not significant as a percentage of the supply voltage for cables rated above 1000 V unless very long route lengths are involved.

In the tables included in this publication and in the IEE Wiring Regulations (BS 7671), voltage drops for individual cables are given in the units millivolts per ampère per metre length of cable. They are derived from the following formulae:

for single-phase circuits $mV/A/m = 2Z$
for 3-phase circuits $mV/A/m = \sqrt{3}Z$

where $mV/A/m$ = volt drop in millivolts per ampère per metre length of cable route
Z = impedance per conductor per kilometre of cable at maximum normal operating temperature (Ω/km)

In a single-phase circuit, two conductors (the phase and neutral conductors) contribute to the circuit impedance and this accounts for the number 2 in the equation. If the voltage drop is to be expressed as a percentage of the supply voltage, for a single-phase circuit it has to be related to the phase-to-neutral voltage U_0, i.e. 240 V when supply is from a 240/415 V system.

In a 3-phase circuit, the voltage drop in the cable is $\sqrt{3}$ times the value for one conductor. Expressed as a percentage of the supply voltage it has to be related to the phase-to-phase voltage U, i.e. 415 V for a 240/415 V system.

The IEE Wiring Regulations (BS 7671: 1992) used to require that the drop in voltage from the origin of the installation to any point in the installation should not exceed 2.5% of the nominal voltage when the conductors are carrying the full load current, disregarding starting conditions. The 2.5% limit has since been modified to a value appropriate to the safe functioning of the equipment in normal service, it being left to the designer to quantify this. However, for final circuits protected by an overcurrent protective device having a nominal current not exceeding 100 A, the requirement is deemed to be satisfied if the voltage drop does not exceed the old limit of 2.5%. It is therefore likely that for such circuits the limit of 2.5% will still apply more often than not in practice.

The reference to starting conditions relates especially to motors, which take a significantly higher current in starting than when running at operating speeds. It may be necessary to determine the size of cable on the basis of restricting the voltage drop at the starting current to a value which allows satisfactory starting, although this may be larger than required to give an acceptable voltage drop at running speeds.

To satisfy the 2.5% limit, if the cable is providing a single-phase 240 V supply, the voltage drop should not exceed 6 V and, if providing a 3-phase 415 V supply, the voltage drop should not exceed 10.4 V. Mostly, in selecting the size of cable for a particular duty, the current rating will be considered first. After choosing a cable size to take account of the current to be carried and the rating and type of overload protective device, the voltage drop then has to be checked. To satisfy the 2.5% limit for a 240 V single-phase or 415 V 3-phase supply the following condition should be met:

for the single-phase condition $mV/A/m \leq 6000/(IL)$
for the 3-phase condition $mV/A/m \leq 10400/(IL)$

where I = full load current to be carried (A)
         L = cable length (m)

The smallest size of cable for which the value of mV/A/m satisfies this relationship is then the minimum size required on the basis of 2.5% maximum voltage drop.

For other limiting percentage voltage drops and/or for voltages other than 240/415 V the values of 6 V (6000 mV) and 10.4 V (10400 mV) are adjusted proportionately.

Calculations on these simple lines are usually adequate. Strictly, however, the reduction in voltage at the terminals of the equipment being supplied will be less than the voltage drop in the cable calculated in this way unless the ratio of inductive reactance to resistance of the cable is the same as for the load, which will not normally apply. If the power factor of the cable in this sense (not to be confused with dielectric power factor) differs substantially from the power factor of the load and if voltage drop is critical in determining the required size of cable, a more precise calculation may be desirable.

In Appendix 4 of BS 7671: 1992, information is provided to enable account to be taken of the phase angle of the load, denoted by the symbol ϕ. For cables with conductor cross-sectional areas of 25 mm^2 and above the volt drops in the tables, expressed as millivolts per ampere per metre of cable (mV/A/m), are given in columns headed r, x and z. The z values are derived from the cable impedances and correspond to those represented by mV/A/m in the previous formulae. These values apply when no account is taken of phase angles. The r and x values are the resistive and reactive components respectively.

Appendix 4 of the Regulations gives the following formula for calculating the approximate voltage drop taking account of the phase angle of the load:

$$\mathrm{mV/A/m} = \cos\phi\,(\text{tabulated } (\mathrm{mV/A/m})_r) + \sin\phi\,(\text{tabulated } (\mathrm{mV/A/m})_x) \qquad (10.2)$$

In this book the voltage drop values given as such in the tables are $(\mathrm{mV/A/m})_z$ values, but resistances at maximum operating temperature and reactances are given under the heading 'electrical characteristics' for some types of cables for which use could appropriately be made of the above formula. These resistance and reactance values are for each conductor and, for 2-, 3- and 4-core cables, can be multiplied by 2 for single-phase or d.c. circuits and by $\sqrt{3}$ for 3-phase circuits to give the $(\mathrm{mV/A/m})_r$ and $(\mathrm{mV/A/m})_x$ values.

For cables with conductor cross-sectional areas of 16 mm^2 or less the resistive and reactive components of voltage drop are not given separately in the tables of appendix 4 of the Regulations. This is because resistance so predominates that the r and z values are almost the same, the phase angle for the cable being close to zero. For these cables the voltage drops, taking account of the phase angle ϕ of the load, can be calculated approximately by multiplying the tabulated value of mV/A/m by $\cos\phi$.

The following example demonstrates the difference that there could be between calculating the voltage drop from the impedance of the cable as an entity compared with the calculation which takes account of the resistive and reactive components and the phase angle of the load. If equipment having a 3-phase reactance of 1.5 Ω and a resistance of 0.75 Ω is to be supplied by a 3- or 4-core 70 mm^2 (copper) XLPE insulated cable, the impedance of the load is $(1.5^2 + 0.75^2)^{1/2}$, i.e. 1.677 Ω. The cosine of the phase angle of the load is 0.75/1.677, i.e. 0.447, and $\sin\phi$ is 1.5/1.677 i.e. 0.894. The a.c. resistance per kilometre of the cable conductor at maximum operating temperature is 0.342 Ω and the reactance is 0.075 Ω. These values can be obtained from table A14.18. The corresponding 3-phase $(\mathrm{mV/A/m})_r$ and $(\mathrm{mV/A/m})_x$ values, obtained by multiplying by $\sqrt{3}$, are 0.592 and 0.130 respectively. The value of $(\mathrm{mV/A/m})_z$ would be the vector sum of these, which is 0.606. The effective value of mV/A/m taking account of the resistive and reactive components for the cable and the phase angle of the load is $0.592 \times 0.447 + 0.130 \times 0.894$, i.e. $0.265 + 0.116 = 0.381$. This is less than two-thirds of the z value of 0.606 which would be used if phase angles were not taken into account, and allows the use of a cable that is 50% longer for a given voltage drop.

Another factor which can be taken into account when the voltage drop is critical is the effect of temperature on the conductor resistance. The tabulated values of voltage drop are based on impedance values in which the resistive component is that applying when the conductor is at the maximum permitted sustained temperature for the type of cable on which the current ratings are based. If the cable size is dictated by voltage drop instead of the thermal rating, the conductor temperature during operation will be less than the full rated value and the conductor resistance lower than allowed in the tabulated voltage drop. On the basis that the temperature rise of the conductor is approximately proportional to the square of the current, it is possible to estimate the reduced temperature rise at a current below the full rated current. This can be used to estimate the reduced conductor temperature and, in turn, from the temperature coefficient of resistance of the conductor material, the reduced conductor resistance. Substitution of this value for the resistance at full rated temperature in the formula for impedance enables the reduced impedance and voltage drop to be calculated.

BS 7671: 1992 gives a generalised formula for taking into account that the load is less than the full current rating. A factor C_t can be derived from the following:

$$C_t = \frac{230 + t_p - (C_a^2 C_g^2 - I_b^2/I_t^2)(t_p - 30)}{230 + t_p} \qquad (10.3)$$

where t_p = maximum permitted normal operating temperature (°C)
       C_a = the rating factor for ambient temperature
       C_g = the rating factor for grouping of cables
       I_b = the current actually to be carried (A)
       I_t = the tabulated current rating for the cable (A)

For convenience the formula is based on a temperature coefficient of resistance of 0.004 per degree celsius at 20°C for both copper and aluminium. This factor is for application to the resistive component of voltage drop only. For cables with conductor sizes up to 16 mm² this is effectively the total mV/A/m value, but for cables with larger conductors it is the $(mV/A/m)_r$ value in formula (10.2).

Cable manufacturers will often be able to provide information on corrected voltage drop values when the current is less than the full current rating of the cable, the necessary calculations having been made on the lines indicated. If the size of cable required to limit the voltage drop is only one size above the size required on the basis of thermal rating, then the exercise is unlikely to yield a benefit. If, however, two or more steps in conductor size are involved, it may prove worthwhile to check whether the lower temperature affects the size of cable required. The effect is likely to be greater at the lower end of the range of sizes, where the impedance is predominantly resistive, than towards the upper end of the range where the reactance becomes a more significant component of the impedance.

The effect of temperature on voltage drop is of particular significance in comparing XLPE insulated cables with PVC insulated cables. From the tabulated values of volt drop it appears that XLPE cables are at a disadvantage in giving greater volt drops than PVC cables, but this is because the tabulated values are based on the assumption that full advantage is taken of the higher current ratings of the XLPE cables, with associated higher permissible operating temperature. For the same current as that for the same size of PVC cable, the voltage drop for the XLPE cable is virtually the same. If a 4-core armoured 70 mm² (copper) 600/1000 V XLPE insulated cable, with a current rating of 251 A in free air with no ambient temperature or grouping factors applicable, were used instead of the corresponding PVC insulated cable to carry 207 A, which is the current rating of the PVC cable under the same conditions, application of formula (10.3) would give

$$C_t = \frac{230 + 90 - (1 - 0.68)60}{230 + 90} = 0.94$$

(the figure 0.68 is 207^2 divided by 251^2). If the $(mV/A/m)_r$ value for the XLPE cable (0.59) is multiplied by 0.94, it gives, to two significant figures, 0.55, which is the same as for the PVC cable. The $(mV/A/m)_x$ value for the XLPE cable is in fact a little lower than that for the PVC cable, 0.13 compared with 0.14, but this has little effect on the $(mV/A/m)_z$ value which, to two significant figures, is 0.57 for both cables.

POSITIVE, NEGATIVE AND ZERO SEQUENCE IMPEDANCE

The values of cable impedance mentioned so far, and listed in the tables, relate to normal operating conditions. Under fault conditions the general symmetry is disturbed and for the calculation of currents which will flow when a fault occurs the impedances of the equipment in the system applying to the normal symmetrical conditions may not be the relevant values. For the calculation of prospective fault currents, use is made of the resistance and reactance to positive, negative and zero phase currents.

In a fault occurring between all three phases in a 3-phase system, or between phase and neutral in a single-phase supply, the conditions are symmetrical and for the cable involved the normal values of a.c. resistance and reactance per conductor are the operative values. However, for a fault between two phases of a 3-phase system or a fault from phase to earth, asymmetrical conditions occur.

For a fault between phases the current is calculated from

$$I_\mathrm{f} = \frac{U}{Z_1 + Z_2} \quad \mathrm{(A)} \tag{10.4}$$

where I_f = fault current (A)
U = voltage between phases (V)
Z_1 = positive phase sequence impedance (Ω)
Z_2 = negative phase sequence impedance (Ω)

For a fault between a phase and earth the current is calculated from

$$I_\mathrm{f} = \frac{3U_0}{Z_1 + Z_2 + Z_0} \quad \mathrm{(A)} \tag{10.5}$$

where U_0 = phase-to-neutral voltage (V)
Z_0 = zero phase sequence impedance (Ω)

Thus the zero phase impedance, which takes into account the path back to the earthed neutral of the system, enters into the equation for an earth fault but not for a phase-to-phase fault.

The impedances here are the total impedances which will contribute to limiting the current at various points in the system in the event of faults at critical places. The cable impedance is only a part of the total. Other equipment and any impedances deliberately introduced to limit the fault current have to be included. The values for individual parts therefore have to be dealt with as resistances and reactances, or as impedances in the form $R + jX$, so that they can be added as vector quantities.

Here only the values for cables are considered. Usually planners and users will ask cable manufacturers to supply the figures and introduce them into their overall calculations for the system. The following methods of calculation are used by manufacturers.

Positive and negative sequence impedances

For cables, the positive and negative sequence impedances are the same and are the values derived for symmetrical conditions in chapter 2.

The resistance is the a.c. conductor resistance, taking account of skin and proximity effects. The values of a.c. resistance listed in the tables are for maximum normal operating conductor temperature. These are not the most appropriate values to use to calculate maximum fault current, as the fault might occur under lightly loaded conditions when the resistance is lower. The a.c. resistance at ambient temperature is therefore required instead. On the other hand, the a.c. resistance at maximum normal operating temperature would be an appropriate value to use in determining what short-circuit capability is required of the cable, because the short-circuit ratings of cables are normally based on a limiting temperature rise from the maximum normal operating temperature (see chapter 9).

The reactance is the reactance per conductor calculated from the inductance, as described in chapter 2.

Zero sequence resistance

The zero sequence resistance of the conductors per phase is the a.c. resistance of one conductor at ambient temperature (normally 20°C) without the increase for proximity effect described in chapter 2. The increase for skin effect is included. To this value is added the following as appropriate:

for 3-core cables:	three times the resistance of the metallic covering;
for single-core cables:	the resistance of the metallic covering;
for SL cables:	the resistance of one metallic sheath in parallel with three times the resistance of the armour.

The metallic covering for 3-core and single-core cables may be a metal sheath, armour or a copper wire and/or tape screen, whichever is the metallic layer which will carry earth fault current. If the cable has a metallic sheath and armour, the resistance of the metallic covering is the resistance at ambient temperature of the two components in parallel.

Some 3-core cables may contain interstitial conductors or copper wire screens on each core to carry earth fault currents. In general, where there is such a component for each core, it is the resistance of one of them which is considered. When the earthing component is a collective one, i.e. a common covering for all three cores, its resistance is multiplied by three. Where both types of earth current carrying components are present, the example given for the SL cable (three sheaths but only one armour) is followed. In effect the calculation is equivalent to determining the resistance of all the earth paths in parallel and multiplying by three for the resistance per phase.

The resistances are calculated from the cross-sectional areas and resistivities of the metallic layers. For a sheath the cross-sectional area can be calculated from the diameter and thickness. For wires, the resistance of a single wire may be divided by the number of wires and a correction made for their lay length as applied. When wires are applied helically or in wave form, their individual lengths exceed the length of the cable by an amount depending upon the ratio of their lay length to their pitch diameter. For example, armour wires applied with a lay of eight times their pitch diameter have a length approximately 7.4% greater than the length of the cable.

Zero sequence reactance

For single-core cables, the zero sequence reactance of the cable per phase can be calculated from the equation

$$X_0 = 2\pi f \times 10^{-3} [0.2 \log_e(D/d) + K] \quad (\Omega/\text{km}) \qquad (10.6)$$

where X_0 = zero sequence reactance
 f = frequency (Hz)
 D = mean diameter of metallic covering (mm)
 d = conductor diameter (mm)
 K = a constant depending on the conductor construction as in chapter 2, equation (2.2)

The similarity between this and the combination of the equations for inductance and reactance in chapter 2 is evident. The value D, which amounts to twice the spacing from the conductor axis to the metallic covering, replaces the $2S$ where S is the axial spacing between conductors in normal operation.

For a frequency of 50 Hz the equation simplifies to

$$X_0 = 0.314[0.46 \log_{10}(D/d) + K] \quad (\Omega/\text{km}) \qquad (10.7)$$

For a 3-core cable the equation is

$$X_0 = 0.434 \log_{10}(D/GMD) \quad (\Omega/\text{km}) \qquad (10.8)$$

where GMD is the geometric mean diameter of the conductors in the laid-up cable. Conventionally, the value of GMD is taken as 0.75 of the diameter of the circle which circumscribes the conductors in the laid-up cores, assuming the conductors to be circular.

The equation for the 3-core cable is analogous with that for single-core cable, the value K now not being relevant and the 0.434 including multiplication by 3 for reactance per phase.

IEE WIRING REGULATIONS

The IEE Wiring Regulations (published as BS 7671: 1992) relate to electrical installations at voltages up to 1000 V a.c. in and around buildings. They do not apply directly to the public supply system, to which statutory regulations apply, but to consumers' installations which, more often than not, take their energy from the public supply system.

The Regulations are not statutory in England and Wales but are generally accepted as providing the basis for a safe and reliable installation. The electricity supply companies would be unlikely to connect to the public supply system an installation which did not comply substantially with them.

The implications of the Regulations for the selection of cables arise from aspects which should be taken into account whether covered by the Regulations or not, e.g. voltage drop; the requirements in this respect have already been covered as being of general importance irrespective of the particular requirements of the Regulations. However, the Regulations give specific guidance on other aspects which are based on international consensus, as well as the advice drawn from wide ranging representation of the electrical engineering interests in the UK, and it is convenient to consider the impact on cables of the requirements of the relevant regulations as specifically stated.

The following sections relate to the sixteenth edition of the IEE Wiring Regulations published in 1992 and Amendment No. 1, 1994. Other relevant publications are those by Jenkins[1] and Whitfield.[2]

Protection against overload current

Chapter 52 of the Regulations, dealing with selection of cables, conductors and wiring equipment, requires by Regulation 523-01-01 that the cable shall have a current carrying capacity adequate for the load to be carried and it refers to Appendix 4, where the detailed information on current rating is given. However, it is Section 433 'Protection against overload current', which effectively decides the current rating required for the circuit cabling. There, in Regulation 433-02-01, it is required that the device protecting a circuit against overload should satisfy the following conditions:

(a) its nominal current or current setting I_n is not less than the design current I_b of the circuit;
(b) its nominal current or current setting I_n does not exceed the lowest of the current carrying capacities I_z of any of the conductors of the circuit;
(c) the current I_2 causing effective operation of the protective device does not exceed 1.45 times the lowest of the current carrying capacities I_z of any of the conductors of the circuit.

These requirements may be expressed as follows:

$$I_b \leq I_n \leq I_z \tag{10.9}$$

$$I_2 \leq 1.45 I_z \tag{10.10}$$

Clearly from conditions (a) and (b), as abbreviated in the relationship (10.9) the required current rating of the cable is not determined directly by the circuit current but has to be not less than the nominal current of the fuse, or other overload protective device, which in turn has to be not less than the circuit current. The cable rating is determined directly by the overload protective device selected for the circuit and, as fuse ratings progress in discrete steps, the cable rating may be more than it would need to be if related directly to the circuit current.

Condition (c), abbreviated in the relationship (10.10), does not constitute an invitation to overload cables by up to 45% of their rated current, but is to safeguard against the worst consequences of overloads occurring through mistake or accident. The intention is that, if an overload of 45%, or preferably less, should occur, the protective device should operate, interrupting the current, within 'conventional time'. Conventional time varies with the device but is not more than 4 hours.

For a cable in air, a 45% overload persisting for several hours would cause a temperature rise of more than double the design value for sustained operation. While the degree of deformation of PVC due to softening in a single excursion to such a temperature is regarded as tolerable on the basis of tests, the chemical deterioration of most insulants, including PVC, is accelerated by increased temperature, and overloading, especially if repetitive, is bound to take a toll of cable life.

It has to be accepted that, if the devices providing overload protection are to pass their full rated currents for an indefinite period without interruption, there must be a reasonable margin between their rated currents and the currents causing them to interrupt the supply. The 1.45 factor is met by most of the standard types of device and also corresponds to an overload temperature close to the limit that can be tolerated for a relatively short time by PVC, which, because of its thermoplasticity, is perhaps the most vulnerable at high temperatures of the commonly used insulants.

Of the standard protective devices it is accepted in the Regulations that fuses to BS 88 or BS 1361, or circuit breakers to BS 3871: Part 1 or BS EN 60898: 1991, if they comply with relationship (10.9), will automatically ensure compliance with (10.10). This may not be obvious from examination of the standards for the protective devices. For example in BS 88 the conventional fusing current is given as 1.6 times the rated current. However, account is taken of the difference in the conditions of the tests used to determine values in the standard and the practical conditions of use, in which, for example, the freedom to dissipate heat differs. These devices are therefore considered to operate in practice at not more than 1.45 times their nominal current or current settings, and compliance with the requirement that the nominal current of the device is not more than the current rating of the cable ensures also that the operating current of the device is not more than 1.45 times the current rating of the cable.

It is when rewirable fuses to BS 3036: 1958 (1992) are chosen as the means of protection that condition (c) takes effect. For these fuses the currents causing operation within 4 hours may be twice their rated currents. Therefore, to comply with the requirement that the operating current of the fuse should not exceed 1.45 times the current rating of the cable, the cable must have a higher rating than when use is made of the other standard types of protective device previously designated.

Referring back to the relationship (10.10) expressing condition (c) of Regulation 433-02-01, i.e. $I_2 \leq 1.45 I_z$, if $I_2 = 2 I_n$ it follows that I_n should not be greater than $0.725 I_z$. Thus Regulation 433-02-01 states that where the device is a semi-enclosed fuse to BS 3036, compliance with condition (c) is afforded if its nominal current I_n does not exceed 0.725 times the current carrying capacity of the lowest rated conductor in the circuit protected.

The effect of this is, of course, that the current rating of the cable, with any correction factors for ambient temperature, grouping or other relevant conditions, must be not less than the fuse rating divided by 0.725, i.e. not less than 1.38 times the fuse rating. Therefore the use of a rewirable fuse imposes a penalty in size of cable required when this is determined by current rating. Of course, if the size of cable is determined by other considerations, such as voltage drop, there may be no such penalty.

These requirements are in line with international (IEC) rules, which are published in sections having the common reference number 364 and in which the relevant section on protection against overcurrent is IEC 364-4-43.

Protection against short-circuit current

Section 434 of the IEE Wiring Regulations ostensibly specifies requirements for the devices used to protect against short-circuit currents, but it will also affect the choice of the cable to be used. The device has to be chosen to effectively protect the cable used, but conversely the cable has to be chosen so that it is effectively protected by the device used. As with Regulation 433 (overload protection) it is concerned with co-ordination between protective devices and conductors.

The basic requirement for protection against short circuit is that all currents caused by a short circuit at any point in the circuit shall be interrupted in a time not exceeding that which brings the cable conductors to the admissible limiting temperatures. The maximum time T for which the protective device should allow the short circuit to persist is derived from the following equation:

$$T = k^2 S^2 / I^2 \quad \text{(s)} \tag{10.11}$$

where S = cross-sectional area of conductor(s) (mm^2)
I = effective short-circuit current, r.m.s. if a.c. (A)
k = a value depending on the conductor metal and the cable insulant

In Regulation 434-03-03 values of k are given for cables with copper and aluminium conductors with each of the usual insulating materials.

The derivation of the formula is essentially as described in chapter 9 of this book, dealing with short-circuit ratings. The important difference between equation (10.11) and equation (9.1) in chapter 9 is that the value k in (10.11) differs from the value K in (9.1). In chapter 9, K is a function of the volumetric specific heat of the conductor metal, its resistivity and its temperature coefficient of resistance. In the IEE Wiring Regulations k embraces all these parameters but also includes the permitted temperature rise between the normal sustained maximum conductor temperature and the maximum short-circuit conductor temperature. The values of k in Regulation 434-03-03 are, in fact, equivalent to the figures by which $ST^{-1/2}$ are multiplied to give the short-circuit current shown in the last column of table 9.5 in chapter 9.

If the same device is used to protect against short-circuit as against overload current, in accordance with the requirements for protection against overload already given, it may generally be assumed to provide the required protection against short circuit and checking against the short-circuit requirements is not necessary. This is confirmed in Regulation 434-03-02. There is a proviso in this regulation, however, warning that for certain types of circuit breaker, especially non-current-limiting types and for conductors in parallel, this assumption may not be valid for the whole range of short-circuit currents. It is valid for the standard types of fuse where parallel conductors are not protected by a single fuse. If the fuse rating is not greater than the cable rating and if the fuse will operate within 4 hours at 1.45 times the cable rating, then the current–operating time characteristics of fuses, and the relationship between sustained current carrying capacities and short-circuit capacities of cables, are such that cable short-circuit capacity will not be exceeded.

In order to serve as the short-circuit protection, the protective device must be placed at the supply end of the circuit so that it protects the whole of the circuit cabling, wherever the short circuit may occur, and it must have a breaking capacity adequate for the prospective short-circuit current which could occur at that point. The overload protective device, on the other hand, can be placed close to the load since an overload, defined as excess current occurring in an electrically sound circuit, can only arise from the load. A common example would be an overload device in a motor starter.

The device providing protection against short circuit may therefore be separate from the overload protection and may have a nominal current in excess of the sustained current rating of the cable. It is then necessary to check that the combination of cable and short circuit protection complies with Regulation 434-03-03. Constraints on the selection of the device for short-circuit protection may result in a magnitude and duration of short-circuit current that dictate the size of cable, instead of the sustained current, overload protection requirements or voltage drop. In this case the required conductor size can be calculated from the following transposed form of equation (10.11):

$$S = IT^{1/2}/k \qquad (\text{mm}^2) \qquad (10.12)$$

The conductor size used should be the next higher standard size to the value calculated as S, unless S happens to be very close to a standard nominal cross-sectional area.

More conveniently, the required conductor size can be obtained from the data for standard cables given in chapter 9.

Where the short-circuit condition is critical, it is necessary to check that the protection is adequate for the minimum short-circuit current it has to protect against, and not only for the maximum current occurring due to a fault at the supply end of the circuit. The characteristics of protective devices are such that the heating effect of a relatively low short-circuit current in the time taken for the device to interrupt the current may be greater than the heating effect of a higher short-circuit current causing the device to operate in a shorter time.

The sixteenth edition of the IEE Wiring Regulations includes in appendix 3 graphs showing the maximum time to operation against current for the standard types of fuse and for miniature circuit breakers. These can be used to derive the value of T for the prospective short-circuit current for checking compliance with Regulation 434-03-03. The Commentary on the Regulations includes the same graphs together with superimposed lines showing the maximum time for which short-circuit currents of varying magnitude may be allowed to persist for compliance with Regulation 434-03-03 when the cable is PVC insulated with copper conductors; these graphs are useful for a quick check when the cable is of this commonly used type.

From the graphs in the Commentary it is evident that for short-circuit protection by fuses to BS 88 and BS 1361, the minimum short-circuit current is the critical value as far as the heating of the conductors is concerned, but table 10.1 based on the characteristics of a 100 A fuse to BS 88: Part 2, taken as a typical example, demonstrates the point. As the heating effect is proportional to I^2T, the lower fault currents, with longer durations, are the more critical. The pattern is not quite the same for fuses to BS 3036: 1958 (1992) for which I^2T may be somewhat higher for small values of T than for some longer periods, but checking against the maximum and minimum prospective currents will cover the intermediate values.

In the Regulations a short circuit is defined as an overcurrent resulting from a fault between live conductors of differing potential. A fault to earth is not a short circuit as

Table 10.1 Comparison of heating effects of short-circuit currents interrupted by a 100 A fuse to BS 88: Part 2

Short-circuit current, I (A)	Maximum time for fuse to operate, T (s)	I^2T (A^2 s)
550	4.5	136×10^4
700	1.5	74×10^4
850	0.7	51×10^4
1000	0.35	35×10^4

defined. Nevertheless it is necessary to check that the conductors will not overheat during a fault to earth. Because the earth fault current is likely to be less than the short-circuit current between conductors, it is likely to persist for longer and therefore cause greater heating, as illustrated in table 10.1.

Protection against indirect contact

In the measures for protection against electric shock detailed in chapter 41 of the Regulations, protection against shock by indirect contact refers to contact with exposed conductive parts made live by a fault, as distinct from direct contact with parts which are live in normal operation. An exposed conductive part is defined as a conductive part of equipment which can be touched and which is not a live part but which may become live under fault conditions.

When the method of protection against shock by indirect contact is by automatic disconnection of supply in the event of a fault to earth, the impedance of the protective conductor included in the cable or of the metal covering over the cable, e.g. metal sheath and/or armour, may be an important parameter.

As with other aspects, much of the terminology in the Regulations follows international documentation: a protective conductor is a conductor used to connect conducting parts (other than live parts) together and/or to earth. The circuit protective conductor is the conductor which generally follows the cable route connecting the exposed conductive parts of the equipment supplied back to the main earthing terminal of the installation.

The circuit protective conductor may be a conductor in the cable or other continuous metallic component of the cable. It may also be separate from the cable, e.g. a separate conductor, metallic conduit or trunking, but for present purposes it is protective conductors which are parts of cable that are considered.

The basic requirement contained in Regulation 413-02-04 is that during an earth fault the magnitude and duration of the voltages between simultaneously accessible exposed and extraneous conductive parts anywhere in the installation shall not cause danger.

In Regulation 413-02-08 the requirement of Regulation 413-02-04 is considered to be satisfied if

(a) for final circuits supplying socket outlets, the earth fault loop impedance at every socket outlet is such that disconnection occurs within 0.4 s;
(b) for final circuits supplying only fixed equipment, the earth fault loop impedance at every point of utilisation is such that disconnection occurs within 5 s.

The requirement for faster disconnection in circuits supplying socket outlets than in circuits supplying only fixed equipment takes account of the use of socket outlets for supplying equipment which may be held in the hand while in operation, when the risk is greater.

When the protection is afforded by an overcurrent protective device, the time for disconnection depends upon the current and the current–time characteristics of the device. Regulations 413-02-10 and 413-02-11 give the maximum values of earth fault loop impedance which, at a voltage to earth U_0 of 240 V, will give currents sufficient to operate various types of fuse and miniature circuit breaker in 0.4 s or in 5 s (tables 41B1 and 41B2). The values are calculated by dividing 240 by the current giving operation in the appropriate time, as derived from the graphs in appendix 3 of the Regulations.

The earth fault loop impedance includes the impedance of the phase conductor and the protective conductor of the circuit and the impedance of the supply to the circuit. For larger installations the consumer may have his own supply transformer and the impedance of this and of any cabling and other equipment between it and the final circuit is within the control of the designer. For smaller installations, where the supply is taken direct from the public distribution system, the earth fault loop impedance includes a part external to the installation. This external impedance includes that of the supply authority's transformer and the phase and protective conductors of the cabling from there to the consumer's installation.

When the installation has been made, the earth fault loop impedance can be measured, but at the planning stage an estimate has to be made if protection against shock by indirect contact is to be by interruption of supply by an overcurrent device. When part of the impedance is in the supply authority's system, it is indicated in the IEE Guidance Notes No. 6 to the Regulations that this impedance external to the installation will not normally exceed 0.35 Ω if the system is one with CNE conductor, as applies to most new parts of the public supply system, and will not normally exceed 0.8 Ω if the earthing is provided via the lead sheath of the older type of public supply cable. In particular cases, the supply authority may be able to give a more precise value and should be asked anyway, if only to confirm that the normal maximum will apply.

Occasionally the supply authority may not be able to supply an earth terminal with reliable connection back to the substation earth and in this event an earth electrode has to be supplied by the consumer and the earth fault loop impedance includes the resistance of this and the path through the ground to the supply earth. The installation and its source of supply then becomes a TT system instead of a TN system and circuits supplying socket outlets are required by Regulation 471-08-06 to be protected by residual current devices (current operated earth leakage circuit breakers). The designations, such as TN and TT, to describe types of system earthing are fully explained in Part 2 of the Wiring Regulations.

In the Regulations, the measure for protection against indirect contact by automatic disconnection of supply is designated more fully as 'protection by earthed equipotential bonding and automatic disconnection of supply'. It is required that extraneous conductive parts should be connected by equipotential bonding conductors to the main earthing terminal of the installation. In the event of a fault to earth, the voltage between exposed conductive parts and simultaneously accessible extraneous conductive parts is therefore the voltage drop in the circuit protective conductor between the exposed conductive part and the main earthing terminal. This is the product of the earth fault current and the impedance of the circuit protective conductor.

In IEC 364-4-41, which is taken into account in the IEE Wiring Regulations, the corresponding measure of protection relates the magnitude of the voltage to maximum disconnecting time for circuits which may be used to supply equipment to be held in the hand. The IEE Wiring Regulations, while adopting the 0.4 s disconnection time as a condition satisfying Regulation 413-02-04, do not preclude other methods of complying with the regulation. For instance, Regulation 413-02-12 makes use of the IEC relationship between the magnitude of the voltage and the time for which it may be allowed to persist.

A critical condition in the IEC relationship between 'touch voltage' and maximum disconnecting time is 50 V for 5 s. It can be shown that, for most of the standard overcurrent protective devices, if the 50 V, 5 s condition is met, then all the shorter disconnection times required for higher voltages will also be met. This does not apply to semi-enclosed fuses BS 3036: 1958 (1992) and from appendix C of the Commentary on the Regulations it is apparent that, for these, the critical condition is 240 V for 0.045 s; if this is satisfied the permissible longer times for lower voltages will also be satisfied.

In Table 41C of the Regulations, impedances of protective conductors are tabulated for the various standard protective devices. These values ensure that the voltage drop in the protective conductor will not exceed 50 V when the disconnection time is 5 s and will not reach excessive voltages when the disconnection time is shorter. If the impedance of the protective conductor does not exceed the appropriate value in this table, the total earth fault loop impedance may be up to the value for disconnection in 5 s, for circuits supplying socket outlets, instead of for disconnection in 0.4 s.

In some circumstances it may be convenient to make use of Regulation 413-02-12, rather than try to arrange for disconnection within 0.4 s. The following figures extracted from the Regulations can be used as examples to demonstrate this when the protective device is a fuse to BS 1361:

Fuse rating (A)	5	15	20	30
Z_s for 0.4 s disconnection (Ω)	10.9	3.43	1.78	1.20
Z_s for 5 s disconnection (Ω)	17.1	5.22	2.93	1.92
Z_2 as in Table 41C (Ω)	3.25	0.96	0.55	0.36

where Z_s = earth fault loop impedance
Z_2 = maximum impedance of protective conductor required by Regulation 413-02-12

If the supply is from the 240/415 V public supply network the earth fault loop impedance Z_e external to the installation may, under adverse circumstances, be 0.8 Ω. Adjusting the previous figures to obtain the maximum impedances Z_c of the cable conductors to meet the two disconnection times gives the following:

Fuse rating (A)	5	15	20	30
Z_c for 0.4 s disconnection (Ω)	10.1	2.63	0.98	0.40
Z_c for 5 s disconnection (Ω)	16.3	4.42	2.13	1.12
Z_2 as in Table 41C (Ω)	3.25	0.96	0.55	0.36

For circuits protected by 5 and 15 A fuses, it is not likely that advantage can be taken of the higher cable impedance for 5 s disconnection because of the associated limitation on the impedance of the protective conductor. The protective conductors in cables are not usually of lower impedances than the phase conductors and, if they are of the same impedances, they may be up to 5.05 Ω and 1.31 Ω respectively to give disconnection in 0.4 s, i.e. higher than they need to be to comply with Regulation 413-02-12. For a circuit protected by a 20 A fuse, the difference is smaller, but with the 30 A fuse there is a clear benefit from the method of Regulation 413-02-12. With the 30 A fuse the impedance of the protective conductor may be 0.36 Ω provided that the total cable impedance, phase plus protective conductor, is not more than 1.12 Ω, whereas for disconnection in 0.4 s the impedance of the phase and protective conductors in series must not exceed 0.40 Ω, i.e. fractionally more than that of the protective conductor alone for compliance with Regulation 413-02-12. In this instance the method of Regulation 413-02-12 is the better proposition.

The cable in the circuit might be a 4 mm² flat twin and earth cable to BS 6004: 1995 in which the protective conductor is the 1.5 mm² earth continuity conductor. The resistance of the phase and earth conductors together, per metre of cable, is 0.0231 Ω. For these small sizes the resistance and impedance are practically the same and so this value can be taken as impedance. It is derived from tables B1 and B2 of IEE Guidance Notes No. 5 to the Regulations which takes account of the increase in temperature and consequently in resistance during the fault period. The maximum length to meet the 0.4 s disconnection time would be 0.40 divided by 0.0231 (0.40 Ω being the maximum value of Z_c to allow disconnection in 0.4s), i.e. 17 m to the nearest metre. For disconnection in 5 s the length could be almost three times this (1.12 divided by 0.0231), but with the method of Regulation 413-02-12 the limit in this case will be set by the resistance of the protective conductor. The 1.5 mm² protective conductor has a resistance per metre at 20°C of 0.0121 Ω, and multiplying this by 1.38 in accordance with table B2 of IEE Guidance Notes No. 5, to take account of the temperature rise during the fault, gives 0.0167 Ω/m. The maximum cable length to comply with the value of Z_2 is therefore 0.36 divided by 0.0167, i.e. 21 m. The tabulated voltage drop for this cable is 0.011 V/A/m so that, at a load of 30 A, the maximum length to meet the 2.5% (6 V) maximum, if this is applied, would be 18 m. This is only a little more than the 17 m to give disconnection in 0.4 s, but the actual current to be carried in normal operation may well be less than 30 A, the rating of the fuse. Given that the next lower standard fuse rating is 20 A, the load to be carried might be only 25 A, for example. The rating of the cable clipped direct or embedded in plaster is 36 A, and for an actual load of 25 A application of formula (10.3) gives a factor of 0.93 which can be applied to the voltage drop. Using this and 25 A instead of 30 A in calculating the cable length would allow 23 m to be used. The adoption of the method of Regulation 413-02-12 therefore permits the cable to be used in the maximum length which still meets voltage drop requirements, whereas disconnection in 0.4 s would set a lower limit. In general, recourse to Regulation 413-02-12 will be helpful when a large part of the earth fault loop impedance to give disconnection in 0.4 s is taken up by the external impedance.

In the example quoted above, reference was made to the need to take account of the heating effect of the fault current on the resistances of the conductors. IEE Guidance Note No. 5 give guidance on this aspect for phase and protective conductors, and combinations thereof, for sizes up to and including 35 mm², for which inductance can be ignored. In that publication, table B1 gives resistance values at 20°C, which are as quoted in BS 6360: 1991 and appendix A4 of this book for individual conductors. The values for combinations of phase and protective conductors are derived by simple addition. Table B2 of the publication gives the factors by which the values of resistance at 20°C can be multiplied to give values applying during the period of the fault. The data are presented here in table 10.2. The multipliers are based on the simplification that the effective resistance during the fault period is that corresponding to a temperature halfway between the maximum normal operating temperature and the

Table 10.2 Multipliers to apply to resistance at 20°C to give resistance during fault

Insulation material	PVC (70°C)	85°C rubber	90°C thermosetting (e.g. XLPE)
Multiplier	1.38	1.53	1.60

maximum permitted temperature which may be reached at the instant of interruption of the fault current, i.e. the maximum short-circuit temperature (see chapter 9). For PVC insulated cables of the smaller sizes, for example, the two temperatures are 70 and 160°C and the mean of these is 115°C. This is 95°C above the 20°C at which the standard resistances are quoted. Thus, using a temperature coefficient of resistance of 0.004 for both copper and aluminium conductors the multiplier for temperature during the fault is $1 + (95 \times 0.004)$, which gives the tabulated figure of 1.38.

These multipliers apply to conductors which are within the cable. For protective conductors external to the cable, lower values are appropriate because the starting temperature is the ambient temperature instead of the cable temperature. Strictly the multipliers are also only applicable to conductor sizes up to and including 35 mm^2. Table 10.2 is applicable to larger conductors in cables operating on direct current but for a.c. operation inductance and the skin effect may become significant for large sizes.

Of the various combinations of phase conductors and protective conductors that might be employed, a common practice in the UK for heavier loads is to use wire armoured PVC or XLPE insulated cables with the wire armour as the circuit protective conductor. These types of cable, with PVC insulation to BS 6346: 1989 or with XLPE insulation to BS 5467: 1989 and BS 6724: 1990, therefore merit particular consideration.

The contribution to earth loop impedance of these cables comprises the impedance of a phase conductor in series with the impedance of the armour, the path of the return fault current via the armour being approximately concentric with the phase conductor. Under these conditions the impedance of the conductor can be taken to be approximately its d.c. resistance at 20°C multiplied by the appropriate factor in table 10.2, to take account of temperature. In theory, with a.c. operation the skin effect will slightly increase the resistance of the larger conductors, but in the cables with the larger conductors the resistance of the armour so far exceeds the resistance of the conductor that a small approximation in the latter is insignificant in the calculation of the total earth fault impedance.

Because the return current is concentric to the phase conductor, inductance is low, but experiments done by ERA Technology for cable makers have indicated that for steel wire armour, which is, of course, magnetic, an allowance for reactance should be made. The conclusion from the measurements is that 0.3 mΩ is an appropriate value for all sizes of steel wire armoured cable.

In the same experiments there was no discernible rise in the temperature of the armour when currents flowed at magnitudes and for times to give disconnection using fuses of ratings equal to or slightly less than the cable ratings. It was concluded that the armour resistance could be taken as the maximum applying during normal operation, i.e. 10°C less than the maximum permissible conductor temperature in normal operation. The armour temperatures applying are therefore 60°C for PVC insulated cables and 80°C for XLPE insulated cables. On the basis of a temperature coefficient of resistance for steel at 20°C of 0.0045, the multipliers to convert resistances at 20°C, as quoted in the cable standards, to resistances at these temperatures are 1.18 for PVC insulated cables and 1.27 for XLPE insulated cables.

The contribution per metre of a steel wire armoured cable to earth fault loop impedance is then the sum of the resistances of phase conductor and armour in milliohms, each adjusted for temperature as above, added vectorially to 0.3 mΩ, the armour reactance. For example, for an XLPE insulated cable with conductor resistance R_c mΩ/m at 20°C and steel wire armour resistance of R_A mΩ/m at 20°C, the contribution per metre of cable to the earth fault loop impedance would be

$$[(R_c \times 1.6 + R_A \times 1.27)^2 + 0.3^2]^{1/2} \qquad (m\Omega)$$

Tables 10.3 and 10.4 give calculated values for 2-, 3- and 4-core steel wire armoured PVC insulated 600/1000 V cables to BS 6346: 1989 and XLPE insulated 600/1000 V cables to BS 5467: 1989 and BS 6724: 1990, all with copper conductors. Comparing tables 10.3 and 10.4, it appears that the earth loop impedances for XLPE insulated cables are, size for size, significantly greater than those for PVC insulated cables. However, this is partly due to the assumption made in calculating the values that the full thermal ratings of the cables are utilised, both in normal operation and under fault conditions, i.e. that in normal operation the conductors of the PVC cables will be at 70°C and those of the XLPE cables will be at 90°C and under fault conditions the conductors will reach 160°C in the PVC cables and 250°C in the XLPE cables. If an XLPE cable is used for the same duty as the PVC cable of the same size and with the same overcurrent protection, its temperatures in normal operation and during an earth fault will be almost the same as for the PVC cable. The multipliers for PVC insulated cables to take account of temperature during the fault can then be applied to the XLPE insulated cable. The earth fault loop impedance of the XLPE cable will generally still be higher at the same temperature because its dimensions and consequently the cross-sectional area of its armour are less, giving a higher armour resistance, but the difference will be smaller than the tables indicate.

Table 10.3 Contribution of steel wire armoured cables to earth fault loop impedances, 600/1000 V PVC insulated cables to BS 6346: 1989, copper conductors

Conductor size (mm²)	Contribution by cable to earth fault loop impedance (mΩ/m)			
	2-core	3-core	4-core (all equal)	4-core including reduced neutral
1.5	29.3	28.7	27.9	–
2.5	21.0	20.6	19.6	–
4	15.2	14.6	11.8	–
6	12.3	9.68	9.09	–
10	7.13	6.90	6.54	–
16	5.72	5.37	4.19	–
25	4.08	3.85	3.49	3.49
35	3.57	3.22	2.98	2.98
50	3.03	2.79	2.09	2.56
70	2.63	2.04	1.81	1.81
95	1.83	1.71	1.45	1.48
120	1.65	1.54	1.09	1.11
150	1.50	1.09	0.98	1.01
185	1.10	0.99	0.89	0.90
240	0.97	0.87	0.78	0.80
300	0.88	0.78	0.70	0.73
400	0.78	0.71	0.55	0.56

Table 10.4 Contribution of steel wire armoured cables to earth fault loop impedances, 600/1000 V XLPE insulated cables to BS 5467: 1989 and BS 6724: 1990, copper conductors

Conductor size (mm^2)	\multicolumn{4}{c}{Contribution by cable to earth fault loop impedance (mΩ/m)}			
	2-core	3-core	4-core (all equal)	4-core including reduced neutral
1.5	31.3	30.9	30.2	–
2.5	23.0	22.3	21.6	–
4	17.4	16.9	16.0	–
6	13.8	13.3	10.4	–
10	10.6	8.01	7.63	–
16	6.67	6.42	5.91	–
25	5.87	4.35	4.10	4.10
35	4.02	3.77	3.39	3.52
50	3.55	3.17	2.92	3.05
70	2.98	2.73	1.98	2.10
95	2.11	1.98	1.73	1.73
120	1.92	1.79	1.25	1.49
150	1.75	1.23	1.10	1.14
185	1.24	1.10	0.98	1.00
240	1.09	0.97	0.86	0.88
300	0.99	0.89	0.78	0.81
400	0.88	0.79	0.6	0.72

For both types of cable the probability is that the temperatures during normal operation and on occurrence of a fault will be less than the maximum permitted values, on which the tabulated data are based. The current actually to be carried in normal operation must not exceed the rating of the overload protective device and the rating of the overload protective device must not exceed the rating of the cable. This makes it very likely that the current to be carried will be less than the cable rating and that the temperature at which the cable operates normally will be less than the permissible maximum on which the rating is based. If the cable is protected by a standard device that gives protection against overload as well as short circuit, i.e. with a rating not higher than the current rating of the cable, the current let-through for a given time under fault conditions will fall short of that which would bring the conductors to their maximum permissible short-circuit temperature, as previously indicated in the section on protection against short-circuit current. The tabulated impedances may therefore be higher than apply in practice. They are useful to provide a check on earth loop impedance, but if, for a particular part of an installation, this is a critical factor in determining the size or length of cable which can be used, it may be possible to justify lower temperatures and associated multipliers for resistance by more accurate calculation of the temperatures applying to the circuit.

For example, a circuit to carry 120 A three phase might be protected by 125 A fuses to BS 88: Part 2. A suitable cable installed in free air with no rating factors to be applied for ambient temperature or grouping would be a 4-core 25 mm^2 (copper) armoured XLPE cable with a current rating of 131 A. Within reasonable limits of accuracy the temperature rise of the conductors is proportional to the square of the current, and since the temperature rise at 131 A is 60°C the temperature rise at 120 A will be $60 \times 120^2/131^2$, i.e. 50.3°C, giving a conductor temperature in normal operation of 80.3°C. Under fault conditions the current which would operate the fuse within 5 s is slightly under 700 A. In formula (10.11) ($T = k^2 S^2/I^2$) the value of k for a copper conductor XLPE insulated cable is 143, based on a maximum temperature rise of 160°C (from 90 to 250°C). With $T = 5$ and $S = 25$, the current giving this temperature rise is $143 \times 25/\sqrt{5}$, i.e. 1599 A. As the fuse will operate within 5 s at 700 A for the circuit being considered, the temperature rise will be $160 \times 700^2/1599^2$, i.e. 30.7°C. For the circuit, then, the conductor temperature can be taken as 80.3°C in normal operation and 111°C at the end of the fault period, giving a mean temperature during the fault of, say, 95.7°C. The armour temperature can be taken as 70.3°C. Multipliers for resistances calculated for these temperatures during the fault are 1.303 for the conductor and 1.226 for the armour. The earth fault loop impedance calculated using these multipliers is 3.78 mΩ/m compared with the value of 4.10 mΩ/m shown in table 10.4 as applying when the temperatures are maximum permissible values.

For paper insulated lead sheathed cable, the lead sheath and the wire armour, if the cable is so armoured, can serve as the protective conductor, BS 6480: 1988 gives the resistances of these components for 600/1000 V cables. The resistance multipliers for calculation of earth loop impedances, based on maximum permissible temperatures, are 1.4 for conductors and 1.225 for armour, and it seems appropriate to allow 0.3 mΩ/m for armour reactance.

Cross-sectional area of protective conductors

The implications for circuit protective conductors of the requirement for interruption of supply in the event of an earth fault have already been considered, making reference to Section 413 of the Regulations. However, it is Section 543 which is actually headed 'Protective conductors'. Regulations 543-01-01 to 543-01-04 contain requirements for the cross-sectional area of protective conductors in general, and therefore for circuit protective conductors, which may be cables or parts of cables.

Regulation 543-01-01 requires, so far as it affects protective conductors which are a part of a cable or in an enclosure formed by a wiring system (e.g. single-core cables in conduit), that the cross-sectional area should either be calculated by a formula given in Regulation 543-01-03 or be in accordance with Regulation 543-01-04, which includes table 54G, relating the cross-sectional area of the protective conductor to that of the associated phase conductors.

After the earlier consideration of the requirements for protection against short circuit the formula in Regulation 543-01-03 should be familiar. The requirement is that the cross-sectional area shall be not less than is given by

$$S = (I^2 T)^{1/2}/k \qquad (\text{mm}^2) \qquad (10.13)$$

where S = cross-sectional area (mm^2)
  I = the value (r.m.s. for a.c.) of fault current, for an earth fault of negligible impedance, which will flow through the associated protective device (A)
  T = the operating time of the disconnecting device(s), corresponding to the fault current I
  k = a factor depending on the material of the protective conductor and the appropriate initial and final temperatures

The initial and final temperatures take account of the insulation and any other materials with which the conductor is in contact.

Values of k are given in the Regulation in tables 54B, 54C, 54D, 54E and 54F and are repeated here in table 10.5. These values are derived by calculation on the lines described in chapter 9 of this book.

As an alternative to calculating the cross-sectional area of the protective conductor in accordance with Regulation 543-01-03, Regulation 543-01-04 allows that it may be selected in accordance with table 10.6, which is copied from table 54G of the Regulations. Where the table requires the minimum cross-sectional area of the protective conductor to be a non-standard size, the nearest larger standard cross-sectional area shall be used.

Use of Regulation 543-01-04 and table 10.6 will often result in a much larger earth conductor than if Regulation 543-01-03 and table 10.5 were used. This is because table 10.6 effectively ensures that the earth conductor has the same fault capability as the phase conductor, even if the latter is far in excess of what is actually required to meet the prospective fault current and the fault duration dictated by the protective device. Therefore, the table 10.5 approach is preferred, especially if the adequacy of an available earth conductor is being confirmed, because the table 10.6 approach may unjustifiably lead to the conclusion that the existing earth conductor is inadequate. This situation arises when assessing the earth fault capacity of the steel wire armour of, say, 600/1000 V XLPE-insulated industrial cables to BS 5467: 1989. In this case, the table 10.6 formulae will indicate that the fault capacity of the armour on larger cable sizes does not match that of the phase conductors, while Regulation 543-01-03 will confirm that the armour is entirely adequate when considering the I^2T let-through energy associated with any normal protective device.

For non-sheathed single-core PVC insulated cables run in conduit or trunking, if a cable is used as a protective conductor it will be similar in construction to the phase and neutral conductor and the cross-sectional area can be chosen to comply with table 10.6; no additional calculation is necessary.

For the flat twin and earth cable (2-core plus protective conductor), also to BS 6004: 1995, only the 1 mm^2 cable has a protective conductor of equal area to the phase conductor. The other sizes, all 16 mm^2 or less, do not comply with table 10.6 and calculation in accordance with Regulation 543-01-03 is required to check the adequacy of the protective conductors.

For these cables the temperature limit for the protective conductors may exercise control over the circuit arrangements. Tables A1 to A3 of Appendix A to IEE Guidance Notes No. 5 to the Regulations give maximum earth fault loop impedances to give compliance with both the disconnecting times required by Regulation 413-02-08 and the thermal constraint on protective conductors, for copper protective conductors from 1 mm^2 up to 16 mm^2. Separate tables are given for application when fuses to BS 3036, BS 88 and BS 1361 provide the overcurrent protection and the tables are further subdivided according to whether the circuits feed socket outlets or fixed equipment. Many of the values of impedance given are the same as those in Regulations 413-02-10 and 413-02-11, governing the disconnecting time, and this implies that for these conditions the requirements of Regulation 413-02-08 are decisive. Some of the values, however, differ from those in Regulations 413-02-10 and 413-02-11 and where this occurs it indicates that the thermal restraints on protective conductors contained in Regulation 543-01-03 are decisive in setting the level of earth fault loop impedance.

Table 10.5 Values of k for protective conductors

(a) Insulated conductors not incorporated in cables or bare or insulated conductors in contact with outside of cable covering (but not bunched with cables)

Material of conductor	Material of insulation or cable covering		
	PVC	85°C rubber	90°C thermoset
Copper	143/133*	166	176
Aluminium	95/88*	110	116
Steel	52	60	64
Assumed initial temperature (°C)	30	30	30
Final temperature (°C)	160/140*	220	250

*Above 300 mm^2

(b) Conductor as a core in the cable (or bunched with cables)

Material of conductor	Insulation material			
	70°C PVC	90°C PVC	85°C rubber	90°C thermoset
Copper	115/103*	100/86*	134	143
Aluminium	76/68*	66/57*	89	94
Assumed initial temperature (°C)	70	90	85	90
Final temperature (°C)	160/140*	160/140*	220	250

*Above 300 mm^2

(c) Conductor as a sheath or armour of a cable

Material of conductor	Insulation material			
	70°C PVC	90°C PVC	85°C rubber	90°C thermoset
Steel	51	46	51	46
Aluminium	93	85	93	85
Lead	26	23	26	23
Assumed initial temperature (°C)	60	80	75	80
Final temperature (°C)	200	200	220	200

(d) Conductor as a steel conduit, ducting and trunking

Material of conductor	Insulation material			
	70°C PVC	90°C PVC	85°C rubber	90°C thermoset
Steel (conduit, ducting or trunking)	47	44	54	58
Assumed initial temperature (°C)	50	60	58	60
Final temperature (°C)	160	160	220	250

(e) Bare conductors where there is no risk of damage to any neighbouring material by the temperatures indicated (assumed initial temperature 30°C)

Material of conductor	Conditions[a]		
	Visible and in restricted areas[b]	Normal	Fire risk
Copper	228 (500°C)	159 (200°C)	138 (150°C)
Aluminium	125 (300°C)	105 (200°C)	91 (150°C)
Steel	82 (500°C)	58 (200°C)	50 (150°C)

[a] The figures in parentheses are the limiting final temperatures.
[b] The temperatures indicated are valid only where they do not impair the quality of connections.

Table 10.6 Minimum cross-sectional area of protective conductor in relation to area of associated phase conductor

| | Minimum cross-sectional area, S_p, of the corresponding protective conductor (mm^2) ||
Cross-sectional area, S, of phase conductor (mm^2)	Protective conductor and phase conductor of same material	Protective conductor and phase conductor of different materials
$S \leq 16$	S	$\dfrac{k_1 \times S}{k_2}$
$16 < S \leq 35$	16	$\dfrac{k_1 \times 16}{k_2}$
$S > 35$	$\dfrac{S}{2}$	$\dfrac{k_1 \times S}{k_2 \times 2}$

k_1 is the value of k for the phase conductor, selected from table 43A in chapter 43 of the IEE Wiring Regulations.
k_2 is the value of k for the protective conductor selected from table 10.5(a)–(e), as appropriate

The levels are reduced compared with those for 5 s and 0.4 s disconnection in order to achieve shorter times and lower values of the product I^2T in equation (10.13).

Table 10.7 shows the fuses of the three types which would be used with each size of flat twin and earth cable when the fuse is to provide protection against overload as well as short-circuit current. Two values of earth fault loop impedance are included against each fuse rating. These are taken from tables A1 to A3 of IEE Guidance Notes No. 5 to the Regulations. The first value is for circuits feeding socket outlets and the second for circuits feeding fixed equipment (the latter is always the higher value). These values are the appropriate ones for the size of protective conductor in the cables.

The values marked by an asterisk are the impedances necessary in order to comply with the thermal restraint (Regulation 543-01-03) and those without an asterisk are necessary to give the required disconnection time (Regulation 413-02-08). In the first

Table 10.7 Maximum earth fault loop impedances for compliance with disconnecting times and temperature limits for protective conductors for flat twin and earth cables

Cable (mm^2)	Cable rating (A)	BS 3036 fuse			BS 88 fuse			BS 1361 fuse		
		Rating (A)	MEFLI (Ω)		Rating (A)	MEFLI (Ω)		Rating (A)	MEFLI (Ω)	
1.5/1	20	15	2.67	5.58	20	1.85	2.18*	20	1.78	2.18*
2.5/1.5	27	20	1.85	4.00	25	1.50	1.63*	20	1.78	2.80*
4/1.5	36	20	1.85	4.00	32	1.09	1.22*	30	1.20	1.50*
6/2.5	46	30	1.14	2.76	40	0.86	1.08*	45	0.60	0.66*
10/4	63	45	0.62	1.66	63	(0.48	0.67*)	60	(0.35	0.37*)
16/6	85	60	(0.44	1.14)	80	(0.33	0.47*)	80	(0.24	0.27*)

*See text

column of the table the cables are designated by two values, the first being the cross-sectional area of the phase conductors and the second the cross-sectional area of the protective conductor, e.g. 1.5/1 signifies phase conductors of $1.5\,\text{mm}^2$ and a protective conductor of $1\,\text{mm}^2$.

The table shows that when the disconnection time has to be 0.4 s the thermal constraint on the protective conductor does not influence the maximum earth fault loop impedance required provided that the rating of the fuse is selected to provide protection against both short-circuit and overload currents. This is not necessarily so if the fuse rating is higher and another device provides the overload protection.

The values in parentheses are not included in the tables of the Regulations and have been calculated to complete the picture for the type of cable. When fuses to BS 3036 are used the thermal restraints do not influence the maximum impedances acceptable for 5 s disconnection, but the other fuses do place a restraint on the impedance. The main reason for this difference is that the BS 3036 fuse has in general to be of a lower rating than the BS 88 and BS 1361 fuses for the same cable in order to provide overload protection (fuse rating ≤ 0.725 times cabling rating). Of course, the normal circuit currents using the BS 3036 fuses would have to be no greater than the fuse ratings, and so the full capacity of the cables would not be utilised. Lower rated fuses to BS 88 or BS 1361 could also be used with similarly reduced circuit currents.

The current ratings given for the cables are those for the usual conditions of installation, at 30°C ambient temperature, as single cables, and without contact with thermal insulating materials. In other conditions the ratings might be reduced and lower rating fuses used accordingly. In that event the earth fault loop impedances both for the disconnection times and for compliance with Regulation 543-01-03 would change, as would their relationship to each other.

For the wire armoured cables to BS 6346: 1989, BS 5467: 1989 and BS 6724: 1990 the armour is more than adequate to meet the requirements of Regulation 543-01-03 provided that the fuse protecting against short-circuit is chosen to give protection against overload current also, in accordance with Regulation 433-02-01.

Cross-sectional areas of armour for the 600/1000 V cables are given in the British Standards. Therefore, Regulation 543-01-04 and table 10.6 could be used to compare the fault capability of the armour with that of the phase conductors. However, this approach will indicate that the armour of many cable sizes, particularly the larger sizes with copper conductors, fail to meet the criterion in Regulation 543-01-04. This merely indicates that those armours do not match the phase conductors for earth fault capability, not that the armours are unsuitable as protective conductors in the particular circuit. If armour does not meet Regulation 543-01-04, then the simple approach has failed and recourse must then be made to Regulation 543-01-03, which constitutes the fundamental requirement. Failure to take this additional step will lead to the installation of unnecessary additional protective conductors.

Tables 10.8 and 10.9 show how the required cross-sectional areas of steel wire armour to comply with Regulation 543-01-03 compare with the areas provided on the cables when fuses to BS 88: Part 2, with ratings not higher than the cable ratings, are used as overcurrent protection. Table 10.8 relates to 2-, 3- and 4-core XLPE cables to BS 5467: 1989 and BS 6724: 1990 and table 10.9 relates to 2-, 3- and 4-core PVC cables to BS 6346: 1989, all with copper conductors.

The required areas in these tables have been calculated from equation (10.13) using $k = 46$ for the XLPE insulated cables and $k = 51$ for the PVC insulated cables. They

Table 10.8 Comparisons of steel wire armour areas required with areas actually provided when fuse is to BS 88 with rating no greater than the cable rating: 600/1000 V XLPE insulated cables with copper conductors

Cable conductor area (mm²)	Cable rating (A)	BS 88 fuse rating (A)	Current for 5 s disconnection (A)	Armour area Required (mm²)	Armour area Actual (mm²) 3-core	Armour area Actual (mm²) 4-core
3- and 4-core cables						
1.5	25	25	99	4.8	16	17
2.5	33	32	130	6.3	19	20
4	44	40	170	8.3	21	23
6	56	50	220	11	23	36
10	78	63	280	14	39	43
16	99	80	400	19	44	49
25	131	125	680	33	62	70
35	162	160	900	44	70	80
50	197	160	900	44	78	90
70	251	250	1500	73	90	131
95	304	250	1500	73	128	147
120	353	315	2100	102	141	206
150	406	400	2600	126	201	230
185	463	400	2600	126	220	255
240	546	500	3500	170	250	289
300	628	500	3500	170	269	319
2-core cables						
1.5	29	25	99	4.8	15	
2.5	39	32	130	6.3	17	
4	52	50	220	11	19	
6	66	63	280	14	22	
10	90	80	400	19	26	
16	115	100	540	26	41	
25	152	125	680	33	42	
35	188	160	900	44	62	
50	228	200	1200	58	68	
70	291	250	1500	73	80	
95	354	315	2100	102	113	
120	410	400	2600	126	125	
150	472	400	2600	126	138	
185	539	500	3500	170	191	
240	636	630	4300	209	215	
300	732	630	4300	209	235	

Table 10.9 Comparisons of steel armour areas required with areas actually provided when fuse is to BS 88 with rating no greater than the cable rating: 600/1000 V PVC insulated cables with copper conductors

Cable conductor area (mm^2)	Cable rating (A)	BS 88 fuse rating (A)	Current for 5 s disconnection (A)	Armour area Required (mm^2)	Armour area Actual 3-core (mm^2)	Armour area Actual 4-core (mm^2)
3- and 4-core cables						
1.5	19	16	54	2.4	16	17
2.5	26	25	99	4.3	19	20
4	35	32	130	5.7	23	35
6	45	40	170	7.5	36	40
10	62	50	220	9.6	44	49
16	83	80	400	18	50	72
25	110	100	540	24	66	76
35	135	125	680	30	74	84
50	163	160	900	39	84	122
70	207	200	1200	53	119	138
95	251	250	1500	66	138	160
120	290	250	1500	66	150	220
150	332	315	2100	92	211	240
185	378	315	2100	92	230	265
240	445	400	2600	114	260	299
300	510	500	3500	153	289	333
400	590	500	3500	153	319	467
2-core cables						
1.5	22	20	78	3.4	15	
2.5	31	25	99	4.3	17	
4	41	40	170	7.5	21	
6	53	50	220	9.6	24	
10	72	63	280	12	41	
16	97	80	400	18	46	
25	128	125	680	30	60	
35	157	125	680	30	66	
50	190	160	900	39	74	
70	241	200	1200	53	84	
95	291	250	1500	66	122	
120	336	315	2100	92	131	
150	386	315	2100	92	144	
185	439	400	2600	114	201	
240	516	500	3500	153	225	
300	592	500	3500	153	250	
400	683	630	4300	189	279	

are the areas required to cater for 5 s disconnection time, when the product I^2T is higher than for short disconnection times. They are therefore the maximum areas required when the fuses are as stipulated. The current ratings of the cables are taken from appendix 4 of the Regulations and are the maximum values listed therein for the various methods of installation and an ambient temperature of 30°C. The table is not intended for use to derive the current ratings of the cables, which are only given to show a step in the compilation of the data.

It is apparent that the actual areas of armour wires as used are all comfortably in excess of those required.

For cables with aluminium conductors the current ratings are lower and accordingly the fuse ratings and the currents to give disconnection in 5 s are lower. The result is that the required armour areas are substantially less than for cables with copper conductors. The actual areas of armour on cables with solid aluminium conductors are a little lower, because the cable diameters are a little less, than on cables with copper conductors, but the overall effect is that the actual areas exceed the required areas by a greater margin than for cables with copper conductors.

2-, 3- and 4-core cables with solid aluminium conductors and aluminium strip armour generally comply with Regulation 543-01-04. The main need is to ensure that connections are good and the armour is kept free from corrosion.

For paper insulated lead sheathed and wire armoured cables, in so far as they may be used where the Regulations apply, the armour alone will provide a cross-sectional area of protective conductor similar to that for the XLPE and PVC cables and the extra capacity of the lead sheath helps to ensure that compliance with Regulation 543-01-03 presents no problems as long as the rating of the overcurrent protective device is not too far removed from the cable ratings.

When single-core cables are used, there will be at least two cable armours in parallel for a d.c. or single-phase circuit. For a 3-phase circuit there will be three or four cables with armours to serve as the protective conductor. The earlier statement that the armour of cables to BS 6346: 1989 and BS 5467: 1989 is adequate to meet Regulation 543-01-03 provided that the fuse provides overload protection as well as short-circuit protection was based on the armour area for the circuit being at least twice the area for a single cable.

Steel wire armour will only be used on single-core cables for d.c. circuits. With a.c. circuits the armour wires will generally be aluminium. As the k values are higher for aluminium than for steel, this increases the fault current carrying capacity. This is largely due to the higher conductivity of aluminium, and for many of the single-core cables the two, three or four cable armours in parallel have a combined area sufficient to meet the requirements of Regulation 543-01-04, making calculation in accordance with Regulation 543-01-03 unnecessary. This applies to the following cables.

Cables to BS 6346 with copper conductors
   2 cables in circuit: sizes up to 185 mm^2 inclusive
   3 cables in circuit: sizes up to 500 mm^2 inclusive
   4 cables in circuit: sizes up to 1000 mm^2 inclusive

Cables to BS 6346 with solid aluminium conductors
   2 cables in circuit: sizes up to 740 mm^2 inclusive
   3 or 4 cables in circuit: all sizes (up to 1200 mm^2)

Cables to BS 5467 or BS 6724 with copper conductors
  2 cables in circuit: sizes up to 185 mm² inclusive
  3 cables in circuit: sizes up to 500 mm² inclusive
  4 cables in circuit: sizes up to 630 mm² inclusive

Cables to BS 5467 or BS 6724 with solid aluminium conductors
  2, 3 or 4 cables in circuit: sizes up to 300 mm² inclusive (whole range)

If the fuse rating has to be higher than the cable rating, with separate protection against overload, for instance to avoid nuisance operation at starting current, then tables 10.8 and 10.9 do not apply and the fault current carrying capacity of the armour may be inadequate to comply with Regulation 543-01-03. The check would have to be made on the lines of the tables taking account of the current for disconnection in the appropriate time. The only final control in the Regulations over the rating of the device used to give short-circuit protection is that it should comply with Regulation 434-03-03, which is concerned with protection of the phase and neutral conductors.

There is no assurance that a device selected on this basis would protect protective conductors complying with Regulation 543-01-04 from earth fault currents. It is recommended that when the short-circuit protective device has an appreciably higher rating than the cable, a check should be made with equation (10.13) for protection against earth fault current for both the protective conductor and, if necessary, the phase conductor.

In the Commentary on the Regulations, tables 3, 4, 5 and 6 give the required areas of protective conductor to comply with equation (10.13) according to the type and rating of device used for short-circuit protection, for various values of k. These would be very useful in any exercise such as illustrated in tables 10.8 and 10.9 but with different protection and/or cable. The answers derived from these tables may differ slightly from those in tables 10.8 and 10.9 for the same fuses and value of k because of differences in reading the graphs of fuse characteristics, which are plotted on logarithmic scales.

In table 10.10 the exercise is carried a step further for 2-, 3- and 4-core wire armoured 600/1000 V XLPE insulated cables with copper conductors. Here the minimum sizes of cable having cross-sectional areas of armour sufficient to comply with Regulation 543-01-03 are listed against each of the standard overcurrent protective devices mentioned in the Regulations. The current ratings of the cables in ampères are shown in parentheses for ready comparison with the ratings of the protective devices. The calculations are based on the I^2T values of the protective devices for 5 s disconnection. For shorter disconnection times the cable sizes providing the required areas would be no larger and in some instances might be smaller.

The table shows that the armour areas are generally adequate even when the rating of the protective device is a step above the cable rating and often adequate when the rating of the protective device is two or more steps above the cable rating.

For steel wire armoured PVC insulated cables the general situation is similar but, of course, not the same in detail.

Correlation of requirements

If any of the regulations exercised the ultimate control of the size of cable and its associated protective conductor and the maximum length in which it could be used,

Table 10.10 Minimum sizes of 600/1000 V XLPE insulated SWA cables with copper conductors to meet Regulation 543-01-03 for each rating of circuit protective device

Protective device designation	Required area of SWA (5 s disconnection) (mm^2)	Smallest cable providing required area of SWA		
		2-core (mm^2)	3-core (mm^2)	4-core (mm^2)
Fuses to BS 88: Part 2				
63A	14	1.5 (29)	1.5 (25)	1.5 (25)
80A	19	4 (52)	2.5 (33)	2.5 (33)
100A	26	10 (90)	10 (78)	6 (56)
125A	33	16 (115)	10 (78)	6 (56)
160A	44	35 (115)	16 (99)	16 (99)
200A	58	35 (188)	25 (131)	25 (131)
250A	73	70 (291)	50 (197)	35 (162)
315A	102	95 (354)	95 (304)	70 (251)
400A	126	150 (472)	95 (304)	70 (251)
500A	170	185 (539)	150 (406)	120 (353), 150 (406)
630A	209	240 (636)	185 (463)	150 (406)
800A	335	–	–	400 (728)
Fuses to BS 1361				
45A	12	1.5 (29)	1.5 (25)	1.5 (25)
60A	16	2.5 (39)	1.5 (25)	1.5 (25)
80A	23	10 (90)	6 (56)	4 (44)
100A	31	16 (115)	10 (78)	6 (56)
Fuses to BS 3036				
60A	10	1.5 (29)	1.5 (25)	1.5 (25)
100A	21	6 (66)	4 (44)	4 (44)
Miniature circuit breakers to BS 3871				
30A type 1	5.8	1.5 (29)	1.5 (25)	1.5 (25)
30A type 2	10.2	1.5 (29)	1.5 (25)	1.5 (25)
30A type 3	15	1.5 (29)	1.5 (25)	1.5 (25)
50A type 1	9.7	1.5 (29)	1.5 (25)	1.5 (25)
50A type 2	17	2.5 (39)	2.5 (33)	1.5 (25)
50A type 3	24	10 (90)	10 (78)	6 (56)

For protective devices smaller than those listed the smallest cable (1.5 mm^2) provides more than the required area of armour.

Where two pairs of figures appear in the column for 4-core cables the first pair apply to cables with all conductors of equal size and the second pair to cables with a reduced neutral conductor.

there would be no need for the others which are directed towards exercising similar controls. It is not possible to say, for example, that, if the voltage drop requirement is met, then the earth fault loop impedance is bound to be low enough to give disconnection of supply in the required time when a particular type of cable is used, or the other way about.

It has only been possible to say that, if the protective device gives protection against overload currents in accordance with Regulation 433-02-01, then, if it has the required breaking capacity and is suitably placed, it can be assumed to provide protection against short-circuit current, and even then an exception has to be noted for certain types of circuit breaker.

In an example quoted when considering protection against indirect contact by interruption of supply, it was noted that disconnection of supply within 0.4 s by a fuse to BS 1361: 1971 (1986) would limit the length to less than that required for 2.5% voltage drop when the impedance external to the installation was high, but if recourse were made to Regulation 413-02-12 to meet Regulation 413-02-04 it was the voltage drop and not the impedance of the protective conductor which set the limit on length.

Table 10.7 indicated that, using fuses which provided overload protection, the earth fault loop impedances with flat twin and earth cables would be set by the disconnection requirement if this had to be effected in 0.4 s, but if the disconnection time could go to 5 s the limit on earth fault loop impedance would often be set by the thermal restraints for the protective conductor (Regulation 543-01-03). It would be possible to extend the exercise by deducing, for a given external earth fault loop impedance, what length of cable could be used without exceeding the remainder of the loop impedance. This could then be compared with the maximum length for compliance with voltage drop requirements. The exercise could be extended further to see how reference to Regulation 413-02-12 affects the outcome when the external impedance takes up a large part of the total impedance for 0.4 s operation.

However, in the Commentary on the Regulations, which is recommended reading, chapter 12 is devoted to smaller installations, in which the smaller cables are used, and this deals thoroughly with the impact of the various regulations on the cables. It indicates that the voltage drop requirement is often the factor restricting the length of the circuit or, for a given length of circuit, determining the size of cable to be used. Of course, this is not invariably so and table 10.7 indicates that, if it is desired to use the flat twin and earth cable, the earth fault loop impedance may sometimes determine the size required, either to achieve the disconnecting time or to avoid overheating of the protective conductor on earth fault.

It may be useful to consider in a little more detail the larger cables, PVC or XLPE insulated with steel wire armour, used in larger industrial installations.

Tables 10.8 and 10.9 show that, if the fuse provides protection against overload current as well as short-circuit current, overheating of the protective conductor by earth fault current is not likely to be a problem. However, the armour as protective conductor cannot by any means be forgotten once compliance with Regulation 543-01-03 has been checked, because it contributes as a major component of the earth fault loop impedance, affecting the disconnection time. If disconnection is effected within 5 s with the fuses indicated in the tables, then Regulation 543-01-03 is met, but this is of no account unless the earth fault loop impedance permits disconnection in that time.

The current rating of the cable obviously sets a limit to the minimum size which can be used and, on the basis that the current carrying capacity is fully utilised, table 10.11 shows the maximum lengths of 600/1000 V copper conductor XLPE insulated wire armoured cables to comply with the requirements for earth fault loop impedance. The voltage drops for these lengths are then shown in the final column. For these cables the assumption is made that the supply will be provided by the consumer's own transformer. There is no provision to be made for external impedance, but, for the

Table 10.11 Maximum lengths of 600/1000 V copper conductor XLPE insulated SWA cables to meet earth fault loop impedance requirements for 5 s disconnection and voltage drops for these lengths

Cable size (mm^2)	Cable rating (A)	Fuse rating (A)	Maximum earth fault loop impedance (table 41D) − 10% (Ω)	Maximum length (m) 3-core	Maximum length (m) 4-core	Voltage drop per metre (mV/A)	Voltage drop for length (%) 3-core	Voltage drop for length (%) 4-core
3- and 4-core								
16	99	80	0.54	84.1	91.4	2.5	4.1	4.4
25	131	125	0.315	72.4	76.8	1.65	3.6	3.8
35	162	160	0.243	64.4	71.7	1.15	2.9	3.2
50	197	160	0.243	76.6	83.2	0.87	2.6	2.8
70	251	250	0.144	52.7	72.7	0.60	1.9	2.6
95	304	250	0.144	72.7	83.2	0.45	2.0	2.3
120	353	315	0.099	55.3	79.2	0.37	1.6	2.2
150	406	400	0.0864	70.2	78.5	0.30	2.0	2.3
185	463	400	0.0864	78.5	88.2	0.26	2.0	2.2
240	546	500	0.0585	60.3	68.0	0.21	1.5	1.7
300	628	500	0.0585	65.7	75.0	0.185	1.5	1.7
2-core								
16	115	100	0.396	59.3		2.9	7.2	
25	152	125	0.315	53.6		1.9	5.3	
35	188	160	0.243	60.4		1.35	5.4	
50	228	200	0.180	50.7		1.00	4.2	
70	291	250	0.144	48.3		0.69	3.5	
95	354	315	0.099	46.9		0.52	3.2	
120	410	400	0.0864	45.0		0.42	3.2	
150	472	400	0.0864	49.4		0.35	2.9	
185	539	500	0.0585	47.2		0.29	2.9	
240	636	630	0.0486	44.6		0.24	2.8	
300	732	630	0.0486	49.1		0.21	2.7	

purpose of this illustration, 10% is subtracted from the maximum permissible earth fault loop impedance given in table 41D of the Regulations to allow for the possible effect of the transformer. The data are based on the use of fuses to BS 88 having a rating not higher than the cable rating and on disconnection in 5 s. The voltage drops are calculated from the mV/A/m values given in appendix 4 of the Regulations assuming loads equal to the fuse ratings. The maximum lengths to satisfy the requirements for earth fault loop impedance are derived from table 10.4 herein.

For the smaller sizes (including those below 16 mm^2, not listed) it is likely that voltage drop will more often impose the limit on cable length than earth fault loop impedance. If the old 2.5% limit on voltage drop still applied it would control the lengths of all the 2-core cables listed and those of the 3- and 4-core cables up to 50 mm^2.

For the larger sizes of 3- and 4-core cable the requirements for earth fault loop impedance appear to exercise the main control and this could also apply to some of the 2-core cables when voltage drops in excess of 2.5% are acceptable. When account is taken of differences in phase angles and that some of the cables would be carrying currents appreciably below their ratings (because of the steps in fuse ratings) and therefore be operating at lower temperatures than those assumed for the tabulated values, the voltage drop limitations might be further alleviated. The earth fault loop impedance could then prove to be the decisive restraint for many of the cables.

The figures in table 10.11 relate to the hypothetical arrangement, which is one to be considered, where a cable of one size runs over the whole distance from the supply to the point of utilisation. However, the larger cables will often carry a supply to a number of smaller cables which each supply a final circuit. There may be intermediate divisions of the current supplied by the large cable, so that there are several sizes of cable in series between the supply and the final circuit.

In these conditions, if each cable carries a current up to the rating of the fuse which protects it, the percentage voltage drops are additive and the total length of cable to meet the acceptable limit will lie between the lowest and highest values for the particular cables involved when they are used alone.

On the other hand, the total length of cable, as limited by earth fault loop impedance, can be substantially increased in these circumstances. The large cable at the supply end makes only a small contribution per unit length to the earth fault loop impedance required to ensure operation of the fuse protecting the smaller cable of the final circuit, and the total route length of cable can therefore be increased.

This can be illustrated simply by consideration of a 4-core $150\,mm^2$ XLPE insulated wire armoured cable supplying a number of circuits of which one comprises 4-core $25\,mm^2$ cable of the same type. Suppose, for convenience of calculation, that the $150\,mm^2$ cable is of the length which just meets requirements for earth fault loop impedance, i.e. 78.5 m. Suppose further that a voltage drop of 3.8% was acceptable for the equipment being supplied by the $25\,mm^2$ cable at a current of 125 A. This is the voltage drop shown in table 10.11 when 76.8 m of this cable is used alone between supply and point of utilisation and the 76.8 m is the maximum length, under these conditions, to satisfy the requirements for earth fault loop impedance. If the $25\,mm^2$ cable was taking its supply from the $150\,mm^2$ cable, the voltage drop in the $150\,mm^2$ cable (carrying 400 A) would be 2.3% and the length of the $25\,mm^2$ cable would be limited to that causing an additional voltage drop of 1.5%, i.e. $76.8 \times 1.5/3.8 = 30.3$ m.

The contribution of the $150\,mm^2$ cable to the earth fault loop impedance for the $25\,mm^2$ cable is $0.0864\,\Omega$, which is a relatively small part of the $0.315\,\Omega$ required for the $25\,mm^2$ cable protected by 125 A fuses. It leaves $0.229\,\Omega$ for the $25\,mm^2$ subcircuit and this would allow $76.8 \times 0.229/0.315$ m, i.e. 55.8 m, of the $25\,mm^2$ cable to be used if no limit were imposed by voltage drop.

It can be concluded that either the earth fault loop impedance or the voltage drop requirement might limit the length of cable or the size required for a given route length, depending on particular circumstances. If a large cable supplies a number of circuits of smaller cables, the earth fault loop impedance is less likely to be the critical factor for the total route length.

Where the earth fault loop impedance is critical, an additional protective conductor separate from the cable can be used to reduce it, or supplementary bonding between exposed conductive parts and simultaneously accessible conductive parts in the same equipotential zone can be employed. When the voltage drop is excessive at the necessary load current it can only be reduced by using a larger cable.

In the calculations in this section protection against short-circuit currents is provided by fuses having ratings no higher than the ratings of the cables they protect. If the fuses have higher ratings, the values calculated will be changed, although the basic methods will be valid. From table 10.10 it can be seen that the armour areas will often comply with the requirements relating to temperature rise on earth fault, even when the fuse rating is one or two or even more steps above that corresponding to the cable rating. However, the use of higher rated protective devices has a profound effect on the requirements for earth fault loop impedance. It is clear that, for the benefit of optimum utilisation of cables, the protection should be as close as possible, consistent with the circuit current and other restraints on the selection of the devices. This is also consistent with optimum safety in minimising energy released under fault conditions.

REFERENCES

(1) Jenkins, B. D. (1981) *Commentary on the 15th Edition of the IEE Wiring Regulations.*
(2) Whitfield, J. F. (1981) *Guide to the 15th Edition of the IEE Wiring Regulations.*
(3) *Guidance Notes to the 16th Edition of the IEE Wiring Regulations.*

PART 2
Wiring Cables, Flexible Cables and Cables for General Industrial Use

CHAPTER 11
Cables for Fixed Installations

HISTORY AND DEVELOPMENT OF WIRING SYSTEM CABLES

In some of the earliest installations, which were in private houses, the current was carried by copper wires covered with cotton yarn, either lapped or braided, stapled to wooden boards and subsequently varnished. By the time the filament lamp was introduced, around 1880, multicore cables were available with wax impregnated cotton and silk insulation and lead sheaths.

Gutta percha and rubber were then being used for telegraph cables and in 1889 Hooper introduced a vulcanised rubber cable with insulation in three layers: (a) pure rubber next to the conductor; (b) a layer, often termed a 'separator layer', also unvulcanised and believed to contain a high proportion of zinc oxide which imparted a white or drab colour, and (c) a final vulcanised layer with optimum physical properties. Sulphur and probably litharge were used as vulcanising agents – hence the black colour of the outer layer.

This construction was probably used to protect the copper conductor from reaction with residual sulphur from the outer layer but tinned copper conductors were later introduced to prevent interaction between copper and sulphur. The tinning also limited oxidation of copper wires exposed at terminals and facilitated soldering of the conductor for joints and terminations. Tinning was also advantageous in reducing oxidation and minimising adhesion between the conductor and insulation, as well as easing the stripping of the insulation when making connections.

Rubber insulated cables were initially expensive and other types continued for many years, e.g. fibrous materials such as cotton or jute and later oil-impregnated paper. Woven cambric tapes coated with linseed oil variants found some use, particularly in North America. Power cable applications of the latter types continued for many decades.

The outer covering for rubber insulated cores was commonly a lapped woven cotton tape, waterproofed on one side by a coating of rubber compound, followed by a braid of cotton or jute impregnated with a preservative compound, usually wax based. In the early 1900s rubber outer sheaths were developed, based on the compounds then in use for cab tyres: hence the term cab tyre sheaths (CTS). As pneumatic tyres were developed this was changed to a tough rubber sheath (TRS), the type of compound being very similar. One type of cable had rubber insulated cores laid side by side and enclosed in the TRS in a flat formation.

The need for concealment and protection of electric lighting wiring was recognised as desirable early in the history of domestic installations. Two basic systems developed, one on walls and building surfaces generally and the other buried in the plaster or otherwise in the decorative or structural parts of the building. One of the earlier versions of the former group consisted of wooden boards with grooves in which the insulated conductors were placed. A wooden cover or capping was then screwed on. In the second type of system, zinc conduit tubes were used into which were drawn the wiring cables. Later zinc was replaced by galvanised iron and then enamelled steel.

These two methods of installation determined the geometrical designs of modern wiring cables. The surface system led to the provision of flat cables with conductors laid side by side, the whole having an approximately rectangular cross-section. Early designs consisted of rubber insulated conductors enclosed in a lead sheath, followed by a similar arrangement in which the lead was replaced by a mechanically robust rubber compound, i.e. CTS or TRS.

In the buried or conduit system of wiring installations a number of individual insulated and protected conductors were drawn into the tubes as required. These were generally of the taped, braided and compounded type of finish, the wax compound on the braid acting as a lubricant during installation as well as serving its original preservative function.

Rubber eventually replaced all the types of insulation in the wiring systems. TRS and lead sheathed rubber insulated cables for surface wiring and the vulcanised rubber insulated taped, braided and compounded conduit cables were in general use until well after the Second World War.

Safety considerations required the bonding and earthing of exposed metal parts of domestic and public wiring systems at a very early stage. The forerunner of the present IEE Wiring Regulations (now BS 7671) was published in 1883. This required provision of conductors specifically for the purpose. Initially the lead sheath, when available, was considered to provide this function, but it was soon realised that separate low resistance earth circuits were necessary because of the difficulty of making reliable contact with such sheaths. The bonding of metallic sheaths and the provision of definite low resistance paths to a common earth at the supply source were also shown to be necessary, to avoid electrolytic corrosion due to stray currents. Consequently, wiring systems were developed in which various means of achieving these ends were featured. One of these consisted of a flat lead sheathed cable with rubber insulated cores which had a bare copper wire placed underneath the lead sheath and in contact with it throughout its length. It was introduced by the Callenders Cable and Construction Co. Ltd, in 1927. This basic design set the pattern for present day surface wiring cables.

Although different insulants and sheaths have been introduced, the emphasis on earthing as a safety feature has increased and hence also the need to provide an integral earth conductor. Consequently the most commonly used surface wiring cable in the UK has been the flat twin with earth (reference type 6242Y) as shown in fig. 11.1. CPC denotes circuit protective conductor, formerly known as earth continuity conductor (ECC).

Cables for Fixed Installations

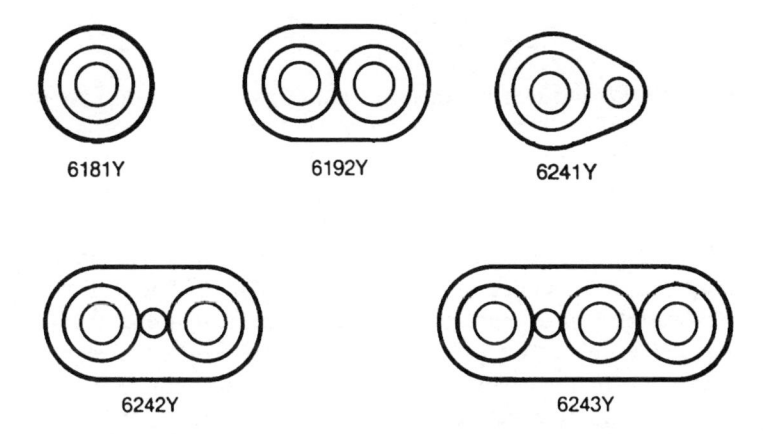

Fig. 11.1 Typical single-core and flat type wiring cables (6241Y, 6242Y and 6243Y with CPC)

The cable reference number as given above and used throughout part 2 of the book is based on a system originally introduced, some decades ago, by UK cablemakers. Although not to be found in any official standards, the system is still considered to be useful to manufacturers and users alike, and remains in general use.

The rubber insulated cable, with a variety of protective coverings, became the dominant type in wiring systems because of its flexibility and ease of handling. It remained virtually unchanged until the late 1930s. At that time a new thermoplastic material, plasticised polyvinyl chloride (PVC), began to be used in Germany and America for both the insulation and sheath of cables. The effect of the following war on world availability of rubber encouraged the use of this alternative material and within twenty years it had all but replaced rubber in wiring system cables.

Today PVC is still the most commonly used material. However, in recent years there has been considerable development of compounds with negligible halogen content and reduced smoke emission under fire conditions. Wiring system cables employing these new low smoke and fume materials are already being installed in public buildings and other areas where performance in fires is of special concern, and their use is increasing. A British standard (BS 7211: 1994) for low smoke single core non-sheathed conduit cable (6491B-Harmonised reference H07Z) is now available. The British standard also includes circular and flat twin and CPC (reference type 6242B) cables, these cables being distinct from BS 6724: 1990 for armoured and sheathed cables with low smoke and fume characteristics.

DESIGN OF MODERN WIRING SYSTEM CABLES

The two forms of cable have changed little over the years, i.e.

(a) insulated and sheathed: circular single-core and 2- or 3-core flat cables for surface wiring or direct burial in plaster;
(b) insulated only: single-core for use in conduit, or bunched on trays, or in trunking.

Voltage designation

BS 7671: 1992 (IEE Wiring Regulations) recognises two voltage designations: extra low voltage and low voltage. The latter usually applies in buildings and is defined as 'exceeding extra low voltage but not exceeding 1000 V (1500 V d.c.) between conductors or 600 V (900 V d.c.) between conductors and earth'. Extra low voltage denotes voltages up to 50 V a.c. (120 V d.c.)

Historically, the voltage designations of wiring system cables, in the form U_0/U defined in chapter 2, have tended to relate to the relative thickness of insulations rather than the required electrical characteristics. To some extent the minimum thickness which can be applied reliably is a determining factor and over the years there has been significant reduction in dimensions as manufacturing techniques and materials have improved. However, the thicknesses of insulation (and sheaths) of wiring system cables are largely governed by consideration of the mechanical hazards encountered during both manufacture and installation. They cannot be determined on a strictly theoretical basis and are the result of practical experience. Whilst such cables are categorised by their mechanical duty, it is still important that they should not be used for applications above their rated voltage.

The system voltage in domestic and public buildings in the UK is rarely other than the standard 230/400 V. IEC 38 notes that voltages in excess of 230/400 V are intended exclusively for heavy industrial applications and large industrial premises.

The BS 7671: 1992 definition given above aligns with the corresponding voltage category (Band II) recognised by both the IEC and CENELEC within the low voltage category. CENELEC Harmonisation Documents for wiring cables provide principally for two voltage classes: 300/500 V and 450/750 V.

In superimposing voltage classifications it is not surprising that apparent anomalies arise. However, it is necessary to match the voltage designation of a cable with that of the range of supply voltages for the intended application, if only as an easy means of eliminating unsuitable cables, e.g. telephone cables from power distribution systems. It is also traditional to discriminate between different degrees of protection applied to the same conductor size by ascribing different voltage ratings. This introduces a further complicating consideration in that it implies a requirement for discrimination based on the risk of mechanical damage in service and the possible consequences of electrical failure, e.g. electric shock and fire.

When the subject of voltage ratings was considered in drawing up the European Commission (EC), now the European Union (EU), cable standards (CENELEC Harmonisation Documents), it was found that in Europe the highest voltage between the conductors in final power distribution circuits was 480 V. 500 V was therefore chosen as the standard. As most wiring cables operate on 3-phase four-wire systems or on single-phase systems derived therefrom, the corresponding voltage to earth is $500/\sqrt{3}$, which becomes 300 V when rounded up to the nearest 100 V – hence the above-mentioned 300/500 V designation as applied to the sheathed wiring system cables only.

The single-core types, with a single covering only and no separate sheath, are intended for use with further protection in the form of metal or plastic conduit, trunking or cable trays, where the risk of damage and contact is less. They are therefore given the higher voltage designation of 450/750 V. The 600/1000 V designation is given to such types when they are used for the internal wiring of switch and control equipment, the thickness of insulation being similar to that of the cores in armoured and sheathed cables.

Materials for wiring cables

Conductors

Conductors are now predominantly of copper as described in chapter 4, which also refers to the problems with aluminium in the commonly used types of wiring fittings.

Elastomeric wiring system cables and wires have tinned copper conductors. Tinning is less necessary now with synthetic rubber compounds than it was when natural rubber insulation was prevalent, but the tin layer has been retained as a safeguard against possible interaction between the copper and the insulation. It also aids removal of the insulation, as previously mentioned.

Insulation

The insulation of wiring cables until recently was almost exclusively PVC. However, as already noted, significant quantities of low smoke and fume (LSF) materials are now being employed. Other exceptions are that ethylene–propylene rubber (EPR) and silicone rubber (SR) are used for cables installed in very hot or very cold situations where the temperature is outside the range of PVC.

PVC insulation is usually of the general purpose type (type TI 1, chapter 3) compounded to have a higher tensile strength, better resistance to deformation and better electrical properties than the sheath.

The insulation for the single-core non-sheathed cable (6491B) with low smoke and fume characteristics, e.g. as described by BICC as LSF (the description also used for cables to BS 6724: 1990), is a crosslinked material having physical and electrical properties comparable to those of PVC at normal ambient temperatures. Under fire conditions it shows reduced flame propagation and smoke emission and minimal acid gas emission. The cross-linked insulating compound is covered by BS 7655: Section 5.1: 1994 (type EI5–Maximum operating temperature 90°C).

Requirements for thermoset insulation are given later, with the descriptions of the cables.

Sheath or protective finish

The PVC compound (BS 7655: Section 4.2: 1993 type 6) used for sheathing is specially formulated for the purpose and is somewhat different from that used in other applications, e.g. for the sheath of PVC insulated armoured power distribution cables (type TM 1). As it is not expected to contribute to the electrical performance of the cable, it can be more easily formulated to assist in its removal at terminations; an important point in installation economics.

A grey colour, with a matt surface finish, has become the standard for the sheath of wiring system cables, after experience with a number of other colours. It represents the best compromise between appearance, economy and technical performance. However, white sheathed cables are available where this is essential and black is usual for circular sheathed cables. To distinguish the product from PVC cables, light grey sheathing is employed on low smoke and fume 6242B (BS 7211: 1994) type cables.

The protective finish applied over EPR insulation consists of a braid of textile yarn treated with a compound which renders the yarn moisture resistant and also behaves as a lubricant when the cable is drawn into conduit. Silicone rubber is usually protected by a braid of glass fibre yarn treated with an appropriate lacquer to prevent fraying.

Identification of cores by colours

While it has proved relatively easy to obtain agreement within Europe on common core colours for flexible cords, the extension into identification of fixed cables is still not fully resolved. Agreement to date covers the use of the bi-colour green/yellow for the insulated circuit protective conductor. The use of light blue for the neutral, which is the practice in the continental European countries, has also been accepted in principle for future standardisation. However, for the UK, blue for the neutral cannot be adopted without changing the UK phase colours, which include blue, with red and yellow for the other phases. This subject is under discussion at the IEE (BS 7671).

A satisfactory scheme for the phase colours has not yet been achieved, and lack of resolution within IEC and CENELEC has been a major obstacle to completion of universal wiring regulations and cable standards. A number of countries wish to use black for one or more of the phases, but because of the long-standing use in the UK of black to indicate the neutral core, there is reluctance, on safety grounds, to changing its use to a phase core, particularly in association with blue for the neutral, which would reverse the significance of the colours for the two cores between old and new cables.

Consideration is being given to other means of resolving the problem, such as having all phase cores brown, with numbers where necessary to identify the phases, but the final resolution is uncertain.

It will be noted that the Harmonised Cable Standards, e.g. BS 6004: 1995, CENELEC HD 21, preclude the use of the single colour yellow for the core identification of single core non-sheathed wires (type H07V). BS 7671 (sixteenth edition of the Wiring Regulations) stipulates the use of yellow to identify the second phase conductor in a 3-phase circuit. The yellow single-core (type 6491X) wire is hence not a harmonised type but is recognised as a UK national standard only. For other individual cores it is advisable to consult the relevant cable standard or wiring regulations.

Standards and references

Details of standard cable types, reference numbers, voltage designations, British Standards and CENELEC Harmonisation Documents are shown in table 11.1.

Current CENELEC Harmonisation Documents are:

HD 21 (current issue HD 21 S2) – poly(vinyl chloride) insulated cables and flexible cords of rated voltage up to and including 450/750 V;

HD 22 (current issue HD 22 S2) – rubber insulated cables and flexible cords of rated voltage up to and including 450/750 V;

HD 22.9 S1 – insulated non-sheathed single-core cables (low smoke and fume) of rated voltages up to and including 450/750 V.

Cables for Fixed Installations

Table 11.1 References and standards for wiring cables

Insulation	Cable type	Voltage (V)	Reference	CENELEC code	British Standard	CENELEC HD
PVC	Non-sheathed general purpose cable, single-core solid conductor	450/750	6491X	H07V-U	BS 6004	HD 21
PVC	Non-sheathed general purpose cable, single-core stranded conductors	450/750	6491X	H07V-R	BS 6004	HD 21
PVC	Non-sheathed single-core solid conductor for internal wiring	300/500		H05V-U	BS 6004	HD 21
PVC	Insulated and sheathed single-core solid and stranded conductor	300/500	6181Y		BS 6004	
PVC	Insulated and sheathed single-core cable, solid conductor	300/500	6241Y[a]		BS 6004	
PVC	Insulated and sheathed flat 2-core,	300/500	6192Y		BS 6004	
	solid and stranded conductors	300/500	6242Y[a]		BS 6004	
PVC	Insulated and sheathed flat 3-core	300/500	6193Y		BS 6004	
	solid and stranded conductors	300/500	6243Y[a]		BS 6004	
PVC	Insulated and sheathed 2-core circular	300/500		[b]	BS 6004	HD 21
	Insulated and sheathed 3-core circular	300/500		[b]	BS 6004	HD 21
	Insulated and sheathed 4-core circular	300/500		[b]	BS 6004	HD 21
	Insulated and sheathed 5-core circular	300/500		[b]	BS 6004	HD 21
EPR	60° rubber insulated OFR sheathed flat 2-core (festoon lighting)	300/500	6192P		BS 6007	
EPR	85° rubber insulated textile braided and compounded single-core	450/750	6101T		BS 6007	
XL-LSF	Thermosetting insulated non-sheathed single-core cable with solid and stranded conductor	450/750	6491B	H07Z-U/R	BS 7211	HD 22.9.S1

(cont.)

Table 11.1 continued

Insulation	Cable type	Voltage (V)	Reference	CENELEC code	British Standard	CENELEC HD
XLPE	Thermosetting insulated and LSF sheathed single-core solid and stranded conductor	450/750	6181B		BS 7211	
XLPE	Thermosetting insulated and LSF sheathed single-core cable, solid conductor	300/500	6241B[a]		BS 7211	
XLPE	Thermosetting insulated and LSF sheathed flat 2-core, solid and stranded conductors	300/500	6242B[a]		BS 7211	
XLPE	Thermosetting insulated and LSF sheathed flat 3-core, solid and stranded conductors	300/500	6243B[a]		BS 7211	
XLPE	Thermosetting insulated and LSF sheathed circular					
	2-core	450/750	6182B		BS 7211	
	3-core	450/750	6183B		BS 7211	
	4-core	450/750	6184B		BS 7211	
	5-core	450/750	6185B		BS 7211	

[a] With CPC
[b] Harmonised designs, but not eligible for ◁HAR▷ marking because of non-standard core colours

These documents embody the definitions that 'wiring cables' (German: Leitungen) are for nominal voltages up to 750 V. 'Mains cables' or 'power cables' (German: Kabel) are regarded as having rated voltages of 1000 V or above.

Differing installation practices and philosophies between the European countries, particularly with respect to the protective system, have made necessary the retention of certain national types of wiring system cable (e.g. flat sheathed cables in the UK and NYIF cables in Germany). These are designated National Types, which recognise national practices. As such they are subject to the national certification procedure (provided in the UK by BASEC) but cannot carry the CENELEC common marking ◁HAR▷ because they are not harmonised designs. Chapter 7 includes further explanation of CENELEC harmonisation, certification and the ◁HAR▷ mark.

Certain cables which are otherwise made in conformity with the Harmonisation Documents deviate in one or more details and cannot therefore have the ◁HAR▷

Table 11.2 Designation of cable type

1st group: Voltage designation	Following letters: Covering materials and construction	Final letter: Type of conductor
H03 ≡ 300/300 V H05 ≡ 300/500 V H07 ≡ 450/750 V	N ≡ PCP R ≡ rubber T ≡ textile V ≡ PVC Z ≡ XL-LSF	-F ≡ flexible -R ≡ strand -U ≡ solid wire -K ≡ flexible for fixed installation

certification applied to them. An example is the PVC light non-flexible circular multicore cable to BS 6004 which has the UK national core colour identification.

A common shorthand code for designating the construction of cables has been formulated in CENELEC, as will be evident from the foregoing. The complete code is given in Harmonisation Document HD 361 (current issue HD 361 S2), 'System for cable designation'. CENELEC countries are obliged to use the code for designating harmonised cables in their national standards, but its use for other cables is optional. In the UK it has been adopted for harmonised cables only. Some parts of the code which are applicable to the simpler designs of wiring cables and flexible cords and cables are given in table 11.2, in which H signifies that the cable is a Harmonised Type. Examples are:

H07V-U: 450/750 V, PVC insulated solid wire conductor;
H07RN-F: 450/750 V, rubber insulated PCP sheathed flexible conductor.

TYPES OF WIRING SYSTEM CABLES

PVC insulated and sheathed cables

In the UK these are 300/500 V to BS 6004. Details of sizes and technical data are given in appendix A6. Once installed they can be operated at temperatures from $-30°C$ to $+70°C$. Precautions to be observed when installing cable at low temperatures are given in chapter 15.

Single-core cables
The single-core cables are normally supplied in a range of conductor sizes from $1\,\text{mm}^2$ to $35\,\text{mm}^2$. Single solid wires are used for conductor sizes of 1.0, 1.5 and $2.5\,\text{mm}^2$ and stranded conductors for the larger sizes. These conductor formations are also used in the flat cable designs described below, which they complement.

Flat cables
The flat form cable has proved to be very convenient for attaching to building surfaces, its small depth facilitating concealment. As described here, it is peculiar to UK practice,

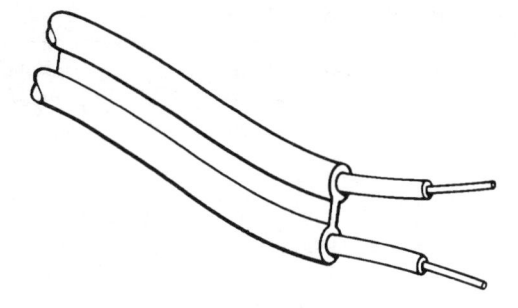

Fig. 11.2 German type of NYIF flat cable

although somewhat similar cables are used in Germany (NYIF cable, fig. 11.2) and the USA. It has won its position relative to competing types as a result of technical performance and economic advantage, both in use and in manufacture, together with its appropriateness to installation conditions and to UK safety practices.

It is available in two forms, i.e. with and without circuit protective conductor (CPC). The CPC is not insulated and lies between two insulated conductors (see fig. 11.1). The size range for 2- and 3-core flat cables is from 1 to 16 mm^2.

Single-, 2- and 3-core cables with CPC have a single solid or stranded protective conductor, the size of which, relative to the area of the current carrying conductors, takes account of the IEE Wiring Regulations (section 543 of BS 7671). With copper conductors there is a minimum cross-sectional area of 1.0 mm^2.

The thicknesses of insulation and sheath for the various sizes of cable have varied over the years. They have been influenced by a blend of practical experience and testing in relation to service requirements, improvement in manufacturing plant, advances in PVC and LSF compound technology and the application of quality assurance techniques to every aspect of cable making.

Circular multicore cables
A range of light PVC insulated and sheathed circular 2- to 5-core cables comes into this category, although they are not in common use in the UK. They may have solid wire conductors up to 10 mm^2 and stranded conductors above that. The total range is from 1.0 mm^2 to 35 mm^2. An additional extruded, soft unvulcanised inner covering (mastic) is used to fill the core interstices and to facilitate stripping. They find application on the continent in the larger installations particularly in damp situations, for prefabricated houses and in business, industrial and agricultural premises.

Although they are generally to an agreed Harmonisation Document, some sizes of the UK range included in BS 6004: 1995 are not included in the harmonised range. As colour coding for the core identification has not yet been agreed in CENELEC, this type of cable, whilst included in HD 21, is not regarded as fully harmonised and cannot carry the ◁HAR▷ mark. LSF versions are included in BS 7211: 1994.

60°C rubber insulated and sheathed cable

This cable is designated 300/500 V and is to BS 6007: 1993. Although strictly not a cable for fixed installations, it is similar in design to the flat PVC insulated and sheathed cables described above. It has rubber insulation (60°C insulation, BS 7655: Section 1.1: 1993 type EI4) and a tough polychloroprene (PCP) sheath (60°C sheath, BS 7655: Section 2.1: 1993 type EM2) with stranded tinned copper conductors.

It is a direct descendant of the early wiring system cables and is retained solely for festoon lighting used as temporary circuits, for instance on building sites or for decorative purposes. The cable is designed to provide self-sealing connections with specially designed lamp holders, whose spike terminals penetrate the sheath and insulation to make contact with the conductor when they are clamped at intervals along the cable length.

Only one size, 2.5 mm^2, 2-core is available and further details are given in appendix A6, tables A6.7 and A6.8.

This cable is not a harmonised type and it is available only as a National Standard Type. The core colours are red and black and the sheath colour is black. The operating temperature range is from $-30°C$ to $+60°C$.

PVC insulated non-sheathed single-core cable

This cable, of 450/750 V designation (reference 6491X), is covered by BS 6004: 1995 and HD 21. It is intended for use with additional mechanical protection as an inherent part of the wiring system, i.e. in conduit, ducts or trunking. It uses the same cores as those already described for flat sheathed cables but has a higher voltage rating, because when installed it is less subject to mechanical risks or hazards.

The conductor size range is considerably larger than for the flat cables as the conduit or trunking wiring system is much more widely used in large commercial and industrial installations. The standard range of conductor sizes is 1.5 mm^2 to 630 mm^2. In HD 21 the sizes from 1.5 to 16 mm^2 may be single solid wires, but stranded conductors provide additional flexibility and are available as an alternative. In the UK it is customary to use stranded conductors, although there is limited use of solid wires in sizes of 1.5 and 2.5 mm^2. Stranded conductors are used exclusively in sizes from 25 to 630 mm^2, ranging from 7-wire at the lower end to 127-wire at the upper limit.

Intended principally for the internal wiring of equipment such as switchgear, an extension of the conductor size range down to 0.5 mm^2 is available in three sizes of single solid wire, i.e. 1.0, 0.75 and 0.5 mm^2. The voltage designation for these three sizes is 300/500 V.

Sizes of 1.0 and 1.5 mm^2 are most commonly used for domestic lighting circuits; 2.5 and 4.0 mm^2 are used to supply socket outlets and the larger sizes are used on power circuits, e.g. lifts, pumps and on main feeds to distribution boards. The largest sizes of 500 and 630 mm^2 are used primarily in short runs at the mains intake positions between items of switchgear on large installations.

Details of dimensions, weights and technical data for PVC cables with copper conductors are given in appendix A6, tables A6.1, A6.2 and A6.4.

These cables are available in a range of colours including green/yellow, yellow, red, black, white, blue, brown and grey. It should be noted that some colours are available only as National Types.

LSF single-core non-sheathed cable

This cable to BS 7211: 1994 (reference 6491B) is rated 450/750 V and is used for fixed power, lighting services and earthing in industrial and other environments where its fire performance (reduced flame propagation, low smoke generation and minimal acid gas emission) can be beneficial. It is primarily intended for installation in conduit or trunking. Sizes 0.5, 0.75 and 1.0 mm^2 have a voltage designation of 300/500 V.

The conductor size range is 0.5–630 mm^2; with stranded conductors commonly used for sizes 1.5 mm^2 and above. The LSF insulation is suitable for operation at continuous conductor temperatures up to 90°C.

Dimensions are similar to the single-core PVC cables (reference 6491X) described above, but 6491B LSF cables are somewhat lighter than the same sizes of PVC cables. Current ratings are the same at 30°C ambient temperature, but the higher operating temperature allowable permits them to be used in ambient temperatures up to 50°C without applying a temperature correction factor.

Crosslinked LSF compound is more flexible than PVC at low temperatures, thus allowing the installation of LSF cables at temperatures below 0°C provided that the precautions given in chapter 15 are followed.

LSF cables can be supplied in a range of colours, including red, black, yellow, blue, green and green/yellow.

Details of dimensions, weights and technical data for LSF cables with copper conductors are given in appendix A.6, tables A6.10–A6.15.

85°C rubber insulated, single-core, textile braided and compounded

These cables are designated 450/750 V and are covered by BS 6007: 1993. They provide a range of cables for conduit installations where ambient temperatures exceed those appropriate to PVC insulation. The insulation is a heat-resistant elastomeric compound (85°C insulation, BS 6899: 1991, type GP 1). Although the particular polymer is not specified in the relevant standard, that most commonly used is ethylene–propylene rubber in one of its forms, i.e. EPM or EPDM.

The conductor sizes range from 1.0 to 95 mm^2, the 1.0 mm^2 size having a single solid wire, the 1.5 and 2.5 mm^2 sizes solid or stranded conductors and the remainder having stranded conductors with 7 or 19 wires, as appropriate.

Sometimes a separator tape, commonly of synthetic film such as polyethylene terephthalate (PETP), is applied over the conductor to facilitate removal of insulation at terminals. It is not common on the smaller sizes as it increases the diameter and is disadvantageous when a number of cables have to be drawn into conduit.

A further occasional feature, at the manufacturer's option, is a woven proofed tape applied over the insulation, mainly dictated by the method of manufacture. This may carry the mandatory printed legend indicating the nature of the insulation and its temperature category, namely Heat Resisting 85. Alternatively the legend may be printed on the surface of the insulation or on a narrow longitudinal tape inserted under the braid expressly for this purpose. Marking of this sort is advantageous in identifying the different types of rubber.

The overall mechanical protection of the insulation is provided by a braid of textile yarn treated with a preservative compound, usually based on a wax. The outer surface of this braid is coloured red or black to provide circuit identification. Further information is provided in appendix A6, tables A6.7 and A6.8.

The operating temperature range is −60 to +85°C. These cables constitute a National Type.

CURRENT RATINGS FOR WIRING SYSTEM CABLES

Sustained ratings

Methods of deriving sustained current ratings and all the parameters associated with them are discussed in chapter 8. Ratings for individual types of cables are given in appendices A5–A11 and where appropriate these ratings are in accordance with BS 7671 (sixteenth edition of the IEE Wiring Regulations).

Short-circuit ratings

The method of deriving short-circuit ratings is given in chapter 9, and table 9.6 in that chapter includes data for calculating ratings for cables with the main types of insulation. A graphical presentation for PVC insulated conduit wires is included as fig. 9.11.

Short-time ratings

For current flows of shorter time than, say, 1 minute, equilibrium conditions are not attained. Consequently for the recognised temperature rise of the conductor the current rating increases as the duration of current flow becomes less. These considerations are important in circuits such as those supplying motors where the starting current is considerably in excess of the full load current. Where frequent stopping and starting occurs, the cumulative effect of the starting current on the temperature rise of the cable needs to be taken into account.

Figure 11.3 shows the short-time ratings for the smaller size of conduit wires to BS 6004, run singly in air, assuming an ambient temperature of 35°C and a maximum conductor temperature of 70°C. The values are calculated on the basis that the conductor temperature starts at the ambient temperature, that the current flow ceases at the end of the defined period and that the conductor returns to ambient temperature before current flows again.

CABLES FOR GENERAL INDUSTRIAL USE

Industrial cables bridge the gap between wiring cables and the power distribution cables described in chapter 18. As discussed in chapter 20, cables with PVC insulation replaced the earlier paper insulated lead sheathed cables. The types now most commonly used are multicore cables with copper conductors insulated with PVC, EPR or XLPE. Following the voltage designation previously discussed, they are usually classified as 300/500 V or 600/1000 V. As indicated by the conductor sizes, the applications include low power and control uses and supplies to small machines, e.g. in heating and ventilation systems. The use of armoured cables and trunking systems has replaced former conduit practice and increasing use is being made of unarmoured cables, particularly in continental Europe.

In terms of mechanical protection, the lower end of the range is similar to that represented by the circular PVC insulated, PVC sheathed 300/500 V, 2- to 5-core cables already described under the heading 'Types of wiring system cables'.

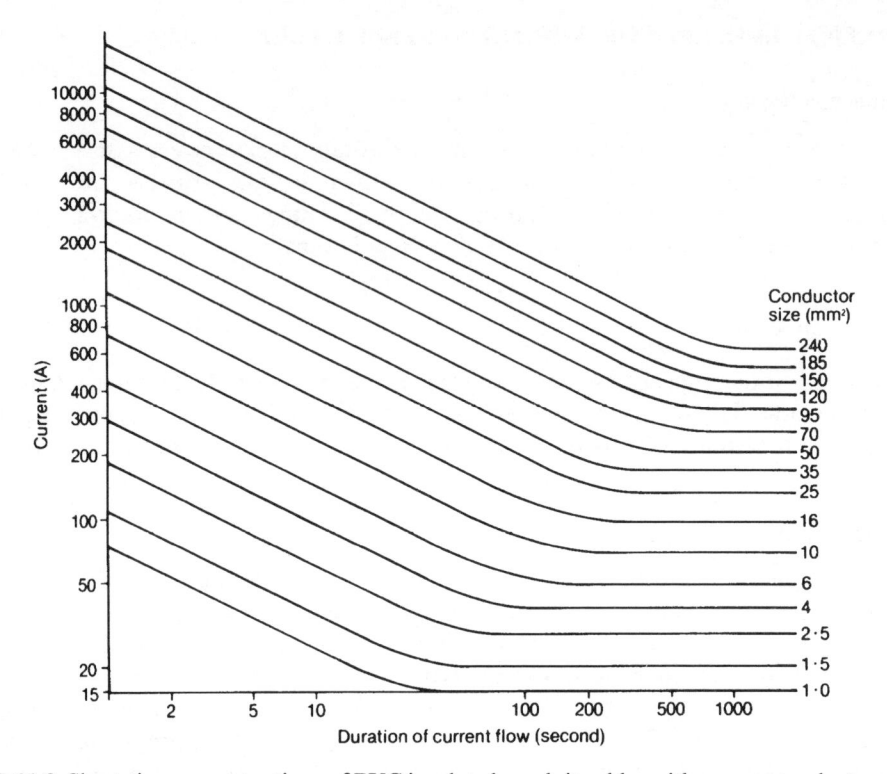

Fig. 11.3 Short-time current ratings of PVC insulated conduit cables with copper conductors, run singly in air at 35°C ambient temperature

PVC, armoured, power/control cables

In cases where significant risk of mechanical damage exists, it is usual to use the smaller sizes (up to about 16 mm^2) of 600/1000 V cables to BS 6346 or BS 5467 (XLPE insulated). This standard covers cables with galvanised steel wire armour, as described in chapters 18 and 20.

Up to 16 mm^2 the conductors are of copper, with circular solid wires for the smaller conductors and stranded construction for the larger sizes. Details are given in appendix A5, tables A5.1–A5.3.

Such cables are used for power supply or control in industrial and other environments, either outdoors, indoors or underground. Appropriate glands as specified in BS 6121 provide the proper termination for armour wires which is essential to provide adequate earthing of the armour and earth continuity.

Cables with thermoset insulation

PVC cables have been pre-eminent in power distribution and control systems in industrial environments. However, the thermoplasticity of PVC is a limitation which influences sustained current rating, overload and short-circuit rating of cables. This limitation is emphasised at high ambient temperatures. XLPE and EPR insulated cables, with the appropriate heat-resisting sheaths, offer advantages in that the conductors can be operated to a maximum continuous temperature of 90°C, with a maximum short-circuit temperature of 250°C.

EPR insulated CSP sheathed cables, as originally used for shipwiring cables (BS 6883: 1991), are often used in applications such as steel manufacturing plants and rolling mills, where ambient temperatures can be high.

XLPE insulated armoured cables to BS 6724: 1990 with LSF bedding and sheath, which emit lower levels of smoke and corrosive products on exposure to fire, are becoming more widely used in industrial and other environments where their fire performance is an advantage, and can also be operated with sustained conductor temperatures up to 90°C.

XLPE insulated armoured cables with PVC bedding and sheath are covered by BS 5467: 1989. From the increased emphasis on co-ordination of overcurrent protection, short-circuit protection and cable ratings, these cables are likely to be of much more interest in the future in installations at normal ambient temperatures. They are the equivalent of PVC types in terms of ease of installation and jointing and of permissible bending radii. The larger cables of this type with XLPE insulation are described in chapter 21.

Cables for the oil and petrochemical industries

Because the PVC insulated, wire armoured, PVC oversheathed cable design, as used in general industry, has good ability to withstand a broad range of hostile environments, it is also the cable mainly used in oil and petrochemical plants.

An important difference, however, is that, as discussed in chapter 5, if such cables are buried in ground containing hydrocarbons, these materials may pass through the oversheath and into the centre of the cable and the hydrocarbons could be transmitted into fire risk areas. It is sometimes necessary, therefore, to incorporate a metallic sheath over the inner sheath. In the UK, a lead sheath is used with steel wire armour over it. In North America, an aluminium sheath with PVC oversheath and no armour is often preferred. Cables with XLPE insulation are also protected similarly.

Specifications for such cables are issued by individual oil companies and in the UK by the Engineering Equipment and Materials Users Association (EEMUA).

Cables with specially fire-resistant or fire-retardant constructions, either to keep important circuits in operation or to restrict the spread of fire, are of importance as also is the use of materials with low emission of acid gas and smoke in fires. This is discussed in chapter 6.

Cables for offshore oil installations generally follow the same pattern as for cables in ships and are described later in the chapter.

Cables for mining and quarrying

Fixed cables for mining and quarrying follow a similar pattern to that for general industrial distribution. For voltages up to 1.9/3.3 kV the most usual construction is with PVC insulation and steel wire armour. Some PVC insulated cable has been used for 3.8/6.6 kV and even a little for 6.35/11 kV, but currently the mining industry specifies EPR or XLPE insulation for these voltages. Usually the armour is a single layer, but double wire armour is used for cables to be installed in mine shafts and is sometimes preferred for roadway cables.

There is also an overlap between what may be considered as 'wiring cables' and the power distribution cables covered in part 3 of this book. As cables of larger conductor size predominate, reference should be made to chapters 18 and 20 for more detailed discussion on cable types. Dimensions, weights and ratings are given in appendix A10 for cables in accordance with British Coal Specifications 295 and 656.

CABLES FOR RAILWAYS AND UNDERGROUND TRANSPORT

Trackside equipment and signalling

Before the electrification of parts of the UK railway system, the use of cables was predominantly for such applications as signalling, power operation of points and mechanical signals and signal lights etc. The cables were installed alongside the track on posts or in ducts. Initially the cable designs and materials represented an extension of domestic wiring practice with natural rubber insulation and sheaths.

Environmental factors such as water, mechanical hazards and fire subsequently encouraged the development of special compounds and the so-called 'oil based type rubbers' became used for insulation. These were vulcanised compounds with natural rubber, bitumen and vegetable or animal oil residues. They continued in use for railway cables long after similar use in general wiring cables because of their resistance to water penetration and mechanical damage. The introduction of PCP permitted a composite insulation consisting of a layer of natural rubber next to the conductor followed initially by two layers of PCP (RNN insulation, i.e. rubber/Neoprene/Neoprene (Du Pont registered trademark)) and later by one layer of PCP (RN insulation). This had significant advantages in flame retardance and resistance to oil.

As PCP sheathing compounds are very tough, have outstanding weather and oil resistance and do not burn easily, the RN insulated type with PCP sheath became the standard for signalling, associated power circuits and the internal wiring of signal boxes. These cables continued to be widely used in the UK until the revision, in the mid-1980s, of BR 872, 'Specification for railway signalling cables'. This retained many of the heavy duty PCP sheathed types but with RN insulation now replaced by EPR to BS 6899: 1991, type GP 1 (plus a maximum permittivity requirement). In addition to these, the new issue of BR 872 contains a range of halogen-free cables with reduced flame propagating properties, made up of non-sheathed single-core and sheathed single-core and multicore types, for wiring in signal boxes and other enclosed areas where fire performance is an important factor. With the increasing adoption by British Rail of the solid state interlock (SSI) signalling system, it can be expected that elastomeric types will eventually be replaced on longer routes by optical fibre data link cables, similar in design to British Telecom trunk telephone cables.

Signalling and communication cables associated with modern 25 kV overhead catenary systems have to be protected by electromagnetic screening because of the large, short duration power surges which are a feature of these systems. The use of copper and/or high permeability magnetic alloy tapes over the laid-up cores reduces the magnitude of induced voltages to tolerable levels.

Track feed cables

Power supply to the point of track feed follows fairly conventional public distribution practice and this also includes the supply to the 25 kV a.c. catenaries. In the case of low voltage live rail systems, heavy duty rubber cables are used for the connection to the live rail.

Early cables had moisture-resistant rubber for the insulation and heavy duty rubber sheaths. PCP was later used for sheathing. Present cable designs take advantage of the higher performance synthetic rubbers using EPR for insulation and CSP for sheaths.

Cables for locomotives and rolling stock

Conductors for cables for locomotives and rolling stock are of multistrand flexible construction because of the tortuous routes involved.

Single-core cables in ducts or conduit are normally used. For many years, EPR/CSP composite coverings on the conductor were dominant but, as in the case of more general wiring types, there has been an increasing demand for limited fire hazard (LFH) cables, often with negligible halogen content (ZH) materials, particularly for use below ground (e.g. in tunnels). Elastomeric cables may typically have a composite flame-retardant EPR/EVA covering. For the more complex wiring of electrical and electronic equipment, high performance 'thin wall' types will probably be employed, commonly insulated with one of the wide-ranging family of polyolefins. There has been a steady move towards materials with lower smoke emission and improved resistance to oils and fluids.

The need for improved communication with trains operating at the higher speeds has also encouraged the use of radio systems involving the use of radio frequency cables which are installed along the track and are designed to radiate and receive radio energy along their length. They are usually coaxial cables whose outer conductor is 'leaky'. This technique has been adopted to a limited extent in various national railway systems, particularly where lengthy tunnels prevent the use of conventional radio equipment, but its main application has been in metropolitan underground systems.

Mass transit underground railways

Avoidance of problems due to fire is particularly important in underground railway systems because of their effect in disrupting services and causing discomfort and panic among passengers, and possible loss of life or injury. The subject is discussed in chapter 6.

As a result, PVC insulated cables, which were widely employed until about ten years ago, both for power distribution and for signalling and information systems, have been virtually superseded by LSF types. Methods of reducing flame propagation and the emission of toxic fumes and smoke to the lowest possible levels have been the subject of major development work in the cable industry. The requirements are generally embodied in performance specifications, which thus allow a variety of materials to be employed, and it is too early to predict which designs of cable will eventually prove most successful.

CABLES FOR SHIPS AND OFFSHORE OIL INSTALLATIONS

Standardisation of cables within the shipbuilding industry has been steady but, with the rapid growth in ships' electrical systems in the last few decades, there has been a greater incentive to adopt high performance materials in efficient designs. The development of the offshore oil industry has extended this development with specialised requirements which, although they follow shipwiring practice, have also introduced new features.

Proliferation of cable types has been encouraged by the increasing complexity of electrical systems and the differing requirements of the classification authorities who register and insure ships throughout the world. The varying views of ship owners and builders, offshore operators and national specifying authorities have further inhibited standardisation. However, IEC Standard 92-3, 'Electrical installations in ships: cables (construction, testing and installation)', has been influential. It is now being completely revised and re-issued in separate parts to cover each aspect of ships' cables, as listed in chapter 7. Within Europe most national shipwiring specifications adopt some, if not all, of the IEC 92 series of specifications. Such cables are also accepted by most classification authorities, including the American Bureau of Shipping (ABS), Bureau Veritas (BV), Det Norske Veritas (DNV), Germanische Lloyd's (GL), Lloyd's Register of Shipping (LRS) and the USSR Register of Shipping.

Materials

Conductors

Copper conductors are used universally, aluminium not having found application because of possible corrosion problems at joints and terminations. Owing to the more tortuous installation routes, most cables have circular stranded conductors of more flexible construction than is used for equivalent cables on land, but some low voltage power distribution cables have shaped conductors in the range from 25 to 185 mm^2.

Insulation

EPR insulation, as specified in BS 6883: 1991, is widely used for its flexibility and good operating and short-circuit characteristics. Special formulations to BS 6899: 1991, type GP 2, are used for the higher voltage cables.

PVC is specified by a few shipyards, but it has a limitation because of a maximum conductor temperature of 60°C in ships. XLPE is growing in popularity, but it results in a stiffer cable, which is unfavourable for bending and handling in the complex routes and confined spaces occurring in ship installations.

Sheaths

The choice of sheathing materials is varied and subject to local preference. BS 6883: 1991 provides for the requirements of offshore installations as well as ships. It includes CSP sheaths with reduced acid gas emission (in fires) and with enhanced oil resistance, together with zero halogen (ZH) sheaths. The latter can be formulated to give significantly lower smoke emission than CSP, but comparable ageing characteristics and oil resistance.

PCP is the preferred sheathing material in French and German yards, although it has an inferior ageing performance to that of CSP. The choice may be influenced by indigenous manufacture. In the Netherlands, Norway, Sweden and Finland, a combination of PCP inner sheath and PVC outer sheath is used.

Construction

Cores are laid up in standard formations with up to 61 cores for the smallest sizes and 4 in the larger ones. Core identification is by the printing of numbers directly on to the insulation surface or on woven proofed textile tapes lapped over the core, as preferred. Screened constructions, with a semiconducting layer over the insulation followed by copper tapes, are used at 6 and 11 kV. For difficult route conditions, a tinned copper wire braid may be used instead of copper tapes.

A sheath of CSP or other material, as discussed above, is applied over the laid-up cores, together with further protection as appropriate for the installation. A braided layer of galvanised steel or phosphor bronze wire is commonly applied over the inner sheath and is a necessary feature for cables where there is a danger of explosion, i.e. in tankers and offshore installations. For single-core cables, the wire for such a braid must be non-magnetic to minimise electromagnetic induction effects. The oversheath may be CSP (HOFR) or PVC (BS 7655: Section 4.2: 1993, type 4).

Sustained current ratings

Ratings are generally more conservative than for land cables. They are tabulated in the IEE 'Regulations for the electrical and electronic equipment of ships' and 'Recommendations for the electrical and electronic equipment of mobile and fixed offshore installations'. They are based on an ambient temperature of 45°C and a maximum conductor temperature of 90°C for PVC insulation.

Short-circuit ratings

Typical short-circuit ratings for EPR insulated ship cables operating at a conductor temperature of 90°C are given in table 11.3. These are based on the conductor achieving a temperature of 250°C.

In view of the large electromagnetic forces developed between the cores of multicore cables during short circuits (see chapter 9), it is recommended that armoured types should be used when the current is likely to exceed 20 kA. This applies also to single-core cables run in trefoil. At currents approaching 70 kA, there is at the moment no firm evidence about the ability of even armoured cables to stand up to the forces involved. Currents in excess of this value have been omitted from the table or placed in parentheses, although they could theoretically be carried by the cables on thermal considerations alone. Recent tests have shown that overheating of the conductor (due to the asymmetrical waveform of the current during the first few cycles of a short circuit) is not a serious problem for EPR cables at durations down to 0.1 s, and ratings for this duration have therefore been included.

Table 11.3 Maximum permissible short-circuit current for EPR insulated cables in ships

Conductor size (mm²)	Current (r.m.s.) (kA)			
	1 s	0.5 s	0.2 s	0.1 s
1	0.15	0.22	0.34	0.48
1.5	0.22	0.31	0.49	0.70
2.5	0.35	0.49	0.79	1.12
4	0.57	0.80	1.27	1.80
6	0.85	1.20	1.90	2.7
10	1.45	2.0	3.2	4.5
16	2.3	3.2	5.1	7.2
25	3.6	5.1	8.1	11.4
35	5.0	7.1	11.2	15.8
50	6.8	9.6	15.1	21
70	9.8	13.6	22	31
95	13.6	19.2	30	43
120	17.1	24	38	54
150	21	30	47	67
185	26	37	59	
240	35	49		
300	44	62		
400	56			
500	70			
630	(91)			

Reactance and voltage drop

These factors are of particular significance for power cables in ships because of the likelihood of proximity to steelwork and the consequent inductive influence. Typical reactance values for certain dispositions of cables remote from steelwork are given in table 11.4. They are based on current BS 6883: 1991 insulation and sheath thicknesses. The method of calculation using these values is given in chapter 10.

Installation

Installation must be carried out in accordance with the appropriate regulations governing each application, as required by the classification authority. In the UK these are the IEE 'Regulations for the electrical and electronic equipment of ships' and the Rules of the Lloyd's Register of Shipping. IEC 92-352, 'Electrical installations in ships', gives guidance which is accepted internationally.

Precautions concerning the temperature of installation of cables are given in chapter 15. The adoption of minimum bending radii is an important feature to be taken into account and figures for guidance are given in table 11.5.

Table 11.4 Reactance per conductor (EPR insulated cables) (mΩ/100 m at 60 Hz)

Conductor size (mm^2)	2-core cable (single-phase) or 3- or 4-core cable (3-phase) Unarmoured or armoured	Two single-core cables touching (single-phase) or three single-core cables touching in trefoil (3-phase)	
		Unarmoured	Armoured
1.5	142	178	222
2.5	133	165	207
4	133	159	196
6	126	150	184
10	118	139	177
16	112	132	161
25	107	124	150
35	104	120	145
50	103	119	141
70	102	113	134
95	99	111	130
120	97	108	127
150	97	108	126
185	97	108	126
240	96	106	123
300	96	105	121
400		104	119
500		103	117
630		101	114

Cables required to maintain circuit integrity under fire conditions

The subject is covered in chapter 6 and special constructions applicable to cables for ships and offshore installations include the following.

Silicone rubber/glass insulation
This was the first successful design for elastomeric cable which could provide substantially increased integrity for circuits involved in fires. The cables were required to withstand the 3 hour fire resistance test which was later adopted internationally as IEC 331 (see chapter 6). A composite of silicone rubber and glass braid was applied to conductor sizes up to 16 mm^2, and silicone rubber coated glass fibre fabric tapes was applied to larger sizes. The cables were sheathed with CSP. The construction was widely adopted for essential power distribution and signalling circuits in naval ships, and was embodied in UK Ministry of Defence (Navy) Specification DGS 211 (later NES 527) and US Navy Specification MIL-C-915. Similar types of cable, usually incorporating a phosphor bronze or galvanised steel wire braid armour, were used to a more limited extent in merchant ships and offshore installations, but there was a swing away from this construction when micaglass/EPR insulated types were introduced during the 1970s (see below).

Table 11.5 Minimum installation bending radius

Cable	Bending radius (× cable outer diameter)
150/250 V, 440/750 V, 600/1000 V	
Up to 10 mm diameter	3
10–25 mm diameter	4
Over 25 mm diameter	6
Any armoured cable	6
Any cable with shaped conductors	8
1.9/3.3 kV and 3.3/3.3 kV	
Without armour or screen	6
1.9/3.3 kV to 6.35/11 kV	
Unarmoured	8
Armoured	12

Silicone/glass has continued to be the preferred insulation system for essential circuits in naval vessels, but the urgent need to eliminate cable materials which produce large quantities of corrosive gases and smoke in fires has brought about the development of the elastomeric sheath to MOD (N) Specification NES 518, now Def. Stan. 61-12 Part 31. It is based on non-halogenated polymer(s) and has outstanding resistance to many types of organic fluids; it has significantly lower smoke emission than CSP, but comparable ageing and flame-retarding characteristics. As already noted, halogen-free ZH sheaths with many of the properties of Def. Stan. 61-12 Part 31 are available for cables to BS 6883: 1991, and can be expected to be in demand for other applications where exceptional performance under fire conditions is required.

Micaglass/EPR insulation
This design has a special halogen-free ZH inner sheath, together with phosphor bronze or galvanised steel wire armour (helical or braid). Outer sheaths are available in either halogen-free ZH material or CSP, the latter having reduced acid gas emission (limited to 5% when decomposed at 800°C or 8% when decomposed at 1000°C). Cable constructions are to a proposed revision of BS 6883: 1991 and are designed to withstand the IEC 331 fire resistance test at the increased temperature of 950 ± 50°C.

Special designs
Constructions complying with BS 6883: 1991 and IEC 502, as appropriate, are available, which in addition to withstanding a 950°C fire resistance test will also withstand other conditions which may occur during a fire, i.e. disturbance of the cable by vibration or falling debris, and application of water (by sprinkler or hose). Details of this test are given in BS 6387: 1994.

FLOOR WARMING AND HEATING CABLES

As the purpose of these cables is not the supply of electricity but its conversion into heat, they are in the category of terminal equipment. Applications include underfloor heating, either as the main source of space heating or as background heating in domestic, public and industrial buildings, together with soil warming in horticulture and sports grounds. They are also used for road heating to prevent freezing and for pipe heating to maintain the temperature of pipes and containers in industry.

'Cold tails', consisting of standard insulated cables, are used to connect the heating cable to the mains supply.

Construction and materials

Heating cables are of single-core construction with conductors of resistive alloy, chosen to give an appropriate resistance per unit length and for resistance to corrosion and stress cracking.

Insulation is of crosslinked polyethylene (XLPE), specially compounded for heat and abrasion resistance, of radial thickness appropriate to the mechanical duty and voltage rating. Subsequent coverings, where used, are designed to provide integrated constructions to meet different degrees of mechanical hazard and the differing electrical protection and earth continuity requirements for the regulating authorities.

Where a PVC sheath is used, the PVC compound is of a heat-resisting grade (85°C, BS 7655: Section 4.2: 1993, type 4). Four constructions are available as indicated in appendix A6, table A6.9.

The resistance values are chosen to give particular heat emission per unit length of cable (W/m) at the commonly used nominal supply voltages of 220, 230, 240 and 380 V. Cables are supplied in standard manufacturing lengths to allow the maximum variety of installation requirements to be covered.

In many heating systems it is more convenient to design an installation to give the required heat output per unit area by using prefabricated heating cable units having a fixed standard total heat output at a particular supply voltage. These units have integral cold tails (supply leads) jointed at the factory, which are usually about 3 m in length. The length of the active part of the unit varies with the total wattage output and the supply voltage.

Performance characteristics

The heat output required is usually defined in terms of power input to unit area of the heated surface (W/m^2). This will depend on the temperature to be attained at the surface, the thermal characteristics of the ground or floor to be heated and the efficiency of any thermal insulation. With heated floors the presence and type of floor covering or carpet will also have an influence. The spacing of floor heating cables is determined by the required power input per unit area and the heat emission per unit length of cable (W/m).

For a given voltage, heating units have a predetermined heat output per total length to simplify selection and installation, and the cables have a range of conductor

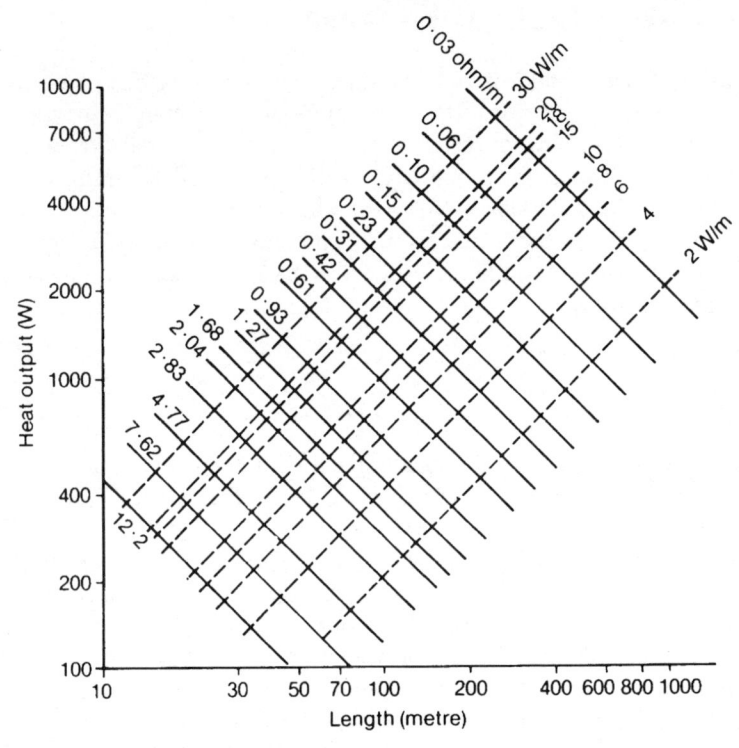

Fig. 11.4 Heat output at 240 V for XLPE insulated heating cable, types X, XB, XJ and XSBV

resistances. Information concerning the length required of a particular cable and the total power input to a given area to be heated is provided by the manufacturer in the form shown in fig. 11.4, which is a typical chart for 240 V supply.

Installation

In the UK the installation of heating cables is subject to the IEE Regulations (BS 7671), which invoke compliance with BS Code of Practice CP 1018, 'Electric floor warming systems for use with off-peak and similar supplies of electricity'. This gives, among other information, guidance on the temperatures of floors in floor warming systems. Other national authorities have corresponding requirements which vary significantly according to the particular applications common in that country and the practices with respect to protection. There are corresponding IEC recommendations. It should be noted, however, that consideration should be given to the installation environment (thickness of screed etc.) to ensure that the conductor temperature does not exceed 80°C. Temperatures in excess of this will seriously reduce the life expectancy of the cable.

In a typical floor heating installation, the cables are embedded in a compact screed forming the upper part of a floor structure which incorporates a damp-proof membrane underneath. These cables should not be installed where permanently damp conditions exist or in constantly wet or frequently washed floors. Care must be taken to avoid

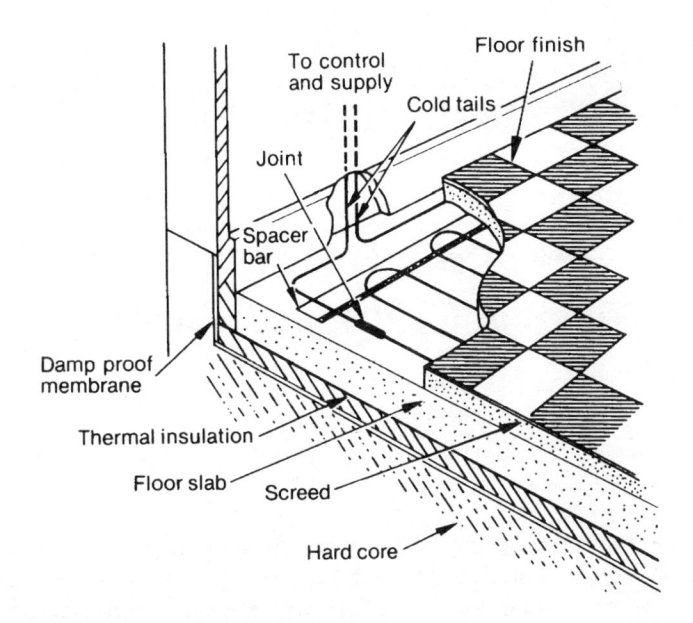

Fig. 11.5 Typical floor warming installation

mechanical damage to the cables during installation when laying floor screeds, during floor repairs and when driving or fixing metal nails, screws etc. for floor fittings. Cables and units should be fixed in position by spacer bars or other means to prevent cable movement or contact. Joints should be embedded in the floor screed along with the cable. A typical installation is shown in fig. 11.5.

In the event of damage occurring, faults can normally be located and repaired without disturbing a floor area greater than 200 mm × 200 mm. Advice on fault location and the specialist equipment used may be obtained from the cable manufacturer.

CHAPTER 12
Flexible Cables and Cords

Wherever a piece of moveable electrical equipment requires to be connected to the fixed wiring system in domestic, public, commercial or industrial premises, an appropriate flexible cable or cord is required. Because of the wide range of equipment using electric power, there is a corresponding wide range of flexible cables and, particularly, flexible cords. Flexible cords are for use with electrical appliances characterised by the use of relatively small amounts of power, while flexible cables are used for the connection of heavier equipment.

The variety of forms and protective finishes is indicative of the range of environmental factors in which electrical equipment is used. In the general wiring field they range from single insulated wires for the internal wiring of electrical appliances, including luminaires, to large multicore cables supplying heavy duty motors. Additionally there are many important special applications such as cables for cranes, hoists and lifts, cables for high temperature environments, for mining and quarrying, and for electrical welding.

The operating temperature range for which there is an appropriate flexible cable is from −55 to 250°C. The majority of general purpose flexible cables and cords are for systems below 1 kV and are designated 300/300 V, 300/500 V or 450/750 V. However, there are also specialised applications, notably power cables for mining and quarrying, with voltage designations of 600/1000 V, 640/1100 V, 1.9/3.3 kV and 3.8/6.6 kV.

In some cases of open cast mining operations, special designs of flexible power cables for 11 kV operation are supplied. Large flexible cables have been used for even higher voltages, but they are not entirely satisfactory as it is difficult to maintain effective dielectric stress control when the cable is subjected to repeated movement.

GENERAL PURPOSE CORDS AND CABLES

Standards

Flexible cords and cables were the first types of cable to be the subject of European harmonisation (CENELEC Harmonisation Documents HD 21 and HD 22). Much preliminary work had been done, however, in the preparation of IEC Standards,

IEC 227 for PVC insulated flexible cables and cords and IEC 245 for rubber insulated flexible cables and cords, many of the requirements of which are embodied in HD 21 and HD 22.

In this work, agreement was reached on a colour code for core identification, an essential factor in complying with the safety aspect of the European Union Low Voltage Directive. The standard colours and sequences for 2-core cables and 3-, 4- and 5-core cables which include a protective conductor are as follows:

Cable	Core colours
2-core	Blue, brown
3-core	Green/yellow, blue, brown
4-core	Green/yellow, black, blue, brown
5-core	Green/yellow, black, blue, brown, black

The bi-colour green/yellow must be used only for the protective or earth core. The colour blue is normally reserved for identification of the neutral core. However, in circuits not having a neutral, blue may be used for other functions, provided that these are identified at the terminations. The colours brown or black are for identification of the live (phase) cores. Colours for single-core non-sheathed cords or cables include green/yellow and blue. Other colours not expressly forbidden by regulations may be used.

Colours expressly forbidden, for reasons of safety, for flexible cords and cables complying with Harmonisation Documents, British Standards and the IEE Regulations are any bi-colour other than green/yellow, green alone and yellow alone.

In the UK the British Standards as given below were amended to align with HD 21 and HD 22.

BS 6500: 1994 Insulated flexible cords and cables
BS 6004: 1995 PVC insulated cables (non-armoured) for electric power and lighting
BS 6007: 1993 Rubber insulated cables for electric power and lighting

BS 6500: 1994 includes both PVC insulated and rubber insulated flexible cords. BS 6004: 1995 includes PVC insulated non-sheathed flexible single-core cables in sizes of 1.5 mm^2 and above in addition to the non-flexible cables referred to in chapter 11. BS 6007: 1993 includes 60°C and 85°C rubber insulated and sheathed flexible cables in single-core and 2-core to 5-core versions and multi-core versions (up to 36 cores). Also included in BS 6007: 1993 are single-core heat resistant (EI3 to 110°C) and silicone-rubber-insulated (EI2 to 180°C) cables. Also included in BS 6500: 1994 are a number of National Standard Types, regarded as representing a particular requirement in the UK and elsewhere but not reflected in other European countries.

Table 12.1 provides a cross-reference between the CENELEC code designations and the conventional UK reference numbers for the main types of harmonised flexible cords. The fourth figure in the UK 'code' indicates the number of cores. For a further discussion on these reference codes, see chapter 11.

Materials

General purpose flexible cables and cords divide into two main groups, those insulated with PVC and those with thermoset (rubber) insulation. A further group for the upper end of the temperature range has silicone rubber or PTFE insulation.

Table 12.1 Designations for harmonised flexible cords

Voltage rating	CENELEC code	UK 'code'	Description of cord
PVC flexible cords			
300/300	H03VV-F	218(N)Y	Light cord, circular
300/300	H03VVH2-F	2192Y	Light cord, parallel twin
300/500	H05V-K	2491X	Insulated only, single
300/500	H05VV-F	318(N)Y	Ordinary cord, circular
300/500	H05VVH2-F	3192Y	Ordinary cord, parallel twin
Elastomeric flexible cords			
300/300	H03RT-F	204(N)	Insulated and braided
300/500	H05SJ-K	2771D	Insulated, glass braided, single
300/500	H05RR-F	318(N)	Insulated and sheathed
300/500	H05RN-F	318(N)P	Insulated and OFR sheathed
450/750	H07RN-F	398(N)P	Insulated and OFR sheathed

Insulation

The PVC insulated types have insulation compounds of type TI 1, TI 2 or TI 3 to BS 7655: Section 3.1: 1993 and HD 21, except that type TI 3 compound is used in cables to BS 6141: 1991, 'Insulated cables and flexible cords for use in high temperature zones'. Type TI 1 is a general purpose insulation compound. Type TI 2 is a special flexible insulation compound which includes a transparent version. They have maximum operating temperatures of 70°C and 60°C respectively. Type TI 3 is a hard grade of PVC with a maximum operating temperature of 90°C.

Rubber insulated cords and cables have insulation compounds of type GP 1 to BS 6899: 1991 and types EI 2, EI 3 and EI 4 to BS 7655: Section 1.1: 1993. Type EI 4 is a natural or synthetic thermoset insulation compound having a maximum operating temperature of 60°C. Type GP 1 is a synthetic thermoset insulation compound having a maximum operating temperature of 85°C. Type EI 3 is formulated to have a maximum operating temperature of 110°C. Type EI 2 is a silicone rubber insulation compound having a maximum operating temperature of 180°C.

60°C rubber insulation was originally based on natural rubber, but both 60°C and 85°C rubber insulations are now generally based on EPR.

PTFE insulation for BICC Intemp 250 (Registered trademark) is covered with a secondary insulating layer of glass/braid. The maximum operating temperature is 250°C.

Conductors

Plain copper wires are almost always used for PVC insulated cores and tinned copper is used for rubber insulated types, but plain copper is now permissible for most elastomeric cables and cords to BS 6007 and BS 6500, provided that a separator is used. The high temperature cords and cables often have copper wires protected by nickel plating or may be of heat-resistant alloy in special cases.

Sheath and coverings

PVC for sheaths is specially compounded for flexibility and is normally type TM 2 to BS 7655: Part 4: 1993 and HD 21. However, a special grade is used for transparent sheaths. Other exceptions are type TM 3 to BS 7655: Part 4: 1993 for heat-resisting cords and a special compound retaining its flexibility at low temperatures for BICC 'Polarflex' cords. The standard colours for the sheath are generally black or white. In special cases, provided that the quantity is sufficient to justify special manufacture, the clear bright colours possible with PVC may be used to advantage. Notable applications are the use of yellow (the standard colour for Polarflex) or orange to make cords supplying gardening power tools more visible and to reduce the risk of damage, and the use of pastel colours to match particular colour schemes on domestic appliances such as vacuum cleaners and food processors.

The variety of protective coverings for rubber insulated flexibles is influenced primarily by the environment. 60°C rubber insulated cords have either a textile braid finish, a tough rubber (TRS) or a PCP sheath. Multicolour patterns are possible with a textile braid to provide an attractive appearance. Examples of those offered as standard are maroon, old gold, white, black with white tracers, blue/green/grey or red/green.

The sheath is provided in two standard versions, type EM 1 (ordinary duty) and type EM 2 (ordinary duty, oil resisting and flame retardant), both to BS 7655: Part 2: 1993 and HD 22. EM 1 is usually based on natural rubber or a synthetic alternative, EM 2 on PCP. 85°C rubber insulated cords have a sheath of ordinary duty heat-, oil- and flame-retardant rubber (HOFR) to BS 7655: Part 2: 1993, type RS 3, which is usually based on CSP or CPE. The corresponding cables have a sheath of heavy duty heat-, oil- and flame-retardant compound (heavy duty HOFR) to BS 7655: Part 2: 1993, type RS 4. As the provision of the necessary mechanical properties for natural rubber and PCP compounds depends largely on the incorporation of carbon black, especially with heavy duty versions, the sheath colour is black. HOFR compounds are less dependent on this type of reinforcement and are produced in white for cords in ordinary duty form. They can also be supplied in other colours for special applications.

The overall covering for silicone rubber cords and cables is a braid of glass fibre yarn treated with a heat-resistant varnish usually based on silicone resin. The outer covering for single-core PTFE insulated 'Intemp 250' cables consists of a glass braid with heat- and abrasion-resistant finish. For multicore types, single-core cables are laid together and then covered with a PTFE binding tape and a further glass braid with heat- and abrasion-resistant finish.

Construction and applications

The design and construction of flexible cords for particular applications is very important. The main factors to be considered are the degree of mechanical hazard, the amount of flexing, the balance between the size of the cord and the restraint it imparts on the appliance which it supplies, the environmental conditions, such as the presence of dampness or oils and solvents, and the operating temperature. Compromise has to be reached between flexibility and handling behaviour on the one hand and protection against mechanical damage on the other. Maximum flexibility and docile handling is achieved by building into the construction as much space and freedom of relative movement as is practical. Materials used to resist mechanical stresses are tough and resilient, and hence impose a corresponding influence on the amount of effort to bend, flex or otherwise move the cord.

The conductor construction involves the use of a larger number of smaller diameter wires relative to those of fixed cables. The maximum diameter of the wires used to obtain a given cross-sectional area of conductor is specified, thus defining the minimum number also. The wires are twisted together in a random bunch or bundle of wires. The number of twists per unit length is carefully chosen to permit reasonable bending and flexing of the conductor without damage, whilst avoiding stiffness. In the larger cables this bunched form of conductor better resists crush damage than the more regularly geometric, mechanically stable, rope stranded conductor which has some advantage in resistance to bending, flexing and twisting. In cases where compromise has to be struck between these factors, a number of bunched assemblies are stranded together in rope formation.

For very light duty applications on very small hand-held appliances such as electric shavers, the conventional forms of cord are not suitable, as a stage is reached in reducing the size of a cord where the smallest mechanically adequate conductor begins to dictate the flexibility. To meet this situation, a tinsel conductor is used, consisting of flattened wires of very small thickness wound round a supporting textile thread, a number of such threads being twisted together to give the required conductivity. This type of cord is available in parallel twin formation in one size only, maximum current rating 0.2 A.

Single-core and 2-core twisted cords are intended for the internal wiring of appliances, particularly luminaires (see below), and are used almost entirely with 90°C or higher temperature forms of insulation. Mechanical stresses are minimal.

The lightest duty general purpose cord is the flat 2-core unsheathed type, which has the two conductors arranged parallel in one plane with a single covering of PVC. The cross-section of this covering is of 'figure-of-eight' (dumb-bell) or 'double D' form, thus providing a groove on each side between the insulated cores whereby they can easily be separated without damage for termination. It is designed for small light duty portable appliances and is subject to a UK Statutory Instrument: The Electrical Equipment (Safety) Regulations. The PVC sheathed version of the flat 2-core cord is also for light duty and achieves its flexibility by the use of a relatively thin sheath of a flexible PVC compound. This point demonstrates the conflict which can arise between desirable characteristics. PVC provides a cord which has a bright colourful easily cleaned smooth surface and is resistant to moisture. It has good appeal to users. However, as the thickness of sheath is increased to obtain better mechanical protection, the stiffness of the cord increases significantly, and this is particularly evident at lower ambient temperatures. The light duty cords, which also include in the standard range 2- and 3-core circular versions with twisted cores, have smaller thicknesses of insulation and sheath than the corresponding ordinary duty cords. They are appropriate for light portable appliances in domestic premises, kitchens, offices etc., where the risk of mechanical damage is small and there is a limited range of ambient temperature. Examples of their use are connections to table and standard lamps, office machines, and radio and television sets. In appropriate cases, the heat-resisting (90°C) version to BS 6141: 1991 can be employed, for instance when a power supply lead is required to operate in a high ambient temperature or is subject to heating by the appliance.

The importance of controlled freedom of relative movement of the components of flexible cords and cables is recognised in the specification requirement that the sheath must not adhere to the cores. This is of consequence both in flexing behaviour and in stripping the sheath when preparing connections, especially when automatic stripping machines are used. In some cases separate core interstice fillers, consisting of threads of textile fibres or extruded sections of plastic or unvulcanised rubber, are used. A tape or film separator may also be used over the laid-up cores to achieve these ends. PVC, with its lower resistance to cut propagation and lower stretch than rubber, is often preferred when automatic machine stripping is of concern.

Ordinary duty PVC cords have the same constructional features as the light duty cords, including flat forms. Because of the greater thickness of insulation and sheath, they are more robust and tend to be stiffer and less docile. This may not be of great consequence with appliances of limited movement, such as washing machines, refrigerators etc., particularly as these applications imply the possibility of a damp environment. Also, as they are designed for medium duty applications, the inference is that they will supply more heavy and powerful appliances than are appropriate to the light duty cords. Therefore their stiffness has less influence on the control of the appliance. Standard ordinary duty cords also include 4- and 5-core assemblies. Nevertheless, because of this stiffness, PVC flexible cables of size $4\,mm^2$ and above have not proved popular.

BS 6141: 1991 makes provision for 90°C versions of the ordinary duty PVC cords, which can be substituted when cables are subject to higher operating temperatures than usual. For lower ambient temperatures, cables from the BICC Polarflex range (which are dimensionally the same as ordinary duty cords but remain flexible down to $-20°C$) can be used for a variety of purposes on outdoor sites provided that they are adequately shielded from mechanical damage (see below).

Rubber insulated flexible cords and cables have, of course, a longer history than the PVC equivalents. The thickness of insulation and sheath which is practical with PVC is not so appropriate to rubber. However, rubber is considerably more flexible and resilient, a factor which becomes more evident as the size of the cable increases. The thermosets cover a wide temperature range and their properties are not as temperature dependent as those of PVC. For these reasons some of the constructions adopted historically to produce attractive light duty cords for normal indoor ambient conditions have persisted, against the competition from PVC. These consist of the use of textile yarn braids or a combination of rubber sheath/textile braid as the final covering. The degree of abrasion resistance afforded by this type of covering is dependent on the form and treatment of the yarn and the construction of the braid. Special performance test techniques have been developed, details of which are given in a later section.

In some circumstances, where considerable movement is involved, the degree of freedom of movement of the cores of a textile braided cord can be so high that it becomes disadvantageous in that the conductors can be bent round such a small radius that permanent uneven elongation of the conductors results and kinks are formed which lead rapidly to fatigue failure of the wires. The use of a rubber sheath controls this behaviour, but for domestic purposes the attractive appearance of the coloured textile braid would be lost if this was used alone. Consequently a long established combination of the two types of protection exists in a National Standard cord, the so-called UDF cord. In this design, a thin covering of rubber, much less than the normal rubber sheath thickness, is applied over the twisted cores with their interstice fillers. The textile braid is then applied so that it is partially embedded in the rubber layer. The combined coverings thus complement each other in controlling the natural bending radius of the cord whilst retaining the attractive appearance of the textile braid. A new Harmonised Document is currently under preparation which specifies flexible cords for extra flexible applications. Three types of cord are specified, including a

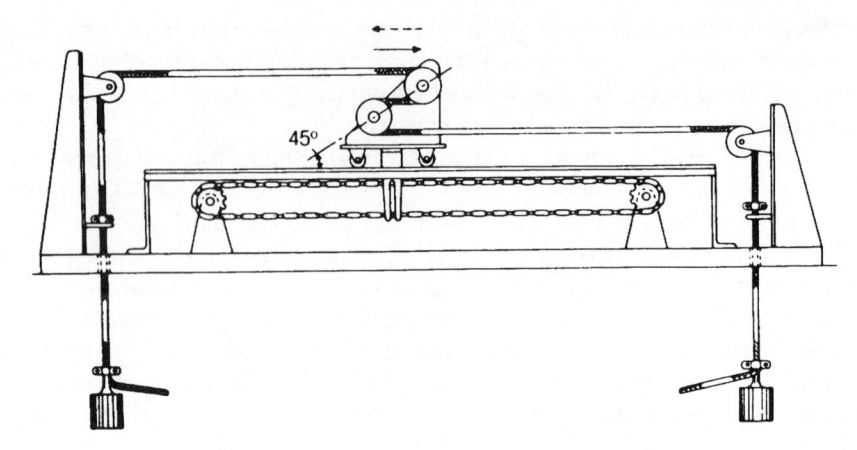

Fig. 12.1 Equipment for flexing test on flexible cords (Courtesy of British Standards Institution)

version utilising crosslinked PVC insulation and sheath. Another type has elastomeric insulation and XLPVC sheath. Common to all types however is the use of extra flexible (class 6) conductors and a tight core lay-length to ensure a long flexing life.

To determine further fitness for purpose, a modified flexing test has been included which, in addition to flexing the cord along its own axis (fig. 12.1), twists the cord to simulate the stresses encountered in use.

An additional test simulates a kink forming in the cord, and requires that circuit integrity is maintained whilst the cord is held under mechanical and electrical loading, as the kink is formed and unwound repetitively.

The upper end of the temperature range for which standard flexibles are available is covered by 300/500 V silicone rubber insulated and glass braided cables (up to 180°C), used mainly for fixed wiring in appliances, and 600/1000 V Intemp 250 PTFE/micaglass insulated and glass braided cables (up to 250°C), mainly for control and supply circuits in industrial environments.

A requirement of wiring regulations is that flexible cords or cables should be protected by a metallic screen or armour when they are exposed to risk of mechanical damage, or in situations where a particular hazard of electric shock exists, or in atmospheres where there is the risk of fire or explosion. Therefore versions of the standard PVC and rubber cords are available with either a tinned copper or galvanised steel wire braid. Cables for control purposes are available in up to 37 cores with or without such protection or electrical screening. Where a visual check of the condition of the protection is required, cables are provided with a transparent PVC oversheath.

Tests

Identical test methods for the abrasion resistance and flexibility of rubber insulated flexible cords are contained in the relevant British and International Standards.

Abrasion resistance
Figure 12.2 shows a diagrammatic representation of the test equipment. A sample of the cords is caused to rub against a similar sample of cord wound on a flanged mandrel. The

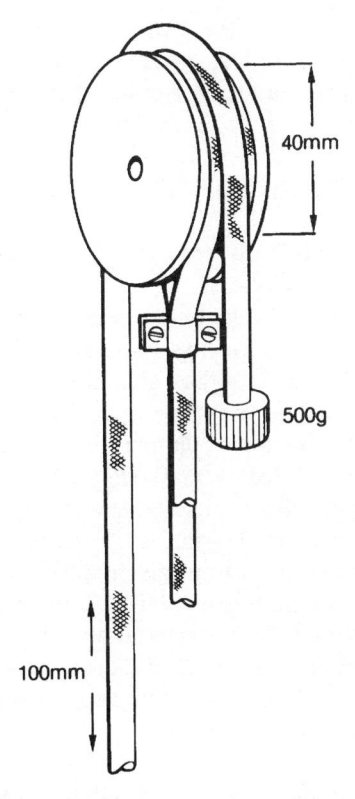

Fig. 12.2 Equipment for abrasion test on textile braided flexible cords

test sample is maintained under tension by a weight fixed to one end and the other end is attached to a mechanical device which imparts movement of a defined distance and rate of travel. After 20 000 single movements, the insulation of the cord must not be exposed by more than a specified amount and the cord must maintain its electrical integrity.

Flexing test
The apparatus used is shown in fig. 12.1. The test is carried out on a sample energised with an a.c. voltage and mechanical loading. It must withstand 60 000 strokes without loss of electrical integrity. The electrical and mechanical loading and the form of the pulleys vary according to the type and size of the cord.

Full details of both tests are given in BS 6500: 1994 which also includes tests for the flexing and impact behaviour of cords with tinsel conductors.

Technical data

General details of the size ranges, dimensions, weights, volt drop and current ratings are given in appendix A8.

Current ratings
It will be noted that the rated currents for flexible cords given in appendix A8 are lower than those given for the corresponding size and type of fixed wiring cable. Although

they are based on the same logic as the current ratings for other cables, they are influenced by subjective considerations, primarily human reaction to the surface temperature. As flexible cords are handled or touched by people, the concept of 'touch temperature' is of overriding importance. Empirically the current ratings for cords have been established in relation to PVC wiring cables protected by overcurrent devices which will not operate within 4 hours at 1.5 times the designed load current of the protected circuit, e.g. the rewirable fuse. From experience and measurement it has been established that, slightly modified, these result in cord surface temperatures, in air, that are acceptable to the majority of users.

Although the temperature designations of materials are retained in flexible cords, all cords have the same basic rating, for the reasons explained above. Temperature designation of the materials of a cord has significance in relation to rating when ambient temperatures higher than normal have to be taken into account. Consequently rating factors are given for ambient temperatures related to the temperature designations of the insulation and sheath materials.

In the case of general purpose flexible cables, the current ratings are derived in the same way as for comparable fixed cables and have no subjective element.

It should be noted that the current ratings given apply only to a single cord or cable in free air, not wound in coils or on a drum. The current ratings which apply in these circumstances are considerably less than those tabulated. The appropriate values vary with the number of coils and the design of the drum. This needs to be emphasised particularly in relation to the reeling drums supplied for use with extension cords for domestic purposes. In some cases the appropriate current rating may be less than half the current rating given in the tables.

The current rating for the parallel twin, PVC insulated, tinsel conductor cord is 0.2 A maximum.

CABLES FOR MINING AND QUARRYING

The difficult geological conditions and particularly the mechanically hostile environment at the coal face in UK mines probably represent the most demanding application for flexible cables in industry. Three main types of flexible cable are in use. One is an unarmoured cable for supplying power to coal cutting machines, often referred to as cutter cables or trailing cables. Another, with flexible steel wire armour or pliable wire armour (PWA), is for supplies to coal conveying and loading machines, remote control gear, coal face lighting and moveable transformers. Also under this heading come auxiliary multicore cables for control and monitoring of large mining machines. The third type is a relatively light unarmoured cable for hand-held shot hole drilling machines and similar equipment requiring a highly flexible yet robust supply cable. These systems are fed by the fixed cables described in chapter 11.

All these cables, or similar ones, find application in other types of mining or rough industrial situations worldwide. Versions of the PWA cables are used for supplies to machinery in open cast mines, quarries and other miscellaneous mines; also to cranes, dredgers, excavators, tunnelling machinery and supplies on civil engineering construction sites.

The relevant specifications are as follows.

BS 6708: 1991 Flexible cables for use at mines and quarries
BCS 188: Flexible trailing cables for use with coal cutters and for similar purposes
BCS 504: Flexible trailing cables with galvanised steel pliable armouring
BCS 505: Flexible trailing cables for use with drills
BCS 653: Flexible multicore screened auxiliary cables with galvanised steel pliable armouring

The use of cables in UK mines is governed by regulations issued by Her Majesty's Inspectorate of Mines.

Unarmoured cables for mining machinery

Historical
The first successful attempts to use electrical coal cutting machines took place in the period 1885–1890. Many devices were proposed to protect the supply cables and they all exhibited degrees of compromise between high flexibility and mechanical protection. They included single rubber insulated cores drawn into hose-pipes enclosed in a leather jacket formed by sewing the layers of a leather strip together, or wrapped in canvas and bound with rope. Eventually, multicore cables with cores twisted together and padded to a circular section with jute yarn and then jute braided were introduced. These were unsatisfactory in wet mines and around 1906 a non-hygroscopic sheath was developed. This consisted, initially, of vulcanised bitumen protected by bitumen impregnated tape, impregnated jute yarn and spirally applied tarred rope. In later versions the rope was replaced by a braid of whipcord, or in some cases leather strips or even metal chain armour. These forms of protection were superseded by the tough rubber sheath, sometimes reinforced by an embedded cord braid or layer of canvas. The introduction of PCP in the late 1930s brought a further step forward because it made practical a cable which was tough, flame retardant and water resistant. By the middle of this century PCP was the only sheathing material for flexible cables used in mines and quarries in the UK and most other industrialised countries. Heavy duty CSP has more recently been introduced as an alternative material.

Materials
There had been little challenge to natural rubber as the insulation for these cables until the designs were metricated in 1970. However, the introduction of butyl rubber, EPR and CSP to cable making indicated that the cores of flexible mining cables could be improved in respect of current rating and resistance to mechanical damage, notably the effects of crushing. CSP was only used for a short period for insulation because electrical problems were experienced with the earth protection system, owing to the high cable capacitance. EPR and composite EPR/CSP are now the only recognised insulants.

Construction
Because of the risk of accidental explosion and fire in coal mines, devices to minimise the contribution of flexible supply cables to such accidents are major design features. Equally important is the protection of operators against electric shock. Salient principles of electrical safety philosophy in UK mines are low impedance earthing of the neutral

point and protective switchgear based on earth fault current, excess current and current unbalance sensing. Consequently, almost all flexible cables have a separate earthing conductor, the only exception being type 11 (see below) in which the four core screens form the earth. Earthing conductors may or may not be insulated. All cables also have a metallic barrier or screen of defined conductance enclosing the cores. In the case of early cables this screen consisted initially of a braid formed of tinned copper wires, or in some cases a layer of seven-strand formations applied spirally. The advantage of the latter arrangement was that it enabled connections for earthing to be made more easily. The essence of these, as with all components, is maximum flexibility with maximum resistance to the mechanical stresses arising from constant movement, often in confined spaces over extremely rough ground. The spiral screen was found to be more prone to damage due to twisting than the braid and it was eventually eliminated.

Further development provided the composite braid screen in which the members are composed of tinned copper wires in one direction and of textile yarn in the other direction. This was originally cotton, but nylon fibre is now used because of its greater strength and stretch. In this context it will be noted that the collectively screened arrangement, i.e. all cores within a common braid, is not a recognised design because it does not provide protection against a phase to phase fault, a common result of crush damage. Additionally it was shown that the volume of combustible material within the screen was sufficient, under fault conditions, to disrupt the cable explosively and to provide a potential source of explosion or fire externally.

Some standard types of trailing cable are shown in fig. 12.3. These constructions comply with all the requirements described earlier for flexibility and provide more positive arrangements for maintaining geometric stability in view of the extreme conditions of bending, flexing and twisting to which they are submitted. The life of this type of cable is determined by its mechanical history and the amount of damage it sustains before it becomes irreparable.

All these types are designed for 3-phase operation and consist of three individually screened power cores, an earthing conductor and a pilot conductor.

The cores are assembled in a geometrically symmetrical and mechanically stable arrangement and the maximum pitch or lay length of conductor wires, cores and screen wires is closely controlled. The three power cores are placed with their screens in contact and the earth conductor is not insulated but is placed centrally in contact with the power core screens. In type 14, the earth conductor is insulated to the same diameter

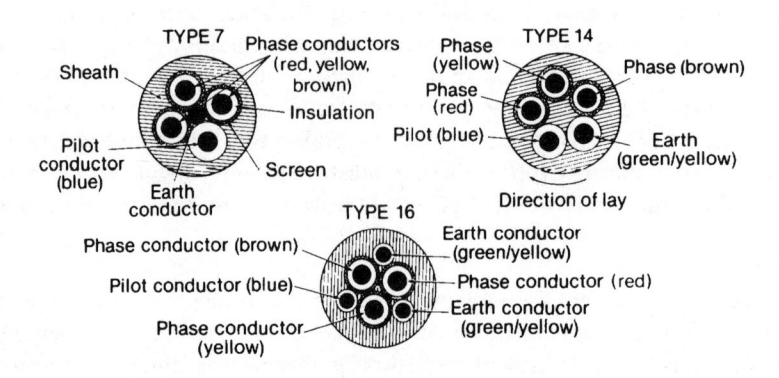

Fig. 12.3 Cross sections of coal cutter cables with metallic and conducting rubber screens

as the power cores thus giving the most mechanically correct arrangement but making the cable larger than types 7 and 16. The earth conductor in type 16 is divided into two parts, each insulated and of the same size as the pilot core. This provides the smallest cable of the three patterns.

Several variants of type 7 are readily available. Type 7M has a screened pilot core of the same conductor size as the power cores, which provides a more balanced construction for conditions in which the cable may be subject to an unusual degree of flexing or tension. In type 7S the pilot core is replaced by a unit made up of three cores, to give additional monitoring or control circuits to the machine being supplied by the cable. In recent years, the voltage ratings of these types have been increased by the introduction of a 1900/3300 V range of cable with the same design philosophy. The voltage rating has been increased to accommodate higher power requirements at the coal face. Type 11, which is made only in the 16 mm^2 size, has a screened pilot core but no separate earth conductor, and is designed to meet special service conditions in which a central conductor (as in the normal type 7) would be likely to break. In UK mines all coal face machines are remotely controlled and the pilot conductor is used to operate the remote control circuit. The core assembly is enclosed in a substantial sheath of black PCP compound complying with BS 7655: Part 2: 1993, 60°C heavy duty OFR sheath, type RS 2, with an additional tear resistance requirement (see BS 6708).

The core insulation complies with type FR 1, 85°C insulation (or GP 4, 90°C insulation) to BS 7655: Part 1: 1993 or is an MEPR compound with special mechanical properties as specified by British Coal. Core identification is by a layer of tape applied over the insulation. Red, yellow and brown colour coding is used for the power cores, and blue for the pilot core or pilot unit, the three pilot cores in type 7S being coded black, white and blue.

As will be evident from the designs discussed above, considerable development has occurred in improving the safety, robustness, flexibility, size and weight of trailing cables. The diversity of conditions has dictated the degree to which the various designs are adopted. However, with the increasing emphasis on mechanisation, the elimination of the more difficult seams, and common ownership, considerable rationalisation of cables has taken place.

Technical data
Details of the cable range together with dimensions, weights and current ratings are given in appendix A10, tables A10.1–A10.3.

Voltage and current rating
The rated voltages for cutter cables are 640/1100 V and 1900/3300 V. Current ratings are based on 25°C ambient temperature and a maximum conductor temperature of 75°C. Both continuous and intermittent ratings are provided, the latter based on the most severe operating cycle normally employed with coal cutting machines. The ratings are given in table A10.2, together with the cycle from which the intermittent ratings are calculated and the rating factors to be applied for higher ambient temperature.

As with other flexible cables the ratings do not apply if the cables are used on drums or in coils or where heat dissipation is otherwise reduced. Nor do they take account of volt drop. This can be calculated from the values of resistance and reactance in table 12.2. The values are for one power conductor.

Table 12.2 D.C. resistance and reactance of coal cutter cables

Nominal area of power conductor (mm²)	Nominal d.c. resistance (mΩ per 100 m at full load current)	Nominal reactance (mΩ per 100 m at 50 Hz)				
		Types 7, 7S and 7M	Type 10	Type 11	Type 14	Type 16
16	150	10.9		10.9		
25	96	10.7	12.2		10.9	9.9
35	69	10.1	11.8		10.4	9.4
50	47	9.8	11.5		10.0	9.0
70	34	9.5	11.1		9.8	8.7
95	25	9.4	10.3		9.7	8.5
120	19	9.2				

Pliable wire armoured (PWA) cables

PWA cables are designed for situations in mines and quarries where the risk of mechanical damage is high but where limited, relatively infrequent, movement is involved. For instance in UK mines they are used as flexible extensions of the fixed power cables in roadways on the primary side of transformers supplying power to face machines. They are also used to connect the secondaries of the transformers to remote control switchgear. These transformers and switchgear are moved periodically as the coal face advances. Except in auxiliary cables to BCS 653 (see below), the materials and constructional features are the same as for cutter cables. 2-, 3- and 4-core assemblies are provided as standard for voltages up to 3.8/6.6 kV with screened cores and at 320/550 V and 1.9/3.3 kV with unscreened cores. 320/550 V auxiliary cables have up to 24 ETFE insulated screened cores and are intended for use on large mining machines (external to the machine), to provide interconnection between machine sections or between machine sections and associated auxiliary equipment.

Some designs are available for supplies to dredgers, draglines and excavators used in alluvial and open cast mining. The cores are normally unscreened except for voltages higher than 6.6 kV.

A summary of the PWA cables available for mining purposes is as follows:

(a) 320/550 V cables with unscreened cores (rubber insulated) and screened cores (ETFE insulated);
(b) 640/1100 V cables with screened cores for remote controlled circuits, coal face lighting, conveyors and loaders;
(c) 1.9/3.3 kV cables with screened cores for roadway distribution, conveyors etc. and with unscreened cores for operating pumps etc.;
(d) 3.8/6.6 kV cables with screened cores for use primarily as roadway extension services.

BS 6708, BS 6116 and BCS 504 detail the standard designs recognised under UK Mining Regulations for use in mines and quarries. They are referred to by type numbers.

Types 62, 63, 64, 70, 71

These types are assumed to carry currents well below their thermal capacity and are standardised in one size only, normally 4 mm². The conductor size is decided primarily by mechanical considerations. With the exceptions of types 70 and 71, all have screened cores. Core insulation is MEPR/CSP to BS 7655: Part 1: 1993, type FR 1, or special MEPR to BCS 504.

Types 201, 211, 321, 331, 631
These are for 3-phase distribution and are standardised within a range of sizes from 10 to 120 mm^2. Type 201 has three screened power cores only, type 321 has three unscreened power cores and an unscreened earth core, and the remaining types have screened power cores plus an unscreened earth core. Core insulation for types 201 and 211 is MEPR/CSP to BS 7655: Part 1: 1993, type FR 1, type GP 4, or special MEPR. For types 321, 331 and 631 the insulation is of composite type FR 2 or type GP 5 to BS 7655: Part 1: 1993 or special MEPR as specified by British Coal.

Construction of above types to BS 6708 and BCS 504
In all types the cores are assembled (round a PCP or other thermoset centre when required) and covered with an inner sheath of PCP, type EM 2 to BS 7655: Part 2: 1993, which provides a bedding for the armour.

The PWA consists essentially of a helical layer of seven-wire strands of galvanised steel wires applied with short lay. In the UK there is a statutory requirement that the armour of mining cables shall have a conductance of not less than 50% of the conductor within the cable having the highest conductance. To meet this requirement it is necessary, in some cases, to include some copper wires in the armour.

The outer sheath of these cables is PCP, heavy duty OFR to BS 7655: Part 2: 1993, type RS 2, with the tear resistance increased to 7.5 N/mm.

Construction of cables to BCS 653
Types 506, 512 and 518 have 6, 12 and 18 cores respectively, with conductors of area 1.34 mm^2 (19/0.3 mm wires). Type 524 has 24 cores, with conductors of area 0.93 mm^2 (19/0.25 mm wires). The cores are insulated with ETFE and then individually screened with copper wire braids. After assembly of the cores, the cables are completed in the same way as the elastomeric types described above.

Types 730 and 830
These are non-armoured cables having individually metallic screened cores primarily for use as trailing cables or large machines at quarries.

Types 20, 21, 321, 621
These cables are intended for use in quarries, open cast mines and other miscellaneous mines in which the 'Regulations for quarries and metalliferous mines' apply.

The standard type references and voltage designations are as follows:

type 20: 600/1000 V, 3-core, 2.5–150 mm^2, single-phase
type 21: 600/1000 V, 4-core, 2.5–150 mm^2, 3-phase
type 321: 1.9/3.3 kV, 4-core, 16–150 mm^2, 3-phase
type 621: 3.8/6.6 kV, 4-core, 16–150 mm^2, 3-phase
type 730: 3.8/6.6 kV, 3-core, 35–150 mm^2, 3-phase
type 830: 6.35/11 kV, 3-core, 50–150 mm^2, 3-phase

All the cables include an earthing conductor for the apparatus fed by the cables, as required by the above Regulations. The power cores do not have a metallic screen for fault protection. Type 621 has a semiconducting, stress control tape over the insulation. The PWA does not have copper wires included. All other details are the same as for PWA cables to BS 6708.

Technical data
Details of dimensions, weights and current ratings are given in appendix A10, tables A10.4–A10.7, A10.9 and A10.10.

Current ratings
The ratings quoted in the appendices do not apply to cables on drums or in coils. Under these conditions the ratings depend on the provision for heat dissipation. Also where cables are used outdoors, in direct sunlight, particularly in tropical zones, a factor must be applied.

Voltage drop
Values of d.c. resistance and reactance for use in calculating voltage drop are given in table 12.3.

Unarmoured flexible cables – drill cables

Originally these cables were designed to supply hand-held shot hole drilling machines in mines. The growth of mechanical mining has resulted in a corresponding reduction in this application. However, they have come to be used in a number of other situations where a highly flexible yet robust supply cable is required.

Table 12.3 D.C. resistance and reactance of PWA cables to BCS 504

Conductor area (mm^2)	Nominal d.c. resistance (mΩ per 100 m at full load current)	Nominal reactance (mΩ per 100 m at 50 Hz)				
		Types 62, 63	Types 64, 70	Type 71		
4	570	12.3	13.0	13.4		
		Type 201	Type 211	Type 321	Type 331	Type 631
10	235	12.0	12.9			
16	150	10.9	11.7			
25	96	10.4	11.2		12.4	
35	69	9.9	10.6	12.0	11.8	
50	47	9.6	10.4	11.4	11.2	12.7
70	34	9.2	10.0	10.9	10.7	12.0
95	25	9.0	9.9	10.5	10.4	
120	20	8.8	9.5	10.1	10.1	

BS 6708: 1991 and BCS 505 formerly covered the designs and requirements for unscreened and screened cables, but on grounds of safety the unscreened design has been withdrawn by British Coal. The present cables are

type 43: 5-core, conducting rubber screen
type 44: 5-core, composite copper/nylon braid screen

Although shot hole drilling machines in UK mines operate at 125 V between phases, with the neutral point earthed so that the voltage to earth is limited to 72 V, the insulation thickness and protection on drill cables are such that they are rated at 600/1000 V and may be used at voltages up to this in installations other than coal mines and where the UK Mining Regulations do not apply. Generally, the construction and materials are the same as for the coal cutter cables but only one conductor size of 6 mm^2 is produced as standard.

To give better flexibility to the type 44 cable, a polyethylene terephthalate film tape is included over the braided core screen. It will be noted that, because of the extreme requirements for these cables, the geometry of the cross-section is symmetrical, with the conductors all in the same pitch circle. The shaped thermoset cradle centre ensures mechanical stability. As an aid to identification, a yellow longitudinal stripe is applied during manufacture to the sheath of the conducting rubber screened type 43 cable.

Technical data
The dimensions and weights of drill cables are given in appendix 10, table A10.8.

Current rating
The size of conductors for drill cables is selected primarily on mechanical considerations and when used in UK coal mines they operate at currents much less than their maximum thermal ratings. When used elsewhere the rating for 6 mm^2 types 43 and 44 is 46 A.

FLEXIBLE CABLES FOR OTHER INDUSTRIAL APPLICATIONS

Welding cables

Welding cables are used to connect the secondary, high current side, of a welding transformer to the welding electrode holder and for the earthing or return lead. As electrical welding involves low voltage, high current conditions, the electrical demands on coverings are not onerous. Prior to the 1960s, when compounds based on natural rubber were the most common coverings for flexible cables, it was considered necessary to use two-layer coverings. The first layer was insulation compound (VR) to provide the electrical performance and the second layer a heavy duty sheath (TRS) to withstand the very rough mechanical conditions in service.

Because of the high current involved, conductor cross-sections tend to be large. With hand-held electrodes, minimum restraint on the operator is essential and flexibility is of paramount importance. These cables are predominantly of single-core construction and conductors consist of multiple bundles of small diameter wires. A paper or polyethylene terephthalate separator tape is applied over the conductor to prevent the coverings keying into the conductor and hence provide controlled relative movement with improved flexibility and mechanical life.

The introduction of the high performance synthetic rubbers in the 1960s enabled improved single- and dual-covering welding cables to be developed. The most successful have been EPR and CSP, used individually or in combination, the choice being dependent on conditions in the particular circumstances. The salient advantage common to both materials is heat resistance. Whereas natural rubber covered cables are limited to a maximum operating temperature of 60°C, both EPR and CSP can be operated at 85°C, with significant improvements in current ratings, ageing and weather resistance. In addition, CSP has resistance to oil and chemical contamination and is flame retardant. Compounds of both materials have more than adequate electrical properties together with combinations of other properties to meet the specific requirements for many situations.

Conductors have traditionally been of copper, but in recent years aluminium (sometimes alloyed with small amounts of other metals to give greater strength) has also been used on a considerable scale, though its popularity has varied to some extent with the fluctuations in world metal prices. Aluminium cables are lighter than copper and offer significant savings in weight when high welding currents are required, and on projects where the distances and complexity involved in welding current distribution call for large amounts of cable. As pointed out in chapter 3, extra care is required when fitting terminations to aluminium conductors.

As aluminium is more difficult to draw reliably to the same small sizes of wire that are common with copper, larger wire sizes are used in the conductors. Also aluminium conductors have a lower fatigue resistance and breakage can occur more frequently, particularly in the cable near to the electrode holder. This can be overcome by connecting a short piece of the appropriate copper conductor cable between the electrode holder and the end of the aluminium cable.

The relevant British Standard is BS 638, 'Arc welding power sources, equipment and accessories', Part 4: 1979, 'Welding cables'; currently being revised to align in part with HD 22.6. For cables rated at 100 V a.c. and 150 V d.c. it includes single-layer coverings EM 5 or RS 5 type compounds to BS 7655: Part 2: 1993; with a two-layer option for type EM 5, and for cables rated at 100 V a.c. and for use up to 450 V a.c., an inner layer of type GP 4 compound to BS 7655: Part 1: 1993, with an outer layer of type EM 5 or type RS 3 compound. The alternatives of plain or tinned copper wires or aluminium wires with each covering are recognised. The coverings are identified by colour as follows:

single-layer RS 5 or EM 5: grey
two layer GP 4 and EM 5 or RS 3: black or orange

Details of dimensions and weights are given in appendix 9, table A9.1.

Voltage designation

The cables discussed here are recognised for welding duty at voltages to earth not exceeding 100 V d.c. or a.c.

This limitation does not apply to the voltage produced by a superimposed high frequency supply or other similar low power device used for starting or stabilising an arc. When required they may be used at voltages above 100 V but not exceeding 450 V. When used under the latter conditions adequate protection against damage to the coverings must be provided.

Current rating

The quoted ratings of welding cables, as given in appendix A9, table A9.3, are based on an ambient temperature of 25°C and a maximum conductor temperature of 85°C. They are assumed to be in free air with unrestricted natural ventilation.

For equivalent current rating, the aluminium cable has a larger conductor diameter and a correspondingly greater overall diameter, although it is appreciably lighter than the copper cable.

Early current ratings for welding cables were established by experience, based on hand welding practice where, in general, cables were energised for short periods and had long rest periods. The introduction of welding machines with appreciably longer periods of energisation necessitated a review of rating procedures. This was also needed because of the introduction of the higher performance materials. As a result, the concept of a 'duty cycle' was introduced which allows currents higher than the continuous rating to be used, dependent on the times for which the cable carries current in a given period. The period chosen is five minutes and the duty cycle is defined as

$$X = \frac{\text{time (minute) of current flow}}{5} \times 100\%$$

The continuous ratings are now determined by means of the methods dealt with in chapter 8 and the ratings for any particular situation are related to them through the use of the duty cycle. Current rating for representative duty cycles are given in appendix 9, table A9.3. For any other duty cycle the following formula can be used:

$$I_1 = I_c(100/X)^{1/2} \qquad (12.1)$$

where I_1 = rating for required duty cycle
I_c = continuous rating (100% duty cycle)
X = required duty cycle (%)

Resistance and voltage drop

Values for the range of both aluminium and copper conductor welding cables are given in appendix 9, table A9.2. When long lengths of cable are used between the welding set and the electrode, larger conductors than are indicated by thermal considerations must be used to avoid excessive voltage drop. These larger cables are more awkward to handle and may cause inconvenience to the operator. This can be avoided by reverting to the normal size of cable indicated by current rating at a point near to the electrode holder.

Flat flexible cables for mobile machines (cranes and hoists)

The electrical supplies to mobile industrial machinery are frequently provided by conventional circular-section cables as described previously. Where the path of the flexible cord or cable needs to be closely defined and controlled, or where space is limited, the geometry of the conventional cord or cable may be inadequate and a special solution would then be required. This can be achieved by restricting the flexing to one plane by arranging the cores side by side in one layer and enclosing them in a rectangular-section sheath. Thus the flexibility of the cable is improved in the required plane compared with a circular-section cable. Such cords and cables have a better space factor and are most appropriate where the supply cable is wound on a drum or reel as the machine is operated over a regular route. Examples are cranes and hoists where the cable is often suspended in festoons from a travelling hanger on a track.

Cores are insulated with EPR type GP 4 to BS 7655: Part 1: 1993 and are the same as those in the corresponding circular flexible cords and cables. The sheath is HOFR type RS 3 to BS 7655: Part 2: 1993. The maximum number of cores available as standard varies with conductor size. Cores are arranged parallel, in a single layer, and are divided into groups of two or three separated by webs of sheathing material to ensure transverse stability. Identification is by means of numbered cores. The range and dimensions of flat flexible cords and cables are given in appendix A8, table A8.21.

Current ratings
Because the rectangular section gives a greater area for heat loss, the ratings of flat-section cords and cables on a thermal basis are equal to or better than the corresponding circular versions. They are given for guidance in appendix A8, table A8.20. These values assume the cable to be in free air and straight, i.e. not coiled or reeled, and that a maximum of three cores are loaded. Requirements for individual installations vary widely and specific consideration has to be given to each application.

Coil leads

BS 6195: 1993, 'Insulated flexible cables and cords for coil leads', defines a coil lead as 'an insulated flexible conductor connected directly and permanently to a coil winding or other component of electrical apparatus and usually connected to some form of terminal'. This description covers a surprising range of cords and cables which have to cater for widely varied environmental conditions. They provide connections to the electromagnetic coils of motors, generators, transformers, relays, circuit breakers, actuators etc.

Although simple in construction, they include a wide range of materials; BS 6195 deals with a comprehensive selection but it is by no means exhaustive. Insulation materials range from general purpose PVC through a selection of rubbers to PETP, glass fibres, polyimide film, ETFE, FEP and PTFE. Increasing use is also being made of irradiation crosslinked polymers for the lower voltage categories. BS 6195 only deals with the more commonly used types and schedules them by operating temperature range and insulation material as shown in table 12.4.

Conductors
The flexible conductors are either simply bunched or multi-bunched, depending on size, and are formed of plain copper (with PVC) or tinned copper wires for operation up to 150°C or nickel plated copper above 150°C.

The standard size range is from 0.22 to 400 mm^2. Sizes above 2.5 mm^2 with thermoset insulation are usually supplied with a PETP film under the insulation to facilitate stripping.

Table 12.4 Types of coil leads to BS 6195

Type	Insulation	Continuous operating temperature (°C) Minimum	Maximum
1a	PVC, general purpose (type TI 2 to BS 7655: Part 3: 1993)	−20	+70
1b	PVC heat resisting (type 4 to BS 7655: Part 3: 1993)	−20	+85
3 and 4	Thermoset polymer and compound dependent on voltage category: CSP (type OR 1), EPR/CSP (type FR 1 and FR 2 to BS 7655: Part 1: 1993)	−30	+90
5	Silicone rubber compound (type EI 2) to BS 7655: Part 1: 1993)	−60	+150
7	PETP fabric tape/PETP braid, synthetic resin varnished	−50	+110
8a	Varnished glass fibre fabric/glass fibre braid, treated with synthetic resin varnish	−50	+130
8b	Varnished glass fibre fabric/glass fibre braid, treated with silicone resin varnish	−50	+180

Insulation
The choice of material is governed by the environmental conditions and the operational temperature. Coil leads are used in air, in inert gases and in oil, water or refrigerating liquids. Various conditions impose limitations, e.g. exposure to oil reduces the maximum operating temperature of thermosets and exposure in an enclosed static atmosphere limits the operating temperature of silicone rubber.

A common requirement in the manufacture of equipment is that the coil lead must withstand the temperatures and chemicals involved in potting, encapsulation or impregnation. The choice of insulation must take these into account by testing under actual conditions. BS 6195: 1993 recognises the situation but, as it is impractical to specify standardised conditions of test, the subject is left for collaboration between cable supplier and user.

Voltage designation
The classification (for which reference should be made to BS 6195: 1993) is unique to the UK and is based on the 1 minute test voltage applied to the equipment. It covers a range of working voltages from 300 V to 11 kV. The higher voltage categories (E and F), with EPR/CSP insulation, have a semiconducting graphite coated fabric tape over the conductor for stress control and the size range is limited for the same reason.

Current rating
As the conditions for heat dissipation vary so widely in different designs of equipment, it is difficult to quote specific ratings.

Standards in Europe and North America

The coil leads used in Europe and North America are generally similar to those in the UK and are controlled by National Standards and associated Approval Schemes. The latter have considerable importance as the key to obtaining import licences for electrical equipment is often the use of approved components. Typical of bodies operating such schemes are the Underwriters Laboratories Inc. (UL) in the USA, the Canadian Standards Association (CSA) and Verband Deutscher Electrotechniker (VDE) in Germany.

Because of the different way of expressing the conditions applicable to operating temperature, the comparison between UK and North American Standards may often be misleading. For example, in North America heat-resisting grade PVC is given a 105°C temperature designation. PVC can degrade if held for long periods at this temperature, and clearly such temperature designations are not intended to indicate continuous operational life, as is UK practice, where similar material would be rated at 90°C.

In addition to silicone rubber and varnished PETP or glass fibre constructions for operation at continuous temperatures above 100°C, use is made of EVA (110°C), ETFE (150°C), FEP and polyimide (200°C) and PTFE (260°C). Some of these materials are also used for lower temperatures because of their resistance to chemicals.

CHAPTER 13
Auxiliary Cables (Pilot and Telephone)

The term auxiliary refers to cables associated with power distribution and transmission systems used for control, protection, signalling and speech, and data transmission purposes. Such systems are mainly operated by the electricity transmission and distribution companies but similar applications occur in many industrial systems.

STANDARDS

The total range of cable types used throughout the world is vast. It may be illustrated, however, by the cables for public supply systems in the UK, which are covered by the Electricity Association Technical Specification (EA TS) 09–6, 'Auxiliary multicore and multipair cables'. This standard includes three types of thermoplastic insulated cable:

(a) PVC insulated multicore cables;
(b) polyethylene insulated multipair cables;
(c) PVC insulated, light current, multipair control cables.

It also defines the general operating conditions for the power systems with which the above cables are associated. Cables to EA TS 09–6 are required to withstand induced voltages caused by surges on the adjacent power system. This is achieved by varying the insulation thicknesses and/or the design of armour, according to the design level of surge. Three levels of disturbing conditions are catered for, defined by the induced voltage which would be anticipated between conductor and earth on an unprotected circuit: not exceeding 5 kV, 5–15 kV, and above 15 kV.

The screening factor of the armour is designed to reduce the induced voltage on the conductors to that appropriate to the thickness of insulation. Three standard armour designs are used to cater for the majority of power system installations. However, instances may occur where, due to exceptional lengths of parallel routes and/or fault current levels, special designs are necessary. All the cables meet the requirements of BS 4066: Part 1: 1980, 'Tests on electric cables under fire conditions', but in some cases cables with reduced flame propagation characteristics are required (chapter 6).

For industrial applications the standard cables described for fixed installations in chapter 11 are generally used, e.g. PVC insulated wire armoured cables to BS 6346: 1989 and, for ships and oil installations, elastomeric insulated cables to IEC 92-3.

PVC INSULATED MULTICORE CABLES

These cables are almost the same as the PVC insulated and sheathed SWA cables to BS 6346 described in chapter 11. The differences are that only one size, $2.5\,mm^2$, is used and, to provide better flexibility in terminal boxes, the conductor is of stranded form. The standard range of core numbers is more limited, namely 2, 3, 4, 7, 12, 19, 27 and 37 cores. Core identification is by means of black numbers on white insulation throughout.

In some cases the galvanised steel wire armour is coated with a waterproof compound, commonly based on bitumen, to inhibit penetration of water along the armour wire interstices. Corresponding to transmission cable practice, when it is necessary to check the integrity of the oversheath by d.c. tests, a graphite coating can be applied to the surface of the oversheath.

PVC multicore cables to EATS 09-6 are used to connect substations and power stations for the remote operation of, for example, tap changers and for protection circuits associated with transformers, switchgear etc. The cable capacitance is often of importance to the circuit design and the nominal equivalent star capacitance is 440 nF/km.

The voltage designation is 600/1000 V and the cores will withstand an induced voltage up to 5 kV.

POLYETHYLENE INSULATED MULTIPAIR CABLES

Intended largely to provide circuits for speech and data transmission, as well as for feeder protection, these cables are essentially telecommunication cables.

Construction

The cores consist of a single plain copper wire 0.8 mm in diameter insulated with polyethylene (type 03 of BS 6234: 1987). There are two thicknesses of insulation which are dependent on the anticipated induced voltage, 0.5 mm for 5 kV and 0.8 mm for higher voltages. The cores are twisted into pairs which are then laid up in combinations of 4, 7, 19, 37 and 61.

All cables, except the four-pair arrangement, have three of the pairs designated as being capable of operation as carrier frequency circuits. Pair and cabling lay lengths are chosen so as to minimise coincidence and hence crosstalk. Carrier pairs are placed in the innermost layer and are separated by audio pairs to minimise mutual coupling. The cores are identified by the use of coloured insulation with the pairs arranged in a specified scheme. The assembled pairs are contained by a plastic binder tape followed by an inner sheath of black polyethylene (type 03C of BS 6234: 1987) and then armour and a PVC outer sheath.

Armour

Three standards of armour are specified to provide differing screening factors corresponding to three levels of disturbing electromagnetic field intensity from adjacent power circuits. They are graded by induced voltage level as quoted previously.

For induced voltages up to 5 kV the armour consists of standard thickness galvanised steel wire SWA to BS 6346. For the range 5–15 kV the armour is similar but the wire is of heavier gauge. For voltage levels which would otherwise exceed 15 kV an improved screening factor is provided by the use of a single layer of aluminium wire armour. The wire is in condition H 68 to BS 2627: 1970.

Because of the susceptibility of aluminium to corrosion in the presence of moisture, coating of the aluminium wire armour with a waterproof compound is mandatory. When specified by the purchaser for SWA cables, and in all cases for aluminium wire armour, an increased thickness of outer sheath is required. This is to be graphite coated to enable d.c. electrical tests on sheath integrity to be carried out.

Filling between cores

To cater for wet situations where water ingress to the cable cores may be a hazard, with consequent disruption of transmission characteristics, the core interstices may be filled with an appropriate compound during laying-up. The compound is formulated to have low mobility in the operating temperature range of the cable and to be compatible with the insulation and sheath materials. Although also chosen to have minimum effect on the transmission characteristics, it does cause some modification relative to those of unfilled cables. Consequently different values of primary transmission characteristics are applicable in the test requirements. A water penetration test is also specified to check the effectiveness of the filling in preventing longitudinal transmission of water.

PVC INSULATED MULTIPAIR LIGHT CURRENT CONTROL CABLES

Although of similar construction to the polyethylene insulated cables in the previous section, these cables are not intended for telecommunication and therefore no transmission characteristic measurements are required. They are designed primarily for use where an independent two-wire circuit is needed for control, indication and alarm equipment associated with switchgear and similar power apparatus. For such circuits the working voltage does not normally exceed 150 V d.c. or 110 V a.c., with currents lower than the thermal rating of the conductor. The voltage designation is 100 V.

Most of the cable is installed indoors and it is not armoured. However, to cater for other situations an SWA version is provided which will withstand induced voltages up to 5 kV.

The standard conductor size is a single wire of plain annealed copper to BS 6360 (class 1), 0.8 mm in diameter.

Pair numbers are standardised as 2, 5, 10, 20, 40, 60, 100 and 200.

A hard grade general purpose PVC compound (type 2 of BS 6746) is used for the insulation and, in view of the electrical duty, a thickness of 0.3 mm is required, thus enabling cable dimensions to be achieved which are appropriate to the equipment with which the cables are used.

Two cores are twisted together to form pairs, except in the case of the two-pair cable which is laid up in quad formation. The twisted pairs are then laid up in layers to give the appropriate total number. Core and pair identification is by self-coloured insulation, two different colours identifying a pair in accordance with a specified colour scheme following conventional telephone cable practice. Plastic binder tapes are applied over the laid-up pairs.

The non-armoured cable is sheathed with a black general purpose PVC compound type TM 1 or type 6 in accordance with BS 6746. A rip-cord may be inserted longitudinally between the outer binder tape and the sheath to facilitate sheath removal.

In the case of the armoured cable, the inner sheath (bedding) is the same as the sheath on non-armoured cable. Over this is applied standard galvanised steel SWA, appropriate to the diameter of the cable, followed by a black PVC sheath (type TM 1 to BS 6746).

CHAPTER 14
Manufacture of General Wiring Cables

GENERAL FEATURES OF MANUFACTURING FACILITIES

The very wide range of applications and the consequent large variety of cable types, and the variety of materials used, create a complex situation for the provision of manufacturing facilities. Because of this the general wiring cable factory employs a larger number of processing techniques than other more specialised cable plants. The considerable size range, which includes cables from a few millimetres to a hundred or more millimetres in diameter, is a further complication because of the range of machine sizes which must be provided to carry out any particular operation.

Volume of production may also vary a good deal, so that manufacturing facilities must on the one hand be versatile and flexible enough to cater for short production runs and on the other hand be capable of dealing with large volume production. Often these requirements result in several types of plant existing in the same factory to carry out a particular process.

However, factories may vary considerably in their degree of specialisation. At one extreme is the purpose designed factory to achieve maximum economic advantage from rationalisation of cable types and materials, and optimum use of mechanical handling, with automatic control and virtually continuous output. At the other is the factory making cables for various applications, using many different materials and a wide range of plant, which enables it to react quickly to changes in pattern of demand. Commonly the manufacturing facilities within and between factories are grouped according to whether the cables produced have thermoset or thermoplastic coverings.

MANUFACTURING PROCESSES

As a consequence of the range and variety of general wiring cables, a large number of processes are used for cable manufacture. These can be grouped as:

(a) conductor forming, i.e. wire drawing and annealing, wire coating (tinning and plating), bunching and stranding;
(b) insulating and sheathing, which consist mainly of the various techniques of extrusion appropriate to the characteristics of the material used, but also include

tape and yarn wrapping and braiding; with thermoset materials there is also, essentially, a vulcanising, i.e. crosslinking, operation;
(c) assembly, including laying-up of cores, taping, braiding and armouring.

Many of the cable manufacturing processes have common principles and are described in chapter 25. Emphasis is therefore placed here on those especially associated with general wiring cables.

Conductor forming

Conductors for general wiring cables are characterised by the high proportion of small diameter wires used. In wire drawing there is a predominance of multi-wire machines using natural and synthetic diamond dies for the smaller sizes. Copper rod is drawn to wire in stages, the wire being annealed continuously.

Tin coating of copper wire is usually carried out in association with wire drawing, the tin being applied by electroplating. In modern high speed production units, which combine drawing and annealing in-line, on an almost continuous basis, the tin may be applied on the input wire. In the case of nickel-coated conductors, used in cables for high temperatures, the nickel is usually electroplated on the final wire size, although it is sometimes applied by cladding in which a composite copper–nickel billet is drawn into wire.

The introduction of multihead wire drawing machinery has taken the conductor forming process a step further. Nineteen- and 24-wire multiwire machines are now available. The purity of the copper rod, in particular the absence of oxide impurities, has to be very high to achieve good results with this configuration. Twelve- and 16-wire machines are more common, the wires being split (2×6, 2×8) on to separate bobbins or directly into high speed double twist machines making unilay strands (bunches).

The conductor assembly operations of bunching and stranding have the common principle of twisting, as used in rope making. The methods used are decided by the small size and large numbers of wires which make up flexible conductors. For the bunching and stranding operations, which are used to produce the majority of conductors for flexible cords and cables, four principal types of machine are used, the elements of which are as follows.

(a) The reels containing the single wires are held in fixed frames or 'creels', the wire being drawn through a fixed guide and a forming die and then passing via a rotating arm or 'bow' to the take-up reel, which rotates about its own axis. The action imparts two twists to the wire for each revolution of the bow. This is the high speed double twist machine. A modified version of this machine is used to manufacture the high quality formation-stranded conductors required for cores with low radial thicknesses of insulation.
(b) The reels are held in fixed frames or 'creels', the wire being drawn through a fixed guide and a forming die by a take-up reel which rotates about its own axis and also about the principal axis of the machine. This machine is the drum twisting buncher.
(c) The wire reels are carried on a rotating cradle, as in a stranding machine (chapter 25). However, no attempt is made to lay the wires in a definite geometric pattern as in stranding, the wires being gathered in a bunch in the forming die.

(d) The reels containing the single wires are supported within a rotating cage. The reels remain stationary in space, the necessary twisting action being imparted by the cage, to which the wires are attached by guides. This is the tubular stranding machine.

There are variations of these machines, imparting differing degrees of twist to the individual wires. The types of machine in (a) and (d) give the higher production speeds but are usually limited to the small and medium conductor sizes. The larger flexible conductors can be made on the machines described in (b) and (c), or by further bunching processes using any of the bunching methods (multi-bunch conductors). Flexible conductors of the 'rope stranded' type (chapter 4) are made on stranding machines.

The above machines differ in scale, depending on the size and length of conductor to be produced. The considerable lengths of conductor necessary for high speed continuous covering operations have resulted in the development of improved process handling techniques. These include the 'spinning off' of wire from stationary reels, where wire is thrown over the flange of a reel by a rotating guide, the reel being placed with its axis parallel to the direction of wire travel. Alternatively the reel may be placed flange down and the wire 'spun off' vertically.

Insulating and sheathing

The fundamental components of an extrusion line for rubber and thermoplastic cable coverings are indicated in fig. 14.1 (see also chapter 25). Over the years, there has been some rationalisation of extruder design. Whereas early machines for use with thermoplastics were distinctly different from those for rubber, those now in use are similar for both types of material, with only minor differences in design and operation. The introduction of high temperature vulcanising techniques and high performance rubbers has contributed to this trend.

For small-scale jobbing work, the smaller and more versatile extruders tend to be used, though larger machines may be required for sheathing. Large extruders with high output are used for insulation on high speed lines for large volume production. Twin or triple extruders with a combined head are now relatively common and are used for high voltage cores requiring the simultaneous application of screens with the insulation, for multilayer insulation, and for applying insulation and sheath on single-core cables such as welding cables.

Extrusion lines, particularly the high speed lines for thermoplastics such as PVC and PE, now have a high degree of instrumentation and automatic control, in respect of both the performance of the line units and the physical characteristics of the product. For example, eccentricity and diameter of insulation can be monitored continuously.

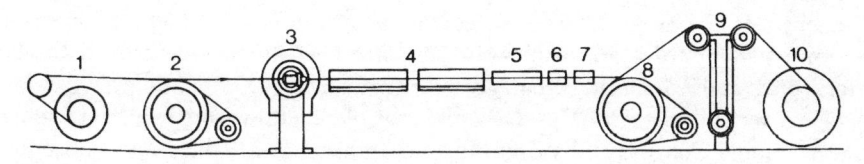

Fig. 14.1 Typical layout of cable extrusion line with (1) input drum, (2) capstan wheel, (3) extruder, (4) cooling trough, (5) spark tester, (6) diameter gauge, (7) eccentricity gauge, (8) capstan wheel, (9) tension controller and (10) output drum

Electrical testing of the insulation, by application of high voltage, is also carried out in-line. Tension control is usually effected automatically, and on bulk production lines there is mechanical handling for input materials and processed wire and cable, together with provision for continuous running by changing input and output drums whilst the line is operating normally. Tandemisation of the extrusion process with other processes is a common feature of the manufacture of wiring cables which enables a more cost effective approach for bulk production. In certain cases it has been possible to combine wire drawing, insulation extrusion, laying-up and overshathing in one continuous operation.

In principle, high temperature polymers are processed in the same way as described above. Silicone rubber is extruded similarly to other rubber compounds, although it can be applied in liquid form.[1]

Fluoropolymers such as FEP and ETFE are thermoplastic and (like polyethylene) have a well defined melting point. They are therefore melt extruded[2] similarly to PE. Because of their relatively high melting temperatures, special heat-resistant alloys are needed for some components of the extruder. PTFE is an exceptional case as it is not truly thermoplastic, owing to its very high molecular weight. Screw extrusion cannot be used because it results in shearing of the polymer particles. Instead, the powdered polymer is blended with a suitable liquid lubricant, formed into a billet, and then applied to the wire by a ram extruder. The lubricant is subsequently evaporated off in an oven, and the temperature of the cable is then raised to 400°C, causing the polymer particles to coalesce. The process is analogous to the forming of powdered metals and ceramics by sintering. The sintering oven is usually placed in line with the extruder.

Curing (crosslinking)

In the chemical process of crosslinking, or vulcanisation as it has usually been termed with rubber, the modification of the polymer matrix in thermoset materials is initiated by heating the covering after extrusion. It may be at atmospheric or high pressure. Originally the treatment was a batch process, with the uncured cores or cables supported in trays containing powdered chalk or talc, or contained in tightly wrapped textile tapes or within a lead sheath and wound on a steel cylinder. The trays or cylinders were heated by pressurised steam in an autoclave. Subsequently the tapes or lead sheath were removed, if not required as a final component of the cable. This practice still survives for short length production of cables with certain rubber polymers, such as PCP and CSP.

A commonly used high temperature method is that of continuous vulcanisation (CV) carried out by means of high pressure steam (14–25 bar) contained in a tube into which the extruded covering passes directly. Introduced in the 1940s in a horizontal arrangement (HCV), as an alternative to the steam autoclave method for the bulk production of small cores and cables, it has since been further developed for larger diameter cores and cables in a category arrangement (CCV), and for the largest products in a vertical form (VCV).[3] The coverings of general wiring cables are generally smaller in volume than, say, distribution and transmission cables and consequently higher extrusion speeds are possible, particularly when insulating.

As crosslinking by chemical means is time and temperature dependent, greater throughput from a CV line involves the use of higher temperatures or a longer steam pressure tube. Steam as a heating medium has a disadvantage in this context, as large increases in steam pressure are required to gain only modest increases in temperature.

Alternative heat transfer fluids have therefore come into use. Molten salt baths (LCM) at atmospheric pressure are now widely used for continuous curing of sheaths. Besides providing a convenient means of high temperature heating, they overcome other disadvantages of steam CV. The LCM system utilises a relatively deep bath of salt into which the cable is dipped, thus eliminating the restrictive CV tube which can cause damage when the cable makes contact with it. Also the cable is accessible as soon as it leaves the extruder, so that there is a considerable reduction in scrap at the beginning and end of each cable run compared with long CV lines.

However, because of operation at atmospheric pressure, porosity of the covering is a problem. Porosity is due to volatile substances, principally moisture, in the compound and is suppressed in the high pressure environment in a steam CV tube. By special formulation this porosity can be almost entirely eliminated in sheath compounds, and the salt bath technique has been found most useful for sheathing, especially for cables which could not be satisfactorily processed by steam CV, e.g. large multicore signalling cables. Insulation compounds cannot be processed in this way, however, as the special formulations render the electrical characteristics unsatisfactory.

More recently, the problems of porosity in the crosslinked material have been tackled by combining some of the features of LCM and high pressure steam CV. This approach uses what is called a 'pressurised liquid salt continuous vulcanisation system' (PLCV).[4,5] In this system molten salt is circulated through a vulcanising tube which is similar in arrangement to the steam HCV tube but much shorter. Because of the density of the molten salt, many cables are relatively buoyant and contact with the CV tube can be eliminated. This reduces the problems with the control of tension in the cable, which occur in all steam CV lines, so that a wider range of cables can be processed. The vulcanising tube is pressurised using air or an inert gas such as nitrogen, depending on whether oxidation of the compound is a hazard. The gas pressure system is independently controlled and is varied to suit the product. The pressure may be up to about 7 bar.

Other 'dry' vulcanising systems (see also chapter 25) have been used, e.g. nitrogen alone as both heat transfer and pressuring medium, with the addition of infrared heating.[6] Many polymers can be crosslinked without the use of chemicals and heat, by means of electron beam irradiation.[7] So far, because of cost, this method has found application mainly for specialised cables for the communication, electronic and aircraft industries.

Silicone rubber coverings are commonly cured in hot air continuous vulcanising tubes. To develop maximum performance they may be subject to a second (or post) cure in a hot air oven as a batch operation, to eliminate the volatile products of the crosslinking reaction.

The introduction of the Sioplas and Monosil processes (described in chapter 3) has seen a more general return to batch curing. As pointed out in chapter 24, the silane crosslinking technique allows greater flexibility and much improved running speeds at the extruder, with a useful reduction in scrap levels (compared with long steam pressure lines). Used initially for curing relatively low-filled compounds, its first major impact in the UK was on the production of XLPE. With improved control of moisture content in filling materials (vital in order to avoid premature crosslinking of compound in the extruder), the range of polymers which can be processed by this technique has increased considerably.

Assembly

The cable assembly operations are similar in principle to conductor stranding and bunching. The laying-up of cores and armouring of general wiring cables are carried out on similar machines to those described above and in chapter 25 for conductor forming.

Although there is a predominance of small diameter cores containing flexible conductors, there must be a sufficient range of machines to handle cores from 2–3 mm in diameter up to sizes overlapping those of distribution cables. The laying-up of cores for large volume products such as 2- and 3-core flexible cords is carried out on high speed machines similar to those used for bunching small conductors. However, cables with a large number of small cores are common, and the laying-up of these cores is very similar to armouring or large conductor stranding, where provision must be made to apply a large number of cores or wires in one pass. These machines usually have more than one cradle containing the cores or wires, the cradles being arranged in tandem. The laying-up machines are usually equipped with taping heads, so that binder tapes can be applied, as necessary, over each layer of cores.

The braiding process may be used for applying materials ranging from yarns of cotton, viscose, rayon, nylon, PETP and glass, through plain, tinned or nickel-plated wires to phosphor bronze and steel wire. Depending on the cable sizes the braiding machines may be set to operate in either a vertical or a horizontal arrangement.

QUALITY ASSURANCE

Over the last two decades, formal quality assurance systems have become an integral part of general wiring cable manufacture. Final test and inspection functions have been combined with a specialised in-process inspection organisation, as previously operated for cables for defence applications, and expanded into a comprehensive system involving stage monitoring, statistical data analysis and auditing. The approval scheme operated by BASEC, which is described in chapter 7, is in most cases interlinked with the factory quality assurance organisation, and provides the link with the CENELEC certification scheme.

In modern factories quality assurance has been extended into all aspects of production, from the examination of incoming materials and the servicing of supplier approval schemes to final inspection and maintenance of product approvals. The overall scheme is recorded and defined in a quality manual which forms the basis of the manufacturer's relationship with his customers. More information is given in chapter 7.

REFERENCES

(1) Fresleigh, R. M. and Kehrer, G. P. (1980) 'Fabricating wire and cable with liquid silicone'. *Rubber World* **181**(5), 33–35.
(2) Edwards, I. C. (1978) 'Fluorinated copolymers – properties and applications in relation to PTFE'. *Plast. Rubber Inst.* **3**(3), 59–63.
(3) Blow, C. M. (1975) *Rubber Technology and Manufacture*. London: Newnes-Butterworths.

(4) Smart, G. (1978) 'PLCV – a progress report'. *48th Annual Convention of the Wire Association*.
(5) Smart, G. I. (Oct. 1977, May 1978) 'Continuous vulcanising systems using liquid salts under pressure'. *47th Annual Convention of the Wire Association*, Boston, USA; also in *Wire J*.
(6) Sequond, D. C. and Kailk, D. (1979) 'Power cable vulcanisation without steam'. *Elastomerics* **111**(3), 32–37; *Elastomers Plast.* **11**, 97–109.
(7) Brandt, E. S. and Berijka, A. J. (1978) 'Electron beam crosslinking of wire and cable insulations'. *Rubber World* **179**(2), 49–51.

CHAPTER 15
Installation of General Wiring Cables

In most electrical systems a major consideration is the installation of the wiring. Well defined practices and conventions in relation to fundamental principles have formed the framework of national codes of practice, with adaptations to suit particular applications. As there are usually a number of engineering interpretations of any particular set of circumstances, it is perhaps not surprising that philosophies vary on some aspects of wiring installation in different countries. In particular, attitudes in other European countries on earthing, and consequently earth conductors and cables containing them, differ appreciably from those in the UK. Protection of circuits is another area where significant differences occur. For instance, the use of circuit breakers in domestic premises has been more widely adopted in France and Germany than in the UK. The use of domestic ring final subcircuits and their associated fused plugs is confined to the UK and certain Commonwealth countries.

NATIONAL AND INTERNATIONAL WIRING REGULATIONS

The legal status of wiring regulations has also differed worldwide. as has the attitude to inspection and approval of both installation and equipment, not least regarding cables. Many of the less industrially developed countries in the world had no wiring regulations or codes of practice but are now seeking to establish them. In the light of this and a greater interest in wider international markets by the electrical engineering industries, a realisation developed that common internationally accepted requirements were necessary. These requirements are defined in general terms and can be interpreted at a national level, the aim being to establish common levels of safety and to eliminate barriers to international trade.

IEC requirements

This realisation resulted in the setting up by the IEC of a committee to deal with recommendations for electrical installations in buildings. The work of this committee has produced an outline of requirements for electrical installations within which most of the detailed considerations are at a sufficiently advanced stage to influence national wiring regulations. It is contained in IEC 364, 'Electrical installations of buildings', which has the following framework:

Part 1: scope, object, fundamental principles
Part 2: definitions
Part 3: assessment of general characteristics
Part 4: protection for safety
Part 5: selection and erection of equipment
Part 6: inspection and testing
Part 7: requirements for specific installations

Chapter 52 of part 5, 'Cables, conductors and wiring materials' covers:

(a) methods of installation;
(b) general rules;
(c) current carrying capacity;
(d) cross-sectional area of neutral conductor;
(e) voltage drop;
(f) terminations and joints;
(g) mechanical and external stresses;
(h) corrosion;
(i) electromechanical stresses;
(j) fire barriers.

Thus the factors to be considered in designing and carrying out an electrical wiring installation are defined. The treatment of each factor is essentially in general terms and the requirements for particular installations are expressed in terms of 'external influences'. A comprehensive framework of installation practice results, within which it is possible to consider the factors relating to wiring system cables. The level of engineering competence is taken to be that of a professional engineer responsible for the design of electrical installations. External influences include, for example, environmental conditions, types of installation and persons using the installation. It is therefore clear that the IEC requirements cannot be used directly as simple rules. In the UK BS 7671 (sixteenth edition of the IEE Wiring Regulations) interprets these IEC requirements. In the case of developing countries it is anticipated that it will be some time before particular requirements for installations emerge and part 7 of IEC 364 caters for installations in special situations, such as bath/shower rooms, swimming pools, saunas, construction sites and agricultural and horticultural premises.

UK wiring regulations

Through membership of IEC, and in Europe through membership of the EU and consequently CENELEC, there is a degree of commitment for participating countries to embody the IEC requirements into their national wiring regulations. Indeed there is an obligation for CENELEC members to follow CENELEC Harmonisation Documents, which are based on IEC Standards. Consequently BS 7671 (sixteenth edition of the IEE Wiring Regulations) adopts the IEC format and concept of expression.

In the case of cables in particular, reference to the standards has been used to make the intent clear in specific cases and to avoid conflict with the fourteenth edition and cable standards. All designated cables are required to comply with the appropriate British Standard.

The format of BS 7671 is the same as that of IEC 364. The section dealing directly with cables is chapter 52. However, chapter 51 is relevant for core identification (clause 514 Identification and notices). Chapter 52 is subdivided into the following.

521: selection of type of wiring system
522: selection and erection in relation to exernal influences
523: current-carrying capacity of conductors
524: cross-sectional area of conductors
525: voltage drop in consumers' installations
526: electrical connections
527: selection and erection to minimise the spread of fire
528: proximity to other services
529: selection and erection to maintainability, including cleaning.

Whereas previous editions of the IEE Wiring Regulations were aimed particularly towards domestic and commercial premises, both IEC 364 and BS 7671 apply to any installation. There are, of course, parallel regulations for specific applications as, for example, IEC 92, 'Electrical installations in ships' together with the regulations for the various Classification (insurance) Authorities referred to in chapter 11.

In the case of UK installations it may also be necessary to take into consideration the regulations and the legal requirements relating to particular classes of electrical installations, as listed in appendix 2 of BS 7671. These cover such areas as factory installations and those on construction sites, cinematograph installations, installations in coal mines, quarries and metalliferous mines, and agricultural and horticultural installations.

The framework of IEC 364 and BS 7671 form a convenient and thorough list of the factors influencing the installation of cables. While no attempt is made to discuss this information in detail it is appropriate to make comments as below.

SELECTION OF TYPES OF CABLES AND CORDS

Appendix 3 of the sixteenth edition (BS 7671) of the IEE On-site Guide contains a comprehensive list of the types of non-flexible and flexible cables and cords, with details of their intended use and any additional precautions where necessary. It considers the hazards of mechanical damage and corrosion, and points out that other limitations may be imposed by the relevant regulations, in particular those concerning maximum permissible operating temperature. Thus attention is drawn to an important point that the factors referred to above cannot, in practice, be taken in isolation. Appendix 3 does not claim to be exhaustive.

The IEE On-site Guide does not apply to flexible cords used as part of a portable appliance or luminaire. In connection with such applications the following publications contain further information and guidance:

Installation of General Wiring Cables

(a) BS 3456: Safety of household and similar electrical appliances;
(b) BS 4533: Electric luminaires (lighting fittings);
(c) For cable to BS 6500, BS 6007, BS 7211, BS 634, BS 5467 and BS 6724, further guidance may be obtained from those standards. Additional advice is given in BS 7540, Guide to use of cables with a rated voltage not exceeding 450/750 V for cables to BS 6004, BS 6007 and BS 7211. Flexible cords and cables to BS 6500 and BS 6141 are also covered in BS 7540;
(d) The Electrical Equipment (Safety) Regulations 1975;
(e) The Electrical Equipment (Safety) Regulations 1976 and the associated booklet;
(f) 'Administrative Guidance on the Electrical Equipment (Safety) Regulations 1975 and the Electrical Equipment (Safety) (Amendment) Regulations 1976' (available from HM Stationery Office).

Armour

It is indicated in the regulations that flexible cables and cords may incorporate a flexible armour of galvanised steel or phosphor bronze, or a screen of tinned copper wire braid. Such armour becomes necessary in a number of situations to comply with the requirements for protection against risk of mechanical damage. Of the British Standards listed in BS 7671, BS 6500, 'Insulated flexible cords', expressly makes provision for the use of a screen of braided tinned copper wire. Cables to BS 6007 can be supplied with suitable armour and reference should be made to the cable manufacturer for details. Cables to BS 6708, 'Specification for flexible cable for use at mines and quarries', referred to in chapter 12, also find applications in these circumstances. Flexible armour includes braids of galvanised steel wire or phosphor bronze wire as applied to cables to BS 6883, 'Elastomer insulated cables for fixed wiring in ships and on mobile and fixed offshore units'. The use of phosphor bronze wire, either as PWA or braid, or the use of aluminium wire armouring, enables mechanical protection to be provided for single-core cables used in a.c. circuits. The use of such protection, with appropriate glands, becomes essential in flame-proof installations in environments where there would be a particular hazard of explosion.

Operating temperature

The tables of current ratings in the appendices include the limiting temperatures on which the current carrying capacities of the various types of cable are based. These are the maximum conductor operating temperatures and it should be appreciated that sheath temperatures can reach values only 10–15°C lower. Where cables reach these temperatures it is necessary to consider the effect of such temperatures on materials which might come into contact with them, such as non-metallic conduit, trunking and ducting, and decorative finishes. The surface temperature of cables is also important where there is the possibility of contact with people or livestock. With humans the threshold of pain occurs at a temperature of about 70°C but temperatures near this may cause some discomfort and involuntary reaction. This situation has been taken into account in determining the rating for flexible cords but special attention needs to be given to flexible cables insulated with silicone rubber or PTFE.

A common mistake with flexible cords and wiring system cables is to apply tabulated current ratings to extension cords which are coiled on reeling drums used in association with portable appliances or to cables which are retained on their despatch reels and used for temporary circuits. In the case of PVC cables, melting of the insulation or even the whole of the covering may occur in such circumstances.

Where cables are installed in roof spaces, or in cavities between floors and ceilings, they should be installed in such a way that they are not covered by thermal insulation material. A cable so covered can have its current carrying capacity reduced by up to 50%. This is particularly important in single-storey buildings where a major part of the wiring installation can be in the roof space. With certain forms of thermal insulation material, precautions are necessary to prevent interaction with PVC sheaths; otherwise loss of plasticiser, affecting the physical properties of the sheaths, can result.

Another important thermal effect may arise when cables with a lower limiting temperature are installed alongside fully loaded power circuits using cables with a higher limiting temperature. A particular example of this can occur in distribution systems where the thermoplastic telecommunication or control cables are installed adjacent to power circuits. In this and similar cases the lower temperature cables must be installed at such a distance that their limiting temperature is not exceeded.

In wiring regulations, heating cables are generally treated as energy-using equipment, and the requirements for cables do not normally apply. National requirements for heating cable installations vary considerably in different countries and for different applications. Some general comments are made in chapter 11 on the installation of floor heating systems and advice is available from the cablemaker. British Standard CP 1018: 1971 (1993), 'Electric floor-warming systems for use with off-peak and similar supplies of electricity', provides general guidance relating to the UK.

ENVIRONMENTAL CONDITIONS

Appendix 5 of BS 7671 (sixteenth edition of the IEE Wiring Regulations) gives an explanation of a code for classification of external influences and lists three main categories:

A – Environment
B – Utilisation
C – Construction of buildings

At present the IEC work is insufficiently complete to be used, but it is intended that particular installations will be described in terms of the relevant external influences, which will then determine the applicable requirements. BS 7671 goes some way towards dealing with the types of environmental conditions identified in the IEC work as indicated below.

Temperature

The influence of ambient temperature on the current rating of cables is dealt with in chapter 8 and typical correction factors for different ambient temperatures are given in the relevant appendix for the particular type of cable.

There is a warning in appendix 4 of BS 7671 dealing with cable current ratings, which refers to the effect of exceeding the limiting temperatures used in calculating full thermal ratings. The need for this may readily be appreciated from the discussion in chapter 3 on the thermal degradation of insulating materials and effects on service life.

A common example of these effects occurs in lampholders where it is sometimes not realised that very high lamp cap temperatures can arise with filament lamps. As a result of conduction and convection of heat, the temperature of the cord in and adjacent to the lampholder can exceed what is tolerable for most general purpose insulants. This can result in a much reduced service life and often a short-circuit fault. The recommendation to use an appropriate heat-resisting flexible cord for pendant and batten lampholders and with tungsten filament lamps cannot be too strongly emphasised.

Although less emphasis is placed in wiring regulations on the lower end of the ambient temperature range, low temperatures can be important both during and after installation. Most insulating and sheathing materials stiffen to some extent as their temperature is progressively reduced. PVC shows this effect more markedly than do the rubbers or polyethylene. Once installed, PVC is quite satisfactory at temperatures down to $-30°C$ if protected from violent impact. However, the general duty compounds will shatter if bent violently, or particularly if struck hard at temperatures below freezing point. PVC compounds specially formulated for low temperatures show a marked improvement in this situation. Overvigorous stripping of the sheath of a flat wiring system cable after it has been stored overnight in frosty conditions has been known to shatter the core insulation. PVC insulated and/or sheathed cables which have been stored at temperatures below freezing point should be allowed to stand for at least 24 hours at a temperature above $10°C$ before they are handled for installation. BICC LSF insulation and sheathing compounds are more flexible than PVC at low temperatures but, to avoid the risk of damage during handling, cable should be installed only when both the cable temperature and the ambient temperature have been above $-10°C$ for the previous 24 hours, or where special precautions have been taken to maintain the cable above this temperature. Most rubber cables also remain flexible at quite low temperatures. For example, EPR and silicone rubber insulated cables can withstand reasonable bending down to $-40°C$ and $-50°C$ respectively. PCP stiffens as the temperature is reduced below $0°C$ but will not shatter from impact above $-35°C$. It is advisable to allow PCP sheathed cables that have been subjected to low temperatures time to warm up before they are handled. CSP and CPE behave in a similar manner to PCP and the same precautions should be taken.

BS 7540, 'Use of cables with a rated voltage not exceeding 450/750 V', gives minimum installation and handling temperatures for cables and flexible cords to BS 6004, BS 6007, BS 6141, BS 6500, BS 6726, and BS 7211.

Presence of water

The IEC code covers eight classes of presence of water, from 'negligible' to 'submersion'. All are of concern to some extent in connection with the use of cables, because even the 'negligible' category does not preclude the occasional presence of water vapour. It may not be generally realised that all organic materials are permeable to water to some extent, and as a result cables which are subjected to prolonged immersion in water will eventually reach the stage where water penetrates through the sheath and insulation to the conductor unless special precautions are taken, such as the incorporation of a metal sheath or other form of water barrier. Whether deterioration of performance occurs in the case of unprotected cables at low voltage depends on the type of insulation and the degree of contamination of the water. Polyethylene is less susceptible than PVC and thermoset materials.

Some circumstances amounting to continuous submersion are not always obvious. The surroundings of swimming pools, both indoors and out, the substrata of roads and paths, paved or concreted outdoor areas and the floors of domestic or industrialised buildings where large amounts of water are regularly used so that they can be almost permanently waterlogged, are all examples of locations where the failure of inappropriate cables has occurred. The need to limit the use of glass braid insulated cables to only the 'negligible' category of location should be self-evident.

Corrosive or polluting substances

The risk of corrosion to cable components incorporating aluminium is specifically dealt with in the regulations. As far as cables are concerned, the most likely cause of contact with active substances arises from accidental spillage or the treatment of surfaces to which they are attached. Painting of cables is usually harmless. However, creosote and preservative liquids based on copper compounds should not be applied to surfaces to which cables are attached. It is almost impossible to prevent substantial contamination with these substances in liquid form, resulting in chemical deterioration of cable coverings. There is no objection to the installation of cables on surfaces pretreated with these substances and allowed to dry.

In some industrial environments hostile substances can be present as an inherent or incidental part of the process being served, such as in chemical and oil installations, electroplating etc. PVC, LSF, PCP and CSP sheaths all have excellent records in providing protection in these cases.

If present in quantity for an appreciable length of time, hydrocarbon fluids and organic solvents will cause embrittlement of PVC compounds by extracting plasticiser. Absorption of certain oils and solvents will cause softening and swelling of thermoset cable components. The permanence of the latter effect will depend on the volatility of the liquid and the degree of contamination. The softening and swelling caused by volatile solvents generally disappears once the liquid has evaporated and little permanent damage may result. The effect on cables of installation in ground containing petrol or other such hydrocarbons is discussed in chapter 5.

Mechanical stresses

The IEC classification recognises impact and vibration as two categories of mechanical stress, with a further class for other mechanical stresses. Bending, flexing, twisting and abrasion are obvious further considerations where cables are concerned. Flexing and twisting relate especially to flexible cords and cables.

Avoidance of excessive strain at permanent bends in field installations is covered by specifying minimum installation radii. Values for specific types of cable are given in the appropriate chapters and are summarised in appendix A19.

During installation it is necessary to avoid excessive bending. Repeated bending can result in irreversible straining of the conductor wires, with resultant kinking. Cables with larger numbers of small cores are particularly vulnerable. Twisting, either during installation or in use, can result in similar damage. If the torsion is in the direction of lay of the cores, so tending to tighten the construction, the forces may be sufficient to kink the cable, even with large sizes, causing permanent damage. Twisting in the opposite direction may appear to give less obvious damage, but buckling of the conductors can occur, especially in large multicore cables. Armoured cables are subject to 'bird-caging' of the armour. Flexible PWA cables are particularly prone to early failure due to twisting arising from the cable being allowed to roll in use. Braid armour overcomes this problem.

During installation, twisting should always be avoided by allowing the despatch reel to rotate as the cable is pulled off. Turns of cable should not be allowed to escape over the drum or reel flange. If long lengths cannot be pulled directly into position they should be coiled down in figure-of-eight formation (chapter 26). This also applies to large flexible cables in service, including those with pliable armour.

Damage by fauna

In the UK the most likely form of damage under this heading is that due to gnawing by rats. It is apparent that rats will gnaw any material or object which crosses the line of a regular path or run. The only cable protection which can be relied upon is a hard metal covering in the form of conduit, sheath or armour. Cases of this type of damage were reported when lead sheathed wiring cables were used.

In countries where termites are common they are a recognised hazard to most organic materials including of course cable coverings. Further information is given in chapter 5.

Solar radiation

Exposure to direct solar radiation. even in the UK, has a pronounced accelerating effect on the degradation of organic cable sheaths. All sheaths for outdoor use should contain carbon black, which if well dispersed and of appropriate type provides the best protection against attack by ultraviolet light (photo-oxidation). CSP, PCP, LSF, PVC and polyethylene sheath compounds formulated with weather resistance in mind have probably the best performance.

The infrared radiation from the sun is absorbed by all bodies directly exposed, and their resultant temperature, relative to their surroundings, is dependent on ventilation and the colour and smoothness of their surface. Rough black bodies, which have the highest coefficient of absorption, can have their temperatures considerably increased by this effect. It is therefore necessary to make allowance for it when determining the ambient temperature for current rating purposes.

JOINTS AND TERMINATIONS

In domestic and similar wiring systems, the consideration of joints and terminations is almost entirely concerned with the selection of appropriate fittings with mechanical connectors. With the increasing use in the last decade of somewhat stiffer cables, and cores having a single solid wire conductor, the need has been recognised for adequate space within the fitting, and for due attention to be given to minimum bending radii to avoid localised pressure on the insulation.

Purpose-designed joints and terminations are required for industrial distribution systems for fixed installations, and these no longer demand the high degree of skill necessary in the past. Conductor joints and terminations can be made simply and readily with compression fittings, and soldering is seldom now required. The enclosure is usually a plastic mould filled with cold-pouring resin, as described for distribution cables in chapter 27.

Many terminations and connectors for large flexible cables, including mining and welding cables, now also rely on compression fittings. Joints in flexible conductors, unless effected by means of a plug and socket cable connector, are strongly deprecated, as any other form of conductor joint leads to a very rapid flexing failure due to the localisation of stress at the inevitable short stiff section of conductor.

Owing to the very arduous mechanical duty of flexible trailing cables, techniques for their repair have developed into a specialised craft, especially in the mining industry. Advice and materials for this purpose are available from the cablemaker.

CHAPTER 16
Mineral Insulated Cables

The trade name 'Pyrotenax' aptly characterises the outstanding feature of mineral insulated cables as the ability to survive in fire. Increasingly this is a requirement of modern installations where continuity of supply, in the event of a fire, is of paramount importance. The beginning of mineral insulated (MI) cable can be traced back to the late nineteenth century when a Swiss engineer, Arnold François Borel, first proposed a cable constructed from inorganic materials completely enclosed in a metallic sheath, patenting this invention in 1896. His intention was to construct a cable that would operate at high temperatures, even in fire, and at the same time would be resistant to severe mechanical stresses. The idea then lay dormant until 1934, when a French company devised a commercially viable process and began to manufacture MI cable at Clichy under the trade name 'Pyrotenax'. In 1936 a British company, Pyrotenax Limited, was established to produce MI cable under licence.

It was soon recognised that, by the introduction of various metallic sheaths and conductor materials, MI cables could be adapted to such diverse applications as temperature measurement and heating. Applications include the design and development of mineral insulated thermocouple cables capable of operating up to 1250°C, instrument/signal cable for the power, aerospace, downwell and petrochemical industries and a wide variety of cables for applications such as ramp heating and pipe tracing.

Its compact dimensions and its freedom from the need for additional mechanical protection make it economically attractive for a wide range of domestic wiring applications and it is ideally suited for neat and unobtrusive wiring in churches and historic buildings. MI cables have been developed and adapted for use in almost every conceivable application, including general lighting, small power circuits, lateral and rising mains, control wiring, emergency services such as fire alarm systems, emergency lighting, standby generator and lift supplies, public address systems, computer supplies, and smoke extraction systems.

CONSTRUCTION

The construction of a typical MI cable is illustrated in fig. 16.1. Current carrying conductors are embedded in highly compacted mineral insulation and the whole is

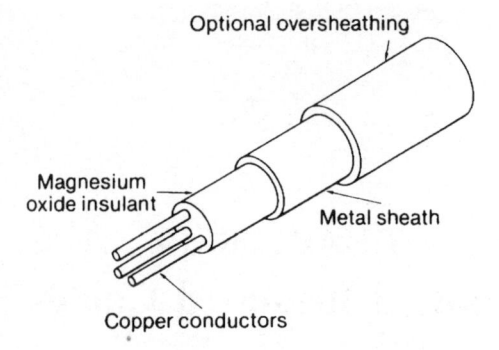

Fig. 16.1 Construction of a typical mineral insulated cable

encased in a metal sheath. Typically, the conductors and sheath are made of copper and the insulation is magnesium oxide. On copper sheathed cables an extruded thermoplastics or other outer covering may be applied when required for corrosion protection, identification or aesthetic appeal.

Wiring cables are made in a number of conductor sizes, ranging from 1 mm^2 to 240 mm^2, and with a number of cores, ranging from 1 to 19. Usually, only single-core cables are made with conductors larger than 25 mm^2. Cables with 7, 12 and 19 cores usually have conductors of small cross-sectional area. The details relating to cable dimensions, weights and number and size of conductors are given in appendix A11.

Cables are designed and manufactured in accordance with the provisions of BS 6207: Part 1: 1995, which implements CENELEC harmonisation document HD 586.1 S1. There is also an equivalent non-identical IEC standard IEC 702.1.

CHARACTERISTICS

Performance at high temperatures

The non-organic materials used in the construction of MI cable give it characteristics significantly different from those of other electric cables. Of special note are its non-ageing properties and its performance at high temperatures. Copper MI cables can operate continuously at up to 250°C and for short periods of time at temperatures up to 1083°C, the melting point of copper. This property enables them to continue operating in a fire, supplying power to essential services and, in many cases, afterwards to remain in working condition.

MI cables comply with the highest category of performance specified in BS 6387 'Specification for performance requirements for cables required to maintain circuit integrity under fire conditions' category CWZ. However, unlike other types of fire resistant cable this rating can be achieved using the same piece of cable for each of the three separate tests: fire alone, fire with water spray, and fire with mechanical shock.

Mechanical deformation

The cable can withstand severe mechanical abuse such as bending, twisting and impact without appreciable deterioration of its electrical properties. Even when heavily

Fig. 16.2 A mineral insulated cable hammered flat and continuing to supply power

deformed, as illustrated in fig. 16.2, it can maintain electrical integrity and continue to function. No additional protection is necessary when it is used under mechanically hazardous conditions, and this feature makes the cable economically attractive for many applications.

Protective conductor

The copper sheath, with its extremely low resistance, provides an excellent circuit protective conductor.

Corrosion

Copper sheathed cables are resistant to most organic chemicals and can operate indefinitely in most industrial environments. A green patina may form on the sheath after long exposure to atmosphere but this does not indicate harmful chemical attack. Copper will corrode in contact with ammonia or mineral acids and if the presence of either of these is suspected then bare copper sheathed cables should not be used. Adequate protection against this kind of attack can be provided by an outer covering made of a thermoplastic material, e.g. PVC or polyethylene, or, now being used increasingly, halogen-free polyolefin coverings containing additives such as alumina trihydrate, which do not readily propagate flames and have very low smoke and acid gas emission under fire conditions (see chapter 6). Alternatively, the metal sheath can be made of another material, such as cupro-nickel or stainless steel. It is worth noting that copper does not suffer significant corrosion when exposed to air bearing sea-water spray and therefore corrosion protection is not usually necessary when copper sheathed cables are used in offshore oil installations.

Resistance to radiation

With the exception of the optional plastic oversheath the constituent materials are practically unaffected by radiation and therefore MI cables can be used in high radiation environments with confidence that no deterioration with time will occur.

Current rating

Since magnesium oxide is a refractory material, the current carrying capacity of MI cables is not determined by the usual criterion of deterioration of the dielectric material with rising temperature but by a combination of the temperature characteristics of the surrounding materials with which cables are in contact and consideration of permissible volt drop. For their size, MI cables have a higher current rating than most other cable types.

In appendix A11, tables A11.1 and A11.2 give current ratings for the 500 V light duty range of cables. Tables A11.4 and A11.5 give similar data for the 750 V heavy duty range. Voltage drops are in Tables A11.6 and A11.7. The values in the tables are in accordance with the values given in the Wiring Regulations, BS 7671: 1992. They cover only the sizes of cable which are readily available (i.e. not all the sizes of cable listed in BS 6207: Part 1: 1995).

Resistance to voltage surges

The characteristics of a mineral insulation such as magnesium oxide, unlike those of organic insulation, do not alter with time and the initial dielectric strength is retained indefinitely. Consequently a smaller safety margin can be used as there is no deterioration to allow for.

In the initial stages of manufacture magnesia powder is compressed until it reaches a stable density beyond which no further compression occurs. In this state the powder density reaches about 75% of the density of magnesium oxide crystals, and there is very little air left in the spaces between the particles. However, the air which does remain may form continuous paths between the conductors or conductors and sheath. The mechanism of electrical breakdown is complex, being influenced by that of air and magnesium oxide crystals, and the dielectric strength of compacted magnesia is between that of a magnesium oxide crystal and that of air.

It follows that, when a breakdown occurs, the discharge path follows a path of ionised air but does not penetrate powder particles. When the discharge ceases, the air deionises and the insulation reverts to its normal state.

Low energy discharges through compacted magnesia powder do not damage the insulation and can be repeated indefinitely without any adverse effect. On the other hand high energy discharges can, and on occasion will, result in insulation breakdown leading to short circuit. A probable explanation of the observed effects is that a low energy discharge ionises the air but leaves the metal conductors unaffected, and, as energy increases, metal in increasing quantities is vaporised from the conductors and on cooling is deposited on magnesia particles. A sufficiently high energy discharge can vaporise enough metal to form a continuous path between conductors.

In summary, compacted magnesia insulation cannot be degraded by heat, by mechanical forces (other than excessive bending) or by being subjected to electrical stress, and thus will not be damaged by low energy voltage surges. If voltage surges are sufficiently energetic to vaporise conductor metal then, as a consequence, the insulation can also be affected.

MATERIALS

Insulation

A wide variety of mineral materials has been proposed, and occasionally used, as the dielectric in MI cable. These include aluminium oxide, beryllium oxide, calcium carbonate, fine clay, sand, and powdered glass, but long experience has shown magnesium oxide to be the most suitable. It is chemically and physically stable, has a high melting temperature, high electrical resistivity combined with high thermal conductivity, is non-toxic and is readily available in a sufficiently pure form.

Magnesia is derived from one of two raw materials, either magnesite (magnesium carbonate) mined in various parts of the world but notably Greece and India, or magnesium hydroxide precipitated from sea-water or subterranean brine by the addition of limestone or dolomite. The carbonate or the hydroxide is heated in kilns at temperatures between 900 and 1800°C to produce magnesia powder whose properties will vary with the type and source of raw material and the calcination temperature. Low calcination temperatures yield soft, amorphous and cohesive powder, whereas higher temperatures produce harder, more crystalline powder with less affinity for water. The type of powder used will depend on the method of manufacture employed and the intended application of the cable.

All types of magnesia powder, even when compacted, have some affinity for water and therefore cables have always to incorporate appropriate terminations which provide a seal against the ingress of moisture.

The resistivity of any magnesium oxide at room temperature is very high but decreases with rising temperature. High temperature properties, particularly resistivity, depend greatly on the kind and amounts of impurities present and studies are continuing aimed at obtaining a better understanding of the effects of impurities. Figure 16.3 shows the relationship between resistivity and temperature for several grades of magnesia.

Because magnesia is hygroscopic it is important to dry the powder thoroughly before it is introduced into the cables. Wide variations in high temperature performance may

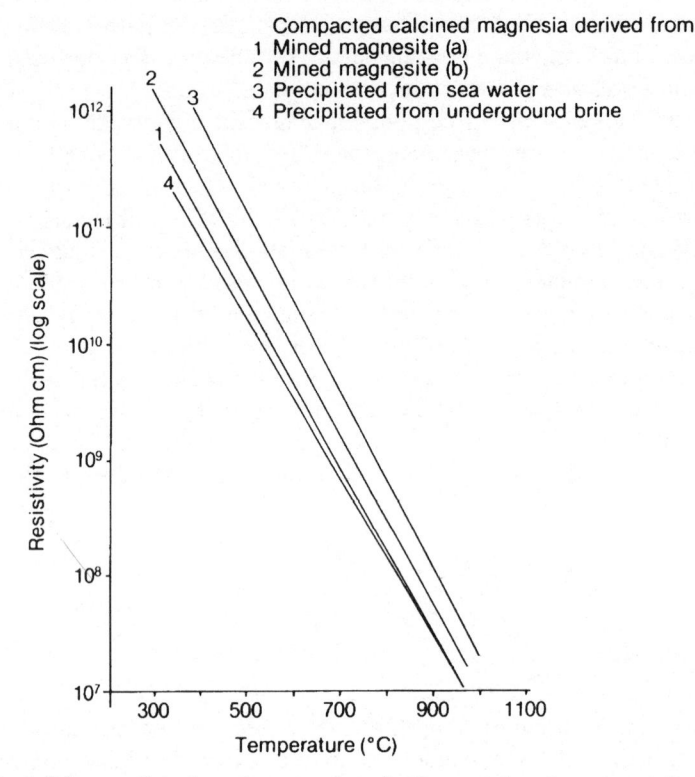

Fig. 16.3 Resistivity as a function of temperature for four grades of compressed magnesia powder used as cable insulants

Compacted calcined magnesia derived from
1 Mined magnesite (a)
2 Mined magnesite (b)
3 Precipitated from sea water
4 Precipitated from underground brine

result from traces of moisture in the powder although very small amounts are unlikely to have a significant effect either on resistivity or dielectric strength.

Table 16.1 shows a typical composition of magnesia produced from mined magnesite.

Conductor material

The two conditions which have to be met by the conductor material are high conductivity, essential for the efficient transfer of electrical energy, and ductility which

Table 16.1 Typical composition of magnesia produced from mined magnesite

		%
Calcium oxide		2.5
Iron oxide		0.26
Silica		5.5
Boron	less than	0.001
Sulphur		0.075
Chlorides (NaCl)		0.002
Magnesium oxide		Remainder

is necessary for satisfactory processing. Tough pitch high conductivity copper satisfies these requirements for most general wiring cables, although in some types of cable it is necessary to employ oxygen-free h.c. copper.

Sheath material

The standard sheath material is phosphorus deoxidised copper with a conductivity of about 80% IACS. Other materials which may be used include mild steel, cupro-nickel and austenitic stainless steel which finds increasing application in aggressive environments.

MANUFACTURE

Several manufacturing methods have been evolved during the period of time spanning about 60 years. Until recently all methods were similar in that in all cases a relatively short, large diameter composite comprising an outer metal tube, conductor and a mineral filler was assembled initially and then compacted and elongated to finished size by repeated drawing through dies. Frequent interstage annealing was necessary in order to restore ductility to the composite cable. On reaching final diameter, the cable was fully annealed to impart maximum conductivity to the conductors and to make it pliable in order to improve handling during installation. In most cases the 'start tube' was about 9–12 m long and 55–60 mm in diameter.

Methods of continuous production of MI cables have now been developed.

Batch process

In the block filling method, illustrated in fig. 16.4, blocks of magnesia are first formed under high pressure to the required density and appropriate dimensions. The blocks are heated in a furnace to expel moisture present either in the free state or in chemical combination and then pushed into the tube with the holes for the conductor rods kept in registration, and the rods are inserted.

The composite 'start assembly' is processed by successive drawing and annealing to final size, during which operation the blocks are crushed and reduced to highly compacted powder.

A similar start assembly can be produced in less time by a ramming process. A start assembly is prepared consisting of a tube and conductors retained by a plug pressed

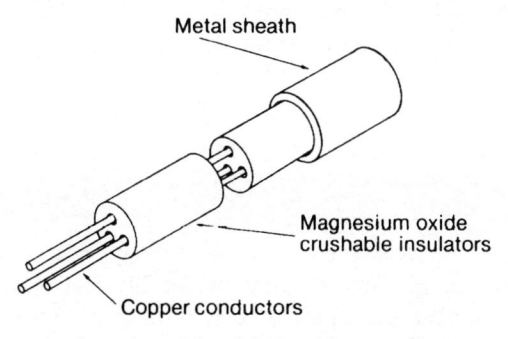

Fig. 16.4 Block filling process for mineral insulated cable manufacture

into the bottom end of the tube. A ram with suitable apertures for the conductor rods is placed in the tube and magnesia in powder form is introduced into the space between the plug and the bottom of the ram, either along the annular space between the tube and the ram or through a tubular passage along the ram axis. The ram reciprocates relative to the tube assembly and the powder is compacted at each down stroke. The reciprocation is continued until the tube is completely filled with packed powder. It is necessary to achieve sufficient packing density to ensure that when the assembly is drawn through a die for the first time the conductor rods are held firmly in place and the required conductor geometry is maintained.

The composite assembly is processed to finished cable in the manner already described.

An alternative method is to fill tubes with powder and to compact by drawing through a die to avoid the need for ramming. A start assembly consisting of a tube and conductor rods is prepared, with the lower end of the tube reduced in diameter so that it can be passed through a die. Powder is introduced through a guide tube usually incorporating shroud tubes which enclose the conductors and maintain them in correct registration. Then the composite assembly is drawn progressively through a die, the cross-sectional area of the bore is reduced and the powder is compacted sufficiently to hold the conductor rods firmly in place for further processing.

Continuous process

For a long time now, cables of other types have been made by continuous process but, until recently, the manufacture of MI cables has remained essentially a batch process. The evident disadvantages of batch processing have been the cause of considerable effort expended on development aimed at making the process continuous.

Essentially the process developed by the manufacturers comprises several steps carried out consecutively. The process begins with metal strip of suitable dimensions which is continuously formed into a tube and seam welded. Conductor rods and powder are continuously introduced through a guide tube. At the lower end of the guide tube there is a die which precisely locates the rods in relation to the outer tube. The space below the die is filled with powder so that the rods are held in the correct position. Before the cable can be handled it must be reduced in cross-sectional area in order to compact the powder to a density sufficiently high to preclude any undesirable conductor movement relative to the outer tube. This can be done by rolling or drawing and both methods have been used.

It has been established that cable produced by this process has mechanical and electrical properties indistinguishable from those of cable produced by conventional processes.

SEALS

As mentioned above it is necessary to exclude atmospheric moisture from MI cables and seal development has perforce accompanied cable development. The earliest seals utilised a bituminous substance applied to a pot, which also did duty as a gland, screwed to the end of the cable sheath. This was superseded by a cold seal which relied on a castor oil based plastic compound contained in a screw-on pot. Similar types of seal are still used with various types of sealing media, including epoxy resins and putties, silicone oil based compounds and ceramics.

The design and manufacture of seals are carried out in accordance with the provision of BS 6081.

PART 3
Supply Distribution Systems and Cables

CHAPTER 17
Supply Distribution Systems

In the UK the total generating capacity of the public electricity supply system at the beginning of 1988 was 63 868 MW and the maximum demand 55 241 MW. From 1948 to 1988 the total number of consumers in England and Wales rose from 10.80 to 21.91 million and the total length of the main circuits energised at voltages below 132 kV increased from about 266 000 km to 597 320 km. The latter figure is made up of 228 124 km of overhead line and 360 196 km of underground cable. From 1948 a vigorous programme of full rural electrification was pursued in which overhead lines were used for economic and practical reasons, but in urban and industrial areas the system has been developed using cables laid directly in the ground and connected to ground-mounted substations. Mains cables operating at voltages up to 11 kV are usually laid in public footpaths or service reservations alongside the carriageways, where space is allocated for each public service, but cables operating at higher voltages often have to be laid in the carriageway.

Since 1980 attention has turned increasingly to refurbishment of the existing system and this is expected to continue over the next ten years. Priority will have to be given to switchgear, transformers and overhead lines, but the underground cables now in service are still generally sound and are unlikely to need to be replaced for many years, with the possible exception of some of the transmission fluid-filled cables which may be considered environmentally unsuitable due to fluid leaks, or fire risk.

While there is a broad similarity among the European public electricity distribution systems, a much wider variation is found in the USA, but one feature common to most of the American utilities has been the use of ducted systems with manholes to accommodate cables in city centres. Another is the high proportion of overhead circuits in urban networks. Objections to the appearance of overhead lines, and mandatory requirements that in certain areas new supplies be provided by underground cable, have led to the development of a system known as underground residential distribution (URD), which is discussed later in this chapter, and there is a commitment to the eventual undergrounding of existing overhead lines in urban areas.

PARAMETERS OF ELECTRICITY SUPPLY SYSTEMS

In the development of early electricity supply systems, many direct and alternating voltages, frequencies, phase numbers and connections were used to achieve the most economical use of capital, but the arrangements discussed below have become predominant in modern systems.

Frequencies and phase numbers

Symmetrical three-phase system

A symmetrical 3-phase a.c. system with an earthed neutral and a frequency of 50 Hz or 60 Hz is now used almost universally for main power distribution systems (fig. 17.1(a)), some of its advantages being the following.

(a) Alternating current may be generated without a commutator and transformed to higher or lower voltages by static transformers.
(b) A 3-phase winding makes efficient use of the armatures of cylindrical machines and of the cores of transformers.
(c) In a symmetrical 3-phase system the phases are mutually displaced by the same angle, 120°, and the magnitudes of the voltages between the phases are all equal, as are the magnitudes of the voltages between the phases and the neutral.
(d) A 3-phase supply will excite a magnetic field rotating in a definite direction which is easily reversed.
(e) The voltage applied to a symmetrical load that is normally connected in delta may be reduced by connecting it in star without the use of a neutral conductor.
(f) A frequency of 50 Hz or 60 Hz relates well to the normal speeds of mechanical drives and is high enough to prevent discernible flicker from electric lamps and low enough to avoid undue interference with telecommunications equipment.
(g) While 3-phase and single-phase loads may be connected between the phases without the use of a neutral conductor, the provision of a neutral conductor allows single-pole-and-neutral supplies to be given to individual loads and if these are distributed between the phases the currents will tend to cancel each other in the neutral.
(h) A 3-phase system with loads connected between phase and neutral may be supplied from a 3-phase system without a neutral conductor by the use of a transformer with a star-connected secondary winding.
(i) A single-phase system may be supplied from a transformer connected between two phases of a 3-phase system.

The symmetrical 3-phase system is used throughout the UK with a frequency of 50 Hz. In some 3-phase systems, one phase is connected to earth instead of the neutral, as shown in fig. 17.1(a). A 3-phase line then has only two live conductors, and no neutral conductor being provided, but their potential to earth is higher than it would be if the neutral were earthed. Only one conductor of a single-phase tapping need be live but if no single-phase tappings are made between the two live phases a balance of loads on the phases will not be obtained.

Single-pole-and-neutral system

In the UK isolated low voltage single-pole-and-neutral systems (fig. 17.1(b)) are used where the total demand is less than 50 kV A and a small number of single-phase loads would be difficult to balance between three phases.

Supply Distribution Systems

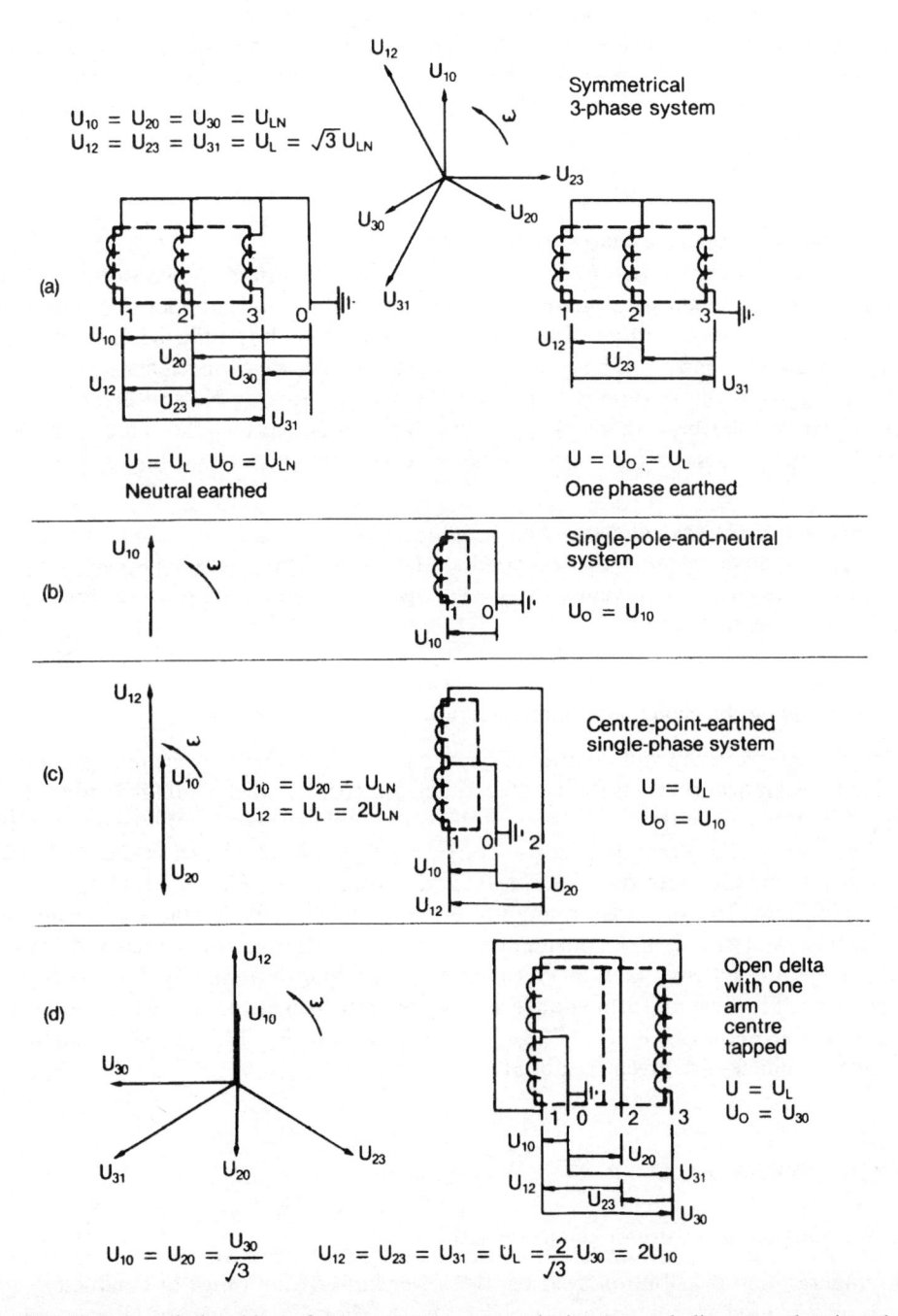

Fig. 17.1 Voltage relationships of some systems currently in use and diagrams showing the arrangements and connections of the secondary windings of the infeeding transformers. The heavy broken lines represent the magnetic cores

Centre-point-earthed single-phase system

Some systems have two live conductors carrying potentials relative to an earthed neutral that are in phase opposition to each other, as shown in fig. 17.1(c). If a neutral conductor is provided, the current from loads connected between opposite poles and the neutral will tend to cancel in the neutral, as in the 3-phase system. Such a system may be supplied from a centre tapped winding on a single-phase transformer. In the USA low voltage systems are commonly of this type.

'V' system with one arm centre tapped

The extension of the centre-point-earthed single-phase system shown in fig. 17.1(d) is sometimes used. It allows 3-phase and single-pole-and-neutral supplies to be given from a 4-wire distribution system. There is a low voltage winding utilising two limbs only of a 3-phase transformer. The windings on the two limbs are connected as two sides of a delta and one is centre tapped to provide the neutral. Single-pole-and-neutral supplies then have half the line to line voltage. A disadvantage is that the load is unevenly shared between the phases. In Tokyo low voltage supplies are given from this type of system.

Importance of the neutral conductor

In all the systems using a neutral conductor to give supply to single-phase loads, its integrity is essential to prevent excessive voltages from being applied to the loads on the less heavily loaded phases.

Constraints on the choice of circuit voltage

For a given frequency and number of phases, it can be shown that the most economical circuit design occurs when the conductor related cost C_c of the circuit is equal to the capitalised cost C_i of the I^2R losses and the voltage related cost C_v of the circuit is equal to the sum of the other two costs, i.e. $C_v = 2C_c = 2C_i$. Conductor size and circuit voltage should therefore rise together as circuit power rating P rises, i.e. $C_v \propto C_i \propto \sqrt{P}$. In fact the high cost of transforming stations, and the need to standardise transformers, switchgear and cables, limits the number of voltage levels that can be used and normally a distribution system takes the form of three or four superimposed networks each operating at its own uniform voltage and fed by transformers connected to the network operating at the next higher voltage level, as shown in fig. 17.2. Typical examples of network voltages are given in table 17.1.

CABLE NETWORKS

Constraints on the choice of conductor size

In underground distribution systems, the lower limit of the range of conductor size S that may be used at each voltage level is set by the short-circuit current I that could flow and the total operating times T of the protection and switchgear in the approximate relationship $S = kI\sqrt{T}$. The result is that a mat of cables in a small range of sizes is operating at each voltage level in most urban and industrial areas. The alternative policies followed in organising these cables into networks are outlined below.

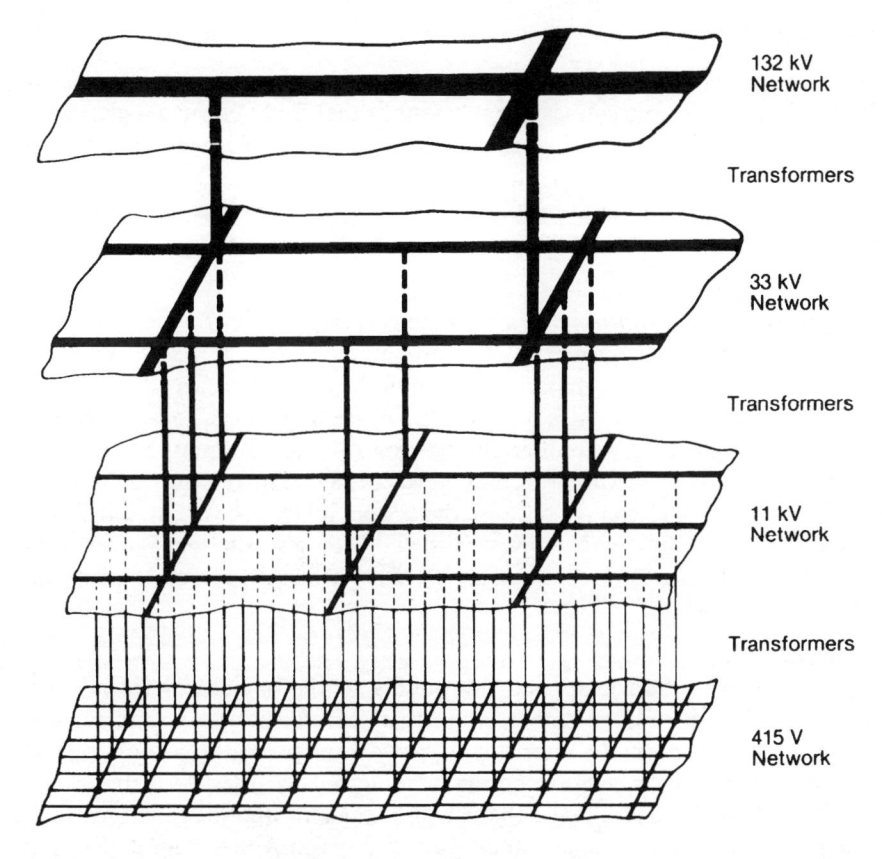

Fig. 17.2 Superimposed networks forming a distribution system

Table 17.1 Typical examples of nework voltages

UK	Germany	USA
132/76.2 kV	110/63.5 kV	138/79.7 kV or 120/69.3 kV
33/19.1 kV	30/17.3 kV	46/26.5 kV or 40/23 kV or 24/13.9 kV or 22/12.7 kV
11/6.35 kV or 6.6/3.81 kV	20/11.5 kV or 10/5.77 kV or 6/3.46 kV	13.2/7.6 kV 4.8/2.8 kV
415/240 V	380/220 V	220/110 V

The second group of voltage levels is sometimes omitted, i.e. 33/19.1 kV etc.

Contiguous networks

The cables operating at each voltage level can be arranged into a contiguous network of distributing interconnectors connecting substations that feed in from a higher voltage as well as feeding loads directly or through transformers stepping down to a lower voltage network. This arrangement allows load normally supplied by a substation that is out of service, or has to be taken out of service, to be transferred to other substations through the interconnectors.

Parallel operation of transformers through the network
If the network and transformer ratings are matched to the load, the in-feeding transformers may be operated in parallel through the interconnecting distributors. This makes for a more even sharing of load by the in-feeding substations, reduced losses, reduced voltage regulation, smoother voltage control and continuity of supply on the loss, owing to a fault, of an in-feeding transformer. Fluctuating loads cause less disturbance to the supply voltage because the system impedance is reduced throughout the network while the short-circuit levels at the in-feeding substations are less than they would be if the same transformer capacity were concentrated at one substation.

Unit protection
Further, if unit protection is applied thoroughly to the higher voltage networks, as shown in fig. 17.3, there is greater freedom to interconnect and a fault on any high voltage cable will not normally interrupt supplies. Unit protection is not applicable to low voltage distributors, which are normally protected by fuses. Distribution systems developed on these principles invite the use of a single rating for transformers feeding into each network and a single circuit rating at each voltage level. Examples of contiguous networks are shown in figs 17.4, 17.5 and 17.6.

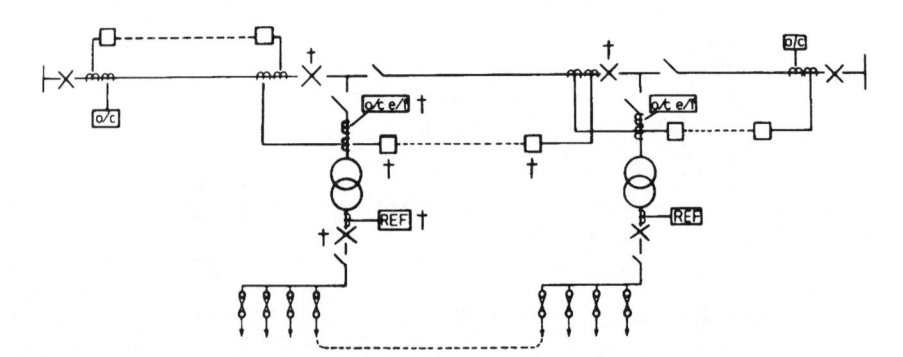

Fig. 17.3 A protection scheme used in an HV interconnected network. The automatic circuit breakers are arranged to form units that include one transformer and one section of cable linking two substations. Unit protection, having unlabelled relays in the diagram, covers the HV cables and switchgear and an overcurrent and earth fault relay covers the transformer and LV busbars. There is also restricted earth fault protection on the low voltage winding of the transformer. Each of the relays marked with a dagger trips all the circuit breakers marked with a dagger to isolate the unit. Any number of substations may be connected into the interconnector. Back-up overcurrent protection covers the whole interconnector

Supply Distribution Systems 289

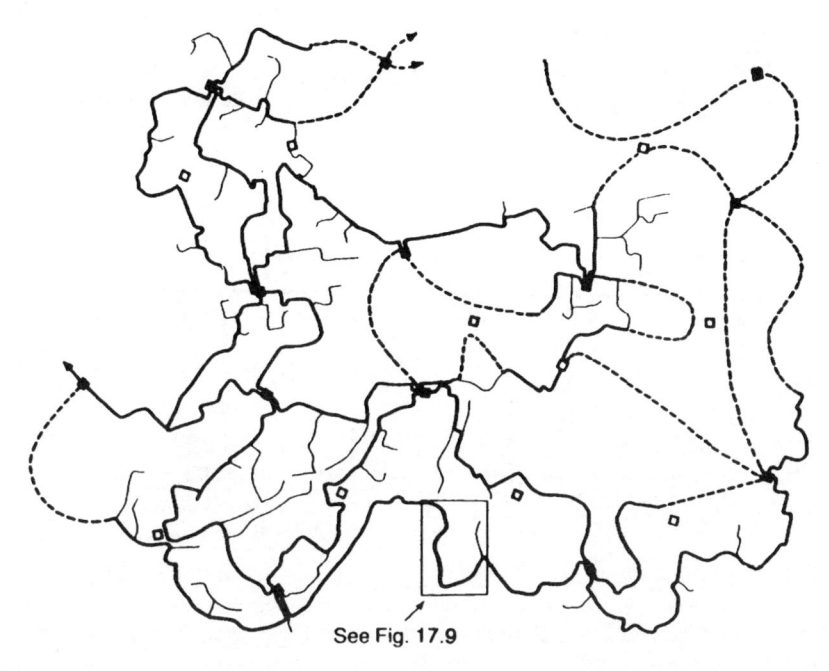

Fig. 17.4 A uniform LV network on a residential development that is in the process of construction. All the cables are 95 mm² waveform. The fine lines represent spurs off the interconnectors which are shown as heavy lines. The broken lines show further interconnectors that will be formed as the development extends. Sites acquired for reinforcement substations are shown by open squares

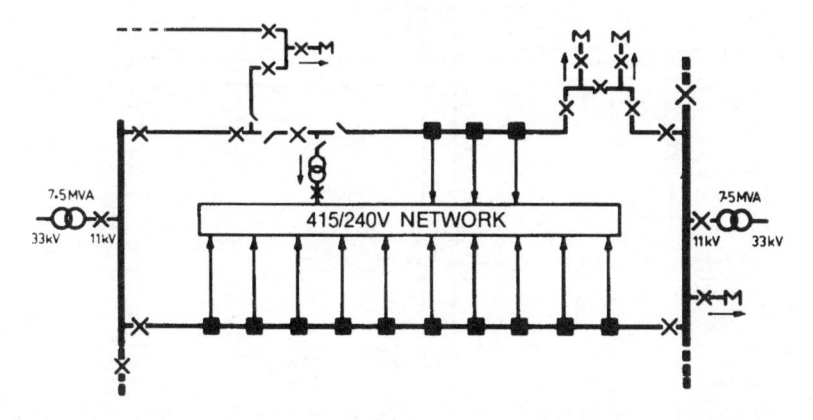

Fig. 17.5 Part of an HV interconnected network on an industrial estate. The cable and busbars extend as shown by the broken lines. Power flows out of the HV network are shown by arrows. M indicates commercial metering. One substation with an HV branch is shown in detail. The solid squares represent substations of the type detailed in fig. 17.3

Radial networks

An alternative policy is to develop radial distribution networks supplied from relatively large substations where a security of supply appropriate to the load is provided, multiple feeders being used to supply a substation in which several transformers are

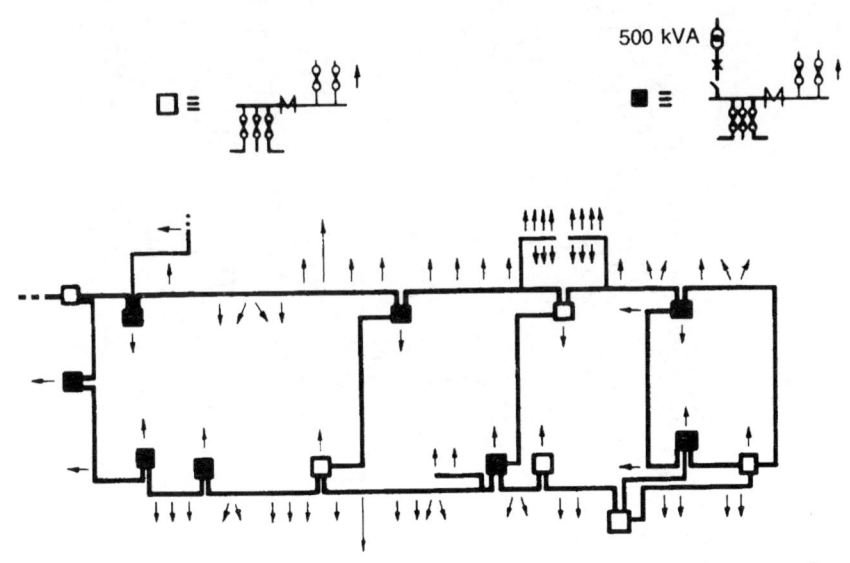

Fig. 17.6 Part of the LV network fed from the HV network shown in fig. 17.5. 185 mm² LV cables are used throughout the industrial estate. Points from which supplies are given are shown by arrows. Substations equipped with LV fuse boards only are shown by open squares. M represents busbar metering

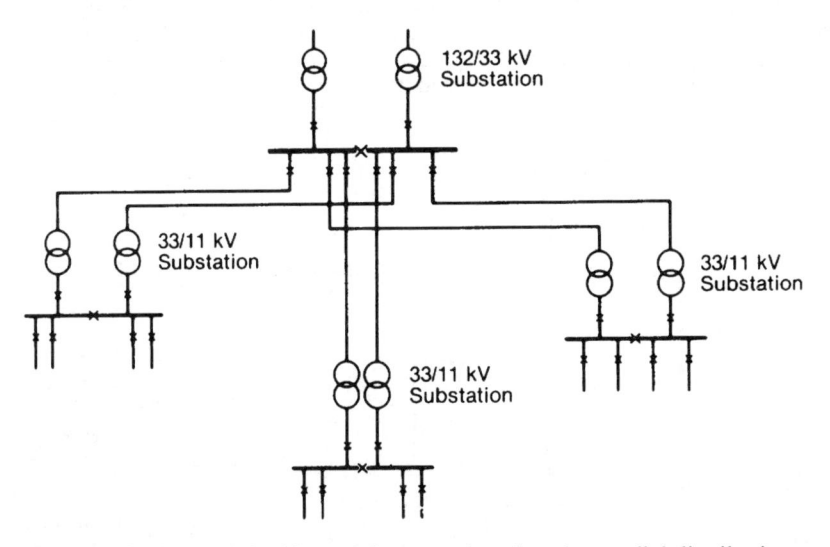

Fig. 17.7 Duplicate feeders and double transformer substations in a radial distribution network

installed (fig. 17.7). Several cable sizes appropriate to the prospective loads that they are likely to carry might be used at each voltage level within the limitations set by fault currents. This policy aims to provide system capacity at the lowest cost per kilovolt amp installed and to match each component to its prospective load.

Provision for load growth

Throughout most of the history of electricity supply there has been a vigorous growth in demand and the new systems that are installed should be able to cater for growing

Uniform network

One policy has been to lay down a uniform contiguous network at each voltage level with one main cable size throughout, matched to a standard transformer rating, the substations having the minimum dimensions to accommodate the transformer and its associated switchgear. The substations are spaced to provide for the estimated early load density and when the load rises above the rating of the network, which is matched to the transformers, the most heavily loaded interconnectors are relieved by being turned into further suitably sited standard substations, as shown in fig. 17.8. In this way the installed transformer and substation capacity are kept as small as possible, the distribution of transformers corresponds to the distribution of load and the capacity of the cable network is not based on a speculative assessment of load growth.

Tapered network

The other strategy has been to construct a 'tapered' distribution network in which the cable size at each part of the network is matched to the load density that is expected to be reached after a number of years. Transformer capacity sufficient for the estimated early load density is provided initially and further capacity is installed in the same substation when the load requires it. Additional cables and switchgear may also be necessary. A tapered network does not lend itself so readily to parallel operation of transformers through the network, nor to reinforcement from additional substations.

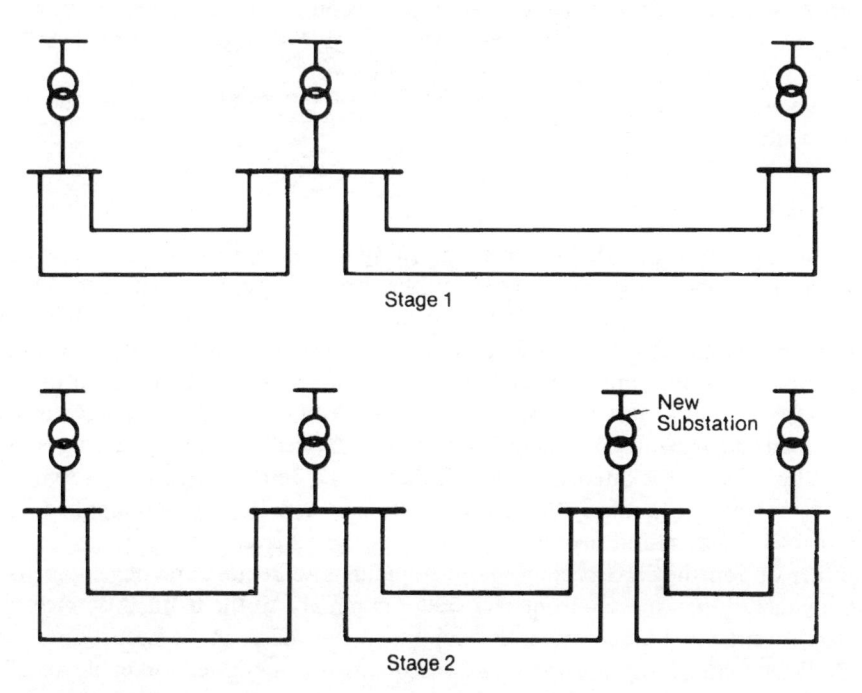

Fig. 17.8 Reinforcement of a uniform contiguous distribution network by the provision of an additional infeeding substation

SUPPLIES TO RESIDENTIAL DEVELOPMENTS

The choice of suitable locations for the substations is most important in the economic design of any distribution network. This applies whether the system comprises a uniform interconnected network with small standard substations or is a tapered network with fewer larger substations.

Design of uniform low voltage networks for residential developments

As only one cable size and one transformer rating are used, the main objectives in the design of a uniform residential network are that the substations and interconnecting distributors be evenly loaded, that the distributors provide adequate interconnection between adjacent substations, that the substations be suitably spaced for the early load density, that sites for additional substations sufficient for the predicted ultimate load density be provided in suitable locations and that the distributors be routed so that they may be turned in to the additional substations. Figure 17.4 shows a uniform network on a residential development that is in the process of construction. All the cables are 95 mm^2 waveform type and the substations all have 500 kVA transformers.

Design of tapered low voltage networks for residential developments

In recent years extensive studies on the design of tapered underground distribution systems for new housing estates have shown the following.

(a) The ratings assigned to low voltage cables should be economic ratings in which the capitalised costs of the losses are taken into account. This requires some additional capital outlay to cover extra cable cost but some saving can be made in overall costs, which include jointing and electrical losses.
(b) A maximum of four cable sizes should be used as the benefits from the use of a larger number are small.

Residential service connections

In the UK supplies to private houses are given by single-phase services consisting of an earthed neutral conductor and a phase conductor carrying a potential to the neutral of 240 V. The service connections account for about half of the total capital cost of the network supplying a new housing estate and much thought has been given to means of simplifying the installations and reducing costs. An arrangement in which up to four single-phase concentric service cables are jointed directly to the 3-phase waveform main in one joint and connections are laid from houses serviced directly to adjacent houses is shown in fig. 17.9. In the interest of phase balance it is preferred that three service cables are connected at one joint and that not more than two houses are supplied by each service cable. This method has been found to be very successful.

Another option that has become more popular is to require the developer to lay a 50 mm diameter duct directly from each meter cupboard to the main. A service cable is threaded through the duct when the cupboard has been made secure and jointed directly to the main. Jointing costs may consequently be increased but cable installation costs and voltage drop are reduced. With this option care must be taken to secure the duct against the ingress of gas.

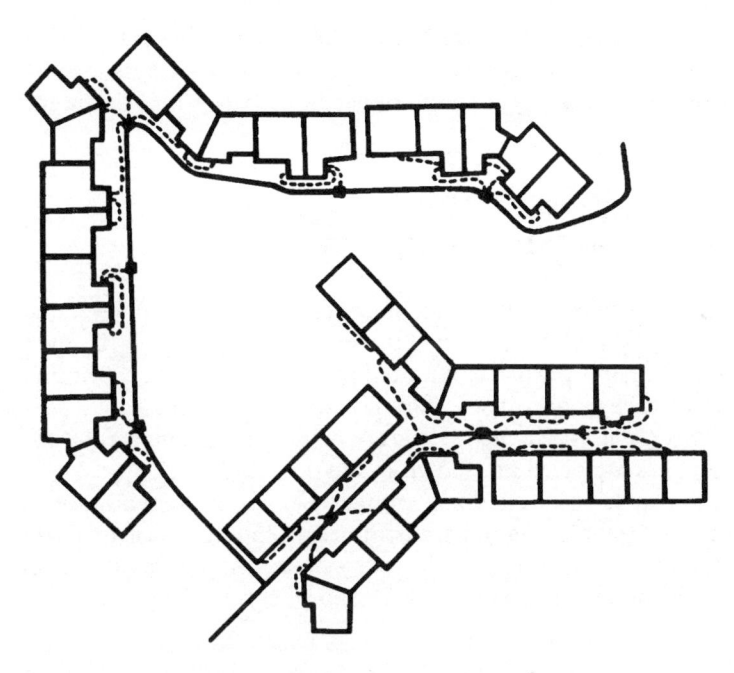

Fig 17.9 A detail of fig. 17.4 showing service connections. 25 mm^2 single-phase waveform service cables are distributed evenly between the three phases of 95 mm^2 main

Underground residential distribution

Underground residential distribution (URD) is a system that was developed in the USA to take the place of the pole-mounted residential distribution systems that were used there until the 1960s. It consists of single-phase high voltage distributors supplying 50 kVA, 100 kVA and 150 kVA transformers which step down to low voltage. Each feeds between eight and sixteen residential consumers by service cables radiating from the transformer or from multiservice pillars. In the USA the transformers are centre tapped and three conductors are taken to each house, giving supplies at 220 V between live conductors and 110 V between each live conductor and the earthed neutral.

In one arrangement studied in the UK, high voltage three-phase-and-neutral cables run from the primary substation in a ring through several distribution points where the three phases are split by switchgear that provides individual control for the incoming cables and individual control and fuse protection for the single-phase distribution cables. The distribution cables have a single live core with a concentric copper neutral conductor directly in contact with the ground. At each transformer the high and low voltage neutral/earth conductors are bonded together. The transformers are installed in small underground vaults. In each transformer there is a high voltage fuse. Cable connections to the high voltage side of the transformer are made by plug connectors which operate satisfactorily under water. The low voltage connections are made off at a sealed busbar.

In the USA high air-conditioning loads and lower housing densities favour this type of system especially in view of the lower value (110 V) of phase–neutral voltage.

In the UK it was estimated in 1970 that a URD system of this type, for a development with an average diversified demand of 6 kW per dwelling, could cost 26%

less than a conventional low voltage distribution system. However, the saving would be reduced to 16% if account were taken of the cost of losses. As the cost of transformers increased with the increasing price of copper in the early 1970s no significant savings could be achieved.

COMMERCIAL AND INDUSTRIAL SUPPLIES

Loads at commercial and industrial premises range from a few hundred watts to hundreds of megawatts, and the voltage at which the supply is given is generally determined by the supply capacity required in comparison with the normal transformer ratings used at each voltage level. The range of supply capacities provided at each voltage level by one electricity board is given in table 17.2. Some of the arrangements by which the supplies are given are shown in figs 17.5 and 17.6.

For supply capacities within the ranges shown in parentheses in table 17.2, accommodation for one or more transformers stepping down from the next higher voltage is normally required by the electricity board. An alternative supply is normally provided for loads greater than 1 MVA and wherever economically practicable for lower loads.

LOADS THAT CAUSE DISTURBANCE ON THE SYSTEM

National regulations set limits to the permissible variation of the voltage at a consumer's supply terminals, but disturbances to the supply voltage within those limits are often unacceptable to the consumer. In general a lower system impedance to any point from which a load is supplied reduces voltage fluctuations caused by that load. On the other hand a lower system impedance probably means that more transformer capacity is operating in parallel and supplying more customers from a more extensive network. This means that more customers will be affected by any disturbance, whether caused by loads or short-circuit faults, and disturbances will be reflected more strongly on to the higher voltage levels. There is therefore a trade-off.

Table 17.2 Typical ranges of supply capacities provided at each voltage level

Network voltage	Range of supply capacities normally provided	Normal ratings of transformers used
415/240 V	35–1000 kVA (100–1000 kVA)	11 kV/415 V, 500 kVA
11 kV	1–20 MVA (2–20 MVA)	33 kV/11 kV, 7.5 MVA
33 kV	Above 15 MVA (Above 30 MVA)	132 kV/33 kV, 60 MVA

Fluctuating load currents

Relatively small fluctuations in voltage, if repeated frequently, may have a quite unacceptable effect on incandescent lighting. A change in voltage of 1% once every second is very irritating. At ten times per second some people will notice even a variation of 0.25%. Such fluctuations in voltage are caused by reactive current surges drawn by motors, welding machines and arc furnaces, a major component of the voltage dips being generated in the reactive impedances of the transformers supplying the loads. Current surges can also cause protection systems to operate. If the surges cannot be reduced sufficiently, e.g. by the use of special motor starting equipment, it may be necessary to supply the offending load from a higher voltage network with a lower impedance and higher protection settings, the supply probably being metered at the higher voltage.

Another solution may be to increase the transformer capacity supplying the affected network, or, if the cable or line from the transformer is contributing greatly to the system impedance to the load, an additional cable may be provided to supply the fluctuating load alone. A subtler approach, which is often very effective, is to join two radial distributors from different transformers so forming an interconnector from which to supply the fluctuating load. This effectively reduces the impedance of the system to the load by increasing both the circuit capacity and the transformer capacity through which it is drawn. In any case, the voltage fluctuation at the 'point of common coupling', i.e. at the point on the system nearest to the disturbance generator from which both the disturbance generator and other consumers are supplied, is the condition to be controlled.

Harmonic currents

Harmonic currents generated by non-linear impedances, rectifiers, thyristor controls etc. can overload power factor correction capacitors and interfere with electronic equipment. Supply authorities set limits for the magnitudes of the harmonic currents that they will allow consumers to inject into the public distribution system and it may sometimes be necessary for the consumers to install resonant by-pass filters to absorb the harmonics generated by his equipment. Again the total solution may require the connection of the load into the system at the higher voltage.

Adjustment of network voltage

In general it is better to operate switchgear and cables at the highest voltage for which they are designed, so as to exploit their full short-circuit and load ratings. Short-circuit power levels on the HV network tend to rise as generation capacity and dynamic loads increase and the short-circuit ratings of switchgear may become inadequate, even if the capacity of the transformers feeding that part of the network is not changed. Even so, after an increase in network voltage, the capacity of the transformers feeding into the network can usually be increased by the ratio of the increase in voltage, and so the capacity of the whole network can be raised and the impedance of the system to the points of common coupling of fluctuating loads or harmonic generators reduced.

In the UK various voltages ranging upwards from 2 kV, have been used for HV distribution systems and some cables and switchgear have been installed which are only suitable for 6.6 kV or less. However, most of the high voltage cables now in use are suitable for 11 kV and so is most of the switchgear installed since 1948. The switchgear that is not insulated for 11 kV does not comply with modern standards for other reasons and will have to be replaced. There is therefore a strong case for the increase of lower HV network voltages to 11 kV.

Before applying the higher voltage it is usual to pressure test each circuit, unless there is a record that the appropriate test voltage has been previously applied. Sometimes old joints will fail on test and will have to be remade. Sometimes it is known that the type of joint used in a certain period was not suitable for 11 kV and those joints are remade before testing.

One of the obstacles to this course is the cost of changing transformers to operate at the new voltage, but there are compensating factors.

(a) Modern transformers have much lower iron losses; for example a typical 500 kVA transformer with a laminated core of grain-oriented cold-rolled steel has an iron loss of 0.7 kW compared with 1.7 kW for a similar transformer with a hot-rolled laminated steel core. The capitalised savings in the cost of losses resulting from the renewal of the transformer, added to the scrap value of the old transformer, help to pay for the new transformer.
(b) Newer transformers recovered as a result of a change in system voltage can often be reused on a part of the network that is to remain at the lower voltage.

A change of voltage will be made much simpler if transformers purchased for use initially at the lower voltage are also suitable for use at the higher voltage. This facility can be provided for about 10% extra cost.

EARTHING

In the UK the 1937 Electricity Supply Regulations require that every distribution system should be connected to earth. The secondary windings of transformers stepping down to 6.6 kV or 11 kV are normally star connected and the star point is connected to earth. The connection may be solid, or if it is necessary to reduce the potential earth fault current an impedance may be inserted. Low voltage distribution systems must be solidly connected to earth to ensure that there is no rise in potential that would cause danger in the case of a fault connecting the high voltage system and the low voltage system. 3-phase transformers feeding low voltage distribution systems have star-connected secondary windings with solidly earthed neutrals, and one pole of the secondary winding of each single-phase transformer stepping down to low voltage is solidly earthed.

The 1937 Regulations allowed transformers to be operated in parallel through a low voltage network with an earth connection at each transformer provided that each transformer had a delta-connected winding, and relaxations of the regulations extended these provisions to high voltage networks. If the delta connection is applied to the secondary winding, an interconnected-star earthing choke may be used to provide the earth connection. Typical system earthing arrangements are shown in fig. 17.10. The 1988 Regulations do not specify the provision of a delta-connected winding but make the 'person concerned' responsible for limiting 'the occurrence and effects of circulating currents' between earthed neutrals.

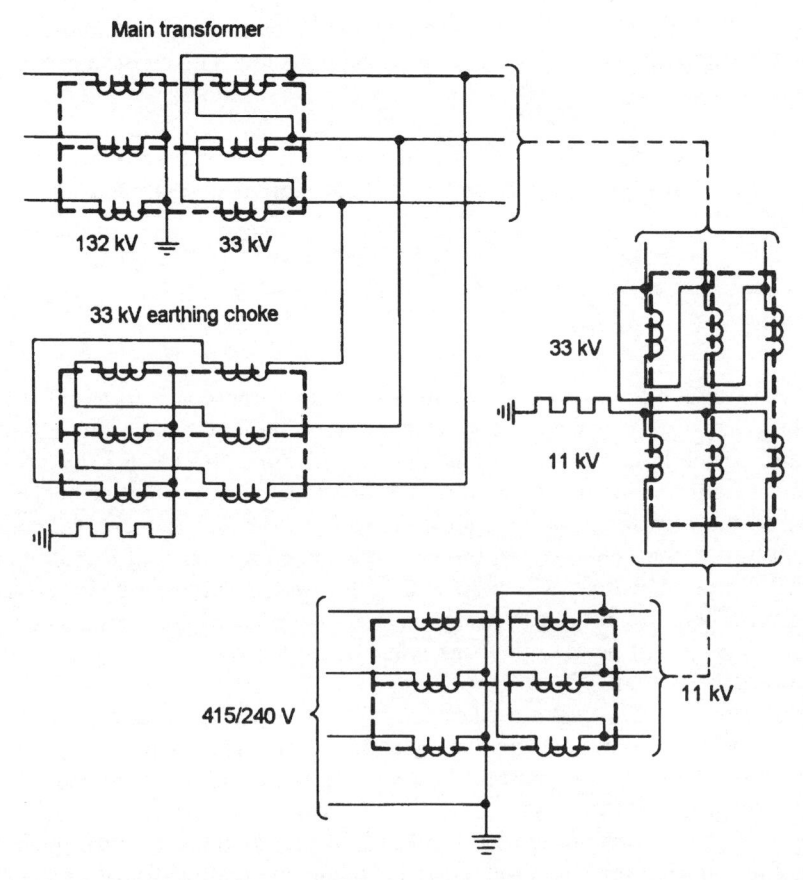

Fig. 17.10 Typical system earthing arrangements. An 11 kV earthing resistor is not always used

Protective multiple earthing
Where underground cables with contiguous metal sheaths bonded to the substation earthing systems were used in low voltage distribution networks, and earth terminals connected to the cable sheath were provided for the consumers' use, a low resistance connection between the system neutral and earth and low resistance return paths from the consumers' earthing systems to the transformer neutral were maintained. At the same time there was a problem on low voltage overhead line networks, where it was more difficult to make a low resistance connection between neutral and earth and consumers' earth fault currents had to return through the ground. One solution was to string an earth-continuity conductor on the overhead line, but it was found that if it was broken or a joint failed a long time could elapse before the fault was reported as supplies did not depend on it. This led to the idea that if the neutral conductor were used to provide a low resistance return path for earth fault currents its failure would be noticed quickly while, with suitable bonding and earth electrodes, dangerous potential differences on consumers' premises could be avoided.

In 1955 approvals were granted for additional connections to be made between neutral and earth, collectively referred to as protective multiple earthing (PME). Under these approvals PME was applied to many overhead low voltage distribution systems, although the imposed conditions made it difficult to apply to existing underground installations.

This experience led to relaxations and clarification of the original conditions and opened the way for the general application of PME to underground low voltage distribution networks and the use of CNE cables for extensions and alterations of existing networks.

SHORT-CIRCUIT CURRENTS

When a short circuit is closed on an inductive source, there is a transient d.c. surge which dies away within a few cycles. This is superimposed on the alternating fault current and so the first peak of current may be 1.7 times the peak value of the a.c. component of current alone. If synchronous or asynchronous rotating machines are connected to the network they will contribute to both the d.c. and a.c. components of the current during the initial surge, but only the a.c. components of the contributions from synchronous machines will persist after the initial surge, and these only at a reduced level. Therefore, if there is a preponderance of motor loads on the network, the ratio of the first current peak to the peak value of the steady state short-circuit current may be even greater than 1.7.

A transformer with a delta-connected primary winding and a star-connected secondary winding will make a greater contribution to the current in a nearby earth fault than it will to each phase current in a 3-phase fault, unless an earthing resistor is used as shown in fig. 17.10.

However, if transformers in separate substations are interconnected through a cable network, the contributions to fault currents made by remote transformers will be reduced by the impedances of the network and, because a cable network offers a higher impedance to earth fault currents than it does to currents flowing between phases, the maximum earth fault current on an interconnected network is often lower than the maximum 3-phase fault current even if earthing resistors are not used.

On high voltage urban networks there is not usually any difficulty in obtaining sufficient fault current to operate the appropriate protection, but on low voltage networks the system voltage and the ohmic impedances of the transformers and network are lower, so the impedance of cable local to the fault and the impedance of the fault itself are more effective in reducing the short-circuit current. Therefore, if low voltage cables, particularly small mains and service cables, are too long, the fault current will not be sufficient to operate the circuit protection at the substation.

PROTECTION

The main objective of power system protection is to maintain the stability and integrity of as much of the power system as possible by disconnecting faulty equipment that endangers the system. Therefore, it is necessary for the protection system not only to detect and disconnect the fault but also to be discriminative in isolating as little of the system as possible in the process.

The distribution system may be thought of as a number of units, each of which can be isolated by automatic circuit breakers. Unit protection compares the currents flowing at the ports of each unit, normally using pilot wires. It will discriminate to isolate only the faulty unit, independently of the level of short-circuit current, and without using time delays, although the protection zone cannot correspond exactly with the protected unit and there has to be either a blind spot or an overlap between the protection zones within the switchgear. In applying unit protection it is necessary, and desirable, so to arrange the automatic circuit breakers that the units are as small as possible, as in the arrangement shown in fig. 17.3.

In other classical protection systems, which may be used as a back-up to unit protection and to cover its blind spots, the basis for the detection of the fault is current level, while the basis for discrimination may be current level, graded time delays or system impedance between the protection relay and the fault. If there is a transformer in the circuit, an instantaneous overcurrent relay, discriminating purely on the basis of current level, may be used to protect the circuit as far as the primary winding of the transformer, but otherwise graded time delays are used to provide discrimination in both overcurrent and impedance protection schemes. This is significant because the size of cable required to carry a particular short-circuit current depends on the time for which the current is allowed to flow.

Allowing for the operating time of the circuit breaker, and for errors and overshoot in the relays, a differential time delay of about 0.4 s is necessary to ensure discrimination between circuit breakers. Within a high voltage distribution network there may be three overcurrent protection levels, one on the out-feeding transformers discriminating with the lower voltage protection, one on the circuit breakers that control the distributors at the in-feeding substations and an intermediate one. To obtain discrimination between these levels on a radial distributor the cables covered by the main distributor protection would therefore be required to carry the short-circuit current for about 1.5 s or longer at the higher voltages. A short circuit on an interconnector is fed from both ends and so the problems of discrimination and stress on cables are not as great.

The advent of digital techniques is making it possible to use criteria other than mere current level or impedance in detecting faults or discriminating in their isolation. For example 'source protection' sees faults not only on a cable distributor but also on the small transformers connected to it. It looks for sudden changes in the phase angles of the currents in the feeder at the source from which it is fed. It is very likely that a further development, 'differential source protection', will allow the same principle to be applied to a distributor fed from more than one source. It may also be possible to transmit digital signals directly over the power cable, so providing what is effectively unit protection without the use of pilot cables.

On LV distribution networks HRC fuses are used which, by their speed of operation, protect cables from damage due to high short-circuit currents. The greater problem is one of ensuring that sufficient short-circuit current will flow to operate the fuse. A 300 A fuse is unlikely to clear a fault that is fed by a 95 mm^2 aluminium cable longer than 800 m. This corresponds to the lengths of the interconnectors on a residential development with a housing density of 25 houses per hectare and a diversified demand of 1 kW per house, supplied from fully loaded 500 kVA substations. At higher load densities or lower transformer loadings, the interconnectors are shorter. The length of a spur cable that can be protected by the fuses at the ends of an interconnector depends on the position at which the spur is connected.

COMPUTER AIDS TO POWER SYSTEM ANALYSIS AND DESIGN

The electrical power system load flow was amongst the earliest of engineering problems to be tackled by digital computer, and it is now possible to perform large power flow, fault level and even transient stability calculations in a matter of seconds. Interactive computing allows a dialogue to take place between the engineer and the analysis program at a computer terminal, while a graphical display unit can present results in a clearly comprehensible form and also offer easy interaction via a 'light pen' or similar facility. These features are exploited in the Interactive Power System Analysis (IPSA) system, developed by the University of Manchester Institute of Science and Technology and Merseyside and North Wales Electricity Board. The engineer simply 'points' to the network diagram to indicate the next configuration to be studied, and results are displayed against the network diagram, or where appropriate as a graph.

Other interactive systems, for instance the ICL DINIS, are being developed to allow cable routes to be digitised graphically and used to calculate impedance and susceptance values that are then used in the system analysis.

For some years the Electricity Council in the UK, and subsequently the Electricity Association following privatisation, has monitored the loads of selected samples of consumers, and EA Technology Ltd, Capenhurst, has developed a computer program (DEBUT) which makes use of the typical daily load curve and variance of certain consumer types in finding the optimum tapering for a specified radial network.

CHAPTER 18
Distribution Cable Types, Design and Applications

In this chapter, distribution cables are regarded as power cables for fixed installation of rated voltages from 600/1000 V up to 19 000/33 000 V. Cables of lower voltage rating, auxiliary cables and flexible wiring cables are covered in part 2, and the types of cables used for systems operating above 33 kV are included in part 4.

VOLTAGE DESIGNATIONS AND EFFECTS ON CABLE DESIGN

Generally, distribution cables are designated according to the nominal system voltage on which they are intended to operate. The standard system voltages used in the UK are 1.9/3.3 kV, 3.8/6.6 kV, 6.35/11 kV, 12.7/22 kV and 19/33 kV. Cables are also manufactured for use on 8.7/15 kV systems for the export market.

The 600/1000 V cables are used in 230/400 V systems in the UK and at similar voltages in other countries. At this voltage level the cables are designated by the highest nominal system voltage for which they are suitable. The design of these cables is dictated by the mechanical rigours associated with their installation conditions and not simply by their electrical duty. Allowing for an accepted 10% upward variation on the nominal voltage, these cables are suitable for a maximum of 1100 V.

Voltage designations are in the form U_0/U, where U_0 relates to voltage between the conductor(s) and earth and U the voltage between conductors. Under normal operating conditions, the ratio of $U_0:U$ is $1:\sqrt{3}$. However, in some systems designed to enable continued operation with a fault to earth on one phase, the voltage on the two sound phases will rise towards the phase-to-phase voltage. It is acceptable, for example, to use 6.35/11 kV cables in an 11 kV system in which immediate isolation of the affected circuit does not occur when there is an earth fault, provided the operation time under fault conditions is limited. IEC Publication 183, 'Guide to the selection of high voltage cables', contains three categories of system A, B and C.

Category A Where earth faults are cleared as rapidly as possible but in any case within one minute.
Category B This category comprises those systems which, under fault conditions, are operated for a short time only with one phase earthed. This period should, in general, not exceed 1 hour, but a longer period can be tolerated as specified in the relevant cable standard.
Category C This category comprises all systems which do not fall into categories A or B.

In addition, the cable standards recommend cables of the same nominal voltage ratings for both category A and B systems. In most cable standards, the period for which cables are allowed to operate under category B conditions is extended up to 8 hours on any single occasion. In the UK there is a further stipulation that for classification as category B, the expected total duration of earth faults should not exceed 125 hours per year. When cables are expected to operate for longer periods with one phase earthed, cables of the next voltage higher should be specified, thus effectively increasing the insulation between phase conductor and earth.

For paper insulated cables of the belted type, it is necessary only to increase the thickness of belt insulation to cater for category C systems. Designs are therefore provided for category C systems with voltage designations of 3.3/3.3 kV, 6.6/6.6 kV and 8.7/11 kV. The value of 8.7 kV for the belted 11 kV category C system is chosen to provide consistency with the U_0 value for screened 8.7/15 kV cables.

So far in this chapter, for simplicity, the designation of voltage rating of cables from 3.3 to 33 kV has been explained in terms of standard British system voltages. In some other European countries, the nominal system voltages are not quite the same, being in general 10% lower, i.e. 3.0, 6.0, 10, 20 and 30 kV. These variations in the nominal system voltage can lead to doubts about the suitability of UK cables for continental applications and vice versa. This situation is clarified in IEC 183 by the designation of a maximum system voltage U_m.

U_m = The maximum r.m.s. power frequency voltage between any two conductors for which cables and accessories are designed. It is the highest voltage that can be sustained under normal operating conditions at any time and at any point in the system. It excludes temporary variations due to fault conditions and the sudden disconnection of large loads.

The relationship between U_0/U and U_m for cables rated from 3 kV up to 33 kV has been extracted from IEC 183 and is shown in table 18.1.

All cables of rated voltages to which a common value of U_m applies have to be suitable for that maximum voltage. 10 kV and 11 kV cables, for example, should be suitable for a maximum sustained voltage of 12 kV.

In North America different standard systems voltages exist (see IEC Publication 38, 'IEC standard voltages'). The standards published by ICEA, NEMA and AEIC for XLPE insulated cables include steps for insulation thickness, based on changes in voltage from 600 V through 2 kV, 5 kV, 8 kV, 15 kV, 25 kV, 28 kV, to 35 kV. These are not necessarily standard system voltages, but for a voltage between any two steps, the insulation thickness associated with the higher value is utilised.

Table 18.1 Relationship between U_0/U and U_m

Rated voltage of cables and accessories, U_0/U (kV)	Highest voltage for equipment, U_m (kV)
1.8/3 and 3/3; 1.9/3.3 and 3.3/3.3	3.6
3.6/6 and 6/6; 3.8/6.6 and 6.6/6.6	7.2
6/10 and 8.7/10; 6.35/11 and 8.7/11	12
8.7/15	17.5
12/20; 12.7/22	24
18/30; 19/33	36

The method of catering for differences between systems is also different in the USA. The cable standards refer to 100%, 133% and 173% insulation levels. The 100% insulation level is suitable for systems which provide immediate isolation when an earth fault occurs, the time limit is 1 minute. For longer periods of up to 1 hour, the 133% insulation level is suitable. If a phase is anticipated to remain earthed indefinitely, or presumably for more than 1 hour, the 173% level applies. Cable standards do not include XLPE insulation thicknesses for the 173% level, but do recommend consulting the cable manufacturer.

APPLICATION OF CABLE TYPES FOR PUBLIC SUPPLY

22 kV and 33 kV cables

In the UK and many other countries, 33 kV cables are widely used for distribution in public supply systems. Historically, pressure assisted types, such as fluid filled or internal gas pressure cables with either reinforced lead sheaths or aluminium sheaths, were utilised. In the UK, 3-core paper insulated solid type cables with steel wire armour have had the greatest use. However, since the 1980s, XLPE insulated cables have had increasing use as an alternative to paper insulated types.

The use of 22 kV cables for distribution in public supply is confined to the north-east of England where they are used in place of the 33 kV cables used elsewhere. However, 20 kV is widely used for distribution in other European countries such as France and Germany.

The traditional paper insulated cable used in the UK incorporates metallic screens on each core, lead sheath and steel wire armour, with either hessian serving or an extruded PVC sheath. In recent years there has been a trend to replace the steel wire armouring and serving with an extruded Medium Density Polyethylene (MDPE) sheath. The use of alternative 3-core, paper insulated constructions, such as the SL type of cable with a lead sheath on each core, has been very small.

In many parts of the world there has been a marked swing towards cables with crosslinked types of insulation, especially XLPE. Skill and care are required in the jointing of all types of cable, but the extruded insulations are less susceptible to moisture pick-up and do not involve the plumbing of metallic sheaths. Moreover, the crosslinked insulations can operate at higher temperatures and this is of greater benefit in hot countries such as those of the Middle East. Armouring is not always required for

XLPE insulated cables and if the magnitude and duration of earth fault currents can be kept reasonably low by the system arrangements, a relatively light copper wire screen will suffice as the surrounding earthed metallic layer.

11 kV and 15 kV cables

In the UK public supply system, 11 kV is the main distribution voltage between 240/415 V and 33 kV. Although 6.6 kV systems were once common they have now largely been converted to 11 kV. As already indicated, the continental system of 10 kV is regarded as equivalent to the British 11 kV.

For several decades up to about 1970 the type of cable used in the UK, Europe and countries influenced by them, was predominantly 3-core paper insulated, lead sheathed, armoured and served. The corresponding single-core cable, generally un-armoured, was used for short interconnectors or for very high currents requiring large conductors. Aluminium sheathed 11 kV cables were introduced into the UK in the early 1970s. A clear cost saving could be achieved by using an aluminium sheath with extruded PVC in place of the traditional lead sheath with steel wire armour and bituminised fibrous bedding and serving. Initially cables with both corrugated sheaths and smooth sheaths were used. However, the consensus of users favoured the corrugated version, mainly because of the greater flexibility and lower mechanical forces at joints.

As mentioned above, for 33 kV cables, in the world at large during the 1970s, the trend was away from paper insulated cables towards cables with crosslinked insulation. This change occurred more rapidly for 10–15 kV cables as the demands on the new insulating materials were less onerous at these lower voltages.

In view of the general world trend and the effect that this would have on the manufacturing plant, the cable makers and the electricity supply industry in the UK gave consideration to preparing a provisional design standard for a cable with crosslinked insulation. The first design from these joint deliberations was a 3-core cable having solid aluminium conductors, extruded semi-conducting screens, either XLPE or HEPR insulation, a semi-conducting extruded rubber bedding as a common core covering over the laid up cores, a concentric screen of copper wires applied in a waveform and a PVC oversheath. This design was used for only a relatively short period of time and the vast majority of UK supply companies continued to use the paper insulated, corrugated aluminium sheathed (PICAS) cable.

In recent years, the UK supply industry has again turned its attention to the potential benefits of cables with crosslinked insulation. Several designs are currently finding favour. In general the conductors are aluminium, either solid or stranded and the XLPE insulation is of the water tree retardant (WTR) type (see chapter 24). The majority of cables are three core in construction with collective metallic screens/armouring and MDPE outer sheath. The choice of collective copper wire screen or steel wire armour is largely based on the level of earth fault current the cable will be required to carry.

600/1000 V cables

The types of 600/1000 V cable used for public supply are covered in chapter 22, which deals with CNE cables. Before these types became established, the commonly used cable was the 4-core paper insulated, lead sheathed and armoured type. The four conductors were the three phase conductors and the neutral, with the lead sheath providing the main earth return path.

When PVC insulated 600/1000 V cable with shaped conductors became an established type at the beginning of the 1960s, it was not adopted for buried installations in the public supply network in the UK mainly because, as a thermoplastic material, the PVC could soften and deform at the overload temperatures which could arise. In the UK, adoption of 600/1000 V cables with extruded insulation for underground public supply systems awaited the development of cables with crosslinked types of insulation, which were first used in the 'Waveconal' type described in chapter 22. These types of insulation are more tolerant of overload than thermoplastic material such as PVC. In accordance with the normal approach of adopting new insulants first at the lower voltages, the Waveconal cable was the first type of distribution cable with XLPE insulation to be put to major use in the UK, early in the 1970s. In France, cable of the same basic design as the Waveconal type is used and there is also substantial use of Districable with XLPE insulation.

CABLES FOR GENERAL INDUSTRIAL APPLICATIONS

For power supplies within the more general types of industrial plants, the same types of cable have been used at the higher end of the voltage range as for the public supply system, but the relative usage of the various types is inclined to be more weighted towards those which are simpler to install and terminate.

11–33 kV cables

At 33 kV, the paper insulated, lead sheathed and armoured cable has predominated, but in recent years there has been a gradual change to the cables with extruded insulations, typically XLPE. For 11 kV systems, the traditional paper insulated lead sheathed and armoured cable of belted construction has, in the main, been replaced by XLPE insulated alternatives.

The use of aluminium conductors favoured for the public supply system in the UK is less general in the industrial sector. This is because, whilst cables with aluminium conductors have been intrinsically more economical than copper, effective terminations are more easily achieved with copper. There are two aspects to this: first the effective connection to copper by soldering or mechanical means, and second, the smaller copper cables are easier to accommodate in the connecting chambers of equipment.

The design of XLPE insulated cables used for industrial applications was initially based on IEC 502, 'Extruded solid dielectric insulated power cables for rated voltages from 1 kV up to 30 kV'. In the main these cables have stranded copper conductors, extruded semi-conducting conductor and core screens, copper tape screens on each core, extruded separation sheaths that act as a bedding for the wire armour, and extruded outer sheaths. Many variations of the above basic design have been developed to suit particular industrial applications. These variations have often been associated with the use of alternative bedding and sheathing materials such as PVC, MDPE and Low Smoke and Fumes (LSF) compounds. The relevant British Standard is BS 6622: 1991, 'Cables with extruded insulation for rated voltages from 3800/6600 V up to 19 000/33 000 V'.

Cables for voltages up to 3.3 kV

At the lower end of the voltage range (up to 3.3 kV), in contrast with the higher voltages, there was a rapid changeover from paper insulated to PVC insulated power cables for industrial use in the late 1950s and early 1960s. There was no need for conductor or insulation screens, thought desirable at higher voltages, and shaped conductors could be used, once the extrusion techniques were established. No metal sheath was required and the simplicity of terminations was a major factor in promoting the swing. The cable was also 'flame retardant' as defined by the current standard test.

The thermal limitations of PVC, which inhibited its use for cables in the underground public supply, are of much less account in industrial installations or in power stations. In these installations the current to be carried by the cable is much better defined and close excess current protection is provided. The international standard giving rules governing electrical installations up to 1 kV in buildings (IEC 364) requires that the overload protective device should operate within a time limit at a current not greater than 1.45 times the current rating of the cable. This requirement is repeated in the IEE Wiring Regulations, now BS 7671: 1992. This requirement ensures that conductors will be limited to a temperature not causing a high degree of deformation of the insulation.

The type of cable most generally used in the UK was the wire armoured construction specified in BS 6346: 1989. More recently this is being superseded by similar cables with XLPE insulation to BS 5467: 1989. In both cable specifications, the insulated cores are laid up and have an extruded PVC inner covering which acts as a bedding for the wire armour, over which is applied a PVC outer sheath. Both BS 6346 and 5467 include cables with stranded copper conductors and with solid aluminium conductors. Cables with stranded aluminium conductors can be supplied, but the concept, when the standards were prepared, was that a user wishing to have the lower cost aluminium conductor would want the most economical form of these, i.e. the solid type.

The XLPE insulated cables have advantages over the PVC versions of higher operating and short-circuit temperatures and are eminently suitable for the same types of installation. Advantage can be taken of the ratings when the current is limited by thermal considerations, but not when the conductor size is determined by voltage drop considerations.

During the 1980s, sheathing materials were developed which evolve very limited amounts of smoke and acidic fumes when involved in fires. This development, as exploited for example in the LSF (low smoke and fume) cable, led to BS 6724: 1990, 'Armoured cables for electricity supply having thermosetting insulation with low emissions of smoke and corrosive gases when affected by fire'.

Cables with low smoke and fume characteristics were first developed in the UK for the London Underground railway system, where of course the avoidance of dense smoke and irritant gases in the event of a fire is a very important consideration. Once the materials and means of production had been developed, their use spread to public buildings, hospitals, installations in tunnels, commercial premises and industrial environments.

Cables used in coal mines

The cables used to carry power into mines towards the working face are regarded essentially as cables for fixed installation, as distinct from the flexible cables used near the coal face which undergo movement and flexing while carrying current. The roadway cables, laid along the underground tunnels, form the large component of the fixed cabling but the cables installed in the shafts are an essential part. The roadway cables may be moved on occasions but are fixed while carrying load.

Until the introduction of PVC insulated mains cables, the type of cable used in the UK was generally insulated with impregnated paper, lead sheathed and armoured with two layers of galvanised steel wires (double wire armour). A typical outer covering was tape and jute braid impregnated with a flame-retardant paint.

When PVC insulated mains cables became available there was a fairly rapid change to them for the mining application. A particular advantage of PVC over insulation consisting of layers of paper tape is its greater mechanical resilience. Failures occasionally occurred in paper cables as a result of internal damage resulting from impact or crushing by rock falls. Tests indicated that PVC insulation was more resistant to such damage.

At that time most of the roadway and shaft cables were rated 1.9/3.3 kV and the design of PVC insulated cable adopted and still used followed fairly closely the design for this voltage covered by the British Standard. The main differences are that the bedding for the armour, which is extruded, is increased to give additional mechanical protection to the cores, a hessian tape is applied over the extruded bedding, i.e. immediately under the armour, and the latter has bitumen compound applied to it between and over the wires. The hessian tape is primarily to act as a holder for the bitumen and the bitumen is regarded as a desirable extra protection against corrosion for the armour. Originally the cables were double wire armoured (DWA), but later some use of single wire armoured (SWA) cables developed. A PVC sheath is extruded over the armour.

At one stage solid sector shaped aluminium conductors were adopted for these 3-core 3.3 kV cables, but they were subsequently abandoned, with reversion to stranded conductors in the interest of greater flexibility. The desire for flexibility also accounted for the adoption of single wire armour as an alternative to double wire armour.

The regulations relating to installations in mines required that the armour on the paper insulated cables should have a conductance not less than 50% of the largest conductor in the cable. Depending on the cable size, this made it necessary to include a number of tinned hard drawn copper wires in the armour of some of the cables. When the PVC insulated cable was adopted, as it had no lead sheath to supplement the armour conductance, a 75% minimum armour conductance was specified for the DWA cables. For the SWA cables, used later, to avoid too high a proportion of the wires being of copper, a 60% level of conductance is specified.

When 6.6 kV came into use in some mines the use of PVC insulated cables was extended to this voltage. This has been the major use of PVC insulated cables at 6.6 kV in the UK. One design of 6.6 kV cable was the same as for 3.3 kV except for an increased thickness of insulation. A second design was similar but had in addition a copper tape applied on a semiconducting tape bedding around each core. The prime purpose of the copper tapes is to provide an earth return path around each phase to carry sufficient current to operate sensitive protection in the event of an earth fault on any phase. The design with the copper tape screens tended to become the standard.

In the 1970s, when a further increase in voltage to 11 kV was contemplated, a small amount of 11 kV PVC cable was supplied. This differed from the 6.6 kV cable in that the conductors were circular and conductors and insulation were fully screened in addition to having greater insulation thickness. However, the UK National Coal Board (now British Coal) later decided to adopt EPR insulation for 11 kV cables and to extend its use downwards to 6.6 kV. The elastomeric properties of EPR, with their benefits for flexibility and resistance to impact and crushing, played an important part in this decision. The designs of the EPR insulated 6.6 kV and 11 kV cables are basically similar to those of the PVC insulated cables, as described above, except for the insulating material. In recent years XLPE has also been approved as an alternative to EPR for 6.6 kV and 11 kV cables.

British Coal have their own cable specifications. These have the numbers 295 for the PVC insulated cables and 656 for the EPR and XLPE insulated cables.

In most countries there are special designs of cable for mining applications. Usually extruded insulations and armouring are features. In the USA, however, an unarmoured mining cable is used which has copper tapes on the screened cores and copper conductors for earth currents laid in the interstices between cores. It has a fairly heavy thermoplastic or rubber sheath. As the mining techniques, traditions and experience vary, differences in emphasis on the characteristics required of the cables are reflected in the designs employed in different countries.

Cables for oil refineries

In those areas in oil refineries where there may be spillage of oil and its products or chemicals used in the industry, the designs of cable used generally incorporate some modification from the normal standard designs. Solvents especially are able to penetrate through the materials used for extruded insulations and sheaths. While this does not have a serious effect on the properties of the materials themselves, it is clearly desirable to prevent inflammable liquid or vapour gaining access to the cable, passing along it, and emerging at terminations. For such situations the cables usually have a metal sheath, not required for extruded insulations in most other installations.

For cables with PVC, XLPE or other extruded insulations which have to meet these conditions, a number of oil companies specify that a lead or lead alloy sheath should be applied over the assembly of cores. Some specifications require an extruded inner PVC sheath under the lead sheath, but more often the lead sheath is applied directly over a binder holding the cores together. The cables are wire armoured and have a PVC oversheath. The bedding for the armour is an extruded layer of PVC. For the conditions of use it is necessary that the covering over the lead should provide adequate protection against mechanical damage and corrosion.

When paper insulated cables are used, they are metal sheathed anyway, but for lead sheathed wire armoured cable, which is the usual type, PVC beddings and oversheath are used.

BASIS OF CABLE DESIGN

Thickness of paper insulation

Table 18.2 shows the thickness of paper insulation specified in the international standard IEC 55-2.

For paper insulation the thicknesses are minimum values; the specification requires that, when measured by the method stipulated, the thickness should not be less than

Table 18.2 Thicknesses of impregnated paper insulation

Conductor size (mm²)	600/1000 V single core (mm)	600/1000 V multicore Conductor/conductor (mm)	600/1000 V multicore Conductor/sheath (mm)	1.8/3 kV single core (mm)
4	–	1.2	1.0	–
6	–	1.2	1.0	–
10	–	1.2	1.0	–
16	–	1.2	1.0	–
25	–	1.4	1.2	–
35	–	1.4	1.2	–
50	1.2	1.4	1.2	1.8
70	1.2	1.4	1.2	1.8
95	1.3	1.4	1.2	1.8
120	1.3	1.4	1.2	1.8
150	1.4	1.8	1.4	1.8
185	1.4	1.8	1.4	1.8
240	1.6	2.0	1.6	1.8
300	1.7	2.0	1.6	1.8
400	1.8	2.0	1.6	1.9
500	2.0	–	–	2.0
630	2.0	–	–	2.0
800	2.0	–	–	2.0
1000	2.0	–	–	2.0

Rated voltage (kV)	Number of cores	Range of conductors (mm²)	Insulation thickness Conductor/conductor (mm)	Insulation thickness Conductor/sheath (mm)
1.8/3	3	16–400	2.4	1.8
3/3	3	16–400	2.4	2.1
3.6/6	1	50–1000	–	2.4
3.6/6	3	16–400	4.2	2.7
6/6	3	16–400	4.2	3.1
6/10	1	50–1000	–	3.0[a]
6/10	3 belted	16–400	5.8[a]	3.5[a]
6/10	3 screened	16–400	–	3.0[b]
8.7/10	3 belted	16–400	5.8[b]	4.3[b]
8.7/15	1	50–1000	–	3.9[b]
8.7/15	3 screened	25–400	–	3.9[b]
12/20	1	50–1000	–	5.0[b]
12/20	3 screened	25–400	–	5.0[b]
18/30	1	50	–	7.3[b]
		70–1000	–	7.0[b]
	3 screened	35	–	7.8[b]
		50	–	7.3[b]
		70–400	–	7.0[b]

[a] May include screening layers
[b] Includes screening layers

that shown in the table. The method of measurement in effect gives the average thickness of the layer around the core or around the cable, but with an insulation comprising layers of tape each of substantially uniform thickness there is no significant variation around the periphery.

As indicated by the notes to table 18.2, the thicknesses given in the tables for rated voltages 6/10 kV and above include allowances for semiconducting or metallised screening layers. Some of these screening layers are required and some are optional, the option of whether to include them being the manufacturer's. It is because some of the screening layers are optional that the insulation thicknesses are given in this way. If the manufacturer includes the screens, he may reduce the thickness of actual insulation by the applied screen thicknesses or by a certain allowed amount, whichever is smaller, the concept being that the improvement in electrical performance achieved by screening permits this reduction in the insulation. If the manufacturer uses a screen of thickness greater than the allowed value for either required or optional screens, then he may only reduce the insulation thickness by the allowed value. If the manufacturer does not apply the optional screens, then the full thickness given in the table has to be applied as insulation.

The following summarises this aspect.

6/10 kV single-core cables
Screening of conductor and over insulation are both optional. The allowance included in the thickness if either or both are used is up to 0.2 mm.

6/10 kV 3-core belted cables
Screening of conductor and over belt insulation are both optional. The allowances included in the thicknesses are as follows:

(a) if conductor screening only is used, up to 0.4 mm between conductors and up to 0.2 mm between conductors and sheath;
(b) if belt screening only is used, up to 0.2 mm between conductors and sheath;
(c) if both conductor and belt screening is used, up to 0.4 m between conductors and between conductors and sheath.

6/10 kV 3-core screened cables
Conductor and core screening are both required. The allowance included in the thickness is up to 0.2 mm.

8.7/10 kV 3-core belted cables
Conductor screening is required, and screening over belt insulation is optional. The allowances included in the thicknesses are as follows:

(a) if conductor screening only is used, up to 0.4 mm between conductors and up to 0.2 mm between conductors and sheath;
(b) if both conductor and belt screening are used, up to 0.4 mm between conductors and between conductors and sheath.

8.7/15 kV, 12/20 kV and 18/30 kV single-core and three-core screened cables
Screening of conductor(s) and over insulation is required. The allowance included in the thickness is up to 0.3 mm.

In BS 6480: 1988 those screening layers which are optional in the IEC standard are specified as required layers. This simplifies the way in which insulation thicknesses can be presented and they are specified as thicknesses of insulation alone, not including screens, appropriate amounts to allow for the latter having been subtracted from the IEC values.

While there are some variations, the thicknesses of paper insulations specified in the national standards of the Western European countries are much in accord with the IEC standard. This is less a matter of the national standards being based on the IEC standard than of the IEC standard taking account of the experience and the national standards of the developed countries, which existed long before the international standardisation took place, with compromises to resolve differences.

Thickness of extruded insulation

Table 18.3 shows the thicknesses of extruded insulation specified in IEC 502.

For this type of insulation the thicknesses given are 'minimum average' values, i.e. the average of a number of measurements around the core is required to be not less than the value in the table. The thickness of an extruded insulation is likely to be less uniform than that of a laminar insulation and the smallest of the set of measurements around the core, often designated the 'minimum thickness at a point', is allowed to fall below the thickness given in the table by an amount not exceeding 10% plus 0.1 mm.

These thicknesses for extruded insulants, given in table 18.3, are all exclusive of screening. The 3.6/6 kV cables with PE or XLPE insulation are required to have semiconducting screens over the conductor and over the insulation, but the 3.6/6 kV PVC and EPR cables need not include such screens. Cables for 6/10 kV and above have semiconducting conductor and insulation screens for all the insulating materials.

Generally these thicknesses are adopted by the European countries. There is a partial exception in the UK where the thicknesses for the material known as HEPR for 600/1000 V and 1.9/3.3 kV cables are those specified in IEC 502 or XLPE instead of those for EPR. The test requirements relating to mechanical properties for HEPR differ, however, from those for EPR in the IEC standards.

It is understood that in Japan the insulation thicknesses specified in national standards for XLPE insulated cables are somewhat greater than the IEC values but for export purposes the IEC values are used.

In the USA, where the rated voltages differ from those taken as the standard values in IEC 502, the IEC insulation thicknesses are not applicable. However, table 18.4 compares the 100% levels for XLPE specified in the ICEA/NEMA publication mentioned previously with IEC thicknesses, and it will be seen that there is not a very wide difference.

The insulation thicknesses specified in the USA for EPR for the voltage range 5–35 kV are the same as for XLPE. For these voltages both types of cable have conductor and insulation screening.

Dimensions of other component layers

Thicknesses of metallic sheaths and extruded non-metallic coverings and the dimensions of armour wires and tapes are generally derived from formulae or tables which relate the thickness to the diameter of the underlying cable. For some layers

Table 18.3 Thicknesses of extruded insulations

Conductor size (mm²)	600/1000 V			1.8/3 kV			3.6/6 kV			
	PVC (mm)	XLPE (mm)	EPR (mm)	PVC (mm)	XLPE (mm)	EPR (mm)	PVC (mm)	PE (mm)	XLPE (mm)	EPR (mm)
1.5 and 2.5	0.8	0.7	1.0							
4 and 6	1.0	0.7	1.0							
10	1.0	0.7	1.0	2.2	2.0	2.2	3.4	2.5	2.5	3.0
16	1.0	0.7	1.0	2.2	2.0	2.2	3.4	2.5	2.5	3.0
25	1.2	0.9	1.2	2.2	2.0	2.2	3.4	2.5	2.5	3.0
35	1.2	0.9	1.2	2.2	2.0	2.2	3.4	2.5	2.5	3.0
50	1.4	1.0	1.4	2.2	2.0	2.2	3.4	2.5	2.5	3.0
70	1.4	1.1	1.4	2.2	2.0	2.2	3.4	2.5	2.5	3.0
95	1.6	1.1	1.6	2.2	2.0	2.4	3.4	2.5	2.5	3.0
120	1.6	1.2	1.6	2.2	2.0	2.4	3.4	2.5	2.5	3.0
150	1.8	1.4	1.8	2.2	2.0	2.4	3.4	2.5	2.5	3.0
185	2.0	1.6	2.0	2.2	2.0	2.4	3.4	2.5	2.5	3.0
240	2.2	1.7	2.2	2.2	2.0	2.4	3.4	2.6	2.6	3.0
300	2.4	1.8	2.4	2.4	2.0	2.4	3.4	2.8	2.9	3.0
400	2.6	2.0	2.6	2.6	2.0	2.6	3.4	3.0	3.0	3.0
500	2.8	2.2	2.8	2.8	2.2	2.8	3.4	3.2	3.2	3.2
630	2.8	2.4	2.8	2.8	2.4	2.8	3.4	3.2	3.2	3.2
800	2.8	2.6	2.8	2.8	2.6	2.8	3.4	3.2	3.2	3.2
1000	3.0	2.8	3.0	3.0	2.8	3.0	3.4	3.2	3.2	3.2

Rated voltage (kV)	Range of conductors (mm²)	Insulation thickness			
		PVC (mm)	PE (mm)	XLPE (mm)	EPR (mm)
6/10	16–1000	4.0	3.4	3.4	3.4
8.7/15	25–1000	5.2	4.5	4.5	4.5
12/20	35–1000	6.4	5.5	5.5	5.5
18/30	50–1000	–	8.0	8.0	8.0

comprising lapped materials, such as tapes, the thickness may be the same for the whole range of diameters. Examples are fibrous beddings and servings for armoured paper insulated lead sheathed cables, which in IEC 55-2 and BS 6480: 1988 have specified approximate thicknesses of 1.5 and 2 mm respectively for the whole range of cable sizes. For other lapped layers there may be only two or three different thicknesses related to quite wide steps in diameter. For example, two thicknesses of steel tape, 0.5 or 0.8 mm, are sufficient for steel tape armouring for the range of cables in IEC 55-2 and BS 6480: 1988. Extruded layers, including metallic sheaths, generally have the most gradual

Table 18.4 Comparison of USA and IEC thicknesses of XLPE insulation

Rated voltage (V)	Insulation thickness USA (mm)	Insulation thickness IEC (mm)	Rated voltage (V)	Insulation thickness USA (mm)	Insulation thickness IEC (mm)
600	0.76–2.03		10 000		3.4
1000		0.7–2.8	15 000	4.44	4.5
2000	1.14–2.29		20 000		5.5
3000		2.0–2.8	25 000	6.6	
5000	2.29		28 000	7.11	
6000		2.5–3.2	30 000		8.0
8000	2.92		35 000	8.76	

increments in thickness corresponding with comparatively small steps in diameter. Wire armour is intermediate, with five different diameters for the same range of cable sizes as accommodated by the two thicknesses of steel tape.

IEC 502 (cables with extruded insulants) and IEC 55-2 (paper insulated lead sheathed cables) both include appendices which give the methods for calculating the thicknesses of the various coverings applied over the core, or laid-up cores, of the cable progressively up to the final oversheath. The diameters to which the thicknesses are related by formulae or tables are termed 'fictitious diameters', because they are calculated by a formalised method and do not represent the real diameters that any particular manufacturer is expected to achieve. Thus there are standard thicknesses required by the specification irrespective of variations in diameter between manufacturers, due, for example, to degree of compaction or methods of shaping conductors, or between batches produced by one manufacturer.

In IEC 55-2 the thicknesses are pre-calculated and presented in comprehensive tables for each standard voltage and number and size of conductors. In this document the inclusion of the method of calculation is therefore mainly a record, but it may be used by manufacturers to calculate values for cables generally in accordance with the standard but not of the standard sizes included in the detailed tables. In IEC 502 such pre-calculated dimensional tables are not included, the standard covering a wide range of possible constructions, but manufacturers follow the method to derive values for the particular construction they quote against individual enquiries.

The practice in British Standards is to give constructional tables with pre-calculated values of thicknesses and armour dimensions even for the cables with extruded insulants, and the calculation methods, which are basically similar to the IEC methods are not included.

Metallic coverings

Metallic sheaths are essential for paper insulated cables to exclude water from the insulation. For the extruded insulants there is not such an obvious need for a surrounding metallic layer for the effective functioning of the cable. However, IEC 502 requires that such cables of rated voltages above 1 kV should have a metallic covering. Unless there is an earthed low resistivity layer around the cable the electric field will extend beyond the limits of the cable and, if the stress is high enough, which may be the case at voltages above 1 kV, this may cause sparking to earthed metal in contact with or close to the cable or may affect any person who comes into contact with it.

If the cable includes a metal layer surrounding the cores individually or collectively for another purpose, for example metal tapes as part of the core screening, an earthed concentric conductor, a metallic sheath or an armour, then this may serve also for the shielding function. Otherwise a shield, which usually consists of copper tape(s) or wires, is applied especially for this purpose.

Cables without any metallic covering are used and are suitable for voltages above 1 kV in particular situations. An example would be an overhead insulated conductor which is out of reach and well spaced from earth. At ground level, cables without metallic coverings might be used within enclosures and spaced from earthed metal. In standards for cables for general purposes, however, the emphasis is on requirements which normally apply, leaving special designs to be dealt with on an individual basis, ensuring full understanding of any limitations on their use.

In addition to fulfilling this shielding function, and perhaps also being a water-impermeable sheath or an armour, the metallic covering is often used as the cable component to carry fault current to earth in the event of an earth fault on the system. For this purpose it needs to have sufficient conductance and thermal capacity to avoid it becoming overheated when carrying the prospective fault current for the time required to operate the overcurrent protective device. Steel tape armour is generally not suitable for this function. It is applied helically with a short lay and a gap between the turns and its resistance per unit length of cable is high compared with wire armour, which is applied with a comparatively long lay. Although there are two layers of steel tape, contact resistance between the layers is likely to be high when carrying heavy currents, especially if, as often applies, the tape surfaces are coated with a preservative compound.

For similar reasons copper wires are preferable to tapes as a shield when it is required to carry heavy earth fault currents.

In British practice standard types of cable with extruded insulation, when armoured, are armoured with wires rather than tapes. The size of the armour wire increases with increase in cable size and usually the wire armour will cope with the earth fault current. However, British manufacturers supply cables armoured with steel tape when this is required by overseas customers. Steel tape might be preferred, being of lower cost, when it is primarily to provide mechanical protection or serve as a shield and not to carry substantial fault current.

The use of wire armour in the UK has been generally more extensive than in most other countries. It was customary in the UK, when using lead sheathed cables in the public supply systems, to have steel tape armoured cables for low voltage and wire armoured cables for the higher voltages. Wire armour was regarded as the superior type appropriate for the more important cables and the different armouring was a convenient means of distinguishing between low and high voltage cables when carrying out excavation after the cables had been installed.

On lead sheathed cables the current carrying capability of the armour is of less importance than on cables not having a metallic sheath to fulfil this function. Wire armour, however, is advantageous in providing longitudinal strength to the cable as well as resistance to impact. The benefits of this are obvious for cables to be installed vertically or supported horizontally in cleats, or for cables to be pulled into ducts or laid elsewhere where pulling tensions are likely to be high. Even for buried cables, it can be an advantage if there is some ground movement after the cables have been laid. The steel wire used is galvanised and this generally preserves it well from corrosion. Steel tape armour is sometimes galvanised but more often has a coating of a bituminous compound.

In continental Europe steel tape is the most common form of armouring for lead sheathed cables over the range of distribution voltages, the tendency being to reserve wire armouring for installations where longitudinal strength is particularly required. Wire armour, when it is used, is often in the form of flat wires, whereas the standard in the UK is round wire. There is sometimes difficulty in inducing flat steel wires to lie flat on the cable when applied and often a steel tape is applied helically with an open lay on top of them to hold them in position. This tape is often called a 'counter helix'. Such a binder tape is sometimes used even on round wire armour, under a serving or oversheath, but the British experience indicates that this is unnecessary.

CHAPTER 19
Paper Insulated Distribution Cables

For distribution and transmission purposes impregnated paper insulated cables have had an impressive record of reliability throughout the 20th century, except for the development phases in the 1920s and 1930s when the use of belted cable was attempted at 33–66 kV (chapter 2). The reliability of paper cables is indicated by the fact that the UK supply industry depreciates paper distribution cables over a 40 year life.

Although the basic construction has changed little since impregnated paper was first introduced, there have been continuous improvements in materials and manufacturing techniques which have led over the years to high quality and reductions in dimensions. Probably the most significant change was the introduction of non-draining impregnants in the 1950s. This overcame the problem of migration of impregnant which left a relatively 'dry' dielectric.

Much information on the properties of all the materials used and of the impregnated paper dielectric has been included in chapter 3, which should be read in conjunction with this chapter.

CONSTRUCTION

Paper cables in the 1–33 kV range are often referred to as 'solid type' as they are designed to operate without internal or external pressure.

The insulation consists of helically applied paper tapes with a small gap between turns (fig. 19.1). The registration of tapes in relation to each other is important to avoid successive butt gaps in a radial direction (chapter 3). When cables are bent for drumming and laying, the paper tapes have to slide over each other without undue creasing, wrinkling or tearing and are therefore applied with a gap between turns. The gap width must be such that, when the cable is bent to the smallest permissible radius, it will not close completely and cause wrinkling of the paper. For bending reasons the mechanical design requirements are as important as electrical aspects in relation to insulation thickness, certainly for low voltage cables. These mechanical requirements cover such features as the angle and lapping tension during paper application, the width of the tape (generally 12–28 mm), the thickness of the paper (0.07–0.19 mm), the paper density and tensile strength.

Paper Insulated Distribution Cables 317

Fig. 19.1 Single-core, 300 mm², 600/1000 V, paper insulated lead sheathed cable with PVC oversheath

The conductors in multicore cables are usually sector shaped up to 11 kV and oval for 33 kV. Solid aluminium is used extensively at 1 kV. Stranded conductors are normally pre-spiralled (chapter 4) to reduce the possibility of damage to the insulation by twisting during the laying-up operation but this is not universal for the smaller sizes of 1 kV cable.

Belted construction

The cable design with a 'belt' of insulation over the laid-up cores (fig. 19.2) is the most economical in terms of total material cost. Such cables are nearly always used up to 6.6 kV and are the most common type at 11 kV.

The spaces between the cable cores under the belt are filled with jute or paper. Whereas the main insulation consists of paper tapes precisely applied, the filler insulation has to be softer and less dense so as to be compressed into the space available and is weaker electrically. Stresses in the fillers have to be limited to an acceptable level and therefore belted cables are not generally used at voltages greater than 11 kV.

Although any discharge between the outside of the insulation and the metallic sheath may produce discoloration of the impregnating compound, it has little effect on the life of the cable. However, when BS 6480 was revised in 1969 it was decided to specify the inclusion of a semiconducting carbon paper tape over the insulation for 11 kV belted cables, optional in IEC 55.

Fig. 19.2 4-core, 70 mm², 600/1000 V, paper insulated lead sheathed cable with STA and bituminous finish

Fig. 19.3 3-core, 150 mm^2, 6.35/11 kV, screened PILS cable with PVC oversheath

Screened construction

The electric strength of impregnated paper is weaker in the tangential direction than in the radial direction and for cables at voltages above 11 kV it is necessary to ensure that the electrical field is radial (chapter 2). As operating voltages were raised with 3-core cables in the 1920s and early 1930s, non-radial fields were the cause of extensive cable failures of belted cables. Hochstadter first pointed to the need for screening and the design is still sometimes referred to as H type. Screening consists of a thin metallic layer in contact with the metallic sheath (fig.19.3). As it carries only a small charging current, the thickness is unimportant but it is necessary to have smooth contact with the insulation together with an ability to withstand cable bending without damage. A paper/aluminium foil laminate, pinholed to allow penetration of compound during impregnation (chapter 25) and applied with an overlap, is normally used but a thin copper or aluminium tape is often preferred because it is less susceptible to damage. Multicore cables usually require a binder over the laid-up cores and this is normally a copper woven fabric (CWF) tape. Fine copper wires are included in the fabric to provide electrical contact between the screen and the metallic sheath.

Where screening is optional, i.e. at 11 kV, the disadvantages concern the extra complexity and cost of providing screened joints. The advantage of screening is improved electrical quality, permitting operation to a higher temperature and hence increased rating. Most users, however, prefer the belted design because of simpler jointing.

At voltage levels where it is necessary to adopt insulation screening a screening layer over the conductor is also required. This provides a smooth interface between the wires of a stranded conductor and the insulation, thus limiting discharge which may arise due to electrical stress enhancement on the strands or voids at the interface. Conventional practice is to apply two semiconducting carbon paper tapes over the conductors.

SL and SA screened cables

These are radial field single-core metallic sheath cables with electrostatic tape acting as the insulation screen. SL and SA refer to sheathing with lead and aluminium respectively. The three corrosion protected cores of SL cables are laid up together, armoured and finished with further corrosion protection (fig. 19.4). SA cables are laid up similarly with a PVC oversheath on each core but are not normally armoured.

Although the amount of metal in the three individual sheaths is little different from that in a cable having three cores within a single sheath, the greater diameter results in

Fig. 19.4 3-core, 150 mm^2, 19/33 kV, SL cable

extra bedding and armouring material, thereby increasing the total cable cost. However, jointing and terminating is more convenient. Some users prefer to install three separate single-core cables rather than a multicore cable, but the multicore SL and SA designs provide some of the advantages of single-core cable and avoid the high magnetic losses which would occur with steel armour.

SL and SA cables are not commonly used in the UK but are extensively employed in continental Europe and in other countries following European practice. A factor which has been associated with individual preference relates to the draining of oil–rosin impregnant on hilly routes, as there is less compound under the sheath with the single-core construction. In the UK the problem was overcome by the use of non-draining (MIND) insulation.

INSULATION CHARACTERISTICS

The characteristics of impregnated paper insulation, including moisture sensitivity and need for metallic sheathing, electric strength and resistance to discharge in butt-gap spaces, are reviewed in chapter 3.

It is not only the electric strength which defines insulation thickness. Particularly at the lower end of the voltage range mechanical requirements predominate and at the upper end allowance has to be made for the condition of the dielectric resulting from impregnant movement during load cycles to defined temperature. An indication of insulation thickness and operating stresses is shown in table 19.1 (see also chapter 18).

Even in the 6–15 kV range the electrical stress is not a primary factor and the operating stress of 2–4 MV/m is well below the capability of impregnated paper. However, because of the possibility of ionisation in voids it is necessary to ensure a high level of filling with impregnant.

The electrical stress becomes important at 33 kV and, partly to reduce stress at the conductor surface, oval conductors are usually used in preference to sector-shaped conductors. The use of oval conductors for multicore cables also improves the bending performance of cable with the thicker insulation. For the same reason, more insulation is required on very small conductor sizes, in contrast with the increase in insulation thickness with conductor size on lower voltage cables.

Table 19.1 Insulation thickness (minimum) and electrical stress (maximum) for paper insulated cables

Voltage (kV)	Conductor size (mm^2)	Belted design insulation thickness Between conductors (mm)	Conductor sheath (mm)	Single-core and screened design Insulation (mm)	Stress (MV/m)
0.6/1	50	1.4	1.2	1.2	0.24
	1000			2.0	0.13
1.9/3.3	50	2.4	1.8	1.8	1.3
	1000			2.0	1.0
3.8/6.6	50	4.2	2.7	2.4	2.0
	1000			2.4	1.7
6.35/11	50	5.6	3.4	2.8	2.9
	1000			2.8	2.4
8.7/15	50			3.6	3.3
	1000			3.6	2.6
12.7/22	50			4.9	3.8
	1000			4.9	2.8
19/33	50			7.3	4.4
	1000			6.8	3.2

MASS-IMPREGNATED NON-DRAINING (MIND) CABLES

Special impregnants are used to prevent drainage of the compound when cables are installed vertically or on slopes. Such cables were pioneered by BICC in 1949, initially at 1 kV. Following extensive development work in collaboration with Dussek Bros Ltd, during the next two decades they subsequently became almost universally standardised in the UK over the whole voltage range.

When cables with oil–rosin impregnant are installed on routes having differences in level, the impregnant migrates towards the lower end and, unless great care is taken in making joints and terminations, there may be leakage of the compound. On steep gradients barrier joints may be necessary and there is also a danger of lead sheath expansion resulting in fracture. More importantly, the loss of compound reduces the electrical strength of the insulation at the higher end, and the creation of a vacuous condition may lead to ingress of moisture if terminations are not completely sealed. At voltages of 11 kV and above, the loss of compound can cause a high level of discharge within the cable and experience has shown that this may be important even for quite modest vertical sections such as at pole terminations. Before the advent of MIND insulation it was necessary for cables on slopes either to be partially drained and to have considerably increased insulation thickness to reduce the stress or, alternatively, to be of the pre-impregnated paper type of non-draining construction.

During impregnation, the MIND compound is very fluid and readily saturates the insulation. On cooling to ambient temperature, the change of state to a soft solid form is accompanied by a slight contraction. With MIND insulation it is even more important than with oil–rosin to ensure that the cooling process is slow, the compound being circulated through a cooling system to obtain temperature uniformity. This also prevents the formation of a skin on the compound surface which could act as a barrier to further compound entry. Even so, the volume impregnation factor is of the order of 93% in comparison with 96% for oil–rosin. Later MIND compounds have been better than the original ones with respect to coefficient of contraction.

In service the MIND compound will remain in position whereas oil–rosin compound may readily drain away and leave more voids in the insulation. However, when cables are tested immediately following manufacture, MIND insulation shows more ionisation, i.e. greater difference in power factor between one-half working voltage and twice working voltage. When BS 480 was amended in 1954, the limit for multicore 33 kV cables was raised from 0.0006 for oil–rosin to 0.006 for MIND to allow for this feature and further revision is detailed in BS 6480: 1988. In addition, the impulse strength of MIND cable is slightly below that of virgin oil–rosin cable. Nevertheless, the BS 6480: 1988 requirements of 194 kV at 33 kV and 95 kV at 11 kV are readily met, the latter being above the IEC 55 requirement of 75 kV.

Initial ionisation is not of importance and MIND cable provides excellent long-term performance, as shown by a stability test with daily load cycles to a temperature 5°C above the maximum permitted continuous limit and a voltage of 1.33 times the working level.

Periodic measurements of the power factor at 1.5 and 2.0 times working voltage show that both oil–rosin and MIND cables reach a peak and then stabilise. MIND cables reach this peak more quickly, generally after a few cycles, and the level may subsequently decrease. At working voltage, however, in contrast with oil–rosin insulation, MIND insulation shows little or no ionisation after a prolonged period of testing.

It has been argued in some countries that, under short-circuit conditions, MIND insulation could be inferior because the conductor temperature permitted is such that the compound may change to the liquid state. This has been difficult to establish experimentally and it is apparent that the temperature of the mass of cable is well below that of the conductor. MIND cables have shown a very satisfactory service performance for over 30 years. Any shortcomings in this and any other respects arising in non-draining cables made elsewhere have no doubt been due to lack of recognition of the expertise necessary in formulating satisfactory compounds and correct processing techniques.

METALLIC SHEATHS

Paper insulated cables are sheathed with either lead or aluminium and the characteristics of these two metallic sheaths have been discussed in chapter 3.

ALUMINIUM SHEATHED PAPER INSULATED CABLES

In the early 1970s, many users throughout the world were considering changing to MV polymeric insulated cables. British manufacturers however, conscious that the service reliability of such cables appeared to be much lower than that of paper cables, drew attention to a further economy which could be obtained with paper cables whilst further development of polymeric insulation was proceeding. Substitution of an aluminium sheath for a lead sheath and armour provided a price saving of around 25%. Service trials were quickly arranged and by 1975 nearly all the UK electricity utilities had changed to this construction for their 11 kV networks.

One problem in the early years related to the multiplicity of designs available, all adding to the cablemakers' stocks. In addition to lead sheathed cables not being superseded for many years, the aluminium sheath could be smooth, which was slightly cheaper, or corrugated (fig. 19.5) and, as with lead sheathed cables, some authorities favoured a belted and others a screened design. A review was carried out by the utilities and in 1978 it was considered that the much better handling characteristics of the cable with corrugated sheath had a distinct advantage. It was decided to standardise the belted cable with corrugated sheath in conductor sizes of 95, 185 and 300 mm^2, though the utilities which had traditionally used screened cable remained with this type, as detailed in EA 09-12.

As cable weight was reduced by 50%, smooth sheath designs up to 185 mm^2 were no more difficult to install than lead sheathed cable but problems could arise with larger sizes in duct installations because of the stiffness of the cable. This extra rigidity was also detrimental in setting cable ends into equipment when space was limited in substations. The corrugated sheathed cable was however easy to install and had a distinct advantage over lead sheathed cable.

The increased rigidity of smooth sheath designs, together with the high coefficient of expansion of aluminium, also caused concern regarding mechanical stresses in joints. Within the joint there can be 'bowing' of the cores in the joint sleeve after heating cycles. Whilst these problems can readily be overcome by the use of joints of the cast resin filled type, the increase in cost is not always considered to be justified. Those utilities which used smooth sheathed cables continued with bitumen filled joints on the basis that the cables were installed in open-ring circuits and very seldom reached full design load.

Manufacturing technique is particularly important for cables with corrugated sheaths because of the considerable increase in space between the insulation and the inside of the sheath. If this is completely filled with impregnating compound, it is possible to generate very high internal pressures at operating temperature. It is consequently necessary to control the amount of compound between that necessary to prevent ionisation within the insulation and that producing maximum pressure on heating. In addition, if the aluminium sheath should be punctured and water entered the cable, all the cable affected would have to be replaced. Sufficient impregnating compound therefore needs to be present to prevent water from penetrating along the cable. The

Fig. 19.5 3-core, 11 kV, belted PIAS cable with corrugated aluminium sheath

annular corrugation design used for these solid type cables (fig. 19.5) has positive advantages. In the event of local puncture of the sheath, the small clearance at the root of the corrugation, containing impregnating compound, is an obstacle to transmission of water along the cable. Furthermore, it is much easier to cut around one crest of a corrugation rib when a piece of sheath has to be removed for jointing or terminating.

The economic advantage of the aluminium sheathed design (EA 09-12) has undoubtedly been the main reason for the continued use of medium voltage distribution paper cables in the UK, whereas XLPE insulation has proceeded to oust paper in many other countries. XLPE cables are becoming competitive in price and their other attraction, that they do not require such highly skilled jointers, is having more impact as the traditional skills become less available.

ARMOUR

The choice between steel tape (STA) and galvanised steel wire (GSW) armour for multicore cables is very much related to local practice. Generally in the UK STA is adopted for 0.6/1 kV and GSW is favoured for higher voltages because of its better resistance to mechanical damage and provision of improved longitudinal cable strength for handling purposes. However, in continental Europe and many other countries, steel tape armour, sometimes galvanised, tends to be used more extensively throughout the voltage range.

Aluminium sheaths are not armoured and are finished with an extruded layer of polymeric material, commonly PVC. Lead sheathed and armoured cables are now also often finished with extruded PVC or MDPE but bituminous corrosion protection is also employed (chapter 5).

OPERATIONAL AND INSTALLATION PARAMETERS

Continuous operating temperature

Conductor temperature limits as a basis for cable ratings are referenced in chapter 8. Table 19.2 summarises the requirements for paper cables.

The temperature of 80°C for 1 kV cables is based on the physical and electrical degradation of the dielectric materials. At the higher voltages the limit is primarily associated with the imposition of conditions to prevent increase in ionisation within the

Table 19.2 Conductor temperature limits for paper cables

Voltage (kV)	Cable design	Maximum conductor temperature (°C)
0.6/1, 1.9/3.3, 3.8/6.6	Belted	80
6.35/11	Belted	65
6.35/11, 8.7/15	Screened	70
12.7/22, 19/33	Screened	65

dielectric. Most distribution cables operate with cyclic loading predominantly on a daily basis. During heating there is expansion of the impregnant and for various reasons the compound may not flow back between the layers of paper during cooling, thus creating voids in which discharge may occur. Lead sheaths readily expand under pressure but do not contract when the pressure is reduced, thus reducing the likelihood of the compound returning into the insulation. All these effects have a bearing on the maximum temperature which can be adopted. The temperature is lower for belted cables, owing to the non-radial field, and reduces with increasing voltage because of the more serious effect of ionisation on cable life.

Technically it has been proved feasible to develop satisfactory impregnating compounds and cable construction to permit a temperature of 85°C up to 33 kV but such cables have not been brought into commercial use.

Bending radii

Materials and methods of manufacture have to be chosen to mitigate against dielectric disturbance on bending. IEC 55 and most national standards prescribe the minimum bending radii quoted in appendix A19.

The stiffness of impregnants changes with temperature to an extent which varies with composition and types of compound used. If cables are to be installed in cold climates, it is usual to prescribe a bending test to be carried out at an appropriate temperature, to ensure that there is no disruption of the insulation.

LIFE AND PERFORMANCE CHARACTERISTICS

The service performance of paper cables has been so good that it is accepted as the standard by which other types are judged. Apart from mechanical damage, one of the main causes of cable replacement is the entry of water and the travel of the water through a considerable length of cable. Electrical failures in the dielectric are rare up to 6 kV. At 11 kV and above they are more usually associated with the discharge and carbon tracking ('treeing') mechanism discussed in chapter 2.

An interesting type of breakdown was known for many years as 'pinhole type failure below terminations on 11 kV belted cable'. It was common in South Africa but also arose elsewhere. Although not necessarily associated with high altitudes the incidence was greater at heights of around 1000 m. The failures invariably occurred in the vertical cable below terminations and at a distance of 1 to 15 m from the termination. There was intense pinhole type erosion of the dielectric papers, with no carbonisation but with green deposits both on the conductor and on adjacent insulating papers. After lengthy investigations it was established that the breakdown started from drainage of oil–rosin impregnant which allowed air to enter the cable through incomplete sealing of the wire interstices of the conductor in the termination. Discharge then occurred in the air-filled voids in the dielectric in the region of highest stress, i.e. where the conductors approach each other most closely and where the radii of the shaped conductors are smallest. The discharges resulted in the formation of oxides of nitrogen and ozone which attacked the paper to produce the pinholes and led to production of moisture. This combined with the oxides of nitrogen to form nitric acid which reacted with the copper to give the green copper nitrate.

The solution was to use hermetically sealed termination boxes. Also the use of non-draining cables mitigates against creation of voids and the suction of air into the cable.

TESTS

Testing procedure and requirements are reviewed and discussed in chapter 28.

CHAPTER 20
PVC Insulated Cables

PVC rapidly replaced rubber for wiring cables in the 1940s and was introduced in Europe for power cable insulation in the late 1950s, initially in significant commercial quantity in Germany. For corrosion protection, particularly with aluminium sheathed transmission cables, extruded PVC oversheaths became firmly established in the mid-1950s.

At that time, paper insulated cables were being used for industrial distribution and as fully impregnated non-draining paper had not then been completely accepted it was necessary to adopt some form of limitation of the amount of impregnating compound in the cable to minimise compound drainage problems. This caused either reduction in quality or increased insulation thickness. The use of PVC provided cables of excellent quality which were clean and much easier to handle. Being little affected by moisture they do not require a metal sheath and this also simplifies jointing and terminating. Consequently they quickly became adopted for industrial power applications.

A particular feature of the early development was associated with the fact that, whereas wiring cables have circular conductors, the conductors of power cables were sector shaped and the larger sizes pre-spiralled. Shaped-solid aluminium conductors had emerged as a strong competitor to copper at the same time. The use of shaped dies to extrude PVC to the profile of the conductor presented great concentricity problems which led to the 'float-down' or 'tubing-on' extrusion technique. This involves extruding the PVC as an oversize circular tube which is drawn to a snug fit on the conductor by a combination of vacuum and controlled conductor/extrusion speeds.

It can be said that PVC power cables truly became established in the UK by the IEE Symposium on Plastic Insulated Mains Cable Systems in November 1962 and the papers presented provide an interesting record.

APPLICATIONS

The fields of use fall into four distinct categories as discussed below. Especially as PVC from the outset had to compete with impregnated paper, the most important factor to be taken into account was the amount of softening at raised temperatures due to its thermoplastic nature. This can result in deformation of the insulation due, for example,

PVC Insulated Cables

to conductor thrust at bends. Paper cables will withstand fairly high short time overloads and consequently the fuse co-ordination does not need to be particularly refined. PVC cables need adequate protection against overload or alternatively a reduced rating has to be assigned to them. The situation in relation to the sixteenth edition of the IEE Wiring Regulations (BS 7671: 1992) is discussed in chapter 10.

Industrial cables

Industrial usage covers distribution in and around factories at voltages up to 3.3 kV. Except possibly in North America, PVC cables have been almost universally used throughout the world since around 1960 and overcurrent protection does not present any serious problem. In the UK the conductors have been stranded copper or solid aluminium. In some other countries, stranded rather than solid aluminium has been preferred, but this has been due to some extent to economic factors relating to conductor manufacture rather than strict technical preference. Standard designs comprise single-core (stranded and solid sectoral) construction (fig. 20.1) and multicore constructions of 2, 3 and 4 cores (fig. 20.2) of equal shaped conductor sizes.

Public supply

Apart from house service cables, PVC has not been used for public supply in the UK because of the overcurrent protection problem. Normal distribution systems provide little or no protection for the cables, e.g. two electricity supply companies operate with

Fig. 20.1 600/1000 V, single-core sectoral aluminium conductor, PVC insulated cable

Fig. 20.2 600/1000 V, 4-core copper conductor, PVC insulated SWA cable with extruded bedding

a solid mesh and many others have a very coarse fuse at the substation at the end of a cable which may be tapered in cross-section along the route. This is not so in many other countries and there has been widespread use of PVC, even sometimes in tropical regions, where its use would be unexpected because of the undesirable effects of derating etc. However, PVC clearly has drawbacks for any public supply system and XLPE is taking over in this field.

Coal mining

Much power cable is used for fixed installation along underground roadways to take the supply from a main colliery shaft to the working face. Because the point of use is constantly changing, the individual cable lengths are fitted with couplers to enable the distribution system to be moved around as required. In the UK the standard voltages are 3.3 kV and 6.6 kV, with some requirement at 11 kV.

Paper cables were conventional until the 1960s but were then replaced by PVC primarily because of the better resilience of PVC in withstanding rock falls. UK requirements for 3.3 kV and 6.6 kV cables in the later 1970s were at the rate of about 650 km/year. Similar use at 11 kV has largely been the only application of PVC at this voltage in the UK, but, as with other high voltage applications, it has now given way to XLPE-insulated versions. At 6.6 kV, the trend from PVC to XLPE is well advanced and a similar movement is expected at 3.3 kV in due course. In recent years, the decline in coal mining activity in the UK has been reflected in a large reduction in cabling needs.

Initially the conductors were copper, but for economic reasons solid aluminium was introduced. After some years there was a further change to stranded aluminium because of flexibility problems in coiling the cables to take them down the mine shafts. Until the time that more attention was paid to flexibility (mid-1970s) it had been traditional for all such roadway distribution cables to be double wire armoured. This provided high earth conductance and a rugged construction. Subsequently both DWA and SWA designs have been used with earth conductances of 75% and 60% respectively. Shaft cables always have DWA and with single-point suspension the top section has quadruple armour. Figure 20.3 shows a typical PVC insulated mining cable.

Fig. 20.3 3.8/6.6 kV, 3-core aluminium conductor, PVC insulated cable for roadway distribution in coal mines

High voltage cables

Reference to the use of PVC for HV cables is made in a later section.

DESIGN AND CONSTRUCTION

PVC cables may be unarmoured or armoured. If unarmoured they have an extruded PVC sheath over the single core or laid-up cores. If armoured the bedding may comprise plastic tapes or extruded PVC and an extruded PVC sheath is applied over the armour. Dimensions, weights, electrical characteristics and ratings of cables to BS 6346: 1989 are given in appendix A13.

Prior to the first issue of IEC 502 in 1975, there were some small differences between National Standards, but since then most have been amended to become almost identical.

Insulation

In the 1–3 kV range the insulation thickness required is largely dependent on mechanical considerations, e.g. the physical loading which might be imposed on the cable. Such loading may result in some deformation of the insulation at raised temperatures, e.g. during emergency conditions in service operation. This is why there is more variation of thickness with conductor size than with voltage. At higher voltages electrical requirements predominate.

The general characteristics and properties of PVC insulation are discussed in chapter 3 and the special formulations for further reduction in flame propagation, together with reduced generation of acid gas and smoke, in chapter 6. For the normal run of cables BS 6346: 1989 stipulates a 'general purpose' compound from the types listed in BS 7655. However, BS 6346: 1989 caters only for cables up to 3.3 kV. For higher voltages the reduction in insulation resistance and increase in dielectric power factor at the top end of the operating temperature range become important to avoid instability (chapter 3). IEC 502 caters for this by stipulating PVC/A up to 3 kV and PVC/B for higher voltages. The essential requirement for PVC/B is that the product of the permittivity and the power factor must not exceed 0.75 between ambient temperature and 85°C; also the power factor at 80°C must not exceed the value at 60°C.

A matter which causes some confusion is that three essential limits dictate the service operation of PVC insulation: (a) a sustained maximum temperature of 70°C, which is determined by the thermal ageing characteristics of the material; (b) a temperature of the order of 120°C which governs the maximum degree of deformation permissible in the time/temperature range required for the circuit protection to operate; and (c) a limit of 160°C used for the calculation of short-circuit ratings. The performance of the material can be assessed by tests on the material but overload and short-circuit limits have to be defined by tests on complete cables because they are at least partially dependent on the performance of the cable as a whole. Much work on these aspects has been carried out by ERA Technology Ltd.[1]

The confusion arises particularly with heat-resisting grades of PVC because, whilst they give some benefits in better resistance to ageing at temperatures above 70°C, they may be little or no better in relation to deformation at these higher temperatures.

BS 7655: Part 3: 1993, 'PVC insulating compounds' gives requirements for heat-resisting PVC compounds for a maximum conductor temperature of 90°C. References may also be found to compounds for 100°C or even 105°C. The latter are generally not for power cables and relate to applications where the service at these temperatures is of relatively short duration and a limitation on cable life is recognised.

The essential point is that although heat-resisting compounds cannot be used to obtain higher sustained ratings, because of the circuit protection issue, they do have a value if cables have to be derated due to operation at higher ambient temperatures. If the ambient temperature is 25°C the permissible temperature rise for standard PVC cable is 45°C and this governs the sustained rating value. With a 90°C compound the same temperature rise and rating can be obtained from an ambient temperature of 45°C. The situation on circuit protection remains the same at 1.45 times the rated current.

Such heat-resisting compounds are, of course, significantly more expensive and when high temperature operation is important, ambient or otherwise, it is now usually more economic to use insulation of the thermoset type (e.g. XLPE or EPR) as this has a sustained conductor temperature limit of 90°C.

Considerations relating to the flexibility of PVC also apply to cables required for installations at low temperatures, i.e. in countries with severe winters. It is not recommended to install normal PVC cables below 0°C because standard PVC compounds become increasingly stiff and brittle at low temperatures. Special compounds with increased amounts of selected plasticisers can be formulated but here again it is now generally preferable on economic and technical grounds to use polyethylene or thermoset materials.

Core identification

The cores are normally identified by the colour of the insulation but there is no international standardisation. British practice is based on table 20.1, with red, yellow and blue indicating phase conductors and black the neutral.

Fillers and armour bedding

Fillers are required mainly with cables having circular cores so as to provide a reasonably circular profile. The armour bedding may comprise either (a) non-hygroscopic tapes of around 0.8 mm thickness or (b) an extruded layer of suitable material (usually PVC) of thickness 1.0–2.0 mm according to cable diameter.

Table 20.1 Core identification in PVC cables

Cable type	Colours
Single-core	Red or black
2-core	Red, black
3-core	Red, yellow, blue
4-core	Red, yellow, blue, black

The extruded bedding provides a more robust construction and is normally recommended for cables buried in the ground. It is also essential if certain types of termination gland are used in flame-proof enclosures.

Armour

Standard specifications cater for a wide range of choice for armour, e.g. IEC 502 covers galvanised steel wire, plain or galvanised steel tape, plain or galvanised steel strip and aluminium strip. Comments on design considerations relevant to the different forms of armour used are given in chapter 18, and much depends on individual customer preferences or what had traditionally become accepted practice with paper cables. This has varied widely in different countries and established practices remain.

In the UK only two types of armour have been adopted and recognised by BS 6346: 1989 – galvanised steel wire and aluminium strip. Steel tape armour, as commonly used with paper cables, is more frequently used in continental Europe, and in some countries it is galvanised when the users have a preference for preserving the armour in aggressive environments. With PVC cables there is a greater tendency to use galvanised tapes, thus avoiding the need to use bitumen. The armour can also act as an earth conductor, though steel tape is not very efficient in this respect. Steel strip armour is also widely used in continental Europe and the strip is normally galvanised.

The use of aluminium strip armour came early in the development of PVC cables and with the backing of the aluminium industry was associated with the introduction of solid aluminium conductors to produce an all-aluminium cable. The incentive at the time was also towards the use of the armour as a concentric neutral conductor. The strip type neutral never came to fruition, but for a time many users adopted aluminium strip armour, either because of the reduced cable price then prevailing or to obtain higher armour conductance in the earthing system. Also in the UK a view prevailed for a time that the IEE Wiring Regulations should be interpreted as requiring that in any cable the armour should have 50% of the conductance of a phase conductor – a situation which did not exist with many sizes of lead sheathed cable or PVC cables having galvanised steel armour. It was eventually established that such a requirement was not necessary. Nevertheless, there are occasions when the extra conductance of aluminium strip armour is beneficial and nowadays it is mainly only in such instances that it is used. The price advantage compared with galvanised steel wire has been lost, although it must be recognised that relative prices vary from time to time and from country to country.

Aluminium in strip or wire form also has an advantage for single-core cables because of the need to use non-magnetic material. A factor to be taken into account with aluminium armoured cables is that the termination glands should also be made from aluminium or be otherwise designed to be compatible.

A final point about the comparison between steel wire and aluminium strip armour is the resistance of the cable to impact blows. The aluminium strip is of much lower thickness and provides less resistance to damage.

Oversheath

IEC 502 allows for galvanised steel armour to be left bare or for cables to be oversheathed with PVC, polyethylene or similar material. Specific test requirements are quoted for PVC and polyethylene. Crosslinked elastomeric materials such as PCP and CSP are allowed in IEC 502, but these materials are not intended for use with PVC cables because of the risk of deformation and degradation during the vulcanisation process.

BS 6346: 1989 excludes bare armoured cables and stipulates only a PVC oversheath of the same grade of material as the standard general purpose insulating compound. In practice bare armoured cables would only be suitable for clean dry indoor conditions and apart from the improved visual appearance the small extra cost for the oversheath is well worthwhile. It is believed that very little bare cable has been used throughout the world.

When PVC cables were first developed it was common practice to apply bitumen over the armour to obtain additional corrosion protection, i.e. following conventional paper cable practice. However, the bitumen extracted plasticiser from the PVC, so becoming a thin mobile liquid. Many industrial cables are installed with terminations below a vertical run of cable and the result was a nasty dripping of black liquid at the termination. The plasticiser migration can be avoided by having a separation layer of suitable tape between the bitumen and oversheath but such a layer is not easy to apply effectively and the value of the bitumen is doubtful. It was subsequently discarded for standard types of cable, although retained for special applications such as mining cables where the use of bitumen has been considered to be beneficial.

HIGH VOLTAGE PVC CABLES

IEC Specification 502 covers designs of PVC cables up to 15 kV but with a few exceptions the amount of cable used above 3.3 kV has been relatively small. These exceptions consist mainly of the mining cables discussed earlier and those used in Germany. Chapter 24 deals with the subject of high voltage polymeric insulated cables and only the main significant features are outlined below.

Table 20.2 Comparative dielectric losses in cables

Voltage (kV)	Conductor size (mm^2)	Insulation	Dielectric loss (kWh/km year)
3.8/6.6	50	PVC	3109
	1000	PVC	11658
	50	XLPE	11
	1000	XLPE	30
	50	Paper	57
	1000	Paper	224
6.35/11	50	PVC	8955
	1000	PVC	29258
	50	XLPE	23
	1000	XLPE	78
	50	Paper	145
	1000	Paper	546

In comparison with the alternative materials, such as impregnated paper and polyethylene, the electrical losses in PVC are high. Table 20.2 gives figures for some typical cables. Nevertheless, the dielectric loss is still only a small proportion of the total losses, the conductor loss being predominant.

However, as PVC is not competitive with XLPE the previous limited use has now almost ceased. A higher operating temperature can be sustained by XLPE, and hence higher ratings, and it is a more universal material in that it can be used up to much higher voltages.

Some interest in PVC has derived from the fact that it has much better resistance than PE or XLPE to partial discharge. Thus whilst PE and XLPE cables need to have conductor and dielectric screens above 3.3 kV, they are only required above 6.6 kV with PVC (and also with EPR).

OPERATING CHARACTERISTICS

Ground containing hydrocarbons (oil refineries)

One of the few situations requiring special attention with PVC cables concerns installation in ground containing a significant quantity of petroleum hydrocarbon. This arises most often in oil refineries but it can happen in any place where such hydrocarbons are being handled. As discussed in chapter 5, PVC oversheaths are comparatively resistant to attack by petrol or similar hydrocarbons. All that happens from occasional spillages is that some plasticiser is leached out, with corresponding hardening, but if the cables are not subsequently subjected to severe bending this has little effect. The loss of plasticiser is not detrimental to the protective properties of the PVC.

However, whilst surrounded by hydrocarbons in this way, some of the hydrocarbon can diffuse through the PVC and will be present in any air spaces within the cable. It can then travel along the cable to leak out at joints or terminations and this could create a hazard. Consequently it is a common practice for PVC insulated cables in oil refineries to be lead sheathed. An alternative solution in some circumstances is to seal the cable and conductors at joints and terminations.

Operational life of PVC cables

As PVC insulation for power cables has only been in general use since the early 1960s it is perhaps still a little early to judge whether the ultimate life of PVC power cables will match that of paper insulated cables. There is no doubt that operating performance has been good and very few failures have occurred due to overheating. Problems due to softening at raised temperatures have already been referred to. A few failures have also occurred due to overheating resulting in thermal degradation. Some may have been due to inadequate attention to formulation to obtain the required stability but inevitably, if the standard PVC compound is heated for long periods much above its rated temperature, degradation is likely. The extent of degradation depends on temperature and time. The PVC becomes hard and brittle and darkens in colour and, due to the liberation of hydrochloric acid, copper conductors will be attacked with the production of green chlorides.

Loss of plasticiser due to heating in normal service or due to contact with other materials, such as bitumen in joint boxes, is rarely a problem. Some hardening of the PVC occurs but the electrical properties do not deteriorate. Providing the cables do not have to be moved subsequently, the loss of flexibility is not usually of importance.

Flame retardance

PVC cables in general, and the standard PVC compound as a material, are, to a degree, intrinsically flame retardant. However, under some conditions they may burn and it is even possible for flames to travel along the cable. Alternative compounds and construction are available when improved flame retardance is required, as described in chapter 6.

REFERENCE

(1) Parr, R. G. and Yap, J. S. (1965) 'Short-circuit ratings for PVC insulated cables'. ERA Report No. 5056.

CHAPTER 21

Thermoset Insulated Cables up to 3.3 kV

This chapter is essentially a supplement to the previous one, but for cable with an alternative insulation material. The two thermoset materials required by most specifications are XLPE and EPR, but the latter is rarely adopted outside Italy and in this chapter the reference is essentially to XLPE. Requirements for EPR cables are identical to those for XLPE except, of course, that specific values for the insulation as a component are slightly different for the individual materials. Cable constructions are basically the same as those for cables with PVC insulation. The applications are also the same, but with some extension because in many countries, for reasons already stated, PVC cables have not found much favour for public supply systems. In general, XLPE insulated cables are a competitive alternative to PVC cables for industrial use and to paper insulated cables or public supply systems.

Cables up to 3.3 kV do not require conductor or insulation screens, whereas for 6.6 kV upwards the position is different in this respect. Cables for 6.6 kV and higher voltages are therefore more appropriately covered in chapter 24 on high voltage cables with thermoset insulation.

Cables with all three types of thermoset insulation are covered by IEC 502, but IEC 502 and BS 5467: 1989 cater for both XLPE and EPR. Dimensions, weights, ratings and electrical characteristics of XLPE insulated cables are given in appendix A14.

COMPARISON BETWEEN THERMOSET INSULATED CABLES AND OTHER TYPES

Comparison with PVC cables

The important difference is the extra toughness of the insulation and, in particular, the ability to withstand much higher temperatures without deformation due to mechanical pressure. The better physical properties of XLPE enable the insulation thickness to be reduced and hence also the overall size of the cable. The continuous rating temperature is increased from 70 to 90°C and the temperature for short-circuit ratings for the cable itself (as distinct from the whole system) from 160 to 250°C. The large increase in continuous ratings indicated in appendix A14 implies that in many cases it will be possible to use a cable one size smaller, but in a majority of industrial applications this benefit cannot be achieved because the dictating factor is voltage drop rather than current rating. However, where derating factors have to be applied because of high ambient temperature, e.g. in tropical countries, the benefit is substantial.

Although XLPE cables have been available for many years, they have been slow in the UK in taking over from PVC power cables. This is largely because the crosslinking process affected manufacturing cost and their use tended to be restricted to applications where the current rating advantage could be fully exploited. With the reduction in cost obtained by developments in the silane process for crosslinking, the situation has now changed. It is to be expected that XLPE will gradually supersede PVC for power cable insulation.

PVC has a degree of flame retardance, though the extent depends on the test method taken as the criterion. XLPE burns readily and particularly important is the fact that drops of molten material fall away. Nevertheless, it is usually the characteristics of the complete cable which are more important than those of the insulation itself, as the bedding, armour and oversheath exert a strong influence on the behaviour of cable in a fire. XLPE has not suffered greatly, therefore, in comparison with PVC in this respect. For cables in the 1 kV range it is possible to compound XLPE with additives to introduce a modest degree of flame retardancy and such a material finds some application in the USA.

However, the additives generally contain halogens and give rise to similar fume and smoke problems as discussed in relation to PVC in chapter 6. When flame retardance is important, therefore, it is still necessary to establish all the factors relating to the test performance required to meet the service conditions and the ability of the complete cable design to meet the test conditions.

XLPE tends to be more stiff and rigid than PVC, but the difference in cable flexibility is small and only noticeable in cores of small conductor size.

Comparison with paper cables

Because paper cable ratings are based on 80°C (up to 3.3 kV) the rating advantage of XLPE is less than with PVC but is still positive. The main advantage compared with paper cables is the availability of a cable without metallic sheath and the consequent advantages in laying and jointing, together with reduced bending radii. XLPE cables are much cleaner, more robust, lighter and easier to handle – important advantages in countries without traditional cable installation skills. Outside the UK they have replaced paper cables in the voltage range under consideration. Meanwhile, in the UK the waveform type of concentric neutral cable with XLPE insulation described in chapter 22 has now completely replaced the Consac cable with paper insulation.

CONSTRUCTION

As the concentric neutral design is covered in chapter 22, we are concerned here only with the conventional designs of armoured and unarmoured cables (figs 21.1 and 21.2) and, as already stated, the constructions are essentially the same as for PVC cables. The main design difference is that, because the insulation thickness is based more on

Fig. 21.1 600/1000 V, single-core XLPE insulated cable with stranded conductor, taped bedding and aluminium wire armour

Fig. 21.2 600/1000 V, 4-core unarmoured cable with solid aluminium conductors

mechanical considerations than on electrical requirements, the tougher nature of XLPE permits a considerable reduction in thickness (chapter 18). Details for individual cable sizes are included in appendix A14.

As with PVC cables, the bedding under armour may consist of either extruded polymeric material or two layers of PVC or other non-hygroscopic tapes. However, when PVC bedding and XLPE insulation are in direct contact with each other it is also necessary to ensure that no contamination occurs by migration of plasticiser from the PVC into the insulation. This may be done by suitable formulation of the PVC or by the inclusion of a separating layer of polyethylene terephthalate or polypropylene film.

The oversheath is normally extruded PVC and this is specified in BS 5467: 1989, the general requirement being for a black oversheath. While IEC 502 appears to allow alternative sheathing materials such as polyethylene, polychloroprene or chlorosulphonated polyethylene, such materials are intended for other types of cable, where their use is more appropriate.

FUTURE PROSPECTS FOR CABLES UP TO 3 kV

XLPE is a material which has not only proved itself to be eminently suitable for this voltage range of 1–3 kV, but is also steadily finding acceptance for the whole field of power cables up to 132 kV and higher voltages. For the 1–3 kV range it is displacing the present main alternatives of PVC and impregnated paper and, although cable users are naturally slow to change from products which have given good service, this process is likely to continue.

Economics constitute a key factor. Another is that cablemakers maintain stocks of the most popular cables for industrial use and are naturally reluctant to hold too many alternative types. Therefore, when a trend based on sound technical and economic considerations is clear, it tends to dictate stocking policy, which in turn gives impetus to the trend. This has already happened with the trend from PVC to XLPE in 1–3 kV industrial cables and, therefore, where PVC cables are required, the customer suffers the disadvantages associated with ordering non-stock items. EPR insulation has largely replaced butyl rubber and has a few devotees, particularly where flexibility is important, but XLPE is likely to dominate in the years immediately ahead.

OPERATIONAL CHARACTERISTICS

The notes below indicate any differences compared with the situation for cables with PVC insulation described in chapter 20.

Ground containing hydrocarbons

As the cables usually have a PVC oversheath, the situation is essentially the same as for cables with PVC insulation. Permeation of hydrocarbon through the oversheath into spaces between the armour wires and between the cores is usually more important than any subsequent effect of further permeation through the insulation into the conductor. Although XLPE has better resistance to hydrocarbons than PVC, this does not have great practical significance.

Ageing and operational life of the insulation

It has been explained that the limits of operational characteristics of PVC insulation are defined by the resistance to deformation at operating temperature, e.g. due to external mechanical pressure or conductor thrust at bends, and by the resistance to thermal degradation. XLPE is superior to PVC in both these respects. For material test purposes the ageing temperature for XLPE is 55°C higher than for PVC.

CHAPTER 22
600/1000 V Cables with Combined Neutral and Earth for Public Supply

Combined neutral and earth (CNE) cables are used in the UK in systems having protective multiple earthing (PME), as discussed in chapter 17. The incentive for development of PME came from the need to retain good earthing and consumer protection when difficulties arose from the introduction of plastic water pipes. If the lead sheath of the paper insulated supply cable was not plumbed at joints it was not suitable for carrying fault currents through a sufficiently low impedance path back to the supply transformer. By using the neutral conductor of the supply cable for the purpose there was no need for a separate earth conductor in the cable and it soon became apparent that this could lead to considerable saving in the cost of distribution and house service cables operating at 240/415 V.

The reduction in cable diameter and hence materials cost is well illustrated in fig. 22.1 which shows a comparison between 600/1000 V 4-core PILS/STA cable with one form of CNE cable (Consac) of equal rating.

Initially PME systems had to be kept separate but when regulations were changed and it was appreciated that there were no problems in converting existing distributors to PME, or in mixing cable types within any single distributor, all new or replacement cable could be of the CNE type. Thus the supply authority did not need to maintain a stock of conventional cable with a separate earth conductor. This has largely been the situation in the UK since around 1970. The saving in cable cost was then of the order of 20–30% against PILS cable with aluminium conductors and the reduction in total installed cost about 10–15%.

In other countries such as France and Germany, the basic network and regulations have been somewhat different, mainly because each consumer has to provide a separate earth. CNE cables were in use earlier than in the UK and there have been differences in the types of cable which have been favoured.

Although, certainly in the UK, the use of CNE cables has been associated with the use of PME, they can be used to advantage on any type of system. They need not be restricted to PME systems and this is the case in many countries of continental Europe.

The integrity of the neutral conductor is a feature to which reference is made later and another aspect is that the system should be so arranged that the neutral may never

Fig. 22.1 Comparative diameters of 600/1000 V, 4-core PILS/STA and CNE cable of Consac type of equal rating

be disconnected. This also applies during operations to provide service joints and all the cable designs allow for joints to be made without severance of the neutral conductor.

TYPES OF CNE CABLE

Four basic designs have been used in Europe as a replacement for lead sheathed paper cables, namely Consac, Waveform, Districable and 4-core polymeric insulated cable without armour. Usually only two of these types have been predominant in any one country at the same time. In some cases a third design has emerged when one of those originally used has dropped out of favour.

General design details

Phase conductors
The use of aluminium conductors is almost universal, mostly of stranded design in France but always solid in the UK. In Germany some copper is still used and with aluminium there has been a progressive change from stranded to solid. In the UK the use of solid shaped conductors emanated from 5 years of successful experience by two of the regional electricity companies (RECs) with conventional PILS cables.

Neutral conductors
The characteristics of the neutral conductor are the key design feature of CNE cables. If for any reason there is a loss of phase conductor there is the inconvenience of loss of supply to a consumer but the fault is known immediately. Corrosion of a neutral conductor means reduction of fault current carrying capacity and possible failure in its function as a protective conductor. This may not be immediately apparent because of secondary earthing paths. Nevertheless a rise in voltage of exposed metal could occur on consumers' premises.

Corrosion protection of the aluminium neutral conductor in CNE cables is a feature to which great attention has been paid in the design of individual cable types. The possibility of loss of conductance due to corrosion is also the reason why little attempt has been made to produce designs with a reduced size neutral conductor, although the Waveconal design has a 185 mm^2 neutral in the 240 and 300 mm^2 sizes.

Cable weight

The weight of aluminium CNE cables is about half that of PILS/STA cable with aluminium conductors. Consac cable is the lightest, with only 35–40% of the weight of paper/lead cable. The low weight reduces the cost of cable laying because fewer workmen are needed.

Consac cable

First introduced in Germany with stranded phase conductors in the early 1960s, Consac retained the well-proven paper insulation but substituted an aluminium sheath for lead and the sheath became a concentric conductor (fig. 22.2). As with all CNE cables, there was a PVC oversheath for corrosion protection and with Consac a bitumen layer was applied between the aluminium and the PVC to prevent passage of water along the interface in the event of damage to the oversheath. Consac was the first CNE cable to be used in the UK and it was adopted extensively from the late 1960s up until the early 1990s.

The application of aluminium for sheathing followed a decade of successful experience with high voltage pressure cables, e.g. of the fluid filled type. However, more care is taken in the installation of expensive transmission cables. They are also installed in deeper trenches and in sand. Hence with Consac there was a greater risk of damage to the oversheath and subsequent corrosion of the aluminium sheath in the early years of cable life.

A feature of Consac was that in the event of damage to the oversheath the corrosion mechanism on the aluminium sheath took the form of pin holes. Water could then enter the insulation and cause cable breakdown before there was sufficient general corrosion of the sheath to result in significant loss of conductance, i.e. unknown loss of the protective conductor was highly unlikely.

Steps were taken during manufacture to limit the clearance between the insulation and the sheath and to control the amount of impregnant in the space so as to limit longitudinal travel of moisture along the cable in the event of sheath damage.

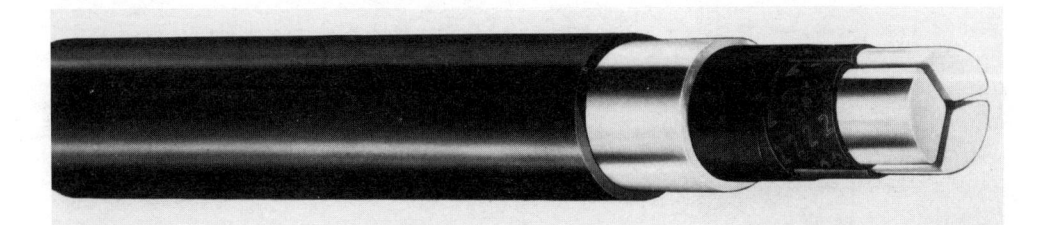

Fig. 22.2 Consac type of CNE cable

Although Consac was available in sizes up to 300 mm^2 it tended to be rather stiff and difficult to handle at that size. In general, therefore, 185 mm^2 was the largest size used.

When first introduced, Consac had a disadvantage in comparison with other CNE cables because only soldering techniques were available for conductor joints and the joint casing was filled with bitumen. The advent of mechanical conductor joints and cast resin filling reduced this drawback considerably but the procedure for cutting the sheath to open a flap has undoubtedly been one important reason why users have preferred to adopt alternative designs of CNE cable.

Waveform type

This type of construction was first introduced in Germany with PVC insulation in the 1960s under the name 'Ceander'. The neutral conductor takes the form of a layer of aluminium or copper wires applied with constantly reversing lay as illustrated in the Waveconal cable shown in fig. 22.3. This makes service jointing very easy, without any cutting of the neutral conductor, because when the oversheath is removed the wires may readily be pulled out to form two bunches for jointing purposes.

In Ceander cable in Germany, copper wires were used for the neutral conductor. It was considered that there was an undue risk in using aluminium because it could not be adequately protected against corrosion, and soil water passing through a damaged oversheath could easily travel along the cable. The design then became used in France with XLPE insulation as an alternative to PVC and with an aluminium wire neutral conductor. For additional corrosion protection of the aluminium, extruded layers of unvulcanised rubber were applied under and over the aluminium wire layer.

In the UK in the mid-1960s there were some trials with a cable similar to the French cable but with a copper wire neutral, under the name 'Wavecon'. However, for maximum economy the aluminium neutral version was soon preferred and widespread use began of 'Waveconal' cable. Compared with the French cable, an important difference of the Waveconal design was that rather larger diameter aluminium wires were used so that they could be spaced apart. This allowed the two rubber layers to meet each other between the wires, so that in effect each wire is encapsulated in rubber. Much work was also carried out on the composition of the rubber type material so that it could be removed cleanly from the wires, had adequate resistance to deformation and good ageing properties. In the event of mechanical damage to the oversheath some wires may be exposed and can corrode, but water does not readily come into contact with neighbouring wires. In laboratory tests it has been shown that the voluminous corrosion products of aluminium may disturb the rubber to permit some further passage of water but the service performance of the cable has been satisfactory.

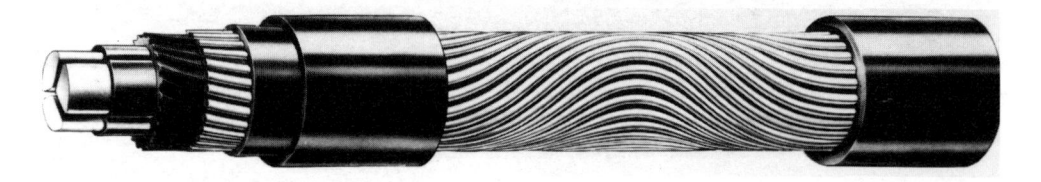

Fig. 22.3 Waveconal CNE cable with XLPE insulation

Because of fears of excessive corrosion of the aluminium neutral some regional electricity companies in the UK have adopted a Wavecon design with a copper wire neutral. This has an unvulcanised rubber bedding under the concentric neutral/earth wires, as with the Waveconal cable, but not necessarily a rubber layer over the wires. When PME systems were first adopted in the UK about half the electricity companies opted for the paper insulated Consac cable with most others using the waveconal design. However, in recent years the Wavecon design with a copper neutral/earth conductor has again found favour. Those regional electricity companies that had used Consac for many years have now converted to Wavecon along with some of the Waveconal users.

In the UK the insulation has been XLPE from the outset and apart from service cables this was the first major departure from impregnated paper for public supply cables. Although the sustained ratings could be somewhat higher than for Consac, as they may be based on a conductor temperature of 90°C instead of 80°C, they are generally taken to be the same, voltage drop usually being a limiting factor.

Waveform has a slightly greater diameter and material content than Consac. A novel feature of manufacture is that in addition to the application of the neutral wires using a reversing lay head, two layers of rubber and the PVC oversheath are applied in the same operation.

Districable

Two UK regional electricity companies used Districable for a period of time but in general it has found little adoption outside France. Construction (fig. 22.4) is of 4-core design with solid aluminium conductors and XLPE insulation (formerly PVC) on the phase conductors. The neutral conductor, which is circular, has a thin lead covering for corrosion protection. A thin soft galvanised or tinned steel binder tape holds the laid-up cores together and there is a PVC oversheath. As the steel binder tape makes contact with the neutral/earth conductor there is an earthed metallic envelope, although the ability of the thin steel tape to convey fault currents to the CNE conductor may be somewhat limited.

Doubts have been expressed about the ability of the thin lead coating to serve as an adequate corrosion protection for the aluminium neutral conductor during installation and service, but the service performance has been satisfactory.

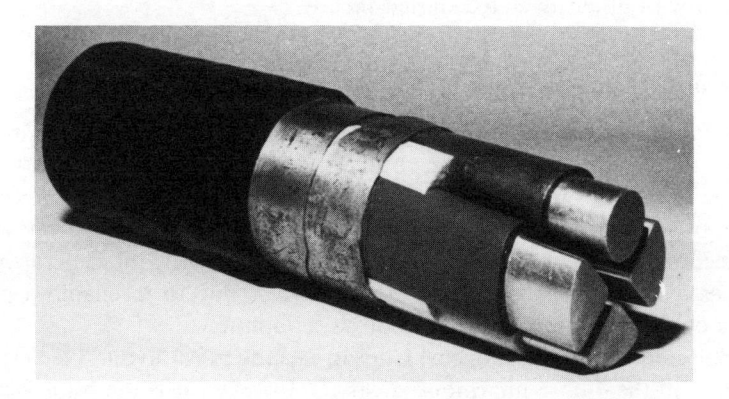

Fig. 22.4 Districable with solid aluminium conductors and XLPE insulation

Fig. 22.5 600/1000 V, 4-core unarmoured cable with solid aluminium conductors and XLPE insulation

4-core unarmoured cable

This design (fig. 22.5) reaches the maximum simplicity in distribution cable design as it merely consists of shaped conductors, generally solid aluminium, polymeric insulation and a PVC oversheath. Originally introduced in Germany with PVC insulation it has now become the standard design in Germany for mains distribution and services. There has been some drift from PVC to XLPE insulation, however.

This unarmoured cable is used in some other European countries and elsewhere but not in the UK or France. Largely on grounds of safety, the authorities in the UK have not given permission for its use for direct burial and extra protection would not be economic. Possibly a desire, in the interest of standardisation, not to extend the multiplicity of cable designs brought into use over a short period of years has had some bearing on the decision. Another factor is that the Electricity Supply Regulations stipulate that cables in the high voltage range (specified as 650 V upwards) are required to have a metallic envelope. Although the cables concerned operate at 230/400 V, they are classified as 600/1000 V design.

However, the most important point concerns effects following third-party mechanical damage to cables and the fact that such damage, particularly from mechanised tools and excavators, has been occurring at an increasing rate during the last two decades. Any damage to the insulation on the neutral conductor could lead to the loss, by corrosion, of the protective conductor and damage to a phase conductor could cause currents to flow in other insulated buried metals.

Comparison of CNE cable designs

All the four types are very easy to handle for installation, especially in comparison with the former PILS armoured cables, but as already mentioned Consac has the disadvantage of stiffness in sizes above 185 mm^2 and has now been discontinued.

They are, of course, less robust than the PILS cables and consideration was given to better protection, such as by a tougher oversheath. However, the general opinion has been that, as a large proportion of cable damage is due to mechanised plant, extra expenditure on cable protection would not be economic.

In materials and manufacturing cost Consac was the most favourable design, even in comparison with the 4-core unarmoured cable. However, it is the total installed cost, including jointing, which is significant. The 4-core unarmoured cable is clearly the most attractive for jointing simplicity and Consac had a disadvantage because of the somewhat more difficult step in opening up the aluminium sheath. There is not a great difference between the overall costs for any of the designs and so once usage had become established for any particular one there has not been much incentive to make a change. Service experience with all of them has been satisfactory.

Knowing that the cables were less robust than former types and that faults due to third-party damage were increasing, consideration was given at an early stage to safety aspects when live cables were damaged by tools such as spikes and spades. One series of tests was carried out by the UK Electricity Council Research Laboratory (unpublished) and one by McAllister and Cox.[1] The results and conclusions were similar in each case.

The aspects investigated were related primarily to effects on personnel due to shock because of the tool becoming live and possible injury due to the flash and explosion. Little was known about the situation with existing cables, except that electrocution was very rare and that cases of severe burning had occasionally occurred.

Contrary to expectation, it was found that with PILS cables the reason that tools did not become live was not due to conductance of fault currents through the sheath and armour. Both the outer metal on the cable in contact with the tool and the end of the tool itself just burned away to leave the tool isolated. Thus cables with a metallic sheath, such as PILS and Consac, showed a greater severity of flash than Waveconal, Districable or 4-core unarmoured cable, but there was less chance of the tool remaining live. It was concluded that greater incidence of severe accidents due to electrocution was unlikely and that overall there was little chance of increasing hazards by adoption of any of the CNE cable designs. A possibly important factor with 4-core unarmoured cable was that, because of the absence of any metal envelope, a workman would not be aware of striking a live cable, as there would not be any flash until the implement bridged between two cores. Because of the geometry there is therefore a somewhat greater chance of the more severe flash associated with a core to core rather than a core to earth fault.

Many different types of fault can be generated by sharp and blunt objects and the overall situation is quite complex. However, one aspect which did come out clearly was that synthetic insulation had some advantage over paper in that faults rarely developed to become of a sustained type which could damage a significant length of cable and/or lead to a 3-phase fault. Another facet was that the test result showed that there was no advantage in using a binder/cross flow tape around the neutral wires in Waveconal cable. Other tests had shown the undesirability of including this tape because it interfered with the corrosion protection capability of the rubber layer. In this context the limited value was also apparent of the thin steel binder tape in Districable to carry fault currents to the neutral earth conductor.

REFERENCE

(1) McAllister, D. and Cox, E. H. (1972) 'Behaviour of m.v. power distribution cables when subjected to external damage'. *Proc. IEE* **119** (4), 479–486.

CHAPTER 23
Service Distribution Cables

APPLICATIONS

As the name implies, the classification of service cable is associated primarily with the provision of an electricity supply from a distribution main, generally underground, to individual houses or commercial premises. The cables described here are designed for direct burial, as distinct from providing a service from an overhead line.

The type of cable is also suitable for any small-scale supply up to around 200 A single- or 3-phase with conductor sizes from 4 to 35 mm^2 in copper and from 6 to 50 mm^2 in aluminium. Apart from dwellings and shops with direct or looped services there are other wide-scale applications. The smaller sizes are used for street lighting and all the multiplicity of street furniture such as traffic lights and signs, and the larger sizes are very convenient for motorway lighting where the lighting needs to be controlled by wiring in specific circuits, i.e. as distinct from the majority of street lights which are now connected individually to a distribution main.

Paper insulated lead sheathed cables

A high proportion of existing houses in the UK are still supplied through a 2-core paper cable with steel tape armour and bituminous hessian serving. Use of this cable in new installations ceased in the 1960s. With this cable, terminations comprise a wiped connection to the lead sheath and usually a bitumen filled cable sealing chamber: altogether rather large and untidy compared with modern installation practice. Nevertheless, although the rubber insulated internal wiring of older houses will no doubt have been replaced, the main feed to the meter to pre-1960 vintage is still probably the original one and will be capable of continuing for a few more decades.

Split concentric cable design

When PVC insulation was developed for power cables in the later 1950s, the UK regional electricity companies carried out trials but came to the conclusion that problems might arise due to thermal deformation at high conductor temperatures which could not readily be prevented by the provision of better circuit protection. With service cables, however, the fuse at the point of supply provides good control, apart from mechanical damage to the cable itself, which is comparatively rare.

At this time it was still necessary to include a separate earth conductor in single-phase service cables, though on the larger sizes this earth conductor did not need to have the full conductance of a phase conductor. Hence the split concentric design shown in fig. 23.1 emerged and was fairly quickly adopted as standard. For copper phase conductors up to 16 mm^2, the size of the earth conductor matches that of the phase conductor. For the larger sizes, the earth conductor is smaller than the phase conductor. For instance, the 25 mm^2 copper cable has a 16 mm^2 earth conductor, while the 35 mm^2 copper cable has a 25 mm^2 earth conductor. A 3-phase design is very similar and has three laid-up insulated conductors in the centre instead of the single one shown.

In order to ensure that no corrosion of the neutral conductor occurs and to eliminate any possibility of faults in connections within the consumer's cut-out, the two conductors in the concentric layer, i.e. the neutral and the earth, each comprise copper rather than aluminium wires. The wires of the earth conductor are bare and are only separated from the lightly PVC covered wires forming the neutral conductor by a PVC string.

Within the building, the clean PVC sheath was much more attractive than the bituminised hessian finish of the paper cable and for the common 16 mm^2 size the overall diameter was reduced from 22.4 mm to 14.7 mm. Furthermore, the termination illustrated in fig. 23.2 is much neater and easier to make than the previous types necessary with a lead sheathed cable.

An armoured version of the single-phase cable is used for motorway lighting. Essentially the armoured design is the cable as already described with two additional layers applied over it, these being a layer of galvanised steel wires and an external PVC oversheath.

When the split concentric design was first introduced, aluminium had already replaced copper for the conductors of the main distribution cable and consideration was given to the use of aluminium for the service cable. For the small conductor sizes involved, the saving with aluminium has never been very great but designs with a circular solid aluminium phase conductor have always been available and some users

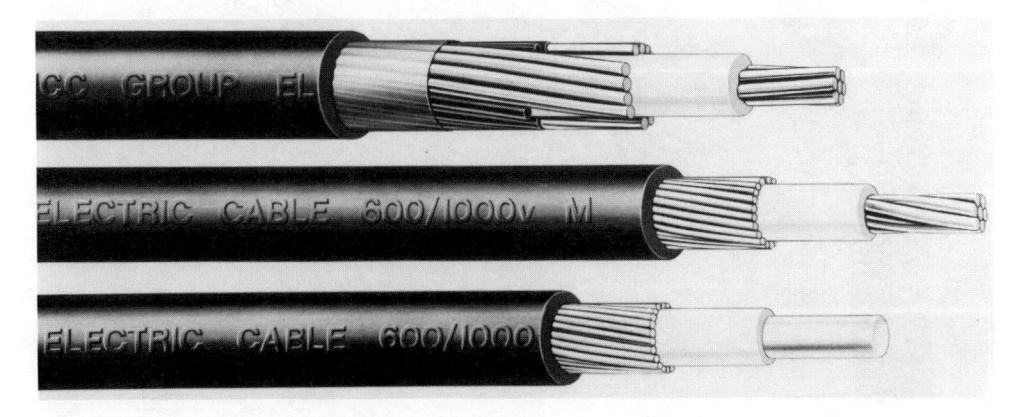

Fig. 23.1 PVC insulated 600/1000 V service cables: split concentric with copper phase conductor (*top*); CNE concentric with copper phase conductor (*middle*); CNE concentric with solid aluminium phase conductor (*bottom*)

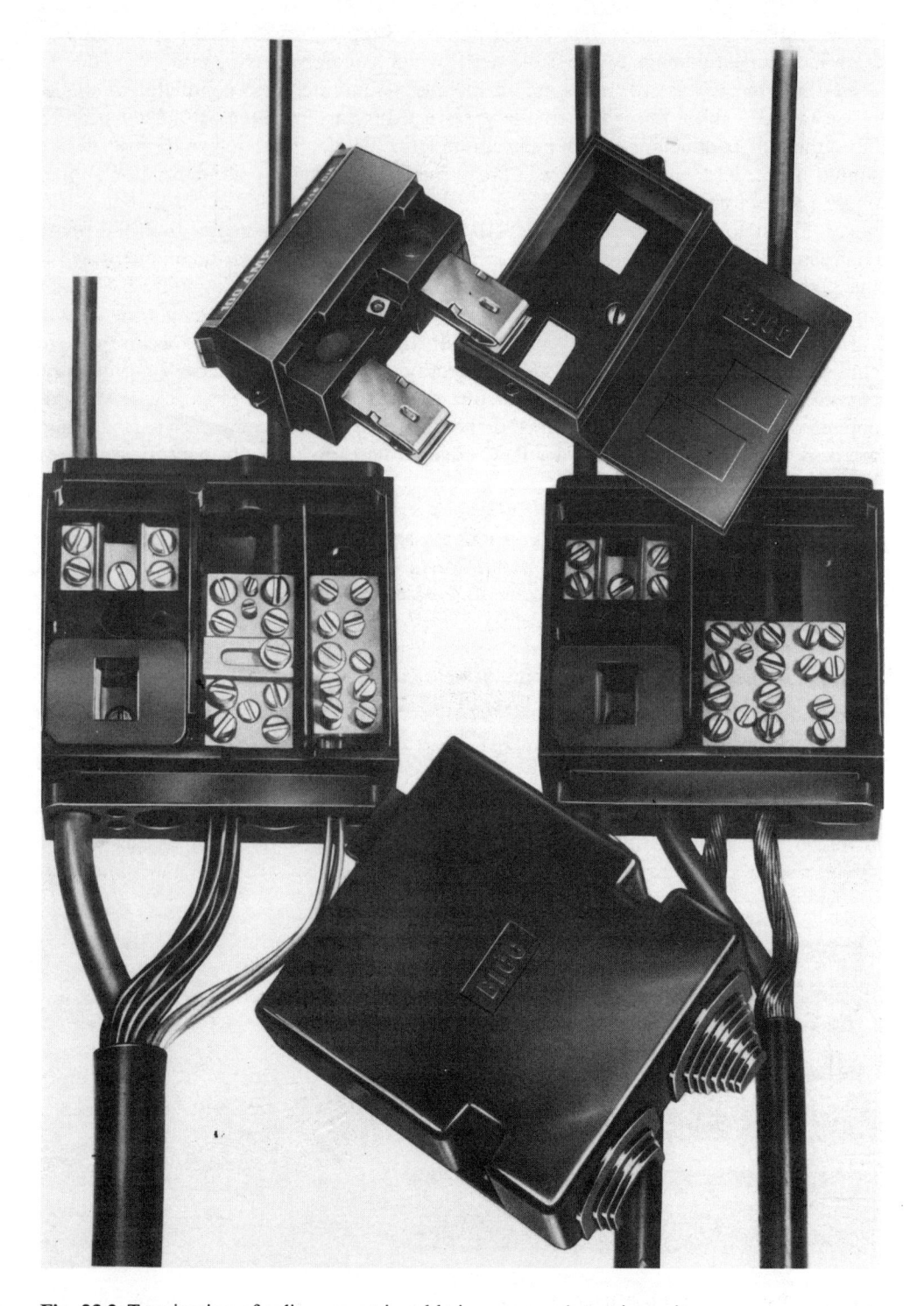

Fig. 23.2 Termination of split concentric cable in consumer's service unit

have preferred them as an alternative to stranded copper. The connection in the cut-out is usually of the pinch screw type and there are not the same problems with a solid conductor as exist with a stranded conductor. Another reason that aluminium wires have not been used in the concentric earth and neutral layer is that, although mechanical damage to service cables was stated above to be rare, it does happen on occasions. Water may then enter the cable and with aluminium there would be a probability of serious corrosion of the important protective earth conductor. The provision of extra corrosion protection, as is done with Waveconal cable, would make the aluminium design uneconomic in comparison with copper.

Combined neutral and earth design

This design is often termed 'CNE concentric' or 'straight concentric' and is a simplification of the split concentric version for use when both the distributor and the consumer can operate with protective multiple earthing. One conductor may be eliminated and the concentric layer then consists of a simple layer of bare copper wires (fig. 23.1). For the copper 16 mm^2 size of around 100 A rating referred to above, the diameter is further reduced from 14.7 mm with the split concentric to 12.1 mm for the straight concentric cable. Both single- and 3-phase designs are available and these types now represent the bulk usage of service cable in the UK.

XLPE insulation

As PVC was the original insulating material for the concentric designs, its use has tended to continue. XLPE is the obvious alternative and has the advantage of being suitable for higher temperatures. The trend is towards replacement of PVC by XLPE, but the process is gradual. XLPE can also be used with a lower insulation thickness, but for some sizes of single-phase cable the diameter over the insulation is dictated by the space required to accommodate the concentric layer of wires.

UK standards

The single-phase split concentric cable with copper conductor and PVC insulation is covered by BS 4553: 1991. Currently, the 3-phase designs, the single-phase designs with aluminium phase conductors and those with XLPE insulation are supplied to manufacturers' own specifications. However, a revision of BS 4553 is being prepared which will incorporate these other single-phase types.

The CNE types are included in the common standards of the electricity supply companies, namely EA TS 09-7, but will soon be published as a part of BS 7870.

Waveconal service cable

There are some users of the Waveform distribution cable described in chapter 22 who have preferred to use small sizes of the same cable for service connections. However, this application is diminishing, largely on economic grounds. If such cable is used it is also important that the design of termination in the consumer's service unit is suitable to accommodate bunched aluminium wires.

Service cables outside UK

The split concentric and straight concentric designs of the type described are peculiar to the UK and are seldom used elsewhere. Practices vary widely from country to country and, in addition to the use of conventional 2-core and 3-core armoured and unarmoured cables, other prominent arrangements include the following.

USA

Underground cable services are less frequently adopted than in Europe and when used are generally associated with the URD system (chapter 17). A common practice is for 6–12 consumers to be fed from a small transformer in the high voltage (10–30 kV) network either directly from the transformer or through a distribution pillar using short lengths of secondary or service cable. The cabling needs to be of 3-wire type to meet the dual voltage supply requirement. The neutral is usually of reduced area. Solid aluminium is now the most popular choice of conductor and XLPE is common for the insulation, although copper conductors and polyethylene insulation are also used. Three single-core cables, merely comprising conductor and insulation, may be installed separately, frequently in a plastic duct, or more usually triplexed (laid-up) together with no outer sheath. There is no armour or earth conductor.

Germany

Changes from PILS cables to both Consac and Waveform cables for PME type systems have been similar to those in the UK and are discussed in chapter 22. Practice has subsequently moved strongly towards the use of 4-core PVC or XLPE insulated distribution cable with a PVC oversheath but no armour. Services to consumers are of 3-phase type and a similar 4-core unarmoured cable is now widely used also for services.

France

The 3-phase services provided in France are as in Germany and 4-conductor service cable of combined-neutral-and-earth type is required. The normal design is now of 4-core construction having PVC or XLPE insulation on the phase conductors, a thin soft steel tape over the laid-up cores, in contact with the neutral conductor, and a PVC oversheath. With copper conductors the neutral conductor is left bare and with aluminium it has a lead sheath, i.e. the design is a small size of the Districable type described in chapter 22.

CHAPTER 24
Polymeric Insulated Distribution Cables for 6–30 kV

Since 1970, in the 6–30 kV cable range there has been a world-wide swing away from paper to polymeric insulation for distribution cables. In the UK, in terms of usage, the scale of change has been smaller and its speed slower than in many other countries but, because much of the UK cable is made for export, the pattern of manufacture has moved substantially towards an increased proportion of cable with polymeric insulation.

Trends in the comparative costs of the cables themselves can be expected to have influenced the change, but another factor is that it has become increasingly difficult to secure, at reasonable cost, the skills required to joint and terminate the traditional paper cable. This is especially the case in the less developed countries. Inevitably, the simpler concept and reduced number of skills involved with the installation of polymeric cable has a significant influence on the swing away from paper cables.

Recently, most of the large distribution networks in the Middle East, in the field of power cables, have predominantly used XLPE insulation. The improved current rating of polymerics at the higher ambient temperatures experienced in this part of the world is also a factor in preferring XLPE to paper.

Against this background, most industrialised nations have seen a reduction in demand for paper cables as the traditional markets have declined and there has been a growth in demand for polymerics from newer markets.

Synthetic polymeric materials were not new to power cables when the move towards them began to gather pace; polyethylene was used as an insulant as early as 1943. The high intrinsic electrical strength and low dielectric loss angle of polyethylene, together with its good resistance to chemicals, ease of processing and low cost, make it an ideal material for use in the manufacture of power cables. Its main disadvantage is the relatively low melting point of the material (105–115°C), which means that the sustained current rating, overload and short-circuit temperatures are limited. By converting the thermoplastic polyethylene into a crosslinked thermoset, the melting point is greatly increased and the material is well able to surpass the thermal capabilities of a paper insulated cable. The crosslinking can be achieved either by radiation or by chemical means and for the manufacture of power cables the latter is by far the most economic. For the manufacture of cables with a voltage exceeding 6 kV the chemical additives are best kept to a minimum; in the case of the thermoplastic insulant only an antioxidant is added to the polyethylene and for crosslinked material the only further additives are those chemicals which are essential to achieve the chemical bonds.

Ethylene–propylene rubber (EPR) consists of approximately equal quantities of ethylene and propylene units and is often referred to as a copolymer. It was first used as an insulant during the early 1960s and is always used in its crosslinked form, which is achieved using similar chemical means to polyethylene.

FIELD OF USE

Initially the thermoplastic polyethylene insulation was the most popular with many kilometres of cable installed in Germany, France and the USA. Mainly low density polyethylene (LDPE) has been used but there is also some experience in the USA with high density polyethylene (HDPE) as an insulant. HDPE has superior electrical properties to LDPE but is more rigid; HDPE is more crystalline than LDPE and thus capable of operating at a higher temperature and it has been assigned a sustained rating of 80°C, in comparison with 70°C for LDPE. Its crystallinity at 70°C is 75% whereas the figures for LDPE and XLPE are 40% and 30% respectively. Its saturation moisture content is approximately a fifth of that of low density materials; nevertheless, its greater stiffness has mitigated against its wider acceptance.

Generally the use of EPR has been confined to those applications where its property of greater flexibility can be used to advantage, but in Italy and Spain EPR has been used widely throughout the voltage range. Although 132 kV cables have been manufactured with EPR, the higher loss angle and poorer thermal resistivity make it a poor competitor against XLPE at higher voltages. While EPR has excellent resistance to ozone and electrical discharges and the material was for this reason preferred to PE during the 1960s, changes in manufacturing techniques have led to a reduction in discontinuities and size of voids within PE cables to such an extent that advantage can no longer be taken of this positive characteristic. EPR is also more expensive than XLPE and it seems likely that applications for it will be confined to voltages of less than 30 kV, especially where flexibility of the core is an important factor, e.g. in the mining industry and in power stations.

Crosslinked polyethylene has become the most favoured insulant. Germany, Japan, the USA and Scandinavian countries have all installed vast quantities of such cables.

SERVICE EXPERIENCE WITH POLYETHYLENE (PE AND XLPE)

In Europe, the move from paper to XLPE insulated cables in the 6–30 kV range has occurred gradually. It is reported[1] that by the end of the 1970s there were approximately 10 000 km of 10 kV and 3000 km of 20 kV circuits in operation with failure rates of 0.23 and 0.35 faults per 100 km per year. By the end of 1991 the installed circuit lengths had increased to approximately 59 000 km (10 kV), 97 000 km (20 kV), and 3000 km (30 kV) with respective failure rates of 0.21, 0.61 and 0.76 per 100 km per year.

XLPE has been installed in Japan since 1965 and today all cables in Japan up to 69 kV are made with this material. The Scandinavian countries swung to XLPE in the early 1970s and have installed many kilometres of cable in the voltage range 12–170 kV.

In both of these countries the service experience has been very good. The operating experience with 11 kV polymeric cable amongst regional electricity companies in the UK is growing and to date is good.

The American experience has been well documented and publicised in the surveys conducted by Thue.[2] A later survey[3] reported that 60 000 km of PE cable and 116 000 km of XLPE cable had been installed in the USA by the end of 1983. The service experience in the USA, which included the early designs, is unsatisfactory compared with that in Europe, the failure rate being five to six times that of Germany.

Inevitably, everyone can derive benefit from mistakes made by others, and the Americans are to be congratulated on their bold and rapid swing to the extruded dielectric. Unfortunately, the sudden change did not provide sufficient time to understand fully the limitations of these otherwise excellent materials. The susceptibility of both PE and XLPE to premature failure arising from electrical discharge and water trees was overlooked. Early American designs utilised a semiconducting tape screen applied helically over the conductor; this was subsequently found to be a potential source of water trees and has since been superseded by an extruded layer of semiconducting material. Also insufficient care was taken in avoiding contamination of the insulant during both compounding and extrusion processes. Finally, many American engineers may rue the decision not to apply an oversheath over the cable. More often than not, the cable was laid in wet ground with the concentric copper wires, applied directly over the core screen, thus being exposed to the elements.

The tremendous technical effort which has been devoted in the USA to investigating and understanding the phenomenon of premature failures has not necessarily produced a clear reason or satisfactory explanation of the cause, but both producer and user have derived considerable advantage from this work.

BREAKDOWN PHENOMENA

Breakdown mechanism

The electrical breakdown of polyethylene as a result of discharge is well understood and much work has been conducted on this subject.[4-6] It was shown that the failure occurs because of discharge activity, the intensity of which erodes fine channels in the polyethylene. The erosion can take three forms: melting, chemical decomposition, and the production of microcracks in the polyethylene. In the case of chemical attack, the decomposition products of polyethylene are conducting and, with certain voids, can lead to a diminution of any subsequent discharges. This has the effect of increasing the life of the dielectric but only appears in unvented voids or channels.

Experiments with fine needles inserted into polyethylene have shown that the tree growth is far more rapid when using a vented needle, i.e. a needle with a hole through its length vented to atmosphere, than an unvented one. In the case of the unvented sample, the discharges arising from the tip of the needle will erode the polyethylene and decomposition products, H_2 and CO_2 will be released which will increase pressure in the channel. Paschen's law states that the breakdown strength of a gas is a function of the product of gas pressure and distance between electrodes. Hence as the pressure in the channel increases, the voltage required to cause the gas to discharge increases, and the discharges are suppressed. If the pressurised gas is given sufficient time it will diffuse

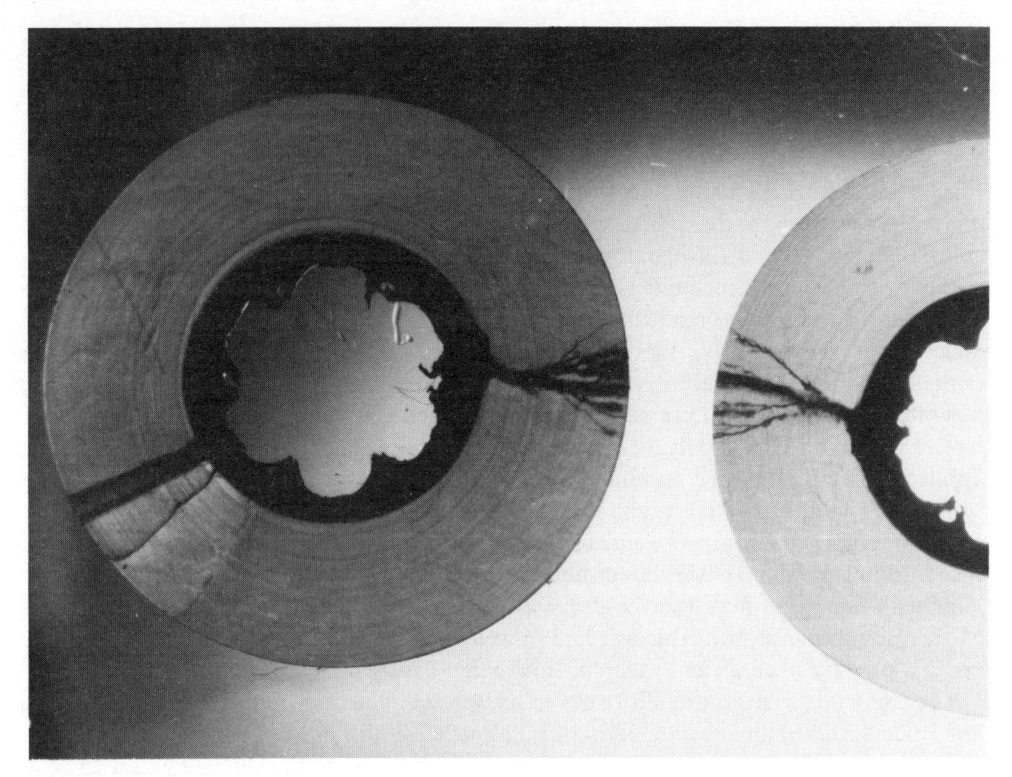

Fig. 24.1 Electrical impulse failure of sample of 6.35/11 kV XLPE cable

through the polyethylene, reducing the pressure in the channel and thus the voltage required to cause further discharges. It is for this reason that accelerated tests on plastics containing unvented voids or channels are not useful. It is far more likely that such a model will have a reduced life at very low frequencies; thus in this case high frequency retards the growth of trees whilst low frequency accelerates their growth.

The opposite is true for vented trees. As the pressure in such a channel or void is constant, the energy from successive electrical discharges is the same and, in the frequency band 50–1200 Hz, it is now well established that the lifetime of a dielectric is inversely proportional to frequency.

All electrical trees are readily visible in polyethylene and fig. 24.1 shows an impulse failure on a sample of 6.35/11 kV XLPE cable.

The thermal breakdown mechanism, generally associated with paper and PVC insulated cables, may well account for some failures in low dielectric loss materials where the presence of contaminants, introduced during compounding and processing, can lead to small localised areas having high loss. It is not possible to detect their presence as any measurement of loss angle records the predominant low figure for the dielectric, thus swamping the very small number of localised high loss sites.

Breakdown strength

The intrinsic breakdown strength of thin film polyethylene is up to 500 kV/mm, whilst the a.c. breakdown stress of a cable is nearer a tenth of this figure. For an 11 kV cable, the r.m.s. breakdown stress at the conductor is typically 50 kV/mm and the maximum peak strength at the same position under an impulse test is 100 kVp/mm. The d.c. strength for polyethylene is nearly 20% higher than this figure at room temperature and 120 kV/mm is typical. The figures for EPR are somewhat lower at 40 kV/mm for a.c., 80 kVp/mm for impulse and 95 kV/mm for d.c. voltage.

The level of breakdown is a function of the rate of application of voltage. As an example, the figure for a cable where the voltage rise is rapid can be twice that for the same cable where a step rise is used. Care must be exercised in the conditioning of samples exposed to such tests. For example the breakdown strength of a freshly extruded cable will be approximately 20% greater than that of the same cable left unenergised for a year.

As well as the effect of the history of the sample on its breakdown strength, the volume of dielectric under test is also a factor. In Schultz[7] reference is made to Leonardo da Vinci's conclusion that the tensile strength of wire decreases as the length of wire tested increases. Obviously, failures were occurring at the sites of flaws and the greater the length of wire tested, the greater were the chances that they would include imperfections. The same is true of plastics. Artbauer and Griac[8] found that, in the case of polyethylene, the breakdown stress was inversely proportional to the area of material under test to the power 0.15.

Again, because the probability of having flaws in samples of equal length is remote, it is necessary to evaluate approximately twenty specimens in order to establish a meaningful breakdown level for a polymeric cable. It has become accepted practice to use the Weibull distribution (appendix A20), which is a mathematical representation to cover reliability. Its first application in the field of engineering was associated with the quality control of ball bearings during the early 1920s.

The most commonly used equation defines the probability of failure P as

$$P = 1 - \exp\left[-\left(\frac{E}{E_0}\right)^b \left(\frac{t}{t_0}\right)^a \left(\frac{Lr^2}{L_0 r_0^2}\right)\right] \quad (24.1)$$

where E_0 = value of the stress corresponding to the breakdown probability of 63.2% of a voltage applied for a time t_0 to a sample having length L_0 and radius r_0 over the conductor screen, i.e. derived on a model cable
E = conductor stress
t = time of application
L = length of cable under examination
r = radius of conductor under examination
a = characteristic constant of the dielectric (sometimes referred to as the time equivalent)
b = stress exponent (a characteristic constant of the dielectric)

It follows that to compare two cables

$$\frac{E}{E_0} = \left(\frac{L_0 r_0^2}{Lr}\right)^{1/b} \quad (24.2)$$

The relationship between life and stress is given by

$$E^n t = constant \quad (24.3)$$

Therefore from the Weibull equation

$$\left(\frac{E}{E_0}\right)^b \left(\frac{t}{t_0}\right)^a = constant$$

$$\left(\frac{E}{E_0}\right)^n \frac{t}{t_0} = \left(\frac{E}{E_0}\right)^{b/a} \frac{t}{t_0}$$

where $n = b/a$. n is often referred to as the life exponent. Typically for polyethylene n is between 9 and 20, b between 10 and 20 and a between 0.5 and 2. The value of b is indicative of the amount of scatter of the breakdown figures. The higher the value of b the better and the more consistent is the polymer and process.

Weibull plots are extremely useful laboratory tools which enable judgements to be made with regard to the selection of polymers and also comparison of products made by different processes. However, use of n, b and a values obtained from short lengths of cable tested in a laboratory to derive the incidence of failure and the life span in service is not recommended. Inevitably, the failure mechanism at high voltage is completely different from the failure mechanism which occurs at the much lower stress in service and therefore the two cannot be compared.

WATER TREES

The growth of tree-like features in PE insulated power cables was widely reported in the early 1970s. It has since been the subject of much discussion with regard to mechanism of growth and effect on long-term cable performance.[9] Water trees have normally been classified by descriptive words such as bow tie and vented. They develop from voids, contaminants and defects which occur within the insulation or on the semiconducting screens. Many of the premature failures which have occurred in the USA have been ascribed to the existence of water trees emanating from the edges and irregular protrusions of semiconducting taped screens which were applied to many of the early cables.

The phenomenon is not confined to highly stressed dielectrics. Many trees have been found in cables operating at stress levels of less than 1 kV/mm. While most of the photographs shown in technical literature cover water trees in polyethylene they are not peculiar to this material and are present in EPR, EVA, polypropylene and PVC. However, they are far more difficult to study in these opaque materials.

Water trees grow in the direction of the electrical field and emanate from imperfections which have the effect of increasing the electrical stress at local sites. The branches of water trees are very narrow, of the order of 0.05 μm, and are not visible if the water is allowed to dry out of the specimen. Staining techniques are used and retention of the tree structure can then be maintained indefinitely. As an example, 0.1 g of rhodamine dye in a litre of water provides excellent definition of trees in polyethylene if the samples are boiled in the solution for 15 minutes.

The growth of a vented water tree with its source at a conducting screen is generally far more rapid than that of one emanating from within the dielectric. In nearly all cases the latter appear as bow-tie type and fig. 24.2 shows a typical example. The trees are generated from contaminants or voids existing within the dielectric and propagate in one or two directions towards the semiconducting screens.

Fig. 24.2 Bow-tie tree emanating from a contaminant in an XLPE insulated cable

The fact that water trees have a deleterious influence on the electrical properties is now well established but the mechanism is not fully understood. Figure 24.3 depicts the reduction in breakdown strength of samples aged in water over a period of time; ageing in air results in only small decreases in strength. Figure 24.4 is a photograph of an electrical tree which developed within a water tree.

Water trees increase in length with time, frequency and increasing voltage. However, in the case of frequency the acceleration factor is less than would be expected. As an example, samples energised at 500 Hz and 50 Hz show an acceleration factor of only 2 at the higher frequency.

Water trees appear to be more profuse in highly strained dielectrics, and the same phenomenon seems to exist as in electrical trees where, on bent samples of cable, there is an excess of trees in the area of the dielectric which is in tension and a paucity in the area which is in compression.

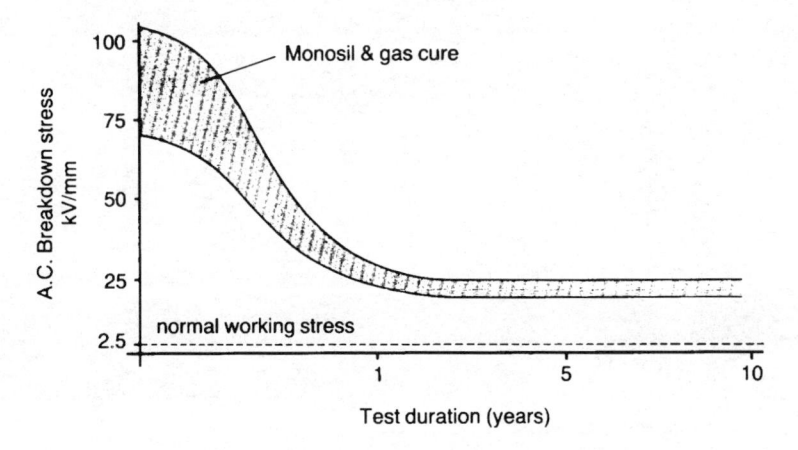

Fig. 24.3 Reduction in a.c. breakdown strength of XLPE cable aged in water at working voltage

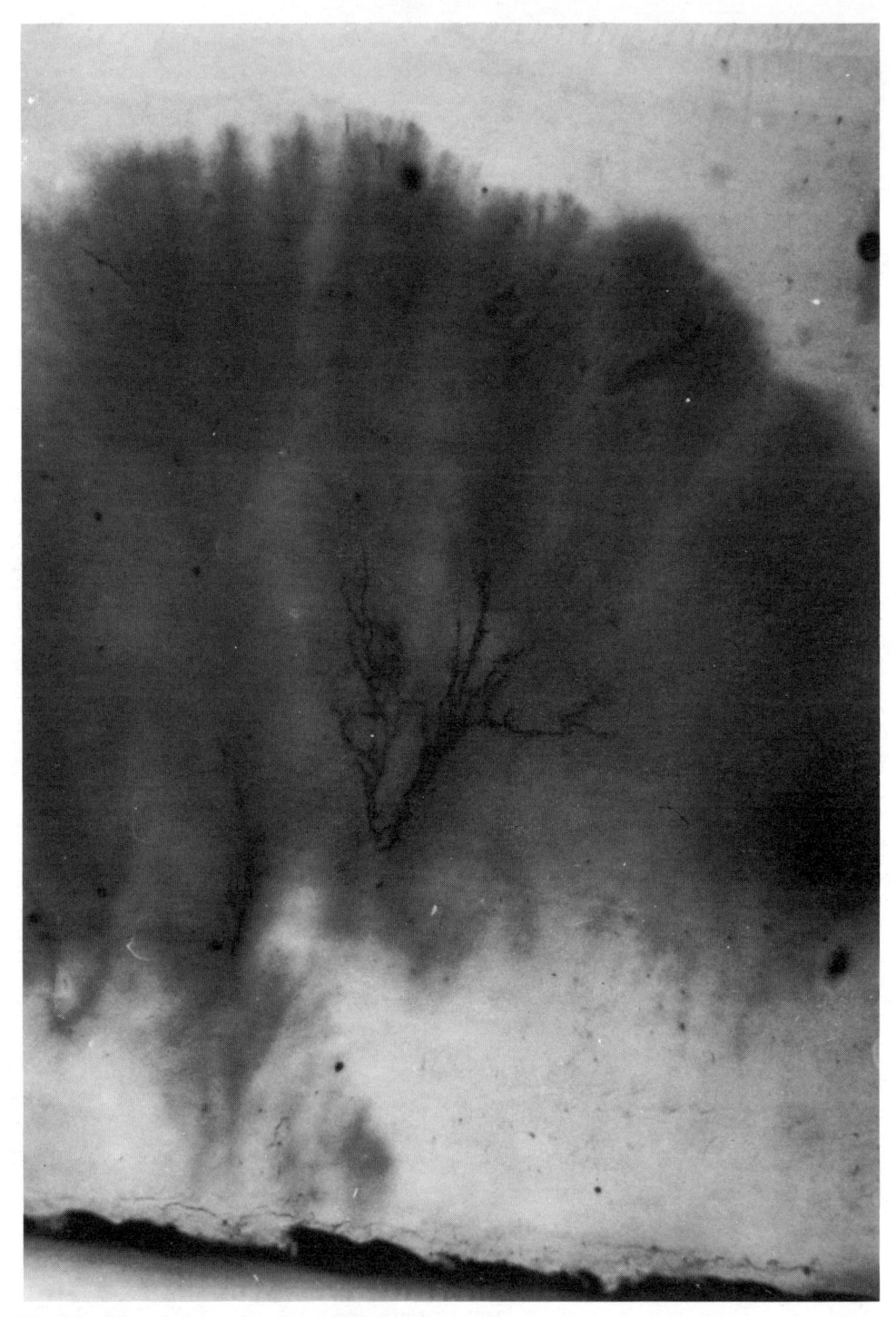

Fig. 24.4 Electrical tree forming within a water tree

The phenomenon of water trees is still not fully understood and a number of mechanisms have been postulated and discussed in terms of chemical, electrical, electrostatic and mechanical effects. It is possible that several mechanisms are involved, the importance of each depending on the conditions and materials used.[5,6,9] Polymer morphology, internal mechanical stress, material purity (insulation and screens), the ionic concentration of the water, ion penetration and polymer oxidation are among the factors considered important in the propagation of water trees. As an example, the failure may be accounted for by dielectrophoresis, i.e. the motion of water in an electric field could serve to answer the reason why water diffuses into a cable against a thermal gradient. The water contained in a void vaporises or expands, thus increasing the pressure in the void and so generating fine cracks. The formation of oxygen during this process may accelerate the oxidation of the polyethylene, which would account for the increase in loss angle. The final electrical breakdown could then be a thermal runaway.

DESIGN OF POLYMERIC INSULATED CABLES

The conductors of polymeric cables are generally circular with either stranded copper, stranded aluminium or solid aluminium. For three-core cable in the range 3.6/6.0 kV to 8.7/15 kV some use has been made of both sector shaped stranded and solid conductors.

At 3.6/6 kV and above, as a means of containing the electrical field within the insulation, semiconducting screens are applied over the conductor and insulation. By this means it is possible to eliminate any electrical discharges arising from air gaps adjacent to the insulation. The coefficient of thermal expansion of polyethylene and EPR is approximately ten times greater than that of either aluminium or copper, and when the conductor is at its maximum operating temperature of 90°C a sufficiently large gap is formed between the insulation and conductor to enable electrical discharges to occur. This discharge site and any others which are formed around a conductor when the cable is bent can be eliminated by applying a semiconducting layer over the conductor. Similarly, any discharges arising from air gaps between laid-up cores can be nullified by the use of a screen over the insulation.

During the early 1960s semiconducting tape screens were applied over the conductor but these have since been superseded by an extruded layer. This has the advantage of providing a smoother finish and, as it can fill the interstices between the wires, a circular envelope around the conductor. By reducing the concentration of flux lines around individual wires, the electrical stress around the conductor is reduced by between 10% and 15%. The semiconducting layer is compatible with, and bonds to, the insulation. A nominal thickness of 0.4 to 1.0 mm is typical.

The insulation thicknesses for the three insulants PE, XLPE and EPR are identical at each voltage level above 3.6/6 kV; at this voltage EPR is thicker. The radial thicknesses and electrical stresses are given in table 24.1.

The outer semiconducting screen is normally an extruded layer of semiconducting material. The extruded screen can be a compatible material which bonds itself to the insulation or a compound, such as ethylene–(vinyl acetate) (EVA), which is strippable from the insulation.

In order for the strippable screen to have sufficient tear strength during removal from the insulation, it is necessary for the thickness to be approximately 1.0 mm, but it may

Table 24.1 Insulation thicknesses and stresses for polymeric cables

Rated voltage (kV)	Insulation thickness (mm)			Electrical stress (kV/mm) 185 mm² conductor	
	PE	XLPE	EPR	Maximum	Minimum
3.6/6[a]	2.5	2.5	3.0	1.63	1.28
6/10	3.4	3.4	3.4	2.07	1.52
8.7/15	4.5	4.5	4.5	2.38	1.60
12/20	5.5	5.5	5.5	2.79	1.74
18/30	8.0	8.0	8.0	3.12	1.67

[a] These figures are true for conductors up to 185 mm². Above this size the thickness increases.

be thinner for harder materials. There are no such constraints with a bonded screen and, because semiconducting materials are very expensive, thickness is kept to a minimum, 0.5 mm being a typical figure.

The manufacture of single core cables is generally completed by the application of a layer of copper wires to provide an earth envelope with a cross-sectional area of 16 to 50 mm², depending upon the phase to earth fault level existing on the network. The cable is finished with an extruded oversheath. For networks with a very much higher fault level, or where increased mechanical protection is required, a copper tape is applied over the semiconducting layer, followed by an extruded bedding, then a helical application of aluminium armour wires and finally an extruded oversheath.

Fig. 24.5 Construction of 3-core 8.7/15 kV XLPE insulated steel wire armoured cable

For 3-core constructions, the application of copper tape around each screened core is more common. By the use of polypropylene strings to fill the gaps between the laid-up cores, the cable is formed into a circular shape over which is extruded a bedding sheath. The cable (fig. 24.5) is completed by the use of either steel wire or steel tape armour and an extruded oversheath. This type of cable construction has found considerable favour for industrial applications around the world.

Some UK regional electricity companies view the 3-core steel wire armoured construction as over-engineered and expensive. These companies are actively investigating alternative 3-core designs of 6.35/11 kV cable, which include collective metallic screens over the laid-up cores and polyethylene oversheaths. The choice between copper wire or galvanised steel wire for the collective screen is largely dependent upon the system earth fault level. Figures 24.6 and 24.7 show two designs of cable with collective wire screens.

There are also variants of the single-core design and at 12/20 kV a triplex construction has been developed in France. The dielectric screen is extruded over the insulation with a surface finish in the form of longitudinal grooves in the semiconducting material. These are filled with a powder immediately over which is laid a thin plastic coated aluminium foil which adheres to the outer oversheath. If water penetrates through the sheath, the powder swells thus preventing longitudinal travel of water along the cable.

At 18/30 kV a number of single core design options are readily available, these include copper wire screens either embedded in semiconducting butyl rubber, applied helically over the semiconducting core screen or sandwiched between layers of water swellable tapes. Alternatively, cables incorporating a full metallic sheath offer the greatest protection against radial moisture penetration.

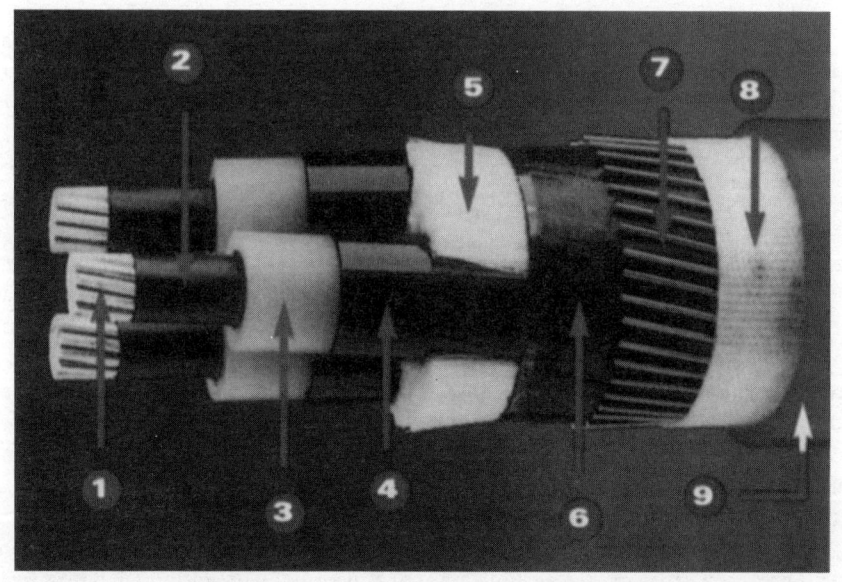

Fig. 24.6 3-core, circular stranded conductors, XLPE insulated, collective copper wire screen, MDPE oversheathed, 6.35/11 kV cable to IEC 502: (1) circular stranded conductor; (2) conductor screen; (3) XLPE insulation; (4) extruded semiconducting screen; (5) non-hygroscopic fillers; (6) semiconducting tapes; (7) copper wire screen; (8) synthetic tape; (9) MDPE oversheath

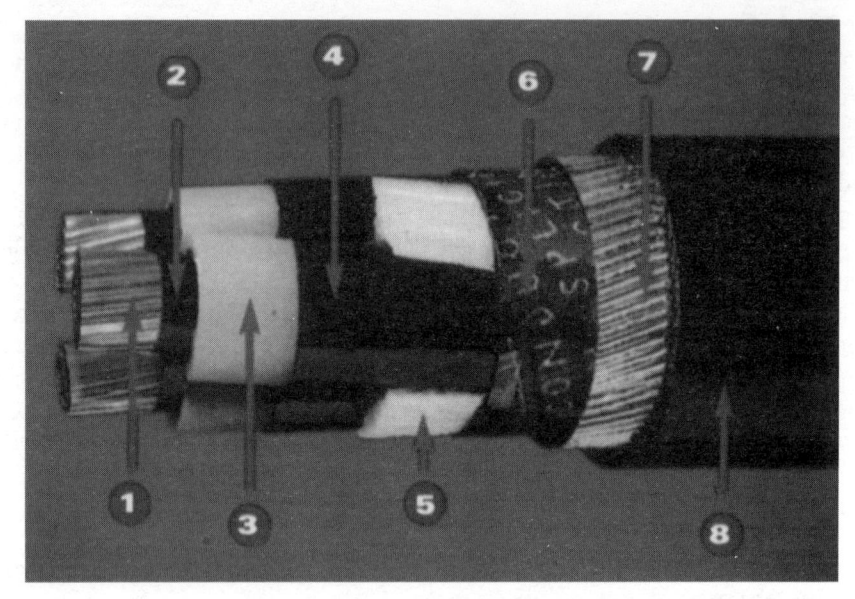

Fig. 24.7 3-core, shaped stranded conductors, XLPE insulated, galvanised steel wire armoured, MDPE oversheathed, 6.35/11 kV cable to IEC 502: (1) shaped stranded conductor; (2) conductor screen; (3) XLPE insulation; (4) extruded semiconducting screen; (5) non-hygroscopic fillers; (6) semiconducting tapes; (7) galvanised steel wire armour; (8) MDPE oversheath

MANUFACTURING PROCESSES USED TO PRODUCE CURED MATERIAL

For the manufacture of power cables the chemical process is adopted to effect the change from thermoplastic to a thermoset material. Until the 1980s, continuous catenary vulcanisation (CCV) was the most popular. However, following the discovery of an alternative chemical reaction by Dow Corning[10], designated Sioplas, simpler processes have gained in popularity.

In the CCV process, the pre-compounded polyethylene, containing an antioxidant and peroxide, is extruded on to a conductor at a temperature below the decomposition temperature of peroxide (130–140°C). The insulated conductor immediately enters a heated vulcanisation zone, located within a tube of typically 250 mm diameter and 50 m in length, which is in the form of a catenary. The now-cured material is cooled in water, or an inert gas, in a continuation of the tube which can extend its length by a further 50–80 m. When the insulation first enters the tube it is still thermoplastic, and if at elevated temperature it touches the wall of the tube, deformation will occur – hence the need for a tube in the shape of a catenary. Obviously, a tube held in a vertical position meets these requirements and such equipment is referred to as a vertical continuous vulcanisation (VCV) line. The cost of installing a VCV far exceeds that for a CCV, hence it is only used for the highest voltages.

The heat required to raise the temperature of the insulation in the curing zone of the tube can be obtained from steam, high temperature nitrogen, radiant heaters fixed on the outside of the tube, or by the injection of heated salts or oils into the tube. The heat provided to the insulation then causes the peroxide (e.g. 2.5% dicumylperoxide in polyethylene) to decompose, generating volatiles such as acetophenone, methane, water vapour, methystyrene and ethane. In order to keep the voids which are formed within the insulation by these gases down to an acceptable size, the tube is pressurised to about 10–15 bar. The insulation has to remain in the heated zone sufficiently long for all the peroxide to decompose and generally the temperature of the conductor has to be raised to 180°C to produce curing of the insulation adjacent to its surface. Therefore, throughput speed is reduced both by increase of conductor size and by thicker insulation.

Traditionally, steam was used to provide both heat and the pressure in the tube, but with the interest in higher voltage cables, the dry cure processes have been looked upon as producing a better dielectric. Long die cure, radiant heat under nitrogen or silicon oil and salt cure are all variants, and no voids other than those formed during the decomposition of the peroxide are produced.

The Sioplas and Monosil[11] processes, which have been used since the mid-1970s, were exciting developments, providing greater flexibility and economy to the producer. Both these processes are a chemical means of effecting the crosslinking and are discussed in more detail in chapter 3. In the Sioplas process an extrusion/compounding operation is required in which 10% of the amount of peroxide used in the CCV process is decomposed and silane is grafted on to a carbon atom. This material, which is still thermoplastic, is then extruded on to a conductor. In the Monosil process there is no separate pre-compounding/grafting operation as all materials are fed into the insulation extruder. In both cases the curing is effected in a water or steam vessel, the lines are cheaper to install and maintain and the production rates are much faster. It should be noted that the Monosil cure can still be considered a dry cure as the resulting water content of the insulation is < 250 ppm compared with a typical CV dry cure water content of 200 ppm. A wet or steam cure system would give a moisture content of several thousand ppm. The simpler technique also allows greater flexibility and the production of less scrap during starting up and closing down of the lines.

FUTURE PROSPECTS

Throughout the whole voltage range of distribution cables, the swing away from the traditional paper insulated types to polymeric insulated designs with advantages in ease of installation has been very significant. The lower capital cost and higher productivity processes associated with XLPE are likely to ensure a continued demand for these cables.

XLPE insulated cables are now being introduced at EHV, and work on XLPE insulation that will operate at increased stresses and reduced thicknesses is progressing.

REFERENCES

(1) UNIPEDE 'Activity Report On Cables'. Distribution Study Committee 50, April 1994.
(2) Thue, W. A. (1977) 'Field performance of polyethylene and crosslinked polyethylene cables'. IEEE Eng. Soc. Insulated Conductors Committee.
(3) Mashikian, M. S. (1986) 'Extruded medium voltage cable materials and practices in the U.S.A., Europe and Japan'. *Conf. Ref. 1986 IEEE Int. Symp. on Electrical Insulation*, pp. 13–22.

(4) Mason, J. H. (1953) 'Breakdown of insulation by discharges'. *Proc. IEE, Part 1* **100**, 149–158.
(5) Ku, C. C. and Liepens, R. (1987) *Electrical Properties of Polymers*. Munich: Hanser.
(6) Bartnikas, R. and Eichhorn, R. M. (1984) *Engineering Dielectrics*, Vol. IIA. ASTM 57P783. Philadelphia.
(7) Schultz, J. M. (ed.) (1977) *Treatise on Materials Science and Technology, Vol. 10B, Properties of Solid Polymeric Materials*. London: Academic Press.
(8) Artbauer, J. and Griac, J. (1970) 'Some factors preventing the attainment of intrinsic electrical strength in polymer insulations'. *IEEE Trans.* **EI-5**, 104–112.
(9) Shaw, M. T. and Shaw, S. H. (1984) 'Water treeing in solid dielectrics'. *IEEE Trans.* **EI-19**(5), 419–452.
(10) British Patent 1 286 460 (Dow Corning 1972).
(11) Swarbrick, P. (1977) 'Developments in the manufacture of XLPE cables'. *Electr. Rev.* **200**(4), 23–25.

CHAPTER 25
Manufacture of Distribution Cables

The essential stages of manufacture are shown in fig. 25.1. The initial processes of wire drawing and stranding are the same for all cable types, as are the later stages of laying up, metal sheathing, armouring, bedding and oversheathing. It is in the materials and processes used for insulation that significant differences are found between cable types. Because of the difficulty of handling the weight of longer cable lengths on site, dispatch cable lengths rarely exceed 400 m, but clearly it is beneficial if longer factory lengths can be made. In factories it is usual to employ mechanical handling to deliver, lift, and load both materials and cables to each process machine; this not only enables longer lengths to be used, but also cuts down loading time and optimises machine usage.

One of the largest items in processing cost is the time to load the individual process plant with part-manufactured cable and materials and it is common for machines to operate for only 25–50% of the total time that they are available. Production in long lengths is helpful but the main factor is the time and labour cost for loading.

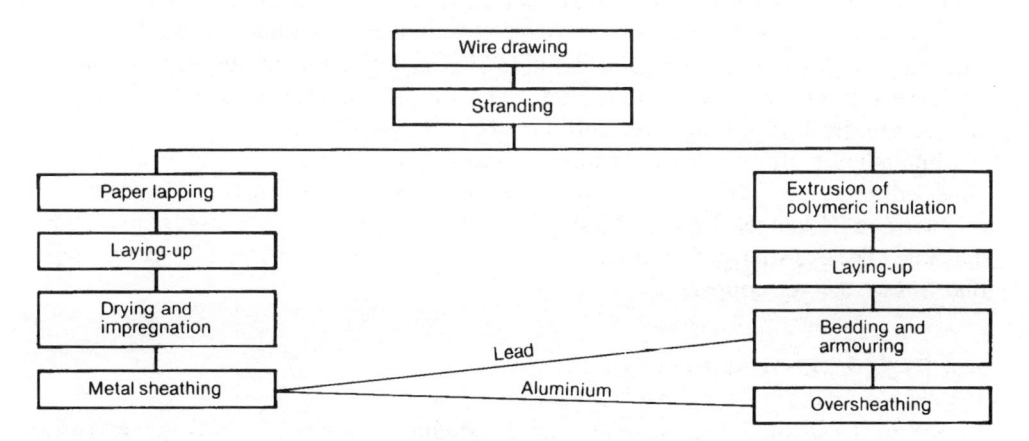

Fig. 25.1 Main processes in the manufacture of paper and polymeric insulated cables

Fig. 25.2 Rod breakdown machine

WIRE DRAWING

For the wire sizes required for distribution cables, i.e. 1.43–3.29 mm, copper wire is drawn through a continuous series of dies from annealed rod of around 8.0 mm diameter. The number of dies normally required is 6–13, according to wire size, and they are copiously lubricated by a stream of suitable compound. In modern plant, speeds up to 1200 m/min are obtained and the finished wire is continuously annealed by electrical heating in a steam atmosphere. During the winding on to steel spools the change to a fresh reel is achieved without stopping the machine.

Aluminium is drawn similarly from continuously cast 9.5 mm rod, equivalent to an annealed temper. For the reasons given in chapter 3 the drawn wire is not annealed and is in a broad $\frac{3}{4}$H temper. Figure 25.2 shows aluminium rod entering a large wire drawing machine. The resistivity of aluminium is affected far less by its metallurgical temper than is the case for copper.

SOLID ALUMINIUM CONDUCTORS

Except for the smallest sizes, solid aluminium conductors are produced by hot extrusion from billets of aluminium. The extrusion may be by an aluminium fabricator, in which case a conventional straight-through extrusion press is used, or by the cablemaker himself using an aluminium sheathing press. The purity is normally 99.7% minimum but with the Schloemann sheathing press a higher temperature is possible and 99.5% minimum purity may be used. With this press it has also been found possible to use as-cast rather than machined billets, i.e. it is not necessary to remove the outer skin by machining. Generally two conductors are extruded together. The hot extrusion practice provides conductors in a soft temper, which although generally advantageous is undesirable for very small sizes because of possible stretching. These sizes are produced by drawing. (See also chapter 3.)

STRANDING

Conventional stranding practice was based around a layer of six wires laid over one wire and then succeeding layers, with lay reversal on each layer, giving a total number of wires increasing as 7, 19, 37, 61, 91, 27, 169. However, for conductors of shaped cross-section and for compacted circular conductors, which have now largely replaced the uncompacted type for distribution cables, these are not always the ideal numbers of wires. For example, six wires in the centre of a 2-core or 3-core shape of conductor conform better to the cross-section to be formed than six wires around one, and in a compacted circular conductor the compacting of the inner layers sometimes produced too small a diameter to accommodate six additional wires in the next layer. The basic practice of separate layers with reversed lays is still the most common construction although unidirectional stranding is sometimes used.

Figure 25.3 shows typical stranding machines, each carriage having sufficient bobbins for one layer. The bobbins in the carriage may be 'fixed' or floating, the latter often

Fig. 25.3 Stranding machines for conductors, with conveyor system for storage and delivery of bobbins of conductor wire

being referred to as 'sun and planet' arrangement. Compacting tools or dies are normally installed for each layer. Shaping and pre-spiralling for multicore conductors is carried out at the same time. (See also chapter 4.)

A more recent technique, which permits higher output for small conductors up to 19 wires, is to dispense with a revolving carriage and feed the wires directly, as a 'bunch', into a shaping head. Shaping of conductors is achieved by passing the strand at each layer between suitably shaped rollers. Fine adjustment of the gap between the rollers permits some trimming of the height aspect of the conductor and also, therefore, some trimming of the resistance ultimately obtained. Shaping rollers are normally driven so as to process as a pair around the axis of the machine. This produces a pre-spiralled strand which then, at the later laying-up stage, requires no twisting to make it take up its place as one of a number of similar cores laid-up in a spiral form. Pre-spiralling greatly reduces the chances of unintentional disruption of the strand at lay-up.

PAPER LAPPING

Insulating paper is delivered in rolls of approximately 100 kg weight and 700 mm width. These are accurately slit into pads of appropriate width and diameter for use on lapping machines as illustrated in fig. 25.4. The angle of lapping, lay length, paper width and gap are all interrelated and dependent on the ratio of the speeds of head rotation and linear travel of the conductor.

Fig. 25.4 Paper lapping machine for 11 kV cables

The pads are mounted on a carriage which usually accommodates eight or more papers and individual carriages are able to rotate in either direction. Reversal of carriage rotation and hence direction of lay every eight or so papers is the most common practice for high voltage cables. It provides a more stable construction both during the lapping operation and in subsequent cable handling but, of course, has some penalty in that at each change there are butt gaps of double depth. A mechanism is provided for tension control for each paper and for the machine to be stopped if a paper should break.

The important factors of lapping in relation to the quality of the insulation are discussed in chapter 3, e.g. tension, gap width and registration of tapes in successive layers.

DRYING AND IMPREGNATION FOR PAPER INSULATION

Air and moisture have to be removed separately by heat and vacuum, both from the impregnant and the paper insulation plus fillers. The degasification of the impregnant is accomplished in a tall tank with the compound falling as droplets or flowing over plates to form a large surface area. The cable, either on steel drums or wound into trays, is placed in tanks which may be up to 4 m in diameter. The tops have special seals and the surfaces are heated by either steam or high pressure water. In modern plant the cable conductors are also heated by passage of direct current, to reduce the time for the whole mass to build up to temperature and to obtain better temperature distribution. Monitoring of the conductor resistance provides an accurate record of the cable temperature. A temperature of 125°C is maintained during the drying process.

During the heating process the tank is evacuated to a level of the order of $13\,N/m^2$ and nowadays the whole process is controlled automatically on a basis of temperature, time and pressure. Adequate dryness may be checked by a pressure drop test, i.e. shutting in the tank and measuring the pressure rise in a given time period.

The impregnating compound, also heated to 125°C and under vacuum, is then drawn slowly into the tank and the pressure is raised to $200\,kN/m^2$. After an appropriate impregnating period the tank is allowed to cool very slowly, and as discussed in chapter 19 it is important, particularly for MIND cables, for the impregnant to be circulated through a heat exchanger to obtain uniform cooling throughout the mass of cable insulation.

For low voltage cables, where complete filling of the insulation with impregnant is not essential, the compound may be pumped out of the tank whilst still hot and the cable left for surplus compound to be drained off. For high voltage cables, however, uniformity of cooling under compound to a temperature below the set-point of the compound is important, and this is most easily accomplished when the cable is in trays rather than on drums. Trays filled with compound to above cable level may be lifted out of the tanks to allow the final cooling to ambient temperature to proceed on the shop floor. Drums have to be removed at a temperature close to the set-point of the compound to avoid a large amount of compound adhering to the drum.

The total drying and impregnation process may take from 10 to 60 hours according to the plant, the amount of cable in the tank and the cable voltage.

EXTRUSION OF THERMOPLASTIC MATERIALS

All thermoplastic materials are applied by an extrusion process and PVC may be taken as a particular example. Pellets of the material are fed into the hopper of the extruder. They may be fully compounded, including colour, or the colour may be added at the same time by a metered second feed of coloured master batch. The hopper feeds by gravity into a long heated barrel through which an Archimedean screw revolves. As the PVC is forced along the barrel the additional frictional heat generated causes it to soften and in the compression process air is forced out backwards. An extrusion head containing male and female dies is located at the end of the barrel and may be at an angle of 45°–90° to it. Uniform flow of the softened material throughout the whole circumferential aperture between the dies is most important to obtain a concentric tube and freedom from overheating, which could lead to local decomposition within the extrudate.

To obtain the required degree of compression, the volume of material between the screw and barrel must be decreased towards the extrusion head, and screw designs to this end vary considerably. Figure 25.5 illustrates the essential principles with the separate zones along the screw. Different materials require a varying ratio of barrel length to diameter, and 15–20 is a typical range for PVC. The depth of flights and amounts of compression also need to be adjusted to suit the material, as do the heating and cooling systems.

Maintenance of optimum temperature for the particular compound being extruded is vital and when the required temperature is reached the barrel heating may need to be reduced or even forced cooling applied. For PVC, a barrel temperature of 150–180°C and a die temperature of 180°C would be typical.

For thermoplastic extrusion, the dies are usually designed with concentric circular tips, i.e. to extrude a tube. If the surface to be coated is non-circular, e.g. a shaped conductor, an oversize tube is reduced in size by a combination of vacuum and controlled line speed to form a perfect fit over the surface – often called a 'float-down' or 'tubing-on' technique. The technique is equally applicable to the provision of a close

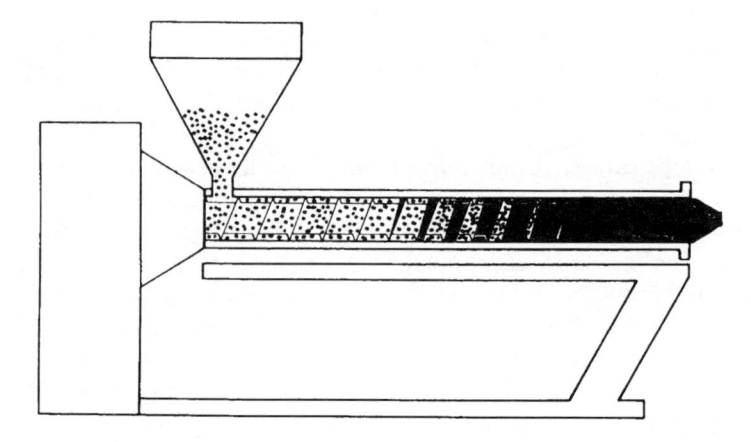

Fig. 25.5 Passage of thermoplastic material along an extrusion screw (Courtesy of Maillefer SA, Ecublens-Lausanne)

fitting oversheath on a corrugated aluminium sheath; it is the only option for insulating shaped spiralled conductors, all of which present a continuously rotating aspect at the die exit.

EXTRUSION AND CURING OF THERMOSETTING MATERIALS

At the time of extrusion, thermosetting materials are still thermoplastic and the only basic difference in technique is that they have subsequently to be cured (vulcanised). Originally this was a separate process but nowadays, except for the silane process discussed later, this operation is almost always carried out in line and known as continuous vulcanising (CV). For distribution cables the materials involved are XLPE and EPR. More use is made of XLPE and more procedures are available, but most of the basic principles apply also to EPR. Curing processes which have been used commercially for XLPE include

(a) continuous catenary vulcanisation (CCV) with heat transfer from either steam or a high temperature compatible fluid, or from thermal radiation, in a nitrogen gas atmosphere
(b) vertical catenary vulcanisation (VCV) which is similar but with a vertical tube
(c) irradiation by high energy electron beams
(d) MDCV: heat transfer by use of a very long land die in the extruder
(e) PLCV: liquid curing medium vulcanisation with heat transfer from a molten mixture of chemical salts under pressure
(f) the silane chemical crosslinking process.

A universal requirement for high voltage cables is that the insulating materials must be extremely clean initially, and during handling provision must be made to exclude any entrapment of dust or contaminants. This requires feeding arrangements from storage containers into the extruders without exposure to the atmosphere.

Many of the curing processes used for power cable insulation, e.g. (a) (b) (c) and (e) above, require the thermal decomposition of peroxides which are either precompounded into the polymer or added directly at the extruder. The temperature required is above 140°C and the rate of cure is temperature dependent. It is important, however, that the curing should not begin during extrusion and hence the extrusion temperature should not exceed 140°C. The curing temperature after extrusion may be up to 300–350°C (somewhat less for elastomers) and to prevent void formation due to decomposition products from the peroxides a pressure above about 7 bar (0.7 MN/m^2) is required.

Vertical and catenary tubes

Initially these processes were developed for the curing of natural rubber with sulphur and involved the passage of the cable through a long tube filled with steam at high pressure. The vertical tube has an advantage in that there is no undue sag of thick-walled soft extrudate as it leaves the die. However, as very long tubes of the order of 150 m became desirable for economic reasons, it became impractical on cost grounds to erect towers of this height. The catenary shape (fig. 25.6) is necessary to match the

Fig. 25.6 CCV line

natural sag of the cable core as it leaves the extruder die at the top end of the tube, and so avoid any scraping along the tube wall. Various tension and other devices are employed to maintain the core centrally in the tube.

When extrusion is commenced, the upper end of the tube is coupled solidly to the extrusion head and high pressure steam is admitted to the tube at a temperature of the order of 210°C. The lower part of the tube is filled with water, to cool the extrudate under pressure and so that it can be handled normally for drumming on leaving the tube. Approximately half the total length may be used for the cooling process but the proportion varies with technique and the thickness of the extruded layer.

Much attention to detail is necessary to prevent unnecessary wastage by the presence of inadequately cured material on starting and stopping extrusion and the process is clearly not suitable for producing short lengths of cable. Once set up it is preferable to extrude for days rather than hours and this, of course, necessitates very long continuous lengths of conductor of one size to be supplied to the extruder. Separate drums of conductor are used and welded together at the input end. An accumulator arrangement allows maintenance of steady conductor feed during the welding process.

Because the steam pressure is in the region of 20 bar ($2 \, \text{MN/m}^2$) the extrudate is pressed firmly against the conductor. This presents no problem with solid or heavily compacted stranded conductors having a relatively smooth surface, but if this is not the case a tape has to be applied longitudinally around the conductor to prevent passage between the wires.

For high voltage cables 'dry curing', as described later, is more often used for XLPE insulation than steam curing but, whichever technique is employed, at least three extruders are required for the insulation and two semiconducting screens. These may feed into a single extrusion head (often referred to as 'triple head') or the conductor screen may be applied from a separate head situated behind a dual head from which the insulation and outer screen are applied. Another variation is to employ two dual heads, the conductor screen and a thin layer of insulation being applied from one, followed by the bulk of the insulation and the outer insulation from the second. One purpose of the thin layer of insulation applied in the same head as the conductor screen is to protect the latter, which preferably should be kept as smooth as possible, from any slight abrasion which might occur on entry to the second head. The surface of the insulation from this first head becomes integrated with the main part of the insulation from the second head.

The output is an important factor in CV extrusion and is mainly related to the degree of cure which can be obtained in the length of heated tube available. Optimum productivity is derived from computer programs based on conductor size, volume of extrudate, temperature and length of tube etc.

In VCV and CCV lines dry curing is usually carried out by thermal radiation from the tube in a pressurised atmosphere of inert gas, usually nitrogen.

The pressure does not need to be as high as with steam as the temperature is independent of pressure and pressurising is mainly to prevent formation of voids from the peroxide decomposition products. The nitrogen in the curing section is not circulated, but is bled away at a low rate to prevent accumulation of gases produced by the curing reaction. The temperature, and hence output, is increased as the surface of the cable can be heated to 300°C. To deal with such higher temperatures the tube may be heated up to 450°C and needs to be fabricated from stainless steel or other suitable metal.

The heating of the tube may be by attachment of electrical elements to the outside of the tube or by d.c. heating of the tube itself. Good temperature control is essential and the core has to be maintained in the centre of the tube to maintain uniform curing. Cooling of the cable may be obtained by water in the lower part of the tube or by nitrogen circulated through external heat exchangers. Plant often provides for either cooling method to be used, depending upon what is preferred for the particular core being processed. Because of the high curing temperature, and hence a soft extrudate, it is important in a CCV line that the 'touch-down' point in the tube should not be reached until adequate cooling has taken place.

High temperature compatible liquids, such as silicones or polyalkylene glycol, may also be used for both curing and cooling.

Irradiation methods

High energy electron beams generated by linear accelerators of Van de Graaff type have been used for curing since the 1960s. In this process no peroxide is necessary and heating is not required. However, the process has mainly been applicable only to small cables, such as equipment wires, and to produce heat shrinkable materials for jointing purposes. For thicker wall insulation it has not proved to be economic.

Long land die process (MDCV)

Mitsubishi Dainichi continuous vulcanisation (MDCV) is named after the companies which developed the process in the early 1970s, again with the objective of obtaining curing in the absence of steam. The equipment is horizontal and very compact, the essence being to extend the die of the extruder to form a heated tube several metres long, to maintain the pressure and prevent void formation. Fast curing is achieved by the use of high temperature and on emergence the cooling is also carried out in a horizontal tube.

Liquid curing baths (PLCV)

Pressurised liquid continuous vulcanising (PLCV) is a technique based on the long established heated salt bath process (LCM) for vulcanising rubber cables. The salts consist of a eutectic mixture of potassium nitrate (53%), sodium nitrite (40%) and sodium nitrate (7%) and the temperature is up to 300°C. Because of the buoyancy effects of the molten salts, an inclined tube can be used and the catenary form is not always necessary: this is dependent on cable weight.

Silane chemical linking (Sioplas and Monosil)

Silane crosslinking was first developed as the Sioplas process by Dow Corning in the 1970s. The progression to the Monosil process was a joint BICC–Maillefer innovation. Sioplas is a two-stage crosslinking process whilst Monosil achieves the same result in a single stage. As a consequence, the basic polyethylene extruder feedstock is cheaper for Monosil and a compounding stage is eliminated. Both processes have been used for low- and medium-voltage insulation, but the greater simplicity of the single-step Monosil process has resulted in it being preferred over Sioplas for medium-voltage insulation.

Sioplas requires the preparation, in a material compounding operation, of a grafted polyethylene. It is sometimes claimed that under the correct storage conditions, grafted material is stable for periods of up to one year; at least in hot humid countries, these claims may prove optimistic. In any event, Sioplas graft co-polymer must be considered as having a limited shelf life, the length of which is strongly dependent on storage conditions.

Once grafted, the polyethylene is fed to the insulating line extruder along with a catalyst masterbatch, usually dibutyl-tin-dilaurate. For MV cables, the extrusion head is fed not only by the main extruder but also by at least two smaller ones, each feeding a carbon loaded semiconducting version of the grafted polymer to form the conductor screen and the core screen. Extrusions can therefore be tubed on to conductors, avoiding the need to apply tapes for the purpose of stopping inter-wire penetration by insulant. Tubing tooling also greatly facilitates the insulation of sector shaped pre-spiralled conductors with XLPE.

Having insulated the conductor with grafted and catalyst-bearing material, crosslinking is achieved by the action of heat and moisture off-line in a curing tank containing water at 80°C. Crosslinking reaction speed is governed by the diffusion rate of water in the polyethylene. However, the ability of polyethylene to absorb water is low and, within the material, water is continually converted to methanol so that the eventual water content of silane crosslinked polyethylene is always comparable to the water content of dry cured CCV produced materials.

A feature which contrasts with CCV is that, for silane crosslinking, there is no requirement to maintain a high pressure during crosslinking. The low temperature during curing reduces the occurrence of voids resulting from evolution of volatiles.

The limited ability to absorb water, which confers the advantage of a low moisture content end-product, also limits speed of cure in inverse relation to the square of the insulation thickness. Typical cure times range from 2 hours for 0.9 mm radial thicknesses at low voltage to 120 hours for 8 mm thicknesses at 33 kV.

The Monosil process uses similar chemistry to Sioplas but in this case the grafting reaction is carried out in the extruder barrel. As a result, the extruder feedstock is a base resin, either LDPE or LLDPE, and the need for the separate compounding process is obviated.

In Monosil, a silane cocktail is mixed – containing the alkoxy-silane, peroxide and catalyst, all in liquid form. Polyethylene granules may enter the hopper in the same way as for thermoplastic insulations, or they may first be mixed with the silane cocktail, which they absorb under the action of heating to about 50°C. This method has certain advantages but is highly complex and requires comparatively large amounts of space and plant costs. As a result, the more usual method is to inject the silane cocktail into the extruder feed port or into the hopper throat in a quantity proportional to the extruder output. Final mixing is achieved in the extruder barrel, where the screw incorporates a mixing section and grafting occurs after the thermally initiated decomposition of the peroxide. Extruder length to diameter ratios as high as 30:1 are used so as to provide a long residence time and an opportunity for all the required stages of mixing, peroxide breakdown and grafting to take place. Melt temperatures are usually in the range 210–260°C, where, in contrast to the CCV process, the initiating of peroxide breakdown by the extra heat is advantageous.

At MV, the screening layers are grafted in a similar fashion. However, the choice of carbon blacks is limited to grades which retain relatively low levels of water. Additionally, it is necessary to dry the screen polymers thoroughly immediately before feeding to the extruder.

As for Sioplas, except for a few cases which rely on ambient moisture and temperature, curing is achieved by immersion in hot water.

LAYING-UP

To lay cores together, either very large diameter machines are necessary so that bobbins of core can be rotated around a common axis, or the cores can be run off horizontally from their drums into a die and then proceed on to a take-up drum which revolves on its own axis (drum-twister machine). The latter (fig. 25.7) is now commonly used because of higher output. Belt insulation for paper cables, or taped bedding for polymeric cables, is conveniently applied in the same operation.

LEAD SHEATHING

The discontinuous type extrusion presses with vertical rams and containers which have to be filled with liquid lead have now largely given way to continuous extrusion machines of the Hansson type. These operate somewhat on the lines of a plastic

Fig. 25.7 Drum-twister laying-up machine

extruder except that the screw is vertical and is fed at its bottom end from a tank of liquid lead. The extrusion temperature is about 300°C and the sheaths are sprayed with water on leaving the dies.

ALUMINIUM SHEATHING

Aluminium sheaths are formed by direct extrusion over the insulation of paper cables or the bedding of polymeric cables. As the temperature is around 500°C, special methods of cooling have to be adopted. Two types of press are used, both being supplied with heated billets of aluminium.

Fig. 25.8 Hydraulik continuous aluminium sheathing press

Fig. 25.9 Schloemann aluminium sheathing press

In the Hydraulik press (fig. 25.8) the sheathing is continuous because a subsidiary ram under the die box maintains extrusion pressure when the main ram is withdrawn for the insertion of a new billet. The Schloemann press (fig. 25.9) is of the horizontally opposed twin-ram type and has an ingenious automatic ratchet spanner arrangement so that the dies move apart when pressure is released. This prevents nipping of the metal at stop-marks, a feature which causes bad bending properties in conventional aluminium tube presses.

For the Schloemann press it is possible to use as-cast billets but surface machining is necessary for the Hydraulik press.

For the corrugated form of sheath an oversize tube is extruded over the cores and a corrugating head is placed between the extruder and the take-up drum.

ARMOURING

Steel tape armour is applied by a conventional taping process to which reference is made in chapter 5, beddings and servings of the bituminous type being applied during the armouring operation.

The conventional method for the application of steel wire armour is for wire on bobbins to be mounted on a large carriage which rotates around a central mandrel through which the cable is drawn, by a large capstan wheel, at an appropriate speed to

Fig. 25.10 Drum-twister design of power cable armouring machine

obtain the appropriate length of lay. Larger cables may require more than 70 wires, so that very large and heavy machines are necessary. For this reason more modern armouring machine design follows a practice originally developed for the high speed production of small wiring cables and based on the 'drum-twister' technique. The armouring wires are drawn from stationary packs to form a ring of wires around the cable and the final drum containing the armoured cable revolves to derive the required lay of the wires. Figure 25.10 illustrates a typical arrangement, the stationary packs being at the far end and the haulage mechanism of 'caterpillar' type in the middle of the machine. Such machines can handle copper wires down to 0.067 mm in diameter and steel wires up to 2.5 mm in diameter. The equipment shown is designed for cables having extruded beddings and oversheaths, i.e. there is no provision for the textile and bitumen bedding and serving applications common on conventional armouring machines. Nevertheless, taping heads for plastic or metal binder or separator layers can be seen on the machine.

OVERSHEATHING

The final operation in the manufacture of distribution cables, prior to final testing, is usually to oversheath in PVC or HDPE. The process uses a conventional 'float-down' extrusion technique, to accommodate small variations in cable diameter.

CHAPTER 26
Installation of Distribution Cables

The majority of public supply distribution cables are buried directly in the ground, possibly with short duct sections, and this chapter deals mainly with such installations. In some areas such as in North America, however, there is a difference in that complete installation in ducts is generally preferred. This has the advantage in busy city areas that there is less disturbance to traffic and cable replacement is easier. Installation above ground is common for industrial sites and for cables associated with railways, the cables being supported on hangers or laid on fabricated steelwork or trays.

Distribution cables are usually supplied on wooden drums. The handling of the drum and suitable support at the optimum position is an important part of the installation operation.

Drum lengths of distribution cables are commonly of 250–500 m. Considerably longer lengths can often be installed satisfactorily but require much more care during running off. The inner end moves backwards and slack turns developing within the drum may result in kinking of the cable. The greater attention necessary and the possibility of damage may outweigh the cost of an extra joint.

For most types of installation the important stages may conveniently be divided into initial site inspection, trial holes, trenching, cable laying, reinstatement and final fixing.

SITE INSPECTION AND TRIAL HOLES

During the visit to site, decisions will be made on such matters as the method of installation, the special equipment required, e.g. pulling winches and any mechanised plant for digging, any duct positions for road crossings etc., and the general items of equipment necessary, including road signs and illumination.

It is then often necessary to dig trial holes to establish the trench route in detail. The trial holes will indicate the position of other services so that smooth bends can be provided to reduce the pulling loads when long lengths of cable are being installed. The trial holes also provide information on the nature of the ground for excavating and timbering purposes and on whether there is any chemical activity which would necessitate a special anticorrosion finish for the cable. Any unusual soil characteristic which might affect the thermal resistivity, and hence cable rating, would also be

DRUM HANDLING

Cable drum trailers

Very often drums of cable have to be transported from a central depot to an installation site and it is when short lengths of cable are required that trailers are particularly valuable, i.e. instead of cutting and re-drumming in the depot, the whole drum length may be taken to site by a single crew vehicle and then returned.

Figure 26.1 shows suitable equipment which makes the loading very simple by use of a manually operated hydraulic pump on the trailer. On site the drum is left in the cradle. On completion of pulling the cable end is capped and the drum is made secure and then returned to the depot. In the case of open routes, where the trench is free from obstruction, it is possible to pay off the cable from the towed trailer directly into the trench.

General handling

Drums of cable should normally be lifted by crane or forklift equipment or alternatively ramped-off using boards and wind off winch. Dropping drums of cable on to the ground from a lorry can cause damage to both the cable and the drum.

The drum should be mounted at the most convenient position for cable pulling and for manual installation. This is normally at the start of a reasonably straight section, preferably near the commencement of trenchwork.

Fig. 26.1 Cable drum trailer in use

It is important that any rolling of the drum to this position should be in accordance with the arrow on the drum wing as loose turns will develop, by unwinding, if the opposite direction is used. The distance of rolling should be kept to a minimum.

Drums are normally mounted so that the cable is pulled from the top of the drum and for very heavy cables it may be necessary to use a ramp to support the cable during passage into the trench. When cables have significant stiffness, e.g. those with non-corrugated aluminium sheaths, it may be preferable to pull from the bottom to reduce the tendency for the cable to come off with a wavy or spiral profile. As the cable is paid off, the drum rotates counter to the arrow which is intended to indicate the direction for rolling the drum into position.

Another factor which may affect the drum position is the presence of any services or obstructions at the trench entry, which could cause abrasion damage to the cable.

For mounting the drum a pair of screw jacks is adequate for relatively light drums but fabricated A-frames containing hydraulic jacks are necessary for heavier drums.

EXCAVATION

Most underground distribution cables are installed in the footpaths of urban areas. In the UK, this work is carried out in accordance with the New Roads and Street Works Act 1991. The purpose of this Act is to ensure safety, minimise inconvenience and protect the structure of the street and the apparatus in it. A Code of Practice is also followed giving instructions for signing, lighting and guarding of the works. Mini excavators are the most common type of mechanical plant employed, supplemented by hand digging for conventional open cut trenches. The advancement of non-dig techniques gives a greater range of options to the cable installation engineer, particularly at major road and rail crossings.

Depth of laying

The standard depth of burial in the UK is 500 mm at 0.6/1 kV, 800 mm from 3.3 to 11 kV, and 900 mm for 33 kV. As indicated in table 8.4 of chapter 8 there are some variations in depths throughout the world but they are not very substantial.

Timbering

As the excavation depth for most distribution cable installations is less than 1 m, there is seldom any need for full close boarded timbering. However, skeleton timbering may frequently be necessary to prevent deterioration of trench sides due to traffic vibration and to protect building foundations, street lamps etc. At a depth of 1 m the use of close timbering is dependent on ground conditions and for safety it is essential at depths below 1.3 m.

However, the key factor is overall safety for personnel and property. Safeguards may be necessary for a shallow trench next to an old building to avoid danger from possible collapse of walls etc.

Vertical timbers are usually 225 mm × 38 mm, horizontal poling boards 225 mm × 75 mm and struts 100 mm × 100 mm. Proprietary stretchers may be used instead of timber struts.

Excavated material

To comply with the 1991 Act, most excavated material is taken away to a licensed disposal site. Care must be taken to ensure drains and ditches are not blocked. Excavated material should be placed clear of the trench to avoid being a hazard to operatives, cable and the public. On completion of the excavation the base of the trench should be inspected to ensure that it is stone free and where specified, a bedding (normally 75 mm) of riddled earth or sand should be used.

PIPES AND DUCTS

RIGIDUCT® has largely replaced all other duct types used for the installation of distribution cables installed by regional electricity companies in the UK. RIGIDUCT® is manufactured by extruding two high density pipes simultaneously one over the other. The outer pipe is corrugated for extra strength and flexibility and the inner pipe is smooth or low friction. The duct is generally used in 3 m and 6 m lengths. Jointing is achieved by means of a push fit coupler and multi-duct systems are also easily achieved. It is normal practice to install RIGIDUCT® surrounded by fine granular material. Only in special circumstances is reinforced concrete used as the surrounding medium.

All pipes and ducts should be installed and jointed in accordance with the manufacturers' instructions. Before installing cable in ducts, a duct brush should be drawn through to remove any debris. This should be followed by pulling through a mandrel slightly smaller than the duct bore in order to ensure correct alignment of the ducts. If ducts are not to be used immediately, then a polypropylene drawcord should be left in the ducts and duct mouths should be sealed with a cap to prevent ingress of debris and moisture. A bell mouth should be placed at the duct entry to ensure smooth and safe passage of the cable into the duct. If the rating of the cable is critical, it may be necessary to pump bentonite or similar grout into the duct to improve heat transfer (see chapter 34).

Thrust boring, headings and tunnels

These constructions have to be considered by the installation engineer as an alternative to 'open-cut' duct installations. They are of particular advantage for road crossings where the normal depth of laying is precluded by existing services and/or the road or railway traffic flow cannot be restricted.

PREPARATION FOR CABLE LAYING

Cable stockings

A pulling rope has to be attached to the leading end of the cable, and a cable stocking is normally used for this purpose. Figure 26.2 shows types of cable stockings with single and double thimbles. The latter are normally preferred because there is less damage to the cap on the cable end if the pulling load is high.

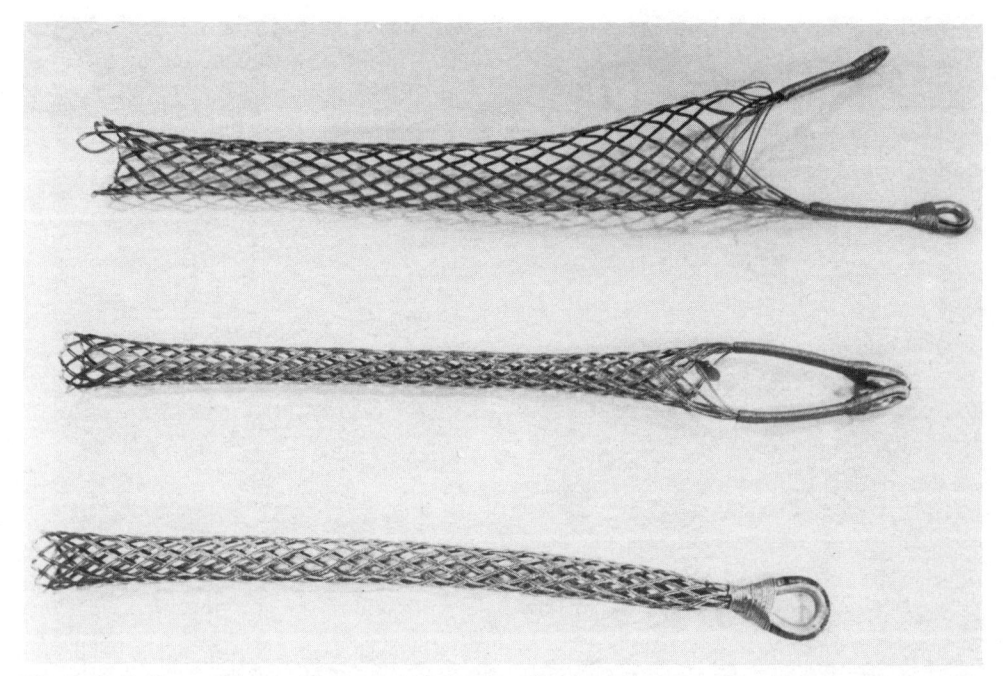

Fig. 26.2 Cable stockings with single and double thimbles (Courtesy of S.E.B. International Ltd)

Pulling eyes

With a stocking, the load is initially taken by the external cable components and is transferred by frictional forces to the conductors. If the load is high it may cause stretching of the outer layers and to avoid this a pulling eye may be plumbed to the armour, sheath and conductors to ensure distribution of the load across the whole section.

Many designs of pulling eye are available for different types of cable. Some are tubular with the conductors being sweated inside the tube and the metal sheath plumbed to the outside. Figure 26.3 shows a pulling eye which can be adapted for most cable designs.

Figure 26.4 shows an arrangement used by the UK regional electricity companies for aluminium sheathed cables. It is not a pulling eye in the strict sense, as it is only a means of anchoring the conductors and sheath in the cable end, and a stocking is necessary. Three holes are drilled through the cable with 60° phasing and threaded pins are screwed into tubes placed in the holes. When using this method it is important, immediately after laying, to cut off the cable end and re-seal the cable with a new cap.

Power winches

Power winches fall into two categories:

(a) compact lightweight designs utilising either a small petrol or compressed air engine as the power unit, which is suitable up to 2 tonnes safe working load with speeds of 5–8 m/min;

Fig. 26.3 Typical pulling eye for distribution cables

Fig. 26.4 Cable end with conductors anchored to aluminium sheath by the insertion of three pins (Courtesy of Southern Electricity Board)

Fig. 26.5 Diesel driven winch for safe working load of 4 tonnes. Transmission is through a gear box coupled to a pair of 'bull wheels'

(b) medium weight designs suitable for 2–4 tonnes, the larger sizes having a diesel power unit. Instead of relying on a direct pull these larger units utilise a pair of 'bull wheels' for wire bond haulage as illustrated in fig. 26.5.

With winch pulling, it is important to take steps to keep the pulling load to a minimum. The drum position should be chosen so that the longest length of straight trench is at the pulling end with any severe bends as close as possible to the drum.

Pulling tension in a wire bond can be reduced by passing the bond through a snatch block where the trench changes direction (fig. 26.6). The pull is stopped just before the cable reaches the snatch block so that the bond can be removed from it. This arrangement also prevents the bond from causing damage by scoring the skid plates on the inside of the bend.

An alternative procedure for pulling by winch is the continuous bond method. Instead of attaching the bond to the leading end of the cable, the cable is lashed to the bond at about 1 m intervals. This method is mainly required for heavy cables and is seldom needed for distribution cables. It is described in chapter 36.

Preparation of the trench

Preparation comprises the installation, as necessary, of skid plates, rollers, etc., and paying out the winch rope if using power assistance. Typical rollers and dual purpose rollers are shown in fig. 26.7. Cable rollers are necessary to prevent the cable from touching the ground and should be spaced a maximum of 2 m apart for normal size cable. With heavy cables this spacing may need to be reduced to 1.2 m. Correct positioning is important to keep the friction load component to a minimum.

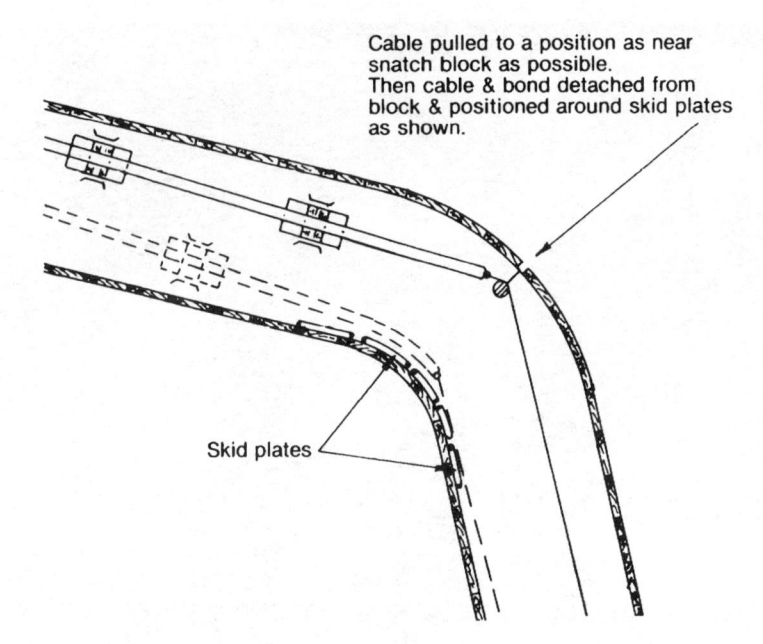

Fig. 26.6 Snatch block arrangement to reduce pulling tension on a bend

Ducts should be clean and smooth and fitted with bell mouths at entry, and also at exit if followed by a bend.

The pulling tension is determined by a summation of the weights of cable up to a point of friction multiplied by the coefficient of friction at that point. Conditions vary widely according to cable type, cable finish and bend in a route, but a general average for the coefficient of friction is around 0.25. Under difficult conditions in ducts it may increase to 1.0, and in such situations graphite lubricants should be applied at duct entries to reduce the friction.

CABLE PULLING

The cable should preferably be drawn to its final position in a continuous manner. During stops, it will settle between rollers and may cause high strain on men and machines during re-starting. Whether the pull is manual or with a winch, it is necessary for one man to be stationed at the drum with a plank wedged against the wing so that over-running of the drum is prevented if pulling stops. Otherwise many loose turns can easily develop on the drum.

Heavy lead sheathed paper cables in long lengths may need very large gangs of men if winch pulling is not used. However, because of the large reduction in cable weight, only four or five men are needed for a 200 m length of 11 kV cable with corrugated aluminium sheath for an average route. Polymeric cables having no metallic sheath are even easier to install.

When pulling by a winch it is advantageous for the cable end to be taken by hand as far as possible before attaching the winch rope. This allows the leading cable rollers, skid plates etc. to take the load and settle under well controlled conditions. The winch

Fig. 26.7 Typical cable rollers, normal and dual purpose, i.e. combined with skid plate (Courtesy of S.E.B. International Ltd)

operator must carefully observe the dynamometer to prevent overloading. On long pulls, good communication is essential, preferably by radio.

Final placing

Prior to disconnecting the pulling rope, the cable is laid-off, i.e. starting at one end it is carefully lifted from the rollers and deposited on the bottom of the trench. About 10 m of cable should be lifted at one time, any slack being carried forward. This exercise is simple if only one cable is being installed but needs careful control if the rollers are to be re-used for further cables in the trench. The cables cannot be positioned until the last one has been pulled.

The end position of a cable may require double handling because it is not possible to draw cable straight into a substation or other building. In this case the cable is overpulled and then manhandled around to the duct entry and fed into the required position. At all times the loops should be kept as large as possible so that the bending radius is always above the minimum permitted. Similarly, at the drum position, the necessary length of cable may be unwound from the drum and laid out, if necessary in a figure eight if space is limited (see next section), prior to cutting to length and placing in position. Immediately after cutting, the cable must be suitably sealed to prevent ingress of moisture. In this respect it is also important to examine the pulling end seal to ensure that it has not been damaged during laying.

Flaking cables

Cables have to be flaked, i.e. a substantial amount laid on the ground, when for some reason the drum cannot be mounted at a favourable position near the joint or termination. Cables of voltages higher than 11 kV must not be flaked. It is most frequently done on long lengths of auxiliary or pilot cables in order to obviate a joint in the middle of the length or, where the cables are laid in the ground, to enable the trench to be backfilled when only half the drum length has been laid. Where drums of pilot cable are twice the length of the accompanying feeder cable, it is common practice to position the pilot cable drum at the mid-position and pull the cable in the first half of the section, which can then be backfilled.

The remaining cable must then be flaked out in the shape of a figure eight in order to avoid twisting when it is carried forward into the trench or route ahead. The size of the figure eight will depend on the amount of space available and the length of cable to be absorbed, but in no circumstances may the diameter of each half of the eight be less than twice the minimum permissible bending radius r of the cable. The distances between the outer layer of cable on the drum and the nearest point of the figure eight must exceed $3r$.

For a drum mounted at the side of a trench or joint bay, the process of flaking is shown in fig. 26.8. It should be noted that in forming the second layer of the figure eight the cable coming up from the trench passes under that coming off the drum to form the half of the eight farthest from the drum as shown at E in fig. 26.8. Care must be taken to ensure that at no point in the flaking process is the cable bent anywhere at less than the minimum permissible bending radius, especially when turning the cable at the top nearest the drum. When flaking has been completed the inner end of the cable on the drum becomes the leading end for laying.

Installation of Distribution Cables

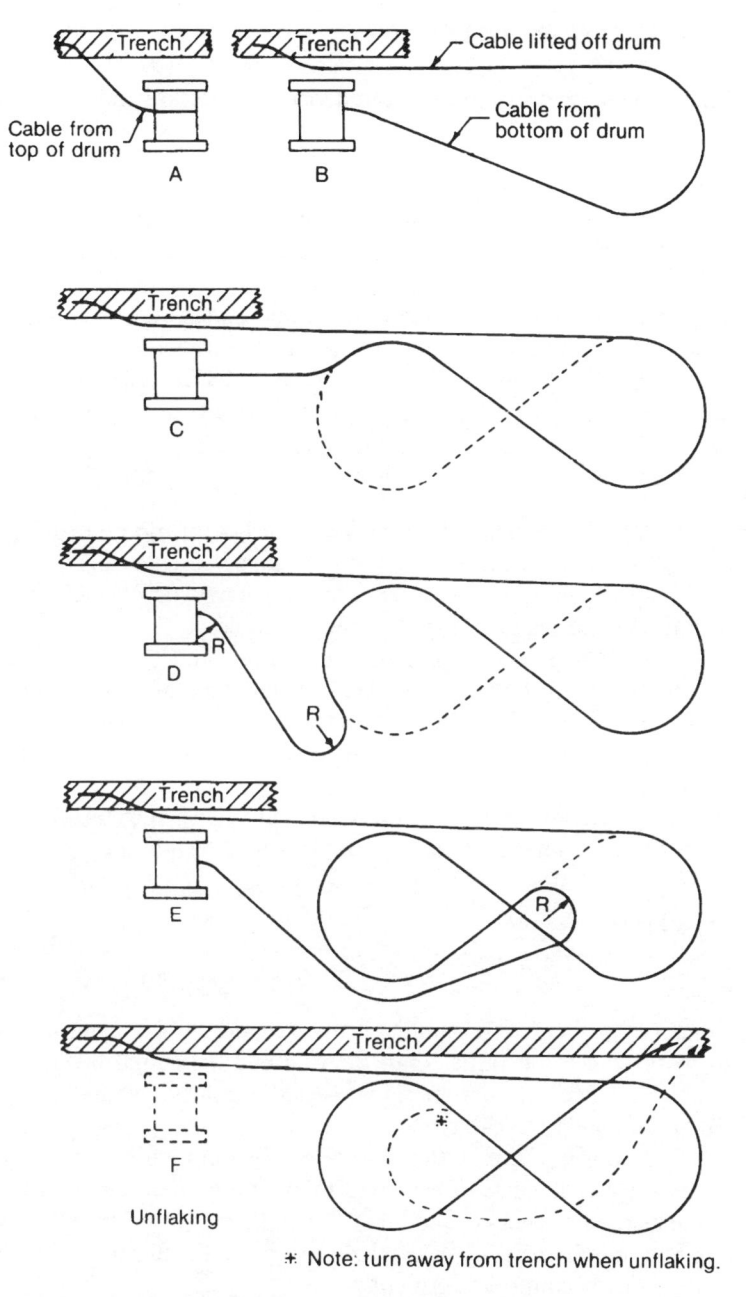

Fig. 26.8 Procedure for flaking cables by the figure eight method

At the commencement of unflaking it is important that the new leading end of the cable is so directed that the cable does not incur a twist or kink as it comes off the top layer. Note particularly that if the end of the cable points towards the drum it must be reversed by turning away from the trench as at F. Care must also be exercised to prevent infringement of the minimum bending radius as layers of cable come off the figure eight during installation.

BACKFILLING AND REINSTATEMENT

Prior to backfilling, it is necessary to carry out a visual inspection and some items which require to be checked are as follows.

(a) The cables have a proper bedding.
(b) The spacing is correct if there is more than one cable in the trench.
(c) Cables at duct mouths are suitably supported and ducts sealed to prevent ingress of moisture or vermin.
(d) Pulling equipment is removed. In the case of skid plates etc., this may need to be combined with backfilling to prevent collapse of the trench wall.
(e) The cable is free from obvious damage caused by installation. A very high proportion (possibly 90%) of cable failures in service is due to such damage.

Backfilling

The cables should first be surrounded and covered with appropriate bedding material, using sand or riddled soil as necessary, to give a compacted cover of 75 mm thickness over the cable. Any required earthenware, concrete cover or plastic tiles or plastic marker sheet may then be placed centrally over the cable.

In urban areas, the trench should be backfilled in layers and with approved materials conforming to the New Street Works Act 1991. The degree of compaction should be verified using, for example, a Clegg impact soil tester.

The use of special backfill materials to improve current ratings are discussed in chapters 34 and 36.

Temporary reinstatement is rarely specified and it is normal for permanent reinstatement to be laid immediately after completion of backfilling.

GENERAL ASPECTS

Bending radii

Cables should never be bent to a small radius. The prescribed minimum should be considered to be the exception rather than the rule and the actual bending radius the largest which circumstances will permit. This eases the task of installation and reduces the possibility of damaging the cable.

It is particularly important to maintain a generous bending radius when cables are to be pulled by a power winch, so as to keep within maximum permissible tension and to prevent the cable being flattened around bends or in ducts. In the immediate vicinity of joints and terminations where the cable can be set in position after the pulling operation is complete, the radius becomes less critical.

The minimum bending radii required by UK specifications for the various types of cable are given in appendix A19. These radii are also generally representative of requirements throughout the world for similar types of cables.

Cold weather precautions

A cable must not be laid or otherwise bent when it is at such a low temperature that damage might be caused to the insulation or serving. With normal paper insulated cables and any cables having a standard PVC oversheath, cable laying should take place only when both cable and ambient temperature have been at or above 0°C for the previous 24 hours, or when special steps have been taken to heat the cable to above this temperature.

If there is reason to suspect that the cable is below 0°C, its temperature must be measured by inserting a standard glass bulb thermometer ($-10°C$ to $+10°C$) between the turns of cable. For this purpose two to three battens should be removed from the drum at 45° and 135° positions. If the cable is below the temperature limit and the drum cannot be taken into a heated building, laying must be deferred until its temperature has been raised by the following method.

(a) Roll or run the drum until the gaps made in the lagging occupy 135° and 225° positions.
(b) Place lighted paraffin operated danger lamps (not hurricane lamps) in a row on the ground beneath the gaps. The number of lamps will depend on the weight of cable, the ambient temperature, and the time available for heating. As a general guide, the lamps should be placed 100–150 mm apart, and the heating period should be at least 24 hours.
(c) Cover the drum to the ground with tarpaulin sheeting fitted close to the drum, but draped at a safe distance from the lamps. A little ventilation between the sheeting and the ground may be found to be necessary, but any large gaps allowing cold air to enter will seriously slow down the heating.
(d) After a heating period of 24 hours a couple of battens should be removed at 45° and 315° positions and the temperature of the cable measured.
(e) To allow for inequalities, and for cooling during laying, the temperatures reached should not be less than 5°C.

Once a cable has become really cold, it will require many hours of continuous heating to bring it up to the required temperature. It is essential that heat is applied for sufficient time to warm the whole of the cable and not merely the outside layer. Drums to be heated should be placed as close as possible to the pulling positions so that the minimum time elapses after heating is stopped until installation takes place.

When pulling a cable which is only a little above the temperature limit, the region of bending must be watched for any cracking, and on detection of such an effect the speed of pulling must be reduced until it ceases. If cracking of a minor nature occurs, the protective finish must be repaired.

As the heating of cables involves a fire risk, it is very important that the heating arrangements are kept under supervision and that a frequent check is made to ensure that the lamps are burning properly.

Cable laying in subsidence areas

In areas subject to subsidence, principally due to mining installations, care must be taken to prevent damage due to uncontrolled movement of the cable. Slight subsidence problems can normally be accommodated by laying the cable with a horizontal wave formation. The cable under this condition is held by the backfill and only minor movement can be achieved. More major movement of the cable can be accommodated utilising wave boxes which keep the cable in a less restrained vertical plane (fig. 26.9).

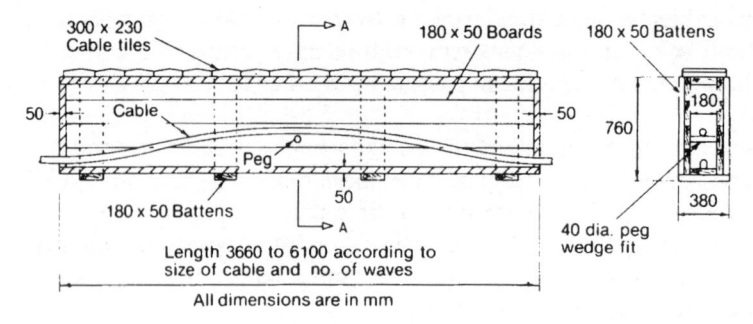

Fig. 26.9 Wave box for use in laying cable in areas subject to mining subsidence

Information regarding subsidence must be made available to cable design engineers at the earliest opportunity in order that wave boxes, if required, can be incorporated in the system.

Repairs to PVC or polyethylene oversheaths

It is very important that any damage to the oversheath observed after installation should be repaired before backfilling is commenced. If the cable has an aluminium sheath it is also essential to make a careful inspection for damage.

Strict attention to detail is necessary in making such repairs and only the approved materials specified should be used. Only a brief outline is given below and full instructions should be obtained from the cable manufacturer. Some of the methods quoted are only suitable for use after installation has been completed, when the cable will not be subjected to significant movement in service.

Superficial damage
The local area of damage is rubbed down with carborundum strip to the depth of the damage and chamfers of 25 mm length are formed at the edges. After cleaning with a suitable solvent, PVC self-adhesive tape of 25 mm width is applied under tension with 50% overlap. The taping is continued up the chamfer until the top is reached. Then another four layers are applied over a length extending 75 mm beyond the chamfer.

Holes or slits in PVC oversheaths
The edges of the hole or slit are chamfered for a distance of 30 mm and the area around this is abraded over a length of 20 mm. The area is then cleaned with a suitable solvent and, if bitumen is present under the oversheath, care must be taken to remove it from the prepared surface. A patch is then applied to fill the area using an approved grade of special putty. This is followed by an overlapped layer of 50 mm wide PIB self-amalgamating tape extending 50 mm from the patch and three overlapped layers of PVC self-adhesive tape extending 100 mm from the edge of the PIB tape. In the case of slits, further strengthening by the addition of an epoxy resin bandage should be made if the cable is likely to be moved. Details are given later.

Removal of a complete ring of oversheath
After removal of the damaged ring a chamfer is formed, unless the cable has a corrugated aluminium sheath. In the latter case the edges are left square. The surface is then

thoroughly cleaned with solvent, taking care to remove the graphite layer, if present. Four overlapped layers of 50 mm wide PIB self-amalgamating tape are then applied at high tension over a length extending to 50 mm beyond the original cut. PVC self-adhesive tape is then applied at one-third overlap to build up to a level corresponding to the original oversheath diameter. For corrugated aluminium sheathed cables the length should be up to the oversheath cut and for other cables up to the end of the chamfer.

Five layers of PVC self-adhesive tape are then applied, each one extending 5 mm further along the cable. The repair is completed with a resin poultice reinforcement consisting of six layers of ribbon gauze or bandage impregnated and painted with an approved grade of freshly mixed epoxy resin. This requires about 12 hours to cure.

PULLING LOADS

Cable stocking over normal cable end

For the majority of installations a cable stocking is applied over the end of the cable as cut and capped, i.e. it grips the outside of the cable and the pulling load may not be fully transmitted to the conductors. For paper insulated cables with lead or aluminium sheaths, unarmoured or steel tape armoured, the maximum pulling load tends to be dictated by the way the metal sheath is pulled forward over the insulation. This could cause damage to the insulation over 1–3 m at the leading end. In the USA the following formula has been suggested for lead sheathed cables without wire armour:

$$T = 3.33t(D - t) \quad \text{(kgf)} \tag{26.1}$$

where T = pulling load (kgf)
  D = diameter over lead sheath (mm)
  t = lead sheath thickness (mm)

Rather higher loads are possible if the cable has wire armour, whether the cable is pulled with a stocking over the armour or, as is sometimes done, the armour is extended over the end of the cable to form an attachment for the pulling rope, i.e. a direct pull on the armour. Whatever method of attachment is used, the limiting load is then governed by the rotation on the end of the cable caused by untwisting of the armour.

Winch pulling with a prepared cable end

When cables are pulled by a power winch, more attention has to be given to the maximum pulling load that is permissible. For such an installation, a pulling eye attached to the conductors is necessary and traditionally it has been general practice to relate the maximum permissible load to the strength of the conductors, e.g. 6 kgf (58.8 N) per square millimetre of total conductor area for copper and 3 kgf (29.4 N) for aluminium. These figures are based on the ultimate tensile strength of the materials with a safety factor of about 2.5. Similar figures may be obtained by taking the 0.1% proof stress and a lower safety factor. It is desirable to specify a maximum load of 2000 kgf as such a load would indicate some unnecessary obstruction.

However, derivation of a maximum load in this way ignores some important factors, in particular the effect of side pressure in passing over skid plates or against duct surfaces at bends in the route. Excessive pressure can cause flattening and damage to the cable insulation or other components.

Muhleman[1] gives a detailed analysis of all the effects which govern pulling loads for duct installations but the approach is by mathematical treatment rather than practical observations.

It is extremely difficult and expensive to carry out effective trials on cables with different constructions and using a typical site installation with a range of tensions up to levels that cause damage. However, a comprehensive series of tests was undertaken in the UK in 1975–76 by the Southern Electricity Board in collaboration with the Electric Cable Manufacturers Confederation. The types of cable investigated included 11 kV paper insulated with smooth and corrugated aluminium sheath, Consac and Waveconal CNE cables. The route included two 120° or 90° bends and a duct section. By connecting the two ends of the cable with a wire rope in which a Tirfor winch was included, it was possible to vary the tension over a wide range. After each pull, samples of cable used were examined visually and also subjected to impulse test. In a separate series of tests, samples of cable were subjected to static pulling loads to investigate effects such as sheath extensibility.

The work highlighted a number of important points.

(a) Vertical rollers should not be used at bends because they cannot be set up with sufficient accuracy to prevent an individual roller from standing proud of the others.
(b) Skid plates combined with horizontal rollers (fig. 26.7) appeared to provide the optimum arrangement and, if high pulling loads are involved, careful setting is necessary to obtain a smooth curve.
(c) Skid plates easily become scored by the pulling hawser and good maintenance is required to prevent damage to PVC oversheaths.
(d) When aluminium sheathed cables were pulled with a stocking applied over a normal cable end, i.e. with a plumbed end cap, it was confirmed, as already mentioned, that one limit to maximum pulling tension was the amount of sheath extension which could be tolerated. By pinning the sheath and conductors together, as described earlier, much higher tensions could be withstood. Ultimately the amount of flattening due to side pressure was found to be the limiting factor, although in the case of cables with smooth sheath the limit was at a level below that at which it is necessary to anchor the sheath to the conductor.

From this work, recommended maximum tensions for 11 kV PIAS cables were as shown in table 26.1.

Table 26.1 Maximum pulling load for 11 kV PIAS cables

Conductor size (mm^2)	Smooth sheath (kN)	Corrugated sheath Without sheath anchor (kN)	Corrugated sheath With sheath anchor (kN)
95	5.9	4.0	6.9
150	7.9	5.4	9.8
185	9.8	6.4	11.8
240	13.7	7.9	14.7
300	19.6	9.8	19.6

At high ambient temperatures above 20°C, there may be a tendency, particularly for cables with smooth metal sheaths, for the PVC oversheath under the stocking to stretch beyond underlying components. If the cable has a metal sheath it may be necessary to remove the PVC oversheath and apply the stocking directly over the metal sheath.

Consac and Waveconal cables

Consac cable is so light that winch pulling is seldom required and even then the loads are low. It is not necessary to anchor the conductors and sheath. Therefore, when pulling with a stocking over the cable end, the limit is imposed by stretching of the aluminium sheath. A reasonable maximum is that which corresponds to 90% of the 0.1% proof stress of annealed aluminium. This can be taken as $30\,\text{N/mm}^2$ of sheath cross-sectional area. The above tests showed that there was no undue deformation by side pressure at this level.

In the case of Waveconal cables, the limit is imposed by deformation of the rubber layer for the corrosion protection of the concentric neutral conductor. Tentative maximum pulling loads vary from $3.0\,\text{kN}$ for $70\,\text{mm}^2$ cable up to $7.0\,\text{kN}$ for $185\,\text{mm}^2$ and above.

SUPPORTS FOR CABLES ERECTED IN AIR

Support spacings for lead sheathed cables

Most distribution cables in air are suspended from open J-hangers or cleats and only a small proportion are installed on trays. Whatever method is used, it is necessary to pay attention to the effect of expansion and contraction, particularly if there are regular operating cycles to maximum load and the cables are lead sheathed. Fracture of lead sheaths due to local bending and fatigue has happened frequently, e.g. in generating stations on the cables connecting the generators to transformers, when the cable supports have been too close together. Expansion causes a thrust in the cable which can result in bodily movement of the cable across or through the support until slack cable builds up at some convenient position, such as a bend in the route. A short span of cable then has to accommodate the expansion and contraction of a long section length and the repeated flexing can lead to failure.

Holttum[2] was probably the first to publish a detailed investigation and draw attention to the benefits from using much longer spacings between supports than had been traditional. He provided a somewhat complicated formula to determine optimum spacing and recommended that in each span between supports the cable should be installed with a sag of 2% of the spacing. The object was that each span should be able to cater for its own expansion and contraction by variation of the sag, with no possibility of movement across the point of support. Such movement cannot be prevented adequately by clamps or cleats.

Prior to Holttum's work, a common conventional spacing was about 750 mm and for lead sheathed paper insulated cables he advocated that it should be increased to around 2–3 m according to cable size and weight. For a long time his theory was a little unpopular because it is much more difficult to obtain satisfactory visual appearance with long spacings and somewhat more care is required to obtain uniform sagging during installation. However, it became accepted that it was quite impractical with short

Table 26.2 Constants for PILS cables

	K		V
Unalloyed lead	0.60×10^4	Solid type, single-core	4.1
Alloy B	1.6×10^4	Solid type, 3-core and SL	0.34
Alloy $\frac{1}{2}$C	0.91×10^4	Oil-filled, single-core	0.46
Alloy E	0.85×10^4	Oil-filled, 3-core	0.38

spacings to cleat cables sufficiently rigidly to prevent bodily movement through supports and that longer spacings were undoubtedly beneficial. In addition to preventing the movement of cable across supports it is also necessary to provide positive restrictions by adequate cleating in sections containing joints and at bends or changes of level.

Subsequent to Holttum's original investigations, a simplified formula has become used more generally:

$$L^3 - \frac{Kd^2tL}{W} = \frac{V(9ne^4 + 11ma^4) \times 10^6}{W} \qquad (26.2)$$

where L = distance between support centres (mm)
K = a constant depending on sheath material
d = lead sheath diameter (mm)
t = lead sheath thickness (mm)
W = weight of cable (kg/km)
V = a constant dependent on number of cores and cable type
n = number of conductor wires
e = diameter of each conductor wire (mm)
m = number of armour wires
a = diameter of each armour wire (mm)

The constants for PILS cables are given in table 26.2.

Whereas Holttum recommended that the spacings from the formula should be minima, it is now taken that they should be maxima with a general average being 30% lower. Values are normally between about 1.5 and 2.5 m according to conductor size.

When cables are installed on trays or flat surfaces, so that it is not possible to obtain 2% sag between points of support, it is often necessary to adopt a similar principle but to arrange for horizontal instead of vertical movement. This situation arises most commonly with large conductor size cables, which are hence of single-core type and often installed in trefoil cleats. A convenient arrangement with such cleats is to form the cable into a uniform wave shape with the deflections in opposite directions on each side of the cleat. The cleats are installed in such a way that they can pivot about their axis. The cables are supported by further trefoil cleats half way between the pivoting cleats and the latter cleats are on a suitable base to allow for horizontal movement. (See also chapter 34). A similar principle is required if the cables are installed in flat formation.

Support spacing for cables without lead sheaths

As polymeric insulated cables such as PVC cables to BS 6346 and XLPE cables to BS 5467 do not have lead sheaths, there is not the same problem of possible cable failure

Table 26.3 Support spacings for polymeric insulated cables

Overall cable diameter (mm)	Solid aluminium conductor		Stranded copper conductor	
	Horizontal spacing (mm)	Vertical spacing (mm)	Horizontal spacing (mm)	Vertical spacing (mm)
Below 15			350	450
15 to 20	1200	550	400	550
20 to 40	2000	600	450	600
40 to 60	3000	900	700	900
Over 60	4000	1300	1100	1300

but similar cable movement may occur and it is still desirable to arrange for 2% sag to be provided between supports. Bodily movement of cable due to thrust is more likely to occur if the cable has solid aluminium conductors and much wider spacings are necessary. Table 26.3 shows suitable spacings for cables in buildings, where compliance with the IEE Wiring Regulations is required. For any other condition of installation the recommendations of the cable manufacturer should be obtained.

Other considerations

Cable cleats provide a satisfactory means of supporting cables and wherever possible their use should be considered in preference to hangers, which if used should have adequate bearing surface free from sharp edges.

Multicore power cables installed in groups in air should have spacing all round for the dissipation of heat and even so some cables may need to be derated by 10%. Exact details vary with the number of cables and the method of installation. Full details are given in ERA Reports 74–27 and 74–28 which respectively deal with the heat emission from cables in air and cables on perforated trays. Where, depending upon conductor size, 75 mm to 150 mm spacings can be arranged between fully loaded cables, no derating should be necessary provided there is free ventilation around each.

When power cables are double or treble tiered they should be cleated so as to provide free ventilation between layers, but with long cable lengths where voltage regulation governs the cable size the cables may not be fully loaded and mutual heating is accordingly alleviated.

As control cables can be accommodated in the spaces between power cables, careful planning of the configuration of the cables should result in an economic layout.

Provided the trays are of the perforated type and permit the circulation of free air, multicore power cables may be installed on trays in single-layer formation without cleats. Control cables can be bunched and uncleated when installed on trays.

Single-core cables are normally installed in trefoil formation, each trefoil cleat containing three different phases, and are spaced as mentioned above.

In order to restrain the forces set up between phases under heavy short-circuit conditions it is sometimes necessary to bind the three phases together with specially designed binders positioned between the supporting cleats.

Another method of installing single-core cables on horizontal runs is on expanded metal trays with binders as mentioned above, but without attachment to trays. Where cables change horizontal or vertical direction they should be cleated to prevent an accumulation of expansion at the bends, the expansion being taken up along the straight sections.

REFERENCES

(1) Muhleman, C. E. (1976) 'Cable pulling'. *IEEE Conf. Rec. Pulp and Paper Industry Tech.*, Boston, USA, 15–21.
(2) Holttum, W. (1975) 'The installation of metal sheathed cables on spaced supports'. *Proc. IEE, A* **102**, 729–42.

CHAPTER 27
Joints and Terminations for Distribution Cables

Joints and terminations are an integral part of a power cable distribution system and must perform the same basic functions as the cable on which they are installed. In addition to providing conductor connections suitable for the full current rating of the cable and adequate insulation, they must also provide stress control for screened cables, mechanical protection for joints, and environmental protection for terminations. Selection of joints and terminations needs to be based on the installation skills available together with the economic, technical and physical constraints of the installation. Joint designs must cater for straight joints, where two identical cables are joined, and branch joints which may be either T or Y formation. Transition joints for connecting cables having different types of insulation are also required. Joint designs for 0.6/1 kV cables must be capable of accommodating live jointing practices, which are widely used for making service connections to energised distribution cables. Straight and branch jointing is carried out in this way by making the cable to be connected open circuit so that large electric arcs are not drawn when the conductors are connected. A number of safety measures are required for this technique, including gloves, insulated tools, shrouds and blankets together with accessories specifically designed for live jointing.

With the reduction of the number of highly skilled jointers and economic pressures to reduce installation times a number of new concepts have evolved which reduce the skills required, shorten jointing times and result in more reliable products.

DESIGN PRINCIPLES

Common requirements for all cable accessories are adequate conductor connection and the reinstatement of insulation. Methods of conductor connection are discussed in the next section. For all medium voltage accessories, i.e. those operating at or above 3.8/6.6 kV, electrical stress control is required at both the dielectric screen termination and the connector position in joints and at the dielectric screen position in terminations. There are also specific requirements for terminations and joints which need to be addressed.

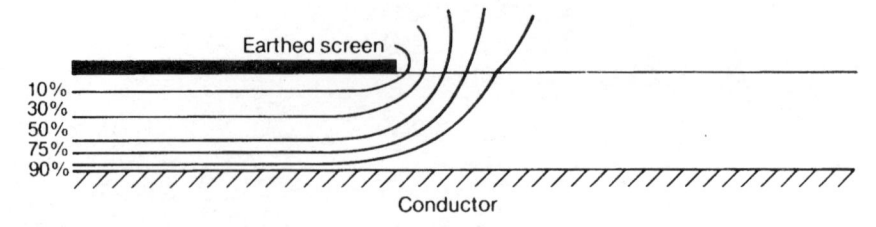

Fig. 27.1 Potential field at the dielectric screen termination

Stress control at the dielectric screen termination

In the preparation of the cable prior to jointing, the termination of the dielectric core screen produces an increase in potential gradient along the interface between the dielectric and the surrounding space. This can be seen in fig. 27.1.

The stress in the dielectric will be much higher than the design stress of the cable, increasing the probability that premature failure will occur. If the medium surrounding the termination is air, or if there is an air gap between the dielectric and the filling compound, then the stress may cause the air to discharge at working voltage. Polymeric materials are less resistant to discharges than paper insulation, hence discharges in the termination region will erode polymeric dielectrics, leading to premature failure.

A number of techniques are used for stress control; the main types are capacitive, high permittivity and non-linear resistive stress control.

Capacitive stress control

The construction of a stress cone, a traditional method of stress control, provides a means of controlling the capacitance in the area of the screen termination. This reduces the stress in the dielectric until, at the actual screen termination, the stress is at an acceptable level. The stress cone is continued beyond the screen termination to reduce the potential gradient at the surface of the dielectric to a level where discharges will not occur. The use of a stress cone is shown schematically in fig. 27.2.

High permittivity stress control

The use of materials with relative permittivities significantly higher than the dielectric is an extremely common method of stress control in polymeric cable accessories.

When materials of dissimilar permittivity are subjected to a potential gradient across their combined thickness, the material which has the lower permittivity will experience

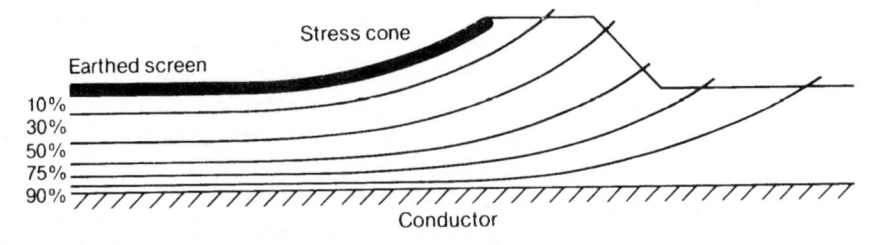

Fig. 27.2 Potential field at a cable termination using a stress cone

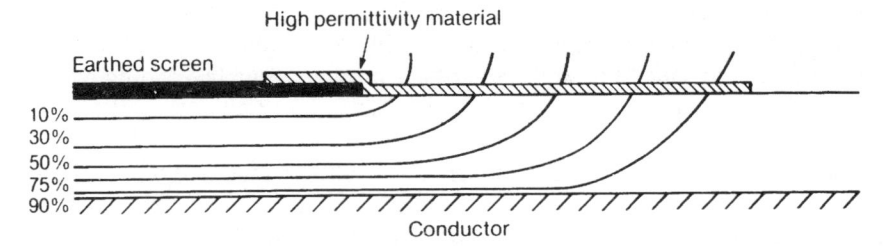

Fig. 27.3 Potential field at a cable termination using a high permittivity stress control layer

the higher stress. By using a material having a higher permittivity the lines of equipotential are spread further apart and emerge more gradually from the dielectric. This can be seen by comparing fig. 27.3 with fig. 27.1.

Non-linear resistive stress control

Another popular method of stress control is to utilise materials with non-linear current versus voltage characteristics. When applied to the screen termination, the material allows small currents to flow through the cable dielectric and then along the layer to the earthed screen. As the current in the layer increases, the resistance of the material drops, resulting in a smooth linear voltage gradient along the layer.

Stress control at the connector position

For medium voltage joints, stresses at the connectors are increasingly important at the higher voltages such as 12/20 kV and 19/33 kV. This is also becoming increasingly important at 6.35/11 kV when bulky fittings such as mechanical connectors are used. The most common method is to incorporate a semiconducting layer into the accessory which is brought into contact with the connector. This can be applied in tape form or be part of the accessory, and has the effect of creating a larger diameter smooth electrode which rises to working voltage. This encloses the sharp edges of connectors and reduces the stresses on external layers.

An alternative technique is to use a high permittivity layer at the connector position. This works in the same manner as described above by placing higher stresses on the layer with the lower relative permittivity.

Terminations

Terminations are classified as either outdoor or indoor types, depending on whether or not they are capable of withstanding the effects of precipitation, pollution and ultraviolet radiation. Indoor terminations are nearly always fitted inside a metallic enclosure, which provides an earth envelope around the termination and in the case of compound filled terminations retains the compound. The only exceptions to this are screened separable connectors (see later) which are designed to be safe to touch when in operation.

Outdoor terminations

For applications outdoors terminations must be protected from weathering and ultraviolet radiation along their entire length. This protective layer must be designed to resist the phenomenon of surface tracking. This occurs because the outer surface has a voltage gradient along it; water and other pollutants reduce the surface resistance, allowing leakage currents to flow. These leakage currents quickly dry out areas or rings to form dry bands which are then forced to withstand almost all of the voltage gradient. Arcs across these dry bands can erode the outer protection or attack the surfaces leaving carbonaceous tracks that can lead to failure. The traditional method of overcoming this is to use shedded porcelain insulators, but an increasingly common approach is to use sophisticated materials designed to resist tracking. (See also chapter 48.)

Indoor terminations

The insulation inside the terminal box may be compound, air or a combined dielectric consisting of air and a shroud applied over the bare connector. Air insulation is achieved by using a box large enough to allow adequate phase to phase and phase to earth spacings for the system voltage. Using shrouds allows the clearances to be reduced, whilst compound filling offers further reductions in clearances. Terminations must be designed to resist humidity, which is present whenever air is either the sole or part dielectric. Table 27.1 gives details of clearances for air insulated and compound insulated boxes for a variety of voltages. Recommendations for clearances on shrouded terminations for inclusion in European standards are still under discussion.

Joints

In addition to the necessary measures for stress control, joints must be able to reinstate the insulation between phases and phase to earth to that of the cables they are connecting. Another important requirement is that the joint design is capable of providing mechanical protection against external damage and prevents moisture entering the joint from the surroundings.

At medium voltage an earth envelope is also required as a safety precaution. Joints are also required to be capable of carrying through fault currents; in some versions cross-bonds are used whilst in others the earth screen is designed to carry both through earth faults as well as faults at the joint position. Connection of cross-bonds to the

Table 27.1 Air and compound clearances for terminations

Rated voltage,[a] U_m (kV)	Insulating medium	Phase to phase clearance (mm)	Phase to earth clearance (mm)
1.1	Compound or air	20	20
3.6	Compound	20	20
3.6	Air	90	65
12	Compound	45	32
12	Air	127	76
24	Compound	100	75
24	Air	242	140
36	Compound	125	100
36	Air	356	222

[a] Maximum voltage for equipment, see chapter 18

metallic sheaths of cables was traditionally done by plumbing with solder. This has been largely replaced with cold connections such as constant force springs, worm drive clips or tinned copper straps.

CONDUCTOR CONNECTION

A number of different methods of conductor connection are available. They vary in the level of skill required and the equipment needed to install them.

Soldered connectors

The traditional method of jointing paper insulated cables is to use soldering. The individual strands of the conductor are tinned using the appropriate flux for the conductor metal, the ferrule is placed around the aligned conductors and filled by basting with molten solder. Grade M solder is used for copper conductors and Alca P for stranded aluminium. The ferrule or lug is then wiped and rubbed down to give a smooth outer surface. For solid aluminium conductors a more reliable tinning technique is the use of abrasion solder (predominantly tin with approximately 10% zinc), and filling the ferrule with H grade solder.

For joints the weak-back ferrule is the most popular, so called because its design allows it to be opened to fit the stranded conductors and permit the introduction of solder. The same technique is used for fitting lugs, where the hot metal is poured in at the end, and for solid ferrules where the solder is introduced through filling holes.

The fitting of lugs and ferrules is a highly skilled operation which relies on the jointer's expertise to produce a solid ferrule or lug with no cavities or badly tinned conductor strands. These defects will result in a high resistance connection which could lead to premature failure of the joint or termination.

Soldered ferrules are limited to a maximum temperature under short-circuit conditions of 160°C. They are therefore unsuitable for cable systems taking full advantage of the 250°C short-circuit rating offered by polymeric cables such as XLPE.

Compression connectors

This system utilises an installation tool equipped with specially designed dies to deform a metallic tube on to the conductor, producing a low resistance electrical contact with good mechanical grip. The tools are either hydraulic or electro-hydraulic in operation (fig. 27.4) and the dies impart an indent, hexagonal, or circumferential form on the connector depending on the conductor and customer preference. Examples of the different types of connectors are shown in fig. 27.5.

The connectors themselves are either made from copper or aluminium tube, which is machined to produce ferrules and flattened to produce lugs. For a number of applications bimetallic connectors are required; these are produced by friction welding copper forgings to an aluminium bar which is then machined into the desired form. An example of a bimetallic connector used in medium voltage separable connectors can be seen in fig. 27.6.

The use of compression tooling provides a more reproducible connection that requires a lower level of operator skill than for soldered connections, which is

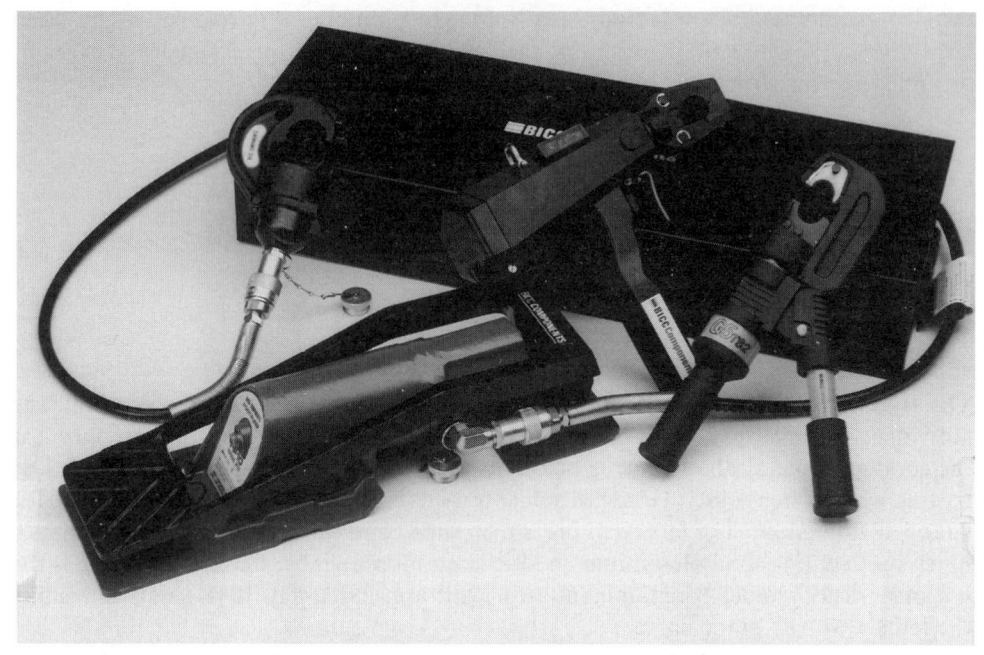

Fig. 27.4 Hydraulic and electro-hydraulic compression tools (Courtesy of BICC)

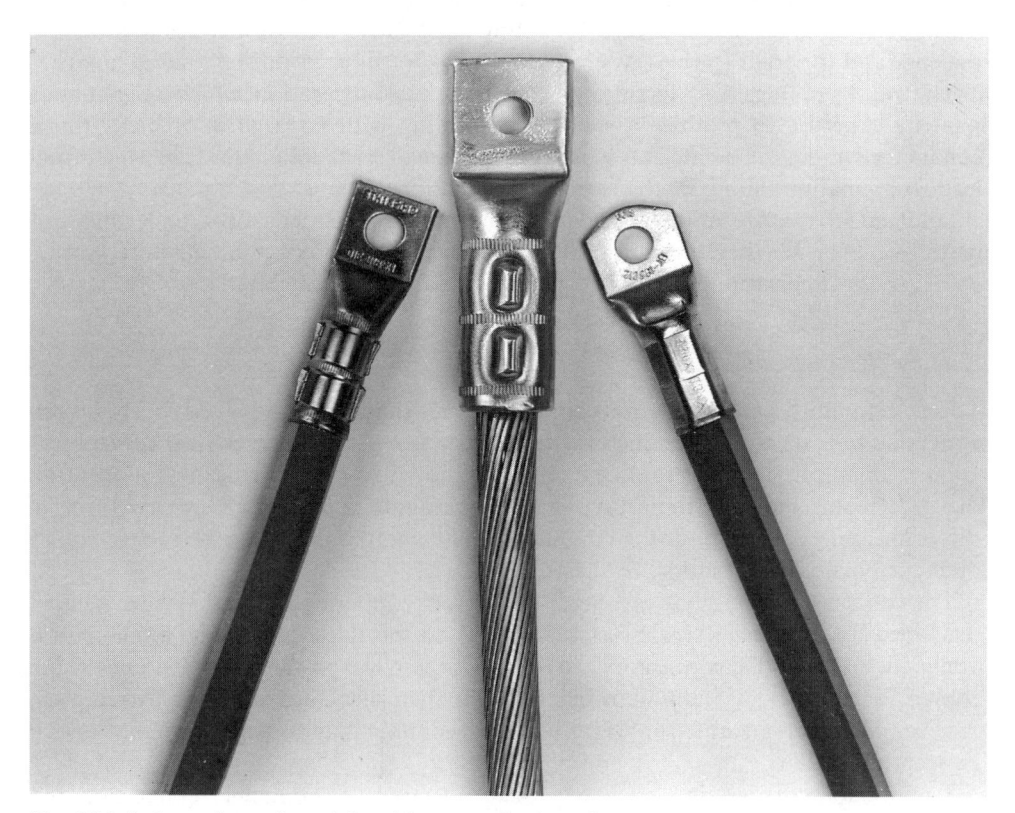

Fig. 27.5 Indent, circumferential and hexagonal crimped connectors

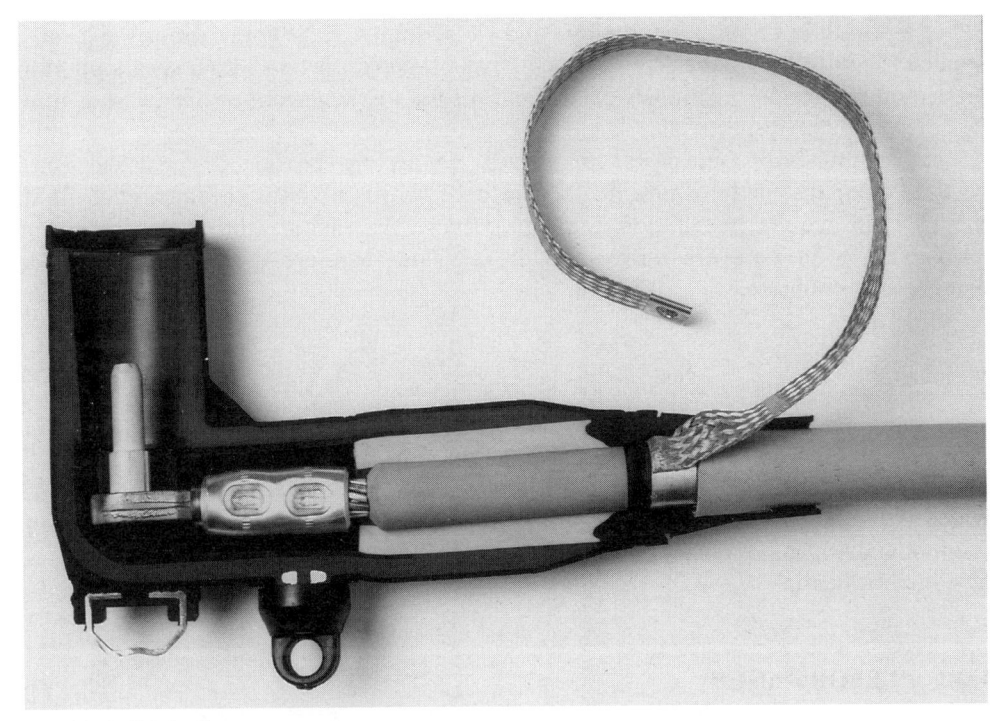

Fig. 27.6 400 A elbow connector

particularly beneficial for connection to aluminium conductors. The use of compression connectors is firmly established as the preferred method of connecting both low and medium voltage terminations and in mainland Europe for medium voltage joints.

Mechanical connectors

The use of mechanical connectors has been almost universally adopted for low voltage jointing. In the UK they are being used increasingly for 6.35/11 kV and 19/33 kV cable jointing.

For low voltage applications, a number of types are available. The most common types for solid conductors are parallel groove types for straight and branch joints and for service connections to aluminium mains cables. Examples of these types are shown in fig. 27.7.

Fig. 27.7 Low voltage mechanical connectors

For stranded low voltage conductors and all medium voltage applications the torque applied is controlled by the screws which have a tapered portion which shears off at a predetermined torque. Examples of low and medium voltage connectors are shown in figs 27.8 and 27.9 respectively.

Like compression connectors, mechanical connectors provide a consistent connection when installed by unskilled operators. They also have the advantage that they do not require installation tooling. Because the manufacturing processes for mechanical connectors are more difficult they are therefore more expensive than compression connectors.

Welding

Conductor welding is normally associated with transmission cables and joints, but is occasionally used on lower voltage cables. The main advantages of welding are that a conductor joint with excellent electrical and mechanical properties is achieved, which is particularly important for submarine cables. The operation requires skilled jointers and expensive ancillary equipment and so it is only used when reliability is absolutely essential.

FIRE PERFORMANCE

With the increasing use of cables designed to have low smoke emission on burning and without toxic gas emissions, complementary accessories have been developed.

Work has concentrated on joints and has involved using a number of specially formulated thermoplastic and thermosetting materials which are flame retardant, halogen free and emit no acid gases on burning. BICC designs have been produced for low and medium voltage joints, which have been tested, approved and installed in London Underground and the Channel Tunnel. These products are also being used in the new airport in Hong Kong.

TRADITIONAL CABLE ACCESSORIES

The traditional methods of jointing and terminating paper insulated cable are relatively inexpensive compared with the more modern systems described later. They do however require the availability of jointers with the relevant skills.

Low voltage joints and terminations

Low voltage joints are either of the sleeve or mechanically bonded type. In the sleeve type the conductors are soldered together, insulated with impregnated cotton or paper tapes and encased with a copper or lead sleeve which is plumbed on to the cable sheath. An outer cast iron or earthenware box is then fitted and the whole box filled with bitumen. In the mechanically bonded version the sleeve is omitted and a bitumen filled cast iron outer box with lead sheath clamps is used.

Indoor terminations consist of a metal box with a gland to which the sheath of the cable is plumbed. Conductor connections are soldered, the cores separated and the box

Joints and Terminations for Distribution Cables 409

Fig. 27.8 Low voltage mechanical connectors for stranded aluminium

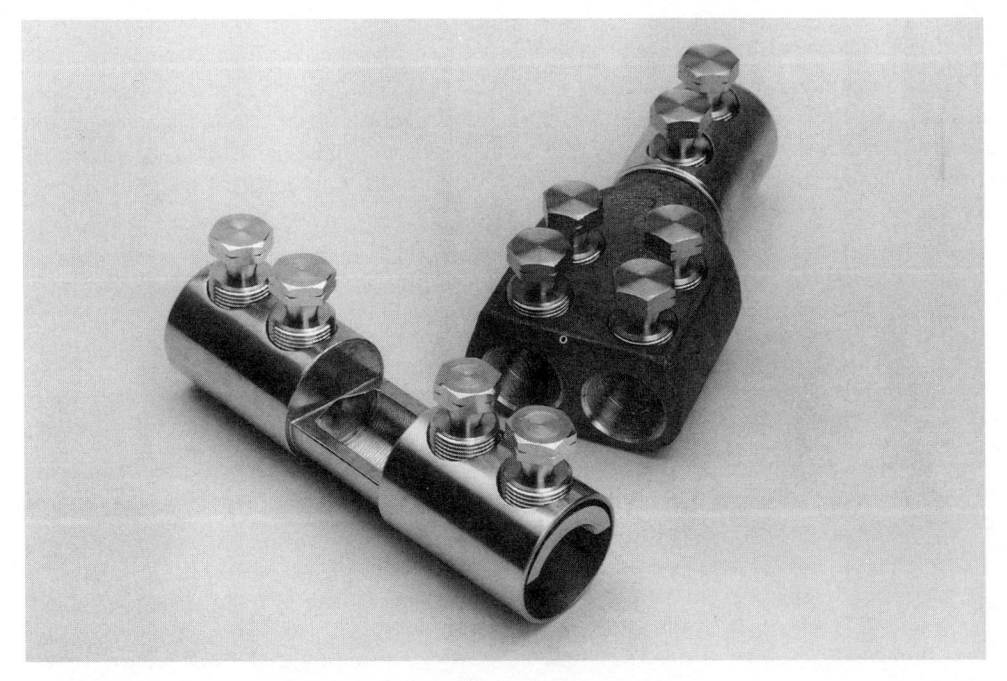

Fig. 27.9 Medium voltage mechanical connectors (Courtesy of B&H (Nottm.) Ltd)

filled with a bituminous compound. Outdoor terminations include porcelain insulated bushings and may be of the inverted type where the cable enters from the top.

Medium voltage joints and terminations

For belted cables indoor MV termination designs are similar to the low voltage designs above, but with larger creepages and clearances. For screened cables stress control is required; either carbon paper or antimonial lead wire is used up to 6.35/11 kV with stress cones being used at higher voltages. These are formed by building impregnated crêpe paper into a cone of specific dimensions and then extending the dielectric screen part way up the cone with lead wire. Outdoor terminations are either 3-core inverted type with porcelain bushings or single-core terminated directly within porcelain sealing ends. The filling medium is bitumen up to 12.7/22 kV and oil–rosin at 19/33 kV.

Joint designs for belted cables are similar to the sleeve type used at low voltage. For 6.35/11 kV screened cables both screened and unscreened joints are available. In the unscreened joint the core screen is terminated and a carbon tape or lead wire stress relief is applied; for screened joints each core is insulated and the screen carried over with knitted copper tape. A lead sleeve is then used together with a cast iron, pitch fibre or earthenware outer box which is bitumen filled.

33 kV joint designs are generally screened with impregnated crêpe paper insulation and copper tinsel screening. Copper sleeves are used filled with an oil–rosin compound. 3-core joint designs are similar but include tie rods connected to armour clamps, the whole assembly being contained within a bitumen-filled glass fibre box.

MODERN CABLE ACCESSORIES

Resin

The use of cold curing resin systems has been adopted for low voltage joints throughout Europe. In the UK the system involves injection moulded or more usually vacuum formed shells which act as a mould into which a polyurethane, acrylic or epoxy resin is poured, having been mixed on site. An example of a low voltage resin straight is shown in fig. 27.10.

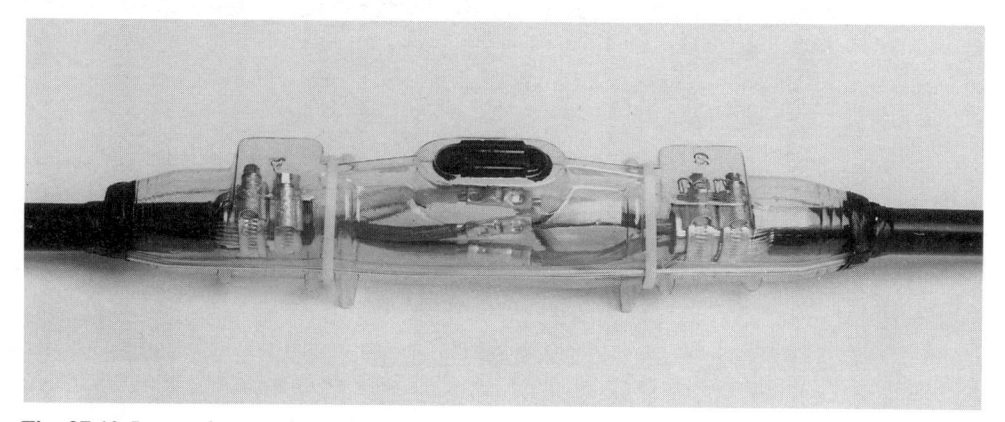

Fig. 27.10 Low voltage resin straight joint

The resin sets solid in approximately 30 minutes depending upon ambient temperature. Resin shelf life in the unmixed state varies from 1 to 2 years according to the resin type and storage conditions. For distribution cables by far the most popular system is polyurethane; acrylic is used for special applications such as zero halogen flame retardant joints. Acrylic resins are the safest resin system to handle, but the health risks associated with the mixing and disposal of polyurethane have been eliminated in the UK by the recent development of enclosed bag mixing systems. A typical enclosed bag mix is shown in fig. 27.11. This consists of a barrier between the resin and hardener components which keeps the two components apart under storage conditions but ruptures when required to mix the two components together. No vapours are emitted at this stage and after pouring there is no free hardener requiring special disposal in the resin pack.

In resin inject systems, which are more common on mainland Europe, the jointed cables are surrounded by a spacer tape and enclosed by a pressure tape. A pump is then used to force the resin into the cavity created and maintain it there until it cures.

In both cases the phase and neutral/earth connections are made by some means of mechanical connector. By bonding to the cable the resin provides a seal against the ingress of moisture and the high impact strength of the resin gives protection against mechanical damage.

The use of resin systems has been extended to 24 kV cables where the resin is the primary insulation between phases and phase to earth. They incorporate one or more of

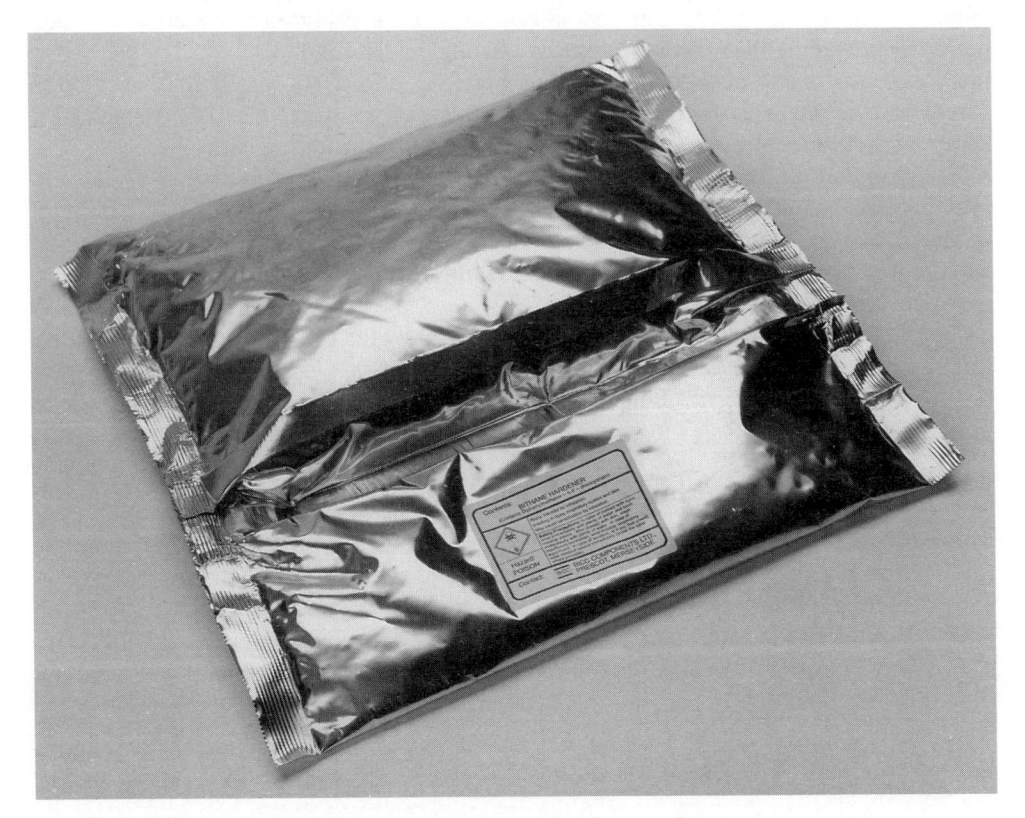

Fig. 27.11 Twin pack enclosed mixing bag for polyurethane resins

the methods of stress control mentioned earlier, together with an earth screen and, where appropriate, cross-bonding to ensure the transmission of through earth faults. At 33 kV a well established technique for polymeric cables again uses the resin to provide mechanical protection and seal against moisture. Hand applied insulating and semiconducting EPR self-amalgamating tapes are used to control stress, provide insulation and reinstate the semiconducting dielectric screen.

Heat shrink

The use of heat shrink polymeric materials has firmly established this technique for low and medium voltage terminations and increasingly for 11 kV joints. The heat shrinkable property is imparted by first extruding or moulding the polymeric material into the required shape and then crosslinking by either irradiation or chemical means. The components are then warmed and expanded and allowed to cool in this expanded state. On the application of heat the memory imparted by crosslinking causes it to return to the shape in which it was crosslinked.

Low voltage terminations consist of a breakout or udder where the cores are splayed out together with tubing to protect the cores themselves. Mastics and hot melt adhesives are used to line the components to ensure watertight seals.

Medium voltage terminations use track resistant weatherproof materials for both tubing and breakouts together with heat shrinkable rain sheds, also in the same material, for outdoor terminations. Stress control is by means of either a high permittivity or non-linear resistive tube or coating, whilst for 11 kV belted paper cables the breakout is made from semiconducting material, effectively converting the cable to a screened design. Heat shrinkable terminations can be used in boxes designed for compound clearances by the addition of a heat shrink shroud or boot.

Early versions of joints for medium voltage cables consisted of a number of tubes to provide stress control, insulation, screening and overall protection. More modern designs combine these layers, for example by combining the insulating and semi-conducting layers as a single coextrusion, making installation somewhat easier.

An increasingly popular method of 11 kV jointing in the UK is to use either a moulding at the connector position and encapsulate with resin, or to use heat shrink tubing to carry out the electrical functions of stress control and insulation with the resin to provide mechanical protection and moisture sealing.

Pre-moulded joints and terminations

These products are made from either EPR or silicone rubber and are only suitable for polymeric cables. Premoulded products, being factory made and tested, are of consistent quality and can be installed very quickly. They are designed to form an interference fit on the cable. They will only accommodate small variations in diameters so it is very important to know the dimensions of the cable to be jointed or terminated.

Indoor terminations designed for air clearances consist of a stress cone and the conductor connector; these are converted to outdoor terminations by adding inter-locking rain sheds together with a cap to seal on to the conductor connector (fig. 27.12).

Joints are also available which use the stress cone in the form of a cable reducer. The inner bore of the cable reducer is sized to suit the cable diameter and the outer diameter

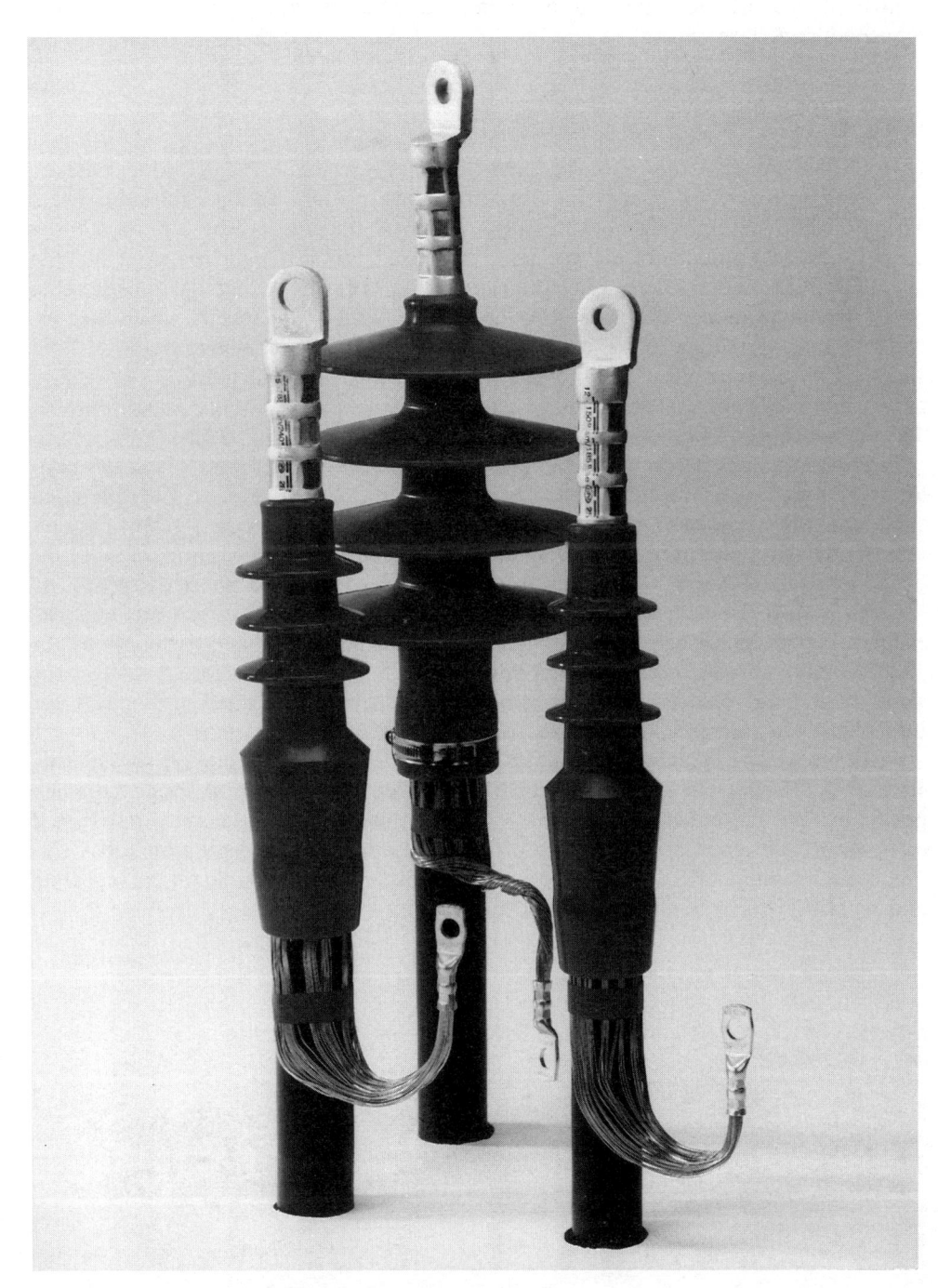

Fig. 27.12 Outdoor premoulded cable termination

forms an interference fit with a housing. This has an inner semiconducting portion to control stresses at the connector position, a middle insulating layer and an outer semiconducting layer.

Cold shrink

Over the last few years the use of elastomeric materials which shrink on to the cable are now rivalling heat shrink as a method of terminating medium voltage cables. Like heat shrink they can accommodate a wide range of cable diameters, but they offer a number of advantages. Equipment such as gas bottles and torches is unnecessary and the installation is less dependent upon the skill of the operator. Being elastomeric, cold shrink continues to grip the cable after installation, unlike heat shrink, which 'freezes' when it is recovered on to the cable and is not capable of further movement. Cold shrink is therefore capable of following any subsequent movement such as resetting cores or the cable operating at a lower current. There are two basic types, stretch rubber and prestretched.

In the first type, stress control is applied separately as a mastic pad which has non-linear resistive properties. Mastic pads readily conform to the contours of the prepared cable and will also accommodate rough edges if the dielectric core screen is poorly terminated. A simple tool is used to inflate the outer protection moulding on to a tube which holds it by partial vacuum until it can be placed over the prepared cable. The vacuum is then released causing the moulding to instantly collapse on to the cable. Examples of indoor single-core and 3-core outdoor stretch rubber terminations for 12/20 kV cables are shown in figs 27.13 and 27.14 respectively. For lower voltages the same outdoor termination is used together with the number of sheds appropriate for the voltage; for example one shed per core is used at 6.35/11 kV.

In the prestretched type the moulding is expanded in the factory on to a plastic support tube. This tube is partially cut in some way to allow it to be removed when positioned over the cable, causing the moulding to conform to the cable. Prestretched components for terminations include plain and shedded outer protection tubes and breakouts for multicore cables. Sleeves and end caps are also available for cable sealing and repairs.

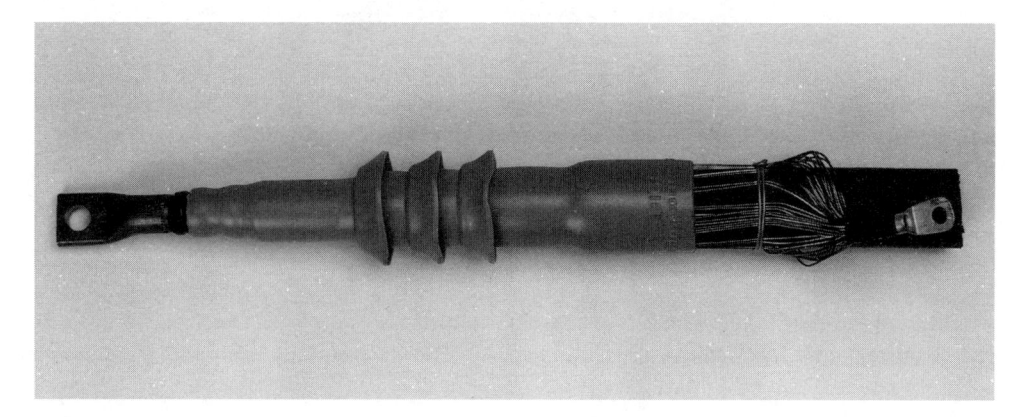

Fig. 27.13 Indoor stretch rubber termination for 12/20 kV cables

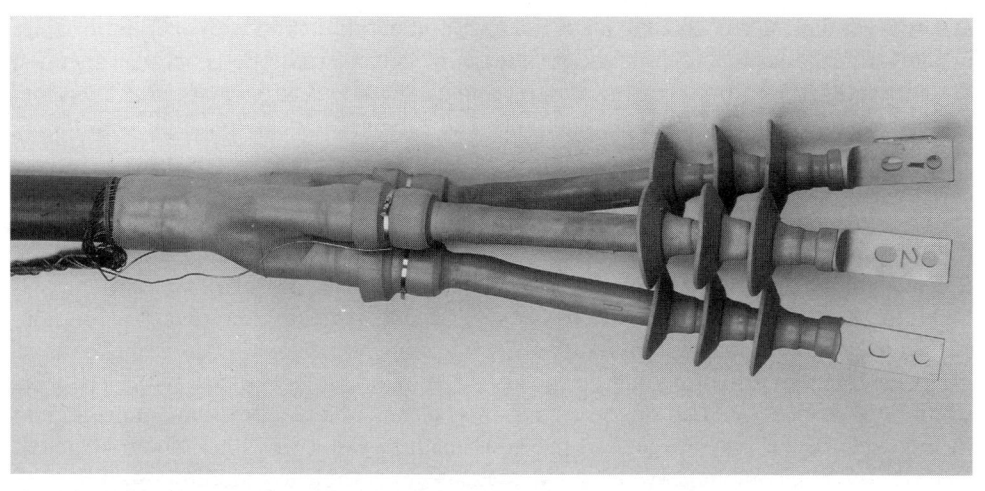

Fig. 27.14 Outdoor stretch rubber termination for 12/20 kV cables

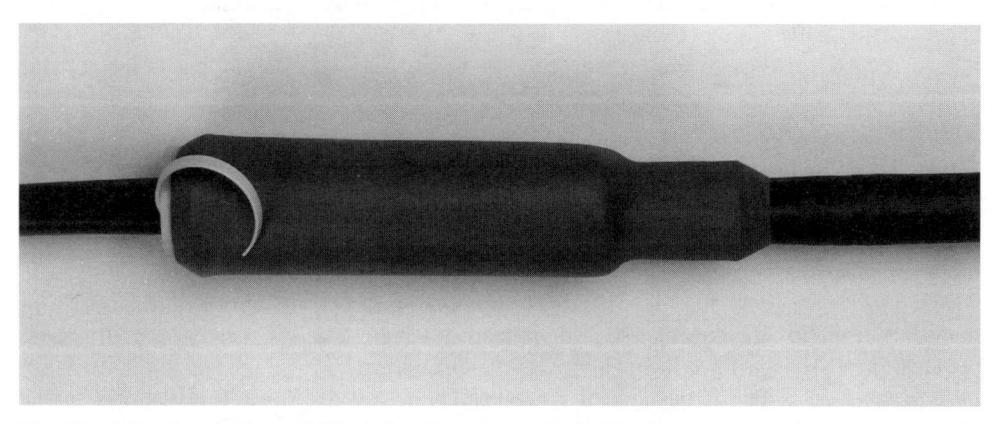

Fig. 27.15 Prestretched moulding being fitted to 11 kV cable core

At 11 kV designs are available consisting of a moulding with an insulating inner layer and an outer insulating portion. Figure 27.15 shows this in the process of being collapsed on to a cable core by the removal of a spiral cut former. Stress control at the dielectric screen is carried out by non-linear resistive mastic and the whole assembly resin encapsulated.

Prestretched joints have recently been made available for 12/20 and 18/30 kV cables. They are made from either EPR or silicone materials and consist of a single moulding or extrusion which incorporates stress control layers for both the connector and dielectric screen positions, insulation and outer screening. Outer protection is provided by either an integral or separate tube which compresses sealing mastic on to the cable outer sheath.

Separable connectors

Separable connectors are a special form of premoulded termination. They are required to mate with a series of standardised equipment bushings and have been used extensively on transformers in France, Spain, Germany, Italy and with polymeric cables in the UK. In addition to providing a means of cable termination, because they are disconnectable they allow sections of plant and cable to be isolated and either energised or earthed, providing operational flexibility and the opportunity to simplify switching arrangements.

There are two main types: the outside cone type in which an elbow or T-connector mates to a male bushing fitted to the equipment, and the inside cone, in which a male rubber moulding plugs into a well type bushing.

The first type has gained the widest use and is based on the same principles as the premoulded accessories described earlier. Cable reducers are again used to control stresses at the dielectric core screen termination. The housing – which again has a semiconducting inside layer, an insulating middle layer and a semiconducting outside layer – is either elbow or T-shaped. Conductor connection is by means of a compression connection on to the cable which is connected to the equipment bushing by either a pin or bolted contact. The semiconducting outer surface encloses the whole termination by an earthed envelope so that the termination is safe to touch; this also means that terminations on adjacent phases can touch thus reducing the size of the 3-phase termination. An example of an outside cone elbow connector is shown in figure 27.6.

Unscreened versions of the outside cone system are also available, but these are not as popular as the two types mentioned above.

Transition joints

When jointing polymeric to paper insulated cables, the design must be such that the impregnants in the paper insulated cable are prevented from coming into contact with the materials used in polymeric cables. In these circumstances, cable dielectrics such as XLPE and EPR swell and lose their mechanical properties, whilst semiconducting materials used in the conductor and dielectric screen layers experience a dramatic increase in resistivity, impairing their ability to function.

A number of different designs of transition joint are available for voltages up to 36 kV including heat shrink, prestretched and tape/resin. Passage of impregnant at the connector position is prevented by having compression or mechanical connectors with blocked or solid centre sections. The materials used in the joint are either specially formulated to give resistance to impregnants, for example silicone rubbers and some heat shrink materials, or in the case of resins and EPR materials protected by a barrier tube or tape. Examples of barrier materials are silicone rubber tubing, terylene tape and PTFE.

DESIGN AND PERFORMANCE STANDARDS

In the UK design standards are limited to spacings for compound and air filled boxes and the dimensions of separable connector interface, with performance standards for connectors and low voltage joints and connector. Both design and performance standards are well established in France, Italy, Spain and Germany.

Over recent years a large amount of effort has been put into creating both IEC and CENELEC standards for medium voltage joints, terminations and separable connectors. Both are close to final publication. An IEC standard on connectors covering performance has been published but is not regarded favourably; a CENELEC

Table 27.2 Summary of design and performance standards for joints and terminations

Subject	Standards	Type
Low voltage termination dimensions	BS 5372	Design
Dimensions for compound and air filled boxes	BS 2562 BS 6345	Design
Interfaces for premoulded	BS 7616	Design
MV separable connectors	BS 7215 EN 50180 EN 50181	Design and performance Design Design
Compression and mechanical connectors	BS 4579: Parts 1 & 3 NF C63-061 IEC 1238: Part 1	Performance
Low voltage joints	BS 6910: Part 1 HD 623 VDE 0278: Parts 1 & 3	Mainly performance Performance Performance
Medium voltage joints, terminations and separable connectors	*HDs 628, 629: Parts 1 & 2* VDE 0278: Parts 2, 4, 5 & 6 CEI 20-24 HN 52-S-61 C33-001 *IEC 55-1* *IEC 502-4* *IEC 1442*	Performance
Cold connections to metallic sheaths	Engineering Recommendation C93	Performance

Draft standards are shown in italics

equivalent is in the early stages of preparation whilst work in IEC on defining envelope dimensions for connectors is continuing. A summary of relevant standards for joints and terminations is given in table 27.2.

CHAPTER 28
Testing of Distribution Cables

Testing is primarily aimed at two aspects: (a) any new design of cable will perform satisfactorily for a reasonable life-time and (b) cables as manufactured meet specification requirements for quality, e.g. conductor resistance and minimum thickness of important components, parameters which may readily be specified and measured. However, testing to determine incipient faults has often had to be developed by experience over a long period, e.g. application of excessive voltage may itself create more incipient faults than it detects.

The main IEC, CENELEC and British Standards concerned with testing are given in chapter 7. The tests to be carried out on each type of cable are included in the appropriate cable standard. For some tests complete detail is given in the cable standard, but for many cross-reference is made to separate standards for tests, often for details of the method of testing and sometimes also for the requirements to be met. Although national requirements still differ in some respects, there is a proposed move towards adoption of CENELEC and IEC standards, especially for polymeric cables. Nevertheless, as CENELEC and IEC standards require a consensus of opinion and are slow to evolve, it is often the case that national requirements are more severe for some specific details. Comparisons are made below but the outline given is based on IEC 55-1 for paper cables and IEC 502 for polymeric cables, rather than CENELEC HD 621 and CENELEC HD 603/ HD 620 respectively

Most standards cater for tests in four categories:

(a) routine tests by the manufacturer on every finished length of cable (or, for spark testing, during manufacture) to ensure compliance with construction requirements and demonstrate the integrity of the cable;
(b) sample tests which are not practicable on every complete length of cable; they are made on samples of cable to represent production batches and provide a periodic check on manufacturing consistency;
(c) type tests to be carried out during the development of a new grade of insulation or cable design to establish performance characteristics; they are not repeated unless changes are made which could alter these characteristics;
(d) site tests after installation made to demonstrate the integrity of the cable and its accessories as installed.

Tests on materials taken from the cables may be included in the type tests but there are standards for the materials as purchased or produced by the manufacturer and methods of tests on materials. Materials testing was referenced in chapter 3 and this chapter is mainly confined to tests on cables.

COMMENTS ON INDIVIDUAL TESTS FOR CABLES

Dimensions

Great care is necessary to obtain accurate and reproducible results and reference should be made to the appropriate specification for details of the technique and equipment necessary, e.g. ordinary or special micrometer, diameter tape, or microscope.

Insulation resistance and capacitance

Numerical values for insulation resistance and capacitance give little guidance on the quality of paper cables because they are predominantly related to the types of material used and the processing conditions. Although no values are included in cable specifications, tests are usually carried out by cable manufacturers on individual cable lengths because they give a good indication of consistency of manufacture.

Similar remarks apply to polymeric cables but there are type test requirements for the insulation resistance of low voltage PVC cables at ambient and working temperature in IEC 502, and BS 6346: 1989 includes routine tests at ambient temperature.

High voltage tests

For paper insulated cables, application of a high voltage provides the most searching test for any defects. With polymeric insulated cables, the measurement of partial discharge above working voltage is possibly more significant. When testing at a cable factory, it is convenient to use a.c. but for testing complete installations on site the use of d.c. is more practicable. Most specifications, e.g. IEC 55-1 and 502, at present allow the use of d.c. for factory tests at a level of 2.4 times the a.c. voltage. The use of d.c. for polymeric cables is currently under review.

The choice of voltage level always poses a problem because (a) if too high it could cause incipient damage which might affect subsequent service life and (b) breakdown under high voltage is time dependent. For example paper cables will withstand around ten times working voltage for short periods but the breakdown level falls to about 65% in 50–100 hours. By the end of this time the level has fallen to an asymptotic value and it used to be common practice to carry out a number of voltage versus time to breakdown (VTB) tests at various voltage levels to determine the 100 hours value. The results have little relevance for assessment of service life because the expansion and contraction of the impregnant during load cycles has an important effect, particularly in the case of non-pressurised cables in the 10–30 kV range. However, a 100 hour a.c. test at four times working voltage has recently been included in BS 6480: 1988 for 19/33 kV cables.

Partial discharge tests

Partial discharges in a cable are caused by the breakdown of the gas contained within voids in the insulation. The voids may be either dielectric bounded or at the interface between dielectric and semiconducting screens. The voltage at which the breakdown first occurs is known as the discharge inception voltage. The stress in the void is directly proportional to the relative permittivity of the insulation and, because the breakdown strength of a gas is much less than for solid insulation, the void can break down, causing discharges at voltages much lower than the operating stress of the cable.

For many years the quality of paper insulated cables has been assessed by measurement of the dielectric loss angle against increasing voltage. This is possible because of the even distribution of the losses due to the lapped construction of the paper dielectric. However, in extruded dielectric the random distribution and sizes of voids no longer allows characterisation of discharge by measurement of dielectric losses.

Most national and international specifications for polymeric cable now require measurement of partial discharge and define the maximum level of discharge acceptable at particular test voltages.

Dielectric power factor (dielectric loss angle)

Although the power factor is quite low for paper cables and is even lower for polymeric insulated cables, it does represent some loss in the energy distributed. Limits are therefore prescribed in routine and type tests for paper cables of 8.7/15 kV and above, and also in type tests for XLPE and EPR cables of 6/10 kV and above. Power factor is a function of both voltage and temperature and limits are specified over a range of conditions. With paper cables, the check of increase with voltage is largely a test of the quality of impregnation as such increase indicates ionisation in voids.

Impulse voltage tests

Switching operations or lightning may cause high transient or surge voltages to appear occasionally on cable systems and the ability to accommodate them has become a normal part of type test procedures. The test is performed on cable heated to maximum conductor temperature (tolerance $-0, +5°C$ for paper and $+5°C, +10°C$ for polymeric cable) and, after application of ten positive and ten negative impulses at the withstand level, the same sample is subjected to a high voltage test. Some specifications require only three positive and negative impulses. In order to ensure that the sample tested is in a condition similar to cable in service, the testing is carried out on a sample which has been subjected to a bending test. Impulse testing is restricted to paper cables of 8.7/15 kV and above and polymeric cables of 1.8/3 kV and above. The IEC withstand voltage requirements are as follows.

Rated voltages (U_0/U)	kV	1.8/3	3.6/6	6/10	8.7/15	12/20	18/30
Impulse voltages	kV	40	60	75	95	125	170

Bending test

The ability of cables to bend during drumming and installation, without undue distortion or damage to any of the components, is an important requirement for any cable. Features of particular significance are damage to the insulation of paper cables and disturbance of the screen on all types of cable.

The test is carried out with a sufficient length of cable to produce a complete turn around a drum and normally comprises three complete reverse bending cycles at a radius which is more severe than the smallest permissible bending radius during installation. The radii are given in appendix A19.

After bending, paper cables are submitted to a high voltage test and semiconducting screened polymeric cables are subjected to a partial discharge test. In addition, there is a requirement for visual examination of the lead sheath, armour and protective finish of paper cables. British practice in BS 6480: 1988 also requires a visual examination of the insulation, with limits for disturbance and damage to individual papers. However, IEC specifications cater for any damage to the insulation and screens on all types of cables to be assessed by passing the electrical high voltage or the partial discharge test as appropriate.

For paper cables the bending test is a separate test in the category of sample tests. For polymeric cables, however, it is part of the type test sequence in which various tests are carried out on the same cable sample.

The general national and international specifications cater for testing at ambient temperature (10–25°C) but special attention may need to be given to the insulation and protective finishes for cables to be installed in geographical regions having very low temperatures. This is covered by user requirements.

Drainage from paper insulation

The test applies to non-draining cables only and the IEC sample test procedure is based on a 300 mm sample open at both ends. After heating in an oven for 8 hours at maximum conductor operating temperature, the amount of compound drainage must not exceed 2% of the interior volume of the metal sheath for single-core and SL-type cables, and 3% for multicore cables. British practice in the sample test/type test for 19/33 kV cable, and the IEC type test require a longer sample of 900 mm, sealed at both ends with sufficient space at the lower end to collect drained compound. The heating period is 7 days. The test requirement is 3% maximum up to and including 3.8/6.6 kV and 2.5% for higher voltage cables. In this test the temperature condition is more severe than that arising in service because the heating is from the outside of the cable.

Tests under fire conditions

The test specified in IEC 332: Part 1 and BS 4066: Part 1: 1980 (1995), which relate to flame propagation on a single vertical cable, may be required on cables having appropriate types of oversheath, i.e. PVC and suitable elastomeric compounds. This form of test has been used for many years, with some variations in the details over that period.

Now, however, IEC 332: Part 3 and BS 4066: Part 3: 1994 recognise that flame propagation in cable installations depends upon the amount of cable in the location and the disposition of the cables with respect to each other, as well as on the properties of the component materials. These specifications include three categories of test conditions, with variation of the density of non-metallic cable components as one of the differences in the test conditions between the three categories.

A test to measure the amount of hydrochloric acid gas evolved during combustion of component materials has also been standardised in IEC 754: Part 1 and BS 6425: Part 1: 1990. BS 6724: 1990 and BS 7835: 1996 also include a test for smoke emission, i.e. BS 7622: 1993 equivalent to IEC 1034. These tests and the test specified in BS 4066: Part 3: 1994 are not applicable to the general range of standard cables but only to those specially designed to meet them. A number of BS cable standards have been drafted to call for these tests under fire conditions. Although there are no IEC standards for distribution cables having the requisite properties, CENELEC standards are in place, e.g. HD 604 and HD 622.

These tests may be invoked in users' specifications for cables outside the scope of existing and draft BSs but then the total cable design and the choice of materials for each component have to be specially considered. London Underground Ltd, for example, specifies a smoke emission test and did so for a number of years before a similar test was specified in BS 6724: 1990.

The background and details of the tests are covered in chapter 6. All the tests are currently under review.

Long duration tests

The influence of water on medium voltage polymeric cables has been investigated for many years, particularly as a result of problems in service with the first generation of such cables, i.e. water treeing.

Recently, a large number of long duration tests of up to 2 years have been designed, not to predict life-time but to evaluate the water treeing resistance of polymeric cables and enable discrimination between good and bad cable designs.

Test methods (14 in number) have been included in CENELEC HD 605 but most suffer from lack of experience. A working group within CENELEC is studying the effects of the various ageing parameters (voltage, temperature, time and water) in an attempt to rationalise these test methods, taking into account that the ageing mechanism in the test method is the same as in service. No conclusions have yet been reached by the working group.

No such tests are present in IEC, but they are beginning to appear in national standards as implementations of CENELEC standards. Consequently, all European countries have long duration tests in progress.

ROUTINE TESTS ON CABLES

Tests during manufacture

Although useful to the manufacturer for quality control purposes, tests during processing are not significant to the user. However, as the size and weight of distribution cables make it impracticable for any tests on complete cables to be carried out with the cable immersed in water, high voltage a.c. spark tests are made on polymeric insulation and sheaths during manufacture.

Conductor resistance test

The resistance of every conductor is measured and corrected to a standard length at 20°C. The main sources of error are inaccurate length and the sensitivity to temperature. Finished cables cool slowly and the conductors tend to be at above ambient temperature.

High voltage test

The test requires a 5 min application of an a.c. voltage of $2.5U_0 + 2\,\text{kV}$ for cables rated up to 3.6/6 kV, $2.5U_0$ and $3.5U_0$ respectively for paper and polymeric cables of 6/10 kV and above. For multicore non-screened cables the voltage is applied between conductors and also between any conductor and sheath. Cables with individually screened cores are tested from conductor to sheath only.

Power factor test for high voltage paper cables

This test applies only to screened cables of 8.7/15 kV and above. The power factor is measured at 0.5, 1.25 and 2.0 times U_0. The value at $0.5U_0$ must not exceed 0.006 and the differences between the values measured at $0.5U_0$ and $1.25U_0$ and between the values at $1.25U_0$ and $2U_0$ are required not to exceed the limits given in table 28.1. British practice in BS 6480: 1988 requires the test only for 19/33 kV cables and the limits are half of the IEC values (IEC 55-1) given in the table.

Partial discharge test for high voltage polymeric cables

The test is required by IEC 502 for semiconducting screened cables insulated with EPR or XLPE of rated voltage 6/10 kV and above. The magnitude of discharge at 1.73 times U_0 must not exceed 10 pC for EPR or XLPE.

There are wide variations in national requirements for this test, together with a progressive tendency towards higher test voltages and lower permissible amounts of discharge. BS 6622: 1991 at present specifies 10 pC as a maximum at $1.5U_0$ (soon to be amended in a revision of this standard) for XLPE and EPR insulated cables in the voltage range from 3.8/6.6 kV to 19/33 kV. In this standard the 3.8/6.6 kV EPR cables are of screened design whereas in IEC 502 screening is not mandatory for 3.6/6 kV EPR cables.

SAMPLE TESTS ON PAPER CABLES

Measurements of dimensions, primarily the thicknesses of insulation, lead sheath, non-metallic sheaths and armour
Bending test, followed by a voltage test and sample examination
Drainage test, for cable with non-draining insulation

The categorisation of sample tests into 'regular' and 'special' is currently being removed from British Standards.

Table 28.1 Maximum increase in power factor between test voltages for paper cables

Test voltage	Mass impregnated		Non-draining	
	$U = 15\,\text{kV}$ or below	U above 15 kV	$U = 15\,\text{kV}$ or below	U above 15 kV
$(0.5\text{–}1.25)U_0$	0.0010	0.0008	0.0050	0.0040
$(1.25\text{–}2.0)U_0$	0.0025	0.0016	0.0100	0.0080

SAMPLE TESTS ON POLYMERIC CABLES

Conductor examination, for compliance with IEC 228
Measurement of dimensions, including insulation, metallic and non-metallic sheaths and armour; also overall diameter, if specified
High voltage test, only applicable to cables of rated voltages above 3.6/6 kV and requires a voltage of $4U_0$ for 4 hours.
Hot set test for EPR and XLPE insulation: conditions are prescribed for a test to check that the material has been properly cured to give the required thermal properties

A current revision of BS 6622: 1991 is removing the categories 'regular' and 'special'. BS 6622: 1991 (6.6–33 kV cables) also provides for measurement of the resistivity of extruded semiconducting screens and the cold strippability of extruded semiconducting insulation screens, when this is required, as sample tests.

TYPE TESTS FOR PAPER CABLES

The type tests for IEC 55-1 are summarised below.

Power factor/temperature test

The test applies to screened cables of 8.7/15 kV and above and measurements are taken up to 10°C above rated operating temperature. The limits are as follows: 20–60°C, 0.006; 70°C, 0.013; 75°C, 0.016; 80°C, 0.019; 85°C, 0.023.

Dielectric security tests

The tests apply only to screened cables of 8.7/15 kV and above and are in two parts.

High voltage a.c. test
This test is carried out at an ambient temperature for a period of 4 hours at $4U_0$ for mass-impregnated cables and $3U_0$ for non-draining cables.

Bending/impulse a.c. test
The impulse test is carried out at maximum permissible operating temperature with a voltage of 95, 125 or 170 kV for cables with U_0 equal to respectively 8.7, 12 and 18 kV. The subsequent a.c. voltage application is at the same voltage as for the routine test.

Other tests

These comprise a drainage test as previously detailed together with non-electrical tests on non-metallic oversheaths.

British requirements

The most important difference in BS 6480: 1988 is that, although type tests are restricted to 19/33 kV cables, there is an onerous requirement for a loading cycle test requiring a minimum of 30 m of cable. A vertical rise of 2 m to the terminations is necessary and non-draining cables must have a vertical loop at least 6 m high. The installation has to undergo at least 100 cycles at 25.4 kV (1.33 times rated voltage) to a temperature of 70–75°C. Each cycle comprises 6 hours heating and 18 hours cooling and not more than five cycles may be carried out in 7 days. Hot and cold power factors at 25.4 kV are measured and, if graphical plotting shows stability, the test may be terminated after 100 cycles; otherwise it is continued up to 250 cycles. In addition, a 100 hour test at $4U_0$ has been included.

TYPE TESTS FOR POLYMERIC CABLES

Type tests for polymeric cables are still in the process of evolution and reference should be made to chapter 24 for an indication of the way in which knowledge of fundamental requirements is leading to a better understanding of the testing to be undertaken. Although the characteristics of IEC 502, as given below, are generally accepted internationally, there are many other additional specific tests which are favoured by different countries, e.g. CENELEC pr HD 620, and are currently under consideration for inclusion in IEC 502.

Electrical tests for screened cables

Applicability
Tests are required for screened cables insulated with EPR or XLPE of rated voltage of 6/10 kV and above.

Test sequence
In general the tests are required successively on the same sample of 10–15 m length between accessories with the following sequence:

(a) partial discharge test;
(b) bending test followed by repeat of (a);
(c) power factor/temperature test;
(d) three heat cycles to a conductor temperature 5°C to 10°C above maximum conductor temperature in normal operation: 8 hour cycle with at least 2 hours at temperature and at least 3 hours of natural cooling;
(e) partial discharge test;
(f) impulse test at 5°C to 10°C above maximum conductor temperature in normal operation;
(g) 15 min a.c. test at the routine test voltage;
(h) a 4 hour a.c. test at $4U_0$.

Measurement of the resistivity of the semiconducting screens is now included.

Electrical tests for unscreened cables

For unscreened cables rated at 3.6/6 kV and below, the sequential testing is limited to:

(a) insulation resistance at room temperature;
(b) insulation resistance at operating temperature;
(c) a 4 hour a.c. test at $4U_0$.

In addition, a separate sample shall be subjected to an impulse test at 5°C to 10°C above maximum conductor temperature in normal operation.

Non-electrical tests

A wide variety of tests is stipulated in IEC 502 as follows:

(a) measurement of dimensions;
(b) measurement of mechanical properties of the insulation and sheath materials before and after ageing;
(c) ageing tests on complete cable samples to test compatibility between materials;
(d) specific tests for the insulation material;
(e) specific tests for the sheath material;
(f) semiconducting screen strippability test;
(g) water penetration test for longitudinally water blocked designs.

There are also proposals for additional tests to limit the amount of contamination in the insulation and for measurement of void content.

TESTS ON CABLES WITH SPECIAL PERFORMANCE IN FIRES

Cables, designed to avoid some of the effects produced by standard cables when they are involved in fires, are subject to the same tests as apply to standard cables of the same voltage rating with the same insulation, including tests on components. However, additional tests and requirements apply to the components and the cables related specifically to the performance in fires.

In part 2 reference is made to requirements related to fire performance applying to cables for applications such as ships installations, railway systems and offshore installations but, within the scope of part 3, the British Standard for cables specially designed for fire performance is BS 6724: 1990, which is for wire armoured cables of rated voltages up to 1.9/3.3 kV with low emission of smoke and corrosive gases when affected by fire. A corresponding standard for MV polymeric cables has been published (BS 7835).

The fire-related tests specified in BS 6724: 1990 and BS 7835: 1996 are as follows:

Corrosive and acid gas emission

This test, in accordance with BS 6425: Part 1: 1990, is made separately on samples of insulation, bedding for armour, any fillers and binders and the oversheath. The test measures the amount of hydrochloric acid gas (HCl) generated when a sample of given mass is burnt under specified conditions. The requirement of BS 6724: 1990 and BS 7835: 1996 is that the HCl generated should not exceed 0.5% of the mass of the sample for each component.

Fire test on single cable

This is the test to BS 4066: Part 1: 1980 (1995) made on a sample of cable 600 mm long clamped vertically. One or two Bunsen or propane gas burners are used to ignite the sample (two burners for cables exceeding 50 mm in diameter), the time of application of the flames depending upon the mass per unit length of the cable. When the burners are removed the flames should extinguish before reaching within 50 mm of the top clamp, which amounts to 425 mm from the position where the flames impinge on the cable.

IEC 502 and British Standards for cables with extruded insulations provide for this test to be made, as a type test, on cables with PVC coverings if it is desired to claim compliance as a feature of the design. Generally PVC oversheathed cables will meet the test, but they would not meet the combination of fire tests specified in BS 6724: 1990 and BS 7835: 1996.

Fire test on multiple cables

This test is to BS 4066: Part 3: 1994. It is made on a set of cables, each 3.5 m long, installed vertically, touching or spaced according to their size, and approaches much more closely a test on a full-scale installation than the test to Part 1 of the standard. The fire source is a multiple jet propane gas burner with controlled fuel and air input and a special test rig is required. BS 6724: 1990 and BS 7835: 1996 requires the test category to be that involving a volume per metre of non-metallic cable materials of 1.5 litres. The requirement is that, after the cables have been exposed to the burner for a specified period, after which the fire source is stopped, the flames should not propagate up the cables to a distance greater than 2.5 m above the burner.

Smoke emission test

This test is to BS 7622: 1993. Samples of cable are arranged over a fire source consisting of a tray of alcohol, which is set alight in a test chamber which has come to be known as a 'three metre cube' because of its dimensions. A specified light source and a photocell at opposite sides of the chamber are used to make measurements from which the light transmittance of the smoke generated is calculated.

BS 6724: 1990 and BS 7835: 1996 includes an appendix which describes how, by performing these tests as type tests on cables of selected sizes, type approval for the range of sizes can be obtained.

SITE TESTS AFTER INSTALLATION

IEC 55 for paper cables requires a d.c. test at a voltage based on 70% of that permissible for factory tests, i.e.

(a) paper cables up to 3.6/6 kV, 5 minutes at $4.2U_0 + 3.36$ kV;
(b) paper cables of 6/10 kV and above, 5 minutes at $4.2U_0$.

IEC 502 for polymeric cables is under revision and a d.c. test equal to $4U_0$ for 15 minutes is being proposed. In some countries doubts have arisen about the effects of d.c. voltage on polymeric cables and IEC 502 allows alternative a.c. testing, if agreed between the parties concerned, consisting of

either   the system voltage U applied between conductors and screens for 5 minutes
or   the normal voltage U_0 applied between conductors and screens for 24 hours.

BS 6622: 1991 does not provide for this alternative for new installations, but does recommend that for installations which have been in use the manufacturers should be consulted for test conditions to take account of the particular circumstances.

These tests are under active review both in the UK and elsewhere, e.g. CIGRE SC21 have working groups studying after laying tests and diagnostic methods with the target of making recommendations for consideration by IEC.

PART 4
Transmission Systems and Cables

CHAPTER 29
Basic Cable Types for A.C. Transmission

PARTIAL DISCHARGE PROBLEMS WITH PAPER INSULATION

The use of the screened cable construction, first conceived in 1914, led to satisfactory cable designs for operation at 33 kV and to a limited extent at 66 kV. However, many failures occurred at 66 kV and higher voltages and were found to be caused by discharges in minute vacuous voids. Such voids are formed in the butt-gap spaces by expansion of the impregnating compound on heating, followed by insufficient contraction on cooling to fill the insulation completely. The voids contain low pressure gas extracted from the impregnant and are electrically weak. Partial discharges occur with sufficient energy for carbonisation of the paper to take place and this leads to the presence of carbon trees and ultimate electrical breakdown. The electrical stress in the insulation at which this breakdown phenomenon operates, i.e. in the region of 5 MV/m, also determines an upper limit of voltage of around 66 kV for such cables. There are no manufacturing problems in producing single-core 66 kV cables having an appropriate design stress, but 3-core cables are generally too large for normal manufacture. Discharge at lower stresses, in the range 2–4 MV/m, will cause some polymerisation of the impregnants with the formation of waxes (commonly called 'cheesing') but have little effect on the serviceability of the cable.

Emmanueli in the early 1920s was the first to find a solution to the problem. The fluid-filled cable which he pioneered has continued ever since to be the most widely used design for very high voltages.

In the fluid-filled cable, void formation is eliminated by maintaining the liquid impregnant inside a metal sheath at a positive pressure, which for this purpose need not be very high. An alternative arrangement is to maintain the insulation under a high gas pressure. If the sheath over the insulation is reasonably flexible the gas pressure can be applied externally. In other designs the metal sheath is made sufficiently strong to withstand a high internal pressure and the gas is admitted into direct contact with the insulation.

TYPES OF PRESSURISED PAPER INSULATED CABLE

Over the years a large variety of cables with pressurised insulation have been developed and put into service. Two designs, both fluid-filled, one at low pressure and one at high pressure, now account for the vast majority of new installations throughout the world. However, others using gas pressures still find some application and it is convenient to divide designs into groups, as shown in table 29.1. Each group may in turn be split into self-contained designs with lead or aluminium sheath and designs in which the cable is pulled into a pre-installed steel pipe, the pipe being subsequently filled with the pressurising medium of insulating liquid or gas.

By the inclusion of operating voltages in table 29.1 it will be seen that fluid-filled cables are suitable up to the highest voltage at present in service (525 kV); they also have potential for use at 750 kV and 1000 kV. Gas-pressurised cables, however, are limited to 132 or 275 kV according to design and this is associated with somewhat inferior electrical breakdown strength. On the whole they are not economically competitive with fluid-filled cables but for some specialised circumstances they may be favoured because of advantages in terms of greater simplicity with the associated equipment and accessories or where steep gradients are involved.

Fluid-filled cables

There can be some confusion about the interpretation of the generic description of a fluid-filled (FF) cable and, in particular, the coupling of the term 'low pressure' with it. In general, when no other description is given, FF cable is taken to mean a self-contained cable which operates at a maximum static sustained pressure of 5.25 bar with transient pressures up to 8 bar. This is the conventional cable which has very wide usage and is discussed in chapter 30.

However, two other designs of cable which may be termed 'low pressure' have found applications in specific countries, although the use has now generally been discontinued.

Unreinforced sheath LPFF cable

In the USA there has been use of a cable which operates at a pressure which is sufficiently low to be withstood by a specially alloyed lead sheath having no reinforcement. The pressure has therefore to be kept below 1 bar, which limits its application to fairly flat routes.

Table 29.1 Pressure cable types and voltage ranges in a.c. commercial service

Fully fluid-impregnated		Gas within insulation	
Design	Existing voltage range (kV)	Design	Existing voltage range (kV)
Lead or aluminium sheath			
Low pressure FF	30–525	Internal gas pressure	30–275
Mollerhoj flat cable	30–132		
Steel pipe			
High pressure FF	30–500	Internal gas pressure	30–132
External gas pressure with diaphragm sheath	30–275		

Mollerhoj cable

Figure 29.1 shows a type of cable which has been used in Denmark at voltages up to 132 kV. It is a truly self-contained cable because after the installation has been completed the ends are sealed and there are no feed tanks. The cores are laid side by side in flat formation under a lead sheath. By providing a longitudinally applied corrugated tape along each flat face of the lead sheath and binding them with helical wires and brass tapes, an elastic diaphragm is created which responds to changes in internal pressure on heating and cooling. The design has been used for submarine transmission links for applications up to 400 kV d.c. but it has found little application outside Denmark. One of the reasons for this is that it is necessary to limit the temperature rise to prevent excessive fatigue of the lead sheath.

TRANSMISSION SYSTEM GROWTH

In most cases, transmission cables are used where sections of overhead line circuits have to be placed underground and so the requirements have been dictated by the needs of the overhead transmission grid system. When pressure cables first became established in the 1930s, the voltage range was from 33 to 132 kV and conductor sizes up to 200 mm^2 were adequate for a rating which did not exceed 110 MVA at 132 kV. In 1938, when bulk power transmission first commenced in the UK, the CEGB generating capacity was 8500 MW, and some idea of the rapid increase in capacity may be obtained from the figure of 12 841 MW in 1949–50 and 62 564 MW in 1973–74, with a maximum demand which subsequently continued at around 50 000 MW. A transmission voltage of 275 kV, with a winter circuit rating of 760 MVA, was required in the late 1950s and 10 years later the voltage had risen to 400 kV with a rating of 2600 MVA. To meet the requirements during the period, the cable designs were constantly being pushed towards their limits so as to match a single circuit of cable with a similar overhead line circuit. Conductor sizes increased to 2500 mm^2 in the 1970s. As the maximum cable diameter consistent with manufacturing and laying practicability was reached, it became necessary to obtain higher rating by more efficient removal of the heat generated in the

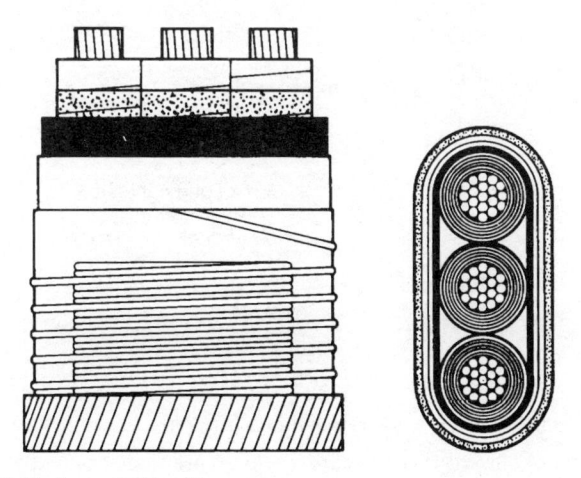

Fig. 29.1 33 kV Mollerhoj type FF cable

Table 29.2 Circuit length of underground cable and overhead line in the UK system in 1990/91 (km)

Voltage	Overhead lines	Cables	Total
33 kV	30 386	15 608	45 994
66 kV	3228	1145	4373
132 kV	21 442	2652	24 094
275 kV	4910	472	5382
400 kV	10 285	161	10 446

cables by control of the properties of the trench backfill material and/or introduction of specific cooling techniques to prevent overheating of the cables.

In 1980 the amount of main transmission line and cable in service in the UK was 14 659 circuit km compared with 7660 circuit km in 1948 (see also table 29.2).

Although the proportion of underground cable in the UK is small, it is still large in comparison with most other countries. In spite of constant endeavours to improve the price ratio between a.c. underground cables and overhead lines very little change has occurred and it remains between about 10:1 and 23:1 according to voltage and installation circumstances.

IMPROVEMENT OF DIELECTRIC BY PRESSURISING

For the reasons explained earlier, relating to partial discharges associated with heat cycles, the conventional maximum operating temperature for 33 kV solid type paper cables is 65°C. For many years pressurised cables were designed for continuous operation at 85°C and more recently this has been increased to 90°C in the case of modern fluid-filled cables. The important effect of this difference on current carrying capacity is shown in fig. 29.2, which illustrates the situation for lead sheathed cables

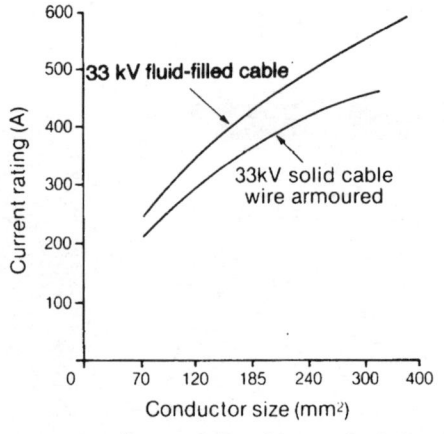

Fig. 29.2 Comparison of current carrying capacity of 3-core lead sheathed SC FF and solid type paper cables

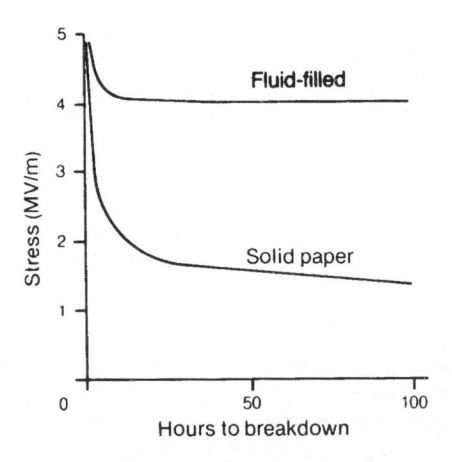

Fig. 29.3 A.C. breakdown strength of SC FF and solid type paper cables

Table 29.3 Comparative a.c. dielectric performance of SC FF and solid cables

Parameter	FF cable (MV/m)	33 kV solid cable (MV/m)
Short time strength	50	50
Maximum long time strength	30–40	12–15
Design stress	7–20	5 max

under British conditions. Part of the improvement is due to lower insulation thickness and hence better thermal conductance.

The improvement in dielectric strength is indicated in fig. 29.3 and table 29.3.

DESIGN STRESS AND PERFORMANCE DATA FOR SCFF PAPER CABLE

One of the paramount features of design stress for transmission cables relates to impulse voltage requirements to deal with transients caused by lightning and switching operations. In the case of FF paper cables, typical maximum stresses are about 100 MV/m for impulse and 40 MV/m for a.c., a ratio of about 2.5:1. Figure 29.4 shows that the service requirement ratio varies between 10:1 and 6:1 according to voltage. Consequently cables must be designed on an impulse breakdown stress basis and this provides a large safety margin for a.c. performance. The reverse would lead to failure by impulse breakdown.

As the ratio decreases with system voltage it is possible to increase the design stress with voltage. For paper cables the maximum design stresses range from about 7.5 MV/m at 33 kV to 15 MV/m at 400 kV.

Attention must also be paid to dielectric loss angle, capacitance and improvement of current carrying capacity. Dielectric loss angle becomes more important with increasing voltage. In the case of solid type cables it also increases with cable operating temperature and may rise sharply with voltage due to the onset of partial discharge in voids. Pressurised cables generally have a lower dielectric loss angle and with fluid-filled cables there is little increase with temperature and voltage.

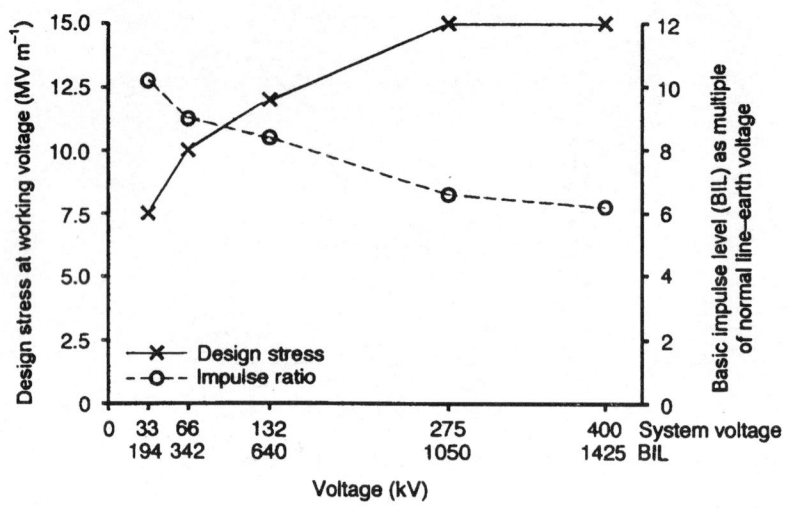

Fig. 29.4 Design stress for SC FF paper cables and relationship with basic impulse level of the UK system

The dielectric loss angle results in a transmission power loss which appears as heat and the dielectric loss D may be expressed as

$$D = U_0^2 2\pi f C \tan \delta \times 10^{-3} \quad \text{(kW/km)} \tag{29.1}$$

where U_0 = voltage to earth (kV)
f = power frequency (Hz)
C = electrostatic capacitance (μF/km)
$\tan \delta$ = dielectric loss angle

The heat generation has an effect on the current carrying capacity if a maximum cable temperature has not to be exceeded. At 33 kV the effect is very small but, as the power losses are proportional to the square of the voltage, the situation is very different at 400 kV and the reduction in rating could be as much as 25% for the same conductor size. Reduction of dielectric loss angle has been obtained by such means as the use of deionised water during paper-making (chapter 3) or the use of polypropylene paper laminate (chapter 41).

It will also be seen that, in addition to keeping the dielectric loss angle to a minimum for the highest voltages, the capacitance should be as low as possible. To achieve this by increasing the dielectric thickness is not practical and reduction of the permittivity of the paper by adoption of lower apparent density is limited as it reduces the impulse strength of the cable. Polypropylene paper laminate is particularly beneficial in reducing both permittivity and dielectric loss angle.

ALTERNATIVE DIELECTRICS TO IMPREGNATED PAPER

Ever since the excellent dielectric properties of polyethylene were recognised in the 1940s it has been predicted that this material would eventually replace paper for the insulation of power cables. It has a high intrinsic electric strength and a low dielectric loss angle and permittivity.

The problems encountered and the solutions adopted before its potential for transmission cables could be realised are described in chapter 33.

Cables with crosslinked polyethylene insulation (XLPE) for operating voltages up to and including 132 kV have now been established for some years. To avoid the incidence of water trees a metallic barrier, usually an extruded sheath, is used. The thickness of the insulation is at present significantly greater than that of a comparable pressure assisted paper cable.

Initially most progress with cables for higher voltages, i.e. 220 kV and above, was made using low density polyethylene. In 1993 it was reported that some 1100 km of 225 kV cable and short lengths of 400 and 500 kV cable insulated with low density polyethylene had been installed worldwide.

The development of XLPE insulation for these voltages followed later but long lengths of 275 kV cable have now been installed together with short lengths of 500 kV cable. Improved manufacturing and polymer handling techniques have enabled design stresses approaching those of pressurised paper cables to be employed. The incentive to use XLPE insulation is that a maximum operating temperature of 90°C can be used compared with 70°C for low density polyethylene and 80°C for high density polyethylene.

Some countries have developed cables with ethylene propylene rubber (EPR) insulation. The term EPR insulation encompasses a range of polymers, fillers, red lead, oils to assist extrusion, antioxidants, and a crosslinking agent. Its dielectric properties are not as good as polyethylene or XLPE. In particular its dielectric loss angle and permittivity are similar to those of paper insulation and its impulse strength is lower than XLPE.

The main advantages of EPR are its flexibility and resistance to electrical partial discharge when compared with polyethylene. The material requires crosslinking in the same manner as XLPE and similar processes are used for extrusion, crosslinking and degassing. Although it is generally accepted that it has better resistance to water treeing than XLPE, the higher voltage cables have usually been provided with a metallic barrier when installed underground.

EPR insulated cables are used in some countries (e.g. Italy) for voltages up to and including 150 kV.

CHOICE OF CABLE DESIGNS AVAILABLE

It is difficult to give a clear summary because choice based on current carrying capacity, economics and reliability tends to be influenced by traditional user practice, by national preferences and by environmental issues but some of the issues are determined by the cable voltage.

The technique of life cycle analysis (LCA) is intended to account systematically for the impact of the raw materials, manufacturing energy, packaging, installation, in-service energy loss, recovery and recycling on the environment. A true comparison cannot yet be made because of the absence of an internationally agreed method of calculation and list of costs. Some general observations are:[1] the metals employed in cables are of high purity and can be recycled; insulating paper is derived from renewable forestry resources; XLPE, EPR and impregnated paper are unsuitable for reuse, but can be burnt as fuel; thermoplastics such as LDPE insulation and oversheaths can be reused for lower grade applications; the choice of high stress cables such as FF paper and polypropylene paper laminate permit material volumes to be reduced; energy loss in service is reduced at voltages of 220 kV and above by the use of FF polypropylene paper laminate, XLPE and LDP cables.

33–132 kV

The choice is between solid type impregnated paper (up to 66 kV), thermosetting insulation (mainly XLPE), gas-filled and fluid-filled. The gas-filled cable tends to be used only where it has specialised advantages (chapter 31). The use of solid type cable above 45 kV is mainly for maintenance of existing systems of this type.

The current carrying capacity of solid paper cables is lower than for the other types but in some cases capitalisation of losses may detract from the advantage of using the smaller conductor size possible with other types. Solid type paper cables are also less attractive for use in countries where the ambient temperature is high, because of the effect of derating factors.

The use of polymeric insulated cables, particularly XLPE, is growing rapidly. A disadvantage is that 3-core designs have hitherto not been practical at voltages above 33 kV, due to their large diameter.

In comparing pressure cable with extruded polymeric cable there are three important factors. One is the effect of route length because for short routes the cost of the ancillary equipment for pressure cables may be a relatively high proportion of total cost. The second is whether the extra skill and expertise necessary for installing and maintaining the system is readily available. The third is concern about the environmental consequences of impregnant leakage from old cables (although the hydraulic integrity of improved designs together with biodegradable impregnant installed since 1980 has been excellent). Largely for these reasons the use of pressure cable is diminishing at these voltages.

200 kV and above

Until recently the choice was between self-contained FF cable and high pressure pipe type FF cable. The latter is mainly confined to North America and whilst having some advantages (chapter 32) is generally limited by current rating and less economically competitive. Both designs have been developed up to 750 kV and the self-contained FF cable up to 1100 kV. Commercial installations are in service at a voltage of 500/525 kV. The self-contained FF cable and high pressure pipe type cables now increasingly employ polypropylene paper laminate insulation, they combine the inherent electrical reliability of the FF cable with reduced losses and increased current rating.[2]

Polymeric insulated cables are now being installed for long circuits at system voltages of 220/275 kV. Initially the cables were insulated with low density polythene but more recently XLPE has been used because of its higher operating temperature. A factor which delayed the introduction was the lack of availability of reliable joints. Short lengths of cable without joints have been installed at 500 kV. Generally design stresses are lower than fluid-filled cables resulting in larger cables but the lower dielectric loss angle and lower permittivity are advantages.

REFERENCE

(1) Endersby, T. M., Galloway, S. J., Gregory, B. and Mohan, N. C. (1993) 'Environmental compatibility of supertension cables'. *3rd IEE Int. Conf. on Power Cables and Accessories 10 kV–500 kV*.

(2) Gregory, B., Jeffs, M. D. and Vail, J. (1993) 'The choice of cable type for application at EHV system voltage'. *3rd IEE Int. Conf. on Power Cables and Accessories 10 kV–500 kV*.

CHAPTER 30
Self-contained Fluid-filled Cables

The fluid-filled cable is the most widely used type of transmission cable throughout the world and for a long time in the UK has been the only design used for new installations at 275 kV and 400 kV (fig. 30.1). Originally known as the oil-filled cable, the name was changed to fluid-filled to take into account the fact that the most widely used impregnants today are synthetic fluids. First introduced in the early 1920s, development has been continuous ever since, to meet progressive demands for increases in voltage and current carrying capacity. This is still so today with a capability to meet future requirements for 750 kV and 1100 kV cables,[1] together with further up-rating of 400 kV overhead lines. The highest voltage cables in service are the 525 kV cables which were first installed in the early 1970s at the Grand Coulee Dam.[2]

CABLE DESIGN FEATURES

Construction

The standard design consists essentially of copper or aluminium conductors, paper insulation and an aluminium or reinforced lead sheath designed to withstand a sustained internal pressure up to 5.25 bar, with transient pressures up to 8 bar. In general, higher pressures might enable economies to be obtained in fluid feeding arrangements but would usually be insufficient to justify the associated increase in the cost of cable and accessories. However, in special cases, such as cables installed on routes which have steep changes in elevation, it can be economic to design up to a maximum pressure of approximately 30 bar.

The basic concept is one of full impregnation of the whole of the insulation at all times by a low viscosity hydrocarbon fluid under pressure. As temperature rises, the surplus fluid due to expansion is forced out of the cable into storage tanks and reverse flow takes place on cooling. The cable is filled with fluid at the time of impregnation and sheathing and is subsequently kept in this condition throughout its life. It is despatched to site with a tank attached to take care of change of fluid volume with ambient temperature. Even during the jointing operations of cable installation, fluid pressure and flow are maintained. The only exception to this is for cables which are installed

Self-contained Fluid-filled Cables

Fig. 30.1 275 kV FF cable with 2000 mm² Milliken conductor and corrugated aluminium sheath

vertically, e.g. cables for hydroelectric generating schemes. At Cruachan in Scotland the generators and transformers are in a cavern at the base of a mountain with 275 kV cables feeding an overhead line termination at a height of 352 m. To avoid the need for numerous stop-joints in such applications, the cable sheath has to be reinforced to withstand the hydrostatic pressure, in the above case 30 bar, and the cable needs to be partially drained under vacuum to control the flow during jointing at the lower end. Careful re-impregnation is then necessary before making the upper sealing end.

Ducts to provide channels for fluid flow have to be incorporated in the cable design and are indicated in fig. 30.2, which also shows typical cable constructions.

For single-core cables a duct is normally included in the centre of the conductor, although for the short cable lengths which are used as terminations for 3-core cables the fluid channel may be on the outside of the insulation.

For 3-core cables having fillers between the cores, the duct is formed by the use of an open helix of steel or aluminium strip incorporated into the filler. When fillers are not necessary, i.e. with corrugated aluminium sheaths, the empty space between the cores produces a low impedance to flow without the necessity for a specific duct (fig. 30.3). Such constructions are known as ductless shaped fluid (DSF) or ductless circular fluid (DCF) according to the conductor shape.

The normal size range comprises single-core cable with conductors from 120 to 2500 mm² and 3-core cables from 120 to 630 mm², although at 33 kV the range extends downwards to 70 mm². Above about 150 kV, only single-core cables are available. This is because of a limit on overall diameter of about 150 mm associated with the maximum size of cable and cable drum which can conveniently be handled for installation.

Compared with distribution cables the use of aluminium for conductors was held back by problems associated with jointing, but following the introduction of the simple metal–inert gas (MIG) welding technique, the proportion of aluminium conductors has grown during periods when the price comparison has been favourable. For very heavy currents, of course, use of aluminium more readily causes increase in cable diameter toward the manufacturing limit. Conductor construction is similar for both metals but with the segmental form a small number of larger segments may be used with aluminium.

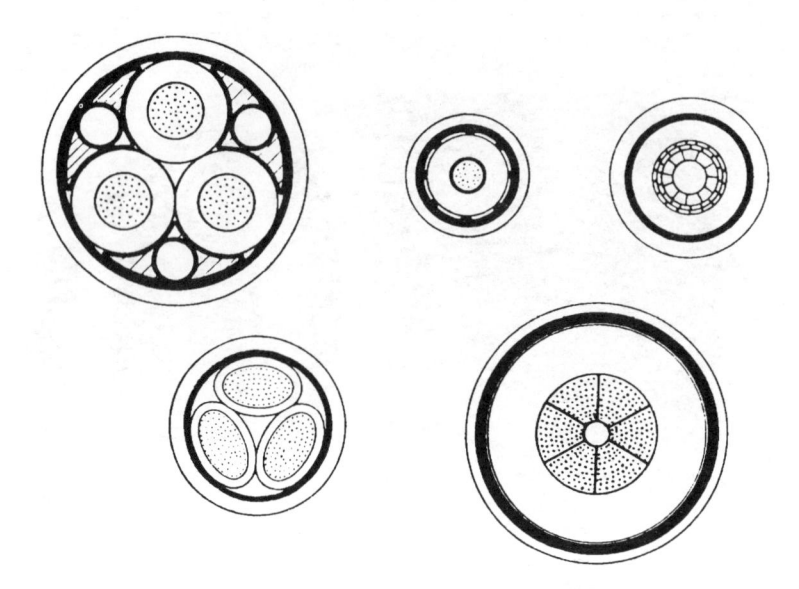

Fig. 30.2 Cross-section of typical self-contained fluid-filled cables

Fig. 30.3 33 kV ductless FF cable with oval conductors

Except for the Milliken construction (fig. 30.1), the former general practice to strand circular wires around an open steel spiral duct to form a hollow centre in single-core conductors now only applies for sizes of 150 mm² and below. Instead, a self-supporting ring is formed from segmental wires and this may form the basis for additional layers of circular wires, flat strips or segments applied with alternating direction of lay. Shaped segments provide better space occupancy than can be obtained by compacting circular wires and also better flexibility. Very great control of the segment dimensions and precision in placing them together in the stranding machine are both vital (fig. 30.4).

Two important factors arise with very large conductors for single-core cables, namely the mechanical thrust on heating and the skin and proximity effects on a.c. resistance. The former is discussed in chapter 37. Reference to skin and proximity effects and the

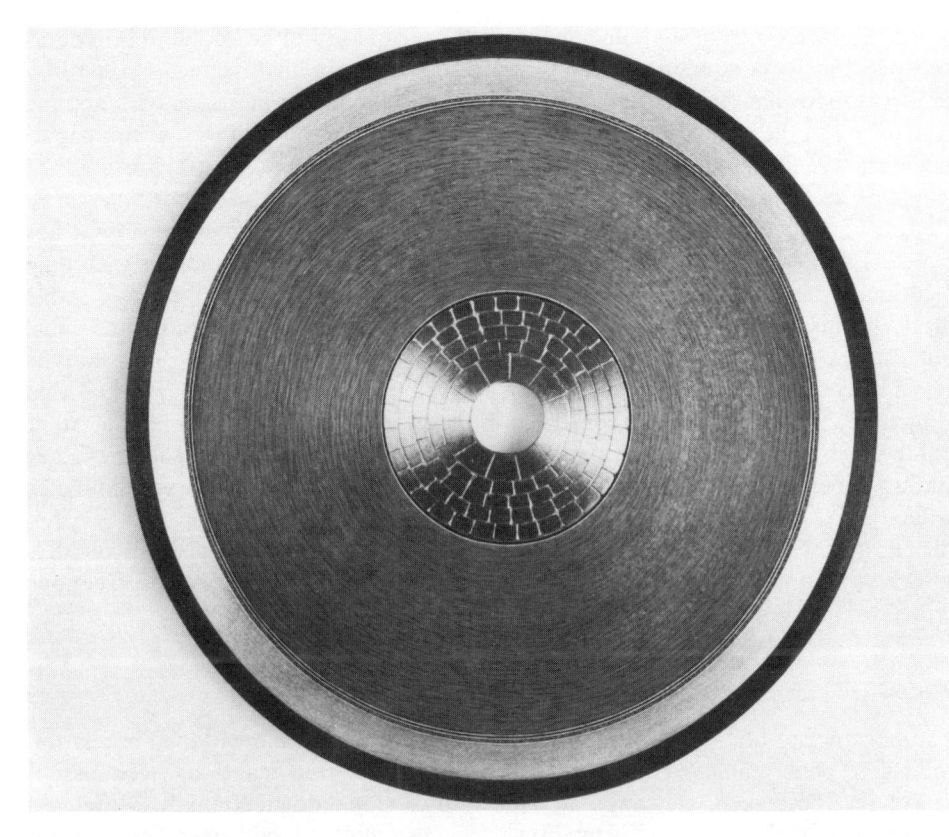

Fig. 30.4 525 kV FF cable with segmental wire conductors

use of Milliken conductors to reduce losses has been made in chapters 2 and 4. Usually the self-supporting duct construction is used up to about 1000 mm^2 but in certain cases it may be economic to extend this size if a larger internal duct is used. For larger sizes, Milliken conductors are generally adopted. They consist of six stranded cores usually laid around a 12 mm duct. Three cores are wrapped with plain paper and the alternate cores with carbon paper. The total number of wires may exceed 500.

The shape of 3-core cable conductors is normally oval at 33 kV and circular for higher voltages. They are stranded from round wires and compacted to provide a smooth surface or stranded from shaped or flat wires. Although solid aluminium conductors have been used extensively for 1 kV paper cable they are unsuitable for the higher voltages, except for the smaller conductor sizes at 33 kV, because of disturbance of the lapped insulation on bending.

Conductor screens

Lapped tape screens are used to give the conductor a smooth surface and so avoid electrical stress concentrations on individual wires. A possible exception exists at 33 kV, where the stresses are low. Two main constructions are: (a) plain carbon papers and (b) metallised plain paper or coated carbon paper. Carbon paper is paper in which carbon is included during manufacture in sufficient quantity and of suitable type to make it electrically conducting. If a plain carbon paper conductor screen is used in a cable, it is found that the dielectric loss angle of the insulation increases with applied voltage owing to the influence of the electrical field on the surface of the carbon paper. As this occurs at the conductor surface the resultant effect on the dielectric loss angle of the cable decreases as the insulation thickness increases, i.e. as cable voltage increases. This effect of carbon paper is not as harmful as may at first be thought because it has been found that, with the constant application of voltage, the effect reduces with time and ultimately becomes negligible. Nevertheless, during the initial routine cable electrical testing it is possible that it may mask other imperfections and some cable manufacturers use screen constructions which eliminate or reduce it. This can be achieved by coating the carbon paper with metal foil or alternatively with paper of high air impermeability. The latter construction is carried out at the time of paper manufacture by producing 2-ply paper, one ply being carbon paper and the other insulating paper. Conductor screening by metallised paper also, of course, eliminates the effect.

The effect of screen construction on dielectric loss angle versus voltage characteristics is recognised in test specifications by allowing different values for the two constructions described.[3]

Insulation

The importance of butt-gap width and depth have already been mentioned in relation to solid type paper cables. The subject is of still greater significance at transmission cable voltages because of effects on a.c. and impulse breakdown strength. Moreover, with the greater thickness of insulation, it is essential to take steps to prevent disturbance during bending. In addition to optimum selection of paper specification in terms of purity, density, impermeability and surface finish for satisfactory coefficient of friction, factors which have to be taken into account during design and manufacture include choice and control of paper width and thickness, lapping tensions and precision of application. Moreover to minimise the change in interlayer pressures due to changes in paper dimensions as a result of drying and impregnation, the moisture content of paper for cables for higher voltage is reduced before application and the lapping is carried out in a controlled humidity atmosphere (fig. 30.5). In the machine illustrated, precision of paper tensioning is obtained by electrical servo control which also maintains uniformity of tension when head rotation starts and stops.

Requirements for the optimum mechanical and electrical condition of the insulation are interrelated and often in opposition, so that compromise is necessary, e.g. thick papers favour mechanical stability but thin papers give the best electrical strength. In general, in the region of the highest electrical stress adjacent to the conductor, thicknesses of 75 µm are used, and at the outside variation from 125 to 200 µm is possible.

In most cases the design stress of FF cables is determined by the lightning voltage specified for the system rather than the a.c. system or test voltage. The effect of a lightning strike on the cable is represented by impulse testing (chapter 39), the impulse strength of the insulation being affected by the constitution of the paper, the thickness of paper and the uniformity of the conductor. In the latter aspect it has been found that cables with Milliken conductors have a somewhat lower impulse strength than cables having conductors with normal stranding. Table 30.1 shows the design stresses used in

Fig. 30.5 Paper lapping machine in a conditioned atmosphere

Table 30.1 Effect of impulse level on design stress

Operating voltage (kV)	Lightning impulse level[a] (kVp)	Ratio of impulse to operating voltage	Maximum design stress (50 Hz) Non-Milliken (MV/m)	Maximum design stress (50 Hz) Milliken (MV/m)
33	194	10.2	[b]	[b]
66	342	9.0	10	9
132	640	8.4	12	10
275	1050	6.6	15	13
400	1425	6.2	15	15

[a] Voltage to earth
[b] Insulation thickness at 33 kV is determined by mechanical requirements

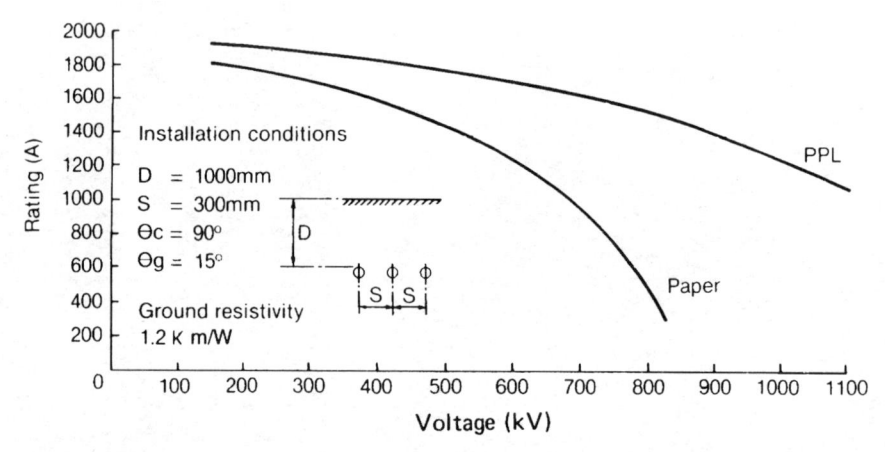

Fig. 30.6 Variation of current rating with operating voltage for paper and PPL insulation; conductor size 2500 mm^2

the UK. It will be seen that, as the operating voltage increases, so does the design stress, because the specified impulse requirements become relatively less onerous.

While it is possible to design cables with paper insulation for voltages up to 1100 kV, reference to equation 29.1 indicates the increasing value of dielectric losses with voltage. This reduces the amount of current dependent losses that can be permitted, thus reducing the current rating of the cable. This is illustrated in fig. 30.6. As will be seen, at a voltage of about 850 kV the paper cable has no rating capability if it is naturally cooled.

To overcome this limitation polypropylene paper laminate has been developed as a replacement for paper.[4] The material is known as PPL, PPP or PPLP depending on the country of origin. It consists of a film of polypropylene coated on both sides with a layer of paper as illustrated in fig. 30.7. The material consists of approximately 50% paper and 50% polypropylene. The important properties of the material compared with paper are given in table 30.2. The physical properties are such that it can readily be substituted for paper using the same lapping machines.

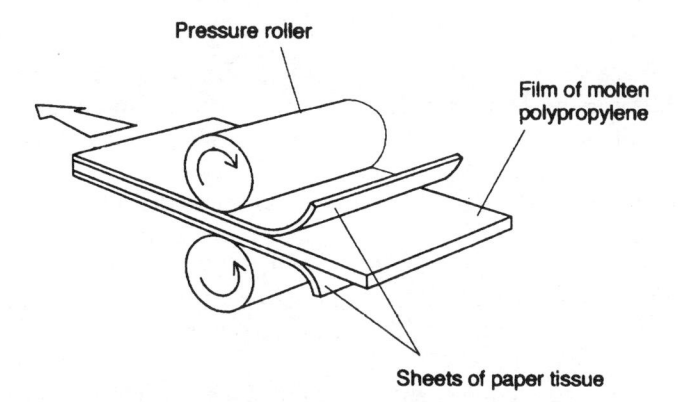

Fig. 30.7 Construction of polypropylene paper laminate

Table 30.2 Typical properties of 100 μm paper and PPL insulation

Property	PPL	Paper
Tensile strength (MN/m^2)	50	110
Elongation at break (%)	2.0	2.5
Air impermeability (G s)	Infinity	15 000
Density (g/cm^3)	0.9	0.9
[a]Relative permittivity at 90°C	2.7	3.4
[a]Dielectric loss angle at 90°C	0.0008	0.0023
[a]Dielectric loss factor at 90°C (permittivity × DLA)	0.0021	0.0078
[a]Impulse strength (MV/m)	160	135
[a]Short time a.c. strength (MV/m)	55	50

[a] Characteristics from fluid-impregnated model cables

Compared with paper, the benefit of the lower dielectric loss angle and lower permittivity on the current rating is shown in fig. 30.6. It can be seen that at a voltage of 1100 kV there is still a useful current capability and that there are significant increases down to about 220 kV. Other advantages are the higher impulse strength and lower permittivity. The latter reduces the capacitive charging current.

The material is more expensive than paper and therefore the extra cost must be set against the benefits given by its use. In general its use is only economic at voltages of 275 kV and above. PPL insulated cables are now being regularly supplied at voltages of 275 kV, 400 kV and 500 kV. In the case of the latter voltage a cable has been supplied designed to a maximum stress of 18.2 MV/m.[5] The previously highest design stress for a paper insulated 500 kV cable was 16.2 MV/m.

Insulation screen

All cables have a screen over the insulation to constrain the electric field to the lapped insulation. The screen must be sufficiently permeable to gases and fluid to permit drying and impregnation of the cable and usually consists of carbon paper, metallised paper or metal tapes, singly or in combination.

Impregnants

Until the early 1960s, low viscosity mineral oil had always been used, the only change from the initial inception being some reduction in viscosity to enable the lengths of individual hydraulic sections to be increased. When synthetic alkylates of dodecylbenzene type became available from other applications in the chemical industry, a progressive change was made to them because it was found that they were less subject to minor changes in quality, this being of particular importance during the turbulent conditions experienced by the oil industry in the 1970s. They are also technically superior at higher temperatures and the higher stresses being reached in service. The preferred type of dodecylbenzene impregnant is readily biodegradable when tested by the OECD Closed Cup Test.[6]

One of the important characteristics required from the impregnant is to absorb hydrogen and moisture. Hydrogen is not formed under normal operating conditions but can be generated, for example, by debris introduced during jointing operations.[7] Alkylates can absorb much more hydrogen and this ability increases rather than decreases with temperature.

Drying and impregnation

As with other types of cables with lapped insulation, FF cables are vacuum dried. It is necessary to achieve very low levels of moisture content to ensure that the lowest dielectric loss angles are obtained.

Several methods are used for impregnation. For many years the standard practice was to remove the cable from the drying tank, apply the lead sheath, place the cable in a steam heated vessel, evacuate the cable from both ends and then finally to vacuum impregnate the cable. Limitations to this process were that the insulation picked up some moisture during sheathing and that, as manufacturing lengths increased, the time for subsequent evacuation became very long. Whilst improvements are possible to reduce moisture regain, two alternative procedures have been developed, namely mass impregnation and vacuum sheathing. In the former the cable is impregnated in the drying vessel. To ensure that no impregnant is lost from the cable during lead or aluminium sheathing the vessel is connected directly to the sheathing extruder and the cable is sheathed whilst still surrounded by impregnant. For this process (fig. 30.8) special drying vessels are required which can be situated directly behind the sheathing extruder and connected by suitable pipework. Techniques had to be developed to keep the cable under fluid pressure throughout processing.

Similar techniques and equipment are needed for vacuum sheathing. At the end of the drying process the vessel is connected to the sheathing extruder and the sheathing is carried out under vacuum. A separate process is then required for filling with impregnating fluid. The main advantage of this process is that only the impregnant required to fill the cable is used, whereas a large quantity is in circulation for the mass impregnation process and special attention has to be given to drum and tank cleaning.

Lead alloy sheaths

Originally all FF cables had lead sheaths but now a significant proportion are sheathed with aluminium for economic reasons. Exceptions are for submarine cables and for

Fig. 30.8 Stainless steel drum being removed from stainless steel mass-impregnation tank

some installations where the higher electrical losses with aluminium exert an excessive influence because small cable spacings are employed. Nevertheless, lead has served well, provided that the cablemaker has recognised its limitations. In spite of the reinforcement which is necessary, hoop stresses may still arise in the sheath and a high quality with freedom from extrusion defects is paramount. The use of a continuous screw press is necessary to secure this quality.

Another important factor is that under the very low creep stresses which may arise due to the internal pressure, some types of lead alloy may fail by cracking after very small extensions. This limits the choice of alloy and the metallurgical structure associated with the type of extrusion press has also to be taken into account. For normal applications UK manufacturers use $\frac{1}{2}$C alloy (0.2% tin, 0.075% cadmium).

Aluminium sheaths

Aluminium sheaths have many advantages because of reduced weight and cost and the elimination of reinforcement. Moreover by using a corrugated seamless aluminium (CSA) sheath, the thickness is reduced by around 50% compared with a smooth sheath and the cable flexibility is as good as or better than with a lead sheath. The extra resistance to crushing and flattening provided by such sheaths also means that fillers may be omitted between the cores of many 3-core cables, so that the interstitial space can serve as the fluid duct. Finally, aluminium has vastly better fatigue strength for installations where vibration or slow high stress bending movement may be involved, e.g. in manholes or on bridge crossings.

The first FF cables introduced in 1952 with aluminium sheaths had smooth sheaths, but after the corrugated version became available in 1959 it soon became standard. Smooth sheaths will withstand higher pressures but for the vast majority of installations the much thinner corrugated sheath is entirely adequate. If necessary the thickness may also be increased to cater for high static pressures resulting from steep gradients or vertical installation. In contrast with the corrugation for solid type paper cables, in which the ribs are of a discrete annular type, i.e. transverse to the cable axis, FF cable sheaths are usually helically corrugated. The pitch varies from approximately 40% of the sheath internal diameter for small cables to about 25% for larger cables. Rib heights are within a range of 4–7% of the internal diameter and 3-core cables have somewhat deeper ribs than single-core cables because the permissible bending radii are smaller.

Reinforcement for lead sheaths

Although steel tapes may be used to reinforce the lead sheaths of 3-core cables it is now more usual to use copper alloy or stainless steel tapes for all cables. The tape is normally applied as a single layer and the thickness varies from 100 to 200 µm according to the cable diameter.

Armour

The only FF cables which require armour are those for submarine or high pressure installations (the former are discussed in chapter 42).

Anticorrosion protection

Extruded sheaths of PVC or medium to high density polyethylene are used. The oversheath can be modified to include chemicals to reduce the propagation of flame in the case of fire, or to prevent attack by termites as appropriate.[6] Details are as given in chapter 5. A thin conducting layer is usually applied over the oversheath to permit d.c. testing to check its integrity both at completion of manufacture and after installation.

HYDRAULIC DESIGN

An important consideration in the design of self-contained fluid-filled cables is to ensure that under any operating conditions the pressure in the cable system is maintained between permitted values. The maximum is usually taken as 5.25 bar for sustained pressures and 8 bar for transient pressure. The minimum design value is generally 0.2 bar. As the pressure in the cable will change depending on elevation, it is necessary to carry out a survey of the cable route to determine its profile. To simplify the analysis of the pressures, it is normal to express the internal pressure in terms of metres of fluid (1 m of fluid at 15°C corresponds to approximately 0.085 bar). Figure 30.9 shows the variation of internal fluid pressure along a route under static temperature conditions.

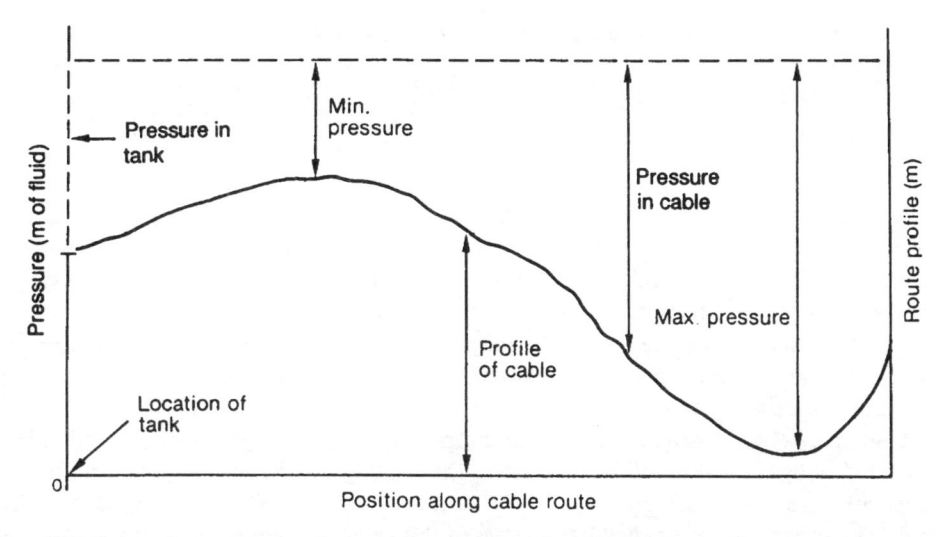

Fig. 30.9 Internal pressure variation along route of self-contained FF cable under static conditions

If the current in the conductor of the cable is increased, the temperature of the cable will rise and will result in thermal expansion of the impregnating fluid. Therefore fluid will flow from the insulation into the duct and then into the tanks. The flow will create a hydraulic pressure drop along the duct, and it is necessary to calculate this pressure in order to ensure that the specified maximum transient pressure of the cable is not exceeded.

For a unit length of cable δx situated at a distance x from the tank and assuming that the average temperature of the cable is rising at the rate of $d\theta/dt$, then the rate of fluid expulsion (l/s) is

$$V \delta x \, \alpha (d\theta/dt)$$

where $V =$ fluid volume of the cable (l/m)
 $\alpha =$ coefficient of expansion of the fluid (°C^{-1})

The quantity of fluid flowing through the sample of cable situated at x in a route length of L metres, assuming that a tank is situated at one end, is

$$V(L-x)\alpha(d\theta/dt)$$

The pressure drop δP along the duct of the section of cable δx in length is

$$\delta P = V(L-x)\alpha(d\theta/dt)b\delta x \qquad (30.1)$$

where b is the hydraulic impedance of the duct per metre. Integrating from $x = 0$ to $x = L$ the pressure drop P along the complete length of cable is given by

$$P = V\alpha(d\theta/dt)b\,L^2/2 \qquad (30.2)$$

It will be noted that the pressure drop is proportional to the square of the cable length, the fluid volume per unit length and the rate of temperature rise. The latter depends on the increase in current loading but will be a maximum when full load is applied suddenly.

It can be shown that for a duct with a circular cross-section

$$b = 25.5\eta/r^4 \tag{30.3}$$

where η = viscosity of the impregnating fluid (centipoise)
 r = internal radius of the duct (mm)

The above assumes streamline flow. It will be seen that the transient pressure can be reduced by increasing the duct diameter and by reducing the viscosity of the impregnant. Except for special installations, the diameter of the central duct of single-core cables is standardised at approximately 12 mm.

In a similar manner to the transient heating pressure, a transient cooling pressure will arise when the temperature of the cable reduces. In this case the fluid flows from the tank into the cable system.

These transient pressures will be superimposed on the static pressures and this is illustrated in fig. 30.10. It will be noted that the cooling transient pressure starts from a higher pressure at the tank position. This is because, during the heating period, fluid has passed into the tank and hence has increased the pressure (see chapter 35).

The transient pressures only occur during changes in temperature of the system. The above analysis is a simplification of the hydraulic conditions as it does not allow for such factors as compressibility of the fluid impregnant, the temperature gradient through the cable insulation, the variation of viscosity with temperature etc. In general the maximum heating transient will occur shortly after switching full load on to a cold cable, while the maximum cooling transient will occur shortly after switching off full load.

The analysis of static and transient pressures is an important part of self-contained fluid-filled cable design. In practice the tanks are connected to a cable system at

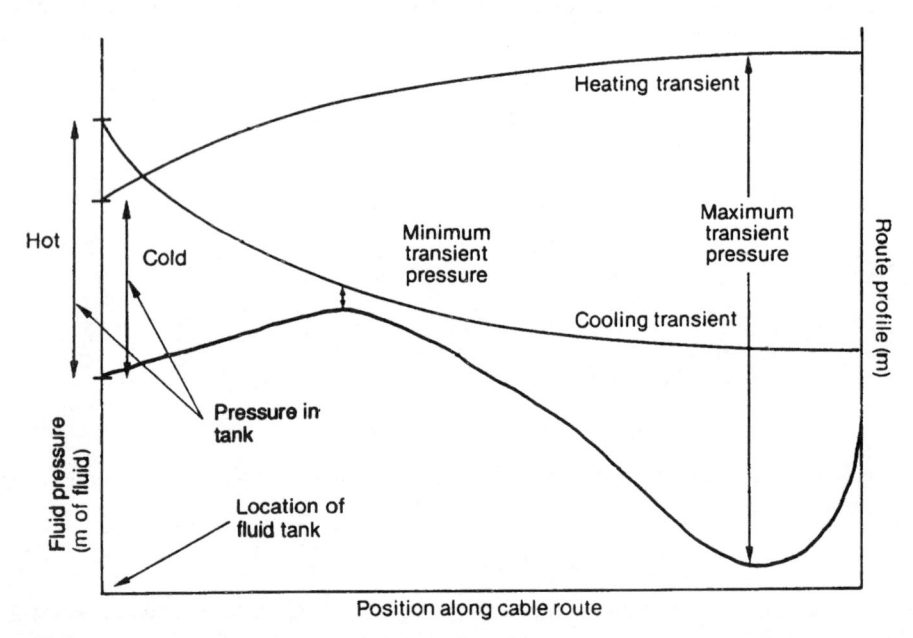

Fig. 30.10 Internal pressure variation along route of self-contained FF cable under transient conditions

terminations or stop joints. In the case of 3-core cables, connection can also be made at straight joints as with these cables the ducts are situated in the interstices between cores.

The fluid feed tanks are available in a number of sizes from 88 to 300 litres to suit the volume of fluid in the hydraulic section. The tanks may be installed above or below ground; in the latter case they are enclosed in a concrete shell. Details of their construction are given in chapter 35.

The tanks may have to be installed at a site which is at a low level relative to the route profile and if the gas-filled elements within the tank had initially been sealed at atmospheric pressure they would be significantly compressed under no-load conditions. This would be wasteful of tank capacity and is avoided by pre-pressurisation of the elements.

Particularly when the route is undulating, the installation may need to be split into a number of separate hydraulic sections and this is done by the use of stop joints, e.g. the inclusion of an epoxy resin barrier section in the centre of joints to prevent any flow of impregnating fluid across the joint. The maximum economic difference in head is about 30 m to 40 m. Individual sections vary very considerably in length according to circumstances and may extend up to 3 km or more in special cases. Figure 30.11 shows a typical arrangement for a 3-core cable system.

Whilst preparation of a preliminary hydraulic scheme which would be quite satisfactory is quite a quick and straightforward matter, the choice of the most economic solution at the planning stage is usually quite complex. It is advantageous not only to concentrate the tankage in large units at the smallest number of locations, but also to use a minimum of the more expensive stop joints. Such factors have also to be taken into account in calculating cable lengths between feed positions, to ensure that no excessively high static pressure, or low transient pressure, will arise at positions remote from the feed points.

Reliability of such a system is extremely high and very little maintenance is involved. Each hydraulic section is normally fitted with a pressure gauge having a low pressure alarm contact which is connected via a supervisory system to a suitable control centre.

SYSTEM DESIGN

As outlined in fig. 30.11, a complete installation will consist of terminations to overhead lines or switchgear or transformers, plus usually straight and/or stop joints. The terminations are always made on single-core cables and with 3-core cables it is necessary to use trifurcating joints for connection to single-core 'tails'.

Use of single or 3-core cables

Up to about 150 kV, 3-core cables are manufactured as standard. Upwards from about 200 kV, size limitations dictate that only single-core cables are practicable. Between these voltages special consideration may be given to the manufacture of 3-core cable of smaller conductor sizes.

In general, a 3-core installation provides a more economic solution unless the route length is very short in which case the saving by eliminating trifurcating joints may

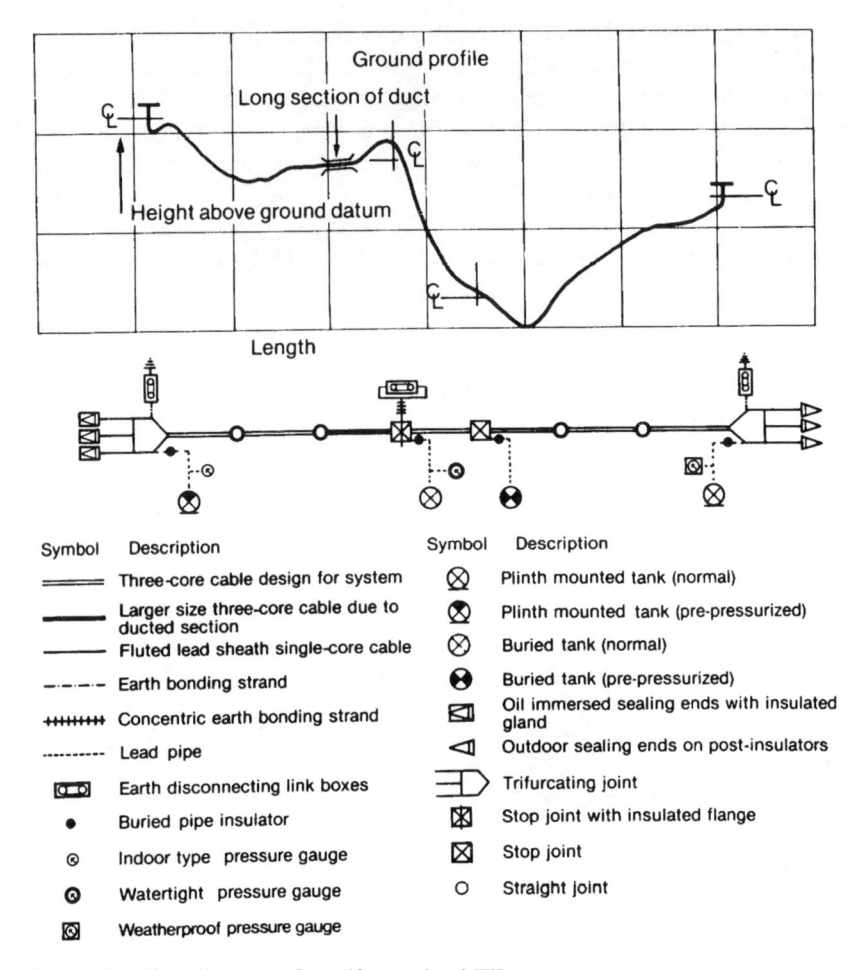

Fig. 30.11 Typical line diagram of a self-contained FF system

be significant. However, the rating required may exert a substantial influence, particularly with cables requiring large conductor sizes, and each scheme needs to be evaluated in detail.

Cables for railway electrification

In the UK and several other countries, railway electrification now uses an overhead catenary at 25 kV a.c., single-phase. A somewhat unusual FF cable is used as a supply cable, unusual because a two-conductor design is required, one being near earth potential. The construction is concentric with a small amount of insulation over the outer conductor. By using aluminium conductors the lightweight cable is particularly suitable for track-side installation and long spans of the order of 4.5 m may be adopted between cable support posts.

GENERAL TECHNICAL AND PERFORMANCE DATA

Appendix A16 contains information on the following:

Technical data: dimensions, weights, charging current, a.c. resistance and reactance for 33–400 kV cables with lead and corrugated aluminium sheaths (tables A16.2–A16.8); d.c. resistances are given in appendix A4 (table A4.1)
Power ratings: for single- and 3-core cables laid direct and in air (figs A16.1–A16.8)
Electric losses: representative values for small and large conductor size cables from 33 to 400 kV with lead and aluminium sheaths (fig. A16.9)

REFERENCES

(1) Rosevear, R. D. and Vecelli, B. (1979) 'Cables for 750/1100 kV transmission', *2nd IEE Int. Conf. on Progress Cables for 220 kV and Above*.
(2) Ray, J. J., Arkell, C. A. and Flack, H. W. (1973) '525 kV self-contained oil-filled cable systems for Grand Coulee third powerplant – design and development'. *IEE Paper No. T73*, pp. 492–496.
(3) IEC 141-1. 'Tests on oil-filled and gas pressure cables and their accessories'.
(4) Endersby, T. M., Gregory, B. and Swingler, S. G. (1992) 'The application of polypropylene paper laminate insulated oil-filled cable to EHV and UHV transmission'. CIGRE Paper 21-307.
(5) Endersby, T. M., Gregory, B. and Swingler, S. G. (1993) 'Polypropylene paper laminate oil filled cable and accessories for EHV application'. *3rd IEE Int. Conf. on Power Cables and Accessories 10 kV–500 kV*.
(6) Endersby, T. M., Galloway, S. J., Gregory, B. and Mohan, N. C. (1993) 'Environmental compatibility of supertension cables'. *3rd IEE Int. Conf. on Power Cables and Accessories 10 kV–500 kV*.
(7) Gibbons, J. A. M., Saunders, A. S. and Stanett, A. W. (1980) 'Role of metal debris in the performance of stop-joints as used in 275 and 400 kV self-contained oil-filled cable circuits'. *Proc. IEE, C* **127** (6), 406–419.

CHAPTER 31

Gas Pressure Cables

There are two basic types of paper insulated gas pressure cables, one type employing gas within the dielectric to suppress ionisation, and other type applying gas pressure external to a diaphragm to maintain the dielectric under compression under all service conditions. Cable types which employ a gas pressure external to the diaphragm are the gas compression cables and the high pressure gas-filled (HPGF) pipe type cable.

This chapter does not deal with SF_6 insulated cables (which are not paper insulated), these being dealt with in chapter 41.

INTERNAL GAS PRESSURE CABLES

In this type of cable the insulation is saturated with nitrogen gas under a nominal pressure of 14 bar, which suppresses discharges which may otherwise have occurred in any cavities within the insulation. Two methods of manufacture are used, i.e.

(a) mass impregnation of the dried paper insulation with a viscous impregnant;
(b) the application of paper tapes pre-impregnated with a petroleum jelly type of compound.

The early designs of mass-impregnated cable were known as the impregnated pressure (IP) cable and the first commercial installation was carried out in 1940.[1] This type of cable was extensively used in the UK for operating voltages up to and including 132 kV during the 1940s and 1950s, first with a reinforced lead sheath and later with a smooth aluminium sheath.

Problems were encountered as a result of gas absorption by the impregnating compound, which caused frothing when the cable was being degassed, sometimes causing a blockage in the gas pipes or in the annulus below the metal sheath. After this occurred it was often difficult to re-pressurise the cable following repairs or diversions, the re-gassing process sometimes lasting many days before the cable could be recommissioned. Because of these problems and the improvements made in the self-contained fluid-filled cable, commercial applications ceased at the end of the early 1960s. The design has continued to be used in North America as a pipe type cable[2] in which migration of impregnating compound does not cause a significant increase in the impedance to gas flow because of the large amount of free space between the cable cores and the pipe.

Another variety of this type of cable, used in North America, is the low pressure gas-filled cable.[3] This is a self-contained cable which operates under a nitrogen gas pressure of 0.6–1 bar. The cable has an unreinforced sheath. After impregnation the cable is drained of surplus compound to avoid the problems associated with compound migration described earlier. Because of the low gas pressure, the maximum design stress of this type of cable is only slightly higher than that of solid type paper insulated cables.

The basic principles of the gas-filled (GF) cable[4,5] are that the conductor is insulated with pre-impregnated paper tapes which are processed to ensure that the butt gaps in the insulation are devoid of impregnant and that, prior to entering service, the dielectric is charged with nitrogen to a nominal pressure of 14 bar to suppress ionisation. Hence the gas is an integral component of the composite dielectric.

Substantial lengths of single- and 3-core gas-filled cables (fig. 31.1) are in use in the voltage range 33–138 kV and a limited amount of single-core 275 kV cable has been installed. The main attraction of this type of cable system is the relative simplicity of the ancillary equipment and jointing procedures.

Conventional stranded copper or aluminium conductors are normally used, but in some cases involving long routes of single-core cable, hollow-core conductors are used. All conductors are screened with metallised or carbon loaded paper tapes.

The insulating paper used for gas-filled cables is dried and impregnated in sheet form prior to slitting into tape widths. This process involves passing the paper over two rollers heated to about 200°C, which reduces the moisture content from about 8% to less than 0.15%, thence into a vacuum vessel, which degasifies the paper, and finally through a tank filled with a paraffinic jelly, which impregnates the paper. Before winding the dried and impregnated paper into roll form it passes over two knives heated to 90°C to scrape off all impregnating compound adhering to both surfaces of the paper. The large rolls of paper are slit into narrow tapes immediately prior to the start of the core insulating process.

The application of the pre-impregnated paper tapes follows standard dry paper lapping techniques, but as the core does not need to be dried and impregnated after the insulating process, the papers are dimensionally stable and the original lapping tensions

Fig. 31.1 3-core 33 kV gas-filled cable

are retained in the finished cable. The core therefore has good bending characteristics. The use of pre-impregnated paper tapes ensures that the butt gaps in the helically applied tapes are free of compound, thus providing a path for gas to permeate the dielectric when the cable is charged with nitrogen following the installation of the cable.

In 3-core lead sheathed gas-filled cables, impregnated and drained jute strings are laid into the interstices to allow gas to penetrate along the cable length. Aluminium sheathed gas-filled 3-core cables may be supplied without ducts or fillers in the interstices.

Initially gas-filled cables were provided with a lead alloy pressure retaining sheath and the application of metallic reinforcing tapes external to the sheath was necessary to enable it to withstand the internal gas pressure. At a later date gas-filled cables were introduced with smooth aluminium sheaths which, unlike the lead sheathed cables, did not require additional reinforcement against the internal gas pressure and these are now generally favoured. The thickness of the aluminium sheath is dictated by the bending requirements of the cable. Both lead and aluminium sheaths are applied with a small annular clearance over the core(s) to provide a gas channel from end to end of each cable length. Modern cables are protected from corrosion by an extruded oversheath of high or medium density polyethylene or PVC.

After installation, the gas pressure is maintained by a cubicle located at one end of the feeder, or, in the case of long cable routes, by a cubicle at each end of the route. The cubicle contains two or three cylinders of nitrogen at a pressure of about 150 bar connected into a manifold, the gas being fed into the cable at the base plate of the sealing end or into the trifurcating joint of 3-core cables. The gas feed is controlled by a two-stage pressure regulator and the pressure is monitored by pressure gauges which are equipped with electrical contacts to signal an alarm at a remote substation should the gas pressure fall below 12 bar. In an emergency the cable can be kept in service for a limited time provided that the pressure does not fall below 9 bar.

The gas-filled dielectric has a low relative permittivity (typically 3.2) but the dielectric loss angle is nearly twice that of an equivalent fluid-filled cable because of the higher moisture content of the papers. Figure 31.2 shows the effect of the thickness of the paper (and therefore the depth of the butt gaps) on the ionisation inception stress when under 14 bar nitrogen pressure at ambient temperature.

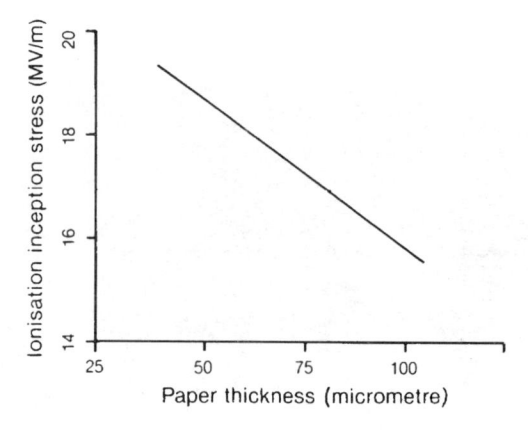

Fig. 31.2 Effect of paper thickness on ionisation inception stress at a gas pressure of 14 bar

The thickness of insulation applied to 33 kV gas-filled cables is 3.3 mm, this being dictated by mechanical considerations.

The minimum lightning impulse withstand stress of gas-filled cables is about 85 MV/m at the maximum permissible conductor temperature of 85°C and the lowest permissible gas pressure. It is this parameter which determines the thickness of insulation required for the higher voltage cables. Taking impulse withstand requirements into account, it is usual to design 66 and 132 kV gas-filled cables to produce a maximum stress at the conductor surface of 8.5–10 MV/m at normal working voltage. When operating at the normal gas pressure, the gas suppresses all discharge activity within the dielectric at all voltages up to about 1.75 times the working voltage.

The cable is substantially non-draining at the maximum conductor temperature and may therefore be laid on severe gradients without adverse effect.

The service record of gas-filled cables has been satisfactory since they were first introduced in 1937 and there is still a demand for this type of cable principally from utilities which already have this type of cable in their system. The relatively simple terminal equipment required, the ease of jointing and the ready availability of bottled nitrogen in most parts of the world are attractive features, particularly where specialised supertension cable jointers are not available at short notice to repair a damaged cable. Often a damaged cable can be left in service by maintaining a gas feed until repairs can be undertaken during a planned outage. Owing to the lower permissible electrical design stress and thicker metal sheaths, gas-filled cables are invariably more expensive than equivalent fluid-filled cables. This type of cable is also facing competition from the non-pressurised extruded polymeric cables.

EXTERNAL GAS PRESSURE CABLES

The gas compression cable[6] is designed to facilitate the use of high electrical stresses by maintaining the dielectric in the fully impregnated state under all service conditions by the application of gas pressure external to a diaphragm sheath to prevent void formation within the dielectric. It was first introduced as a self-contained cable, the gas pressure being applied between a diaphragm lead sheath applied directly over the cable and an outer reinforced gas retaining lead sheath. In recent years, gas compression cables are only used in pipe type cable systems, three individual cores or a 3-core cable being pulled into a pre-laid steel pipe which is then charged with nitrogen to a pressure of about 14 bar.

Oval shaped stranded conductors are used for single cores. The conductor is usually screened with carbon loaded paper tapes and insulated with paper tapes and the core is screened with metallised paper or non-ferrous metal tapes. The single-core or 3-core cable is dried and impregnated in a conventional cable vessel, the impregnant being a viscous compound. Metallic and fabric tapes are applied over the diaphragm sheath to reinforce it against the internal pressure created by thermal expansion of the compound and by pressure due to compound migration if the cable is laid on a gradient. For installation in pipes single cores are wrapped with D-shaped skid wires and 3-core cables with flat armour wires.

The original design of gas compression cable was excessively heavy as both diaphragm and pressure retaining sheaths were of lead alloy. In later designs, the use of polyethylene diaphragm sheaths in both self-contained and pipe type cables introduced other problems. As polyethylene is slightly permeable, gas slowly penetrates the diaphragm sheath and goes into solution in the impregnant. When depressurising the cable the gas in the dielectric creates a residual pressure which can result in distortion of the polyethylene sheath.

Because the self-contained gas compression cable requires both a diaphragm and a pressure-resistant sheath it is generally uncompetitive with other types of pressure-assisted cables. The gas compression cable is therefore now in rather limited use, being mainly restricted to use in Germany. Incidences of electrical failures have occurred at terminations and adjacent to joints. These failures have been attributed to the action of compound drainage.

The maximum design stresses used in the UK were as follows: for 66 kV cables, 8.5 MV/m; for 132 kV cables, 12 MV/m and for 275 kV cables, 15 MV/m. The maximum conductor size was 800 mm^2.

REFERENCES

(1) Brazier, L. G., Hollingsworth, D. T. and Williams, A. L. (1953) 'An assessment of the impregnated-pressure cable'. *Proc. IEE, Part 2* **100** (78).
(2) Association of Edison Illuminating Companies (AEIC) (Oct. 1982) 'Specifications for impregnated paper-insulated cable, high pressure pipe-type'. New York: AEIC CS2–82.
(3) Association of Edison Illuminating Companies (AEIC) (1969) 'Specifications for impregnated-paper-insulated, lead covered cable, low pressure gas-filled type'. New York: AEIC CS3–69.
(4) Beaver, C. J. and Davey, E. L. (1944) 'The high pressure gas-filled cable'. *J. IEE, Part 2* **91**.
(5) Thornton, E. P. G. and Booth, D. H. (1959) 'The design and performance of the gas-filled cable system'. *Proc. IEE A* **106** (27).
(6) Sutton, C. T. W. (1952) 'The compression cable'. Paris: CIGRE Paper No. 202.

CHAPTER 32
High Pressure Fluid-filled Pipe Cables

GENERAL DESCRIPTION

High pressure fluid-filled (HP FF) pipe type cables are used mainly in North America and the former USSR. This type of cable was pioneered in the USA and was known in its early days as 'Oilostatic' cable. Figure 32.1 illustrates samples of HP FF cables. The three insulated conductors are drawn into a steel pipe which is subsequently filled with liquid and maintained under a high pressure. Except at terminations, all cables are of 3-core construction, no ducts being required within the conductors or between cores.

The route has to be carefully chosen to keep bends to a large radius to prevent any damage to the cores, during pulling-in, arising from side pressure against the pipe wall. Joint bays may be spaced at longer intervals than for self-contained cables.

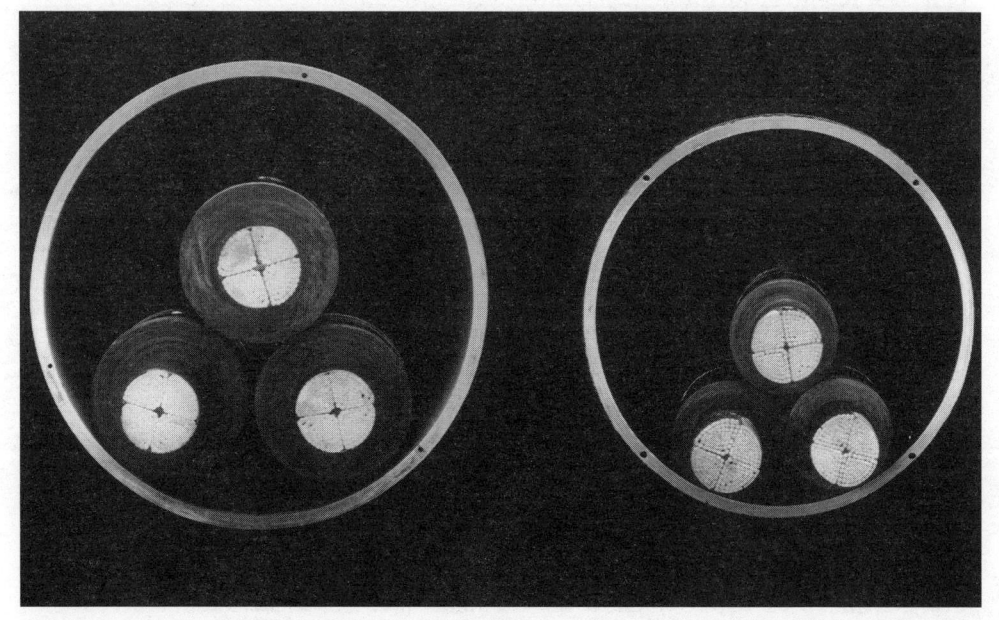

Fig. 32.1 Samples of 345 kV HP FF pipe type cable, paper insulated (*left*) and PPL insulated (*right*)

The basic principle of the HP FF cable is the same as that of the self-contained fluid-filled (SC FF) cable, i.e the insulation is kept fully impregnated at all times. To reduce drainage of fluid from the insulation during transit from the factory to site, the cable is impregnated with an impregnant which has a much higher viscosity than that used for the SC FF cable. To ensure that full impregnation of the insulation is maintained during cooling cycles, it is necessary to use high fluid pressures, the nominal pressure of present day systems being approximately 14 bar.

HP FF cables are at present in operation at voltages up to 550 kV (former USSR),[1] but the amount at the maximum voltage is very limited and the highest voltage at which significant quantities of cable have been installed is 345 kV (New York, USA).[2] Cables have been developed for 765 kV operation.[3]

Conductors

Both aluminium and copper conductors are used up to a maximum conductor size of approximately 1250 mm^2. Larger sizes are generally not economic because the closeness of the cores in the pipe gives rise to a high proximity loss in the conductor which significantly increases the a.c. resistance[3] (chapter 8). For the same reason, the change to the Milliken conductor construction is usually made at a lower conductor size than for SC FF cables. Normally the Milliken conductor is constructed from four stranded segments although conductors consisting of five segments have been used for experimental cables.

Insulation

The requirements for the insulating papers and lapping are generally similar to those already described for SC FF cables. However, in the case of HP FF cables there is an additional requirement for a very firm insulation to minimise deformation due to the side pressures which occur during installation, as previously mentioned.

The impregnant used for the cores would typically have a viscosity of about 3000 centistokes at 20°C and for the higher voltage cables is usually of the synthetic type.

An important requirement is that the cable cores are transported to the installation site without significant moisture pick-up. In the early days this was achieved by applying a temporary lead sheath on each core and this was stripped off as the cores were pulled into the steel pipe. This technique has been replaced by the application of a taped moisture barrier around the individual cores and the despatch of the cores in an atmosphere of a dried gas on sealed drums. The barrier is applied to the impregnated core and usually consists of several layers of metal and plastic tapes. To produce a low coefficient of friction between cores and pipe and hence minimise pulling tensions, a D-shaped wire, usually of bronze, is applied helically with a short lay over each core with the rounded surface outwards.

The permittivity, dielectric loss angle and thermal resistivity of the dielectric are essentially the same as for SC FF cables. The higher operating pressure gives an advantage in a.c. strength and the somewhat higher viscosity of the impregnant at maximum operating temperature gives a slightly higher impulse strength. In practice, however, paper insulated HP FF cables tend to be designed to slightly lower maximum stresses than SC FF cables to allow for the higher mechanical stresses imposed during installation.

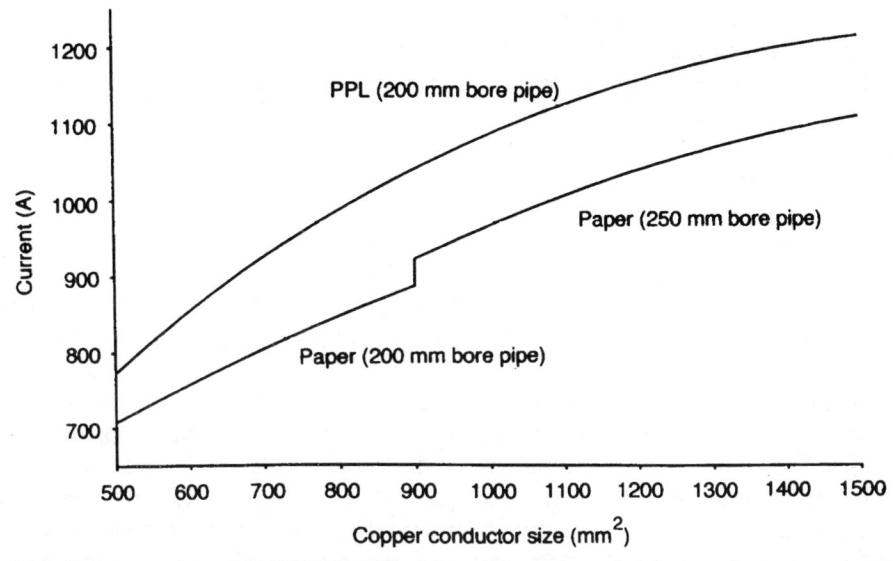

Fig. 32.2 Current rating of 345 kV HP FF cables with paper and polypropylene paper laminate insulation

HP FF cables using polypropylene paper laminate insulation have been developed in the USA.[4] The electrical characteristics are similar to those of SC FF cable insulation. Because pipes are made with diameters in discrete steps there is a significant advantage if the insulation can be reduced sufficiently for a smaller diameter pipe to be used. This not only arises from a lower pipe cost but also because the amount of pipe filling fluid is reduced. Therefore in development attention was directed to reducing insulation thickness to the practical minimum. The result of this is illustrated in fig. 32.1 in which a comparison is made between 345 kV cables with paper and polypropylene paper laminate. The insulation thickness of the polypropylene paper laminate cable is approximately 60% of that of the paper insulated cable.

Another benefit of using polypropylene paper laminate insulation is the reduction of dielectric losses and lower charging current. Reduction of dielectric losses is particularly important for pipe type cables because these are usually a greater proportion of the total losses than in self-contained cables. The effect of this reduction on the current rating is shown in fig. 32.2.

For the reasons given above the commercial use of polypropylene paper laminate insulation in pipe type cables preceded that in self-contained cables and is now widely used at 345 kV.[5]

STEEL PIPE AND INSTALLATION

The pipes are made of carbon steel complying with appropriate national standards for thickness and testing requirements. The size is chosen to permit adequate clearance between the three cores and the pipe. It is important to avoid an internal diameter of the pipe of about three times the diameter of the individual cores as it is possible for the cores to align across a diameter and jam during the pulling-in operation. A requirement of great importance is a need to have the internal surface smooth and clean so that on final filling with insulating fluid no serious contamination is introduced. Hence in manufacture the pipe must be cleaned and in some cases the internal surface is coated with a resin which is compatible with the pipe filling fluid. Corrosion protection follows general pipe line practice and usually consists of an asphaltic mastic applied to a thickness of about 12 mm. This is supplemented by the application of cathodic protection to the pipes on completion of the installation. The principles of cathodic protection are similar to those adopted for other steel pipe applications. However, the cathodic protection system must be capable of handling the high currents flowing to ground under fault conditions. This is achieved by using devices such as cadmium–nickel batteries or polarisation cells which have a high impedance to d.c. but a very low impedance to a.c.

Compared with the installation of self-contained cables, a major advantage in heavily congested city areas is that only a short stretch of roadway needs to be opened up at a time. Installation practice for the pipe is similar to general procedures for oil and gas pipelines in that sections are joined together before being lowered into the trench. A difference is that simple butt welding between pipes cannot be adopted because of the need for a smooth internal surface with no step or projections which might cause damage to the cores during pulling-in. The pipe ends may be flared or a bell and spigot type of joint may be used. The individual welds are tested as the pipe laying proceeds and then the whole length between jointing manholes is checked with dry nitrogen gas at high pressure.

Before cable installation proceeds, any moisture is removed by evacuation of the pipe. For the pulling-in operation the ends of the three cores are connected together and they are pulled into the pipe in one unit with close monitoring of the pulling tension. Depending on the ratio of core to pipe diameter, the cores may be in triangular or cradle formation.

When the laying, jointing and terminating have been completed, the pipes have to be filled with insulating fluid. This is carried out by evacuating the pipe to a low vacuum level and then filling with degasified fluid. Because of the large quantity of fluid contained in the pipe system, the unit used to degasify the fluid must have a large output. The fluid used to fill the pipe has a lower viscosity than that used to impregnate the insulation, a value of 160 centistokes at 20°C being typical. This viscosity, associated with the large cross-sectional area of the pipe, results in low longitudinal transient pressures and these can usually be ignored in HP FF cable system design. However when fluid circulation is employed for forced cooling, a low viscosity synthetic fluid similar to that used in SC FF cables may be used to reduce pumping pressures.

After the fluid filling operation the system is connected to a reservoir and a pumping module is installed, either at one end or both ends according to the length. The purpose of the module is to maintain the required fluid pressure in the system at all times. During a period of increasing cable temperatures, a pressure relief valve vents excess fluid from the cable system into the reservoir. When the temperature of the cable system falls, a pressure switch set at a pressure below that of the relief valve comes into operation to activate the pumps to maintain the required pressure in the cable system. The reservoir is of sufficient volume to accommodate fluid expansion from the cable system plus an additional amount to cater for fluid leaks. The reservoir has either a blanket of dry gas or is operated under vacuum to prevent deterioration or contamination of the insulating fluid.

SYSTEM DETAILS

In the USA, large underground manholes are prepared at the positions where the cores are pulled into the pipes and these subsequently serve as positions for jointing. They are usually of very large construction produced from precast concrete sections and are of sufficient size for jointers to work comfortably, 5.5 × 2 × 2 m being fairly typical. Although manholes are somewhat costly, the distance apart of jointing positions is usually longer than for self-contained cables.

Jointing practice is generally similar to that with other paper insulated cables, compression ferrules being used to joint the conductors and crêpe paper for the insulation. The three joints are accommodated in a single steel sleeve which bridges the main pipes. An important difference from self-contained fluid-filled cables is that, because of the higher strength of the steel pipe, no stop joints are required to limit internal pressure due to differences in route profile.

Termination is achieved by an accessory which connects the main pipes to three smaller pipes. These three separate pipes need to be of non-magnetic material, e.g. stainless steel, to prevent magnetic hysteresis losses. The porcelains used to terminate the single cables have to be designed to cater for the high internal pressure.

The pumping equipment includes duplicated pumps, automatic switching of pumps, low pressure alarms etc. and is connected through an appropriate monitoring system to a control centre.

CURRENT CARRYING CAPACITY

The current carrying capacity of HP FF cable is slightly inferior to that of 3-core self-contained fluid-filled cables and significantly lower than that of single-core SC FF

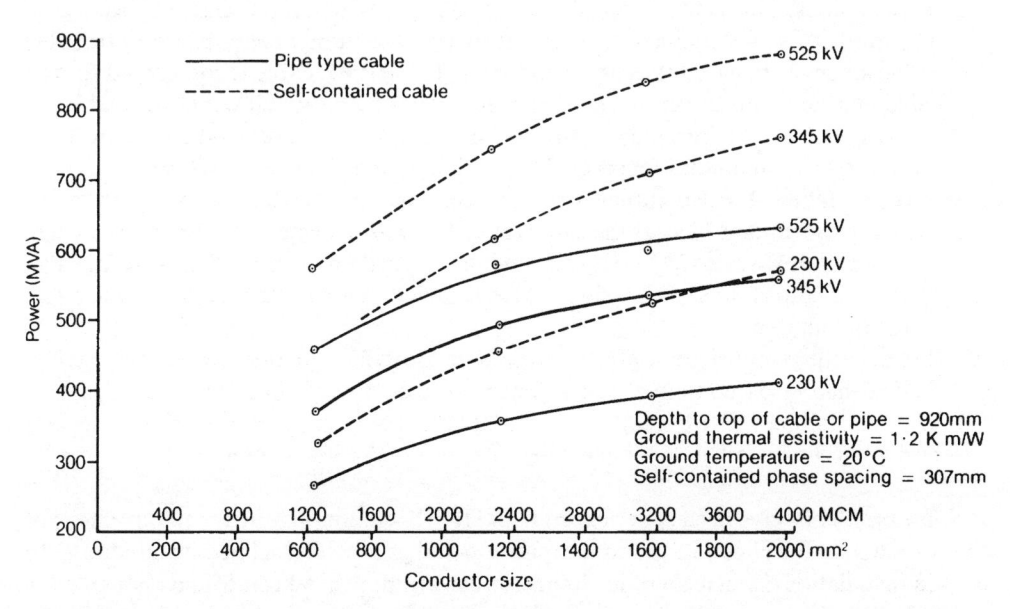

Fig. 32.3 Comparative ratings of paper insulated HP FF and single-core self-contained fluid-filled cables

cable. The latter aspect is illustrated in fig. 32.3. The lower current rating compared with single-core self-contained cable arises primarily because of two reasons.

(a) The proximity of the cores in the pipe causes the HP FF cable to have a higher a.c. conductor resistance and a higher thermal resistance to ground. In the case of SC FF cables both these effects can be reduced by increasing the spacings between the single-core cables.
(b) There are hysteresis losses in the steel pipe.

As with other types of cable, the ratings of the HP FF cable can be increased by forced cooling. The most widely used form of cooling is forced movement of insulating fluid along the pipe. In the case of a single feeder, a return pipe is laid alongside the cable and the fluid is circulated along the cable and through the return pipe and a heat exchanger before being pumped back into the cable. In the case of a double feeder, a return pipe can be eliminated by using the second cable for the return fluid.

Although not strictly forced cooling, fluid oscillation has been used to overcome localised thermal limitations.

APPLICATIONS

On strictly economic grounds, it is usually difficult to justify the use of HP FF cable systems in place of directly buried SC FF systems for the majority of applications. However, there are certain situations when customers prefer this type of cable. These arise as follows.

(a) In certain countries it is standard practice to install all self-contained power cables in duct banks. This is to permit easy removal in case cable replacement is necessary. The rating of self-contained cables in ducts is lower than when directly buried due to the air space between the cables and duct. The HP FF cable is considered to be a cable in a duct and therefore the cost comparison with the self-contained cables in this case is more favourable. A further point is that self-contained cables in ducts require special manholes (chapter 36) which can affect the comparison.
(b) When installing HP FF cables it is not necessary to excavate the complete cable trench between joint bays at the same time. This has considerable advantages when a cable circuit is required to pass through a crowded city centre. The principle also applies, of course, to self-contained cable installed in ducts, but the economic effect is less favourable.
(c) Some Utilities consider that the steel pipe provides greater protection against dig-ins than a self-contained cable system protected with concrete cover tiles.
(d) When installing HP FF cables it is possible to lay a spare pipe so that a second feeder can be installed at a later date without excavating a trench.

Some problems have been experienced with HP FF systems arising from movement of the cores due to the effects of thermal expansion and gravity on inclined installations. In certain installations, when there is change of elevation, it has been found that there is progressive movement of the cores towards the bottom of the installation. This can result in localised bending of the cores or disturbance at joints. Attempts have been made to overcome the problem by stretching the cores after installation to take up slack but it is probably best to avoid the use of HP FF cable on such routes. Somewhat similar difficulties have been experienced due to thermomechanical movement of the cores, particularly adjacent to joints. Most of the problems have been associated with 345 kV cables where localised flexing of the cable has caused the gaps between papers to increase, with resultant soft spots.[6] To overcome the problem the joints and adjacent cable are now reinforced to prevent the thermal expansion movement from being localised.

REFERENCES

(1) Gleizer, S. E., Goldobin, D. A., Kadomskaja, K. P., Khanukov, M. G., Obraztsov, Y. V. and Peshkov, I. B. (1982) 'Insulation development of oil-filled cables for heavy load transmission'. Paris: CIGRE Paper No. 21–08.

(2) Association of Edison Illuminating Companies (AEIC) (May 1990) 'Specification for impregnated paper and laminated paper polypropylene insulated cable high-pressure pipe-type' (5th edition). Birmingham, Alabama: AEIC CS2-90.

(3) *Underground Transmission Systems Reference Book* (1992) Electric Power Research Institute.

(4) Allam, E. M., Cooper, J. H. and Shimshock, J. F. (1986) 'Development and long term testing of a low-loss 765 kV high pressure oil filled pipe cable'. CIGRE Paper No. 21–06.

(5) Allam, E. M., McKean, A. L. and Teti, F. A. (1988) 'Optimised PPP-insulated pipe-type cable system'. *IEEE/PES Transmission and Distribution Conf.* Anaheim, California.

(6) McIllveen, E. E., Waldron, R. C., Bankoske, J. W., Matthews, H. G. and Dietrich, F. M. (1978) 'Mechanical effects of load cycling on pipe type cable'. *IEEE Trans.* **PAS-97** (3).

CHAPTER 33
Polymeric Insulated Cables for Transmission Voltages

Nearly all the 66 kV cable and much of the 132 kV cable for new installations is now of the extruded polymeric type, with XLPE as the usual insulation. In the UK, where there is long experience with pressure-assisted cable systems, there has been less incentive to use polymeric cables in the 66–132 kV range, but some substantial installations have been completed and the polymeric cable is regarded as an alternative to the pressure-assisted types for use when the circumstances of the particular installation favour it.

The operating electrical stresses for polymeric insulation are lower than for FF cables and the insulation thicknesses are therefore greater. This leads to the use of a greater volume of material in the cable and the absence of 3-core designs above 33 kV. One advantage is that there is no need for the auxiliary equipment required for pressure-assisted cable systems or for the same arrangements for system maintenance. Dispensing with insulating fluid may also be a factor in situations where the effects of fire have to be considered, or may influence users who have had leaks on old cable systems.

INSULATING MATERIALS

Thermoplastic (linear) polyethylene (PE) and crosslinked polyethylene (XLPE) have been the most commonly used polymeric insulants, while ethylene propylene rubber (EPR) has had limited use up to 150 kV. The low dielectric losses of polyethylene, both PE and XLPE, make it an attractive proposition at 220 kV and above. The importance of dielectic losses in high voltage cables is discussed in chapter 41. With a dielectric loss angle less than 0.001 and a relative permittivity of 2.3, polyethylene has a distinct advantage over impregnated paper and EPR.

XLPE has a major advantage over PE as the cables can be operated at higher temperatures. 90°C is the maximum sustained operating temperature for XLPE, compared with 70°C for LDPE, low density polyethylene. This has a significant effect on the relative current ratings. In France, some use has been made of HDPE, high density polyethylene, with a maximum sustained temperature of 80°C allocated; this reduces the difference with XLPE without achieving equality.

XLPE insulated cables are therefore preferred in most countries and are by far the most widely used type of polymeric cable. In France where there has been substantial experience with LDPE its use has continued because a degree of confidence has been developed. The manufacturing process for PE is simpler, without the complications of crosslinking. However, even in France there is now a move to the use of XLPE insulation. EPR is used in a few countries as it has the same sustained operating temperature (90°C) as XLPE.

DESIGN OF POLYMERIC CABLES

As XLPE insulation is by far the most widely used polymeric material it is intended to concentrate on the design aspects of this type of cable. However, many of the areas covered apply equally to cables with LDPE, HDPE and EPR insulation. Figure 33.1 shows an example of a 400 kV cable.

Conductors

Conductors are stranded using aluminium or copper wires. Sizes range from 150 mm^2 for the lower voltages to 3000 mm^2 at the higher voltages. Sizes up to 1000 mm^2 are usually of the compacted circular design and 1200 mm^2 and above of the Milliken type. These latter designs consist of four or five compacted insulated segments.

Fig. 33.1 400 kV XLPE insulated cable with a CSA sheath

Increasing use is being made of water blocked conductors. If a cable is damaged in service there is a possibility that, if the damage is severe enough, the conductor could be exposed and groundwater enter the cable. Although conductors are compacted during manufacture there are still channels between wires through which water can pass. In a normal conductor under particularly adverse conditions water could travel many hundreds of metres. To prevent the growth of water trees this length of cable would need replacing before the circuit could be returned to service.

To overcome this problem designs of conductor have been developed which contain special powders or tapes which swell on contact with water and so prevent the movement of water along the conductor. The materials used for this purpose must be specially selected so that they do not have an adverse effect on other materials in the cable during manufacture or in service. To ensure that the conductor connections in the accessories are effective the water blocking materials have to be removed at the jointing position when assembling the accessories.

Conductor screen

To ensure that the conductor presents a smooth surface to the insulation, a layer of semiconducting polymer is extruded over the conductor. It is essential that this is fully bonded to the insulation so that during the temperature changes that occur in service, no spaces develop between the screen and the insulation in which harmful partial discharges could take place. Particularly for the higher voltage cables which operate at the highest electrical stresses, the screen must be as smooth as possible to minimise any local high stresses; specially formulated 'super smooth' semiconducting compounds are used for these cables. As the screen is in intimate contact with the insulation it is important that no harmful chemicals can migrate from the screen into the insulation.

Insulation

Unlike the situation below 33 kV there are no international constructional standards which dictate the thickness of the insulation required at the various voltages. Nevertheless there is some reasonable correlation between the thicknesses used in most countries. Table 33.1 shows typical figures for the insulation thicknesses and maximum stresses used.[1]

There is no international agreement on the design basis for this class of cable. Some countries, notably Japan, believe that cables should be designed on the basis of mean stress in the insulation. This leads to the use of a constant insulation thickness for all conductor sizes at a given voltage. Other countries mainly design on the basis of

Table 33.1 Relationship between cable voltage and insulation thicknesses

Voltage (kV)	Insulation thickness (mm)	Maximum stress (MV/m)
66	12	3–6
132	18–22	5–8
275	22–32	8–10
400	25–32	13–14

maximum stress at the conductor screen which results in different insulation thicknesses for different conductor diameters. In certain cases limits are set for the stress at the outside of the insulation as this has an important influence on the design of accessories (see chapter 35). The criteria for the design of these cables are discussed in reference 2.

There is however universal agreement that to achieve the highest design stresses it is necessary to have:

(a) no cavities within the insulation or at the interfaces with the screens;
(b) the lowest possible level of contamination within the insulation and screens;
(c) the smoothest interfaces between the screens and insulation to minimise local stresses;
(d) the best insulation circularity and concentricity to minimise stress variations and to facilitate the precise assembly of accessories.

While some of the reasons for the selection of higher design stresses at the higher system voltages are the less onerous lightning impulse voltage requirements (see chapter 30), much is due to the improved manufacturing and material handling techniques developed for this class of cable.

Design stresses are based on the results of short time a.c. tests, impulse tests and long term tests. Statistical analysis is used because the mechanisms responsible for the reduction of XLPE strength with life have not been fully identified or characterised. First, a 'threshold' or minimum value of a.c. breakdown stress is identified by the short-term voltage ramp testing of many cable samples. Second, the reduction factors due to operation at 90°C are determined and applied. Third, this stress is extrapolated to the required 30 to 40 year cable design life using data from accelerated long term endurance testing. Fourth, the residual stress at the design life must be greater than the cable design stress. The Weibull distribution method is described in chapter 25. As the short-term breakdown voltages of 275 and 400 kV cables are extremely high, results are extrapolated from tests on model cables with thinner insulation, made using the same materials and manufacturing plant.

Insulation screen

To ensure that the electrical field is confined to the polymeric insulation an extruded semiconducting polymer screen is applied overall. The material used is similar to that used for the conductor screen. In contrast to the strippable screens used on lower voltage cables, the screens on this class of cable are fully bonded to the insulation to avoid any cavities forming between the insulation and the screen.

Outer covering

It is now generally agreed that transmission voltage cables should be provided with a moisture impermeable metallic barrier to prevent the possibility of water trees developing in service, the growth of trees will be accelerated by the higher stresses used. Extruded metallic sheaths have been widely used for this purpose as they also provide the essential earth return path along the cable for both a.c. capacitive currents and short-circuit fault currents. These sheaths have the excellent mechanical robustness necessary to withstand side wall loads during installation, thermomechanical fatigue strain, and impact damage in service. The sheath can be either a smooth lead alloy or corrugated aluminium. Discrete annular corrugations are preferred, rather than helical ones, to restrict the passage of water in case the sheath is penetrated in service by external damage. As a further longitudinal barrier to the passage of water it is now normal practice for cables with both types of sheath to be provided with water swellable tapes applied over the insulation. In the case of water accidentally being introduced into the cable the tapes swell and so prevent water moving along the cable. The tapes, which are semiconducting to ensure that the insulation screen is at earth potential, also provide mechanical protection for the cable core by absorbing the radial thermal expansion of the insulation and by cushioning side wall loading at bends.

For cables up to 132 kV the use of a metallic foil for the impermeable barrier is beginning to be used instead of an extruded sheath, particularly if the cable is not to be buried directly in the ground. The foil is coated with a special polymer so that when it is wrapped around the cable and then covered with an extruded polymeric sheath, the foil bonds both to itself and to the extruded sheath. To achieve these requirements the foil is applied longitudinally to the core immediately prior to the extrusion of the polymeric sheath. The foil is not usually of sufficient thickness to provide a low resistance path for fault currents and it may be necessary to apply a layer of copper wires to meet this requirement. While this type of covering may have a lower cost than an extruded metallic sheath, it does not provide the same degree of protection and so extensive mechanical and electrical type test procedures are recommended.[3]

In the case of an extruded metallic sheath, an oversheath of PVC or medium to high density polyethylene is used. The oversheath can be modified to include chemicals to reduce the propagation of flame in the case of fire, or to prevent attack by termites as appropriate.

MANUFACTURING PROCESSES

The main difference in the manufacture of polymeric transmission cables compared with other types is in the application of the insulation and screens. Unlike paper insulated cable, extruded polymeric cable is inaccessible to visual in-process inspection whilst in the die head and crosslinking tube; thereafter, it is covered by the outer black insulation screen. The specified routine electrical test performed on each completed cable is only intended to confirm that a minimum level of performance has been achieved during the short-term over voltage test (chapter 39). Thus the quality of the resulting cable is largely dependent upon the type and condition of the manufacturing plant and in particular upon the quality system in the cable factory. The key requirements are:

(a) manufacturing plant designed without compromise for supertension cable;
(b) precise and predictable process control;
(c) routine laboratory inspection of incoming and in-process insulation and screening materials;
(d) routine laboratory dissection and microscopic analysis of samples as a routine from each cable;
(e) stringent routine HV test regime on each cable;
(f) periodic sample testing of cable for statistical HV breakdown tests;
(g) type approval and prequalification test certification.

Considering the manufacture of XLPE cables first, the operation consists of five main stages: preparation of the insulation and screening compounds, the extrusion of the insulation and screens, heating the cable to initiate the crosslinking of the polymer, cooling the cable to near ambient temperature, and degassing the cable.

It is essential that the insulation and screening compounds are prepared with the minimum level of contamination. Therefore the materials are fed to the extruder hoppers via a sealed pipe system. For the lower voltage low stress cables, suitable quality pre-compounded 'superclean' insulating and 'supersmooth' semiconducting materials can be purchased. They are transported comparatively long distances to the factory and then unloaded and sampled under clean room conditions prior to entering the enclosed transportation system to the extruder. For the higher voltage cables which operate at much higher stresses the key factor is to achieve the highest purity of materials. To obtain this standard both the insulation and screening material are filtered by extrusion through the finest possible mesh before the addition of the crosslinking agent. After filtering, the materials must remain within a fully enclosed storage system; the crosslinking agent is then added. The fully compounded ultra clean material is then conveyed through a fully enclosed storage and transportation system connected to the extruders. The compounding process is best carried out in the cable factory immediately adjacent to the extrusion plant. For both low and high voltage cables it is good practice to verify material cleanliness with either an insulation pellet checker in-line with the extruder and an automatic contamination counter on an extruded tape, or on the melt at the exit from the main extruder.

The next three manufacturing stages are carried out in one operation. For the highest voltage cables the conductor is drawn through a triple extrusion die box where the semiconducting conductor screen, the insulation and the outer semiconducting screen are applied together, each layer being injected from a different extruder. For the lower voltages, tandem extrusion can be used in which the conductor screen and a thin layer of insulation are applied some 2–3 m from the second die box where the bulk of the insulation and the outer screen is applied. It is usual for the insulation 'melt' to be filtered through a 'strainer pack' upon leaving the extruder and before entering the die box. To avoid the crosslinking reaction being initiated by shear heating of the compound as it passes through the mesh, this filter has a much larger mesh size than that used during compound preparation.

Dicumyl peroxide is used as the crosslinking agent. To activate the process the cable must be heated to above 200°C. As explained in chapter 3 the crosslinking process generates volatile by-products and, to avoid these creating cavities in the insulation, external pressure must be applied during the heating and subsequent cooling operations to ensure that the gaseous by-products remain dissolved in the insulation.

In the early days of XLPE cables, steam was used as both the pressurising and heating medium, however it was soon realised that this resulted in an insulation with a high moisture content. In consequence, dry curing processes were developed: three basic types of plant are outlined below.

Continuous catenary vulcanisation (CCV)

Figure 33.2 shows the arrangement for a typical CCV line.

The conductor passes through the extrusion equipment directly into a large diameter tube in the shape of a catenary which contains nitrogen gas at a pressure of about 8 bar.

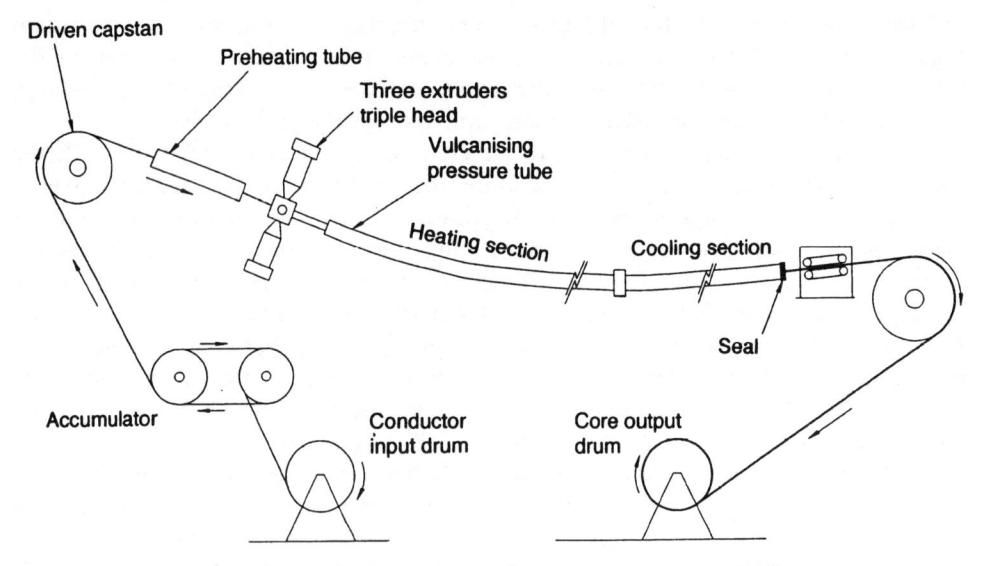

Fig. 33.2 Arrangement of a continuous catenary vulcanisation (CCV) line

To achieve the temperature for activating the crosslinking agent, the first section of the tube is electrically heated. The core is prevented from touching the sides of the tube by accurately controlling the tension in the conductor. On completion of the crosslinking operation the cable passes into the cooling section of the tube in which water or nitrogen gas is circulated. Immediately after extrusion and in the initial stages of crosslinking the polymer is a viscous liquid. Gravity causes it to sag giving the tendency for an oval, eccentric shaped insulation. Methods of reducing this effect are for the cable to be slowly rotated during its passage through the tube and for a more viscous polymer to be used. Alternatively the tube can be filled with silicone oil which gives buoyancy to the insulation thereby counteracting the effect of gravity.

The CCV process has the advantage that a long line can be installed and is usually optimised for a limited range of cable sizes for which a high output speed can be achieved. It is widely used for cables up to 150 kV with comparatively small conductor sizes. For higher voltage cables, where conductor sizes and insulation thicknesses are greater, a vertical line is generally preferred.

Horizontal line (MDCV)

Figure 33.3 shows the arrangement of the horizontal line initially developed by Mitsubishi Dainichi and known as the MDCV process.

The conductor passes through a triple extrusion die head. The insulated conductor then passes into a closely fitting tube approximately 10 m in length. The tube is heated to approximately 200°C to transfer heat directly to the insulation and initiate crosslinking. To ease the passage of the core through the tube, a special lubricant is injected into the tube after the extrusion die head. Each cable diameter requires a special die tube. After crosslinking the cable passes into a horizontal tube through which cooling water is circulated.

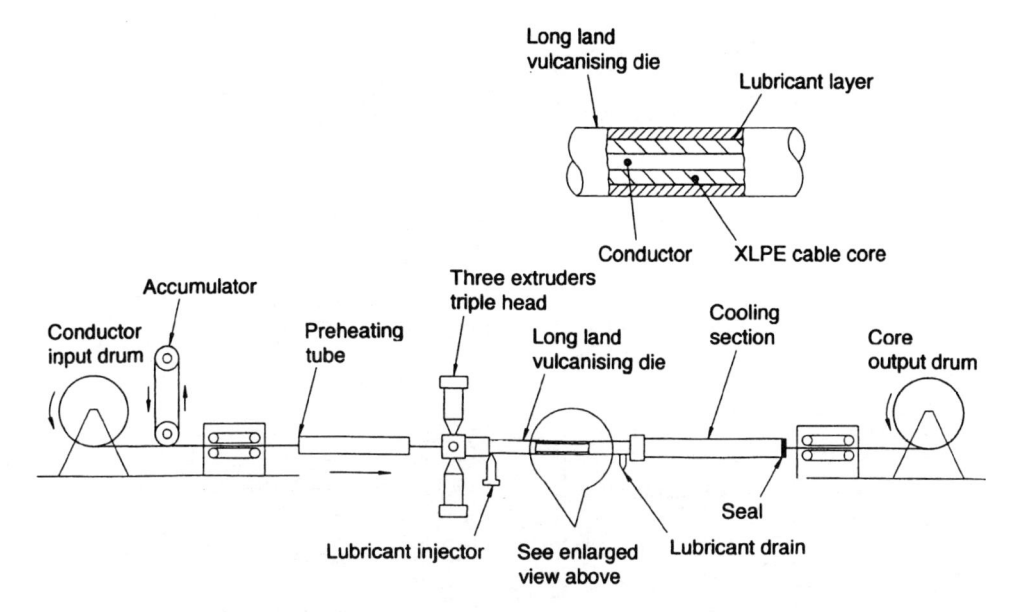

Fig. 33.3 Arrangement of an MDCV extrusion line

The capital cost of the line is relatively low compared with other processes, but suffers from a problem similar to CCV. In this case it is the tendency for the conductor to droop through the insulation. To minimise eccentricity, similar techniques are used as for CCV, i.e. conductor twisting and the use of high viscosity polymers.

Vertical continuous vulcanisation (VCV)

The plant is illustrated in fig. 33.4.

The conductor passes vertically downwards through the triple extrusion die head into a nitrogen gas pressurised tube. Crosslinking is activated by electrically heating the tube and also by circulating hot gas. On completion of crosslinking, the cable passes into the cooling section of the plant.

There is no tendency for the insulation to droop because the cable moves vertically through the plant. It is thus possible to achieve a cable with the high level of concentricity and circularity which is essential for cables operating at high electrical stress. A greater range of polymers can be used because it is not necessary to select a high viscosity material as used for CCV and MDCV extrusion. It is particularly suitable for the manufacture of cables with large conductor sizes and is the preferred process for EHV cables. The capital cost is high due to the necessity of providing a vertical tower to house the plant as illustrated in fig. 33.5.

The final stage of manufacture associated with extrusion is the degassing of the extruded core. The gaseous by-products of crosslinking are dissolved in the insulation during the extrusion process and to obtain optimum performance it is necessary for them to be removed. This is carried out by heating the drum of extruded cable in a ventilated oven for a process time period dependent on the insulation thickness.

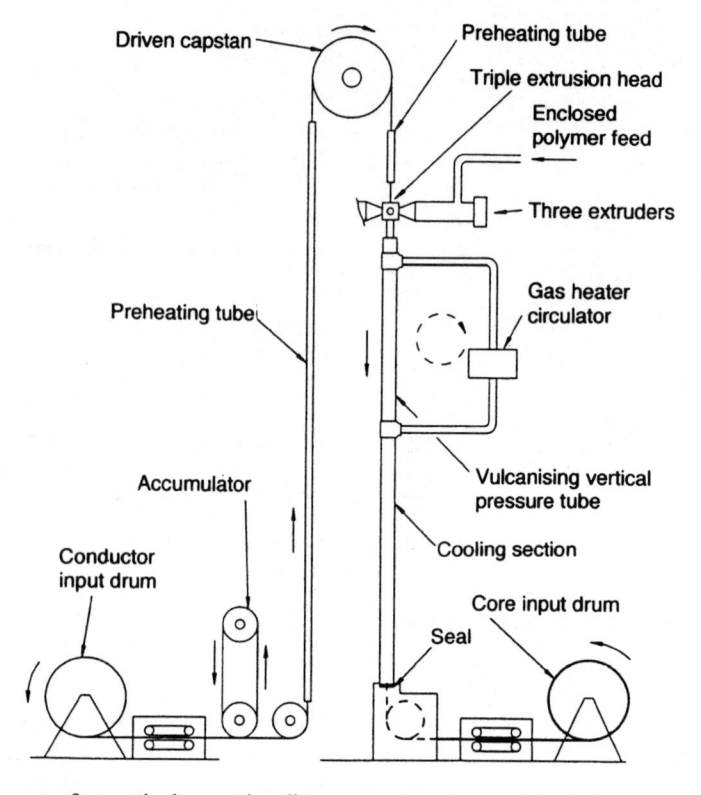

Fig. 33.4 Layout of a vertical extrusion line

PRESENT POSITION OF POLYMERIC INSULATED CABLES

Polymeric insulated cables are now widely used at voltages up to and including 150 kV. In most countries XLPE insulation is preferred, but a few use cables with EPR insulation.

At higher voltages the initial developments used LDPE. This is a much simpler system than XLPE and enabled high degrees of cleanliness and void free insulation to be obtained more rapidly. This design of cable was pioneered in France and has been developed up to a voltage of 500 kV. In 1993 it was reported that 1100 km of 225 kV cable together with short lengths of 400 and 500 kV cables had been supplied.[4] Generally conductor sizes are below 1000 mm^2.

Development of XLPE cables followed later than that of LDPE cables due to the more complex manufacturing process. A further complication was the requirement in many countries for conductor sizes of 2000 mm^2 and greater. Much of the initial development was carried out in Japan where there is emphasis on the need for cables with reduced fire risk due to the installation of multiple circuits in tunnels. Long lengths of 275 kV 1400 mm^2 cable with joints were installed in the late 1980s and 275 kV 2500 mm^2 cable in the early 1990s.[5] Short lengths of 500 kV cable without joints were first installed in 1986[6] and with joints in 1996. The advanced material and manufacturing technology developed to make EHV XLPE cables and accessories possible is

Fig. 33.5 Vertical extrusion tower and compounding plant

now being applied to HV XLPE cables. At 132 kV, significant reductions in insulation thickness are being made, making possible smaller single core cable, the introduction of three-core cable and longer drum lengths.[7]

In the development of polymeric cable systems, the availability of cable designs has usually preceded that of proven joints. The development of joints suitable for site assembly, as well as meeting the technical standards, has proved to be a difficult task and in certain cases has delayed the application of major polymeric cable systems. The position has now largely been resolved, this being dealt with in chapter 35.

The manufacture of cables with EPR insulation follows the same stages as that described for XLPE cables. As this type of cable is not usually used for voltages in excess of 150 kV, they are normally manufactured on CCV plants.

Cables with linear polyethylene (LDPE and HDPE) do not require crosslinking or degassifying processes. For EHV cables with large conductors, vertical extrusion is the preferred process for similar reasons to those given for the manufacture of XLPE cables.

REFERENCES

(1) CIGRE WG 21-09 (1991) 'Working gradient of HV and EHV cables with extruded insulation and its effect'. *Electra* (139), 63.
(2) CIGRE WG 21-04 (Not yet published) 'Criteria for electrical stress design of HV cables'.
(3) CIGRE WG 21-14 (1992) 'Guidelines for tests on high voltage cables with extruded insulation and laminated protective coverings'. *Electra* (141), 53–65.
(4) Argaut, P., Auclair, H. and Favrie, E. (1993) 'Development of 500 kV low density polyethylene insulated cable'. *3rd IEE Int. Conf. on Power Cables and Accessories 10 kV–500 kV*. pp. 77–81.
(5) Kaminaga, K., Asakura, T., Ohashi, Y. and Mukaiyama, Y. (1986) 'Development and installation of long distance 275 kV XLPE cable lines in Japan'. Paris: CIGRE Paper No. 21-103.
(6) Sato, T., Muraki, K., Sato, N. and Sekii, Y. (1993) 'Recent technical trends of 500 kV XLPE cable'. *3rd IEE Int. Conf. on Power Cables and Accessories 10 kV–500 kV*. pp. 59–63.
(7) Bartlett, A. D., Attwood, J. R. and Gregory, B. (1997) 'High standards for modern HV power cable system environments'. CIRED, 14th Int. Conf. and Exhib. on Electricity Distribution, *IEE Conf. Pub 438*, Paper No. 3.5.1.

CHAPTER 34

Techniques for Increasing Current Carrying Capacity

The high cost of supertension cables relative to overhead lines means that underground cables account for less than 10% of the UK's transmission circuits.[1] Hence transmission voltages and power ratings are selected to give the most economic overhead rather than underground transmission system. Cable ratings are normally dictated by the current rating of the overhead line to which they are connected.

In the UK the overhead line ratings have been progressively increased and ratings have been specified for 50°C, 65°C and 75°C operation. A selection of the resulting overhead line continuous pre-fault ratings for 275 and 400 kV lines is shown in table 34.1.

UK overhead lines also have a post-fault rating which they are required to maintain for a period of up to 24 hours, these being 17% higher than the pre-fault ratings. It has

Table 34.1 Transmission capability of 275 and 400 kV UK overhead lines

		Rating (MVA) at three operating temperatures		
		50°C	65°C	75°C
275 kV, 2 × 400 mm² conductors	Cold weather	795	900	960
	Normal weather	740	855	920
	Hot weather	640	775	850
400 kV, 2 × 400 mm² conductors	Cold weather	1150	1310	1400
	Normal weather	1070	1240	1340
	Hot weather	930	1130	1240
400 kV, 4 × 400 mm² conductors	Cold weather	2310	2610	2790
	Normal weather	2150	2480	2680
	Hot weather	1860	2250	2470

Ambient temperatures: cold weather <5°C; normal weather 5–18°C; hot weather >18°C

become practice to match the continuous cable ratings with the short term post-fault rating of the overhead line. It should be noted that this is a safe situation because the cables have a higher thermal capacity and can also sustain an increased short term loading dependent upon the particular circuit conditions.

Many developments in cable design, system design and installation practices have been made to enable these increases in continuous ratings to be met (where possible with one cable per phase).

The four most important aspects which have enabled cable systems today to meet the wide range of rating requirements at voltages from 132 to 550 kV have been (a) the use of special backfills, (b) attention to sheath bonding on single-core cables, (c) the use of low dielectric loss insulating materials, and (d) artificial cooling of the cable by the use of external water pipes or internal fluid circulation.

SPECIAL BACKFILLS

Background

It was customary in the UK, up to the early 1960s, to assume a constant value for the soil thermal resistivity along the whole length of the cable route, and to calculate the conductor size required to carry the load continuously without exceeding the maximum design conductor temperature. Experience had shown that for most installations an assumed value of thermal resistivity of 1.2 k m/W had resulted in perfectly reliable cable systems. However, in July 1962 there were two successive failures of 132 kV transmission cables in the London area. Both circuits had been operating at or near full load almost continuously during the summer and the backfill in the region adjacent to the hot cables had dried out. The consequent increase in soil thermal resistivity and resultant overheating of the cables led to thermal runaway conditions and cable failure.

Because of these failures, considerable work was carried out to develop backfills for the cable surround which have a low value of thermal resistivity under fully dried conditions. These backfills are now grouped into two classes as follows.

Selected sands

These are sands obtained from approved selected sources so that the *in situ* thermal resistivity in the dried out state is not greater than 2.7 K m/W.

The category of selected sand has been arrived at from consideration of the availability in the UK of naturally occurring sands with good cohesion properties and of the general relationship between thermal resistivity and density of such sands in the dry state. The sand needs to be of a coarse type with a mixture of particle sizes.

Stabilised backfills

Stabilised backfills are composite materials specially selected so that the *in situ* thermal resistivity in the dried out state is not greater than 1.2 K m/W.

At present two main types of stabilised backfill are used: cement-bound sand which consists of selected sand mixed with cement in the proportions 14:1, and a sand–gravel mixture consisting of a 1:1 mixture of selected sand and gravel. The latter should have a particle size nominally of 10 mm, preferably rounded, with not more than 50% crushed material.

UK practice

Where required by rating and installation conditions, allowance for migration of moisture is today made by assuming that the thermal resistivity of all materials surrounding the cable within the 50°C isotherm will have a value corresponding with a dry condition.[2] However, the cost of stabilised backfills is approximately three times that of selected sands, and therefore special backfills are only used where they can be justified economically. In practice, therefore, for 132 kV and below, stabilised backfills are not normally used unless heavy loading is envisaged during summer conditions. At 132 kV the cable bedding is generally a selected sand and below 132 kV traditional locally available sands and riddled soil are usually used for backfilling. For 275 kV and higher voltage cables, where the dielectric losses tend to maintain elevated temperatures during periods of low load, stabilised backfills are now the normally accepted materials for the cable surround.

The typical improvement in rating of a 132 kV circuit, which results from the use of special backfills, can be seen in fig. 34.1. This shows the comparison between placing selected sand and stabilised backfills within the 50°C isotherm, in an indigenous soil having a normal thermal resistivity of 1.2 K m/W and dried out thermal resistivity of 3.5 K m/W. It will be seen that in this case the use of selected sand results in an increase

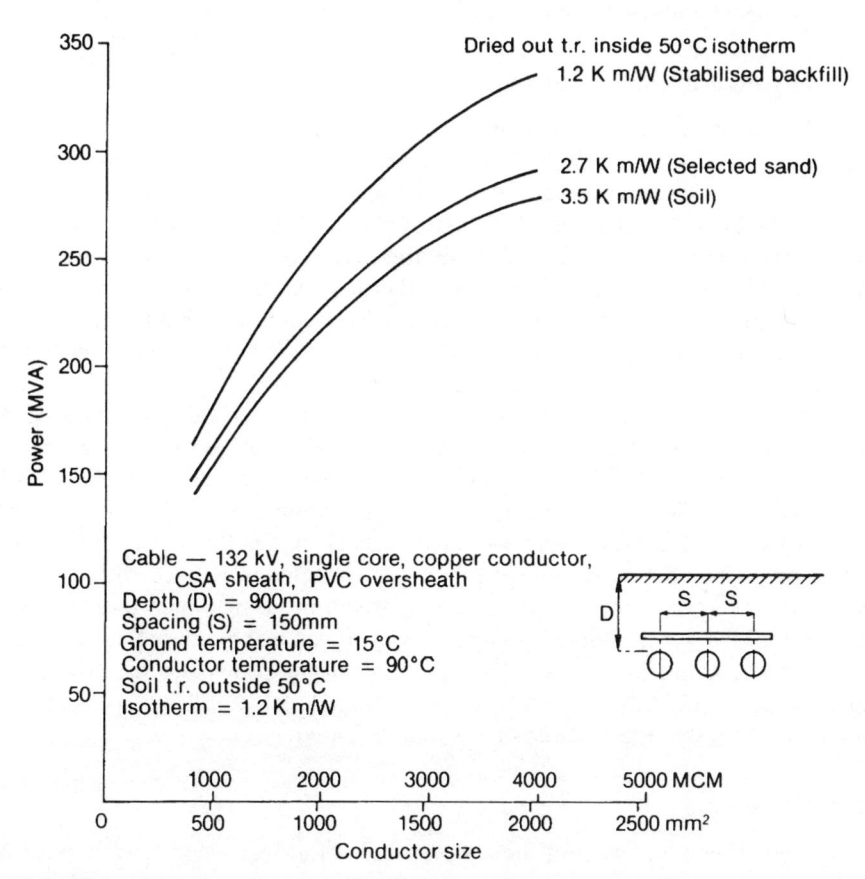

Fig. 34.1 Effect of dried-out thermal resistivity (t.r.) on the power rating of laid-direct 132 kV FF paper insulated cable installation

in rating of approximately 3%, and the use of stabilised backfill in an increase in rating of between approximately 15% and 20%, depending on conductor size. The method of calculating these ratings is based on that given by Cox and Coates.[3]

Extension of UK practices overseas

Around the world a wide variation in soil thermal resistivities may be encountered, from about 0.6 to 3.5 K m/W. The value for any particular installation has to be based on local measurements, and the calculation of current ratings normally assumes that this thermal resistivity remains constant. For instance, in Hong Kong the accepted value for rating calculations is 0.9 K m/W, which has been determined from a considerable amount of work by the local supply authority. However, for their first 400 kV installation it was considered that the effects of soil moisture migration should be allowed for. Thermal resistivity measurements along the route showed that the ground where the trench would be dug had a thermal resistivity of no greater than 0.7 K m/W and measurements of fully dried soil samples showed these to have thermal resistivities as low as 1.2 K m/W. The rating was therefore based on the soil within the 50°C isotherm being fully dried out but with a general ground thermal resistivity outside this region not greater than 0.7 K m/W. Thus, in this case, the soil was used for direct backfilling of a 400 kV circuit.

Reduction of ground thermal resistivity

Where high resistivity soils are encountered, such as in areas of the Middle East, replacing the cable surround with low thermal resistivity material can enable the required rating to be met with a reduced conductor or trench size, resulting in an overall cost saving. The lower thermal resistivity material placed in the trench and the thermal resistivity of the ground will give an 'effective' thermal resistivity value between the two. It is not possible, however, to calculate the thermal resistance of the elements in a multiple thermal resistance backfill arrangement by means of a simple mathematical formula. A number of computerised methods of solution are possible; one developed by Winders[4] enables the effective ground thermal resistivity to be deduced. This value can then be used to calculate the thermal resistances by the method given in IEC 287.[5] Another is to model the cable and trench using finite element analysis.

A typical example of the effect of placing lower thermal resistivity material in the trench is shown in fig. 34.2 where, for the 132 kV cable configuration shown, the effective thermal resistivity of the middle cable to ground is given for the cables buried in soil of thermal resistivity 3.5 K m/W and with the trench progressively filled with material of 1.2 K m/W. Also shown in fig. 34.2 is the improvement in rating which results from the reduction in the effective ground thermal resistivity. It can be seen here that the effect of placing lower thermal resistivity materials in the trench is to increase the power rating from approximately 125 MVA to a maximum of 170 MVA ($\approx$36% increase) depending on the amount of lower thermal resistivity material used.

Bentonite filled ducts

The air space between a cable and the inner surface of a duct has relatively poor heat transmission properties and this results in a reduced rating for cables installed within ducts compared with laid-direct systems.

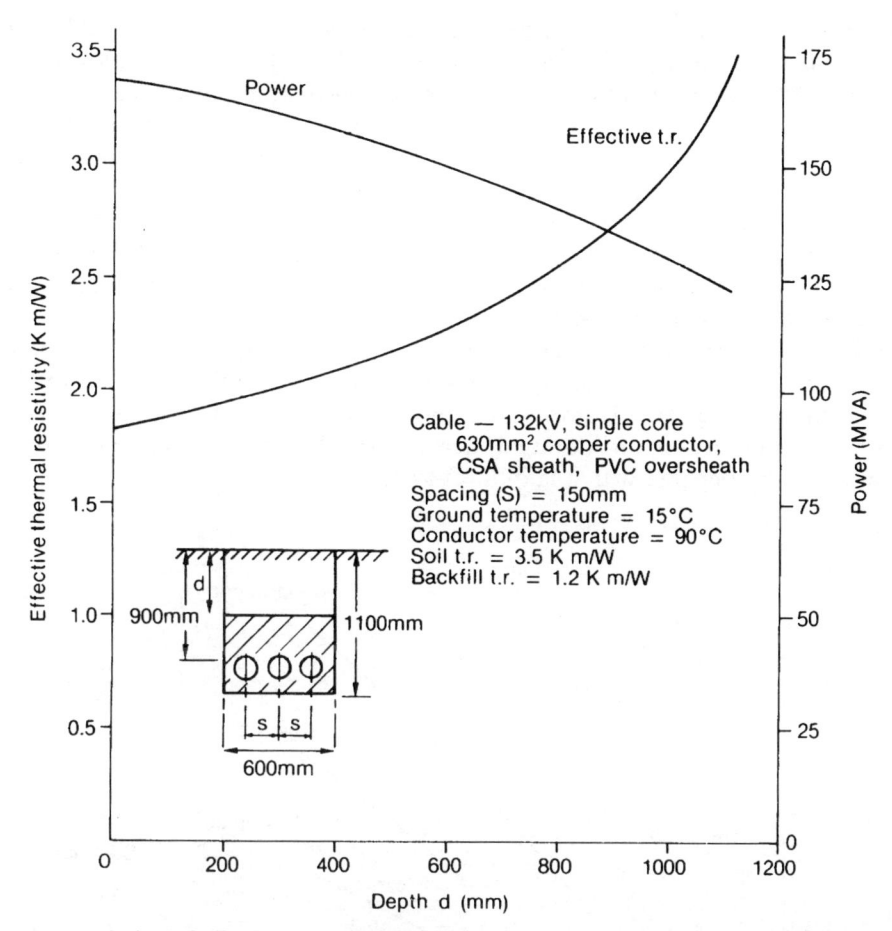

Fig. 34.2 Variation of effective ground thermal resistivity and power rating with depth of special backfill for a 132 kV FF paper insulated cable

Today this thermal problem has been eliminated on relatively short duct runs (less than about 70 m) by replacing the air with a mixture of bentonite, cement, sand and water. This mixture can be pumped into the duct using standard pressure grouting techniques. When sealed within the duct it remains a gel which can be flushed out using water jets.

The grout is sufficiently stiff to provide a constraint against thermomechanical movement of the cable and it was for this reason that it was first adopted.

The ducts must be effectively sealed to prevent loss of the filling medium and also to preserve its moisture content under service conditions. Provided that the moisture is retained, the thermal resistivity of the bentonite grout will be less than 1.2 K m/W. The use of bentonite results in an increase in rating of approximately 10%.

Troughs

The availability of stabilised backfills has encouraged the adoption of an alternative method of installing cables in filled surface troughs. Considerable savings in excavation costs are possible and troughs have been used extensively in the UK for cable routes alongside railways, canals, in switchyards and in rural situations. The shallow troughs are formed from concrete with thick reinforced concrete lids to afford mechanical protection. The filling can consist of cement-bound sand or a sand–gravel mix.

Thermally, a trough system can be regarded as a shallow laid-direct system. However, conventional current rating procedures are not directly applicable, because the trough lid is not isothermal due to the close proximity of the cables. The rating is also much more influenced by changes in climatic conditions than a deeply buried cable. Experimental work has shown that these effects are best allowed for by adopting a higher effective ground temperature.

For a trough system with an effective ground temperature of 40°C and a thermal resistivity of 1.2 K m/W, the resulting ratings are very similar to those of the laid-direct system as shown in fig. 34.3. However, if a lower effective ground temperature of 10°C can be assumed (such as occurs during winter) the trough installation generally gives an improved rating compared with a laid-direct system.

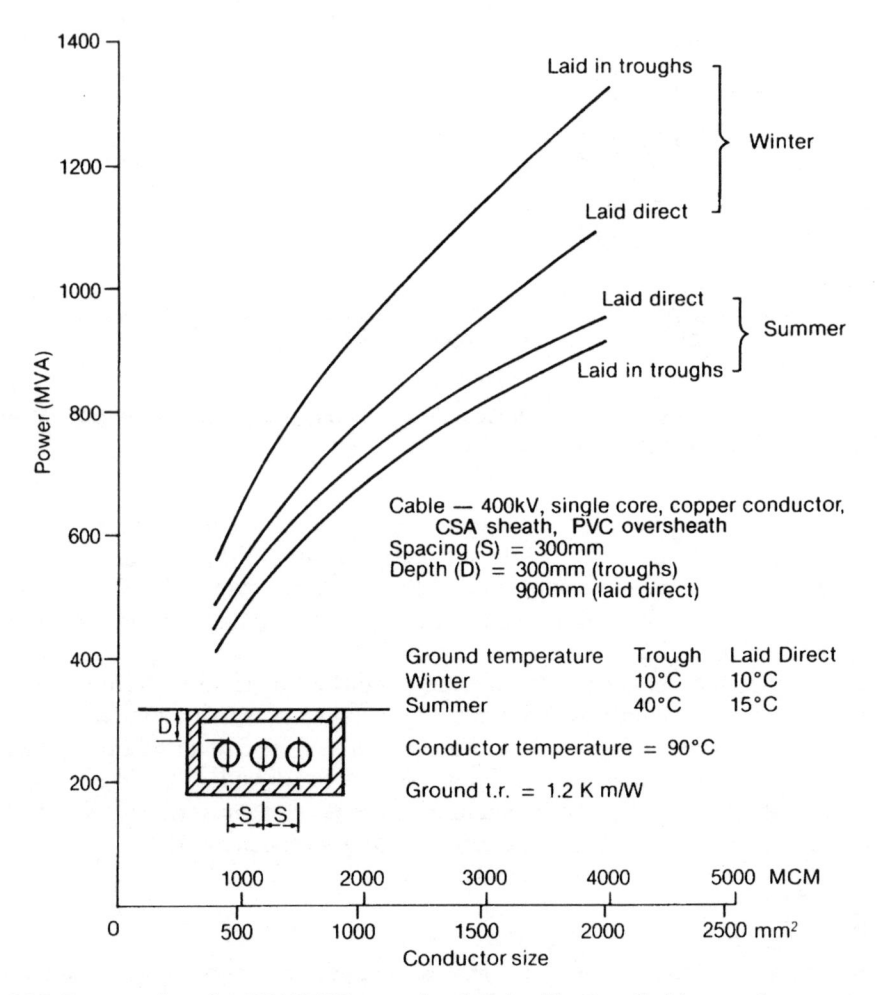

Fig. 34.3 Power ratings for 400 kV FF paper insulated cables installed in troughs

SPECIALLY BONDED CABLE SYSTEMS

Distribution voltage cables are normally installed with solidly bonded sheaths and, in order to minimise the sheath circulating currents on single-core cables produced by the magnetic flux linking the conductors and sheaths, they are nearly always laid in close touching trefoil formation. However, trefoil formation is poor for heat dissipation, as the three cables have a considerable heating effect upon one another. This is generally not a limitation for cable systems at 33 kV but with larger conductor sizes and higher voltages alternative 'specially bonded' systems[6] are more economic. At 275 and 400 kV they often provide the only practical means of meeting the required ratings with natural cooling.

Special bonding involves earthing the single-core cable sheaths at one point only and insulating all other points of the sheath from earth, so that the circulating sheath losses are eliminated and the phase cables can be spaced apart to reduce their mutual heating effect without increasing sheath losses.

When the sheaths are solidly bonded at both terminations, the circulating currents balance the induced voltages so that the whole of the sheath (assuming zero earth impedance) is at ground potential. If one termination only is grounded, the sheaths are subjected to a standing voltage of zero at the ground connection and maximum at the point furthest from this connection. This voltage is proportional to the conductor current and in fig. 34.4 the manner in which the sheath voltage is influenced by cable spacing is shown. To protect the sheath insulation against transient voltages arising from lightning or switching transients it is therefore necessary to fit sheath voltage limiters (SVLs) at all joint and sealing end positions where the sheath is insulated from earth.[7]

Three basic variations of specially bonded systems are commonly used: end-point bonding, mid-point bonding and cross-bonding.

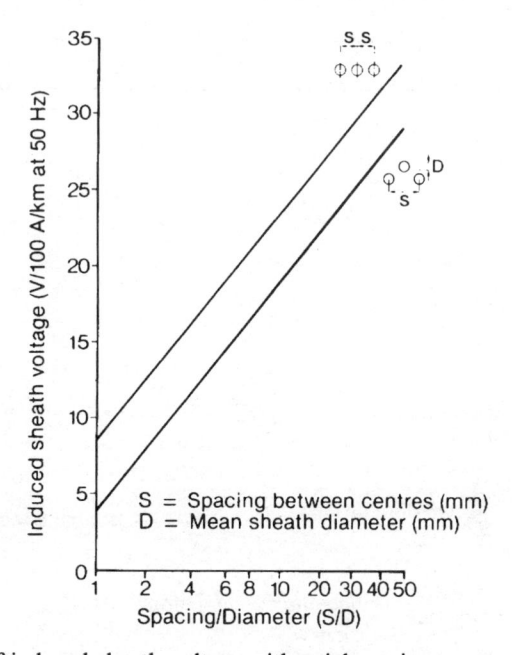

Fig. 34.4 Variation of induced sheath voltage with axial spacing

End-point bonded system

In this system the sheaths at one termination are earthed and at the other termination are insulated from ground and fitted with SVLs as shown in fig. 34.5. It is necessary to provide a separate earth continuity conductor for fault currents which would normally return via the cable sheaths. In the UK the standing voltage at the sealing end is usually limited to 65 V for 66 kV and 132 kV systems and 150 V for 275 kV and 400 kV systems, and protection from contact with the exposed metalwork at the terminations is provided by resin bonded glass fibre shrouds. The standing voltage is proportional to the cable length and therefore the voltage limitation imposes a limitation on the length of the cable that may be bonded in this manner (about 1.2 km).

Mid-point bonded system

Bonding of the mid-point is used where the route length is too long to employ an end-point bonded system. In this system the cable is earthed at the mid-point of the route and is insulated from ground and provided with SVLs at each termination. It can be seen that this doubles the possible route length, as the maximum allowable standing voltage can be tolerated at each sealing end.

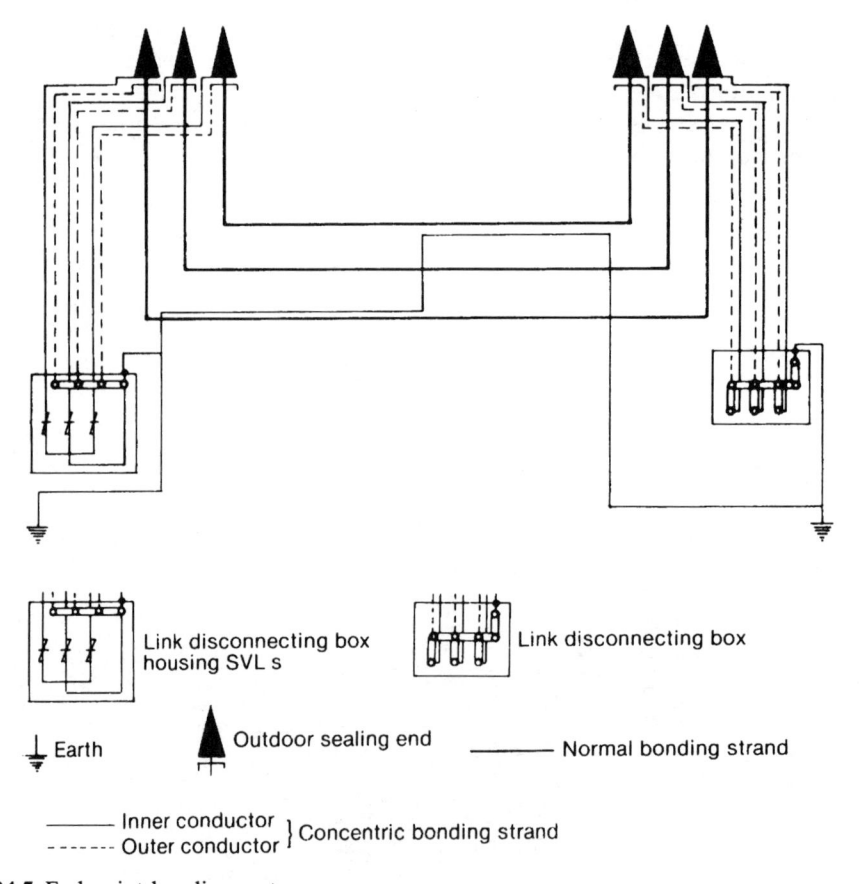

Fig. 34.5 End-point bonding system

Cross-bonded system

Either of the foregoing methods of bonding is suitable for comparatively short routes, but for longer routes the cross-bonded system must be used, as shown in fig. 34.6.

In this system the route is split up into 'major' sections, each comprised of three drum lengths and all joints are fitted with insulated flanges. At each third joint position the sheaths are connected together and at all other positions they are connected so that all sheaths occupying the same position in the cable trench are connected in series. The sheaths at the intermediate positions are also connected to SVLs.

The three sheaths connected in series are associated with conductors of different phases and when the cables are installed in trefoil formation their currents, and hence the sheath voltages, have equal magnitude but phase displacements of 120°. The overall effect is that the resultant voltage and current in the three sheaths are zero.

When cables are laid in flat formation the voltages induced in the sheaths of the outer cable are greater than that induced on the sheath of the middle cable and the phasor sum is not zero. The cables are therefore transposed at every joint position and the cross-connections are made with a phase rotation opposite to that of transposition so that the sheaths are effectively straight connected, i.e. the sheath of the middle cable of section 1 connects on to the middle cable of section 2 etc. By this method the phasor sum of the sheath voltages over three successive elementary sections is again zero.

The voltages of the sheaths to earth for a 132 kV system are as shown in fig. 34.7. It can be seen that the maximum voltage is the maximum voltage on each cable length, i.e. 65 V, but the voltage over the three-drum-length section is zero. This pattern can be repeated continuously for any route length without the sheath voltage exceeding 65 V.

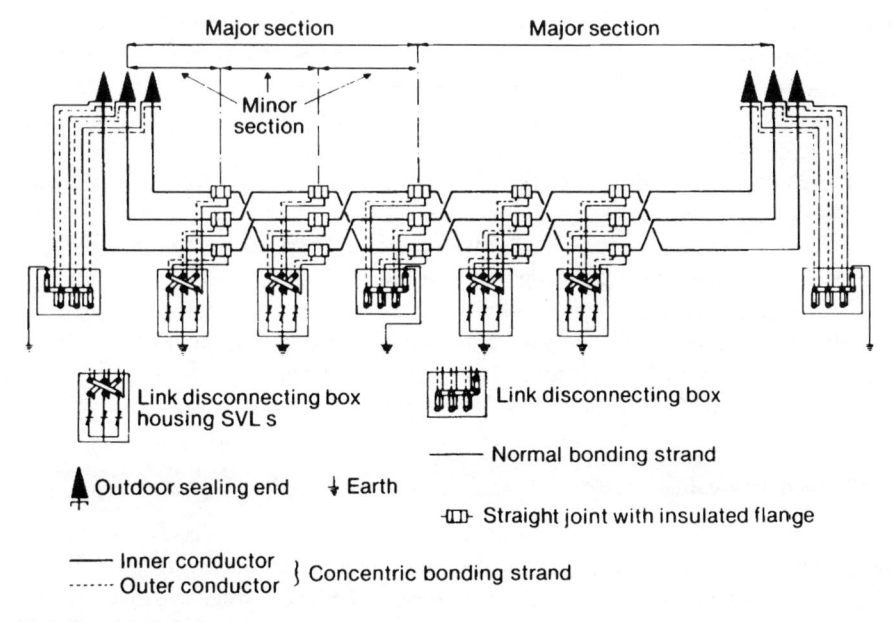

Fig. 34.6 Cross-bonded system

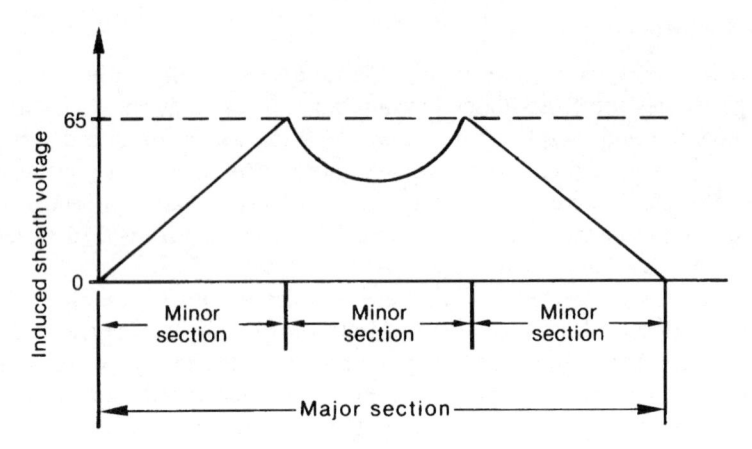

Fig. 34.7 Variation of sheath voltage to earth on a 132 kV cross-bonded system

The system does not require an earth continuity conductor as the sheaths are continuously connected and it is only necessary to earth the sheaths at the ends of each major section.

Ratings of specially bonded systems

It is apparent from the foregoing that it is possible to increase the current rating of a cable system by improving the heat dissipation through laying the cables in flat spaced formation and eliminating the sheath circulating current losses by employing a special bonding system. This is illustrated in fig. 34.8 where comparison is made between the ratings applicable to 400 kV solid bonded and cross-bonded cable systems having lead and corrugated aluminium (CSA) sheaths.

In this case the effect of special bonding is to increase the current rating of the lead sheathed cables by between approximately 15% and 50% and the CSA sheathed cables by between approximately 25% and 80%, depending on conductor size. The bonding system does not, of course, eliminate the sheath eddy current losses but these are reduced by the wider spacing which would be employed.

An improvement in rating is achieved but this is at the expense of having a more complex and costly system due to the provision of SVLs, link boxes and in some cases an earth continuity conductor. It is only economic, therefore, to provide specially bonded systems where the circulating losses are large. Today, specially bonded systems are generally only utilised when conductor sizes greater than 630 mm^2 are required. Although they are cheaper for smaller sizes than this, the saving involved is not usually considered to be worth the additional system complication.

All specially bonded systems must of necessity be fully insulated sheath systems.

The fully insulated cable system

To take advantage of the specially bonded cable systems it is necessary to insulate the cable sheath from earth. This is achieved by having an extruded serving of PVC or PE on the cables and housing the joints in compound filled fibreglass boxes to insulate them from the surrounding soil. The sealing ends are also mounted on pedestal

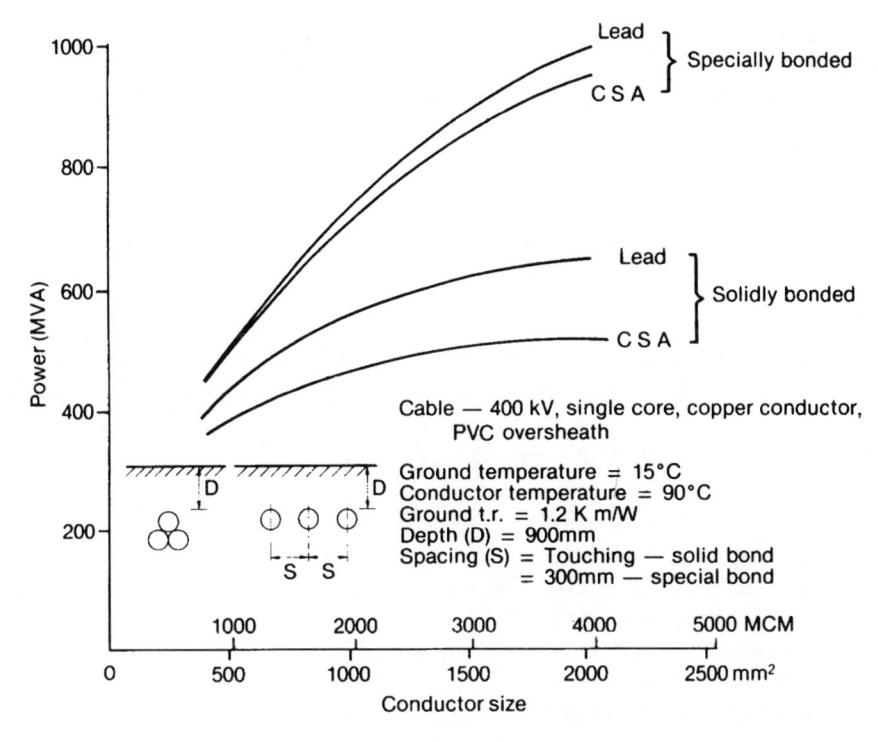

Fig. 34.8 Power ratings for specially bonded 400 kV cable systems with FF paper insulation

insulators to isolate them from their supporting steelwork. The cables are provided with a conducting graphite coating on the external surface so that periodically the serving can be tested for any damage by applying a voltage (usually 10 kV d.c. for 1 minute initially and 5 kV for subsequent tests) between the cable sheath and the graphite conducting coating. For ease of testing, bonding and earthing connections are via removable links in lockable boxes provided at all bonding positions.

Sheath voltage limiters (SVLs)

To limit high voltages appearing on the open-circuited sheaths of specially bonded cable systems under transient voltages, it is necessary to fit SVLs across the interruption of the metallic sheaths. The type of SVL used in the UK is a non-linear resistor having characteristics of a high impedance to current flow when the voltage across it is low but an extremely low resistance when the voltage is high. The SVLs are housed in the associated link box. Concentric bonding leads are used to connect the accessories to the link boxes, the concentric design being used to reduce the voltage across the sheath interruption by minimising the surge impedance and pulse propagation time.

LOW DIELECTRIC LOSS CABLES

Low dielectric loss cables offer significant rating advantage over conventional paper insulated fluid-filled cables. The heat generated in the cable dielectric is directly

Table 34.2 Comparison of insulation properties

	Paper cable	PPL cable	XLPE cable
Dielectric loss angle (DLA)	0.0024	0.0010	0.0005
Relative permittivity (ε_r)	3.5	2.8	2.3
Dielectric loss factor (DLF)	0.0084	0.0028	0.00115
Thermal resistivity	5.0	5.5	3.5

proportional to the dielectric loss factor. Table 34.2 shows a comparison of the dielectric loss factors for 400 kV paper insulated, PPL insulated and XLPE insulated cables.

The equation for dielectric loss can be represented as:

$$W_d = 0.35(DLF f V E r) \tag{34.1}$$

where W_d = dielectric loss (W/m)
DLF = dielectric loss factor ($DLA \times \varepsilon_r$)
f = power frequency (Hz)
V = phase voltage, U_0 (kV)
E = insulation design stress at U_0 (MV/m)
r = radius of conductor screen (m)

It can be seen from the above equation that, for a given design stress, dielectric loss is proportional to voltage and therefore becomes increasingly significant for EHV cables. In addition, it can be seen that, for a given voltage, the conductor radius and insulation design stress have a direct effect on the magnitude of the dielectric loss, and hence on the current rating.

XLPE cables have a lower design stress and hence greater insulation thickness and diameter than either paper insulated or PPL insulated cables, and this offsets the benefits gained by the lower dielectric loss and lower insulation thermal resistivity of XLPE. In addition, XLPE cables have a higher a.c. to d.c. resistance ratio of the conductor (due to the increased compaction which occurs during manufacture). They also have an increased thermal resistance at the space between the XLPE core and the metallic sheath, included to permit thermal expansion of the core. PPL cables have a lower conductor a.c. resistance as a result of the lower compaction and the presence of the fluid duct. They also benefit from a smaller insulation thickness and a fluid-filled space of low thermal resistivity between the insulation and the sheath.

These factors combine to give PPL cables at 275 kV and 400 kV a superior rating to both paper and XLPE insulated cables.[8] This is illustrated in fig. 34.9. XLPE is included for comparison purposes although it has still to gain service experience for long circuit lengths containing joints.

The use of low loss cables in a transmission system increases ratings for a given installation condition and can produce savings in both cable and installation costs. Under standard UK conditions a 400 kV PPL cable with a 2500 mm² conductor has a 14% increased current rating compared with a paper insulated cable.[9] Ratings for low loss cable are even more attractive in tropical regions because the percentage of dielectric losses to total losses increases under high ambient conditions. For example, a 400 kV

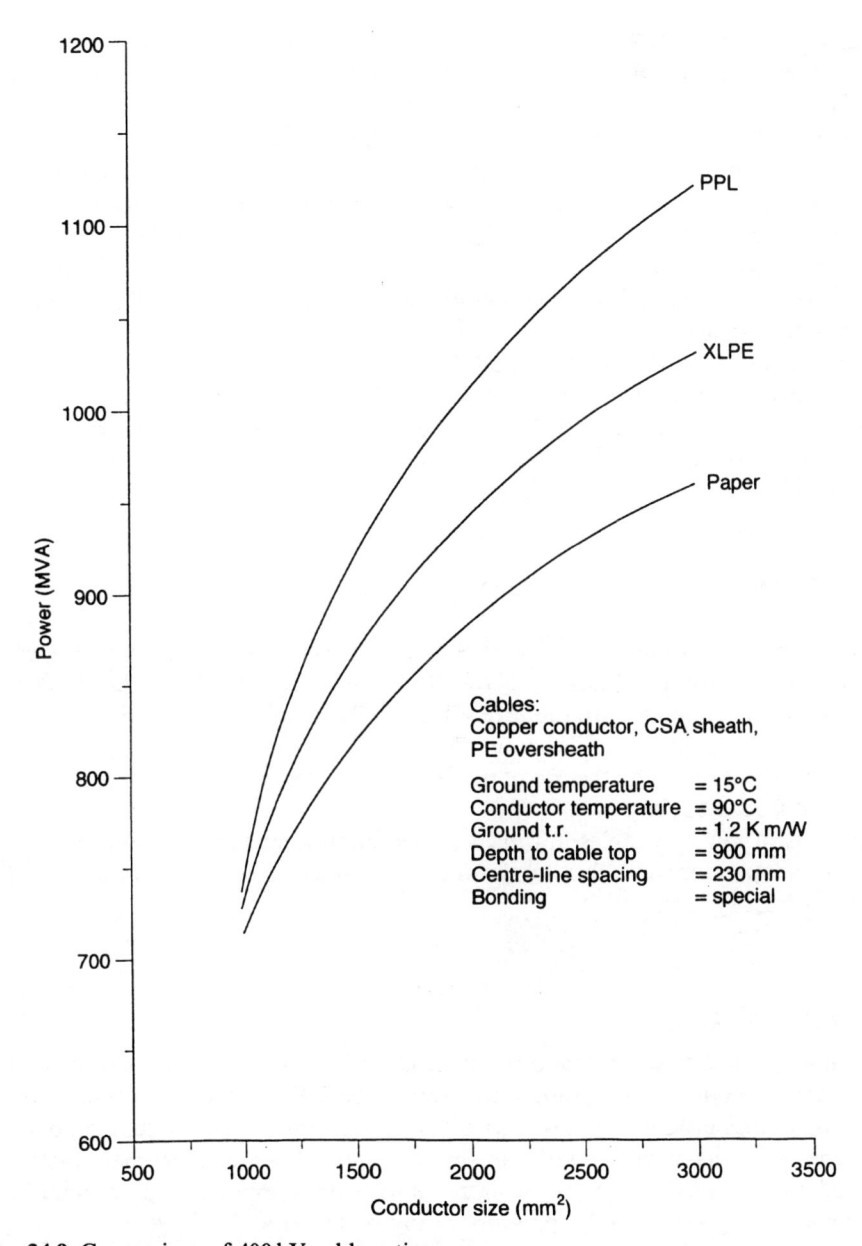

Fig. 34.9 Comparison of 400 kV cable ratings

2500 mm² PPL cable installed at a depth of 1400 mm in a region with an ambient temperature of 30°C exhibits a 30% increase in rating compared with a paper cable.[10]

Where it is necessary to install cables at increased depth, the reduction in rating with low loss cables is significantly less than with a paper cable. For example, for the same cables and conditions as for the previous example, PPL cable at 1000 mm depth of burial exhibits a 25% higher rating compared with a paper cable and at 3000 mm depth the advantage is increased to 44%.[10] For a given rating, the use of PPL insulation at

400 kV can permit the conductor size to be reduced by 40% compared with a paper insulated cable. Alternatively, a change to PPL will enable a reduction in trench width of 40%.[9]

Advantages of low loss cables also accrue to those Utilities which employ higher transmission voltages. At 500 kV, for a burial depth of 900 mm, spacing of 450 mm and ambient temperature of 15°C, a PPL cable with a stress of 18.8 MV/m exhibits a 25% increase in rating compared with an equivalent paper cable. At 765 kV, operating at a stress of 20 MV/m, the increased rating with PPL is 243%.[11] The increased ratings of naturally cooled low loss cable can also eliminate the need for forced cooling equipment for highly rated circuits.

The low permittivity of both PPL and XLPE cables compared with paper cables reduces the capacitive current and capacitive MVAr. For a 400 kV PPL cable the capacitive MVAr is reduced by 20% thus permitting longer circuit lengths and an increase in their useful load current, or a reduction in reactive compensation.[9]

FORCED COOLED CABLE SYSTEMS

To meet the highest rating requirements it is not always possible to design a cable system with one cable per phase, where all the heat generated can be dissipated naturally by the surrounding ground, even if specially bonded systems with very wide phase spacings and special backfills are used. In such circumstances it is only possible to achieve the required rating by the use of multiple cables per phase, or by making use of some form of forced cooling.

The benefits of low loss cables in forced cooled applications are not so pronounced because the temperature rise due to the dielectric losses is less than in a naturally cooled, laid direct system.

Separate pipe cooling

For laid-direct installations the simplest method of forced cooling for self-contained fluid-filled cables, and the most commonly used in the UK, has been to pump cooling water through pipes installed in close proximity to the cables. In the UK, high density polyethylene pipes have been used for this purpose. They can be manufactured, transported and installed in the same lengths as the cables. Pipes with a nominal bore up to approximately 85 mm and a pressure rating of 11 bar at 25°C are used.

A typical arrangement of cables and cooling pipes is shown in fig. 34.10 which shows two cooling pipes installed between the cables and two positioned above the outer cables. Cooling water is normally passed down two of the pipes and returned via the others to cooling stations where the water temperature is reduced by either water–water or water–air heat exchangers before being returned to the inlet pipes. To achieve the highest ratings, the heat extracted by the pipes must be maximised but at the same time cooling section lengths must be made as long as possible to minimise problems in siting heat exchangers. The cooling stations are therefore constructed at suitable positions along the cable route, taking account of the environmental as well as technical requirements. Many such stations have been installed in the suburbs of London and

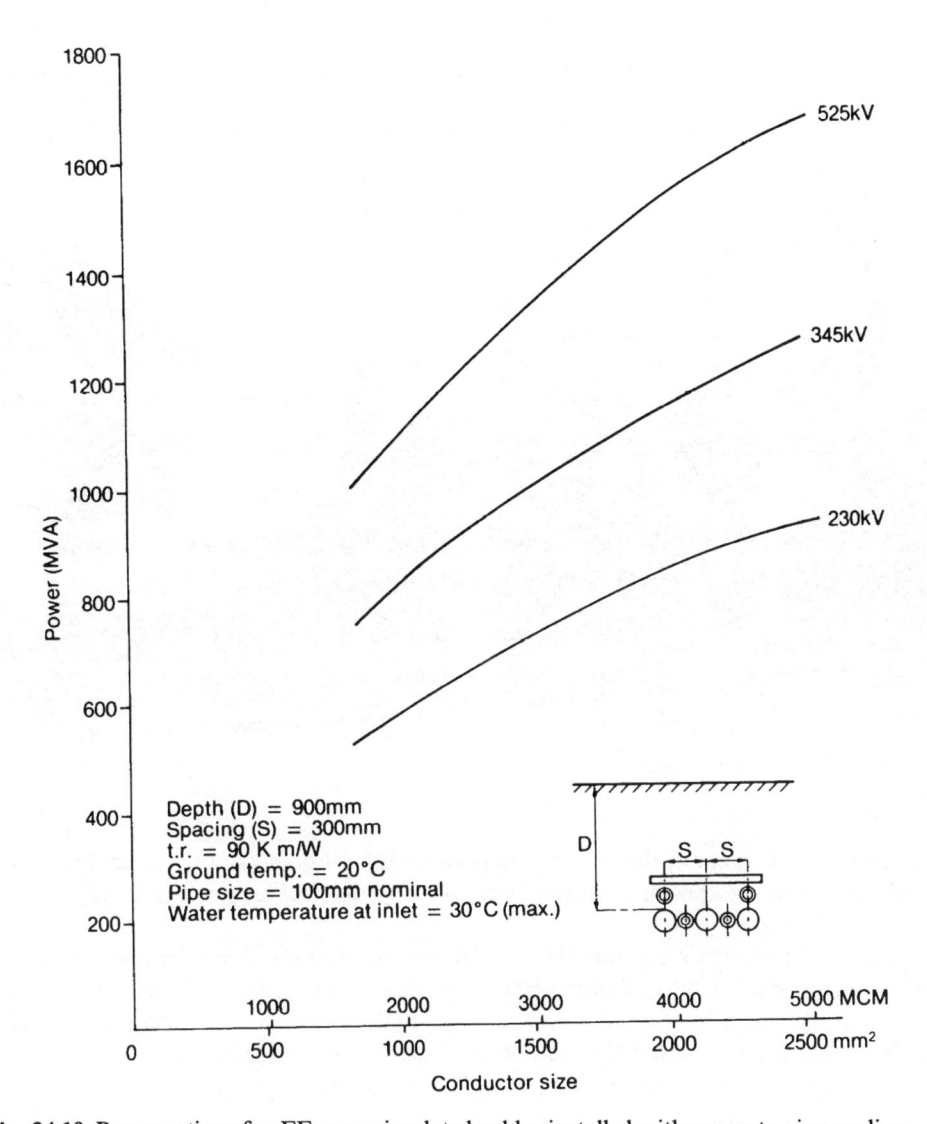

Fig. 34.10 Power ratings for FF paper insulated cables installed with separate pipe cooling

fig. 34.11 shows a typical example. A considerable number of separate pipe cooled systems have been installed in the UK[12,13] and subsequently in Vienna.[14] The thermal design of such systems is complex[15] and, in addition to the normal environmental parameters, several other variables have to be allowed for, e.g.[16]

(a) cable–pipe configurations;
(b) water flow patterns;
(c) number of pipes;
(d) pipe sizes;
(e) pipe operating pressure and life;
(f) coolant temperature;
(g) heat exchanger cooling temperatures.

Fig. 34.11 Typical cooling station in an urban area

The hydraulic design of the cooling pipe system takes into account such factors as the profile of the route, cooling characteristics of the heat exchangers and availability of cooling station sites.

The problem of obtaining suitable sites for cooling stations along the route (a typical size for an air–water heat exchanger station is 28 m × 11 m × 4 m high and for a water–water heat exchanger station 24 m × 8 m × 4 m high) will frequently limit the possible number of such stations and this can be the limiting design criterion.

Heat transfer relationships in separate pipe cooled systems
If it is assumed that the ground surface is effectively isothermal and that thermal resistivities are not temperature dependent, it is possible to apply the principle of superposition. Thus, at any given position, the cable sheath and water temperature rises ($\theta °C$) and heat emissions (H W/m) are governed by a matrix equation of the form

$$[\theta] = [T][H] \tag{34.2}$$

where the square matrix $[T]$ (K m/W), consisting of 'self' and 'mutual' heating coefficients, is dependent on the geometry of the cables and pipes and on the thermal resistivities of the ground, cable oversheaths and pipes. In this equation, whenever the pipe is cooling the system, the pipe heat emission will be negative.

The above, however, is concerned only with radial heat transfer. If water is flowing along the pipes at a rate of q l/s, then heat will be absorbed to raise the temperature of the water along the route of length L km. The longitudinal heat transfer is governed by the relationship

$$\frac{d\theta}{dL} = \frac{-H}{4.18q} \qquad (°C/km) \qquad (34.3)$$

The factors which affect the matrix elements $[\theta]$, $[T]$ and $[H]$ and the influence of water flow rate are many and have to be fully considered in order to design the most economic scheme for any particular installation.

Ratings of separate pipe cooled systems
Typical ratings achievable with 230 kV, 345 kV and 525 kV systems for the conditions specified and a typical cooling section of 3.2 km are included in fig. 34.10. Two cooling sections are normally fed from each heat exchanger and this results in heat exchangers sited every 6.4 km along the route. For comparison, the ratings applicable to the same installation but without forced cooling are shown in fig. 34.12. It will be seen that separate pipe cooling provides very significant improvements in rating, particularly at the higher voltages. For instance at 525 kV the increase in current rating of a 2500 mm² cable is from 1120 MVA to 1700 MVA, an increase of over 50%.

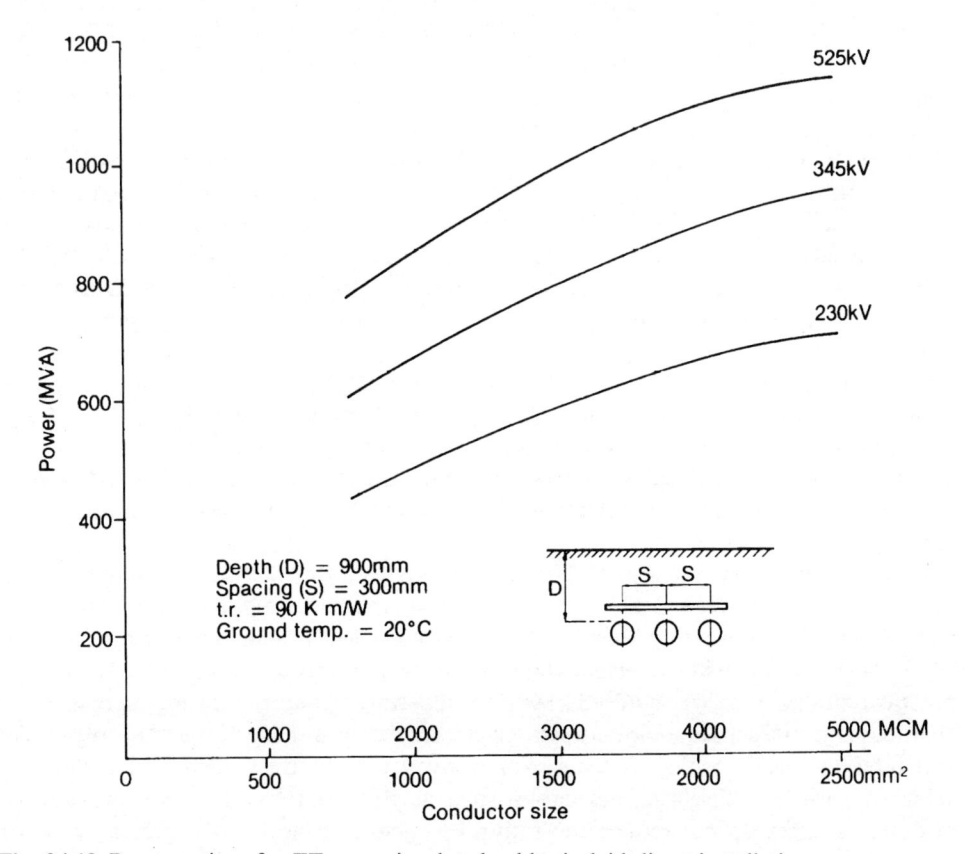

Fig. 34.12 Power ratings for FF paper insulated cables in laid-direct installations

Integral sheath cooling

Integral sheath cooling is an extension of separate water pipe cooling with the difference that, instead of the pipe being laid alongside the cable, the cables are pulled into the water pipes. The pipes are larger than those used for separate pipe cooling and to date systems with one cable per pipe have been used. A circulatory cooling system is normally used, the cooling water being pumped out along one pipe and returned via the other two. The pipe with the outward water flow, which has a flow rate twice that of the other two pipes, can have a larger diameter to reduce pumping pressures. To avoid current induced losses and corrosion problems, non-metallic pipes are preferred. Suitable materials are unplasticised PVC, glass-fibre reinforced PVC or asbestos cement. A corrugated aluminium sheath cable protected with a high density polyethylene oversheath is essential for this application, because this type of cable sheath has a high resistance to fatigue strains arising from thermomechanical movements. Where long cooling sections are employed it may be necessary to increase the diameter of the conductor duct to allow cooling of the accessories by oscillation or circulation of the fluid impregnant.[17]

Ratings of integral cooled cable systems

As the water is in direct contact with the oversheath of the cable and the flow is turbulent, the external thermal resistance is reduced effectively to zero. The rating therefore depends only on the cable characteristics and the maximum water temperature in the system. The latter depends on the inlet water temperature, cable losses, water flow rate and heat exchanger characteristics.

Typical ratings achievable with 230 kV, 345 kV and 525 kV integral sheath cooled systems for the conditions stated are shown in fig. 34.13. The lower thermal resistance between the cable and the cooling water, compared with the separate pipe cooled system, produces a significant improvement in rating and at 525 kV a rating of over 2000 MVA can be achieved, an increase of approximately 80% on the laid-direct rating.

With this type of cable cooling, and to a lesser extent with separate pipe cooling, significant increases in rating are possible by increasing the insulation design stress. This is illustrated in fig. 34.14 for a 525 kV 2000 mm^2 cable. In the case of naturally cooled FF paper insulated cables, the maximum rating is achieved with a design maximum stress of about 13 MV/m for the assumed conditions. For higher design stresses, the reduction in cable thermal resistance is not sufficient to offset the increase in dielectric losses, the net effect being a reduction in the current rating of the cable system.

In the case of the integral sheath cooled system, the thermal resistance of the insulation is virtually the total thermal resistance, and hence any reduction will have a far greater effect on rating than in a naturally cooled system. Hence, as shown in fig. 37.14, the rating will increase with design stress and it is desirable to use the highest practical design stress. Figure 34.14 also illustrates that the rating versus stress characteristic of a separate pipe cooled system lies between those of the naturally cooled and integral sheath cooled systems. For a given design stress, increased ratings are achieved with low dielectric loss cables such as PPL and XLPE, however increases in design stresses do not reduce the rating by such a pronounced degree as with FF paper insulation.

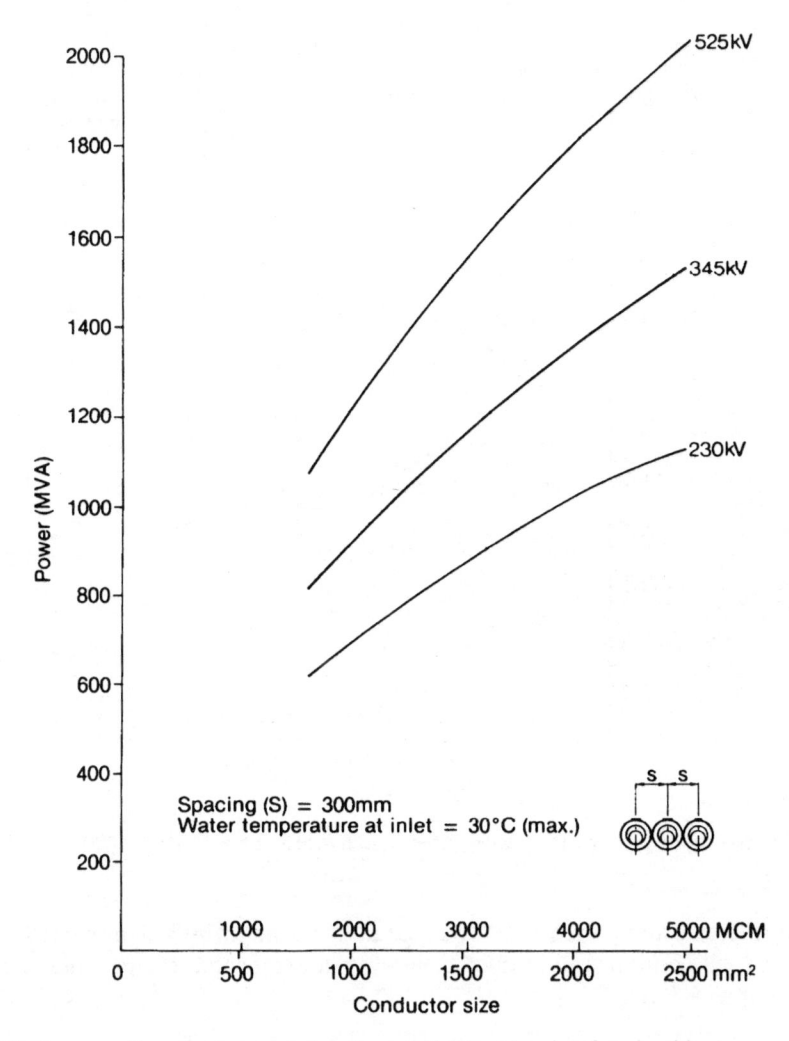

Fig. 34.13 Power ratings for integral sheath cooled FF paper insulated cables

The 400 kV cable system for the Severn Tunnel

An integral sheath cooled cable system has been installed in the UK.[17] It consists of a 400 kV 2600 mm² fluid-filled cable with a design stress of 15 MV/m installed in resin bonded glass-fibre-reinforced PVC pipes. The installation is in a 3670 m tunnel beneath the River Severn and the River Wye. The design maximum rating is 2600 MVA at a maximum conductor temperature of 95°C.

Internal fluid cooling

The most efficient method of obtaining increased rating is to remove the losses at source. This is the principle of the internally fluid-cooled cable in which the hydrocarbon impregnating fluid is circulated through the conductor, which is itself the main source of the losses. To obtain adequate flow rates the central duct is enlarged from the standard 12 mm diameter and fluid is fed in and extracted at feed joints

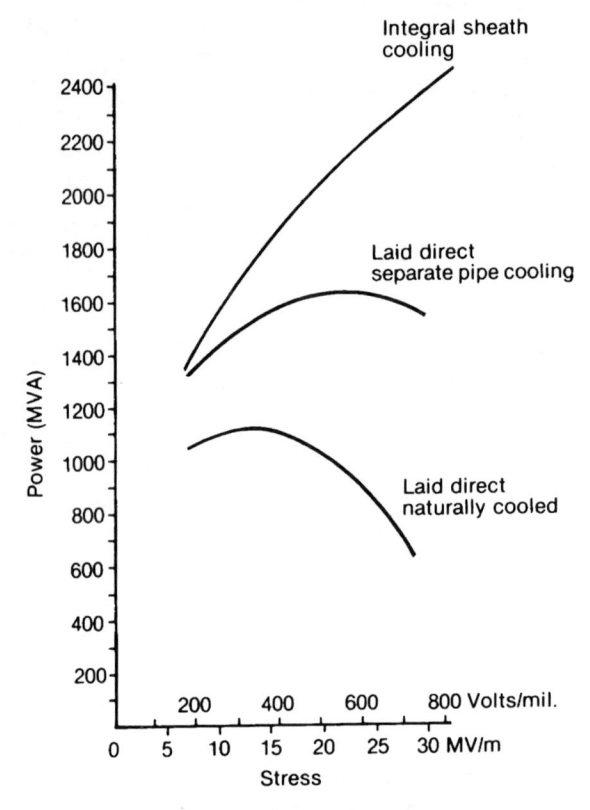

Fig. 34.14 Variation of power rating with cable design stress for a 525 kV, 2000 mm² FF paper insulated cable

situated at intervals along the route. The feed joint is similar in design to the stop joint used with the self-contained fluid-filled cable system except that the passages are larger to take the greater flow.

Ratings of internally cooled cable systems
With internal fluid cooling there is virtually no inherent limit on current rating. The practical limitations become the maximum duct size that can be incorporated in the conductor, the spacing of feed joints and the maximum pumping pressure. Figure 34.15 illustrates the variation in rating of a 525 kV cable system with spacing between feed joints, using a copper conductor of 2000 mm² cross-section having a 50 mm duct.

A rating of just over 4000 MVA is possible with a spacing between feed joints of 1 km. An advantage of the system is that, because the losses are extracted at the conductor, it can be made virtually independent of its environment. Therefore, factors such as depth of burial and the thermal properties of the ground can be ignored.

Field trials of 400 kV internally fluid-cooled cables
To date, although trial installations of internally cooled 400 kV cables have been subjected to extensive tests in the UK,[18] no commercial internally fluid-cooled systems appear to have been completed. One of the test installations consisted of 3100 mm² aluminium conductor cable with a 35 mm duct and the other of 2000 mm² copper

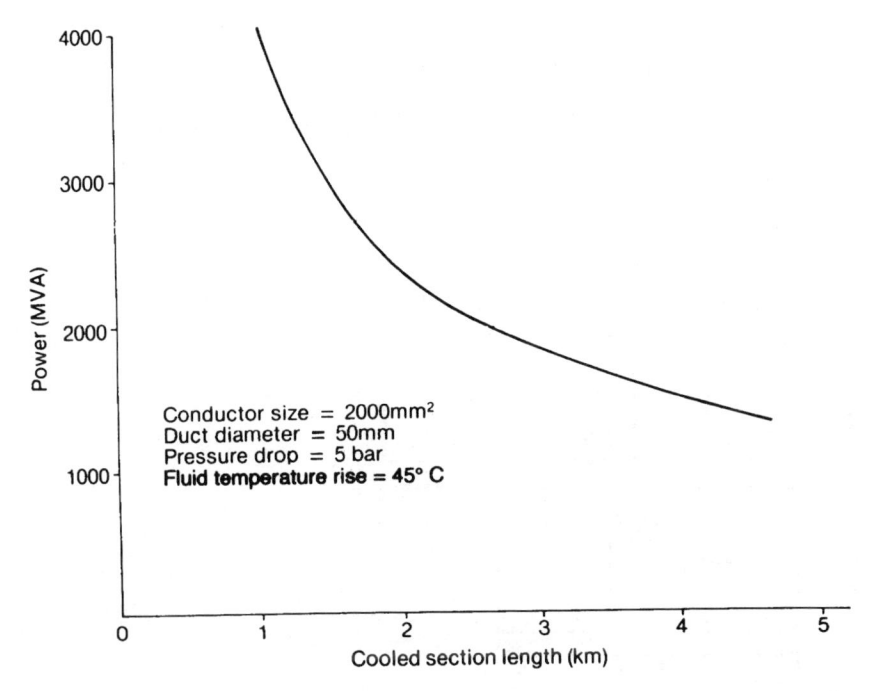

Fig. 34.15 Variation of power rating with cooled section length for 525 kV, 2000 mm^2 FF paper insulated cable

conductor cable with a 50 mm duct. The installations included a feed joint, straight joint and sealing ends, together with associated pumps and cooling equipment. The tests on the aluminium and copper conductor cables extended over 100 and 200 load cycles respectively, while energised at 1.1 times working voltage. The power factor measurements on the cable system and the circulating fluid were stable throughout the trials and examination showed that the condition of the cable after the trial was satisfactory.

SUMMARY

At the beginning of this chapter it was pointed out that many of the advances in supertension cable technology have resulted from the requirement to meet the overhead line ratings given in table 34.1. As a summary, some details are given below to assess how four major developments have contributed towards meeting these ratings and the effect on the economics of underground cable transmission in the UK.

These economic factors are often complex and difficult to quantify[9,19] but an appreciation can be obtained by considering the cable size and number of circuits required to meet a nominal 4000 amp (2270 MVA at 400 kV) overhead line rating.

The effects of the developments can be illustrated by considering them in seven separate stages, as shown in fig. 34.16. The first case assumes that special backfills, special bonding, low loss cables and forced cooling had not been developed, but that conductor sizes up to 2000 mm^2 are available. For this case five thermally independent trefoil circuits would be required to meet the rating.

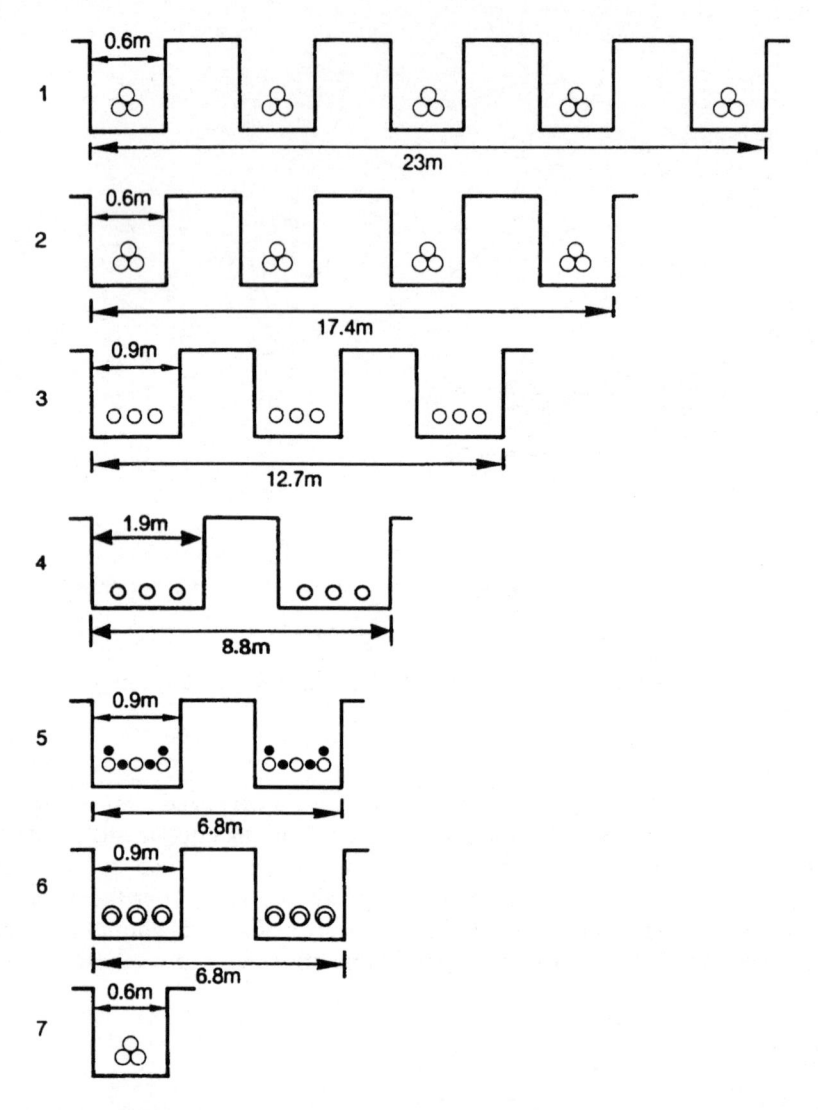

Fig. 34.16 Comparative space requirements for 400 kV FF circuits of similar rating (copper conductor cable with lead sheath and polyethylene protective finish):

(1) five circuits, 1600 mm^2 – paper insulated solid bonding, dried-out thermal resistivity of 3.0 K m/W within 50°C isotherm;
(2) four circuits, 2000 mm^2 – paper insulated solid bonding, stabilised backfill of dried-out thermal resistivity of 1.2 K m/W;
(3) three circuits, 1600 mm^2 – paper insulated 300 mm spacing, cross-bonding, stabilised backfill;
(4) two circuits, 2500 mm^2 PPL insulated 650 mm spacing, cross-bonding, stabilised backfill;
(5) two circuits, 2000 mm^2 – paper insulated separate pipe cooling, cross-bonding, stabilised backfill;
(6) two circuits, 1300 mm^2 – paper insulated integral sheath cooling, cross-bonding, stabilised backfill;
(7) one circuit, 2000 mm^2 – paper insulated internal fluid-cooling, cross-bonding

Case 2 shows the effect of using stabilised backfill within the 50°C isotherm and this results in the number of trefoil cable circuits required being reduced to four. Case 3 illustrates the use of special bonding, with the effect that the number of flat spaced cable circuits required is reduced to three. Case 4 demonstrates that the use of low loss cables can reduce the required circuits to two. Cases 5, 6, and 7 demonstrate that, by forced cooling using water pipes, the trench widths and conductor sizes can be reduced and internal fluid cooling can reduce the required circuits to one.

Though quoted in a somewhat arbitrary manner, the above indicates how the reduced costs from the successive developments provide savings resulting from using fewer circuits which are considerably greater than the increased cost due to the extra complexity of the installation.

REFERENCES

(1) Banks, J. (1974) 'Electric power transmission: the elegant alternative'. *Proc. IEE* **121** (1), 419–58.
(2) Cox, H. N., Holdup, H. W. and Skipper, D. J. (1975) 'Developments in U.K. cable-installation techniques to take account of environmental thermal resistivities'. *Proc. IEE* **122** (11).
(3) Cox, H. N. and Coates, R. (1965) 'Thermal analysis of power cables in soils of temperature-responsive thermal resistivity'. *Proc. IEE* **112** (12).
(4) Winders, J. J. (1973) 'Computer program analyzes heat flow cables buried in regions of discontinuous thermal resistivity'. Vancouver: *IEEE PES Summer Meeting* and EHV/UHV Conf. Paper T 73 502–2.
(5) IEC 287 (1969) 'Calculation of the continuous current rating of cables (100% load factor)'.
(6) CIGRE Study Committee 21, Working Group 07 (1973) 'The design of specially bonded cable systems'. *Electra* (28).
(7) Engineering Recommendation C55/4 (1989) 'Insulated sheath power cable systems'. *Electricity Council Chief Engineers Conf.*, Mains Committee.
(8) Gregory, B., Jeffs, M. D. and Vail, J. (1993) 'The choice of cable type for application at EHV system voltage'. *3rd Int. Conf. on Power Cables and Accessories 10 kV–500 kV*, IEE Conference Publication 382, pp. 64–70.
(9) Endersby, T. M., Gregory, B. and Williams, D. E. (1991) 'Low loss self contained oil filled cables insulated with PPL for UK 400 kV transmission circuits'. *5th Int. Conf. on AC and DC Power Transmission*, IEE Conf. Publication 345, pp. 337–343.
(10) Gregory, B., Jeffs, M. D. and Smee, G. J. (1992) 'Self-contained oil filled cable insulated with polypropylene paper laminate for EHV transmission'. *9th CEPSI*, 3–29.
(11) Endersby, T. M., Gregory, B. and Swingler, S. G. (1992) 'The application of polypropylene paper laminate insulated oil filled cable to EHV and UHV transmission'. CIGRE 21-307.
(12) Alexander, S. M., Smee, G. J., Stevens, D. F. and Williams, D. E. (1979) 'Rating aspects of the 400 kV West Ham–St. John's Wood cable circuits'. *IEE 2nd Int. Conf. on Progress in Cables and Overhead Lines for 220 kV and Above.*

(13) Arkell, C. A., Doughty, D. F. and Skipper, D. J. (1979) '400 kV self-contained oil-filled cable installation in South London, UK (Rowdown–Beddington)'. *7th IEEE/PES Transmission and Distribution Conf.*
(14) Arkell, C. A., Bazzi, G., Ernst, G., Schuppe, W.-D. and Traunsteiner, W. (1980) 'First 380 kV bulk power transmission system with lateral pipe external cable cooling in Austria'. Paris: CIGRE Paper No. 21–09.
(15) CIGRE Study Committee 21, Working Group 08 (1979) 'The calculation of continuous ratings for forced cooled cables'. *Electra* (66).
(16) Alexander, S. M. and Smee, G. J. (1979) 'Future possibilities for separate pipe cooled 400 kV cable circuits;'. *IEE 2nd Int. Conf. on Progress in Cables and Overhead Lines for 220 kV and Above.*
(17) Arkell, C. A., Blake, W. E., Brealey, A. D. R., Hacke, K. J. H. and Hance, G. E. A. (1977) 'Design and construction of the 400 kV cable system for the Severn Tunnel'. *Proc. IEE* **124** (3).
(18) Brotherton, W., Cox, H. N., Frost, R. F. and Selves, J. (1977) 'Field trials of 400 kV internally oil-cooled cables'. *Proc. IEE* **124** (3).
(19) Cherry, D. M. (1975) 'Containing the cost of undergrounding'. *Proc. IEE* **122** (3).

CHAPTER 35

Transmission Cable Accessories and Jointing for Pressure-assisted and Polymeric Cables

The basic accessories are the terminations at the ends of the cable circuit, which seal the cable and provide a connection to other items of transmission plant, and the joints which connect lengths of cable together. Accessories are an integral part of the cable system and are required to exhibit an equal performance to the cable to achieve the highest system rating at the lowest cost.

Unlike the cable, the accessories are assembled on site, often in adverse weather conditions and in confined locations such as joint bays excavated in the ground or located within tunnels and subterranean chambers. The joint insulation is applied by hand without the factory benefits of controlled tape tension, superclean extruders and controlled temperature and humidity. Therefore accessory designs and jointing techniques have been developed to reduce the jointing time to a minimum, both to reduce costs and particularly to reduce the contamination of the insulation by the absorption of moisture, air and airborne debris.

GENERAL CRITERIA

Electrical considerations

The concept of designing joint insulation is similar for all types of cable. Compared with the cable, a greater thickness of insulation is applied over the conductor connection so as to reduce the electric stress and thus compensate for the reduction in strength caused by the presence of an interface between the cable and joint and by enlarged gaps between adjacent and overlying insulating tapes. The difference in diameter of insulation between the cable and accessory introduces a component of electric stress along the surface of the insulation, i.e. its weakest direction. The magnitude of this longitudinal stress is an important design parameter, particularly for paper insulated cable, which determines the profiled shape of the conductor connection, the dimension of the cable insulation pencil and the profiled shape of the earth screen, termed the stress control profile. The increase in diameter over the insulation also increases the thermal resistance of the accessory and, unless alleviating steps are taken, it increases the operating temperature of the joint in comparison with that of the adjacent cable. Each accessory has to be thermally designed by limiting the insulation diameter to improve radial heat dissipation and by minimising the length to improve longitudinal dissipation through the conductor.

The type of cable insulation is the determining factor in the design of the accessories and in the techniques of jointing. Fluid-impregnated paper insulated cable has been developed to operate at the highest electrical design stress and is established at the highest transmission voltage and ratings. Developments to extend the performance of this dielectric further continue to be made to the impregnant, in the form of synthetic hydrocarbon and silicone oils, and to the tape, in the form of low loss laminates of paper and polypropylene.[1] The basis for reliable performance remains unaltered, that the degasified low viscosity fluid is both the principal insulant and the means of eliminating gaseous voids. Jointing on FF cables is facilitated by the ease of removal of discrete layers of screening and insulating tapes. The performance of the accessory is tolerant to variations in jointing accuracy because of the inherent resistance of paper to electrical discharging and because the fluid is formulated to quench a discharge by the absorption of gas.

Polymeric cable has been made available at transmission voltages up to 550 kV by the development of superclean insulation, super smooth screens and a sophisticated triple extrusion process which employs dry curing under high nitrogen pressure.[2] The quality of the insulation depends on the complete exclusion of sources of electrical discharge such as imperfect screen interfaces, gaseous voids and particulate contamination. The potential for the presence of these features results in a lower design stress for polymeric cables thereby significantly increasing the cable diameter.[3] Polymeric cables are preferred because of the absence of a fluid impregnant, simplifying the ancillary equipment and maintenance.

XLPE is the commonly chosen insulant because of its inherent low dielectric loss, although particular EPR compounds are selected for some applications because of increased elasticity and tolerance to moisture. The accessories are similar for both XLPE and EPR cable. The absence of a fluid impregnant is a disadvantage to the accessory and requires that the jointer uses a high standard of accuracy and cleanliness and that the designs incorporate other means of eliminating the voids which are inherently associated with the preparation of the surface of the cable dielectric.

There are three methods of forming the interface on a polymeric cable. The first is to melt and mould the accessory insulation on to the cable insulation *in situ*. This is a complex process for use on site particularly for crosslinked cables because a second heating process is required. The second is to employ a gaseous or liquid filling as used in the design of outdoor and GIS terminations, however this is generally not preferred for this because of the large size of the accessory and because of the problem of pressure/level monitoring and of maintenance. The third method is to apply elastomeric accessory insulation under pressure, which is then maintained during the service life of the accessory. Pressure can be applied by elastomeric tape or by prefabricated elastomeric mouldings. The three design principles are described later in the chapter.

Mechanical considerations

The accessories experience increased mechanical loads compared with the cable due to their presence as a mechanical discontinuity in the system. As the cable conductor and sheath generate both tensile and compressive thermomechanical forces in service[4-6] the conductor connection is designed to withstand such loads and the insulated core is reinforced to prevent excessive lateral movement. Polymeric insulation has a memory of the extrusion process in the form of elastic strain, which is released by cutting, and of locked-in strain, which is relieved by heating.[7,8] Retraction of the insulation is confined to be local to the cut end because of the distributed nature of adhesion to the stranded conductor. To prevent retraction from the joint it is important to constrain the insulation additionally and to transfer the tensile loads from one cable to the other.

Straight joints do not experience a differential longitudinal load if the adjacent conductors are of equal area and if the cables are equally constrained. When thermomechanical imbalance is known to be present, an anchor joint can be employed to transmit the imbalanced load through the insulated barrier to the joint shell and then through a stanchion to the ground.[5]

Adjacent to cable terminations it is preferred to continue the rigid constraint of the buried cable by close cleating and thus prevent fatigue of the metallic sheath. A rigid insulator is provided for both paper insulated and polymeric supertension cables to constrain conductor movement and to withstand the full thermomechanical conductor load plus the combined effects of busbar loading, short-circuit forces and wind and ice loading. In special circumstances the termination is required to withstand vibrations due to earthquakes and to the operation of the circuit breakers in metal enclosed switchgear.

Pressurised cable systems

The pressure in fluid-filled cable systems is maintained by permanently installed tanks of pressurised impregnant. In gas pressure cable systems, cylinders of compressed gas are connected either to the terminations or to special 'feed joints' installed at predetermined positions along the cable route. In both types of system the function of separating the pressure in adjacent sections of cable is performed by a joint with an insulating barrier which is sealed to the conductor and to the joint shell. This is termed a 'stop joint' in fluid-filled cable systems and a 'sectionalising joint' in gas-filled systems. Sectionalising joints are required much less frequently than stop joints. The stop joint and sectionalising joint are also used as feed joints for the insulating fluid. In fluid-filled systems straight joints predominate and are installed at intervals which are typically 400 m long, whereas stop/feed joints are installed at intervals of 2–5 km, depending on the variation of the static hydraulic pressure along the route. Transition joints are used to connect different types of cable and are similar to stop joints and sectionalising joints in design.

Following manufacture, the joint shells and terminations are submitted to twice the maximum operating pressure, i.e. 10.5 bar for fluid-filled cable and up to 34 bar for gas-filled cable.[9] Special accessories have been designed to withstand operating fluid pressures of up to 35 bar.[10,11]

Cable ancillaries

The ancillaries are those items of equipment which are permanently installed together with the cables and joints. They provide facilities for maintaining and monitoring pressure and for providing access to the sheath bonding connections to test the cable/accessory sheath insulation.[9] Metal sheaths are fitted to polymeric transmission cables to exclude moisture and as a return conductor for short-circuit current. Single-core cable systems in transmission circuits often employ special sheath bonding to eliminate induced sheath currents and hence increase the current rating.[12,13] This requires that, as with the cable sheath, each accessory shell or sleeve is insulated from ground potential, joint shells being either taped, sleeved or surrounded by bituminous compound and the terminations being mounted on insulated supports. The metallic sheaths of adjacent cables are electrically separated at the joint by an insulated flange in the joint shell and by an insulated gap in the earth screen which covers the joint insulation.

ACCESSORIES FOR SINGLE-CORE FLUID-FILLED CABLES (33–550 kV)

Fluid flow in single-core cable systems

Fluid pressure is maintained in the single-core cable system by a duct located at the centre of the conductor. Termination and stop joints contain a narrow channel which permits the flow of fluid between a port in the conductor connection at high voltage to a union at low voltage in either the termination sleeve or joint shell and hence to a fluid pressure tank. Further details of these are given later. The conductor connection in a straight joint contains a hollow pin which permits fluid flow between the adjacent conductor ducts. Additional unions are located in the joint shell and termination to permit the accessory to be evacuated from the top union and impregnated with fluid through the bottom union. Fluid flow from the space between the sheath and cable insulation is prevented during evacuation by a synthetic rubber glove or taped poultice to seal the cable sheath to the core. Fluid flow from the conductor connection is sealed during evacuation by an impregnation pin in outdoor terminations and by a drop-out valve in stop joints. These are opened following impregnation, manually in the former case and by fluid pressure in the latter.

Outdoor terminations

A 525 kV termination and a 132 kV termination are shown in figs 35.1(a) and 35.1(b) respectively and overall dimensions are given in table 35.1.

Insulator

The insulator is made from glazed electrical grade porcelain. The clearance height is determined by the basic impulse level voltage (BIL) specified by standards.[14] Alternate long–short (ALS) or anti-fog sheds are provided in the outer surface to give protection from pollution.[15] The surface creepage length is commonly based on a stress of 0.023 MV/m and the protected to total surface creepage length is based on a ratio of 2. The flashover voltage reduces at low barometric air pressure and should be calculated for installations at high altitude.[16] Similarly the creepage distance may require to be increased in special circumstances for installations adjacent to the sea which are subjected to excessive salt spray or at high altitude when subjected to heavy industrial pollution.

Insulation co-ordination

Arcing horns are provided when required by the user. At 400 kV the gap is formed between two fixed toroids, one suspended from the corona shield and the other

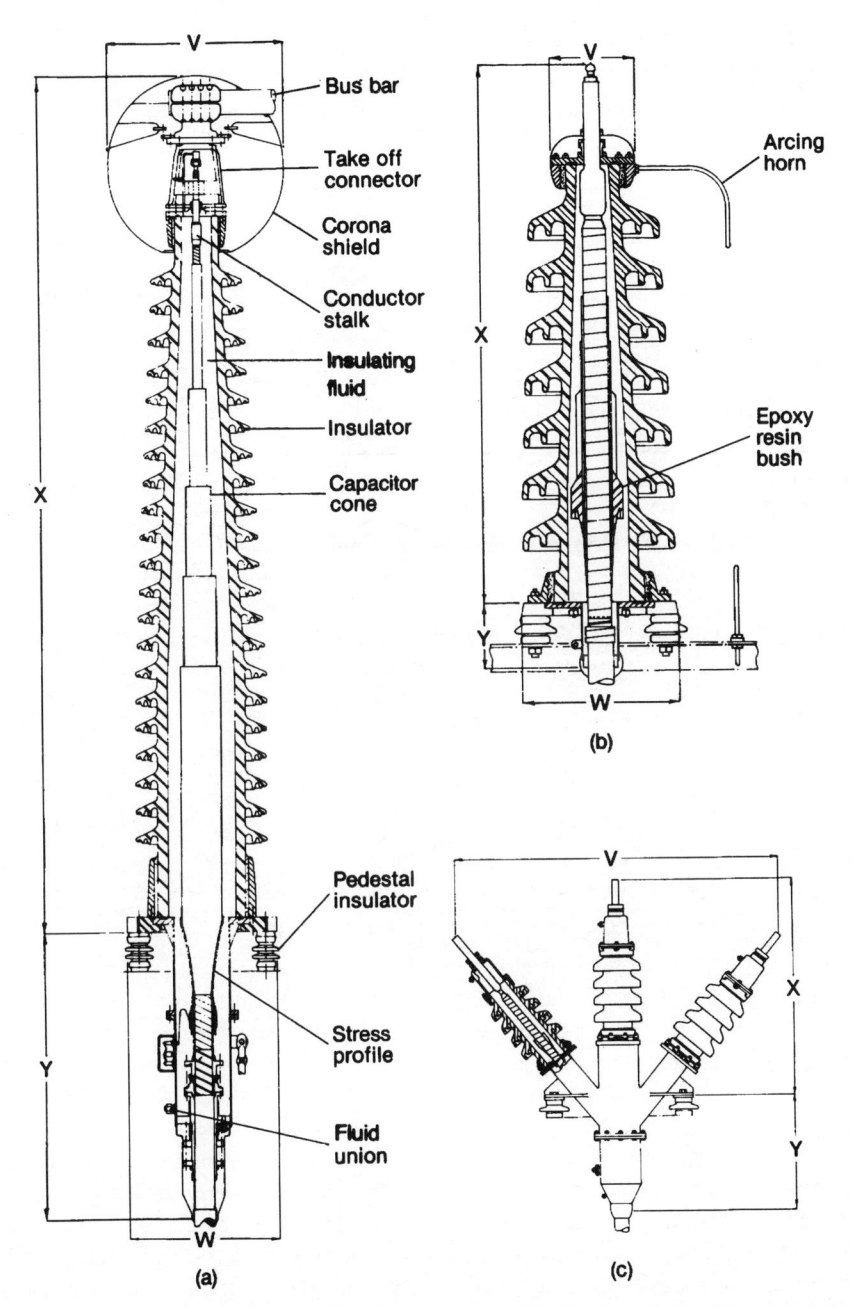

Fig. 35.1 Outdoor terminations for fluid-filled cables: (a) 525 kV single-core; (b) 132 kV single-core; (c) 33 kV 3-core

supported on the baseplate. For lower voltages an adjustable rod type is used (fig. 35.1(b)). The gap may be reduced by the user to be compatible with the insulation co-ordination policy of the adjacent transmission plant, e.g. IEC 71-1[17] recommends that the gap be set to flash over at 80% of the BIL. Table 35.2 gives typical gaps.

Table 35.1 Dimensions of outdoor terminations

U (kV)	BIL (kVp)	X^a (mm)	Y^a (mm)	Total (mm)	V^a (mm)	W^a (mm)
		Height			Width	
525	1800	5080	1525	6605	1015	840
400	1425	3810	325	4135	585	710
275	1050	2840	290	3130	415	510
161	750	2110	200	2310	255	405
132	650	1710	165	1875	265	405
66	342	1100	310	1410	185	375
33	194	770	185	955	205	230

[a] See fig. 35.1

Table 35.2 Horn gaps in outdoor terminations

U (kV)	BIL (kVp)	At BIL (mm)	At 80% BIL (mm)
400	1425	2540	2080
275	1050	1905	1550
161	750	1370	1090
132	650	990	790
66	342	533	410
33	194	320	220

Account should be taken of the climatic conditions of humidity and barometric pressure.[16,18] The arcing horns should not be set to flash over frequently as power arcs may damage and rupture the porcelain.

Take-off connector

The connection to other manufacturers' equipment is made by air insulated flexible or rigid busbars of either aluminium or copper. Copper busbars are connected by a simple clamp directly to the copper stalk joined to the cable conductor and protruding through the top plate of the termination. Aluminium busbars are connected indirectly by a weather protected aluminium/brass take-off connector to avoid the problems of bimetallic corrosion of the conductor stalk (fig. 35.1(a)).

Stress control

Stress control is provided to achieve as uniform a voltage gradient as possible in the air external to the insulator and in the fluid channel between the core and the inside of the termination, both of which are electrically weaker than the paper insulation. At voltages of up to 132 kV a simple epoxy resin bush (fig. 35.1(b)) is positioned at the termination of the cable earth screen. A metallic re-entrant of large curvature is set into the bush so that the stress concentration is reduced and contained within the isotropic epoxy resin insulation. At voltages of 161 kV and above, a capacitor cone stress control has advantages (fig. 35.1(a)). This is applied on site and consists of a number of equally spaced cylinders of overlapping aluminium foil insulated from each other and embedded within the paper roll insulation. There is an equal voltage drop between each foil so as to achieve a uniform stress distribution along the insulator. The positions of the outer earth potential foil and the inner conductor potential foil are carefully chosen to achieve optimum flashover withstand level on both impulse polarities. A corona shield is provided to screen the top metal work, to reduce corona at working voltage and to improve the impulse performance.

Some authorities specify additional electrical proving tests, especially for the cable termination.[18] These can consist of a short time a.c. voltage withstand, both dry and wet under simulated conditions of heavy rainfall, and of radio interference tests. The 525 kV termination shown in fig. 35.1(a), for example, was required to withstand 875 kV r.m.s. for 1 minute dry and 690 kV r.m.s. for 10 s wet.[19] The most onerous proving test is the cable system lightning impulse BIL.

Terminations into SF_6 insulated metal-clad equipment

Metal-clad equipment is significantly more compact than that in air insulated substations and outdoor switching yards, thus enabling savings to be made in land utilisation and cost, particularly in urban areas. The equipment has been increasingly employed, to the extent that at some voltages the supply of metal-clad cable terminations has exceeded that of the outdoor type.[20]

A single-phase 400 kV SF_6 termination is shown in fig. 35.2(a) and a 132 kV 3-phase termination in fig. 35.2(b). The leading dimensions are given in table 35.3. These comply with an IEC publication[21] which was issued to harmonise the dimensional compatibility at the interface between the cable termination and switchgear, thus ensuring interchangeability and elimination of special designs.

Insulator

The external clearance length compared with an outdoor termination (table 35.1) is typically reduced by a factor of 3. A precision cast epoxy resin insulator (fig. 35.2(a)) is employed to withstand the increased radial and longitudinal stresses that occur, both on the outer insulator surface in contact with the SF_6 and on the inner surface in contact with the fluid channel. Figure 35.3 shows a field plot of a 400 kV termination. The SF_6 pressure and chamber diameter are determined by the manufacturer of the metal-clad equipment. The SF_6 working pressure can be up to 8 bar,[21] which is greater than the maximum pressure within the fluid-filled cable (5.2 bar) and poses the risk of gas ingress into the fluid-filled cable system with the possibility of electrical failure of the termination.

This risk has been eliminated for the designs shown in figs 35.2(a) and 35.2(b) by the introduction of an insulator with a solid embedded electrode (fig. 35.2(c)) which, being unpierced by seals, completely segregates the SF_6 pressurised switchgear from the cable.[22] The possibility of SF_6 leakage into the cable is eliminated and the need to set the cable fluid pressure above the switchgear SF_6 pressure is removed. The solid electrode has been made possible by the development of a plug-in connector with a mechanical lock. At the earthed end of the insulator double O-ring seals are employed with the intermediate gap vented to atmosphere.

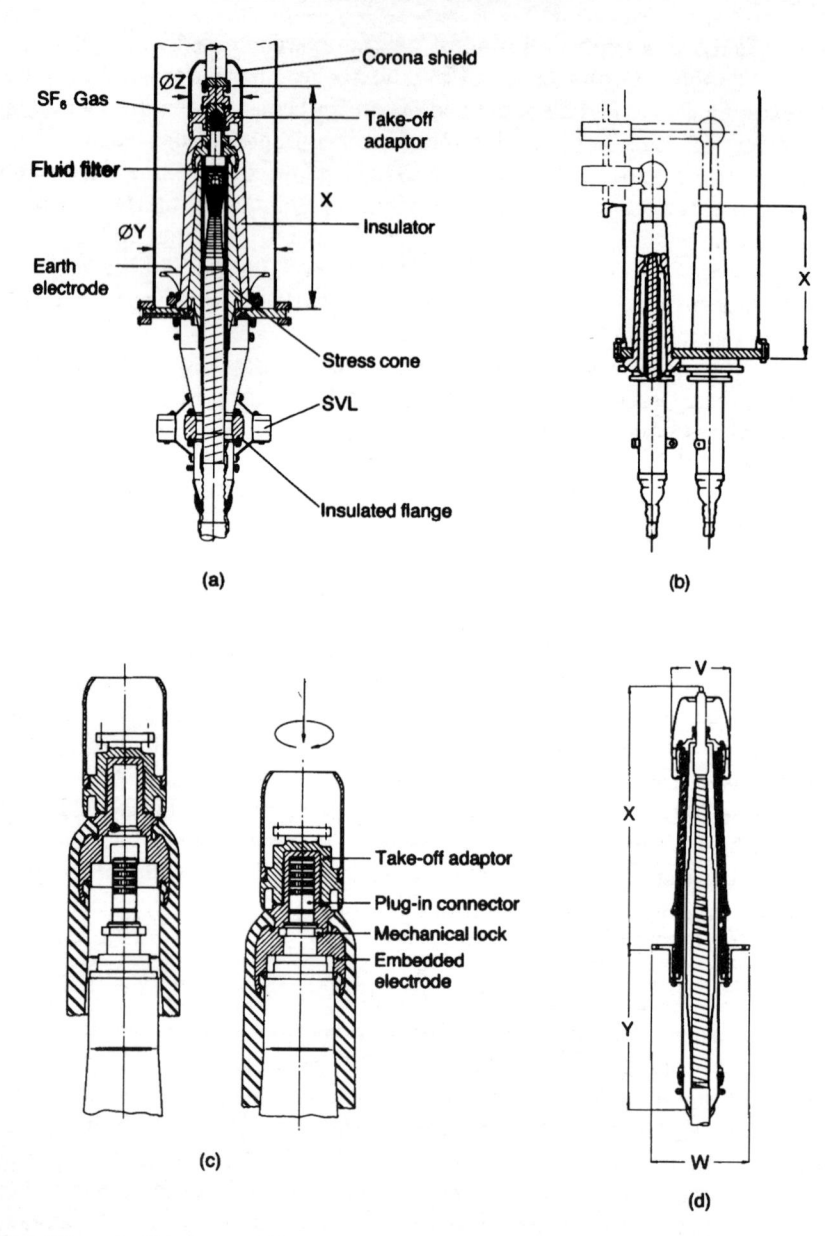

Fig. 35.2 SF$_6$ and oil-immersed terminations for fluid-filled cables: (a) 400 kV SF$_6$; (b) 132 kV 3-phase SF$_6$; (c) assembly of plug-in connector; (d) 275 kV fluid immersed

Take-off connector
The take-off connector is designed to be compatible with the particular type of busbar and its current rating and is therefore designed in conjunction with the manufacturer of the metal-clad equipment. The design shown in fig. 35.2(a) matches the 4000 A rating of the busbar and consists of a plug-in connection to the embedded electrode and a bolted palm connection to the busbar.

Table 35.3 Dimensions of terminations into SF_6 insulated equipment

U (kV)	BIL (kVp)	Length X (mm)	Diameter (mm) Trunking, Y	Diameter (mm) Shield, Z
66	342	583	300	110
132	650	757	300	110
275	1050	960	480	200
400	1425	1400	540	280
500	1550	1400	630	280

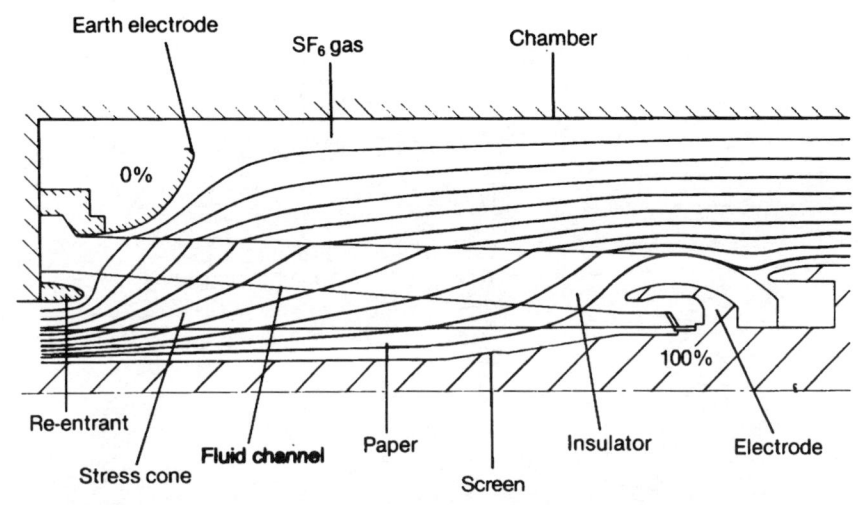

Fig. 35.3 Field plot of 400 kV SF_6 termination showing 10% equipotentials

Stress control

At 275 and 400 kV particularly short insulator lengths are achieved by employing a close fitting epoxy resin stress cone to reduce the fluid channel depth to 1 mm, thus increasing its electric strength. Stainless steel mesh filters are positioned at the ends of the fluid channel to prevent the introduction of fluid-borne particles. Stress control is achieved by the relative positions of the earth electrode, the stress cone re-entrant, the embedded connector screen and the corona shield (fig. 35.2(a)) which are determined from a computer field plot. At 66 and 132 kV, either cast resin stress cones or capacitor cones can be fitted.

Special performance requirements

The termination is required to withstand lateral thrust from the busbar and vibration from the switchgear breakers in addition to the thermomechanical cable loads and the fluid pressure.[20,22] The termination must not exceed either the cable or the busbar operating temperature. Fortunately the dissipation of heat is enhanced by convection in the SF_6 gas and by longitudinal conduction in the conductor and the insulator. The insulated flange in the metal sleeve (fig. 35.2(a)) experiences exceptionally steep fronted voltage transients of the order of 10^{-1}–$10^{-2}\,\mu s$ due to the close proximity of the switchgear.[23,24] Flashover is prevented by connecting two sheath voltage limiters directly across the flange to present a low surge impedance (figs 35.4(a) and 35.4(c)). The earthing arrangement in fig. 35.4(b) is not recommended as circulating current of high magnitude can flow in the earth loop.[12]

Oil-immersed terminations

These terminations are in the tanks of oil-filled transformers or switchgear and are used less frequently than outdoor or SF_6 metal clad types. Figure 35.2(d) shows a 275 kV termination and table 35.4 gives typical dimensions.

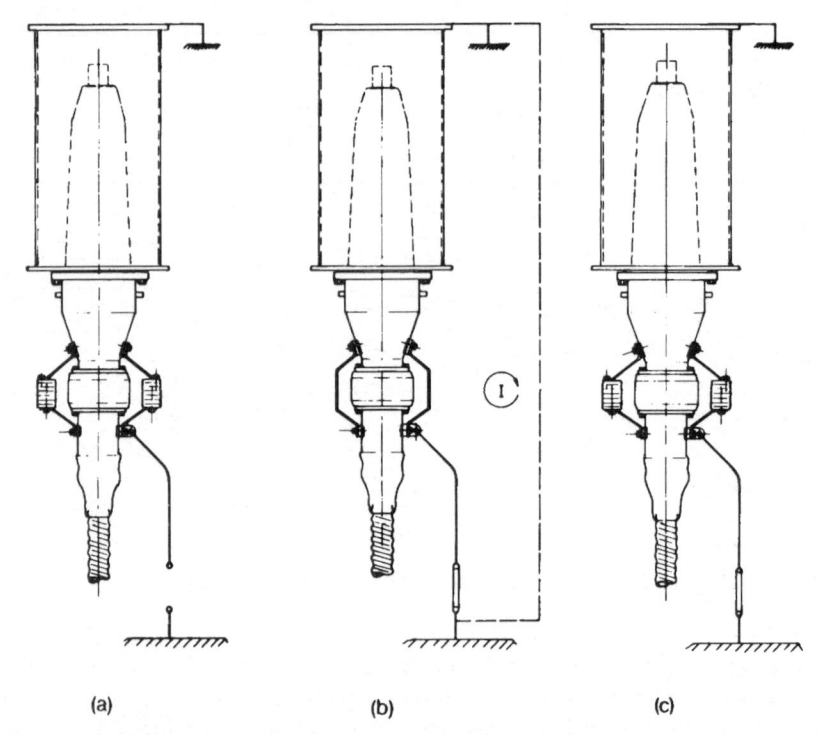

(a) (b) (c)

Fig. 35.4 Sheath bonding arrangements: (a) unearthed, end-point bonded; (b) direct earthed sheath; (c) direct earthed sheath with SVLs

Table 35.4 Dimensions of fluid-immersed terminations

U (kV)	BIL (kVp)	X^a (mm)	Y^a (mm)	Total (mm)	V^a (mm)	W^a (mm)
400	1425	1640	1210	2850	385	565
275	1050	1345	890	2235	335	530
132	650	1010	905	1915	270	460
66	342	730	450	1180	205	320
33	194	540	320	860	200	300

Length (mm) and Diameter (mm) columns as shown.

[a] See fig. 35.2(d)

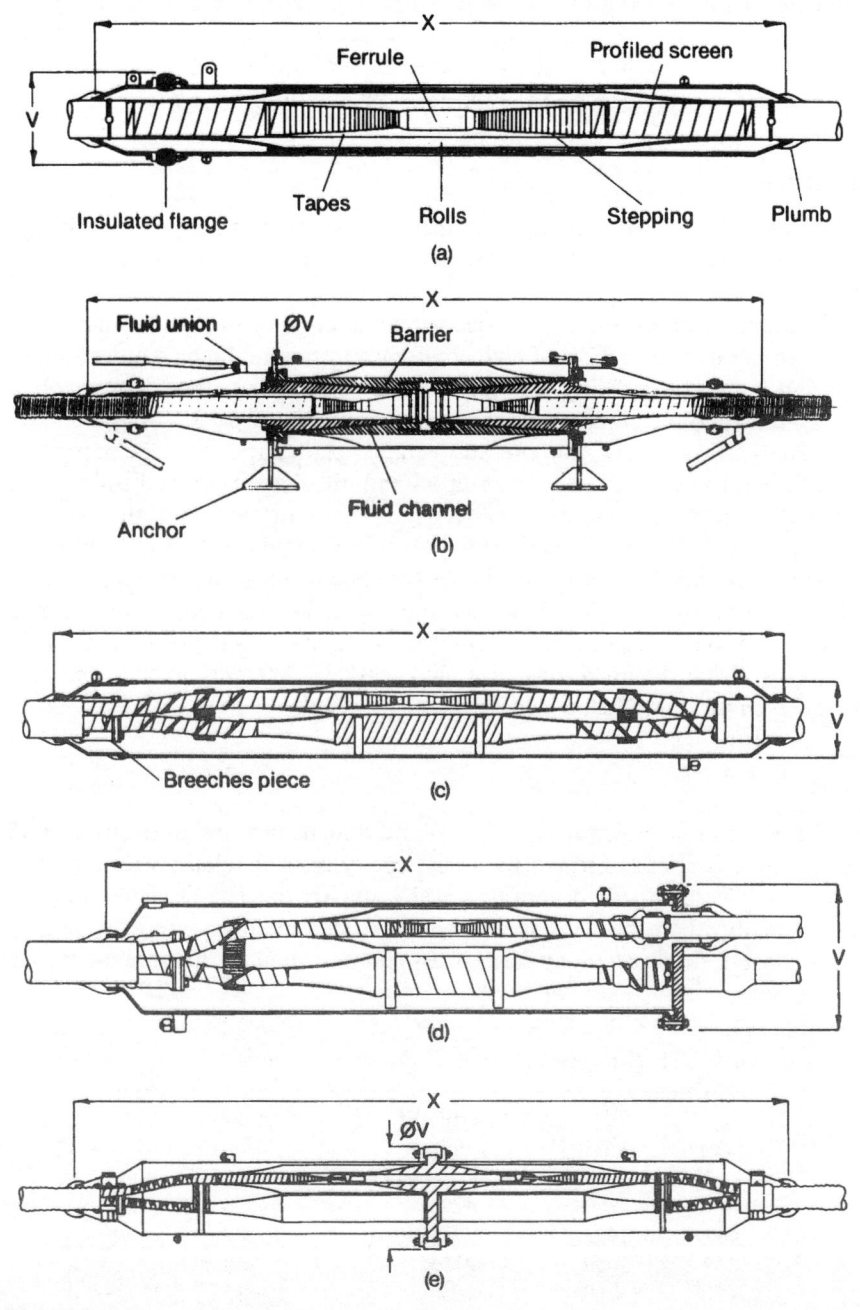

Fig. 35.5 Cable joints for fluid-filled cables: (a) 400 kV single-core straight joints; (b) 400 kV single-core stop joint; (c) 132 kV 3-core straight joint; (d) 66 kV 3-core trifurcating joint; (e) 132 kV 3-core stop joint

Transformer oil exhibits a lower electrical strength than cable fluid because of its application as a bulk insulator and because it is non-degasified and subject to contamination from debris and oxidation. To prevent flashover during the d.c. voltage commissioning tests on the cable, care must be taken to ensure that the transformer oil is dry and free from fibres. Similar materials and stress control are used in both the oil immersed termination and the outdoor termination.

Straight joints

A typical 400 kV joint without its joint shell protection is shown in fig. 35.5(a) and dimensions are given in table 35.5. Different types of joint shell protection are shown in fig. 35.6.

The conductor connections are suitable for conductors of up to 3000 mm^2. The high operating temperature of 90°C and high conductor retraction loads require a high creep strength and long-term stability. Compression ferrules are used on copper conductors and MIG welded connections on aluminium conductors (figs 35.7(b) and 35.7(c)). To achieve high electric strength, thin pre-impregnated plain paper or crêpe paper tapes are applied adjacent to the conductor connection and pre-shaped profiled rolls are applied thereafter. Crêpe carbon tapes are applied over the conductor connection and over the paper roll profiles to achieve a smooth conducting boundary, free of fluid-filled gaps and stress raisers. Slim joints are available for special applications and these employ flush bronze welded and MIG welded connections for copper and aluminium conductors and hand taped insulation, the joint shell being insulated with heat shrink sleeves. Typical dimensions of a slim 400 kV 2000 mm^2 joint are 2800 mm in length and 146 mm in diameter over the shell.

Stop joints

A 400 kV stop joint is shown in fig. 35.5(b) and dimensions are given in table 35.6.

The hydraulic barrier is formed by a cast epoxy resin insulator with an embedded electrode which is clamped to and electrically shields the ferrule. The flange on the insulator is clamped to the joint shell. Two fluid feed channels are provided and are formed by close-fitting cast epoxy resin stress cones located in the bore on each side of

Table 35.5 Dimensions in single-core straight joints

U (kV)	BIL (kVp)	Length, X[a] (mm)	Maximum diameter, V[a] (mm)
525	1800	4293	300
400	1425	2718	280
275	1050	2156	265
161	750	1650	215
132	650	1562	205
66	342	1422	190
33	194	600	165

[a] See fig. 35.5

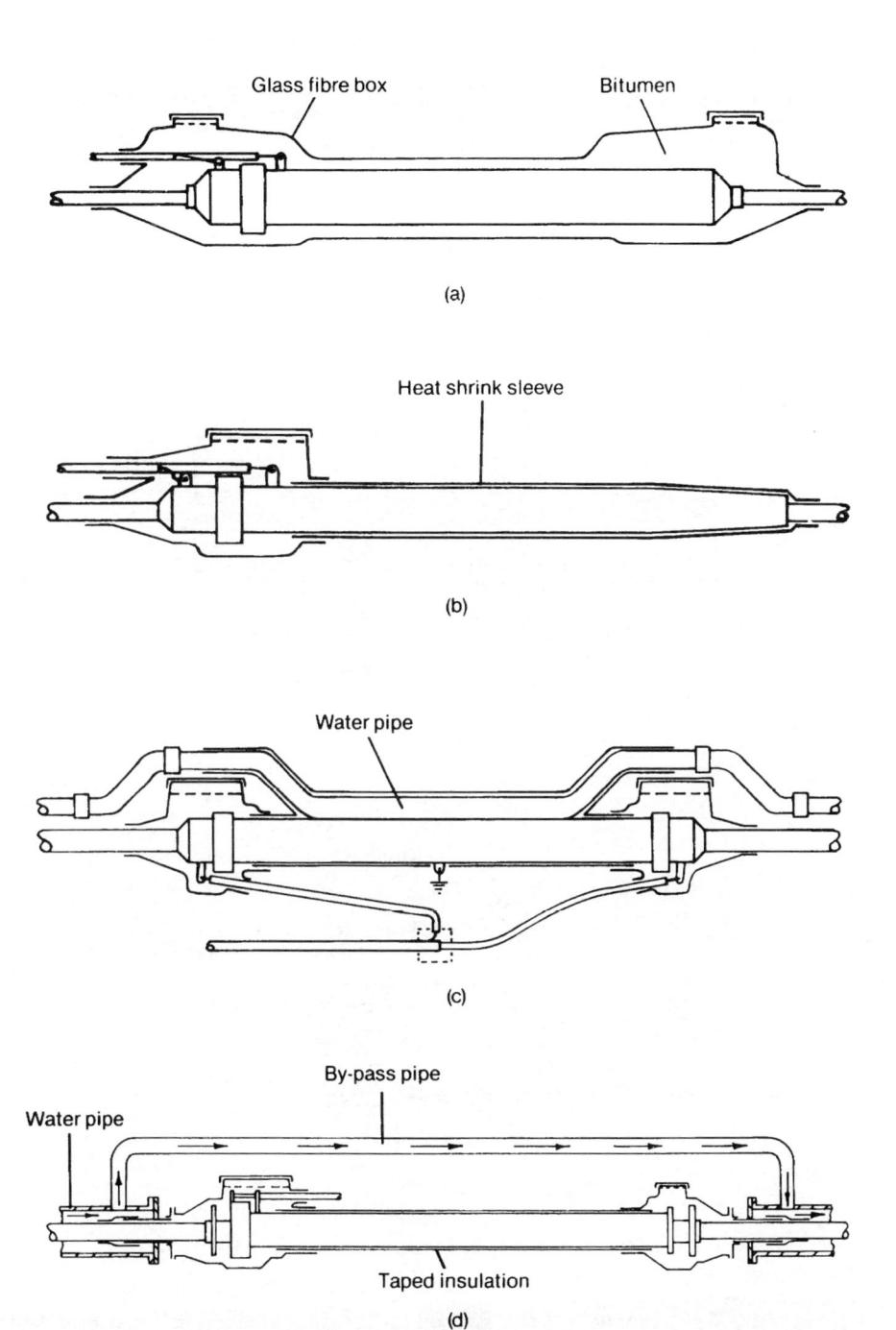

Fig. 35.6 Arrangements for straight joints: (a) 400 kV conventional protection; (b) 275 kV low thermal resistance protection; (c) 400 kV water cooled joint; (d) 400 kV joint in integrally cooled cable

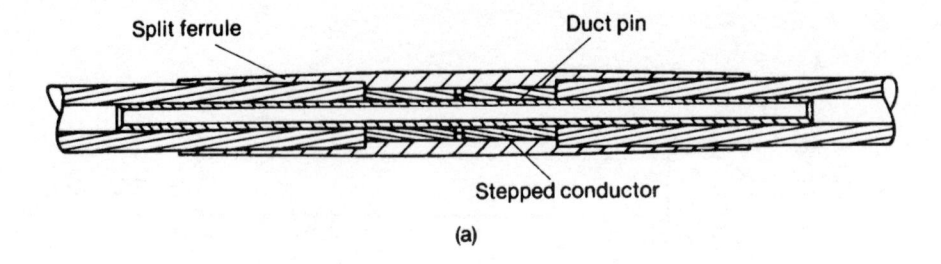

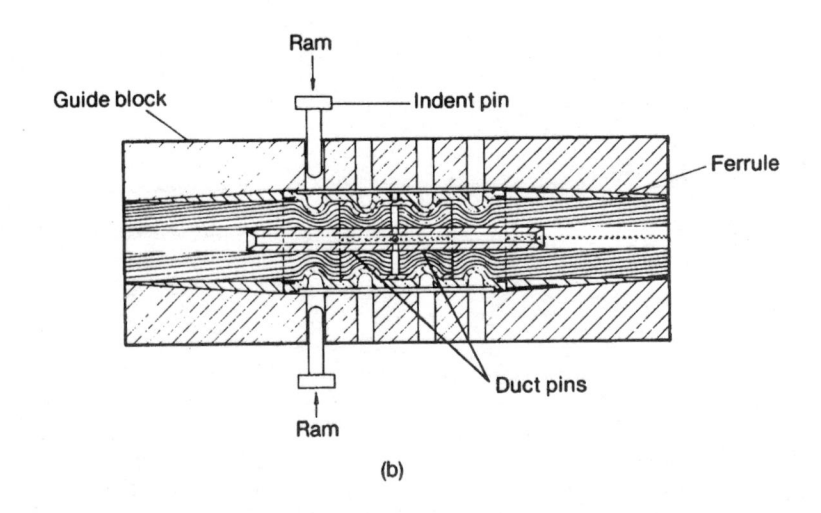

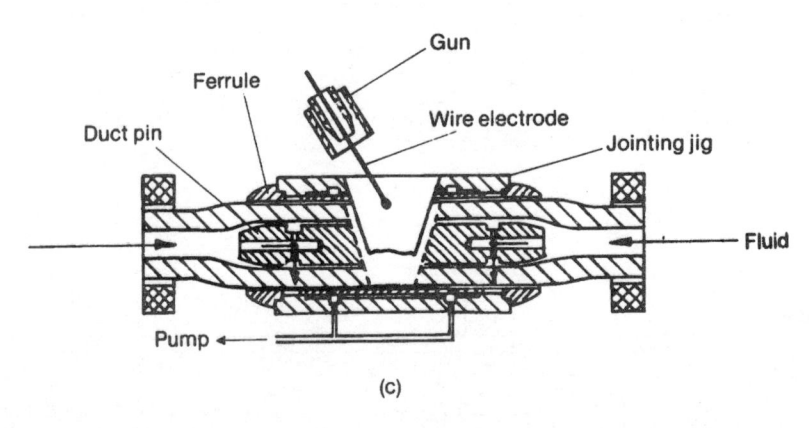

Fig. 35.7 Conductor connectors: (a) soldered; (b) compression; (c) MIG welded

the barrier. At 275 kV and 400 kV the fluid channels are sealed at each end by stainless steel filters. Computer field plotting is employed to determine the electrical geometry of the joint, primarily to achieve a uniform stress and dielectrophoretic force distribution within the fluid channel.[25] The latter is the force experienced by an uncharged particle in a non-uniform electric field. A uniform distribution minimises the possibility of the accumulation of microscopic fluid-borne particles in the fluid channel.

Table 35.6 Dimensions in single-core stop joints

U (kV)	BIL (kVp)	Length, X^a (mm)	Maximum diameter, V^a (mm)
400	1425	4370	530
275	1050	3600	455
161	750	3600	455
132	650	2110	320
66	342	1780	255
33	194	1780	255

[a] See fig. 35.5(b)

ACCESSORIES FOR 3-CORE FLUID-FILLED CABLE SYSTEMS (33–132 kV)

Design principles for conductor connection, insulation and stress control are similar to those for single-core accessories. The 3-core cable differs in that fluid pressure is maintained by ducts located in the spaces between the three cores. Unlike a single-core cable, the conductor contains no duct and the fluid flow is under the sheath at earth potential. The straight joints are required to transmit the fluid flow from one cable sheath to the other and the stop joints to form a hydraulic barrier.

Terminations

Figure 35.1(c) shows a 33 kV pole-mounted 3-phase termination which is unusual in that it permits the 3-core cable to be made off directly into the termination. This termination would be prohibitively large at higher voltages and instead the 3-core cable is divided into three single-core cables at either a trifurcating or a splitter joint. The single-core cable has a corrugated, fluted, or loose sheath to permit the flow of fluid between the 3-core cable and the terminations. Standard single-phase terminations are employed with the addition of a special sheath seal containing a drop-out impregnation valve to prevent the ingress of impregnant during evacuation.

3-core joints

Figure 35.5(c) illustrates a straight joint. The three insulated and screened cores are bound together and bandaged to form a tight fit in the joint shell and prevent lateral movement and buckling due to thermomechanical expansion of the cable cores. The cores are sealed during jointing by a breeches piece containing a synthetic rubber seal. Following evacuation and impregnation, communication for fluid flow between the sheaths is made by inserting a key through a port in the joint shell to operate a valve in each breeches piece. The sheath cut and breeches piece are designed to permit a small cyclic movement of the cores, of up to 6 mm, without abrasive damage.

Figure 35.5(d) shows a 66 kV trifurcating joint. This is essentially the same as a straight joint but with the addition of drop-out impregnation valves in each of the single-core tails. A stop joint in which the barrier is formed by an integral epoxy resin casting with three solid conductor rods embedded in separate insulated bushings is illustrated in fig. 35.5(e). Typical dimensions are given in table 35.7.

Table 35.7 Dimensions of 3-core FF joints

U (kV)	Straight X[a] (mm)	Straight V[a] (mm)	Trifurcating X[a] (mm)	Trifurcating V[a] (mm)	Stop X[a] (mm)	Stop V[a] (mm)
132	2335	265	2030	315	3200	440
66	1645	145	1855	300	2130	290
33	1205	140	1415	300	2060	290

[a] See fig. 35.5

A slim 132 kV 3-core straight joint with a length of 2185 mm and a joint shell diameter of 140 mm has been employed for submarine crossings.

ACCESSORIES FOR GAS-FILLED CABLES (33–132 kV)

The predominant type of cable is the pre-impregnated paper type (chapter 31). The accessories are similar to those for fluid-filled cables. Gas pressure is maintained by a central duct in single-core cables of long length and by a duct between the cores in 3-core cables. 3-core cables are manufactured up to 66 kV and single-core cables up to 132 kV. Although the operating temperature is 85°C the maximum conductor size of 630 mm^2 is comparatively small compared with fluid-filled cables and permits the use of soldered conductor connections which are specially elongated to increase the creep strength (fig. 35.7(a)).

Terminations

Figure 35.8(a) shows a 132 kV outdoor termination. The gas pressure is contained via a high pressure porcelain insulator. The stress is controlled by a pre-shaped stress control profile of pre-impregnated paper rolls, surmounted by a brass toroid insulated with impregnated crêpe paper. A gas entry chamber is sealed to the sheath cut, to which is connected an insulated pressure equalisation pipe of PTFE, which terminates at the conductor stalk. The gas entry chamber permits pressurisation and depressurisation of the cable through the conductor and also under the sheath on both single and 3-core cables. The porcelain is filled with a viscous oil/polyisobutylene compound with a gas space to allow for expansion. At 33 kV an alternative termination is available in which the cable is terminated to a conductor rod embedded in a cast epoxy resin pneumatic barrier. This is sealed to the base of a low pressure porcelain partly filled with compound.

Terminations into SF_6 switchgear employ the same insulator as the fluid-filled design (fig. 35.2(b)) and fluid immersed terminations are also similar to the fluid-filled design (fig. 35.2(d)).

Joints

Figure 35.8(b) shows a 132 kV single-core straight joint of simple construction. The insulation consists of hand applied pre-impregnated tapes. The joint is not filled with

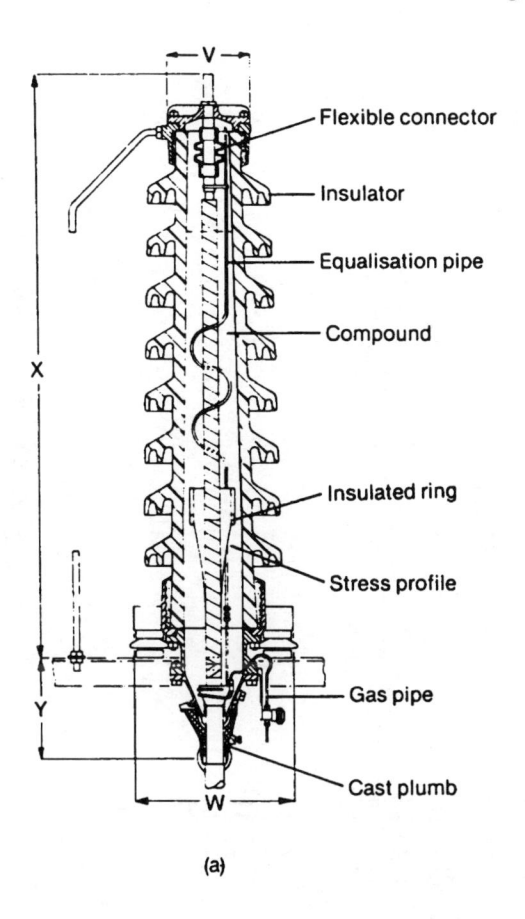

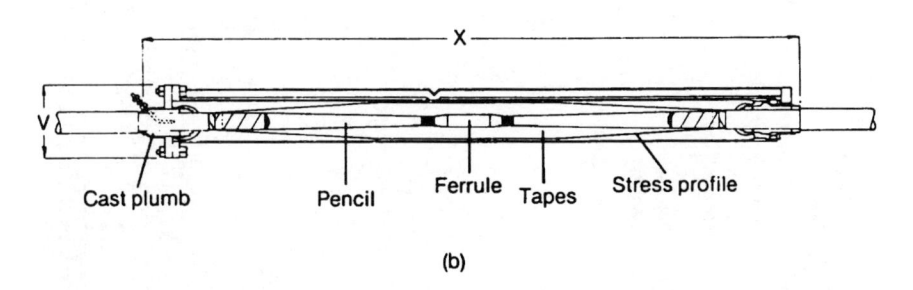

Fig. 35.8 Accessories for gas-filled cable: (a) 132 kV outdoor termination; (b) 132 kV straight joint

compound and sealing of the sheath cut is not required. Gas-filled accessories are not evacuated and the cable can be pressurised from the terminations, thus reducing the need for gas unions in the joint shell. The 3-core straight and trifurcating joints are also similar to simplified versions of the fluid-filled joints shown in figs 35.5(c) and 35.5(d). The terminating joints to other type of cable employ a pneumatic barrier and are similar to the stop joint shown in fig. 35.5(e) for 3-core applications up to 66 kV and for single-core applications up to 132 kV (see fig. 35.10(c) later). A low cost single-core

520 Electric Cables Handbook

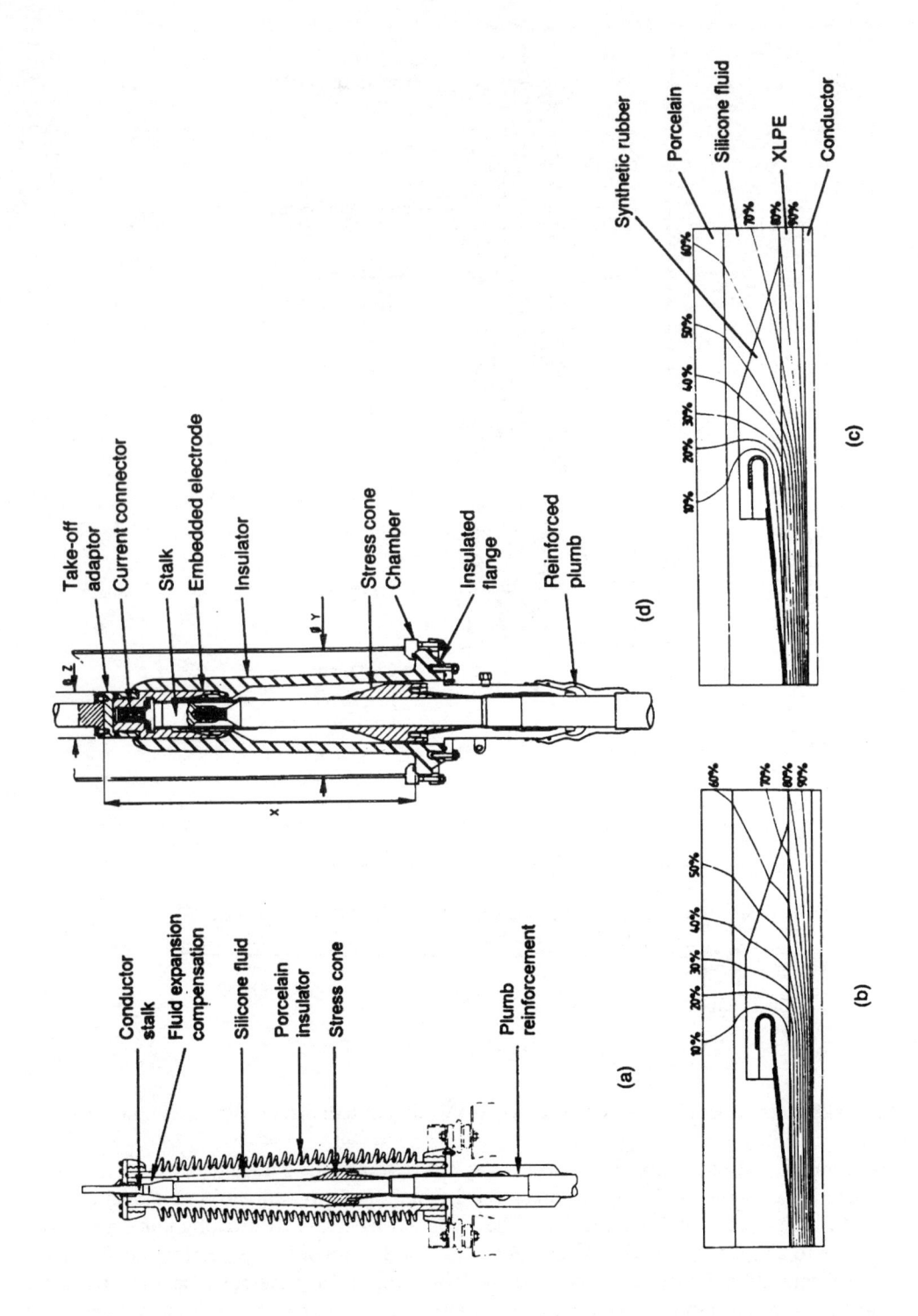

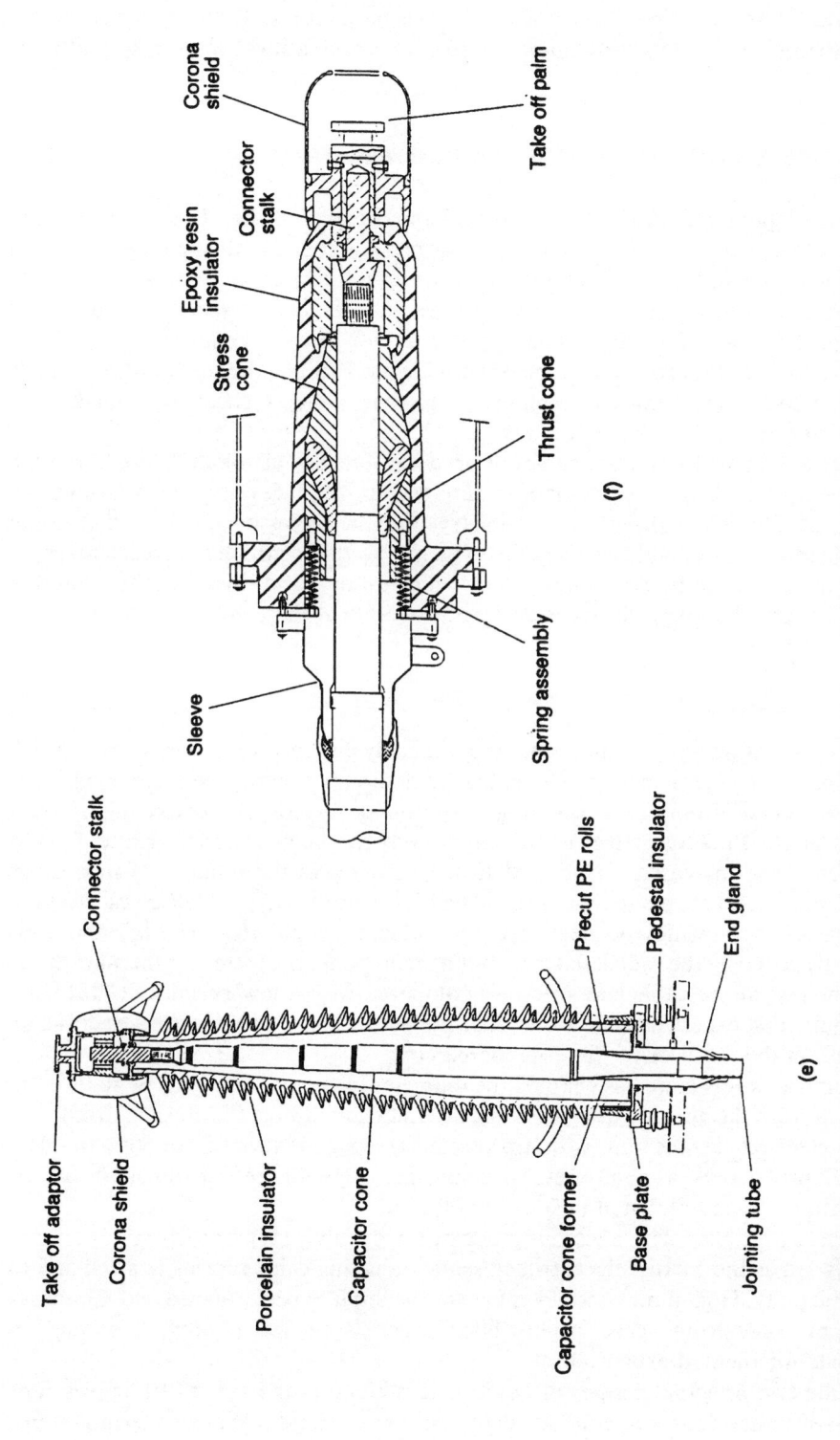

Fig. 35.9 XLPE cable terminations: (a) 132 kV outdoor termination; (b) high permittivity stress cone; (c) low permittivity stress cone; (d) 132 kV SF$_6$ termination; (e) 400 kV prefabricated capacitor cone type outdoor termination; (f) 400 kV prefabricated dry type SF$_6$ termination

straight joint is used and it is contained within a lead sleeve reinforced with epoxy resin impregnated glass fibre tape. A low cost 3-core splitter joint is available which separates and sheaths the cores, thus avoiding the need for a trifurcating joint and special tails.

ACCESSORIES FOR POLYMERIC CABLES (66–550 kV)

The cables are predominantly XLPE insulated and, because of their large diameter, are of single-core construction. A metallic sheath over a water swellable tape is preferred to prevent the radial and longitudinal ingress of water into the cable. Where appropriate the accessories incorporate the service proven features of accessories on pressure-assisted cables, such as compression connectors for copper conductors and MIG connectors for aluminium conductors. Sheath closures to metal glands on joint shells and terminations are similarly reinforced to withstand vertical soil loads and longitudinal oversheath retraction loads.

In order to be able to produce cables of practical dimensions for EHV operation the design stress at both the conductor and core screen interfaces must be substantially higher than for HV cables. The geometry for the accessory is designed using sophisticated computer aided techniques which, together with the correct choice of suitable materials, in-house manufacture and testing, and high quality jointing techniques, can reliably produce accessories to operate at these stresses.

Outdoor terminations

Figure 35.9(a) shows a porcelain insulator specifically designed to terminate HV XLPE cables. The cable core screen is terminated by a slip-on rubber stress cone, which is factory moulded from an insulating compound based on chlorosulphonated polyethylene (CSP). Alternative rubber compounds can be considered such as EDPM and silicone. The increased elastic stretch of CSP reduces the number of mouldings required to cover the cable design range. The high permittivity of CSP gives excellent stress control, depressing the field into the cable dielectric (fig. 35.9(b)) and thus significantly reducing the radial and longitudinal components of stress at the start of the stress cone and at the cable core interface compared with a low permittivity rubber[26] (fig. 35.9(c)). Test experience has shown that the performance of these areas is critically dependent on the finish of the hand-prepared core.

The compact stress cone has enabled the diameter of the porcelains to be reduced in comparison with the anti-fog designs.[14] The alternate long–short (ALS) shed profile has the same creepage distance but with a protected creepage reduced from 50% to 40%. The ALS profile has a good wet pollution performance and is resistant to the accumulation of wind blown atmospheric pollution.[15]

The insulator is filled with a viscous grade of silicone fluid. Hydrocarbon oils are not recommended as these cause the extruded semiconducting cable screens to swell and to lose conductivity. High fluid viscosity enhances the impulse performance and minimises the risk of leaks from seals. An air-filled space is provided above the fluid to compensate for thermal expansion.

There are two principal designs of outdoor termination for EHV XLPE cables. The first is based on the capacitor cone stress control used in fluid-filled cable terminations.

The thermal expansion of the XLPE core is significantly greater than that of the paper capacitor cone. To prevent damage in service the capacitor cone is prefabricated in the factory by wrapping on to an epoxy resin impregnated paper former. The former is designed such that, when assembled on to the stripped XLPE core on site, a small radial channel is formed to allow the core to expand. The termination is impregnated with a silicone fluid which is maintained under pressure using a pressure tank. A 400 kV prefabricated capacitor cone outdoor termination is shown in fig. 35.9(e). Alternatively a rubber stress cone can be employed as for HV terminations.

Terminations into SF_6 insulated metal enclosed equipment

A 132 kV insulator which complies with IEC Publication 859 dimensions[21] is shown in fig. 35.9(d) and dimensions of the 66 kV and 132 kV designs are given in table 35.3. The insulators are cast epoxy resin and are the same as those employed for fluid-filled cables with a solid embedded electrode to prevent SF_6 leakage into the insulator and thence to overpressurise the metal glands and enter the cable sheath and conductor. Stress control is achieved by high permittivity CSP rubber stress cones which fit inside the insulators. The insulators can be filled with viscous silicone fluid or low pressure SF_6 gas, the former being preferred because of the reduced risk of leakage. Compensation for fluid thermal expansion is provided by either a small fluid-filled pressure tank or a gravity fed reservoir.

An alternative is the prefabricated dry design which does not need to be filled with insulating fluid. A 400 kV termination is shown in fig. 35.9(f) and the dimensions of the prefabricated dry designs from 66 kV to 400 kV are given in table 35.8. All other interface dimensions comply with IEC Publication 859. The rubber stress cone is designed to be an intimate fit in the bore of the epoxy resin insulator. A spring loaded thrust ring ensures that air is completely excluded from the electrically stressed interfaces during expansion and contraction in service. This design requires the cable dimensions to be tightly toleranced in manufacture and the jointing to be precise. The dimensions of the range of stress cone bores are required to be compatible with the integral steps in the range of cable diameters.

Terminations into oil immersed terminations

These terminations are seldom required. At 66 kV to 400 kV the terminations for SF_6 insulated metal enclosed equipment are employed as these have an adequate external surface creepage distance for operation in transformer oil.

Table 35.8 Dimensions of dry type terminations into SF_6 insulated equipment

U (kV)	BIL (kVp)	Length, X (mm)
66	342	480
132	650	530
275	1050	650
400	1425	1150

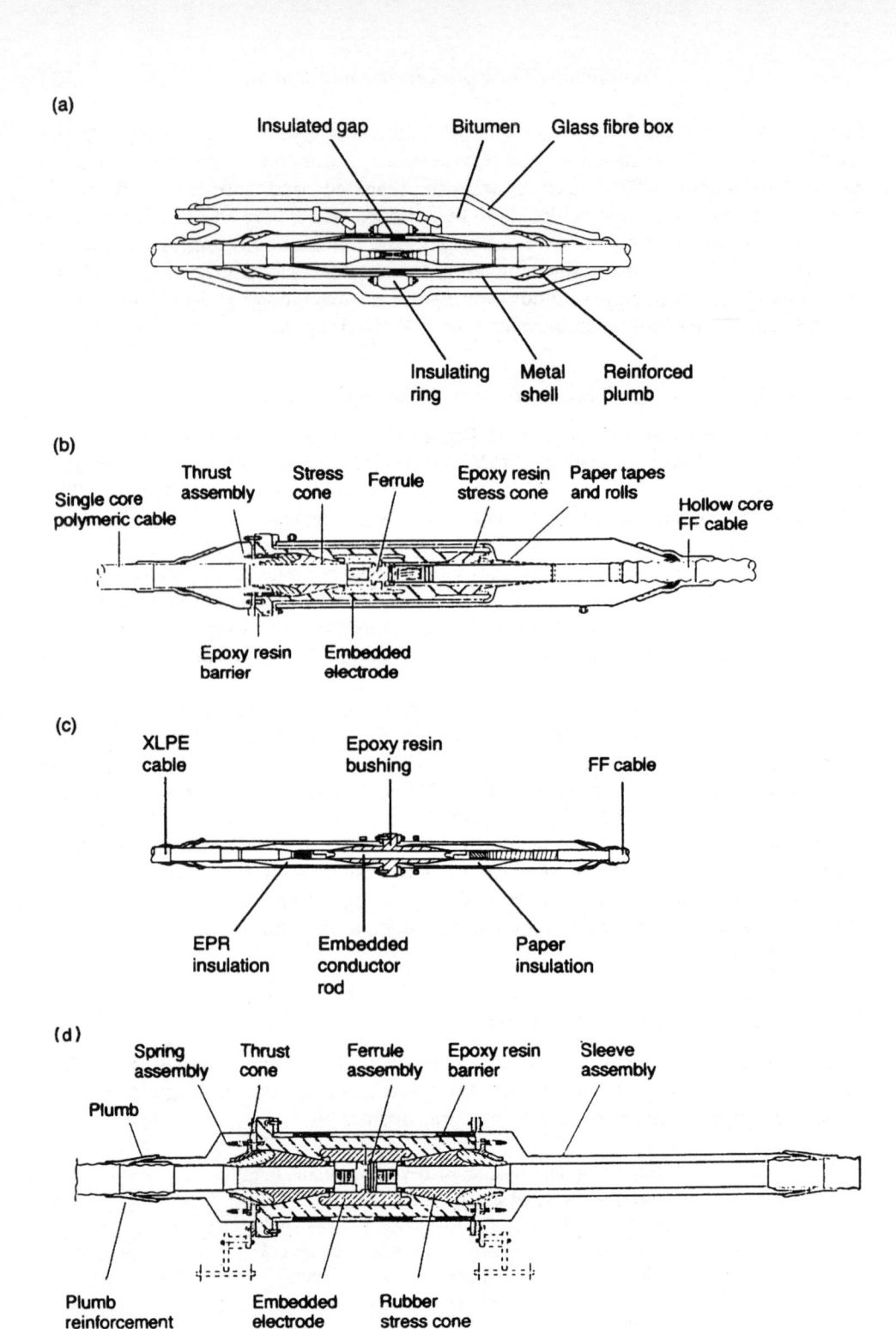

Fig. 35.10 Cable joints for XLPE cable: (a) 132 kV joint with metal sheath and bitumen protection; (b) 132 kV transition joint; (c) 66 kV transition joint; (d) 400 kV prefabricated joint; (e) 132 kV prefabricted one piece rubber moulded joint

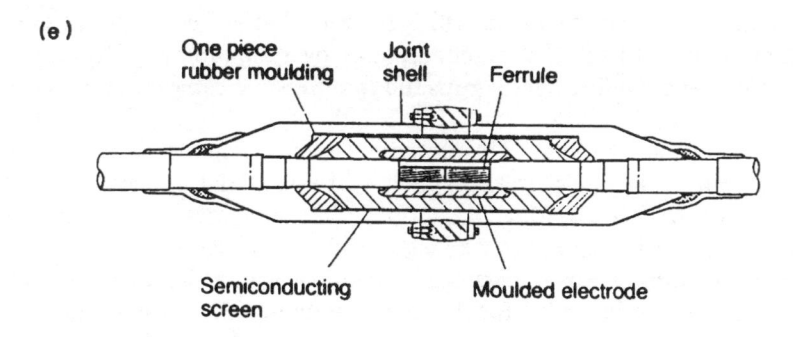

(e) One piece rubber moulding / Joint shell / Ferrule / Semiconducting screen / Moulded electrode

Fig. 35.10 Continued

Straight joints

Three main types of joint are in use at 66 kV and above, being characterised by the method of re-insulating the cable. Compression ferrules are used to connect copper conductors and MIG welds to connect aluminium conductors. The joint design either must be tolerant to longitudinal retraction of the cable insulation[7,8] or must constrain movement. Most designs of joint are encased within a metal shell (fig. 35.10(a)), insulated from earth using the techniques evolved for specially bonded fluid-filled cable. It is preferred to fill the shell with a thermosetting resin to constrain movement of the joint and cable insulation and to improve the radial dissipation of heat.[26]

Taped joint
This is the most versatile design and is preferred for general application. The joint in fig. 35.10(a) has been insulated with self-amalgamating EPR tape and screened with a similar semiconducting grade. The act of stretching the tape activates amalgamation to the underlying layers, this being enhanced by thinning of the tape and by cumulative compressive force. Although joints can be hand insulated it is preferred to use a taping machine to improve speed and quality. This design of joint readily accommodates variations in cable dimensions and can be used to connect different sizes of cable.

Pre-fabricated joints
This category employs factory moulded and tested insulation, permitting the jointing processes of core pencilling and taping to be eliminated. The same high level of skill is required in the removal of the cable core screen and in the smoothing of the core insulation. The joint tends to employ more complex components with special tooling needed to fit them. Dimensional accuracy of manufacture of the cable and accessories and of hand preparation of the cable is paramount. Prefabricated designs are best suited to applications in which the ambient jointing conditions are adverse and time is limited, such as in a confined tunnel under a roadway.

The simplest design is a one-piece elastomeric moulding (fig. 35.10(e)),[27] usually EPDM rubber, which incorporates an embedded semiconducting rubber electrode to screen the ferrule and the core cuts. The large size of the moulding limits the radial stretch that can be achieved by hand, which requires that a special tool be employed. The tool stretches it on to a hollow mandrel placed over the cable and then extracts the mandrel, so that the moulding tightly grips the two prepared cable cores. It is necessary to remove the cable sheath for twice the length of the moulded insulation, thus increasing the length of the joint. A more complex design employs one large EPDM centre moulding to fit all cable sizes.[28] This is part pulled and part floated on to two constant-diameter stress cones, which are termed 'cable adaptors'. It is not necessary to remove extra cable sheath, and thus the length of the joint shell is reduced. A range of cable adaptor mouldings needs to be stocked. Another variant with design advantages employs a rigid epoxy resin casting for the centre moulding, shown in fig. 35.10(d). This permits smaller and more flexible rubber stress cones to be fitted by hand with no special tooling. Spring loaded thrust rings maintain the stress cones in intimate contact with the bore of the epoxy resin casting irrespective of thermal expansion of the cable and of compression set of the rubber.[29]

Field moulded joints

The cable factory extrusion and crosslinking process (curing) is emulated on site to reconstitute, consolidate and crosslink the polyethylene joint insulation.[27] This is a sophisticated process which requires a regime of accurately controlled temperature and pressure for in excess of 12 hours, irrespective of conductor size, ambient temperature and variability of the power supply. A disadvantage is that the moulded insulation needs to be shaped, smoothed and screened by hand; this produces a screen interface significantly inferior in quality to the extruded and bonded factory screen. The joint is most suitable for solidly bonded systems because of the problem of forming an insulated gap in the outer screen. The field moulded design of joint is suited to applications in which a semi-flush diameter is required, such as for submarine cables or trough installations. This design of joint has been employed in significant numbers in Japan on 275 kV XLPE cables installed in tunnels where there is limited space at the joint position. The design has also been successfully developed for 500 kV application.

The insulation can be directly injected by a small screw extrusion press bolted to a mould tool around the joint.[30] A simpler method is to apply the polyethylene in tape form and then to constrain it in a mould. The insulation is first heated to melt and consolidate the tapes to each other and to the cable. The temperature is further raised to initiate degradation of the dicumyl peroxide within the tape such that crosslinking is initiated. The temperature is progressively elevated to accelerate the process. Gas is evolved during the reaction and it is essential to keep it in solution by controlling the temperature accurately and by the application of pressure. Pressure can be applied by gas, by liquid or by springs. It is similarly important to control the rate of cooling to prevent the formation of bubbles and of contraction voids.

Transition joints

A prime requirement of the transition joint is that the pressurising medium in a FF or GF cable must not enter the polymeric cable. The low viscosity cable fluid will rapidly degrade the polyethylene insulation and screens. The polymeric cable system is not designed to withstand internal pressure. The designs of FF cable stop joint and the GF cable terminating joint are ideal for this application. They also have the capability of withstanding the unbalanced thermomechanical loads arising from the differences in physical size between FF and XLPE cable and in particular between 3-core FF and single-core XLPE cable.[3,31]

Figure 35.10(b) shows a 132 kV transition joint. This is based on a FF hollow-core cable stop joint which has the capability of feeding fluid to the conductor duct. Hydraulic segregation is formed by rubber seals, which are clamped between the embedded electrode in the epoxy resin casting and the ferrule. On the XLPE cable side an elastomeric stress cone is held in intimate contact in the bore of the barrier to form a dry dielectric interface.

3-core FF and GF cables do not have conductor ducts and thus a simple through-bushing casting can be employed of the type shown in fig. 35.10(c). It is necessary to increase the creepage length and the diameter of the bushing to match the design parameters of XLPE straight joints; thus transition joints tend to be larger than FF stop joints. On the XLPE side the epoxy resin bushing and cable are insulated with self-amalgamating EPR tape. It is preferred to have three individual bushings which can be insulated without physical interference between cores. The three cores are mounted on a common metal plate. The FF or GF cable side is then insulated with paper in the normal way. For large conductor sizes it is preferable to incorporate the bushings into separate single core joints which connect to the 3-core cable by single-core tails and a splitter joint.

ANCILLARY EQUIPMENT

Sheath bonding equipment

The connections between the cable sheaths (chapter 34, fig. 34.6) are housed within an accessible link box to enable commissioning tests to be conducted.[12] These tests consist of the application of 10 kV d.c. for 1 minute to each cable oversheath and the measurement of the imbalanced alternating circulating current through the links. The links may be housed in a street pillar or within a manhole, either in a diving bell box, which is suitable for occasional immersion in water at 1 m depth, or in a watertight sealed box.

Figure 35.11 illustrates cross-bonding link boxes with the links arranged to transpose the cable sheaths at a minor bonding section. A sheath voltage limiter (SVL) unit is housed in the box to permit disconnection and hence safeguard it from damage during the 10 kV d.c. sheath test. The SVL assembly comprises three non-linear resistors in star connection with each resistor connected to one link so that the insulated flange in each joint shell is bridged by two resistors in series.

The purpose of the SVL is to limit the voltage rise caused by an incident surge on the joint screen gap, insulated flange, cable sheath and bonding lead. Such surges are initiated by normal circuit switching operations, by lightning impulse and, on rare occasions, by flashover of the cable terminations. They travel along the cable in cylindrical mode between the conductor, sheath and earth. At the instant of arrival at the flange, the sheath current is interrupted, thus causing a transfer of energy to the electrostatic field by an instantaneous increase in voltage. Across the flange this can reach approximately 40% of the power cable BIL if the SVLs are not connected. The impedance of the two SVLs in series falls to approximately 1–2 Ω and limits the voltage

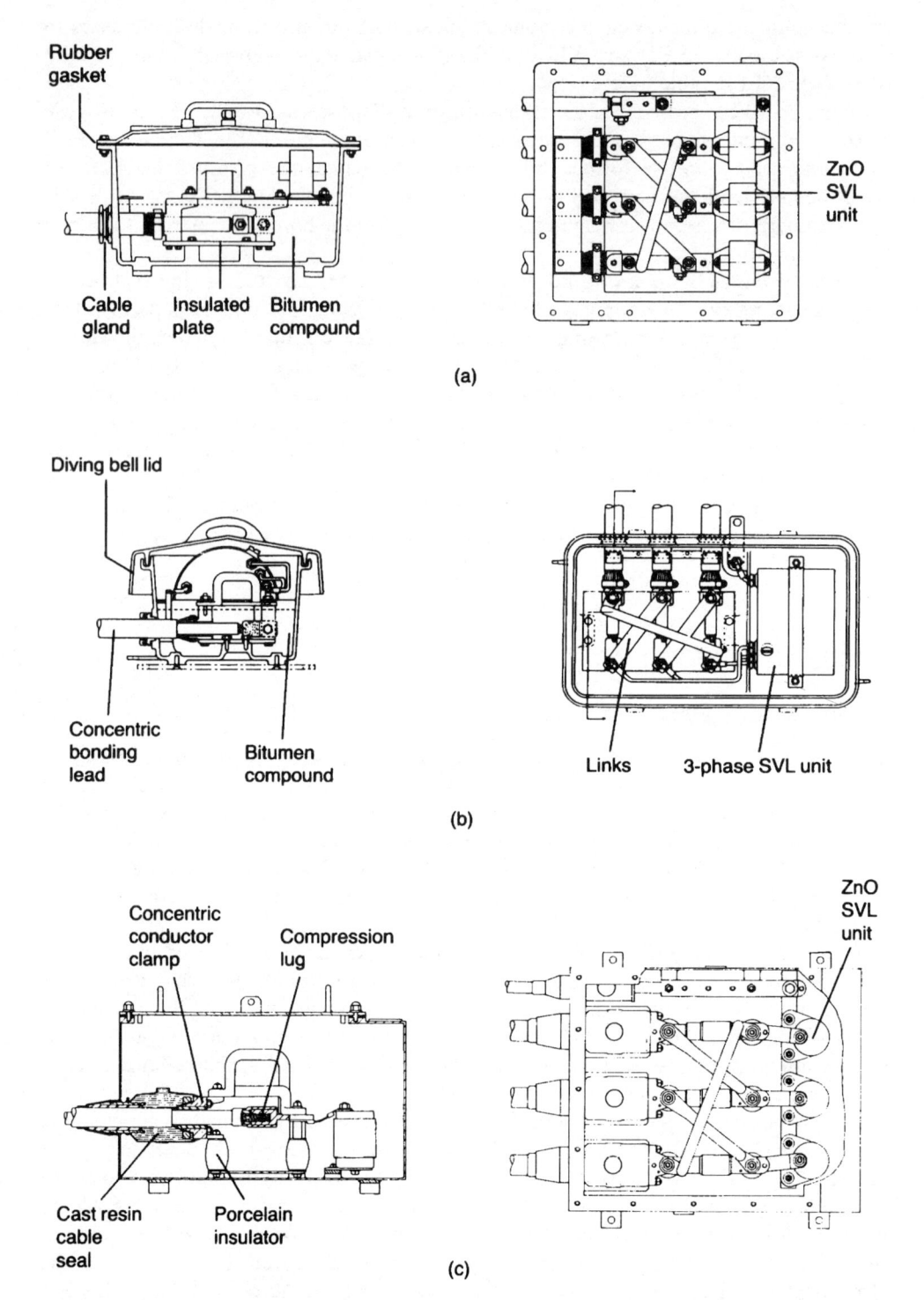

Fig. 35.11 Sheath bonding equipment: (a) cast iron bolted lid design; (b) cast iron diving bell lid design; (c) stainless steel bolted lid design

rise across the links to approximately twice the peak residual voltage (PRV) of each resistor, e.g. 40 kVp at 40 kAp for two SVL 60s. The bonding lead surge impedance and transmission time significantly reduce the effectiveness of the protection to the joint. It is normal practice therefore to minimise the impedance by employing concentric leads and the time by restricting the length to 10 m. Such arrangements limit the flange voltage to approximately 8% of the cable BIL, e.g. 125 kVp, for a 400 kV system.[12,13]

Typical SVL voltage versus current characteristics are shown in fig. 35.12. In normal service the maximum induced a.c. sheath voltage is less than 150 V r.m.s. to which the resistor acts as an open circuit. The SVL rating is based upon withstanding the calculated r.m.s. induced sheath voltage, which arises from an external through-fault, for two periods of 1 s each, this being the assumed maximum clearance time of the secondary circuit protection. To prevent damage to the SVL and to the link housing it is important that the user ensures that the sheath does not experience additional voltage rises, e.g. due to a high value of substation earth resistance. Zinc oxide (ZnO) resistors have now superseded silicon carbide (SiC). They exhibit the advantages of smaller size, increased rating for a single disc and more repeatable voltage versus current characteristics. The characteristics of SVL 20, 40 and 60 resistors (2, 4 and 6 kV r.m.s.) are shown in fig. 35.12 and are compared with an SVL 28 SiC resistor. At 10 kAp the impedances of the ZnO resistors tend to be less than that of SiC, whilst at rated voltage the ZnO resistors exhibit a much higher impedance. The SVL 60 ZnO unit (fig. 35.13) can withstand the 5 kV d.c. routine test on the oversheath, thus avoiding the need to open two-thirds of the link boxes in a cross-bonded circuit during circuit maintenance work.

Arresters with a spark gap in series with an SiC disc have occasionally been employed, with the advantage that the gap remains open circuit until a surge is seen. A disadvantage is that the effectiveness of protection is reduced by an increase in the PRV and in the time to flash over. For these reasons they are not recommended. Capacitors have been employed for special applications but they tend to be prohibitively large and expensive.

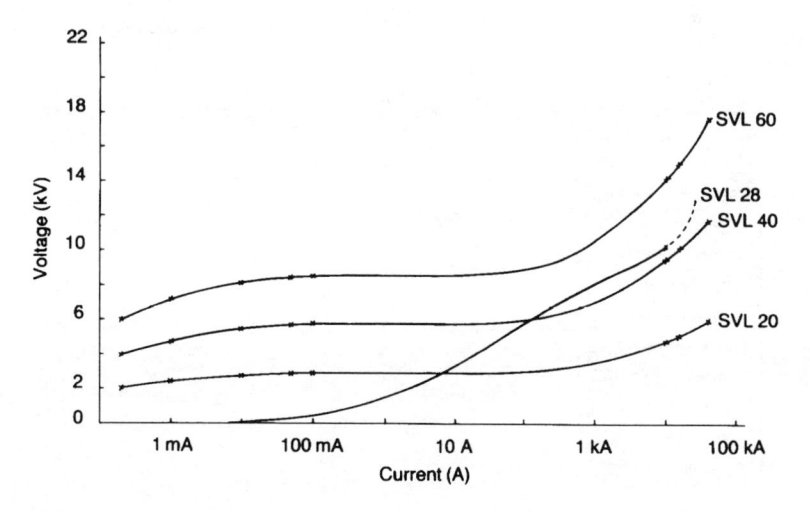

Fig. 35.12 Characteristics of sheath voltage limiters

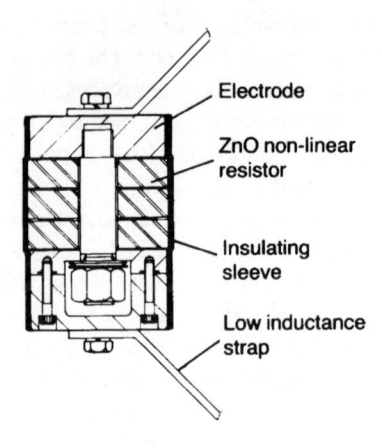

Fig. 35.13 SVL 60 ZnO unit

Pressure equipment

The pressure in the gas-filled cable is maintained by conventional gas cylinders (BS 5045) pressurised with dry nitrogen to 170 bar. The pressure is reduced to that of the cable by a regulator. The equipment is housed within a surface mounted street pillar.[9]

The pressure in the fluid-filled cable is maintained by pressure tanks. Figure 35.14 shows the equipment diagrammatically. The tank contains a number of sealed elements each containing approximately 4.5 litres of CO_2 gas, this gas being used because of its high solubility in cable fluid (120% by volume). The faces of the elements are flexible diaphragms either of corrugated tinned mild steel or of stainless steel. The volume around the elements is filled with degasified cable fluid. The tank is connected to the cable termination or stop joint by a pipe which contains an insulated link to permit

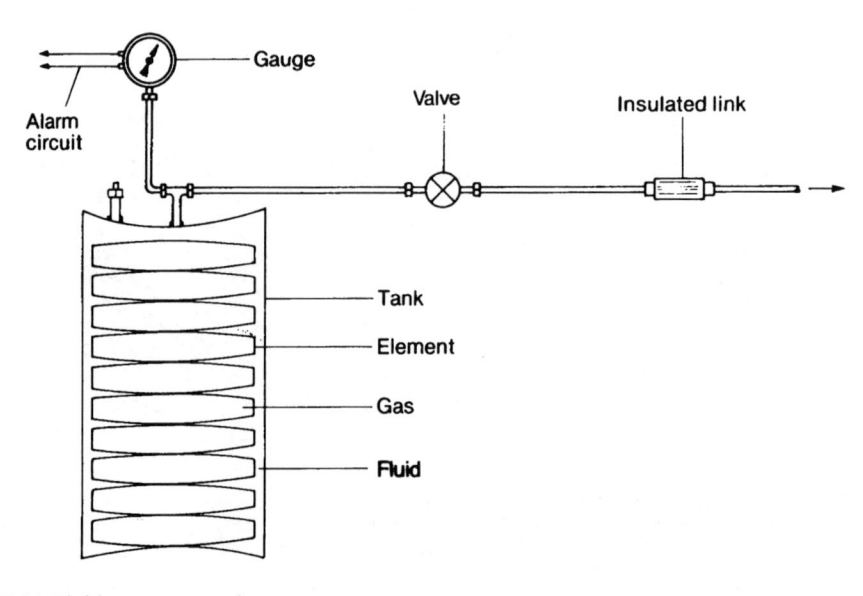

Fig. 35.14 Fluid pressure equipment

electrical testing of the cable sheath system. The gas in the elements is the source of pressure in the cable system. When the cable carries current, the heated fluid expands and flows into the pressure tank and acts upon the diaphragms to compress the gas. The fluid in the cable contracts upon cooling and is forced back from the pressure tank by the compressed gas. The pressure tank size is specified by its gas volume in litres and by the pre-pressurisation index (PP), which is the absolute pressure in atmospheres to which the element is pressurised in the factory. Table 35.9 gives typical dimensions of the tanks.

The pressure versus volume characteristic follows the combined gas law and details are given in fig. 35.15 for a 300 litre tank with a range of PP values up to 4. The minimum pressure is the PP level and the maximum pressure is that at which the diaphragms cease to be fully flexible. Mounting of the tanks is discussed in chapter 30.

Table 35.9 Dimensions of FF pressure tanks

Gas volume (litre)	Height (mm)	Diameter (mm)
44	640	565
88	970	565
135	1230	565
180	1500	565
225	1730	565
300	2200	565

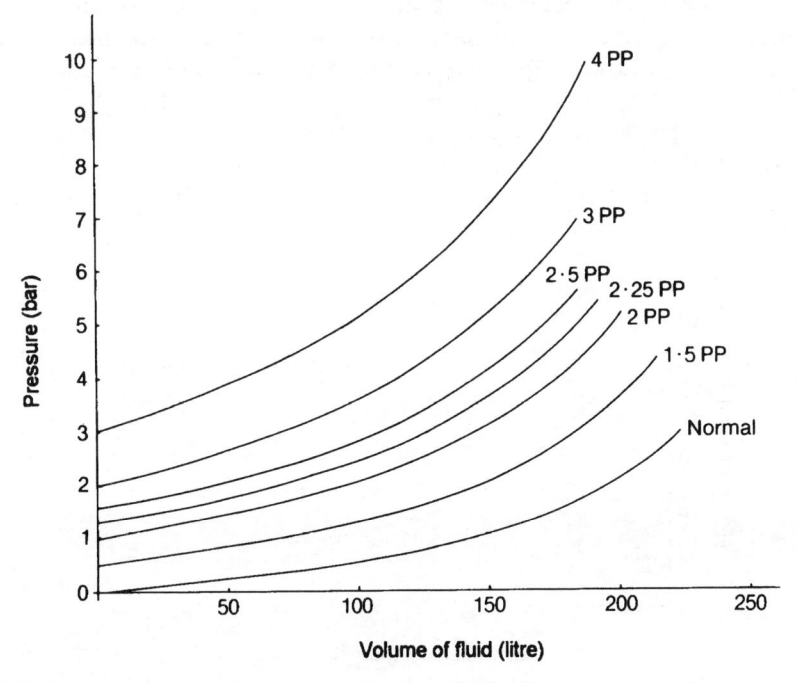

Fig. 35.15 Pressure versus volume characteristics of FF cable pressure tanks

532   Electric Cables Handbook

Figure 35.14 also shows a pressure gauge which enables the performance of the system to be checked visually. Two pressure-sensitive contact switches in the gauge sound an alarm if the fluid pressure falls, as a result of damage to the cable system, to either the minimum operating pressure or the prescribed emergency minimum pressure. It is usual to locate the pressure gauge and fluid pipe connections within manholes in a container or in a street pillar.

JOINTING METHODS

Conductor connections

Figure 35.7 includes the three main methods of conductor connection. Soldered connections are the simplest but exhibit an inferior hot creep strength as indicated in fig. 35.16, so that they are prone to failure due to the tensile load developed in heavily loaded conductors. Strength can be increased by lengthening the ferrule and by achieving a tight fit to encourage capillary penetration. Such ferrules are employed for gas-filled cables of conductor area up to 630 mm^2. Adequate capillary action is not achieved in larger conductors and for this reason soldered ferrules have been largely superseded for fluid-filled cable by compression ferrules and fusion welded connectors. Soldered ferrules are not acceptable for XLPE cable because the conductor is permitted to operate up to 250°C during short-circuit conditions (FF cable up to 160°C) and this is significantly higher than the eutectic temperature of solder. The tensile strength of compression and welded connectors approaches that of the adjacent conductor and is less dependent on jointer proficiency.

The compression ferrule is preferred for copper conductors. For high compaction conductors up to 1000 mm^2, circumferential dies are forced into the ferrule by a hydraulic press, the dies locating either on a circumferential ridge (bump-type compression) or in a circumferential groove (knife-edge compression). For low compaction conductors of the Milliken type, a pin indent compression is used. The ferrule (fig. 35.7(b)) is contained within a guide block to prevent distortion and two diametrically opposed steel pins are forced into the ferrule by 300 kN hydraulic rams. Two rows of indents are made on each side of the ferrule with six indent pins per row. This technique has been applied for ratings up to 4000 A.

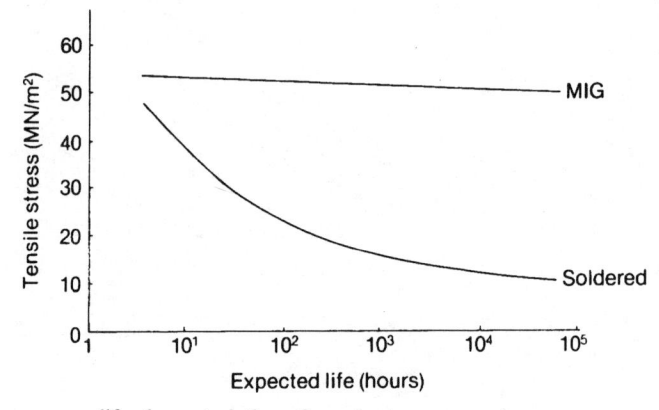

Fig. 35.16 Stress versus life characteristics of conductor connections

Fusion welded connectors are preferred for aluminium conductors and in this case are considered superior to compression ferrules because of the difficulty of breaking the oxide layer and achieving a stable resistance. Figure 35.7(c) shows a connector for metal–inert gas (MIG) welding. In this process aluminium wire is constantly fed into a welding gun and is used as a consumable electrode. A d.c. arc is struck between the wire and the conductor and continuously propels a fine spray of molten aluminium into the conductor face. The inert gas, argon, is used as a shield to prevent oxidation and to dictate the arc energy. Cable impregnant is prevented from reaching the weld by capwelding the conductor faces and applying a vacuum. A voltmeter and ammeter are connected to the welding gun to ensure that the optimum welding parameters are achieved independently of the length of connecting lead and type of welding set. The arc is self-regulating and the method gives a good tolerance for variations in jointer proficiency. A flush connection is produced for conductors in 3-core cables. A ferrule is used for hollow-core conductors and fluid communication is formed along a channel in the outer surface of the ferrule and then radially inwards through the conductor to the central duct. Another method uses pulsed MIG welding, which permits the weld to be built up around a central hollow duct pin by horizontal spray transfer. However, this method requires greater jointer proficiency and involves an additional compression process which seals each conductor to the pin to prevent cable impregnant from entering during welding.

MIG welding is used to connect aluminium conductors in XLPE cable. However, the weld temperature is above the crystalline melting point at which XLPE behaves as a soft elastomer. It is necessary to cool the conductor to prevent distortion and degradation of the insulation. The temperature of the conductor is monitored during welding and heat is extracted by the use of water cooled welding jigs.

Because of its simplicity, thermit welding offers possible attractions for small conductors, but still requires good conductor preparation and fluid control. Flush bronze welded connections have been developed for slim 400 kV FF cable joints on copper conductors up to 2000 mm^2 and have been employed in single-core submarine cable joints at 100 and 250 kV d.c. and in 3-core submarine joints at 132 and 150 kV a.c.

Jointing paper insulated cables

Following the completion of the conductor connection the cable insulation is prepared by tearing the paper tapes to form a conical shape, often termed 'pencilling', for gas-filled cable (fig. 35.8(b)) or in a series of steps, i.e. 'stepping', for fluid-filled cable (fig. 35.17). The ferrule is screened and the insulation is applied. It is important to minimise moisture absorption and the size of gaps and creases during application of the hand applied insulation. Moisture absorption is minimised by using pre-shaped and pre-impregnated paper rolls which are supplied to site in sealed containers and are stored under hot cable fluid until required by the jointer. Whenever possible, the joint is insulated in a continuous operation. The insulation is basted frequently with hot fluid or compound and fluid-filled joints are insulated under fluid pressure to ensure an outward flow of fluid. At voltages above 220 kV it has become practice to employ humidity control in the joint bay to a relative humidity of nominally 50% at 20°C. This is because moisture significantly increases the dielectric loss angle (DLA) and hence the dielectric heating of the paper insulation and also reduces the insulating properties of fluid channel surfaces.

Fig. 35.17 Joint bay for 275 kV fluid-filled cables

Fluid channel cleanliness

The fluid channels in accessories with electrically stressed short fluid channels, such as stop joints, SF_6 terminations and oil-immersed terminations, are carefully washed with cable fluid to remove jointing debris. At 220 kV and above, a high pressure flushing probe is used to clean the channels vigorously. These are then visually inspected using an endoprobe and the seepage fluid is sampled and checked using either an automatic particle counter or a filter membrane. The cable duct is flushed with filtered cable fluid before jointing and the joint is flushed after impregnation. Hydraulic shears are used to cut the conductor to avoid the generation of swarf. Cleanliness is facilitated by jointing in a well prepared joint bay with good lighting (fig. 35.18).

Fig. 35.18 Joint bay with 275 kV stop joint for fluid-filled cable

Jointing XLPE insulated cables

The metal cable sheath is first vented to release gaseous by-products of the crosslinking process, some of which are flammable. A set of specialised tools is required to prepare the insulation. The exposed core is clamped and heated in a cylindrical jig to remove curvature and to encourage the insulation to retract[7,8] before insulation is applied. A 'screen stripping' tool removes the bonded core screen (fig. 35.19),[26] it being important to ensure that the tool compensates for the eccentricity of the core, thus avoiding the formation of a stress-raising step into the insulation at the screen termination. An 'end stripping tool' (fig. 35.20) exposes the conductor. For joints which are to be taped or injection moulded a similar 'pencilling tool' is used to form the conical insulation pencil and to expose the conductor screen. The core is smoothed and polished by hand using successive grades of fine abrasive cloth. All traces of indentations and scratches must be removed as these form air-filled voids which will electrically discharge in service.

The conductors are joined using a compression ferrule or a MIG welded connection. A stream of ionised air is directed over the insulation to prevent the electrostatic accumulation of airborne debris on the core. The core is finally solvent cleaned and carefully inspected prior to the application of insulation.

For a prefabricated joint the insulating components are pushed back over the prepared core before making the conductor connection. These are pulled over the ferrule using the specialised tooling appropriate to the particular joint, taking care not to contaminate the bore of the insulation by contact with the semiconducting core screen

Fig. 35.19 Core screen stripping tool

(fig. 35.21). For an EPR tape joint the screen is hand reconstituted over the conductor connection and a powered taping machine applies the insulation under conditions of pre-set tension and stretch. The core screen and insulated screen gap are formed by hand. The joint shells are slid over, plumbed to the cable sheath, filled with a thermosetting resin compound and encased in the particular type of anticorrosion protection.

Jointing and terminating polymeric cable is a highly skilled process despite the apparent simplicity of the cable. The joint bays are double lined to ensure cleanliness and are well lit to aid visual examination of the prepared core. In temperate climates the bays are heated to achieve consistent properties of the elastomeric components. In tropical climates the bays are air conditioned and ventilated to prevent contamination of the core by perspiration from the jointer. Jointers are required to pass a rigorous training programme culminating in tests of proficiency. For polymeric cable jointing visual adjudication alone is inadequate. A direct way to confirm quality is for the jointer to assemble a joint and termination in the high voltage laboratory (fig. 35.22) and to require the accessories to withstand a 4 hour test at $3U_0$ and to pass a partial discharge test at $1.5U_0$.

Sheath closure

The joint shell is sealed to the sheath to contain the maximum design pressure within the FF or GF cable.[9] The seal on to an XLPE cable is required to be non-porous so that water is not sucked into the sheath during a cable cooling cycle. Each design of seal is required to conduct the return current during a system short circuit and to withstand the sheath mechanical loads in normal service.

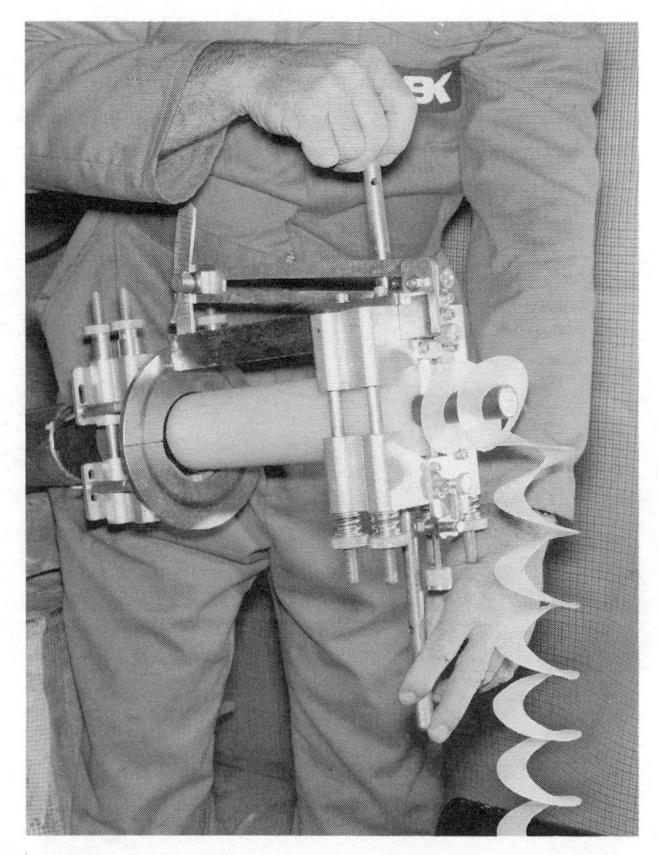

Fig. 35.20 Cutting back XLPE insulation to prepare for conductor jointing

External plumbed wipe
A plumbed wipe is the normal method used with lead or aluminium sheaths for FF cable, for XLPE cable and for some types of GF cable. The cable sheath is tinned and the joint shell sealed with lead strip to exclude cable fluid during plumbing. A lead based alloy of H metal in stick form is softened to a plastic state using a gas torch and is patted and wiped into position on the sheath until a thickness of 15 mm above the gland is achieved (fig. 35.23). The same method is used on aluminium sheathed cables, with the addition of a friction tinning process to break the oxide layer. The sheath is vigorously brushed and heated and a thin stick of metal is rubbed into the surface (fig. 35.24). Aluminium sheaths, unlike lead sheaths, do not readily creep and thus mechanical loads are concentrated on the plumbs which, being of lead alloy, exhibit a low creep strength at the operating temperature of typically 70°C. For this reason it is now the practice to reinforce all supertension cable plumbs to withstand sheath loads of up to 10 kN using epoxy resin impregnated glass fibre tape (fig. 35.25). Reinforced external plumbs have been used satisfactorily at pressures of up to 17 bar.[32]

Cast plumb
A mould is incorporated into the joint shell (fig. 35.26) or formed around the sheath of pressure cables and is filled with hot wax or oil. Molten lead alloy is poured into the

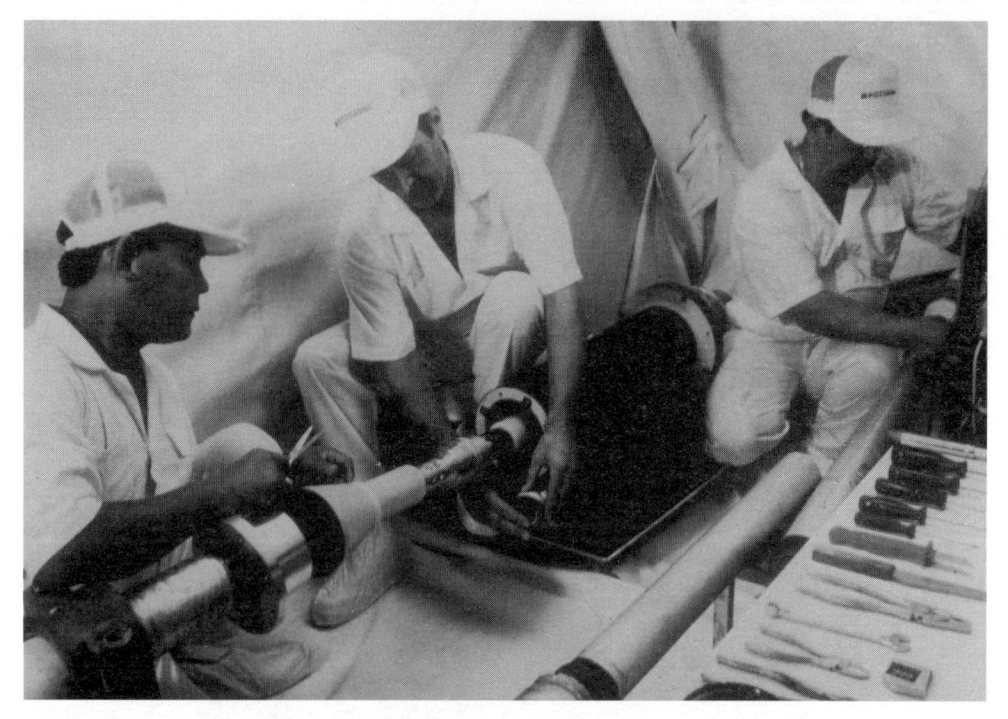

Fig. 35.21 Joint bay with a 66 kV prefabricated joint

mould to displace the liquid. This method requires careful control of the alloy temperature to avoid porosity and is now employed for high pressure applications in conjunction with a wiped pressure seal. An externally wiped plumb is included in fig. 35.8(a).

Internal plumbed wipe
This is made using the same technique as the external wipe but is reversed in direction so that it is located inside the joint on a plumbing gland (fig. 35.26) and experiences the pressure loads in the stronger mode of compression. The joint shell is mechanically sealed to the plumbing gland. The internal plumb has been employed for fluid, and gas pressures up to 31 bar,[11] usually in conjunction with external mechanical reinforcement in the form of a cast plumb, a wiped plumb, cast epoxy resin, or a mechanical lock.

Welded closures
Figures 35.27 and 35.1(a) show welded closures for a special application[10] on a smooth (non-corrugated) thick wall aluminium sheathed 525 kV cable for operation at a pressure of 25 bar. The sheath is belled out using hydraulic tools and sealed with an O-ring to prevent cable fluid from contaminating the weld. Tungsten–inert gas (TIG) welding, with a shield of helium, is used to fuse the sheath to an aluminium closure plate. Welding has also been employed on non-corrugated aluminium sheathed gas cables at 17 bar. Pulsed MIG welding has been developed for fluid-filled cables but has not yet been adopted for cables with thin wall corrugated sheaths because of the risk of puncture of the sheath by the weld pool. MIG welding is used on EHV XLPE cables with corrugated aluminium sheaths. A tube known as a plumbing platform is welded to

Fig. 35.22 Preparing cable termination for electrical tests in laboratory

the sheath following the crest of the corrugations. The sheath closure is made by plumbing on to this platform, preventing damage to the core by avoiding direct heating of the cable sheath for an extended period.

Lead burning
Lead burning is a highly skilled technique used principally for submarine cables to achieve a butted or overlapped flush connection between a lead alloy sleeved joint and the lead alloy sheath of the cable. Oxygen and hydrogen are burnt in a fine flame to melt and fuse the alloy locally. This technique has been successfully employed with submarine 150 kV 3-core fluid-filled cable joints.

Impregnation and pressurisation

Air is prevented from entering the fluid-filled cable during jointing by raising the joint above the level of the adjacent cable and by jointing under fluid flow. A vacuum drain

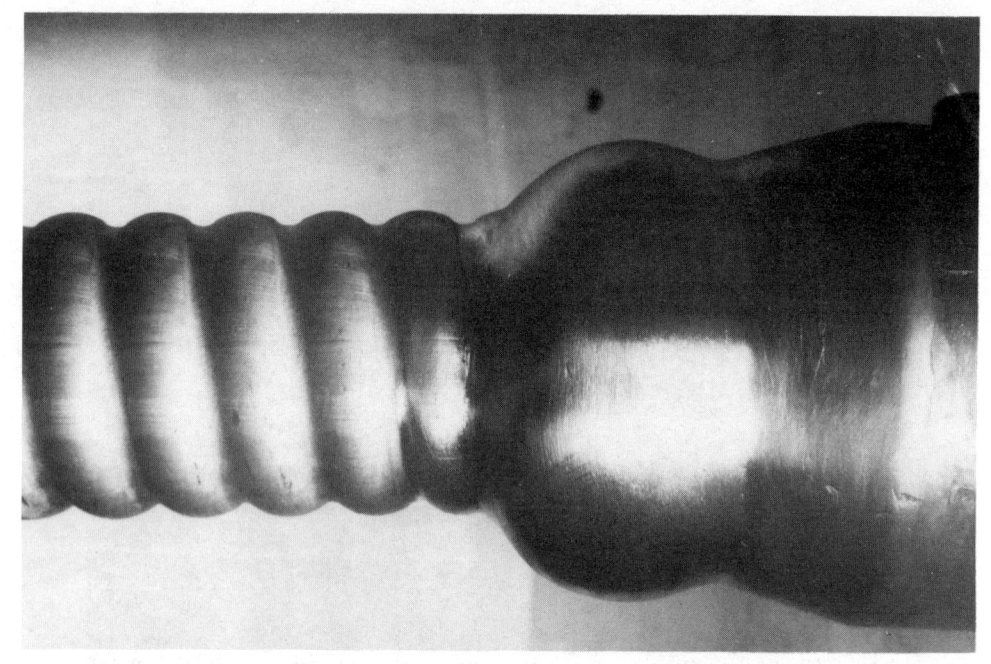

Fig. 35.23 Sheath closure with plumbed wipe

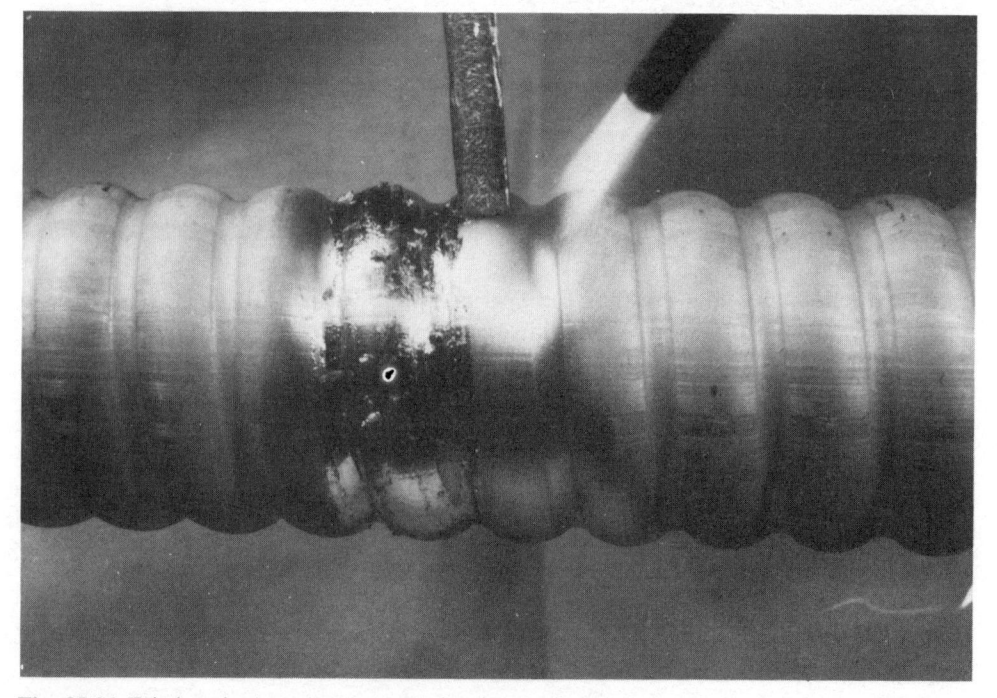

Fig. 35.24 Friction tinning a corrugated aluminium sheath

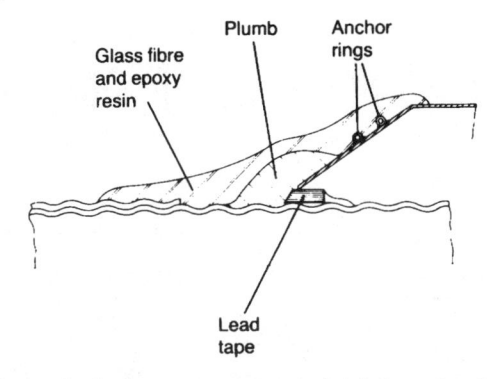

Fig. 35.25 Reinforced plumbed wipe on corrugated aluminium sheath

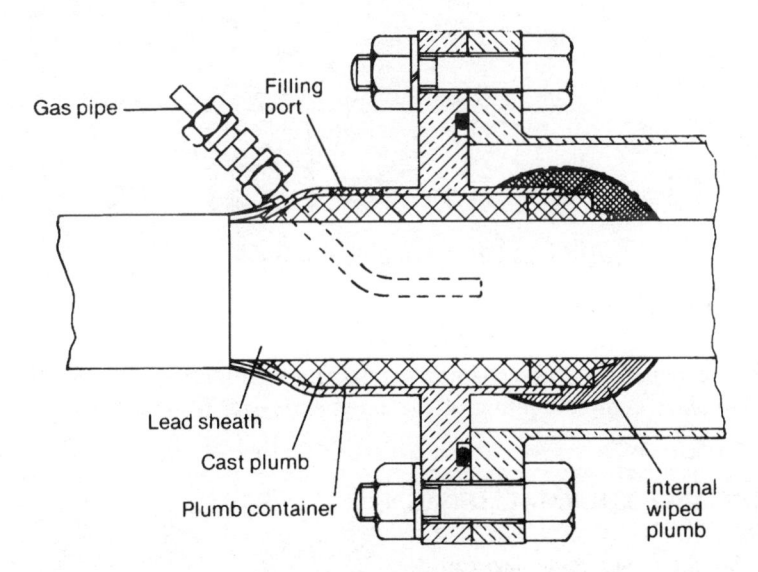

Fig. 35.26 Sheath closure with cast plumb and internal plumbed wipe

jar is connected to the bottom of the completed joint to remove seepage fluid and a vacuum pump is connected to the top. Evacuation is conducted for a typical period of up to 6 hours for accessories up to 132 kV. At higher voltages, periods of up to 24 hours are required together with a 'pressure rise' test in which the vacuum pump is disconnected and the joint pressure is measured to ensure that it does not exceed typically 0.1 to 0.15 torr in 30 min. The joint is impregnated under vacuum by admitting degasified fluid through a bottom union. At 400 kV the residual gas pressure (RGP) of the fluid is measured after impregnation to ensure that it is below 5 torr. A pressure test at the maximum system pressure (i.e. up to 5.2 bar) is conducted with the unions, seals and sheath closures whitewashed to locate possible leaks.

Accessories on gas-filled cable are not evacuated but are simply purged and pressurised with dry oxygen-free nitrogen together with the cable.

Joints on XLPE cable up to 132 kV are not liquid filled. Non-degasified silicone fluid is used to fill porcelain insulators on outdoor terminations up to 132 kV. A simplified

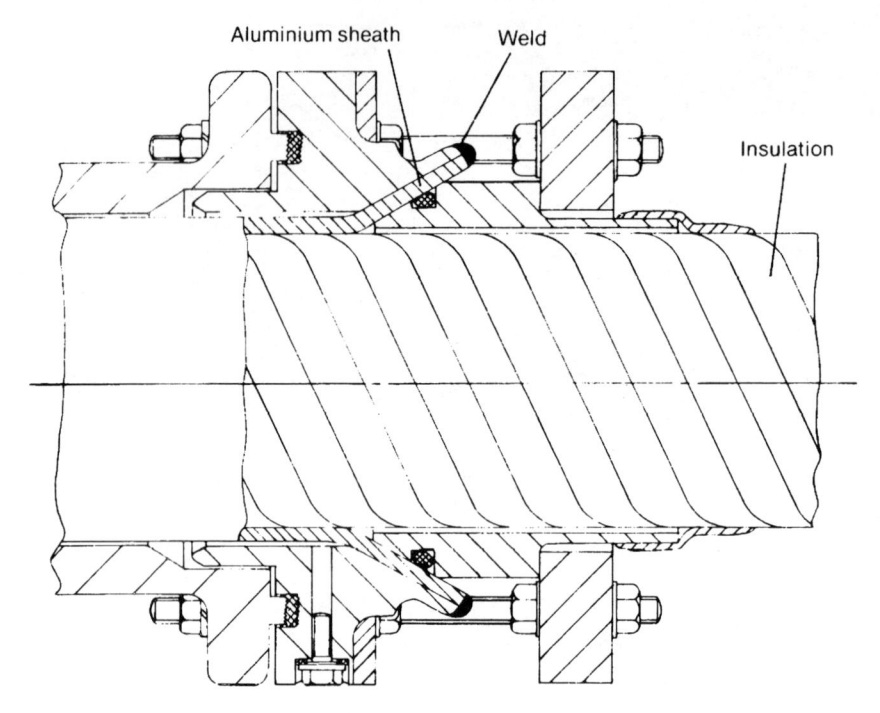

Fig. 35.27 Welded sheath closure

degasifying and evacuation process is necessary for a termination into metal-enclosed equipment to achieve complete filling of the short epoxy resin insulator.

INSULATION AND THERMAL DESIGN

Insulation design for paper cable accessories

The limiting performance of the insulation occurs at the lightning impulse BIL.[33] The maximum radial stress occurs at or adjacent to the ferrule in a straight joint and is chosen in conjunction with the thermal rating and method of insulation. For example, a high radial stress would require the joint to be insulated with paper tapes, thus increasing the jointing time. The reduced diameter would increase the dielectric losses but reduce the radial thermal resistance. Typically the maximum radial stress is 40–80% of that of the cable. The stress and outer diameter are determined using the formula

$$E = V \left\{ r_x \epsilon_x \left[\frac{1}{\epsilon_a} \log_e \left(\frac{r_{oa}}{r_{ia}} \right) + \frac{1}{\epsilon_b} \log_e \left(\frac{r_{ob}}{r_{ib}} \right) + \cdots \right] \right\}^{-1} \qquad (35.1)$$

where r_x = radius at which the stress is required (m)
E = stress (MV/m)
V = voltage (MV)
r_{oa} = outer radius of the material (a, b etc.) (m)
r_{ib} = inner radius of the material (a, b etc.) (m)
ϵ_a = relative permittivity of the material (a, b etc.)

Typical values for relative permittivity are as follows:

Fluid impregnated paper	3.5
Fluid impregnated PPL	2.8
Cable fluid	2.2
Pre-impregnated paper (gas cable)	3.4
Epoxy resin	4.2
Porcelain	6.0

The maximum radial stress in complex accessories (e.g. stop joints and resin bush terminations) is determined by computer field plotting (fig. 35.3).

The shape of the stress control profile, cable stepping and ferrule is calculated to control the longitudinal component of stress along the surface of the paper insulation. The longitudinal impulse strength of the paper is 0.5–5% of the radial strength and is influenced by the anisotropic nature of the paper, the presence of fluid-filled gaps and creases and the magnitude of the radial stress. For the purpose of design, the longitudinal stress parameters are made a function either of the radial stress or of the longitudinal distance along the profile and incorporate a design margin to allow for jointing variations.

For simple accessories such as straight joints the angle of the stress control profile may be approximated from

$$\theta = \arctan\left(\frac{E_1}{E}\right) \tag{35.2}$$

where θ = angle of stress control profile
E_1 = limiting value of longitudinal stress (MV/m)
E = radial stress at the profile from equation (35.1) (MV/m)

For complex accessories such as SF_6 terminations and stop joints with large profile angles, the above approximation is invalid and longitudinal stress should be measured from a computer field plot. Finite element solutions are employed to allow complex geometries to be analysed. Powerful computer workstations running sophisticated suites of software can calculate the electrical field distribution throughout an accessory.

Using modern analysis, models containing up to 100 000 nodes can be solved to obtain greater accuracy in regions where divergent fields are expected. A computer printout of the equipotential distribution allows the solution to be verified visually. Stress and dielectrophoretic force[25] distributions along each interface can be graphically printed and compared with the design parameters.

Insulation design for polymeric cable accessories

The limiting performance of accessories on polymeric cable can be either the hot impulse test or the $3U_0$ 4 hour a.c. withstand test at ambient temperature.[34] The a.c. test is particularly searching in those aspects of design, material and jointing which introduce stress raisers and voids and hence promote electrical discharging. As with pressurised cables the insulation geometry must be designed to control the radial and longitudinal components of stress along the interface with the cable core. Elastomeric insulating materials have the advantages of isotropic design parameters and of greater choice of permittivity (table 35.10) but the disadvantages of a less well defined geometry and a greater temperature-dependent permittivity.

Table 35.10 Relative permittivity of materials in accessories for polymeric cable

Material	Relative permittivity[a]
Extruded XLPE	2.5
Extruded EPR	3.0
Tape EPR	2.8
Moulded EPDM	3.0
Moulded CSP	8–10
Stress control elastomer	15–25
Silicone fluid (viscous)	2.8

[a] Values are for the average of the ambient and operating temperatures

Thermal rating

The rating formulae and parameters are the same as those for the cable (chapter 34) but with the addition of longitudinal heat flow along the conductor and cable sheath. Typical thermal resistivities are given in table 35.11 and typical values of DLA in table 35.12.

To calculate the maximum operating temperature of the joint, the centre joint in the bay is chosen and is assumed to be thermally symmetrical about its centre line. The joint and joint bay are divided transversely into sections of similar radial and longitudinal geometry and these are further subdivided to obtain accuracy of solution. An equivalent thermal network is constructed and solved using a computer. The finite element representation typically requires 50–150 transverse sections and 5–10 radial

Table 35.11 Thermal resistivity of materials in accessories

Material	Thermal resistivity (Km/W)
Cable fluid paper	5.0
GF paper	5.5
Cable fluid PPL	5.5
Cable fluid: annular gap	3.5[a]
Extruded XLPE cable	3.5
Extruded EPR	5.0
EPR tape	4.8
Acrylic resin compound	2.0
Polyurethane compound	4.8
Filled epoxy resin	0.9
Porcelain	0.95
Bitumen	6.2
Stabilised backfill	1.2
Copper conductor	0.0026
Aluminium conductor	0.0043

[a] 50% of static value to allow for convection in typical applications

Table 35.12 Dielectric loss angles of materials in accessories used for thermal ratings

Material	DLA[a]
FF paper cable	0.0024
FF paper joint	0.003
FF PPL cable	0.001
GF paper insulation	0.0045
Extruded XLPE	0.001
Extruded EPR	0.005
EPR/EPDM joint insulation	0.005

[a] Values are for the operating temperature

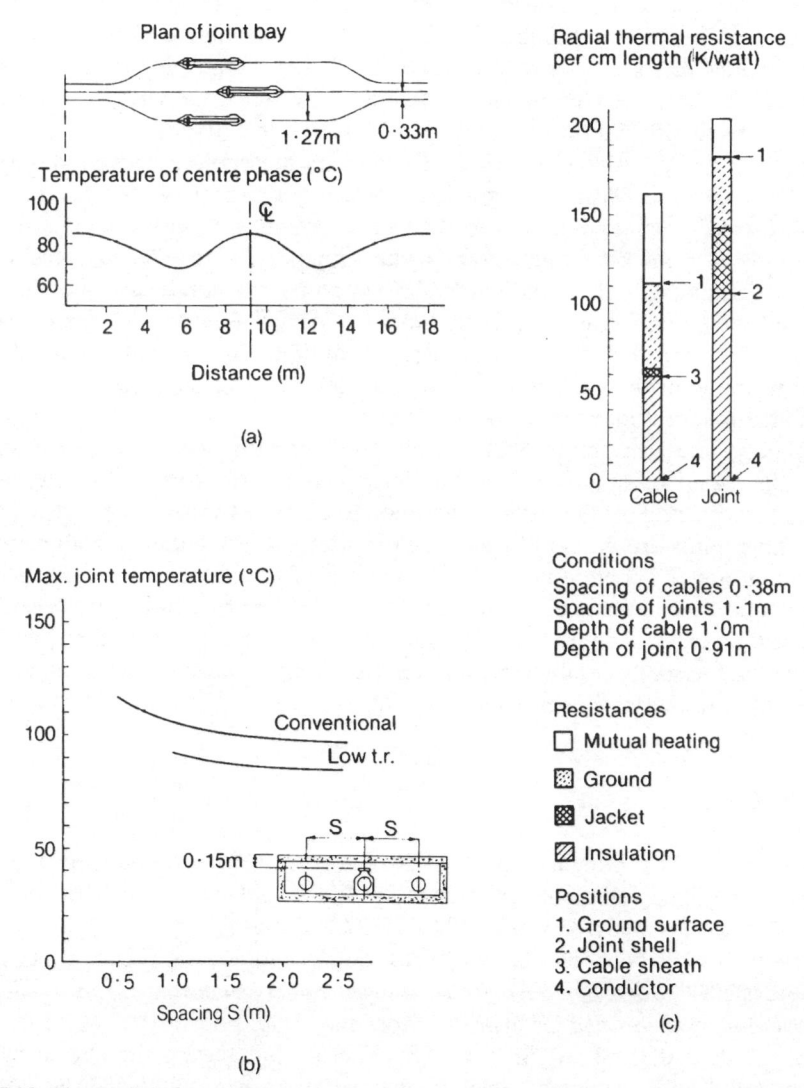

Fig. 35.28 Thermal design characteristics: (a) conductor temperature in a 275 kV FF straight joint bay; (b) 275 kV joint temperature versus spacing characteristic; (c) comparison of 275 kV cable and joint thermal resistances

sections. A graphical printout is obtained of the temperature distribution along the conductor (fig. 35.28(a)). If the temperature exceeds the maximum cable design temperature (typically 90°C) the spacing between the joints is increased to reduce mutual heating and the solution is repeated to obtain a graph of the variation of maximum temperature with spacing (fig. 35.28(b)).

For directly buried naturally cooled cables, an increase in the joint spacing is usually adequate to limit the temperature. However, this is seldom possible with special cable systems designed with enhanced heat dissipation to achieve increased current density (e.g. shallow trough and water cooled systems). Figure 35.28(c) shows a comparison of the cable and joint thermal resistances to the ground surface for a 275 kV FF system. It will be seen that if either the external thermal resistance of the ground around the cable or the mutual heating is significantly reduced, a comparative reduction in resistance can only be achieved by altering the joint design, e.g. by replacing the bitumen filled glass fibre box by low thermal resistivity insulation of either tape or heat shrink sleeve. Figure 35.28(b) shows the reduction of temperature achieved with a low thermal resistivity joint shell insulation for a 275 kV FF trough installation.

Efficient force cooled cable systems require that superior cooling is given to the joints. Figure 35.6(c) shows a 400 kV straight joint with a water pipe welded to the copper joint shell, the pipe being connected to the cable cooling pipe with the coolest water. Figure 35.6(d) is a 400 kV straight joint in an integrally water cooled cable system. The heat is dissipated by longitudinal heat flow along the conductor and fluid duct to the water cooled cable and to a lesser extent by radial flow to air. The cable fluid within the cable central duct is reciprocated at laminar flow velocity. Turbulence is promoted in the joint duct to improve heat transfer by increasing the velocity over a flexible ridged mandrel inserted during jointing.[32]

Terminations and stop joints on FF cable are thermally less critical than straight joints, as the radial dissipation of heat is improved for the former by internal and external convection and for the latter by conduction through the cast epoxy resin barrier.

XLPE cable joints are potentially at a greater thermal disadvantage compared with the cable than are FF cable joints. This is because of the absence of convective heat transfer by fluid to the joint shell, the reduced level of dielectric heating in the cable and the increased diameter of some types of joint. It is important to ensure that joint shells are filled with a thermally conductive material, that joint dimensions are minimised and that joint bays are thermally designed.[26]

REFERENCES

(1) Endersby, T. M., Gregory, B. and Swingler, S. (1993) 'Polypropylene paper laminate oil filled cable and accessories for EHV application'. *3rd IEE Int. Conf. on Power Cables and Accessories 10 kV–500 kV*, London.

(2) Gregory, B. and Nicholls, A. W. (1986) '66 kV and 132 kV XLPE supertension cable systems'. *6th Conf. on Electric Power Supply Industry. Transmission and Distribution Systems and Equipment.* Paper No. 3.08.

(3) Smee, G. J. and West, R. S. V. (1986) 'Factors influencing the choice between paper and XLPE insulated cables in the voltage range 66 kV–132 kV'. *2nd Int. Conf. on Power Cables and Accessories 10 kV to 180 kV*. IEE Publication No. 270, pp. 193–197.

(4) Arkell, C. A., Arnaud, U. C. and Skipper, D. J. (1974) 'The thermomechanical design of high power self-contained cable systems'. Paris: CIGRE Paper No. 21–05.
(5) Ball, E. H. and Holdup, H. W. (1984) 'Development of cross-linked polyethylene insulation for high voltage cables'. Paris: CIGRE.
(6) Nakagawa, H., Nakabasami, T., Sugiyana, K. and Shimada, A. (1984) 'Development of various snaking installation methods of cables in Japan'. Jicable '84 Conference Publication, pp. 413–420.
(7) Aalst, R. J., Laar, A. M. F. J. and Leufkens, P. P. (1986) 'Thermomechanical stresses in extruded HV cables'. Paris: CIGRE Paper No. 21–07.
(8) Asada, Y. and Maruyama, Y. (1987) 'A study on insulation shrinkback in crosslinked polyethylene cables'. Jicable '87 Conference Publication, pp. 264–269.
(9) ESI Standard 09–4 (1979) '66 and 132 kV impregnated paper insulated oil-filled and gas-pressure type power cable systems'. Electricity Supply Industry.
(10) Ray, J. J., Arkell, C. A. and Flack, H. W. (1974) '525 kV self-contained oil-filled cable systems for Grand Coulee third powerplant: design and development'. *IEEE Trans.* **PAS-93**.
(11) Williams, A. L., Davey, E. L. and Gibson, J. N. (1966) 'The 250 kV d.c. submarine power cable interconnection between the North and South Islands of New Zealand'. *Proc. IEE* **113** (1).
(12) Engineering Recommendation C55/4 draft (1988) 'Insulated sheath power cable systems'. Electricity Council Chief Engineers Conference, Mains Committee.
(13) CIGRE Study Committee 21, Working Group 07 (1976) 'The design of specially bonded cable circuits', part II. *Electra* (47), second report.
(14) ESI Standard 09–10 (1976) 'Porcelain insulators for 33, 66, 132, 275 and 400 kV pressure assisted cable outdoor sealing ends'. Electricity Supply Industry.
(15) Looms, J. S. T. (1988) *Insulators for High Voltages. IEE Power Engineering Series 7*. Peter Peregrinus.
(16) IEC 137 (1973) 'Bushings for alternating voltages above 1000 V' (2nd edition).
(17) IEC 71–1 (1976) 'Insulation coordination'. Part 1: 'terms, definitions, principles and rules' (6th edition).
(18) IEEE (1975) 'Standard test procedures and requirements for high voltage a.c. cable terminations'. *IEEE* **48**.
(19) Arkell, C. A., Johnson, D. F. and Ray, J. J. (1974) '525 kV self-contained oil-filled cable systems for Grand Coulee third powerplant: design proving tests'. *IEEE Trans.* **PAS-93**.
(20) Arkell, C. A., Galloway, S. J. and Gregory, B. (1981) 'Supertension cable terminations for metalclad SF_6 insulated substations'. *8th IEEE/PES Conf. and Exposition on Overhead and Underground Transmissions and Distribution*.
(21) IEC 859 (1986) 'Cable connections for gas-insulated metal-enclosed switchgear for rated voltages of 72.5 kV and above'.
(22) Gregory, B. and Lindsey, G. P. (1988) 'Improved accessories for supertension cable'. Paris: CIGRE paper No. 21–03.
(23) Sütterlin, K. H. (1972) 'Cable lead-ins into 110 kV metal-clad switchgear'. IEE Conference Publication No. 83, pp. 45–52.
(24) Ishikawa, M. *et al.* (1981) 'An approach to the suppression of sheath surge involved by switching surges in a GIS power cable connection system'. *IEEE Trans.* **PAS-100** (2).

(25) Gibbons, J. A. M., Saunders, B. L. and Stannett, A. W. (1980) 'Role of metal debris in the performance of stop joints as used in 27 kV and 400 kV self-contained oil-filled cable circuits'. *Proc. IEE C* **127** (6).
(26) Gregory, B. and Vail, J. (1986) 'Accessories for 66 kV and 132 kV XLPE cables'. *Second Int. Conf. on Power Cables and Accessories 10 kV–180 kV*. IEE Publication No. 270, pp. 248–256.
(27) Rosevear, R. D., Williams, G. and Parmigiani, B. (1986) 'High voltage XLPE cable and accessories'. *2nd Int. Conf. on Power Cables and Accessories 10 kV–180 kV*. IEE Publication No. 270, pp. 232–237.
(28) Stepniak, F. M., Burghardt, R. R., Boliver, V. J. and Shimshock, J. F. (1987) 'Effects of aging on premoulded elastomeric splices for 138 kV XLPE cable'. *IEEE Trans.* **PWRD-2** (3), 632–637. Paper 86 T & D 585–4.
(29) Attwood, J. R., Gregory, B., Lindsey, G. P. and Sutcliffe, W. J. (1993) 'Prefabricated accessories for extruded polymeric cable at 66 kV to 150 kV'. *IEE 3rd Int. Conf. on Power Cables and Accessories 10 kV–500 kV*, London.
(30) Nakabasami, T. *et al.* (1985) 'Investigations for commercial use of 275 kV XLPE cables and development of extrusion type moulded joint'. *IEEE/PES 1985 Winter Meeting*, Paper 85 WM 007–0.
(31) Attwood, J. R., Gregory, B. and Svoma, R. (1991) 'A range of transition joints for 33 kV to 132 kV polymeric cables'. Jicable '91 *3rd Int. Conf. on Polymer Insulated Power Cables*, Paris.
(32) Arkell, C. A. *et al.* (1977) 'Design and construction of the 400 kV cable system for the Severn Tunnel'. *Proc. IEE* **124** (3), 303–316.
(33) Engineering Recommendation C47/1 (1975) 'Type approval tests for single core impregnated paper insulated gas pressure and oil filled power cable systems for 275 kV and 400 kV'. Electricity Council Chief Engineers Conference, Mains Committee.
(34) ESI Standard 09–16 (1983) 'Testing specification for metallic sheathed power cables with extruded crosslinked polyethylene insulation and accessories for system voltages of 66 kV and 132 kV'. Electricity Supply Industry, Issue 1.

CHAPTER 36
Installation of Transmission Cables

PROJECTION PLANNING

The installation situations in which cables are used at transmission voltages fall into five main categories:

(a) for the interconnection of substations within urban areas where the use of overhead transmission is neither environmentally acceptable nor practical;
(b) to form part of a circuit in rural areas where the use of an overhead line is environmentally unacceptable;
(c) to span obstructions within a circuit, such as bridges, rivers, estuaries and in some cases towns;
(d) to replace overhead line connections in the vicinity of a new substation or power station, thereby improving the overall environmental acceptability;
(e) inside power stations or substations to provide more compact and less obtrusive connections than busbar or overhead lines.

The first category is the most common for transmission cable, although most of the considerations discussed below apply to each group.

Some interesting trends in installation techniques are emerging. In cities these are: (a) to install cables in specialised tunnels (e.g. 132 kV cable tunnels under London and a 380 kV cable tunnel under Berlin), and (b) to install cables in pre-laid pipe runs, thereby avoiding the problems of route planning and of disruption to vehicular traffic. In rural areas the trend is to develop methods of mechanical trenching and cable laying, thereby reducing the cost disadvantage of cables to overhead lines.

Owing to the high cost per unit length of the transmission cable and the great diversity of types and sizes manufactured, it is relatively rare for a manufacturer to be able to supply any new requirements from stock. In general, cables at these voltages are 'made to measure' project by project. In a limited number of applications the standard drum length approach is adopted, where one or more standard drum lengths are selected from experience as typical of the optimum length for installation in such a situation, and manufacture proceeds on this basis.

Such an approach to the installation will yield faster project mobilisation times even where the routes are not cleared and will enable a more flexible approach to the site work, but suffers from the drawback that it inevitably involves a higher level of surplus cable at the end of the project. Such a method is not practicable where joint bay location is in any way difficult due to the size of the joint bays, lack of available space in the roadways, or the need to match cable lengths for cross-bonding. It is rarely encountered, therefore, at voltages above 132 kV or in heavily developed urban environments.

The more commonly used approach is first to determine and clear a substantial part or preferably all of the cable route. Once the line of the route is set the position of every joint bay is determined. In principle it is normal to use as long a length of cable between joint bays as possible, i.e. the minimum number of joints practicable. In practice the maximum section length (distance between consecutive joint bays) can be limited by any one or more of a number of different constraints:

(a) local restrictions on the length of continuous trench that can be opened;
(b) manufacturing length;
(c) transportable length on one drum;
(d) handling limitations at the site of installation;
(e) induced voltage in the cable sheath;
(f) balancing of minor section lengths on cross-bonded systems;
(g) positioning of joint bays.

In fluid-filled cable installations, the need to install stop joints instead of simple straight through joints is identified and all the materials are then put into manufacture, each drum of cable being manufactured to a specific predetermined length. In this type of approach to the installation work, the planning and manufacturing lead times are much greater but wastage is limited and the whole project is fully detailed and programmed prior to work starting at site.

TRANSPORTATION OF CABLES

Although transportation systems are now quite highly developed in most parts of the world, problems are still encountered, even in the most developed countries. The difficulties are usually limited to the road transportation of large cable drums with regard to size rather than weight, as weight limitations are overcome by increasing the number of axles utilised.

As the rated voltage of the cable increases, the diameter and the minimum bending radius become greater. Thus the minimum hub diameter of the drum increases and also the volume of the cable to be added to it. In order to comply with the width limitations contained in most sets of highways regulations, the increase in the bulk of the drums results in larger diameter drums.

A drum carrying 400 m of 400 kV cable will be of the order of 4 m in diameter over the battens. When this is placed on the back of a conventional low loader, bridge clearances in excess of 4.5 m are required and these are not always available. Specialist vehicles have had to be developed to meet these situations, both in the form of heavy duty drum carriers for short haul applications and specialised trailer units for overland transportation, in which the drums are carried very close to the ground.

CABLE TRENCHES

The line of the cable trench and position of the joint bays will have been selected during the planning stage and basically proved at that time by sample trial holes. The extent of the trial holing is determined by experience and moderated in accordance with the extent and accuracy of the available records of existing services and obstructions along the proposed route. A subsurface electronic survey can be particularly accurate in locating existing underground electric services and other metallic services which have an induced 50–60 Hz 'hum'. With major services, water mains for example, where the exact intersection with the cable route is not known, a signal can be injected on to the pipe at a known reference point and its direction and depth can be plotted.

The final line of trench should be selected to have as few changes of line and direction as possible and corners should be taken at a radius at or greater than the minimum installation radius of the cable. This enables the most efficient cable laying operation to be followed because cables of this size and weight tend to be difficult to bend. An indication of minimum bending radii for cables 33 kV and above may be obtained from table 36.1, where d_o is the overall cable diameter.

Transmission cables can be laid in four different ways:

(a) by the traditional open-cut method, buried directly in the ground;
(b) by pulling into pre-installed plastic ducts, with break trenches at bends;
(c) in buried concrete troughs which are either pre-cast or cast *in situ*;
(d) by pulling into existing pipes with direct buried joints or in underground chambers.

Table 36.1 Minimum bending radii (summary)

Voltage (kV)	Number of cores	Minimum installation bending radii			
		Adjacent to accessories		Laid dried	Laid in ducts
		with former	without former		
Non-pressure-assisted					
33	1	$15d_o$	$20d_o$	$21d_o$	$35d_o$
	3	$12d_o$	$15d_o$	$18d_o$	$30d_o$
Pressure-assisted					
33–132	1	$15d_o$	$20d_o$	$30d_o$	$35d_o$
	3	$12d_o$	$15d_o$	$20d_o$	$30d_o$
275–400	1	$20d_o$	$20d_o$	$30d_o$	$35d_o$
Extruded dielectric (metallic sheath)					
33–200	1	$12d_o$	$12d_o$	$15d_o$	$25d_o$
200–400	1	$15d_o$	$20d_o$	$30d_o$	$35d_o$
Extruded dielectric (foil laminate)					
33–200	1	$15d_o$	$20d_o$	$20d_o$	$30d_o$

The method to be employed will depend upon a number of factors, which include economics, inconvenience to the public and the need to provide for the security of the system against external influences.

The traditional open-cut method is by far the most economical and allows great flexibility in installation. This method, however, can cause considerable inconvenience to the public when the whole trench is excavated for the cable to be installed, and security of the trench and cable to third party damage is low.

Inconvenience is localised with cables being pulled into pre-installed ducts, the installation of which can be carried out well in advance of cabling, a small portion at a time, by a cut and fill operation. This method of construction makes for an expensive installation although the security of the system is good. If the ducts are filled with a bentonite grout, no system derating due to air in the duct is necessary.

The third method, to lay cables in buried reinforced concrete troughing, has become increasingly popular, particularly with the installation of cable systems for the transmission of bulk power to load centres and to major industrial users where security of the system is a priority. Although as inconvenient to the public as the open-cut method, it has major advantages in that it is particularly safe from damage by third party activity and it also retains the selected low thermal resistivity backfill material which otherwise would normally be lost when work is carried out under or adjacent to the cables so that the system would be put at risk of derating. Whilst this method of installation is most secure and retains substantial flexibility, it is comparatively expensive.

The final method is finding increasing use when existing cables installed in pipelines approach the end of their life. In these cases it is often possible to replace existing cables without replacing the pipe. This is achieved by pulling out the old cables and replacing them with modern designs of the same or higher system voltage. This allows use of existing pipes which are regarded as an asset and also minimises disruption to the environment during installation.

The line of trench must also allow for the necessary clearance from other services (usually 300 mm) and other parallel cable circuits (preferably 5 m). Due regard should also be taken of statutory regulations and the rights and interests of landowners, which may also affect the line of trench.

Transmission cables are usually laid deeper than lower voltage cables and standard installation depths of 1.0–2.5 m are quite normal. These cables are frequently of the single-core type and the system design will then specify a minimum spacing of phase centres at standard laying depth to ensure that the cables do not overheat at the design rating. The cables are usually laid at this spacing to minimise induced sheath voltage and enable the narrowest (cheapest) trench to be excavated. When transverse obstructions are met, it may be necessary for the cables to be laid deeper, in which case the need to widen the phase spacing must be considered to ensure that the cables do not overheat. Increased ratings may be achieved by the installation of polyethylene water cooling pipes alongside the cables. By passing water through these pipes, heat may be removed from the cable environment. The pipes are usually installed at a set distance from the cables, the distance being controlled by the use of plastic spacers.

During the design stage, work will have been carried out to measure the ground thermal resistivity in the area. Once the cable trench is excavated, sample tests should be carried out in the trench wall to ensure that the surrounding ground is at least as good as the standard figure used for determining the cable rating.

Open-cut trenches are usually excavated to a level about 75 mm below the final position of the bottom of the cable, to allow for a bedding layer of selected backfill to be placed in the trench below the cables.

For buried trough installations the trough is either cast *in situ* or constructed from pre-cast elements which are hoisted into position on to a bedding layer of weak mix concrete. The speed of installation of the cast *in situ* trough can be rapidly increased by the use of plasticiser or quick curing additives in the concrete. After the cables are installed the trough is filled with a selected backfill and a reinforced concrete lid, sufficiently waterproofed, is placed on top.

Selected backfill materials may be divided into three groups: sand, cement bound sand and sand–gravel mix. The importance of using a selected backfill is to achieve the required thermal resistivity demanded by the system design and local conditions (chapter 34). Thermal resistivities are difficult to measure accurately on site, and as the measurement is time consuming it may delay installation work. On site quality control measurements are made of the dry density of the backfill, which is related to the thermal resistivity figure. As a precaution to eliminate lumps of foreign material from the cable environment, it is quite normal to have transmission cable trenches close timbered where unstable ground is encountered. Open poling is used in moderately firm ground.

ROAD CROSSINGS

The most frequent obstacle a trench encounters, other than drains, water or gas pipes, cables etc., is the crossing of side roads where they join the road in which the trench is being excavated. These road junctions must usually be kept open and operational throughout the cable installation period. Thus the trench cannot simply continue across the side road and some special arrangement is required. The most useful method is to install ducts, with one duct for each cable to be installed. The diameter of the inside of the duct is about twice the cable diameter. The ducts are set in a block of concrete at predetermined spacings, dependent on the depth of laying, to ensure that cables do not overheat (fig. 36.1). The road way, however, is often the location for large services such as high pressure sewer lines and stormwater drainage systems, and it is not always possible for a rigid block of ducts to pass between these services.

A duct block requires careful planning; a trial trench should therefore be excavated across the road to the exact dimensions of the proposed installation. This trench can be either temporarily backfilled or covered with heavy steel plates to allow continuity of traffic flow. Once the trench has been proved any temporary backfill can be removed and the duct block installed, taking account of any adjustments to the trench dimensions which may be necessary due to changes in depth to overcome obstructions.

The ducts themselves are usually made from rigid PVC or equivalent material and are laid in short sections. They require adequate wall thickness to ensure that they do not deform to an oval shape under the pressures exerted during installation and should be installed in such a manner as to ensure a smooth internal surface to prevent damage to the cable oversheath during installation. With the increasing variety of modern materials that are becoming available in the market, an excellent alternative to rigid PVC ducts is the flexible high density polyethylene pipe. This has advantages for various applications in that its coefficient of friction is much less than rigid PVC pipe and hence the pulling force on the cable is reduced. The flexibility of the pipe also

Fig. 36.1 A triple circuit of 132 kV XLPE insulated cables installed in a duct block

allows for a more flexible installation where the pipe can easily follow the contour of obstructions and return to normal laying depth over a shorter distance, economising on both the dimensions of the trench and the concrete used in constructing the block. This type of pipe is also well suited to undercrossings of rivers and canals that can be dredged, the installation of the pipes being made in one continuous length.

After installing the cable in the ducts they should be filled with a mixture of bentonite sand and cement. This is kept in position by sealing the annular gap around the cable at each end of the duct. The bentonite mixture improves the conduction of heat away from the cable and supports it thermomechanically in the duct. When duct runs exceeding 50 m are to be filled with bentonite, it is normal to fit a vent pipe near the centre of the duct during installation to facilitate filling from both ends.

An alternative method of road crossing is flush decking. Although mainly superseded by duct blocks this is still used quite satisfactorily in some major cities. This method is based on cutting the trench as usual but going straight across side roads with the trench. As the trench opens across the road, the top 0.3 m is filled in with heavy timbers supported by cross-bearers in such a manner as to reinstate the surface of the road flush. The traffic can then drive on the timbers over the trench. Sufficient room must be left, of course, for men to work under the flush decking which may cause the cables to be laid deeper than otherwise necessary. Plating is another technique to keep traffic flowing. This usually necessitates timbering of the trench prior to the laying of heavy duty steel plates of two to three times the width of the trench.

Thrust boring (pipe jacking) and conventional tunnelling, although certainly more costly, are often practical solutions in crossing roads, canals, rivers etc. where other methods are unsuitable.

JOINT BAYS

It is often a requirement that the top of the joint is as deep as the top of the cable at the standard depth of laying, where the top of the joint is taken to be the lid of the fibre glass box. The axis of the joint is thus below that of the cable. In addition there must be adequate clearance under the cable to allow jointing operations to be carried out without difficulty. The joint bay floor tends to be quite deep as a result and certainly deeper than the trench bottom at normal laying depth.

Apart from thermal considerations, the spacing between cables in joint bays may need to be increased to enable the jointers to get between them with their equipment. Joint centre spacing is typically two to four times the cable phase centre spacing at normal depth of laying. In any case, this amount of joint spacing is usually necessary to limit the operating temperature of the joints. Since space is also required between the outside joints and the joint bay wall, it is often found that the joint bay is two or three times the normal trench width and sometimes greater.

The transition between trench spacing and joint spacing must not take place too sharply after leaving the joints as some straight cable is required during jointing for passing back the joint sleeve and bending radii must meet the specified requirement. Thus although the finished length of the joint may be 1.5–2.0 m, the length over which the trench must be widened may be up to about 16 m.

Certainly at the higher end of the voltage range the joint bays are sizeable constructions, particularly at stop joint positions in fluid-filled systems, for which it can be difficult to find space in congested urban areas. The increasing widths and depths commonly required at transmission voltages have led to the adoption of reinforced concrete floors and side retaining walls as part of the construction. This also ensures a firm flat base for jointing upon, with adequate anchorage for cable cleats should they be required.

The most economic configuration of joints is usually to have all three abreast, but for special requirements, to produce a narrower joint bay, they may be staggered in relation to each other. An arrowhead layout is usually preferable, although an echelon arrangement is not unusual in very confined situations.

SPECIAL CONSTRUCTIONS

Unless route lengths are very short, special constructions may be required. Typical examples are river crossings, rail crossings, cable tunnels, cable bridges and troughs.

While it is not possible to consider the detailed design of each of these on a general basis, certain aspects of the installation require special attention:

(a) the thermal environment must be satisfactory;
(b) any thermomechanical forces that could be experienced must be adequately constrained or dissipated;
(c) where cables are in the open, due account must be taken of risks due to fire, vandalism, accidental third party damage and solar radiation;
(d) protection from vibrations;
(e) specified installation radii must be observed.

CABLE LAYING

The most basic method of laying cable involves pulling the cable from the drum by hand. This technique is still used in many parts of the world today, particularly where labour is cheap and plentiful and the cable is relatively short and lightweight. The drum is mounted to rotate freely and a cable pulling crew is spread along the cable trench at intervals which are determined by the weight per unit length of the cable and the route complexity. The foreman or supervisor of the crew co-ordinates the physical effort made by each individual to move the cable forward, through the use of a whistle or by a shout. This system of cable laying is crude but effective under the right circumstances and is termed 'hand pulling'.

Nose pulling As the need to lay longer cable lengths of heavier cable designs developed, it became clear that some form of motive power should be introduced to act as the prime mover for the cable in the trench. This brought about the use of a power winch at the other end of the trench from the cable drum, with a steel rope, of at least the same length as the cable, paid out from the winch through the trench to connect to the end of the cable. When the rope is drawn back by the winch the cable is pulled into the trench. In this method the cable drum must be mounted by a spindle or rim system to allow the cable to be pulled off the drum freely. The frictional forces on the trench bottom are minimised by running the cable over free running rollers spaced at specified intervals of a few metres. Skid plates are usually installed at bends. If rollers mounted vertically are used the curvature of the roller must match that of the cable to ensure that the cable is not deformed. This method of cable laying is termed 'nose pulling' and probably in one form or another is the most common practice throughout the world. However, it should be noted that there are limitations in nose pulling cables. The two principal limitations are:

(a) the maximum tensile load that can safely be applied to the cable conductor or to the pulling eye fitted to the cable end;
(b) the maximum side wall thrust developed on the cable as it traverses directional changes along the cable trench.

These limitations are directly proportional to the complexity of the cable pull and the size and weight etc. of the cable involved; it is therefore necessary to calculate the anticipated loads before installing a cable by the nose pulling method. A load cell or dynamometer should be utilised to indicate actual tensions developed whilst pulling in cables by mechanically aided means. Some modern winches are fitted with a tension recording device and an automatic shut-down facility to respond to a pre-set maximum permissible tension.

However, there are many projects which involve extremely heavy cables, very long lengths between joints, tortuous routes involving continuous changes in direction and level, or a mixture of these difficulties in varying degrees. In such circumstances it is often found that the tension or the side thrust would reach an unacceptable level and possibly lead to damage of the cable during installation. To deal with these situations a range of more sophisticated cable pulling techniques has been developed which distribute the pulling force more evenly throughout the cable length. Three of the more widely used methods are described below.

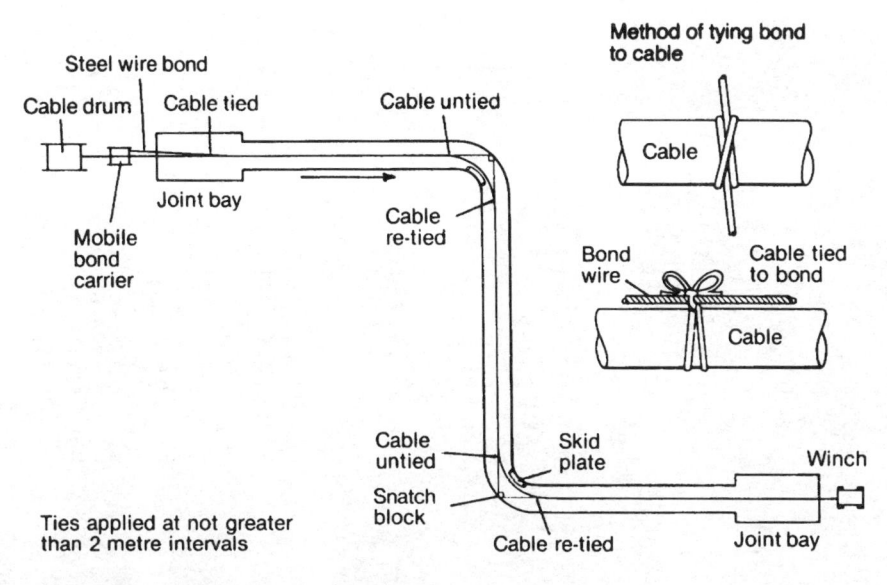

Fig. 36.2 Principles of bond pulling technique

Bond pulling The first method is called 'bond pulling' and is illustrated in fig. 36.2. A wire bond of more than twice the length of the cable route is coiled on a drum mounted on a suitable mobile trailer unit equipped with a braking device to maintain the bond in tension, and this is placed near the cable drum. The bond is run out through the whole length of the trench over cable rollers and attached to the pulling winch. At each change of direction of the route the bond is taken through a snatch block anchored to the side of the trench. These snatch blocks take the full side force on the bond from the change in direction. Initially, some 20–25 m of cable is hand pulled off the drum to allow 10–12 ties of jute yarn and cable. Before applying any ties, the bond wire must be back-tensioned to a load value approximately of 12% of the total weight of cable to be pulled. As the cable installation progresses the back tension may be relaxed to hold the load value essentially constant throughout the pull. The bond must not be allowed to relax totally until the cable is in its final position when the ties can be removed. Once the winch has started, the cable is tied to the bond at intervals of approximately 2 m as it is drawn off the cable drum. Prior to each change of direction at snatch block positions the ties are removed whilst the cable is taken round the bend, and the ties are re-attached to the bond immediately after the change of direction. Care must be taken to guide the nose of the cable over the rollers to avoid jamming or displacement of rollers and possible consequential damage to the cable. Once the cable has been pulled into position and the jute ties have been removed the bond is rewound on to the bond carrier ready for the next cable pull. An example of a bond pull route is shown in fig. 36.3.

Driven rollers The second method of cable laying is the driven roller technique. At intervals of 10–20 m along the trench, power roller units are installed. Each of these comprises a pair of wheels between which the cable passes and from which it gets the motive force to push it as far as the next roller unit. One wheel is a powered wheel driven by an electric motor through a gear box. It transmits the driving force to the cable by the frictional force between the cable sheath and the pneumatic tyre on

Fig. 36.3 A bond pull route for 400 kV PPL cable which has a very steep gradient in addition to onerous bends at termination positions

the driving wheel. The frictional force is maintained by the physical spacing of the two wheels. The distance between roller units in a straight trench is determined by the weight per unit length of the cable. At corners, ducts etc., extra units are employed to provide the increased power requirements involved in bending the cable. The nose of the cable is guided from one unit to the next by a pulling rope run back from a light capstan winch at the opposite end of the trench to the drum. The forces applied to the cable nose in this system are a small fraction of those involved in the nose pulling method. The roller units and the winch are synchronised by electrical interconnection. Free-running rollers are used to support the cable between driven roller units.

Caterpillar machines The third method is similar to the driven roller technique except that caterpillar machines are used instead of driven rollers. The caterpillar machines, of similar design to those used in cable manufacturing, have the advantage over motor driven rollers that their surface area in contact with the cable sheath is much greater and hence there is less likelihood of the cable slipping. Both the driven roller and the caterpillar machine have the advantage of being able to position the cable in the trench accurately with both forward and limited backward movement of the cable. This is something which is not easily achieved with the bond pulling system. The caterpillar machine is also recommended when cables are installed in confined spaces such as tunnels and substations where accurate control is required to minimise the risk of damage to the cable from protruding steelwork, concrete beams etc., and also damage to essential equipment already commissioned. The use of pulling blocks, winches, rollers and wire bonds is not always recommended in confined spaces.

Cable oversheath protection and checking The most economic and straightforward method of cable pulling in most situations is the nose pulling method. The basis of safe cable pulling irrespective of the method used is good communication, usually by two-way radio, and a disciplined workforce. Good co-ordination between operators will ensure a satisfactory cable installation. Laying of cables must only take place when cable and ambient temperature have been at or above 0°C for the previous 24 hours.

Transmission voltage cables are invariably of the insulated sheath type, the cable being covered with an oversheath of either medium density polyethylene or reduced flame propagating PVC depending on its location. Over this sheath a graphite coating is provided to form a conducting layer. For such cables it is common practice to check the integrity of the cable oversheath electrically after cable laying to ensure that the oversheath has not sustained damage. This may be caused by sharp stones or other deleterious matter which had not been removed from the trench, thus necessitating oversheath repairs (for repair methods see chapter 26). The optimum time to carry out this test is immediately following the placing of primary backfill and cable protection tiles. Any faults can then be rectified without re-excavation of the road. The test usually involves applying 25 kV d.c. between the metallic cable sheath and earth where the cable is in ducts or 12 kV d.c. if no ducts are present. The integrity of the oversheath of the auxiliary cables laid with the main feeder cables is also checked, albeit at lower voltages.

After pulling in it is recommended that all non-pressurised cable ends at joint bays be raised up from the joint bay floor, if possible to above ground level, to prevent the possible ingress of water.

JOINTING

At transmission voltages, jointing operations take days in total rather than hours and it is essential to use some form of weather/security protection in the form of a tent over the joint bay and dewatering pumps. At the highest voltages, on the more sophisticated system designs, it is not unusual for joint bay occupation to spread over 5–8 weeks in total and for jointing procedures to demand a clean, humidity controlled environment. In these cases complete houses are erected over joint bays. Special air-conditioned enclosures are constructed inside and substantial quantities of electricity have to be provided to power the equipment used. A typical 400 kV straight joint enclosure is shown in fig. 36.4 where the internal tubular frame supports a fire-retardant PVC cover to contain the dehumidified atmosphere required for jointing. Overall protection is provided by a fabricated steel building which also allows adequate space for the dehumidification, jointing and ancillary equipment. The joint bay floor and walls are coated with a special non-slip epoxy paint which seals the concrete and prevents build-up of dust.

Fluid-filled cables For fluid-filled cables, cleanliness during jointing is of paramount importance and it is necessary to replace the fluid lost during the jointing operations. The required quality is achieved by utilising a degasification plant, which filters and heats the replacement cable fluid under vacuous conditions within defined limits, thus ensuring that the fluid is up to the specified standard.

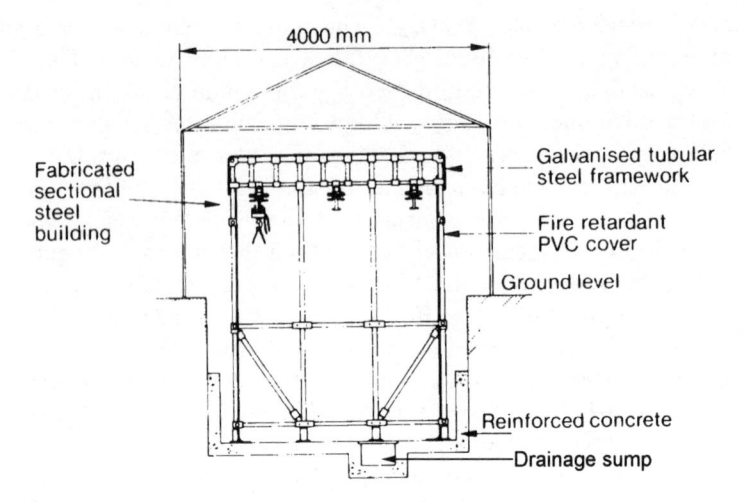

Fig. 36.4 Enclosure for joint bays to obtain atmospheric humidity control

The sequence of jointing is considered at the planning stage to ensure that the effects of profiles and fluid feeding requirements on jointing sequence are adequately catered for. This of course affects the cable laying sequence.

Fluid-filled cables should be jointed with the cables maintained under a positive internal pressure; however, in some unusual situations where this is not possible, e.g. for cables installed vertically in shafts, special techniques have to be adopted.

On completion of jointing, the joint is evacuated for a specified period to remove all traces of air before it can be impregnated. After satisfactory testing to prove the efficacy of this operation, the joint is filled with clean processed fluid and the jointed cable system is returned to its normal pressure. A whitewash test is applied to the plumbs for a period of no less than three days before the outer parts of the joint are placed in position as a final test to prove their integrity.

XLPE cables Joint bays for XLPE cable systems also need to achieve a basic standard of cleanliness although dehumidification is not normally a requirement. Cleanliness is important while the insulation of the joints is being applied; this needs a small clean tented area immediately surrounding the joint being completed. This tented area may be moved from joint to joint within the bay and represents savings in costs compared with close sheeting the entire bay (fig. 36.5). Few savings in terms of bay dimensions over other systems are possible due to the similarity of some of the tooling and to the larger cable size. For example an identical hydraulic indent press for the copper conductor connection ferrule is used on all systems; this press needs to rotate about the cable axis. Additionally a thermal rating calculation of the bay may prevent spacings being reduced.

Joint bays for all systems need to restrain cable thermomechanical movement and it is customary to anchor the cable and joints against thermal expansion and contraction forces using a cement bound sand backfill at least up to the centre line of the joints. Cement bound sand offers mechanical restraint as well as a stable, low thermal resistivity.

Fig. 36.5 Completed joints in a 132 kV XLPE tunnel installation

SITE TESTS

Tests after installation are made in order to demonstrate the integrity of the cable system and its accessories. Typical tests (other than hydraulic tests on fluid-filled cables and optical time domain reflectometry tests on optical fibre distributed temperature sensing cores) are conductor resistance, 10 kV d.c. oversheath tests (on insulated sheath systems), cross-bonding circulating sheath currents (on cross-bonded systems) and high voltage tests.

The high voltage d.c. test is carried out on fluid-filled cables at a multiple of the a.c. cable voltage varying between 1.7 and 2.6, depending on the cable voltage and test specification used. Mobile a.c. HV test sets are under development for after-laying tests on XLPE cable systems at test voltages of up to $1.7U_0$ for 30 minutes. The cross-bonding check is designed to measure that the sheath currents flowing under simulated load conditions in cross-bonded cable systems are minimal. This is achieved by applying a low voltage short-circuit current to the three cable conductors and measuring the induced currents in the cable sheaths at the link box locations, from which the full load sheath circulating currents may be calculated to determine the efficacy of the cross-bonding system.

CHAPTER 37
Thermomechanical Design

A change in the load being carried by a power cable causes a variation in the temperature of the various components and this results in thermal expansion or contraction of these components. The effects of these thermal changes on the insulation have been discussed in chapters 2 and 29. In this chapter the effects of temperature on the metal components of the cable are considered. These effects have a considerable influence on the design of cable installations, bearing in mind that cables installed in air can be subjected to temperature variations of approaching 100°C under normal loading conditions and even greater excursions when short circuits are taken into account. Thermomechanical design must be considered for all types of power cable installations but it is of special importance in the case of transmission cables where maximum operating temperatures are high and conductor cross-sectional areas up to 2500 mm^2 are used.

Theoretically there are two extremes in installation practice as far as thermomechanical effects are concerned:

(a) where the cable is completely unrestricted and the changes in temperature result in the full expected thermal expansion; in these situations no compressive or tensile forces develop in the cable;
(b) when the cable is fully restrained and no movement of the cable is permitted; in these cases, the thermal expansion or contraction is fully absorbed by internal compressive or tensile forces.

In practice it is not possible to achieve these two extremes as the cable has more than one metallic component. These will be at different temperatures and have different coefficients of thermal expansion. Therefore, if no differential movement of the cable components is permitted, it is impossible to obtain free expansion, i.e. without tensile or compression forces developing over a range of temperature. In practice, the conductor is usually the dominant factor in thermomechanical effects. This is because it experiences the greatest temperature change and has a high elastic modulus. An exception to this is the case of smooth (i.e. non-corrugated) aluminium sheaths, particularly when associated with small conductor sizes and thin insulation (i.e. a small difference in conductor and sheath temperatures).

Similarly it is not possible to restrain all the cable components fully. In practical installations only the sheath or pipe is restrained and therefore the conductors are not rigidly held and will have some freedom for movement. This freedom of movement will be extremely small in the case of a single-core self-contained cable but will be of significant proportions in the case of the cores in a pipe type cable.

Thermomechanical design of installations must take into account the movements and forces which develop and ensure that these can safely be withstood by the cable and its accessories. As mentioned earlier, the effects of the thermal expansion depend on how the cable is installed and therefore it is convenient to review the design under the headings of methods of installation.

BURIED CABLES

Considering a single-core cable installed in a well compacted backfill, longitudinal and lateral movement of the complete cable is virtually eliminated, and the only possible movement is longitudinal displacement of the conductor and insulation within the sheath. This movement is resisted by friction between the core and the sheath but can be eliminated if the conductor is firmly held at the cable ends. If it is assumed that no longitudinal movement occurs, then the force developed in the conductor will be

$$F = ES\alpha\theta \qquad (37.1)$$

where F = force in conductor (N)
E = effective modulus of elasticity of the conductor (N/m²)
S = cross-sectional area of the conductor (m²)
α = coefficient of expansion of the conductor (°C⁻¹)
θ = temperature rise of the conductor (°C)

In this equation it is assumed that the conductor acts as an elastic member. The effective modulus of the conductor will depend on its material, the method of construction and the state of temper of the material. Figure 37.1 shows a typical initial force versus temperature characteristic of a single-core Milliken conductor cable where no longitudinal movement is permitted. It will be seen that, up to about half the maximum temperature rise, the relationship between temperature rise and force is linear, but for higher temperatures the rate of rise with force decreases.

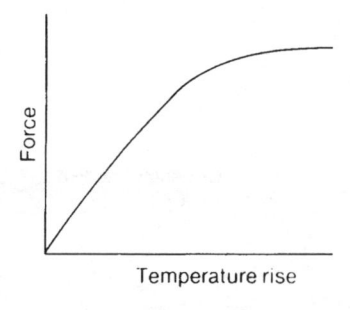

Fig. 37.1 Force versus temperature characteristic for a restrained single-core cable with Milliken conductor

The reason for this is that the conductor no longer acts as an elastic member and some permanent set occurs. Tested under conditions similar to those experienced in service, a 2000 mm² Milliken conductor cable can produce a maximum thrust of some 60 kN.[1] If end movement is permitted, the maximum force will be reduced, the magnitude of the reduction depending on the amount of movement permitted and the frictional force between the core and the sheath. In the case of a fully restrained cable, if the cable is allowed to cool to its initial temperature a tensile force will develop in the conductor. This is due to the fact that the conductor has effectively been shortened at the maximum temperature, as a result of creep of the conductor. The effect of repeated heat cycles is illustrated in fig. 37.2 where it will be seen that the cable system progresses to a position where compressive and tensile forces of approximately equal magnitude develop during the respective heating and cooling periods.

When a buried cable is laid round a bend the tensile or compressive force in the conductor will subject the insulation to a side force which is given by the following formula:

$$F_s = \frac{T}{r} \qquad (37.2)$$

where F_s = side force per unit length (N/m)
        T = conductor force (N)
        r = radius of bend (m)

With conductors of large cross-sectional area the insulation can be subjected to very high side forces. The compressibility of impregnated paper insulation is not greatly affected by temperature and a design force of approximately 20 kN/m is generally used.[1]

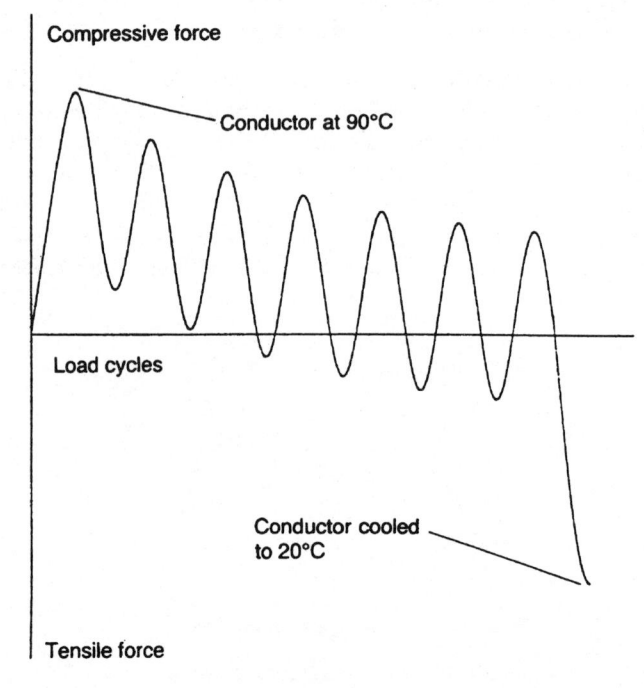

Fig. 37.2 Tensile and compressive forces generated in a single-core cable during loading cycles

In the case of a polymeric insulated cable, as the temperature of the cable rises, the force developed in the conductor increases and at the same time the ability of the insulation to withstand the pressure decreases. Much work has therefore been carried out to determine suitable maximum overload temperatures associated with installation radii. In the case of XLPE cables a maximum overload temperature of 105°C is now taken when associated with the installation radii given in chapter 36. Effects due to short circuits are not considered significant because of the short times involved and the modest temperature rises that occur with the large conductors generally associated with transmission cables.[2]

CABLES INSTALLED IN AIR

Cables are often installed in air, outdoors and indoors, and it is usually necessary to provide a cleating system to support the cables. Basically two design philosophies are adopted.

Rigid system
With this system, the cables are supported in such a manner that no longitudinal or lateral movement is permitted, i.e. it is similar to a buried cable system.

Flexible system
Lateral expansion of the cable between supports is permitted but it is controlled in such a manner that the cyclic strains imposed on the cable components are kept within acceptable limits. The component most affected by these strains is the cable sheath.

Because expansion is not permitted in the rigid system, the compressive forces will be high. It is therefore necessary to determine the spacing of the cleats so that buckling of the cable will not occur.

The spacing between supports can be calculated using a development of Euler's buckling theory.[3] It should be noted that the theory assumes longitudinal uniformity and therefore is not strictly applicable to corrugated sheaths. However, experimental work has shown that it can be applied to corrugated aluminium sheaths if the values given in the following sections are used.

$$L = \frac{2\pi}{s}\left(\frac{E_{\text{eff}}I}{1000\,F_{\text{p}}}\right)^{1/2} \tag{37.3}$$

where L = maximum spacing between supports (m)
   s = factor of safety
   E_{eff} = effective modulus of sheath material (GN/m^2)
   I = second moment of area of sheath (mm^4)
   F_{p} = thermomechanical force developed in conductor and sheath (N)

It is usual to take a factor of safety s between 2 and 4. For corrugated aluminium sheath cables, E_{eff} is taken as 25% of that of aluminium. The second moment of area is calculated as follows:

$$I = \frac{\pi(d_o^4 - d_i^4)}{64} \quad (\text{mm}^4)$$

where d_o = the outside diameter for smooth aluminium and lead sheaths, or the mean of the outside diameter of the crest and trough for corrugated aluminium sheaths (mm)

$d_i = (d_o - 2t)$ (mm), where t = sheath thickness (mm)

Where cables are installed around bends, it is necessary to reduce the spacing between cleats to 30–60% of the values used for straight sections.

Fig. 37.3 Railway electrification feed cables suspended on posts

The most usual forms of flexible system are where cable movement is permitted at right angles to the longitudinal axis of the cable. To ensure that the cyclic strains in the cable are within acceptable limits it is necessary to set up the cable initially with offsets. Two different systems are used: those in which the cable moves in a horizontal plane and those in which the cable is permitted to move in the vertical plane. An example of the latter is shown in fig. 37.3, which shows a 25 kV corrugated aluminium sheathed cable installed along a railway. In the case of the horizontally trained cable it is usually necessary to provide intermediate sliding supports between cleats. This is illustrated in fig. 37.4.

Systems which permit vertical movement usually take the form of widely spaced supports with the cable sagged between them in the manner advocated by Holttum.[4] The initial work has been extended to smooth and corrugated sheath cable and the following formulae may be used.

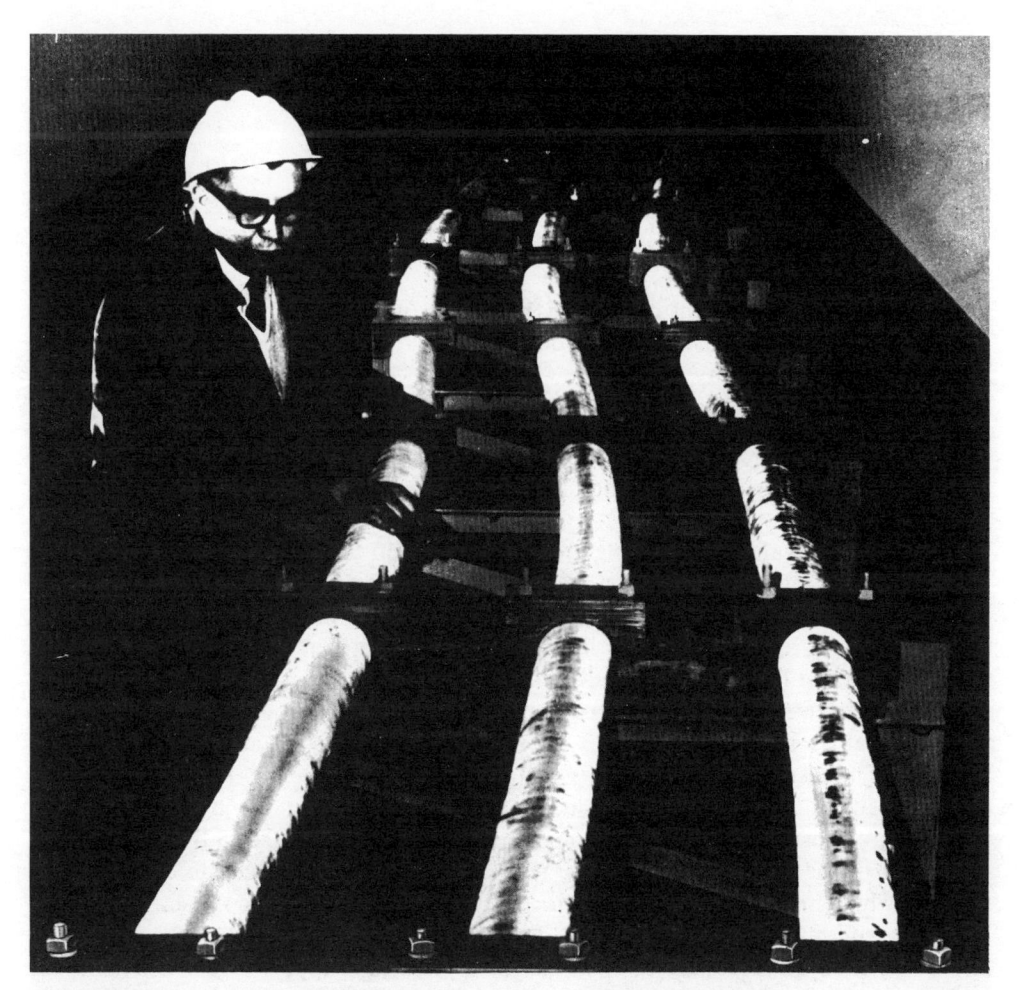

Fig. 37.4 Cables cleated horizontally with allowance for cable movement by sliding at alternate cleat positions

Lead sheath cable

$$L = \left(\frac{K d_m^2 t}{W \times 10^3}\right)^{1/2} + 0.2 \quad \text{(m)} \tag{37.4}$$

where L = spacing (m)
d_m = mean diameter of sheath (mm)
t = sheath thickness (mm)
W = cable weight (kg/m)
K = factor as given below:

Material	K
Pure lead	5.98
Alloy E	9.5
Alloy $\frac{1}{2}$C	9.15
Alloy B	16.19

Aluminium sheath cables

$$L = \left(\frac{0.00244 \, Y I}{d_r W}\right)^{1/2} \quad \text{(m)} \tag{37.5}$$

where L = spacing between supports (m)
Y = yield stress of aluminium or effective yield stress in the case of aluminium sheaths (MN/m^2)
I = second moment of area of the sheath (mm^4)
d_r = outside diameter for smooth sheaths or root diameter for corrugated sheaths (mm)
W = cable weight (kg/m)

The value for I is calculated in the same manner as for equation (37.3). The effective yield stress for corrugated aluminium sheaths is taken as 25–30% of the yield stress of aluminium.

The calculated values for spacing are not critical and the distance between supports in a vertically sagged system can be adjusted within limits of 20% to suit local conditions. The initial sag in the vertical direction is chosen so that the bending strain in the sheath, as a result of changes in length due to temperature variations, does not exceed the design limits for the sheath.

In cable systems that accommodate expansion in the horizontal plane, it is usual to train the cables in approximately a sine wave with swivel cleats at the points of inflexion. The distance between cleats and the offset are chosen so that the change in sheath strain due to temperature variations does not exceed the design limit. For simplicity, it is usual to assume that movement of the cable is equal to the unrestrained expansion of the conductor. However, when required, a more accurate calculation is possible which makes allowances for the compressive force in the conductor.

Compared with the rigid system, the flexible system uses fewer cleats and supports and the mechanical forces imposed on the accessories and support system are relatively low. Against this, the cable has to be carefully positioned during installation and allowance must be made for the space occupied by the offsets over the full temperature

Table 37.1 Comparison of spacings between cleats for rigid and vertically sagged flexible systems (FF cable with CSA sheath)

Cable type and size	Spacing between cleats on straight sections (m)	
	Rigid	Vertical sagged
Single-core 400 kV 2000 mm^2	1.0	4.0
Single-core 132 kV 300 mm^2	0.55	3.15
3-core 33 kV 185 mm^2	0.7	3.45

range. Table 37.1 makes a comparison of the spacing between supports for rigid and vertical sagged systems with self-contained fluid-filled cables having a corrugated aluminium sheath.

Cleats used for cables installed in air can be made of aluminium alloy or of a filled resin. When the cleat is metallic, an elastomeric liner is usually used. In situations where special restraint is required, e.g. when cables are installed in long vertical runs, long cleats manufactured from hardwood are normally used. For cleating cables underground, cleats manufactured from a filled resin are now preferred to the hardwood cleats previously employed.

Polymeric insulated cables without a metallic sheath, or with a lead sheath, or in particular a foil laminate sheath require special consideration. With this type of cable the thermal expansion of the polymeric insulation causes a significant increase in cable diameter. If a normal type of cleat is used it can restrain the insulation from expanding and result in deformation of the insulation, particularly if a stranded earth return conductor (shield wires) has been applied on to the extruded core. To prevent this, the cleats for this type of cable are designed to permit a limited expansion of the cable and at the same time provide the necessary restraint. This can be achieved by providing a specially resilient elastomeric liner or by using a bolt assembly which allows a degree of movement, e.g. by the use of specially designed spring washers.

CABLE INSTALLED IN DUCTS

The duct method of installation of self-contained cables is widely adopted in North America and Japan. Traditionally, the thermal expansion of the cables is accommodated by permitting movement of the cable into the manholes in which the joints are situated. Because of the friction between the cable and the duct, the full expansion of the cable is not experienced in the manhole. To cater for the thermal expansion, the joints are offset from the centre line of the ducts as illustrated in fig. 37.5. However, the movement of the cable into the manhole imposes bending strains on the cable adjacent to the joints. Before the mechanism was fully understood, considerable trouble was experienced with fatigue cracks developing in the cables due to cyclic bending. To keep

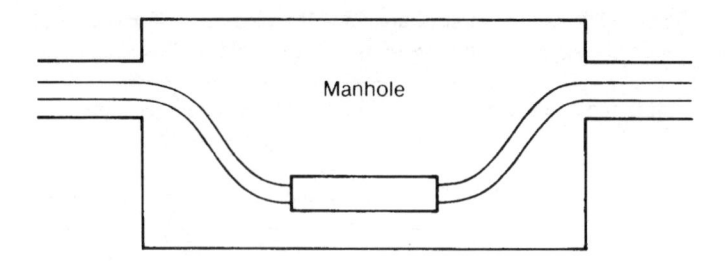

Fig. 37.5 Cable joint installed in a manhole

cyclic strains within acceptable limits, it is necessary to adopt generous offsets for the joints and in some cases to allow the joints to move. More information on the layout of cables in manholes is given elsewhere.[5-7] A manhole for a 230 kV cable system is shown in fig. 37.6 in which it will be noted that movement of the joints is catered for by suspending the joints from the manhole ceiling.

There are advantages in using aluminium sheathed cables for duct systems because a cyclic strain some 2.5 times greater than for lead alloy sheath can be tolerated. As a result of this, smaller offsets can be employed, and hence smaller manholes.

Even for cables with aluminium sheaths, the manholes for duct installations take up a considerable amount of space and hence are expensive. Attempts have been made to overcome this problem by rigidly cleating the cable and thus preventing movement, instead of permitting movement of cable into the manholes. Thermal expansion of the cable has then to be absorbed by lateral movement of the cable within the duct and the development of compressive forces.

Fig. 37.6 Manhole for 230 kV system with suspended joints to cater for cable expansion

INSTALLATION DESIGN

The analysis given in the preceding sections assumes that the method of installation is uniform throughout the cable route. In many cases this is not the case. For example short duct runs are often used for road crossings in directly buried systems and terminations are invariably associated with at least short lengths of cable in air. If this transition involves a change from a fully restrained system to a flexible system, it is possible that core movement from the fully restrained system to the flexible portion will occur on heating and vice versa during cooling. If this movement is excessive there is a possibility of damage to the insulation and its screen and also that excessive sheath strains occur in the flexible part of the system. The importance of these effects increases with conductor size and with maximum conductor operating temperature.

If possible it is best to avoid such transitions. For example where portions of a directly buried system come to the surface, a cleating arrangement which produces a rigid system should be used. Similarly for short duct runs it is possible, after installation of the cable, to fill up the duct with a special pumpable grout which provides similar support to that which the cable experiences in the ground. If it is required to remove the cable at a later stage, the grout can be removed using high pressure water hoses.

If a transition is unavoidable, movement can be restricted by increasing the frictional force between the core and sheath by snaking the cable at the transition. For FF cables with extremely large conductor sizes and where it is necessary to avoid all movement, an anchor stop joint can be used at the transition position. With this type of accessory, the conductor is effectively connected to the joint sleeve by the epoxy resin moulding and the joint sleeve is anchored, by insulated metalwork, to the concrete pad of the joint bay.

JOINTS AND TERMINATIONS

The thermomechanical forces developed by the cable have a significant effect on the design of accessories. In fully restrained systems, the conductor connections must be capable of withstanding the maximum compressive and tensile forces. Soldered connections, which are widely used for small conductor cables, are susceptible to slow creep failure under tensile forces. For the larger conductor size used in transmission cables, it is now established practice to use connections with a superior long-term tensile performance, i.e. compression ferrules or welded connections.

The thrust developed by the cable conductors is also important with regard to the electrical insulation of the joint. If the joint is not sufficiently strong in compression there is a possibility of collapse due to buckling. Generally there is no significant problem with single-core joints but 3-core cable joints are significantly weakened by the need to offset the cores within the joint shell due to the requirement to build up the insulation over the ferrule. This makes the pitch circle diameter of the conductors in the joint greater than that in the cable and the offset of the cores encourages movement of cable cores into the joint.

The performance of joints is illustrated in fig. 37.7. The full line shows the relationship between movement of the cores from the cable and the force in the conductors for the maximum operating temperature. The broken lines are the movement versus force characteristics of two joints. The latter curves are established by subjecting a joint in the laboratory to an increasing compressive load and noting the movement of the conductor

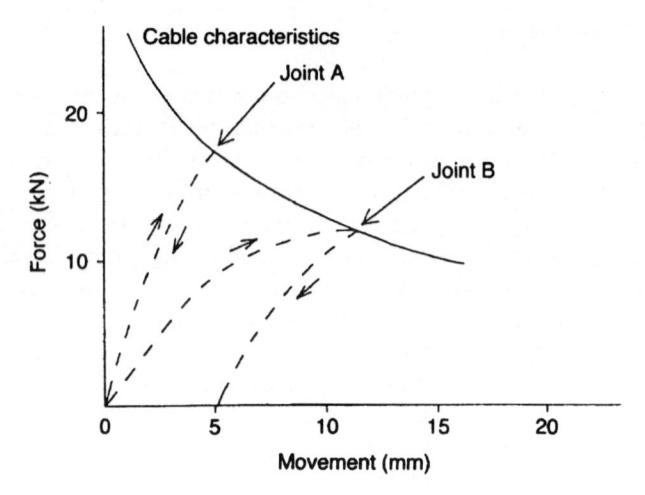

Fig. 37.7 Cable and joint compressive force versus movement characteristics

into the joint. The position where the joint characteristics cut the curve of the cable versus force characteristics is the situation which would occur if the joint was installed in a long length of cable. In the case of joint A, it will be seen that relatively little core will move from the cable into the joint. In the case of joint B, the two curves do not intersect until significantly more cable core has moved into the joint. To confirm that the joint design is suitable, the compressive conductor force is progressively reduced. If the cores have deflected in an elastic manner then, at zero force, the residual movement is small and the joint is judged to have performed satisfactorily. If the residual movement is large then the cores are judged to have buckled, i.e. deformed inelastically (joint B).

The joint A characteristic is important not only from an initial thrust point of view but also when consideration is given to the performance of the joint under the compressive and tensile forces developed as a result of load cycles. Repeated longitudinal core movement and local bending of the cable in the joint can cause soft spots to develop in the paper insulation. These soft spots are caused by the butt gaps between papers enlarging and accumulating at one position. In an extreme form they can lead to cable failure. It is also necessary to ensure that any contact between the cable cores and joint fittings does not cause cable damage as a result of the cyclic movement. These considerations apply both to self-contained and pipe type cables.

Terminations must be capable of withstanding the full conductor thrusts. Generally this presents no problem but in some cases, when large conductor cables are associated with a high internal cable pressure, this can become an important design criterion for the porcelain or epoxy resin moulding.

Terminations into SF_6 insulated metal-clad equipment require special consideration. It is normal practice for the equipment manufacturer to allow small movements of the cable termination chamber to accommodate thermal expansion of the equipment. If the cable adjacent to the equipment is flexibly cleated, it is comparatively simple to arrange the cleating to allow for this movement. The situation is more complicated if the cable is rigidly cleated – this is often the case when most of the cable length is directly buried. In this situation, a semi-rigid member such as an I beam can be fixed from the cable termination chamber to the cleat support structure. The length and cross-section of the I beam is chosen to allow the chamber to move without imposing too great a load on it.

The cable is then close cleated to the I beam. With this arrangement the cable is prevented from buckling by the close cleating and the flexibility of the I beam allows for the small movements of the metal-clad equipment.

Finally, the connection between the cable sheath and the accessory must be capable of withstanding the thermomechanical forces developed by the cable in addition to any other forces such as those arising from any internal pressure. Generally plumbing is satisfactory for lead and corrugated aluminium sheaths, backed up by resin and glass fibre reinforcement. In the case of non-corrugated aluminium sheaths, the thermomechanical forces are much higher. When these are associated with the high internal pressures used in some designs of pressure-assisted cables, it is necessary to provide a connection with increased mechanical strength, e.g. reinforced plumbs or welded connections.

REFERENCES

(1) Arkell, C. A., Arnaud, U. C. and Skipper, D. J. (1974) 'The thermomechnical design of high power self-contained cable systems'. Paris: CIGRE Paper No. 21–05.
(2) Head, J. G., Crockett, A. E., Wilson, A. and Williams, D. E. (1991) 'Thermomechanical behaviour of XLPE cables under normal & short-circuit conditions'. Versailles, *Jicable '91*, Paper No. A.4.4.
(3) Cavalli, M., Guaktien, G. and Lanfranconi, G. M. (1973) '330 kV oil-filled cable laid in 1600 ft vertical shaft at Kafue Gorge hydro electric plant'. *IEEE* Paper No. T. 73. 126.
(4) Holttum, W. (1955) 'The installation of metal sheathed cables on spaced support'. *Proc. IEE, A* **102**, 729–742.
(5) Schifreen, C. S. (1951) 'Thermal expansion effects in power cables'. *Proc. AIEE* **70**. Paper No. 51–22.
(6) Hata, H. (1967) 'On the design of cable offsets in manholes'. *Sumitomo Electr. Tech. Rev.* (10), 32–40.
(7) Mochlinski, K. (1961) 'Cables in ducts: arrangement of unarmoured lead covered cables at duct ends and in manholes'. *ERA* Report No. F/T 201.

CHAPTER 38
D.C. Cables

Although many of the first electrical supply systems were based on d.c. distribution, these were rapidly superseded by a.c. systems which had the desirable feature of easy transformation between generation, transmission and distribution voltages. The development of modern electrical supply systems in the first half of this century was based exclusively on the a.c. transmission system. However, by the 1950s there was a growing demand for long transmission schemes and it became clear that in certain circumstances there could be benefits by adopting a d.c. voltage. These include reduction of system stability problems, more effective use of equipment because the power factor of the system is always unity and the ability to use a given insulation thickness or clearance at a higher operating voltage. Against these very significant advantages has to be weighed the high cost of the terminal equipment to convert the a.c. to d.c. and to invert the d.c. back again to a.c. For a given transmission power, the terminal costs are constant and therefore, for d.c. transmission to be economic, the system must be greater than a certain length so that the saving in the transmission equipment exceeds the cost of the terminal plant.

In the case of submarine transmission cables there is a further factor of importance. The high capacitance of cables results in a comparatively high charging current in a.c. transmission. As capacitance is proportional to length, there is a critical length at which the charging current equals the thermal current rating of the cable and hence the cable system has no capability for useful power. The charging current increases with voltage and hence the critical length reduces as operating voltage increases (see chapter 41). In the case of underground cables, the problem can be overcome, albeit at the penalty of significant cost, by the use of shunt reactors at intervals along the route. However, this is not practical for submarine cables and hence a.c. submarine transmission systems have significant length limitations. In d.c. transmission, a charging current only occurs during switching on or off and therefore has no effect on the continuous current rating of the system. Thus the length limitation is eliminated and this explains why d.c. transmission has been so widely used for long submarine crossings.

To date, the same basic types of cable that have been developed for a.c. transmission have been adopted for d.c. transmission. Hence many of the design features are similar and in the following sections it is intended only to discuss features where differences in design occur.

D.C. Cables

CONDUCTORS

As there are no electromagnetic induction effects, except when switching on and off, there are no skin and proximity effects to be taken into account in the determination of the continuous current rating. As the a.c. resistances of a conductor due to these effects can in practice be up to 20% greater than the d.c. resistance, the use of d.c. results in lower conductor losses. Furthermore there is no need, from a rating point of view, to use the more complex Milliken construction for the larger conductor sizes.

INSULATION

As might be expected, there are considerable differences in the design of the dielectric for d.c. cables in comparison with a.c. operation.[1]

The long-time electrical strength of dielectrics under d.c. conditions is significantly higher than under a.c. This arises from the reduced discharge activity under d.c. As was explained in chapter 2, an important deterioration phenomenon in paper insulated cables is the effect of discharge activity in the butt gaps between adjacent turns of paper. In the case of a.c., multiple discharges can occur during both positive and negative half cycles because, following a discharge, voltage conditions are rapidly re-established by the capacitive current flowing in the insulation. In a d.c. cable, the build-up of voltage following a discharge is much slower, being controlled by the leakage current through the insulation, which is several orders of magnitude lower than the a.c. charging current. Therefore the discharge repetition rate in d.c. cables is very much lower than with a.c. cables and, for equal life, the d.c. cable can be operated at much higher electrical stresses.

This is particularly evident in the case of mass-impregnated solid type paper insulated cables where the maximum stress under a.c. operation is limited to about 4 MV/m (r.m.s.) but for d.c. operation stresses up to about 32 MV/m have been employed.[2]

The dielectric design of d.c. cables is more complex than that of a.c. cables because of the different stress distribution. In chapter 2 the stress distribution in a.c. cables was derived and the assumption was made that the insulation has a uniform permittivity. This characteristic of the insulation is affected only to a very minor extent by changes in cable temperature and hence the stress distribution does not change significantly as a result of current loading conditions. In the case of a d.c. cable, the stress in the insulation depends on the geometry of the cable and the resistivity of the insulation. It can be shown that, if the resistivity is uniform throughout the insulation, then the stress distribution is the same as that in an a.c. cable. However, the resistivity of the insulation is highly dependent on temperature and to a lesser extent on the electrical stress because these increase the mobility of the charge carriers.

The relationship is given by the following formula:

$$\rho = \rho_0 \exp(-\alpha\theta) \exp(-\beta E) \tag{38.1}$$

where ρ_0 = resistivity at reference temperature (Ω m)
 θ = difference in temperature between the actual and reference temperatures (°C)
 α = temperature coefficient of electrical resistivity (per °C)
 β = stress coefficient of electrical resistivity (per MV/m)
 E = electrical stress in the insulation (MV/m)

In the case of hydrocarbon fluid impregnated paper, a typical value for α is 0.1 and for β 0.03. Hence temperature has the greater influence on resistivity.

The following equation has been derived for the stress distribution in a d.c. cable:[3]

$$E_r = \frac{\delta V(r/r_s)^{\delta-1}}{r_s[1 - (r_c/r_s)^\delta]} \tag{38.2}$$

where E_r = stress at radius r (MV/m)
V = working voltage (MV)
r_c = conductor screen radius (m)
r_s = radius over insulation (m)
W_c = conductor loss (W/m)
T = thermal resistivity of insulation (Km/W)

and

$$\delta = \frac{\alpha W_c T/2\pi + \beta V/(r_s - r_c)}{\beta V/(r_s - r_c) + 1}$$

When the cable is carrying load, there will be a temperature gradient across the insulation and the influence of this is illustrated in fig. 38.1. It will be seen that the effect of the gradient, compared with the isothermal case, is to reduce the stress nearest the conductor and increase it at the outside of the insulation. For each case, by definition, the area under the curve remains constant and is equal to the applied voltage. Increasing the conductor loading will increase the temperature gradient and increase the stress at the outside of the insulation. The current rating of the cable is usually limited so that the stress at the outside of the cable under full load conditions does not exceed that at the conductor under no load.

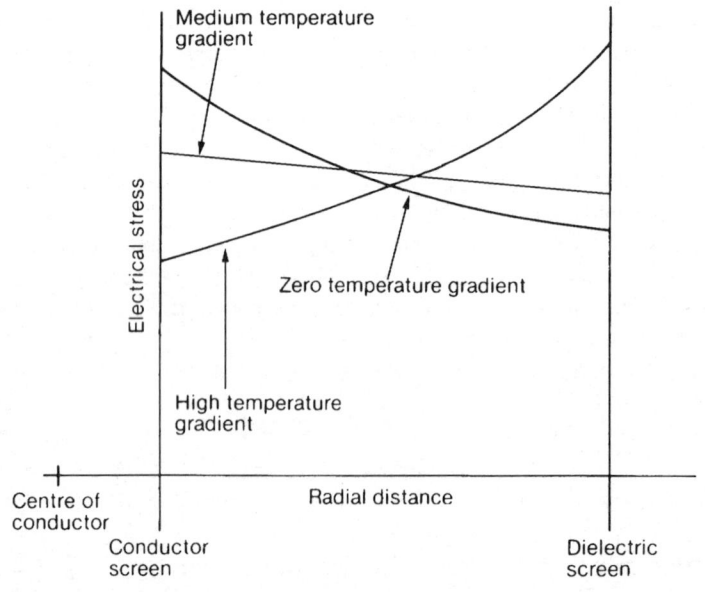

Fig. 38.1 Electrical stress in d.c. paper insulated cable showing dependence on temperature gradient

D.C. Cables

As in the case of a.c. transmission cables, transient voltages can determine the insulation thickness of d.c. cables.[4] It has been found that the most onerous condition occurs when a transient voltage of opposite polarity to the operating voltage is imposed on the system when the cable is carrying full load. If the cable is connected to an overhead line system, this condition usually occurs as a result of lightning transients. The condition is illustrated in fig. 38.2. Its effect on the stress distribution within the cable can be determined using the principle of stress superposition, i.e. the resultant stress at any point in the insulation is that resulting from the summation of the stress due to the operating voltage U_o and that due to a transient voltage equal to the impulse level of the system U_p plus U_o. The overall effect of this is illustrated in fig. 38.3, and it will be seen that the maximum stress appears next to the conductor and decreases through the insulation. The magnitude of the maximum stress will depend on the stress next to the conductor immediately before the transient occurs. When the d.c. stress is at its minimum (full load conditions) the resultant stress will be at its maximum, and when the d.c. stress is at its maximum (zero load conditions) the resultant stress will be at its minimum.

It is this requirement which usually determines the insulation thickness necessary and it is an important test requirement which has now been included in international test specifications.[5]

Fluid-filled, gas-filled and paper insulated solid type cables impregnated with a fluid compound have been used for d.c. transmission projects. As nearly all the recent installations have involved long submarine crossings, solid type cables predominate; because as there are no requirements for maintaining pressure there are no restrictions on circuit length (see chapter 42).

As stated earlier, the low repetition discharge rate associated with d.c. operation results in the ability to design solid type mass impregnated cables with design stresses (32 MV/m) approaching those used for fluid-filled cables (40 MV/m). However, the discharge performance is still of importance and to restrict the formation of cavities the operating temperature is restricted to 50°C. One of the most critical operating stages with this design of cable is when full load is switched off. Under these conditions there is a high cooling rate and the compound is at its lowest viscosity. These conditions are most conducive to the formation of cavities and hence discharge activity. To overcome

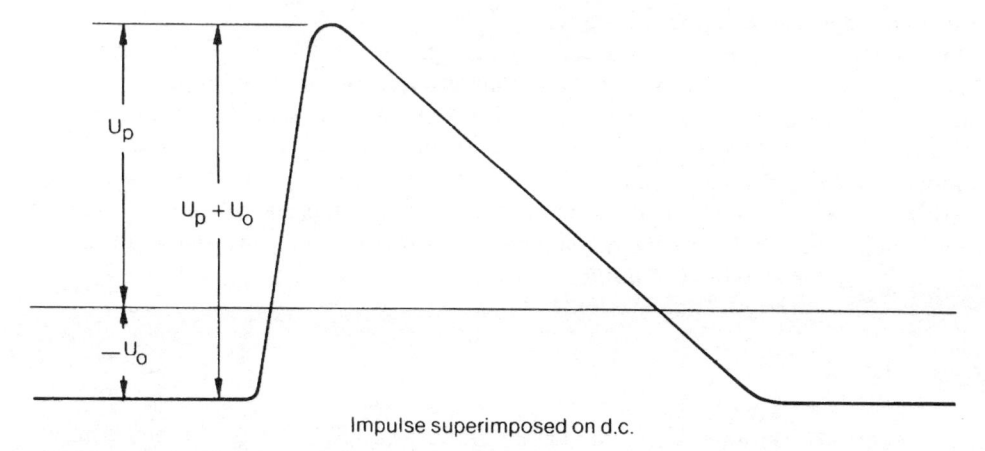

Fig. 38.2 Impulse transient on a d.c. cable

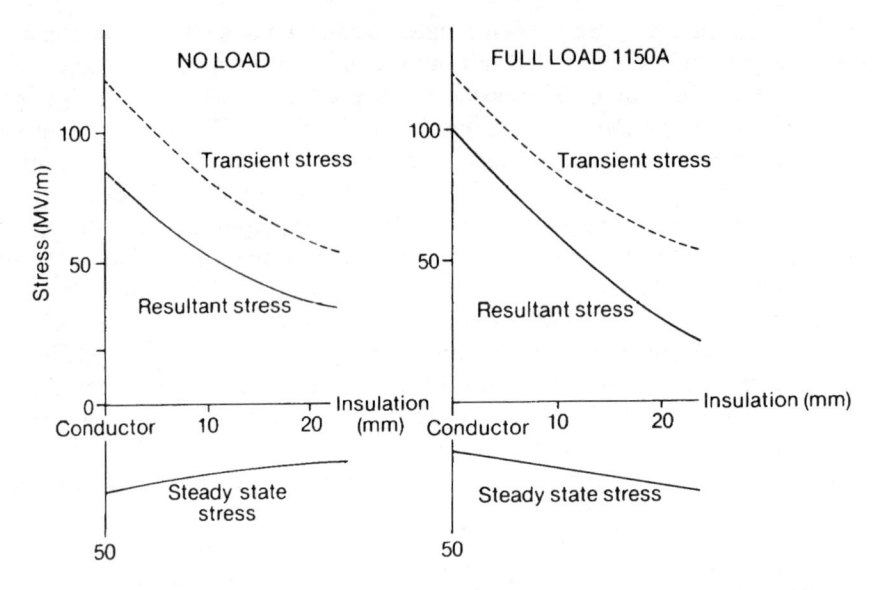

Fig. 38.3 Stress distribution in a 600 kV, 1000 mm², paper insulated d.c. cable with a positive impulse superimposed on negative operating voltage

this situation Cable Dependent Voltage Control (CDVC) is being introduced. CDVC operates to reduce the load by reducing the voltage rather than the current and hence lowers the cable cooling rate and the formation of cavities, while at the same time reducing the electrical stress in the insulation at this critical period. This change in operating conditions together with the general increases in system voltage have resulted in CIGRE initiating a review of the previous test requirements.[5]

Trials with polymeric insulated cables have been disappointing; a wide scatter in breakdown voltages being obtained. It is suspected that this is due to the development of space charges that affect the stress distribution. Most polymers have a very high insulation resistance and any space charges that occur will persist for long periods. Research has been directed to the use of additives to reduce the insulation resistance without seriously affecting the other properties but to date no commercial polymeric insulated d.c. cables have been installed.

An important benefit of d.c. operation is the virtual elimination of dielectric losses. The d.c. leakage current is of such small magnitude that it can be ignored in current rating calculations, whereas in a.c. cables dielectric losses cause a significant reduction in current rating. This is of considerable importance for higher system voltages. Similarly, high capacitance is not a penalty in d.c. cables. This permits the use of high density papers which have a high electric strength and low thermal resistivity. The latter is of importance in reducing the temperature gradient, which in turn reduces the effect of loading on the stress distribution.

SHEATHS

The absence of continuous electromagnetic induction effects results in the complete absence of sheath and armour losses. This is of particular importance for submarine cables, where in the case of single-core a.c. cables the wide separations required between cables for maintenance purposes can result in high sheath and armour losses.

ACCESSORIES

Designs are very similar to those used for a.c. cable systems. In the case of accessories using a range of materials in the electric field, e.g. stop joints, the electric field analysis under varying load conditions is most complex. Special attention has to be paid to the pollution of terminations where the d.c. field attracts dust particles to the porcelain.

SYSTEMS

The net effect of the differences in design of a.c. and d.c. cables is that, for equal power ratings, d.c. cable systems can be installed at an appreciably lower cost than a.c. cable systems. This is illustrated in fig. 38.4 in which a comparison is made between a.c. and d.c. cable systems. For the a.c. case a naturally cooled cable system using 275 kV 2000 mm^2 fluid-filled cables has been considered. This cable, with suitable modification to the dielectric design, can be considered as a d.c. cable rated at 500 kV. For equal power ratings it is obvious that the d.c. cable scheme will be significantly cheaper with respect to both cable and installation costs, as two cables in a single trench can be used for the d.c. scheme whilst six a.c. cables are required in two trenches.

As mentioned earlier, the cost of cable is only part of the total cost and the cost of the converter equipment must be included. The latter is very expensive and at present it is unlikely that a d.c. submarine cable system would be economic for a route length of less than approximately 30 km and a d.c. underground scheme for less than approximately 80 km.

It is for long submarine connections that d.c. cables have by far their greatest use and most land cables are associated with these projects. Circuits with operating

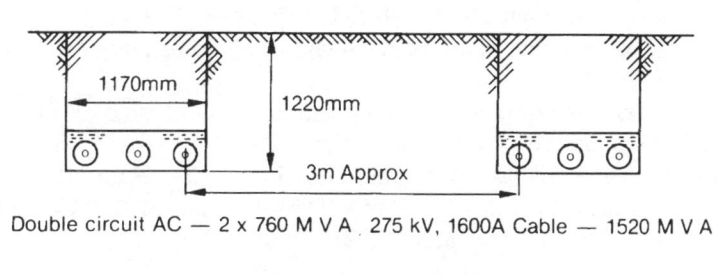

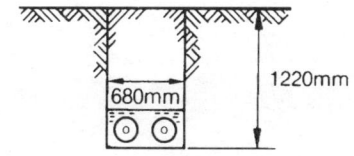

Double circuit AC — 2 x 760 M V A 275 kV, 1600A Cable — 1520 M V A

Two pole DC — $\mp$ 500kV, 1600A Cable — 1600 M V A

Fig. 38.4 Comparison of a.c. and d.c. transmission capacities, taking as a basis 275 kV, 2000 mm^2, a.c. fluid-filled paper cable

voltages up to 450 kV and lengths of 250 km are in operation. More information is provided in chapter 42. Notable exceptions are the UK 266 kV Kingsnorth–Willesden underground circuit which had a route length of approximately 80 km[6] and the 450 kV cables which were installed in a tunnel beneath the St Lawrence river in Canada.[7] Self-contained fluid-filled cables were used for both these circuits.

Research has indicated that paper insulated cables can be designed to operate at system voltages well in excess of 1000 kV.[8,9]

A significant development has been the design and testing of a PPL insulated, fluid-filled d.c. cable for the submarine cable to link the Honshu and Shikoku islands in Japan. PPL was selected because its high impulse and d.c. voltage strength permitted the diameter of the cable to be reduced. The circuit is a 500 kV bi-pole, rated at 2800 MW, with a total length of 50.7 km, of which 48 km is submarine.[10] The conductor size is 3000 mm^2.

REFERENCES

(1) Arkell, C. A. and Parsons, A. F. (1982) 'Insulation design of self-contained oil-filled cables for D.C. operation'. *IEEE Trans.* **PAS-101**, 1805–1814.

(2) Ekenstierna, B. and Nyman, A. (1993) 'Baltic HVDC interconnection'. *CIGRE Colloquium*, New Zealand. 29 September–1 October 1993.

(3) Eoll, K. (1975) 'Theory of stress distribution in insulation of high voltage d.c. cables', Part II. *IEEE Trans.* **EI-10** (1).

(4) Povh, D. and Luoni, G. (1992) 'Impact of overvoltages on design of HVDC cables'. Paris CIGRE Paper No. 14-104.

(5) (1980) 'Recommendations for tests on power transmission d.c. cables for a rated voltage up to 600 kV'. *Electra* (72), 105–114.

(6) Casson, W. (1966) 'Kingsnorth–London d.c. transmission interconnector'. *IEE Conf. H.V.D.C. Transmission*, Paper 9. Manchester.

(7) Bell, N., Bui-Van, Q., Couderc, D., Ludasi, G., Meyere, P. and Picard, C. (1992) '±450 kV D.C. underwater crossing of the St. Lawrence river of a 1500 km overhead line with five terminals'. CIGRE Paper No. 21-301.

(8) (July 1985) 'Design limits of oil/paper high voltage direct-current cables'. EPRI report EL-3973.

(9) Arkell, C. A. and Gregory, B. (1984) 'Design of self contained oil-filled cables for UHV D.C. transmission'. CIGRE Paper No. 21-07.

(10) Fujimori, A., Tanaka, T., Takashima, H., Imajo, T., Hata, R., Tanabe, T., Yoshida, S. and Kakihana, T. (1966) 'Development of 500 kV D.C. PPLP-insulated oil-filled submarine cable'. *IEEE Trans.* **PD-11**.

CHAPTER 39
Testing of Transmission Cable Systems

The testing of transmission cable systems is dictated primarily by international, national and user specifications with procedures grouped into five categories:

(a) routine tests on cables and accessories;
(b) special (sample) tests on cables requested by the purchaser;
(c) prequalification tests for XLPE cable and accessories;
(d) type tests on cables and accessories;
(e) site tests on systems after installation.

The information given in this chapter covers the essential tests in IEC specifications and those of the British Electricity Boards for cables of the paper and PPL insulated type, primarily self-contained fluid-filled cables and cables with polymeric insulation for system voltages of 33 kV and above for non-submarine use. Since the privatisation of the UK electricity industry the BEB specifications are gradually being superseded by individual electricity company specifications. Electricity transmission is now the responsibility of the National Grid Company (NGC). The test requirements for polymeric insulated cables are similar in many respects to those required for paper and PPL insulated cables, but there are significant differences and the main ones are included herein.

The most important IEC, BEB and NGC specifications are as follows:

IEC 141 Tests on oil-filled and gas pressure cables and their accessories up to and including 400 kV

Engineering Recommendation C28/4 Type approval tests for impregnated paper insulated gas pressure and oil-filled power cable systems from 33 kV to 132 kV inclusive

Engineering Recommendation C47/1 Type approval tests for single core impregnated paper insulated gas pressure and oil-filled cable systems for 275 and 400 kV

National Grid Company NGTS 3.5.1 275 kV and 400 kV oil-filled cable

National Grid Company NGTS 3.5.2 132 kV, 275 kV and 400 kV XLPE cable

ESI Standard 09-3   33 kV impregnated paper insulated oil-filled and gas pressure type power cable systems
ESI Standard 09-4   66 and 132 kV impregnated paper insulated oil-filled and gas compression type power cable systems
ESI Standard 09-5   275 and 400 kV impregnated paper insulated oil-filled and gas compression type power cable systems
IEC 840   Tests for power cables with extruded insulation for rated voltages above 30 kV ($U_m = 36$ kV) up to 150 kV ($U_m = 170$ kV)
ESI Standard 09-16   Testing specification for metallic sheathed power cables with extruded cross-linked polyethylene insulation and accessories for system voltages of 66 kV and 132 kV
IEC 233   Tests on hollow insulators for use in electrical equipment

Other important specifications include the following:

Electra (72), 105–114, October 1980   Recommendations for tests on power transmission d.c. cables for rated voltages up to 600 kV
Electra (151), 15–19, December 1993   Recommendations for electrical tests, prequalification, and development on extruded cables and accessories at voltages >150 (170) kV and ≤400 (420) kV
Electra (151), 21–27, December 1993   Recommendations for electrical tests type, sample, and routine on extruded cables and accessories at voltages >150 (170) kV and ≤400 (420) kV
AEIC CS2-90   Specification for impregnated paper and laminated paper polypropylene insulated cable high pressure pipe type
AEIC CS4-93   Specification for impregnated paper insulated low and medium pressure self contained liquid filled cable
AEIC CS7-93   Specification for crosslinked polyethylene insulated shielded power cables rated 69 kV through 138 kV
IEEE 48-90   Standard test procedures and requirements for high voltage a.c. cable terminations

Many principles are the same as for the testing of distribution cables, as given in chapter 28, and are not repeated here.

PAPER AND POLYPROPYLENE PAPER LAMINATE (PPL) CABLES

Routine tests

Most specifications require electrical tests along the lines illustrated below for fluid-filled (FF) cables, in accordance with IEC 141-1. The examples given vary to some extent in other specifications but are typical of the requirements.

Dielectric loss angle (DLA) and high voltage tests
The DLA is measured at ambient temperature and corrected to 20°C. The voltage range is from U_0 to $2U_0$ for cables up to a U_0 of 87 kV and $1.67U_0$ for higher voltage cables. Requirements are shown in table 39.1 for cables not screened with carbon-black paper.

Table 39.1 Test voltage and DLA for FF cables

Cable voltage (U_0/U) (kV)	Highest voltage for DLA test (kV)	Maximum DLA $U_0 \times 10^{-4}$ Paper	PPL	Maximum DLA Highest voltage $\times 10^{-4}$ Paper	PPL	Maximum difference in DLA from U_0 to highest voltage $\times 10^{-4}$ Paper	PPL	A.C. withstand test (kV)
19/33	38	35	–	43	–	10	–	53
38/66	76	35	–	43	–	10	–	86
76/132	152	33	–	40	–	8	–	162
160/275	230	30	14	34	16	5	4	275
230/400	385	28	14	31	16	4	4	395

Slightly higher values are permitted when carbon-black paper screens are used. The test is followed by application of high voltage for 15 minutes at the level shown in table 39.1.

The DLA and high voltage tests for gas-compression and gas-filled cables are carried out at any gas pressure up to 2 bar. These cables normally operate at a pressure of approximately 14 bar and, as the reduced pressure is not representative of normal service conditions, the test voltages are accordingly reduced.

In the case of d.c. cables the high voltage test comprises the application of $2U_0$, with negative polarity, for 15 minutes, and as with a.c. cables the high voltage test ensures freedom from mechanical defects. For d.c. cables, however, no convenient method is available which is comparable with the a.c. measurement of DLA for monitoring cable quality. It is therefore usual to measure the DLA at reduced test voltage and the results give a good indication of freedom from excessive moisture content and introduction of contamination during processing. For fluid-filled d.c. cable with U_0 from 100 to 600 kV, the DLA is measured over a voltage range equivalent to a.c. maximum stress values of 10–20 MV/m. For gas pressure cable and mass-impregnated cable the stress values are reduced.

The application of these high voltages between conductor and sheath necessitates special attention to the provision of suitable cable terminations. Permanent installations require expensive porcelain containers and the services of skilled personnel. It is beneficial to have more simple procedures for temporary terminations for routine testing. For cables of 33 kV and below a test termination may be made by removing the metal sheath and dielectric screen over a length of approximately 1 m and applying a stress cone or other means of electric field control.

For cables above 33 kV and up to 132 kV a re-usable cast resin stress cone is used completely enclosed in an insulating material chamber filled with insulating fluid. For higher voltage cables a part gas insulated termination has been developed. This consists essentially of a horizontal gas insulated metal chamber with a vertical gas insulated high voltage bushing at one end and a cable entry chamber at the opposite end. The cable entry chamber is equipped with a cast resin insulator with a solid embedded electrode which separates the gas insulated side from the cable. The cable termination employs a re-usable resin stress cone and a conductor connection which allows the terminated cable to be plugged into the embedded electrode in the cast resin insulator.

The assembled termination is filled with either insulating fluid for fluid-filled cable tests or with insulating gas when used with a rubber stress cone for polymeric cable tests. The opposite end of the cable is terminated in a similar design of termination which does not include the high voltage bushing.

Other electrical tests

Conductor resistance and capacitance tests are mandatory. Capacitance is conveniently calculated from Schering bridge or transformer ratio bridge measurements made during the DLA test and is a useful check on cable quality. The figure should not be greater than 8% above the declared value.

Particularly when the cable contains an aluminium sheath and when it is to be installed in an insulated sheath system, it is very important that there are no defects in the corrosion protective finish. This is checked by the application of d.c. with a voltage corresponding to a stress of 8 kV/mm of average thickness of the finish (maximum value 25 kV). To make this test practicable the extruded oversheath is coated with a conducting layer of colloidal graphite during the manufacture of the cable.

Routine tests on accessories for paper, PPL and polymeric cables

Cable joints and terminations form an integral part of a cable transmission system and to ensure their reliability in service are subjected to rigid inspection and routine test procedures before despatch. Most cable joints and terminations are manufactured from several individual components and to ensure correct assembly and compliance with appropriate standards each item is visually examined and dimensionally checked before being subjected to any further tests. The additional test requirements can be considered in two main groups: (a) hydraulic and pneumatic pressure tests and (b) electrical tests.

Pressure tests

Whenever possible, the various pressure tests required are carried out with all the individual components of the accessory completely assembled, except for the cable. Cable entry positions are sealed with a plate soldered on or by a specially manufactured cover with rubber or similar seals. All accessories for use in fluid-filled and gas-filled cable systems, i.e. sealing ends, transformer or switchgear terminations and joints, together with any epoxy resin or other insulating components which form part of the pressure-retaining envelope, are subjected to pressure withstand tests. The following are examples of routine pressure tests carried out on various components.

(a) Porcelain and epoxy resin terminations: each porcelain or epoxy resin termination complete with base plate, bottom extension gland, sheath insulating ring and top plate fitted with the conductor terminal for use on fluid-filled cable systems is subjected to a hydraulic pressure test at 11 bar (just over twice the maximum system operating pressure) for a minimum time of 15 min. The same assembly for use on gas-filled cable systems is subjected to a hydraulic pressure test at 34 bar (twice maximum system operating pressure) for a minimum time of 15 minutes followed by a pneumatic pressure test for 24 hours at the maximum design pressure of 17 bar. To comply with the test specification there should be no evidence of leakage or unacceptable distortion of the metal components during or at the end of the test.

(b) Joint sleeves complete with end bells, insulating rings or resin barriers are subjected to the same pressure test procedures as for porcelain and epoxy resin terminations.
(c) Pressure tanks for all fluid-filled cable systems: each pressure tank is subjected to a hydraulic pressure test at a pressure equal to 1.1 times the design pressure for 8 hours. There should be no evidence of leakage. Following this test the pressure versus volume characteristic of the tank is checked.
(d) Auxiliary components for gas-filled cable systems: gas cylinder regulators, safety valves, pressure gauges and gas control cubicles are subjected to pressure tests to ensure their safe and correct operation.

Electrical tests
(a) Cast epoxy resin components incorporating cast-in electrodes or conductors are now widely used in the construction of joints and terminations for paper, PPL and polymeric cables. Moulded elastomeric components are widely used in accessories for polymeric cables, either alone or in combination with cast epoxy resin components. These components require electrical tests to ensure integrity and material quality. The above components are formed of a homogeneous material which is extremely sensitive to electrical discharge and when subjected to such discharge rapidly erodes, leading to breakdown of the component. If voids are present within the casting or moulded component, electrical discharge will occur in the void when the voltage across the void reaches a critical value.

Voids can be caused by cracks or splits in the resin, lack of adhesion to the embedded electrodes or conductors, or gas bubbles. Two high voltage tests are applied to these components: partial discharge tests to indicate whether voids are present in the casting, and DLA measurements to indicate the quality of the material forming the casting. For both these tests an outer earth screen, if not already present, is applied and the component is installed in an insulated tank filled with either insulating fluid or gas. The high voltage is connected to either the embedded electrode or conductor(s). DLA and partial discharge measurements are made at a test voltage which ensures that the stress imposed on the critical areas of the component is the same as, or higher than, the stress encountered in service.
(b) Joint sleeves: metal joint sleeves are sometimes provided with factory applied external insulation to enable the d.c. voltage site test on the anticorrosion covering of the cable to be carried out. To ensure that the insulation is free from pinholes or mechanical damage the insulation is subjected to a voltage of 10 kV d.c. for 5 minutes.
(c) Porcelain insulators: to verify the electrical integrity of the wall of porcelain insulators, each insulator is subjected to a high voltage a.c. test, equivalent to a stress of 1.5 kV/mm of wall thickness with a minimum requirement of 35 kV for 5 minutes. For this test the insulator is filled with water to form the inner electrode and wire or chains are placed around the barrel between the sheds to form the outer electrode. Any insulator which punctures during the test is rejected.

Non-destructive tests
In addition to the electrical tests described above it is good practice to perform X-radiography on elastomer mouldings and either X-radiography or ultrasonic tests on cast resin components.

Table 39.2 Number of samples taken from contract cables

Cable length				Number of samples
3-core cables		Single-core cables		
Above (km)	Up to and including (km)	Above (km)	Up to and including (km)	
2	10	4	20	1
10	20	20	40	2
20	30	40	60	3

Special tests on cables

Because of the importance of the integrity of expensive transmission cables, careful attention is given to a visual examination of a sample cut from each length and to a check of dimensions. As an additional check on cable quality the customer may request tests to be carried out on cable samples: for example the frequency of taking samples as stipulated in IEC 141-1 is shown in table 39.2.

In this category the most usual test is a mechanical test, which comprises a bending test followed by the application of high voltage a.c. for 15 minutes and subsequent detailed examination of the cable.

Bending test

A sample of cable is subjected to severe cycles of bending to demonstrate that it will not suffer from normal cable laying operations.

The bending test is carried out at ambient temperature and consists of winding the cable on to a test drum, unwinding, rotating the cable through 180°, rewinding and unwinding. This cycle is repeated three times, precautions being taken to prevent the sample from twisting. The hub diameter of the test cylinder is given in table 39.3.

After a 15 minute application of a.c. at the voltage specified for the routine test, a 1 m sample is cut from the centre of the test length and is examined in detail. IEC specifications prescribe limits for damage and displacement of component parts, including the insulation papers.

Table 39.3 Diameter of cylinder for bend test on fluid-filled cables

Type of cable	Hub diameter
Single-core cables with lead or corrugated aluminium sheath	$25(D+d)$
Three-core cables with sheaths as above	$20(D+d)$
All cables with smooth aluminium sheath	$36(D+d)$

D = measured overall diameter of the metal sheath
d = measured diameter over the conductor or the equivalent diameter for shaped conductors

Special tests on accessories

IEC 233 requires tests on porcelain insulators and details the rate of sampling.

Temperature cycle test
For this test the insulator, at ambient temperature, is immersed for 30 minutes in a bath of hot water and then withdrawn and immediately immersed in cold water for 30 minutes. The temperature difference between the two baths is related to the dimensions of the insulator and is generally between 35 and 50°C. The cycle of operations is repeated three times. The result is satisfactory if there are no cracks or damage to the glaze or loosening of the cemented top and bottom rings.

Porosity tests
For this test freshly broken pieces of porcelain are required which in production have been fired adjacent to the insulator. The samples are immersed in a 1% solution of fuchsine dye in alcohol at a pressure of 150 bar for not less than 12 hours. The samples are then further broken and examined. No sign of dye penetration should be found at the freshly broken surfaces.

Type approval tests

Type tests are tests made in order to demonstrate satisfactory performance characteristics to meet the intended application. They are of such a nature that, after they have been made, they need not be repeated unless changes are made in either the material or the design. It is important to carry out type tests on the complete transmission system, i.e. to include the particular design of cable and the accessories designed for use on the cable. Approval of a range of conductor sizes of a specific design and voltage, together with the associated accessories, is obtained by successfully completing all the required type tests on the smallest and largest conductor size of the range. Alternatively if the cables are designed to have the same conductor design stress throughout the range, it is necessary to complete only the required type tests successfully on the largest conductor size within the range. A summary of the various forms of test in accordance with IEC 141-1, British Electricity Boards Engineering Recommendations, and National Grid Specifications is given below.

Mechanical tests
A bending test is carried out as previously described and it may be required that a further sample of the bent cable is submitted to a mechanical integrity test of the metal sheath. For this test an internal pressure equal to twice the maximum design pressure is applied for seven days. No leakage should occur.

Loading cycle test
(a) A.C. systems: In this test a cable system is subjected to conditions more severe than encountered in normal operational service. Three miniature cable installations, each consisting of 30 m of cable previously submitted to the conditions of the bending test, and including the accessories designed for use with the cable, are subjected to a minimum of 20 loading cycles with a continuously applied voltage between $1.33U_0$ (275 and 400 kV) and $1.5U_0$ (33–132 kV). During each load cycle,

circulating current is passed through the conductor and/or sheath for 8 hours so that the conductor temperature is maintained between 5 and 10°C above the maximum design value for the last 3 hours. There is then a cooling period of 16 hours. The DLA and capacitance of each installation is measured hot and cold during each load cycle. Circulating current is induced in the cable by the use of ring core transformers, independently of the high voltage. The conductor and/or extreme ends of the sheath are connected together to make closed loops which form single-turn secondary windings of the ring core transformers. The magnitude of the induced current is controlled by voltage regulators connected to the transformer primary windings. The result is considered to be satisfactory if there is no breakdown and the measured values of DLA remain stable.

(b) D.C. systems: in service, d.c. systems may be energised with either positive or negative voltage and under certain conditions may be subjected to rapid polarity reversal due to failure of termination or converter equipment. To simulate this condition the test requirements are similar to those for an a.c. system, with 30 loading cycles, but for the first ten cycles the cable is energised at a voltage of $2U_0$ with positive polarity. The next ten cycles are with $2U_0$ negative polarity and during the last ten cycles the voltage is $1.5U_0$ with polarity reversal every 4 hours. No breakdown should occur in any of the components during this test.

DLA versus temperature test

IEC 141-1 does not include the loading cycle test but specifies requirements for DLA measurement at U_0 on the cable at ambient temperature, 5°C above maximum operating temperature, 60°C, 40°C and again at ambient temperature. The DLA must not exceed specified values.

Thermal stability test

This test comprises an extension to the a.c. loading cycle test to demonstrate stability over a prolonged period. Following the last cycle, current loading is applied continuously to each installation at a voltage of $1.33U_0$ (275 and 400 kV) or $1.5U_0$ (132 kV) until the conductor temperature is steady at a value between 5 and 10°C above the design operating temperature. The current loading is then held constant for 6–12 hours and the variation of conductor and sheath temperature should not exceed 2°C after making allowance for any change of ambient temperature. As a further check the DLA is measured at test voltage immediately following the test period.

Impulse tests

(a) A.C. systems: in most cases, underground power transmission cables are connected to overhead lines and must withstand the high transient voltages generated by lightning strokes on or near the overhead line. Switching operations occurring in other parts of the transmission system may also generate high transient voltages. The impulse test is specified to ensure that the cable system will operate satisfactorily under these severe service conditions. For test purposes an impulse voltage can be defined as either positive or negative with respect to earth, rising rapidly to a maximum voltage and decaying less rapidly to zero without any appreciable oscillations. The majority of impulse tests are made using the standard lightning impulse in which the voltage rises to a crest value in 1.2–5 µs and decays to half the crest value in approximately 50 µs.

Table 39.4 Comparison of impulse test and a.c. voltages

Cable voltage (U_0/U) (kV)	Lightning impulse voltage (kV) (crest)	Switching impulse voltage (kV) (crest)
19/33	194	–
38/66	342	–
76/132	640	380
160/275	1050	850
230/400	1425	1050

For system operating voltages of 132 kV and above, an additional impulse test is made using standard switching impulse. In this case the voltage rises to a crest value in 250 μs and decays to half the crest value in approximately 2500 μs.

The recommended method of measuring the voltage wave shapes for impulse tests is given in IEC 60-1: 1989 and BS 923: Part 1: 1990. For the lightning or switching impulse test, twenty impulses, ten positive and ten negative, are applied to the test installations submitted previously to the load cycle test or dielectric loss angle versus voltage test. The test is carried out with the cable conductor temperature 5°C higher than the maximum design operating temperature. Table 39.4 shows the values of lightning impulse and where appropriate switching impulse voltages, for cable system voltages from 33 to 400 kV, specified in British Electricity Boards Engineering Recommendations.

(b) D.C. systems: lightning can also impose high transient voltages on d.c. cable transmission systems and the most severe condition occurs when the polarity of the lightning impulse is of opposite polarity to the cable voltage. This condition is recognised in the type test procedure by the inclusion of a lightning impulse superimposed on the d.c. test. For this test the cable conductor is heated to 5°C above its maximum operating temperature and energised at a d.c. voltage of U_0 negative polarity for a minimum time of 2 hours, and then subjected to ten impulses of positive polarity. The test is then repeated with the cable energised at U_0 positive polarity and ten impulses of negative polarity are applied. Specialised equipment is required to enable these tests to be carried out.

Dielectric security test
To ensure that the cable system has an ample safety margin under abnormal a.c. operating conditions, each installation must be submitted successfully to a high voltage a.c. dielectric security test of up to $2.5U_0$ for 24 hours at ambient temperature. A high voltage d.c. test may also be carried out to ensure than cable accessories will withstand the d.c. voltage applied during the tests after installation. As an alternative to the dielectric security test, a switching impulse test may be carried out.

Thermal resistivity of dielectric
The thermal resistivity is important in the calculation of current ratings and some specifications require measurement on a minimum length of 11 m of cable at maximum operating temperature. The value obtained should not exceed the design value by more than 5% for 275 kV and 400 kV or 10% for 33–132 kV.

Tests on cable corrosion protective finishes

To demonstrate that cable finishes have adequate properties to meet installation and laying conditions, tests are specified in IEC 229 and BEB C48/1.

A sample of cable is first submitted to the bending test conditions already described, and then abraded by 50 passages along its surface by a length of steel angle with its point at right angles to the cable and loaded with a mass of $0.018D^{1.7}$ kg where D is the overall diameter of the cable. The maximum mass is 55 kg.

To represent possible damage to the surface by stones, a preloaded cycle chain wheel is then passed along the cable diametrically opposite the length abraded, the loading being a mass of $(0.2d + 2)$ kg (maximum load 18 kg). In this case d is the diameter under the oversheath.

The sample is then placed in a 0.5% saline solution and submitted to 100 daily temperature cycles with 10 V d.c. applied between the metal sheath and the saline solution. The saline solution is heated to a temperature of 75–80°C for 5 hours and allowed to cool for 19 hours. The leakage current is measured daily for 100 days and the sample is finally subjected for 1 minute to a d.c. voltage equivalent to a stress of 2 kV/mm of oversheath thickness (maximum 5 kV).

The finish is deemed to be satisfactory if no failure occurs and an examination of the centre 1 m section reveals no sign of corrosion.

Reference to another test sometimes requested for protective finishes for aluminium sheaths is made in chapter 5. This involves removal of four circular pieces of oversheath 10 mm in diameter and immersion in 1% sodium sulphate solution.

Additional tests for accessories

Additional type tests for outdoor sealing ends may include power frequency voltage withstand – wet and dry – and measurement of radio interference.

Tests after installation

Electrical and hydraulic tests are required on the complete installation to ensure that no damage has occurred during cable laying and assembly of the accessories. In addition to measurement of conductor resistance, typical tests in accordance with IEC 141-1 comprise the following.

High voltage test

For a.c. cables the application for 15 minutes of a d.c. voltage which is the lower value of either 50% of the specified lightning impulse withstand voltage or $4.5U_0$ for cables with U_0 not exceeding 64 kV, $4U_0$ for U_0 not exceeding 130 kV and $3.5U_0$ for U_0 exceeding 130 kV.

For d.c. cables a voltage of $1.8U_0$ is typical.

Sheath insulation d.c. test

The fully insulated sheath system including cable oversheath, terminal base insulation, joint external, and sectionalising insulation if present, together with the insulation of bonding leads and link boxes or pillars, must withstand a voltage of 10 kV d.c. applied between sheath and earth for 1 minute.

Oil flow test (for FF cables)
Each cable section is subjected to a fluid flow test to ensure that no abnormal restriction is present in the cable or accessories. The test is carried out by measuring the pressure drop in the section under measured fluid flow.

Pneumatic test on gas pressure cables
Each complete circuit, including joints and terminations, is pressurised to 17 bar for seven days, followed by a further seven days at normal operating pressure to 12 bar. The gas tightness of the circuit is considered satisfactory if there is no leakage of gas during the seven day period at normal pressure.

POLYMERIC CABLES

Many of the routine tests, sample (special) tests and type tests required for cables insulated with polymeric materials (PE, XLPE or EPR) are similar to those required for paper insulated cables, but there are important differences, as shown below. In addition to the electrical tests on complete cable, physical tests are required on samples of the component parts of the cable to check that the materials used have satisfactory properties.
Prequalification tests are required for 275 kV and 400 kV XLPE cable systems.

Various countries have produced their own specifications for tests on polymeric cables and the examples given below are those required by ESI Standard 09-16, IEC Publication 840 for power cables with XLPE insulation for 66 and 132 kV transmission systems, and NGTS 3.5.2 for power cables with XLPE insulation for 275 kV and 400 kV transmission systems.

CIGRE has published in *Electra* (151), December 1993, recommendations (a) for electrical tests, type, sample and routine, on extruded cables and accessories at voltages above 150 kV and up to 400 kV, and (b) for electrical tests, prequalification and development.

Tests on accessories for polymeric cables are described earlier in this chapter and are similar to those on fluid-filled cables.

Routine tests

Partial discharge test and high voltage test
For polymeric cables the partial discharge test has replaced the traditional method of assessing insulation quality applied to paper insulated cables by measurement of the DLA against increasing voltage. This is explained in chapter 28. The partial discharge test is made at ambient temperature. The test voltage is raised initially to $1.73U_0$ for 10 s and then reduced to $1.5U_0$ and the partial discharge is measured. The magnitude of the partial discharge at $1.5U_0$ is required not to exceed 10 pC. The test is followed by the application of a voltage of $2.5U_0$ for 30 minutes; for 275 kV and 400 kV cable the test voltages are reduced to $2.3U_0$ and $1.9U_0$ respectively and the time increased to 60 minutes.

Some countries specify a maximum partial discharge level of 5 pC at $1.5U_0$. This has led to the continued development of test methods to obtain higher test sensitivities, e.g. screened rooms, tuned active filters. The general expectation is that as cables for higher transmission voltages are introduced it will be necessary to review the a.c. withstand voltage to avoid overstressing the dielectric and to reduce the partial discharge magnitude pass level below 5 pC.

Special tests

The frequency of tests is similar to that required for paper insulated cables. The tests include the following.

(a) Conductor examination for compliance with IEC 228.
(b) Measurement of the electrical resistance of the conductor for compliance with IEC 228 or BS 6360: 1991 (while classed as a special test in IEC 840, with measurement not required on every length of cable, this is a routine test in ESI Standard 09-16).
(c) Measurement of cable dimensions, including thicknesses of insulation and metallic and non-metallic sheaths and overall diameters.
(d) Hot set tests for XLPE insulation to check that the material has been properly cured to give the required thermal properties (test conditions are given in IEC Publication 811-2-1).
(e) Volume resistivity of semiconducting screens: the values for the inner and outer semiconducting screens are required not to exceed $500\,\Omega\,m$ at 90°C.
(f) Test for shrinkage of insulation. Extruded synthetic materials exhibit a degree of shrinkage when heated, which is associated with residual strain in the material resulting from the heating and cooling cycles that take place during manufacture. If excessive, this might cause retraction of the insulation, when it becomes warm, at joints and terminations. For this test a 200 mm length of cable, with the sheath and metallic screens removed, is subjected to a temperature of 130°C for 1 hour; after the sample has cooled to ambient temperature, the shrinkage is required not to exceed 4%.
(g) Measurement of capacitance: the value of the capacitance is required to be not greater than 8% above the value declared by the manufacturer in his tender.

Prequalification tests

This test (described in *Electra* (151), December 1993) is to indicate the long-term performance of the cable system. For this test a cable assembly containing a minimum of 100 m of cable including accessories is subjected to a continuously applied test voltage of $1.7U_0$ (275 kV and 400 kV) for a period of 1 year. During this period the cable is subjected to repeated loading cycles to obtain the maximum conductor design operating temperature. The conductor must be maintained at this temperature for 2–4 hours before being allowed to cool. The test is considered satisfactory if no failure occurs. On completion, impulse and a.c. tests are performed on samples of the cable including accessories.

Type tests

Electrical tests on complete cable
ESI 09-16 and NGTS 3.5.2 require electrical tests on two miniature cable installations, one being of cable only and the second being of cable together with accessories designed for use with the cable. In both cases each installation must include a minimum of 10 m of cable previously submitted to a bending test. The bending test conditions are the same as those required for single-core paper insulated cables with lead or corrugated aluminium sheaths. IEC 840 is being amended to include tests on cables and accessories together. Figure 39.1 shows an installation of a 400 kV XLPE cable under test.

Fig. 39.1 A 400 kV XLPE cable installation under test

Sequence of tests
The electrical tests are required successively on the same miniature installations generally in the sequence:

(a) partial discharge test;
(b) DLA measurement as a function of the voltage and cable capacitance (cable installation only);
(c) DLA measurement as a function of temperature (cable installation only);
(d) loading cycle test;
(e) lightning and/or switching impulse test;
(f) power frequency voltage test;
(g) d.c. voltage test (accessory installation only);
(h) examination of cable and accessories.

Partial discharge test
The partial discharge test is made at ambient temperature. The voltage is initially raised to $1.73U_0$ for 10 s and then reduced slowly to $1.5U_0$ for the cable installation and to $1.25U_0$ for the accessory installation and partial discharge measurements are made. The magnitude of the partial discharge at $1.5U_0$ and $1.25U_0$ must not exceed 5 pC.

DLA measurement as a function of voltage and capacitance
The DLA and capacitance of the cable are measured at $0.5U_0$, U_0 and $2U_0$ at ambient temperature. The DLA requirements given in ESI 09-16 for XLPE insulated cables are that the value at U_0 should not exceed 0.001 and the difference between the values at $0.5U_0$ and $2U_0$ should not exceed 0.001.

DLA measurement as a function of temperature
The DLA of the cable is measured at U_0 at ambient temperature and the measurement is repeated with the conductor temperature 5°C higher than the maximum design operating temperature. The maximum DLA at ambient temperature must not exceed 0.001. The maximum DLA at a conductor temperature 5°C higher than the maximum design operating temperature must not exceed 0.002 (ESI 09-16) or 0.001 (IEC 840 and NGTS 3.5.2).

Loading cycle test
The conditions for the loading test are similar to those for paper and PPL insulated cables with the addition of partial discharge measurements. Both miniature test installations must be subjected to 20 loading cycles with a continuously applied test voltage of between $1.5U_0$ and $2U_0$. During the last 2 hours of each loading cycle the conductor temperature must be maintained at temperatures not less than 5–10°C and not greater than 10–15°C above the maximum design operating temperature. The test is considered to be satisfactory if there is no breakdown and the measured values of DLA or partial discharge magnitude remain stable.

Impulse test
The test conditions and withstand levels are the same as those required for paper and PPL insulated cables.

Power frequency voltage test
This test is made at ambient temperature. Both installations must be submitted successfully to the application of a voltage of $3U_0$ for 4 hours. IEC 840 and NGTS 3.5.2 do not include this test but specify the application of an a.c. voltage between $2.0U_0$ and $2.5U_0$.

D.C. voltage test
If an after-installation d.c. voltage test is specified, then performance should be confirmed in the laboratory on the accessory installations loop, the test level being 3 to $4U_0$ for 15 minutes. However, after-installation d.c. voltage tests are no longer recommended and are being superseded by a.c. voltage tests.

Physical tests on cable components

These tests are required to check that the materials used for the component parts of the cable have satisfactory properties. In addition to those already described under sample tests, IEC 840, ESI 09-16, and NGTS 3.5.2 include the following:

(a) mechanical properties of the insulation before and after ageing;
(b) ageing tests on complete cable samples to test compatibility of materials during operation;

(c) insulation/screen moisture content;
(d) effect of material compatibility on semiconducting screen resistivity;
(e) swell height measurements on water blocking tapes.

Non-electrical tests on complete cable

These tests are required to check that the thermal resistivity of the insulation and resistance to longitudinal water penetration are satisfactory.

Thermal resistivity of the dielectric
This test is similar to that specified for paper and PPL insulated cables.

Water penetration test
This test is applicable to buried polymeric cables which include a barrier to prevent longitudinal water penetration along the gap between the outer surface of the insulation screen and the water impermeable layer and/or a barrier which prevents longitudinal water penetration along the conductor. The test is carried out on a 3 m test sample of the cable. The barriers are exposed to water with a 1 m head by cutting a ring approximately 50 mm long at the centre of the cable to expose the barriers. The sample is subjected to ten heating cycles at a conductor temperature between 5 and 10°C higher than the maximum design operating temperature. The test is considered satisfactory if no water emerges from the ends of the test sample.

Tests after installation

The electrical tests required on complete installations are similar to those required for paper insulated cables. The tests described are designed to confirm the quality and performance of the cable and accessories. They are in addition to the quality assurance system of the supplier. The quality assurance system should be formally recognised and certified to an international specification such as BS EN ISO 9001 for manufacture and installation.

High voltage tests

There is growing concern that d.c. testing may not be effective in giving confidence in the integrity of polymetric cable systems. Experience has shown that d.c. testing does not find faults that manifest themselves on a.c. system voltage after a few weeks in service. The voltage distribution in accessories is very different under d.c. energisation to that under a.c. energisation and therefore the application of d.c. may cause damage to the accessories, also d.c. trapped charge (space charge) may accumulate in the cable insulation and persist into the service life such that the cable may experience electrical stress higher than the maximum design operating stress of the cable.

D.C. testing is still in use in some countries, however in other countries it has been the practice to introduce prototype testing methods based on variants of an a.c. test voltage. A disadvantage of a.c. testing is that the test current is much higher than conventional d.c. testing and therefore the test equipment is physically larger at the preferred voltage test frequency of at or near to the power supply frequency (50 Hz).

The most robust type of test equipment to date uses the principle of series resonance with a fixed inductor and variable frequency (30–250 Hz). The variable frequency allows a wider range of cable lengths with different capacitance to be tested. Very low frequency (0.1 Hz) test sets have the advantage of being physically smaller but there is doubt that the stress distribution in accessories is representative of that at the power supply frequency. The partial discharge repetition rate is much lower, for example to be equivalent to the repetition rate at the power supply frequency, the test duration would have to be increased by a factor of 500. CIGRE Working Group 21-09 has agreed that a power frequency a.c. test at either $1.7U_0$ for 1 hour or U_0 for 7 days would give better confidence in the integrity of the complete installation. In France the acceptance test is to energise the cable in the transmission system at U_0 for 7 days.

Partial discharge site acceptance and monitoring tests

There is interest in the possibility of performing partial discharge tests as site acceptance tests. The most commonly used partial discharge detectors operate at a relatively low frequency bandwidth (20–500 kHz) and are prone to high noise interference from radio transmissions and electrical equipment. The development of partial discharge detectors operating at a higher frequency bandwidth (up to 10 MHz) has made it possible to select a quieter frequency and make possible site measurements particularly on accessories using electrodes either applied internal or external to the joint shell or termination metalwork. At these frequencies the attenuation is high in the cable thus localising the detection at the accessories or terminations. It is considered that the cable routine test ensures the cable quality and providing there is no damage during the laying operation the integrity of the cable is maintained. Similarly if prefabricated routine tested accessories are employed the main variability that might occur is during on site accessory assembly. Localised partial discharge detection offers the prospect of testing each accessory separately after completion of the installation, however some modification to the accessory design is required and the connections to accessory electrodes have to be accessible for test purposes and be protected against damage from the type of transient voltages that occur during system operation. The problems associated with accessibility are especially significant for laid direct installations.

Sheath insulation test

The requirements for the sheath insulation test are similar to those for paper and PPL transmission systems. It is recognised as the most important test available for monitoring the integrity of the cable during the life of the cable. This is true for paper, PPL and polymeric insulated cables; in each case damage to the oversheath can lead to rapid deterioration of the main cable insulation.

CHAPTER 40
Fault Location for Transmission Cables

The relative importance of a cable circuit within a power system increases with increasing power transmission capability. Thus transmission class cables (which for the purpose of the present chapter include 132 kV cables) require and achieve a higher level of reliability than distribution class cables. It is equally important that, should a transmission cable experience damage, an efficient fault location service is available such that it can be quickly repaired and returned to service.

Transmission cable systems, including cables, joints and terminations, are distinct from distribution cables in the following ways:

(a) they are larger and heavier and thus are designed to be more robust against external contact;
(b) they may employ substantially different technology in the main insulation systems, with PPL or paper insulated fluid-filled systems predominating, although XLPE insulated systems are beginning to be used at up to 500 kV;
(c) they always possess an outer insulation system which provides corrosion protection and insulates the metallic sheath or other outer metallic layer from earth;
(d) their installation methods give additional protection, with the cables and joints laid at greater depths, often encased in cement bound sand;
(e) they are installed and jointed under more controlled conditions.

These factors are reflected in relatively lower levels of internal and externally induced faults. They also have consequences for the fault finding techniques employed.

MAIN INSULATION FAULTS

Transmission class cables always consist of individually screened insulated cores providing a purely radial electric field within the insulation. In the majority of cases, a 3-phase a.c. circuit comprises three (or a multiple of three) single-phase cables, although small-conductor 132 kV fluid-filled or gas compression cables are often 3-core. In both cases, and in contrast to the situation in lower voltage systems, faults in the main insulation are from phase to earth only, not from phase to phase. This reduces the range of applicable fault finding techniques. However, even after a fault has caused circuit protection to operate, the fault may still have high enough resistance and residual electric strength to make it impossible to locate by the bridge or low-voltage time-domain methods applicable to lower voltage systems.[1]

The basic technique for main fault location is the impulse current method.[2] A high voltage surge generator is used to apply an impulse voltage to the faulted core, and possibly to a second core in parallel. All the cable sheaths must be temporarily bonded together for the purpose of the test both at the cable terminals and at any cross-bonding points. The impulse current is measured at the cable end using an impulse current coupler and displayed on a digital storage oscilloscope. Where two cores are connected in parallel, the coupler is connected differentially between them as shown in fig. 40.1, reducing the effect of any joints between the equipment and the fault position.

The high voltage impulse travels along the core and on reaching the fault causes it to break down. Following the breakdown, impulses travel along the cable in both directions from the fault. These pulses are then reflected repeatedly between the ends of the core and the fault position which has the characteristics of a short circuit for the duration of the impulse. The time between successive reflections between the near end of the cable and the fault gives the fault position. The time from the transmission of the original impulse and the arrival of the first reflection cannot be used directly because this also includes an unknown and variable time delay between the arrival of the pulse at the fault position and the breakdown of the fault. Where the fault is near the far end of the cable, a difficulty can arise where reflections from the far end are received before those from the fault, causing possible confusion. Where sufficient output voltage is available from the impulse generator, the impulse voltage can be increased until this delay becomes very short which may make the problem clearer.

Many of the problems of the impulse current method are alleviated by combining the technique with conventional low-voltage time-domain reflectometry (TDR) resulting in the secondary impulse technique. As stated above, the breakdown created by the impulse generator at the fault position has the characteristics of a short circuit for the duration of the impulse. This short circuit is then relatively easy to detect and locate by a conventional TDR system which injects a fast low voltage pulse into the cable core and detects reflections from discontinuities. The low voltage TDR set is automatically triggered to launch its pulse shortly after the breakdown of the fault is detected. A TDR set can equally well be used with a high voltage d.c. supply instead of an impulse generator. In both cases, the TDR trace is particularly straightforward to interpret because the

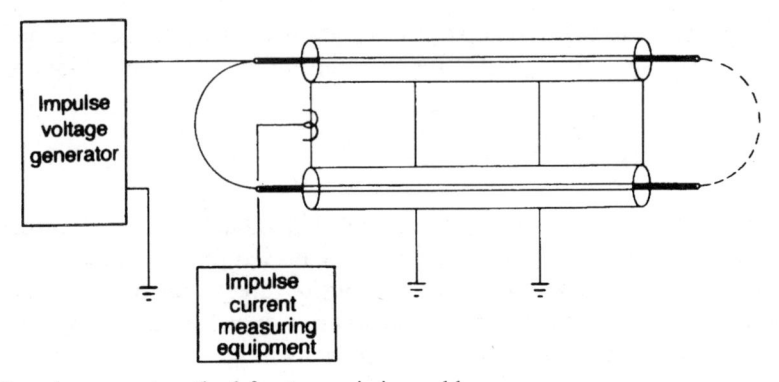

Fig. 40.1 Impulse current method for transmission cables

waveform with the high voltage present can be compared with the waveform with no high voltage; the traces are identical up to the fault position at which point they diverge.

Time-domain methods are rarely precise enough to determine fault locations with sufficient accuracy to permit immediate excavation. The exception may be where the fault is due to gross external damage. In other cases pinpoint location techniques must be applied. The impulse generator can be used to generate an audible 'thump' at the fault position which is detected by sensitive microphones at the ground surface. The use of microphones is normally combined with detection of the magnetic pulse from the impulse using a magnetometer; this gives a less precise location but allows the operator to be sure that the acoustic pulses he is tracking down are actually due to the impulse generator.

PARTIAL DISCHARGE DETECTION AND LOCATION

Incipient faults in extruded cable systems may manifest themselves as partial discharge (PD) activity. PD is defined as an electric discharge which does not completely bridge the insulation between conductors. PD can occur at defects in the insulation where sufficiently high electric fields exist. The cable itself will have been subjected to a PD test in the factory and so is unlikely to give problems. However, damage may occur during installation and jointing.

PD does not immediately result in a full insulation fault. However, the localised but intense electrical, thermal and chemical effects of the discharge are liable to degrade the adjacent insulation materials. This leads to a progressive deterioration of the insulation and eventually to a fault.

Conventional techniques used in the high voltage laboratory and factory environments for detecting and locating PD rely on measurement of the fast electrical pulse which propagates away from the PD site. A typical PD detector in accordance with IEC Publication 270[3] is sensitive within a specified bandwidth, which normally does not extend above a frequency of 1 MHz; frequencies in this range travel freely along the entire length of a cable installation. It is possible to locate the PD source within the cable by time-of-flight techniques, either by comparing the time of arrival of the pulse at two positions or by comparing the time of arrival of the pulse with that of its reflection from the far end. Systems of this type have been developed and used in trials on service cables. The cable requires to be energised by an alternating voltage which can either be the system voltage or can be derived from a separate test voltage source. However, there are problems because of reflections from discontinuities in the cable at joints. The resolution is also poor due to the limited bandwidth employed resulting in the detection of oscillating pulses which are hard to time exactly. The sensitivity is limited by the presence of a band of extraneous electrical noise including radio transmissions and noise from the operation of electrical equipment such as switchgear.

Newer techniques for PD detection in extruded polymeric cables rely on equipment sensitive to much higher frequencies up to hundreds of Megahertz. These techniques may permit relatively accurate location of the PD source due to the wider bandwidths.[4] Higher detection sensitivity is also obtainable by tuning the detector to a part of the frequency band which is quiet for the particular circuit.[5] In the highest frequency range, the PD pulse is rapidly absorbed by the cable semiconducting screens and cannot travel far along the cable. Because the cable itself is almost certain to be PD free, at least initially, it is sufficient to detect PD by suitable sensors incorporated in the joints and

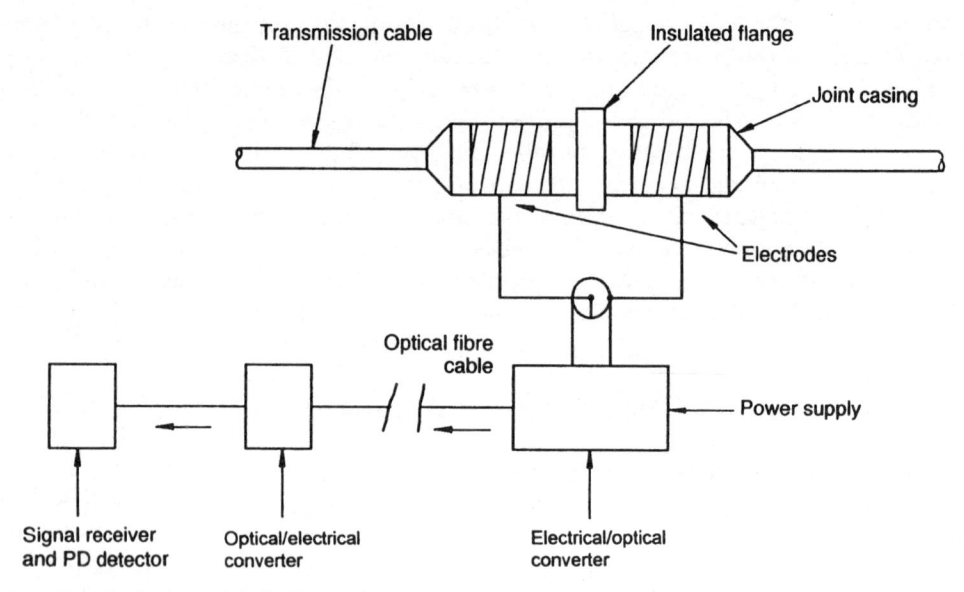

Fig. 40.2 Typical partial discharge detection system

terminations within the cable system. A typical system is shown schematically in fig. 40.2. Such prototype schemes are intended to be suitable for monitoring cable installations in service with the detector connected to each sensor in turn at regular intervals. Detection of PD activity will give an indication of impending problems and an identification of the joint or termination involved. Such schemes have been introduced in Japan as part of the overall system design to monitor joints in 275 kV XLPE cable circuits installed in tunnels.[6] Significant engineering obstacles have to be overcome before this type of PD monitoring can be applied to cable systems which are laid direct in the ground. Laid-direct systems usually employ special sheath bonding requiring a high standard of electrical insulation and corrosion protection. Reliable energy sources to power the PD detection equipment are not usually available at each joint bay position.

FAULTS IN THE PROTECTIVE COVERING

Transmission cable systems almost invariably employ special bonding methods to reduce or eliminate circulating currents in the metallic cable sheath, requiring the sheath to be insulated from earth except at specific points. This insulation is provided along the bulk of the cable route by a layer normally of PE or PVC over the cable metallic sheath. The insulating layer is variously known as the serving, oversheath or corrosion protective covering. The external metalwork of joints is similarly covered typically by a GRP housing filled with an insulating compound. Damage to this insulation system may allow water penetration and consequent corrosion of the metallic sheath, accelerated by the induced sheath voltage. Low resistance faults may allow significant circulating sheath currents to flow, so increasing the temperature rise of the cable. For the longevity of the cable system it is essential to maintain the integrity of this outer insulation, thus suitable fault-finding techniques are required.

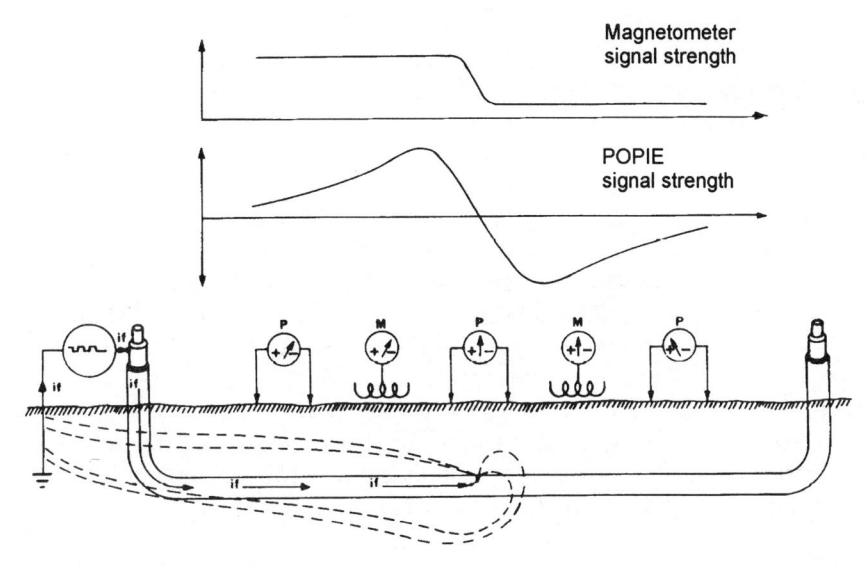

Fig. 40.3 Location of faults in protective finish using the magnetometer and POPIE methods

Since the insulated sheath of a transmission cable is in effect a single unscreened buried conductor, it is not possible to use any of the prelocation or pinpointing methods which can be applied to faults in the main insulation. More recently use has been made of very sensitive magnetometers to trace the current flowing along the sheath of a cable from a transmitter to the fault point. The transmitter used is a modified version of the unit originally developed for pinpointing serving faults by the POPIE method.[7] The principle of operation of both the magnetometer and POPIE instruments is shown in fig. 40.3. The d.c. generator applies an easily recognised characteristic signal between the metallic sheath of the cable and the general mass of earth. The current leaves the cable sheath at the point of fault and returns to the generator by a distributed path creating, around the fault, a voltage gradient which appears at the ground surface as a series of circular equipotential lines. The POPIE detector consists of a sensitive high impedance millivoltmeter connected to a pair of probes which are moved along the cable route until they straddle the fault.

On long cables it would be extremely tedious to survey the whole of the route with the POPIE and so the initial pre-location, or sectionalising, is done using the magnetometer which responds to the magnetic field produced by the sheath current up to the fault point. The magnetometer does not give as precise a location as the POPIE method but it is extremely quick and easy to use and the two methods are therefore complementary.

FLUID AND GAS LEAKS

Extruded cable technology has not yet supplanted the proven insulation systems in the transmission cable class. Transmission cables continue to give excellent service using pressurised fluid or gas. Environmental considerations make leaks of cable fluid unacceptable and the loss of fluid or gas leads to expense and operational difficulty. Techniques must therefore be available for the location and repair of leaks of fluid or gas.

Leaks can frequently be accurately prelocated by isolating sections of the system and measuring pressure drops. The highest percentage of leaks will be found to be from pipework, control equipment or plumbed sheath connections at accessories. If such a leak is in an accessible position, it is relatively easy to pinpoint visually (for fluid), aurally, or with the aid of tracer gas or soap bubbles (for gas).

Gas leaks from the cable itself or from joints and other underground equipment can often be located by introducing a tracer gas which can be detected above ground. This is not possible for fluid leaks. In high pressure pipe-type cables there can be very high fluid flow rates, the direction of which can sometimes be determined at an accessible point by heating a portion of the pipe and subsequently measuring the longitudinal temperature profile. In self-contained low-pressure fluid-filled cables accurate leak location can often be achieved quickly by the use of highly sensitive pressure and flow measuring instruments.[8] However, if the leak rate is of low magnitude, this pressure drop method becomes inaccurate and it is necessary to subdivide the cable hydraulic sections by use of liquid nitrogen filled freezing sleeves. Ultimately with each method, it is necessary to excavate the cable to confirm the location and make repairs.

REFERENCES

(1) Tanaka, T. and Greenwood, A. (1983) *Advanced Power Cable Technology Volume I: Basic Concepts and Testing*. Boca Raton: CRC Press.
(2) Dean, R. J. (1993) 'Users' guide to power cable fault location'. ERA Report 93-0233R, ERA Technology Ltd.
(3) IEC Publication 270 (1981) 'Partial Discharge Measurements'.
(4) Braun, J.-M., Horrocks, D. J., Levine, J. P. and Sedding, H. G. (1993) 'Development of on-site partial discharge testing for transmission class cables'. IEE Conference Publication 382, pp. 233–237.
(5) Pultrum, E. (1995) 'Site testing of cable systems after laying, monitoring with HF partial discharge detection'. *IEE Colloquium on Supertension (66–500 kV) Polymeric Cables and their Accessories*, p. 21.
(6) Toya, A., Goto, T., Endoh, T., Suzuki, H. and Takahashi, K. (1995) 'Development of a partial-discharge automatic-monitoring system for EHV XLPE insulated cable lines'. *Jicable '95 4th Int. Conf. on Insulated Power Cables*, D.3.2.
(7) Gooding, H. T. and Briant, T. A. (1962) 'Location of serving defects in buried cables'. *Proc. IEE, A* **109**, 124–125.
(8) Burgess, V., Brailsford, J. and Dennis, G. (1987) 'Hydraulic method for the location of oil leaks from pressure-assisted 3-core and single-core cables'. *IEE Proc., C* **134** (2), 170.

CHAPTER 41
Recent Improvements and Development of Transmission Cables

ESSENTIAL REQUIREMENTS

Higher power transmission capability

While in many parts of the industrialised world the late 1970s and early 1980s saw little growth in the demand for electricity, the late 1980s and early 1990s have seen a return to growth, although not at the same rate as in previous decades. It is reasonable to assume that this modest growth will continue and that in the future there will be a requirement for higher power circuits than are currently used. The improvements in space occupancy achieved by intensively cooling conventional cables is illustrated in fig. 41.1. With space beneath city streets becoming scarcer, the incentive to obtain greater transmission capability through one cable circuit is expected to persist. An interesting trend is the construction of cable tunnels underneath cities to avoid street congestion. Tunnels have recently been constructed in London to house 132 kV cables, in Tokyo to house 275 kV and 500 kV cables, and in Berlin to house 380 kV cables. Provision is usually made to cool the tunnel by air circulation and, for highly rated circuits, to provide water cooling.

Most underground cable systems are part of longer overhead line transmission circuits which determine the operating voltage and load transmission capability. Operating voltages for overhead line systems have already reached 1000/1100 kV. It is reasonable to suppose that, just as has happened in the past, requirements for undergrounding will occur, initially at line terminations and in rural areas of beauty, but later extending to load centres in urban areas. Identified projects have initiated the development of 750/800 kV and 1000/1100 kV cable systems.[1,2]

Reduced running losses

With increasing environmental pressures and energy costs, there is a greater incentive to develop cable systems with lower losses, the two basic forms of which are current-dependent and voltage-dependent losses. To make significant changes in current-dependent losses it is necessary to use a conductor which has a much lower resistance

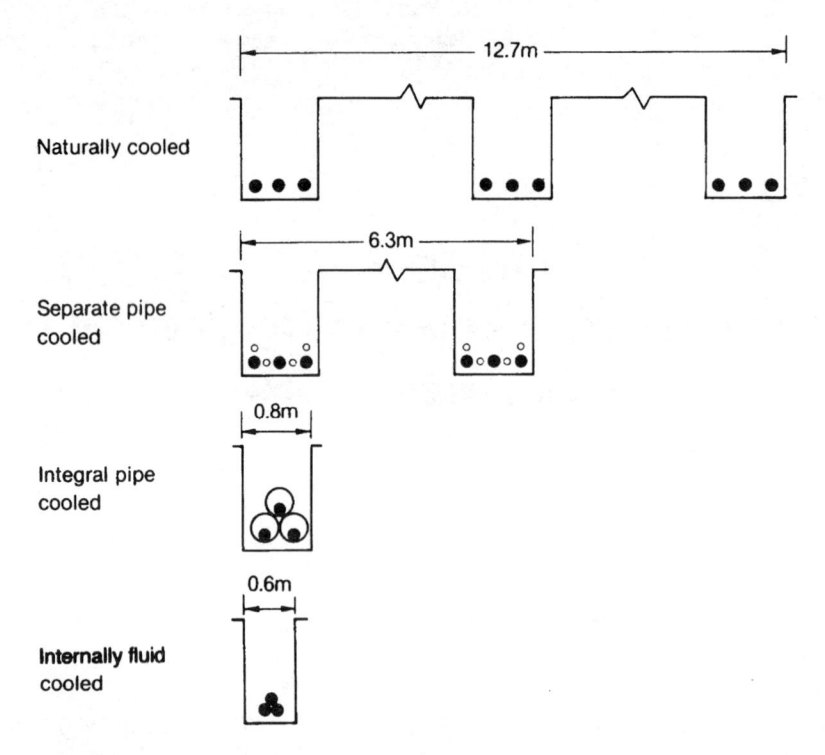

Fig. 41.1 Land requirements for a 400 kV FF paper cable circuit of 2600 MVA capacity

than the copper or aluminium conductor used at present, e.g. a superconductor. However, in the immediate future, there is far more scope for a reduction of the voltage-dependent loss, i.e. the dielectric loss. These losses are far more predictable than current-dependent losses as they exist whenever the circuit is energised. Calculation of current-dependent losses requires assumptions to be made for the daily and seasonal load variations over the complete life of the cable

The calculation of dielectric losses is dealt with in chapter 2, equation 2.12. However to understand the influence of the various factors on the dielectric loss, the equation can be rewritten as follows:

$$\text{dielectric loss} = 34.9 f S_r r U_0 \epsilon_r \tan \delta \times 10^{-5} \qquad \text{(kW/km per phase)} \qquad (41.1)$$

where f = frequency (Hz)
S_r = stress at conductor (MV/m)
r = conductor screen radius (mm)
U_0 = voltage to ground (kV)
ϵ_r = relative permittivity of insulation
$\tan \delta$ = dielectric loss angle

It can be seen that the dielectric loss is proportional to frequency, conductor screen stress, conductor screen radius, system voltage, relative permittivity and dielectric loss angle. The two latter terms are properties of the insulation. The magnitude of the losses increases with operating voltage and this is illustrated in table 41.1. It will be seen that at the higher voltages the losses become very significant. The values given in the table are

Table 41.1 Dielectric losses for 33–750 kV paper insulated 2000 mm^2 fluid-filled cables[a]

Voltage (kV)	Design stress (MV/m)	Dielectric loss (W/m)
33	4.5	0.4
66	9	1.5
132	10	3.3
275	13	9.0
400	15	15.2
750	20	38.0

[a] Assumptions: $DLA = 0.0024$; relative permittivity, 3.5; operating frequency, 50 Hz

based on paper insulation. Assuming the same design parameters for the cable but changing only the insulation, the use of polypropylene/paper laminate (ϵ_r 2.8, $\tan \delta$ 0.001) would reduce the losses by 66% and XLPE (ϵ_r 2.3, $\tan \delta$ 0.001) by 73%.

There is a growing tendency to take these losses into account when assessing costs of cable systems. This is done by adding to the initial cost a notional capital sum of money which would pay for the cost of the losses throughout the life of the cable.[1] The calculation takes into account the magnitude of the losses, the cost of the losses, the life of the cable and interest rates. The latter is important as it is assumed that the sum of money would be invested and that the interest would assist in paying for the losses as well as the capital, which would reduce to zero at the end of the assumed cable life. The same calculation can be used for the current-dependent losses, but as indicated earlier there is some difficulty in predicting the current loading over the life of the cable. A further complication is that the losses are proportional to the square of the current and this makes accurate calculation more difficult.

In addition to the cost effects, the presence of dielectric losses reduces the current rating of the cable. This is illustrated in fig. 41.2, which shows the variation of

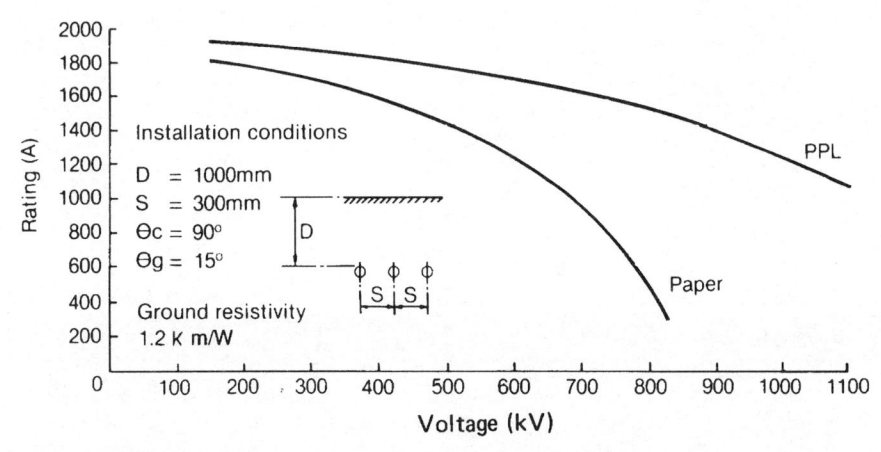

Fig. 41.2 Variation of current rating with operating voltage for paper and PPL insulation; conductor size 2500 mm^2

current rating of a 2500 mm² fluid-filled cable with operating voltage. It will be seen that the current rating reduces with voltage until at a voltage of about 850 kV the paper insulated cable system has no rating capability. PPL insulation is referred to later in this chapter.

It should be noted that forced cooling can have an effect on the dielectric losses of a cable system. For example, if it is possible to use one cable in place of two in parallel by using cooling, the single cable will have a significantly lower capacitance than the two-cable circuit and the dielectric losses, which are proportional to capacitance, will therefore be reduced.

Reduction of capacitance

Any reduction in capacitance, provided that the dielectric loss angle is not increased, will reduce the dielectric losses. However, capacitance is important in other respects. By re-arranging the formulae given in chapter 2 it is possible to express the charging current, that is the current flowing into the cable system due to its capacitance, as follows:

$$I = 3.49 f \epsilon_r S_r r l \times 10^{-4} \quad \text{(A)} \quad (41.2)$$

where l = cable length (km).

It will be seen that for a given cable design the charging current is proportional to cable length. The charging current is not directly affected by operating voltage but in practice as the voltage of the cable increases so does the design stress. There is a critical length at which the charging current equals the thermal rating of the cable and therefore the cable circuit can carry no useful load. Figure 41.3 shows the effect of feeder length on the useful current rating of paper insulated FF cables. By changing to a low loss dielectric such as polypropylene paper laminate (PPL) the critical length benefits in two ways: first the charging current is reduced in the ratio of the relative permittivities and second the thermal rating of the cable is higher, as shown in fig. 41.2.

In the case of land cables the length effect can be overcome by splitting the cable into sections and fitting shunt reactors. In some situations shunt reactors are used in any case to compensate for the effect of the cable capacitance on the transmission system.

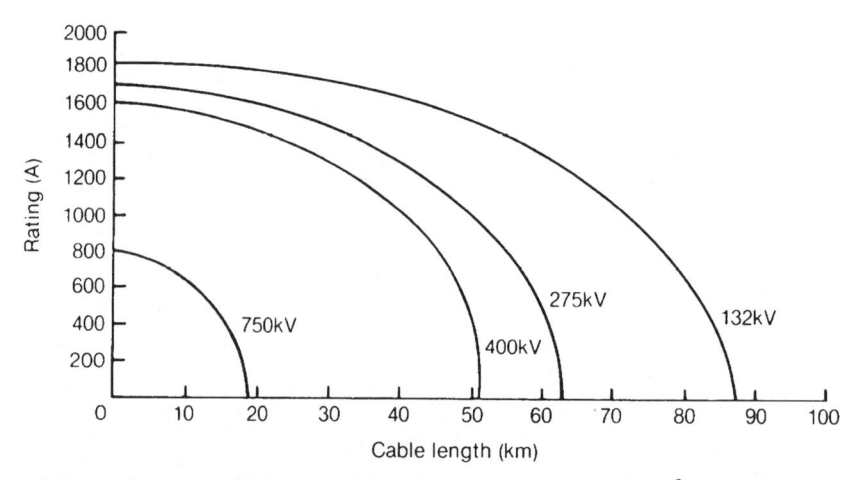

Fig. 41.3 Effect of feeder length on useful current rating for 2500 mm² FF cable with paper insulation

Use of a reactor involves a capital cost and a running cost due to its losses. These can be taken into account financially in a similar manner to that explained for dielectric losses. Therefore there are economic advantages in lowering the capacitances of cable systems.

Reactors can be used to compensate but not to overcome the problem for submarine systems and the reduction of cable capacitance is of great importance on long installations.

Reduced cost of installation and maintenance

A modern pressurised cable system requires sophisticated installation techniques. For example, both self-contained and high pressure pipe type systems need specialised treatment of the impregnating fluid for the installation of the cables and the jointing requires the employment of skilled jointers working in air-conditioned enclosures for the higher voltages. While this equipment and expertise is usually readily available in the industrialised nations, it can present problems in other areas. Perhaps of greater significance, is the need for this equipment and expertise for possible maintenance during the life of the cable. With pressurised systems, there is inevitably a possibility of leakage in the event of third party damage. Although the systems are designed to continue in operation with minor leaks, these should be repaired in a timely manner.

Much work has been carried out to simplify existing systems and to improve performance in relation to leakage.[3] This has resulted in the elimination of fluid loss in modern FF cable systems. Nevertheless the fact that polymeric cable systems do not require pressurisation has been an important incentive in their development.

DEVELOPMENT OF EXISTING CABLE SYSTEMS

The information given in the previous sections outlines the objectives of the developments currently being carried out. In several cases the development covers more than one of the objectives.

Pressurised cable systems

Cooling of conductors
Much work has already been carried out to improve the current rating of low pressure and high pressure fluid-filled cables. The area where further developments can be expected is in conductor cooling. In chapter 34 reference was made to the use of fluid circulation through the conductor to improve rating and fig. 34.15 illustrated that it is only this type of cooling that would permit the use of one cable per phase to meet a required rating of 4000 MVA at 525 kV. Fluid circulation has been used in trials carried out in Italy on a 1100 kV cable system.[2] It is therefore to be expected that development of this system will continue further. Similar developments, but using different fluids, have been carried out in Germany and Japan.

In Germany, water has been used for conductor cooling.[4] Compared with hydrocarbon fluid, water has several attractions, e.g. its heat capacity is about 2.5 times higher and its viscosity is only about one-tenth of that of fluid. Its major disadvantages are that it has to be kept isolated from the fluid in the cable and that injection and removal must be through long insulators at the terminations. For the latter purpose the water must be deionised to reduce the losses. Because the injection of water cannot be made at intermediate feed joints along the route, the cooling section must be the full length of the circuit. To achieve this a stainless steel tube approximately 55 mm in internal diameter is incorporated into the conductor and pumping pressures up to 30 bar are employed. The cable system has a very high rating, but the presence of high pressure water inside the conductor of a fluid-filled cable means that the integrity of the cooling system must be exceptionally high. A minute leakage could irreparably damage considerable quantities of cable and it is from this aspect that doubts about its practicability arise. In Japan, trials have been made using evaporative conductor cooling.[5] This type of cooling is most efficient and has the advantage that coolant pumps can be eliminated. However, the coolant must be separated from the cable insulation and the coolant duct in the cable must be sufficiently large to permit 2-phase flow, i.e. liquid and gas. Whilst elimination of pumping equipment is a positive advantage, gravity distribution of coolant restricts the length and profile of the cable system. It therefore appears that applications will be restricted to relatively short lengths.

Alternative tape materials for insulation
From dielectric considerations, work with conventional fluid-filled paper insulated cables has indicated that it is possible to extend the operating voltage to about 1100 kV a.c.[3]

To achieve the highest voltages, an increase in minimum operating pressure to 15 bar is envisaged. While from an electrical strength point of view, it is clear that conventional paper insulation can be extended to these voltages, the problem of high dielectric losses remains. For example, at 1100 kV the dielectric losses of the cable would be so high that unless some form of cooling is employed the cable would be thermally unstable. Thus it would be necessary to employ some form of cooling whenever the cable is energised. At 750 kV in certain favourable installation conditions, it would be possible to operate with natural cooling but the rating would be severely restricted.

It is with these problems in mind that designers have looked for alternatives to paper insulation. Initially much work was carried out on the use of plastic films instead of paper but various restrictions have prevented successful development, e.g. (a) compatibility of the plastic and impregnant: (b) difficulty in obtaining 100% impregnation with fluid; (c) relatively low impulse strength; (d) poor bending performance; (e) high cost. A compromise which has found favour is a composite tape consisting of a plastic film sandwiched between layers of Kraft paper. Various forms of laminate have been considered and some of these are described in references 6–10. The preferred construction is now considered to be a laminate consisting of a film of polypropylene between two layers of Kraft paper.[11–12] The proportion of paper to polypropylene is approximately 50:50. It has been found that this type of laminate overcomes many of the problems associated with the plastic films. The paper layer provides a good path for impregnation of the insulation and mechanically reinforces the plastic, which results in an improved bending performance and restricted swelling. It also protects the plastic from local discharges, which increases the a.c. electrical strength of the insulation. The reduction in relative permittivity of the laminate compared with that of the fluid filled butt gap reduces the stress in the impregnant and significantly increases the impulse strength compared with a paper cable.

As described in chapters 30 and 32 the use of polypropylene paper laminate (PPL) results in a significant reduction in the dielectric losses of both FF and HPFF cables and it is now being used in increasing amounts for commercial installations at 275 kV and above.

Cost of maintenance

As mentioned previously, work is being carried out to minimise maintenance associated with pressure-assisted cables. In self-contained fluid-filled cables, the main source of leaks is in accessories. Although welding has been employed on some installations,[13] plumbing remains the most widely used method for making the connection between the cable sheath and the line accessory. Methods have been developed to provide mechanical reinforcement of the plumbs to avoid problems that have been experienced due to slow creep of the plumb under adverse conditions of pressure and longitudinal stress in the sheath. Improved methods of leak detection and location are also being pursued.

Polymeric insulated cables

As stated in chapter 33 polymeric insulated cables are now firmly established for voltages up to 150 kV. PE and XLPE insulated cables have been developed up to voltages of 500 kV. Now that cables with XLPE insulation are at voltage parity with those with PE insulation it is probable that PE insulation will drop out of favour due to its lower operating temperature and hence lower power rating.

Commercial lengths of 275 kV large conductor XLPE cable have recently been installed together with short lengths of 500 kV. The main problem is the provision of reliable joints. Initial joint designs for the highest voltages were based on injection moulded insulation. However this requires the same high standards of cleanliness and process control in the field that are necessary in the cable factory without the ability of carrying out effective routine tests on completion. Recent work has therefore been directed at the development of prefabricated joints as described in chapter 35. Viewed from the requirement of a 40 year life, experience with commercial installations is too short to know whether the same high level of joint reliability will be obtained as with the pressurised laminar cable systems.

Although electrical design stresses are approaching those of laminar cables, they are still some 15-20% lower at the highest voltages. This, together with the use of thicker screens and the necessity for a space beneath the sheath to allow for thermal expansion, results in significantly larger cables. Apart from direct material cost considerations, for a given despatch drum size only shorter lengths can be transported. A further factor of larger cables is that sheath eddy losses will be higher. This is of particular significance with aluminium sheathed cables which are laid close together. Reference to IEC 287 shows that the loss increases with sheath diameter and sheath conductance. To reduce eddy losses cables have been developed with stainless steel sheaths.[14] Stainless steel has a much higher resistivity than aluminium and can also be used in much lower thicknesses. It is applied longitudinally in the form of a strip which is wrapped around the cable and the two edges welded together. The sheath is then corrugated. This construction results in much reduced losses, however such a sheath would be totally inadequate for carrying fault currents. To provide a low fault resistance path, copper wires are applied in a layer with a long lay over the core. The copper wires do not contribute significantly to the eddy loss because the circumferential resistance of the construction is high.

A further aspect of the current rating of the higher voltage XLPE cables is the a.c. resistance of the conductor. Compared on a size to size basis with self-contained FF cables, the a.c. resistance of the XLPE cable is higher because of: (a) the absence of a central duct; and (b) lower inter-strand resistance due to increased compaction during the extrusion and crosslinking process. To improve the situation XLPE cables have been developed with Milliken conductors constructed with individually insulated copper wires. The insulation prevents the flow of current between adjacent wires and therefore reduces skin and proximity effects. The normal method of making conductor connections in accessories is by the use of compression connections. These rely on current flowing between wires and it is therefore necessary to remove the insulation from the wires locally in the joints and terminations.

500 kV appears to be the highest voltage possible with present technology. To make practical designs at 750 kV an increase of 30–40% in design stress will be required. Work is being directed to obtain a better understanding of the mechanisms of ageing and deterioration in these types of cable, with particular attention being paid to the role of insulation formulations with significantly higher electrical strength including additives to enhance service life and to inhibit tree initiation. Increases in the design stress of XLPE cable will place greater emphasis upon the development of specialised designs of joints and of assembly technology. An interesting result should be that high stress designs of XLPE cables and joints may eventually become available at the lower system voltages.

Cables for d.c. systems

Almost all the d.c. cable systems supplied so far have been for submarine crossings or the land cable associated with them. For long crossings the mass-impregnated solid paper insulated type cable is chosen because there are no restrictions on length due to pressurising requirements. It has been supplied for operating voltages of 450 kV and research is being carried out for use at even higher voltages. To date paper insulation has been used but the application of using polypropylene paper laminate is being pursued at voltages up to 500 kV to gain advantage of the increased impulse strength and reduced diameter.

The possibility of the use of XLPE and PE insulation has been under investigation for many years. Cables with these insulations have the same advantage as the mass-impregnated cable in that for d.c. transmission there are no restrictions on circuit length but they have the potential of operating at a higher temperature. In the case of XLPE, 90°C instead of 50°C. However it has not been possible to obtain the full potential of these materials in tests on full size cables. It is believed that one of the main reasons for this is the development of space charges in the dielectric which distort the stress distribution. These space charges persist for long periods because of the high resistivity of the polymers. Attempts have been made to improve the situation by the use of additives to reduce the resistivity but it has not been possible to date to match the electrical performance achieved with paper insulated cables. Successful tests have been reported on a 250 kV cable with a maximum stress of 20 MV/m using XLPE insulation with a mineral filler.[15] This stress value compares with 32 MV/m which has been used for mass-impregnated paper cables.

To date the use of underground d.c. cables has been very limited. There is a 120 km length of underground 400 kV paper insulated fluid-filled cable installed in Denmark but this is a direct consequence of the submarine crossing between Denmark and Germany.[16] However this is at present the longest underground cable in the world and this length probably would not have been considered practical for an a.c. system. This illustrated the advantages described in chapter 38 of operating insulated cables in d.c. systems. For land cables the FF and HPFF designs use their higher electrical design stress and operating temperature, when compared with a mass-impregnated cable. Research work has shown that operating voltages in excess of 1000 kV should be possible together with power ratings exceeding 4000 MW per bipole.[17] Such systems are at present the only practical solutions if a requirement occurred to transmit high power reliably underground over long distances as an alternative to a.c. transmission by overhead lines.

NEW DESIGNS OF TRANSMISSION CABLE

Compressed gas insulated cables

Compressed gas insulated cables may also be termed gas insulated transmission lines (GIL). It may be argued by some of the supporters of compressed gas insulation that cable designs have already been established. However, although many installations of this type of cable have been completed and are continuing to be installed, they are all of relatively short length (30–3000 m) and have not been used for main buried direct transmission links through cities.

Compressed gas insulated (CGI) cables have developed from the metal-clad substations and the initial designs were in effect an extension of the gas-insulated busbar, many of the early installations being associated with switching stations. Figure 41.4 shows a sample of a typical single-phase rigid CGI cable.

The inner conductor consists of a rigid aluminium tube and is supported in a larger aluminium pipe by means of rigid spacers, usually of filled epoxy resin, which are situated at intervals along the pipe. For burial in the ground, an anticorrosion layer would be essential to protect the pressurised enclosure from corrosion. The space between the conductor and the outer pipe is filled with pressurised SF_6 gas. This is an electronegative gas and at atmospheric pressure has an electrical breakdown strength some 60% greater than nitrogen. To improve the electrical strength of the gas further, it is used under a pressure of 3–5 bar.

Lengths of up to 20 m are pre-assembled in the factory and transported to site. The amount of work in jointing these lengths together on site depends on the precise design. A typical design would include plug-in connections for the conductor connection and a welded joint for the outer pipe. The plug-in connector permits expansion of the conductor to take place and avoids high thermomechanical forces in the conductor

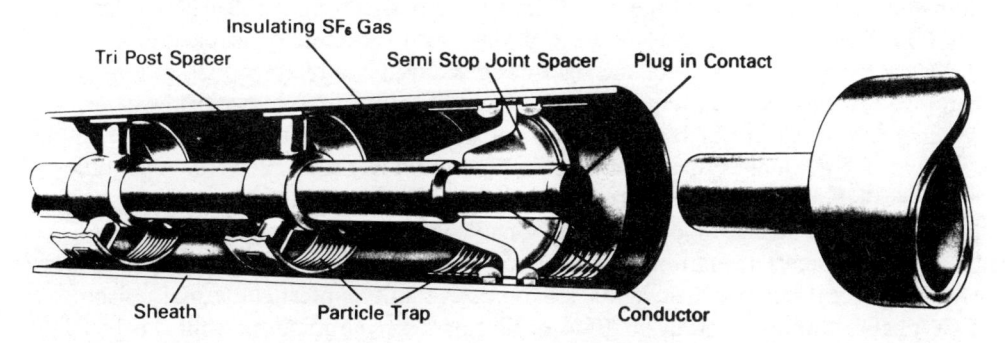

Fig. 41.4 Typical 230 kV single-phase rigid CGI cable (Courtesy of Westinghouse Electric Corporation)

system. A key feature of the cable is the design of spacer. The most usual forms are discs, cones or post type insulators. They are designed to minimise their effect on the flashover voltage between the conductor and outer pipe. Metal inserts are usually cast in the spacer to eliminate high electric fields between the spacer and the pipe conductor. The spacer system must be designed to allow longitudinal gas flow during pressurisation although sectionalisation is necessary to prevent complete depressurisation and contamination of the system in the event of a fault and to inhibit the longitudinal travel of a power arc in the event of a flash-over.

The electrical strength of the gas–spacer system is very much influenced by the presence of conducting particles and it is of greatest importance that the highest cleanliness standards are maintained during assembly of lengths in the factory and during installation at site. Some designs of cable include special low electrical field regions at the earthed end of the spacer which are designed to trap conducting particles.

At present, most single-phase CGI cables operate with the outer pipes fully bonded and earthed. The dimensions of the cables are such that the sheath currents are very high and an increase in the thickness of the pipe results in a lowering of the losses. The precise thickness is a balance between the cost of the losses and the initial cost of the pipe. An advantage of the fully bonded system is that the electromechanical forces arising on the conductor from short circuits are greatly reduced by the opposing currents induced in the outer pipes and the external magnetic field is negligible.

Terminations for single-phase CGI cables are relatively simple, stress control being achieved by corona rings or shields placed at the top and bottom of the insulator. To cater for changes in direction, factory manufactured elbows are provided. T-joints are also a possibility.

Following the trends of the insulated busbar, 3-phase CGI cables are being developed. The reasons for this development are similar to those which apply to more conventional cable systems, i.e. economy in sheath materials, reduction of space requirements and simpler installation. For this system, post type insulators are used. Systems are in commercial service at voltages up to and including 550 kV. Trials have been made on a 1200 kV cable installation.[18]

In order to reduce the costs of CGI cables further, development of long length flexible designs is being carried out. If possible, this would greatly reduce installation costs as the lengths between joints could be increased by a factor of about 10. The cable consists of a flexible conductor, rigid spacers and a flexible corrugated aluminium sheath.[19] The pressure of SF_6 is similar to that used in rigid systems and hence overall diameters are also similar. Figure 41.5 shows a cross-section of a 230 kV design.

CGI cables have attractions in terms of high current rating, low capacitance and low dielectric losses. In the case of the rigid designs, large conductor cross-sections can be employed. A further advantage is the low internal thermal resistance of the cable. This results from heat transfer by convection and to a lesser extent by radiation. The high density of SF_6 makes heat transfer by convection most effective. The different mechanisms of heat transfer make a general comparison of the thermal resistance of traditional solid insulation cable and CGI cable insulations difficult. Furthermore, heat transfer in gaseous insulation is not simply proportional to temperature difference. However, specific comparison at 400 kV indicates that the internal thermal resistance of a CGI cable is approximately 20–25% of that of a conventional fluid-filled cable. Thus the CGI cable will have a superior rating in terms of both the possibility of larger conductor sizes and a lower thermal resistance. In the case of surface cooling, the effect

Fig. 41.5 Example of flexible CGI cable with rigid spacers (Courtesy of Electrical Power Research Institute, USA)

of internal thermal resistance on rating is greater than for natural cooling, as the internal thermal resistance forms the major part of the total thermal resistance. The increase in rating obtained by surface cooling is therefore much greater for CGI cables than for conventional cables.

The relative permittivity of SF_6 is virtually unity and the power factor is virtually zero. Dielectric losses are effectively zero. The low capacitance of the cable arising from the low relative permittivity and large dimensions results in a much longer critical length before the cable system becomes self-loading due to capacitive current.

The CGI cable therefore has many attractive features from an electrical point of view. However, the rigid system is very expensive in terms of material and installation costs. There are considerable practical difficulties in installing such a system in urban areas, owing to the space it occupies and the difficulties that would be encountered at obstructions. Additionally, its characteristic impedance lies between that of a conventional cable and that of an overhead line; thus if it were to be installed in parallel with overhead lines in a grid system it would tend to attract a higher proportion of transmitted current (the funnel effect). The flexible cable is more practical for installation and has potential for a cheaper system, but it has significant problems of engineering and reliability to overcome and has not been proven in normal service.

More recently consideration has been given to the use of CGI cable for high power a.c. transmission over long distances.[20] Because of the very large quantities of SF_6 gas that would be involved and because SF_6 is a 'greenhouse' gas which accumulates in the atmosphere, nitrogen is being proposed to dilute the SF_6. To obtain an electrical performance approaching that of SF_6 it will be necessary to use a significantly higher gas pressure. The development of this type of system is at a very early stage.

Superconducting cables

Superconducting cables have been under consideration for many years, initially based on low temperature superconductors but more recently using higher temperature materials.

Considering first the low temperature design, as originally conceived this was to be a low voltage cable capable of carrying a current of many hundred kiloamps, so eliminating the need for voltage transformation. Unfortunately, research work quickly showed that this dream was not possible. This arises from the fact that in addition to a critical temperature, i.e. a temperature above which the material cannot be superconducting, there is also a critical magnetic field. If the superconductor is placed in a magnetic field above this critical value, the conductor becomes normal, i.e. possesses resistance. The critical field is related to the temperature of the conductor: the lower the temperature, the higher is the critical magnetic field. Conductors which have the above characteristics are known as type 1 superconductors and include pure metals such as lead, mercury and tin. As current produces a magnetic field, for a particular conductor geometry the critical magnetic field can be expressed as a critical current, and hence a current limitation exists for a particular design of superconducting cable.

Further research into the behaviour of conductors at near absolute zero identified a class of materials known as type 2 conductors. These materials were found to have two critical fields: H_{C1}, the field below which the material is truly a superconductor, and a higher field H_{C2} above which the material has normal resistance. In the region between fields H_{C1} and H_{C2} the material is in a mixed state and has some losses. However, these losses are of very small magnitude and type 2 conductors can be used in higher critical fields than type 1 conductors.

Two such type 2 materials which have been considered for superconducting plant are niobium alloys containing either tin or titanium.

The design of superconducting cables is so different from that of conventional cables that the principles will be best understood by describing a design that has been taken to the field trial stage.[21] Figure 41.6 shows the experimental 138 kV cable designed for a rating of 1000 MVA. It comprises two flexible cores installed in a rigid pipe (commercial cables would have three cores). Each core consists of an inner and outer conductor separated by the main insulation which consists of lapped polypropylene tapes. The conductor is made from niobium–tin strips. The magnetic field from the inner conductor induces a circulating current in the outer conductor almost equal in magnitude but in the opposite direction to the conductor current. This effectively confines the magnetic field to the cable core, thus avoiding eddy currents anywhere in the cable system. The cable was installed in a thermally insulated enclosure to minimise the entry of heat into the system, i.e. a completely opposite situation to conventional cables where the heat generated is dissipated to the environment. The enclosure consists of two concentric pipes with diameters of 215 and 405 mm. The inner pipe is held inside the outer by spoke type supports and is thermally insulated by pumping the air out of the space between the two pipes to produce a vacuum. The inner pipe is wrapped with multilayer aluminised plastic sheets to minimise the transfer of radiated heat. The cable core is cooled to less than 10 K by the circulation of liquid helium.

Fig. 41.6 Experimental 138 kV superconducting cable (© 1986 IEEE)

The test installation has been operated for extensive periods in the superconducting mode. However, much more development work would be required to make it a practical proposition. Other studies of low temperature superconducting cables suggest that they could only become economic compared with conventional 400 kV cables when circuit ratings of 5000 MVA or more are required.[22]

The discovery of high temperature superconductors (HTS) provided a boost to the development of superconducting cables. HTS materials have critical temperatures in the range 90–120 K as opposed to 8–20 K for low temperature superconductors. The significance of the higher temperature is that it leads to the possibility of using liquid nitrogen as a coolant in place of helium. This results in a much lower cost for the coolant, refrigerators and the thermal insulation of the cable.

One basic cable design under investigation follows the same principles as developed for the low temperature conductors, i.e. layers of superconductor inside the insulation and on the outside with the three cores in a pipe held at the low temperature.[23] Another design has thermal insulation applied between the superconductor and the insulation so that the rest of the cable is at ambient temperature. This type of cable is being developed so that it can be installed in existing ducts or pipes and so replace an existing cable with one that can carry a much higher power.[24]

Development of both designs is at an early stage but initial economic studies suggest that cables based on HTS conductors could be economic at lower powers than those using low temperature superconductors. HTS is described in detail in part 6.

CONCLUSIONS

For load requirements up to about 3 GVA it appears that developments of existing designs of cable will probably be the most economic for the foreseeable future. For ratings in the range 3–5 GVA, compressed gas insulated cables or superconducting cables based on HTS materials are theoretically capable of meeting the requirements. Compressed gas insulated cables for these ratings will be extremely large and rigid and it remains to be seen whether it is practical to install such systems in the centre of large urban conurbations or even in long lengths in rural areas. A further aspect is the necessity for a high degree of cleanliness both in installation and maintenance. Much more development is required before superconducting cables can be considered as viable alternatives to present designs.

For higher voltage applications, impregnated paper insulation is technically feasible for voltages up to 1100 kV a.c. and d.c. In the case of a.c. systems, the dielectric losses cause severe reductions in current ratings at the higher system voltages and polypropylene paper laminate insulation is being increasingly used in FF and HPFF pipe cables.

In the case of polymeric cables for a.c. transmission, XLPE is the preferred insulation. It is becoming established for long length high power circuits at 275 kV with long length applications planned for 500 kV. It is attractive for application at 500 kV because of its low dielectric loss and low permittivity. Its general use at this voltage is dependent on demonstrating a reliable service performance.

Compressed gas insulated cables offer prospects for longer lengths of a.c. transmission underground but the engineering problems are as formidable as are those of superconducting cable. The best prospect for high power underground transmission using conventional methods of heat dissipation and small trenches is the paper insulated d.c. cable, which has no restriction on circuit length.

REFERENCES

(1) Endersby, T. M., Gregory, B. and Williams, D. E. (1991) 'Low loss self contained oil filled cables insulated with PPL for U.K. 400 kV transmission circuits'. IEE Conf. Publication 345, pp. 337–343.

(2) Donazzi, F., Fameti, F., Luoni, G. and Mosca, W. (1984) 'Power transmission of a self-contained oil-filled 1100 kV cable system. Full scale tests and design criteria'. Paris: CIGRE paper No. 21-09.

(3) Endersby, T. M., Galloway, S. J., Gregory, B. and Mohan, N. C. (1993) 'Environmental compatibility of supertension cables'. *IEE Conf. on Power Cables and Accessories 10 kV–500 kV*, pp. 71–76.

(4) Blasius, P., Marjes, B., Henschel, M., Kunisch, H. J. and Martin, W. (1982) 'Testing a 110 kV low-pressure oil-filled cable with a water cooled conductor in Berin (West)'. Paris: CIGRE Paper No. 21-01.

(5) Kojima, K. and Kubo, H. (1976) 'A study of internally cooled cable systems for bulk power underground transmission'. *IEEE Transmission and Distribution Conf.*

(6) Itoh, H., Nakagawa, M. and Ichino, T. (1979) 'EHV self-contained oil-filled cable insulated with composite paper DCLP'. *IEE Conf. on Progress in Cables and Overhead Lines for 220 kV and Above*.

(7) Sakurai, T., Iwata, Z., Shimizu, M., Fujisaki, Y. and Furisawa, H. (1980) '275 kV self-contained oil-filled cable insulated with polymethylpentane laminated paper'. IEEE Paper No. 80 SM 555-3.

(8) Kusano, T., Soda, S., Fujiwara, Y. and Kinoshita, S. (1981) 'Practical use of "Siolap" insulated oil-filled cables'. IEEE Paper No. 81 WM 114-8.

(9) Matsuura, K., Kubo, H. and Miyazaki, T. (1976) 'Development of polypropylene laminated paper insulated EHV power cables'. *IEEE Underground Transmission and Distribution Conf.*
(10) Soda, S., Kojima, T., Fujiwara, Y., Kinoshita, S. and Takeuchi, K. (1979) 'Development of "Siograthene" laminated paper insulated oil-field cables'. *IEE Conf. on Progress in Cables and Overhead Lines for 220 kV and Above.*
(11) Arkell, C. A., Edwards, D. R., Skipper, D. J. and Stannett, A. W. (1980) 'Development of polypropylene paper laminate (PPL) oil-filled cable UHV systems'. Paris: CIGRE Paper No. 21-04.
(12) Allam, E. M., Cooper, J. H. and Shimshock, J. F. (1986) 'Development and long-term testing of a low-loss 765 kV high-pressure oil-filled pipe cable'. Paris: CIGRE Paper No. 21-06.
(13) Ray, J. J., Arkell, C. A. and Flack, H. W. (1974) '525 kV self-contained oil-filled cable systems for Grand Coulee. Third power plant: design and development'. *IEEE Trans.* **PAS-93**, 630–639.
(14) Sato, T., Muraki, K., Sato, N. and Sekii, Y. (1993) 'Recent technical trends of 500 kV XLPE cable'. *IEE Conf. on Power Cables and Accessories 10 kV–500 kV*, pp. 59–63.
(15) Maekawa, Y., Yamaguchi, A., Ikeda, C., Sekii, Y. and Hara, M. (1991) 'Research and development of DC XLPE cables'. *Jicable '91*, pp. 562–569.
(16) Poulsen, S. H., Svarrer Hansen, B., Herrman, B. and Kjaer Mielsen, O. (1994) '400 kV flat type oil-filled cable for KOMTEC HVDC interconnection Denmark/Germany'. Paris: CIGRE Paper No. 21-204.
(17) 'Design limits of oil/paper high voltage direct current cable'. (1985) EPRI Report No. EL-3973.
(18) Bolin, P. C., Cookson, A. M., Corbett, J., Garitty, T. F. and Shimshock, J. F. (1982) 'Development and test installation of a three-conductor and UHV compressed gas insulated transmission line for heavy load transmission'. Paris: CIGRE Paper No. 21-04.
(19) Spencer, E. M., Samm, R. W., Artbauer, J. and Schatz, F. (1980) 'Research and development of a flexible 362 kV compressed gas insulated transmission cable'. Paris: CIGRE Paper No. 21-02.
(20) Aucount, C., Boisseau, C. and Feldmann, D. (1995) 'Gas insulated cables: from the state of the art to feasibility for 400 kV transmission'. *Jicable '95*, Paper No. A.5.4.
(21) Forsyth, E. B. and Thomas, R. A. (1986) 'Operational test results of a prototype superconducting power transmission system and their extrapolation to the performance of a large system. *IEEE Trans.* **PWRD-1** (1).
(22) Maddock, B. J., Cairns, D. H. H., Sutten, J., Swift, D. A., Cotrill, J. E. J., Humphries, M. B. and Williams, D. E. (1976) 'Superconducting a.c. power cables and their application'. Paris: CIGRE Paper No. 21-05.
(23) Metra, P., Ashworth, S., Slaughter, R. J. and Hughes, E. M. (1993) 'Preliminary analysis of performance and cost of high temperature superconducting power transmission cables'. *IEE Conf. on Power Cables and Accessories 10 kV–500 kV*, pp. 248–252.
(24) Von Dollen, D. W., Hingerani, N. G. and Summ, R. W. (1993) 'High temperature superconducting cable technology'. *IEE Conf. on Power Cables and Accessories 10 kV–500 kV*, pp. 253–257.

PART 5
Submarine Distribution and Transmission

CHAPTER 42
Submarine Cables and Systems

Submarine cables are used in three basic types of installation:

(a) river or short route crossings which are generally relatively shallow water installations;
(b) major submarine cable installations, coast to coast and island to mainland often laid in deep water and crossing shipping routes and fishing zones; these cables are generally required for bulk power transfer in a high voltage either a.c. or d.c. transmission scheme;
(c) between platforms, platforms and sea-bed modules or between shore and a platform in an offshore oil or gas field; these cables are currently laid in depths up to about 200 m but it is anticipated that much deeper installations will be required in the future.

Submarine cables are usually subject to much more onerous installation and service conditions than equivalent land cables and it is necessary to design each cable to withstand the environmental conditions prevailing on the specific route. Subject to certain restrictions, mass impregnated solid type paper insulated cables, fluid-filled cables, gas-filled cables and polymeric cables are all suitable for submarine power cable installations. Polymeric and thermoplastic insulated cables are used for control and instrumentation applications. There is an increasing tendency to include optical fibre cores, where possible, in power cable constructions, for purposes of control, communication and measurement of cable strain and temperature.

Cables for river crossings

Cable routes for river crossings are generally only a few kilometres long and cross relatively shallow water. The length of cable required can often be delivered to site as a continuous length on a despatch drum. Normal methods of installation include laying the cable from a barge into a pre-cut trench and mounting the drum on jacks on one shore and floating the cable across the river on inflatable bags. Installation of the cable is therefore a relatively simple operation which does not involve excessive bending or tension. The cable design is the most similar to land cable practice of all the different types of submarine cable. However, it is considered prudent to improve the mechanical security of the cable by applying slightly thicker lead and anticorrosion sheaths and to use armour. If the cable is to be laid across the river at the entrance to a port, it is recommended that the cable be buried to a depth of at least 1 m. A cheaper but less effective alternative is to protect a surface-laid cable by laying bags of concrete around it.

Should the cable be considered liable to damage due to shipping activities, an alternative solution to direct burial of the cable is to entrench a suitable pipe into the river bed and then pull in the selected type of cable.

Requirements for long cable lengths laid in deep water

Any cable to be laid on the sea bed should have the characteristics given below, the relevant importance of each particular characteristic being dependent on the depth of water and length of cable route.

(a) The cable must have a high electrical factor of safety as repair operations are generally expensive and the loss of service before repairs can be completed is often a serious embarrassment to the utility concerned.
(b) The cable should be designed to reduce transmission losses to a minimum, as submarine cable routes are generally long and the operating power losses are therefore significant in the overall economics of the system.
(c) The cable should preferably be supplied in the continuous length necessary to permit a continuous laying operation without the need to insert joints while at sea. Proven designs of flexible joints are available to permit drum lengths of cable to be joined together, either during manufacture or prior to loading the continuous cable length onto the laying vessel.
(d) The cable must withstand, without deterioration, the severe bending under tension, twisting and coiling which may occur during the manufacture and installation programmes.
(e) The cable must also withstand, without significant deformation, the external water pressure at the deepest part of the route.
(f) The cable, and where appropriate the terminal equipment, must be designed to ensure that only a limited length of cable is affected by water ingress if the metal sheath is damaged when in service.
(g) The armour must be sufficiently robust to resist impact damage and severance of the cable if fouled by a ship's anchor or fishing gear.
(h) For deep water installations, the cable must be reasonably torque balanced to avoid uncontrolled twisting as it is lowered to the sea bed.
(i) The weight of the cable in water must be sufficient to inhibit movement on the sea bed under the influence of tidal currents. Movement would cause abrasion and fatigue damage to the cable.
(j) The cable must be adequately protected from all corrosion hazards.
(k) All cable components must have adequate flexural fatigue life.
(l) All paper insulated and some polymeric insulated cables are required to be watertight along their complete lengths. Water ingress impairs the electric strength of these cables.

The requirements for cable laid on the sea bed also largely apply to cables which are to be buried. The bending characteristics of the cable as it passes through the burial device may need further consideration, and the friction of the serving against rollers and skid plates has to be taken into account. It is essential that information be provided on the length of the proposed route, the nature and contour of the sea bed, tidal currents, temperatures etc. before a provisional cable design can be prepared for the proposed installation. Sufficient information can often be obtained from Admiralty charts to enable a tentative cable design to be prepared to complete a feasibility study but in most cases it is necessary to carry out a hydrographic survey before the cable design can be finalised.

A.C. cable schemes

Where practical, a.c. submarine cable schemes are the first choice. As in the case of land cable circuits, the use of 3-core cables up to and including 150 kV is preferred to single-core cables provided that they can meet the required rating. 3-core cables also offer savings in both cable and installation costs as only two cables compared with four for a 3-phase scheme need be installed when security of supply is required if one cable is damaged.

If a 3-core submarine cable is damaged externally, e.g. by a ship's anchor or trawling gear, all three cores are liable to be affected. It would therefore be necessary to install two 3-core cables from the outset, preferably separated by 250 m or more to obtain reasonable security of supply. In 3-core solid type cable installations (i.e. for circuits up to 33 kV rating) single lead type cables (i.e. HSL cables) are sometimes preferred, particularly for deep water installations.

For major a.c. power schemes it will probably be necessary to use single-core cables as 3-core cables will be unable to meet the rating. In this case the cables are spaced far apart so that the risk of more than one cable being damaged in a single incident is minimised. The installation of four single-core cables for one circuit, or one spare cable for two or three circuits, would be expected to provide reasonable assurance of continuity of supply. However, widely spaced single-core magnetically armoured cables give rise to high sheath losses. These losses can be reduced substantially by the use of an outer concentric conductor underneath the armour.

D.C. cable schemes

For major cable installations requiring bulk transfer of large quantities of power, the choice has to be made between an a.c. or a d.c. transmission scheme. The longest submarine transmission schemes are invariably d.c. as their length is not limited by the necessity of supplying the charging current inherent in the a.c. system. Where it is intended to connect two separate power systems, e.g. between different countries, d.c. is again chosen as it is possible to keep the two systems independent, thereby preventing risk of instability. A d.c. system also allows for a greater degree of control of power flowing through the cables. Where the link is relatively short, the a.c. system will be more attractive than the d.c. because of the high capital cost of the converter stations required at both ends of the d.c. route. In a small number of routes it may be possible to position some reactive compensation for the a.c. system on conveniently placed islands, although this solution may affect the economics of the scheme. One additional advantage of a d.c. scheme is that, for major links incorporating single-core cables, only two cables are necessary whereas a minimum of three cables is invariably required for a.c. schemes.

D.C. transmission schemes may consist of a one polarity cable (monopole) carrying full circuit power with sea return, or preferably two cables of positive and negative polarity (bipole) each carrying half circuit power. The magnetic field created by a monopole cable causes compass errors near the cable route which may be unacceptable to the relevant Admiralty authorities as it would create a potential hazard to shipping.[1] The monopole scheme suffers the further disadvantage that if the submarine cable is damaged there would be no transmission capability until the cable was repaired. In a bipole transmission scheme, if only one pole cable is damaged, the remaining cable will continue to carry half circuit power with sea return. The installation of three single-core cables from the outset provides reasonable assurance of full transmission capability even if one cable is damaged.

ELECTRICAL DESIGN FEATURES OF CABLES

Conductors

The conductor design will be influenced by the choice of transmission scheme in which it is required to operate. Long submarine routes are generally d.c. schemes, which allow for the use of concentrically stranded conductors. As most land cables operate on a.c. there will be no difference in the design of conductor used in a.c. submarine cables. Large a.c. conductors will be of the Milliken type.

Copper is generally preferred to aluminium for the conductors of all submarine cables as its use permits a higher current density, thereby reducing the overall diameter of the cable. In the event of the cable being damaged and sea-water entering it, a copper conductor is much more resistant to corrosion than an aluminium conductor.

Details of the design of conductors are given in a later section dealing with specific types of cable.

Insulation thickness

The insulation thickness is designed on the same basis as for land cables and for the higher voltages is usually determined by the lightning impulse test requirement. It is considered prudent, however, to employ slightly lower maximum design stresses for submarine cables than would be adopted for land cables, to compensate for the more severe bending and tension which a submarine cable may need to withstand during the laying operation. The resultant increase in the electric strength of the cable is considered to justify the small increase in cost. For d.c. operation, design stresses up to 32 MV/m have been used for solid type cables and 35 MV/m for FF cables (paper and PPL insulated). Therefore both types of cable can be considered for the highest voltage schemes.

All d.c. submarine cables are expected to meet the electrical test requirements specified in the latest CIGRE 'Recommendation for tests for power transmission d.c. cables for a rated voltage up to 600 kV' (see chapter 38), following mechanical tests carried out in accordance with the CIGRE 'Recommendations for mechanical tests on submarine cables'.[2]

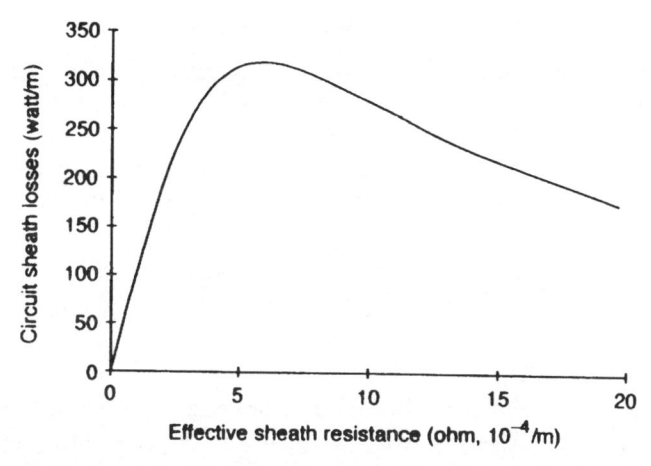

Fig. 42.1 Relationship between effective sheath resistance and sheath losses in 630 mm^2 single-core fluid-filled submarine cable

Current ratings

The current rating of submarine cables is mainly dependent on the maximum recommended conductor temperature, the thermal resistivity of the dielectric and the environment in which the cable is laid, and in the case of single-core a.c. cable schemes the axial spacing of the cables on the sea bed and the choice of armouring material.

The thermal resistivity of the environment in which the cables operate is dependent on site conditions and the method of laying. If the cables are laid on the surface of the sea bed, thermal values of 0.3 K m/W have sometimes been assumed. If the cable is to be buried in the sea bed, however, values of up to 1.0 are usually used for cables buried no deeper than 2.0 m.

In single-core a.c. submarine cable schemes, the cables are normally widely spaced on the sea bed and the sheath circuits are bonded at both ends. The wide cable spacing increases the sheath circulating current and reduces the current rating of the cables. Whereas the use of aluminium alloy armour permits the highest current density in the phase conductors at minimum cost, the mechanical properties and corrosion susceptibilities of the aluminium alloy are often inadequate to protect the cables from external hazards. Galvanised steel wire armour has excellent mechanical characteristics but when applied to single-core a.c. cables gives rise to eddy currents, hysteresis and circulating currents in the magnetic material.

The current rating of single-core a.c. cables is dependent, *inter alia*, on the effective resistance of the sheath circuit (i.e. the metallic sheath, armour wires and any reinforcing tapes in parallel). The currently favoured method of reducing the losses and increasing the current rating is to reduce the effective resistance of the sheath circuit. This may be achieved by using an outer concentric conductor which usually consists of a layer or layers of copper wires. This reduces the external magnetic field and permits the use of steel wire armour. In certain cases to achieve a low conductance, the steel wire is replaced with hard drawn copper strips.

Figure 42.1 shows the relationship between the effective sheath circuit resistance and the sheath losses in a typical single-core submarine a.c. cable laid at wide spacing.

Protection of anticorrosion sheath against voltage transients

On long submarine cable routes it is necessary to take steps to ensure that the voltage appearing across the anticorrosion sheath, due to voltage transients in the transmission circuit, does not approach the electrical breakdown level of the sheathing material. It is standard practice, therefore, to bond the lead sheath and any associated reinforcing tapes electrically to the armour wires at regular intervals along the cable length.

MECHANICAL DESIGN FEATURES OF CABLES

Effect of method of installation

Two alternative techniques are available: laying from large coils formed in the hold of a suitable vessel (fig. 42.2), or laying from a large turntable or drum. The turntable or drum technique is the technically preferred option as the cable has to withstand only bending with no twisting action, apart from tension considerations. Provided that turntables are used in all stages of cable manufacture the angle of lay and tension of application of all helically applied cable components need differ little from those used for land cables.

Winding of the cable into a coil imparts a 360° twist in each complete turn of cable in the coil. In a typical case, coiling a cable of 125 mm overall diameter to a minimum eye of 7.5 m (60 times the overall diameter) causes a twist of 15° per metre, the twist per metre decreasing slightly with each turn coiled outwards. For long cable routes the cable coil may be several metres high when coiled into the holds of the laying vessel.

Fig. 42.2 Coiled lengths of cable awaiting transfer to the laying vessel

Cable is generally coiled clockwise when viewed from above. This causes all components which have been applied with right-hand lay, i.e. clockwise, to tighten and all components applied with left-hand lay to loosen when the cable is coiled down. It follows that at the interface between a right-hand component applied over a left-hand component the former is under abnormal tension while the latter is slack. These conditions are conducive to the formation of creases in the slack components unless the effect is controlled by careful selection of the angle of lay and the application tension of every helically applied component of the cable. Lifting of the cable from the coil immediately prior to laying removes the twist from the cable so that it is in the twist-free condition as it passes outboard from the laying vessel. In a well designed cable every component should be free of creases when uncoiled.

A single wire armoured cable will tolerate coiling provided that the armour wires are applied so that they loosen as the cable is coiled down. It is virtually impossible to coil a cable with a double-wire reverse lay armour as there is no combination of lays in the two layers of armour which will avoid either crushing the cable or creating such interfacial pressure between the two layers of armour that it is physically impossible to cause the cable to lay flat in a coil. When site conditions necessitate the use of a double wire reverse lay armoured cable, the only laying technique available for long continuous cable lengths is from a turntable or pipe laying drum barge which obviates the need for coiling. Double wire armoured cable with both layers applied in the same direction can be coiled and handled in a similar manner to single-wire armoured cable.

Resistance of cable to water pressure

The ability of a submarine cable to withstand the external water pressure is determined by the internal pressure of pressure-assisted cables (i.e. fluid- or gas-filled cables), the hardness and coefficient of expansion of non-pressure-assisted cables and the thickness and composition of any metal sheath and metal tapes external to the sheath. The external pressure in sea-water increases by approximately 0.1 bar per metre depth. Distortion of a conventionally designed circular non-pressurised cable does not usually occur at depths less than about 150 m. At greater depths the metal sheath is liable to suffer distortion following the first heating cycle unless the cable is specially reinforced against the external water pressure. Figure 42.3 shows the cross-section of a solid type cable which withstood an external pressure of 27.5 bar at ambient temperature but which became misshapen when cooling to ambient temperature following a heating cycle to 60°C. Thermal expansion of the impregnant had distended the lead alloy sheath. On cooling the external pressure did not contract the sheath uniformly, causing it to become distorted. An adjacent sample of cable withstood an external pressure of 65 bar and repeated heating and cooling when a steel tape 1.5 mm thick was lapped over the lead sheath.

Because of the high coefficient of expansion of XLPE insulation, it may be necessary to restrict the temperature rise and hence the current loading when lead-sheathed XLPE cables are laid in deep water, so as to avoid excessive distension of the lead sheath.

The efficacy of any metal tapes applied external to the lead sheath to reinforce the cable against the external water pressure is proportional to $t^3 n/d^3$ where t is the thickness of one tape, d is the pitch circle diameter of the tapes when applied to the cable and n is the number of tapes applied. A single tape of 1 mm thickness will therefore provide as much resistance to the external water pressure as eight tapes each 0.5 mm thickness.

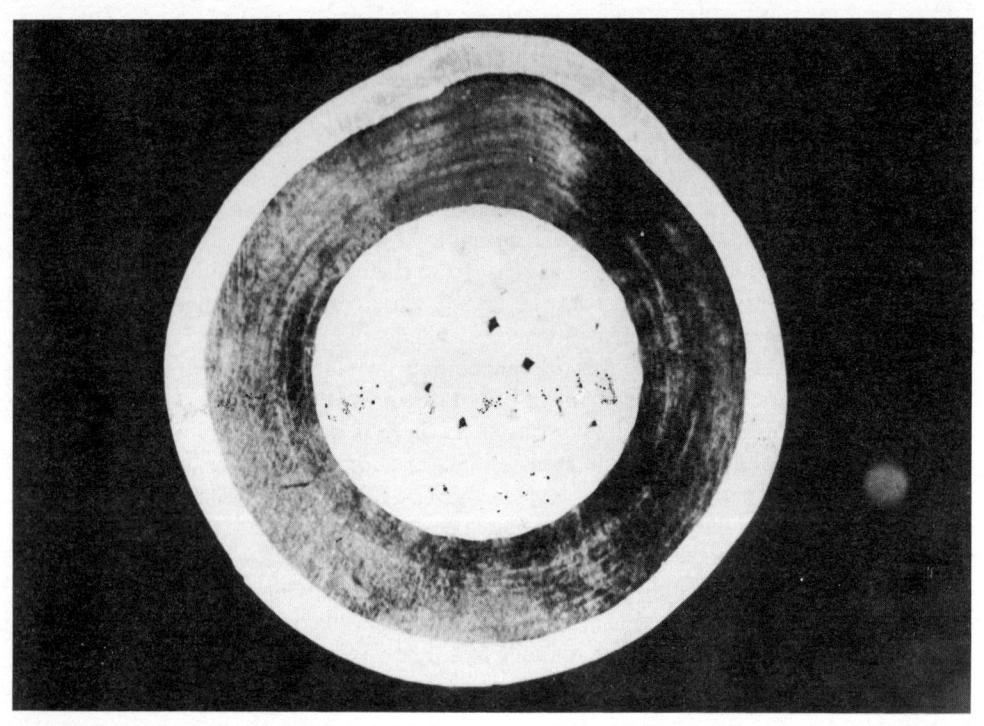

Fig. 42.3 Deformation of solid type PILS cable resulting from a heating cycle to 60°C with hydrostatic pressure of 27.5 bar

Development tests have shown that optimum results are achieved if a polyethylene sheath is applied over the lead sheath followed by one or more layers of reinforcing tape.

In the case of FF cables, the internal pressure of the impregnant will increase as the laying depth increases. However because the density of the impregnant is lower than that of the sea-water a differential pressure will exist which increases with depth by approximately 0.16 bar per metre, the exact value depending on the actual densities of the impregnant and the sea-water. The reinforcement provides some resistance to deformation but cables to be installed in deep water must be designed to withstand this external pressure. This aspect is covered in the CIGRE test recommendations.[2] Gas pressure cables are usually laid without pressurisation and therefore experience the full external pressure of the water during laying.

Choice of lead alloy sheath

The lead alloy sheath of a submarine cable has a very onerous duty imposed upon it compared with that of land cables. It will probably be subjected to strains and vibrations when the cable is laid, when in service if the cable is not buried and due to the motion of the ship if the cable has to be recovered for repair. Tests have shown that lead alloys E (0.4% Sn and 0.2% Sb) and F3 (0.15% As, 0.1% Bi, 0.1% Sn) have adequate fatigue life for this duty.[3] Submarine cables have also given satisfactory service, however, when sheathed with other alloys, e.g. $\frac{1}{2}$B (0.45% Sb) and $\frac{1}{2}$C (0.2% Sn, 0.1% Cd).

Choice of armouring

Submarine cables need to be armoured to withstand the highest tensile loading likely to be encountered when laying and the residual tension left in the cable after laying (typically 1–3 tonnes) and to provide reasonable resistance to impact and abrasion damage from trailing ships' anchors or fishing gear. The armour must be resistant to corrosion as failure of individual wires in service may cause kinks to form followed by electrical failure of the cable. Galvanised steel wires meet the mechanical and anticorrosion requirements at lowest material cost but, as described earlier, the use of magnetic armour on single-core cables may cause an unacceptable reduction in the current rating of the cable.

Resistance to impact and abrasion damage is improved by applying double wire armour. If the cables are to be coiled when transported, it is necessary to apply both layers with the same direction of lay. As explained in the following section, this may not be acceptable for cables to be laid in deep water. Additionally, the added protection may only be considered necessary over part of the cable route such as the shore ends. This raises manufacturing difficulties, e.g. in order to produce a cable with acceptable handling and bending performance it is necessary to 'let in' individual armours by welding them to adjacent wires over a distance of a few metres in the transition between single and double wire armour.

When the economics of steel wire armoured a.c. single-core cable schemes is unfavourable, hard drawn copper wires or strips can be used.[4] Aluminium alloy armour wires were applied to the 420 kV a.c. submarine cables laid between Denmark and Sweden.[5] However, there are published reports of chemical and electrolytic corrosion failure of the aluminium alloy armour wires on the cables laid between Connecticut and Long Island.[6]

Torque balance in cable construction

When a single wire armoured cable is suspended from the bow sheave of the laying vessel, a proportion of the tensile load is carried by the helically applied armour wires. This loading produces a torque in the armour wires which, unless appropriate precautions are taken in the design of the cable, tends to cause the cable to twist so that the lay of the armour wires straightens towards the axis of the cable and thereby transfers strain to the core(s). The twisting action cannot pass backwards through the brakes of the cable laying gear to the cable yet to be laid, nor forward to the cable already laid on the sea bed. The twisting action therefore tends to concentrate in the suspended cable between the bow sheave and the sea bed. The problems become more severe with increasing immersed weight per unit length and increasing depth of laying. Figure 42.4 shows a submarine cable which has developed a kink due to lack of torque balance.

The twisting action can be greatly reduced by applying a second reverse layer of armour wires which under tensile loading conditions produces an approximately equal and opposite torque to that of the inner layer of wires. This construction is now generally used for cables to be installed in water depths greater than about 500 m. The twisting action can be reduced to acceptable levels for cables to be installed at lower depths by the application of metal anti-twist tapes below a single layer of armour wires to produce a counter torque, as shown in fig. 42.5. The requirements may be expressed mathematically.

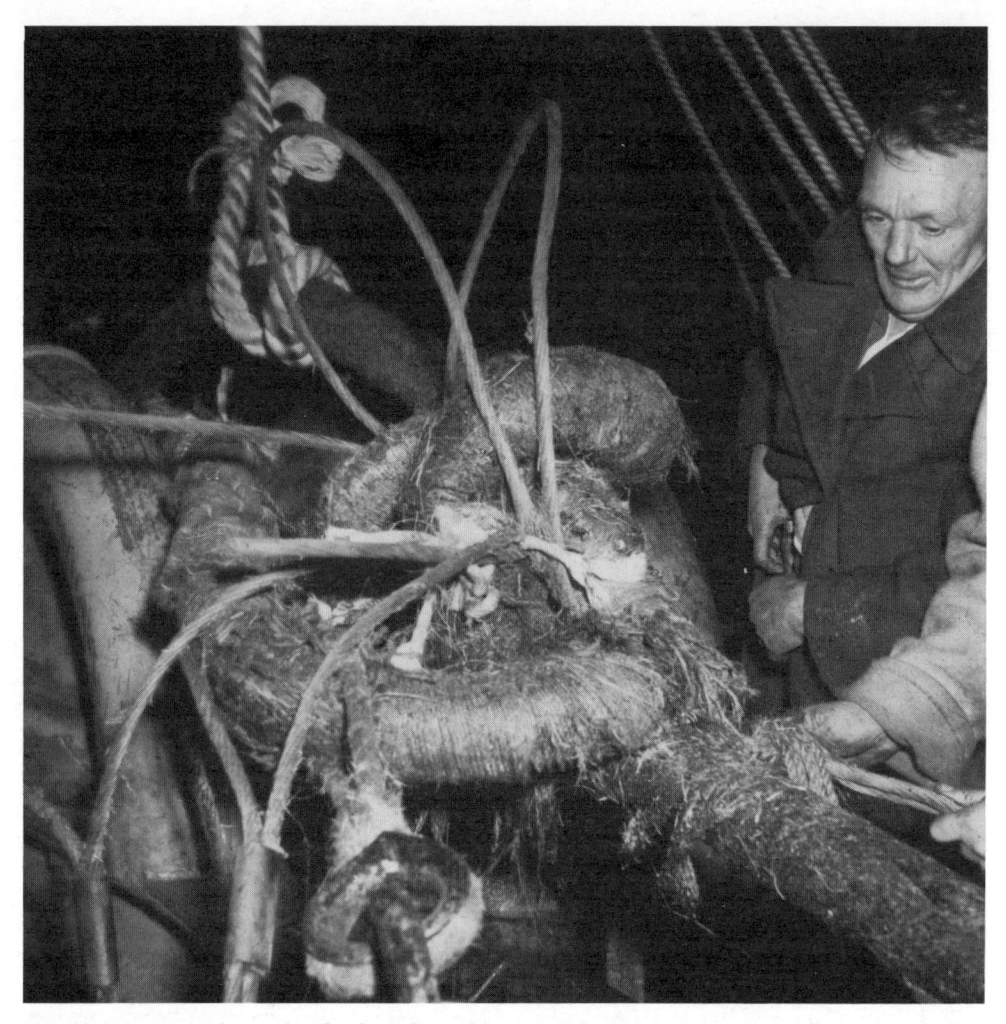

Fig. 42.4 Recovered sample of submarine cable containing kink due to lack of torque balance

The torque produced by the armour is

$$Tr_a \sin \theta \quad \text{(kg m)}$$

and the torque produced by the anti-twist tapes is

$$Tr_{at} \sin \phi \quad \text{(kg m)}$$

where T = tension in cable (kg)
θ = angle of armour wires to cable axis
ϕ = angle of anti-twist tapes
r_a = pitch circle radius of armour wires (m)
r_{at} = pitch circle radius of anti-twist tapes (m)

The torque is balanced when

$$Tr_a \sin \theta = Tr_{at} \sin \phi$$

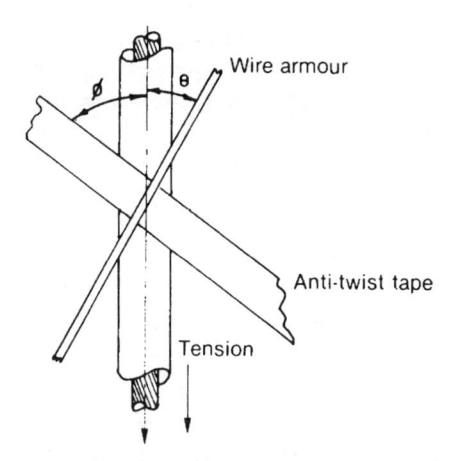

Fig. 42.5 Torque balance in single-core cable by using single wire armour and anti-twist tapes

i.e. when

$$r_a \sin \phi = r_{at} \sin \phi$$

Although it is difficult to achieve complete torque balance with a single layer of armour and anti-twist tapes, the solution is economically attractive for installation in depths up to about 200 m as the anti-twist tapes make a significant contribution to reinforcing the cable against internal and hydrostatic pressures.

Avoidance of cable movement on sea bed

Any cable which is subject to movement across the sea bed due to tidal currents is liable to premature failure due to abrasion damage. It is therefore important that the submerged weight of the cable is adequate to resist the maximum tidal sea-bed currents expected, even under storm conditions. The resistance to movement is proportional to the square root of the coefficient of friction of the cable with respect to the sea bed multiplied by the W/D ratio of the cable, where W is the submerged weight of the cable and D is the overall diameter. The coefficient of friction may be as low as 0.2 if the cable is laid across smooth rock but a value of 0.5 is typical of a cable laid on sand or shingle.

Although in many cases the cable may ultimately be silted over, so inhibiting cable movement, it would be unwise to rely on this occurring uniformly along the complete length of the cable. Therefore, standard practice when designing a submarine cable is to calculate the water velocity likely to cause movement. Should this velocity be less than the maximum expected tidal current at the sea bed, including the shore approaches where tidal currents may be the highest, it would be necessary to increase the weight of the cable. For lead sheathed cables an increase in the weight of the cable is achieved at minimum cost by increasing the thickness of the sheath. For non-metallic sheathed cables it may be necessary to apply lead tapes or to insert lead fillers in the cable to attain an adequate weight-to-diameter ratio.

Mechanical performance of submarine cables

All a.c. and d.c. submarine cables are required to comply with the requirements of the CIGRE 'Recommendations for mechanical tests on submarine cables'.[2] This specification requires that the cable, including flexible joints, should be subjected to tension tests, bending under tension, external pressure withstand tests and, in the case of pressure-assisted cables, an internal pressure withstand test. If the cable is to be coiled, either during manufacture or for transporting to site, all the tests listed above are preceded by a coiling test. The cable is required to withstand a voltage test following the mechanical tests. The cable should be free of damage or untoward features when subjected to visual examination following the mechanical and electrical tests.

PROTECTION AGAINST CORROSION

The service life of a submarine cable will be dependent, *inter alia*, on the efficacy of the anticorrosion protection, as sea-water is an aggressive environment in which to operate. Should the metal sheath of a non-pressure-assisted cable suffer corrosion damage, water would enter the cable and a voltage failure would ultimately occur. In the case of internal pressure-assisted cable, i.e. gas- and fluid-filled cables, a leak in the pressure-retaining sheath would cause the pressure alarms to operate. Small leaks can be tolerated for a short time by maintaining the gas or fluid feed until it is convenient to undertake cable repairs. Corrosion or failure of the sheath reinforcing tapes would cause the sheath to burst. This would necessitate taking the cable out of service and undertaking emergency action to avoid water entering the cable.

Lead alloy sheaths are preferred to aluminium sheaths for submarine cables because of the vastly superior resistance of lead to corrosion when immersed in sea-water. If an aluminium sheath is used, any defect or damage to the anticorrosion sheath would be liable to cause rapid corrosion failure of the metal sheath.

Extruded polyethylene provides the most impermeable barrier to moisture ingress and is used practically universally for the anticorrosion sheath. It is applied directly above the lead alloy sheath or above the reinforcing tapes in the case of pressure-assisted cables. If protection against teredo attack is considered necessary, one or two layers of brass or copper tapes are applied over the polyethylene sheath.

The provision of an anticorrosion sheath external to the armour wires of an a.c. submarine cable is generally unnecessary and technically undesirable. In all but the simplest cable installations it would be impossible to guarantee that any extruded thermoplastic or elastomeric oversheath would be undamaged during the cable laying operation, particularly if laying over rocks. Should the oversheath survive the laying operation without damage, it would still be subject to damage while in service from marine borers, sometimes by fish and invariably by shipping activity. If the oversheath were damaged, some of the current in the cable sheath/armour circuit would flow to the sea at the point of damage to the oversheath, causing electrolytic corrosion of the armour wires.

If the cable is served overall with textile tapes instead of an extruded oversheath, the armour wires are uniformly in contact with the sea-water along the complete cable route and the resultant distributed electrolytic action on the armour wires is very small. Some manufacturers recommend that each individual armour wire should be sheathed with a thermoplastic anticorrosion covering. Severe problems have sometimes been experienced when this solution was adopted on single-core a.c. cables because the corrosion by-products formed following corrosion of one damaged wire caused corrosion damage to adjacent but previously undamaged wires. The use of individually sheathed armour wires reduces the number of armour wires which can be applied to the cable, renders the cable subject to damage between adjacent wires and reduces the ultimate tensile strength of the cable.

Experience has shown that the zinc coating of galvanised steel armour wires may deteriorate after a few years' immersion, thus displaying small areas of non-protected steel. Slight corrosion pitting has been observed in such areas, but in a typical 138 kV a.c. single-core cable installation the diameter of individual armour wires merely decreased from 5.89 to 5.86 mm after 24 years' service.

Overall finishes

When laying a submarine cable it is necessary to brake the cable as it is paid out over the bow or stern skid in order to maintain control of the disposition of the cable on the sea bed (chapter 43). The amount of compression applied to the cable for braking purposes is dependent on the coefficient of friction between the finish of the cable and the braking gear. Apart from the disadvantages of a plastic oversheath on an a.c. cable, to which reference has been made, plastic materials have a low coefficient of friction which necessitates undesirably high pressure on the cable to attain the required braking effort, particularly for deep water installations. For these reasons a bitumen coated jute or polypropylene string serving is preferred, both of which have a high friction coefficient with respect to the braking gear. A string serving also has the advantage of being sufficiently elastic to permit an increase in the pitch circle diameter of the armour wires resultant on coiling the cable without adverse effect on the serving. If a plastic oversheath were applied and the bursting force of the armour wires or other mechanical incident caused rupture of the plastic oversheath, bird-caging of the armour wires would be expected at that position when coiling the cable.

SOLID TYPE MASS IMPREGNATED CABLES

Solid type mass impregnated cables have been used for submarine installations up to 34.5 kV a.c. or 450 kV d.c.

A.C. cables

These are virtually all of the 3-core construction. The 3-core HSL type of cable is generally preferred to the 3-core H type cable, particularly for deep water installations as the construction is inherently more resistant to external water pressure.

The conductors of 3-core HSL and H type cables are normally circular, of compacted construction. The paper insulation, conductor and core screens are applied in a similar manner to that of comparable land cables except that if the cable is to be coiled down, either during manufacture or for transport to site, it may be necessary to modify the paper lapping tensions to attain satisfactory coiling characteristics. The cables are usually impregnated with a viscous fluid compound as this results in more complete filling of the cable than a non-draining compound. This helps to restrict the passage of water along the cable in the case of damage in service. The individual impregnated cores for HSL cable, or laid-up cores for H type cable, are sheathed with a lead alloy and served with an anticorrosion sheath. The three served cores for HSL cables are laid up and padded circular with bitumen impregnated jute or polypropylene string fillers. Multiple lengths of non-armoured cable are jointed together prior to armouring into a continuous length. After armouring and serving, the cable is coiled down or wound onto a turntable to await pre-despatch tests and transfer to the laying vessel.

D.C. cables

The development of the mass-impregnated solid type cable for operation at high d.c. stresses has led to the possibility of high power long length d.c. connections. The important advantage of the solid type d.c. cable over the pressure assisted types is that as there is no requirement to maintain pressure in the cable, there is no limit on the length of cable that can be used. The practical maximum length for a FF cable is about 60 km. The Baltic connection between Sweden and Germany consists of a solid type cable operating at a voltage of 450 kV with a length of 250 km.[7] Proposals have been made for an Iceland–UK connection which would have a route length of over 1000 km.[8] There are now numerous submarine interconnections between the power systems of Europe.

D.C. cables are all single-core construction. The conductors of d.c. cables are circular and of high occupancy. A high occupancy may be achieved by stranding layers of segments around a central circular rod. An occupancy of 96% is claimed for the ± 270 kV d.c. cross channel interconnection.[9] High occupancy in the conductor has the advantage of presenting excellent resistance to water penetration in the event of cable damage especially when used in conjunction with a viscous impregnating compound. Although the construction of the single-core solid cable is intrinsically very resistant to external pressure it is necessary to control the sheath movement due to load cycles. This is done by limiting the maximum conductor temperature to 50°C.

The cables are insulated on high precision lapping machines similar to those used for FF cables. To minimise the number of flexible joints required, the cables are insulated and impregnated in the longest lengths possible. Some cable manufacturers have special drying and impregnating vessels capable of processing 40 or 50 km lengths of cable. This process requires the core to be wound onto a turntable within the drying vessel as it leaves the insulating machine. Flexible joints may be inserted between lengths of core usually at the lead or anticorrosion sheath stage in order to create even longer lengths. This type of joint may be used without any reduction in cable electrical or mechanical strength which makes the maximum length attainable limited only by the size of the storage space or the capacity of the laying vessel.

The lead sheathing, anticorrosion sheathing, armouring and serving processes differ little from standard production techniques for land cable. However, the lead sheathing and plastic sheathing of very long cable lengths necessitate continuous processes, often lasting several days, and require the highest standards of extruder performance to achieve satisfactory product quality.

Water penetration through damaged sheath

If the lead sheath of a single-core or 3-core HSL paper insulated solid type submarine cable is damaged, or the cable is severed, the sea-water causes the paper insulation to swell rapidly, so forming a partial blockage to further water ingress. Further cable beyond the blockage, however, may be affected by moisture owing to the vacuous conditions in the cooling cable and by wick action in the insulating papers. Because of its high occupancy, the conductor has excellent resistance to water penetration.

The interstitial fillers in 3-core H type cable form a low resistance path for water to enter the cable following damage to the sheath. It may be necessary to replace a substantial amount of cable to eliminate moisture should this type of cable be damaged after laying.

FLUID-FILLED CABLES

Fluid-filled paper insulated cables are suitable for use up to the highest voltages and they were the type of cable chosen for the 525 kV a.c. link between British Columbia mainland and Vancouver Island installed in 1984.[10] Fluid-filled cables have the advantage of requiring the smallest insulation thickness for a given voltage so that it is possible in some cases to use a 3-core cable where three single-core cables would otherwise have been necessary. This presents advantages from a manufacturing and installation point of view as well as making the installation less prone to damage when in service.

FF cables are suitable for d.c. operation. However hydraulic considerations limit the maximum circuit length to about 50 km for paper cables. This is too short for many applications and therefore this type of cable is not widely used for d.c. submarine connections. A recent development is the application of fluid-filled PPL cable for a 500 kV d.c. 50.7 km circuit in Japan.[11] PPL offers the possibility of reduced insulation thickness, lower charging current for a.c. schemes, and of permitting longer circuits to be considered before hydraulic limitations become important.

Hydraulic design

It is an essential feature of any fluid-filled submarine cable installation that, if the pressure-retaining sheath is damaged, the internal fluid pressure under leak conditions should exceed the external water pressure at that position, so that water does not enter the cable. The worst possible condition would be for the cable to be severed near one terminal during a period when the cable was carrying full load current, as the rate of cooling would then be at a maximum. Whereas under normal operating conditions the fluid reservoirs at each end of the route effectively feed fluid to the mid-point of the route, if the cable is severed near one terminal the seaward length of cable would be dependent on the fluid reservoir at the remote end of the route. This reservoir would be required to feed fluid at a sufficient rate to compensate for the thermal contraction of the fluid within the complete cable length and also to maintain a positive pressure with respect to the external water pressure where the cable was severed. The fluid feed length would therefore be nearly twice the normal length and the pressure drop nearly four times that which would occur if the cable was severed at the mid-point. It is therefore necessary to incorporate fluid channels within the cable having sufficiently low hydraulic impedance that a positive pressure can be maintained under fluid leak conditions, without the need to operate the cable at an excessive internal pressure.

The calculation of the hydraulic parameters of the cable system is similar to that detailed in chapter 30 for fluid-filled cables installed on land, but for submarine cable installations it may be necessary to calculate many alternative solutions to obtain the optimum size of the fluid duct(s) and the feed pressure of the fluid pumps. There are practical limits to the size of the fluid channels which can be incorporated in a cable and this limits the maximum length of route over which a fluid-filled cable can be operated to about 50 km. On long cable routes and in deep water installations fluid-filled cables are required to operate at higher internal pressure than a conventional land cable and it is therefore necessary to apply additional reinforcing tapes to enable the lead alloy sheath to withstand the internal fluid pressure.

Further details regarding calculation of fluid pressure transients may be obtained by reference to the CIGRE report 'Transient pressure variations in submarine cables of the oil-filled type'.[12] This document also explains how to allow for pressure transients encountered during the laying operation.

Figure 42.6 shows the internal fluid pressures in a typical 1000 mm^2 275 kV single-core cable operating to a maximum conductor temperature of 80°C on a cable route having a maximum depth of 90 m. The cable concerned has a fluid duct of 23 mm bore. The graph shows the variation in maximum fluid pressure along the cable route when the cable is heating, the minimum pressure when cooling and the transient pressures if the cable is severed near one end of the route following a period of full current loading. It will be noted that, in order to maintain an internal fluid pressure of 1 bar at the severed end of the cable, it is necessary to operate the fluid pumps at 11 bar, the pressure drop along the complete route from the remote fluid pump being approximately 10 bar. The graph also shows that the cable would need to be reinforced to withstand an internal fluid pressure of 30 bar.

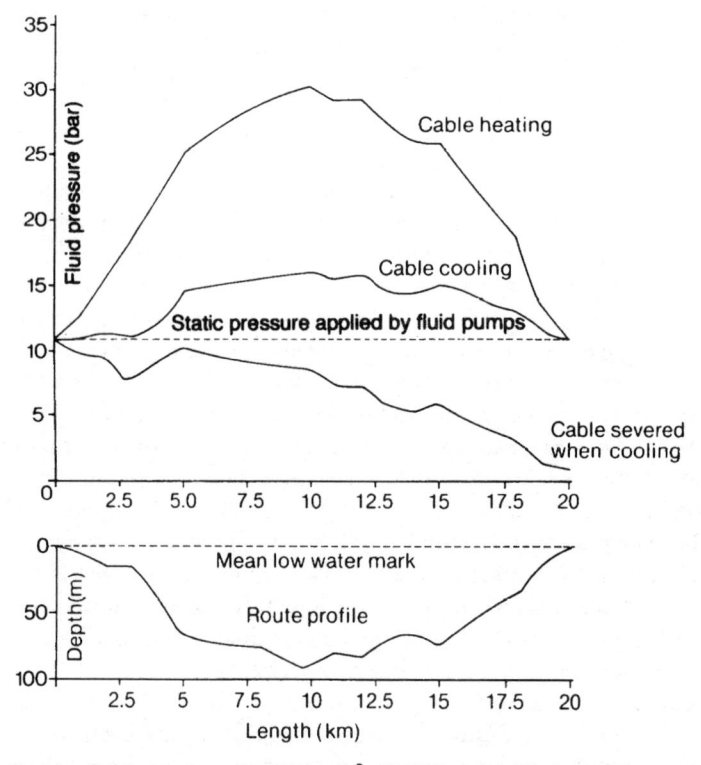

Fig. 42.6 Variation in fluid pressure in 1000 mm^2 275 kV fluid-filled single-core cable under operational and fault conditions

It is sometimes necessary to restrict the temperature rise of fluid-filled submarine cables to reduce the fluid pressure transients to acceptable levels. This necessitates the use of larger conductors than would otherwise be justified by the current loading requirement.

3-core cables

3-core fluid-filled submarine cables can be manufactured for voltages up to 150 kV a.c. subject to certain restrictions on the maximum cross-sectional area of the conductors. As with all types of 3-core cable the capacity of the laying-up machine limits the maximum length which can be manufactured without the need to insert flexible joints in at least the conductors and insulation. Alternatively, drum lengths of completed cable may have flexible joints inserted between drum lengths before they are installed.

Flexible joints are technically acceptable in fluid-filled cables. Comprehensive laboratory testing by several cable manufacturers has shown that the electrical characteristics of a well designed flexible joint can be as good as the machine-made cable and the mechanical performance completely adequate. Flexible joints have additionally given satisfactory service in 150 kV a.c. fluid-filled cables for many years.

The design and construction of the conductors, insulation and screens is similar to that for land cables except that the angle of lay and application tension of the paper tapes may need to be modified if the cable is to be coiled down either during manufacture or on the laying vessel. When laying-up the cores, fluid ducts are usually included in the interstitial spaces between the cores. The maximum bore of the ducts which can be accommodated within the spaces available may impose a restriction on the length of route over which a 3-core cable can be laid.

The laid-up cable is generally sheathed with a lead alloy and reinforced against the internal fluid pressure by two or more layers of metal tape. Tin-bronze and non-magnetic steel tapes have been used for reinforcement. As the transient fluid pressure in submarine cables is often much greater than is normally permitted in land cables, the total thickness of reinforcement is correspondingly increased. The finished cable is either coiled down or wound onto a drum or turntable for pre-despatch testing prior to loading onto the cable laying vessel.

Fluid feed tanks are connected to the cable ends immediately after sheathing and a positive fluid flow is maintained from the cable ends on every occasion that the cable is cut, so that no air or moisture can enter the cable. The fluid tanks remain connected to the cable throughout all subsequent manufacturing and laying operations.

Single-core cables

Single-core fluid-filled cables are not subject to the same restrictions on maximum unjointed manufacturing lengths as 3-core cables are, because the maximum lengths attainable are only dependent on the size of the impregnating and drying vessels available. Some cable manufacturers are equipped with very large vessels capable of drying and impregnating 40–50 km lengths of core.

Single-core cables are required to have a central fluid duct in the conductor and the maximum length which can be installed is limited by the impedance to fluid flow which this duct presents. Significant improvements can be made by increasing the bore of the duct as its impedance varies with the fourth power of the radius. This improvement has to be balanced against the impact of having to increase the diameter of the cable. Additionally it may be possible to employ a low viscosity impregnant. The single-core conductor may be stranded around a helically wound steel duct or alternatively be formed of interlocking segments which form a self-supporting fluid channel with shaped wires stranded over it. If the cable is to be coiled at any stage during manufacture or installation, special attention needs to be given to the length and direction of wire lays to achieve satisfactory coiling characteristics.

All manufacturing processes are carried out on conventional plant, apart from the use of turntables or coiling down processes, if the length of cable required exceeds the capacity of the largest drums which can be handled. If flexible joints are to be made prior to despatch, these are normally inserted after lead sheathing or after any subsequent process, as convenient. Fluid pressure tanks are connected to the cable after lead sheathing and at all subsequent stages of manufacture.

The sheath reinforcing tapes are non-ferrous for a.c. cables but steel tapes offer the cheapest reinforcement for d.c. cables. The anticorrosion sheath, armour and servings generally comply with the details previously quoted.

Fluid pressure supply systems

For short length fluid-filled submarine cable installations, conventional fluid pressure tanks and appropriate alarm equipment, as used for land cables, are often adequate for maintaining the fluid pressure within the cable. On longer cable routes it is often necessary to provide pumping plant and fluid storage tanks at both ends of the route because of the much larger fluid flow and high pressure. The pumps are normally duplicated and alarm circuits are provided to signal any malfunction of the pumping equipment. As these pumps are usually energised from the public electricity supply it may be necessary to install auxiliary power generators to ensure the security of the fluid-filled cables in the event of a failure of the electricity supply.

Prevention of water penetration through a damaged sheath

If the lead sheath of a fluid-filled cable is perforated when in service, the internal pressure causes fluid to flow out so that water does not enter the cable. The fluid tanks or pumps at the cable terminals replace the fluid lost and operate alarms to indicate an abnormal rate of fluid flow. In the event of a major fluid leak the standard procedure is then for the cable to be taken out of service and a restrictor actuated in the hydraulic circuit to limit the fluid feed to the minimum necessary to ensure a positive fluid leak to the sea at all states of the tide. Ideally, the fluid leak should be located at an early stage, the cable cut and lifted and the ends checked for moisture and cut back if necessary. End caps would be fitted and the capped ends lowered to the sea bed with marker buoys attached. Experience has shown that after removing the lengths of cable damaged during the lifting operation the remaining cable is generally free of moisture and suitable for jointing to the new piece of cable required to link the cut ends.

GAS-FILLED CABLE

Cable of the pre-impregnated paper gas-filled type has many advantages from a manufacturing and system design point of view. As manufacture of this type of cable requires no large tank in which to impregnate the core, the possible length of joint-free cable is entirely dependent on the capacity of the laying-up machine or, for single-core cables, the amount of storage space available. The general construction of the conductor, screens and insulation follows solid cable practice except that for the longest lengths of single-core cable a gas duct is required in the centre of the conductor.

The metal sheath is of lead alloy which is heavily reinforced with steel tapes followed by a conventional anticorrosion sheath, beddings, armour and serving.

The static pressure in the gas-filled cable system is maintained higher than the external water pressure to prevent ingress of moisture should the cable be damaged. The minimum design operating pressure is 14 bar but in one installation the pressure had to be raised to 28 bar to fulfil this condition.[13] This places an onerous duty on the cable reinforcement and the accessories. Once the cable is charged with gas there is no longitudinal gas flow within the cable which means that there is not the same pressure restriction on cable length inherent in fluid-filled systems, but gas feed and alarm equipment is still required at the cable ends.

Factory flexible joints have been installed into gas-filled cables and have given satisfactory service for many years.

POLYMERIC INSULATED CABLES

Design limitations

Polymeric insulated submarine cables are in increasing demand for use in offshore oil fields and inter-island links up to a voltage of about 33 kV. Considerable experience in land cable installations in many parts of the world has indicated that the service life of XLPE cables may be limited unless steps are taken to prevent moisture ingress into the insulation. It is therefore necessary to provide an impermeable moisture barrier over the insulation, and for submarine installations this is best achieved by the application of a lead alloy sheath.

The immersed weight-to-diameter ratio of non-sheathed polymeric submarine cables is generally low and the cable is therefore liable to movement over the sea bed when subjected to even modest tidal currents. The application of a lead sheath of appropriate thickness inhibits cable movement, apart from ensuring that the core insulation is maintained in a dry condition.

Submarine cables insulated with EPR require no metal sheath and are installed as a 'wet' construction. This type of cable is suitable for use up to 33 kV providing selection is made of a special formulation of EPR and of a low insulation design stress. Some 3-core designs incorporate optical fibres or other communication cables in the interstices.

To date, polymeric cables have not generally been used for submarine connections above 66 kV. This is because they are generally larger than the equivalent pressure-assisted cables and the concern about the ability to water block the cable adequately in the event of sheath damage. Polymeric insulation is not chosen for d.c. use. The electric strength is reduced by the generation and concentration of trapped space charge near to the conductor and insulation screens.

Cable design and manufacture

The thickness of the screens and insulation applied to polymeric cables up to 30 kV rating should not be less than the appropriate value quoted in IEC 502, although, as with all submarine cables, a slight increase in the insulation thickness may be justified to enhance the factor of safety. The thickness of insulation for cables above 30 kV is subject to agreement between the customer and the cable manufacturer, there being as yet no internationally agreed standard for the insulation thicknesses for the higher transmission voltages.

The economics of the insulating procedure for polymeric cores favour the production of the longest continuous core lengths which can be accommodated on a turntable or core drum. However, the need to regularly clean the insulation extruders limits the length that can be continuously extruded, thus introducing factory joints of a more complex moulded type. As with all types of 3-core cable the capacity of the laying-up machine limits the maximum length of cable which can be manufactured without inserting flexible joints.

When a lead sheath is applied over a single- or 3-core cable the external protection normally comprises a polyethylene anticorrosion sheath, armour bedding, one or two layers of armour wires and a bituminised textile serving overall.

HV routine testing of polymeric cables and factory joints is more problematic, as d.c. testing is regarded as less effective for polymeric insulation and introduces the risk of stress increase by space charge generation.

Effects of damage to lead sheath

Polymeric materials have a relatively high coefficient of expansion which may cause a permanent distension of the lead sheath after an XLPE or PE insulated cable has been on load. Should the lead sheath be damaged after it has been distended, water would enter the annulus between the core(s) and the inside of the lead sheath. The penetration can be minimised by the use of water-swellable tapes over the cable insulation. If the damage is sufficiently severe to reach the conductor, then water penetration can be inhibited by filling the conductor with a blocking compound. While these techniques have been developed for land cables, in the case of submarine cables the external water pressures are much higher.

PILOT, CONTROL AND COMPOSITE CABLES

There is an increasing demand in offshore oil and gas fields for submarine pilot, control and composite cables which are laid between platforms or between a platform and a seabed unit for the remote control and monitoring of equipment. In each case the cable needs to be designed for the specific project, the number and type of cores varying considerably. Generally there are requirements for some pilot cores, speech circuits or TV cores together sometimes with optical fibres, power cores and hydraulic hoses within the same cable envelope. The cable needs to be designed and manufactured to have a high factor of safety, as the failure of a control cable may necessitate closing down an oil well with the consequent loss of production until the cable is repaired or replaced.

PE or sometimes XLPE is favoured for the insulation of speech circuits, PE for TV cores and either PE, XLPE or EPR for pilot and power cores. Flexible hydraulic hoses normally comprise a polyester elastomer inner core which is reinforced with aramid fibre and sheathed with a polyester elastomer jacket.

The individual cable pairs, cores and hoses are laid up with a relatively short lay to reduce the risk that these components are subjected to a significant strain when the cable is under tension. The optical fibre is generally laid up near the centre of the assembly to give maximum protection against damage. Fillers are normally laid up with the cable components to achieve a compact circular shaped cable. Providing there is no severe restriction on the mutual capacitance of the speech pairs it is considered advantageous to inject a thixotropic compound into the remaining interstices of the laid-up cable to restrict water penetration in the event of sheath damage after installation. In most cases a polyethylene sheath is applied over the laid-up cores followed by the appropriate armour and serving, the finish being generally similar to that applied to lead sheathed submarine power cables.

If the cable is required to hang in a catenary from a platform to the sea bed, it is essential that the cable be torque balanced. In these circumstances it is sometimes advantageous to lay up the cable around a central tensile member and to dispense with the external armour wires. If the cable is to be laid on the sea bed it may be necessary to apply a lead sheath to attain an adequate weight to diameter ratio to inhibit movement on the sea bed under the influence of tidal currents. Alternatively, in order to avoid the need for a lead sheath, it may be possible to bury the cable using a 'remote operated vehicle'. Such vehicles are usually equipped with sonar or TV systems and bury the cable using a plough technique.

FLEXIBLE JOINTS FOR CABLES

Techniques have been developed for constructing flexible joints in all types of cable commonly used in submarine cable installations. Most existing major submarine cable circuits contain one or more flexible joints. These joints have generally given trouble-free service. Individual cable manufacturers have developed different techniques for constructing flexible joints but all are expected to meet the mechanical test requirements

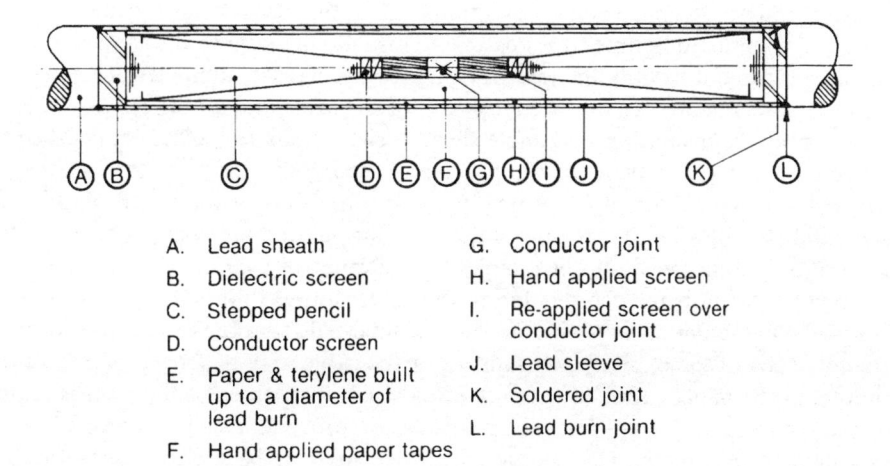

A. Lead sheath
B. Dielectric screen
C. Stepped pencil
D. Conductor screen
E. Paper & terylene built up to a diameter of lead burn
F. Hand applied paper tapes
G. Conductor joint
H. Hand applied screen
I. Re-applied screen over conductor joint
J. Lead sleeve
K. Soldered joint
L. Lead burn joint

Fig. 42.7 Construction of flexible joint in single-core solid type submarine cable up to the lead sheath stage

recommended by CIGRE Committee 21-06.[2] Figure 42.7 shows a typical design of flexible joint in a solid type paper insulated submarine cable.

Conductor connection

Submarine cables are generally laid with a certain amount of tension in the system. This is either unavoidable because of the length of cable suspended underneath the laying ship or is deliberate in order to prevent kinking. The tensile strength of the conductor connection is therefore of paramount importance. Two alternative methods are used for the jointing of conductors. The individual wires of the conductor may be butt-brazed in a staggered configuration so that the conductor retains its original diameter and flexibility. Alternatively, the multiwire conductor may be butt-brazed to form a solid joint which complies with the original diameter but restricts the flexibility of the conductor over a short distance.

Paper insulation and semiconductive paper screens

In all types of paper insulated cable the conductor screen and insulation are profiled on both sides of the joint in the conductor. The length of the stepped profile is designed to ensure a controlled electrical stress when the cable is energised. The insulation is then reconstituted over the jointed conductor, pre-impregnated paper tapes being applied by hand or machine under very carefully controlled conditions often in a humidity-controlled atmosphere. Each paper tape is accurately terminated on the steps of the profiles of the two cables being jointed. The re-application by hand can be a fairly lengthy process and it is possible to design a joint with shorter profiles which are quicker to re-insulate but it is then necessary to restrict the number of bends to which the joints will be subjected. This extra speed in completing a joint is very useful where a cable is being repaired at sea.

The outside diameter of the joint is determined by service performance and electrical test requirements applicable to the system. The use of a flush brazed conductor connection and graded paper tape insulation enables the diameter of the dielectric to be kept to a minimum. Sometimes it is possible to insulate the joint to the same diameter as the adjacent core but usually the joint is insulated to a greater diameter. This diameter, however, is rarely greater than the lead sheath of the cable and allows the lead sleeve to fit closely over the insulation and cable sheath. This enhances the mechanical strength of the joint and is convenient for the process of sealing the sleeve to the sheath.

For solid type cables it is usual to baste the reconstituted insulation with impregnating compound at regular intervals during the insulating process. When making flexible joints in fluid-filled cables two alternative techniques are available for preventing air or moisture from entering the cable during the jointing operation. The joint may be made under a continuous flow of fluid fed from pressure tanks located at the remote ends of the cables being jointed, or the cable may be frozen on each side of the joint and the papers applied without a fluid feed. In both cases the joint is required to be impregnated with fluid after the re-insulation process. This is achieved either via special fluid fittings in the lead sleeve which will need to be sealed afterwards or by impregnating using a temporary sleeve and then swaging down a lead sleeve over the insulation with continuous fluid flow.

Polymeric insulation

The insulation of PE, XLPE and EPR cores is pencilled down to a smooth controlled profile using special tools. The cable core is then reconstituted with special insulating and screening tapes which can also be applied with specially designed machines. Application by machine ensures even tension and correct geometric shape. Temporary binder tapes are applied to the reconstituted joint to compress the hand applied tapes while the joint is pressurised and heated under carefully controlled conditions to bond the PE tapes and to vulcanise the insulation in the case of XLPE or EPR cores.

An alternative procedure is to use an injection moulding machine for all types of core.

Lead sheaths

Prior to the start of jointing lead sheathed cables, a lead sleeve is passed over the lead sheath of one of the cable lengths being jointed. After completing the application of the core insulation and screens, the lead sheath is positioned over the core joint and progressively swaged down until it is a close fit over the insulated core. If the diameter of the reconstituted core insulation is equal to that of the original core, the lead sleeve is swaged down and cut to length to permit butt joints to be made to the cut ends of the lead sheath of the cables being jointed. However, if excess insulation has been applied over the joint, the lead sleeve is swaged down to be a close fit over the original lead sheath.

Table 42.1 Major submarine transmission interconnections

Installation date	Site	Voltage (kV)	Cable type	Maximum depth (m)	Length (km)
1954	Gotland	100 d.c.	MI	160	100
1956	Vancouver	132 a.c.	GF	183	25
1965	Sardinia–Corsica	200 d.c.	MI	500	119
1965	Cook Strait	250 d.c.	GF	245	40
1965	Sweden–Denmark	285 d.c.	MI	80	64
1969	Vancouver	300 d.c.	MI	200	27
1973	Sweden–Denmark	400 a.c.	FF	37	7
1976	Skagerrak	260 d.c.	MI	570	125
1980	Hokkaido–Honshu	250 d.c.	FF	300	42
1984	Vancouver	525 a.c.	FF	400	30
1985	Jersey–France	90 a.c.	FF	25	27
1986	England–France	266 d.c.	MI	55	50
1989	Finland–Sweden	400 d.c.	MI	117	200
1993	Skagerrak	350 d.c.	MI	540	125
1994	Baltic	450 d.c.	MI	30	265
1995	Denmark–Germany	400 d.c.	FFF	21	45
In progress	Spain–Morocco	400 a.c.	FF	610	27

MI Mass impregnated solid type
FF Fluid-filled
GF Gas-filled
FFF Flat fluid-filled

A fusion technique known as lead burning is the conventional method of sealing the lead sleeve to the cable sheath and lead burns have given excellent service over many years. The technique requires highly skilled operators and is very time consuming. More modern methods based on a TIG welding technique are under development.

Other components

Metal tapes such as reinforcing tapes or anti-teredo tapes are jointed by brazing or welding. Polymeric anticorrosion sheaths are reconstituted using polyolefin heat-shrinkable sleeves passed over one of the cable cores prior to jointing.

Armour wires in joints between drum lengths of completed cable or in repair joints may be jointed by welding or by the use of threaded turnbuckles. Both these methods give more than adequate mechanical strength. For factory-made joints it is possible to pass the joint through the armouring machine and armour the cable and joint as one continuous process. This solution is adopted where continuous lengths of cable are specifically required.

MAJOR TRANSMISSION INTERCONNECTIONS

There is a growing number of major high power submarine interconnections. Table 42.1 gives details of some of the more important installations.

REFERENCES

(1) Buseman, F. (1963) 'The magnetic compass errors caused by d.c. single-core sea cables'. ERA Report No. B/T 116.
(2) CIGRE Study Committee 21, Working Group 06 (1980) 'Recommendations for mechanical tests on submarine cables'. *Electra* (68), 31–36.
(3) Anelli, P., Donazzi, F. and Lawson, W. C. (1988) 'The fatigue life of alloy E as a sheathing material for submarine power cables'. *IEEE Trans.* **PD-3** (1).
(4) Dominguez Miguel, M., Ruiz Urbieta, M., Benchekroun, A., El-Kindi, O. and Gallango Faraco, E. (1994) 'Technical requirements for the submarine electrical interconnection Spain–Morocco'. Paris: CIGRE Paper No. 21-205.
(5) Gazzana Prioroggia, P. and Maschio, G. (1973) 'Continuous long length a.c. and d.c. submarine h.v. power cables'. *IEEE Trans.* **PAS-92** (5), 1744–1749.
(6) Chamberland, D. M., Margolin, S. and Shelley, M. (1979) 'The Long Island Sound submarine cable interconnection'. *7th IEEE Transmission and Distribution Conf.*
(7) Ekenstierna, B. and Nyman, A. (1993) 'Baltic HVDC interconnection'. New Zealand: CIGRE SC 14 Colloquium Paper No. 63.
(8) Guonason, E. and Henje, J. (1993) 'A 550 MW HVDC submarine cable link: Iceland–UK–Continental Europe'. *IEE Conf. on Power Cables and Accessories 10 kV–500 kV*, pp. 220–224.
(9) Arkell, C. A., Ball, E. H., Hacke, K. J. H., Waterhouse, N. H. and Yates, J. B. (1986) 'Design and installation of the U.K. part of the 270 kV d.c. cable connection between England and France, including reliability aspects'. Paris: CIGRE Paper No. 21-02.

(10) Foxhall, R. G., Bjorlow Larsen, K. and Bazzi, G. (1984) 'Design, manufacture and installation of a 525 kV alternating current submarine cable link from mainland Canada to Vancouver Island'. Paris: CIGRE Paper No. 21-04.
(11) Fujimori, A., Janaka, T., Takashima, H., Imajo, T., Hata, R., Tanabe, T., Yoshida, S. and Kakihana, T. (1995) 'Development of 500 kV d.c. PPLP-insulated oil-filled submarine cable'. *IEEE/PES Summer Meeting*, July 23–27, 1995.
(12) CIGRE Study Committee 21, Working Group 02 (1983) 'Transient pressure variations in submarine cables of the oil-filled type'. *Electra* (89), 23–29.
(13) Crabtree, I. M. and O'Brian, M. T. (1986) 'Performance of the Cook Strait ±250 kV d.c. submarine cables, 1964–1985'. Paris: CIGRE Paper No. 21-01.

CHAPTER 43
Submarine Cable Installation

A submarine power cable project includes a series of stages:

(a) feasibility study;
(b) survey;
(c) cable laying vessel selection;
(d) cable handling equipment design or selection;
(e) shore end site work;
(f) provision of moorings;
(g) cable laying;
(h) terminating the cable;
(i) testing;
(j) maintenance;
(k) repair.

Although all the above are not necessarily carried out by the same organisation, this is very frequently the case. In any case, it is essential that there should be close co-operation between the cable supplier, the survey team and the installation organisation so that the work can be carried out in the most effective and efficient way. Although it is often thought that difficulties in defining responsibility are likely to arise if separate organisations are employed for the cable supply and for the installation work, this is not borne out in practice. The best installation results from a combination of the most experienced cable manufacturer and the most competent installation organisation working in close liaison or as a joint venture.

FEASIBILITY STUDY

The initial stages of the investigation into the feasibility of a submarine power cable interconnection will normally rely on published survey data given in such publications as UK Admiralty Charts and Pilots and other similar sources of information.

In order to estimate the cost of the installation work it is necessary to work with provisional cable designs which specify the cable diameter, weight per unit length, bending radii, any special cable handling requirements, and the number of cables required. Taking into account alternative operating voltages, conductor sizes, cable construction, types of insulation and possible route variations, this may give rise to a large number of combinations which have to be evaluated in terms of shipping and cable handling requirements.

The feasibility study should result in a positive proposal which defines the type of system to be installed and indicates the preferred location of the interconnection. The precise route of the cables is then determined from the data collected by a full survey of the area.

SURVEY

A preliminary site survey may be included in the feasibility study or as the first stage of the full survey. The objectives are to obtain information concerning local conditions and to confirm the choice of areas to be surveyed in detail. The following features will be investigated:

(a) possible landing points for cables;
(b) local support facilities;
(c) possible practical cable routes;
(d) hazards in the vicinity;
(e) land access to landing points and identity of landowners who may be involved;
(f) a suitable site for placing a navigation correction transmitter based on differential global positioning by satellite (DGPS).

The extent of the full survey will depend upon the sea bed width required to accommodate the cables at a separation adequate to enable one cable to be repaired without a risk of damage to adjacent cables of the interconnection. If the cables are to be buried in the sea bed to protect them from fishing activities, it is essential to include a sub-bottom investigation using seismic methods and core sampling at enough locations to enable the seismic recordings to be interpreted correctly.

A close-up visual survey may also be required to identify the minor sea-bed features which would be indistinguishable by other means but would influence the design of the trench cutting and cable burying machinery. The physical and mechanical properties of the sea-bed material can become the most significant features in the determination of the best route for a buried cable system.

If the cable is to be laid on the sea bed, the survey will be planned to obtain data principally concerned with the water currents and the surface of the sea bed in the area. In particular, it is essential to obtain accurate bathymetric data and to identify areas of exposed rock, wrecks and other hazards so that either the cable route can be planned to avoid the hazards or alternatively the sections of cable requiring additional armour to resist these conditions can be specified.

A substantial amount of the basic data required can be obtained from official sources, and the Hydrographic Department of the country concerned can usually provide assistance with more detailed information concerning the area than that shown on published charts. Available information may include adequate figures concerning tidal conditions, weather statistics and water temperature so that only a limited number of current and tidal measurements are needed to verify that the data apply to the precise location of the selected route. It is normal for power cables to link island to island or islands to a mainland by as short a route as is practical and consequently the crossing of a narrow channel can also involve working in an area of relatively high tidal currents, which could cause movement of the cable with consequential damage, and so it is essential to know the maximum velocity close to the sea bed.

When the prospective routes have been selected it is necessary to prepare profiles, and if possible it is advantageous to make a recording of an echo sounding run along each of the specific routes to confirm the correct interpretation of the survey data obtained. A survey of the landing points, to determine the conditions and distance between the underwater section of the route and the actual cable terminal on land, completes the overall picture required.

From the full survey data it is possible to determine the exact length of each route and other features which enable an accurate estimate of the cable requirements to be made, which will include allowances for contouring and navigational deviations etc. The method of cable installation can also be determined together with the requirements for the cable laying vessel and other equipment needed to carry out the installation. It is preferable to aim at the installation of single continuous lengths of cable to reach from shore to shore without joints, but it may be necessary to allow the inclusion of flexible reconstitutions in long route lengths.

CABLE LAYING VESSEL AND CABLE HANDLING EQUIPMENT

As submarine cable laying projects vary from small river crossings to major international system interconnections, the cable laying vessel has to be selected to suit the particular circumstance of each individual contract. In general the cable laying and repair vessels designed for handling telecommunication cables are only suited to the installation of power cables on comparatively rare occasions. This is because their cable handling machinery is normally designed with smaller bending radii than are required for high voltage cables and their high operating costs make them uneconomic for low voltage installations, which generally tend to be of short route length. The vessel is selected such that it is as small as is compatible with all the requirements for the installation work. The principal features to be considered are:

(a) suitable hold dimensions for the storage of cable coils, drums or a turntable;
(b) suitability of the deck layout for fitting cable handling equipment;
(c) overall dimensions including minimum operating draught;
(d) adequate manoeuvrability;
(e) navigational and communications equipment and facilities for fitting additional items as necessary;
(f) power supplies available for additional equipment;
(g) accommodation and messing facilities and space for fitting temporary additions if necessary;
(h) age and general condition of the vessel such that specially constructed items may be of use for future operations;
(i) charter terms and conditions.

The manoeuvrability of the vessel at relatively low speeds is of considerable importance because this may determine the accuracy with which the cable can be laid on the selected route. A vessel fitted with a bow thruster is an asset when leaving or entering moorings but a full dynamic positioning (DP) system enables the vessel to be controlled at very low speeds and, if necessary, to remain stationary without the use of moorings. The use of DP to avoid the use of mooring anchors is virtually essential for vessels employed in installing cables between offshore platforms. This type of vessel is not suitable for work in shallow water as a minimum operating depth of 10–15 metres is required.

Cable can be laid over the stern, over the side, or over the bow of the vessel and the selection of the direction of pay-out depends upon the proposed method of handling the shore ends of the cable as well as upon the navigational problems of the main crossing. Although opinions differ about the best arrangement, cable repair work is almost invariably carried out over the bow of the vessel. Naturally, the deck layout of the vessel selected for an operation needs to be compatible with the planned scope of the work, which may include the capability for repairing any accidental damage which may occur during installation. Figure 43.1 shows the arrangement of the MV *Photinia* as adapted for the installation of the Cook Strait cables in 1964.

The cable handling equipment includes all the items necessary for the transfer of cable from its stowage in the vessel to its planned position on the sea bed. The main items included in this are the cable storage facility, tension control equipment, tension measuring instrumentation and cable deployment system.

Single wire armoured cables can be coiled down into the hold or on to the deck of the vessel and very long lengths of cable can be stowed in this way with the minimum amount of equipment. Although coiling may require a relatively large labour force to stow the cable, this can be minimised by the use of a suitable rotating mechanical arm which guides the cable into position. Coiled cable may be laid at any speed up to about 8 knots although it is usual to lay at an average of not more than about 3 knots.

Cables with double wire armouring, where the two layers of wire have the same direction of lay, may be coiled in the same way as single wire armoured cables. If the two layers of wires have the opposite hand lay then the cable cannot be coiled and it is necessary to use either a turntable or a drum on to which the cable can be wound. The former is usually preferred for handling long lengths. The use of either arrangement, however, involves the requirement for a system to control the rotation of the mass of the stored cable plus that of the turntable or drum which must be braked or accelerated to match the laying speed. Nevertheless, for short cable lengths the use of a drum, on which cable has been supplied from the factory, reduces the transport and loading costs and only requires the provision of secure drum stands and a braking system to enable cable to be paid out under control at modest speeds.

The cable tension must be changed from the low value at which it leaves the storage system to the laying tension which supports the cable suspended between the vessel and the sea bed. This must be done by a braking system which subjects the insulation of the cable to negligible damage from creasing, crushing or other disturbance. To achieve this, the radius of bends round which the cable passes must be as large as practical, consistent with the size and weight of the cable and the plant which can be accommodated on the vessel.

Cable engines, which may be of the capstan, caterpillar or multiple-wheel linear type, grip the cable and enable the energy to be dissipated in a mechanical, hydraulic or

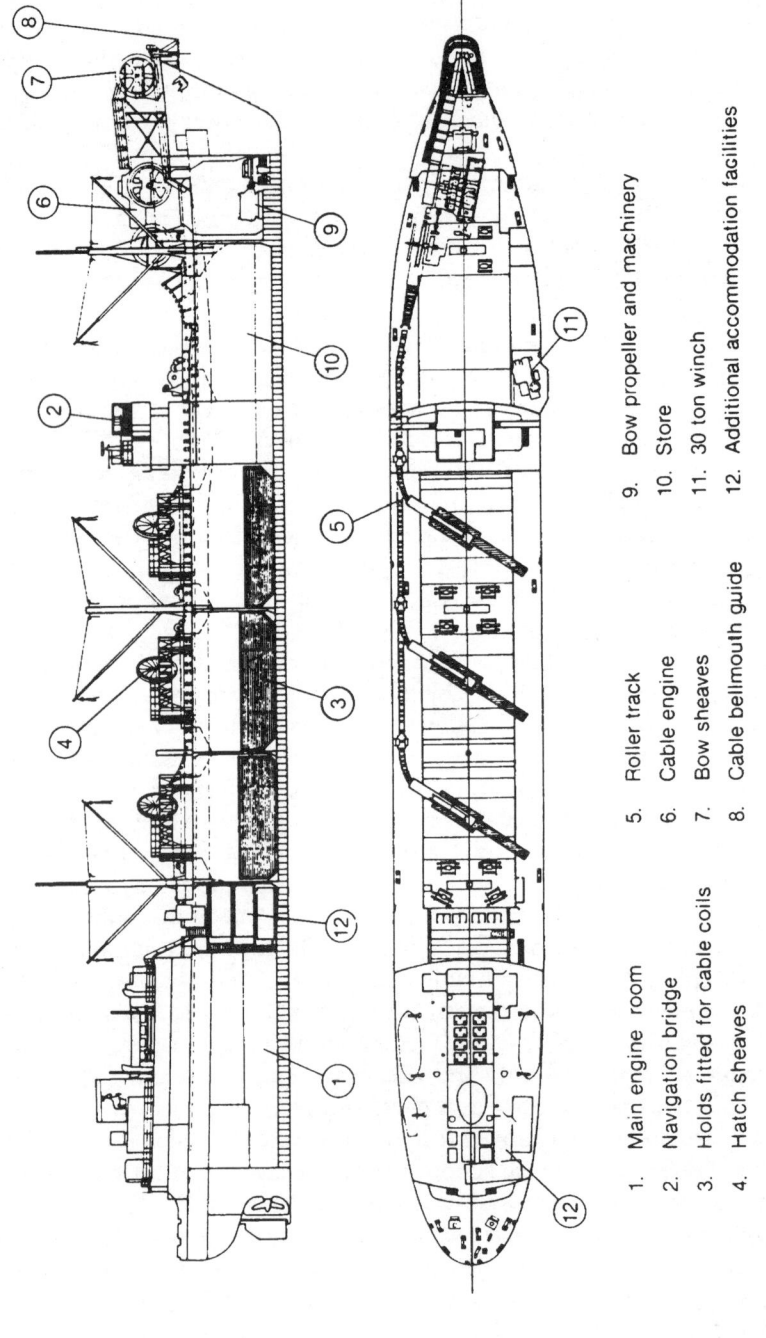

1. Main engine room
2. Navigation bridge
3. Holds fitted for cable coils
4. Hatch sheaves
5. Roller track
6. Cable engine
7. Bow sheaves
8. Cable bellmouth guide
9. Bow propeller and machinery
10. Store
11. 30 ton winch
12. Additional accommodation facilities

Fig. 43.1 MV *Photinia* equipped for laying the Cook Strait cables

electrical braking system. They enable cable to be paid out with much more positive control over a wide tension and speed range and may provide the means of recovering cable for repair work.

Although slack cable coming from storage to the tension control equipment may be passed round bends formed from groups of relatively small diameter rollers to reduce friction, cable under tension must bend round the smooth curved surface of a wheel or skid. All guides on the cable engine and along the cable run to the bow or stern gear should meet the minimum bending radius of the cable.

The degree of sophistication required from the cable tension measuring instrumentation depends upon a number of factors. For slow cable laying in water less than 80 m deep and with tidal currents of, say, less than 1 knot, it may be adequate to observe the angle at which the cable enters the water in order to estimate the tension and no direct tension measurement is essential. In deeper water and in conditions where the relative velocity of the water to the cable is significant, the laying angle becomes inadequate to determine whether cable is being laid slack or with the proper residual tension, and so it is essential to use an instrument to measure the tension at the cable engine, at the bow or stern gear, or at an intermediate point along the cable run. If cable is laid over a skid, it is necessary to know the angle of wrap round the curved surface and the coefficient of friction of the cable servings on this surface in order to calculate the required inboard cable tension.

The cable tension is sensed by either hydraulic or electrical load cells which need to be robust and resistant to salt water, vibration and shock loads. Similarly, the associated amplifiers, recorders and indicators need to be suitable for use in a marine environment and robust reliability is more appropriate than sophistication and high sensitivity.

Although the indication of tension is of prime importance, the measurement of cable speed and of the length paid out are also valuable aids to the control and provision of records of the operation.

The final equipment over which the cable passes on its way out of the vessel is the bow or stern gear which may be in the form of a large sheave wheel or a skid. Although the latter is cheaper, lighter and easier to fit to a vessel it has the inherent disadvantage of causing a change in the cable tension by a factor of $\exp(\mu\theta)$, which makes tension control less sensitive during laying and increases the tension required to recover cable during repair operations. One of the difficulties is in compensating for variations in the angle of wrap (θ) and the other is in achieving a stable known value for the coefficient of friction (μ).

The basic navigational instrumentation for cable laying is that required for measuring the water depth and that for plotting the position of the vessel along the route. Additional features which may usefully be provided are measurements of the speed of the vessel through the water and of speed over the ground, which can be used to assist and verify the correctness of the control being applied during the cable laying operation.

The use of modern position fixing systems, together with a track plotter or a visual display unit, make navigation along the main part of the route a simple matter and enable records of the operation to be obtained with minimal effort. If the cable tension, speed and length instruments are suitably grouped, it is possible to record these on the same time base so that they can be readily related to the navigational record and the trace from the echo sounder which shows the profile of the sea bed along the route actually traversed by the vessel while laying the cable.

PRELIMINARY SITE WORK

Bow and stern moorings are usually required close to the shore at each end of the route to hold the vessel securely whilst the ends of the cable are pulled ashore. Their distance offshore should be the minimum compatible with the safety of the vessel and their spacing must be adequate to enable all the cables of the system to be handled without the need to reposition anchors between laying operations. As the handling of the shore ends is the most hazardous part of a cable laying operation, it is desirable to minimise the length of cable involved and the time taken in effecting this at each end of the route.

To prevent cable from being damaged by surf action or by impact from floating logs, small boats etc., it is normal to bury or protect the cable from some distance below the low water mark across the beach to the terminal position, which may be at a joint pit or at the terminal of an overhead line. The trench should be prepared before cable laying is commenced so that the cable can be protected as soon as possible after it has been laid.

If the cable is landed over a rocky area where it is difficult to cut a trench of adequate depth, protection may be provided by fitting cable protectors of polyurethane or cast iron around the cable. Alternatively mattresses formed from shaped concrete blocks may be laid over the cable. Further protection in sandy areas can be achieved by applying artificial 'sea weed' made from polyurethane fibres. These cause suspended sand grains to be deposited forming a sand wave proportional to the length of the strands applied.

On arrival of the cable vessel on site, the normal procedure is for the ship to enter moorings at the starting end and to practise leaving moorings, navigation along the cable route at cable laying speed and entering moorings at the finishing end. During this practice useful records of the sea-bed profile can be obtained and the Master of the vessel can become familiar with the special requirements for handling his vessel during cabling operations.

CABLE LAYING

The cable laying operation is not commenced until an adequate period of favourable weather is predicted and the sea conditions are seen to be suitable for the laying of the complete length of cable in one uninterrupted operation. A discontinuity in cable laying can readily result in damage to the insulation of a high voltage cable and result in the need to insert one or more repair joints to remedy the defect. On rare occasions the weather conditions may change unexpectedly and prevent the vessel from entering moorings to land the finishing end of the cable. In this event the balance of the length of cable is laid off on an escape route near the shore in moderately shallow water where it can remain safely on the sea bed until conditions are favourable for its recovery and re-laying on the planned route. Only in the last resort would a power cable be cut to enable the laying vessel to escape from some disastrous situation because this makes the expense of a repair or of the replacement of a length of cable inevitable.

Landing the starting end of the cable by pulling it ashore on floats is a simple and fairly short operation but landing the finishing end is more complex and also more vulnerable if the weather deteriorates. It is therefore usual to arrange the direction of cable laying so that the finishing end is at the more sheltered and secure location.

Nevertheless, there may be other overriding considerations, such as the requirement for landing a very long length of cable to reach the terminal, which then makes that site the better location for starting.

With the ship in the moorings at the starting end of the route, the required length of cable is pulled ashore. The end of the cable is secured and the floats are removed from the cable progressively from the shore towards the ship. This operation is timed to take place when the floating cable is as directly over the route as possible with minimal deviation caused by tidal currents or wind. If the cable is floated on twin inflatable floats which are linked together into a continuous string by tack lines between them and all the filler plugs on one side are secured to a rip cord, it is possible to deflate the floats in sequence on that side so that the inflated floats pull the deflated ones clear as the cable sinks to the sea bed. By this means it is possible to release floats very quickly, even though some will be pulled beneath the surface in the deeper water near the ship.

During this operation it is essential to maintain adequate tension control so that the cable is held straight without an excessive pull being exerted as the last floats are released. If the released floats are connected together by tack lines, they are readily recovered for subsequent operations.

As the vessel commences cable laying on the main part of the crossing, it is normal for control of the cable laying operation to be initially by the visual judgement of the Cable Officer while the vessel's moorings are slipped and the ship manoeuvres onto her course along the cable route. Once the vessel is clear of obstructions, the responsibility for handling the cable is handed over to the cable laying control centre and the responsibility for the navigation of the vessel along the desired route and for the maintenance of a steady speed reverts to the bridge.

In the cable laying control centre, the correct cable tension appropriate to the conditions is determined from observation of the ship's position, ship's speed, the water depth and the distance traversed. Instructions are given directly to the cable engine driver or brake operator to adjust the cable tension accordingly. Laying control is simplified if tables relating the ship's position and the required tension are prepared in advance for a range of speeds adequate to cover all likely contingencies.

In shallow water and at relatively low speeds, it is possible to effect laying control adequately by observation of the angle at which cable leaves the vessel and this is frequently the method used when cable is paid out over a stern skid.

The objectives of laying control are twofold. The first is to ensure that no slack cable is laid because this could form a loop on the sea bed which might subsequently be pulled into a kink and damage the insulation. The second is to ensure that the residual tension in the cable on the sea bed is as low as practical to permit it to follow the contours without bridging hollows and this will also ensure that the tension applied to the cable as it leaves the ship is minimised. The value of residual tension employed in the calculation of the laying control figures will vary with the weight per unit length of the cable and also with the sensitivity of the tension measuring equipment. Its value is usually in the range of 2.5 kN for a light cable and 10 kN for a heavy cable in deep water.

The speed at which power cables are laid is usually set by the lowest speed at which the cable laying vessel can safely maintain steady progress accurately along the desired route, which in turn is related to the maximum tidal currents predicted for the duration of the cable laying operation. A cable laying speed of between 2 and 3 knots is usually aimed at, with a maximum of about 5 knots. If it is necessary to maintain the speed at less than 2 knots, as for example when laying from a turntable, it is generally considered essential to use a vessel with a mooring system or with a DP system or to have tugs available to assist in controlling the movement of the vessel. The DP system with its centralised controls is much the more effective arrangement.

If the cable is being buried as it is laid and the laying speed has to be reduced to below 1 knot, a DP system or a system of moorings moved as required by one or more anchor handling vessels is necessary to control the vessel.

Having traversed the main part of the cable route, the cable laying vessel arrives at moorings which are designed to hold her securely while the finishing end of the cable is landed.

The method used for landing the cable end depends upon the type of cable and to some extent upon the length of cable to be handled. Cables operating at 33 kV or below are frequently 'turned over' by taking a bight of cable over to an adjacent coiling area and transferring the balance of the length into a coil or figure eight formation in which the end becomes available for pulling ashore on floats. The operation is completed by lowering the final bight of cable to the sea bed while the shore winch takes away the slack and finally the floats are removed between the ship and the shore.

Other methods which may be used for handling long finishing shore ends employ either a barge on to which cable is transferred[1] or a floating head which enables a bight of cable to be pulled out and laid without necessitating the carrying out of a coiling operation. Figure 43.2 illustrates diagrammatically the floating head, which, being amphibious, enables the cable to be landed on the beach in one continuous operation. This was the method employed in handling the Cook Strait 250 kV d.c. cables.[2]

As it is essential to ensure that the length of cable supplied is certain to reach the terminal position, the allowances for contouring on the sea bed and for navigational deviations, plus a small safety margin, tend to be generous and are seldom totally utilised. During the shore end handling operation it is preferable to cut the cable on the ship to the correct length with only a small safety margin and seal the ends with temporary end caps. The surplus cable can then remain in the vessel until it is convenient to offload it for storage as maintenance cable for possible future repairs.

With the cable end landed, the cable is moved into its final position and secured by a ground anchor before jointing to a section of land cable or terminating on a suitable structure. On completion of the jointing work the cable is finally subjected to the

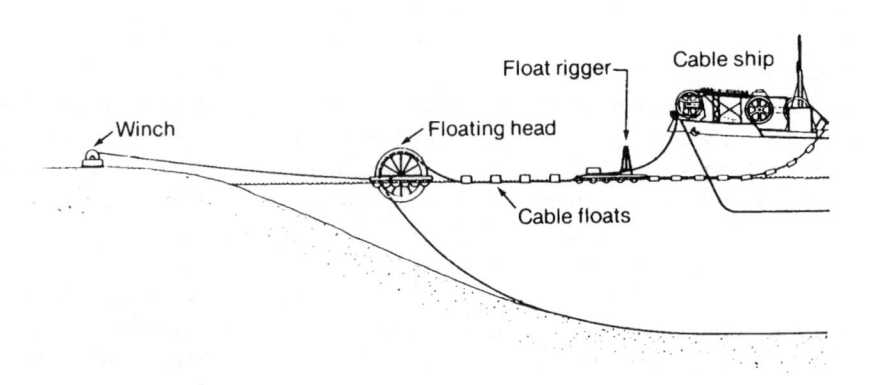

Fig. 43.2 Floating head landing a cable end

appropriate commissioning tests. The installation is completed by fitting cast iron cable protectors or other means of securing the cable against damage in the surf zone if it is not adequately protected by trenching.

MAINTENANCE AND REPAIR

The most frequent cause of failure in the submarine sections of power cable systems is mechanical damage by ships' anchors or by fishing gear. Other faults may arise from abrasion, sometimes coupled with fatigue of the lead sheath, which is the result of movement of the cable on the sea bed caused by tidal currents. More rarely, faults may be associated with joints in the cable because these tend to involve a discontinuity in the mechanical or electrical properties of the cable.

In the event of an electrical fault, a series of tests is carried out which may include measurements of resistance and capacitance and testing by pulse-echo methods. The position of the fault is identified by reference to the installation records of the actual route and length installed in conjunction with technical data of the cable characteristics. However, if the cable has been hooked by an anchor it may have been dragged a considerable distance away from its original route. In this case it may be necessary to find the cable either by using a search coil or towed electrodes to detect a signal injected into the cable or by using side-scan sonar or divers to search for it. If the fault is not electrical and is only leakage of gas or fluid from the cable, the position of the leak is located by pneumatic flow, hydrostatic pressure or hydraulic measurements with reference to the route profile. Leaks may be difficult to locate accurately although, on occasions, gas leaks have been found by the use of an echo sounder and should be identifiable by side-scan sonar.

Having found the fault it can be marked by a float attached to a sinker and a pattern of moorings is then laid round the fault position to secure the vessel during subsequent jointing operations. Access to the faulty section of cable is obtained either by using a grapnel to hook the cable or by picking up cable from one end and recovering it until the fault is reached. In the former case if the cable was not severed at the time of the fault, it may be necessary to cut it by a mechanical cutter to enable it to be lifted to the surface without subjecting it to excessive tension. The damaged section can then be replaced with a new length by using two joints to connect it to the original cable. If the cable is picked up from one end, the damaged section of cable is cut out and it is only necessary to lay a single joint in the fault area, with new cable being jointed in near the shore in order to complete the length required to reach the terminal position.

The method adopted depends upon the location of the fault, the depth of water, tidal currents, weather conditions, equipment and vessels available and also upon the age of the cable. Although recovery of the cable from one end out to the fault area may appear simpler, it nevertheless may result in damage to a considerable length of cable if the cable is old or if weather and tidal conditions make it difficult to control the repair vessel during a long slow recovery operation. However, the two joint method not only involves finding and lifting the cable out at sea, which generally involves cutting it well below the surface, but also involves lowering the final joint and a bight of cable to the sea bed without the formation of kinks. The latter method does have the advantage of minimising the disturbance caused to the original cable, although it also involves staying in position at sea for a longer period.

Fig. 43.3 Ploughing trials on the foreshore for simultaneous embedding of 132 kV three-core FF cable and an optical fibre cable

Whichever method is adopted, it is essential to have a mooring system which enables the vessel to be held with minimal variations of position during the jointing operations and also provides a method of controlling movement of the vessel as the joint or joints are laid. Since the expansion of the offshore oil industry, the development of deep water moorings, DP vessels and the availability of anchor handling vessels make it possible to effect cable repairs which were impractical at the time of the 250 kV d.c. Cook Strait installation in 1964–5.

PRESENT DEVELOPMENTS

Because of the increasing interest in large submarine power cable interconnections, and for their greater security against mechanical damage, more emphasis is being placed on their burial in the sea bed. In relatively soft or non-cohesive material it is possible to use either a plough (see fig. 43.3) or a jetting device to create a trench into which the cable can be introduced in a simultaneous operation, or subsequently after the cable laying operation.

In rock areas, however, it is necessary to use either a mechanical trench cutting machine or explosives. Both these methods are slow, and although it may be possible to cut a trench after cable laying or simultaneously with that activity, it is preferable to complete the trench cutting operation before cable laying is commenced.

REFERENCES

(1) Ingledow, T., Fairfield, R. M., Davey, E. L., Brazier, K. S. and Gibson, J. N. (1957) 'British Columbia–Vancouver Island 138 kV submarine power cable'. *Proc. IEE, A* **104** (18).
(2) Williams, A. L., Davey, E. L. and Gibson, J. N. (1966) 'The 250 d.c. submarine power cable interconnection between the North and South Islands of New Zealand'. *Proc. IEE* **113** (1).

PART 6
High Temperature Superconductivity

CHAPTER 44
Introduction to Superconductivity

BACKGROUND

Many modern industrial countries continue to face increasing demands for power and any major advance in the generation and transmission of energy could yield huge cost savings in the future. Superconducting cables offer the potential to transmit currents of several kiloamps at voltages low enough to eliminate transformation stages. Since the 1960s, many countries have studied and built superconducting cable demonstrators. The motivation behind the development work are the predicted advantages of a superconducting transmission cable over a conventional copper power cable.

Copper cables suffer from resistive losses arising from scattered electrons as they are transmitted through the material (approximate 0.2% for each 10 km). This loss factor effectively determines the length and dimensions of a conventional cable. Typically ohmic losses in the transmission and distribution of energy at present are about 5% of the total. For example, in the European Union this amounts to about 2.5 billion ecu (£1.75 billion) or 70 trillion watt hours; much of this could be saved with the widespread use of superconductivity in electrical engineering. Superconducting cables were quickly identified as having a potential impact in the underground high power cables market sector. Here, the high installation cost of underground force-cooled cables limits their use to densely populated cities. These cables are technically complex and require elaborate cooling often using fluid or gas down a central duct to minimise losses and thermal expansion forces arising from heat generated by ohmic resistance.

For economic reasons, it is unlikely that superconducting cables will be able to compete with copper cables at power levels below 300 MVA, but at power levels of 1 GVA or more, where the heat generated in a conventional cable may be as high as 2×10^5 W km^{-1}, a superconducting cable which must be force-cooled can provide an alternative. In addition, for higher power conventional cables operating at voltages of 132 kV and above, superconducting cables can potentially deliver the equivalent power level at much lower voltages. Several nations embarked on research programmes into superconducting cables from the mid-1960s, but due to the high cost of refrigeration these cables proved to be uneconomical compared with conventional technology, except at power levels much higher than needed (in the range 3–5 GVA).[1,2] The arrival of a new generation of superconductors that could operate at much higher cryogenic temperatures has caused a renewal of work into superconducting cables in the hope that the reduced refrigeration costs may alter the economics of utilising superconducting cables.

Superconducting cables also have potential environmental advantages:

(a) in having no soil contamination from fluid leaks;
(b) thermally insulated superconducting cables will not affect surface vegetation;
(c) for the same power, a superconducting cable is smaller than a conventional cable thereby occupying less space;
(d) external magnetic fields can in principle be eliminated in a superconducting cable;
(e) there is a low fire risk at cryogenic temperatures.

This chapter aims to provide an overall background into the history of superconducting cables using liquid helium technology in the period 1966–1985 and in the recent rapid development of the next generation of superconducting cables using liquid nitrogen technology from 1990 onwards.

THE SUPERCONDUCTING STATE

A superconductor is a material that undergoes a catastrophic spontaneous change in its physical properties when cryogenically-cooled below a critical temperature, T_c. The most widely known feature of a superconductor is that it has zero electrical resistance to the passage of a direct current. This is an obvious attraction in electrical cables where extensive engineering solutions are often required to minimise ohmic heat-generating losses.

The superconducting state is not permanent and increasing the temperature above T_c returns the material to a resistive state. In addition to temperature, two other physical parameters determine the superconducting state and need to be defined to appreciate the properties of superconductors over normal metals. If a superconductor is kept below T_c and the applied current in the sample is gradually increased, a critical value is found above which the material returns to the resistive state despite being cold. This property is important in engineering applications and is usually expressed as the critical current density, J_c. If the superconductor is kept below the critical temperature and below the critical current value for the material and an external magnetic field is applied, then a critical magnetic field is observed, H_c, above which the material is resistive. In an operational situation one would not wish to operate the superconductor at or near one of these phase boundaries and it is found that both J_c and H_c are related to T_c: both increase significantly as the sample is cooled further and further below T_c. For this reason superconductors tend to be used between 50 and 75% of their T_c value. Figure 44.1 is an illustration of the relationship between T_c, J_c and H_c for the superconductor Nb_3Sn. Inside the shaded area in fig. 44.1, Nb_3Sn is in the superconducting state. For any temperature, current density and field outside the triangular area, Nb_3Sn behaves as a normal metal.

Practical superconductors operate in a so-called 'mixed state' between an upper and lower critical field. Below the lower critical field H_{c1} the material behaves as a perfect diamagnet by expelling all the magnetic flux (the Meissner effect) and above the upper critical field H_{c2} magnetic flux lines totally penetrate the material, killing the

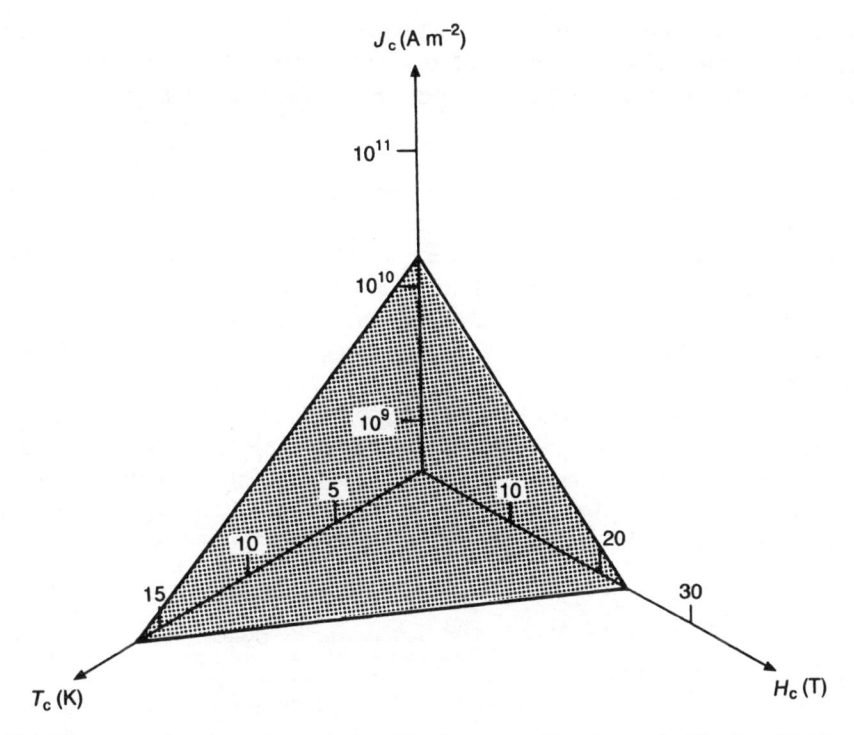

Fig. 44.1 Diagram showing the relationship between T_c, J_c and H_c for Nb$_3$Sn. The superconducting state is contained within the shaded triangular area

superconductivity. In between these values magnetic flux is pinned at specific sites in the material, either crystallographic or mechanical, allowing the supercurrent to flow around the pinning sites, giving superconductors their unique ability to transport high electrical currents whilst generating high magnetic fields. These type 2 superconductors are either metallic alloys with low T_c values (LTS) known since the 1940s and need to be cooled with liquid helium or they are complex copper oxide ceramics (HTS) first discovered in 1987 with T_c values above the boiling point of liquid nitrogen (−196°C, 77 K).

LOW TEMPERATURE SUPERCONDUCTORS

Superconductivity was first discovered at the University of Leiden in Holland by one of the pioneers of low temperature research, Heike Kamerlingh Onnes. He and his co-workers discovered that metallic mercury had zero resistance at 4.2 K. This discovery was completely unexpected and was only made possible by the fact that just 3 years previously Kamerlingh Onnes invented a method of liquefying gaseous helium. In the years that followed, researchers found that several other metals, such as lead, also exhibited superconductivity but research and applications in the phenomenon were dependent on the use of the expensive and unreliable helium liquefaction technology. It was the latter that was the spur for researchers to try and discover new materials with higher T_c values. With no theoretical guide as to the causes of superconductivity, there was no model that could be used to predict where new compounds could be found or what properties they would have even if they could be made. Inevitably research tended to be concentrated on metals (but not exclusively) and a large number of superconductors were discovered, but all had T_c values near to absolute zero and never circumvented the dependence on liquid helium cryogenics. These materials for the purpose of this chapter that are dependent on liquid helium cooling are called low temperature superconductors (LTSs) and can take many forms.

A survey of the literature shows that there are no hard and fast rules as to what type of materials can be LTSs. Many forms are known from ceramics (e.g. InO_x), covalent elements (e.g. P), inorganic compounds (e.g. NbN, La_3Se_4, InTe and PdH-Ag), organic complexes (e.g. $(TMTSF)_2ClO_4$) and even covalent polymers (e.g. $(SN)_x$). These compounds are very interesting in themselves but have no technological interest. However, new metallic alloys yielded compounds that have been (and still are) used in power electrical engineering. An added advantage of these materials, being metals, is that they are amenable to conventional wire manufacturing methods and are widely available in wire, tape and foil forms. These materials are often referred to in terms of their crystal structure. The simplest materials are the B1 materials which have a simple

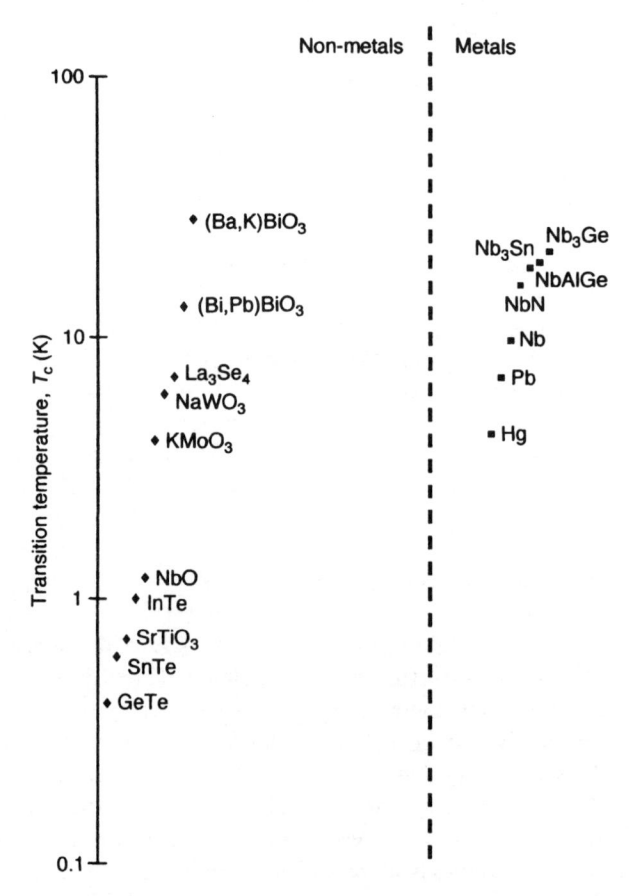

Fig. 44.2 Illustration of the range of known LTS materials. T_c values are indicated on the left-hand axis

rock-salt (or NaCl) structure. The commonest contain niobium with another element, e.g. NbN (with the highest B1 material LTS T_c of 18 K). A further set of alloys have the A15 (or β-tungsten) crystal structure, and it is these materials that have the highest T_c in LTS compounds. A15 alloys have the general formula of A_3B where the A atom occupies a body centred cubic lattice with the B atom arranged as chains on the cubic faces. Examples of A15 materials are V_3Si, Nb_3Sn, V_3Ga and Nb_3Ge (the highest LTS T_c at 23 K). Figure 44.2 shows a sample of the range of LTS materials that are known to exist, along with their corresponding T_c values shown on the left-hand side of the diagram. It is especially interesting to note the much narrower T_c range for the metallic elements compared with that displayed by non-metallic (co-covalent) materials.

The LTS conductor which is most used in cables is the solid solution NbTi. It can be formed from simply melting the two elements in the appropriate proportions into extrusion billets. It is very ductile and can easily be formed into wires and tapes. Nb_3Sn has also been used in conductors but is very brittle and it is difficult to handle without damaging it. It is mostly available as a multifilamentary form embedded within a copper matrix, formed *in situ* by the so-called bronze process.

LOW TEMPERATURE SUPERCONDUCTING CABLES

With the arrival of the A15 alloys, large-scale electrical engineering demonstrator devices utilising superconductors at high current loads began to appear. The world's first model demonstrator to prove that a large a.c. current could be transmitted at liquid helium temperatures was constructed at BICC's Central Research Laboratories,

Fig. 44.3 BICC Cables 138 kV experimental LTS cable system

London in 1968. This is shown in fig. 44.3. It was an experimental 138 kV cable designed to operate at 1000 MVA. It consisted of a coaxial pair of niobium foil conductors, approximately 3 m in length, and approximately 100 μm thick. At each end of the link was a superconducting transformer, housed in a cylindrical container. The cable was designed such that the magnetic field generated by the inner superconductor layer induced an equivalent circulating current in the opposite direction in the outer superconducting layer. This confined the generated magnetic field in the cable core and eliminated any stray eddy currents in the cable system. The conductors were placed in an inner pipe which acted as a cryogenic envelope. The inner pipe was contained within an outer pipe by spoke supports and the air space between the two evacuated to provide thermal insulation. Additional insulation was provided by wrapping the inner pipe in superinsulation (multilayer aluminium-coated plastic sheeting). Liquid helium was circulated around the two conductors and the assembly could be cooled to below 10 K and passed an a.c. current of 2080 A r.m.s. at less than 1 V.

This model formed the basis for further work and the Central Electricity Generating Board in the UK built a more advanced 3-phase version[3,4,5] shown in fig. 44.4. Due to the specific requirements of liquid helium, special engineering design features were employed such as extra superinsulation, the use of liquid nitrogen as a heat shield, extra

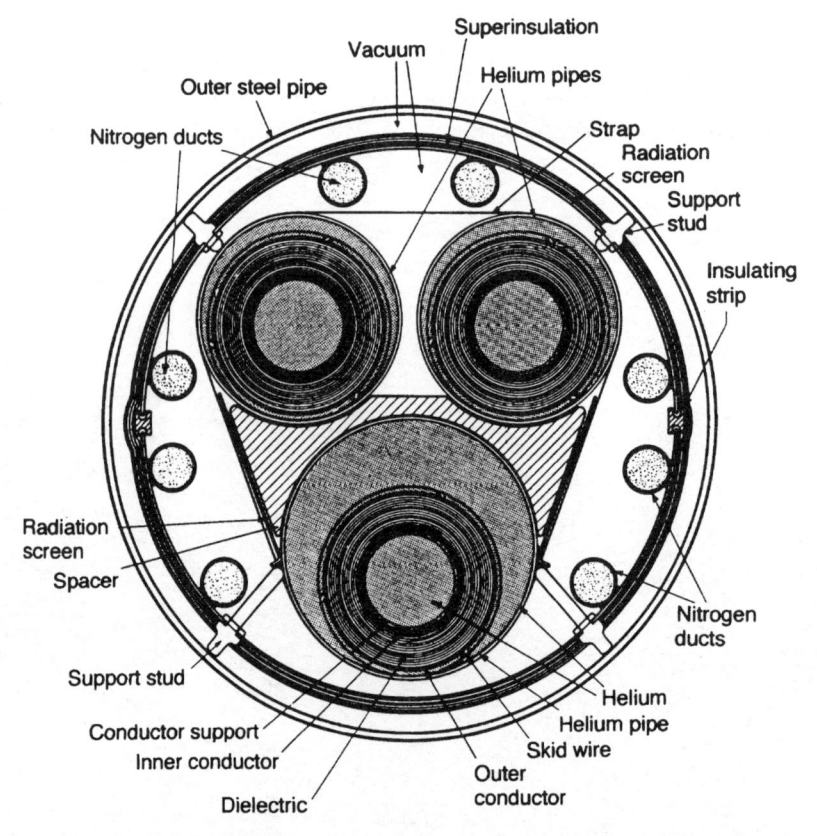

Fig. 44.4 A trefoil of the Central Electricity Generating Board 275 kV, 4 GVA LTS cable. Each of the three phases is flexible and contained inside a rigid outer pipe. (Reproduced with the permission of The National Grid Company plc.)

layers of radiation screening, and vacuum shields. This cable also advanced conductor technology by proposing co-wound niobium superconductor and copper stabiliser helical tapes.

Even further advanced models were constructed outside Britain, the best known was at Brookhaven National Laboratory, Long Island in the USA where a single-phase test rig equivalent to one phase of a 3-phase 1000 MVA cable using a NbTi conductor underwent successful field trials.[6] In Austria, a group under P. Klaudy successfully tested a 50 m working model LTS cable which formed part of a normal power transmission cable attached to the national grid system proving the compatibility of the technology.[7] Trials were also conducted in other countries, e.g. in Germany.[2]

Despite considerable technical success, the complexity of the thermal insulation required to preserve the liquid helium combined with the difficulties in handling pressurised liquid helium almost single-handedly prevented LTS cables competing on economic grounds with underground copper cables. Today, the largest use of LTS cables is in supplying large superconducting magnets such as the one at CERN in Switzerland. They are normally made from NbTi and are of the cable-in-conduit type.

REFERENCES

(1) Edwards, D. R. (1988) *Proc. IEE* **135**, 9.
(2) Bogner, G. (1994) *Superconducting Machines and Device* (eds S. Fonar and B. B. Schwartz), New York: Plenum, p. 401.
(3) Bayliss, J. A. (1973) *Electrical Times (London)*. 24 May.
(4) Carter, C. N. (1973) *Cryogenics* **13**, 207.
(5) Rogers, E. C., Slaughter, R. J. and Swift, D. A. (1971) *Proc. IEE* **118**(10), 1493.
(6) Forsyth, E. B. (1984) *Electricity and Power* **30**, 383.
(7) Klaudy, P. A. and Gerhold, J. (1983) *IEEE Trans. Mag.* **MAG-19**(3), 656.

CHAPTER 45
High Temperature Superconductors

INTRODUCTION

In addition to occurring in metals and metallic alloys, superconductivity was known to exist in covalent compounds such as in mixed oxides, e.g. $Ba_{0.25}Pb_{0.75}BiO_3$ or $SrTiO_3$ and in sulphides, e.g. $PbMo_6S_8$, but all were LTS materials. This fact, along with the difficulty in processing ceramics, meant that these materials were unsuitable for cable applications. In 1986, Alex Müller and Georg Bednorz, two researchers at an IBM Laboratory in Switzerland, found that a mixed oxide $(La_{1-x}Sr_x)CuO_4$ showed an apparent zero resistivity value at the unprecedented high temperature of 35 K.[1] Very soon afterwards, Paul Chu and co-workers found by elemental substitution that $YBa_2Cu_3O_{7-\delta}$ had a T_c of 92 K (−181°C).[2] This was above the technologically significant temperature milestone of 77 K (−196°C), the boiling point of nitrogen, and the discovery of $YBa_2Cu_3O_{7-\delta}$ started a massive worldwide effort to search for new superconducting materials and understanding how to process them into useful artefacts. Materials that are dependent on liquid nitrogen cooling analogous to the helium cooling for LTS are called high temperature superconductors (HTS).

It was quickly established that all the new materials with T_c values above 77 K were complex layered oxides containing layers of square pyramidal copper oxide units (CuO_5) structurally related to the CuO_6 octahedra found in perovskite-type materials, and that these generated and transmitted the supercurrent. Other elements in the crystal provide the charge carriers needed for superconductivity either by the phenomenon known as metavalence or from crystal defects. In $YBa_2Cu_3O_{7-\delta}$, for example, the charge carriers are generated by the deviation of oxygen from the 'ideal' stoichiometry of 7 by a small amount, δ. The consequences of this are that the charge carriers in high temperature superconductors are not electrons (e^-) as in metals but 'holes' which can be envisaged as positive electrons (e^+) and the supercurrent does not flow equally along all the crystallographic axes but is essentially localised in the a–b planes.

A number of HTS oxides were discovered in rapid succession and at the time of writing, the maximum T_c is 133 K in the compound $Hg_{1-x}Pb_xBa_2Ca_2Cu_3O_{8-\delta}$. Figure 45.1 shows the chronological rise in T_c since the discovery of superconductivity.

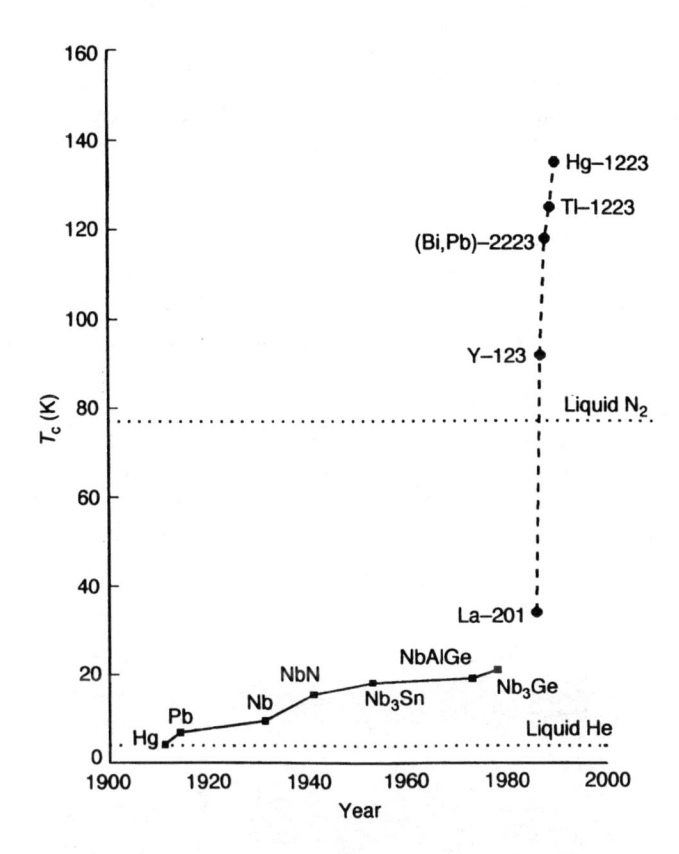

Fig. 45.1 Chronological progress of T_c. Note the enormous discontinuity at 1987 when HTS materials were discovered

Because of the many chemical elements contained in them, HTS materials are usually referred to by a shorthand, e.g. yttrium barium copper oxide ($YBa_2Cu_3O_{7-\delta}$) is called Y-123, the numbers correlating to the cation stoichiometry (with oxygen being ignored in this case). Even more complicated superconducting chemical structures can be built up by exchanging and adding more elements around the basic CuO_5 unit. These complicated formula HTS materials can be classified according to crystal structure as for LTS materials. In addition to the 123 structure exhibited by $YBa_2Cu_3O_{7-\delta}$, there are the 1212 (e.g. $TlSr_2CaCu_2O_7$), 2212 (e.g. $Bi_2Sr_2CaCu_2O_{8-\delta}$ and $Tl_2Ba_2CaCu_2O_{8-\delta}$), 1223 (e.g. $TlBa_2Ca_2Cu_3O_{8-\delta}$) and 2223 (e.g. $(Bi_{2-x}Pb_x)Sr_2Ca_2Cu_3O_{10-\delta}$) structures. Other stoichiometries are also known but are not of technological importance to cables. The bismuth-, thallium- and mercury-containing HTS compounds all have highly anisotropic layered crystal structures which form macroscopic plate-like crystals and have T_c values above 77 K. Although no theoretical basis exists to explain why, T_c is dependent on the number of copper oxide layers and the chemical formula of the material. J_c is also related to the ability of each compound to maintain a mixed state via intrinsic flux pinning. In layered HTS compounds the magnetic flux lines can be envisaged as a series of cylinders (with radius equal to the superconducting coherence length and height equal to the width of CuO_5 units in the crystal). Below a temperature

T^*, these cylinders are locked together through the non-superconducting layers of the material and magnetic flux is pinned allowing high J_c values. Above T^*, the flux cylinders are able to de-couple and can begin to move through the copper oxide planes. The higher intrinsic T_c and J_c compounds have at least two CuO_5 layers sandwiched between a thin non-superconducting oxide layer (e.g. Tl-1223 and Tl-1212 phases with a single Tl-O insulating layer, show stronger flux pinning at 77 K than the double Tl-O layered Tl-2223 phase, while the latter has the highest T_c). Figure 45.2 is a drawing of the crystal structure of $(Bi_{2-x}Pb_x)Sr_2Ca_2Cu_3O_{10-\delta}$. The supercurrent can be imagined to be confined to two dimensions in the layers containing Cu and O (essentially the middle $Sr_2Ca_2Cu_3O_8$ section of fig. 45.2. CuO_5 layers are shown as square-based pyramids surrounding CuO_2 planes and sandwiched between rock-salt structure (Bi,Pb)O layers.

A T_c value significantly above the operating temperature of 77 K is also desirable for engineering considerations. Experience with LTS materials in superconducting machines shows that operation is ideally performed at 50–60% below T_c. For HTS materials operating at 77 K, this would require an ideal T_c of 154 K, and this has not yet been achieved with HTS materials.

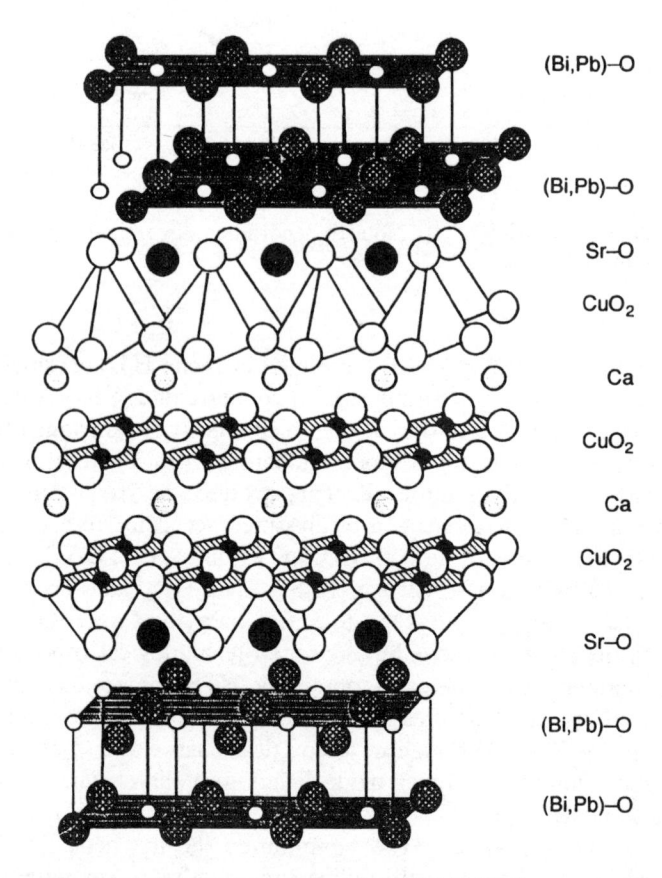

Fig. 45.2 The complex layered crystal structure of the 108 K HTS superconductor $(Bi_{2-x}Pb_x)Sr_2Ca_2Cu_3O_{10-\delta}$

MANUFACTURE OF HTS CONDUCTORS

The technical success of HTS cables relies on the ability to manufacture HTS conductors as cheaply and reliably as LTS conductors despite the fact that the former are hard, brittle ceramics and the latter are malleable metals. To some extent, this is the greatest challenge to HTS as the past work on helium-cooled cables has answered many of the questions associated with cable design, cooling systems and dielectric suitability at cryogenic temperatures. HTS materials need some form of metallic substrate for mechanical support. This can be in the form of a flat strip onto which the ceramic is coated or as a sheath or tube. Being oxides, HTS materials react with common metals such as steel and in so doing become poisoned. Also the high processing temperature excludes aluminium, and the need to process in air or oxygen excludes copper. The one common metal that neither poisons HTS compounds, nor melts or oxidises is silver and its use as a mechanical support for HTS materials is widespread.

Each HTS material has specific advantages and drawbacks for wire production. Y-123 is a granular material with a high melting point above that of silver which is a major drawback. It is also very unstable with respect to its oxygen stoichiometry but it does have a T_c of 92 K and can be processed via thin film deposition techniques to yield high critical current densities at 77 K ($>1 \times 10^9 \, A \, m^{-2}$) with strong magnetic flux pinning. However, for bulk samples, the processing route requires repeated high temperature melting and quenching, which is not a viable industrial production route.[3] Wire fabrication routes have included electrophoretic deposition onto inert metallic round wire substrates, powder-in-tube processing, Doctor Blade ribbons, and vapour deposition. Ultimately Y-123 suffers from having an orthorhombic crystal structure and crystallographic axial alignment at the grain boundaries required to produce a high J_c is difficult to achieve. The result is that, barring a new process development, it is now highly unlikely that Y-123 will be used for producing wire for any high power application by an economically viable industrial route.

The bismuth-containing HTS materials (and the structurally-related thallium-containing phases) have an advantage over Y-123 in that they have melting points below the melting point of silver around 850°C, are slightly more stable in oxygen stoichiometry and a higher T_c. The highest series member, (Bi,Pb)-2223, has a T_c of 108 K and with a $\Delta T = 31$ K it can be used easily at 77 K. (ΔT is the difference between T_c and 77 K.) (Bi,Pb)-2223 decomposes on melting but it was quickly found that a coherent structure could be made using a series of complicated deformation and temperature treatment to sinter the powder. T^* for (Bi,Pb)-2223 is about 30 K, which gives it a low capability to pin magnetic flux at 77 K. Fortunately in the magnetic fields experienced in a lower cable (*ca.* $0.2T$) (Bi,Pb)-2223 is still viable. The bismuth cuprates have also been synthesised using a variety of methods including MOCVD, Doctor Blade and especially powder-in-tube (PIT).[4] (Bi,Pb)-2223 is most suited to PIT manufacture, an industrially well-known technique and relatively cheap for such a high-tech operation; this forms the backbone of all present HTS cable development.

The thallium- and mercury-containing superconductors show even higher T_c values, and in the powder state, Tl-1223 has been shown to have good flux pinning properties.[5] Thallium-containing HTS materials do not seem suited to the PIT process, which produces conductors with only moderate J_c values and poor resilience to applied magnetic fields arising from grain boundary problems similar to that found for Y-123. At the time of writing, the most promising route for fabricating thallium-containing

HTS conductors is a mist pyrolysis of a vapour onto a silver ribbon substrate, and J_c values on short samples of $9 \times 10^8 \, A \, m^{-2}$ at 77 K in self-field have been reported. The most recently discovered HTS materials are the mercury-containing materials. Almost nothing is known about these materials at the time of writing apart from the fact that they are very difficult to prepare phase pure and seem to be very similar in behaviour to the thallium-containing materials. The perceived toxicity of both thallium and mercury makes these unattractive materials to work with and, as a consequence, in the absence of any new breakthrough in HTS, the material most likely to be used in a cable system is (Bi,Pb)-2223 processed in a silver sheath.

The PIT process (shown schematically in fig. 45.3) is in principle very simple and is widely used in the manufacture of mineral insulated cables. The first stage in the process is the loading and packing of loose HTS powder into a thin-walled silver or silver alloy tube, one end of which has been sealed. After loading, the packing end is closed, swaged and the assembly is subjected to a series of drawing reductions to form a ceramic–metal composite wire with an inner ceramic core and outer metallic sheath. Light reductions of around 10% per pass are preferred on all the mechanical operations in the PIT process. The wire is then compressed into a thin strip, typically 0.2 mm thick with a ceramic core thickness of around 0.07 mm. The tape is then annealed just below the melting point of (Bi,Pb)-2223 (820–850°C), followed by a further rolling stage. This procedure can be repeated several times. The thermo-mechanical deformation treatment is designed to induce maximised microstructural alignment, optimised grain growth and suppress the growth of secondary (non-superconducting) phases. Figure 45.4 shows a cross-section of (Bi,Pb)-2223/Ag composite monofilamentary

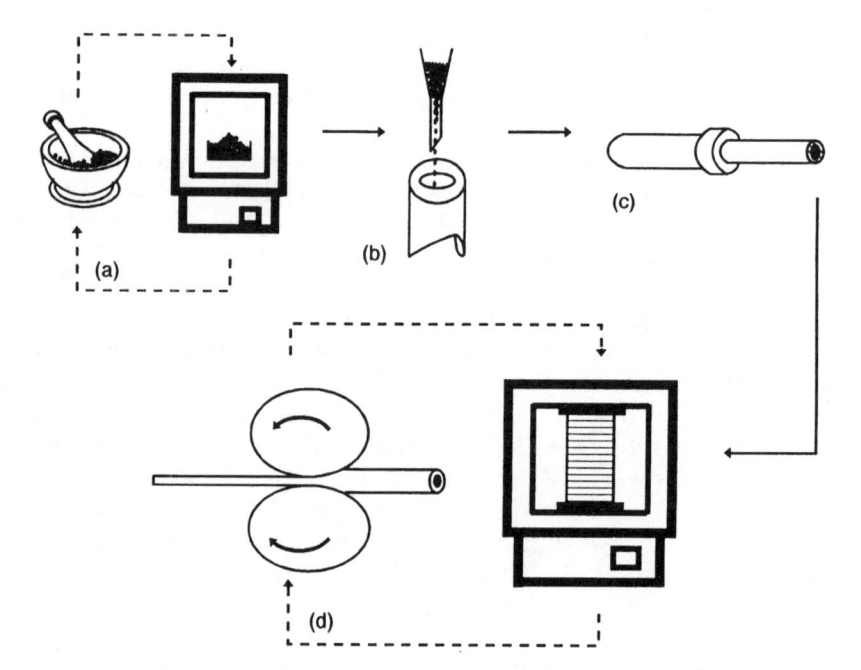

Fig. 45.3 Flow diagram of the powder-in-tube process used to make HTS conductors. The superconducting ceramic powder is first manufactured (a), loaded into a silver tube (b), drawn (c) and then undergoes a complex thermo-mechanical treatment to form a superconducting composite (d)

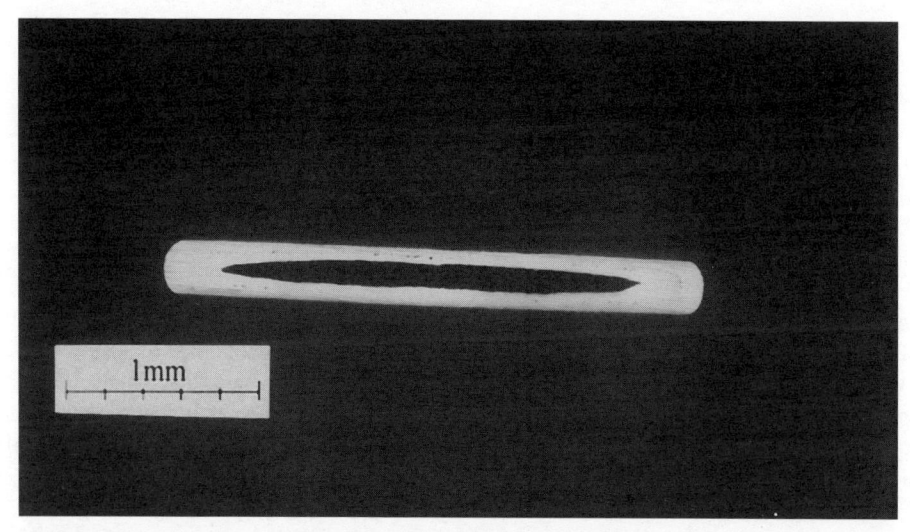

Fig. 45.4 Cross-section of a monofilament tape after thermo-mechanical processing of (Bi,Pb)-2223 (black) in a silver sheath (white) made using the PIT process

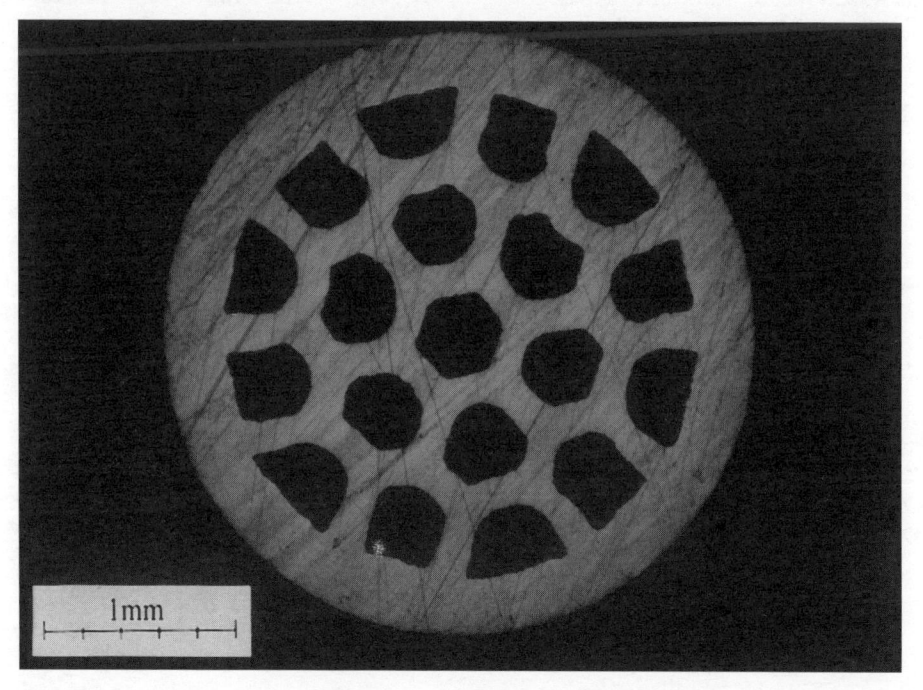

Fig. 45.5 Cross-section of a multifilament (Bi,Pb)-2223/Ag PIT conductor before thermo-mechanical processing

conductor after being flattened and heat treated. Multifilamentary conductors can be made by simply packing the required number of conductors into an appropriately-sized extra silver tube prior to the first mechanical deformation stage. An example of this is shown in fig. 45.5 which shows a photograph of a 19 filament round wire before the thermo-mechanical deformation stage.

PERFORMANCE CHARACTERISTICS OF HTS CONDUCTORS

The key factors that determine the performance of a superconductor are

(a) the magnitude of its transition temperature;
(b) the value of its critical current density;
(c) its electrical performance whilst subjected to an external magnetic field;
(d) its electrical performance under thermal cycling and induced mechanical strain.

The magnitude of T_c has been addressed previously as the most likely HTS conductor is (Bi,Pb)-2223/Ag tapes with a T_c of 108 K. As for J_c, its increase is dependent on advances in improved processing. For LTS conductors J_c values in self-field at 4 K are typically 1×10^9 A m^{-2}. The same value has been achieved with HTS materials using thin film deposition techniques measured at 77 K in self-field, but the highest values currently reported for PIT samples are in the range $5-7 \times 10^8$ A m^{-2} over short lengths of a few centimetres.[6,7] These were produced using static pressing, which is unfortunately not realistic as a large-scale industrial technique. The method of manufacturing long lengths of HTS conductor most used is continuous high pressure rolling. Using this technique, J_c values in the range of 2×10^8 A m^{-2} over a few hundred metres and 4×10^7 A m^{-2} over 1000 m have been reported. Progress in increasing J_c is showing

Table 45.1 Technical requirements for HTS materials

Application	Demonstration Operating field (T)	J_c (A m^{-2})	Production Operating field (T)	J_c (A m^{-2})
D.C. motor				
Heteropolar	0.4–0.8	1.5×10^7	1.5	$>3 \times 10^7$
Homopolar	3	2×10^7	>6	$>3 \times 10^7$
A.C. generator				
Rotating field	2	2×10^7	6	$>3 \times 10^7$
Stationary field	0.8	1.5×10^7	2	$>3 \times 10^7$
Rotating armature (50 Hz)	0.8	2×10^7	>2	$>4 \times 10^7$
Stationary armature (50 Hz)	1	1.5×10^7	2	$>4 \times 10^7$
Transformer (50 Hz)	0.2	1.5×10^7	4	$>4 \times 10^7$
Energy storage	0.6	2×10^7	>6	$>5 \times 10^7$
Magnetic separation	1	2×10^7	2–7	$>2 \times 10^7$
Fault current limiter				
Saturated iron type	0.6	1×10^7	1.5	$>2 \times 10^7$
Quench type (50 Hz)	0.1	2×10^7	0.1	$>4 \times 10^7$
Transmission cable				
1000 MVA (50 Hz)	0.2	1×10^8	0.3	$>1 \times 10^9$
500 MVA (50 Hz)	0.1	1×10^8	0.2	$>2 \times 10^8$
4000 MVA (d.c.)	0–0.1	1×10^8	0.3	$>2 \times 10^8$

continuous iterative improvement. Table 45.1 illustrates the necessary technical requirements for HTS materials to have widespread use in a variety of electrical engineering applications.

Electrical characterisation experiments on HTS materials are relatively easy. Low resistance voltage electrical contacts below $10^{-8}\,\Omega$ need to be made to the outer metallic sheath (indium alloys are suitable for this) and then a second pair of contacts is used to supply an applied current. The sample resistance is monitored using a nanovoltmeter and the applied current gradually increased until a resistance can be detected. The most common voltage criterion used is $10\,\mu\text{V}\,\text{m}^{-1}$. Figure 45.6 shows a typical J_c measurement at 77 K on a PIT-processed (Bi,Pb)-2223/Ag sample. The I_c value is approximately 20 A in this case, and fig. 45.7 shows a corresponding T_c measurement on the same sample determined using a magnetic technique and relies on the Meissner effect (exemplified by the negative value of the magnetic susceptibility). The large transition at 108 K determines the T_c value.

For transmission cable applications, the highest external magnetic field generated in the cable is anticipated to be approximately 0.2 T and engineering solutions exist to minimise this value. Therefore for cable applications the performance of HTS conductors in only moderate fields is required. One of the drawbacks of the ceramic superconductors, owing to their granular nature, is a sharp fall in J_c with applied fields. For PIT conductors, this value is in the range of 20–50% of the zero field J_c. Most of the drop occurs in the first few millitesla and is a result of flux penetrating weak links at

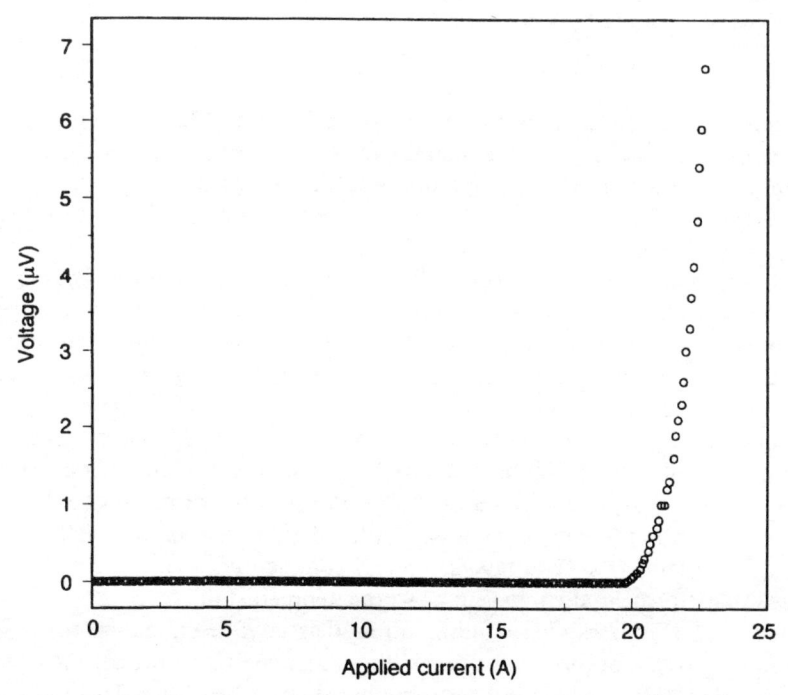

Fig. 45.6 Graph showing the results of a J_c test measurement on a (Bi,Pb)-2223/Ag PIT conductor made at 77 K. The I_c value is approximately 20 A, indicated by the sharp rise in resistivity at this value

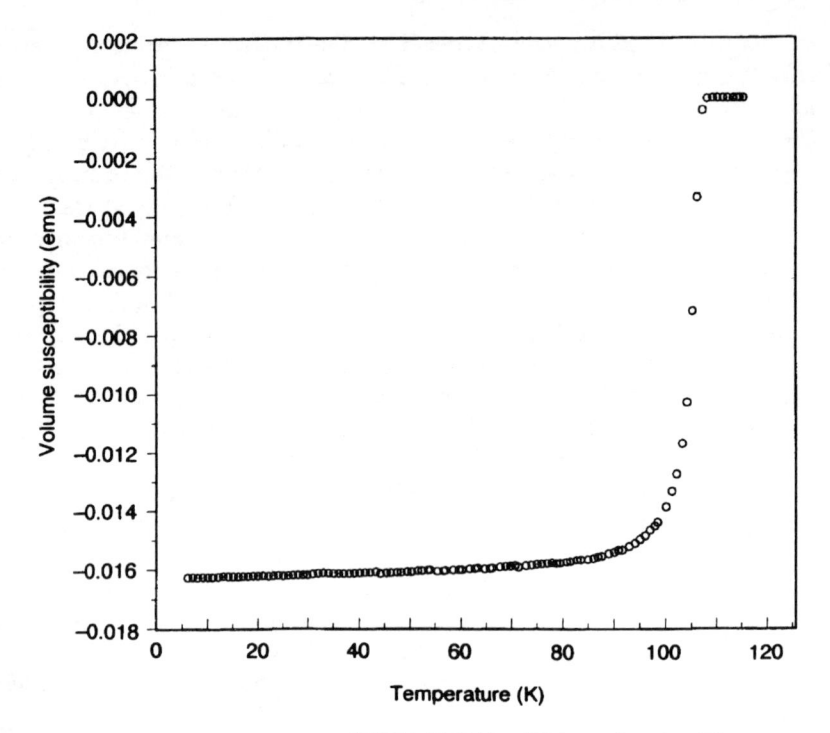

Fig. 45.7 T_c test measurement on a (Bi,Pb)-2223/Ag PIT conductor. The superconducting transition occurs at 108 K and the sharp fall in magnetic susceptibility at this temperature is indicative of the onset of the Meissner effect

grain boundaries, thereby preventing a transport current. Measurements using a.c. currents are carried out in a similar manner to that described above. The value and mechanism of a.c. losses in PIT conductors is still a subject of some debate, however values in the region of $10\,\mu\text{W m}^{-1}$ per phase are usually reported, which are in the range acceptable for a.c. cable operation at 50–60 Hz.

Mechanical measurements show that PIT HTS conductors are fairly weak materials as all their mechanical support derives from the silver sheath. This means that bending strains of only 0.1% can begin to degrade their J_c values and, by 0.2%, this can fall to below 50% of the original J_c value. This is not a devastating problem in itself, as solutions to cable mechanically weak materials already exist. One method of increasing the mechanical strength of PIT conductors is to increase the number of filaments and already examples with over 1000 filaments per conductor have been made. Also silver alloys such as AgMg, AgZr and AgSb can extend the bend strain out to 0.4%. HTS materials seem very resilient compared with their LTS counterparts. Stranded prototype cables made of several layers of HTS conductors have withstood hundreds of thermal cycles from liquid nitrogen to room temperature. Also these early cable models show that HTS stranded conductors models have been shown to be able to withstand fault currents of up to ten times without any change in properties.[8] This is in contrast to LTS materials which fail catastrophically on quenching. This is due partly to the different thermal properties of liquid nitrogen and partly to the highly conducting silver acting as a current and/or thermal shunt.

REFERENCES

(1) Bednorz, J. G. and Müller, K. A. (1986) *Z. Phys* **B64**, 189.
(2) Wu M.-K., Ashburn, J. R., Torng, C. J., Hor, P. H., Meng, V., Gao, L., Huang, Z. J., Wang, Y. Q. and Chu, C. W. (1987) *Phys. Rev. Lett.* **58**, 908.
(3) Murakami, M., *Supercond. Sci. Technol.* **5**, 185.
(4) Sandhage, K. H., Riley, G. N. Jr and Carter, W. L. (1991) *J. Met.* **43**, 21.
(5) Dou, S. X. and Liu, H. K. (1993) *Supercond. Sci. Technol.* **6**, 297.
(6) Sato, K., Hikata, T., Ueyama, M., Mukai, H., Shibuta, N., Kato, T. and Masuda, T. (1991) *Cryogenics*, **31**, 687.
(7) Li, Q., Brodersen, K., Hjuler, H. A. and Freltoft, T. (1993) *Physica C* **217**, 360.
(8) Beales, T. P., Friend, C. M., Le Lay, L., Mölgg, M., Dineen, C., Jacobson, D. M., Hall, S. R., Harrison, M. R., Hermann, P. S., Petitbon, A., Caracino, P., Gherardi, L., Metra, P., Bogner, G. and Neumüller, H.-W. (1995) *Supercond. Sci. Technol.* **8**, 909.

CHAPTER 46
High Temperature Superconducting Power Cables

HIGH TEMPERATURE SUPERCONDUCTING CABLES

The basic arrangement of an HTS phase forming part of a power cable, has a concentric structure. A typical HTS 3-phase a.c. cable assembled from three such units and designed to operate at 77 K and an artist's impression of what such a cable would look like is shown in fig. 46.1 (any design will need at least two conductors to transport the current to and from the load). The basic HTS cable design incorporates the following features. Each conductor has forced liquid nitrogen pumped from a cooling station flowing down the inside of corrugated copper pipe acting as both former and stabiliser called the 'go' line. HTS tapes are non-inductively helically-wound on top of the copper former. This assembly is then insulated using a dielectric. Research suggests

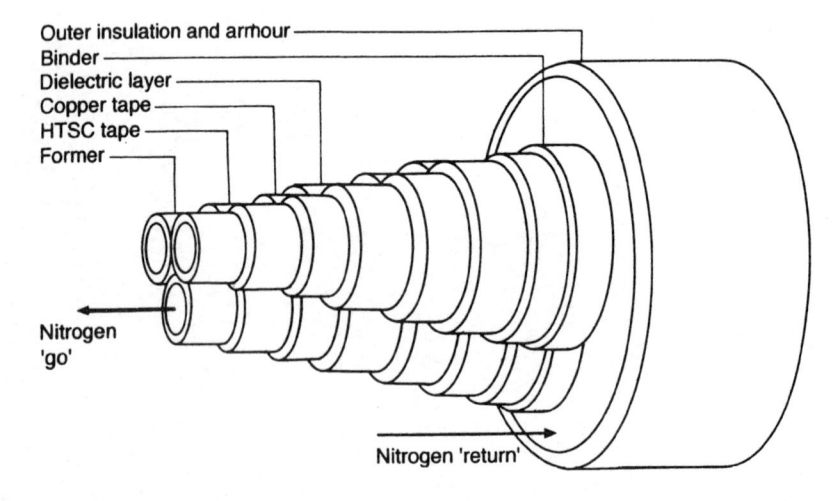

Fig. 46.1 3-phase 1 GVA superconducting a.c. cable operating at 77 K utilising PIT processed (Bi,Pb)-2223/Ag HTS conductors

this could be traditional, e.g. fluid impregnated paper, polypropylene paper laminate or polymeric, e.g. EPR or XLPE. A second conductor layer is then repeated on top of this assembly. The liquid nitrogen is returned along the outside of the conductor assembly after having been collected and re-cooled in a second cooling substation some distance away from the first. The cryogenic envelope is thermally insulated by a vacuum space and encased by an outer shield of stainless steel armour. Placing all the conductors in a single cryogenic envelope is necessary for reasons of refrigeration economy.

Because a superconductor offers no resistance to the passing of an electric current it is plausible to ask if a superconducting cable would also show no loss. In reality it will, as an alternating current travelling down a superconducting cable will generate a small loss arising from the generated alternating magnetic field and from induced eddy currents in the surrounding dielectric and copper in the cable. This problem would not arise in the case of a direct current superconducting cable, which would have only very small losses arising from the unavoidable superconductor–metal joints and terminations. However, for d.c. cables, there would still be a need to have at each end of the cable an a.c./d.c. converter. This adds considerably to the economics of running a superconducting cable. A two-step a.c./d.c. and d.c./a.c. conversion would incorporate an unavoidable 2% loss in the system. This means that there is a minimum break-even economic length for a d.c. superconducting cable of approximately 100 km. Several HTS d.c. cable prototypes have been built in Europe and Japan. Figure 46.2 shows a 1.5 m HTS d.c. transmission cable prototype (for a 1 km GVA d.c. cable) designed and built by BICC Ceat Cavi to operate using cooled helium gas at 40 K. At a test operating temperature of 31 K the prototype cable had a current capacity of 11 067 A (a ten-fold increase in current over a conventional 1000 mm^2 copper cable).

TECHNO-ECONOMIC IMPACT OF HTS CABLES

Analysis of the emergence of high temperature superconductivity in the copper power cable industry has strong parallels with that already seen for optical fibres in the copper telecommunication cable industry.[1,2] Both innovations represent a discontinuity in which a known technology can be displaced by a new process which is superior in at least one performance criterion. This presents new opportunities for business but is also a threat to established markets and careful management of the technological change with adoption of appropriate technological strategies is required for commercial success.[3,4]

The above fact has been widely recognised in many countries and there have been a number of substantial techno-economic assessment studies of the performance requirements for HTS materials to be economical in transmission cables. These were carried out in the early 1990s in each of the three main industrialised blocks of Western Europe,[5] Japan,[6] and the USA.[7]

The Japanese study was carried out by the Tokyo Electric Power Company and suggested that superconducting cables could utilise existing 150 mm diameter ducts under Tokyo as a method of increasing capacity to satisfy a predicted annual growth demand of 2.5%. The spacial limitations of Tokyo make a superconducting cable an especially attractive solution. The study concluded that a superconducting cable operating at 4 K with either LTS or HTS conductors would need a minimum overall diameter of 360 mm, and this is too large to fit into existing ducts. However, if the cable

Fig. 46.2 10 000 A, 40 kV d.c. cable prototype incorporating HTS tapes designed to operate at a temperature of 40 K, achieved by helium gas cooling

could operate at 77 K, i.e. using only HTS materials, the simpler cryogenics allowed a minimum diameter of 130 mm, which could easily fit into existing tunnels. The TEPCO study designed a cable system around this concept and also incorporated a reduced operating voltage of 66 kV to match the HTS cable to the existing distribution voltage network in Tokyo.

The EC study[8] was carried out by the major European cablemakers of ABB, Alcatel, BICC, Pirelli and Siemens along with GEC. This study estimated that a newly installed cryogenically force-cooled superconducting underground a.c. power cable operating at 77 K will have an economic advantage over a conventional force-cooled underground cable for high power transmission with ratings above 500 MVA. For example, a superconducting 10 km 1 GVA cable was calculated to yield cost savings of 15–20% (10 Mecu/installation) and loss savings of 50% (10 GWh/year/cable) compared with a conventional underground cable. Techno-economic studies have been carried out in the USA by both the Electric Power Research Institute and Argonne National Laboratory (on behalf of the International Energy Agency)[9] and identified retro-fitting short high power a.c. links at the 500 MVA power level into existing ducts, in a similar way to the present method of replacing copper telecommunication cables with new optical fibre cables.

All the above studies have arrived at the same conclusions for both technical requirements and economic impact of superconducting cables. They agree, for example,

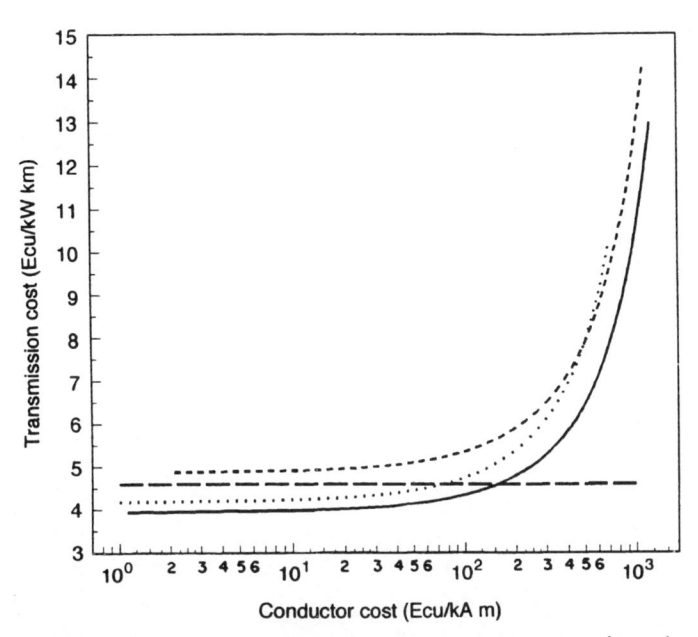

Fig. 46.3 Graph of the calculated transmission cost in units of ecu kW^{-1} km^{-1} against the cost of producing HTS conductor for cables in units of ecu kA^{-1} m^{-1}. Conductor J_c values of 1×10^9 A m^{-2} (——), 1.5×10^8 A m^{-2} (····) and 0.5×10^8 A m^{-2} (----) respectively are represented along with the results for a 400 kV single fluid oil-filled 1000 MVA cable (– – –). Data taken from Ashworth et al. 1994[8]

that a newly installed 1 GVA cable operating at 77 K would need an economic break-even critical current density in the superconductor to be near to 1×10^9 A m^{-2} at 77 K in self-field. However, the EC study also identified a 500 MVA rating retrofit cable as an economic solution utilising J_c values as low as 1×10^8 A m^{-2}. Figure 46.3 shows a graph of the calculated transmission cost in units of ecu kW^{-1} km^{-1} against the cost of producing the conductor for cables in units of ecu kA^{-1} m^{-1}.

The most interesting fact is that an economical break-even J_c can be estimated for a range of HTS conductor properties and operating conditions. These results show that for HTS conductor costs below 100 ecu kA^{-1} m^{-1} the transmission costs are dominated by the ancillary equipment such as installation and cryogenics etc. For values near to 1000 ecu kA^{-1} m^{-1} the HTS conductor would constitute near to 60–70% of the cost of the cable, while at values near to 10 ecu kA^{-1} m^{-1} the HTS conductor would be less than 5% of the total cost of the cable.

Figure 46.4 shows a plot of operating J_c in A m^{-2} against the cost of the conductor in ecu kA^{-1} m^{-1} for an optimised high power cable at a rating of 1000 MVA and a replacement medium power superconducting cable operating at 400 MVA. There are three factors that can influence the shape of the graphs in Figure 46.4. If J_c is increased in the PIT process toward the target value of 10^9 A m^{-2}, both curves would shift upwards parallel to the y-axis. Another factor that can influence the curves is a reduction in the overall cost of production for the HTS conductors and this would have the effect of moving the curves to lower cost values along the x-axis. A third factor that can influence the curves in fig. 46.4 is the identification of new application areas such as retrofitting and this can have the effect of moving the two curves closer together.

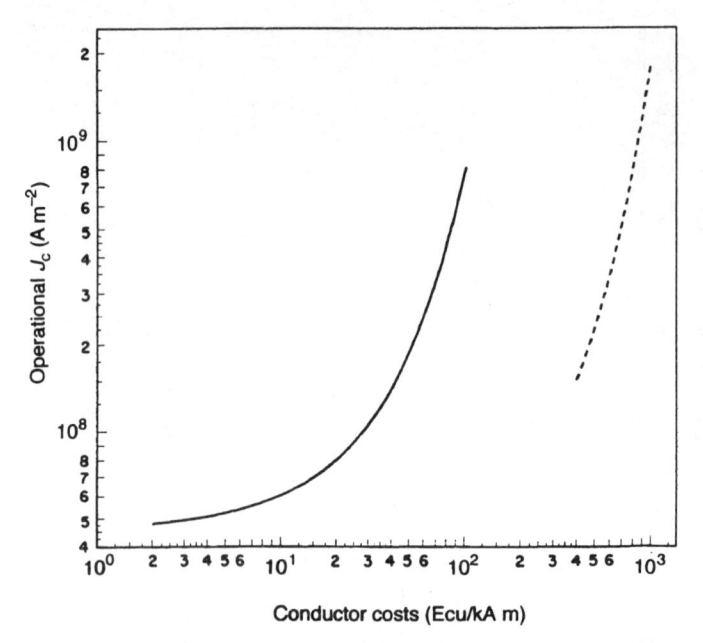

Fig. 46.4 Modified data from fig. 46.3 comparing the economics of the operational J_c in $A\,m^{-2}$ against the cost of the conductor in ecu $kA^{-1}\,m^{-1}$ for a newly installed optimised 1000 MVA high power HTS cable (——) and a replacement 400 MVA medium power HTS cable (- - - -) compared with the equivalent copper cable system

Figure 46.5 is a further representative breakdown which shows the conductor J_c in $A\,m^{-2}$ versus the absolute cost per unit length of the conductor in ecu m^{-1}. These curves will change slightly for different manufacturing methods and for physical properties such as homogeneity of J_c, but on the whole these curves give a good idea of the actual manufacturing costs needed for HTS conductors to be economical in lower cable systems. A retro-fit medium power fixed diameter cable for example could be achieved for an HTS conductor with a J_c of $2 \times 10^8\,A\,m^{-2}$ produced at $400\,\text{ecu}\,m^{-1}$. For this same HTS conductor to be used in a new optimised high power cable it would need to be manufactured at a break-even cost of $60\,\text{ecu}\,m^{-1}$ or less.

TOWARDS AN HTS TRANSMISSION CABLE

Most of the work on superconducting transmission cable models today is performed using HTS conductors. Japan is at the forefront of the development of HTS cables. Several research and development programmes between Japanese cable companies, electric utilities and MITI are already in progress. In December 1994, Sumitomo Electric announced the fabrication and testing of two 1 m a.c. cable models.[10] The first was a 12-layer cable carrying 5800 A at 77 K and the second a 14-layer cable carrying an estimated 12 000 A at 77 K. Other Japanese companies such as Furukawa Electric have demonstrated a.c. prototype cables carrying 3000 A d.c. and Fujikura and Chubu Electric Power also recently demonstrated a 1 metre, 3000 A d.c. power cable prototype.

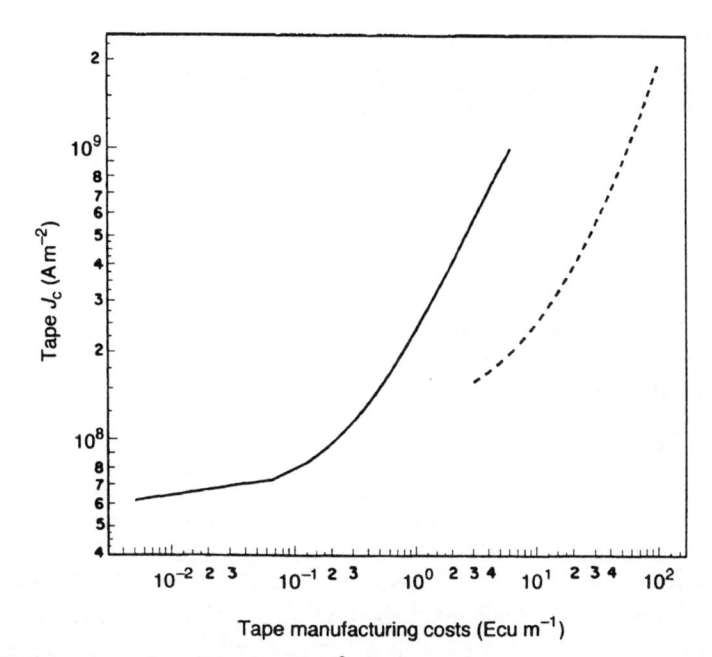

Fig. 46.5 Graph of conductor J_c in $A m^{-2}$ versus the absolute cost per unit length of the conductor in ecu m^{-1} for a newly installed optimised 1000 MVA high power HTS cable (———) and a replacement 400 MVA medium power HTS cable (- - - -). Realistic expected HTS conductor costs are in the range 100–500 ecu m^{-1}

Latest results at the time of writing show that several 50 m prototypes with terminations and carrying up to 2000 A have been built by Sumitomo and Furukawa.

In the EU, HTS superconducting cable research programmes were carried out in partnership between the power cable companies of BICC, Pirelli, Alcatel, Siemens, ABB and GEC which resulted in techno-economic assessment and the production of cable prototypes.[5,11]

In 1991 the US Department of Energy (DoE) initiated a substantial research programme on power applications which has culminated in a $5M part-DoE funded collaborative programme between Pirelli Corp, EPRI, American Superconductor Corporation and the US Federal superconductivity pilot centres at Argonne and Los Alamos National Laboratories to produce a 30 m, 115 kV underground transmission cable prototype. Metre length prototypes have been produced by Pirelli using ASC tapes.[12] The eventual aim of the EPRI programme is to produce a 100 m prototype for testing at the Brookhaven Laboratories, site of the testing of the US Nb-Ti transmission cable. A programme between Southwire, Intermagnetics General Corp and Oak Ridge National Laboratories to manufacture a 1 m HTS cable prototype is also underway.[13]

Despite the manufacturing of working prototypes, there is a long way to go before the viability of HTS cables can be shown beyond doubt. The technical properties which still need to be evaluated or improved are:

(a) higher J_c values at 77 K;
(b) absolute measurement of the total a.c. losses and their reduction to the lowest possible values;

(c) development and appraisal of operational HTS cables;
(d) determination of the high voltage properties of the dielectric materials at 77 K;
(e) development of joints and terminations;
(f) assessment of fault current performance;
(g) research into stability and ageing effects.

REFERENCES

(1) Beales, T. P., Friend, C. M., Segir, W., Ferrero, E., Vivaldi, F. and Ottonello, L. (1996) *Supercond. Sci. Technol.* **9**, 43.
(2) Beales, T. P. and McCormack, J. S. (1995) *Proc. 4th World Congress on Superconductivity* (eds K. Krishen and C. Burnham) Pub NASA Conference Publication 3290, p. 650.
(3) Senior, J. M. and Ray, T. E. (1990) *Int. J. Technol. Man.* **5** (1), 71.
(4) Alic, J. A. and Miller, R. R. (1989) *Int. J. Technol. Man.* **4** (6), 653.
(5) JOULE Final Report, No. JOUE 0050-GB (1992) *'Applications of HTS materials with particular reference to power cables'*. European Community DGXII, Brussels.
(6) Hara, T., Okaniwa, K., Ichiyanagi, N. and Tanaka, S. (1992) *IEEE Trans. PD*, **7** (4), 1745.
(7) Engelhardt, J. S., Von Dollen, D. and Samm, R. (1992) *Proc. American Inst. Phys. Conf.* **251**, 692.
(8) Ashworth, S. P., Metra, P. and Slaughter, R. J. (1994) *Eur. Trans. Elect. Power Eng.* **4**, 243.
(9) Giese, R. F. (1992) *Superconducting Transmission Lines*, International Energy Agency.
(10) (1994) *Superconductor Week*, **8** (40).
(11) Beales, T. P., Friend, C. M., Le Lay, L., Mölgg, M., Dineen, C., Jacobson, D. M., Hall, S. R., Harrison, M. R., Hermann, P. S., Petitbon, A., Caracino, P., Gherardi, L., Metra, P., Bogner, G. and Neumüller, H.-W. (1995b) *Supercond. Sci. Technol.* **8**, 909.
(12) Gannon, J. J., Malozemoff, A. P., Minot, M. J., Barenghi, F., Metra, P., Vellego, G., Orehotsky, J. and Suenaga, M. (1994) *Advances in Cryogenic Engineering* (ed. R. P. Reed) **40**, p. 45, New York: Plenum.
(13) (Winter 1995) *Superconductor Industry* 60.

PART 7
Optical Fibres in Power Transmission Systems

Live line installation of optical cable

CHAPTER 47

Introduction to Part 7

Profound global changes in the commercial structure of both the electricity supply industry and the telecommunications industry have led to a rapidly expanding market for cables which enable the installation of optical fibres on overhead lines. Originally these products met an increasing demand for improved communication networks within power utilities. In this context there has been extensive exploitation of optical fibres, which have the important advantages of high bandwidth, low attenuation, reliability and immunity to electromagnetic interference.[1,2] Deregulation of the telecommunication carrier markets and a more commercial focus in the ESI over recent years have led to a massive increase in the number of fibres required by the electricity utilities. These fibres are now being leased directly or through new commercial organisations to third party telecom carriers and can be expected to contribute significantly to the profitability of an electricity distribution company. Thus, for the transmission line suppliers, the main drive over the last few years and for the foreseeable future is to a more diverse product range and to cables and lines accommodating more fibres. At present cables are regularly installed which contain up to 48 fibres, with 96 and higher fibre counts being considered. This is in contrast to the original thoughts when such cables were first developed for use within the supply industry only. Cables were then designed with six fibres although it was expected that often four or even fewer would eventually suffice.[3]

A number of optical communication cable system options are open to operators of long span overhead line power distribution networks. Optical Ground Wire (OPGW) is used almost exclusively where a communication system is required on a system being re-strung or newly commissioned.[4] However for existing power systems, which do not otherwise require maintenance, the cost and inconvenience of this method leads to a retro-fit system becoming more attractive. Two such systems dominate this post-fit market. The first employs a small and flexible cable (wrap cable) wrapped on to the existing ground wire using specialist installation equipment. This can often be achieved without interruption to the power supply. In the second system an all dielectic self-supporting (ADSS) cable is suspended directly from the tower structures. This widely used technique can also be deployed without circuit outages. In some cases wrap cable can also be used on phase conductors, or composite phase

Table 47.1 Summary of optical cables available for long span overhead power lines

Cable type	Advantages	Disadvantages
Composite earth wire (OPGW)	Well proven, high reliability High fibre count (over 48) Many suppliers Low cost if new system No phase voltage limitations	Difficult to install live High cost if on an existing system High fault condition temperatures
Composite phase conductor (OPPW)	Well established robust cable Can be used on pole routes Cost efficient if on new lines	High cost if on an existing system Additional components needed Live line installation not possible Continuous elevated temperatures
Wrap on earth cables	Fast installation Live line possible sometimes No voltage limitations	Live line installation not always possible High fault condition temperatures
Wrap on conductors	Fast installation Possible on all routes with easy crossing of roads, railways, rivers etc.	Live line installation not possible Vulnerable to gun shot Additional components needed
All-dielectric self-supporting (ADSS)	Many suppliers Well proven Live line installation possible	Reliability on high voltages uncertain Ice loading increases sag Ground clearance
Metal aerial self-supporting (MASS)	No voltage limitation	No live line installation Clashing with conductors may limit application Corona (RF emission) may restrict use

conductors (OPPW) can be used. This chapter describes all of these systems, from cable and component design through to installation techniques. A brief comparison of them all is given in table 47.1.

Optical fibres are dielectric and can withstand substantial voltage gradients,[5] and are immune to electromagnetic interference. Fibres, therefore, present opportunities to improve measurements such as temperature and current on high voltage conductors. There has been little commercial activity in this area, however an example of the potential for fault location is given by Matsuoka.[6]

In the remainder of Part 7 we first introduce the subject of optical fibres, relating their common methods of production and the principles of light transmission in them. This is followed by an outline of common approaches to the cabling of optical fibre and then the application of fibres and cables in overhead lines.

REFERENCES

(1) Martin, J. R. (1993) 'Optical fibres for power utilities'. *CIGRE* Paper No. 5.2.
(2) Sharma, S. C. (1994) 'Solution for fibre optic cables installed on overhead power transmission lines. A review'. *IETE Technical Review* **11**, 215–222.
(3) Dey, P., Gaylard, B., Holden, G., Taylor, J. E., Smith, P., Carter, C. N., Maddock, B. J. and Kent, A. J. (November 1984) 'Optical communication using overhead power transmission lines'. *Elektron*, 38–44.
(4) Bronsdon, J. A. N., Carter, C. N., Kent, A. H. and Iddin, A. (1986) 'The successful evaluation of a 400 kV composite ground wire optical communication system'. *36th Proc. Int. Wire and Cable Symp.*, 496–504.
(5) Looms, J. S. T. (1988) *High Voltage Insulation*. Peter Perigrinus.
(6) Matsuoka, N., Suzuka, T., Ishimaru, H., Hashimoto, H. and Taga, H. (1993) 'Supervisory information system of overhead power transmission lines for CLP'. *Sumitomo Electric Technical Review* **36**, 43–49.

CHAPTER 48
Principles of Optical Fibre Transmission and Manufacture

Optical fibre has moved over a twenty year period from the research phase to now being employed in the long-haul segments of every major telecommunication network in the world. There is little doubt that over the next ten years optical fibre transmission will play a part in every telephone call that the reader of this book makes and every television picture that the reader watches.

As is explained elsewhere, optical fibre is also an important medium for advanced sensing systems with new applications in, for example, temperature sensing of power cables, and, therefore, we expect that most of our readers will have access to optical fibres in their professional lives. In this chapter a brief introduction to the topic of optical fibre and optical fibre transmission is given. The reader who requires a detailed account of the topic is referred to the extensive literature on the subject which is becoming available, some of which is given in appendix A21.

OPTICAL FIBRE WAVEGUIDES

An optical fibre is a thin glass strand, usually a fraction of a millimetre in diameter, which, at its centre, has a region which is able to guide light along its length. This is achieved by the centre (or core) region having a higher refractive index than its surroundings. Light is totally reflected at this boundary and thus guided along the length of the fibre.

In a step index fibre, the refractive index is constant across the core, whereas in a graded index fibre the refractive index decreases gradually away from the centre of the core. Simple geometrical optics can be applied effectively to such structures to gain a good explanation of their waveguiding properties.

Step index fibres

Consider a ray making an angle θ_i, with the end of a step index fibre, as shown in fig. 48.1. Refraction at the air–glass interface causes the ray's path to be deflected

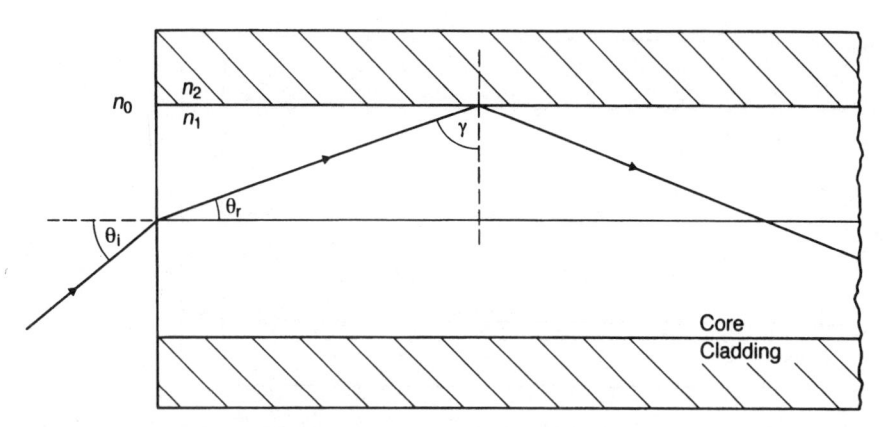

Fig. 48.1 Ray paths in a step index fibre

towards the normal. Snell's law gives the refracted angle inside the glass, θ_r, as follows:

$$n_0 \sin \theta_i = n_1 \sin \theta_r \tag{48.1}$$

where n_0 and n_1 are the refractive indices of air and the glass core respectively. This ray is incident at the core/cladding interface at an angle γ where

$$\gamma + \theta_r = 90° \tag{48.2}$$

n_2 is the refractive index of the glass cladding. For angles larger than the critical angle, γ_c, defined by

$$\sin \gamma_c = \frac{n_2}{n_1} \tag{48.3}$$

the ray experiences total internal reflection at the core/cladding interface. This effect occurs along the length of the fibre and is the basic mechanism by which light is guided along optical fibres.

Graded index fibres

The refractive index in graded index fibre decreases gradually in approximately quadratic fashion radially from a maximum value at the fibre centre. Figure 48.2 illustrates the ray paths in such a refractive index profile.

Fibre rays follow a curved path through the fibre core due to the constant variation in refractive index. The ray is gradually curved with an ever-increasing angle of incidence onto the next generation of refractive index until the condition for total internal reflection is met and the ray then begins to travel back towards the core axis.

Multimode fibres

When the diameter of the fibre core is large compared with the wavelength of light, this simple ray picture is a good approximation and rays can travel within the core of the fibre at essentially any angle between θ_c and $90°$. Rays travelling parallel to the axis of

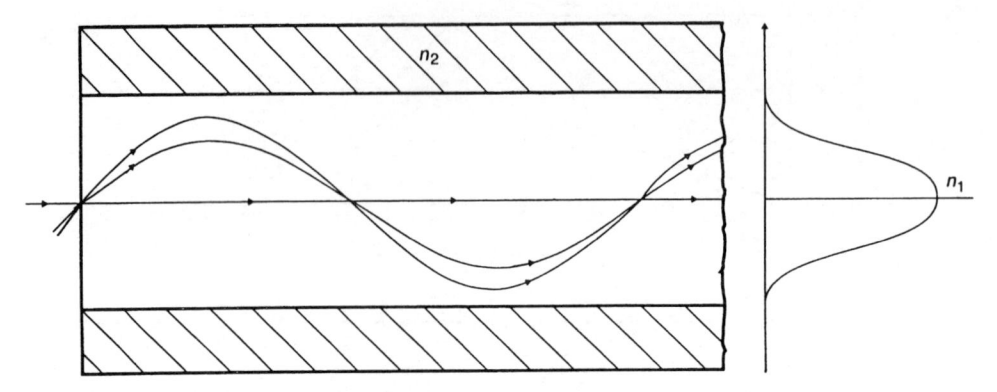

Fig. 48.2 Ray paths in a graded index fibre

the fibre have a shorter physical path length than those travelling at higher angles. In step index fibre these rays travel at the same velocity because the refractive index is constant. Any practical light source connected to the end of the fibre launches rays at a range of angles and so a short pulse of light will arrive at the far end extended in time by virtue of the different paths followed by the light emitted at different angles. This effect is called modal dispersion and is discussed in more detail later.

In graded index fibre, however, the refractive index is lower at the edge of the core, and rays travelling here have a higher velocity. In a fibre with a parabolic refractive index profile light emitted at an angle to the fibre axis will follow a curved oscillatory path. Light travels faster when its path takes it away from the fibre axis and this compensates for the longer path length along which it has to travel. It can be shown in this case that all rays have nearly the same velocity along the fibre axis and thus modal dispersion and, therefore, pulse broadening is minimised.

Single mode fibres

In the foregoing treatment we have assumed a simple model of light propagation. Readers familiar with waveguide theory will appreciate immediately that not all angles of ray propagation can be supported. Angles at which rays are allowed to propagate are equivalent to the electromagnetic modes of the waveguide which have different group velocities. When the core diameter of the fibre is reduced to a critical value, typically in the region of 10 µm for commercial silica fibres, only one mode of propagation is allowed. Such fibres are termed single mode (or monomode), and constitute the standard grade of fibre used in long-haul telecommunications applications. Because pulse spreading is minimised in these fibres they have very large bandwidths, and for practical purposes the bandwidth of a single mode fibre system is constrained by the terminal equipment.

OPTICAL FIBRE MANUFACTURE

Optical fibre is normally made by drawing the fibre down from a large diameter ingot or 'preform'. There are three basic manufacturing steps: preform manufacture, sometimes referred to as laydown; fibre drawing, and final measurements. In the case of silica fibre there are several different processes for making the preforms: Outside Vapour Deposition (OVD), Vapour Axial Deposition (VAD) and Inside Vapour Deposition (IVD).

The largest proportion of fibre produced worldwide is made using the OVD process, and an explanation of the process is given below. It should be noted that once the preform has been completed, the rest of the fibre making process (i.e. drawing, measurements etc.) is the same for each of the three preform production techniques.

Preform manufacture

The initial stage of fibre manufacture is preform deposition which occurs in enclosed thes. Oxides from the combustion of various chemical vapours are deposited as 'soot' a rotating mandrel. This process is known as 'Outside Vapour Deposition'.

Regulated lateral movement of the mandrel relative to the burner ensures uniformity deposition and concentricity with each successive layer. The vapour mix determines composition of each layer, hence the refractive index profile can be established e fig. 48.3).

The deposition process occurs simultaneously in a series of lathes, each individually trolled and monitored by computer in an environmentally controlled clean area. ically the soot is formed by a mixture of silica (SiO_2) and oxides of dopant materials ch raise (using germanium) or lower (using fluorine) the refractive index of the erial. By varying the level of germanium and the number of passes through burner, the refractive index and the ultimate diameter and refractive index profile of the core is determined and thus whether the fibre is single or multimode and step or graded index, as discussed previously.

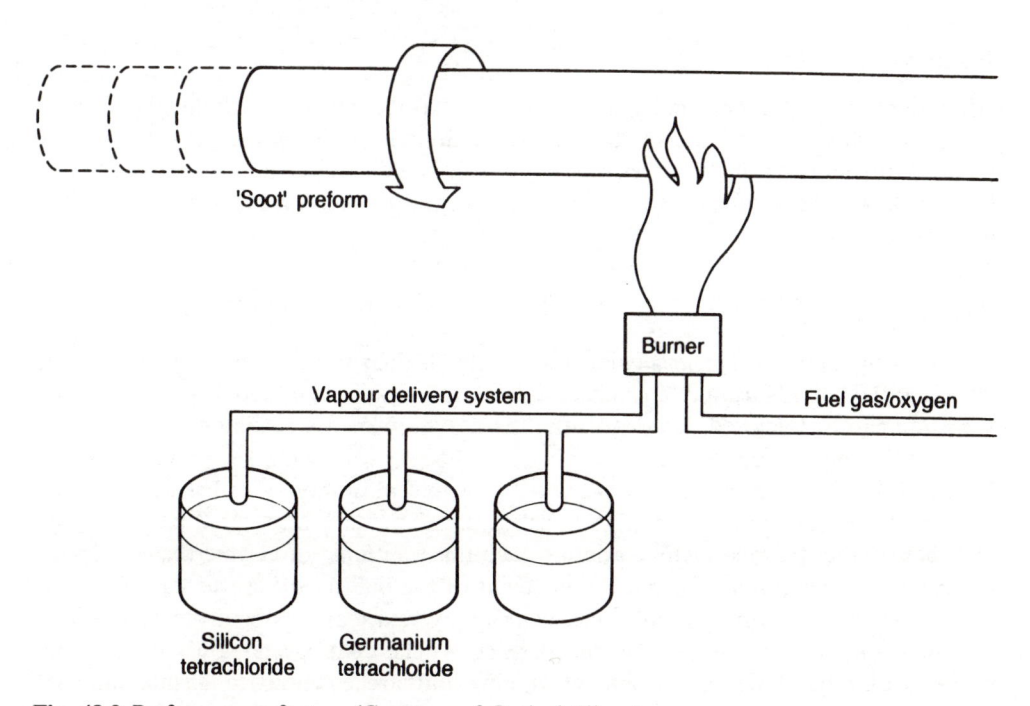

Fig. 48.3 Preform manufacture (Courtesy of Optical Fibres)

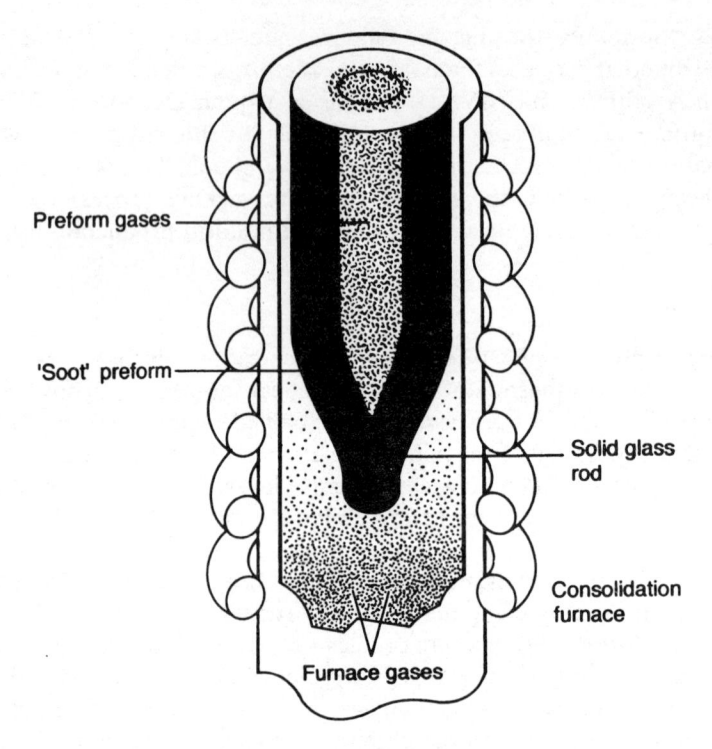

Fig. 48.4 Preform consolidation (Courtesy of Optical Fibres)

Preform consolidation

After deposition is completed the porous 'soot' preform is passed through a two-stage consolidation process. In the first stage drying gas is blown into the furnace reducing the moisture in the pores of the preform to a residue of a few parts per billion (see fig. 48.4). The high temperature, second stage induces preform collapse into a solid glass rod.

Fibre drawing

The drawing facility is constructed as a tower, with the furnace at the top and pulling wheels at the base. Naturally, in a process demanding extreme dimensional stability, no vibration can be tolerated. Consequently, the draw towers are mounted on substantial reinforced concrete foundations.

The glass rod is fed into the furnace, its tip heated until molten and the fibre drawn from below. Once a fibre has been established the diameter is monitored by laser beam just below the furnace. Any variation, within fractions of a micron, results in instantaneous computer controlled adjustment of the pulling speed (see fig. 48.5).

As the fibre is drawn, two polymeric coating layers are applied and the coated fibre is wound on to drums under controlled tension. The coatings not only protect and permit easier handling of the fine glass fibre but are essential to ensure ultimate performance.

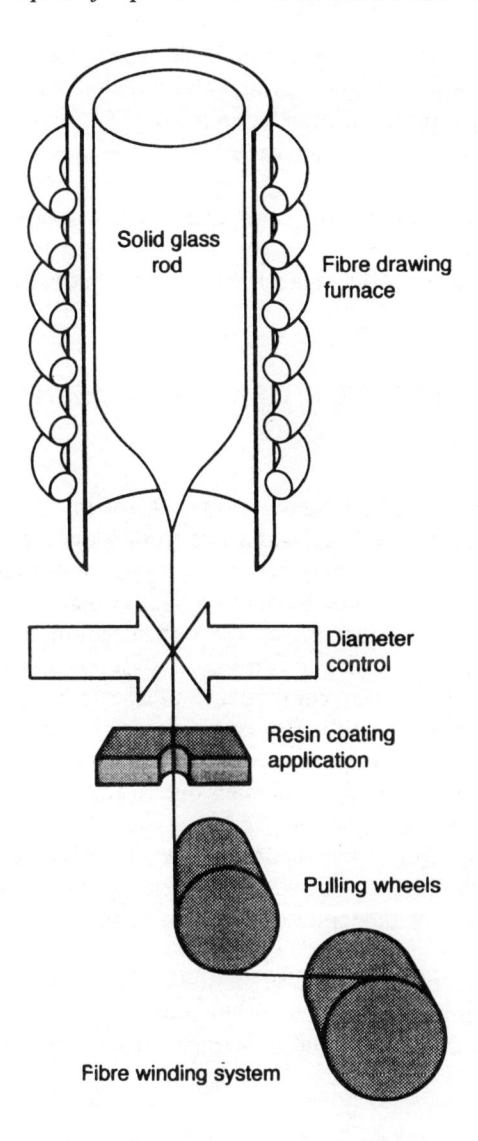

Fig. 48.5 Fibre drawing process (Courtesy of Optical Fibres)

Fibre measurement

Every fibre drawn is subjected to a series of optical and mechanical measurements. Prior to measurement the fibre is rewound on to measurement drums under very low tension and is proof tested at a fixed strain level for a set time. For multimode fibre the measurements include attenuation, bandwidth at two wavelengths, core diameter, and numerical aperture (a measure of the range of angles through which a fibre will accept light and sustain its transmission). For single mode fibre the related measurements include mode field radius (the effective size of the fibre core in which light is propagated), core concentricity (how accurately centred the core is relative to the surrounding cladding layer) and cut-off wavelength (the shortest wavelength at which propagation is single mode, and below which multimode propagation is supported) in addition to attenuation at several wavelengths. Other mechanical and environmental properties are continuously monitored on a sample basis.

All measurement equipment is designed for accuracy and speed of operation. Once the fibre is established in the measurement equipment the measurements are controlled by, and all results stored in, a computer. These measurements can be recalled at any time for any fibre produced.

Finally, the fibre is wound from the measurement drums on to shipping reels.

PROPAGATION CHARACTERISTICS OF OPTICAL FIBRES

Dispersion

Dispersion of the transmitted optical signal causes distortion for both digital and analogue transmission along optical fibre and arises from a number of causes.

Intramodal or chromatic dispersion may occur in all types of optical fibre and results from the intrinsic variation of the refractive index of the material with wavelength, together with the finite spectral linewidth of the optical source. As different wavelengths travel with different speeds in the fibre core, a propagation delay difference occurs between the different spectral components of the transmitted signal. More accurately, intramodal dispersion may be caused by material dispersion resulting from the different group velocities of the various spectral components launched into the fibre from the optical source, and waveguide dispersion due to the variation of the angle of the ray with the fibre axis with wavelength.

Pulse broadening resulting from the propagation delay difference between different modes within the multimode fibre is referred to as intermodal dispersion. As has been discussed, the effect is significant in step index multimode fibres, and much reduced in graded index fibre.

In purely single mode operation, however, it follows that there is no intermodal dispersion and pulse broadening is due solely to the intramodal dispersion mechanisms. As a result, single mode fibre exhibits the least pulse broadening and, therefore, the greatest bandwidth.

Fibre loss

We have discussed how fibre dispersion limits performance of communication systems. Fibre loss also limits this as the performance of optical receivers is degraded as the input signal is reduced. The following sections discuss the various loss mechanisms in optical fibres.

Generally, the power attenuation inside an optical fibre is governed by

$$\frac{dP}{dz} = -\alpha P \qquad (48.4)$$

where α is the attenuation coefficient and P is the optical power; α includes several sources of power attenuation, two important ones being material absorption and Rayleigh scattering. Table 48.1 illustrates typical optical losses for several different

Table 48.1 Typical fibre attenuation values

Fibre type	Attenuation at 850 nm (dB/km)	Attenuation at 1310 nm (dB/km)	Attenuation at 1550 nm (dB/km)
Single mode (8/125 μm)[a]	–	0.35	0.25
Multimode (50/125 μm)	2.4	0.5	–
Multimode (62.5/125 μm)	2.8	0.6	–

[a] Dimensions are core/cladding diameter

types of optical fibres. Note that single mode fibres are often produced with attenuation values as low as 0.18 dB/km at a wavelength of 1550 nm (1.55 μm) and this means that only 4.2% of optical signal is lost per kilometre.

Material absorption

Material absorption can be considered to have two elements. Intrinsic material absorption arising from the structure of pure silica and extrinsic absorption caused by the loss due to presence of impurities. For silica, electronic and vibrational resonances occur in the ultraviolet and infrared regions of the optical spectrum respectively. The main impurity present in modern silica fibres is pure water vapour, albeit reduced to very low concentrations (typically less than 1 part in 10^8), causing vibrational harmonics at 1.39, 1.24 and 09.4 μm, which appear as peaks in the loss versus wavelength spectrum

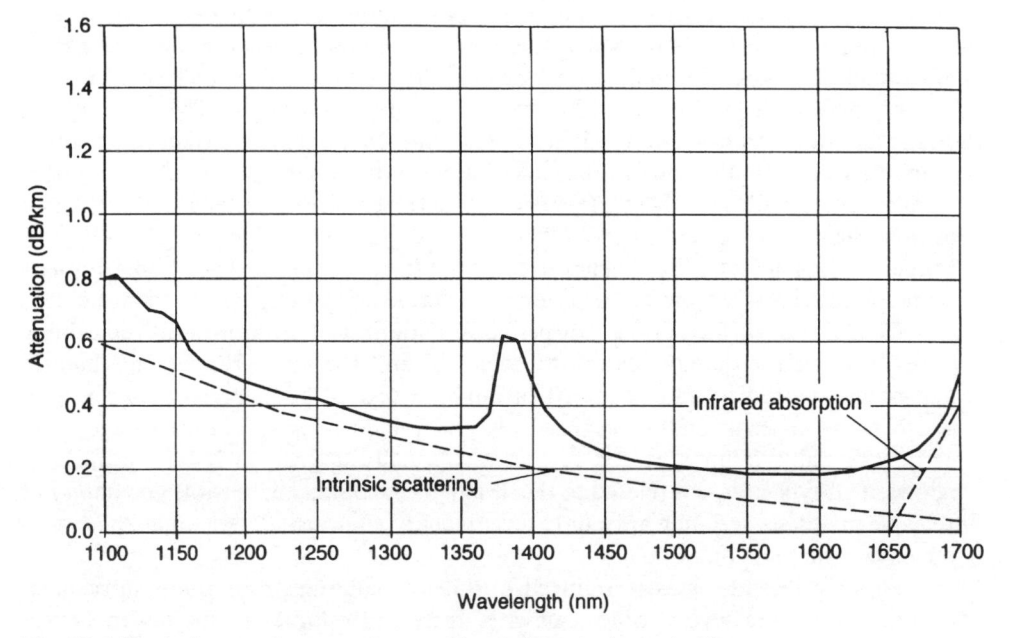

Fig. 48.6 Typical spectral attenuation profile of silica single-mode fibre

as shown in fig. 48.6. Other dopants used in the fabrication process to produce the desired refractive index profile can give rise to other losses.

Rayleigh scattering

Rayleigh scattering is caused by small variations in material density which cause local changes in the refractive index on a scale smaller than the optical wavelength. The scattering cross-section varies as λ^{-4} and this explains why the fibre attenuation normally decreases with wavelength up to about 1.6 μm (apart from the noticeable water peak previously mentioned at 1.39 μm).

As Rayleigh scattering reduces beyond 1.6 μm, infrared absorption begins to dominate, explaining why attenuation then starts to increase with wavelength. Scattering can also be caused by the presence of particulate impurities, but in modern silica fibre, this is negligible. In commercial silica fibres, fibre loss is at a minimum at a wavelength of around 1.55 μm, and this area of the spectrum is termed the third transmission window. Optical transmission at this wavelength is not always practicable for every system, and sometimes dispersion constraints lead engineers to use another wavelength. There are two other low loss regions with the so-called first transmission window at 0.85 μm (850 nm) and the second transmission window at 1.31 μm (1310 nm).

SOURCES OF FAULTS IN OPTICAL CABLES

Bend losses

A significant power loss mechanism for optical fibre is associated with the fact that the optical signal propagating in the fibre can be attenuated whenever the fibre encounters bends which have a radius of curvature below a critical value. These bending losses can be classified in terms of micro-bending or macro-bending, depending on the nature of the bend imposed by the fibre. Macro-bending losses arise whenever the fibre axis is bent by radii of the order of millimetres or more; the attenuation will be localised at the bend and increases significantly with increasing source wavelength. This type of loss impacts the third transmission window (1.55 μm) to a greater extent than other transmission wavelengths and is often introduced by fibre handling upon installation. Macro-bending losses are, therefore, normally detected directly after installation and cable jointing.

Micro-bending losses arise whenever the fibre axis is subjected to a small quasi-periodic perturbation (of the order of tens of microns). This can occur when the fibre is pressed against a surface that is not perfectly smooth, usually as a result the cabling process. In order for the attenuation to be significant, the micro-bending mechanism would normally need to be distributed throughout several kilometres of fibre length. Micro-bending losses normally increase slowly with increasing source wavelength, and consequently cable failures caused in this way usually impact on both transmission windows. This type of loss is related to the design of the optical fibre cable and choice of fibre/cable materials and may arise only several years after installation and exposure to harsh environmental conditions.

In addition to bend losses introduced during the cabling process and subsequent installation, there are several other causes of increased optical attenuation in optical fibre cables.

Ingress of hydrogen

As outlined earlier, hydrogen can impair the optical transmittance of optical fibre by forming absorption bands. First, molecular hydrogen can permeate into the silica leading to infrared absorption at 2.42 μm. Both the 1.31 μm and the 1.55 μm transmission windows can be affected by the flanks of these absorption peaks. The actual attenuation increase is dependent on the hydrogen pressure, and a saturation level is reached after about 500 hours exposure. The attenuation saturation level is found to decrease with increasing temperature. On removal of the hydrogen the attenuation reduces to its original level, hence this effect is sometimes termed the reversible effect.

A permanent attenuation increase results from OH-ions reacting with germanium and phosphorous dopants within the fibre. In this case the loss is manifested as a growth in the already existent OH absorption peak at about 1.4 μm (Ge-OH), and an increase at about 1.6 μm (P-OH).

For hydrogen concentration to build up in the vicinity of fibres there has to be some local generator of hydrogen and some means of containing it in the cable. Many conventional optical cables use a central steel strength member which may corrode with time thus generating hydrogen. There have been instances when a peripheral metallic water barrier applied to the cable has prevented hydrogen from escaping allowing it to build up to levels sufficient to cause measurable losses. The use of non-galvanised steel wire reduces this risk. In some countries the use of a hydrogen scavenging material in the interstices of the cable has proved popular, although the absorbing capacity and longevity of such materials is not well understood.

Non-metallic cable designs, using only polymeric materials, also generate hydrogen during their natural decomposition process. However, the quantity of the hydrogen produced and its retention in the structure is very small compared with that emanating from metals, and consequently associated optical losses are undetectable.

Material compatibility

Optical fibres are manufactured with a variety of protective coatings to preserve the integrity of the underlying glass strand. Changes in properties of the fibre coating can give rise to localised coating delamination resulting in significant microbending losses. It is, therefore, important that the characteristics of fibre coatings and their adhesion to the glass surfaces do not change as a result of ageing, contact with other materials, such as tube filling gel, or materials from the environment in which the cable is installed, such as water, cleaning fluids or fuels. These concerns are reflected in specifications addressing issues of mutual compatibility of cable materials and the exposure of the optical fibre to contaminants (including water).

OPTICAL FIBRE STRENGTH

Contrary to popular perceptions, silica glass is a very strong material which in its pure form without surface flaws results in strains at break of about 5%; the effect of surface flaws is that failure of the fibre can occur at much lower strains. Much effort is taken in fibre manufacture to minimise these flaws and to protect the bare fibre surface from damage by applying a polymeric surface coating immediately after drawing the fibre on the tower. The fibre is then proof-tested by straining it to a pre-determined level for a short time to screen out any small number of larger flaws which would cause the fibre to break at low strains.

It is important to appreciate that surface flaws (or cracks) will grow if the fibre is under strain, a phenomenon known as static fatigue. Eventually the flaws will grow to a point when growth becomes very rapid and the fibre will break. This is a vitally important design criterion for optical cables. It must be ensured that the level of strain which fibres experience over their lifetime through cable manufacture, installation and service in the field is such that the probability of crack growth is acceptably low.

Much theoretical and experimental work has been done to understand fully the mechanisms of crack growth leading to usable design rules for the incorporation of fibre into cables.[1,2] For applications in extremely adverse environments, so-called 'fatigue resistant' and hermetically coated fibres have been developed in which advanced coatings are applied to the fibre to inhibit crack growth.

MEASUREMENT OF OPTICAL TRANSMISSION FAULTS

Optical time domain reflectometry

Optical time domain reflectometry (OTDR) is the most commonly used technique for testing long and short haul cable links. One of the main practical advantages of the technique is that it only requires access to one end of the fibre under test. OTDR may be used to test attenuation (or length) of cable sections or of various joints and demountable connections. The OTDR operates by launching a fast pulse of laser light into the optical cable to be measured. The light is scattered by Rayleigh scattering along the length of the fibre and a small fraction is scattered and guided back toward the OTDR. This backscattered light is analysed to produce an attenuation profile of the fibre along its length as illustrated in fig. 48.7. The distance travelled by the transmitted laser pulse to a certain point along the fibre can be calculated by multiplying half the time taken for the pulse's back-scattered reflection to be received back at the OTDR by the velocity of light

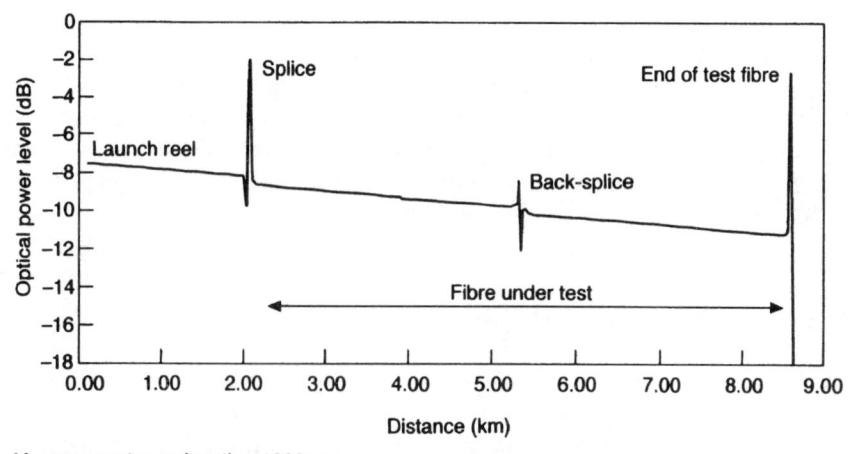

Fig. 48.7 A typical OTDR trace

in the glass core. If the appropriate refractive index is keyed into the OTDR, the time sampling can be converted into distance and the results observed represent the optical power level, and, therefore, loss, as a function of distance along the fibre.

Cutback

A more accurate, but more time consuming, method for determining the optical attenuation of cables is the cutback or insertion loss technique. Here the cable to be measured is connected between a light source operating at the desired wavelength and an optical detector. Herein lies a practical drawback to the method – access is required to both ends of the fibre. The light source should be specified for single mode or multimode requirements, whichever is required. The method of connecting the cable to the source will depend on whether a connectorised source is used. If so, then a connectorised pigtail may be spliced to the launch end of the cable and attached to the source. The output end of the fibre can be coupled into the detector using a bare fibre adaptor. In this configuration, the power meter is set at zero, the cable is cut at the launch end and then the fibre is reinserted into the detector. The resulting change in attenuation equates to the insertion loss of the cable. Great care must be taken to establish suitable launch conditions prior to the measurement to represent transmission accurately over reasonable distances. This may involve ensuring that the cladding is mode stripped (i.e. higher order modes which may be weakly propagated within the cladding layer, rather than the fibre core, are eliminated) and that the core is mode scrambled (artificially evening out the power propagated in the various transmitted modes as they naturally become over a long length of fibre). Such detail is beyond the scope of this section and the reader is referred to other texts for a comprehensive explanation.

REFERENCES

(1) Glaesemann, G. S. and Gulati, S. T. (1991) 'Design methodology for the mechanical reliability of optical fiber'. *Optical Engineering* **30** (6), 709–715.
(2) Hodge, K. G. *et al.* (1992) 'Predicting the lifetime of optical fibre cables using applied stress histories and reliability models'. *41st Proc. Int. Wire and Cable Symp.*, pp. 713–724.

CHAPTER 49
Optical Fibre Cable Construction

Earlier chapters emphasised the need for protection of optical fibres in cable constructions to guarantee their reliable operation.[1] Principally there are two approaches to protecting optical fibres in cables. Fibres may be housed in a loose cavity to ensure that external loadings are not transferred directly to them, or the cable may incorporate cushioning materials which are designed to limit the fibres' exposure to these extraneous influences. Furthermore, the finished cable has to protect the fibre throughout the required envelope of environmental conditions.

GENERAL DESIGN CONSIDERATIONS

Mechanical loading

In order to calculate the degree of protection which must be afforded to the optical fibres it is imperative to assess the effect on the cable of the envisaged mechanical loadings. The tensile loadings expected are highly dependent on the type of cable installation concerned and are usually specified by the customer. It is usual to employ the 'rule of mixtures' approach to evaluate the mechanical response of the cable to a load:

$$\text{cable strain} = \frac{F}{\sum_{i=1}^{n} A_i E_i} \quad \text{(unitless)} \quad (49.1)$$

This expression implies that the tensile force, F (in N), is opposed collectively by n components of the cable structure, each of which have areas and moduli of A_i (m^2) and E_i (Pa). If the different cable elements do not act collectively by locking together under tension, then it may be more accurate to consider only the known cable strengthening elements, such as a central strength member. In small tight cable constructions, internal datacom leads for example, it may be appropriate to include the strength contribution of the fibres themselves in this calculation.

Temperature

It is equally important to assess the effects of changes in temperature on optical cables. As in the case of mechanical loading this requires an assessment of the contributions made to the overall cable performance by each cable component. The composite thermal expansion coefficient of a cable is also provided by the application of the 'rule of mixtures':[2]

$$\text{cable thermal expansion coefficient} = \frac{\sum_{i=1}^{n} A_i E_i \alpha_i}{\sum_{i=1}^{n} A_i E_i} \quad (°C^{-1}) \qquad (49.2)$$

where α_i are the individual cable components' expansion coefficients.

Of particular importance is the relative expansion of the cable with respect to the packaged optical fibres. The fibres, by virtue of their high glass content, have a very low expansion coefficient (in the case of silica fibres, typically $10^{-7}\,°C^{-1}$), and so their length can be considered to be essentially invariant with temperature. This makes it difficult to package them in such a way that they do not suffer any adverse effects when the rest of the cable moves relative to them.

CABLE CONSTRUCTION

Most optical cables have, at their centre, a strength member of one kind or other, around which the fibres, protected in various ways, are stranded as shown in fig. 49.1.

When calculating the necessary size for the central strength member of a cable it is advisable to use a value for the elements' diameters which takes account of manufacturing tolerances. For a generic cable with N elements stranded up with a lay length, L, the minimum size of central member needed to provide support for the stranded fibre units with diameters, d, is given by:

$$D = d \left\{ \left[\frac{1 + \tan^2\left(\frac{\pi}{N}\right)}{\tan^2\left(\frac{\pi}{N}\right) - \left(\frac{\pi d}{L}\right)^2} \right]^{1/2} - 1 \right\} \qquad (49.3)$$

Similarly, it is usual to incorporate peripheral interstitial filling members to provide a more circular cross-section on which to apply subsequent cable layers. The diameter of these fillers, d', if D is the cable centre size and d is the diameter of the stranded fibre elements, is given by:

$$d' = \frac{x^2}{\frac{d}{2} + x} \qquad (49.4)$$

where

$$x = \frac{D}{2} + d - \left[\left(\frac{D}{2}\right)^2 + \frac{Dd}{2} \right]^{1/2} \qquad (49.5)$$

All cables in which the fibres are stranded around another element impose a degree of bend on the fibres. In the case of a helical construction the bending radius of the fibres,

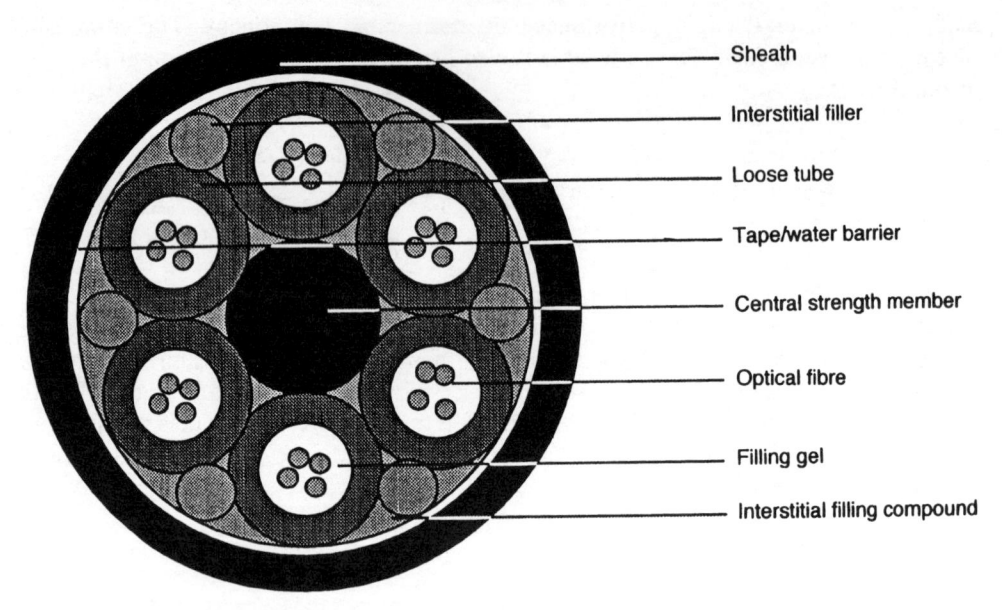

Fig. 49.1 Typical construction of a stranded optical cable

r, is constant along the length of the cable. Given a stranding radius for the fibres around the centre line of the cable, R, and a lay length, L, we can evaluate r thus:

$$r = R\left[1 + \left(\frac{L}{2\pi R}\right)\right]^2 \quad \text{(mm)} \qquad (49.6)$$

If reverse lay (SZ) stranding is employed then the bending radius of the stranded elements is not constant, varying from a minimum value, at points where the elements are effectively helically laid up, to infinity at the reversal points where the elements are momentarily parallel with the cable.

It is important that the bend radius of the fibre elements is maintained above a minimum value for two principal reasons. First, because the fibres are bent to the calculated level along the entire length of cable, there are typically several thousand such turns per kilometre. Together these may be sufficient to cause some degree of optical loss[3] through a combination of the macro-bend and micro-bend mechanisms described earlier. The second potentially damaging effect of having too small a fibre bending radius is that the outer edge of the fibre may experience sufficient bending stress to promote fatiguing of the glass,[4] particularly when combined with severe bending of the cable itself. For standard 125 μm diameter silica fibres a 60 mm minimum bend radius is typically specified.

Loose tube cables

In this type of cable design a polymeric tube (usually polybutylene-terephthalate, or PBT) is extruded loosely over several optical fibres. The maximum number of fibres which can be accommodated in a particular tube depends on the tube size, but typically there may be up to about twelve. A thixotropic, chemically neutral filling compound is usually pumped into the tube as part of the manufacturing process and this serves to prevent longitudinal migration of potentially harmful moisture along the tube during service. The tubes and optical fibres are colour coded to allow easy field identification of individual fibres. By virtue of its loose fit the buffer tube provides considerable crush and impact resistance to the fibres.[5,6] During the manufacturing process it is normal to incorporate a longer length of fibre in the tube than that of the tube itself. This 'excess length' usually amounts to about 0.05–0.10% and its purpose is to prevent the fibres from experiencing any tensile strain during subsequent manufacturing operations, such as stranding, when a back tension may be applied to the tube.

Several of these so-called 'loose tubes' may be stranded up around a central member which usually constitutes the cable's major strength element. Thus the tubes assume a helical path around the central member. This type of cable core structure is common to many different application areas, but there is considerable variation possible in the rest of the cable structure.

Figure 49.2 illustrates a typical cable design that may be used in a conventional long-haul duct installation. Polyester or paper tapes are applied over the loose tubes, and the interstices between the tubes are filled with more water-blocking filling compound. Above these tapes a further tape is applied which is intended to act as a barrier to the radial ingress of water. Typically a laminated aluminium and polyethylene tape is applied at an elevated temperature which causes the polyethylene to melt and form a bond with the tape overlap above it. A suitable sheathing material is added to complete the cable.

In directly-buried applications a more robust corrugated steel tape may be used as a form of mechanical protection and also as a water barrier. Indoor cables generally have no need for either filling compounds or water barriers, but may require special cable sheathing materials to be used to limit the potential fire hazard posed by the cable.

The purpose of the strengthening materials in all of these application areas is to minimise the amount of tensile strain to which the optical fibres are exposed. The most cost-effective form of reinforcing material is steel which for flexibility usually takes the form of a stranded wire. In applications where non-metallic materials are preferred this can be replaced by aramid or glass yarns or pultruded rods.

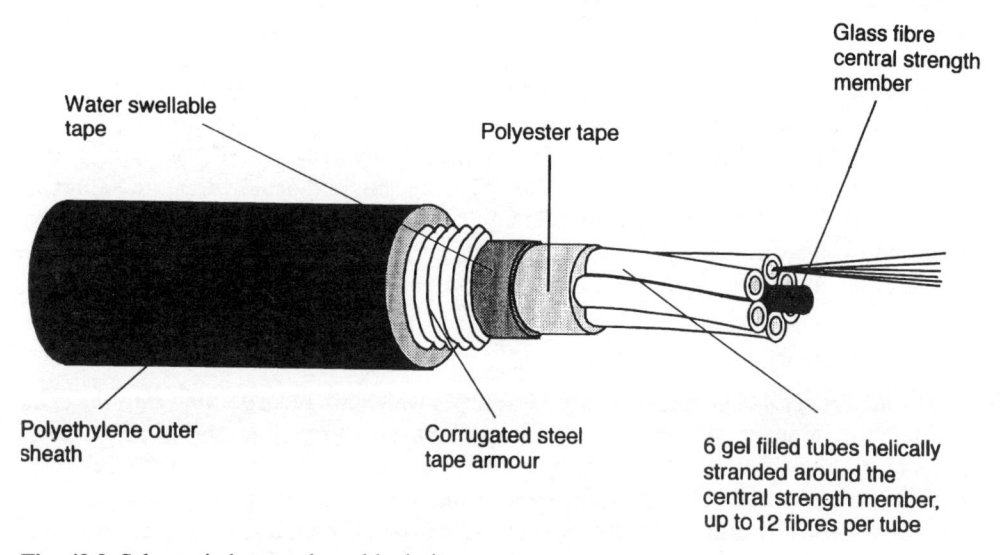

Fig. 49.2 Schematic loose tube cable design

Strain relief of the optical fibres is achieved by their ability to move radially through the loose tube filling compound as the cable is strained. Under tensile load the pitch length of the loose tube, and hence the fibre in a helically laid cable, is extended and this is compensated for by the reduction in the effective stranding diameter of the fibres. When the fibres move to the inner wall of their tubes they will then commence to strain at the same rate as the cable. Mathematically we can define the strain margin thus:

$$\text{strain margin} = 100 \times \frac{\pi^2}{L^2}\left[(b - d_{\text{eff}})(D + d_{\text{t}}) - \frac{(b - d_{\text{eff}})^2}{2}\right] \quad (\%) \qquad (49.7)$$

where L = lay length of the tube
 b = bore of tube
 d_{eff} = effective diameter of bundle of fibres
 d_{t} = diameter of tube
 D = diameter of cable central member

The precise geometrical form of the fibre bundle cannot be precisely determined, but an empirical relationship that yields satisfactory results takes the form:

$$d_{\text{eff}} \approx d_{\text{f}}\left(\frac{5n - 2}{3}\right)^{1/2} \qquad (49.8)$$

where d_{f} is the diameter of an individual fibre (approximately 250 μm) and n is the number of fibres in the bundle.

For a typical cable structure the strain margin normally built into the cable is in the range 0–0.5%, depending on the application. Cables intended for aerial applications where they are likely to be exposed to large dynamically-varying loads generally have larger strain margins than those intended for more benign installations such as ducts. The intention of the strain margin in all cables is that the optical fibres should be isolated from any tensile strain during their lifetime ensuring the maximum possible reliability of the fibres.

Also important in loose tube designs is the so-called compressive margin. This is effectively the opposite of the strain margin, and represents the amount by which the cable can shrink thermally relative to the fibres before they are forced against the outer walls of their tubes. If the cable contracts beyond this point then the fibres may be pushed against the outer wall of the tubes with excessive pressure, potentially leading to micro-bend losses. The mathematical approach to calculation of the compressive margin of a cable is similar to that for the strain margin, and usually the two values are quite similar. Together, the strain margin and compressive margin constitute the strain-free window. The tensile behaviour of a cable with strain and compressive margins is shown in fig. 49.3.

Tight constructions

An alternative class of cable is composed of fibres which are incorporated into the cable without any over-length. When individual fibres are housed in this type of cable mechanical protection is provided by multiple additional coating layers on the relatively fragile primary-coated fibres, these normally consisting of a soft inner layer and a hard outer layer. The purpose of this type of structure is exactly analogous to that employed in loose tube cables – the outer layer protects the fibres from extraneous loadings whilst

Optical Fibre Cable Construction

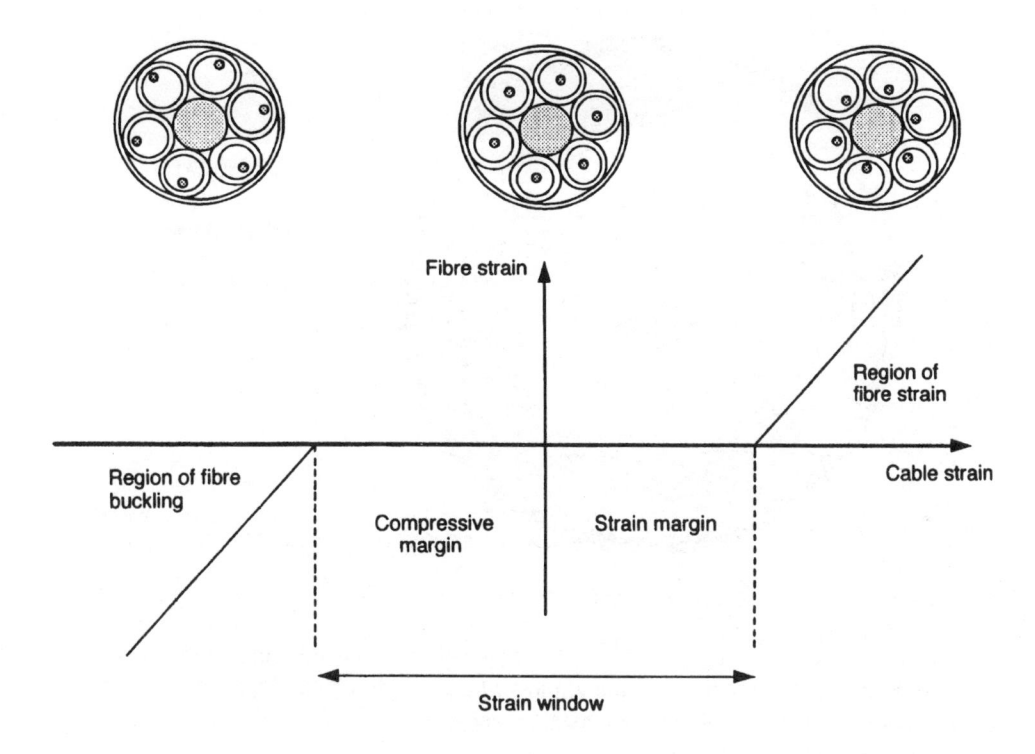

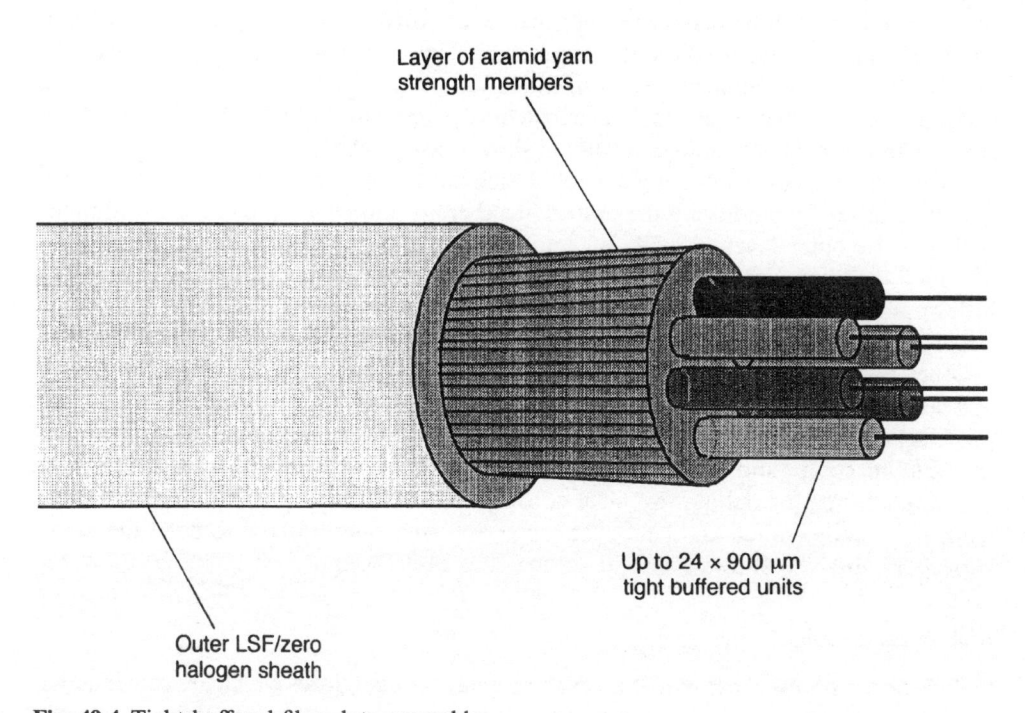

Fig. 49.4 Tight buffered fibre datacom cable

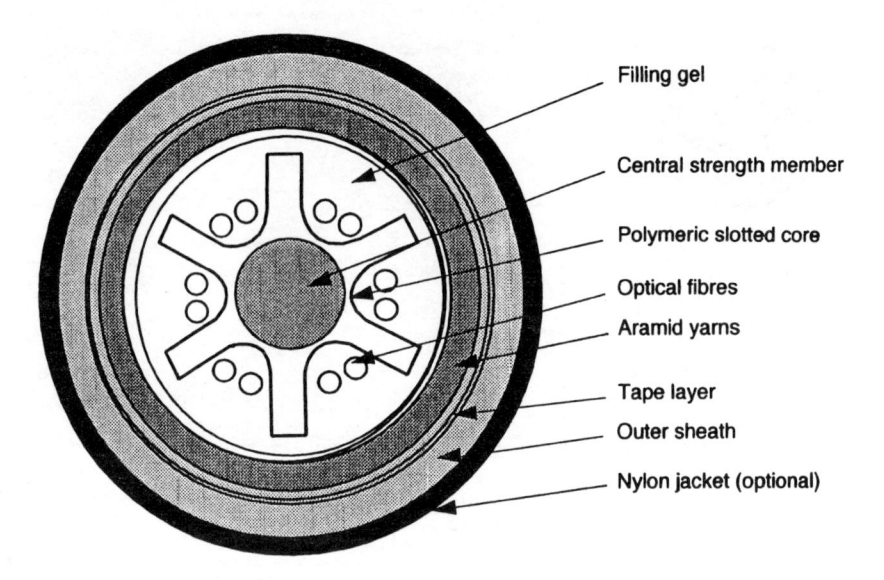

Fig. 49.5 Slotted core optical cable design

the soft inner layer uncouples the fibre to some degree from the outer layer, allowing it to accommodate differential thermal expansion[7] and lateral loadings. One such typical construction consists of a 400 μm diameter layer of silicone and a 900 μm diameter layer of nylon which can be colour coded to allow for easy identification.

Cables employing this type of fibre package tend to be larger than cables housing the same number of fibres in a loose tube configuration when any more than a handful of fibres are required. For this reason these cables are often used for internal applications near terminal equipment where the requirement is for individual fibres which may be quickly and easily terminated. Their lack of strain margin makes this type of cable less suitable for the overhead line environment. A typical tight buffered fibre cable construction for internal datacoms use is shown in fig. 49.4.

It is possible to construct a 'tight' cable which has a degree of strain relief built into it. This is achieved by modifying the central member around which the fibres are stranded such that its outer layer is made of a soft compressible material. When the cable is extended the fibres are allowed to bind down on and compress this soft layer, exerting little strain on the fibres themselves. In cables in which this approach has been adopted strain relief up to a level of about 0.3% has been realised.

Other tight cable constructions involve the packaging of groups of primary coated optical fibres within helical slots in what is termed a 'slotted core'. Typically the slotted rod is made of a tough polymeric material and the fibres are accompanied in the slots with a filling compound. The fibres may be laid into these slots in such a way as to allow a certain latitude for radial movement constituting a strain margin, but this movement tends to be quite limited and hence the cables are usually considered to be of the 'tight' variety. Figure 49.5 shows a typical slotted core cable design.

Ribbon cable designs

Optical fibre ribbons[8,9] are planar arrays of primary coated fibres which are embedded in a polymeric matrix. In theory any number of fibres may be packaged in this way, but the array has typically been limited to between two and twelve fibres. Whilst the approach allows a higher density of fibre packaging within a cable, thus reducing cable size and materials usage, the main advantage to be gained is in the accelerated termination of the fibres afforded by the use of mass fusion splicing and preparation equipment.

Until recently optical fibre ribbons have tended to be incorporated into slotted core cable designs in which the helical slots have a rectangular shape to accommodate a stack of ribbons. Again, it is normal to consider this type of cable design as tight, having negligible strain margin. Further, any packaging of optical ribbons involving their twisting into helical slots involves the imparting of differential bending strains[10] across the stack of ribbons which need to be evaluated before the design can be used with

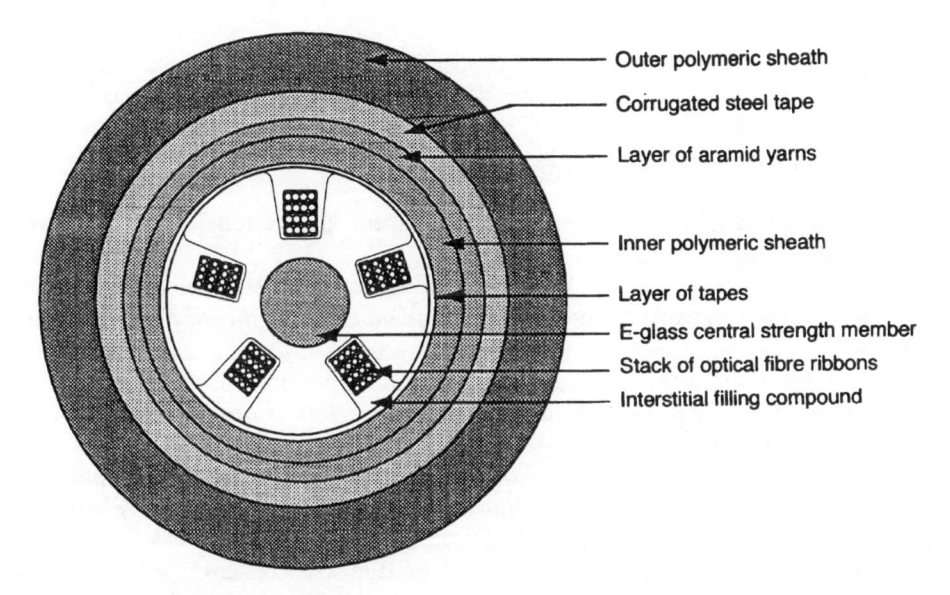

Fig. 49.6 Slotted core ribbon cable

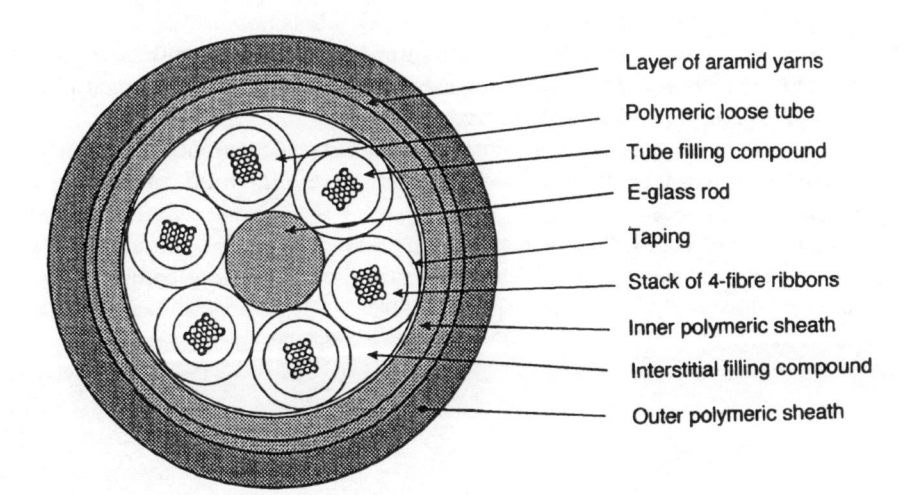

Fig. 49.7 Ribbon in loose tube design

confidence. Whilst some ribbon cable constructions have found application in the aerial cable environment, these tend to have housed relatively few fibres in a single longitudinal slot. A more conventional terrestrial variant is shown in fig. 49.6.

More recently it has become evident that a cable design combining the installation speed and cost-effectiveness of a ribbon-in-slot design with the ready accessibility which can be obtained with loose tube designs may be possible. These cables, known as ribbon in loose tube designs,[11] offer significant advantages in the local loop environment and a representative design is shown in fig. 49.7.

REFERENCES

(1) Bark, P. R., Oestreich, U. and Zeidler, G. (1979) 'Stress-strain behaviour of optical fibre cables'. *Proc. 28th Int. Wire and Cable Symp.*, pp. 385–390.
(2) Cooper, S. M. *et al.* (1986) 'The effect of temperature dependent materials properties on fiber optic cable design'. *Proc. 35th Int. Wire and Cable Symp.*, pp. 148–158.
(3) Stueflotten, S. (1982) 'Low temperature excess loss of loose tube fiber cables'. *Applied Optics* **21** (23), 4300–4307.
(4) Hodge, K., Vinson, J., Haigh, N. and Knight, I. (1994) 'Reliability of optical fibres: impact on cable design'. *IEE Colloquium on Reliability of Fibre Optic Cable*, pp. 1–6.
(5) Hogg, C. and Huang, P. (1991) 'Crush resistance parameters of loose tube optical fibre cables and the application to cable design'. *16th Australian Conf. on Optical Fibre Technology*, pp. 390–393.
(6) Fischer, L., Huber, C. and Rydskov, L. (1990) 'Designing loose tube optical cable for increased crush resistance'. *Proc. 8th Annual European Fibre Optic Communications and Local Area Networks Conf.*, pp. 86–91.
(7) Lenahan, T. (1985) 'Thermal buckling of dual-coated fibre'. *AT&T Technical Journal* **64** (7), 1565–1584.
(8) Greco, F. and Ragni, A. (1992) 'High performance optical fiber ribbon by UV-curing method'. *Wire Journal Int.*, **25** (12), 63–72.
(9) Donazzi, F. *et al.* (1991) 'Problems related to the design and manufacture of optical fibre ribbons'. *Proc. 9th Annual European Fibre Optic Communications and Local Area Networks Conference*, pp. 116–121.
(10) Hatano, S. *et al.* (1986) 'Multi-hundred fiber cable composed of optical fiber ribbons inserted tightly into slots'. *Proc. 35th Int. Wire and Cable Symp.*, pp. 17–23.
(11) Baguer, L. *et al.* (1994) 'Study, manufacture, and test on a S–Z stranded loose tube optical fibre ribbon cable with no pretwisting and no filling compound'. *Proc. 43rd Int. Wire and Cable Symp.*, pp. 499–508.

CHAPTER 50
Composite Overhead Conductors

BARE OVERHEAD CONDUCTORS

Bare overhead conductors form an integral part of almost every electrical power transmission and distribution network. This chapter describes the construction and principal design aspects of these conductors.

In addition to the power carrying phase conductors, single or double circuits on steel supporting towers are normally protected against lightning by at least one earth wire which also passes fault currents. The conductor design for a given application is based on the cost effective satisfaction of the electrical, mechanical and environmental requirements.

Construction

The majority of conductors are configured in a concentric lay stranded form – that is, a single straight core wire surrounded by one or more layers of helically stranded wires. The direction of twist is reversed in adjacent layers, with the direction of lay of the outermost layer conventionally being right-handed.

Types

In addition to stranded conductors constructed from copper or cadmium copper wire, a variety of constructions based on aluminium are available to give an optimum solution to the line requirements. These are as follows.

AAC *A*ll *A*luminium *C*onductor is the lowest cost conductor for a given current rating but its low strength:weight ratio makes it suitable for relatively short span lengths only.

ACSR *A*luminium *C*onductor *S*teel *R*einforced has a higher strength:weight ratio than AAC and because of this can be used on longer spans and can withstand more severe weather conditions. It also possesses a higher modulus of elasticity and a lower coefficient of thermal expansion, both of which enhance its mechanical performance. These properties can be varied by altering the aluminium:steel ratio in the stranding geometry. Higher strength steel can also be used which improves the strength:weight ratio without affecting other properties. The central steel core normally consists of galvanised steel wires but may be of aluminium clad steel wires which give additional conductivity and corrosion protection in exchange for greater cost and some compromise of the mechanical properties. This latter conductor type is abbreviated to ACSAR.

AAAC *A*ll *A*luminium *A*lloy *C*onductor is a homogeneous construction of aluminium alloy wires. The alloy is a heat-treated magnesium silicon aluminium type and the resulting conductor offers several advantages over the AAC and ACSR types. This has led to the increasing use of AAAC in modern lines. These conductors have a superior strength:weight ratio, allowing scope for reduced sags, greater support tower spacing or increased current carrying capacity on existing tower structures, lower a.c. resistance leading to lower line losses, absence of galvanic corrosion, less complex installation fittings, and reduced vulnerability to surface damage.

ACAR *A*luminium *C*onductor *A*lloy *R*einforced is constructed with a central core of aluminium alloy wires and provides a conductor option between those of AAC and ACSR.

AACSR *A*luminium *A*lloy *C*onductor *S*teel *R*einforced has a high strength:weight ratio at the expense of conductivity and is used where these properties are advantageous such as for earth wire applications.

Although wires used in the conductors are most commonly circular in section, two methods, outlined below, are used to maintain the cross-sectional area of a conductor at the same time as reducing the diameter. This helps to minimise ice and wind loading but requires the sacrificing of some conductor flexibility.

Compacted conductors
Conductors stranded from up to two layers of round wires can be manufactured so that at least the outer of these layers is compacted during the stranding process using either a set of shaping rollers or compacting dies.

Segmented conductors
These are conductors stranded with at least one layer of specially drawn wires having the appropriate segmental shape.

Materials

Table 50.1 gives the typical characteristics of metallic materials used in the construction of the common conductor types available.

Of the materials listed in table 50.1, aluminium alloy is worthy of further explanation because of the conductivity/strength balance it offers, and its steadily increasing use. The objective in moving to an alloy is to improve the mechanical properties without having too much effect on the conductivity. The alloy most commonly used in this application is

Table 50.1 Characteristics of conductor materials

	Units	Hard drawn copper	Cadmium copper	Hard drawn aluminium	Aluminium alloy	Galvanised steel	Aluminium clad steel[a]
Conductivity	%IACS	97	79.2	61	53	9	20
Resistance	$\Omega\,mm^2/km$	17.71	21.769	28.264	32.5	192	84.8
Temperature coefficient of resistance per °C		0.00381	0.0031	0.00403	0.0036	0.0054	0.0051
Coefficient of linear expansion per °C	$\times 10^{-6}$	17	17	23	23	11.5	12.96
Linear mass	$kg/mm^2\,km$	8.89	8.945	2.703	2.7	7.8	6.59
Ultimate tensile stress	MPa	414	621	160–200	295	1320–1700	1100–1344
Modulus of elasticity	GPa	125	125	70	70	200	162

[a] 10% by radius aluminium

an aluminium–magnesium–silicon alloy with the BS EN 573 designation EN AW-6101 or EN AW-6201. The main alloying elements of EN AW-6101A, for instance, are limited to the ranges Mg 0.4–0.9%, Si 0.3–0.7%, Fe <0.4% and Cu <0.5%.

Careful control of the composition and processing of the alloy must be maintained throughout the production process to realise the required combination of strength and conductivity. The alloying elements must first be dissolved into solid solution in the base aluminium by a solution heat treatment which involves raising the temperature of the metal to near its melting point, holding the temperature until the elements are in solution, then rapidly quenching so that the magnesium and silicon remain in solid solution in a metastable state. At this stage the metal (which is conventionally 9.5 mm rod with a comparable strength but significantly reduced conductivity compared with 'electrical grade' aluminium) is normally drawn through a set of dies to its final diameter, a process which increases the strength by 'work hardening'. A further heat treatment or artificial ageing process, at a temperature in the range 150–200°C, accelerates the onset of precipitation of the magnesium and silicon as Mg_2Si. As this precipitation begins to occur the conductivity of the metal increases as the impurities come out of solution and the strength initially increases to a maximum before falling away as Mg_2Si recombination occurs.

Typical precipitation heat treatment curves are shown in fig. 50.1. These demonstrate the degree of control over the 9.5 mm rod composition, processing, subsequent drawing, and heat treatment of the final wire necessary to achieve the desired end result in terms of tensile strength and conductivity.

A variety of combinations of strength and conductivity are specified, reflecting the user's requirements, but in general terms the range of final properties that can be obtained is illustrated in fig. 50.2. The dashed line represents approximately the best balance between mean tensile strength and conductivity that can be obtained; the limiting values of some of the alloys currently in use in Europe are also shown.

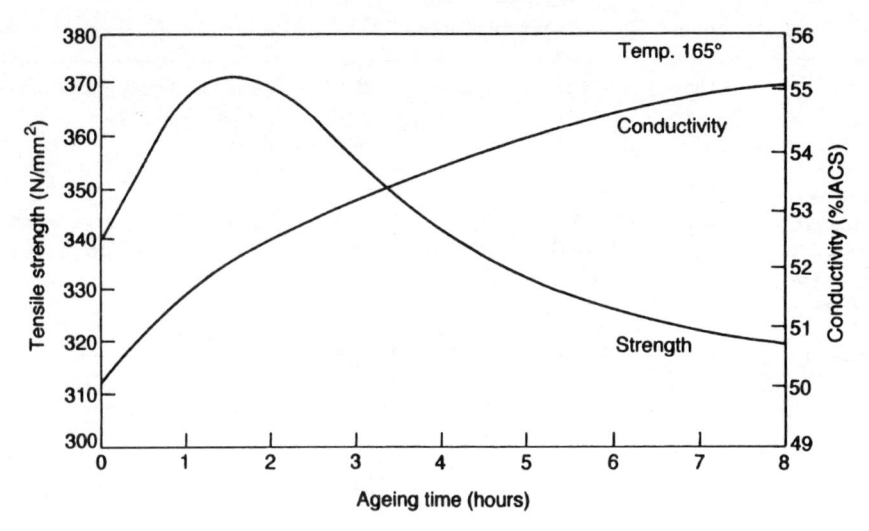

Fig. 50.1 Typical heat treatment curves for aluminium alloy wires

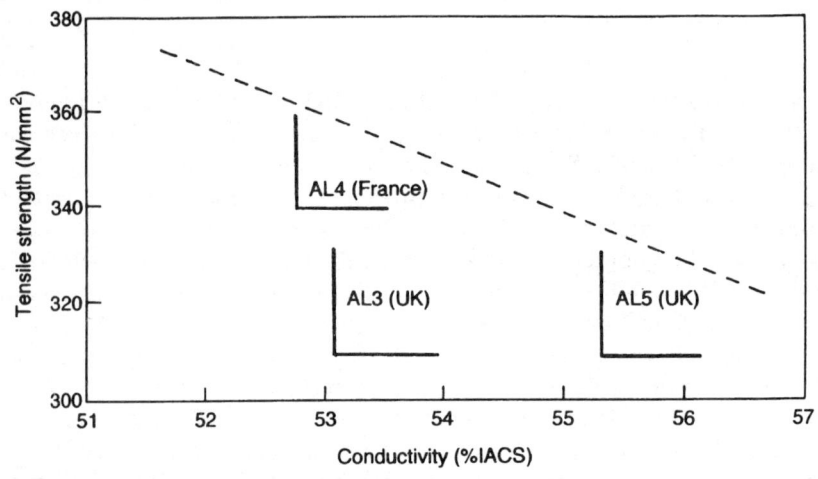

Fig. 50.2 Relationship between tensile strength and conductivity of heat-treatable aluminium–magnesium–silicon alloys

Standards

International, European, and UK standards for bare overhead conductors at the time of publication are as follows.

IEC 1089	Round wire concentric lay overhead electrical stranded conductors
prEN 50182	Round wire concentric lay overhead electrical stranded conductors
BS 215: 1970	Aluminium conductors and aluminium conductors, steel-reinforced
	Part 1: Aluminium stranded conductors
	Part 2: Aluminium conductors, steel-reinforced
	Note: This standard will be replaced by BS EN 50182 on adoption of the European standard

BS 3242: 1970   Aluminium alloy stranded conductors for overhead power transmission
                Note: This standard will be replaced by BS EN 50182 on adoption of the European standard
BS 7884: 1997   Specification for copper and copper–cadmium stranded conductors for overhead electric traction and power transmission systems.

Electrical requirements

The key requirements are for adequate current handling capacity, including under short-circuit fault conditions, and acceptable corona discharge performance.

Current carrying capacity

The prime requirement for the phase conductors, to carry current with acceptable power loss or voltage drop, is satisfied by the use of the appropriate cross-sectional area of high conductivity metal such as aluminium, aluminium alloy or copper.

The electrical constraint on an earth wire is its ability to carry safely the maximum expected short circuit current and is normally expressed in $kA^2 s$ for a given allowable conductor temperature rise. This short circuit rating is also known as the '$I^2 t$ rating'.

The fault current carrying capacity of a conductor can be determined by consideration of the following:

(a) the maximum temperature reached during current transients;
(b) the mechanical degradation of the cable caused by the temperature and duration of the fault;
(c) the amplitude, rate of change and frequency of the electrodynamic forces and bending moments during the fault.

It is conventional for the first criterion to be used when designing a conductor or earth wire although consideration of the second and third criteria is normally addressed during the later testing stages of the cable's development.

Fault current calculations[1] reflect the conversion of the electrical energy dissipated during a fault event into thermal energy with the result that there is an attendant temperature rise in the conductor. In the simplest method of calculation the resistance, heat capacities, and densities of the cable's constituent materials, and the cable dimensions are assumed to remain constant throughout the fault and temperature rise. More exact methods assume quadratic dependencies of these quantities and there are instances recorded of hybrid calculation methods using mixtures of invariants and linear and quadratic dependencies. In general it is assumed that the metal content in conductors will almost instantly reach a temperature peak, and that the non-metallic parts will be heated subsequently by conduction.

An acceptable level of accuracy whilst not being over-complicated may be achieved by assuming temperature-invariant densities, heat capacities, and dimensions, and a linearly varying resistance with temperature. Thus for the conducting material in the cable we define:

A = cross-sectional area   (m^2)
w = weight   (kg/km)
s = specific heat capacity   (J/°C/kg)
R_1 = resistance at initial temperature   (Ω/km)

α = temperature coefficient of resistance   (°C^{-1})
T = duration of fault current   (s)
t_1 = initial temperature before fault   (°C)
t_2 = maximum allowable final temperature after fault   (°C)
I = fault current (r.m.s. if a.c.)   (A)

Additionally

$$R(t) = R_1[1 + \alpha(t - t_1)] \tag{50.1}$$

The cable fault current rating (usually expressed in units of kA^2 s), or the 'action integral' of the fault may be shown to be given by:

$$I^2 T = \frac{ws}{R_1} \int_{t_1}^{t_2} \frac{dt}{1 + \alpha(t - t_1)}$$

$$= \frac{ws}{R_1} \frac{1}{\alpha} \log_e[1 + \alpha(t_2 - t_1)] \tag{50.2}$$

Cables which comprise different conductive materials may be catered for quite simply. The values for the different material parameters can be combined for each of the different materials in the cable. If there are n components in the construction then the following expressions hold:

$$A = \sum_{i=1}^{n} A_i \quad \text{(cross-sectional area)} \tag{50.3}$$

$$W = \sum_{i=1}^{n} w_i \quad \text{(weight)} \tag{50.4}$$

$$S = \sum_{i=1}^{n} w_i s_i \quad \text{(heat capacity)} \tag{50.5}$$

$$\alpha = \sum_{i=1}^{n} \frac{A_i \alpha_i}{A} \quad \text{(thermal coefficient of resistance)} \tag{50.6}$$

$$R = \frac{1}{\sum_{i=1}^{n} \frac{1}{R_i}} \quad \text{(resistance)} \tag{50.7}$$

European utilities tend to limit the maximum fault current temperature to around 200°C primarily to avoid a reduction in tensile strength of the load-bearing elements. In a typical fault the line will be de-energised in a fraction of a second hence most specifications cite a worst-case fault duration of one second. It is important to realise that the fault carrying capacity of any conductor, and therefore the temperature it reaches during the fault, is primarily dependent on the conductivity of the cable.

Lightning
Lightning is an associated problem which tends to affect the outer layer of conductor strands. The peak current which may flow during a lightning strike may be up to 200 kA

although its duration tends to be many orders of magnitude shorter than a system fault. The heating effect tends to be localised near the strike site and damage only rarely occurs to individual strands and then only to those less than about 2.5 mm in diameter. Care should be exercised by the line designer to ensure that the conductor supplied is capable of withstanding the severity and frequency of lightning strikes expected for a given area.[2] If an individual strand of the cable is damaged then a preformed repair element can be applied which should remedy the situation and ensure the continued integrity of the cable.

Corona
If the electrical field at the surface of a conductor exceeds the dielectric strength of the air then the air will continuously ionise and recombine, a phenomenon known as corona discharge. The discharge is a source of radio interference, audible noise and power loss.

Corona performance of a line is affected by the line-neutral voltage, the geometry of the line in terms of conductor spacing and the use of bundled conductors for each phase, the overall conductor diameter, the freedom of imperfections on the conductor surface, and the atmospheric conditions (temperature, pressure, humidity, pollution).

Mechanical requirements

Conductors must have the mechanical attributes that allow them to be installed efficiently and, once erected, satisfy given limits covering minimum ground clearances, tensile load and vibration under all foreseeable weather conditions.

When a new line is designed the size and strength of the supporting poles or towers can enter the optimisation equation. However, when refurbishing or up-grading lines, the task of the conductor designer is often to provide conductors which can deliver the required electrical and mechanical performance when installed on existing support structures.

Tensile strength
The rated tensile strength (RTS) of a conductor is an estimate of the conductor breaking load and is calculated by summing the strength of the component wires. The strength of each component wire is taken as the product of its nominal area and the minimum tensile stress specified for a wire of the given diameter and material. This tensile stress is the ultimate tensile stress (UTS) for all materials except when steel forms a component of a composite conductor. In this case, the strength of the steel, which can only contribute strength corresponding to an elongation compatible with that of the other material(s) of the conductor, is taken to be the specified stress at 1% extension. The calculation of these ratings is best illustrated by some examples.

Example 1
7/3.30 mm AAAC (UK Code 'Hazel')
Minimum UTS of wire after stranding = 295 N/mm^2
RTS = (no. of wires) × (nom. wire area) × 295
   = 17.7 kN

Example 2
54/7/3.18 mm ACSR (UK Code 'Zebra')
Minimum UTS of aluminium wire = 165 N/mm^2
Minimum stress @ 1% extn. of steel = 1100 N/mm^2
Al. strength = (no. of wires) × (nom. wire area) × 165
          = 70.7 kN
Steel strength = (no. of wires) × (nom. wire area) × 1100
             = 61.2 kN
RTS = 131.9 kN

Vibration

Wind blowing across overhead conductors leads to the formation of vortices downstream of the conductor which in turn cause a number of modes of conductor vibration. In steady, low velocity winds of the order of 1 to 15 m/s and with the wind approximately at right angles to the line, so-called aeolian vibration may occur at a frequency in the range 3–100 Hz with an amplitude up to the conductor diameter.

When conductors are arranged in bundles, sub-conductor vibration may be induced at a frequency in the range 0.15–3 Hz with an amplitude up to the individual conductor spacing. This vibration is a function of conductor diameter, conductor spacing, wind velocity and angle.

Where there are very long spans, or when snow or ice accretion has modified the conductor profile, right-angle winds of moderate to high speed may cause aerodynamic lift conditions which can lead to low frequency oscillation of several metres amplitude, known as galloping.

Vibration dampers fitted to the line, either close to the supporting structures or incorporated in the bundle spacers, are used to reduce the threat of metal fatigue at suspension and tension fittings. In order to minimise instability, such as susceptibility to 'galloping', a rule of thumb used by the conductor designer is to limit the number of wires in the outer layer to about 24.

Sags and tensions

After a conductor is erected the sag between supporting structures must satisfy given ground clearance limits and the tension must remain within given limits to reduce the effects of vibration and to contain the probability of failure for a range of weather conditions.

The equation below links together the effects of temperature, wind and ice loading in a single model. Clearly, however, there are significantly shortcomings in the base assumptions for this model, most critically, the actual characteristics of wind, and the effective drag coefficient of both bare and iced conductors.

$$T - \frac{W^2 L^2 EA}{24T^2} = T_i - \frac{W_i^2 L^2 EA}{24T_i^2} - EA\alpha(t - t_i) \qquad (50.8)$$

where T = tension in loaded cable (N)
     T_i = installation tension (N)
     W = effective cable weight per unit length (N/m)
     L = span length (m)
     E = cable modulus (Pa or N/m^2)

A = cable cross-sectional area (m^2)
α = cable thermal expansion coefficient (°C^{-1})
t = temperature (°C)
t_i = installation temperature (°C)

The effective cable weight is given by the resultant load acting on the cable. Normally this is purely due to the effect of gravity. However, if the conductor is iced we must remember to include any additional weight of ice in order to define W_{grav}. If the cable is being loaded by cross-winds then we must incorporate this loading by using vector addition. Hence

$$W = (W_{grav}^2 + W_{wind}^2)^{1/2} \qquad (50.9)$$

where

$$W_{wind} = \tfrac{1}{2} c_D \rho v^2 D \qquad (50.10)$$

and c_D = cable drag coefficient (no units)
ρ = density of air (kg/m^3)
v = wind speed (m/s)
D = cable diameter (m)

To a first approximation, the sag of a conductor is given by

$$\text{Sag} = \frac{WL^2}{8T} \qquad (50.11)$$

where W = conductor weight per unit length (N/m)
L = span length (m)
T = horizontal conductor tension (N)

As the tension will be a function of the rated tensile strength of the conductor under given conditions to provide a safety factor, this equation demonstrates that for a given span length, the strength:weight ratio of the conductor will be a key factor in determining the sag.

Once erected, the conductor will be subject to changing conditions. These will range from the maximum operating temperature at one extreme, to the lowest temperature and with the worst conditions of ice and wind loading envisaged, at the other extreme. Using values for the modulus of elasticity and coefficient of linear expansion for the conductor, the resulting sags and tensions between these extremes can be modelled. Such a model is used at the design stage to ensure that safety and vibration requirements are met, and during installation to give the initial sag value corresponding to the temperature at that time.

Creep

If a single strand of wire is subjected to a constant tensile load, within its elastic limit, an immediate elastic extension will occur. If the load is maintained the extension will continue, although at a slow and ever decreasing rate. This extension, which is inelastic, is called creep.

A high rate of primary creep gives way to a lower rate of secondary creep, which tends to produce a straight line on a log/log scale. In addition to time, the level of creep of a given wire material is dependent on temperature and stress.

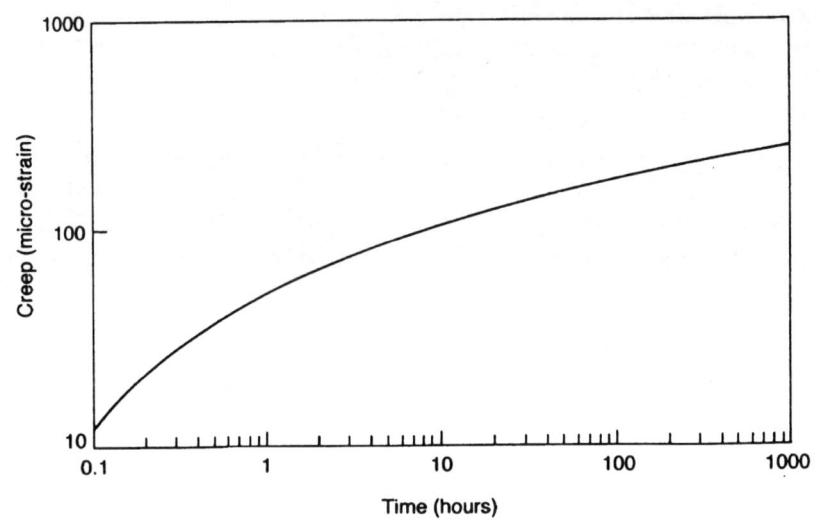

Fig. 50.3 Typical conductor creep characteristics

A stranded conductor exhibits the same form of creep, but the primary component is increased by the effect of settlement and bedding down of the wires within the structure. A typical graph of creep against time is shown in fig. 50.3.

As the effect of creep on a conductor in service will be to reduce the tension and ground clearance, allowance must be made for it at the time of erection. Pre-stressing the conductor at a relatively high tension for several hours before adjusting the sag can remove a significant amount of the potential creep, but a more common method is to raise the value of the maximum temperature at the design sag and tension modelling stage by an amount equivalent to the creep expected, ensuring that design clearances will be maintained after creep has occurred.

Environmental requirements

Corrosion
An overhead conductor in service is exposed to the corrosive effects of the atmosphere which may eventually lead to a reduction in tensile strength and an increase in electrical resistance. Corrosion is worst in locations where sea-salt aerosol and/or industrial halides are present in the air.

In multi-metal conductors galvanic cells may be formed between differing metals. In ACSR conductors, for example, this causes an initial loss of the zinc coating over the steel wires, and is followed by the rapid corrosion of aluminium once it is in direct electrochemical contact with steel. Homogeneous conductors such as all aluminium alloy (AAAC) avoid this form of corrosion but are subject to the formation of crevice corrosion cells. In the presence of moisture these cells can form in the wire interstices where differentials in aeration can exist.

Filling the interstices of the inner layers with a grease compound with a temperature performance which exceeds that of the conductor in service has been found to give effective protection from the above corrosive effects.

COMPOSITE OVERHEAD CONDUCTORS

The concept of incorporating communication transmission media within overhead electrical conductors is a long-standing one, and started with the introduction of copper wires into earth wires[3] long before the widespread use of optical fibres. However, optical fibres offer the dual benefits of immunity from electrical interference and an almost infinite bandwidth which makes them ideal for this application.[4] An Optical Ground Wire (OPGW) is an electrical and mechanical analogue to a conventional high voltage earth wire, but incorporates a number of optical fibres for communications within it. At present typically 24 or 48 fibres are housed in such cables. On lower voltage lines (rated up to 150 kV) it has proven possible to substitute the same type of cable for a phase conductor (OPPW), which is particularly useful in the case of lines not equipped with earth wires, although this type of system is rendered more complex by the difficulties associated with earthing the optical fibres. Other methods of installing fibre optics on power transmission lines have since been developed[5,6] but the OPGW concept remains the solution of choice for most utilities.

OPGW constructions

Different manufacturers supply OPGW employing a variety of ways of packaging the optical elements. Several of these methods have been discussed earlier, but the peculiarities of the OPGW environment have given rise to a number of preferred designs. The first design variant was patented by BICC[7] in 1977 and was based on a conventional loose tube cable of the time. A fully-filled optical core consisting of a number of polymeric loose tubes stranded around a central supporting member and having a layer of tape and a polyethylene sheath was used. This was run into a C-section aluminium alloy tube. Around this tube one or more layers of aluminium, aluminium alloy, or aluminium-clad steel wires were stranded to form the finished OPGW. Figure 50.4 shows the BICC FIBRAL® OPGW of this type which has been extensively used by the UK's National Grid Company (NGC).

This type of loose tube OPGW has many desirable features. The fibres are fully protected from longitudinal strain up to the strain margin of the cable, which at a value of around 0.6% exceeds all foreseeable extensions of the earth wire during service. Also, the fairly substantial aluminium alloy tube provides a high degree of lateral crush resistance for when the cable passes over pulleys and is clamped during installation. It is currently possible to house up to 48 fibres in this type of cable structure although in theory the number of fibres may be arbitrarily increased either by changing the number of fibres per tube or the number of loose tubes. However, it then becomes increasingly difficult to keep the cable diameter down to a realistic size whilst still maintaining a sufficient strain margin.

Other manufacturers have chosen to produce cables which offer less direct protection from tensile loadings to the optical fibres. One such solution has been to combine several fibres together into a discrete bundle and house this in a helical slot along an extruded aluminium rod which is located at the centre of the cable.[8] An aluminium alloy tube is welded around this central former and then conductor wires are stranded around in a conventional manner. Usually there are six fibres per bundle, and it is possible to have up to six slots, although there is little reason, other than practicality, preventing further increase in the number of fibres.

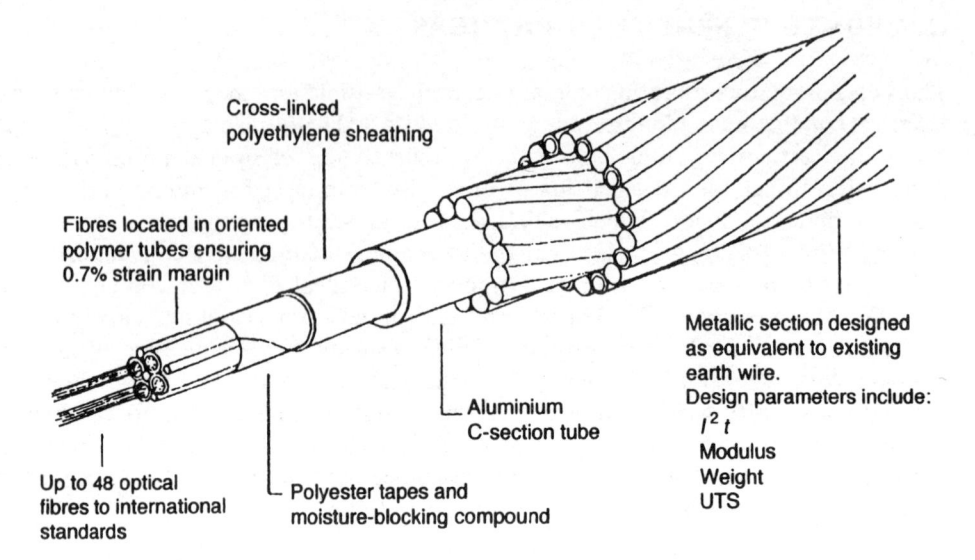

Fig. 50.4 Loose tube OPGW FIBRAL®

Generally the fibre bundles are laid tightly into their respective slots yielding negligible strain margin, and so it is important to design the cable such that the maximum anticipated service extension will not cause undue levels of crack growth on the optical fibre surface. In order to be confident that this will be so, the use of fibres which have been tested to a higher than normal proof level may be required. Optical attenuation penalties arising from lateral pressures produced when the cable is subject to strain may be avoided by using upcoated buffered fibres which ensure that small-scale imperfections in the surface of the central spacer are not transferred to the fibres. Some manufacturers use fibre which has special heat resistive coatings to allow their cables to withstand temperatures of up to 300°C under fault conditions,[9–11] although in these circumstances the effects of such high temperatures on the cable's load bearing elements should be considered.[12] Figure 50.5 illustrates a typical example of an OPGW employing fibre bundles.

A compromise in terms of the mechanical protection of fibres in an OPGW is one in which the fibres, or fibre units, are stranded around a central member which incorporates an outer layer of soft-compressible material.[13] In such a cable, when the cable comes under tension the fibres are allowed to move radially inwards by deforming the soft layer beneath them. This type of cable is less widespread than the foregoing options and generally has been limited to cables accommodating relatively few fibres.

More recently a new type of OPGW structure has become available. One of the metallic strands in an otherwise conventional earthwire may be replaced by a hollow metal tube containing up to twelve optical fibres.[14] In an earth wire consisting of a significant number of individual strands the reduction in strength and electrical conductivity caused by the elimination of one of the solid strands is minimal, and this means that the OPGW's properties are very similar to the earth wire which it replaces. Figure 50.6 shows an example of such an OPGW.

Usually it is one of the inner layer of steel strands in an ACSR conductor which is replaced by a steel tube. This means that the tube is stranded around the cable ensuring

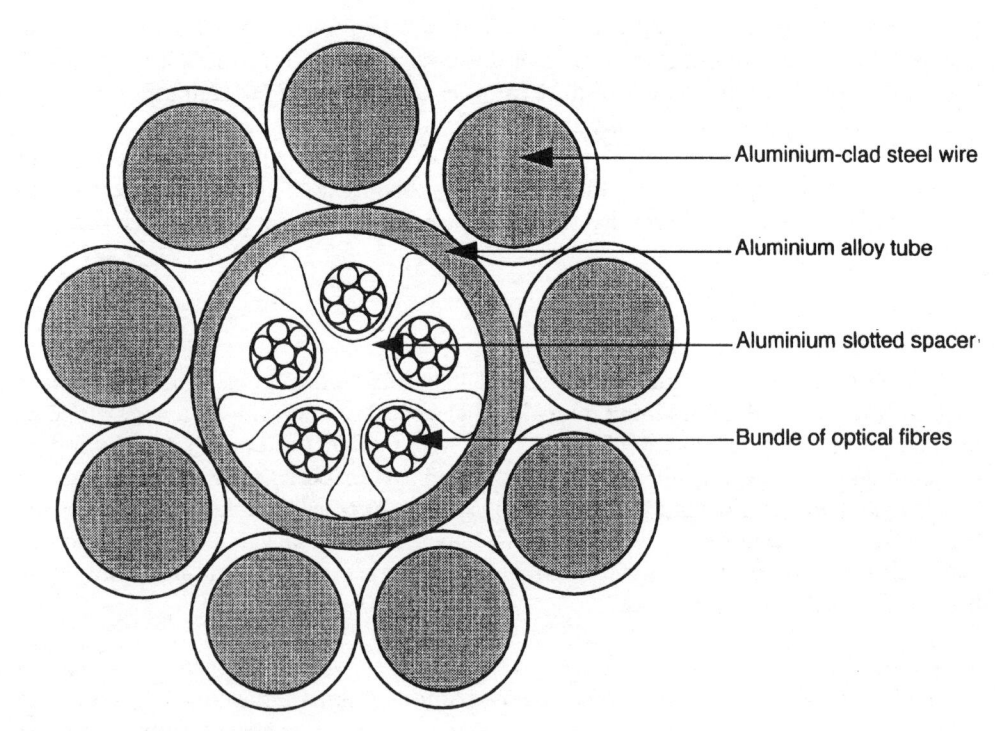

Fig. 50.5 Slotted rod OPGW

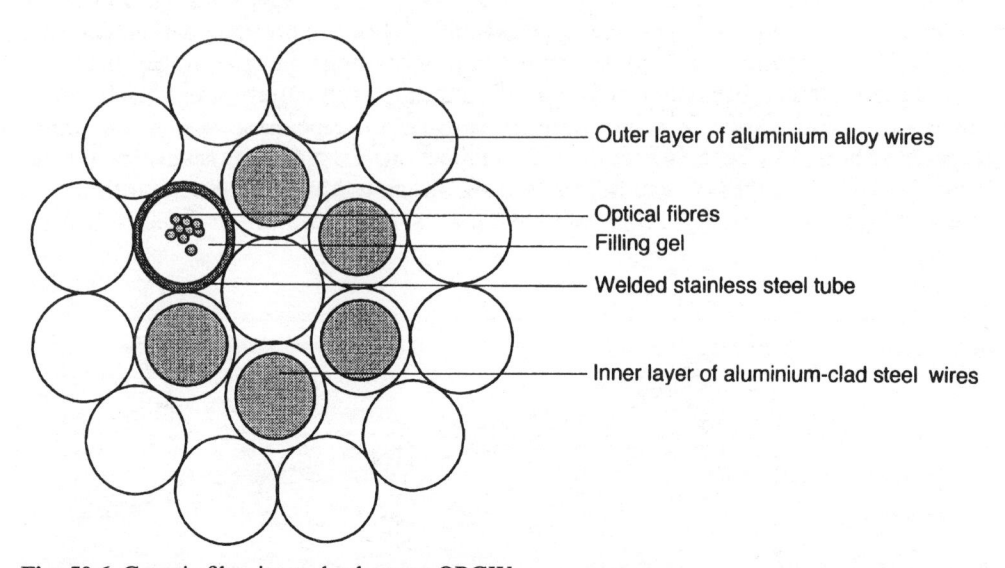

Fig. 50.6 Generic fibre in steel tube type OPGW

that the fibres are endowed with an adequate strain margin as in a normal loose tube optical cable. Typically eight or twelve conventional optical fibres may be inserted into a typical 2.5 mm diameter tube during manufacture, and a number of metallic strands may be substituted by such tubes. The outer layers of strands are applied as normal, thereby ensuring that the metal tube is not subjected to the effects of lightning strikes.

Occasionally the central metal tube in an ACSR may be replaced by what may be quite a large tube (up to 5 mm diameter), but in this case an extra complication results from having to provide an overlength of fibre in order to ensure a significant strain margin.

Such metal tubes are commonly made by forming a strip of metal (usually about 0.1 mm thick) into a tube shape, and then laser-welding the seam. Fibres and a filling compound are incorporated into the tube during this process, and special precautions are taken to protect the fibres from the localised heating effects introduced by the welding operation.

Design considerations

OPGWs are amongst the most technically demanding optical cables to design. They have to be capable of maintaining optical and electrical operation over a wide range of temperatures, typically from −50°C to +80°C, whilst subject to ever-changing wind or ice loadings. In addition, all cable components have to be capable of surviving the effects of system electrical faults and lightning strikes. Furthermore, because such cables are frequently retro-fitted on to existing power lines it is necessary that they have similar characteristics to the conventional earth wires which they replace, and preferably may be installed without a line outage.

Fault currents

For a given level of fault protection it is possible to manipulate the construction of the cable to yield a desired peak temperature and it is thus incorrect to assert that an OPGW which can withstand a temperature of 300°C is any better than one which can withstand 200°C. It is likely that the cable capable of withstanding a higher temperature may be slightly smaller and may exert lower tower loadings on an overhead line.

The exact thermal behaviour of an OPGW during a fault current event has been the topic of several papers[15-17] but until recently the temperature which the fibres experience have only been ascertainable by indirect means. Not surprisingly the precise construction of the OPGW will have a bearing on the temperature experienced by the fibres. For example, in an OPGW employing a dielectric core of stranded polymeric

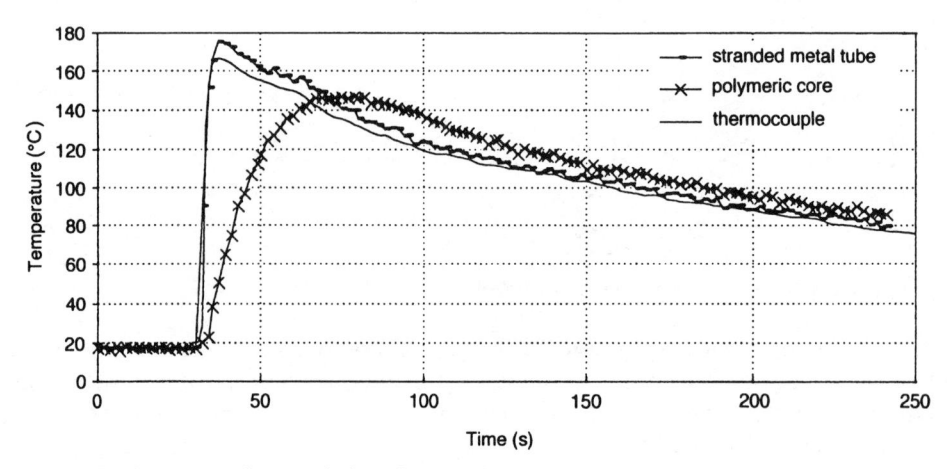

Fig. 50.7 Fault current characteristics of an OPGW

tubes we might expect the heat from the energised conductor strands to be conducted relatively slowly to the fibres, whilst within a metal tube an essentially unhindered thermal pulse might reasonably be expected.

Recently, direct verification of the fibre temperature along its length has become possible by using the Raman optical sensing technique. Figure 50.7 shows a comparison between conventional measurements of temperature by thermocouples inserted between conductor strands and temperatures measured directly by the optical fibre using the distributed sensing method.

OPGWs are not appreciably more prone to lightning damage than ordinary conductors.[18] The normal design constraints governing the minimum conductor strand diameter to withstand given lightning severities apply.

Mechanical loads

The area of environmental conductor loading is extremely complex and a large body of work on the subject exists.[19] However, there is no easy to use, universally applicable model of conductor behaviour which can be employed. Consequently, a simple model of an aerial conductor's responses to loadings is usually adopted by most cable companies and this is usually designed to take into account the combined effects of temperature, wind and ice loadings on the cable.

Under normal conditions an OPGW is suspended between towers with a tension such that the sag of its catenary is similar to that of an ordinary overhead conductor. Typically this tension is about 15% of the ultimate strength of the cable and the resultant cable strain is of the order of 0.05%. Changes in temperature of the cable result in changes in its length (calculable using the simple composite thermal expansion coefficient described earlier) and associated fluctuations in cable tension. A temperature increase will extend the cable, leading to an associated tension decrease and an increase in sag which may be significant in the event of a fault current. An almost instantaneous temperature rise of 150°C will result in a correspondingly sudden sagging of the cable (up to 0.3%) which places significant demands on the mechanical integrity of the cable and the fibres contained therein. Severe low temperatures, such as those experienced in arctic climes, impose not only increased tensions on the OPGW, but also place great demands on the fibre coatings and other protective materials in the cable. It should be clear that in the case of loose tube cables the tube filling compounds should allow movement of the fibres at both these extreme low temperatures and during high temperature excursions.

Ice loading is usually modelled by effectively altering the weight of the cable to reflect the weight of ice accreted along its length. The effect of such a loading is to increase both the sag and the tension of the cable. Typically cable specifications drawn up by utilities require that a cable be able to operate under the influence of a specified radial thickness of ice. However, seldom is the density and exact form of the ice accretion taken into account. Only rarely will the accretion have a density approaching the 'textbook' figure for ice of 900 kg/m^3 and it is similarly rare for it to assume a uniform thickness around the cable circumference. Consequently, the weight of a real ice accretion approaching the specified radial loading is usually less than the theoretical value expected and so the cable will actually not experience the anticipated increase in sag and tension.

Similarly inadequate is the way in which wind loadings on overhead conductors are specified. This normally takes the form of a single wind speed figure under the influence

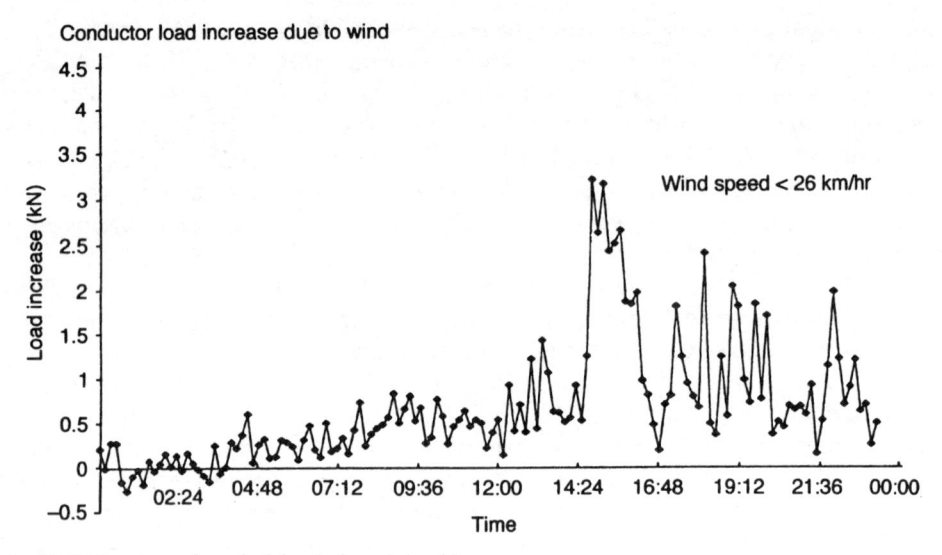

Fig. 50.8 Tension of a wind-loaded aerial cable

of which the cable must be able to operate satisfactorily. Design conservatism ensures that this wind speed is assumed to act uniformly along the length of a complete span and at an angle normal to it. This results in a horizontal force (related via the drag coefficient of the cable to the wind speed) acting at right angles to the plane of the original catenary of the cable. This in turn produces cable 'blow-off' which assumes the form of a catenary projected in the direction of the wind. Any wind incident on an aerial conductor will result in an extension and increase in tension of the cable. Some typical tension data for a cable exposed to wind loading is shown in fig. 50.8.

In addition to these long-term loading effects there exist additional causes of short-term transient loadings which are harder to model. These include aeolian vibration (high frequency, low amplitude oscillations), galloping (low frequency, high amplitude movements), and sleet jump (the sudden relaxation of tension when ice falls off a conductor). Simulation tests of cables help to ensure that these phenomena can be accommodated safely and some mechanical countermeasures to the oscillations have been developed.

Under the worst case combination of environmental conditions anticipated it would be expected that a conductor or OPGW would experience an extension of up to about 0.5%.[20] Thus a typical design strain margin of 0.6% provides ample protection for an OPGW's fibres under the worst conditions envisaged and provides some defence against phenomena such as long term conductor creep.

Historically in the UK the worst case conditions under which an aerial conductor was expected to operate consisted of a combination of a 22 m/s wind speed and a 12.5 mm radial thickness of ice. To apply this condition universally over the country is clearly unrealistic and so considerable effort has been expended in attempting to generate a probabilistic approach to conductor loading. The current state of the art of this approach is embodied in BS 8100 which constitutes a code of practice for lattice towers and masts. Other standards are applicable in this area, e.g. IEC 826.[21]

Installation and accessories

OPGWs may be installed in exactly the same way as conventional earth wires although some designs may require slightly more caution. For instance, designs where the optical fibres are housed in a central tube may be prone to damage if excessive lateral pressure is exerted as the cable is passed under tension over a pulley which is too small. If a cable employs aluminium-clad steel wires in its outer layer of wires then it is prudent to avoid excessive scratching of the strand surfaces which in turn might compromise the integrity of the aluminium layer.

It is especially critical to ensure that precautions are taken to prevent the unwinding of wire strands during installation in designs with a single layer of conductor strands. Torsioning of an underlying fibre optic core due to this effect may unacceptably compromise the strain margin of the cable, and so the conventional preventative means are adopted. In multi-layer cables detorsioning can be built into the design.

Most installations of OPGW are carried out as part of the construction of new transmission lines, or as a planned refurbishment of an existing line, and therefore outages are usually available, allowing normal tension stringing to be carried out. However, if there is no other requirement to de-energise the line than to install the new fibre optic link, then the utility may prefer an installation method that can be carried out live-line. Cradle-block stringing has been developed to allow the replacement of an earth wire with an OPGW whilst maintaining full operation and functionality of the transmission system.

When terminating OPGWs at towers it is usual to use preformed deadend grips which minimise the radial crushing forces on the cable which otherwise may damage the fibre optics housed inside. Further protection is afforded by the use of armour rods between the OPGW and the actual pre-formed deadends.

Standards

Despite having been installed for more than 15 years there are still no published standards relating specifically to OPGW, although both IEC and IEEE are nearing the publication of their deliberations. The draft IEEE standard, P1138, concerns itself with the construction of composite fibre optic groundwires for use on electric utility power lines and it is intended to help assure the ground wire functionality and maintenance, whilst preserving the integrity of the fibre optics and optical transmission. Suitable standards for the optical and mechanical characteristics of the cable are cited within it, and a range of tests specific to OPGW are described.[22] The IEC draft specification, 1396, is similar, and draws on a range of IEC standards to cover the optical, electrical and mechanical aspects of OPGW design. Various countries and individual utilities have their own relevant standards, for example the UK National Grid Company's standard NGTS 3.8.9. CIGRE have also taken a keen interest in the development of OPGWs, and other means of installing fibre optics on power transmission networks, and have issued a comprehensive planning guide.[23]

Optical fibres on energised conductors (OPPW)

The technology of design and manufacture of OPGW has been further developed to enable fibres to be installed within phase conductors (optical phase wire or OPPW). This may be particularly advantageous on towers or wood poles which do not have earth wires. In addition, as overhead lines invariably have more than one phase conductor the possibility exists of installing many more fibres than if one OPGW or self-supporting cable were used. Live line installation of such systems is not possible and the added cost incurred for outages and the phase-to-earth transition equipment for the fibre at terminations makes these systems less attractive than those employing OPGW, wrap cable or ADSS cable if other considerations are equal. However such cables do share the advantage with OPGW of providing the optical elements with heavy armouring, thereby providing more resistance to gunshot and other abuses than wrap cables and all-dielectric self-supporting cables. Of all aerial cables the OPPW is likely to

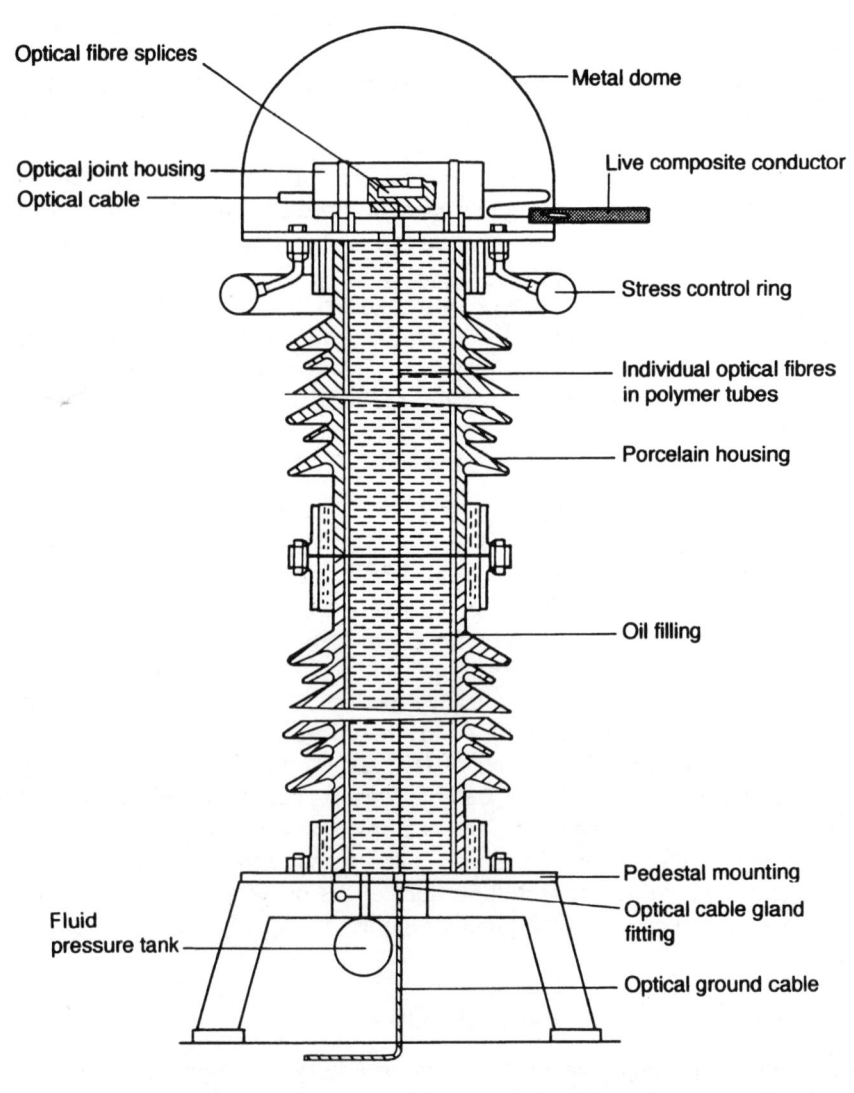

Fig. 50.9 An early method which allows the optical core of an OPPW to pass to ground potential[24]

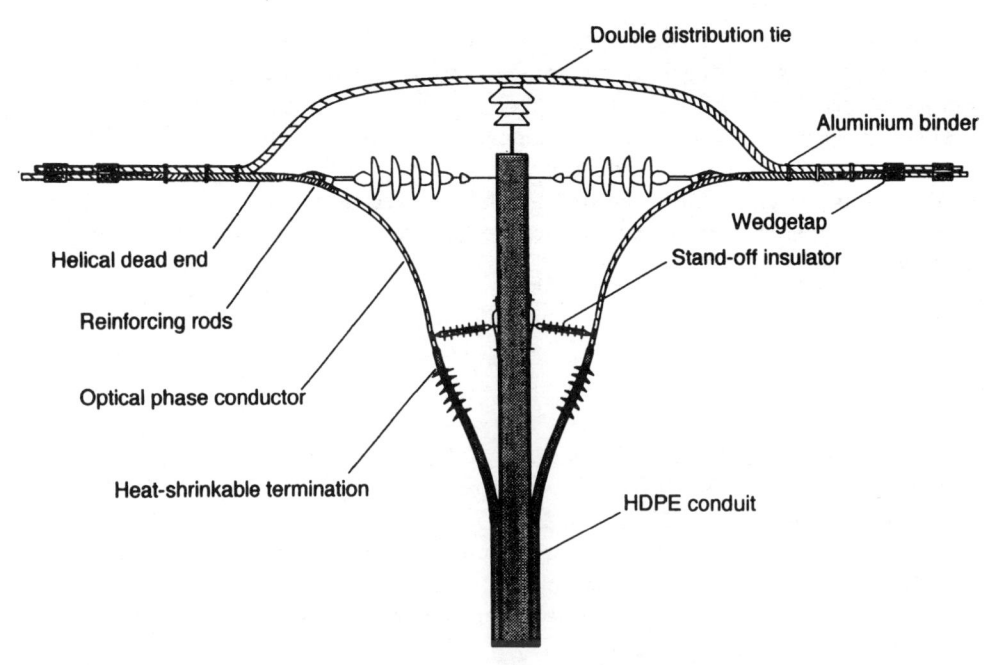

Fig. 50.10 Two phase to earth transitions employed in the Isle of Man on a 33 kV wood pole route[25]

run the hottest. This varies from site to site, but careful consideration must be given to the choice of materials in the optical core if the cable is to run at 65°C or above for many years or decades.

In addition to their use for providing telecoms, embedded fibres have been used directly to monitor the temperature of overhead line conductors much as they have for underground cables. Knowledge of temperature allows the conductors to be driven closer to their thermal limits than they would otherwise be thereby allowing the distribution network to be driven closer to its operation limits. Interest in such measurement for overhead lines has principally be confined to urban Japan where population densities are very high. In general there is currently little interest in this application in Europe.

The principal challenge in designing a system based on OPPW is the need to bring the optical fibres to earth potential for connection to other cables, terminal equipment and so forth. The technical complexity of this increases considerably with system voltage. BICC were the first company to develop such systems for their FIBRAL® cable.[24] In this system an oil-filled porcelain housing was used to effect the transition for systems for up to 400 kV. One such design is shown in fig. 50.9. More modern designs tend to be specifically engineered for a particular system voltage. Figure 50.10 shows a design (necessary for jointing two lengths of cable at ground potential) for a 33 kV wood pole route installed on the Isle of Man. This design used heat-shrink polymeric fittings and is more typical of modern systems.[25] An alternative design using more traditional techniques for a 26 kV line is described by Mercier et al.[26]

Joints between the fibres of two lengths of OPPW are made at phase voltage wherever possible. This reduces the cost and complexity of the joints,[27–29] but is only possible on large structures where the joint box can be satisfactorily mounted and supported.

This essentially limits phase-voltage jointing to lattice towers and high voltage networks. In other cases back-to-back phase/earth transitions are made and the cable is jointed underground.

There is clearly a need for the optical phase wire to have similar electrical characteristics to the phase conductor it is replacing. This is readily achieved using combinations of metals and strand sizes available to design engineers. Issues of lightning resistance and fault current ratings are addressed in exactly the same manner as for OPGW.

REFERENCES

(1) Morgan, V. (1991) *Thermal Behaviour of Electrical Conductors*. John Wiley.
(2) Bonicel, J. et al. (1995) 'Lightning strike resistance of OPGW'. *Proc. 44th Int. Wire and Cable Symp.*, pp. 800–806.
(3) Dageförde, H. et al. (1972) 'ASCR-Aluminium steel communication rope'. *Proc. 21st Int. Wire and Cable Symp.*, pp. 253–268.
(4) Maddock, B. et al. (1980) 'Optical fibre communication using overhead transmission lines'. *CIGRE* paper 35-01.
(5) Coulson, A. (1992) 'High power communications'. *Electrical Power Engineering*, 222–223.
(6) Haywood, B. and Knight, I. (1995) 'Aerial fibre optic systems'. *Distribution 2000*, Brisbane.
(7) GB patent 1598438 filed 13 May 1977.
(8) Ghannoum, E. et al. (1995) 'Optical ground wire for Hydro-Quebec's telecommunication network'. *Proc. IEEE/Power Engineering Society Winter Meeting*, New York.
(9) Okuyama, S. et al. (1986) 'High heat-resistant optical fiber coated with thermal-cured type silicone and fluorine polymer'. *Proc. 35th Int. Wire and Cable Symp.*, pp. 183–188.
(10) Araki, S., Shimomichi, T. and Suzuki, H. (1988) 'A new heat resistant optical fiber with special coating', *Proc. 37th Int. Wire and Cable Symp.*, pp. 745–750.
(11) Biswas, D. (1990) 'High temperature optical and mechanical properties of polyimide coated fibres'. *Proc. 39th Int. Wire and Cable Symp.*, pp. 722–725.
(12) Ash, D. et al. (1979) 'Conductor systems for overhead lines: some considerations in their selection'. *Proc. IEE* **126** (4), 333–341.
(13) Sato, T. and Kawasaki, M. (1987) 'Optimum design of composite fiber-optic overhead ground wire (OPGW)'. *IEEE/CSEE Joint Conf. on High Voltage Transmission Systems in China*, pp. 393–398.
(14) Schneider, J., Schmelter, J. and Herff, R. (1988) 'Optical ground wire design with a minimum of dielectrics'. *Proc. 37th Int. Wire and Cable Symp.*, pp. 83–92.
(15) Madge, R., Barett, J. and Grad, H. (1989) 'Performance of optical ground wires during fault current tests'. *Proc. IEEE/Power Engineering Society Winter Meting*, New York, pp. 1552–1559.
(16) Madge, R., Barett, J. and Maurice, C. (1992) 'Considerations for fault current testing of optical ground wire'. *Proc. IEEE/Power Engineering Society Winter Meeting*, New York, pp. 1786–1792.

(17) Tanaka, S. et al. (1992) 'Analysis of OPGW short-circuit characteristics'. *Sumitomo Electrical Technical Review* **33**, 147–153.
(18) Carter, C., Baldwin, R. and Jones, C. (1984) 'Lightning simulation tests on power transmission conductors carrying embedded optical communications cable'. *IEE Conf. on Lightning and Power Systems*, IEE Conf. Pub. No. 236, 207–209.
(19) ASCE (1984) *Guidelines for Transmission Line Structural Loading*.
(20) Dey, P. et al. (1982) 'Optical communication using overhead power transmission lines'. *CIGRE Conf. on Large High Voltage Electric Systems*, Paris.
(21) Orawski, G. (Sept. 1991) 'Overhead lines – loading and strength: the probabilistic approach viewed internationally'. *Power Engineering Journal*, 221–232.
(22) Dawson, J. (1991) 'Development of test methods for OPGW'. *Wire Journal Int.*, 58–60.
(23) Martin, J. (1993) 'Optical fibre planning guide for power utilities'. CIGRE Paper No. 35-04.
(24) Dey, P., Gaylard, B., Holden, G., Taylor, J. E., Smith, P., Carter, C. N., Maddock, B. J. and Kent, A. J. (1984) 'Optical communication using overhead power transmission lines'. *Elektron*, 38–44.
(25) Friday, A., Smart, T. J., Evans, J. and Bevan, J. (1995) 'Design, development and installation of an optical phase conductor (OPPC) on a 33 kV wood pole line'. *CIRED 13th Int. Conf. on Electricity Distribution*, 11/1-6.
(26) Mercier, E., Rosset, C. and Reber, H. (1991) 'Phase conductor of an overhead line with optical fibres'.*CIRED 11th Int. Conf.*, paper 3.15.
(27) Amerpohl, U. and Bausch, J. (1992) 'Phasenseile mit LWL'. *TEZ* **113**, 368–371.
(28) Amerpohl, U. and Bausch, J. (1993) 'Erstes Phasenseilprojekt mit Lichtwellenleitern in Ostereich – ein Erfahrungsbericht.' *OZE* **46**, 373–377.
(29) Znoyek, G. (1993) 'LWL-Phasenseile fur 20 kV und 110 kV Freileitungen'. *ETZ* **114**, 636–640.

CHAPTER 51
All-dielectric Self-supporting Cables

Optical ground wires (OPGW) have been established as remarkably reliable cable systems since BICC first installed such a cable in 1979.[1] However the cost of installing such cables on to existing lines had been considered prohibitive particularly before the possibility of leasing fibres to third parties was considered. The first generic cable which could be installed with circuits live was developed to provide an economic alternative for existing lines. The all-dielectric self-supporting (ADSS) cables which resulted have been successfully installed over many years on long span overhead power lines and are now automatically considered for any new telecommunication route to be based on existing overhead line structures.

When ADSS cables were originally designed it was assumed that they would be applicable irrespective of system voltage. After having been installed on 132 kV and 150 kV systems for many years without any problems a number of failures occurred when cables were installed on towers carrying higher system voltages particularly in coastal sites.[2,3] This has led to some confusion and disagreement, particularly between cable suppliers, as to the limits of cable application. The ageing mechanisms by which damage is inflicted upon the cables are now better understood than even three years ago; these are reviewed in this section and one engineering analysis of the limitations of the standard cable is made. A new solution to this problem is also reviewed.

The requirements set for an ADSS cable are similar to those for any telecommunication trunk cable. Optical operation is required at 1310 nm and 1550 nm and cables normally contain at least 24 fibres. Optical fibres should be strain free during the cable life or as near to this condition as possible to maximise their life expectancy. The cable structure should prevent moisture coming into contact with the optical fibres to maintain their performance. Although this may seem trivial it is inevitable that during installation or as a result of gun shot damage or the like at some time a cable sheath will be cut through. This will then allow moisture ingress into the cable. If moisture wicks or diffuses into the cable, damage to the fibres may result and electrical damage might also ensue. To allow the cable to be used on most routes, it should be suitable for installation on spans up to 700 m and preferably 1000 m. Ultimate tensile strength (UTS) of the cable and cable/clamp system should be high enough to give a large safety margin. To reduce costs of components and retraining labour the cable should be mechanically robust and crush resistant so that standard installation/suspension fittings can be used. The cable must be lightweight and of small diameter to minimise loading on supporting towers. In addition the cable should be non-metallic to facilitate installation on towers carrying live lines. Cable operation should not be impaired by severe weather conditions such as high wind or ice loading and the cable should never clash with the phase conductors.

Complete cable systems are usually supplied to the customer so that all planning, accessories and installation is supplied as an integral package.

ADSS CABLE DESIGN

A large number of designs are available which meet the criteria discussed above. Essentially these all use traditional loose tube or ribbon fibre optical structures and bonded glass fibre or aramid yarn strength members. The designs are limited by the number of dielectric materials with high modulus, high strength to weight and volume, and low creep, and also by the need to prevent significant strain on optical fibres during the cable's life. Four typical ADSS designs are shown in fig. 51.1.

One BICC design[4] (fig. 51.1(a)) utilises its well proven OPGW (FIBRAL®) optical core.[1] This has four or six polymer tubes which can each contain up to eight optical fibres suspended in hydrocarbon gel. The tubes are stranded together with polymer rods to provide the required strain margin. The interstices are filled with a specially selected gel which gives minimal gaseous products during the process of encapsulation by the strength member. The whole structure is wrapped in polyester tapes and provides a completely blocked cable core impervious to the ingress of moisture. It also provides a large, controlled strain margin throughout the length of the cable enabling guaranteed performance of the fibres at 1300 nm and 1550 nm. A rigid, pultruded glass reinforced plastic cylinder encapsulates the optical core to provide the strength member. This structure has a number of advantages such as providing a torsionally stable product with no preferred bend direction and very good mechanical and environmental protection for the optical core. The whole package is sheathed with either a standard polyethylene or dry-band arc resistant material if appropriate.

An alternative design uses a specially shaped pultruded glass reinforced plastic rod to house optical fibres within ribbons (fig. 51.1(b)).[5] Excess fibre length is provided by feeding more fibre into the slot than the cable length. The excess length is accommodated by a sinusoidal undulation within the core and in this way 0.5% excess length can be provided. This is not enough to provide strain free fibres for all conditions but provides a predicted life of 25 years. Two 12-fibre ribbons can be used to provide a 24-fibre cable. A gel material fills the slot and it is capped by a plastic extrusion. The sheath material encapsulates the resulting circular structure. This design does not offer as much protection to the fibres as the annular pultrusion and does have a preferred bend direction, which may not be helpful during installation, but it does provide good crush resistance and allows easy access to the fibres for jointing.

The third design of cable uses aramid fibre to provide strength to the cable (fig. 51.1(c)).[6] Unlike the pultruded glass products such fibres can allow moisture to wick if the sheath is damaged. To avoid this problem aramid yarns are used which are coated with a water swellable polymer which prevents such wicking. This design provides a very flexible cable which simplifies installation although there is limited protection against crush forces.

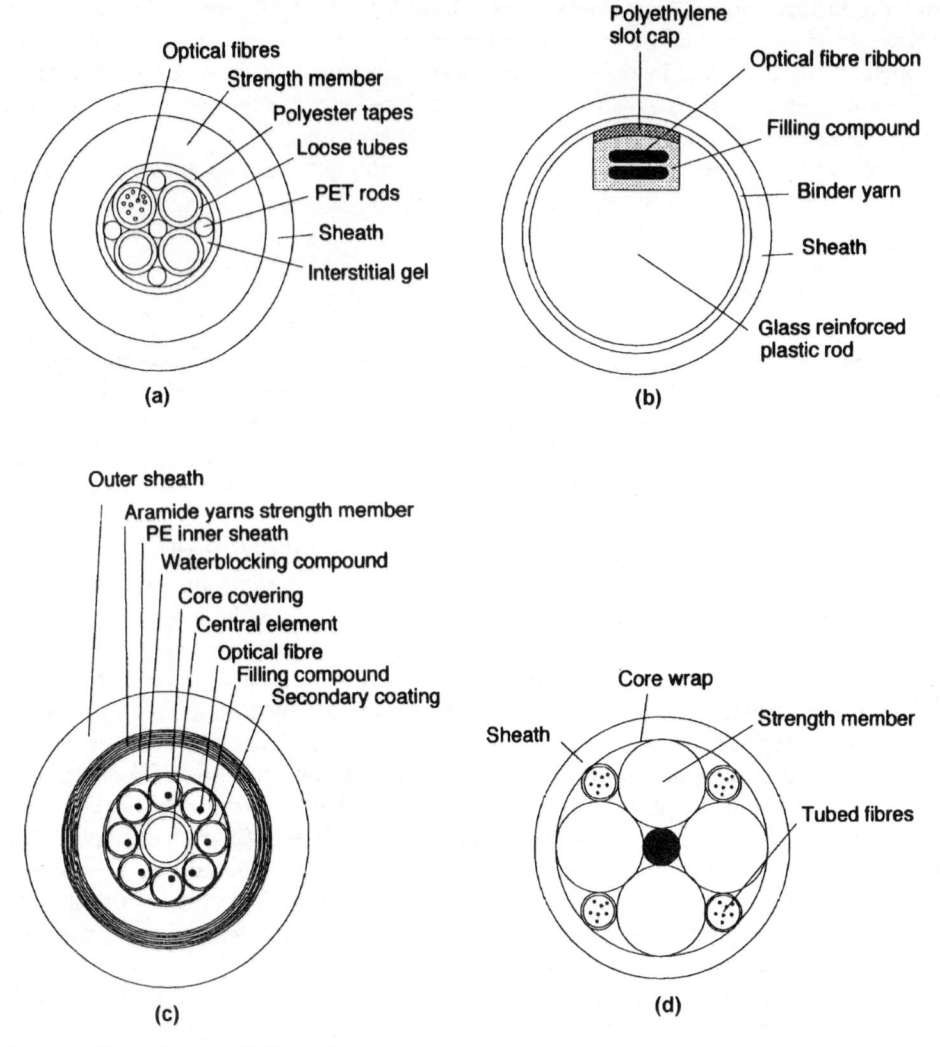

Fig. 51.1 Examples of ADSS designs

The fourth cable employs four separate pultruded strength members which are stranded with the optical tubes laid up between them and the sheath (fig. 51.1(d)).[7] The polymer coating on a central strength member is deformed sufficiently to fill the interstices and prevent moisture migration in the core. This design gives the cable greater flexibility than a single pultruded strength member, and provides easy access to the fibres for jointing. However there is not a great deal of protection of the fibres in this design, and the whole structure is not as crush proof as some of the others. This last feature may be important when designing clamping methods suitable for service conditions.

A number of other designs of long span dielectric cable exist, and no doubt more will be developed in the future, but at present they all tend to have features which are some combination of those described above.[8-13] A summary of typical cable performance characteristics is given in table 51.1.

Table 51.1 A comparison of some typical ADSS cable designs

	24 fibre Annular pultrusion	48 fibre Annular pultrusion	Central pultrusion	Annulus of aramid fibre	Multiple pultrusions
Weight (kg/km)	240	284	220	265	240
Outer diameter (mm)	14.0	15.3	13.0	17.6	16
Min. bend radius (mm)	600	600	500	250	300
Max. working tension (kN)	23	23	22.5	20	25
Ultimate tensile strength (kN)	60	70	65	>50	75
Strength member tensile modulus (kN/mm^2)	50	50	47.5	100	50
Thermal coefficient of expansion (°C^{-1})	7.0E−6	7E−6	5.2E−6	0.7E−6	8E−6
Excess fibre length (%)	0.7	0.7	0.45	0.7	0.9
Fibre strain at max. working tension (%)	0	0	0.25	0	0
Number of fibres	24	48	24	24	24

Note: The figures should only be taken as examples as they change with specific design.

COMPLEMENTARY EQUIPMENT AND INSTALLATION

At tension points a pre-formed galvanised steel helical tension fitting (dead-end) is used to fix the cable to the tower. This fitting is about 1200 mm long and is applied over the top of 1800 mm long reinforcing rods which are applied directly to the cable. These fittings are rated to at least 95% of the cable's ultimate tensile strength. At intermediate towers, where the cable only requires support, an armour grip suspension unit is used.[4] The cables are subject to considerable levels of aeolian vibration because they have almost no self-damping and are installed at relatively high tensions (typically over 10 kN). Each span therefore requires vibration dampers such as the spiral impact or Stockbridge types. Care must be taken to affix such dampers to the reinforcing rods of the clamps, rather than direct to the cable. Direct clamping on to the cable may damage the polymer sheath and may impact on the occurrence of dry-band arcing.

Joint boxes are designed so that they can be installed at any point on the tower body and are vandal resistant. The joint enclosure is environmentally sealed to prevent the ingress of moisture and is re-enterable to facilitate re-splicing in the event of a spur cable being added. Procedures and equipment for jointing ADSS are very similar to those for OPGW.

Cables are usually strung between towers at the level of the bottom or middle cross arm. In general this is mid-way between the bottom four conductors on a standard twin circuit tower, fig. 51.2. This is convenient for a number of reasons; the towers are strong and need no extra reinforcement at these locations, ground clearance is large (see ice accretion) and clearance between conductors minimises the likelihood of clashing.

One of the principal attractions of the ADSS cable is that installation is relatively straightforward even under live-line conditions. Conventional tension stringing techniques are used for these cables. Because the cables are light and installed at tensions of

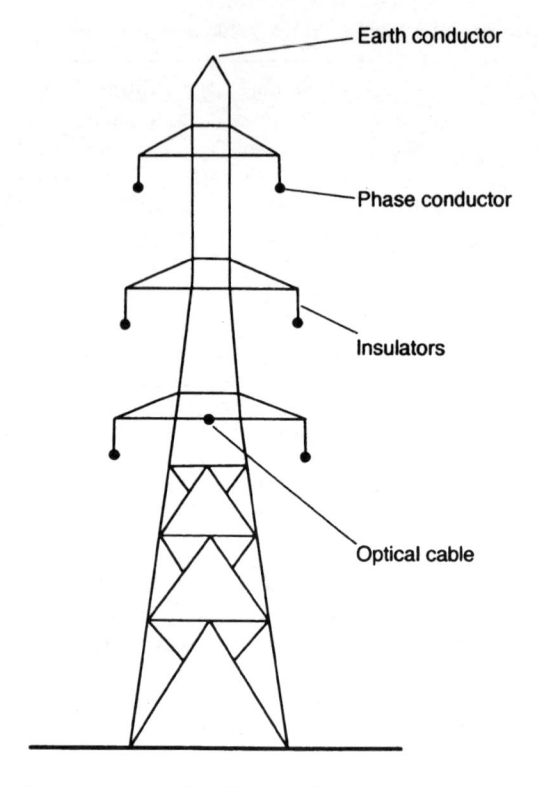

Fig. 51.2 Typical installation position for ADSS cables

typically 12 kN lightweight hydraulic tensioners and pullers can be employed. During installation the minimum bend radius of cables must be respected. This may require the use of sheaves of over 1 m diameter for the pultruded glass products. In addition installation engineers must be aware that the ADSS cables are not as resistant to surface damage as are conductors and so must be treated appropriately.

Once running out blocks have been positioned near clamp locations on towers, a pilot rope is walked into position along the route. This rope is connected to the main bond and the main bond is connected to the cable. The cable can thus be winched into position. Typically cables are supplied in 2–4 km lengths. In such lengths there may be many tension sections; these are progressively clamped off starting at the end furthest from the winch. Conventional sighting techniques are used to determine and set sag. Unlike conductors, ADSS cables do not bed-in or creep and so no allowance is made for this. Keller et al.[11] have described installation by helicopter over particularly rocky terrain. Such techniques are expensive and slow, but may be considered for short sections of a route.

SAG, ICE LOADING AND BLOW OUT

The physical characteristics of each cable are critical in designing the system. Calculations of cable sag must be carried out for installation conditions and for variations in ambient temperature, conditions of ice loading and the effect of wind.[14] These cables behave in a very different manner from conductors stretching in high sidewinds and under ice loading to a considerable degree. The local climate is therefore a significant issue.

When requesting an aerial cable, a utility usually supplies the designer with 'worst case' conditions of combinations of ice and wind loading. The cable designer will then calculate the tensions generated in such conditions and design a cable to withstand them and determine what installation tension must be applied to the cable. A major limitation of this approach is that the conditions specified are not usually realistic worst cases. In an event in the UK in 1990[15,16] an ADSS cable suffered radial accretion of over 45 mm. This resulted in the cable sagging to, and resting on the ground. The cable did not fail as a result of direct ice loading during this event; however numerous conductors and earth wires were broken, one earth wire breaking the ADSS cable by falling on to it. That the ADSS cable is so resilient is very positive, although caution must be expressed because of the infringement of ground clearance. Careful consideration of the effect of reduced ground clearance by icing must be given during the system design.

In service, gusting winds are likely to induce different effects on ADSS cables and conductors. This is because they will each tend to average out the effects of gusts in both time and along their length but because the cables are so different mechanically these averaging processes will not be the same. To be sure transient effects are allowed for, loci of each cable must be drawn for the conditions expected and adequate separation maintained between them.[17]

Galloping is a low frequency, large amplitude oscillation, principally in the vertical plane. It nearly always results from strong steady crosswinds acting on an asymmetrically iced cable. The amplitude of the oscillation can approach the sag of the cable, but because ADSS cables are not normally mounted directly beneath conductors it is unlikely to lead to clashing. For the same reason sleet jumping, whereby the cable rises rapidly after ice has fallen off, does not normally present a problem.

ENVIRONMENTAL STABILITY

One major threat to any aerial installation is gun shot. Damage can be as extreme as complete failure of the strength member or it may just be sheath puncture which allows ingress of moisture or exposure to ultraviolet radiation. Some of the designs shown in fig. 51.1 provide excellent protection to the optical package, notably those in which the package is surrounded by the strength element. The package must also be water blocked to prevent moisture ingress, either with a gel or with a water swellable material. The principal difficulty however is to protect the strength member from secondary damage. For example aramid yarns should not be exposed to ultraviolet radiation. Additional protection by the use of ballistic shields can be achieved,[18] however this usually adds too much weight and diameter to long span cables to be practical. Any associated risks from gun shot need to be taken into the reliability calculations for the system.

Acid pollution might be thought to present an unusual threat to the cables which employ glass strength members. It has long been known that glass rapidly loses its strength in the presence of high values of strain and low values of pH. The same property is true of glass reinforced plastics and composites. Unfortunately the matrix material and type of glass used both impact on the conditions under which this rapid fall-off in tensile strength occurs. The phenomenon is known as stress corrosion[19] and classic work by Metcalfe and Schmitz[20] showed that the cause is ion exchange in the glass matrix. A large body of literature is available on the effects of acid and strain on fibres[19,20] and on mechanical stress on pultrusions alone.[21] There is some literature on composites.[22-24] However little information is available on unidirectional composites with sheaths. Work at BICC has shown that the sheath makes a considerable difference to the properties of the cable, although this is not important at acidities seen in the ADSS environment (where pH is not usually less than 4 and strain levels are not normally in excess of 0.7%). If pH falls below 3.5 and the strain exceeds 0.35% then rapid failure is to be expected. Similar findings are reported by Peacock and Wheeler[25] in three-point-bend tests.

To evaluate the effectiveness of vibration dampers and the cable's optical performance it is typical for a 50 m sample of cable to be subjected to an aeolian vibration test. The cable under test is fitted with dead-end grips and appropriate vibration dampers fitted. It should be noted that spiral vibration dampers are very efficient, but that if Stockbridge type dampers are used, they have to be carefully chosen because they damp a much narrower frequency of vibration. Normally the optical attenuation of the cable is continuously monitored.[4,5,7] Figure 51.3 shows a set of results obtained during testing of the BICC design of ADSS cable Fibras®. The vibrated cable, tensioned to 12 kN, was excited from one end at a frequency of 57 Hz, the resonant frequency of the undamped cable. Once SVDs were attached the amplitude of vibration decreased from 10 mm to 1 mm; these damped vibrations were continued for 150×10^6 cycles. This is considered to be equivalent to 40 years of ageing in service. Throughout the test there were only minor variations (±0.1 dB) of the optical signal. After the test, the cable surface was inspected for damage in the location of the SVD. Only minor polishing of the sheath had resulted. After the test was completed the cable was cut into four pieces

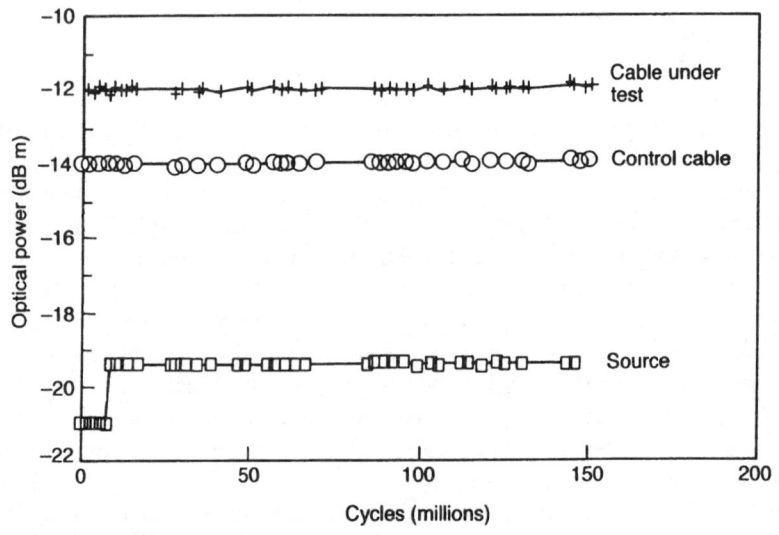

Fig. 51.3 Results of vibration tests on Fibras®. Note: the sensor on the source was changed after 9 million cycles causing the shift in power

and the UTS of each part determined. There was no deterioration in tensile strength when the cable was compared with the control samples.

The cables are designed to operate over wide temperature ranges and so optical attenuation measurements are made on fibres during temperature cycling, both on the drum and under tension laid out straight. This is particularly important because the gel filling compounds should not become so stiff as to prevent free movement of fibres at low temperatures, nor should any of the polymeric materials become brittle.

ELECTRICAL DEGRADATION

The all dielectric self-supporting cable is suspended from the same structure as the phase conductors. There is no direct contact between the conductors and the optical cable but there is a longitudinal field along its length. The cable is at earth potential at the tower by virtue of the metallic clamps and the voltage increases towards mid-span. This field exists irrespective of the presence of the optical cable and in itself presents no problem. If the cable is dry the field has no effect. Examples of field calculations are presented by Berkers and Wetzer[26], Carter and Waldron[2] and Peacock and Wheeler[25]. If the cable surface becomes wet a current will be drawn along its surface as a result of the potential gradient. Again, in itself this current does not damage the cable. However at some time the cable will dry, and the conductive path along its length will be broken. This dry-point on the otherwise wet cable is called a dry-band because it must encircle the cable and it presents a high resistance in the otherwise low resistance path to ground. Thus the majority of the voltage which was previously distributed along the cable surface will be dropped across the dry-band. If the voltage is sufficient then the air can be broken down resulting in an arc across the dry-band.

The voltages and currents involved depend upon the system voltage, the geometry of the overhead line, the age of the optical cable and the conductivity of the pollutant upon its surface. Figure 51.4 shows space potential calculations carried out by using a computer program written in the National Grid Company[27] which has become widely used in the industry. The current on the cable surface increases towards the towers. This can be understood by considering the cumulative effect of distributed capacitances between the optical cable and the phase conductors. That the highest current is adjacent to the clamps means that as a result of Joule heating, dry-bands are most likely to be formed close to the clamps. Once a dry-band has been formed, the current will only flow when an arc is struck. Figure 51.5 shows how the current and potential vary along a cable length as a function of the conductivity of the cable surface.

When ADSS cables were first installed, the threat of dry-band arc activity was not appreciated by design engineers. As will be seen, no threat generally exists until installations are above 150 kV, so that it was not until the mid-1980s when such installations took place that the problem began to be appreciated. Failures do not tend to be reported in the literature but first reports of failure came from Europe where cables were hung on towers carrying system voltages of 220 kV and above. That dry-band arcing activity can result in cable failures on 400 kV lines has been confirmed by experiments in Fawley, England and Hunterston, Scotland.[3,28] In these experiments on 400 kV lines, no damage was inflicted on the cables for considerable periods (up to 3 years) before substantial damage occurred in extreme weather conditions. A number of trials have also been reported in which no damage has yet been seen on cable

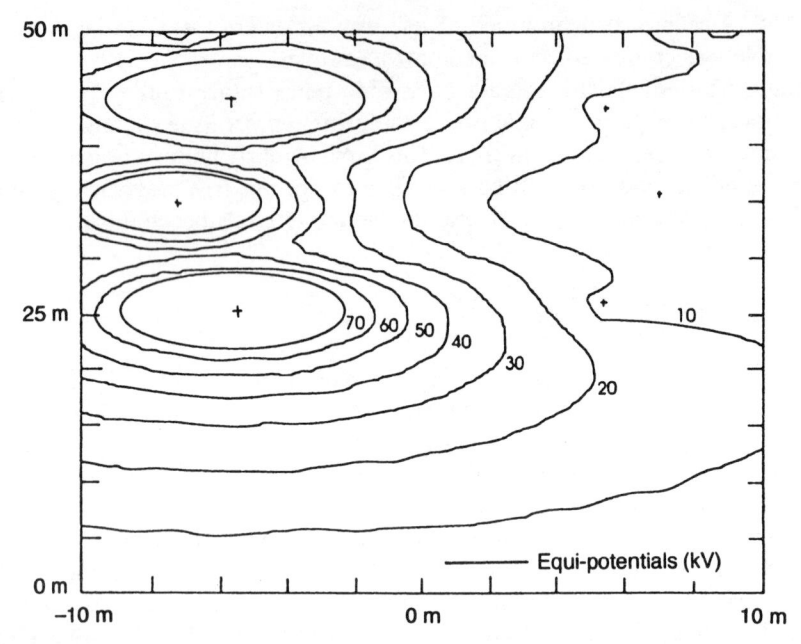

Fig. 51.4 Space potential contour plots on a plane perpendicular to the conductors in mid-span. One of the two circuits is earthed and the other energised

installations.[11,29] This has led to conflicting views in the industry as to where it is safe or unsafe to hang cables. Haag et al.[8] suggest that wherever possible OPGW should be used on systems above 150 kV. We are also of that opinion and that standard ADSS cables are likely to fail on some 400 kV and even 275 kV lines.[30,31] There is also a body of opinion that cables are safe to hang in this environment.[32,33] In truth, the application depends upon commercial decisions on the reliability requirements of the network. Insufficient information is available to make engineering predictions of cable life expectancy depending as it does so strongly on local environments. The remainder of this section describes how some scientific foundation can begin to be put behind these decisions.

Damage to the cable sheath depends upon its composition. Materials which tend to track, such as PVC, will age rapidly, the resulting damage taking the form of continuous carbonaceous tracks on the sheath surface. Another form of rapid failure is simply by melting a sheath, such as polyethylene, which splits on the cable as a result of residual stresses and surface tension in the extruded sheath or stresses imposed by the clamps. The manufacturers' first attempts to solve the problem were therefore to develop superior dry-band arc resistant materials which are resistant to these ageing mechanisms.[25,27,33–37] This has certainly improved performance of the cables. However such sheaths can still fail[30,31] and prime concern in the years between 1990 and 1995 has been to understand the ageing mechanisms better so that suitable testing techniques can be devised.

Once a dry-band has become established and an arc has been struck it will only exist as long as sufficient voltage and current is available to sustain it. If there is no continuous wetting of the cable (by rain or fog, for example) the dry-band will gradually extend until activity ceases. Precipitation tends to land on the top surface and in its presence arcing will continue, the arc largely playing on the top surface with the drying process balancing the wetting action. If moisture deposition is too fast, in cases

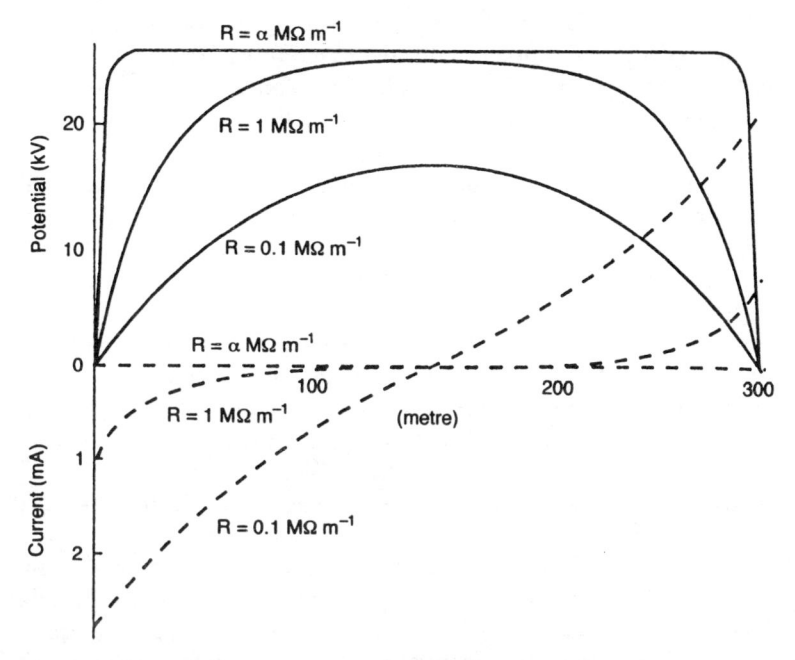

Fig. 51.5 The potential and current on a cable as a function of cable resistance per unit length[27]

of heavy rain, the dry-band will be extinguished altogether. It is the heat generated by the roots of the arc which are in close proximity to the cable surface (they are in contact with the moisture on the cable surface) which degrade the cable sheath.

The nature of an arc depends on the currents drawn. Traditionally arcs have been studied with very high currents (1–1000 A) or at intermediate values (10 mA–1 A). The latter work has been used to consider ageing of external insulators. They differ markedly from the ADSS situation because these insulators are directly coupled to high voltage, high current sources, whereas the ADSS cable has a high impedance source of high voltage. Thus the currents seen will be considerably lower (10 μA–5 mA). It has previously been shown that currents in the region of a few mA are the most deleterious for polymeric degradation;[38,39] unfortunately this is the range of values which can be seen on an ADSS cable in service.

There are two main consequences of the current being low. First, even if there is sufficient voltage available to strike an arc across a dry-band, there may not be a sufficient current to maintain it. This is because once the arc is struck, a voltage is dropped across the impedance of the rest of the circuit (i.e. the moisture on the cable) and insufficient voltage may be left to maintain the arc. Furthermore low current arcs have a very high resistance per unit length, requiring 1000 V/cm at 5 mA as compared with 100 V/cm at 300 mA[40] and so need a higher voltage to maintain them. This has been studied thoroughly by Carter and Waldron[2] and has been shown to lead to arc instability. It has therefore been argued that situations in which the arc is unstable are 'safe' for cable deployment. However it has been shown that even unstable arcs are capable of damaging the cable surface.[31]

The second consequence of low arc current is that the resistance of the arc can be higher than that of the moisture on the cable.[36] In this case the arc will be confined to the dry-band arc length (flashover cannot occur in a test with these conditions). By contrast, a high current arc is not so confined and the roots can move over the moisture surface (flashover can occur in a test with these conditions).[30,41] When the arc is confined, the roots of the arc can only move on the two boundaries between the moisture and the dry-band, and so the roots are restricted to a small area of the cable surface. This increases the rate of ageing of the cable. However the confinement of arcs is even more important when considering that the moisture on a cable is not static. Cables are installed on a (catenary) gradient and will generally be exposed to wind. In addition, the surface of the moisture on which an arc root is present has reduced surface tension (its temperature is above boiling point) and so can move towards the dry-band centre relatively easily. If the dry-band is compressed by movement of moisture, and the arc is of low current, dry-band arc compression can occur resulting in significant damage to the cable surface. One such compression event takes only seconds but can cut through over 1 mm of polymer. Consequently few such events are required in the cable's lifetime to cause severe problems. Even the best polymeric sheaths are severely damaged by this form of ageing, which results in deep carbonised troughs along the cable sheath.[30]

From the considerations above it can be seen that it is not the mid-span potential which should be considered when assessing the threat of damage to the cable but the current which can be drawn to earth. The current clearly depends upon the cable's environment, the conductivity of precipitation, and the sheath condition, whereas the mid-span voltage does not. A large part of the literature still mistakenly regards the magnitude of the space potential in which the cable sits under dry conditions to be the best measure of severity of an installation.

A further result of this thinking, that it is the voltage which threatens the cable rather than the current, is that most laboratory testing is misconceived being based on the erroneous view that high voltages in salt-fog tests, and thus high currents, give rise to more onerous tests. That a sheath which can withstand 1000 hours at 25 kV can also be made to fail after 30 hours at 25 kV just by limiting the current to realistic values[30] should make the user cautious about interpretation of such results.[12,25,42] All test results should specify the source impedance, the conductivity of the salt-spray, and whether arcs were confined to their dry-bands or whether flashover was possible. It is also critical to note that the mechanism of dry-band arc compression suggested by Rowland and Easthope[30] is not a gradual ageing process. Damage occurs very rapidly when conditions are adventitious (i.e. a period of sea spray or heavy pollution), thus predictions made from laboratory tests should not assume standard statistical techniques used for gradual ageing processes.[35] A first attempt at clearly defining conditions for safe installation has given rise to a limit of 1 mA on the cable surface.[31] This is more restrictive in coastal and industrial areas than in inland rural environments where the conductivity of precipitation will be much lower. As higher conductivity pollution will lead to higher currents on otherwise equivalent electrical circumstances it is critical therefore to take into account the local environment.

Some direct measurements have been made of the conductivity of cables exposed to different pollutants. These levels are surprisingly high and are dependent on the 'age' of the cable surface. In table 51.2 the aged cable had been exposed to arcing for 200 hours in a salt fog chamber with a 20 kV supply limited to 8 mA (50 Hz) by an external capacitor. The new cable was not pristine but had not been exposed to ultraviolet or arcing degradation. The sea-water was obtained from open coast line. The cement was applied

Table 51.2 The effect of wet pollutants on the resistance per unit length of new and old ADSS cable

Sample	Resistance			
	Sea-water	Cement	Rain-water	De-ionised water
New cable	130 kΩ/m	600 kΩ/m	48 MΩ/m	>60 MΩ/m
Aged cable	100 kΩ/m	600 kΩ/m	12 MΩ/m	>60 MΩ/m

to damp cable and then gently wetted with rain water until maximum conductivity was achieved. The rain-water was collected in rural Cheshire. Comparison with the literature suggests these laboratory values are worse than would be seen in service.[28]

If the local conditions of the cable are unknown and a worst case of 100 kΩ/m is assumed, then calculations based on standard tower geometries lead to a mid-span potential limit of 12 kV being equivalent to a 1 mA current. This voltage must be recalculated for the specific tower geometry and phase arrangement but will not differ significantly. If however sensible predictions can be made as to the level of pollution, higher space potentials can be acceptable. It should be stressed that the levels of conductivity that need to be considered are the worst case, not the average, as degradation is not principally a gradual effect, it is a rare adventitious phenomenon only occurring when the conditions are severe enough. Similarly fields need to be calculated for conditions of single circuit outages for twin circuit towers to ensure the worst cases are considered.

HIGH VOLTAGE ADSS SYSTEMS

A number of solutions have been proposed to enable high voltage application of ADSS cables, and this suggests a wider appreciation of the limitations of the standard cable than generally acknowledged. Most of these ideas concern grading the field on the cable by using a semiconductive element within the cable.[26,27,43,44] In particular Peacock and Wheeler[25] analysed the requirements and concluded that such a product, though possible in principle would be too difficult to make in practice. The cable would have to be conductive enough to prevent voltage gradients leading to arcing but resistive enough to prevent over-heating. Carter[45] has listed a range of ideas of devices which might prevent damage by arcing but these may not be practical and have not yet been tested.

A solution is presently being developed[46,47] which deploys a semiconducting rod over the ADSS cable adjacent to the towers. This rod is manually pushed out over the first 50 m of each half span, thereby protecting the cable in the region where arcs might otherwise have occurred. By separating the resistive element from the cable, greater heat dissipation is possible and more control over the manufacturing and quality of the rod is possible. This product, in association with any standard ADSS cable, should allow installation on to any voltage system irrespective of local environment. At present two installations of the rod have been completed, and a live line trial with monitoring is underway.[48]

REFERENCES

(1) Bronsdon, J. A. N., Carter, C. N., Kent, A. H. and Iddin, A. (1986) 'The successful evaluation of a 400 kV composite ground wire optical communication system'. *36th Proc. Int. Wire and Cable Symp.*, pp. 496–504.

(2) Carter, C. N. and Waldron, M. A. (1992) 'Mathematical model of dry band arcing on self-supporting, all dielectric optical cables strung on overhead power lines'. *IEE Proc., C* **139** (1992), 185–196.

(3) Carlton, G., Bartlett, A., Carter, C. N. and Parkin, A. (February 1995) 'UK power utilities' experience with optical telecommunications cabling systems'. *Power Eng. Journal*, 7–14.

(4) Davies, A. J., Radage, P., Rowland, S. M. and Walker, D. J. (September 1992) 'The selection of materials for a long span, dielectric, aerial, self-supporting cable for optical communications'. *Plastics in Telecomms VI*, 27/1–27/10.

(5) Rowland, S. M., Craddock, K., Carter, C. N., Houghton, I. and Delme-Jones, D. (1987) 'The development of a metal-free, self-supporting optical cable for use on long span, high voltage overhead power lines'. *36th Proc. Int. Wire and Cable Symp.*, pp. 449–456.

(6) Grooten, A. T. M., Bresser, E. J. and Berkers, A. G. W. M. (1987) 'Practical experience with metal free self supporting aerial optical fibre cable in high voltage networks'. *36th Proc. Int. Wire and Cable Symp.*, pp. 426–437.

(7) Hall, T. and Peacock, A. J. (1989) 'Development and field performance of a self supporting cable for use on high voltage power lines'. *2nd IEE Nat. Conf. on Telecoms*, York, pp. 327–323.

(8) Haag, H. G., Hog, G., Jansen, U., Schulte, J. and Zamzow, P. E. (1994) 'Optical ground wire and all dielectric self-supporting cable – a technical comparison'. *43rd Proc. Int. Wire and Cable Symp.*, pp. 380–387.

(9) Jurdens, C., Haag, H. G. and Buchwald, R. (1988) 'Experience with optical fibre aerial cables on high tension power lines'. *CIGRE* Paper 22-11.

(10) Bonicel, J. P. (1994) 'Optical ground wires and self supporting all dielectric fiber optic cables'. *Electrical Communication*, 1st Quarter, 45–51.

(11) Keller, D. A., Tatat, O., Girbig, R., Adams, M., Bohme, R. and Larsson, C. (1995) 'Design and reliability considerations for long span high voltage, ADSS cables'. *44th Proc. Int. Wire and Cable Symp.*, pp. 786–792.

(12) Kloepper, S., Menze, B., Schulte, J., Maltz, G. and Teucher, F. (1991) 'Short and long span self-supporting, non metallic aerial fibre optic cable'. *40th Proc. Int. Wire and Cable Symp.*, pp. 186–194.

(13) Chung, S.-V. and Ding, L. (1991) 'Stress strain characteristics of self supporting aerial optical fibre cables'. *40th Proc. Int. Wire and Cable Symp.*, pp. 178–185.

(14) Ishihata, Y., Saito, K., Nakadate, K., Horima, H., Niikura, K., Kurosawa, A. and Ohmori, T. (1988) 'Movement of non-metallic self supporting optical fibre cable under wind pressure'. *37th Proc. Int. Wire and Cable Symp.*, pp. 100–108.

(15) Bartlett, A. D., Craddock, K. and Hearnshaw, D. (1991) 'Integrating optical communications within the power network – the practice'. *A.C. and D.C. Power Transmission*, IEE Conference Publication No. 345, 211–219.

(16) Bartlett, A. D., Roberts, J. L. L. and Hearnshaw, D. (1993) 'Overhead power lines designed to contain failure'. *12th Int. Conf. on Electrical Distribution*, 3.1/1–9.

(17) Martin, J. (April 1993) 'Optical fibre planning guide for power utilities'. *CIGRE* Paper 35-04.
(18) Bensink, S. J. B. and Dekker, W. W. J. (1990) 'Aramid tapes as antiballistic protection of aerial optical fibre cables'. *40th Proc. Int. Wire and Cable Symp.*, pp. 362–367.
(19) Metcalfe, A. G., Gulden, M. E. and Schmitz, G. K. (1971) 'Spontaneous cracking of glass filaments'. *Glass Technology* **12**, 15–23.
(20) Metcalfe, A. G. and Schmitz, G. K. (1972) 'Mechanism of stress corrosion in E glass filaments'. *Glass Technology* **13**, 5–16.
(21) Newaz, G. M. (1985) 'A quantitative assessment of debonding in unidirectional composites under long term loading'. *J. Reinforced Plastics and Composites* **4**, 354–364.
(22) Hogg, P. J. and Hull, D. (August–September 1990) 'Micromechanisms of crack growth in composite materials under corrosive environments'. *Metal Science*, 441–449.
(23) Tsui, S.-W. and Jones, F. R. (1989) 'Evaluation of the performance of the barrier layer of sand-filled GRP sewer linings'. *Plastics and Rubber Processing and Applications* **11**, 114–146.
(24) Tsui, S.-W. and Jones, F. R. (1992) 'The effect of damage on the durability of a sandfilled GRP sewer linings under acidic stress corrosion conditions'. *Composites Science and Technology* **44**, 137–143.
(25) Peacock, A. J. and Wheeler, J. C. G. (1992) 'The development of aerial fibre optic cables for operation on 400 kV power lines'. *IEE Proc., A* **139**, 304–313.
(26) Berkers, A. G. W. M. and Wetzer, J. M. (1988) 'Electrical stresses on a self-supporting metal free cable on high voltage networks'. *IEE Conf. on Dielectric Materials, Measurements and Applications*, pp. 69–72.
(27) Rowland, S. M. and Carter, C. N. (1988) 'The evaluation of sheathing materials for an all-dielectric, self supporting communication cable for use on long span overhead power lines'. *IEE Conf. on Dielectric Materials, Measurements and Applications*, pp. 77–80.
(28) Carlton, G., Carter, C. N., Peacock, A. J. and Sutehall, R. (1992) 'Monitoring trials on all-dielectric, self-supporting, optical cable for power line use'. *41st Proc. Int. Wire and Cable Symp.*, pp. 59–63.
(29) Kaczmarski, A. (August 1995) 'Australian trial of ADSS optical cables'. *Electrical Engineer*, 34.
(30) Rowland, S. M. and Easthope, F. (1993) 'Electrical ageing and testing of dielectric, self-supporting cables for overhead power lines'. *Proc. IEE, A* **140**, 351–356.
(31) Rowland, S. M. and Nichols, I. V. (1996) 'The effects of dry-band arc current on ageing of self-supporting dielectric cables in high fields'. *IEE Proc. Sci. Meas. Technol.*, **143**, 10–14.
(32) Dissado, L. A., Parry, M. J. and Wolfe, S. V. (1989) 'Sheath materials for aerial optical cables'. *Plastics in Telecomms V*, London, September.
(33) Lennartsson, H. and Perrett, B. (September 1992) 'Track resistant jacketing material for optical cable'. *Plastics in Telecomms*, TIT **VI**, 27/1–27/10.
(34) Wheeler, J. C. G., Lissenburg, M. J., Hinchecliffe, J. D. S. and Slevin, M. E. (1988) 'The development and testing of a track resistant sheathing material for aerial optical fibre cables'. *IEE Conf. on Dielectric Materials, Measurements and Applications*, pp. 73–76.

(35) Dissado, L. A., Parry, M. J., Wolfe, S. V., Summers, A. T. and Carter, C. N. (1990) 'A new sheath evaluation technique for self-supporting optical fibre cables on overhead lines'. *39th Proc. Int. Wire and Cable Symp.*, pp. 743–751.
(36) Rowland, S. M. (1992) 'Sheathing materials for dielectric, aerial, self supporting cables for application on high voltage power lines'. *6th IEE Conf. DMMA*, IEE Conference Publication No. 363, 53–56.
(37) Daneshvar, O., Hill J. and Mann, X. (1995) 'Development of an all-dielectric self supporting cable for use in high voltage environments' *44th Proc. Int. Wire and Cable Symp.*, pp. 763–769.
(38) Billings, M. J. and Humphreys, K. W. (1968) 'An outdoor tracking and erosion test of some epoxide resins'. *IEEE Trans.* **EI-3**, 62–70.
(39) Bradwell, A. and Wheeler, J. C. G. (1982) 'Evaluation of plastic insulators for use on British Railways 25 kV overhead line electrification'. *IEE Proc., B* **129**, 101–110.
(40) King, L. A. (1961) 'The voltage gradient of the free burning arc in air or nitrogen'. ERA report G/XT172.
(41) Hampton, B. F. (1964) 'Flashover mechanism of polluted insulation'. *Proc. IEE* **111**, 985–990.
(42) Brewer, D. A., Dissado, L. A. and Parry, M. J. (1992) 'Limitations on the damage mechanism in dry band arcing on all dielectric self supporting cables'. *IEE Conf. on Dielectric Materials Measurements and Applications*, pp. 49–52.
(43) Oestreich, U. H. P. and Nassar, H. M. (1988) 'Self-supporting dielectric fiber optical cables in high voltage lines'. *37th Proc. Int. Wire and Cable Symp.*, pp. 79–82.
(44) Wheeler, J. C. G., Lissenburg, M. J. and Hinchecliffe, J. D. S. (1990) 'Advances in the development of aerial optic fibre cables'. *6th BEAMA Int. Electrical Insulation Conf.*, pp. 76–80.
(45) Carter, C. N. (1993) 'Arc controlling devices for use on self-supporting, optical cables'. *IEE Proc., A* **140**, 357–361.
(46) Taha, A. J., Nichols, I. V., Platt, C. A. and Rowland, S. M. (1995) 'A novel system for the installation of all-dielectric self supporting optical cables on high voltage overhead power lines'. *44th Proc. Int. Wire and Cable Symp.*, pp. 171–177.
(47) Nichols, I. V., Platt, C. A., Taha, A. J., Neve, S. and Carter, C. N. (1996) 'The development of a post-fit system for all-dielectric self-supporting optical cable'. To be published CIGRE.
(48) Haigh, H. R., Rowland, S. M., Taha, A. J. and Carter, C. N. (1966) 'A fully instrumented installation and trial of a novel all-dielectric self-supporting cable system for very high voltage overhead power lines'. *45th Proc. Int. Wire and Cable Symp.*, pp. 60–67.

CHAPTER 52

Wrap Cable

An alternative method of installing fibre optics on overhead power lines is to attach a small dielectric optical cable to an existing earth wire. Early attempts to achieve this relied on lashing the optical cable to the earth wire. However, such systems suffer from markedly increased wind loadings, and a propensity to 'gallop' under certain conditions.[1] Furthermore, because the attached cable tends not to be in contact with the supporting wire along its entire length, there is a tendency for loops of loose cable to appear over the course of time.

Early systems involving the helical application of optical cables to earth wires under a controlled tension were generally applied to spans where other means of installing fibre optics were problematical.[2] The cables in these cases were of the 'tight' variety and consequently provided no strain margin. It has been realised, however, that wrap cable technology may have some advantages over other cable systems in certain circumstances. As a result, loose tube cables have been developed which can be quickly and efficiently applied to existing earth wires.[3,4] On the UK National Grid Company's network, installation rates of 20 km/week have been achieved, which is almost a factor of two better than other methods of fibre optic installation.

CABLE DESIGN

When designing a cable for a wrap-on system there are several factors that must be considered as a result of the complex demands placed on the system. Primary requirements are that the cable be small and lightweight to minimise the additional environmental loadings effective on the earth wire. The wrapped cable additionally has to be strong enough to withstand the rigours of processing and installation but must also be flexible enough to allow easy wrapping on to the earth wire and handling at tower tops. It is also important to ensure that the wrap cable does not become loose at any time. To this end, the cable is installed at a tension which is sufficient to stretch it by an amount that exceeds the worst-case total contraction of the earth wire relative to it throughout the operating environmental envelope. A controlled tension is maintained by the wrapping machine along the whole length of the installation.

When the earth wire is exposed to ice or wind loadings, or experiences an electrical fault, it is extended and this results in a similar extension of the wrap cable. Thus the strain margin of the wrap cable has to be sufficient to accommodate these in-service extensions in addition to the small extension applied (typically about 0.15%) during installation. This ensures that the optical fibres are free from longitudinal strain at all times thus maximising the cable's lifetime. Wind tunnel experiments suggest that the presence of the wrap cable does not significantly affect the wrapped earth wire's drag coefficient and the only increase in wind loading results solely from the slightly increased frontal area. Field tests have confirmed these findings and show that a wrapped cable is no more likely to exhibit galloping behaviour than an unwrapped earth wire. It has also been found that the presence of the wrap cable does not affect the degree of icing which occurs on the earth wire in severe icing episodes.[5]

Cable weight is an important system parameter. In order that conductor clearances are maintained at safe levels during installation it is necessary to impose a maximum allowable earth-wire sag. This places an upper limit on the weight of wrap cable that can be carried by a wrapping machine. An optimum cable design is one that minimises weight and diameter thereby maximising the length of cable that can be installed in a continuous length and minimising the number of intermediate fibre joints that have to be made. It has proved possible to install up to 4 km continuous lengths of 24-fibre wrap cable using a cassette containing two drums of cable which are each installed in opposite directions from a central tower.

Finite element analysis of the effects of point loadings on the earth wire during wrap cable installation showed that the resulting extension is dependent on the earth-wire type and span length, the position of the wrapping machine along the span, and the amount of wrap cable left on the drum. Under the worst foreseeable combination of conditions the maximum extension of the earth wire due to the weight of the wrapping equipment is calculated to be about 0.05%.

Ideally the wrapping pitch should be as long as possible because short pitches use up more of the wrap cable and reduce the installed bend diameter of the cable. However, for practical purposes the pitch is set to about 750 mm as this is found to yield sufficient radial pressure on the earth wire to hold the wrap cable in place thereby preventing the creation of loops of slack cable.

Optical performance is enhanced by the adoption of the loose tube structure, eliminating any possibility of distributed microbend losses. The number of splices needed has been minimised by reducing the weight of all system components and hence increasing the distance between splices. This obviously improves the potential system power budget, and at the same time reduces costs. A 48-fibre cable variant has been developed offering improved information-carrying capacity and this is illustrated in fig. 52.1. This cable has a diameter of 8.4 mm and weighs 63 kg/km.

MATERIALS SELECTION

Wrap cables have to be able to withstand the same environment as the earth wires around which they are applied. The normal temperature range of operation for these cables is approximately −30°C to 60°C, although they also have to maintain fault-free operation when an electrical fault occurs on the earth wire. Lightning strikes may

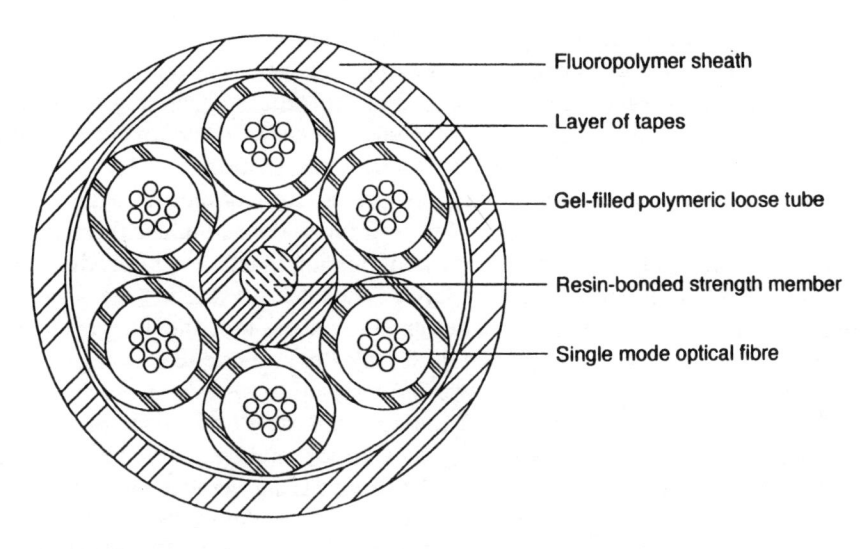

Fig. 52.1 48-fibre wrap-on cable

Labels: Fluoropolymer sheath; Layer of tapes; Gel-filled polymeric loose tube; Resin-bonded strength member; Single mode optical fibre

locally heat the cable even more. To afford extra thermal protection the outer sheath of the cable is made from a fluoropolymer capable of withstanding high temperatures.

In addition to the extreme temperatures which the cable experiences, it must also withstand ultraviolet exposure, prolonged contact with grease applied to the earth wire, possible long term abrasion, and possible exposure to airborne pollutants. Tests have shown that the cable sheath material is eminently suitable for these conditions, and also provides some shot-gun damage protection.

INSTALLATION AND WRAPPING MACHINE DESIGN

The pitch length and tension of the wrap cable must, at all times, be held within the required limits. Uniform pitch length is essential as this is the least length of wrapped cable needed for a given number of turns. Tension in the wrap cable must be maintained throughout all of the operations involved in its installation. Even momentary loss of tension is unacceptable as this is irrecoverable. When calculating the wrapping tension, account must be taken of the elastic and thermal properties of the host earth wire throughout the full range of conditions encountered both in service and, particularly, during installation.

The combination of pitch length and cable tension must be sufficient to generate enough radial force between the host earth wire and the cable to ensure continuous positive contact under all expected service conditions. Performance of the optical cable fittings should be matched to the earth wire so that, provided the host earth wire does not fail, the optical system should remain operational.

The equipment used to install the wrap cable on the earth wire consists of two distinct parts. There is a machine upon which a drum of wrap cable is mounted and which allows rotation around the earth wire of the drum, balanced by a counterweight. Gearing is provided to relate the distance moved along the earth wire with the degree of rotation around it. Additionally there is a motorised tug which provides motive power for the wrapping machine. During installation the wrap cable's tension must be sensibly constant across the whole range of likely wrapping speeds. A tension adjustment is required to compensate for the changing load of cable on the drum during installation, and this has been made as simple and foolproof as possible. A moving counterweight is used to ensure stability as the load reduces during installation.

The combination of drive wheel material, profile and contact pressure is of prime importance to ensure adequate traction for both tug and wrap under the wide range of conditions encountered. Any slippage of the wrapping machine will produce extended and erratic pitch length which is unacceptable. The fairleads guiding the wrap cable off the drum and on to the earth wire are designed to ensure that the wrap cable does not encounter severe bending and to prevent any cable damage due to rubbing and friction.

When the machine arrives at the tower on completion of a span, it is essential that before it is transferred to the next span, the wrapped cable and earth wire are securely clamped together to avoid the possibility of even the slightest cable tension loss. The point at which such a temporary clamp can be applied is inaccessible by normal means so a special remotely operated device has to be designed to facilitate easy access. Figure 52.2 shows this arrangement along with other key elements of the wrap cable system.

Safety and reliability are closely related. As there is usually no secondary protection to the public during the wrapping process, such as netted scaffolds over roads, the design and manufacture of the tug and wrapping machine have to be to a very high standard, backed up by a rigorous series of validation tests prior to entry into service.

Gross weight, payload, and rotational envelope diameter are very much inter-related as there is a counter-weight to the cable load which forms a significant part of the total.

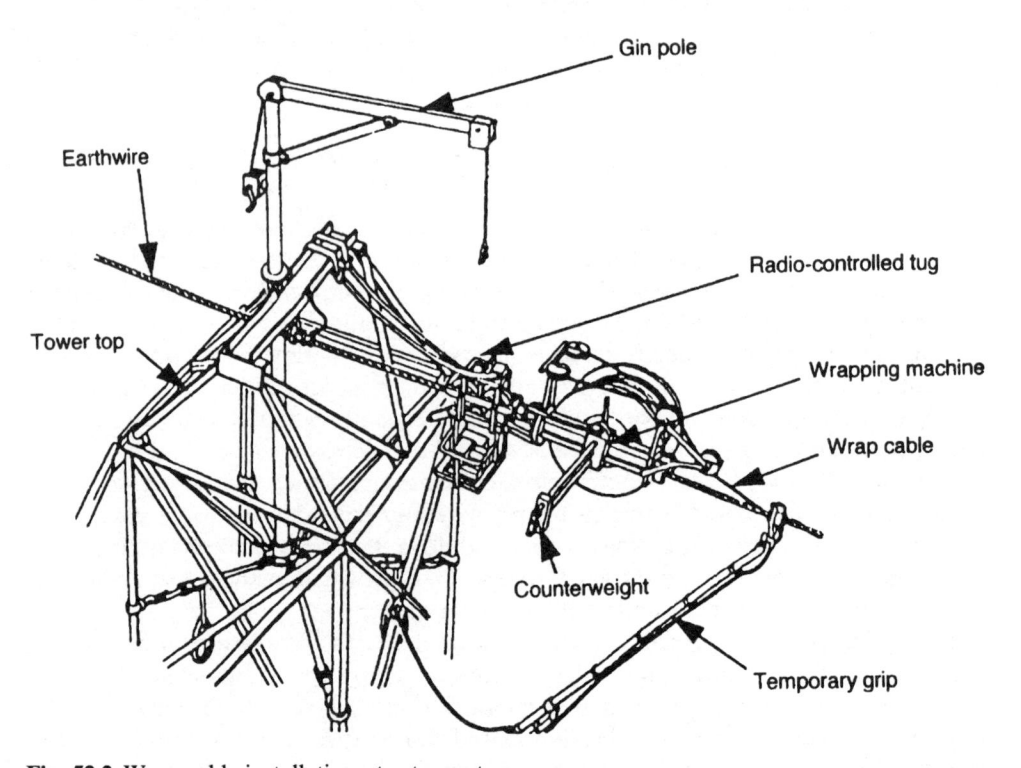

Fig. 52.2 Wrap cable installation at a tower top

Therefore considerable efforts were expended in trying to minimise the weight of the wrapping machine, cable drum, and the wrap cable itself. To maximise installation lengths a carbon fibre composite cable drum is used and high performance metal alloys are used in the machine. Currently the payload: weight ratio is a little over 0.5 and it is unlikely that this could be improved whilst still maintaining all of the built-in safety features. However, from the line owner's point of view, it is only the gross weight that matters. This must be taken in the worse case condition and occurs during recovery of the wrapping machine system with a full drum of optical cable on board. The requirement for the UK National Grid Company's network is a maximum of 250 kg.

Risk assessment indicated that there was a low but significant probability of traction failure from a number of causes. It was thus required to have a recovery system available to enable the retrieval of a broken down wrap machine. It was decided that this could be achieved by sending out a recovery tug unit from the adjacent tower. The recovery tug latches on to the broken down tug unit and then the whole system is pulled back via a pulling rope. This can be achieved either manually or by the use of a winch. Both of the tugs and the wrapping machine are pulled to the adjacent tower and wrapping continues whilst this is done. The recovery tug can then be removed and a repair to either the original tug unit or wrap machine effected before progressing to the next span.

Removal of wrapped cable can be carried out under full live line conditions without the formation of cable loops and without damage to the removed cable even in the case of a mid-span failure. The unwrapping machine is designed to be capable of removing any pitch length of cable that may be present on any conductor on the UK National Grid Company's network. The wrap cable is stored on a drum which processes along the span.

Installation work may be hampered by weather conditions which prevent safe climbing of the towers. High electrical fields are present in the vicinity of the earth wire, their magnitude dependent upon line voltage, phase configuration and circuit outage conditions. Other difficulties encountered during installation may include corroded surfaces of the earth wire or associated fittings.

The installation machines have been designed to be able to negotiate the earth-wire gradients found near tower tops under the combined effects of both rain and earth-wire surface pollution. This necessitates the wrapping equipment to be able to negotiate earth-wire gradients as high as 22°.

Several types of obstructions can be encountered during an installation, most of which have been overcome by specially developed solutions. Mid-span earth-wire joints or repair sleeves are readily negotiated by virtue of the equipment's spring-loaded pressure wheels. These were designed to allow operation on a range of earth-wire diameters, and hence allow the capability to successfully traverse in-line joints. Bird flight deviators and aircraft warning spheres must be removed prior to installation, but specially-designed clamps allow their refitting after wrap cable installation. End of span cable grips and installation ancillaries make it easy to continue installation past transmission towers. Only broken earth-wire strands may prevent successful wrap installation and under these circumstances remedial attention to the earth wire would be advised prior to continuation.

Of prime importance during installation is the safety of the installation crew and of the public at large and to this end all relevant national health and safety legislation must be complied with based on a comprehensive risk assessment of the activities involved and equipment used. In the UK there are safety rules and codes of practice together with specific procedures with which the constructor's method, statements, and work instructions must comply.

Consideration must be also given to the well-being of the earth wire during the installation. The maximum point load on the span due to the weight of the loaded wrapping machine must not exceed the permissible maximum earth-wire tension. Within practical limits the whole of the installation process can be carried out without interruption to the power system.

WRAP CABLES FOR PHASE CONDUCTORS

Whilst wrap cables on earth wires present a very economical method of installing fibres on to power lines, this method is not viable on lines which have no earth wire or whose construction is not sufficiently strong to allow a wrap machine to traverse the ground wire. This last consideration is not unusual for ageing tower constructions the condition and mechanical strength of which are uncertain. One solution to these problems is to wrap on to the phase conductors. This cannot be carried out on live conductors and has the problem shared by OPPW of necessitating complex transition systems between phase voltage and ground, where fibres are taken to underground cables or joint boxes at earth potential, however rapid installation is possible.

Cable design

Cable design issues are principally the same as for earth-wire wrap cables. The fibre in the cable must have sufficient strain margin that the cable can be installed under a range of thermal conditions and must allow for expansion and contraction of the cable during normal operation and fault conditions. In reality, commercial offerings for phase wrap installations have been developed from earth-wire products, so it is not surprising that the cables offered are similar. Those described in the previous section are typical. Of all the cables installed, phase wrap should be seen as the most vulnerable to damage by gun shot, as it is in practice suspended lower than either the ADSS or earth-wrap cable. The phase-wrap cable is particularly close to the ground when installed on lower voltage short span routes. Little can be done about this, as ballistic protection normally involves increasing the diameter of the cable to unacceptable sizes.[6] Thus the possibility of damage by shot gun must be part of the system reliability calculation. As a result of this feature and the need for special termination equipment, the phase-wrap cable is not yet as widely deployed as OPGW, earth-wire wrap or ADSS cable, but it has been in service since 1982.[5]

The effects of snow and ice accretion and lightning strikes are covered in the previous section on wrap cables on earth wires. In summary, the likelihood of a conductor galloping is reduced by the application of a wrap cable and ice accretion is not changed markedly.[5] The effect on drag depends upon the respective cable sizes but essentially drag coefficients are not changed.[1,5,7]

System design

In essence the problems of taking the wrap cable from phase potential to ground are the same as for the core of the OPPW; practically, however, the situation is much simpler.

This is because the optical cable can readily be separated from the conductor. This allows joints between wrap cables to be held at phase potential on the conductor thus removing the need for two expensive phase–earth transition kits or even insulator stacks. Such joints are carried out on the ground and then the joints and cables are wound up into a metal box which is fixed to the phase conductor.[1]

A typical phase to earth joint is shown in fig. 52.3. Essentially it consists of a method of maintaining the wrap cable tension on the span side of the joint (so that it does not unwind), a corona shield for the optical cable where it leaves the conductor and a mechanically protective tube through which the optical cable travels to earth potential at the support.[8] Once the cable has been fed into the tube, the tube is filled with an epoxy based material to prevent internal discharges. The outer surface of the tube is sheathed with a material which is resistant to damage by arcing such as a silicone rubber or EPDM, and usually includes a number of sheds to increase the effective length of the rod as in standard practice.[9] This protective tube arrangement also offers additional shot-gun protection to the cable but must itself be designed with this in mind.

In high voltage applications (over 100 kV for example) the fibres and rod which provide the voltage transition can be factory assembled and spliced to the cables at ground and phase voltage. Excess cable is then stored at low and high potential on the tower or on the phase conductor respectively.[1,7] This construction minimises the risks of discharges within the voltage transition structure due to adventitious voids but can be cumbersome if the cables contain a large number of fibres as many splices take up a substantial volume of space.

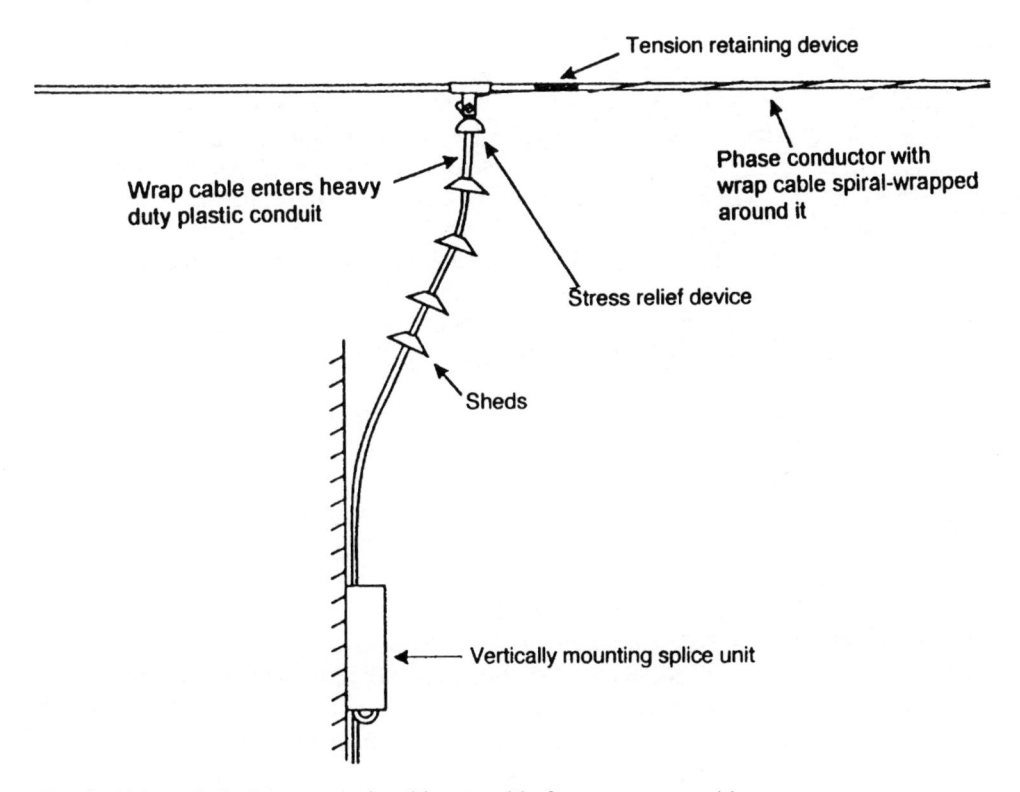

Fig. 52.3 A typical phase-to-earth cable assembly for a wrap-on cable

At each tower the wrap cable must leave the conductor to negotiate either a suspension fitting or tension fittings. In these instances a bale hanger is fitted to the conductor and the wrap cable attached to it. The optical cable should never be separated from the conductor unless it is maintained at the phase voltage by a conductive element also attached to the conductor, otherwise a voltage gradient is likely along its length resulting in dry-band arcing and degradation of the cable.

Installation

There are two substantial differences between wrapping on phase conductors and wrapping on earth wires. Most important is that phase conductors are closer to the ground and are generally more accessible. This means that it is often appropriate to pull the wrap machine from the ground with a rope rather than use a remote controlled engine. This reduces the weight of the machine, an important consideration when installing on weaker structures. The second principal difference is that a smaller wrapping machine has to be used both because of weight allowances and because of physical clearance to other conductors. If conductors are suspended from a tower with a number of cross-arms, the wrap cable is normally installed on the lowest. This allows the wrapping machine to be pulled along from the ground rather than using a motorised tug, and also allows the second cross-arm to be used to support the machine when it is swung from one span to the next.

REFERENCES

(1) Sperduto, R. and Mitchell, M. (1988) 'Installation experience with a fiber optic cable helically wrapped around a Rochester Gas & Electric Corporation 115 kV transmission phase conductor'. *IEEE Trans. Power Delivery* 3 (2), 463–468.
(2) Yoshida, K. *et al.* (1986) 'Winding of optical fiber cable onto existing ground wire'. *Proc. 35th Int. Wire and Cable Symp.*, pp. 472–477.
(3) Carter, C. *et al.* (September 1988) 'The development of wrap-on cables for use on the earthwires of high voltage power lines'. Paris: CIGRE 22-08.
(4) Knight, I. *et al.* (1995) 'Design and development of helically-attached optical fibre cables for use with existing high voltage earthwires'. *Proc. Wires and Cables Conf.*, Florence.
(5) Carson, M. (1985) 'The use of fibre optics and the existing system for the extension and improvement of communication capabilities within public utilities'. *Proc. CIRED Conf.*, Paper 4.06, 270–276.
(6) Bensink, S. J. B. and Dekker, W. W. J. (1990) 'Aramid tapes as antiballistic protection of aerial optical fibre cables'. *40th Proc. Int. Wire and Cable Symp.*, pp. 362–367.
(7) Kourliouros, P. C. and DeSarro, J. F. (1987) 'Innovative helical wrap installation of fibre optic cable'. *Proc. Southern TIER Tech.*, pp. 36–45.
(8) Mikli, N. and Grimwood, R. (1985) 'Ein neues Lichtwellenleitersystem fur die Installation auf Hochspannungs-Freileitungen'. *Elektrizitatswirtschaft* 84, 865–867.
(9) Looms, J. S. T. (1988) *High Voltage Insulation*. Peter Perigrinus.

PART 8
Cables for Communications Applications

CHAPTER 53
Communication Systems

INTRODUCTION

In part 7 we introduced the concept of optical fibres in power transmission lines, described how optical fibres are an increasingly attractive solution to the transmission of information, and gave an introduction to the principles of optical fibre transmission and optical cable design. In this part we review other cables for communication applications and in particular optical and metallic data communication cables in chapter 54 and metallic twisted pair telecommunication cables in chapter 55.

The speedy and accurate transfer of data is an essential requirement of a modern society. The behaviour of the transmission medium is of basic importance, and must produce at its distant receiving end a change of some kind which will be correctly interpreted as meaning a definite event at the transmission point. Communication systems which can use cable as the transmission medium include telephone, radio, television and data, including a whole new host of so-called multi-media services providing the user with access to services such as home banking and shopping, video on demand, and home- or tele-working. The chapters of this part of the Handbook provide an introduction to cables used for these applications. Optical cables which carry long-haul telecommunication services are not covered here as they have already been described in chapter 49.

TRANSMISSION SYSTEMS

Baseband and broadband

In a baseband system, data from different users is combined in a common path in a digital stream. Thus the cable carries only one channel, which by multiplexing in the time domain can accommodate many users. Baseband systems use cables ranging from complex coaxials to simple twisted pairs.

In a broadband system, data from different users are allocated different frequency channels and simultaneously share a common cable. A high bandwidth cable is therefore

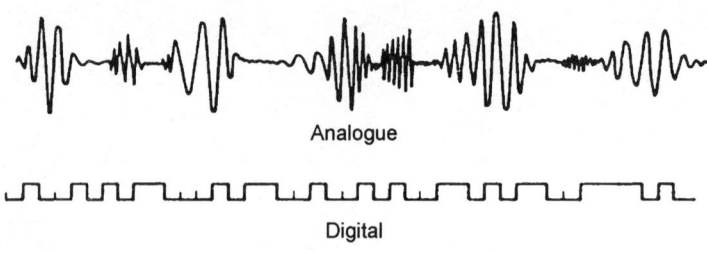

Fig. 53.1 Analogue and digital systems

required, such as the coaxials developed for cable television (CATV) systems. Broadband systems can transmit both digital and analogue signals and can accommodate multiple video, audio and data channels.

Analogue and digital

Analogue systems use a controllable smoothly varying property of electricity such as current or voltage amplitude or frequency to represent information. In digital systems information is manipulated in digital form as a stream of on–off or high–low pulses or binary digits (bits). Figure 53.1 shows an analogue and a digital signal.

With the continual need to increase data communication capacity, digital transmission is appealing because, with relatively low cost electronics, it can substantially increase the capacity of low cost cables. In analogue transmission, video and data are more demanding on the fidelity of transmission, and whenever the signal is amplified the noise is also amplified. With digital transmission, each repeater regenerates the pulses so that a pulse train can travel through a noisy medium being repeatedly reconstructed and remaining impervious to much of the noise-inducing sources.

Digital transmission, however, requires a greater frequency bandwidth than analogue, but can operate at a lower signal-to-noise ratio. The main factors in favour of digital transmission are ease of system design, the potential to increase capacity by the use of digital repeaters at frequent intervals, decreasing costs of circuitry and, last but not least, the rapidly increasing need to transmit digital data on networks.

It should be noted that, for purposes of cable design, digital pulses can be treated as the sum of a series of harmonically related sine waves, with short rise time pulses being in effect wide band radio frequency signals.

Balanced and unbalanced

For balanced systems, twisted pair cables are used and the output of the line driver is balanced to earth so that, when one terminal is positive, the other is negative by the same amount. The receiver responds to the differential voltage between the lines and rejects any signal which changes the voltage on both lines. Provided that the conductors in the pair have identical transmission characteristics, the effects of outside interference are reduced.

For unbalanced systems, one conductor of the pair is connected to earth, and although twisted pair cables can be used, multiconductor and coaxial cables are

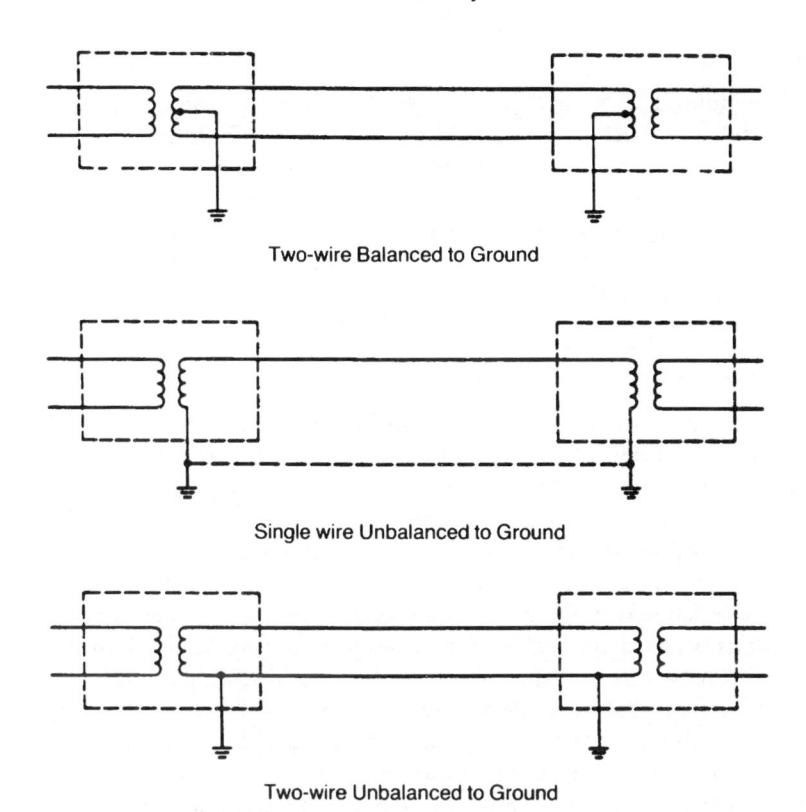

Fig. 53.2 Balanced and unbalanced systems

favoured with less expensive components when interference is not a problem. Examples of balanced and unbalanced systems are shown in fig. 53.2.

Bandwidth and data speeds

Analogue transmission is generally divided into three bands. In telephony the voice band is defined as 300–3400 Hz, and frequency bands wider than voice bands are termed wide band. A communication channel with a bandwidth less than the voice band is narrow band.

In radio engineering the bandwidth of a tuned amplifier is commonly defined as the width of the band of frequencies over which the power amplification does not drop to less than an assigned fraction of the power amplification at resonance. This fraction is usually one-half, corresponding to 3 dB loss in amplification. Transmission lines in general have an attenuation which is a function of frequency and can be constructed to have bands in the frequency spectrum where transmission is good, whilst attenuating other frequencies. In telephone carrier systems such lines or filters are used to separate the various channels.

In general, bandwidth is the frequency range between the lowest and highest frequencies that are passed through a transmission system with acceptable attenuation. For economic reasons, most data communications systems seek to maximise the amount of data that can be sent on a channel.

Serial and parallel transmission

Data are commonly transmitted by changes in current or voltage on a cable. Such transfers are called parallel if a group of bits move over several lines at the same time, or serial if the group of bits move one by one over a single line.

In parallel transmission each bit travels on its own line, and a clock signal on an additional line controls the receiver sampling. Parallel transmission is used over short distances because it is faster than serial which is preferred over long distances when multiple lines become costly.

Local area networks

Computing equipment represents a significant investment and must be used efficiently and effectively. The desire of users to connect computers, terminals, word processing and related equipment to allow discrete units to operate as part of an integrated information system has led to the development of networking.

Local area networks (LANs) are a means of providing universal data communication within a single premises with shared data transmission, storage and peripheral resources. Similarly metropolitan area networks (MANs) provide facilities for data communications between sites within a neighbourhood, for distances up to say 40 km, and wide area networks (WANs) describe the worldwide systems which provide national and international data telecommunications.

Three basic topologies are commonly used for LANs: star, bus and ring, as shown in fig. 53.3. From the user's point of view, they lead to differences in the amount of cable needed to provide a given number of access points, the ease of adding extra connections and the effect of failures. For interconnecting a scatter of points on a site, a bus will generally use the least cable, followed by a ring and then a star. It is generally easier to add connections to a bus than to a star, which may require an extra arm, or to a ring, which must be broken. A star network is least affected by failures of the devices or cables attached to it, but is likely to fail entirely if its central point, or 'hub', fails. A bus will fail if cut; but the whole network will generally not fail if one of the access points, or 'nodes', is faulty, because the node is merely a connection to the highway and can be designed to switch itself out in the event of a fault. Rings, however, are the least reliable because each node is an active repeater, and the failure of one node breaks the ring unless expensive redundancy is built into the system to overcome this.

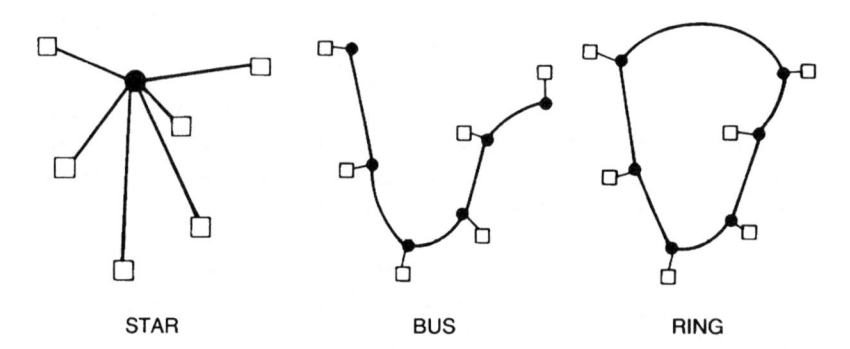

Fig. 53.3 Basic local area network topologies

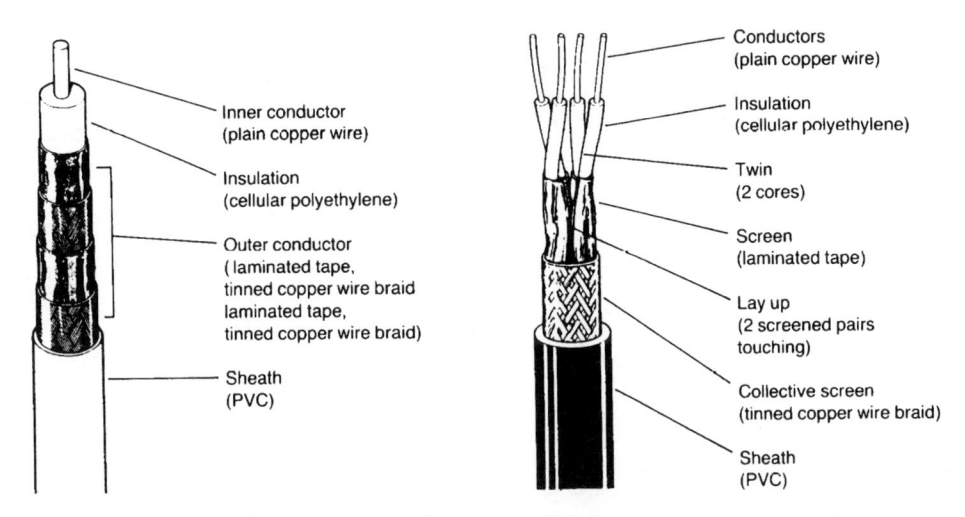

Fig. 53.4 Typical local area network cables showing large coaxial cable for bus systems (*left*) and ring system cable (*right*)

LANs also differ in the type of cables that they use. They may be implemented in coaxial or twisted pair copper cable, or in fibre optics. With some types it is possible to intermix copper and fibre optic sections in the same network, to obtain the specific benefits of each where required. Figure 53.4 shows typical LAN cables.

LANs vary considerably in their size capability, from a few tens of metres to several kilometres. They also differ in the number of devices that can be connected to them, from as few as ten up to several thousand. The rate at which they can transfer information varies widely, from 300 bits per second to 100 megabits per second or more.

OPEN SYSTEMS AND STRUCTURED WIRING

By the mid-1980s, equipment manufacturers were marketing a wide range of products covering a variety of cable types. LAN systems have proliferated, often incompatible and many using dedicated cables, so that when new computing equipment was installed, new cables were generally needed as well, leading to overcrowded cable ducts and trays, complex cable routes and difficulties in re-routing services to accommodate personnel movements.

A growing user demand for rationalisation led to the concept of an open systems interconnection whereby a system could be planned and installed without detailed knowledge of the end equipment. This freedom of choice and independence in the procurement of equipment was a significant advantage when building large multi-storey office blocks where information technology services were required at every desk. Further advantages would result from integrating voice, data, text, image and video services, and from treating these services as a utility, installed as the building is constructed, to provide access points within easy reach of any user who may change equipment and location frequently.

These requirements can be supported by a structured wiring system based on a hierarchical star network (fig. 53.5). Groups of information or telecommunication outlets, which may cover an entire floor of a high rise building, are connected via

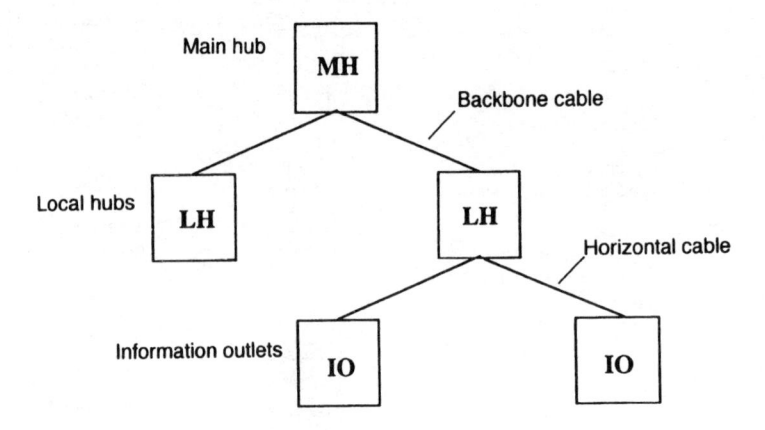

Fig. 53.5 Hierarchical star network

horizontal cabling to a local hub on the backbone cabling. The use of patch cords in the local hub allows a rapid and economic re-routeing of services to the information outlets.

Such a network could also support the interconnection of different systems using increasingly powerful personal computers for more sophisticated applications. An analysis of the requirements for these applications, many of which were the subject of international standards, together with the design and installation practices of systems in different countries, showed that the hierarchical star network could be developed into a general universal cabling system.

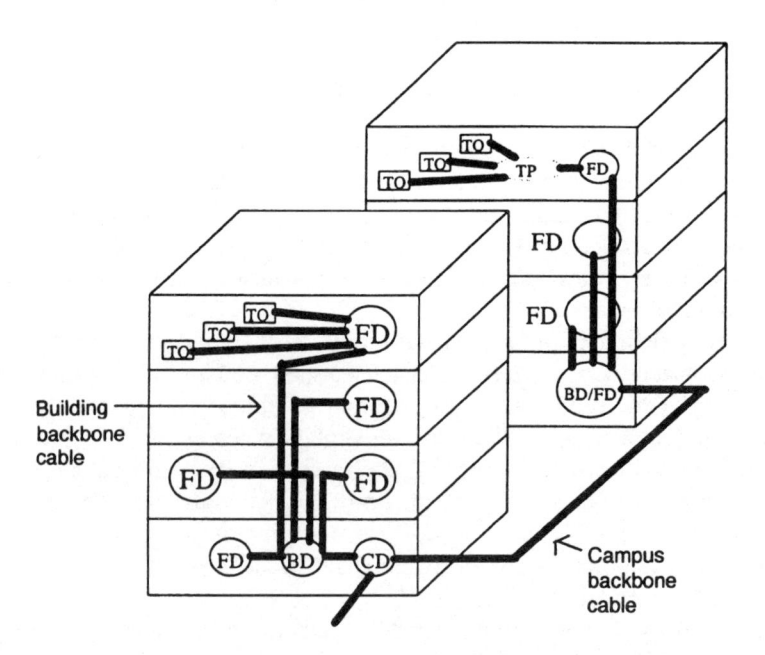

Fig. 53.6 Generic structured wiring cabling system: CD, campus distributor; BD, building distributor; FD, floor distributor; TP, transition point (optional); TO, telecommunications outlet

In 1995, two new generic standards were published for Information Technology Cabling Systems. These are ISO/IEC 11801 – a global international standard drawn up jointly by the International Standardisation Organisation (ISO) and the International Electrotechnical Commission (IEC), and EN 50173 – a European standard drawn up by the European Committee for Elecrotechnical Standardisation (CENELEC). These similar standards specify a generic cabling system for use within commercial premises which may comprise single or multiple buildings on a campus. They cover balanced copper conductor and optical fibre cabling, optimised for premises having a geographical span up to 3000 m, with up to 1 000 000 m^2 of office space and a population of between 50 and 50 000 people.

An example of a generic structured wiring cabling system is shown in fig. 53.6. It may comprise up to three cabling subsystems, a campus backbone, a building backbone and horizontal cabling. The distributors provide a means of configuring the network to support different topologies such as bus, star or ring.

The campus backbone subsystem connects the campus distributor to the building distributor, usually located in separate buidlings. The building backbone subsystem connects the building distributor to the floor distributor and includes backbone cables, sometimes known as riser cables, terminations and cross-connects. As shown in fig. 53.7, the horizontal subsystem connects the floor distributor to the individual telecommunication outlets.

The generic system allows for typical campus backbone cable route lengths up to 1500 m, building backbone cables up to 500 m and horizontal cables up to 90 m. For the horizontal subsystem the maximum combined length of work area cable, patch cords and equipment cables is assumed to be 10 m.

Both ISO/IEC 11801 and ENN 50173 show five classes of applications and links. These are as follows.

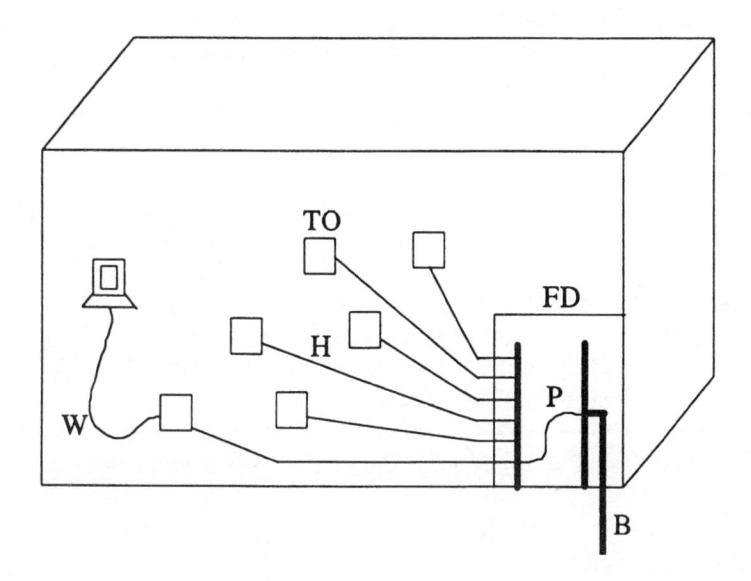

Fig. 53.7 Horizontal subsystem: B, building backbone (riser) cable; P, patch cable; H, horizontal cable; W, work area cable

Class A: includes voice and low frequency applications, with links specified up to 100 kHz
Class B: includes medium bit rate data applications, with links specified up to 1 MHz
Class C: includes high bit rate data applications, with links specified up to 16 MHz
Class D: includes very high bit data applications, with links specified up to 100 MHz
Optical: includes high and very high bit rate data applications, with links specified to support links at 10 MHz and above. Bandwidth is generally not a limiting factor with optical links.

For structured wiring systems, a variety of cables may be used. For backbone cables, balanced copper conductor cables have impedances of 100 Ω, 120 Ω or 150 Ω, and can be constructed in pair or quad configuration with or without screening. Backbone optical cables may be used, with single mode or multimode fibres, and for the latter, 62.5/125 μm fibres are recommended. Similar cables are employed for horizontal links, except that single mode fibre cables are not specified, and 50/125 μm fibres are a permitted alternative to 62.5/125 μm. The balanced copper multipair cables have a much superior transmission performance to those used for telephone or low and medium speed data applications.

TELECOMMUNICATION NETWORKS

A telecommunication system traditionally consists of a trunk network between major conurbations, a junction network between main exchanges and local exchanges, and the local network between the local exchange and the individual customers; it includes the cable used internally within the exchanges and within customer's premises. Cable being installed in the trunk and junction networks is now almost exclusively optical, but cable for the local network, within exchanges and within customers' premises remains metallic twisted pair with the exception of large business customers who in some countries have direct fibre connections. Fibre can be expected to penetrate further into the local network as broadband services emerge accompanied by digital subscriber loop (DSL) technologies, which can carry broadband digital signals over conventional twisted pair telecommunication cables. Further discussion of this is, however, outside the scope of this book.

The external local network is illustrated in fig. 53.8. The network is generally considered in two halves, that between the exchange and the cabinet being the 'primary' network (or 'E' side or 'main-side'), and that between the cabinet and the subscriber the 'D' side, distribution, or 'secondary' network. The local network generally has a star topology with high pair count (typically 100 to 2000 pairs) primary cables branching out to the cabinets, medium pair count (typically 10 to 100 pairs) secondary cables branching out to distribution points, and finally, low pair count (typically 1 to 2 pairs) dropwire cable branching out to the subscriber.

In a conventional local network each subscriber is connected to the local exchange by a dedicated twisted pair. The function of the twisted pair is two fold:

(a) to provide a path for the current required to set up calls, supervise calls, and power the phone (signal function);
(b) to provide a transmission line for voice frequency communication (transmission function).

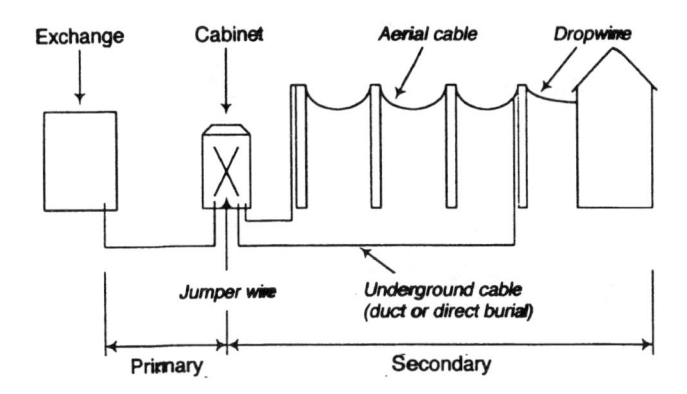

Fig. 53.8 Typical local cable network

The signal function requires that the impedance of the twisted pair be low enough so as not to reduce the current below the system limit. The transmission function requires that the transmission characteristics of the line are such that the signal is received with attenuation and interference which are within the system limits.

TRANSMISSION CHARACTERISTICS OF METALLIC COMMUNICATION CABLES

A transmission line is a network with four fundamental properties of resistance, inductance, capacitance and conductance. These are called the 'primary parameters', and they are defined by the cross-sectional geometry of the line and the materials used in the conductor and dielectric. They are denoted by:

R = loop resistance (ohm)   per unit length
L = inductance   (henry)   per unit length
C = capacitance   (farad)   per unit length
G = conductance   (siemens) per unit length

and are related to the 'secondary parameters'

Z_0 = characteristic impedance
α = attenuation coefficient
β = phase coefficient

by the following relationships:

$$Z_0 = \left(\frac{R + j\omega L}{G + j\omega C}\right)^{1/2} \quad (\Omega) \tag{53.1}$$

and

$$\alpha + j\beta = [(R + j\omega L)(G + j\omega C)]^{1/2} \tag{53.2}$$

where $\omega/2\pi = f$, the frequency in hertz, and the expression $\alpha + j\beta$ is often called the propagation constant.

Expanding and equating the real and imaginary parts of this equation gives

$$\alpha = \left\{ \frac{[(R^2 + \omega^2 L^2)(G^2 + \omega^2 C^2)]^{1/2} + RG - \omega^2 LC}{2} \right\}^{1/2} \quad (53.3)$$

(neper per unit length)

which determines the rate of change in amplitude with cable length and

$$\beta = \left\{ \frac{[(R^2 + \omega^2 L^2)(G^2 + \omega^2 C^2)]^{1/2} - RG + \omega^2 LC}{2} \right\}^{1/2} \quad (53.4)$$

(radian per unit length)

which determines the rate of change of phase with cable length.

Primary parameters

Resistance
A wide variety of conductor materials is used in communication cables. As well as copper in solid or stranded form, copper-covered aluminium and copper-covered steel are used, the latter for small conductors where strength is required. These may also be tinned or silver plated. Aluminium conductors have also been used in the past for telephone cables but they are no longer installed in the light of corrosion problems experienced in the field.

Skin and proximity effects must be taken into account when considering resistance. The ratio that the effective a.c. resistance has to the d.c. resistance of a conductor is called the resistance ratio, which increases with frequency, with the conductor material conductivity and with the size of the conductor.

In a cylindrical plain copper wire at 20°C, when the frequency is sufficiently high, substantially all the current in the conductor is confined to a region very close to the surface, with a depth given by the relation

$$d = \frac{0.066}{f^{1/2}} \quad \text{(mm)} \quad (53.5)$$

where f is the frequency in MHz.

Inductance
Both self-inductance and mutual inductance are manifestations of the interaction of currents and fields. If the current in a circuit changes, the flux is altered and an electromotive force is induced. This effect is known as self-induction.

If part of the flux in a circuit is linked with a second circuit, there is said to be mutual inductance, and the circuits are inductively coupled. Mutual inductance is defined as the ratio of the flux linkages in the second circuit produced by current in the first to the current in the first circuit. It follows that skin and proximity effects which change the flux pattern will have an effect on inductance.

Capacitance and dielectric materials
A capacitor is formed whenever a dielectric separates two conductors between which a potential difference can exist. Capacitance is the ratio of the charge which can be stored

in a capacitor to the potential applied to it ($C = Q/V$, where C is the capacitance in farads, Q the charge in coulombs, and V the applied voltage in volts). If the capacitance between given electrodes with a certain dielectric is compared with that between the same electrodes *in vacuo*, the ratio is a figure characteristic of the dielectric material and is known as the permittivity or dielectric constant, ϵ.

A perfect capacitor, when discharged, gives up all the electrical energy that was supplied to it in charging. In practice some of the energy is dissipated, mainly as dielectric losses.

In electronics cables the materials used for the insulation or dielectric are therefore critical to the transmission performance. Most plastics are considered as electrical insulators, and when assessing their potential use in cables the dielectric constant and power factor are considered over the range of operating temperatures and frequencies. Suitable materials for high performance cable would have low dielectric constant and power factor over a wide range of temperature and frequency and these parameters should not be adversely affected by high humidity.

Molecules of a dielectric may be either polar or non-polar. In the case of polar molecules the dielectric constant is higher as a result of the rotation of the polar molecules under the influence of the applied voltage. The extent to which this polar action is effective, however, depends on the frequency and the temperature. The result is that polar dielectrics have a certain characteristic behaviour with respect to temperature and frequency. If the temperature is lowered sufficiently, polar rotations are prevented, causing the dielectric constant of the material to decrease. Similarly, if the frequency is made sufficiently high, the polar molecules are not able to follow the alternations of the applied field and the dielectric constant decreases. Hence the power factor of a polar dielectric becomes quite large for certain combinations of temperature and frequency. The variation in dielectric constant and power factor for a polar dielectric as a function of frequency for two temperatures is shown in fig. 53.9.[1]

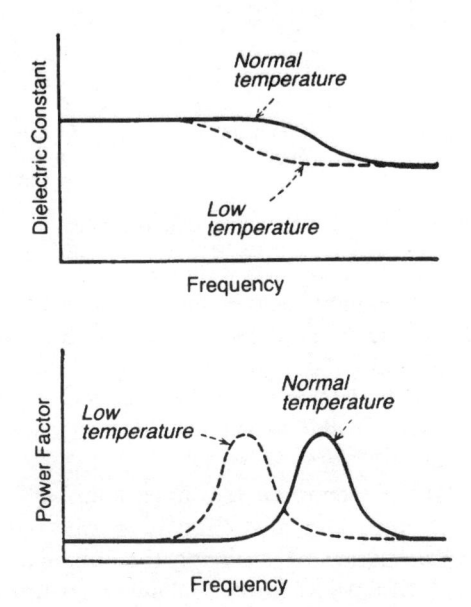

Fig. 53.9 Variation of dielectric constant and power factor of polar materials

Table 53.1 Electrical properties of insulating materials

Material	Dielectric constant		Power factor	
	50 Hz	1 MHz	50 Hz	1 MHz
Polyethylene	2.30	2.30	0.0003	0.003
Polypropylene	2.15	2.15	0.0008	0.0004
PVC	6.9	3.6	0.08	0.09
Nylon	4.0	3.4	0.014	0.04

(a) Effects of mutual capacitance

amplitude

→ time

Original signal — Small capacitance — Larger capacitance

(b) Effect of attenuation

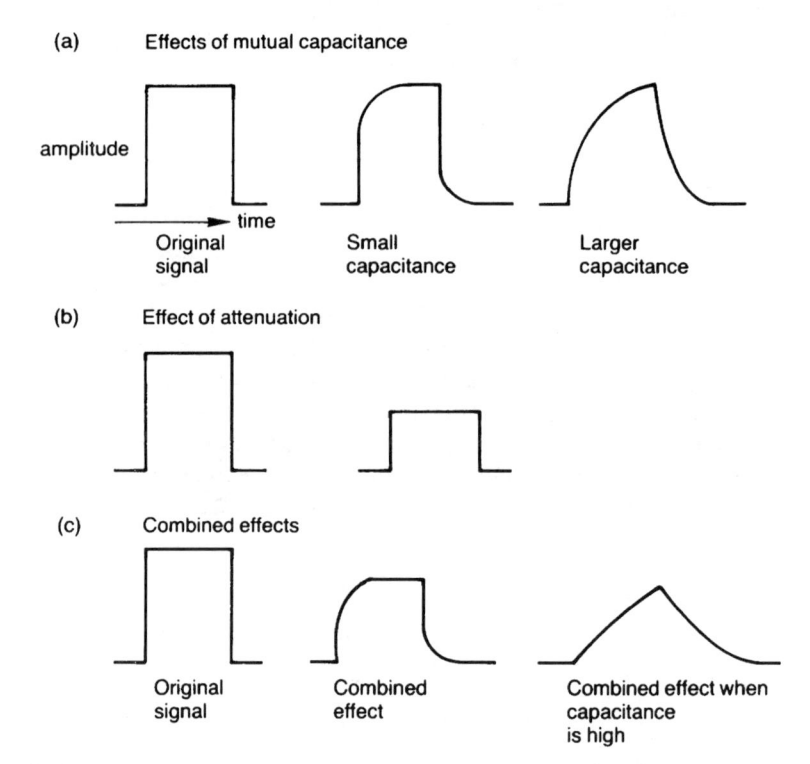

(c) Combined effects

Original signal — Combined effect — Combined effect when capacitance is high

Fig. 53.10 Effects of capacitance and attenuation on digital signals

Non-polar molecules do not exhibit these changes in dielectric constant or peaks of power factor, and materials such as polyethylene are therefore used for higher performance cables. Typical values of dielectric constant and power factor for materials commonly used in electronics cables are shown in table 53.1. In these examples PVC and nylon are polar molecules exhibiting a higher dielectric constant reducing with increased frequency.

The capacitance of a cable depends on the dimensions and configuration of the conductors and screen, if any, and the effective permittivity of the dielectric. Capacitance can therefore be varied by changing the conductor size, the insulation thickness or the permittivity of the insulation. In high frequency digital transmission, capacitance rounds or distorts the pulse shape as shown in fig. 53.10(a), causing errors.

With a low capacitance, cables can cope with faster rise times, increased bit rate capacity and longer transmission distances.

Conductance
When two parallel conductors have a potential difference between them, a certain amount of current will flow because of the finite resistance of the insulation. For an air dielectric the conductance across the gap is negligible, but when other dielectrics are used, as is the case with cables, the conductance is determined by the power factor, capacitance and frequency according to the relationship

$$G = \omega C \tan \delta \tag{53.6}$$

where $\tan \delta$ is the dielectric power factor.

Secondary parameters

Impedance
If a signal is fed into an infinitely long uniform line the ratio of voltage to current at any point along the line is constant and is called the 'characteristic impedance' of the line, usually denoted Z_0. If the line is now cut to a finite length and a load impedance equal to Z_0 is connected across the far end, it behaves as if it were still infinitely long. Power fed into the line will be absorbed by the load and none will be reflected back. The line is then correctly terminated or matched.

Not all the power which is fed into the line will reach the load at the far end. In practice power losses occur in the line itself, because of resistive dissipation in the conductors and dielectric losses in the insulation. For a correctly terminated cable, the ratio of output power to input power, in decibels, is the attenuation of the cable.

The condition in a terminated line can be represented by a wave travelling forward towards the load. If the load is not a correct termination to the line a certain proportion of the forward travelling signal will be reflected back towards the input. Similar reflections can occur from points along the cable which for some reasons may differ from the mean impedance. These reflections cause increased line losses, because both the forward and backward waves are attenuated and each contributes to the overall loss. So for minimum line power losses the amplitude of the reflected wave must be reduced to zero, in other words there must be no standing waves on the line. A standing wave results from the combination of reflected and forward waves travelling in opposite directions. Under conditions of zero standing wave the load is said to be matched to the line.

The ratio of the maximum to minimum values of voltage which occur at points half a wavelength apart on a mismatched line is called the voltage standing wave ratio (VSWR). On a lossless line the VSWR is constant throughout the length and is equal to the ratio of the load impedance Z_L to the characteristic impedance of the line Z_0.

Another term in common use is the voltage reflection coefficient, which is defined as

$$\text{voltage reflection coefficient} = \frac{Z_0 - Z_L}{Z_0 + Z_L} \quad \text{or} \quad \frac{VSWR - 1}{VSWR + 1} \tag{53.7}$$

This is often expressed as a percentage, and it follows that for an open or short-circuited line the reflection coefficient is 100%; it is zero for a correctly terminated line.

From the reflection coefficient follows another common expression, the return loss ratio, which is measured in decibels and is numerically given by

$$\text{loss ratio} = 20 \log_{10} \left(\frac{1}{\text{voltage reflection coefficient}} \right) \tag{53.8}$$

In a perfect cable the dimensions and effective permittivity on which the characteristic impedance depends will be constant throughout the length. Such an ideal cable cannot be achieved in practice because of manufacturing tolerances. For example a change in the diameter of the insulation will result in a corresponding alteration of characteristic impedance at that point. Whenever an impedance change occurs, a small reflection of the forward travelling signal results. If these impedance changes happen in a random manner the effects tend to cancel out, but the manufacture of insulated core involves rotating machinery, and there is always the possibility of dimensional changes recurring at regular intervals along the cable rather than at random. The term 'structural return loss' is used as a measure of the losses which arise from internal reflections within the cable, as distinct from return losses arising from mismatched terminations.

Impedance matching is important in data communications because, as well as reducing the energy reaching the receiver, depending on their location mismatches can add to or subtract from the desired signal level and give rise to data errors.

The relationship between the characteristic impedance and the primary parameters is

$$Z_0 = \left(\frac{R + j\omega L}{G + j\omega C} \right)^{1/2} \quad (\Omega) \tag{53.9}$$

and for most cable insulations G is neglected as it is small in comparison with ωC.

At low frequencies ωL can be neglected as it is small in comparison with ωC and the relationship becomes

$$Z_0 = \left(\frac{R}{j\omega C} \right)^{1/2} \quad (\Omega) \tag{53.10}$$

At high frequencies R can be neglected as it is small in comparison with ωL, and if G can be neglected as it is small in comparison with ωC then

$$Z_0 = \left(\frac{L}{C} \right)^{1/2} \quad (\Omega) \tag{53.11}$$

Attenuation

The voltage amplitude of a signal decreases as it travels along a cable owing to the resistance of the conductor and dielectric losses, as shown in fig. 53.10(b). In the relationship between attenuation and the primary parameters, at low frequencies ωL is small compared with R and if G is small compared with ωC then

$$\alpha = \left(\frac{\omega C R}{2} \right)^{1/2} \quad \text{(neper/length)} \tag{53.12}$$

whilst at high frequencies when R is small compared with ωL and G is small compared with ωC then

$$\alpha = \frac{R}{2Z_0} + \frac{GZ_0}{2} \quad \text{(neper/length)} \tag{53.13}$$

In this case $R/2Z_0$ represents the attenuation due to ohmic losses in the conductors, which varies as the square root of the frequency owing to the skin effect. The term $GZ_0/2$ is the attenuation due to the dielectric losses and is directly proportional to frequency.

The relative variation of these two causes of attenuation is of significance when designing high frequency cables. The permittivity of the insulation should be kept as low as possible for low attenuation, and it should also be noted that the use of lower permittivity insulation in a cable of given impedance and overall size means that larger conductors are required to maintain the given impedance which in turn reduces the ohmic losses.

As the rise time of digital signals can be related to high frequencies, they are attenuated more than the lower frequency components, resulting in rise time degradation and the rouding of pulse edges; this effect, combined with the effects of capacitance, is illustrated in fig. 53.10(c). Excessive attenuation can prevent the signal reaching the amplitude required by the receiver, resulting in data errors.

Phase coefficient

As with other relationships with the primary parameters, phase coefficient can be simplified at low and high frequencies to

$$\beta = \left(\frac{\omega CR}{2}\right)^{1/2} \quad \text{(radian/length)} \tag{53.14}$$

at low frequencies and

$$\beta = (\omega LC)^{1/2} \quad \text{(radian/length)} \tag{53.15}$$

at high frequencies. Of more practical interest at high frequencies is a further relationship given by

$$v = \frac{\omega}{\beta} = \frac{1}{(LC)^{1/2}} \quad \text{(m/s)} \tag{53.16}$$

where v is the velocity of propagation in the cable. In parallel data transmission a uniform velocity of propagation is important to ensure that bits arrive at their destination at the same time and skew errors are avoided.

The velocity ratio is the velocity of propagation in a cable compared with its free-space velocity; it is dependent on the square root of the insulation permittivity and so the lower the permittivity is the higher the velocity of propagation. For polyethylene insulation the velocity ratio is about 0.66.

Ideal balanced shielded pair model

The relationship between the electrical primary parameters and the physical properties of a twisted pair at low frequencies can be represented by the ideal balanced shielded pair model,[2,3] illustrated in fig. 53.11.

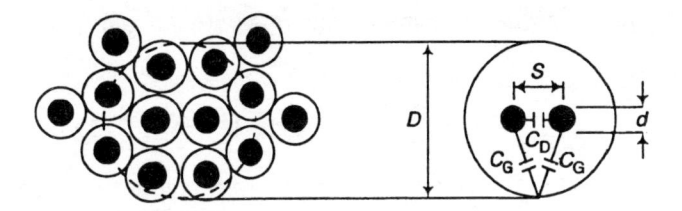

Fig. 53.11 Ideal balanced shielded pair model:
 D = effective diameter of the electrostatic shield seen by the twisted pair due to the other twisted pairs in the cable (mm)
 S = interaxial spacing of the two conductors (mm)
 d = diameter of the conductors (mm)
 C_D = direct capacitance between two conductors of the twisted pair (nF/km)
 C_G = capacitance seen by each conductor between it and the effective screen, equal for a balance pair (nF/km)

Resistance
The resistance at voice frequency is simply the d.c. resistance of the conductors given by:

$$R = \frac{8 \times 10^3 \rho}{\pi d^2} \quad (\Omega/\text{km}) \tag{53.17}$$

where ρ = resistivity of the conductor ($\Omega\,\text{m}$).

Capacitance
The capacitance of the twisted pair (mutual capacitance) is the combination of the individual capacitances given by:

$$C = C_D + \frac{C_G}{2} \quad (\text{nF/km}) \tag{53.18}$$

An approximate expression for the mutual capacitance of a twisted pair in a multipair cable[3,4] is given by:

$$C = \frac{36\epsilon}{\ln\left[1.346\left(\frac{2S}{d} - 1\right)\right]} \quad (\text{nF/km}) \tag{53.19}$$

where ϵ = relative dielectric constant of the cable.

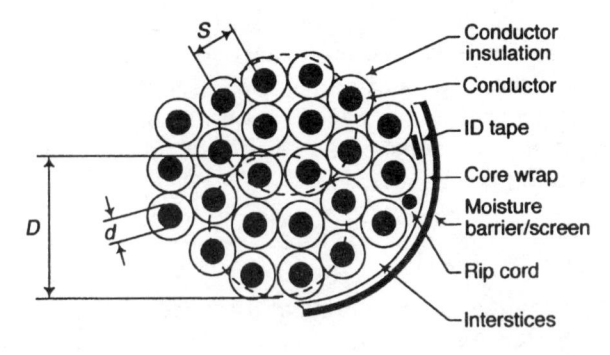

Fig. 53.12 Telephone cable dielectric insulation system

Table 53.2 Relative dielectric constants of telecommunication cables

Insulation	Filling compound	
	Air (1)	Petroleum jelly (2.23)
Solid polyethylene (2.3)	1.8	2.23
Cellular polyethylene (1.78)	1.45	1.9
Foam skin polyethylene (1.81)	1.5	1.92
Paper (1.95)	1.53	N/A
PVC ($\approx$3.4)	5.5	N/A

The relative dielectric constant of the cable is an empirical value dependent on the combination of the relative dielectric constants of all the insulating materials within the cable. Figure 53.12 illustrates the various dielectrics in a multipair cable. To ensure the cable dielectric remains stable, all dielectrics should be frequency stable and have similar dielectric constants.

Typical relative dielectric constants for twisted pair telephone cables at low frequencies are illustrated in table 53.2 for various conductor insulations and interstitial filling materials.

Cable insulation

PVC insulation is used for low speed applications but for high speeds materials such as polyethylene or polypropylene are used because of their better electrical characteristics. In addition to being used in solid form, they can be applied as a cellular material with a permittivity of about 1.45. One method of producing cellular polyethylene is by mixing a chemical additive with the raw material in the extruder. At the extrusion temperature this 'blowing agent' decomposes to produce small cells of nitrogen gas which give the cellular structure, and hence the required low permittivity. There is a tendency for residual blowing agent to give a slightly higher power factor than desirable but on balance the use of cellular material offers a useful advantage. A second technique for producing cellular polyethylene is direct gas injection at extrusion, with the advantage that the reduced chemical residue results in lower dielectric losses.

Although cellular dielectrics can have low permittivities, they can be mechanically weak. A method used to strengthen them is to crosslink them by irradiation.

Screening

An important requirement for many electronic and data cables is adequate screening against various kinds of electromagnetic influences and the need to preserve the integrity of the signal from interference or noise either inside or outside the transmission line.

Many screen designs have been developed to combat the effects of interference. The ideal screen would be a homogeneous thick walled copper tube. Such screens are used on RF and CATV cables, sometimes corrugated to improve flexibility. Sometimes a lighter aluminium tube is used.

Another form of low loss outer conductor is a copper tape applied longitudinally to the insulated core with an overlap. Because current travels longitudinally, a gap or overlap parallel to the axis of a coaxial cable does not affect the attenuation. A typical construction has a copper tape which is corrugated to give better flexibility. It is regarded as a semi-flexible cable, suitable for fixed installations.

Another cheaper version uses a much thinner copper tape. This is not corrugated but has an additional copper wire braid. It is suitable for fixed installations but is not as mechanically robust as the corrugated thick tape construction.

Copper wire braids are used extensively on many types of electronics cables. They are applied with a machine in which two sets of bobbins called spindles or carriers rotate in opposite directions round the cable core with an interweaving chain motion. Each bobbin contains a number of separate wires, referred to as the number of ends. By varying the number of spindles and ends and the lay length of the braid different degrees of coverage can be obtained.

The theoretical understanding of the a.c. resistance of braids and the way in which current flows in them is based mainly on the results of experimental measurements. It is thought that at lower frequencies the current flows mainly along the individual braid wires. The fact that there are two sets of wires spiralling in opposite directions cancels out any inductive effect that would otherwise be present. At higher frequencies the currents tend to flow parallel to the cable axis, so that the contact resistance between crossing braid wires becomes important. This can have an effect on the stability of the attenuation of the cable with time because if the contact resistance increases then the attenuation at microwave frequencies will increase. It has been found that silver-plated braid wires with an appropriate braid design give better stability of contact resistance than plain copper.

Many data cables employ screens manufactured from very thin metallic foils backed with a polyester film to provide strength. Such laminated tapes are applied longitudinally or helically and have the advantage of resulting in only small increases in cable diameter. To optimise screening efficiency, a copper wire braid can be applied over the foil screen or double-sided laminates can be used.

The efficiencies of cable screens are described by the parameter 'coupling impedance' or 'surface transfer impedance', which relates the induced current flowing on one side of a screen to the longitudinal voltage appearing on the other side. For a homogeneous tubular screen it is given by

$$Z_T = \frac{\rho}{2\pi t (ab)^{1/2}} F(u) \qquad (53.20)$$

where

$$F(u) = \frac{u}{(\cosh u - \cos u)^{1/2}}$$

and

$$u^2 = \frac{2t^2 \mu \omega}{\rho} \qquad (53.21)$$

where ρ = screen resistivity
t = screen thickness
$(ab)^{1/2}$ = screen geometric mean diameter
μ = screen permeability
$\omega = 2\pi f$

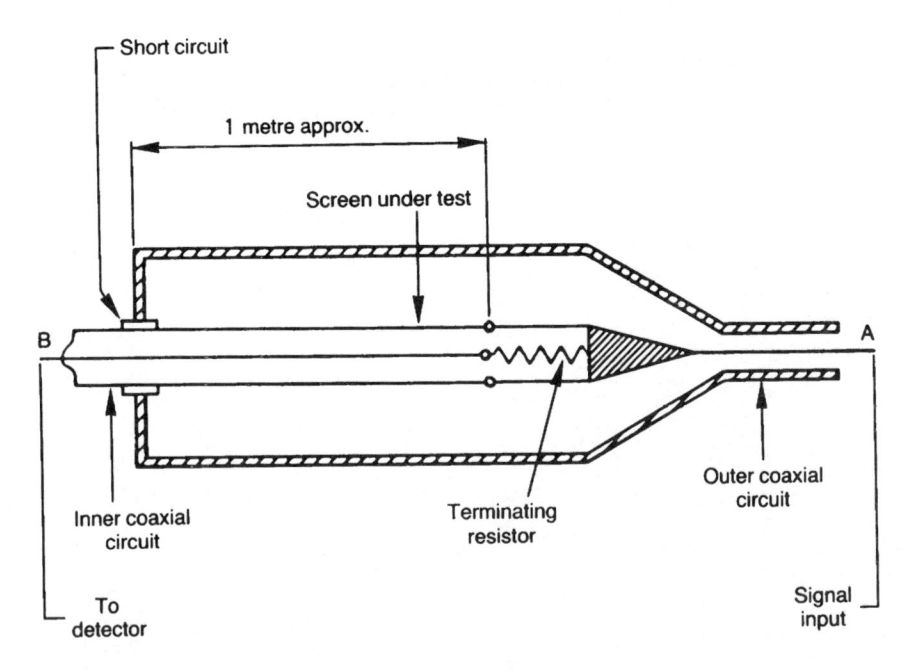

Fig. 53.13 Schematic diagram of triple coaxial apparatus

It follows that the lower the surface transfer impedance is, the better the screen is. For screens other than homogeneous tubes, the surface transfer impedance is usually measured using the triaxial arrangement shown in fig. 53.13.

Surface transfer impedances are shown in fig. 53.14 for various screen constructions at frequencies up to 30 MHz. Curves A and B show the typical drooping characteristic of the better screens. Curve C shows a popular type of screen, a single copper wire braid, which has a rising characteristic with a typical value of 200 mΩ/m at 30 MHz. To improve screening performance a second braid may be applied either directly in contact with the first (curve D) or with an intersheath (curve E).

The ratio of the resistance of a wire braid to the resistance of an equivalent tubular conductor is called the braiding factor and is given by

$$K_\text{B} = \frac{K_\text{L}}{2K_\text{F}} + \frac{\pi K_\text{L} K_\text{F}}{4} \tag{53.22}$$

where $K_\text{L} = 1 + \pi^2 D^2 / L^2$ the lay factor
$K_\text{F} = mndK_\text{L}^{1/2}/2\pi D$, the filling factor
D = mean diameter of braid
m = number of spindles
n = number of wires (ends) per spindle
d = braid wire diameter
L = braid lay length

The cover provided by a braid is $100K_\text{F}(2 - K_\text{F})\%$, and the braid angle is given by $\tan\phi = \pi D/L$.

Typical wire braid screens have a braid angle of about 40° and a cover of just over 90% which is equivalent to a filling factor of 0.7. Braids can be optimised for surface

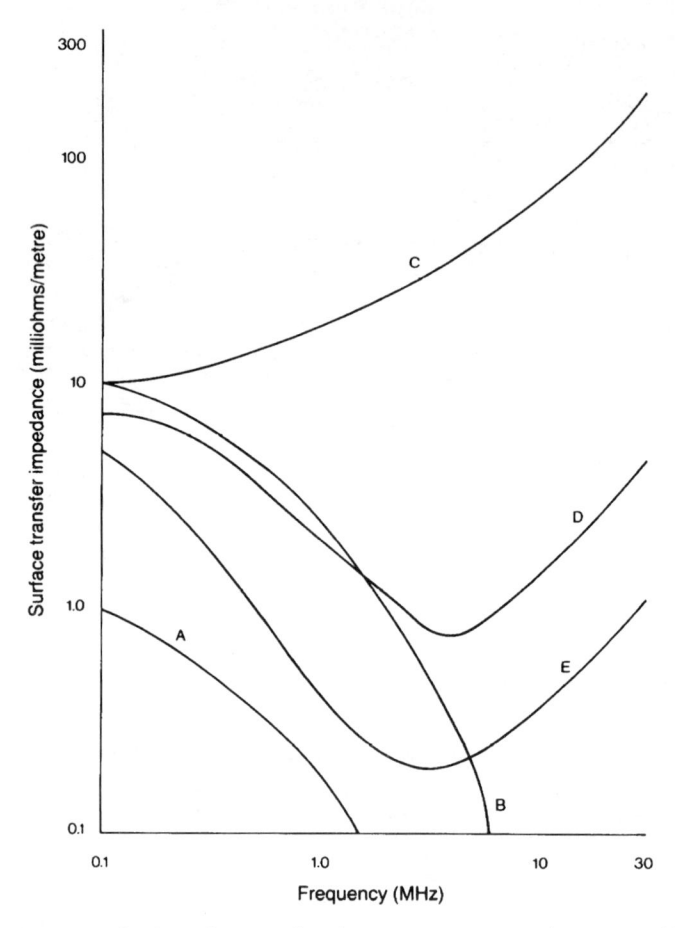

Fig. 53.14 Surface transfer impedances of various screens: curve A, seam welded copper tape 0.25 mm thick; curve B, overlapped corrugated copper tape 0.18 mm thick; curve C, single-wire braid; curve D, double-wire braids in contact; curve E, double-wire braids with intersheath

transfer impedance in terms of the numbers of spindles and ends, wire size and lay length, but generally a compromise is achieved between screening efficiency, manufacturing costs, flexibility and ease of termination.

Laminated foil tape screens provide 100% cover, and their screening efficiency is determined by the material and thickness of their metallic component. Improved efficiencies can be achieved by using double-sided foils in the form of a metal/polyester/metal sandwich, combinations of foils or combinations of foils and copper wire braids. Figure 53.15 shows the surface transfer impedance for various screens incorporating foils. Curve A is for a small coaxial cable with a single tape containing aluminium foil 0.05 mm thick. Curve B is for a multipair cable, with individual pairs screened with a single tape with an aluminium thickness of 0.025 mm, and an overall screen comprising a double-sided foil, each side 0.01 mm aluminium, and a tinned copper wire braid. Curve C shows the surface transfer impedance of the large coaxial cable illustrated in fig. 53.4 (left) which has a tape/braid/tape/braid screen, the tapes being double-sided aluminium/polyester.

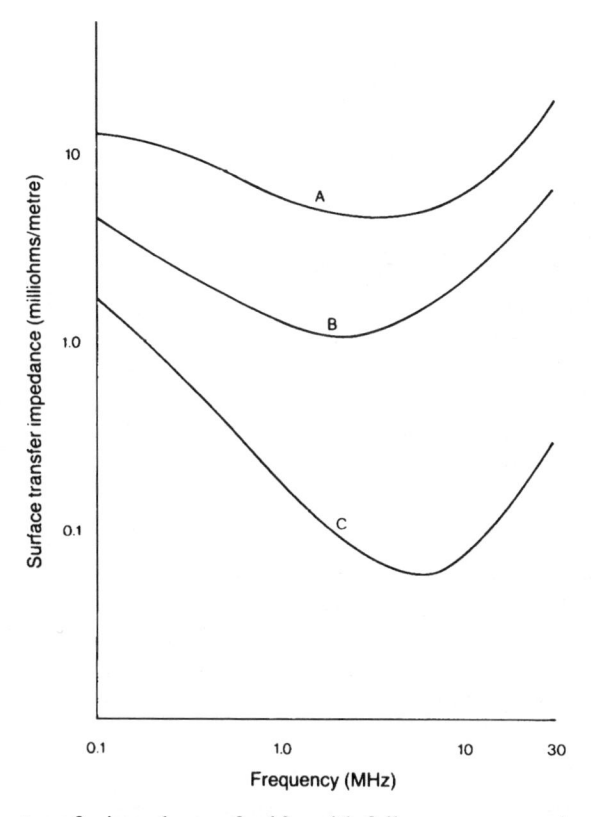

Fig. 53.15 Surface transfer impedance of cables with foil screens: curve A, single aluminium foil 0.05 mm thick; curve B, single- plus double-sided foil with copper braid; curve C, foil/braid/foil/braid

Another type of screen has been developed for applications in nuclear plant installations where very low limits are set on leakage from one control circuit to another. These 'super screened' cables have a mu-metal tape lapped between two braids and show surface transfer impedance values at 30 MHz as low as 1 μΩ/m.

COAXIAL CABLES

In general coaxial cables perform better than their pair counterparts at higher frequencies. For a given attenuation, the size and cost of a coaxial is less than that of a twin or pair cable of comparable performance. They do not display the instability of characteristic impedance, attenuation and electrical length typical of screened twins, but at lower frequencies the screening efficiency of a balanced twisted pair can be superior.

Design features

The basic coaxial cable design is an inner conductor surrounded by a dielectric and a concentric outer conductor.

The general theory concerning transmission characteristics applies to coaxial cables, but in addition consideration should be given to particular aspects of design related to the application and methods of manufacture.

Inner conductor

The preferred form is solid plain annealed copper. For high frequency use, because of the skin effect copper-covered steel may be used, but at frequencies below 10 MHz some current will flow in the steel core. For larger cables, copper tubes or copper-covered aluminium are used.

Stranded conductors can increase the high frequency resistance by up to 25%, and the resistance of tinned copper wire is higher than that of plain copper wire at frequencies above 10 MHz.

Silver-plated conductors are used at very high frequencies because of their smoother surface finish as well as their good conductivity.

Dielectric

Insulation with low and stable permittivity and power factor is preferred. This can be solid polyethylene or cellular polyethylene.

Alternatively, a number of other methods can be employed to fabricate low permittivity dielectrics for coaxial cables by using semi-airspaced constructions with solid polyethylene as the main insulating material. In one method, discs of polyethylene spaced at intervals give an effective permittivity of about 1.07. This type can be made to very close impedance limits and is used extensively for wide band carrier telephone and video trunk routes.

Another construction is the helical membrane type. A continuous tape of polyethylene is applied in the form of a helix along the inner conductor. This gives an effective permittivity of 1.1 which, as in the case of the disc spaced cable, is approaching the lowest practical limit.

A further type shown in fig. 53.16 uses a thread and tube construction in which a thread of polyethylene is lapped round the inner conductor and a close fitting polyethylene tube is extruded over this, giving a permittivity of about 1.3.

Outer conductor

The best form of outer conductor giving low attenuation and good screening is a solid copper tube. For flexible applications copper wire braids are used or combinations of thin tapes and copper wire braids.

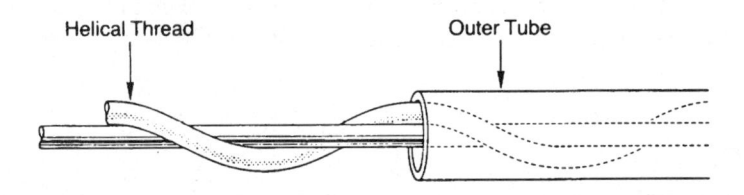

Fig. 53.16 Thread and tube dielectric construction

Outer protection

For indoor and general applications, a PVC sheath may be applied overall. For more arduous environments or outdoor use a steel armour and further PVC sheath can be applied.

Low and high impedance coaxial cables

Some special cables have been made with a characteristic impedance as low as 14 Ω. Low impedance is obtained by the use of a relatively large inner conductor formed by a wire braid over a plastic core. The characteristics of the cable are very sensitive to small variations in conductor diameter.

High impedance cables can be made by increasing the inductance per unit length by the use of a helically wound inner conductor. The increase in inductance reduces the velocity ratio and such cables can be used as delay lines. By the use of a plastic core loaded with magnetic material as a former for winding the inner conductor, impedances as high as 4000 Ω can be achieved together with velocity ratios as low as 0.001. For this low velocity ratio it takes a signal 3.0 μs to travel 1 m of cable. The delay and high impedance are obtained only at the price of high attenuation. At high frequencies the capacitance between adjacent turns of the helix resonates with the inductance at a cut-off frequency above which there is no transmission.

Radiating cables

There is an increasing demand for rapid communication services in the interests of efficiency, safety and convenience as witnessed by the expanding use of two-way mobile radio systems, radio paging and radio control. However, there are certain situations where conventional free-space radio transmission from aerials is not practical, e.g. in mines, tunnels, power stations, industrial areas and linear situations such as railways and motorways. The range of natural radio propagation in tunnels is very poor, only a few hundred metres at the most, so that communication with a distant receiver inside a tunnel is impossible using normal techniques. One way of overcoming this problem is to use a radiating cable or leaky-feeder technique in which the required radio signals radiate from a cable rather than from a conventional aerial.

The basic principle of this technique is shown in fig. 53.17. A two-way base radio station is connected to a long radiating cable transmission line which runs throughout the area where communication is required and transfers signals to and from the mobile stations through the radiation field from the cable. The special design of cable is sometimes called a leaky feeder because when it is fed with radio signals at one end it leaks small amounts of energy all along its length to be picked up by nearby mobile aerials.

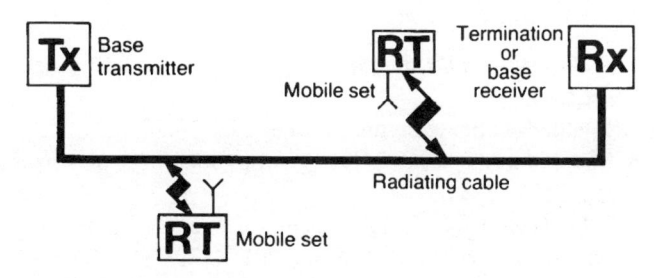

Fig. 53.17 Radiating coaxial cable system

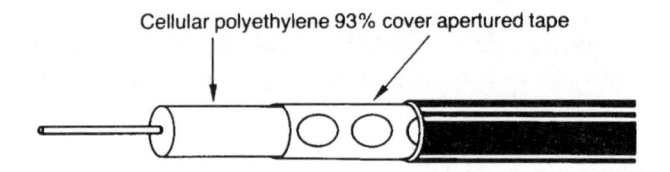

Fig. 53.18 Apertured tape screened radiating cable

The bifilar or unscreened twin cable is economic and radiates well, although the attenuation is very unstable. Even at the lower frequencies a large increase in attenuation occurs when the cable is laid on the floor or mounted close to walls. Experience in coal mines has shown that a serious increase in attenuation occurs when a layer of wet coal dust accumulates on the cable surface. However, with specially designed coaxial cables, good radiation and stable attenuation can be achieved when the cable is wet or dirty or mounted on walls. In these cables the outer screen is a corrugated copper tape perforated with a series of holes, as shown in fig. 53.18, to provide and control the radiation.

Experimental work has shown that radiation from an apertured tape cable depends on the dielectric permittivity of the insulation, and the stability of attenuation under wet and dirty conditions depends on the surface transfer impedance.

A theoretical treatment of the field leakage through a row of holes in an apertured screen has shown that at 30 MHz the surface impedance can be calculated using the formula

$$Z_T = \frac{4nd^3}{D^2} \quad (m\Omega/m) \tag{53.23}$$

where d = hole diameter (mm)
D = screen diameter (mm)
n = number of holes per metre

The optical coverage of apertured or braided screens is sometimes quoted as a characteristic. However, if the surface transfer impedance depends on nd^3/D^2 and the optical cover depends on nd^2/D, it can be seen that optical coverage is not a direct measure of screening efficiency.

A general purpose radiating coaxial will have a permittivity around 1.4 and a surface transfer impedance of 500 mΩ/m.

REFERENCES

(1) Terman, F. E. (1951) *Radio Engineering*, 3rd edn, New York: McGraw-Hill Publishing Co. Ltd.
(2) Mead, S. P. 'Shielded Cable Systems'. Patent No. 2,086,629, 14 April 1936.
(3) Windeler, A. S. (January 1960) 'Design of polyethylene insulated multipair telephone cable'. *AIEE Trans.*
(4) Spencer, H. J. C. (1969) 'Optimum design of local twin telephone cables with aluminium conductors'. *Proc. IEE* **116** (4), 481–488.

CHAPTER 54
Datacommunication Cables

TWISTED PAIR DATA CABLES

Design features

A variety of twisted pair designs is used as shown in fig. 54.1. Conductors are generally in the range 0.2–0.8 mm² (24–18 AWG), solid or stranded, and tinned for ease of termination. PVC insulations are used for short distances and low speed data but for higher speeds and longer distances, polyethylene and polypropylene in solid or cellular form are used. Individual pair screens to reduce crosstalk are generally laminated foils

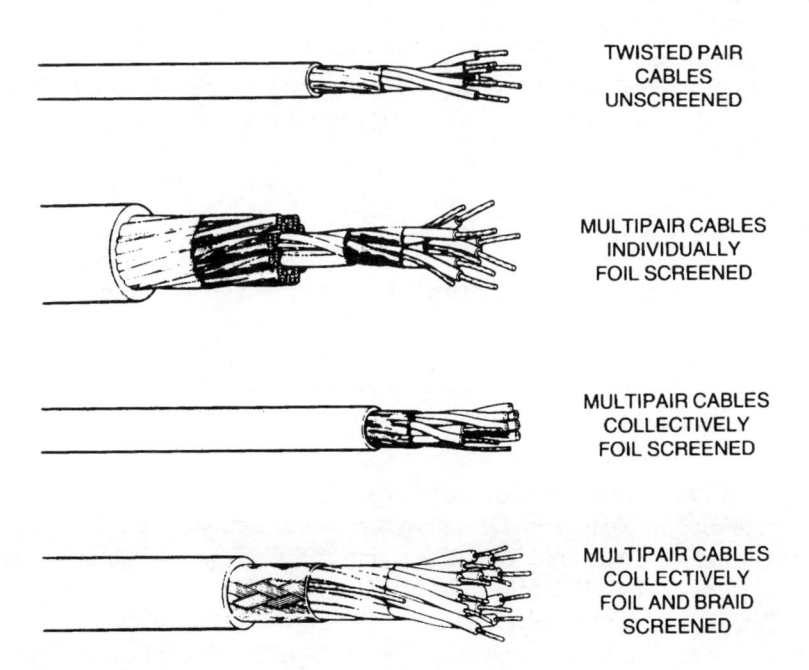

Fig. 54.1 Twisted pair cable

and incorporate a drain wire to achieve satisfactory electrical connections. Overall screens provide protection from outside interference and can be laminated foils with or without an overall braid.

Outer sheaths are generally PVC formulated to suit the particular application, including high and low service temperatures, and to impart flame retardance. A zero halogen low smoke material is used when the emission of smoke, toxic, or acidic gases is considered a hazard in a fire.

High performance twisted pair cables

Many of the early local area network systems were introduced in the USA, and through the efforts of the American Institute of Electrical and Electronics Engineers (IEEE), and particularly Project 802, a family of standards was drawn up for local and metropolitan area networks. Having gained worldwide acceptance many of these systems standards were adopted by ISO.

The earliest standards activity for Structured Wiring was also initiated in USA in 1985 by the Electronics Industries Association (EIA) who established their Engineering Committee TR-41.8 to draw up a Commercial Building Wiring standard to provide a general purpose cabling system to accommodate all communications services. Work commenced to embrace the main features of commercially available networks and cabling schemes; particularly influential was the approval of IEEE 802.3 10 Base T, a system which promoted 10 Mbps transmission over 100 metres of a four pair unscreened cable.

After amalgamating with the Telecommunications Industries Association (TIA) the first Structured Wiring standard was published in July 1991 as EIA/TIA 568. It covered a hierarchical star network with four pair unscreened horizontal cables specified for frequencies up to 16 MHz in terms of impedance, attenuation, crosstalk, mutual capacitance, and capacitance and resistance unbalance.

As systems developed to higher data speeds, attempts were made to classify cables by performance into Levels or Categories. Five copper conductor cable categories were defined as follows.

Category 1: Basic telephone cable for voice frequencies
Category 2: Data cable for use up to 4 Mbits^{-1}
Category 3: LAN cables characterised up to 16 MHz. These cables will support Class C links
Category 4: LAN cables characterised up to 20 MHz
Category 5: LAN cables characterised up to 100 MHz. These cables will support Class D links

These categories were confirmed by the EIA/TIA in their Technical Systems Bulletin TSB-36, published in November 1991. Category 1 and 2 cables are not suitable for the higher performance Structured Wiring systems and Category 4 cables are becoming obsolete. Although ISO/IEC 11801 details Category 3, 4 and 5 cables, EN 50173 specifies Categories 3 and 5 only.

The requirements for balanced twisted pair cables in horizontal subsystems apply to materials, mechanical performance and dimensions, and electrical and transmission

Table 54.1 Electrical characteristics of 100 Ω twisted pair cables to ISO/IEC 11801 and EN 50173

Cable characteristics			Cable category	
Electrical characteristics at 20°C	Units	MHz	3	5
Maximum d.c. loop resistance	Ω/100 m	d.c.	30	30
Maximum resistance unbalance	%	d.c.	3	3
Maximum capacitance unbalance	pF/km	0.0008 or 0.001	1600[a]	1600[a]
Maximum transfer impedance (for screened cables)	mΩ/m	10	100	100
Minimum insulation resistance	mΩ km	d.c.	150	150

[a] These values are specified in EN 50173. ISO/IEC allows up to 3400 pF/km.

characteristics. The physical requirements relate to the operating environment, installation practices and compatibility with connecting hardware, and ISO/IEC 11801 and EN 50173 lay down the permissible range of conductor diameters, the maximum diameter of the conductor insulation, the operational and installation temperature ranges, the minimum bending radius and pulling strength.

The electrical and transmission requirements for Category 3 and Category 5 100 Ω twisted pair cables in horizontal subsystems are shown in tables 54.1 and 54.2. When measuring impedance, structural return loss, attenuation and near end crosstalk, swept frequency techniques are used and it is a requirement that compliance is maintained at every measurement frequency as well as at the nominated frequencies.

The requirements for horizontal subsystem cables can be met by a four pair cable, either unscreened or collectively foil screened as shown in fig. 54.1. Typically these cables have a 0.5 mm (24 AWG) conductor with a polyolefin insulation, either solid or cellular. Because of the demanding performance requirements, control of materials and processing techniques must be at a significantly higher level than when manufacturing basic telephone pair cables.

Balanced cable transmission

The main parameters influencing the performance of a balanced transmission system are characteristic impedance, crosstalk, loop resistance, attenuation, velocity of propagation and the immunity to electromagnetic radiation. For some installations, electromagnetic radiation from the cable is also important.

Constructional features of the cable such as conductor size, variation in conductor diameter, insulation permittivity, variation in insulation diameter, twinning lays, cabling lays and screen foil thickness can have an effect on one or many of these parameters.

Crosstalk

Cables with twisted pairs are used where balanced signal transmission is required, together with low crosstalk. Crosstalk refers to interference on one channel caused by a signal in another channel and occurs between cable pairs carrying separate signals. This definition embraces both near end crosstalk and far end crosstalk.

Table 54.2 Transmission characteristics of 100 Ω twisted pair cables to ISO/IEC 11801 and EN 50173

Cable characteristics			Cable category	
Transmission characteristics at 20°C	Units	MHz	3	5
Velocity of propagation	–	10	0.6c	0.65c
		100	–	0.65c
Minimum longitudinal conversion loss[a]	dB	0.064	–	43
Minimum structural return loss[a]	dB @ 100 m	1–10	12	23
		10–16	10	23
		16–20	–	23
		20–100	–	[b]
Characteristic impedance	Ω	1–100	100 ± 15	100 ± 15
Maximum attenuation	dB/100 m	1	2.6	2.1
		4	5.6	4.3
		10	9.8	6.6
		16	13.1	8.2
		20	–	9.2
		31.25	–	11.8
		62.5	–	17.1
		100	–	22.0
Minimum near end crosstalk	dB @ 100 m	1	41	62
		4	32	53
		10	26	47
		16	23	44
		20	–	42
		31.25	–	40
		62.5	–	35
		100	–	32

[a] Values are not given in EN 50173.
[b] In accordance with the relationship $23 - 10\log(f/20)$ where f is the frequency in MHz.

Near end crosstalk is the crosstalk which is propagated in a disturbed channel in the opposite direction to the current in the disturbing channel, and far end crosstalk is propagated in a disturbed channel in the same direction as the current in the disturbing channel. Crosstalk is fundamentally the resultant of two factors, or couplings, capacitance unbalance and mutual inductance.

Capacitance unbalance gives rise to electrostatically induced potential differences between the wires of the disturbed pair; this coupling is important at all frequencies. When the capacitance of one conductor against the sheath and all of the other conductors combined is not the same as the capacitance of its mate (similarly measured) the pair is said to have a capacitance unbalance; this is measured by the difference between these two capacitances. When the capacitances are equal, the pair is said to be balanced.

Mutual inductance gives rise to electromagnetically induced potentials along the wires of the disturbed pair. This coupling exerts little influence at audio frequencies but becomes increasingly important as the frequency increases.

Both couplings are due to departures from balance between the circuits concerned. In design and manufacture, attempts are made to ensure that the distances between the wires of one pair and the wires of another are equal, as are the distances between each wire and earth. This is done by careful choice and control of materials, insulation thicknesses, twinning lays, and tensions, but during manufacture unavoidable small differences occur which result in residual values of the couplings.

As capacitance unbalance is an electrostatic coupling, it can be reduced by enclosing either or both of the interacting pairs in a screen. The screen need only be thin, and laminated foil screens are generally used. At higher frequencies, mutual inductance becomes important and differential twinning lays in adjacent pairs are used to minimise parallel paths and reduce the effect.

Impedance

The impedance match between a cable and its terminal equipment is important in controlling signal reflections. For twisted pair cables, random structural variations along the cable length occur in addition to periodic structural variations common to manufacturing processes involving rotating machinery. It is therefore usual to specify and measure structural return loss as well as impedance. Maintenance of machinery and control of processes are essential to meet the transmission requirements for Category 5 cables.

Attenuation to crosstalk ratio

Attenuation and crosstalk are important in determining the received signal-to-noise ratio. In ISO/IEC 11801 and EN 50173, alternative limits for attenuation and crosstalk are allowed provided the required attenuation to crosstalk ratio (ACR) is maintained. The ACR of a given length of cable is calculated as:

$$ACR = a_n - a \qquad \text{(dB)} \qquad (54.1)$$

where a_n = near end crosstalk in dB between any two pairs
a = attenuation in dB

Table 54.3 shows the minimum values of ACR specified in EN 50173. The introduction of this parameter allows some flexibility in cable design, for example attenuation can be reduced by increasing the conductor size, reducing the insulation permittivity, increasing pair twinning lays and for screened cables, increasing the screen metal content to reduce its resistance. Crosstalk can be improved by optimising the pair twinning lay scheme or by introducing individual pair screens. Variation in crosstalk performance is reduced by shortening pair twinning lays.

Electromagnetic compatibility

Structured wiring systems can be subject to local regulations regarding electromagnetic emission and immunity. Although the cable as a product is considered to be a passive component of the system, and therefore is not required to be tested for electromagnetic compatibility (EMC), the operational system may be required to comply with, for example, European EMC Directive 89/336/EEC. These requirements should be taken

Table 54.3 Minimum ACR for 100 Ω Category 5 twisted pair cables to EN 50173

Frequency (MHz)	ACR (dB)
1	59.9
4	48.7
10	40.4
16	35.8
20	32.8
31.25	28.2
62.5	17.9
100	10.0

into account when designing a generic cabling system, and it may be advisable to optimise the emission and immunity characteristic of the cable, by optimising the balance of the cable and by the application of screens.

In a balanced system, the receiver responds to the differential voltage between the lines and, if the pair and cable are perfectly uniform with each leg identical, electromagnetic fields generated by the two signals flowing in the twisted pair cancel with the result that the net cable radiation is zero. If the pair is not perfectly uniform, imbalances result as the two signals do not cancel; a net current flows longitudinally which produces electromagnetic radiation. Similarly, imbalances lead to a coupling with external electromagnetic fields making the cable more sensitive to interference. Discontinuities in the cable, which may be produced during installation, can cause a part of the differential mode signals to be converted to longitudinal signals.

Two types of measurements are used to characterise the balance of a cable. These are the longitudinal conversion loss (LCL) and the longitudinal conversion transfer loss (LCTL). LCL measurements are made at the near end of the cable and identify imbalances near to a transmitter which could create radiation; LCTL measurements are made at the far end of the cable and identify the effects on the received signal of imbalances throughout the cable length.

The EMC performance of a cable can be improved by applying a collective screen, and surface transfer impedance is used as a guide to its effectiveness.

Installation

For high performance twisted pair cables operating at the higher frequencies, the cable design and installation practices must ensure that adequate cable balance is maintained after installation and termination. Both ISO/IEC 11801 and EN 50173 specify the performance of cables, connecting hardware and installed links.

Manufacturers recommendations regarding bending radii, pulling tensions and cable spacing should be observed to avoid cable stress due to sharp bends, crushing, excessive tension and tightly bunched cables. When terminating cables in connecting hardware, the untwisted length of pair and the length of removed sheath should be as short as possible.

For screened cables, earthing and bonding procedures should address the EMC performance of the link as well as electrical safety. Screens intended to improve

EMC performance are applied to both cables and connecting hardware and must be terminated in such a way that the screening effectiveness is maintained over the complete link.

OPTICAL DATA CABLES

Cable designs

Initially optical fibre cables were developed for long distance telecommunications, but as the technology advanced, numerous applications arose in data systems where the advantages of optical fibre transmission could be employed. Cables designed to meet the requirements of IEC 794 and EN 187000 will provide a satisfactory performance for the majority of data applications. The basic design principles are the same as those described in chapter 49.

For metropolitan and wide area networks where data transmission over long distances is required, cables designed for aerial, duct and buried routes on public and private telecommunications systems are used with single mode fibres (see chapter 48). For local area networks similar cable designs are used for both campus and building backbones, in some cases modified to be completely non-metallic. On the shorter route lengths, multimode fibres are often used when power budget and coupling parameters are as important as attenuation and bandwidth. IEC 793-1 (Optical Fibres) specifies a variety of step, quasi-step and graded index multimode fibres and includes a guide for fibres for short distance links, with 62.5/125 µm and 50/125 µm graded index fibres gaining popularity for local area networks.

For structured wiring, ISO/IEC 11801 and EN 50173 recommend optical fibre cables for most campus backbone cabling to overcome problems due to earth potential differences and other interfering sources as well as to provide extended route lengths up

Table 54.4 Optical fibre cable performance recommended in ISO/IEC 11801 and EN 50173

Fibre type	Multimode 50/125 or 62.5/125	Single mode
Maximum attenuation (dB/km) at:		
850 nm	3.5	–
1300 nm	1.0	–
1310–1550 nm	–	1.0
Minimum bandwidth (MHz km) at:		
850 nm	200	–
1300 nm	500	–
Maximum cut-off wavelength (nm)	–	1280

Note: Applicable standards are:
Fibre: Type A1a (50/125), A1b (62.5/125) and B1 (single mode) to IEC 793-2, EN 188201, EN 188202, EN 188100, EN 188101.
Cable: IEC 794-2, IEC 794-3, EN 187000, EN 60794-3.

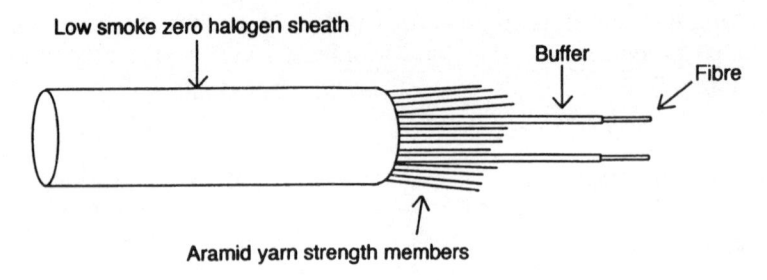

Fig. 54.2 Duplex optical cable for horizontal subsystems

to 3000 m with single mode fibres and 2000 m with multimode fibres. Optical cables are recommended for building backbones on medium to high speed data links, for horizontal data cabling with some very high bit rate systems, and where the electromagnetic environment and security considerations could cause problems with metallic conductor cables.

Table 54.4 shows the optical cable transmission requirements of ISO/IEC 11801 and EN 50173. For the horizontal cables, indoor cables are used, and preferably they should have handling characteristics similar to or better than the four pair Category 5 cables. Figure 54.2 shows a popular compact non-metallic cable featuring individually buffered fibres surrounded by a non-metallic strength member such as aramid yarn, with an overall sheath of low smoke zero halogen material.

In these tight buffered cables the fibre mechanical protection is provided by a layer of material applied directly over the coating of the fibre. This may be a dual layer, with a softer inner layer to provide a cushion against stresses which could cause an increase in attenuation, and a harder outer layer to provide mechanical protection. A typical dual layer buffer has a 140 μm thick inner layer of silicone covered by a 250 μm layer of polyamide. This form of buffer facilitates stripping, a force of only 10 N being required to strip the entire buffer over a 20 mm length.

Blown fibre systems

Of particular interest for local area networks is the development of the technique of blowing fibres into ducts or tubes. Blown fibre systems have been developed as a method of avoiding fibre overstrain for complex route installations. They allow easy upgrading and future proofing with low initial capital investment and the distribution of subsequent costs.

The network infrastructure is created by installing, using the most appropriate cabling method, one or a group of empty plastic tubes. Subsequently as and when circuit provision is required, one or more fibres can be blown by compressed air into the tubes. Individual tubes can, by means of connectors, be extended within buildings up to the fibre terminating equipment.

To enhance the blowing characteristics and to facilitate handling special coatings can be applied to the fibres. Figure 54.3 shows a typical blowable fibre construction whereby fibres can be blown singly or in groups into the same tube. Fibres with different transmission characteristics, or different types, e.g., multimode and single mode, can be blown into the same tube to accommodate particular network requirements.

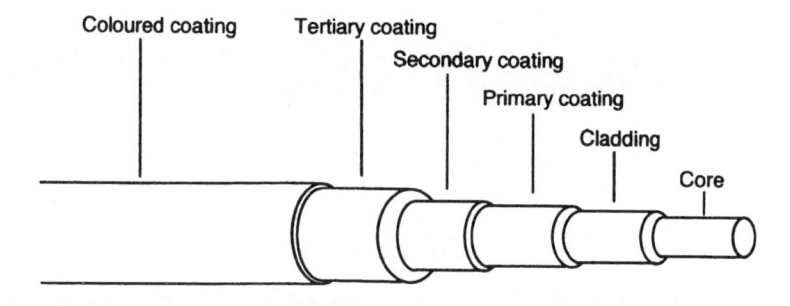

Fig. 54.3 Blowable fibre

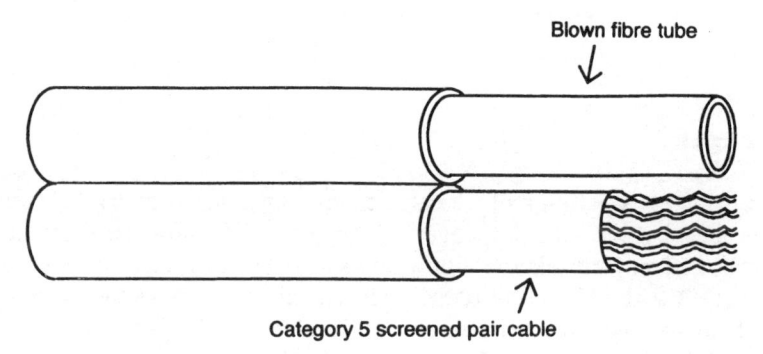

Fig. 54.4 Composite blown fibre and Category 5 pair cable

This technique can also be used to blow fibre ribbons into the tubes to take advantage of multifibre array connectors.

Another cable for local area networks combines a blown fibre tube with a Category 5 structured wiring cable, as shown in fig. 54.4. This flexible approach to the future-proofing of networks provides for immediate requirements to be met with the Category 5 four pair copper conductor cable and future optical up-grading by blowing in fibres as and when required without further cable installation costs.

Pre-connectorised cabling

To save on-site costs, loose tube, tight buffered or ribbon optical cables can be factory terminated with multifibre array connectors, and with suitable protection and end fittings, the terminated cable can be pulled-in and installed using conventional techniques.

CHAPTER 55
Twisted Pair Telecommunication Cables

SYSTEM LIMITS

To ensure satisfactory operation of a local telecommunication network the twisted pair transmission lines between the exchange and the subscriber must be planned to meet both a signal and a transmission limit. Typically the transmission limit is 10 dB (at 1600 Hz) and the signal limit is 1000 Ω. The signal limit defines the maximum loop resistance of an installed twisted pair in the local network. For a known distance between the exchange and the subscriber the maximum resistance can be used to calculate the twisted pair resistance per kilometre and hence the conductor diameter. The transmission limit defines the maximum attenuation for an installed twisted pair in the local network. For a known distance between the exchange and the subscriber (and hence a defined conductor diameter) the ideal mutual capacitance can be calculated.

In practice the twisted pair resistance and mutual capacitance are optimised[1] to:

(a) minimise cable and ducting costs;
(b) rationalise cable stock;
(c) minimise impedance mismatching at connections.

Table 55.1 summarises the conductor resistance, mutual capacitance and attenuation values of typical conductor diameters within the UK's local network.

Table 55.1 Electrical properties of local network cables

Nominal conductor diameter (mm)	Conductor resistance max. av. (Ω/km)	Mutual capacitance max. av. (nF/km)	Attenuation @ 1600 Hz (dB/km)
0.32	223	53	2.99
0.4	143	53	2.4
0.5	91	53	1.91
0.63	58	56	1.57
0.9	28	59	1.12

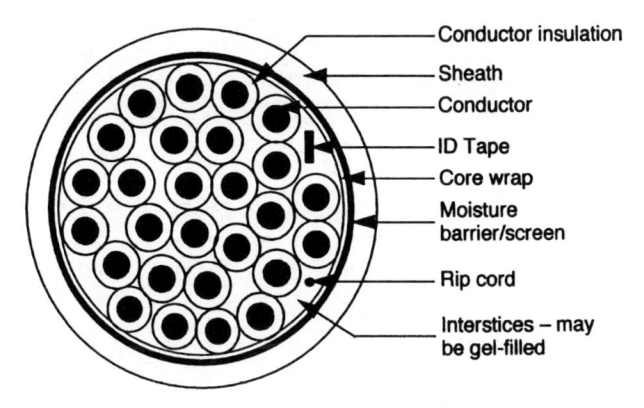

Fig. 55.1 Typical telephone cable layout

EXTERNAL TWISTED PAIR CABLES

Figure 55.1 shows a typical constructional layout for an external duct cable. In the following sections we briefly describe the main constructional elements. (See chapter 53 for a discussion of transmission theory for these cables.)

Conductor

The conductor is typically solid plain copper. In the past aluminium was considered when copper became relatively expensive, however the use of aluminium ceased when copper once again became economically viable; also corrosion problems were encountered in the field.

The gauge of conductor is dependent on the system limits, as discussed earlier. Typical rationalised gauges are 0.32 mm, 0.4 mm, 0.5 mm, 0.6 mm, 0.63 mm and 0.9 mm.

Insulation

Polyethylene has many desirable properties including high insulation resistance, low relative dielectric constant, voice frequency stability and low cost which make it especially suitable for use as the insulation material for external telecommunication cables.

Polyethylene insulation can be either solid, cellular or foam skin. The choice of insulation arises from the development of filled cables. The dielectric constant of a filled polyethylene cable is greater than that of a corresponding air core cable and therefore, for the same mutual capacitance and resistance, the solid insulation of a filled cable must be thicker than for an unfilled cable (a consequence of maintaining the $2S/d$ ratio of equation (53.19)). This disadvantage can be countered in a filled cable by introducing air into the insulation to form a cellular structure. Foam skin is an alternative to cellular structure in which an inner layer of cellular insulation is covered by a coating of solid insulation. If a cellular insulation is used in an air core cable a further reduction in insulation thickness and hence cable diameter, is possible.

Cellular insulation has allowed the introduction of filled cable without an increase in diameter compared with the solid polyethylene air core cable. Whilst cellular insulation is not as robust as solid insulation it has proven over many years to be more than adequate.

Cable assembly

Two or four insulated conductors are twisted around each other to form a twin or a quad respectively. Typically 10 pairs, 25 pairs or 5 quads are grouped together to form a unit. The colour of the insulation, and ink banding if applicable, within the unit are such that each pair can be individually identified. The units are grouped to form multiple units which can be grouped further to produce a cable core with the required number of pairs. The units and multiple units have coloured binder tapes helically applied around them which enable each unit to be individually identified.

Filling

The ingress of water into a cable can increase the mutual capacitance of affected twisted pairs by up to 125%, increasing the transmission loss, or attenuation, by up to 60%. Furthermore pin holes in the insulation can also lead to shorts between the wires under such circumstances. Filling the interstitial spaces within the cable limits both the migration of water into the cable, either through a damaged sheath or joint closure, and permeation of water through a non-hermetic sheath.

Alternatively by pressurisation the cables can be protected against the ingress of water. The cable is sealed at the far end and a constant air pressure is applied to the cable interstitial spacing at the near end. The pressure at the near end is monitored; a drop in pressure would indicate a fault in the cabling system such as damage to the sheath or closures. Pressurisation is usually confined to the primary network with the equipment being housed in the exchange.

Cables to be installed below ground in a non-pressurised network should be filled. Filling compounds are typically petroleum jelly based and are formulated to be compatible with the insulation at temperatures encountered in storage and after installation in the country of use.

Core wrap

One or more core wraps are usually applied to the cable core primarily to provide a heat barrier preventing the softening of insulation due to the heat of extrusion of the sheath. Other desirable properties of the core wraps are to provide a voltage withstand between the conductors and a metallic moisture barrier and to prevent water penetration within the region between the sheath and the cable core. The tapes can be plastic, dry paper, impregnated paper and/or water swelling material depending on the installation.

Moisture barrier/screen

The normally chosen material for external cable sheaths is polyethylene, however polyethylene alone will not block the permeation of water vapour into the cable core. Metallic moisture barriers are bonded to the inner face of the sheath and hence block the permeation of water from the sheath into the cable core. The moisture barrier is typically an aluminium tape with a polyethylene film laminated to one or both sides. The metallic moisture barrier also acts as a limited screen from external electromagnetic interference.

Within smaller cables the presence of a metallic moisture barrier can cause a small increase in the overall mutual capacitance of the cable. The pairs deviate slightly from the ideal balanced shielded pair model due to the proximity of the metallic moisture barrier.

Sheath

The grade of polyethylene used is defined by the preference of the user, the most popular grade being LDPE. Anti-oxidants are compounded into the polyethylene to prevent breakdown with age or during extrusion; also the polyethylene is impregnated with carbon black which prevents degradation due to solar radiation.

Ducted cable

External cable is most commonly installed in ducts, in which case the cable described thus far is sufficient.

Aerial cable

Aerial cable may be the same as duct cable and lashed to a pre-installed strength member (or support wire) or the cable may be self-supporting, that is the sheath will incorporate the strength member within the cable in a 'figure of eight' cross-section. The strength member must be selected to support the weight of the cable for the prescribed span and to withstand any additional loading factors such as wind, ice and perching birds.

Directly buried cable

Directly buried cable requires additional protection above that of duct cable. Depending on the level of additional protection required this can be supplied either by one or more layers of helically or longitudinally applied galvanised steel tapes with an oversheath, or by a layer of helically applied galvanised steel wire and an oversheath. The longitudinally applied galvanised steel tape is typically corrugated to give flexibility. The oversheath may be polyethylene, or PVC, or for protection against particularly voracious insects, nylon may be used.

DROPWIRE

Dropwire provides the final link, from the Distribution Point, typically a 'telegraph' pole to the individual subscriber's premises. Typically the final link is a single span but in some networks a dropwire may follow a route of several aerial spans. The length of the transmission line provided by dropwire within the local network is relatively small and hence the transmission characteristics of dropwire can be relaxed allowing more cost effective designs to be utilised.

In the more traditional double-D or figure of eight designs, as illustrated in fig. 55.2, the conductor combines both transmission performance and strength whereas in the more recent designs the conductors are conventional copper and the strength is provided by additional strength member(s).

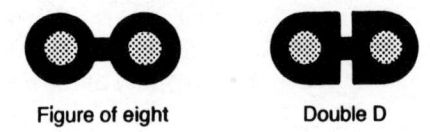

Fig. 55.2 Traditional dropwire designs

JUMPER WIRE

Jumper wire is used in cabinets to link the primary and secondary halves of the network, and for jumpering on distribution frames. The confined spaces require a tough abrasion resistant insulation for which PVC is ideally suited. Increased abrasion resistance can be achieved by irradiating the jumper wire with an electron beam, thus crosslinking the insulation.

INTERNAL CABLE

Internal, switchboard, and equipment cables are all used within buildings with typical cable lengths of 30 m. These small lengths allow the cable transmission characteristics to be less demanding. Internal cables are also likely to be subject to increased manual handling and an increased threat of fire. PVC is commonly used as the insulation and sheath material because it has good abrasion resistance and will not propagate fire.

Where it is necessary, in cases of fire, to limit the smoke emission and acid and toxic gas emission, alternative materials and cable constructions are required. Rigorous testing is required to assure, demonstrate and validate the 'Limited Fire Hazard' performance of these cables.

PVC has a tendency to stick to bare copper conductors making termination difficult, and to prevent this tinned copper conductors are typically used.

REFERENCE

(1) Spencer, H. J. C. (1970) 'A new method of local line transmission planning'. *P.O.E.E.J.* **63**, 84.

Appendices

APPENDIX A1
Abbreviations

MATERIALS

CGI	compressed gas insulation
CPE	chlorinated polyethylene
CR	*see* PCP
CSM	*see* CSP
CSP	chlorosulphonated polyethylene
EPDM	ethylene–propylene terpolymer rubber
EPM	*see* EPR
EPR	ethylene–propylene rubber
ETFE	ethylene–tetra fluroethylene
EVA	ethylene–(vinyl acetate)
FEP	fluorinated ethylene propylene
HDPE	high density polyethylene
HEPR	hard EPR
HNBR	hydrogenated nitrile–butadiene rubber
HTS	high temperature superconductors
IIR	isoprene isobutylene (butyl) rubber
IVD	inside vapour deposition
LCM	liquid curing medium
LDPE	low density polyethylene
LTS	low temperature superconductors
MEPR	medium hardness EPR
MOCVD	metal organic chemical vapour deposition
NBR	acrylonitrile–butadiene copolymer rubber
NR	natural rubber
OVD	outside vapour deposition
PC	polycarbonate
PCP	polychloroprene rubber
PE	polyethylene
PETP	polyethylene terephthalate
PIB	polyisobutylene

PIT	powder in tube
PP	polypropylene
PPL	polypropylene paper laminate
PTFE	polytetrafluoroethylene
PVA	poly(vinyl acetate)
PVC	poly(vinyl chloride)
RTS	rated tensile strength
SBR	styrene–butadiene rubber
SR	silicone rubber
TMTSF	tetramethyltetraselenafulvalene
TPE	thermoplastic elastomer
TPR	thermoplastic rubber
UTS	ultimate tensile stress
VAD	vapour axial deposition
VR	vulcanised rubber
WTR	water tree resistant
XLPE	crosslinked polyethylene

ASSOCIATIONS, INSTITUTIONS AND COMPANIES

ABB	ASEA Brown Boveri
ABS	American Bureau of Shipping
AEIC	Association of Edison Illuminating Companies (USA)
ASTM	American Society for Testing Materials
BASEC	British Approvals Service for Electric Cables
BASEEFA	British Approvals Service for Electrical Equipment for Flammable Atmospheres
BC	British Coal
BCMC	British Cable Makers' Confederation
BCS	British Coal Specification
BR	British Rail
BRB	British Railways Board
BS	British Standard
BSI	British Standards Institution
BT	British Telecom
BV	Bureau Veritas
CEE	International Commission for Conformity Certification of Electrical Equipment
CEGB	Central Electricity Generating Board (UK)
CENELEC	European Committee for Electrotechnical Standardisation
CERN	Centre Européenne pour la Recherche Nucléaire
CIGRE	Congrès International des Grands Réseaux Electriques (International Conference on Electric Transmission Systems)
CIRED	Congrès International des Réseaux Electriques de Distribution (International Conference on Electric Distribution Systems)
CMA	Cable Makers' Association (UK)
DNV	Det Norske Veritas (Norway)

ECMC	Electric Cable Makers' Confederation (UK)
EEC	European Economic Community
EEMUA	Engineering Equipment and Materials User's Association
EFTA	European Free Trade Association
EIA	Electronics Industries Association
EPRI	Electric Power Research Institute (USA)
ERA	Electrical Research Association (now ERA Technology Ltd)
GEC	General Electric Company
GL	Germanische Lloyds (Germany)
ICEA	Insulated Cable Engineers' Association (USA)
IEC	International Electrotechnical Commission
IEE	Institution of Electrical Engineers (UK)
IEEE	Institute of Electrical and Electronics Engineers (USA)
IPCEA	Insulated Power Cable Engineers' Association (USA) (now ICEA)
ISO	International Standards Organisation
LRS	Lloyd's Register of Shipping
LTE	London Transport Executive
LUL	London Underground Limited
MOD(N)	Ministry of Defence (Navy)
NAC	National Accreditation Council
NAO	National Approval Organisation (EEC)
NCB	National Coal Board (UK)
NEMA	National Electrical Manufacturers' Association (USA)
NGC	National Grid Company
OCMA	Oil Companies Materials Association
TEPCO	Tokyo Electric Power Company
TIA	Telecommunications Industries Association
UL	Underwriter's Laboratories (USA)
VDE	Verband Deutscher Electrotechniker

CABLES AND CONSTRUCTIONS

AAAC	all-aluminium alloy conductor
AAC	all-aluminium conductor
AACSR	aluminium alloy conductor steel reinforced
ACAR	aluminium conductor alloy reinforced
ACSR	aluminium conductor steel reinforced
ADSS	all-dielectric self-supporting
ALS	alternate long–short
CNE	combined neutral and earth
CPC	circuit protective conductor
CSA	corrugated seamless aluminium (sheath)
CTS	cab tyre sheath
DCF	ductless circular (conductor) FF cable
DSF	ductless shaped (conductor) FF cable
DWA	double wire armour
ECC	earth continuity conductor

EMC	electromagnetic compatibility
FF	fluid-filled
GSW	galvanised steel wire
HOFR	heat and oil resisting-and flame retardant
HP FF	high pressure fluid-filled
HR	heat retardant
HSA	Hochstadter (screened) separately aluminium sheathed
HSL	Hochstadter (screened) separately lead sheathed
LCL	longitudinal conversion loss
LCTL	longitudinal conversion transfer loss
LFH	limited fire hazard
LSF	low smoke and fume
MASS	metal aerial self-supporting
MI	mineral insulated
MIND	mass-impregnated non-draining
NR	nitrile rubber
OFR	oil resisting and flame retardant
OPGW	optical power ground wire
OPPW	optical power phase wire
PIAS	paper insulated aluminium sheathed
PILS	paper insulated lead sheathed
PME	protective multiple earthed
PWA	pliable wire armour
RNN	rubber/neoprene/neoprene
RP	reduced flame propagation
RPS	reduced flame propagation sheath
SAC	solid aluminium conductor
SC FF	self-contained fluid-filled
SL	separately lead sheathed
SNE	separate neutral and earth
STA	steel tape armour
SWA	single wire armour
TRS	tough rubber sheath
VSWR	voltage standing wave ratio
ZH	zero (negligible) halogen

MISCELLANEOUS

AWG	American wire gauge
BIL	basic impulse level
CATV	community access television (cable)
CCV	continuous catenary vulcanising
CRT	cathode ray tube
CV	continuous vulcanising
DLA	dielectric loss angle
DSL	digital subscriber loop
EHV	extra high voltage (imprecise)

EN	Euro Norme
HCV	horizontal continuous vulcanising
HD	harmonisation document (CENELEC)
HV	high voltage (imprecise)
IACS	International Annealed Copper Standard
IO	information outlet
LAN	local area network
LH	local hub
LV	low voltage (imprecise)
MAN	metropolitan area network
MCM	thousand circular mils (also kc mil)
MH	main hub
MFI	melt flow index
MIG	metal–inert gas
MV	medium voltage (imprecise)
NGTS	National Grid Technical Specification
OTDR	optical time domain reflectometry
PD	partial discharge
PLCV	pressurised liquid continuous vulcanising
PP	pre-pressurisation index
RF	radio frequency
SSI	solid state interlock (signalling)
SVD	spiral vibration damper
SVL	sheath voltage limiter
TO	telecommunications outlet
TP	transition point
TSB	Technical Systems Bulletin
UD	underground distribution (USA)
UHV	ultra high voltage (imprecise)
URD	underground residential distribution (USA)
VCV	vertical continuous vulcanising
VTB	voltage time breakdown
WAN	wide area network

APPENDIX A2
Symbols Used

ENERGY CABLES

A	= volume resistivity
C	= electrostatic capacitance (F/m)
D	= dielectric loss (W/m)
E	= Young's modulus of elasticity (Pa or N/m^2)
E	= electric stress in insulation (MV/m)
F	= force (N)
I	= current (A)
I	= second moment of area of sheath (mm^4)
L	= length (m)
L	= distance between supports (mm)
L	= depth of burial (mm)
L	= inductance (mH/m)
M	= mutual inductance (mH/km)
N	= electrical loss in cable component (W/m)
P	= thermomechanical force (N)
Q	= discharge magnitude (pC)
R	= resistance (Ω)
S	= axial spacing between conductors (mm)
S	= eddy current loss (W/m)
S	= conductor area (mm^2)
T	= torque in cable (kg m)
T	= thermal resistance (K m/W)
T	= time (s)
T	= tension (N)
U	= rated voltage btween any two conductors (V)
U_m	= maximum voltage for equipment (V)
U_o	= rated voltage between conductor and earth (V)
V	= volume of fluid in FF cable (litre)
W	= weight of cable (kg/km)
X	= reactance (Ω)

Y	= yield stress (N/m^2)
Z	= impedance (Ω/m)
a	= diameter of armour wire (mm)
b	= hydraulic impedance (FF cable duct)
d	= diameter (mm)
f	= supply frequency (Hz)
h	= heat dissipation coefficient
k	= conductor/insulant factor for protection calculation
m	= number of armour wires
n	= number of conductors in cable
n	= number of wires in conductor
q	= rate of flow (1/s)
r	= radius (mm)
t	= thickness (mm)
t	= temperature
v	= velocity (m/s)
x	= duty cycle (welding) (%)
y	= skin or proximity effect constant
α	= coefficient of expansion
α	= temperature coefficient of resistance
β	= stress coefficient of electrical resistivity
δ	= dielectric loss angle (tan δ)
ϵ	= relative permittivity
η	= viscosity (cP)
θ	= temperature difference (°C)
θ	= angle of armour wire to cable axis (°)
λ	= ratio of loss in component to loss in conductor
μ	= coefficient of friction
ρ	= resistivity (Ω m)
ρ	= thermal resistivity (K m/W)
ϕ	= angle of anti-twist tape to cable axis (°C)

Note
The unit 'bar' is used for pressure rather than 'bar g'. Gauge pressure is always quoted and not absolute pressure.

SUPERCONDUCTIVITY

A15	= low temperature superconductor crystal structure general formula A_3B
B	= magnetic flux density (T)
B1	= 6:6 co-ordinated NaCl structure
H	= magnetic flux (Wb)
H_c	= critical field (T)
H_{c1}	= lower critical field (T)
H_{c2}	= higher critical field (T)

I_c = critical current (A)
J_c = critical current density (A m^{-2})
T_c = critical temperature (K)

e = electron

μ = magnetic permeability (H/m)

OPTICAL FIBRES AND COMMUNICATIONS

A = cross-sectional area (mm^2 or m^2)
C_D = direct capacitance between two conductors of a twisted pair (nF/km)
D = diameter of optical cable strength member (mm^2 or m^2)
D = effective diameter of electrostatic shield seen by one twisted pair due to another (mm)
D = mean braid diameter
D = screen diameter (mm)
G = conductance (S)
K_L, K_F = lay/filling factor (dimensionless)
L = lay length (m)
L = span length (m)
N = number of loose tubes stranded together
P = optical power (dB)
Q = charge (C)
R = loop resistance (Ω/m)
S = heat capacity (J/°C)
T = time duration of electrical fault (s)
T_i = installation tension (N)
W = effective cable loading per unit length (N/m)

a = attenuation (dB)
b = bore of loose tube (mm or m)
c_D = cable drag coefficient (dimensionless)
d_f = fibre diameter (mm or m)
n = refractive index (dimensionless)
s = specific heat capacity (J/°C/kg)
t = operational temperature (°C)
t = screen thickness (mm or m)
v = wind speed (m/s)
w = weight of cable (kg/km)

α = attenuation (dB/km)
γ_c = critical angle (degrees)
$\tan \delta$ = dielectric power factor (dimensionless)
ϕ = braid lay angle (degrees)
λ = wavelength of light (μm)
μ = screen permeability (H/m)
ρ = density of air (kg/m^3)

APPENDIX A3
Conversion Factors and Multiple Metric Units

CONVERSION FACTORS

From	To	Factor	Reciprocal
mm	in	0.0394	25.4
m	ft	3.2808	0.3048
m	yard	1.0936	0.9144
km	mile	0.6214	1.6093
mm^2	in^2	0.00155	645.16
mm^2	circular mil	1973.5	5.0671×10^{-4}
N	lbf	0.2248	4.4482
N	kgf	0.1020	9.8067
bar	N/m^2	10^5	10^{-5}
bar	lbf/in^2	14.5	68.9476×10^{-3}
N/m^2	torr (mmHg)	7.501×10^{-3}	133.32
N/m^2	lbf/in^2	1.450×10^{-4}	6894.76
kgf/cm^2	lbf/in^2	14.223	0.07031
kg	ton	9.8421×10^{-4}	1016.05
kg	lb	2.2046	0.4536
t (tonne)	lb	2204.6	0.4536×10^{-3}
l (litre)	gal (UK)	0.2202	4.541
l (litre)	gal (US)	0.2642	3.785

MULTIPLE AND SUB-MULTIPLE METRIC UNITS

Multiple	Prefix	Symbol	Sub-multiple	Prefix	Symbol
10^{12}	tera	T	10^{-1}	deci	d
10^9	giga	G	10^{-2}	centi	c
10^6	mega	M	10^{-3}	milli	m
10^3	kilo	k	10^{-6}	micro	μ
10^2	hecto	h	10^{-9}	nano	n
10	deca	da	10^{-12}	pico	p

APPENDIX A4
Conductor Data

Table A4.1 Metric conductor sizes and resistances (20°C) for fixed wiring

Conductor size (mm^2)	Plain copper (Ω/km)	Metal-coated copper (Ω/km)	Aluminium[a] (Ω/km)	Conductor size (mm^2)	Plain copper (Ω/km)	Metal-coated copper (Ω/km)	Aluminium[a] (Ω/km)
0.5	36.0	36.7		500	0.0366	0.0369	0.0605
0.75	24.5	24.8		630	0.0283	0.0286	0.0469
1	18.1	18.2		800	0.0221	0.0224	0.0367
1.5	12.1	12.2		1000	0.0176	0.0177	0.0291
2.5	7.41	7.56		1200	0.0151	0.0151	0.0247
4	4.61	4.70	7.41	1400[b]	0.0129	0.0129	0.0212
6	3.08	3.11	4.61	1600[b]	0.0113	0.0113	0.0186
10	1.83	1.84	3.08	1800[c]	0.0101	0.0101	0.0165
16	1.15	1.16	1.91	2000[c]	0.0090	0.0090	0.0149
25	0.727	0.734	1.20	1150[c]	0.0156		0.0258
35	0.524	0.529	0.868	1300[c]	0.0138		0.0228
50	0.387	0.391	0.641	380[d]			0.0800
70	0.268	0.270	0.443	480[d]			0.0633
95	0.193	0.195	0.320	600[d]			0.0515
120	0.153	0.154	0.253	740[d]			0.0410
150	0.124	0.126	0.206	960[d]			0.0313
185	0.0991	0.100	0.164	1200[d]			0.0250
240	0.0754	0.0762	0.125				
300	0.0601	0.0607	0.100				
400	0.0470	0.0475	0.0778				

[a] Aluminium or aluminium alloy
[b] Non-preferred sizes in IEC 228
[c] Sizes used for fluid-filled cables (not in IEC 228)
[d] Solid sectoral conductors (not in IEC 228 but standard for British practice, BS 6360: 1991)
Except where stated, the data are in accordance with IEC 228 and British Standards.

Table A4.2 Metric sizes and resistances (20°C) for flexible conductors

Conductor size (mm^2)	Maximum d.c. resistance Plain copper (Ω/km)	Maximum d.c. resistance Metal-coated copper (Ω/km)	Conductor size (mm^2)	Maximum d.c. resistance Plain copper (Ω/km)	Maximum d.c. resistance Metal-coated copper (Ω/km)
0.22	92.0	92.4			
0.5	39.0	40.1	70	0.272	0.277
0.75	26.0	26.7	95	0.206	0.210
1	19.5	20.0	120	0.161	0.164
1.5	13.3	13.7	150	0.129	0.132
2.5	7.98	8.21	185	0.106	0.108
4	4.95	5.09	240	0.0801	0.0817
6	3.30	3.39	300	0.0641	0.0654
10	1.91	1.95	400	0.0486	0.0495
16	1.21	1.24	500	0.0384	0.0391
25	0.780	0.795	630	0.0287	0.0292
35	0.554	0.565			
50	0.386	0.393			

The data are in accordance with IEC 228 and British Standards.

Table A4.3 USA stranded conductor sizes and resistances (20°C) for fixed wiring

Nominal area (AWG or kcmil)[a]	Equivalent metric area[b] (mm²)	Nominal d.c. resistance Copper (Ω/km)	Nominal d.c. resistance Coated copper (Ω/km)	Nominal d.c. resistance Aluminium (Ω/km)	Maximum d.c. resistance[c] Copper single-core (Ω/km)	Maximum d.c. resistance[c] Aluminium single-core (Ω/km)	Maximum d.c. resistance[c] Copper multicore (Ω/km)	Maximum d.c. resistance[c] Aluminium multicore (Ω/km)
20	0.519	33.9	36.0		34.6		35.3	
18	0.823	21.4	22.7		21.8		22.2	
16	1.31	13.4	14.3		13.7		14.0	
14	2.08	8.45	8.78		8.62		8.79	
13	2.63	6.69	6.96		6.82		6.96	
12	3.31	5.32	5.53	8.71	5.43	8.88	5.54	9.06
11	4.17	4.22	4.39	6.92	4.30	7.06	4.39	7.20
10	5.26	3.34	3.48	5.48	3.41	5.59	3.48	5.70
9	6.63	2.65	2.76	4.35	2.70	4.44	2.75	4.53
8	8.37	2.10	2.19	3.45	2.14	3.52	2.18	3.59
7	10.6	1.67	1.73	2.73	1.70	2.78	1.73	2.84
6	13.3	1.32	1.38	2.17	1.35	2.21	1.38	2.25
5	16.8	1.05	1.09	1.72	1.07	1.75	1.09	1.79
4	21.2	0.832	0.865	1.36	0.849	1.39	0.866	1.42
3	26.7	0.660	0.686	1.08	0.673	1.10	0.686	1.12
2	33.6	0.523	0.544	0.857	0.533	0.874	0.544	0.891
1	42.4	0.415	0.431	0.680	0.423	0.694	0.431	0.708
1/0	53.5	0.329	0.342	0.539	0.336	0.550	0.343	0.561
2/0	67.4	0.261	0.271	0.428	0.266	0.437	0.271	0.446
3/0	85.0	0.207	0.215	0.339	0.211	0.346	0.215	0.353
4/0	107	0.164	0.169	0.269	0.167	0.274	0.170	0.279
250	127	0.139	0.144	0.228	0.142	0.233	0.145	0.238
300	152	0.116	0.120	0.190	0.118	0.194	0.120	0.198
350	177	0.0992	0.103	0.163	0.101	0.166	0.103	0.169
400	203	0.0868	0.0893	0.142	0.0885	0.145	0.0903	0.148
450	228	0.0771	0.0794	0.126	0.0786	0.129	0.0802	0.132

500	253	0.0694	0.0714	0.0708	0.116	0.118
550	279	0.0631	0.0656	0.0644	0.105	0.107
600	304	0.0578	0.0602	0.0590	0.0967	0.0986
650	329	0.0534	0.0550	0.0545	0.0893	0.0911
700	355	0.0496	0.0510	0.0506	0.0829	0.0846
750	380	0.0463	0.0476	0.0472	0.0774	0.0789
800	405	0.0434	0.0447	0.0443	0.0725	0.0740
900	456	0.0386	0.0397	0.0394	0.0645	0.0658
1000	507	0.0347	0.0357	0.0354	0.0580	0.0592
1100	557	0.0316	0.0325	0.0322	0.0527	0.0538
1200	608	0.0289	0.0298	0.0295	0.0483	0.0493
1250	633	0.0278	0.0286	0.0284	0.0464	0.0473
1300	659	0.0267	0.0275	0.0272	0.0447	0.0456
1400	709	0.0248	0.0255	0.0253	0.0414	0.0422
1500	760	0.0231	0.0238	0.0236	0.0387	0.0395
1600	811	0.0217	0.0223	0.0221	0.0363	0.0370
1700	861	0.0204	0.0210	0.0208	0.0342	0.0349
1750	887	0.0198	0.0204	0.0202	0.0332	0.0339
1800	912	0.0193	0.0198	0.0197	0.0322	0.0328
1900	963	0.0183	0.0188	0.0187	0.0306	0.0312
2000	1013	0.0174	0.0179	0.0177	0.0291	0.0297
2500	1267	0.0140	0.0144	0.0143	0.0235	0.0240
3000	1520	0.0117	0.0120	0.0119	0.0196	0.0200
3500	1773	0.0101	0.0104	0.0103	0.0169	0.0172
4000	2027	0.00885	0.00911	0.00903	0.0148	0.0151
4500	2280	0.00794	0.00817	0.00810	0.0133	0.0136
5000	2534	0.00715	0.00735	0.00729	0.0119	0.0121

[a] AWG up to 4/0, kcmil from 250 upwards
[b] Based on nominal area and incorrect for equivalent resistance
[c] Taken as 2% above nominal resistance for single-core cable and 2% above single-core cable for multicore cable
The data are based on ICEA S–66–524, NEMA WC7–1993, and ICEA S–19–1981, NEMA WC3–1986 for concentric stranded and compact stranded conductors. Different values apply to solid conductors.

Table A4.4 Temperature correction factors for conductor resistance

Temperature of conductor (°C)	Factor to convert to 20°C	Reciprocal to convert from 20°C
5	1.064	0.940
6	1.059	0.944
7	1.055	0.948
8	1.050	0.952
9	1.046	0.956
10	1.042	0.960
11	1.037	0.964
12	1.033	0.968
13	1.029	0.972
14	1.025	0.976
15	1.020	0.980
16	1.016	0.984
17	1.012	0.988
18	1.008	0.992
19	1.004	0.996
20	1.000	1.000
21	0.996	1.004
22	0.992	1.008
23	0.988	1.012
24	0.984	1.016
25	0.980	1.020
26	0.977	1.024
27	0.973	1.028
28	0.969	1.032
29	0.965	1.036
30	0.962	1.040
31	0.958	1.044
32	0.954	1.048
33	0.951	1.052
34	0.947	1.056
35	0.943	1.060
40	0.926	1.080
45	0.909	1.100
50	0.893	1.120
55	0.877	1.140
60	0.862	1.160
65	0.847	1.180
70	0.833	1.200
75	0.820	1.220
80	0.806	1.240
85	0.794	1.260
90	0.781	1.280

Table A4.5 Maximum diameters (mm) of circular copper conductors (BS 6360: 1991)

Cross-sectional area (mm^2)	Conductors in cables for fixed installations		
	Solid (class 1)	Stranded (class 2)	Flexible conductors (classes 5 and 6)
0.5	0.9	1.1	1.1
0.75	1.0	1.2	1.3
1	1.2	1.4	1.5
1.5	1.5	1.7	1.8
2.5	1.9	2.2	2.6
4	2.4	2.7	3.2
6	2.9	3.3	3.9
10	3.7	4.2	5.1
16	4.6	5.3	6.3
25	5.7	6.6	7.8
35	6.7	7.9	9.2
50	7.8	9.1	11.0
70	9.4	11.0	13.1
95	11.0	12.9	15.1
120	12.4	14.5	17.0
150	13.8	16.2	19.0
185		18.0	21.0
240		20.6	24.0
300		23.1	27.0
400		26.1	31.0
500		29.2	35.0
630		33.2	39.0
800		37.6	
1000		42.2	

Notes
(a) For circular copper conductors, maximum diameters only are given and for the stranded (class 2) conductors these are based on uncompacted conductors. The reason for this is that connectors will cope with a wider range of diameters with copper than with aluminium and therefore with copper it is generally only necessary to recommend the maximum diameters to be accommodated. Moreover, circular stranded copper conductors are more frequently used in the uncompacted form than are aluminium conductors.
(b) If minimum diameters for circular copper conductors class 1 and class 2 are needed, reference can be made to the minimum diameters for solid and stranded compacted circular aluminium conductors indicated in table A4.6.

Table A4.6 Minimum and maximum diameters (mm) of circular aluminium conductors (BS 6360: 1991)

Cross-sectional area (mm^2)	Solid conductors (class 1) Minimum diameter	Solid conductors (class 1) Maximum diameter	Stranded compacted conductors (class 2) Minimum diameter	Stranded compacted conductors (class 2) Maximum diameter
16	4.1	4.6	4.6	5.2
25	5.2	5.7	5.6	6.5
35	6.1	6.7	6.6	7.5
50	7.2	7.8	7.7	8.6
70	8.7	9.4	9.3	10.2
95	10.3	11.0	11.0	12.0
120	11.6	12.4	12.5	13.5
150	12.9	13.8	13.9	15.0
185	14.5	15.4	15.5	16.8
240	16.7	17.6	17.8	19.2
300	18.8	19.8	20.0	21.6
400	–	–	22.9	24.6
500	–	–	25.7	27.6
630	–	–	29.3	32.5

Notes
(a) In the exceptional case of uncompacted circular stranded aluminium conductors the maximum diameters should not exceed the corresponding values for copper conductors given in column 3 of table A4.5.
(b) The dimensional limits of aluminium conductors with cross-sectional areas smaller than 16 mm^2 are not given because of the variations in dimensions that exist depending on the wide range of materials and combinations of materials used.
(c) The dimensional limits of aluminium conductors with cross-sectional areas above 630 mm^2 are not given as the compaction technology is not generally established.

APPENDIX A5
Industrial Cables for Fixed Supply

PVC ARMOURED POWER OR CONTROL CABLES
600/1000 V to BS 6346: 1989

Main application: factory wiring

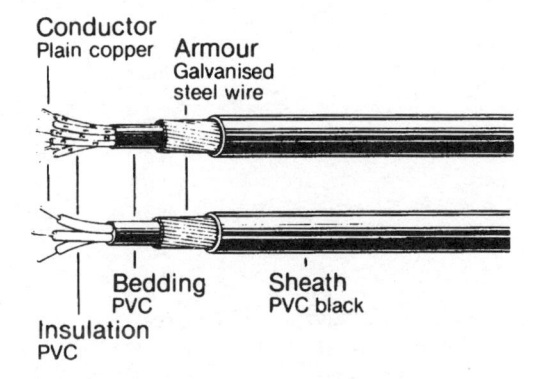

Conductor — Plain copper
Armour — Galvanised steel wire
Bedding — PVC
Insulation — PVC
Sheath — PVC black

For sizes above 16 mm^2 see appendix A13. The current ratings tabulated are in accordance with the sixteenth edition of the IEE Regulations BS 7671. If these do not apply, reference should be made to ERA Report 69–30, Part 3. The ratings are for a single circuit only. Rating factors and other details are given in chapters 8 and 10.

Table A5.1 Current ratings and volt drop

Conductor size (mm^2)	2-core cable, d.c. or single-phase a.c. Rating (A)	2-core cable, d.c. or single-phase a.c. Volt drop per A/m (mV)	3- or 4-core cable, 3-phase a.c. Rating (A)	3- or 4-core cable, 3-phase a.c. Volt drop per A/m (mV)
In free air (ambient temperature 30°C, maximum conductor temperature 70°C)				
1.5	22	29	19	25
2.5	31	18	26	15
4	41	11	35	9.5
6	53	7.3	45	6.4
10	72	4.4	62	3.8
16	97	2.8	83	2.4
Direct in ground at 0.5 m depth (ground temperature 15°C, conductor temperature 70°C)				
1.5	32	29	27	25
2.5	41	18	35	15
4	55	11	47	9.5
6	69	7.3	59	6.4
10	92	4.4	78	3.8
16	119	2.8	101	2.4

PVC ARMOURED POWER OR CONTROL CABLES
600/1000 V to BS 6346: 1989

Table A5.2 Dimensions and weights (2-, 3- and 4-core cables)

Conductor size (mm^2)	Armour wire diameter (mm)	Approximate diameter (mm)	Approximate weight (kg/km)
2-core (ref. 6942X)			
1.5[a]	0.9	11.7	280
1.5	0.9	12.3	298
2.5[a]	0.9	13.1	350
2.5	0.9	13.6	355
4	0.9	15.1	460
6	0.9	16.5	550
10	1.25	20.1	880
16	1.25	21.9	990
3-core (ref. 6943X)			
1.5[a]	0.9	12.3	310
1.5	0.9	12.8	344
2.5[a]	0.9	13.6	390
2.5	0.9	14.1	413
4	0.9	15.8	520
6	1.25	18.3	730
10	1.25	21.2	1010
16	1.25	23.1	1220
4-core (ref. 6944X)			
1.5[a]	0.9	13.0	350
1.5	0.9	13.5	383
2.5[a]	0.9	14.5	440
2.5	0.9	15.0	470
4	1.25	17.8	710
6	1.25	19.2	850
10	1.25	22.8	1200
16	1.6	26.3	1700

[a] Solid conductor

PVC ARMOURED AUXILIARY MULTICORE CABLES
600/1000 V to BS 6346: 1989

Table A5.3 Dimensions and weights (5- to 48-core cables)

Number of cores	Solid conductors			Stranded conductors		
	Armour wire diameter (mm)	Approximate diameter (mm)	Approximate weight (kg/km)	Armour wire diameter (mm)	Approximate diameter (mm)	Approximate weight (kg/km)
1.5 mm² conductor						
5	0.9	13.8	390	0.9	14.3	
7	0.9	14.5	445	0.9	15.2	470
8	0.9	15.4	490	0.9	16.5	
10	1.25	18.1	685	1.25	19.0	
12	1.25	18.6	740	1.25	19.4	784
14	1.25	19.2	800	1.25	20.3	
16	1.25	20.1	920	1.25	21.3	
19	1.25	21.1	965	1.25	22.2	1023
22	1.6	25.0	1330	1.6	26.2	
27	1.6	25.4	1420	1.6	26.7	1506
30	1.6	26.1	1510	1.6	27.4	
37	1.6	27.8	1730	1.6	29.2	1834
40	1.6	28.7	1970	1.6	30.2	
48	1.6	30.8	2080	1.6	32.9	
2.5 mm² conductor						
5	0.9	15.4	495	0.9	16.3	
7	0.9	16.6	590	1.25	18.0	729
8	1.25	18.5	760	1.25	19.2	
10	1.25	20.9	895	1.25	21.9	
12	1.25	21.4	975	1.25	22.4	1049
14	1.6	22.3	1070	1.6	24.6	
16	1.6	23.2	1170	1.6	25.5	
19	1.6	25.4	1500	1.6	26.6	1584
22	1.6	28.8	1770	1.6	30.2	
27	1.6	29.3	1910	1.6	30.7	2043
30	1.6	30.1	2050	1.6	32.0	
37	1.6	32.4	2360	1.6	34.0	2517
40	1.6	33.5	2520	2.0	36.6	
48	2.0	37.5	3200	2.0	39.5	

(*cont.*)

PVC ARMOURED AUXILIARY MULTICORE CABLES
600/1000 V to BS 6346: 1989

Table A5.3 continued

Number of cores	Solid conductors			Stranded conductors		
	Armour wire diameter (mm)	Approximate diameter (mm)	Approximate weight (kg/km)	Armour wire diameter (mm)	Approximate diameter (mm)	Approximate weight (kg/km)
Multicore: 4 mm² conductor (stranded)						
5				1.25	19.0	795
7				1.25	20.5	940
8				1.25	22.1	1050
10				1.6	26.1	1420
12				1.6	26.8	1560
14				1.6	28.0	1730
16				1.6	29.2	1880
19				1.6	30.5	2080
22				1.6	35.0	2490
27				2.0	37.1	3040
30				2.0	38.2	
37				2.0	40.8	
40				2.0	42.5	
48				2.0	46.0	
Multicore: 6 mm² conductor (stranded)						
5				1.25	20.3	980
7				1.25	21.7	1160

Industrial Cables for Fixed Supply 815

> **XLPE ARMOURED POWER
> OR CONTROL CABLES**
> 600/1000 V to BS 5467: 1989 and BS 6724: 1990

Cable design is as shown above table A5.1 for PVC cables but with XLPE insulation. 'Low smoke and fire cable' to BS 6724: 1990 has special material for the bedding and oversheath. For sizes above 16 mm^2 see appendix A14.

The ratings are for a single circuit only. Rating factors and other details are given in chapters 8 and 9.

Table A5.4 Current ratings and volt drop

Conductor size (mm^2)	2-core cable, d.c. or single-phase a.c. Rating (A)	Volt drop per A/m (mV)	3- or 4-core cable, 3-phase a.c. Rating (A)	Volt drop per A/m (mV)
Clipped direct (ambient temperature 30°C, conductor temperature 90°C)				
1.5	27	31	23	27
2.5	36	19	31	16
4	49	12	42	10
6	62	7.9	53	6.8
10	85	4.7	73	4.0
16	110	2.9	94	2.5
In free air				
1.5	29	31	25	27
2.5	39	19	33	16
4	52	12	44	10
6	66	7.9	56	6.8
10	90	4.7	78	4.0
16	115	2.9	99	2.5

XLPE ARMOURED POWER OR CONTROL CABLES
600/1000 V to BS 5467: 1989 and BS 6724: 1990

Table A5.5 Dimensions and weights (2-, 3- and 4-core cables)

Conductor size (mm^2)	Armour wire diameter (mm)	Approximate diameter (mm)	Approximate weight (kg/km)
2-core (stranded conductor)			
1.5	0.9	12.3	300
2.5	0.9	13.6	360
4	0.9	14.7	430
6	0.9	15.9	500
10	0.9	18.0	800
16	1.25	20.0	940
3-core (stranded conductor)			
1.5	0.9	12.8	335
2.5	0.9	14.1	410
4	0.9	15.3	500
6	0.9	16.6	770
10	1.25	19.5	900
16	1.25	21.2	1180
4-core (stranded conductor)			
1.5	0.9	13.5	380
2.5	0.9	15.0	470
4	0.9	16.4	580
6	1.25	18.7	820
10	1.25	21.1	1060
16	1.25	22.9	1410

**XLPE ARMOURED POWER
OR CONTROL CABLES**
600/1000 V to BS 5467: 1989 and BS 6724: 1990

Table A5.6 Dimensions and weights (multicore auxiliary cables)

Number of cores	Armour wire diameter (mm)	Approximate diameter (mm)	Approximate weight (kg/km)
1.5 mm² conductor (stranded)			
7	0.9	15.2	480
12	1.25	19.4	800
19	1.25	22.2	1035
27	1.6	26.7	1525
37	1.6	29.2	1820
2.5 mm² conductor (stranded)			
7	0.9	17.1	600
12	1.25	22.4	1020
19	1.6	26.6	1530
27	1.6	30.7	1960
37	1.6	33.8	2370

APPENDIX A6
Cables for Fixed Installation in Buildings

PVC WIRING CABLES
300/500 V and 450/750 V to BS 6004: 1995

Main application: wiring in buildings

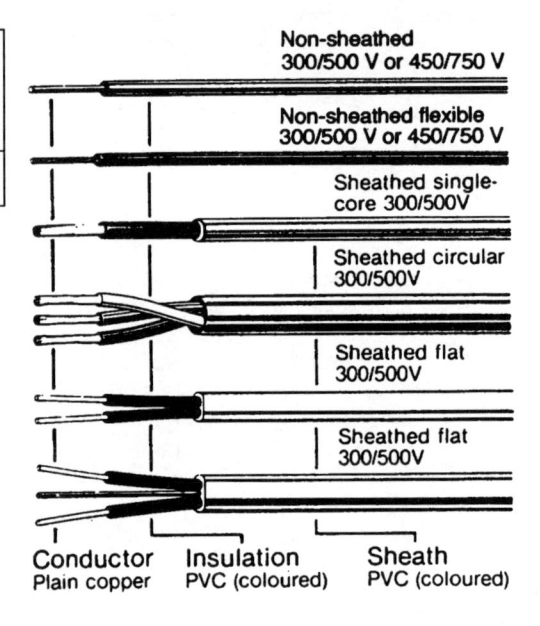

Non-sheathed
300/500 V or 450/750 V

Non-sheathed flexible
300/500 V or 450/750 V

Sheathed single-core 300/500V

Sheathed circular 300/500V

Sheathed flat 300/500V

Sheathed flat 300/500V

Conductor — Plain copper
Insulation — PVC (coloured)
Sheath — PVC (coloured)

Current rating conditions

The ratings given in tables A6.1 and A6.2 are for a single circuit and are in accordance with BS 7671: 1992, 'Requirements for electrical installations', the sixteenth edition of the IEE Wiring Regulations. They correspond to continuous loading at a maximum conductor temperature of 70°C and are based on an ambient temperature of 30°C. When the protection is by semi-enclosed fuse the cable should be selected with a rating not less than 1.38 times the rating of the fuse.

Further information on current ratings is given in chapters 8 and 10. Rating correction factors are included in chapter 8.

Details of the methods of installation applicable to the column headings are in appendix 4 of BS 7671: 1992.

> **PVC WIRING CABLES**
> 300/500 V and 450/750 V to
> BS 6004: 1995

Table A6.1 Current ratings for single-core non-sheathed conduit cables and single-core sheathed cables

Conductor size (mm^2)	Reference method 3 Bunched and enclosed in conduit or trunking[a]		Reference method 1 Clipped direct		Reference method 11 On perforated tray (horizontal or vertical)	
	2 cables, single-phase a.c. or d.c. (A)	3 or 4 cables, 3-phase a.c. (A)	2 cables, single-phase a.c. or d.c. (A)	3 or 4 cables, 3-phase a.c. (A)	2 cables, single-phase a.c. or d.c. (A)	3 or 4 cables, 3-phase a.c. (A)
1.0	13.5	12	15.5	14		
1.5	17.5	15.5	20	18		
2.5	24	21	27	25		
4	32	28	37	33		
6	41	36	47	43		
10	57	50	65	59		
16	76	68	87	79		
25	101	89	114	104	126	112
35	125	110	141	129	156	141
50	151	134	182	167	191	172
70	192	171	234	214	246	223
95	232	207	284	261	300	273
120	269	239	330	303	349	318
150	300	262	381	349	404	369
185	341	296	436	400	463	424
240	400	346	515	472	549	504
300	458	394	594	545	635	584

[a] Non-sheathed cables are not suitable for conduits or trunking etc. buried underground
Note
$\geq 50\,\text{mm}^2$ single core insulated and sheathed (600/1000 V) cables are covered by BS 6346: 1989.

PVC WIRING CABLES
300/500 V and 450/750 V to
BS 6004: 1995

Table A6.2 Voltage drop (per A/m) for single-core non-sheathed conduit cables and single-core sheathed cables

Conductor size (mm^2)	Reference method 3 Bunched and enclosed in conduit or trunking[a]		Reference method 1 Clipped direct		Reference method 11 On perforated tray (horizontal or vertical)	
	2 cables, single-phase a.c. or d.c. (mV)	3 or 4 cables, 3-phase a.c. (mV)	2 cables, single-phase a.c. or d.c. (mV)[b]	3 or 4 cables, 3-phase a.c. (mV)[b]	2 cables, single-phase a.c. or d.c. (mV)[b]	3 or 4 cables, 3-phase a.c. (mV)[b]
1.0	44	38	44	38		
1.5	29	25	29	25		
2.5	18	15	18	15		
4	11	9.5	11	9.5		
6	7.3	6.4	7.3	6.4		
10	4.4	3.8	4.4	3.8		
16	2.8	2.4	2.8	2.4		
25	1.8	1.55	1.75	1.5	1.75	1.5
35	1.3	1.10	1.25	1.1	1.25	1.1
	a.c. d.c.		a.c. d.c.		a.c. d.c.	
50	1.00 0.93	0.85	0.95 0.93	0.84	0.95 0.93	0.84
70	0.72 0.63	0.61	0.66 0.63	0.60	0.66 0.63	0.60
95	0.56 0.46	0.48	0.50 0.46	0.47	0.50 0.46	0.47
120	0.47 0.36	0.41	0.41 0.36	0.40	0.41 0.36	0.40
150	0.41 0.29	0.36	0.34 0.29	0.34	0.34 0.29	0.34
185	0.37 0.23	0.32	0.29 0.23	0.31	0.29 0.23	0.31
240	0.33 0.18	0.29	0.25 0.18	0.27	0.25 0.18	0.27
300	0.31 0.145	0.27	0.22 0.145	0.25	0.22 0.145	0.25

[a] Non-sheathed cables are not suitable for conduits or trunking etc. buried underground
[b] Flat and touching

PVC WIRING CABLES
300/500 V to BS 6004: 1995

Table A6.3 Current ratings and volt drop for flat cables

Conductor size (mm^2)	Reference method 3 Enclosed in conduit on a wall or ceiling or in trunking		Reference method 1 Clipped direct		Reference method 11 On a perforated cable tray or Reference method 13 (free air)	
	2-core cable, single-phase a.c. or d.c	3- or 4-core cable, 3-phase a.c.	2-core cable, single-phase a.c. or d.c.	3- or 4-core cable, 3-phase a.c.	2-core cable, single-phase a.c. or d.c.	3- or 4-core cable, 3-phase a.c.
Current ratings (A)						
1.0	13	11.5	15	13.5	17	14.5
1.5	16.5	15	19.5	17.5	22	18.5
2.5	23	20	27	24	30	25
4	30	27	36	32	40	34
6	38	34	46	41	51	43
10	52	46	63	57	70	60
16	69	62	85	76	94	80
Volt drop per A/m (mV)						
1.0	44	38	44	38	44	38
1.5	29	25	29	25	29	25
2.5	18	15	18	15	18	15
4	11	9.5	11	9.5	11	9.5
6	7.3	6.4	7.3	6.4	7.3	6.4
10	4.4	3.8	4.4	3.8	4.4	3.8
16	2.8	2.4	2.8	2.4	2.8	2.4

PVC WIRING CABLES
300/500 V and 450/750 V to
BS 6004: 1995

Table A6.4 Dimensions and weights (non-sheathed single-core cables)

Conductor size (mm²)	300/500 V solid conductor (ref. H05V-U)		450/750 V solid or stranded conductor (ref. 6491X, H07V)		450/750 V flexible conductor (ref. H07V-K)	
	Mean overall diameter (upper limit) (mm)	Approximate weight (kg/km)	Mean overall diameter (upper limit) (mm)	Approximate weight (kg/km)	Mean overall diameter (upper limit) (mm)	Approximate weight (kg/km)
1.0	2.7[a]	17[a]				
1.5			3.2[a]	21[a]		
1.5			3.3	21	3.4	21
2.5			3.9[a]	33[a]		
2.5			4.0	35	4.1	33
4			4.6	50	4.8	50
6			5.2	71	5.3	70
10			6.7	120	6.8	120
16			7.8	180	8.1	180
25			9.7	280	10.2	290
35			10.9	380	11.7	400
50			12.8	510	13.9	570
70			14.6	720	16.0	770
95			17.1	990	18.2	1000
120			18.8	1200	20.2	1300
150			20.9	1500	22.5	1600
185			23.3	1900	24.9	1900
240			26.6	2500	28.4	2500
300			29.6	3000		

[a] Solid conductor

PVC WIRING CABLES
300/500 V to BS 6004: 1995

Table A6.5 Dimensions and weights (300/500 V, single-core sheathed, flat 2-core and 3-core cables without CPC)

Conductor size (mm^2)	Mean overall diameter Lower limit (mm)	Mean overall diameter Upper limit (mm)	Approximate weight (kg/km)
Single-core sheathed (ref. 6181Y)			
1.0[a]	3.8	4.5	28
1.5[a]	4.2	4.9	36
2.5[a]	4.8	5.8	51
4	5.4	6.8	75
6	6.0	7.4	98
10	7.2	8.8	150
16	8.4	10.5	220
25	10.0	12.5	340
35	11.0	13.5	440
2-core flat (ref. 6192Y)			
1.0[a]	4.0 × 6.2	4.7 × 7.4	53
1.5[a]	4.4 × 7.0	5.4 × 8.4	71
1.5	4.5 × 7.2	5.6 × 8.8	78
2.5[a]	5.2 × 8.4	6.2 × 9.8	100
2.5	5.2 × 8.6	6.6 × 10.5	112
4	5.6 × 9.6	7.2 × 11.5	150
6	6.4 × 10.5	8.0 × 13.0	204
10	7.8 × 13.0	9.6 × 16.0	325
16	9.0 × 15.5	11.0 × 18.5	469
3-core flat (ref. 6193Y)			
1.0[a]	4.0 × 8.4	4.7 × 9.8	78
1.5[a]	4.4 × 9.8	5.4 × 11.5	105
2.5[a]	5.2 × 11.5	6.2 × 13.5	155
4	5.8 × 13.5	7.4 × 16.5	230
5	6.4 × 15.0	8.0 × 18.0	300
10	7.8 × 19.0	9.6 × 22.5	480
16	9.0 × 22.0	11.0 × 26.5	570

[a] Solid conductor

PVC WIRING CABLES
300/500 V to BS 6004: 1995

Table A6.6 Dimensions and weights (300/500 V, single-core, flat 2-core and 3-core with protective conductor)

Conductor size (mm^2)	Protective conductor (mm^2)	Mean overall diameter Lower limit (mm)	Mean overall diameter Upper limit (mm)	Approximate weight (kg/km)
Single core (ref. 6241Y)				
1.0[a]	1.0	4.0 × 5.1	5.2 × 6.4	39
1.5[a]	1.0	4.4 × 5.4	5.8 × 7.0	50
2-core flat (ref. 6242Y)				
1.0[a]	1.0	4.0 × 7.2	4.7 × 8.6	69
1.5[a]	1.0	4.4 × 8.2	5.4 × 9.6	88
1.5	1.0	4.5 × 8.4	5.6 × 10.0	95
2.5[a]	1.5	5.2 × 9.8	6.2 × 11.5	130
2.5	1.5	5.2 × 9.8	6.6 × 12.0	136
4	1.5	5.6 × 10.5	7.2 × 13.0	176
6	2.5	6.4 × 12.5	8.0 × 15.0	243
10	4	7.8 × 15.5	9.6 × 19.0	390
16	6	9.0 × 18.0	11.0 × 22.5	567
3-core flat (ref. 6243Y)				
1.0[a]	1.0	4.0 × 9.6	4.7 × 11.0	92
1.5[a]	1.0	4.4 × 10.5	5.4 × 12.5	120
2.5[a]	1.0	5.2 × 12.5	6.2 × 14.5	173
4	1.5	5.8 × 14.5	7.4 × 18.0	255
6	2.5	6.4 × 16.5	8.0 × 20.0	340
10	4	7.8 × 21.0	9.6 × 25.5	550
16	6	9.0 × 24.5	11.0 × 29.5	730

[a] Solid conductor

Cables for Fixed Installation in Buildings

ELASTOMERIC INSULATED CABLES
300/500 V and 450/750 V to BS 6007: 1993

All tinned copper conductors 450/750 V
85°C rubber insulated, textile braided and compounded

300/500 V
60°C rubber insulated, sheathed (festoon lighting)

Main application: single-core for wiring in buildings and 2-core for festoon lighting

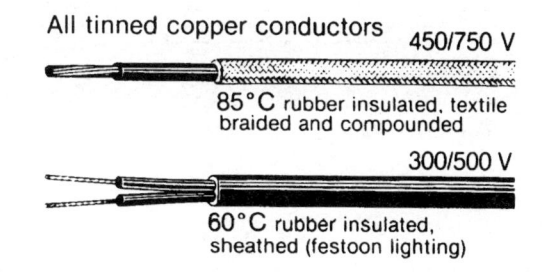

Table A6.7 Current ratings and volt drop

Conductor size (mm^2)	Reference method 3 Enclosed in conduit etc. in, or on, a wall				Reference method 1 Clipped direct			
	2 cables d.c. or single-phase a.c.		3 or 4 cables, 3-phase a.c.		2 cables d.c. or single-phase a.c.		3 or 4 cables, 3-phase a.c.	
	Rating (A)	Volt drop per A/m[a] (mV)	Rating (A)	Volt drop per A/m (mV)	Rating (A)	Volt drop per A/m[a] (mV)	Rating (A)	Volt drop per A/m (mV)
Single-core cables								
1.0	17	46	15	40	19	46	17.5	40
1.5	22	31	19.5	26	25	31	23	26
2.5	30	18	27	16	34	18	31	16
4	40	12	36	10	45	12	42	10
6	52	7.7	46	6.7	59	7.7	54	6.7
10	72	4.6	63	4.0	81	4.6	75	4.0
16	96	2.9	85	2.5	108	2.9	100	2.5
25	127	1.9	112	1.65	143	1.85	133	1.6
35	157	1.4	138	1.2	177	1.35	164	1.15
50	190	1.05	167	0.91	215	0.99	199	0.88
70	242	0.74	213	0.65	274	0.69	254	0.62
95	293	0.58	258	0.51	332	0.52	308	0.48
Flat twin cable								
2.5					24	18		

[a] A.C. only

Note

The ratings are based on an ambient temperature of 30°C and a maximum conductor temperature of 85°C for single-core cable and 60°C for twin cable. They apply to a single circuit only.

The ratings for single-core cables are in accordance with the IEE Wiring Regulations (BS 7671: 1992) and those for twin cable are given for guidance. When the protection is by a semi-enclosed fuse the cable should be selected with a rating not less than 1.38 times the rating of the fuse. Rating factors are given in chapter 8.

ELASTOMERIC INSULATED CABLES
300/500 V and 450/750 V to BS 6007: 1993

Table A6.8 Dimensions and weights

Conductor size (mm^2)	Maximum diameter (mm)	Approximate weight (kg/km)
Single-core, braided 450/750 V (ref. 6101T)		
1[a]	4.3	19
1.5[b]	4.6	24
2.5[b]	5.0	35
4	6.4	57
6	7.0	84
10	8.6	133
16	9.6	194
25	11.5	292
35	13.0	396
50	14.5	530
70	16.5	741
95	19.0	1030
Flat twin 300/500 V (festoon lighting)		
2.5	6.8 × 11.0	116

[a] Solid conductor
[b] Solid or stranded conductor

Cables for Fixed Installation in Buildings

XLPE INSULATED FLOORWARMING CABLES
230 to 380 V (BICC designs)

Main application: heating of buildings and soil warming

Type X
Conductor — Resistive alloy
Insulation — XLPE

Type XS
Insulation — XLPE
Conductor — Resistive alloy
Sheath — PVC

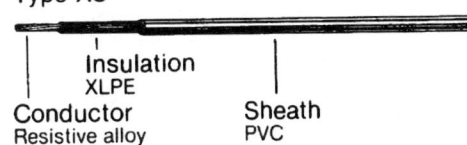

Type XSBV
Insulation — XLPE
Conductor — Resistive alloy
Screen — Galvanised steel wire braid with 1·0 mm² drain wires laid flat
Sheath — Heat resisting PVC (black)

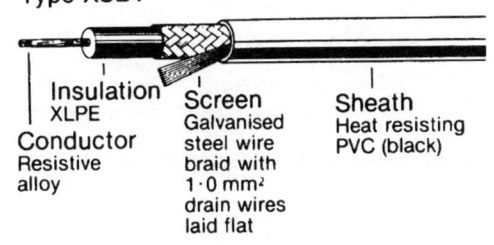

Type XB
Insulation — XLPE
Conductor — Resistive alloy
Screen — Tinned copper wire braid
Sheath — Heat resisting PVC (black)

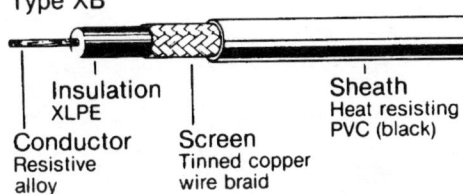

Table A6.9 General data

Types	X: XLPE core, insulated only
	XJ: with heat resisting PVC sheath
	XB: with tinned copper wire braid and sheath as XJ
	XSBV: with galvanised steel wire braid, 1.0 mm² drain wire and sheath as XJ
Cable sizes	18 standard conductor resistances from 12.2 to 0.013 Ω/m
Voltages	220, 230, 240 and 380 V
Lengths and output of heating units	16 standard wattages covering:

Voltage (V)	Heat output (W) Minimum	Heat output (W) Maximum	Length (m) Minimum	Length (m) Maximum
220	250	3500	15	230.5
230	260	3750	16.5	235
240	275	4000	17	236
380	450	6300	27	382

XL-LSF WIRING CABLES
300/500 V and 450/750 V to BS 7211: 1994
Main application: wiring in buildings

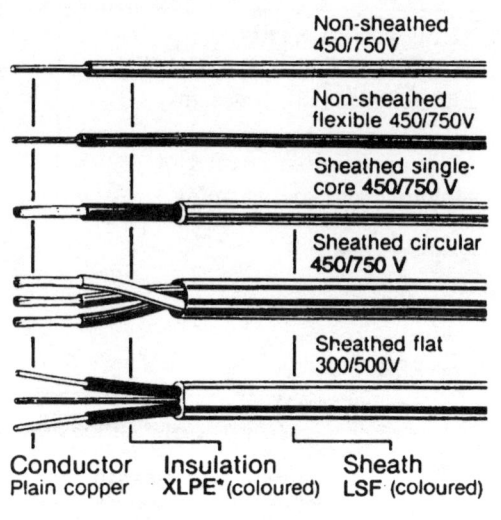

Non-sheathed 450/750V

Non-sheathed flexible 450/750V

Sheathed single-core 450/750 V

Sheathed circular 450/750 V

Sheathed flat 300/500V

Conductor — Plain copper
Insulation — XLPE*(coloured)
Sheath — LSF (coloured)

*XL–LSF insulation on non-sheathed cables

Current rating conditions

The ratings given in tables A6.10 and A6.11 are for a single circuit and are in accordance with BS 7671: 1992, 'Requirements for electrical installations'. IEE Wiring Regulations, sixteenth edition. They correspond to continuous loading at a maximum conductor temperature of 90°C and are based on an ambient temperature of 30°C. When the protection is by semi-enclosed fuse the cable should be selected with a rating not less than 1.38 times the rating of the fuse.

Further information on current ratings is given in chapters 8 and 10. Rating correction factors are included in chapter 8.

Details of the methods of installation applicable to the column headings are in appendix 9 of the IEE Regulations for Electrical Installations.

Cables for Fixed Installation in Buildings

XL-LSF WIRING CABLES
300/500 V and 450/750 V to
BS 7211: 1994

Table A6.10 Current ratings for single-core non-sheathed conduit cables and single-core sheathed cables[b]

Conductor size (mm^2)	Reference method 3 Bunched and enclosed in conduit or trunking[a]		Reference method 1 Clipped direct		Reference method 11 On perforated tray (horizontal or vertical)	
	2 cables, single-phase a.c. or d.c. (A)	3 or 4 cables, 3-phase a.c. (A)	2 cables, single-phase a.c. or d.c. (A)	3 or 4 cables, 3-phase a.c. (A)	2 cables, single-phase a.c. or d.c. (A)	3 or 4 cables, 3-phase a.c. (A)
1.0	17	15	19	17.5		
1.5	22	19	25	23		
2.5	30	26	34	31		
4	40	35	46	41		
6	51	45	59	54		
10	71	63	81	74		
16	95	85	109	99		
25	126	111	143	130	158	140
35	156	138	176	161	195	176
50	189	168				
70	240	214				
95	290	259				
120	336	299				
150	375	328				
185	426	370				
240	500	433				
300	573	493				

[a] Non-sheathed cables are not suitable for conduits or trunking etc. buried underground
[b] Sheathed cables are in sizes up to and including 35 mm^2

XL-LSF WIRING CABLES
300/500 V and 450/750 V to
BS 7211: 1994

Table A6.11 Voltage drop (per A/m) for single-core non-sheathed conduit cables and single-core sheathed cables

Conductor size (mm^2)	Reference method 3 Bunched and enclosed in conduit or trunking[a]		Reference method 1 Clipped direct		Reference method 11 On perforated tray (horizontal or vertical)	
	2 cables, single-phase a.c. or d.c. (mV)	3 or 4 cables, 3-phase a.c. (mV)	2 cables, single-phase a.c. or d.c. (mV)[b]	3 or 4 cables, 3-phase a.c. (mV)[b]	2 cables, single-phase a.c. or d.c. (mV)[b]	3 or 4 cables, 3-phase a.c. (mV)[b]
1.0	46	40	46	40		
1.5	31	27	31	27		
2.5	19	16	19	16		
4	12	10	12	10		
6	7.9	6.8	7.9	6.8		
10	4.7	4.0	4.7	4.0		
16	2.9	2.5	2.9	2.5		
	a.c. d.c.		a.c. d.c.			
25	1.90 1.85	1.65	1.85 1.85	1.60	1.85	1.60
35	1.35 1.35	1.15	1.35 1.35	1.15	1.35	1.15
50	1.05 0.99	0.90	1.00 0.99	0.87		
70	0.75 0.68	0.65	0.71 0.68	0.61		
95	0.58 0.49	0.50	0.52 0.49	0.45		
120	0.48 0.39	0.42	0.43 0.39	0.37		
150	0.43 0.32	0.37	0.36 0.32	0.31		
185	0.37 0.25	0.32	0.30 0.25	0.26		
240	0.33 0.190	0.29	0.25 0.190	0.22		
300	0.31 0.155	0.27	0.22 0.155	0.195		

[a] Non-sheathed cables are not suitable for conduits or trunking etc. which are buried underground
[b] Flat and touching

Cables for Fixed Installation in Buildings

XL-LSF WIRING CABLES
300/500 V to BS 7211: 1994

Table A6.12 Current ratings and volt drop for flat cables

Conductor size (mm^2)	Reference method 3 Enclosed in conduit on a wall or ceiling or in trunking		Reference method 1 Clipped direct		Reference method 11 On a perforated cable tray or Reference method 13 (free air)	
	2-core cable, single-phase a.c. or d.c.	3- or 4-core cable, 3-phase a.c.	2-core cable, single-phase a.c. or d.c.	3- or 4-core cable, 3-phase a.c.	2-core cable, single-phase a.c. or d.c.	3- or 4-core cable, 3-phase a.c.
Current ratings (A)						
1.0	17	15	19	17	21	18
1.5	22	19.5	24	22	26	23
2.5	30	26	33	30	36	32
4	40	35	45	40	49	42
6	51	44	58	52	63	54
10	69	60	80	71	86	75
16	91	80	107	96	115	100
Volt drop per A/m (mV)						
1.0	46	40	46	40	46	40
1.5	31	27	31	27	31	27
2.5	19	16	19	16	19	16
4	12	10	12	10	12	10
6	7.9	6.8	7.9	6.8	7.9	6.8
10	4.7	4.0	4.7	4.0	4.7	4.0
16	2.9	2.5	2.9	2.5	2.9	2.5

The data apply to one cable, with a protective conductor.

XL-LSF WIRING CABLES
300/500 V and 450/750 V to BS 7211: 1994

Table A6.13 Dimensions and weights (non-sheathed single-core cables)

Conductor size (mm^2)	300/500 V solid conductor (ref. H05Z-U) Mean overall diameter (upper limit) (mm)	Approximate weight (kg/km)	450/750 V solid or stranded conductor (ref. 6491B, H07Z-R) Mean overall diameter (upper limit) (mm)	Approximate weight (kg/km)	450/750 V flexible conductor (ref. H07Z-K) Mean overall diameter (upper limit) (mm)	Approximate weight (kg/km)
1.0	3.0[a]	17[a]				
1.5			3.4	22	3.5	21
2.5			4.2	33	4.2	33
4			4.8	50	4.8	50
6			5.4	70	6.4	70
10			6.8	115	7.6	120
16			8.0	175	8.8	180
25			9.8	275	11.0	290
35			11.0	370	12.5	400
50			13.0	500	14.5	570
70			15.0	700	17.0	770
95			17.0	970	19.0	1000
120			19.0	1165		
150			21.0		21.0	1200
185			23.5		23.5	1500
240			26.5		26.0	1900
300			29.5	3000	27.5	2500

[a] Solid conductor

XL-LSF WIRING CABLES
450/750 V to BS 7211: 1994

Table A6.14 Dimensions and weights (450/750 V) single-core sheathed

Conductor size (mm^2)	Mean overall diameter Lower limit (mm)	Upper limit (mm)	Approximate weight (kg/km)
Single-core sheathed (ref. 6181B)			
1.0[a]	3.9	4.8	25
1.5[a]	4.2	5.0	33
2.5[a]	4.6	5.5	43
4	5.3	6.4	65
6	5.9	7.1	90
10	6.7	8.1	130
16	7.6	9.2	190
25	9.4	11.4	300
35	10.6	12.8	440

[a] Solid conductor

XL-LSF WIRING CABLES
300/500 V to BS 7211: 1994

Table A6.15 Dimensions and weights (300/500 V, single-core, flat 2-core, and 3-core with protective conductor)

Conductor size (mm²)	Protective conductor (mm²)	Mean overall diameter Lower limit (mm)	Mean overall diameter Upper limit (mm)	Approximate weight (kg/km)
Single-core (ref. 6241B)				
1.0[a]	1.0	4.1 × 5.2	5.0 × 6.3	
1.5[a]	1.0	4.4 × 5.4	5.3 × 6.6	
2-core flat (ref. 6242B)				
1.0[a]	1.0	4.1 × 7.6	5.0 × 9.1	65
1.5[a]	1.0	4.4 × 8.1	5.3 × 9.7	75
1.5	1.0	4.5 × 8.3	5.4 × 10.0	80
2.5[a]	1.5	4.9 × 9.3	6.0 × 11.2	110
2.5	1.5	5.0 × 9.5	6.1 × 11.4	120
4	1.5	5.5 × 10.4	6.7 × 12.6	150
6	2.5	6.2 × 12.0	7.5 × 14.6	215
10	4	7.3 × 14.5	8.8 × 17.6	330
16	6	8.4 × 17.0	10.1 × 20.5	490
3-core flat (ref. 6243B)				
1.0[a]	1.0	4.1 × 10.0	5.1 × 12.1	90
1.5[a]	1.0	4.4 × 10.7	5.3 × 12.9	110
2.5[a]	1.0	4.9 × 12.0	6.0 × 14.6	150
4	1.5	5.5 × 14.0	6.7 × 16.9	255
6	2.5	6.2 × 16.2	7.5 × 19.5	340
10	4	7.3 × 19.5	8.8 × 23.6	550
16	6	8.4 × 22.8	10.1 × 27.6	730

[a] Solid conductor

APPENDIX A7
Cables for Fixed Installation (Shipwiring and Offshore)

ELASTOMERIC SHIPWIRING POWER CABLES
600/1000 V to BS 6883: 1991

Main application: shipwiring and offshore

Table A7.1 Maximum a.c. continuous current ratings in ships and offshore

Conductor size (mm^2)	Single core (A)	2-core (A)	3- or 4-core (A)	Conductor size (mm^2)	Single core (A)	2-core (A)	3- or 4-core (A)
1.0	17	14	12	70	240	200	170
1.5	21	18	15	95	290	250	205
2.5	30	25	21	120	340	290	240
4	40	34	29	150	385	330	270
6	51	43	36	185	440	370	305
10	71	60	50	240	520	445	365
16	95	81	67	300	590	505	415
					d.c. a.c.		
25	125	105	89	400	690 670		
35	155	135	105	500	780 720		
50	190	165	135	630	890 780		

Note

The ratings are based on an ambient temperature of 45°C and a maximum conductor operating temperature of 90°C. They apply to a single circuit in which up to six cables may be bunched together. If there are more than six cables a derating factor of 0.85 should be applied. Considerably higher ratings may be possible for specific installation conditions (see table A7.2).

Rating factor for ambient temperature

°C	30	35	40	45	50	55	60	65	70	75	80
Rating factor	1.15	1.11	1.05	1.00	0.94	0.88	0.82	0.75	0.67	0.58	0.47

ELASTOMERIC SHIPWIRING POWER CABLES
600/1000 V to BS 6883: 1991

Table A7.2 Maximum a.c. continuous current ratings for specific installation conditions

Conductor size (mm^2)	In free air or on a perforated cable tray[a]					Clipped to a non-metallic surface[a]				
	Single-core cable		Multicore cable			Single-core cable		Multicore cable	
	Two cables single-phase or three or four cables 3-phase spaced (A)	Three cables 3-phase, flat and touching or trefoil (A)	One 2-core cable, single-phase (A)	One 3- or 4-core cable, 3-phase (A)		Two cables, single-phase (A)	3 or 4 cables, 3-phase (A)	One 2-core cable, single-phase (A)	One 3-core cable, 3-phase (A)
1	20	18	20	18		20	18	19	17
1.5	26	24	26	23		26	24	24	21
2.5	36	33	35	31		35	32	33	29
4	49	44	48	42		47	44	45	38
6	64	58	62	54		62	56	58	50
10	88	80	84	74		84	78	80	68
16	120	105	115	98		110	105	105	92
25	160	140	150	130		150	140	140	120
35	200	175	185	155		185	170	175	150
50	245	215	220	195		225	205	210	180
70	315	275	285	245		285	265	265	230
95	385	335	340	300		345	320	320	280
120	450	390	395	345		400	370	375	325
150	520	450	455	400		460	425	430	375
185	595	515	520	455		540	490	490	425

240	710	610		630	575		
300	820	700		720	660		
400	970			860	790	575	
500	1120		535	990	900	660	500
630	1310		620	1130	1040		575
800	1490			1270	1170		
1000	1680			1420	1300		

[a] See below for conditions applicable

Rating factor for ambient temperature

Ambient temperature (°C)	25	30	35	40	45	50	60	70	80
Rating factor	1.04	1	0.96	0.91	0.87	0.82	0.71	0.58	0.41

NOTES ON CURRENT RATINGS FOR SPECIFIC INSTALLATION CONDITIONS

Ambient and maximum cable temperature

Certain authorities do not regard the regulations governing shipwiring cables, e.g. the IEE Regulations for the Electrical and Electronic Equipment of Ships, and the current ratings so prescribed, as applicable to offshore applications. Consequently, higher ratings may be adopted which exploit the full thermal capabilities of the cable materials. Instead of an ambient temperature of 45°C, the ratings in table A7.2 are based on a temperature of 30°C.

General installation conditions

The current ratings in table A7.2 are based on the following installation conditions.

Metallic protection
As cables on platforms at sea are chiefly used in flameproof installations, they have a metallic protection layer which assists glanding and earthing arrangements. The ratings assume that the metallic covering is bonded at both ends, as required in hazardous locations. The covering is usually a galvanised steel wire braid but single-core cables have a non-magnetic (phosphor bronze) wire braid.

Iron or steel surfaces
The ratings assume that single-core cables are installed remotely from iron or ferrous concrete (other than cable supports) and that cables which carry 250 A or more are spaced from any steel deck or bulk head by a distance of at least 50 mm.

Isolation of circuits
One circuit is isolated from other circuits or groups of cables as follows:

(a) horizontal clearances of not less than the overall width of an individual circuit, except that the distance need not exceed 150 mm;
(b) horizontal clearances of not less than six times the overall diameter of an individual cable;
(c) vertical distance between circuits not less than 150 mm;
(d) if the number of circuits exceeds four they are installed in a horizontal plane.

Cables in free air or on a cable tray

The ratings given in columns 2–5 of table A7.2 are calculated assuming the following.

Single-core cables flat spaced (column 2)
(a) Cables are fixed by supports to the vertical surface of a wall, any supporting metalwork under the cables occupying less than 10% of the plan area.
(b) They are installed either vertically one above the other, or horizontally, the distance between cables being not less than the cable diameter D_e and the distance from the wall to the nearest cable not less than $0.5D_e$.*

Single-core cables touching (column 3)

Three single-core cables are installed in contact with one another in flat formation on a perforated cable tray or in trefoil formation either on a perforated cable tray or fixed to a vertical wall. When fixed to a wall condition (a) above for spaced cables applies. The distance from the wall to the nearest cable should be not less than $0.5D_e$ or, where two cables are equidistant from the wall, not less than $0.75D_e$.*

Multicore cables (columns 4 and 5)

Cables are installed singly either on a perforated cable tray or fixed to a vertical wall. When fixed to a wall, condition (a) above for spaced single-core cables applies, and the minimum distance between cable and wall is $0.3D_e$.*

Cables clipped to a surface (columns 6–9)

Cables are clipped direct to or lying on a non-metallic surface. Single-core cables are in contact with one another, and may be installed in trefoil formation.

Group rating factors

For rating factors of groups of cables other than those covered in table A7.2, reference should be made to other published data or to the cable manufacturer.

---

* See general installation conditions above relating to iron or steel surfaces.

ELASTOMERIC INSTRUMENTATION CABLES
150/250 V to BS 6883: 1991
Individually screened pairs and triples

Main application: shipwiring and offshore

Table A7.3 Dimensions and weights

Numbers of pairs	Conductor size (mm²)	Unarmoured Approximate diameter (mm)	Unarmoured Approximate weight (kg/km)	Armoured Approximate diameter (mm)	Armoured Approximate weight (kg/km)
1	0.75	8.6	110	12.9	250
2	0.75	13.3	240	18.3	490
3	0.75	14.1	300	19.1	560
7	0.75	18.8	560	24.4	930
12	0.75	25.1	880	31.1	1390
19	0.75	29.8	1350	37.3	2120
27	0.75	35.8	1890	43.9	2840
37	0.75	40.6	2510	48.8	3590

(cont.)

1	1.0	9.2	120	13.4	270
2	1.0	14.0	270	19.0	530
3	1.0	15.0	350	20.1	620
7	1.0	20.2	630	25.8	1030
12	1.0	26.8	1010	34.1	1660
19	1.0	34.7	1550	39.5	2380
27	1.0	38.2	2170	46.4	3200
37	1.0	43.2	2880	50.2	4060
1	1.5	9.9	140	14.1	310
2	1.5	15.2	320	20.4	600
3	1.5	16.2	410	21.6	710
7	1.5	21.7	750	27.4	1190
12	1.5	28.9	1200	36.5	1920
19	1.5	34.7	1840	42.5	2760
27	1.5	41.6	2570	50.2	3700
37	1.5	47.0	3410	56.1	4710
1	2.5	10.8	170	15.1	350
2	2.5	16.6	390	22.0	710
3	2.5	17.9	510	23.5	850
7	2.5	24.2	960	30.0	1460
12	2.5	32.4	1530	39.9	2350
19	2.5	38.4	2340	46.6	3410
27	2.5	46.5	3300	55.3	4650
37	2.5	52.4	4380	61.9	5920

A similar range of cables is available with collective screens

ELASTOMERIC INSTRUMENTATION CABLES
150/250 V to BS 6883: 1991
Individually screened pairs and triples

Main application: shipwiring and offshore

Table A7.3 continued

Numbers of triples	Conductor size (mm^2)	Unarmoured Approximate diameter (mm)	Unarmoured Approximate weight (kg/km)	Armoured Approximate diameter (mm)	Armoured Approximate weight (kg/km)
1	0.75	8.9	110	13.5	250
2	0.75	14.1	340	19.0	620
3	0.75	15.2	390	20.4	700
7	0.75	20.2	730	26.0	1170
12	0.75	27.1	1220	34.6	2000
1	1.0	9.7	120	14.0	280
2	1.0	15.1	390	20.2	690
3	1.0	16.0	450	21.4	770
7	1.0	21.5	840	27.2	1300
12	1.0	29.2	1420	36.5	2220
1	1.5	10.4	140	14.8	310
2	1.5	16.1	450	21.5	780
3	1.5	17.4	530	22.8	880
7	1.5	23.3	1010	29.2	1530
12	1.5	31.6	1700	39.1	2560
1	2.5	11.4	180	15.8	370
2	2.5	18.0	570	23.6	950
3	2.5	19.1	680	24.7	1080
7	2.5	26.0	1300	33.2	2000
12	2.5	35.2	2200	43.3	3230

A similar range of cables is available with collective screens

ELASTOMERIC SHIPWIRING POWER CABLES
600/1000 V to BS 6883: 1991

Table A7.4 Dimensions and weights

Numbers of cores × conductor size (mm^2)	Circular conductors			
	Unarmoured		Armoured	
	Approximate diameter (mm)	Approximate weight (kg/km)	Approximate diameter (mm)	Approximate weight (kg/km)
1 × 1.0	5.3	40	9.1	135
1 × 1.5	5.5	47	9.3	145
1 × 2.5	6.0	58	9.8	165
1 × 4	6.9	84	10.8	195
1 × 6	7.5	110	11.4	225
1 × 10	9.0	170	13.1	320
1 × 16	10.1	235	14.2	395
1 × 25	12.4	375	16.7	565
1 × 35	13.3	455	17.8	670
1 × 50	15.2	670	19.7	855
1 × 70	17.2	825	22.1	1120
1 × 95	19.2	1100	24.4	1440
1 × 120	21.4	1410	26.6	1770
1 × 150	23.5	1890	28.9	2120
1 × 185	26.0	2350	32.5	2700
1 × 240	29.2	3050	36.0	3410
1 × 300	32.3	3700	39.4	4160
1 × 400	36.4	4470	43.9	5280
1 × 500	40.2	5360	48.1	6480
1 × 630	44.1	6680	52.3	7830
2 × 1.0	8.2	86	12.3	240
2 × 1.5	8.9	105	13.0	260
2 × 2.5	9.8	147	13.9	320
2 × 4	12.0	225	16.3	440
2 × 6	13.1	290	17.6	530
2 × 10	16.0	440	20.7	750
2 × 16	18.3	645	23.4	990
2 × 25	22.9	1020	28.1	1480
2 × 35	24.7	1230	30.1	1710
2 × 50	28.4	1640	35.1	2400

(cont.)

ELASTOMERIC SHIPWIRING POWER CABLES
600/1000 V to BS 6883: 1991

Table A7.4 continued

Numbers of cores × conductor size (mm^2)	Circular conductors			
	Unarmoured		Armoured	
	Approximate diameter (mm)	Approximate weight (kg/km)	Approximate diameter (mm)	Approximate weight (kg/km)
2 × 70	32.3	2200	39.3	3110
2 × 95	37.0	2940	44.4	4000
2 × 120	40.8	3680	48.5	4860
2 × 150	45.1	4500	53.4	5870
2 × 185	50.0	5600	58.8	7170
2 × 240	56.6	7260	65.8	9090
2 × 300	62.7	9050	72.7	11140
3 × 1.0	8.9	100	13.0	270
3 × 1.5	9.4	130	13.5	300
3 × 2.5	10.4	180	14.7	370
3 × 4	12.7	270	17.0	500
3 × 6	13.9	355	18.4	610
3 × 10	17.2	560	22.1	890
3 × 16	19.5	790	24.6	1630
3 × 25	24.4	1290	29.8	1790
3 × 35	26.5	1590	33.1	2290
3 × 50	30.3	2120	37.2	2930
3 × 70	34.7	2900	41.9	3860
3 × 95	39.7	3790	47.6	5000
3 × 120	43.8	4790	52.0	6150
3 × 150	48.4	5860	57.0	7410
3 × 185	53.9	7310	62.9	9110
3 × 240	61.1	9500	70.6	11630
3 × 300	67.6	11960	77.9	14290
4 × 1.0	9.6	130	13.7	300
4 × 1.5	10.3	160	14.6	340
4 × 2.5	11.4	200	15.7	430
4 × 4	13.9	340	18.4	600
4 × 6	15.4	450	20.1	750

(*cont.*)

ELASTOMERIC SHIPWIRING POWER CABLES
600/1000 V to BS 6883: 1991

Table A7.4 continued

Numbers of cores × conductor size (mm^2)	Circular conductors			
	Unarmoured		Armoured	
	Approximate diameter (mm)	Approximate weight (kg/km)	Approximate diameter (mm)	Approximate weight (kg/km)
4 × 10	18.9	710	24.0	1080
4 × 16	21.7	1020	26.9	1450
4 × 25	27.1	1630	33.6	2380
4 × 35	29.4	2058	36.2	2880
4 × 50	33.9	2740	41.0	3670
4 × 70	38.5	3730	46.1	4840
4 × 95	44.4	4946	52.6	6300
4 × 120	48.9	6240	57.4	7770
4 × 150	54.1	7640	63.1	9410
4 × 185	60.0	9510	69.6	11520
4 × 240	68.0	12410	78.3	14840
4 × 300	75.5	15000	86.6	18260
5 × 1.0	10.5	150	14.7	300
5 × 1.5	11.3	190	15.5	370
5 × 2.5	12.6	275	16.9	470
5 × 4	15.3	425	20.0	680
6 × 1.0	11.4	180	15.7	360
6 × 1.5	12.4	235	16.7	420
6 × 2.5	13.7	330	18.2	545
6 × 4	16.7	505	21.6	765

ELASTOMERIC SHIPWIRING POWER CABLES
1.9/3.3 kV and 3.3/3.3 kV to BS 6883: 1991
Main application: shipwiring and offshore

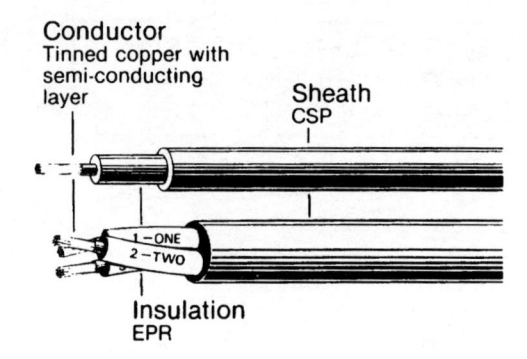

Conductor
Tinned copper with semi-conducting layer

Sheath
CSP

Insulation
EPR

Current ratings

All the relevant data for current ratings, conditions applicable to the ratings and rating factors are the same as for the lower voltage cables as given in table A7.1.

ELASTOMERIC POWER CABLES
1.9/3.3 kV 3.3/3.3 kV, 3.8/6.6 kV, 6.6/6.6 kV and 6.35/11 kV Non-radial field to BS 6883

Main application: shipwiring and offshore

Table A7.5 Dimensions and weights

Conductor size (mm²)	Single-core Approximate diameter (mm)	Single-core Approximate weight (kg/km)	Single-core Approximate diameter (mm)	Single-core Approximate weight (kg/km)	3-core Approximate diameter (mm)	3-core Approximate weight (kg/km)	3-core Approximate diameter (mm)	3-core Approximate weight (kg/km)
	1.9/3.3 kV		*3.3/3.3 kV*		*1.9/3.3 kV*		*3.3/3.3 kV*	
10	15.9	430	17.9	520	27.9	1270	33.1	1720
16	17.1	530	18.9	610	30.7	1600	35.7	2090
25	19.0	690	21.2	790	35.8	2310	39.7	2670
35	20.1	800	22.1	900	37.9	2660	42.2	3070
50	21.7	970	23.7	1080	41.3	3220	45.3	3660
70	23.9	1250	25.7	1340	46.0	4280	49.8	4650
95	26.5	1570	27.7	1650	51.6	5370	54.6	5760
120	28.4	1900	29.8	2000	55.8	6510	58.8	6950
150	30.2	2220	32.7	2450	59.9	7730	63.1	8170
185	33.4	2780	34.8	2900	64.6	9200	67.8	9720
240	36.4	3490	37.8	3600	71.7	11600	74.7	12200
300	39.2	4180	40.6	4330	77.7	14100	80.7	14700
400	43.7	5320	44.7	5420	—	—	—	—
500	48.1	6560	48.9	6660	—	—	—	—
630	52.3	8030	53.3	8170	—	—	—	—
	3.8/6.6 kV		*6.6/6.6 kV*		*3.8/6.6 kV*		*6.6/6.6 kV*	
16	19.9	660	—	—	37.7	2250	—	—
25	22.0	830	27.0	1140	42.0	2880	52.7	4110
35	23.1	950	27.9	1250	44.2	3270	55.0	4570
50	24.5	1120	29.5	1460	47.5	3890	58.1	5280
70	26.7	1410	32.7	1880	52.2	4910	63.0	6420

(cont.)

ELASTOMERIC POWER CABLES
1.9/3.3 kV 3.3/3.3 kV, 3.8/6.6 kV, 6.6/6.6 kV and 6.35/11 kV Non-radial field to BS 6883

Main application: shipwiring and offshore

Table A7.5 continued

Conductor size (mm²)	Single-core Approximate diameter (mm)	Single-core Approximate weight (kg/km)	Single-core Approximate diameter (mm)	Single-core Approximate weight (kg/km)	3-core Approximate diameter (mm)	3-core Approximate weight (kg/km)	3-core Approximate diameter (mm)	3-core Approximate weight (kg/km)
	3.8/6.6 kV		*6.6/6.6 kV*		*3.8/6.6 kV*		*6.6/6.6 kV*	
95	28.9	1730	34.6	2210	56.8	6070	67.2	7600
120	30.6	2060	36.6	2590	60.9	7270	72.0	8960
150	33.5	2510	38.5	2960	65.1	8440	75.7	10280
185	35.8	3000	40.6	3440	69.7	9990	80.6	12000
240	38.8	3690	43.8	4170	76.6	12600	87.5	14620
300	41.8	4420	46.6	4920	82.8	15100	93.7	17260
400	45.7	5540	50.7	6110	—	—	—	—
500	49.3	6700	54.5	7320	—	—	—	—
630	53.9	8240	58.5	8860	—	—	—	—
	6.35/11 kV				*6.35/11 kV*			
35	30.3	1430			60.2	5320		
50	32.8	1750			63.5	6030		
70	34.8	2060			67.9	7190		
95	37.0	2430			72.5	8460		
120	39.0	2800			76.9	9900		
150	40.7	3170			80.8	11230		
185	43.2	3680			85.7	13000		
240	46.2	4420			92.8	15700		
300	49.0	5190			98.8	18400		
400	53.1	6390			—	—		
500	56.9	7670			—	—		
630	60.9	9180			—	—		

Cables for Fixed Installation (Shipwiring and Offshore)

ELASTOMERIC HIGH VOLTAGE CABLES
1.8/3 kV to 8.7/15 kV to IEC 502

Main application: shipwiring and offshore

Conductor — Tinned Copper with semi-conducting layer *

Armour — Galvanised steel wire braid or phosphor bronze

* not at 1·8/3·3 kV

Core Screen *

Insulation — EPR

Sheath — CSP

Protection — CSP or PVC

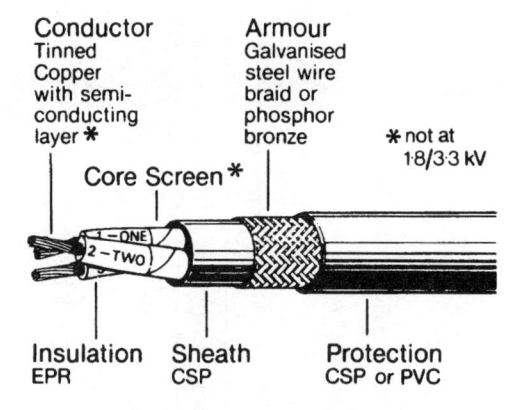

Current ratings

All the relevant data for current ratings, conditions applicable to the ratings and rating factors are the same as for the lower voltage cables as given in table A7.1.

ELASTOMERIC POWER CABLES
3.8/6.6 kV, 6.35/11 kV, 8.7/15 kV Radial field cables to BS 6883

Main application: shipwiring and offshore

Table A7.6 Dimensions and weights

Conductor size (mm²)	Single-core Unarmoured Approximate diameter (mm)	Single-core Unarmoured Approximate weight (kg/km)	Single-core Armoured Approximate diameter (mm)	Single-core Armoured Approximate weight (kg/km)	3-core Unarmoured Approximate diameter (mm)	3-core Unarmoured Approximate weight (kg/km)	3-core Armoured Approximate diameter (mm)	3-core Armoured Approximate weight (kg/km)
	3.8/6.6 kV		*3.8/6.6 kV*		*3.8/6.6 kV*		*3.8/6.6 kV*	
16	16.5	450	21.2	760	33.2	1590	40.2	2500
25	18.2	590	23.1	930	37.2	2050	44.4	3070
35	19.3	690	24.2	1050	39.3	2330	46.5	3420
50	20.6	870	25.6	1270	42.4	3060	50.0	4350
70	22.6	1120	27.8	1580	46.5	3820	54.5	5260
95	24.6	1400	30.2	1910	50.8	4850	59.3	6460
120	26.3	1710	31.9	2260	54.8	5910	63.4	7670
150	28.1	2010	34.6	2730	58.5	6930	67.4	9020
185	30.3	2420	36.9	3230	63.0	8340	72.3	10600
240	33.2	3070	40.1	3940	69.1	10500	78.9	13000
300	35.8	3720	42.9	4690	74.8	12500	85.2	15300
400	39.3	4730	46.8	5830	—	—	—	—
500	43.4	5860	51.2	7180	—	—	—	—
630	47.2	7230	55.4	8700	—	—	—	—
	6.35/11 kV		*6.35/11 kV*		*6.35/11 kV*		*6.35/11 kV*	
16	17.3	480	22.2	810	35.2	1700	42.2	2650
25	19.2	630	24.1	990	39.1	2150	46.4	3240
35	20.2	720	25.2	1120	41.2	2600	48.9	3790
50	21.6	920	26.6	1340	44.1	3180	52.0	4540
70	23.4	1160	28.6	1640	48.5	4140	56.7	5660

(cont.)

ELASTOMERIC POWER CABLES
3.8/6.6 kV, 6.35/11 kV, 8.7/15 kV Radial field cables to BS 6883

Main application: shipwiring and offshore

Table A7.6 continued

Conductor size (mm²)	Single-core Unarmoured Approximate diameter (mm)	Single-core Unarmoured Approximate weight (kg/km)	Single-core Armoured Approximate diameter (mm)	Single-core Armoured Approximate weight (kg/km)	3-core Unarmoured Approximate diameter (mm)	3-core Unarmoured Approximate weight (kg/km)	3-core Armoured Approximate diameter (mm)	3-core Armoured Approximate weight (kg/km)
	6.35/11 kV		*6.35/11 kV*		*6.35/11 kV*		*6.35/11 kV*	
95	25.4	1450	31.0	1980	52.6	5100	61.1	6760
120	27.3	1770	33.9	2470	56.7	6190	65.7	8030
150	28.9	2060	35.6	2820	60.4	7230	69.6	9360
185	31.1	2500	37.7	3310	64.9	8710	74.5	11100
240	34.0	3130	41.1	4050	71.1	11100	81.3	13700
300	36.6	3800	43.9	4810	76.7	13100	87.4	16100
400	40.2	4800	47.6	5920	—	—	—	—
500	43.8	5900	51.6	7230	—	—	—	—
630	47.6	7270	55.8	8740	—	—	—	—
	8.7/15 kV		*8.7/15 kV*		*8.7/15 kV*		*8.7/15 kV*	
25	21.7	740	26.7	1170	44.3	2650	52.1	2650
35	22.6	840	27.8	1300	46.4	2980	54.4	2980
50	23.8	1030	29.0	1510	49.4	3740	57.5	3740
70	25.8	1300	31.4	1850	53.7	4680	62.4	4680
95	27.8	1600	34.3	2310	57.7	5660	66.6	5660
120	29.9	1920	36.5	2720	62.1	6820	71.4	6820
150	31.5	2240	38.2	3080	65.6	7940	75.1	7940
185	33.5	2660	40.5	3540	70.2	9260	80.2	9260
240	36.4	3320	43.5	4310	76.2	11700	86.9	11700
300	39.0	3990	46.5	5100	82.0	13800	93.1	13800
400	42.6	5020	50.4	6320	—	—	—	—
500	46.2	6130	54.2	7560	—	—	—	—
630	50.2	7550	58.4	9110	—	—	—	—

APPENDIX A8
Flexible Cords and Cables

PVC INSULATED FLEXIBLE CORDS
300/300 V and 300/500 V to
BS 6500: 1994 and BS 6141: 1991

Main application: internal wiring and external connection of appliances

Insulated only 300/300 volt parallel twin

Conductors — Plain/tinned copper
Insulation — PVC transparent white, cream

Insulated only 300/500 volt single core

Conductor — Plain/tinned copper
Insulation — PVC coloured

Sheathed 300/300 volt, light cord

Insulation — PVC coloured
Conductors — Plain/tinned copper
Sheath — PVC coloured

Sheathed 300/500 volt ordinary cord

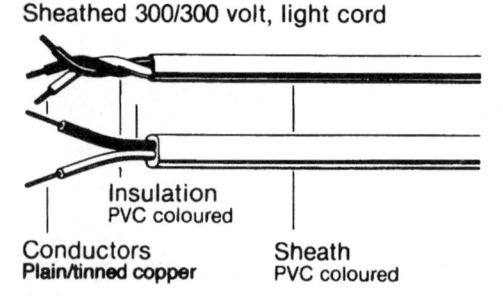

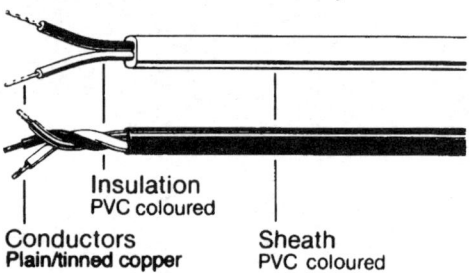

Insulation — PVC coloured
Conductors — Plain/tinned copper
Sheath — PVC coloured

Note
The 300/300 V parallel twin cable is not to a British Standard.

Table A8.1 Current ratings

Conductor size (mm^2)	Rating D.C. or single-phase a.c. (A)	3-phase a.c. (A)
0.5	3	3
0.75	6	6
1.0	10	10
1.25	13	–
1.5	16	16
2.5	25	20
4.0	32	25

Note
The ratings are based on an ambient temperature of 30°C and are in accordance with IEE Regulations. They are only applicable to cords laid straight, not to coils.

Rating factors for ambient temperature

Ambient temperature (°C)	30	35	40	45	50	55	60	70
General purpose cords	1.00	0.91	0.82	0.71	0.58	0.41		
Heat-resisting cords	1.00	1.00	1.00	1.00	1.00	0.96	0.83	0.47

> **PVC INSULATED FLEXIBLE CORDS**
> 300/300 V and 300/500 V to
> BS 6500: 1994 and BS 6141: 1991

Table A8.2 Dimensions and weights (PVC insulated only – 300/300 V and 300/500 V – BS 6500)

Conductor		300/300 V parallel twin (ref. 2812X) (not to BS)			HAR ref. HO5V-K[a] 300/500 V circular single-core (ref. 2491X)	
Size (mm^2)	Maximum diameter of wires (mm)	Mean dimensions minimum (mm)	maximum (mm)	Approximate weight (kg/km)	Maximum diameter (mm)	Approximate weight (kg/km)
0.5	0.21	2.5 × 5.0	3.0 × 6.0	22		
0.5	0.21				2.6	9
0.75	0.21	2.7 × 5.4	3.2 × 6.4	28		
0.75	0.21				2.8	11
1.0	0.21				3.0	14

[a] HO5V2-K Heat-resistant type

> **PVC INSULATED FLEXIBLE CORDS**
> 300/300 V and 300/500 V to
> BS 6500: 1994 and BS 6141: 1991

Table A8.3 Dimensions and weights (PVC insulated, PVC sheathed, 300/300 V, light cord – BS 6500 and BS 6141[a])

Conductor Size (mm^2)	Maximum diameter of wires	Mean dimensions Minimum (mm)	Mean dimensions Maximum (mm)	Approximate weight (kg/km)	
Flat 2-core (ref. 2192Y or HO3VVH2-F)					
0.5	0.21	3.0 × 4.8	3.6 × 6.0	27	
0.75	0.21	3.2 × 5.2	3.9 × 6.4	33	
Circular 2-core (ref. 2182Y or HO3VV-F2 and 2092Y)				2182Y	2092Y[a]
0.5	0.21	4.8	6.0	36	34
0.75	0.21	5.2	6.4	44	41
Circular 3-core (ref. 2183Y or HO3VV-F3 and 2093Y)				2183Y	2093Y[a]
0.5	0.21	5.0	6.2	44	39
0.75	0.21	5.4	6.8	54	49
Circular 4-core (ref. 2184Y or HO3VV-F4)					
0.5	0.21	5.6	6.8	53	
0.75	0.21	6.0	7.4	64	

[a] Heat-resisting circular to BS 6141; others to BS 6500

PVC INSULATED FLEXIBLE CORDS
300/300 V and 300/500 V to
BS 6500: 1994 and BS 6141: 1991

Table A8.4 Dimensions and weights (PVC insulated, PVC sheathed, 300/500 V ordinary cord) BS 6500, BS 6141[a] and BICC Polarflex[b]

Conductor		Dimensions		Approximate weight (kg/km)		
Size (mm^2)	Maximum diameter of wires (mm)	Minimum (mm)	Maximum (mm)			
Flat 2-core (ref. 3192Y or HO5VVH2-F)						
0.75	0.21	3.8 × 6.0	5.2 × 7.6	46		
1.0	0.21	3.9 × 6.4	5.2 × 8.0	48		
Circular 2-core (ref. 3182Y or HO5VV-F2, 3092Y and 3182Y AG[b])						
				3182Y	3092Y[a]	3182Y AG[b]
0.5	0.21	5.6	7.0	47	44	–
0.75	0.21	6.0	7.6	55	52	50
1.0	0.21	6.4	8.0	63	60	58
1.25	0.21	7.0	8.6	78	73	–
1.5	0.26	7.4	9.0	85	81	80
2.5	0.26	8.9	11.0	130	121	119
4.0	0.31	10.1	12.0	182	–	–
Circular 3-core (ref. 3183Y or HO5VV-F3, 3093Y and 3183Y AG[b])						
				3183Y	3093Y[a]	3183Y AG[b]
0.75	0.21	6.4	8.0	65	61	60
1.0	0.21	6.8	8.4	75	70	68
1.25	0.21	7.6	9.4	97	90	–
1.5	0.26	8.0	9.8	104	97	95
2.5	0.26	9.6	12.0	161	150	148
4.0	0.31	11.0	13.0	228	–	–
Circular 4-core (ref. 3184Y or HO5VV-F4, 3094Y and 3184Y AG[b])						
				3184Y	3094Y	3184Y AG[b]
0.75	0.21	6.8	8.6	79	74	72
1.0	0.21	7.6	9.4	95	88	86
1.5	0.26	9.0	11.0	133	124	121
2.5	0.26	10.5	13.0	196	182	177
4.0	0.31	12.0	14.0	280	–	–
Circular 5-core (ref. 3185Y or HO5VV-F5)						
0.75	0.21	7.8	9.6	101	–	–
1.0	0.21	8.3	10.0	118	–	–
1.5	0.26	10.0	12.0	171	–	–
2.5	0.26	11.5	14.0	258	–	–
4.0	0.31	13.5	15.5	359	–	–

[a] Heat resisting to BS 6141
[b] BICC Polarflex

Flexible Cords and Cables

ELASTOMERIC INSULATED FLEXIBLE CORDS
300/300 V, 300/500 V and 450/750 V to BS 6500: 1994

Sheathed and braided designs

Main application: mains supply or extension leads to appliances

Braided 300/300 volt
All tinned copper conductors

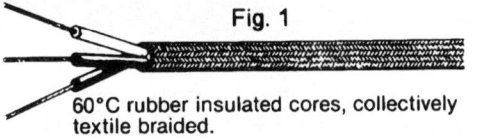

Fig. 1
60°C rubber insulated cores, collectively textile braided.

Fig. 2
60°C rubber insulated cores, rubber layer, collectively textile braided (UDF) semi-embedded

Braided 300/500 volt

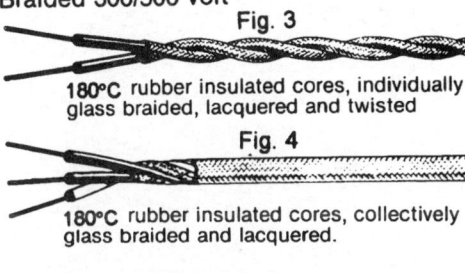

Fig. 3
180°C rubber insulated cores, individually glass braided, lacquered and twisted

Fig. 4
180°C rubber insulated cores, collectively glass braided and lacquered.

Sheathed 300/500 volt

Fig. 5
60°C rubber insulated cores, TR sheath

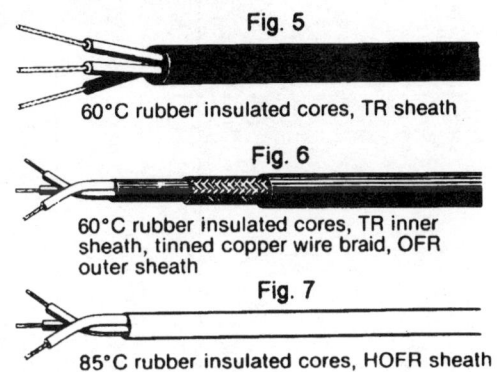

Fig. 6
60°C rubber insulated cores, TR inner sheath, tinned copper wire braid, OFR outer sheath

Sheathed 450/750 volt

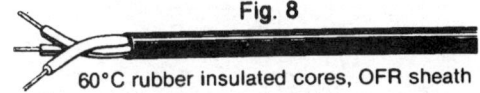

Fig. 8
60°C rubber insulated cores, OFR sheath

Fig. 7
85°C rubber insulated cores, HOFR sheath

Cords to fig. 6 are not supplied to BS 6500.

Current ratings

The ratings for PVC insulated cords in table A8.1 are applicable.

Rating factors for ambient temperature

Ambient temperature (°C)	30	35	40	45	50	55	60	70
60°C rubber	1.00	0.91	0.82	0.71	0.58	0.41		
85°C rubber	1.00	1.00	1.00	1.00	1.00	0.96	0.83	0.47

Ambient temperature (°C)	35–120	125	130	135	140
180°C rubber	1.0	0.96	0.85	0.74	0.60

ELASTOMERIC INSULATED FLEXIBLE CORDS
300/300 V, 300/500 V and 450/750 V to BS 6500: 1994

Table A8.5 Dimensions and weights (60°C insulation, braided, 300/300 V)

Conductor size (mm^2)	Circular (fig. 1) 2-core (ref. 2042) HO3RT-F2 Maximum diameter (mm)	Circular (fig. 1) 2-core (ref. 2042) HO3RT-F2 Approximate weight (kg/km)	Circular (fig. 1) 3-core (ref. 2043) HO3RT-F2 Maximum diameter (mm)	Circular (fig. 1) 3-core (ref. 2043) HO3RT-F2 Approximate weight (kg/km)	Circular (fig. 2) 2-core (ref. 2212) Maximum diameter (mm)	Circular (fig. 2) 2-core (ref. 2212) Approximate weight (kg/km)	Circular (fig. 2) 3-core (ref. 2213) Maximum diameter (mm)	Circular (fig. 2) 3-core (ref. 2213) Approximate weight (kg/km)
0.5	7.6	33	8.0	37	6.4	41	6.8	51
0.75	8.0	44	8.6	58	6.8	49	7.2	59
1.0	8.4	49	9.0	69	7.2	57	7.6	70
1.5	9.0	62	9.6	88	8.6	78	9.2	101

[a] For each braided core

ELASTOMERIC INSULATED FLEXIBLE CORDS
300/300 V, 300/500 V and 450/750 V to BS 6500:1994

Table A8.6 Dimensions and weights (180°C insulation, glass braided, 300/500 V)

Conductor size (mm^2)	Circular single-core (ref. 2771D) HO5SJ-K1		Twisted twin (fig. 3) (ref. 2782D) National type		Circular (fig. 4)[a]			
					2-core (ref. 2792D)		3-core (ref. 2793D)	
	Maximum diameter (mm)	Approximate weight (kg/km)	Maximum diameter (mm)	Approximate weight (kg/km)	Maximum diameter (mm)	Approximate weight (kg/km)	Maximum diameter (mm)	Approximate weight (kg/km)
0.5	3.4	10	6.8	22	5.3	29	5.8	37
0.75	3.6	13	7.2	28	5.8	37	6.2	47
1.0	3.8	16	7.4	34	6.2	45	6.6	58
1.5	4.3	23	8.6	50	7.6	60	8.1	83
2.5	5.0	35	10.0	76	8.8	85	9.4	135

[a] Not included in BS 6500

ELASTOMERIC INSULATED FLEXIBLE CORDS
300/300 V, 300/500 V and 450/750 V to BS 6500: 1994

Table A8.7 Dimensions and weights (60°C insulation, TRS sheathed ordinary cord, 300/500 V)

Conductor size (mm^2)	Circular (fig. 5)							
	2-core (ref. 3182) HO5RR-F2		3-core (ref. 3183) HO5RR-F3		4-core (ref. 3184) HO5RR-F4		5-core (ref. 3185) HO5RR-F5	
	Maximum diameter (mm)	Approximate weight (kg/km)	Maximum diameter (mm)	Approximate weight (kg/km)	Maximum diameter (mm)	Approximate weight (kg/km)	Maximum diameter (mm)	Approximate weight (kg/km)
0.5	7.8	50	8.2	59				
0.75[a]	8.2	61	8.8	75				
1.0[a]	8.8	74	9.2	87	9.6	90	11.0	112
1.5	10.5	109	11.0	130	10.0	106	11.5	131
2.5	12.5	155	13.0	187	12.5	163	13.5	197
					14.0	236	15.5	280

[a] Also available (2- and 3-core) with OFR sheath (cable reference 3182P, 3183P)

> **ELASTOMERIC INSULATED FLEXIBLE CORDS**
> 300/300 V, 300/500 V and 450/750 V to BS 6500: 1994

Table A8.8 Dimensions and weights (60°C insulation, sheathed, screened, OFR sheathed, 300/500 V)

| Conductor size (mm^2) | Circular, screened (fig. 6) ||||||
| | 2-core (ref. 3802P) || 3-core (ref. 3803P) || 4-core (ref. 3804P) ||
	Maximum diameter (mm)	Approximate weight (kg/km)	Maximum diameter (mm)	Approximate weight (kg/km)	Maximum diameter (mm)	Approximate weight (kg/km)
0.5	11.0	139	11.5	155	12.5	189
0.75	12.0	169	12.5	178	13.0	202
1.0	12.5	188	13.0	193	14.0	239
1.5	14.0	248	15.0	275	16.0	324
2.5	16.0	314	16.5	352	18.0	417

Table A8.9 Dimensions and weights (85°C insulation, HOFR sheathed, 300/500 V)

| Conductor size (mm^2) | Circular (fig. 7) ||||||
| | 2-core (ref. 3182TQ) || 3-core (ref. 3183TQ) || 4-core (ref. 3184TQ) ||
	Maximum diameter (mm)	Approximate weight (kg/km)	Maximum diameter (mm)	Approximate weight (kg/km)	Maximum diameter (mm)	Approximate weight (kg/km)
0.5	7.8	40	8.2	57		
0.75	8.2	50	8.8	74	9.6	80
1.0	8.8	61	9.2	88	10.0	107
1.5	10.5	89	11.0	113	12.5	151
2.5	12.5	129	13.0	165	14.0	217

ELASTOMERIC INSULATED FLEXIBLE CORDS
300/300 V, 300/500 V and 450/750 V to BS 6500: 1994

Table A8.10 Dimensions and weights (60°C insulation, OFR sheathed, 450/750 V)

Conductor size (mm^2)	Single-core (ref. 3981) HO7RN-F Maximum diameter (mm)	Single-core Approximate weight (kg/km)	2-core (ref. 3982) HO7RN-F2 Maximum diameter (mm)	2-core Approximate weight (kg/km)	3-core (ref. 3983) HO7RN-F3 Maximum diameter (mm)	3-core Approximate weight (kg/km)	4-core (ref. 3984) HO7RN-F4 Maximum diameter (mm)	4-core Approximate weight (kg/km)	5-core (ref. 3985) HO7RN-F5 Maximum diameter (mm)	5-core Approximate weight (kg/km)
1.0	—	—	10.5	109	11.5	125	12.5	163	13.5	188
1.5	7.2	60	11.5	138	12.5	168	13.5	206	15.0	249
2.5	8.0	78	13.5	194	14.5	233	15.5	280	17.0	355

Circular (fig. 8)

PVC INSULATED FLEXIBLE CORDS
300/500 V multicore control cables
Main application: power and control signal transmission where electrical screening or mechanical protection required

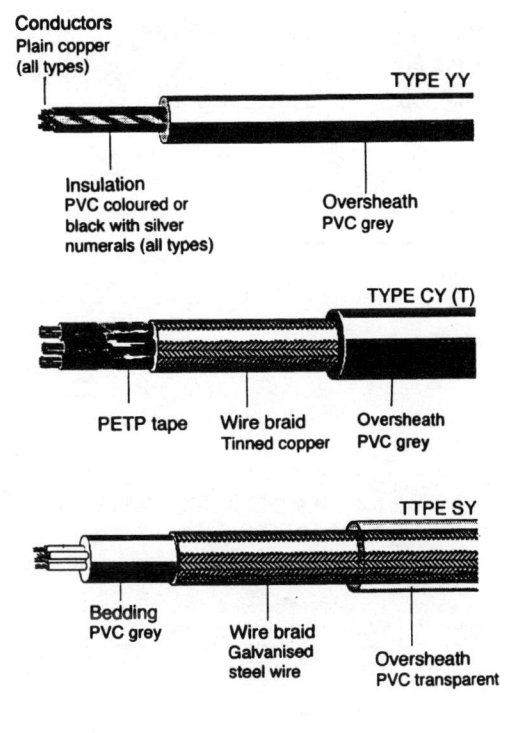

TYPE YY — Conductors: Plain copper (all types); Insulation: PVC coloured or black with silver numerals (all types); Oversheath: PVC grey

TYPE CY (T) — PETP tape; Wire braid Tinned copper; Oversheath PVC grey

TTPE SY — Bedding PVC grey; Wire braid Galvanised steel wire; Oversheath PVC transparent

Table A8.11 Conductors and ratings

Conductor details

Size (mm^2)	0.75	1.0	1.5	2.5
Maximum diameter (mm)	0.21	0.21	0.26	0.26

Current ratings
All the data given in table A8.1 are applicable for cables with up to five cores, or for cables in which the number of loaded conductors does not exceed the square-root of the total number of conductors in the cable. For other situations refer to the manufacturer.

PVC INSULATED FLEXIBLE CORDS
300/500 V multicore control cables

Table A8.12 Dimensions and weights

Number of cores	\multicolumn{3}{c}{0.5 mm^2}			\multicolumn{3}{c}{0.75 mm^2}			\multicolumn{3}{c}{1.0 mm^2}			\multicolumn{3}{c}{1.5 mm^2}			\multicolumn{3}{c}{2.5 mm^2}		
	A	B	C	A	B	C	A	B	C	A	B	C	A	B	C
Type SY															
2	5.7	9.1	104	6.1	9.5	115	6.5	9.8	127	7.0	10.4	150	8.9	12.5	212
3	6.0	9.4	113	6.5	9.9	127	6.8	10.2	150	7.4	10.8	168	9.4	13.1	249
4	6.5	9.9	126	7.0	10.4	150	7.4	10.8	168	8.1	11.7	196	10.3	14.2	293
5	7.1	10.5	150	7.6	11.2	174	8.1	11.7	197	9.0	12.7	241	11.5	15.4	353
Type CY (T)															
2	3.9	6.8	64	4.3	7.2	71	4.6	7.6	78	5.2	8.1	90	6.8	10.0	135
3	4.2	7.1	73	4.6	7.6	83	5.0	7.9	92	5.6	8.8	121	7.3	10.5	169
4	4.7	7.7	83	5.2	8.1	96	5.6	8.7	120	6.3	9.4	142	8.2	11.6	222
5	5.2	8.2	97	5.8	9.0	125	6.2	9.4	141	7.0	10.2	169	9.1	12.6	267
Type YY															
2	–	5.7	43	–	6.1	51	–	6.5	59	–	7.0	74	–	8.9	120
3	–	6.0	50	–	6.5	61	–	6.8	71	–	7.4	90	–	9.4	147
4	–	6.5	60	–	7.0	74	–	7.4	87	–	8.1	115	–	10.3	181
5	–	7.1	73	–	7.6	91	–	8.1	107	–	9.0	141	–	11.5	231

A: Nominal diameter under braid (mm)
B: Nominal overall diameter (mm)
C: Approximate cable weight (kg/km)

Flexible Cords and Cables

ELASTOMERIC INSULATED FLEXIBLE CABLES
450/750 V to BS 6007: 1993

Main application: services or mains supply where flexibility is required

60°C rubber insulated, PCP sheathed flexible cable and 85°C rubber insulated, HOFR sheathed flexible cable

180°C rubber insulated, glass braided, lacquered

Table A8.13 Current ratings and volt drop (60°C rubber insulated)

Conductor		Ratings			Volt drop per A/m						
Size (mm^2)	Maximum diameter of wires (mm)	One 2- or 3-core cable single-phase (A)	One 3-, 4- or 5- core cable, 3-phase (A)	Two single-core cables, single-phase (A)	D.C. (mV)	Single-phase a.c. (mV)	3-phase a.c. (mV)	Two single-core cables, touching			
								D.C. (mV)		Single-phase a.c. (mV)	
4	0.31	30	26		12	12	10				
6	0.31	39	34		7.8	7.8	6.7				
10	0.41	51	47		4.6	4.6	4.0				
16	0.41	73	63		2.9	2.9	2.5				
25	0.41	97	83		1.80	1.85	1.55				
35	0.41		102	140			1.15	1.31		1.32	
50	0.41		124	175			0.84	0.91		0.93	
70	0.51		158	216			0.58	0.64		0.67	
95	0.51		192	258			0.44	0.49		0.53	
120	0.51		222	302			0.36	0.38		0.43	
150	0.51		255	347			0.30	0.31		0.36	
185	0.51		291	394			0.26	0.25		0.32	
240	0.51		343	471			0.21	0.19		0.27	
300	0.51		394	541			0.19	0.15		0.24	
400	0.51			644				0.115		0.21	
500	0.61			738				0.09		0.20	
630	0.61			861				0.068		0.185	

Guidance on the ratings of single core 60°C rubber insulated cables may be obtained by multiplying the ratings in columns 6 and 7 of table A7.2 by a factor of 0.67.

Note: For details of the conditions on which the ratings are based and the rating factors see under table A8.14.

ELASTOMERIC INSULATED FLEXIBLE CABLES
450/750 V to BS 6007: 1993

Table A8.14 Current ratings and volt drop (85°C and 180°C rubber insulated)

Conductor		Ratings			Volt drop per A/m*					
Size (mm²)	Maximum diameter of wires (mm)	One 2- or 3-core cable single-phase (A)	One 3-, 4- or 5-core cable, 3-phase (A)	Two single-core cables, touching, single-phase a.c./d.c. (A)	D.C. (mV)	Single-phase a.c. (mV)	3-phase a.c. (mV)	Two single-core cables, touching, single phase		
								d.c. (mV)	a.c. (mV)	
4	0.31	41	36		13	13	11			
6	0.31	53	47		8.4	8.4	7.3			
10	0.41	73	64		5.0	5.0	4.3			
16	0.41	99	86		3.1	3.1	2.7			
25	0.41	131	114		2.0	2.0	1.7			
35	0.41		140	192			1.2	1.42	1.43	
50	0.41		170	240			0.91	0.99	1.01	
70	0.51		216	297			0.63	0.70	0.72	
95	0.51		262	354			0.48	0.53	0.56	
120	0.51		303	414			0.39	0.41	0.46	
150	0.51		348	476			0.32	0.33	0.38	
185	0.51		397	540			0.27	0.27	0.33	
240	0.51		467	645			0.22	0.21	0.28	
300	0.51		537	741			0.20	0.165	0.25	
400	0.51			885				0.125	0.22	
500	0.61			1017				0.098	0.20	
630	0.61			1190				0.073	0.190	

Guidance on the ratings of single core 85°C and 180°C rubber insulated cables may be obtained by multiplying the ratings in columns 6 and 7 of table A7.2 by a factor of 0.95.

Note: The rating given in tables A8.13 and A8.14 are for a single circuit and are based on an ambient temperature of 30°C. They are in accordance with the sixteenth edition of the IEE Wiring Regulations.

The ratings are for cables in free air but may also be used for cables resting on a surface. They are not applicable to cables in coils or wound on a drum.
* Only applicable to 85°C rubber insulated cables.

Ratings factors for ambient temperature

Ambient temperature (°C)	35	40	45	50	55	60	70
Factor for 60°C rubber insulation	0.91	0.82	0.71	0.58	0.41		
Factor for 85°C rubber insulation	0.95	0.90	0.85	0.80	0.74	0.67	0.52

Ambient temperature (°C)	35–85	90	100	110	120	130	140
Factor for 180°C rubber insulation	1.0	0.96	0.88	0.78	0.68	0.55	0.39

Group rating factors

For the rating factors of groups of cables other than those given in the tables, reference should be made to the cable manufacturer or to other published data.

ELASTOMERIC INSULATED FLEXIBLE CABLES
450/750 V to BS 6007: 1993

Table A8.15 Dimensions and weights (60°C rubber insulated, circular, OFR sheathed cables, 450/750 V)

Conductor size (mm^2)	\multicolumn{2}{c}{One (ref. 6381P) HO7RN-F1}		\multicolumn{2}{c}{Two (ref. 6382P) HO7RN-F1}	
	Maximum diameter (mm)	Approximate weight (kg/km)	Maximum diameter (mm)	Approximate weight (kg/km)
4	9.0	101	15.0	264
6	11.0	142	18.5	367
10	12.5	202	24.0	559
16	14.5	284	27.5	760
25	16.5	406	31.5	1160
35	18.5	553		
50	21.0	754		
70	23.5	957		
95	26.0	1308		
120	28.5	1610		
150	31.5	1960		
185	34.5	2370		
240	38.0	3078		
300	41.5	3775		
400	46.5	4875		
500	51.5	5980		
630[a]	56.5	7450		

(*cont.*)

ELASTOMERIC INSULATED FLEXIBLE CABLES
450/750 V to BS 6007: 1993

Table A8.15 continued

Conductor size (mm²)	Three (ref. 6383P) HO7RN-F3 Maximum diameter (mm)	Three (ref. 6383P) HO7RN-F3 Approximate weight (kg/km)	Four (ref. 6384P) HO7RN-F4 Maximum diameter (mm)	Four (ref. 6384P) HO7RN-F4 Approximate weight (kg/km)	Five (ref. 6385P) HO7RN-F5 Maximum diameter (mm)	Five (ref. 6385P) HO7RN-F5 Approximate weight (kg/km)
4	16.0	330	18.0	406	19.5	509
6	20.0	470	22.0	560	24.5	673
10	25.5	764	28.0	947	30.5	1119
16	29.5	1035	32.0	1292	35.5	1568
25	34.0	1491	37.5	1872	41.5	2420
35	38.0	1940	42.0	2415		
50	44.0	2680	48.5	3419		
70	49.5	3564	54.5	4462		
95	54.0	4860	60.5	6066		
120	59.0	5832	65.6	7367		
150	66.5	7223	74.0	9470		
185	71.5	8667	79.5	11261		
240	81.0	11448	90.0	14756		
300	89.5	14148	99.5	18367		

Notes
[a] National type.
(a) For flexible cords below 4 mm² see tables A8.7 and A8.10.
(b) For conductor details see table A8.13.

ELASTOMERIC INSULATED FLEXIBLE CABLES
450/750 V to BS 6007: 1993

Table A8.16 Dimensions and weights (90°C rubber insulated, circular, HOFR sheathed cables, 450/750 V)

Conductor size (mm^2)	One (ref. 6381TQ) Maximum diameter (mm)	One (ref. 6381TQ) Approximate weight (kg/km)	Two (ref. 6382TQ) Maximum diameter (mm)	Two (ref. 6382TQ) Approximate weight (kg/km)
4	9.0	99	15.0	262
6	11.0	140	18.5	360
10	12.5	200	24.0	550
16	14.5	280	27.5	750
25	16.5	400	31.5	1150
35	18.5	550		
50	21.0	750		
70	23.5	950		
95	26.0	1300		
120	28.5	1600		
150	31.5	1950		
185	34.5	2350		
240	38.0	3050		
300	41.5	3750		
400	46.5	4850		
500	51.5	5950		
630	56.5	7400		

(*cont.*)

> **ELASTOMERIC INSULATED**
> **FLEXIBLE CABLES**
> 450/750 V to BS 6007: 1993

Table A8.16 continued

Conductor size (mm^2)	Three (ref. 6383TQ) Maximum diameter (mm)	Three (ref. 6383TQ) Approximate weight (kg/km)	Four (ref. 6384TQ) Maximum diameter (mm)	Four (ref. 6384TQ) Approximate weight (kg/km)	Five (ref. 6385TQ) Maximum diameter (mm)	Five (ref. 6385TQ) Approximate weight (kg/km)
4	16.0	330	18.0	390	19.5	500
6	20.0	440	22.0	550	24.5	660
10	25.5	700	28.0	900	30.5	1086
16	29.5	950	32.0	1250	35.5	1508
25	34.0	1400	37.5	1800	41.5	2350
35	38.0	1800	42.0	2300		
50	44.0	2500	48.5	3200		
70	49.5	3300	54.5	4250		
95	54.0	4500	60.5	5750		
120	59.0	5400	65.5	6950		
150	66.5	6750	74.0	8850		
185	71.5	8100	79.5	10400		
240	81.0	10600	90.0	13600		
300	89.5	13100	99.5	16850		

Notes
(a) For flexible cords below 4 mm^2 see table A8.9.
(b) For conductor details see table A8.13.

ELASTOMERIC INSULATED FLEXIBLE CABLES
450/750 V to BS 6007: 1993

Table A8.17 Dimensions and weights (180°C rubber insulated, glass braided, 300/500 V)

Conductor size (mm^2)	Single-core circular HO5SJ-K Maximum diameter (mm)	Approximate weight (kg/km)	Two-core twisted National type Maximum diameter (mm)	Approximate weight (kg/km)
4	5.6	51	11.2	105
6	6.2	77	12.4	159
10	8.2	126	16.4	260
16	9.6	188	19.2	388

Notes
(a) For sizes of 0.5 to 2.5 mm^2 see table A8.6.
(b) For conductor details see table A8.13.

HEAT RESISTING FLEXIBLE CABLES FOR 250°C OPERATION
(BICC Intemp 250*)
600/100 V
* Registered trade mark

Main application: flexible cables for industrial control and supply circuits

SINGLE CORE

Conductor: Nickel plated copper
Primary Insulation: PTFE
Secondary Insulation: Glass/mica tape
Covering: Glass braid with heat and abrasion resistant finish

MULTICORE

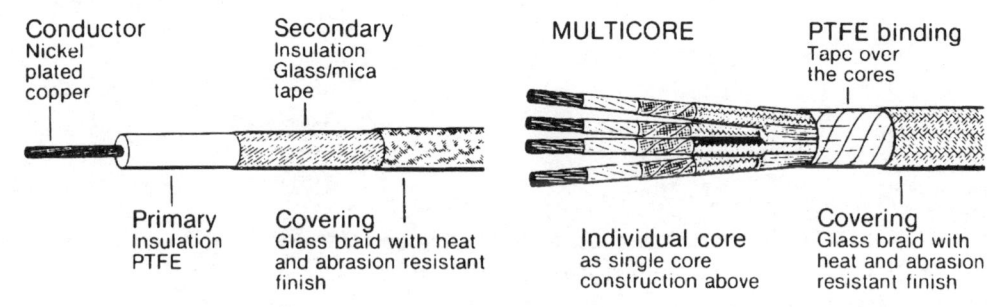

Individual core as single core construction above
PTFE binding: Tape over the cores
Covering: Glass braid with heat and abrasion resistant finish

Table A8.18 Range and dimensions

Conductor		Single-core cable			Maximum number of cores in multicore cable
Size (mm²)	Maximum diameter of wires (mm)	Approximate diameter (mm)	Approximate weight (kg/km)	Minimum bending radius (mm)	
1.0	0.21	3.91	25	39	28
1.5	0.26	4.19	32	42	25
2.5	0.26	4.59	44	46	21
4	0.31	5.19	59	52	16
6	0.31	6.30	84	63	12
10	0.41	7.20	123	72	9
16	0.41	9.24	207	92	5
25	0.41	10.64	299	106	4

The above are standard sizes and other sizes up to 150 mm² are also produced.

For situations where mechanical protection is required, a range of stainless steel wire braided Intemp single and multicore cables is available.

> **HEAT RESISTING
> FLEXIBLE CABLES FOR
> 250°C OPERATION**

Table A8.19 Current ratings

Conductor size (mm^2)	Resistance at 20°C (Ω/km)	Maximum continuous current (A)
1.0	20.0	25
1.5	13.7	40
2.5	8.21	54
4	5.09	74
6	3.39	98
10	1.95	135
16	1.24	180
25	0.795	240

Notes
(a) The above ratings are based on an ambient temperature of up to 80°C and a maximum continuous conductor temperature of 250°C.
(b) It is assumed that the cables cannot be touched and are not in contact with heat sensitive materials.
(c) Because the current rating is high, volt drop and power loss in the cable will be high and special fusing may be required.
(d) The ratings apply to single cables freely ventilated in air. See below for group rating factors.
(e) If the high current approach is not desirable, the ratings for a temperature rise of 50°C may be taken as 54% of the figures in table A8.19. Such ratings need not be corrected for ambient temperature up to 200°C.

Rating factors for ambient temperature

Ambient temperature (°C)	100	120	140	160	180	200	220	240
Rating factor	0.94	0.87	0.80	0.73	0.64	0.54	0.42	0.24

Group rating factors

Number of cables	2	5	10	15	20	25
Rating factor	0.8	0.6	0.45	0.4	0.36	0.33

Flexible Cords and Cables

EPR INSULATED LINAFLEX FLAT FLEXIBLE CORDS (300/500 V) AND CABLES (450/750 V)
(BICC designs)

Main appliction: industrial installations and supply to mobile equipment particularly if required to be wound on reeling drums or festooned

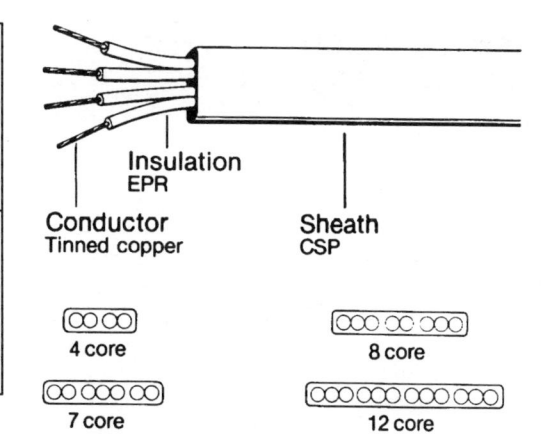

Conductor — Tinned copper
Insulation — EPR
Sheath — CSP

4 core
7 core
8 core
12 core

Table A8.20 Current ratings

Conductor size (mm^2)	Current rating (A)
Flexible cords	
1.5	21
2.5	28
4	39
Flexible cables	
6	50
10	70
16	92
25	125
35	150
50	185
70	235
95	280
120	325

The ratings quoted are based on an ambient temperature of 30°C and a maximum operating temperature of 85°C with three conductors loaded. They are for cables in free air, but may also be used for cables resting on a surface. They are not applicable to cable in coils or wound on a drum. Rating factors for ambient temperature other than 30°C are given under table A8.14 (85°C rubber).

EPR INSULATED LINAFLEX FLAT FLEXIBLE CORDS (300/500 V) AND CABLES (450/750 V)
(BICC designs)

Table 8.21 Range and dimensions

Conductor		Dimensions			
Size (mm^2)	Maximum diameter of wires (mm)	Minor (mm)	Major (mm)	Minor (mm)	Major (mm)

Flexible cords

		4-core		7-core	
1.5	0.26			6.4	31.2
2.5	0.26	7.8	22.3	8.0	37.0
4	0.31	9.0	25.6	9.2	42.3

		8-core		12-core	
1.5	0.26			7.4	51.4
2.5	0.26	8.0	41.2	9.0	60.8

Flexible cables

		4-core		7-core	
6	0.31	9.8	28.7	10.2	48.1
10	0.41	12.2	35.9	14.0	61.4
16	0.41	13.5	41.2	16.1	71.4
25	0.41	15.6	47.8		
35	0.41	17.7	53.8		
50	0.41	20.9	64.0		
70	0.51	23.0	71.6		
95	0.51	26.5	81.6		
120	0.51	29.5	91.6		

APPENDIX A9
Industrial Cables for Special Applications

WELDING CABLES
to BS 638, Part 4: 1979

Main application: low voltage leads for connection to automatic or hand-held metal arc welding electrodes and for the earthing return leads

Conductor
Tinned copper or aluminium (multi-bunched)

Insulation
85°C rubber

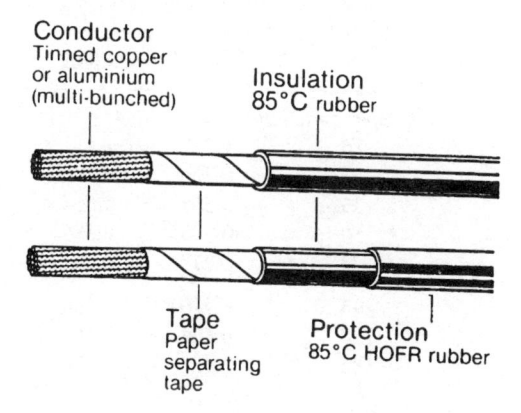

Tape
Paper separating tape

Protection
85°C HOFR rubber

Table A9.1 Range and dimensions

Conductor size (mm^2)	Diameter Minimum (mm)	Diameter Maximum (mm)	Approximate weight Design[a] (kg/km)	Approximate weight Design[a] (kg/km)
Copper conductor			*0361T*	*0361TQ*
16	8.8	11.5	225	230
25	10.0	13.0	320	325
35	11.0	14.5	430	440
50	13.0	17.0	590	600
70	15.0	19.5	820	830
95	17.0	22.0	1090	1110
120	19.0	24.0	1350	1370
185	23.0	29.0	2030	2050
Aluminium conductor			*0361T (A1)*	*0361TQ (A1)*
25	10.0	13.0	165	170
35	11.0	14.5	210	215
50	13.0	17.0	285	295
70	15.0	19.0	370	385
95	17.0	21.5	480	500
120	19.0	24.5	620	640
150	21.0	26.5	740	760
240	26.0	33.0	1150	1180

[a] T signifies 85°C rubber insulation; TQ signifies 85°C rubber insulation and 85°C HOFR covering.

WELDING CABLES
to BS 638, Part 4: 1979

Table A9.2 Resistance and volt drop

Conductor size (mm²)	Maximum resistance (Ω/km) at 20°C — Tinned	Maximum resistance (Ω/km) at 20°C — Plain	Voltage drop – d.c. (V/100 A/10 m of cable) 20°C	60°C	85°C
Copper conductor					
16	1.24	1.21	1.24	1.43	1.56
25	0.795	0.780	0.80	0.92	1.00
35	0.565	0.554	0.57	0.65	0.71
50	0.393	0.386	0.39	0.46	0.49
70	0.277	0.272	0.28	0.32	0.35
95	0.210	0.206	0.21	0.24	0.26
120	0.164	0.161	0.16	0.19	0.21
185	0.108	0.106	0.11	0.13	0.14
Aluminium conductor					
25		1.248	1.25	1.45	1.58
35		0.886	0.89	1.03	1.12
50		0.616	0.62	0.72	0.78
70		0.440	0.44	0.51	0.56
95		0.326	0.33	0.38	0.41
120		0.254	0.25	0.30	0.32
150		0.208	0.21	0.24	0.26
240		0.126	0.13	0.15	0.16

Notes
(a) Excessive voltage drop may occur if long lengths of cable are required. Under such circumstances much larger conductors are required than dictated by rating but for flexibility the final length to the electrode may revert to the size appropriate to rating.
(b) The corresponding voltage drops when using a.c. may be much higher depending on the configuration of the cables.

WELDING CABLES
to BS 638, Part 4: 1979

Table A9.3 Current ratings for repeat cycle operation based on 5 minute period

Conductor size (mm^2)	Duty cycle 100% (A)	85% (A)	60% (A)	20% (A)
Copper conductor (tinned)				
10	100	101	106	143
16	135	138	148	212
25	180	186	204	305
35	225	235	260	400
50	285	299	336	529
70	355	375	426	682
95	430	456	523	850
120	500	532	613	1006
185	665	619	826	1374
Aluminium conductor				
25	140	143	153	218
35	175	180	196	289
50	225	234	257	389
70	275	288	322	502
95	335	353	399	633
120	390	412	469	755
150	455	482	552	869
240	600	640	742	1228

Notes
(a) The above ratings are based on an ambient temperature of 25°C and a maximum conductor operating temperature of 85°C.
(b) High operating temperatures make the cable too hot to handle and reduce the expected service life. Under severe conditions, if a long service life is not expected, e.g. because of the possibility of mechanical damage, or where a high surface temperature can be tolerated, the current ratings for 25°C ambient may be used up to 40°C.

Rating factors

Ambient temperature (°C)	25	30	35	40	45
Rating factor	1.0	0.96	0.91	0.87	0.82

Duty cycle

Current ratings have to be related to the period of the operating cycle, defined as the percentage time per 5 minute period that the cable is operated. The classification of duty cycles adopted is as follows:

Automatic	up to 100%
Semi-automatic	30%–85%
Manual	30%–60%
Intermittent or occasional	up to 30%

APPENDIX A10
Mining Cables

UNARMOURED FLEXIBLE TRAILING CABLES
640/1100 V

Main application: supplies to machinery at underground coal faces (British Coal Specification 188)

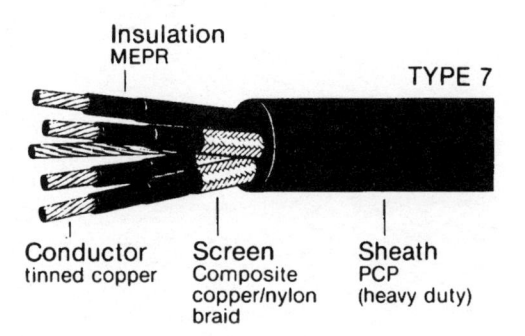

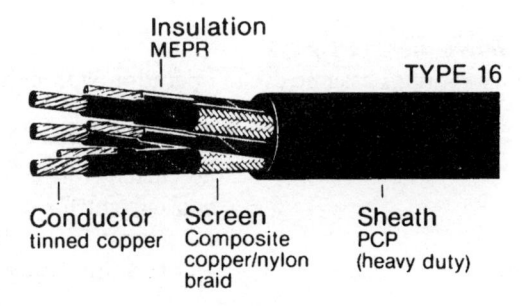

TYPE 7
- Insulation MEPR
- Conductor tinned copper
- Screen Composite copper/nylon braid
- Sheath PCP (heavy duty)

TYPE 16
- Insulation MEPR
- Conductor tinned copper
- Screen Composite copper/nylon braid
- Sheath PCP (heavy duty)

Table A10.1 Construction

Type	Power cores Number	Power cores Screen	Pilot cores Number	Pilot cores Screen	Earth conductor Number	Earth conductor Screen
7	3	Copper/nylon	1	Unscreened	1	Bare (in centre)
7M	3	Copper/nylon	1	Copper/nylon	1	Bare (in centre)
7S	3	Copper/nylon	1 unit[a]	Unscreened	1	Bare (in centre)
10	3	Condg rubber	1	Condg rubber	1	In cradle
11[b]	3	Copper/nylon	1	Copper/nylon	–	–
14	3	Copper/nylon	1	Unscreened	1	Insulated
16	3	Copper/nylon	1	Unscreened	2	Insulated

[a] This sheathed pilot unit contains three unscreened cores
[b] A PETP tape is applied over the conductors

> **UNARMOURED FLEXIBLE
> TRAILING CABLES
> 640/1100 V**

Table A10.2 Current ratings

Power conductor size (mm^2)	Continuous rating (A)	Intermittent rating (A)
16	85	90
25	110	120
35	135	150
50	170	190
70	205	235
95	250	295
120	295	350

Ambient temperature
The ratings are based on an ambient temperature of 25°C.

Intermittent rating
The ratings assume cyclic operation with conditions not more severe than

full current for   40 minutes
no current for   10–15 minutes
half current for   40 minutes
no current for   10–15 minutes
and repetitive cycles of the above

Rating factors for ambient temperature

Ambient temperature (°C)	30	35	40	45	50	55	60
Rating factor	0.93	0.87	0.80	0.73	0.66	0.57	0.48

Note
The above ratings are not applicable to cables in coils or on drums or where circumstances reduce heat dissipation from the cable.

UNARMOURED FLEXIBLE TRAILING CABLES
640/1100 V, 3800/6600 V and 6350/11000 V

Table A10.3 Range and dimensions

Power conductor size (mm^2)	Power (number/diameter)	Pilot (number/diameter)	Earth (number/diameter)	Minimum (mm)	Maximum (mm)	Approximate weight (kg/km)
640/1100 V						
Type 7						
16	126/0.40	126/0.40	126/0.40	35.8	38.6	2310
25	196/0.40	126/0.40	126/0.40	39.7	42.6	2960
35	276/0.40	126/0.40	147/0.40	43.1	46.3	3550
50	396/0.40	196/0.40	196/0.40	48.5	51.8	4600
70	360/0.40	276/0.40	276/0.40	55.1	58.8	6040
95	475/0.50	396/0.40	396/0.40	62.4	66.1	7880
120	608/0.50	360/0.50	396/0.40	68.0	72.5	10500
Type 7M						
16	126/0.40	126/0.40	126/0.40	35.8	38.6	2548
25	196/0.40	196/0.40	126/0.40	39.7	42.9	3163
35	276/0.40	276/0.40	147/0.40	43.1	46.3	3841
50	396/0.40	396/0.40	196/0.40	48.5	51.8	5019
70	360/0.50	360/0.50	276/0.40	55.1	58.8	6498
95	475/0.50	475/0.50	396/0.40	62.4	66.1	8222
120	608/0.50	608/0.50	396.0.50	68.0	72.5	9951
Type 7S						
50	396/0.40	56/0.30	196/0.40	48.5	51.8	4551
70	360/0.50	84/0.30	276/0.40	55.1	58.8	5916
95	475/0.50	84/0.30	396/0.40	62.4	66.1	7576
120	608/0.50	80/0.40	396/0.40	68.0	72.5	9099

(*cont.*)

UNARMOURED FLEXIBLE TRAILING CABLES
640/1100 V, 3800/6600 V and 6350/11000 V

Table A10.3 continued

Power conductor size (mm²)	Conductor formation Power (number/diameter)	Pilot (number/diameter)	Earth (number/diameter)	Diameter Minimum (mm)	Maximum (mm)	Approximate weight (kg/km)
Type 11						
16	126/0.40	126/0.40	—	30.9	33.0	1958
Type 14						
25	196/0.40	126/0.40	196/0.40	43.2	46.5	3450
35	276/0.40	126/0.40	276/0.40	47.3	50.7	4210
50	396/0.40	196/0.40	396/0.40	53.7	57.6	5550
70	360/0.50	276/0.40	360/0.50	61.2	65.0	7270
95	475/0.50	396/0.40	475/0.50	69.3	73.9	9560
Type 16						
25	196/0.40	80/0.40	80/0.40	36.5	39.2	2700
35	276/0.40	80/0.40	80/0.40	39.5	42.7	3170
50	396/0.40	126/0.40	126/0.40	43.9	47.0	4120
70	360/0.50	196/0.40	196/0.40	49.6	53.3	5490
95	475/0.50	276/0.40	276/0.40	56.2	63.0	8200
3800/6600 V						
Type 730						
35	276/0.40	—	—	57.5	60.5	—
50	396/0.40	—	—	62.0	65.0	—
70	360/0.50	—	—	67.6	71.4	—
95	475/0.50	—	—	73.1	76.9	—
120	608/0.50	—	—	76.9	80.9	—
150	758/0.50	—	—	82.1	86.1	—

6350/11000 V Type 830					
50	396/0.40	—	—	75.0	83.0
70	360/0.50	—	—	80.6	86.0
95	475/0.50	—	—	85.9	91.0
120	608/0.50	—	—	89.9	96.5
150	756/0.50	—	—	93.3	103.0

886   Electric Cables Handbook

PLIABLE ARMOURED FLEXIBLE TRAILING CABLES
320/550 V to 3800/6600 V

Main application: supplies to underground services and equipment moved occasionally (British Coal Specification 504)

TYPE 71

- Conductor tinned copper
- Inner sheath PCP
- Insulation MEPR
- Pliable armour Galvanised steel strands (includes some copper strands).

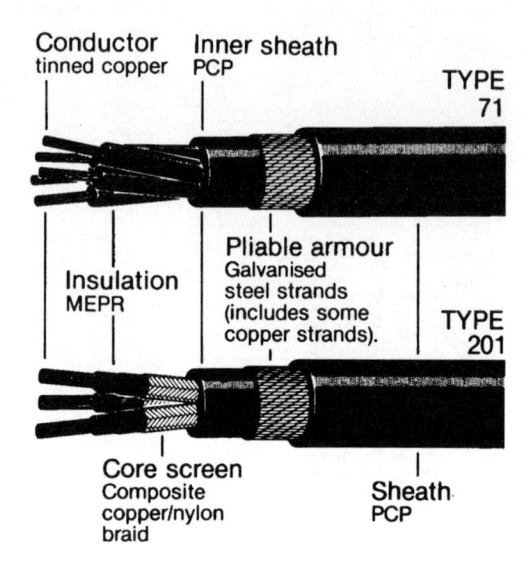

TYPE 201

- Core screen Composite copper/nylon braid
- Sheath PCP

Table A10.4 Construction

Type	Voltage	Insulation	Number of cores	Number and screen on power cables	Earth core
62	640/1100	MEPR	2	2, copper/nylon	–
63	640/1100	MEPR	3	3, copper/nylon	–
64	640/1100	MEPR	4	4, copper/nylon	–
70	320/550	MEPR	4	3, unscreened	Unscreened
71	320/550	MEPR	5	4, unscreened	Unscreened
201	640/1100	MEPR	3	3, copper/nylon	–
211	640/1100	MEPR	4	3, copper/nylon	Unscreened
321	1900/3300	MEPR	4	3, unscreened	Unscreened
331	1900/3300	MEPR	4	3, copper/nylon	Unscreened
631	3800/6600	MEPR	4	3, copper/nylon	Unscreened

Table A10.5 Current ratings

Conductor size (mm^2)	10	16	25	35	50	70	95	120
Continuous rating (A)	63	85	110	135	170	205	250	295

Note

These ratings do not apply to types 62–71 which are control and lighting cables with a rating of 28 A. For rating factors for ambient temperature see table A10.2.

PLIABLE ARMOURED FLEXIBLE TRAILING CABLES
320/550 V to 3800/6600 V

Table A10.6 Range and dimensions

Power conductor size (mm^2)	Conductor formation Power (number/mm)	Conductor formation Earth (number/mm)	Armour[a] (number/mm)	Diameter Minimum (mm)	Diameter Maximum (mm)	Approximate weight (kg/km)
Type 62						
4	56/0.30	–	7/0.45	23.9	26.4	1070
Type 63						
4	56/0.30	–	7/0.45	24.8	27.3	1230
Type 64						
4	56/0.30	–	7/0.45	26.4	28.9	1430
Type 70						
4	56/0.30	–	7/0.45	23.5	26.0	1060
Type 71						
4	56/0.30	–	7/0.45	24.9	27.4	1220
Type 201						
10	80/0.40	–	7/0.71	37.3	39.8	2780
16	126/0.40	–	7/0.71	40.6	43.4	3310
25	196/0.40	–	7/0.71	44.6	47.4	4050
35	276/0.40	–	7/0.71	48.3	51.1	4780
50	396/0.40	–	7/0.71	53.6	57.4	6080
70	360/0.50	–	7/0.90	60.7	64.5	7740
95	475/0.50	–	7/1.25	71.8	75.8	10940
120	608/0.50	–	7/1.25	76.7	81.0	12690
Type 211						
10	80/0.40	80/0.40	7/0.71	39.8	42.3	3160
16	126/0.40	126/0.40	7/0.71	44.0	46.8	3860
25	196/0.40	126/0.40	7/0.71	48.5	51.3	4680
35	276/0.40	196/0.40	7/0.71	52.7	56.5	5590
50	396/0.40	276/0.40	7/0.90	59.7	63.5	7330
70	360/0.50	396/0.40	7/0.90	68.8	72.8	9620
95	475/0.50	360/0.50	7/1.25	80.6	84.9	13450
120	608/0.50	360/0.50	7/1.25	86.2	90.5	16460
Type 321						
35	276/0.40	196/0.40	7/0.90	60.3	64.1	6860
50	396/0.40	276/0.40	7/0.90	68.3	72.3	8700
70	360/0.50	396/0.40	7/0.90	74.4	78.4	10400
95	475/0.50	360/0.50	7/1.25	84.9	89.2	14950
120	608/0.50	360/0.50	7/1.25	89.4	93.7	15950

(*cont.*)

**PLIABLE ARMOURED
FLEXIBLE TRAILING CABLES**
320/550 V to 3800/6600 V

Table A10.6 continued

Power conductor size (mm^2)	Conductor formation Power (number/mm)	Earth (number/mm)	Armour[a] (number/mm)	Diameter Minimum (mm)	Maximum (mm)	Approximate weight (kg/km)
Type 331						
25	196/0.40	126/0.40	7/0.90	57.8	61.6	6440
35	276/0.40	196/0.40	7/0.90	64.6	68.4	8010
50	396/9.40	276/0.40	7/0.90	69.8	73.8	9170
70	360/0.50	396/0.40	7/1.25	80.3	84.6	12740
95	475/0.50	360/0.50	7/1.25	86.4	90.7	14800
120	608/0.50	360/0.50	7/1.25	90.9	95.2	16850
Type 631						
50	396/0.40	276/0.40	7/1.25	86.7	91.0	13830
70	360/0.50	396/0.40	7/1.25	92.8	97.1	16220

[a] Some sizes have copper wires in the armour.

PLIABLE ARMOURED FLEXIBLE MULTICORE AUXILIARY CABLES
320/550 V

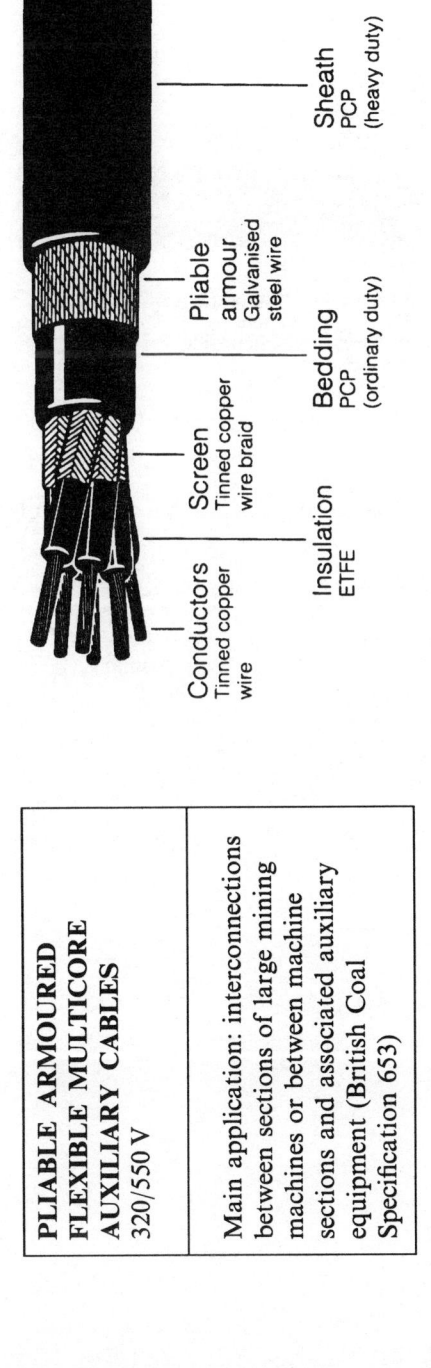

Main application: interconnections between sections of large mining machines or between machine sections and associated auxiliary equipment (British Coal Specification 653)

Cable construction:
- Conductors: Tinned copper wire
- Insulation: ETFE
- Screen: Tinned copper wire braid
- Bedding: PCP (ordinary duty)
- Pliable armour: Galvanised steel wire
- Sheath: PCP (heavy duty)

Table A10.7 Range and dimensions

Cable type	Conductor Area (mm²)	Formation (number/mm)	Number of cores	Armour (number/mm)	Diameter Minimum (mm)	Diameter Maximum (mm)	Approximate weight (kg/km)
506	1.34	19/0.30	6	7/0.45	21.1	23.6	876
512	1.34	19/0.30	12	7/0.45	21.1	23.6	960
518	1.34	19/0.30	18	7/0.45	22.8	25.3	1136
524	0.93	19/0.25	24	7/0.45	24.2	26.7	1219

UNARMOURED FLEXIBLE CABLES 600/1000 V
Main application: highly flexible and robust cables for supply to portable equipment (British Coal Specification 505)

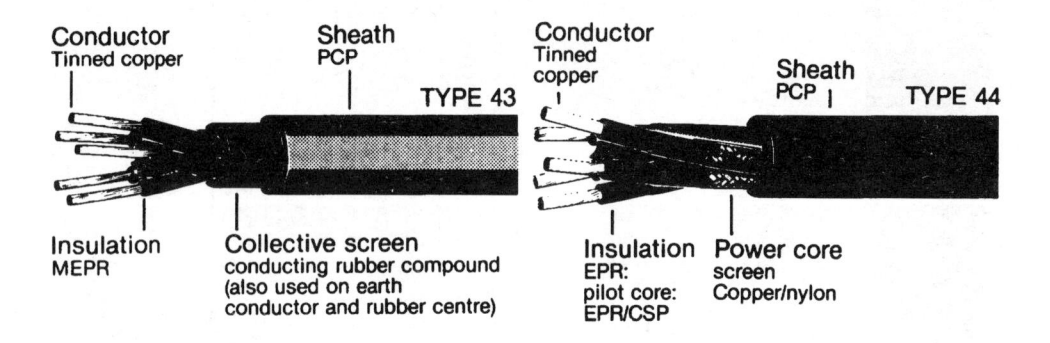

TYPE 43
- Conductor: Tinned copper
- Sheath: PCP
- Insulation: MEPR
- Collective screen: conducting rubber compound (also used on earth conductor and rubber centre)

TYPE 44
- Conductor: Tinned copper
- Sheath: PCP
- Insulation EPR: pilot core: EPR/CSP
- Power core screen: Copper/nylon

CURRENT RATINGS

These cables normally carry currents below their thermal capacity. Where applicable the current rating for types 43 and 44 is 46 A.

Table A10.8 Range and dimensions

Conductor Size (mm^2)	Formation (number/mm)	Diameter Minimum (mm)	Diameter Maximum (mm)	Approximate weight (kg/km)
Type 43 (conducting rubber screen)				
6	84/0.30	25.6	27.6	967
Type 44 (copper/nylon screen)				
6	84/0.30	24.7	26.7	1180

**PLIABLE ARMOURED
FLEXIBLE CABLES**
600/1000 V to 3800/6600 V to
BS 6708: 1991

Main application: trailing cables for power supply in open-cast and miscellaneous mines and quarries

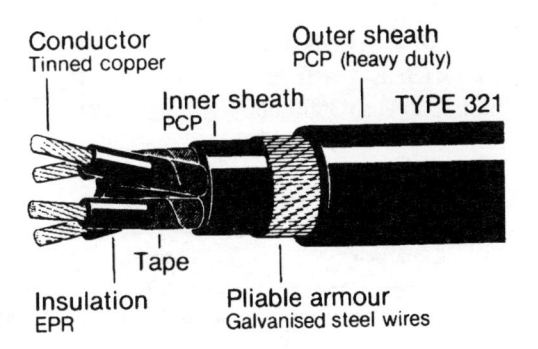

Conductor: Tinned copper
Outer sheath: PCP (heavy duty)
Inner sheath: PCP
TYPE 321
Insulation: EPR
Tape
Pliable armour: Galvanised steel wires

CONSTRUCTION

PETP tape is applied over the conductors for all sizes of type 321 cable and all sizes of types 20 and 21 except 2.5 and 4 mm^2 which are not taped. Type 621 cables have a semiconducting tape. The armour consists of bunches of seven wires.

Table A10.9 Current ratings

Conductor size (mm^2)	Continuous rating[a] D.C. or single-phase a.c. (A)	3-phase a.c. (A)	Conductor size (mm^2)	Continuous rating[a] D.C. or single-phase a.c. (A)	3-phase a.c. (A)
2.5	31	27	35	145	125
4	40	35	50	185	160
6	51	44	70	225	195
10	70	60	95	270	235
16	93	81	120	305	270
25	120	105	150	355	305

[a] Not applicable to cables in coils or on drums

Rating factors for ambient temperature

Ambient temperature (°C)	30	35	40	45	50	60	70
Rating factor	1.0	0.93	0.86	0.80	0.72	0.54	0.31

> **PLIABLE ARMOURED
> FLEXIBLE CABLES**
> 600/1000 V to 3800/6600 V

Table A10.10 Range and dimensions

Conductor Size (mm²)	Conductor Formation (number/mm)	Armour wires (number/mm)	Diameter Minimum (mm)	Diameter Maximum (mm)	Approximate weight (kg/km)
Type 20, 600/1000 V 3-core					
2.5	50/0.25	7/0.45	24.5	26.9	1060
4	56/0.30	7/0.45	25.8	28.2	1200
6	84/0.30	7/0.71	34.3	36.5	2130
10	80/0.40	7/0.71	36.2	38.4	2450
16	126/0.40	7/0.71	38.6	40.8	2930
25	196/0.40	7/0.90	43.6	48.4	3960
35	276/0.40	7/0.90	46.6	50.6	4660
50	396/0.40	7/0.90	51.3	56.7	5740
70	360/0.50	7/0.90	56.6	62.7	7060
95	475/0.50	7/0.90	68.4	72.1	9200
120	608/0.50	7/0.90	72.2	77.0	10750
150	756/0.50	7/1.25	83.3	87.3	14240
Type 21, 600/1000 V 4-core					
2.5	50/0.25	7/0.45	26.2	28.6	1210
4	56/0.30	7/0.45	27.7	30.1	1380
6	84/0.30	7/0.71	36.7	38.8	2440
10	80/0.40	7/0.71	38.8	41.0	2820
16	126/0.40	7/0.71	41.5	44.0	3430
25	196/0.40	7/0.71	46.1	49.8	4310
35	276/0.40	7/0.90	51.0	56.1	5560
50	396/0.40	7/0.90	56.1	61.8	6900
70	360/0.50	7/0.90	64.8	71.4	9150
95	475/0.50	7/0.90	74.9	78.6	11090
120	608/0.50	7/1.25	83.8	89.0	14880
150	756/0.50	7/1.25	91.4	95.3	17230
Type 321, 1900/3300 V 4-core					
16	126/0.40	7/0.71	51.5	55.8	4640
25	196/0.40	7/0.90	56.7	60.1	5810
35	276/0.40	7/0.90	60.7	64.1	6852
50	396/0.40	7/0.90	68.6	72.3	8790
70	360/0.50	7/0.90	74.7	78.4	10390
95	475/0.50	7/1.25	85.3	89.2	14040
120	608/0.50	7/1.25	88.5	93.7	15950
150	756/0.50	7/1.25	95.4	99.3	18280

(cont.)

> **PLIABLE ARMOURED**
> **FLEXIBLE CABLES**
> 600/1000 V to 3800/6600 V

Table A10.10 continued

Conductor		Armour wires (number/mm)	Diameter		Approximate weight (kg/km)
Size (mm^2)	Formation (number/mm)		Minimum (mm)	Maximum (mm)	
Type 621, 3800/6600 V 4-core					
16	126/0.40	7/0.90	67.2	70.4	7460
25	196/0.40	7/0.90	71.4	74.6	8410
35	276/0.40	7/0.90	75.4	78.6	9570
50	396/0.40	7/1.25	85.1	88.6	12920
70	360/0.50	7/1.25	91.2	94.6	14880
95	475/0.50	7/1.25	97.4	100.9	17120
120	608/0.50	7/1.25	100.5	105.2	19000
150	756/0.50	7/1.25	107.5	111.0	21390

PVC INSULATED ARMOURED MINING POWER CABLES
635/1000 V to 6.35/11 kV

Main application: fixed cables for power distribution in coal mines (British Coal Specification 295)

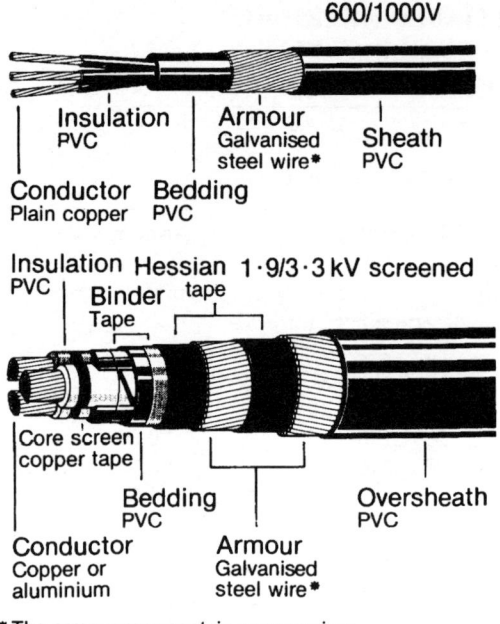

600/1000V

Insulation — PVC
Conductor — Plain copper
Bedding — PVC
Armour — Galvanised steel wire*
Sheath — PVC

1.9/3.3 kV screened

Insulation — PVC
Binder Tape
Hessian tape
Core screen copper tape
Conductor — Copper or aluminium
Bedding — PVC
Armour — Galvanised steel wire*
Oversheath — PVC

*The armour may contain copper wires and may be SWA or DWA

Specification
British Coal Specifications: 295 for 635/1000 V and 1.9/3.3 kV cables and 656 for 3.8/6.6 kV and 6.35/11 kV cables.

Number of cores
635/1000 V: 2-, 3- and 4-core.
Other voltages: 3-core only.

Conductors
All conductors are stranded. For voltages up to and including 3.8/6.6 kV conductors are shaped except as indicated in the tables. For 6.35/11 kV all conductors are circular.

Insulation
PVC for 635/1000 V and 1.9/3.3 kV cables.
EPR for 3.8/6.6 kV and 6.35/11 kV cables.

Core screens
1.9/3.3 kV cables may be screened or unscreened.

3.8/6.6 kV and 6.35/11 kV cables
These cables have conductor and core screens. Conductor screens consist of extruded semiconducting compound. Insulation screens comprise a layer of semiconducting material plus a copper tape.

> **PVC INSULATED ARMOURED MINING POWER CABLES**
> 635/1000 V to 6.35/11 kV

Table A10.11 Current ratings

Conductor size (mm^2)	635/1100 V (copper) 2-core (A)	635/1100 V (copper) 3- and 4-core (A)	1.9/3.3 kV[a] Copper (A)	1.9/3.3 kV[a] Aluminium (A)	3.8/6.6 kV and 6.35/11 kV Copper (A)	3.8/6.6 kV and 6.35/11 kV Aluminium (A)
1.5	24	21				
2.5	33	28				
4	44	38				
6	56	48				
10	77	66				
16	102	87	90		94	73
25		116	118		125	96
35		142	143		150	115
50		172	173		175	135
70		218	217	155	220	170
95		268	266	190	265	205
120			308		305	240
150			351	250	345	270
185			403	290	390	310
240			474		460	365
300			540		520	415
400			620			

[a] Applies to screened and unscreened

Note
These ratings are based on an ambient temperature of 25°C and a maximum continuous conductor temperature of 70°C.

Rating factors for ambient temperature

Ambient temperature (°C)	25	30	35	40	45
Rating factor	1.0	0.94	0.88	0.82	0.75

PVC INSULATED ARMOURED MINING POWER CABLES 635/1000 V

Table A10.12 Range and dimensions (635/1000 V cables, SWA and DWA with copper conductors)

Conductor size (mm^2)	Armour Wire diameter (mm)	Copper wires (number)	Approximate diameter (mm)	Approximate weight (kg/km)
2-core SWA				
1.5	0.9	0	11.7	280
2.5	0.9	0	13.1	350
4	0.9	0	15.1	450
6	0.9	2	16.5	545
10	1.25	1	20.1	870
16	1.25	4	21.9	980
3-core SWA				
1.5	0.9	0	12.3	310
2.5	0.9	0	13.6	390
4	0.9	0	15.8	510
6	1.25	0	18.0	720
10	1.25	1	21.2	1000
16	1.25	4	23.1	1130
25[a]	1.6	4	25.0	1650
35[a]	1.6	7	27.3	2050
50[a]	1.6	11	30.5	2600
70[a]	2.0	10	35.0	3610
95[a]	2.0	15	39.3	4750
4-core SWA				
1.5	0.9	0	13.0	350
2.5	0.9	0	14.5	445
4	1.25	0	17.8	715
6	1.25	0	19.2	855
10	1.25	1	22.8	1190
16	1.6	1	26.3	1520
25[a]	1.6	4	27.8	2050
35[a]	1.6	6	30.5	2540
50[a]	2.0	5	35.4	3480
70[a]	2.0	9	39.2	4490
95[a]	2.0	14	44.3	5890

(*cont.*)

> **PVC INSULATED
> ARMOURED MINING POWER
> CABLES
> 635/1000 V**

Table A10.12 continued

Conductor size (mm^2)	Armour Wire diameter (mm)	Armour Copper wires (number)	Approximate diameter (mm)	Approximate weight (kg/km)
3-core DWA				
1.5	0.9	0	17.0	690
2.5	0.9	0	18.3	795
4	0.9	0	20.6	985
6	1.25	0	23.4	1380
10	1.25	0	26.5	1760
4-core DWA				
1.5	0.9	0	17.6	745
2.5	0.9	0	19.1	870
4	1.25	0	23.2	1360
6	1.25	0	24.8	1550
10	1.25	0	28.2	2110

[a] Shaped conductors

PVC INSULATED ARMOURED MINING POWER CABLES 1.9/3.3 V

Table A10.13 Range and dimensions (3-core 1.9/3.3 kV cables)

Conductor size (mm^2)	Armour Wire diameter (mm)	Armour Copper wires (number)	Approximate diameter (mm)	Approximate weight (kg/km)
Unscreened, copper conductor, DWA				
16[a]	1.6	0	37.2	3210
25[a]	1.6	0	40.1	3420
35	1.6	0	39.1	3900
50	2.0	1	43.8	5090
70	2.0	5	47.0	6140
95	2.0	11	50.4	7220
120	2.5	7	56.4	9370
150	2.5	12	59.2	10570
185	2.5	17	62.5	12020
240	2.5	25	67.9	14540
300	2.5	35	73.4	17060
400	2.5	48	79.7	20790
Unscreened, aluminium conductor, DWA				
70	2.0	0	47.0	4780
95	2.0	0	50.4	5410
150	2.5	1	59.2	7710
185	2.5	4	62.5	8450
Unscreened, aluminium conductor, SWA				
70	2.0	3	41.8	3270
95	2.0	6	45.2	3710
150	2.5	6	52.8	5230
185	2.5	9	56.1	5480
Screened, copper conductor, DWA				
16[a]	1.6	0	38.1	3400
25[a]	1.6	0	40.9	3630
35	1.6	0	39.9	4140
50	2.0	0	44.6	5400
70	2.0	5	47.9	6390
95	2.0	10	51.3	7520
120	2.5	7	57.2	9720
150	2.5	11	60.0	10900
185	2.5	17	63.4	12370
240	2.5	25	68.8	14920

(*cont.*)

> **PVC INSULATED ARMOURED MINING POWER CABLES**
> 1.9/3.3 V

Table A10.13 continued

Conductor size (mm^2)	Armour Wire diameter (mm)	Armour Copper wires (number)	Approximate diameter (mm)	Approximate weight (kg/km)
300	2.5	35	74.2	17520
400	2.5	48	80.6	21170
Screened, aluminium conductor, DWA				
70	2.0	0	47.9	5090
95	2.0	0	51.3	5720
150	2.5	1	60.0	8100
185	2.5	4	63.4	8860
Screened, aluminium conductor, SWA				
70	2.0	2	42.7	3380
95	2.0	5	46.1	3900
150	2.5	6	53.6	5470
185	2.5	9	57.0	6090

[a] Circular conductor; remainder shaped

EPR INSULATED ARMOURED MINING POWER CABLES 3.8/6.6 V

Table A10.14 Range and dimensions (3.8/6.6 kV cables)

Conductor size (mm^2)	Armour wire diameter (mm)	Copper wires in armour (number)	Approximate diameter (mm)	Approximate weight (kg/km)
Copper conductor, DWA				
16[a]	2.0	0	47.7	4830
25[a]	2.0	0	50.8	5450
35[a]	2.0	0	53.4	6030
50[a]	2.0	0	57.0	6660
70	2.5	0	58.4	8580
95	2.5	1	61.6	9790
120	2.5	5	65.3	11150
150	2.5	9	68.1	12570
185	2.5	14	71.4	14120
Aluminium conductor, DWA				
50[a]	2.0	0	57.0	5780
70	2.5	0	58.4	7230
95	2.5	0	61.6	8660
120	2.5	0	65.3	8890
150	2.5	0	68.1	9760
185	2.5	1	71.4	10580
Copper conductor, SWA				
16[a]	2.0	0	44.0	3130
25[a]	2.0	0	47.0	3620
35[a]	2.0	0	50.0	4140
50	2.0	2	53.0	4660
70	2.5	3	52.0	5970
95	2.5	6	55.2	7050
120	2.5	10	58.9	8080
150	2.5	13	61.7	9440
185	2.5	18	65.0	10810
Aluminium conductor, SWA				
50[a]	2.0	0	53.0	3760
70	2.5	0	52.0	4660
95	2.5	1	55.2	5230
120	2.5	3	58.9	5760
150	2.5	5	61.7	6610
185	2.5	7	65.0	7260

[a] Circular conductors; remainder shaped

EPR INSULATED ARMOURED MINING POWER CABLES 6.35/11 kV

Table A10.15 Range and dimensions (6.35/11 kV cables)

Conductor size[a] (mm^2)	Armour wire diameter (mm)	Copper wires in armour (number)	Approximate diameter (mm)	Approximate weight (kg/km)
Copper conductor, DWA				
25	2.0	0	53.1	5600
35	2.5	0	58.6	7340
50	2.5	0	60.9	7995
70	2.5	0	64.5	9090
95	2.5	0	68.5	10365
120	2.5	2	72.7	11770
150	2.5	6	75.9	13240
185	2.5	12	79.9	15090
Aluminium conductor, DWA				
25	2.0	0	52.8	5120
35	2.5	0	58.3	6580
50	2.5	0	60.9	7100
70	2.5	0	64.5	7805
95	2.5	0	68.5	8580
120	2.5	0	72.7	9515
150	2.5	0	75.9	10460
185	2.5	0	79.9	11610
Copper conductor, SWA				
25	2.0	0	47.9	3755
35	2.5	0	52.2	4785
50	2.5	0	54.7	5380
70	2.5	2	58.3	6315
95	2.5	5	62.3	7440
120	2.5	8	66.5	8600
150	2.5	12	69.5	9715
185	2.5	16	73.5	11145
Aluminium conductor, SWA				
25	2.0	0	47.6	3275
35	2.5	0	51.9	4125
50	2.5	0	54.7	4490
70	2.5	0	58.3	5025
95	2.5	0	62.2	5650
120	2.5	2	66.5	6345
150	2.5	3	69.5	6935
185	2.5	6	73.5	7665

[a] Conductors are circular

APPENDIX A11
Mineral Insulated Wiring Cables

Cable design

Copper conductor, copper sheathed, 500 V light duty and 750 V heavy duty cables to BS 6207: Part 1: 1995.

Main application

Fixed installations in industrial, commercial and public buildings.

Data in tables

The tables give data for the more common installation methods and for the sizes of cables which are readily available in the UK. For other installation methods or other sizes of cable consult the cable manufacturer.

(a) Current ratings, volt drop and rating factors for cables with up to 7 cores:
  (i) exposed to touch or covered
  (ii) with bare sheath, not exposed to touch or in contact with combustible material
(b) Dimensions and weights

Notes on current ratings

(a) The current ratings are based on an ambient temperature of 30°C with a sheath temperature rise of 40°C for cables exposed to touch or covered and 75°C for cables with bare sheath not exposed to touch or in contact with combustible material.
(b) Where protection is provided by means of semi-enclosed (rewireable) fuses to BS 3036: 1958, the cable should be selected with a current rating not less than 1.38 times the rating of the fuse (see chapter 10).
(c) The 'free air' ratings apply to a cable or trefoil group spaced at least half of a cable diameter from the wall or other surface to which it is attached. In addition, for single-core cables in flat formation the spacing between cables is one cable diameter (i.e. two diameters between centres).

Notes on volt drop

(a) The volt drops for 500 V light duty cables with conductor sizes up to 4 mm^2 are the same as for the corresponding 750 V cables, as given in tables A11.6 and A11.7.
(b) The values of volt drop for 'spaced' single-core cables apply to cables spaced as for the 'free air' current ratings.

500 V LIGHT DUTY M.I. CABLE

Table A11.1 Current ratings for cables exposed to touch and with outer covering[a]

Conductor size (mm^2)	Two single-core cables or one 2-core cable, single phase a.c. or d.c.		Three single-core cables in trefoil or one 3-core cable, 3-phase a.c.	
	Clipped direct (A)	Free air (A)	Clipped direct (A)	Free air (A)
1	18.5	19.5	15	16.5
1.5	23	25	19	21
2.5	31	33	26	28
4	40	44	35	37

Conductor size (mm^2)	One 4-core cable, three cores loaded, 3-phase a.c.		One 4-core cable, all cores loaded		One 7-core cable, all cores loaded	
	Clipped direct (A)	Free air (A)	Clipped direct (A)	Free air (A)	Clipped direct (A)	Free air (A)
1	15	16	13	14	10	11
1.5	19.5	21	16.5	18	13	14
2.5	26	28	22	24	17.5	19

Rating factors for ambient temperature and for groups of cables are given in chapter 8.
[a] For bare cables multiply current rating by 0.9.

500 V LIGHT DUTY M.I. CABLE

Table A11.2 Current ratings for cables with bare sheaths not exposed to touch or in contact with combustible material

Conductor size (mm^2)	Two single-core cables or one 2-core cable, single phase a.c. or d.c.		Three single-core cables in trefoil or one 3-core cable, 3-phase a.c.	
	Clipped direct (A)	Free air (A)	Clipped direct (A)	Free air (A)
1	22	24	19	21
1.5	28	31	24	26
2.5	38	41	33	35
4	51	54	44	46

Conductor size (mm^2)	One 4-core cable, three cores loaded, 3-phase a.c.		One 4-core cable, all cores loaded		One 7-core cable, all cores loaded	
	Clipped direct (A)	Free air (A)	Clipped direct (A)	Free air (A)	Clipped direct (A)	Free air (A)
1	18.5	20	16.5	18	13	14
1.5	24	26	21	22	16.5	18
2.5	33	35	28	30	22	24

Rating factors for ambient temperature are given in chapter 8.
No rating factors for groups need be applied.

500 V LIGHT DUTY M.I. CABLE

Table A11.3 Dimensions and weights

Conductor size (mm²)	Diameter of bare cable (mm)	Diameter of covered cable (mm)	Conductor diameter (nominal) (mm)	Bare cable (kg/km)	Covered cable (kg/km)
2 × 1.0	5.1	6.6	1.13	104	125
2 × 1.5	5.7	7.2	1.39	136	159
2 × 2.5	6.6	8.1	1.77	187	213
2 × 4.0	7.7	9.4	2.25	248	282
3 × 1.0	5.8	7.3	1.13	136	159
3 × 1.5	6.4	7.9	1.39	176	201
3 × 2.5	7.3	9.0	1.77	256	223
4 × 1.0	6.3	7.8	1.13	162	187
4 × 1.5	7.0	8.5	1.39	203	231
4 × 2.5	8.1	9.8	1.77	277	315
7 × 1.0	7.6	9.3	1.13	236	269
7 × 1.5	8.4	10.1	1.39	295	332
7 × 2.5	9.7	11.4	1.77	411	454

The above are the most commonly used sizes; other sizes are available.

750 V HEAVY DUTY M.I. CABLE

Table A11.4 Current ratings for cables exposed to touch and with outer covering[a]

Conductor size (mm^2)	Two single-core cables or one 2-core cable, single-phase a.c. or d.c.		Three single-core cables in trefoil or one 3-core cable, 3-phase a.c.		Three single-core cables in flat formation, 3-phase a.c.		
	Clipped direct (A)	Free air (A)	Clipped direct (A)	Free air (A)	Clipped direct (A)	Free air Horizontal (A)	Vertical (A)
1.5	25	26	21	22	–	–	–
2.5	34	36	28	30	–	–	–
4	45	47	37	40	–	–	–
6	57	60	48	51	–	–	–
10	77	82	65	69	70	95	84
16	102	109	86	92	92	125	110
25	133	142	112	120	120	162	142
35	163	174	137	147	147	197	173
50	202	215	169	182	181	242	213
70	247	264	207	223	221	294	259
95	296	317	249	267	264	351	309
120	340	364	286	308	303	402	353
150	388	416	327	352	346	454	400
185	440	472	371	399	392	507	446
240	514	552	434	466	457	565	497

(*cont.*)

750 V HEAVY DUTY M.I. CABLE

Table A11.4 continued

Conductor size (mm^2)	One 4-core cable, three cores loaded, 3-phase a.c.		One 4-core cable, all cores loaded		One 7-core cable, all cores loaded	
	Clipped direct (A)	Free air (A)	Clipped direct (A)	Free air (A)	Clipped direct (A)	Free air (A)
1.5	21	23	18	20	14.5	15.5
2.5	28	30	25	27	19.5	21
4	37	40	32	35	–	–
6	47	51	41	44	–	–
10	64	68	55	59	–	–
16	85	89	72	78	–	–
25	110	116	94	101	–	–

Conductor size (mm^2)	One 12-core cable, all cores loaded		One 19-core cable, all cores loaded	
	Clipped direct (A)	Free air (A)	Clipped direct (A)	Free air (A)
1.5	12	13	10	11
2.5	16	17	–	–

Rating factors for ambient temperature and for groups of cables are given in chapter 8.
[a] For bare cables multiply current rating by 0.9.

750 V HEAVY DUTY M.I. CABLE

Table A11.5 Current ratings for cables with bare sheaths not exposed to touch or in contact with combustible material

Conductor size (mm^2)	Two single-core cables or one 2-core cable, single-phase a.c. or d.c.		Three single-core cables in trefoil or one 3-core cable, 3-phase a.c.		Three single-core cables in flat formation, 3-phase a.c.		
	Clipped direct (A)	Free air (A)	Clipped direct (A)	Free air (A)	Clipped direct (A)	Free air Horizontal (A)	Vertical (A)
1.5	31	33	26	28	–	–	–
2.5	42	45	35	38	–	–	–
4	55	60	47	50	–	–	–
6	70	76	59	64	–	–	–
10	96	104	81	87	91	120	105
16	127	137	107	115	119	157	137
25	166	179	140	150	154	204	178
35	203	220	171	184	187	248	216
50	251	272	212	228	230	304	266
70	307	333	260	279	280	370	323
95	369	400	312	335	334	441	385
120	424	460	359	385	383	505	441
150	485	526	410	441	435	565	498
185	550	596	465	500	492	629	557
240	643	697	544	584	572	704	624

(*cont.*)

750 V HEAVY DUTY M.I. CABLE

Table A11.5 continued

Conductor size (mm^2)	One 4-core cable, three cores loaded, 3-phase a.c.		One 4-core cable, all cores loaded		One 7-core cable, all cores loaded	
	Clipped direct (A)	Free air (A)	Clipped direct (A)	Free air (A)	Clipped direct (A)	Free air (A)
1.5	26	28	22	24	17.5	19
2.5	35	37	30	32	24	26
4	46	49	40	43	–	–
6	58	63	50	54	–	–
10	78	85	68	73	–	–
16	103	112	90	97	–	–
25	134	146	117	126	–	–

Conductor size (mm^2)	One 12-core cable, all cores loaded		One 19-core cable, all cores loaded	
	Clipped direct (A)	Free air (A)	Clipped direct (A)	Free air (A)
1.5	15.5	16.5	13	14
2.5	20	22	–	–

Rating factors for ambient temperature are given in chapter 8.
No rating factor for groups need be applied.

750 V HEAVY DUTY M.I. CABLE

Table A11.6 Volt drops for cables exposed to touch or covered (sheath operating temperature 70°C)

Conductor size (mm^2)	Single-phase operation — Two single-core cables touching (mV/A/m)	Single-phase operation — Multicore cables (mV/A/m)	3-phase operation — Three single-core cables Trefoil (mV/A/m)	3-phase operation — Three single-core cables Flat touching (mV/A/m)	3-phase operation — Three single-core cables Flat spaced (mV/A/m)	3-phase operation — Multicore cables (mV/A/m)
1	–	42	–	–	–	36
1.5	–	28	–	–	–	24
2.5	–	17	–	–	–	14
4	–	10	–	–	–	9.1
6	–	7	–	–	–	6.0
10	4.2	4.2	3.6	3.6	3.6	3.6
16	2.6	2.6	2.3	2.3	2.3	2.3
25	1.65	1.65	1.45	1.45	1.5	1.45
35	1.2	–	1.05	1.1	1.1	–
50	0.91	–	0.80	0.83	0.87	–
70	0.64	–	0.56	0.60	0.65	–
95	0.49	–	0.43	0.47	0.53	–
120	0.41	–	0.36	0.40	0.46	–
150	0.34	–	0.30	0.36	0.42	–
185	0.29	–	0.26	0.32	0.39	–
240	0.25	–	0.22	0.29	0.36	–

750 V HEAVY DUTY M.I. CABLE

Table A11.7 Volt drops for cables with bare sheaths not exposed to touch or in contact with combustible material (sheath operating temperature 105°C)

Conductor size (mm^2)	Single-phase operation		3-phase operation			
	Two single-core cables touching (mV/A/m)	Multicore cables (mV/A/m)	Three single-core cables			Multicore cables (mV/A/m)
			Trefoil (mV/A/m)	Flat touching (mV/A/m)	Flat spaced (mV/A/m)	
1	–	47	–	–	–	40
1.5	–	31	–	–	–	27
2.5	–	19	–	–	–	16
4	–	12	–	–	–	10
6	–	7.8	–	–	–	6.8
10	4.7	4.7	4.1	4.1	4.1	4.1
16	3.0	3.0	2.6	2.6	2.6	2.6
25	1.85	1.85	1.65	1.65	1.65	1.6
35	1.35	–	1.2	1.2	1.25	–
50	1.0	–	0.88	0.91	0.95	–
70	0.71	–	0.62	0.65	0.70	–
95	0.54	–	0.47	0.50	0.56	–
120	0.44	–	0.38	0.42	0.48	–
150	0.36	–	0.32	0.37	0.43	–
185	0.31	–	0.27	0.33	0.39	–
240	0.26	–	0.22	0.29	0.36	–

750 V HEAVY DUTY M.I. CABLE

Table A11.8 Dimensions and weights

Conductor size (mm^2)	Diameter of bare cable (mm)	Diameter of covered cable (mm)	Conductor diameter (nominal) (mm)	Approximate weight Bare cable (kg/km)	Approximate weight Covered cable (kg/km)
1 × 10	7.3	9.0	3.57	241	274
1 × 16	8.3	10.0	4.50	327	364
1 × 25	9.6	11.3	5.66	458	500
1 × 35	10.7	12.4	6.66	587	634
1 × 50	12.1	13.8	7.75	760	812
1 × 70	13.7	15.4	9.32	1019	1078
1 × 95	15.4	17.7	10.98	1326	1416
1 × 120	16.8	19.1	12.33	1615	1713
1 × 150	18.4	20.7	13.70	1952	2059
1 × 185	20.4	23.2	15.18	2370	2514
1 × 240	23.3	26.1	17.33	3035	3213
2 × 1.5	7.9	9.6	1.39	225	260
2 × 2.5	8.7	10.4	1.77	278	317
2 × 4	9.8	11.5	2.25	357	400
2 × 6	10.9	12.6	2.75	449	497
2 × 10	12.7	14.4	3.57	623	677
2 × 16	14.7	16.4	4.50	855	918
2 × 25	17.1	19.4	5.66	1178	1277
3 × 1.5	8.3	10.0	1.39	256	293
3 × 2.5	9.3	11.0	1.77	325	366
3 × 4	10.4	12.1	2.25	418	463
3 × 6	11.5	13.2	2.75	529	579
3 × 10	13.6	15.3	3.57	758	817
3 × 16	15.6	17.9	4.50	1039	1130
3 × 25	18.2	20.5	5.66	1451	1557
4 × 1.5	9.1	10.8	1.39	307	348
4 × 2.5	10.1	11.8	1.77	387	432
4 × 4	11.4	13.1	2.25	510	560
4 × 6	12.7	14.4	2.75	648	702
4 × 10	14.8	16.5	3.57	916	979
4 × 16	17.3	19.6	4.50	1292	1393
4 × 25	20.1	22.9	5.66	1813	1956
7 × 1.5	10.8	12.5	1.39	435	482
7 × 2.5	12.1	13.8	1.77	563	616
12 × 1.5	14.1	15.8	1.39	712	772
12 × 2.5	15.6	17.9	1.77	910	1001
19 × 1.5	16.6	18.9	1.39	989	1086

The above are the most commonly used sizes, other sizes are available.

APPENDIX A12
Paper Insulated Distribution Cables

CONTENTS

Tables are included for dimensions and weights, sustained ratings and electrical characteristics. Notes on the cable designs and conditions applicable are given before the tables.

CABLE DESIGNS

Typical constructions are shown in figs. A12.1–A12.4.

Tabulated figures are included for unarmoured single-core cables together with unarmoured and armoured multicore cables to BS 6480: 1988 (HD 621 for 3.8/6.6 kV cables and above) plus corrugated aluminium sheathed cables to EA 09-12 (HD 621).

Conductors
Stranded copper and aluminium, single-core circular and multicore sector shaped. Conductors for 12.7/22 kV and 19/33 kV 3-core cables are oval and for SL cables are circular.

Armour bedding
Bituminised textiles.

Armour
Steel tape and galvanised steel wire.

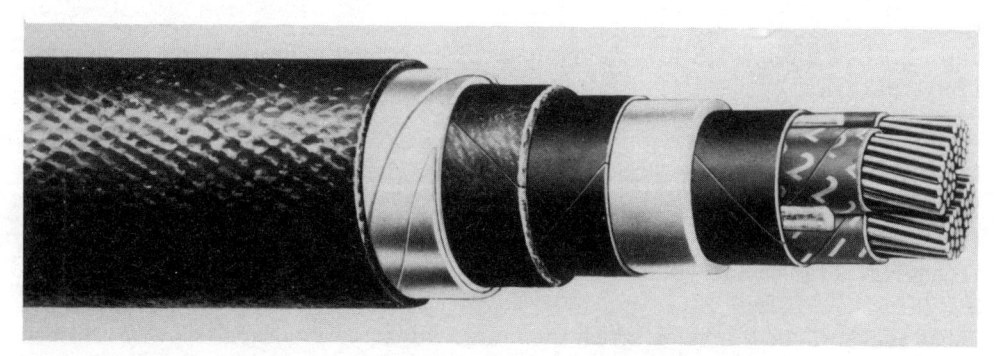

Fig. A12.1 Paper insulated 4-core 600/1000 V steel tape armoured cable

Fig. A12.2 Paper insulated 3-core 6.35/11 kV belted wire armoured cable

Fig. A12.3 Paper insulated 3-core 19/33 kV screened SL cable

Fig. A12.4 3-core, 11 kV, belted PIAS cable with corrugated aluminium sheath

Finish
Extruded PVC oversheath for unarmoured cables and bituminised textile serving for armoured cables.

SUSTAINED RATINGS

The maximum ratings in the tables are for a single circuit and are based on ERA Technology Ltd Report ERA 69-30, Part 1, which is in general conformity with IEC 287.

The standard conditions on which the ratings are based are given below. For other conditions the rating factors included in chapter 8 should be applied.

Maximum conductor temperature

80°C	0.6/1, 1.9/3.3 and 3.8/6.6 kV
65°C	6.35/11 kV belted
70°C	6.35/11 and 8.7/15 kV screened
65°C	12.7/22 and 19/33 kV screened

Cables laid direct in ground

Ground temperature
15°C.

Ground thermal resistivity
1.2 K m/W.

Adjacent circuits
At least 1.8 m apart.

Depth of laying
For voltages up to 1000 V, 0.5 m and for higher voltages 0.8 m (measured from ground surface to centre of cable or trefoil group).

Single-core cables

The data apply to three or four single-core cables operating 3-phase.

Bonding
It is assumed that the sheaths will be solidly bonded, i.e. bonded at both ends of the run. For very short runs it may be found possible to bond at one end only but consideration must be given to the value of the standing voltage which can occur along the cable length under both normal and fault conditions.

Trefoil
A close trefoil is assumed with the cables touching.

Flat formation
The ratings and technical data are based on horizontal installation with a spacing between cable centres of twice the overall diameter for sizes up to and including 185 mm^2 and 90 mm for 240 mm^2 and above.

Installation in air

An ambient air temperature of 25°C.

It is assumed that the cables are shielded from direct rays from the sun and that air circulation is not restricted significantly, e.g. if fastened to a wall the cables are spaced at least 20 mm from it; if in a trench they are not covered.

The data cover adjacent circuits which are spaced apart (chapter 8) and suitably disposed to prevent mutual heating.

ELECTRICAL CHARACTERISTICS

The standard conditions given for sustained ratings are equally applicable to the electrical characteristics. A.C. resistances are tabulated. D.C. resistances are given in table 4.1 of chapter 4.

PAPER INSULATED DISTRIBUTION CABLES (BS 6480: 1988) 600/1000 V

Table A12.1 Dimensions and weights

Conductor size (mm²)	Approximate diameter			Approximate weight, stranded copper conductors			Approximate weight, stranded aluminium conductors		
	PVC sheath unarmoured (mm)	STA and textile finish (mm)	SWA and textile finish (mm)	PVC sheath unarmoured (kg/m)	STA and textile finish (kg/m)	SWA and textile finish (kg/m)	PVC sheath unarmoured (kg/m)	STA and textile finish (kg/m)	SWA and textile finish (kg/m)
Single-core cables									
50	17.0			1.12			0.83		
70	18.8			1.42			0.99		
95	20.9			1.77			1.18		
120	22.7			2.15			1.41		
150	24.4			2.51			1.60		
185	26.5			3.03			1.89		
240	29.5			3.76			2.26		
300	32.4			4.59			2.71		
400	35.7			5.68			3.27		
500	39.7			7.01			3.97		
630	44.1			8.75			4.82		
800	48.8			10.86			5.84		
1000	53.8			13.33			7.00		
2-core cables									
25	19.0	24.9	26.1	1.24	1.92	2.13	0.93	1.61	1.82
35	20.7	26.7	27.9	1.51	2.25	2.48	1.08	1.82	2.05
50	22.7	28.6	29.8	1.91	2.70	2.95	1.32	2.11	2.36

70	25.5	31.4	33.4	2.51	3.35	3.87	1.67	2.51	3.03
95	28.4	34.3	36.3	3.24	4.16	4.74	2.07	2.99	3.57
120	31.1	36.8	38.8	3.98	4.98	5.61	2.50	3.50	4.13
150	34.5	40.0	43.0	4.83	5.91	6.95	3.02	4.10	5.14
185	37.6	42.8	45.8	5.86	7.00	8.11	3.58	4.72	5.83
240	42.3	48.6	50.4	7.49	9.19	10.02	4.50	6.20	7.03
300	46.4	52.2	54.0	9.10	10.89	11.79	5.35	7.14	8.04
400	51.5	56.8	58.6	11.41	13.32	14.30	6.61	8.52	9.50

3-core cables

25	20.9	26.8	28.0	1.62	2.38	2.61	1.15	1.91	2.14
35	23.2	29.1	30.3	2.07	2.91	3.16	1.42	2.26	2.51
50	25.5	31.4	32.6	2.61	3.47	3.79	1.73	2.59	2.91
70	28.5	34.4	36.4	3.37	4.33	4.92	2.10	3.06	3.65
95	32.1	37.8	39.8	4.40	5.45	6.10	2.65	3.70	4.35
120	35.3	40.8	42.8	5.39	6.51	7.21	3.18	4.30	5.00
150	39.3	44.6	47.6	6.69	7.90	9.08	3.97	5.18	6.36
185	43.0	49.2	51.0	8.12	9.83	10.67	4.71	6.42	7.26
240	48.4	54.4	56.2	10.37	12.27	13.21	5.88	7.78	8.72
300	53.1	58.7	60.5	12.64	14.64	15.66	7.01	9.01	10.03
400	59.0	64.1	65.9	15.85	17.98	19.10	8.65	10.78	11.90

4-core cables

25	23.5	29.4	30.6	2.04	2.87	3.13	1.41	2.24	2.50
35	26.0	31.9	33.1	2.53	3.39	3.71	1.67	2.53	2.85
50	28.7	34.6	36.6	3.21	4.16	4.74	2.05	3.00	3.58
70	32.7	38.4	40.4	4.31	5.35	6.01	2.63	3.67	4.33
95	37.0	42.5	44.5	5.65	6.79	7.51	3.31	4.45	5.17
120	40.6	45.9	48.9	6.94	8.16	9.35	3.99	5.21	6.40
150	45.3	51.5	53.3	8.60	10.37	11.24	4.97	6.74	7.61
185	49.5	55.5	57.3	10.44	12.33	13.27	5.89	7.78	8.72
240	56.3	61.8	63.6	13.51	15.57	16.62	7.54	9.60	10.65
300	61.6	66.9	68.7	16.43	18.64	19.78	8.92	11.13	12.27
400	68.6	73.3	76.4	20.61	22.93	24.99	11.02	13.34	15.40

(*cont.*)

PAPER INSULATED DISTRIBUTION CABLES (BS 6480: 1988) 600/1000 V

Table A12.1 continued

<table>
<thead>
<tr><th rowspan="3">Conductor size (mm²)</th><th colspan="3">Approximate diameter</th><th colspan="3">Approximate weight, stranded copper conductors</th><th colspan="3">Approximate weight, stranded aluminium conductors</th></tr>
<tr><th>PVC sheath unarmoured (mm)</th><th>STA and textile finish (mm)</th><th>SWA and textile finish (mm)</th><th>PVC sheath unarmoured (kg/m)</th><th>STA and textile finish (kg/m)</th><th>SWA and textile finish (kg/m)</th><th>PVC sheath unarmoured (kg/m)</th><th>STA and textile finish (kg/m)</th><th>SWA and textile finish (kg/m)</th></tr>
</thead>
<tbody>
<tr><td colspan="10">4-core with reduced neutral[a]</td></tr>
<tr><td>25</td><td>23.3</td><td>29.2</td><td>30.4</td><td>1.88</td><td>2.70</td><td>3.00</td><td>1.31</td><td>2.13</td><td>2.43</td></tr>
<tr><td>35</td><td>25.3</td><td>31.2</td><td>32.4</td><td>2.37</td><td>3.25</td><td>3.57</td><td>1.62</td><td>2.50</td><td>2.82</td></tr>
<tr><td>50</td><td>27.9</td><td>33.8</td><td>35.8</td><td>2.98</td><td>3.94</td><td>4.53</td><td>1.95</td><td>2.91</td><td>3.50</td></tr>
<tr><td>70</td><td>31.6</td><td>37.6</td><td>39.6</td><td>3.95</td><td>5.04</td><td>5.70</td><td>2.47</td><td>3.56</td><td>4.22</td></tr>
<tr><td>95</td><td>36.2</td><td>42.0</td><td>44.0</td><td>5.14</td><td>6.35</td><td>7.09</td><td>3.10</td><td>4.31</td><td>5.05</td></tr>
<tr><td>120</td><td>40.3</td><td>45.8</td><td>48.8</td><td>6.35</td><td>7.66</td><td>8.90</td><td>3.72</td><td>5.03</td><td>6.27</td></tr>
<tr><td>150</td><td>43.6</td><td>48.8</td><td>51.8</td><td>7.62</td><td>8.98</td><td>10.31</td><td>4.47</td><td>5.83</td><td>7.16</td></tr>
<tr><td>185</td><td>48.0</td><td>54.3</td><td>56.1</td><td>9.30</td><td>11.27</td><td>12.22</td><td>5.31</td><td>7.28</td><td>8.23</td></tr>
<tr><td>240</td><td>54.2</td><td>60.0</td><td>61.8</td><td>12.03</td><td>14.16</td><td>15.22</td><td>6.80</td><td>8.93</td><td>9.99</td></tr>
<tr><td>300</td><td>59.4</td><td>65.0</td><td>66.8</td><td>14.60</td><td>16.89</td><td>18.04</td><td>8.05</td><td>10.34</td><td>11.49</td></tr>
<tr><td>300</td><td>61.4</td><td>66.7</td><td>68.5</td><td>14.91</td><td>17.25</td><td>18.43</td><td>8.16</td><td>10.50</td><td>11.68</td></tr>
<tr><td>400</td><td>66.0</td><td>71.1</td><td>74.2</td><td>18.25</td><td>20.70</td><td>21.97</td><td>9.91</td><td>12.36</td><td>13.63</td></tr>
</tbody>
</table>

[a] Size of reduced neutral

Phase (mm²)	25	35	50	70	95	120	150	185	240	300	400
Neutral (mm²)	16	16	25	35	50	70	70	95	120	185	185

PAPER INSULATED DISTRIBUTION CABLES (BS 6480: 1988) 1.9/3.3 kV

Table A12.2 Dimensions and weights

Conductor size (mm^2)	Approximate diameter			Approximate weight, stranded copper conductors			Approximate weight, stranded aluminium conductors		
	PVC sheath unarmoured (mm)	STA and textile finish (mm)	SWA and textile finish (mm)	PVC sheath unarmoured (kg/m)	STA and textile finish (kg/m)	SWA and textile finish (kg/m)	PVC sheath unarmoured (kg/m)	STA and textile finish (kg/m)	SWA and textile finish (kg/m)

Single-core cables

50	18.2			1.22			0.93		
70	20.0			1.52			1.10		
95	21.9			1.87			1.28		
120	23.7			2.25			1.51		
150	25.2			2.59			1.68		
185	27.3			3.13			1.99		
240	29.9			3.82			2.31		
300	32.6			4.63			2.75		
400	35.9			5.73			3.32		
500	39.7			7.04			4.00		
630	44.1			8.78			4.86		
800	48.8			10.90			5.88		
1000	53.8			13.37			7.04		

(*cont.*)

PAPER INSULATED DISTRIBUTION CABLES (BS 6480: 1988) 1.9/3.3 kV

Table A12.2 continued

Conductor size (mm²)	Approximate diameter			Approximate weight, stranded copper conductors			Approximate weight, stranded aluminium conductors		
	PVC sheath unarmoured (mm)	STA and textile finish (mm)	SWA and textile finish (mm)	PVC sheath unarmoured (kg/m)	STA and textile finish (kg/m)	SWA and textile finish (kg/m)	PVC sheath unarmoured (kg/m)	STA and textile finish (kg/m)	SWA and textile finish (kg/m)
3-core cables									
25	23.4	29.3	30.5	1.90	2.74	3.00	1.43	2.27	2.53
35	25.5	31.4	32.6	2.29	3.16	3.48	1.65	2.52	2.84
50	27.8	33.7	35.7	2.86	3.80	4.38	1.98	2.92	3.50
70	31.0	37.0	39.0	3.73	4.77	5.41	2.47	3.51	4.15
95	34.7	40.4	42.4	4.81	5.94	6.64	3.05	4.18	4.88
120	37.8	43.3	46.3	5.82	7.01	8.16	3.61	4.80	5.95
150	40.8	47.3	49.1	6.93	8.60	9.40	4.21	5.88	6.68
185	44.4	50.7	52.5	8.38	10.15	11.02	4.97	6.74	7.61
240	49.5	55.3	57.1	10.58	12.48	13.44	6.10	8.00	8.96
300	54.2	59.8	61.6	13.01	15.05	16.09	7.39	9.43	10.47
400	60.0	65.2	68.3	16.26	18.43	20.30	9.07	11.24	13.11

**PAPER INSULATED
DISTRIBUTION CABLES
(BS 6480: 1988/HD 621 3J1 & 4J2)
3.8/6.6 kV**

Table A12.3 Dimensions and weights

Conductor size (mm²)	Approximate diameter			Approximate weight, stranded copper conductors			Approximate weight, stranded aluminium conductors		
	PVC sheath unarmoured (mm)	STA and textile finish (mm)	SWA and textile finish (mm)	PVC sheath unarmoured (kg/m)	STA and textile finish (kg/m)	SWA and textile finish (kg/m)	PVC sheath unarmoured (kg/m)	STA and textile finish (kg/m)	SWA and textile finish (kg/m)
Single-core cables									
50	19.6			1.33			1.04		
70	21.4			1.64			1.21		
95	23.5			2.06			1.47		
120	25.1			3.38			1.64		
150	26.8			2.82			1.91		
185	28.7			3.28			2.13		
240	31.5			4.07			2.57		
300	34.0			4.80			2.91		
400	37.4			5.91			3.49		
500	40.7			7.19			4.15		
630	45.1			8.94			5.01		
800	49.8			11.07			6.04		
1000	55.0			13.57			7.24		

(*cont.*)

PAPER INSULATED DISTRIBUTION CABLES (BS 6480: 1988/HD 621 3J1 & 4J2) 3.8/6.6 kV

Table A12.3 continued

Conductor size (mm^2)	Approximate diameter			Approximate weight, stranded copper conductors			Approximate weight, stranded aluminium conductors		
	PVC sheath unarmoured (mm)	STA and textile finish (mm)	SWA and textile finish (mm)	PVC sheath unarmoured (kg/m)	STA and textile finish (kg/m)	SWA and textile finish (kg/m)	PVC sheath unarmoured (kg/m)	STA and textile finish (kg/m)	SWA and textile finish (kg/m)
3-core cables									
25[a]	31.1	37.0	39.0	2.39	3.61	4.25	1.92	3.10	3.74
35	30.1	36.0	38.0	2.91	3.93	4.56	2.27	3.29	3.92
50	32.4	38.1	40.1	3.44	4.51	5.18	2.56	3.63	4.30
70	35.9	41.4	43.4	4.39	5.54	6.27	3.12	4.27	5.00
95	39.5	44.8	47.8	5.53	6.75	7.95	3.77	4.99	6.19
120	42.7	48.9	50.7	6.61	8.33	9.17	4.39	6.11	6.95
150	45.4	51.7	53.5	7.73	9.56	10.45	5.01	6.84	7.73
185	49.1	55.1	56.9	9.24	11.16	12.12	5.83	7.75	8.71
240	54.3	59.9	61.7	11.70	13.76	14.81	7.21	9.27	10.32
300	58.8	64.2	66.0	14.05	16.23	17.39	8.42	10.60	11.76
400	64.7	69.6	72.7	17.40	19.71	21.72	10.21	12.52	14.53

[a] Circular conductors

PAPER INSULATED DISTRIBUTION CABLES (BS 6480: 1988/HD 621 3J1 & 4J2) 6.35/11 kV

Table A12.4 Dimensions and weights

Conductor size (mm^2)	Approximate diameter PVC sheath unarmoured (mm)	STA and textile finish (mm)	SWA and textile finish (mm)	Approximate weight, stranded copper conductors PVC sheath unarmoured (kg/m)	STA and textile finish (kg/m)	SWA and textile finish (kg/m)	Approximate weight, stranded aluminium conductors PVC sheath unarmoured (kg/m)	STA and textile finish (kg/m)	SWA and textile finish (kg/m)
Single-core cables									
50	21.1			1.47			1.18		
70	23.1			1.85			1.43		
95	25.0			2.21			1.63		
120	26.8			2.64			1.90		
150	28.3			2.99			2.08		
185	30.4			3.55			2.41		
240	33.0			4.27			2.76		
300	35.7			5.12			3.23		
400	39.0			6.26			3.84		
500	43.3			7.53			4.49		
630	47.7			9.32			5.40		
800	52.4			11.48			6.45		
1000	57.6			14.18			7.85		
960[a]	58.2			13.82					
1200[a]	63.7			16.98					

(cont.)

PAPER INSULATED DISTRIBUTION CABLES
(BS 6480: 1988/HD 621 3J1 & 4J2)
6.35/11 kV

Table A12.4 continued

Conductor size (mm^2)	Approximate diameter			Approximate weight, stranded copper conductors			Approximate weight, stranded aluminium conductors		
	PVC sheath unarmoured (mm)	STA and textile finish (mm)	SWA and textile finish (mm)	PVC sheath unarmoured (kg/m)	STA and textile finish (kg/m)	SWA and textile finish (kg/m)	PVC sheath unarmoured (kg/m)	STA and textile finish (kg/m)	SWA and textile finish (kg/m)
3-core cables, belted									
25[b]	35.5	41.4	43.4	3.25	4.44	5.11	2.78	3.97	4.64
35[b]	38.6	44.3	46.3	3.51	5.02	5.73	3.07	4.33	5.04
50	37.2	42.6	45.6	4.18	5.37	6.45	3.31	4.50	5.58
70	40.6	45.9	48.9	5.20	6.46	7.62	3.94	5.20	6.36
95	44.2	50.5	52.3	6.41	8.19	9.00	4.66	6.44	7.25
120	47.2	53.2	55.0	7.40	9.26	10.12	5.19	7.05	7.91
150	50.2	56.0	57.8	8.61	10.54	11.45	5.89	7.82	8.73
185	53.8	59.4	61.2	10.18	12.21	13.18	6.77	8.80	9.77
240	58.8	64.2	66.0	12.70	14.89	15.95	8.22	10.41	11.47
300	63.3	68.5	71.6	15.13	17.44	19.36	9.50	11.81	13.73
400	69.2	73.9	77.0	18.57	21.00	23.08	11.38	13.81	15.89
3-core cables, screened									
25[b]	35.3	41.2	43.2	3.24	4.43	5.12	2.76	3.95	4.64
35[b]	38.2	43.9	45.9	3.39	4.89	5.60	3.10	4.19	4.90

50	37.1	42.5	44.5	4.15	5.24	5.94	3.17	4.36	5.06
70	40.3	45.8	48.8	5.04	6.32	7.48	3.78	5.06	6.22
95	44.0	50.4	52.2	6.23	8.03	8.84	4.48	6.28	7.09
120	47.2	53.4	55.2	7.36	9.25	10.11	5.15	7.04	7.90
150	50.2	56.2	58.0	8.56	10.53	11.44	5.84	7.81	8.72
185	53.8	59.6	61.4	10.13	12.20	13.17	6.72	8.79	9.76
240	58.9	64.3	66.1	12.48	14.67	13.23	8.00	10.19	11.25
300	63.7	68.8	71.9	15.10	17.42	19.35	9.48	11.80	13.73
400	69.6	74.3	77.4	18.55	20.97	23.08	11.35	13.77	15.88

[a] Milliken conductors
[b] Circular conductors

PAPER INSULATED
DISTRIBUTION CABLES
(EA 09-12/HD 621 4J3)
6.35/11 kV

Table A12.5 Corrugated aluminium sheathed cables. Dimensions and weights

Conductor size (mm^2)	Approximate diameter PVC sheathed (mm)	Approximate weight, stranded aluminium conductors PVC sheathed (kg/m)
3-core cables, belted		
95	49.2	3.40
185	60.8	5.20
300	72.5	7.50
3-core cables, screened		
95	49.0	3.40
185	60.7	5.20
300	72.5	7.50

PAPER INSULATED DISTRIBUTION CABLES (BS 6480: 1988/HD 621 3J1 & 4J2) 8.7/15 kV

Table A12.6 Dimensions and weights

Conductor size (mm²)	Approximate diameter			Approximate weight, stranded copper conductors			Approximate weight, stranded aluminium conductors		
	PVC sheath unarmoured (mm)	STA and textile finish (mm)	SWA and textile finish (mm)	PVC sheath unarmoured (kg/m)	STA and textile finish (kg/m)	SWA and textile finish (kg/m)	PVC sheath unarmoured (kg/m)	STA and textile finish (kg/m)	SWA and textile finish (kg/m)

Single-core cables

50	22.9			1.68			1.39		
70	24.7			2.00			1.57		
95	26.8			2.46			1.87		
120	28.4			2.80			2.06		
150	30.1			3.26			2.35		
185	32.0			3.74			2.59		
240	34.8			4.57			3.07		
300	37.5			5.35			3.46		
400	40.9			6.50			4.09		
500	45.1			7.91			4.87		
630	49.5			9.74			5.81		
800	54.2			11.93			6.91		
1000	59.4			14.52			8.19		
960[a]	60.0			14.32					
1200[a]	65.5			17.34					

(*cont.*)

PAPER INSULATED DISTRIBUTION CABLES (BS 6480: 1988/HD 621 3J1 & 4J2) 8.7/15 kV

Table A12.6 continued

Conductor size (mm²)	Approximate diameter				Approximate weight, stranded copper conductors				Approximate weight, stranded aluminium conductors			
	PVC sheath unarmoured (mm)	STA and textile finish (mm)	SWA and textile finish (mm)		PVC sheath unarmoured (kg/m)	STA and textile finish (kg/m)	SWA and textile finish (kg/m)		PVC sheath unarmoured (kg/m)	STA and textile finish (kg/m)	SWA and textile finish (kg/m)	
3-core cables, screened
25[b]	39.2	44.9	46.9		3.79	5.07	5.82		3.32	4.60	5.35	
35[b]	42.3	47.8	50.8		4.50	5.85	7.08		3.84	5.19	6.42	
50	42.2	47.4	50.4		4.77	6.05	7.23		3.90	5.18	6.36	
70	45.6	51.8	53.6		5.70	7.49	8.31		4.43	6.22	7.04	
95	49.5	55.5	57.3		7.10	9.00	9.88		5.35	7.25	8.13	
120	52.6	58.4	60.2		8.12	10.09	11.02		5.91	7.88	8.81	
150	55.7	61.3	63.1		9.36	11.42	12.40		6.64	8.70	9.68	
185	59.4	64.8	66.6		10.98	13.12	14.16		7.57	9.71	10.75	
240	64.7	69.8	72.9		13.59	15.89	17.80		9.10	11.40	13.31	
300	69.4	74.3	77.4		16.07	18.48	20.52		10.44	12.85	14.89	
400	75.4	79.9	83.0		19.60	22.13	24.33		12.41	14.94	17.14	

[a] Milliken conductors
[b] Circular conductors

PAPER INSULATED DISTRIBUTION CABLES (BS 6480: 1988/HD 621 3J1 & 4J2) 12.7/22 kV

Table A12.7 Dimensions and weights

Conductor size (mm^2)	Approximate diameter			Approximate weight, stranded copper conductors			Approximate weight, stranded aluminium conductors		
	PVC sheath unarmoured (mm)	STA and textile finish (mm)	SWA and textile finish (mm)	PVC sheath unarmoured (kg/m)	STA and textile finish (kg/m)	SWA and textile finish (kg/m)	PVC sheath unarmoured (kg/m)	STA and textile finish (kg/m)	SWA and textile finish (kg/m)
Single-core cables									
50	26.2			2.03			1.74		
70	28.0			2.36			1.94		
95	30.1			2.85			2.26		
120	31.7			3.21			2.47		
150	33.4			3.69			2.78		
185	35.3			4.19			3.05		
240	38.4			5.07			3.56		
300	41.0			5.86			3.98		
400	44.2			7.05			4.64		
500	48.3			8.52			5.48		
630	52.7			10.38			6.46		
800	57.4			12.63			7.61		
1000	62.4			15.28			8.95		
960[a]	63.2			15.07					
1200[a]	68.5			18.14					

(*cont.*)

PAPER INSULATED DISTRIBUTION CABLES (BS 6480: 1988/HD 621 3J1 & 4J2) 12.7/22 kV

Table A12.7 continued

Conductor size (mm²)	Approximate diameter			Approximate weight, stranded copper conductors			Approximate weight, stranded aluminium conductors		
	PVC sheath unarmoured (mm)	STA and textile finish (mm)	SWA and textile finish (mm)	PVC sheath unarmoured (kg/m)	STA and textile finish (kg/m)	SWA and textile finish (kg/m)	PVC sheath unarmoured (kg/m)	STA and textile finish (kg/m)	SWA and textile finish (kg/m)
3-core cables, screened								
25[b]	46.1	51.3	54.4	4.93	6.37	7.70	4.45	5.89	7.22
35[b]	49.2	55.4	57.2	5.72	7.71	8.61	5.07	7.06	7.96
50[b]	52.1	58.1	59.9	6.44	8.51	9.46	5.55	7.62	8.57
70	52.6	58.4	60.2	7.00	9.00	9.94	5.74	7.74	8.68
95	56.3	61.9	63.7	8.29	10.38	11.38	6.54	8.63	9.63
120	59.4	64.9	66.7	9.49	11.68	12.73	7.29	9.43	10.53
150	62.5	67.8	69.6	10.79	13.04	14.13	8.09	10.34	11.43
185	66.2	71.4	74.5	12.45	14.79	16.74	9.05	11.39	13.34
240	71.6	76.4	79.4	15.14	17.57	19.65	10.68	13.11	15.19
300	76.4	80.8	83.9	17.65	20.17	22.37	12.06	14.53	16.78
400	82.4	86.5	89.6	21.25	23.86	26.21	14.10	16.71	19.06
3-core cables, screened SL type								
25[b]	58.2	60.0			7.46	8.41		6.98	7.93
35[b]	60.8	62.6			8.17	9.16		7.52	8.51

50[b]	64.0	65.8	9.26	10.31		8.37	9.42
70[b]	67.9	69.7	10.52	11.64		9.23	10.35
95[b]	72.4	74.2	12.30	13.51		10.52	11.73
120[b]	75.9	77.7	13.63	14.90		11.39	12.66
150[b]	79.6	81.4	15.37	16.70		12.61	13.94
185[b]	83.7	86.8	17.18	19.55		13.71	16.08
240[b]	89.8	92.9	20.28	22.84		15.73	18.29
300[b]	95.0	98.1	23.02	25.73		17.30	20.01
400[b]	101.9	105.0	27.14	30.06		19.83	22.75

[a] Milliken conductors
[b] Circular conductors

PAPER INSULATED DISTRIBUTION CABLES (BS 6480: 1988/HD 621 3J2 & 4J1) 19/33 kV

Table A12.8 Dimensions and weights

Conductor size (mm^2)	Approximate diameter			Approximate weight, copper conductors			Approximate weight, stranded aluminium conductors		
	PVC sheath unarmoured (mm)	STA and textile finish (mm)	SWA and textile finish (mm)	PVC sheath unarmoured (kg/m)	STA and textile finish (kg/m)	SWA and textile finish (kg/m)	PVC sheath unarmoured (kg/m)	STA and textile finish (kg/m)	SWA and textile finish (kg/m)
Single-core cables									
50	31.4			2.73			2.44		
70	33.2			3.04			2.64		
95	34.9			3.49			2.94		
120	36.7			3.89			3.18		
150	37.8			4.29			3.44		
185	39.7			4.82			3.75		
240	42.8			5.77			4.31		
300	45.4			6.57			4.75		
400	48.8			7.81			5.49		
500	52.8			9.33			6.35		
630	57.1			11.31			7.38		
800	61.8			13.75			8.79		
1000	66.8			16.49			10.17		
960[a]	67.6			15.85					
1200[a]	72.9			18.98					

3-core cables, screened								
50[b]	63.7	69.3	71.1	9.07	11.53	12.53	8.16	10.82	11.62
70	64.4	69.7	71.5	10.05	12.57	13.49	8.74	11.40	12.18
95	67.2	72.4	75.5	11.36	14.03	15.75	9.55	12.29	13.94
120	70.5	75.4	78.5	12.76	15.60	17.31	10.47	13.29	15.03
150	72.3	77.0	80.1	14.08	16.90	18.71	11.27	14.14	15.90
185	76.0	80.5	83.6	16.04	18.93	20.85	12.51	15.46	17.32
240	81.1	85.3	88.4	18.76	21.86	23.83	14.12	17.22	19.19
300	85.8	89.8	92.9	21.67	24.81	26.96	15.85	19.08	21.14
400	91.8	95.4	98.5	25.95	29.10	31.49	18.34	21.68	23.88
3-core cables, screened SL type								
50	75.2	77.0			11.87	13.04		11.00	12.16
70	79.1	80.9			13.01	14.20		11.86	13.07
95	82.8	85.9			14.62	16.82		13.01	15.23
120	86.2	89.4			16.00	18.29		13.90	16.21
150	88.7	91.8			17.38	19.72		14.83	17.20
185	92.8	95.9			19.22	21.66		16.09	18.58
240	98.9	102.0			22.50	25.19		18.15	20.79
300	104.1	107.2			25.30	28.04		19.81	22.59
400	111.0	114.1			29.51	32.43		22.50	25.46

[a] Milliken conductors
[b] Circular conductors

PAPER INSULATED DISTRIBUTION CABLES (BS 6480: 1988) 600/1000 V

Table A12.9 Sustained ratings

Conductor size (mm^2)	In air Single-core Trefoil (A)	In air Single-core Flat (A)	In air 2-core (A)	In air 3- or 4-core (A)	In ground Single-core Trefoil (A)	In ground Single-core Flat (A)	In ground 2-core (A)	In ground 3- or 4-core (A)
Copper conductors								
16			105	91			120	105
25			145	120			160	135
35			175	150			195	165
50	215	235	210	180	220	230	230	195
70	275	300	265	230	270	280	285	240
95	335	370	325	280	320	335	345	290
120	390	430	380	325	365	380	395	335
150	445	490	430	375	410	430	445	375
185	520	570	495	430	460	485	500	425
240	620	710	590	510	530	560	580	490
300	710	810	670	590	600	620	650	550
400	820	930	780	680	680	700	740	620
500	940	1040			760	770		
630	1080	1170			850	860		
800	1220	1300			940	930		
1000	1350	1410			1010	980		
Aluminium conductors								
16			81	70			94	80
25			110	93			125	105
35			135	115			150	125
50	165	180	165	140	170	175	180	150
70	210	230	205	175	210	220	220	185
95	260	285	255	220	250	260	270	225
120	305	335	295	255	285	300	305	255
150	350	385	335	290	320	335	345	290
185	405	445	390	335	360	380	395	330
240	485	560	460	400	420	440	455	380

(cont.)

PAPER INSULATED
DISTRIBUTION CABLES
(BS 6480: 1988)
600/1000 V

Table A12.9 continued

Conductor size (mm^2)	In air				In ground			
	Single-core		2-core (A)	3- or 4-core (A)	Single-core		2-core (A)	3- or 4-core (A)
	Trefoil (A)	Flat (A)			Trefoil (A)	Flat (A)		
300	560	640	530	460	475	495	520	430
400	650	740	620	540	540	560	590	500
500	760	840			610	630		
630	880	970			690	710		
800	1020	1090			780	780		
1000	1150	1220			860	850		

For relevant conditions see the notes at the beginning of this appendix.
Single-core cables are PVC sheathed unarmoured. Multicore cables are armoured.

**PAPER INSULATED
DISTRIBUTION CABLES
(BS 6480: 1988/HD 621 3J1 & 4J2)
1.9/3.3 and 3.8/6.6 kV**

Table A12.10 Sustained ratings

Conductor size (mm^2)	In air Single-core Trefoil (A)	In air Single-core Flat (A)	In air 3-core (A)	In ground Single-core Trefoil (A)	In ground Single-core Flat (A)	In ground 3-core (A)
Copper conductors						
16			93			99
25			125			130
35			150			155
50	215	235	180	205	215	185
70	275	295	230	255	265	230
95	335	365	280	305	315	275
120	390	425	325	345	360	315
150	445	490	370	385	400	355
185	520	570	430	435	455	400
240	610	700	510	510	520	460
300	710	800	590	570	590	520
400	820	920	680	640	660	580
500	940	1030		720	720	
630	1080	1170		800	800	
800	1220	1290		880	860	
1000	1350	1410		950	910	
Aluminium conductors						
16			72			78
25			95			100
35			115			120
50	170	185	140	160	165	145
70	210	230	175	195	205	180
95	260	285	215	235	245	215
120	305	335	255	270	280	250
150	350	380	290	300	315	280
185	405	440	335	340	355	315
240	480	560	400	395	410	370

(*cont.*)

**PAPER INSULATED
DISTRIBUTION CABLES
(BS 6480: 1988/HD 621 3J1 & 4J2)
1.9/3.3 and 3.8/6.6 kV**

Table A12.10 continued

Conductor size (mm^2)	In air			In ground		
	Single-core		3-core (A)	Single-core		3-core (A)
	Trefoil (A)	Flat (A)		Trefoil (A)	Flat (A)	
300	560	640	460	445	465	415
400	650	740	540	510	530	470
500	750	840		580	590	
630	880	960		650	660	
800	1020	1090		730	730	
1000	1150	1210		810	790	

For the relevant conditions see the notes at the beginning of this appendix. Single-core cables are PVC sheathed, unarmoured. 3-core cables are armoured.

**PAPER INSULATED DISTRIBUTION CABLES
(BS 6480: 1988/HD 621 3J1 & 4J2)
6.35/11 and 8.7/15 kV**

Table A12.11 Sustained ratings

Conductor size (mm^2)	In air Single-core Trefoil (A)	In air Single-core Flat (A)	In air 3-core Belted[a] (A)	In air 3-core Screened (A)	In ground Single-core Trefoil (A)	In ground Single-core Flat (A)	In ground 3-core Belted[a] (A)	In ground 3-core Screened (A)
Copper conductors								
16			80	95			89	100
25			105	120			115	125
35			130	145			140	150
50	200	215	155	175	190	200	165	180
70	250	270	195	220	235	245	205	220
95	305	330	240	265	280	290	245	265
120	355	385	275	310	320	330	280	300
150	405	440	315	350	360	370	315	340
185	465	510	360	400	405	420	355	380
240	550	620	425	475	470	480	410	435
300	630	700	490	540	530	540	460	485
400	730	800	560	620	600	600	520	550
500	840	900			660	660		
630	960	1010			740	730		
800	1080	1120			810	780		
1000	1190	1220			880	820		
960[b]	1270	1260			930	840		
1000[b]	1420	1370			1010	880		
Aluminium conductors								
16			62	74			69	79
25			82	94			89	99
35			100	110			110	115
50	155	165	120	135	150	155	130	140
70	195	210	150	170	180	190	160	170
95	235	260	185	205	220	225	190	205
120	275	300	215	240	250	260	220	235
150	315	340	245	270	280	290	245	265
185	360	395	280	315	315	330	280	300
240	435	490	335	370	370	380	325	345

(*cont.*)

> **PAPER INSULATED
> DISTRIBUTION CABLES
> (BS 6480: 1988/HD 621 3J1 & 4J2)**
> 6.35/11 and 8.7/15 kV

Table A12.11 continued

Conductor size (mm^2)	In air				In ground			
	Single-core		3-core		Single-core		3-core	
	Trefoil (A)	Flat (A)	Belted[a] (A)	Screened (A)	Trefoil (A)	Flat (A)	Belted[a] (A)	Screened (A)
300	500	560	385	430	415	430	365	390
400	580	640	450	495	475	485	415	440
500	670	730			530	540		
630	780	840			600	610		
800	900	950			680	670		
1000	1020	1050			750	720		

[a] 6.35/11 kV only
[b] Milliken conductors

For the relevant conditions see the notes at the beginning of this appendix.
Single-core cables are PVC sheathed unarmoured. 3-core cables are armoured.

**PAPER INSULATED
DISTRIBUTION CABLES
(EA 09-12/HD 621 4J3)
6.35/11 kV**

Table A12.12 Corrugated aluminium sheathed cables. Sustained ratings

Conductor size (mm^2)	In air 3-core Belted (A)	In air 3-core Screened (A)	In ground 3-core Belted (A)	In ground 3-core Screened (A)
Aluminium conductors				
95	180	200	185	205
185	270	305	270	295
300	370	410	355	380

**PAPER INSULATED
DISTRIBUTION CABLES
(BS 6480: 1988/HD 621 3J1 & 4J2)
12.7/22 kV**

Table A12.13 Sustained ratings

Conductor size (mm^2)	In air				In ground			
	Single-core		3-core		Single-core		3-core	
	Trefoil (A)	Flat (A)	Screened (A)	SL (A)	Trefoil (A)	Flat (A)	Screened (A)	SL (A)
Copper conductors								
25			115	115			120	120
35			135	140			140	145
50	185	200	165	170	180	190	165	170
70	235	255	210	210	225	230	210	210
95	290	310	255	255	265	275	250	250
120	330	360	290	295	305	315	285	285
150	380	410	330	335	340	350	320	315
185	435	470	380	380	385	395	360	355
240	510	560	445	445	445	455	415	410
300	590	640	510	510	500	510	465	455
400	680	730	580	580	570	570	520	510
500	780	820			630	620		
630	890	920			710	680		
800	1000	1020			780	730		
1000	1110	1110			840	770		
960[a]	1170	1150			880	780		
1200[a]	1310	1250			950	820		
Aluminium conductors								
25			88	91			92	93
35			105	110			110	110
50	145	155	125	130	140	145	130	130
70	180	195	160	165	175	180	165	160
95	225	240	195	200	210	215	195	195
120	260	280	225	230	235	245	225	220
150	295	320	255	260	265	275	250	250
185	340	370	295	300	300	310	285	280
240	405	445	350	355	350	360	330	325
300	465	510	400	405	395	405	370	365

(*cont.*)

> **PAPER INSULATED
> DISTRIBUTION CABLES
> (BS 6480: 1988/HD 621 3J1 & 4J2)**
> 12.7/22 kV

Table A12.13 continued

Conductor size (mm^2)	In air				In ground			
	Single-core		3-core		Single-core		3-core	
	Trefoil (A)	Flat (A)	Screened (A)	SL (A)	Trefoil (A)	Flat (A)	Screened (A)	SL (A)
400	540	590	465	465	450	460	420	415
500	620	670			510	510		
630	730	760			580	570		
800	830	860			650	630		
1000	940	960			710	670		

[a] Milliken conductors

For the relevant conditions see the notes at the beginning of this appendix.
Single-core cables are PVC sheathed unarmoured. 3-core cables are armoured.

PAPER INSULATED DISTRIBUTION CABLES (BS 6480: 1988/HD 621 3J2 & 4J1) 19/33 kV

Table A12.14 Sustained ratings

Conductor size (mm^2)	In air Single-core Trefoil (A)	In air Single-core Flat (A)	In air 3-core Screened (A)	In air 3-core SL (A)	In ground Single-core Trefoil (A)	In ground Single-core Flat (A)	In ground 3-core Screened (A)	In ground 3-core SL (A)
Copper conductors								
50	190	205	165	175	180	190	165	170
70	240	255	215	215	225	230	210	210
95	290	315	260	260	265	275	250	250
120	335	360	295	300	305	315	285	285
150	380	410	335	340	340	350	320	320
185	440	470	385	385	385	395	360	355
240	520	560	450	455	445	450	415	410
300	590	630	510	515	500	500	465	455
400	690	720	580	580	570	560	520	510
500	780	810			630	620		
630	890	910			710	670		
800	1010	1010			780	720		
1000	1110	1100			840	760		
960[a]	1170	1130			870	770		
1200[a]	1300	1240			950	800		
Aluminium conductors								
50	150	160	130	135	140	145	130	135
70	185	200	165	170	175	180	165	165
95	225	245	200	205	205	215	195	195
120	260	280	230	235	235	245	225	225
150	300	320	260	265	265	275	250	250
185	340	370	300	305	300	310	285	280
240	405	440	355	360	350	360	330	325
300	465	500	405	410	395	405	370	365
400	540	580	470	470	450	455	420	415
500	620	660			510	510		
630	730	750			580	570		
800	830	850			650	620		
1000	940	950			710	670		

[a] Milliken conductors

For the relevant conditions see the notes at the beginning of this appendix.
Single-core cables are PVC sheathed, unarmoured. 3-core cables are armoured.

**PAPER INSULATED
DISTRIBUTION CABLES
(BS 6480: 1988)
600/1000 V**

Table A12.15 Voltage drop (50 Hz, per A/m)

Conductor size (mm²)	Copper Single-core[a] Trefoil (mV)	Copper Single-core[a] Flat (mV)	Copper 2-core (mV)	Copper 3- or 4-core (mV)	Aluminium Single-core[a] Trefoil (mV)	Aluminium Single-core[a] Flat (mV)	Aluminium 2-core (mV)	Aluminium 3- or 4-core (mV)
16			2.8	2.5			4.7	4.1
25			1.8	1.6			3.0	2.6
35			1.3	1.1			2.2	1.9
50	0.85	0.89	0.95	0.82	1.4	1.4	1.6	1.3
70	0.60	0.66	0.66	0.58	0.97	1.0	1.1	0.95
95	0.44	0.52	0.49	0.43	0.71	0.76	0.81	0.70
120	0.36	0.45	0.40	0.35	0.57	0.62	0.64	0.55
150	0.31	0.40	0.33	0.28	0.47	0.54	0.53	0.46
185	0.26	0.37	0.28	0.24	0.38	0.46	0.43	0.38
240	0.22	0.34	0.24	0.20	0.31	0.40	0.34	0.31
300	0.20	0.32	0.21	0.18	0.26	0.36	0.29	0.26
400	0.18	0.31	0.20	0.17	0.22	0.34	0.24	0.22
500	0.16	0.30			0.19	0.32		
630	0.15	0.29			0.17	0.30		
800	0.15	0.29			0.16	0.30		
1000	0.14	0.29			0.15	0.29		

[a] Data for three unarmoured cables

For the relevant conditions see the notes at the beginning of this appendix.

PAPER INSULATED DISTRIBUTION CABLES
(BS 6480: 1988)
600/1000 V and 1.9/3.3 kV

Table A12.16 Electrical characteristics

Conductor size (mm^2)	Single-core[a] A.C. resistance at 80°C Copper (Ω/km)	Single-core[a] A.C. resistance at 80°C Aluminium (Ω/km)	Reactance (50 Hz) Trefoil (Ω/km)	Reactance (50 Hz) Flat (Ω/km)	Capacitance (μF/km)	Multicore A.C. resistance at 80°C Copper (Ω/km)	Multicore A.C. resistance at 80°C Aluminium (Ω/km)	Reactance (50 Hz) (Ω/km)	Capacitance (μF/km)
600/1000 cables									
16						1.42	2.37	0.080	0.56
25						0.898	1.49	0.076	0.62
35						0.648	1.08	0.074	0.69
50	0.478	0.796	0.102	0.189	0.77	0.479	0.796	0.073	0.79
70	0.330	0.552	0.097	0.183	0.91	0.332	0.551	0.071	0.93
95	0.239	0.397	0.093	0.180	0.98	0.239	0.398	0.069	1.06
120	0.190	0.315	0.090	0.177	1.10	0.190	0.315	0.068	1.20
150	0.155	0.257	0.088	0.175	1.13	0.155	0.257	0.069	1.09
185	0.124	0.205	0.086	0.173	1.25	0.124	0.205	0.068	1.21
240	0.0954	0.156	0.084	0.197	1.26	0.0954	0.157	0.068	1.23
300	0.0771	0.125	0.083	0.190	1.32	0.0771	0.126	0.067	1.37
400	0.0616	0.0988	0.081	0.182	1.40	0.0617	0.0989	0.067	1.54
500	0.0495	0.0778	0.080	0.175	1.42				
630	0.0404	0.0617	0.079	0.167	1.60				
800	0.0340	0.0501	0.077	0.159	1.80				
1000	0.0295	0.0417	0.075	0.152	2.01				

(*cont.*)

PAPER INSULATED DISTRIBUTION CABLES (BS 6480: 1988) 600/1000 V and 1.9/3.3 kV

Table A12.16 continued

Conductor size (mm²)	A.C. resistance at 80°C Copper (Ω/km)	A.C. resistance at 80°C Aluminium (Ω/km)	Single-core[a] Reactance (50 Hz) Trefoil (Ω/km)	Single-core[a] Reactance (50 Hz) Flat (Ω/km)	Capacitance (μF/km)	Multicore A.C. resistance at 80°C Copper (Ω/km)	Multicore A.C. resistance at 80°C Aluminium (Ω/km)	Reactance (50 Hz) (Ω/km)	Capacitance (μF/km)
1.9/3.3 kV cables									
16						1.42	2.37	0.091	0.38
25						0.898	1.49	0.083	0.47
35						0.648	1.08	0.081	0.52
50	0.478	0.796	0.106	0.164	0.61	0.479	0.796	0.079	0.59
70	0.330	0.552	0.100	0.158	0.71	0.332	0.551	0.076	0.68
95	0.239	0.397	0.096	0.154	0.82	0.239	0.398	0.073	0.78
120	0.190	0.315	0.092	0.150	0.92	0.190	0.315	0.072	0.88
150	0.155	0.257	0.090	0.148	1.01	0.155	0.257	0.071	0.96
185	0.124	0.205	0.088	0.146	1.12	0.124	0.205	0.070	1.06
240	0.0954	0.156	0.085	0.168	1.27	0.0954	0.157	0.069	1.20
300	0.0771	0.125	0.083	0.161	1.41	0.0771	0.126	0.068	1.33
400	0.0616	0.0988	0.081	0.153	1.50	0.0616	0.0988	0.068	1.49
500	0.0495	0.0778	0.080	0.146	1.59				
630	0.0404	0.0617	0.078	0.137	1.80				
800	0.0340	0.0501	0.077	0.130	2.02				
1000	0.0295	0.0417	0.076	0.123	2.26				

[a] Three unarmoured cables

For the relevant conditions see the notes at the beginning of this appendix.

PAPER INSULATED DISTRIBUTION CABLES (BS 6480: 1988/HD 621 3J1 & 4J2) 3.8/6.6 kV

Table A12.17 Electrical characteristics

Conductor size (mm²)	Single-core[a] A.C. resistance at 80°C Copper (Ω/km)	Single-core[a] A.C. resistance at 80°C Aluminium (Ω/km)	Reactance (50 Hz) Trefoil (Ω/km)	Reactance (50 Hz) Flat (Ω/km)	Capacitance (μF/km)	Multicore A.C. resistance at 80°C Copper (Ω/km)	Multicore A.C. resistance at 80°C Aluminium (Ω/km)	Reactance (50 Hz) (Ω/km)	Capacitance (μF/km)
16						1.42	2.37	0.106	0.26
25						0.899	1.49	0.096	0.32
35						0.648	1.08	0.092	0.35
50	0.478	0.796	0.111	0.169	0.48	0.479	0.796	0.089	0.39
70	0.330	0.551	0.105	0.163	0.56	0.332	0.550	0.085	0.46
95	0.239	0.397	0.100	0.158	0.64	0.239	0.398	0.081	0.51
120	0.190	0.315	0.096	0.154	0.71	0.190	0.315	0.079	0.57
150	0.155	0.257	0.094	0.151	0.78	0.154	0.257	0.078	0.62
185	0.124	0.205	0.091	0.149	0.86	0.124	0.205	0.076	0.69
240	0.0953	0.156	0.088	0.168	0.97	0.0952	0.156	0.074	0.77
300	0.0769	0.125	0.086	0.161	1.08	0.0768	0.126	0.073	0.85
400	0.0614	0.0987	0.084	0.153	1.21	0.0613	0.0986	0.072	0.95
500	0.0504	0.0793	0.082	0.146	1.34				
630	0.0410	0.0629	0.080	0.137	1.52				
800	0.0345	0.0509	0.078	0.130	1.70				
1000	0.0299	0.0423	0.077	0.123	1.90				

[a] Three unarmoured cables

For the relevant conditions see the notes at the beginning of this appendix.

PAPER INSULATED DISTRIBUTION CABLES (BS 6480: 1988/HD 621 3J1 & 4J2) 6.35/11 kV

Table A12.18 Electrical characteristics

Conductor size (mm^2)	Single-core[a] A.C. resistance at 70°C Copper (Ω/km)	Aluminium (Ω/km)	Reactance (50 Hz) Trefoil (Ω/km)	Flat (Ω/km)	Capacitance (μF/km)	Multicore A.C. resistance at operating temperature[b] Copper (Ω/km)	Aluminium (Ω/km)	Reactance (50 Hz) (Ω/km)	Capacitance (μF/km)
							Belted cables		
16						1.35	2.26	0.112	0.24
25						0.856	1.42	0.105	0.28
35						0.617	1.03	0.098	0.31
50	0.463	0.770	0.113	0.173	0.37	0.456	0.757	0.094	0.34
70	0.320	0.533	0.107	0.167	0.42	0.316	0.524	0.090	0.39
95	0.232	0.384	0.102	0.162	0.48	0.228	0.378	0.086	0.44
120	0.184	0.305	0.098	0.158	0.53	0.181	0.299	0.083	0.48
150	0.150	0.249	0.095	0.155	0.57	0.147	0.244	0.082	0.52
185	0.120	0.198	0.093	0.152	0.63	0.118	0.195	0.080	0.57
240	0.0992	0.151	0.090	0.168	0.71	0.0907	0.149	0.078	0.64
300	0.0745	0.121	0.088	0.161	0.78	0.0732	0.120	0.076	0.70
400	0.0595	0.0956	0.085	0.153	0.87	0.0585	0.0939	0.075	0.78
500	0.0477	0.0752	0.085	0.146	0.96				
630	0.0389	0.0596	0.083	0.137	1.00				
800	0.0329	0.0483	0.081	0.130	1.20				

						Screened cables		
1000	0.0284	0.0400	0.080	0.123				
960[c]	0.0243		0.079	0.121	1.34			
1200[c]	0.0201		0.078	0.114	1.36			
					1.51			
16						1.38	2.29	0.113	0.31
25						0.870	1.44	0.106	0.36
35						0.627	1.04	0.099	0.41
50						0.463	0.770	0.095	0.46
70						0.321	0.533	0.091	0.53
95						0.232	0.385	0.087	0.60
120						0.184	0.305	0.084	0.67
150						0.149	0.248	0.082	0.73
185						0.120	0.198	0.081	0.80
240						0.0921	0.151	0.078	0.90
300						0.0743	0.122	0.077	1.00
400						0.0593	0.0954	0.075	1.11

[a] Three unarmoured cables
[b] 65°C for belted and 70°C for screened cables
[c] Milliken conductors

For the relevant conditions see the notes at the beginning of this appendix.

**PAPER INSULATED
DISTRIBUTION CABLES
(EA 09-12/HD 621 HJ3)
6.35/11 kV**

Table A12.19 Corrugated aluminium sheathed cables. Electrical characteristics

Conductor size (mm^2)	3-core					
	Belted			Screened		
	A.C. resistance at operating temperature[a] (Ω/km)	Reactance (50 Hz) (Ω/km)	Capacitance (μF/km)	A.C. resistance at operating temperature[a] (Ω/km)	Reactance (50 Hz) (Ω/km)	Capacitance (μF/km)
Aluminium conductors						
95	0.379	0.087	0.45	0.384	0.087	0.60
185	0.195	0.080	0.58	0.198	0.081	0.81
300	0.120	0.077	0.71	0.122	0.077	1.00

[a] 65°C for belted and 70°C for screened cables

PAPER INSULATED DISTRIBUTION CABLES (BS 6480: 1988/HD 621 3J1 & 4J2) 8.7/15 and 12.7/22 kV

Table A12.20 Electrical characteristics

Conductor size (mm²)	Single-core[a] A.C. resistance at operating temperature[b] Copper (Ω/km)	Aluminium (Ω/km)	Reactance (50 Hz) Trefoil (Ω/km)	Flat (Ω/km)	Capacitance (µF/km)	Multicore A.C. resistance at operating temperature[b] Copper (Ω/km)	Aluminium (Ω/km)	Reactance (50 Hz) (Ω/km)	Capacitance (µF/km)
8.17/15 kV cables									
25						0.870	1.44	0.114	0.30
35						0.627	1.04	0.106	0.34
50	0.463	0.770	0.118	0.178	0.37	0.463	0.770	0.102	0.38
70	0.320	0.533	0.111	0.172	0.42	0.321	0.533	0.097	0.43
95	0.232	0.384	0.107	0.166	0.48	0.232	0.385	0.092	0.49
120	0.184	0.305	0.102	0.162	0.53	0.184	0.305	0.089	0.54
150	0.150	0.248	0.099	0.159	0.57	0.149	0.248	0.087	0.59
185	0.120	0.198	0.096	0.156	0.63	0.120	0.198	0.085	0.64
240	0.0921	0.151	0.093	0.168	0.71	0.0920	0.151	0.082	0.72
300	0.0744	0.121	0.091	0.161	0.78	0.0742	0.122	0.080	0.80
400	0.0594	0.0955	0.088	0.153	0.87	0.0592	0.0953	0.078	0.89
500	0.0476	0.0751	0.087	0.146	0.96				
630	0.0387	0.0595	0.085	0.137	1.08				
800	0.0324	0.0482	0.083	0.130	1.20				
1000	0.0281	0.0400	0.082	0.123	1.34				

(*cont.*)

PAPER INSULATED DISTRIBUTION CABLES (BS 6480: 1988/HD 621 3J1 & 4J2) 8.7/15 and 12.7/22 kV

Table A12.20 continued

Conductor size (mm²)	Single-core[a] A.C. resistance at operating temperature[b] Copper (Ω/km)	Aluminium (Ω/km)	Reactance (50 Hz) Trefoil (Ω/km)	Flat (Ω/km)	Capacitance (μF/km)	Multicore A.C. resistance at operating temperature[b] Copper (Ω/km)	Aluminium (Ω/km)	Reactance (50 Hz) (Ω/km)	Capacitance (μF/km)
960[c]	0.0243		0.081	0.121	1.36				
1200[c]	0.0202		0.080	0.114	1.51				
12.7/22 kV cables									
25						0.856	1.42	0.125	0.25
35						0.617	1.03	0.116	0.28
50	0.455	0.757	0.126	0.187	0.30	0.456	0.757	0.111	0.31
70	0.315	0.524	0.119	0.180	0.34	0.316	0.524	0.106	0.35
95	0.227	0.377	0.114	0.174	0.39	0.228	0.378	0.100	0.39
120	0.181	0.299	0.109	0.169	0.42	0.181	0.299	0.097	0.43
150	0.148	0.244	0.106	0.165	0.46	0.147	0.244	0.094	0.47
185	0.118	0.195	0.102	0.162	0.50	0.118	0.195	0.091	0.51
240	0.0906	0.148	0.099	0.168	0.56	0.0904	0.149	0.088	0.57
300	0.0730	0.119	0.096	0.161	0.61	0.0728	0.119	0.086	0.62
400	0.0582	0.0937	0.093	0.153	0.68	0.0581	0.0936	0.083	0.69
500	0.0466	0.0737	0.092	0.146	0.75				

630	0.0379	0.0584	0.089	0.137	0.84
800	0.0317	0.0472	0.087	0.130	0.93
1000	0.0275	0.0392	0.085	0.123	1.03
960[c]	0.0239		0.084	0.121	1.05
1200[c]	0.0197		0.083	0.114	1.16

[a] Three unarmoured cables
[b] 70°C for 8.7/15 kV and 65°C for 12.7/22 kV
[c] Milliken conductors

For the relevant conditions see the notes at the beginning of this appendix.

PAPER INSULATED DISTRIBUTION CABLES (BS 6480: 1988/HD 621 3J2 & 4J1) 19/33 kV

Table A12.21 Electrical characteristics

Conductor size (mm^2)	Single-core[a] A.C. resistance at 65°C Copper (Ω/km)	A.C. resistance at 65°C Aluminium (Ω/km)	Reactance (50 Hz) Trefoil (Ω/km)	Reactance (50 Hz) Flat (Ω/km)	Capacitance (μF/km)	Multicore A.C. resistance at 65°C Copper (Ω/km)	A.C. resistance at 65°C Aluminium (Ω/km)	Reactance (50 Hz) (Ω/km)	Capacitance (μF/km)
50[a]	0.447	0.742	0.138	0.200	0.23	0.456	0.758	0.126	0.24
70	0.309	0.514	0.130	0.193	0.26	0.316	0.524	0.119	0.27
95	0.223	0.370	0.124	0.185	0.30	0.228	0.379	0.112	0.30
120	0.178	0.294	0.118	0.180	0.32	0.181	0.300	0.107	0.33
150	0.145	0.240	0.114	0.175	0.36	0.147	0.244	0.103	0.37
185	0.116	0.191	0.114	0.171	0.39	0.118	0.195	0.101	0.40
240	0.0886	0.145	0.110	0.168	0.43	0.0902	0.149	0.097	0.44
300	0.0714	0.117	0.106	0.161	0.47	0.0727	0.120	0.094	0.48
400	0.0569	0.0918	0.103	0.153	0.52	0.0582	0.0935	0.090	0.53
500	0.0464	0.0736	0.098	0.146	0.57				
630	0.0376	0.0582	0.097	0.138	0.63				
800	0.0314	0.0470	0.092	0.130	0.70				
1000	0.0271	0.0389	0.089	0.123	0.78				
960[b]	0.0238		0.089	0.121	0.79				
1200[b]	0.0196		0.087	0.114	0.87				

[a] Circular conductors in multicore cables
[b] Milliken conductors

For the relevant conditions see the notes at the beginning of this appendix.
Except for 50 mm^2, multicore cables have oval conductors.

APPENDIX A13
PVC Insulated Distribution Cables

CONTENTS

Tables are included for dimensions and weights, sustained ratings and electrical characteristics. Notes on the cable designs and conditions applicable are given before the tables.

The data given apply to cables to BS 6346: 1989 in sizes from 16 mm^2 upwards. For smaller sizes see appendix A5.

CABLE DESIGNS

Typical designs are shown in figs A13.1 and A13.2.

Fig. A13.1 PVC insulated 3-core 600/1000 V wire armoured cable with solid aluminium conductor

Fig. A13.2 PVC insulated single-core 600/1000 V, unarmoured cable with sectoral aluminium conductor

Values are provided for unarmoured and armoured cables to BS 6346: 1989 with the following characteristics:

Conductors
Stranded copper and solid aluminium circular conductors for single-core cables and shaped for multicore cables. Cables with stranded aluminium conductors are also available and the dimensions are the same as for cables with copper conductors.

Armour bedding
Extruded PVC for single-core cables, extruded or taped PVC for multicore cables.

Armour
Armoured cables are assumed to have SWA, aluminium for single-core cables and galvanised steel for multicore cables.

Oversheath
Extruded PVC.

SUSTAINED RATINGS

The maximum ratings in the tables are for a single circuit and are based on ERA Technology Ltd Report 69-30, Part III, which is in general conformity with IEC 287. However, the ratings tabulated in ERA Report 69-30 are based on an ambient air temperature of 25°C whereas the ratings given in this appendix are based on a temperature of 30°C to align with the ratings for 600/1000 V cables tabulated in the IEE Regulations for Electrical Installations. When temperature correction factors are applied, the ratings become identical.

The standard conditions on which the tabulated ratings are based are given below. For other conditions the rating factors included in chapter 8 should be applied.

Maximum conductor temperature
70°C.

Circuit protection
The cable should be selected with a rating of not less than the nominal current of the device providing protection against overload, or not less than 1.38 times this value if the device will not operate within 4 hours at 1.45 times its nominal current.

Installation in air
An ambient air temperature of 30°C.

The cables are shielded from the direct rays of the sun. Air circulation is not restricted significantly, e.g. if fastened to a wall the cables are spaced at least 20 mm from it; if in a concrete trough they are not covered.

Adjacent circuits are spaced apart (chapter 8) and suitably disposed to prevent mutual heating.

Cables laid direct in ground

Ground temperature
15°C.

Ground thermal resistivity
1.2 K m/W.

Adjacent circuits
At least 1.8 m apart.

Depth of laying
0.5 m for voltages up to 1000 V and 0.8 m for 1.9/3.3 kV cables (measured from ground surface to centre of cable or trefoil group).

Single-core cables
The data apply to three or four single-core cables operating 3-phase.

Bonding
It is assumed that the armour will be solidly bonded, i.e. bonded at both ends of the run. For very short runs it may be found possible to bond at one end only but consideration must be given to the value of the standing voltage which can occur along the cable length under normal and fault conditions.

Trefoil
A close trefoil is assumed with the cables touching.

Flat formation
The ratings and technical data are based on horizontal installation with a spacing between cable centres of twice the overall diameter. Cables installed vertically will have somewhat lower ratings.

ELECTRICAL CHARACTERISTICS

The standard conditions given for sustained ratings also apply to the electrical characteristics. A.C. resistances are tabulated. D.C. resistances are given in table 4.1 of chapter 4.

PVC INSULATED DISTRIBUTION CABLES (BS 6346: 1989)
600/1000 V – copper

Table A13.1 Dimensions and weights

Conductor size (mm^2)	Approximate diameter			Approximate weight		
	Unarmoured (mm)	Armoured (SWA)		Unarmoured (kg/m)	Armoured (SWA)	
		Tape bedding (mm)	Extruded bedding (mm)		Tape bedding (kg/m)	Extruded bedding (kg/m)
Single-core (aluminium wire armoured)						
50	15.1		19.1	0.60		0.78
70	16.9		21.1	0.81		1.03
95	19.4		23.4	1.10		1.33
120	21.0		26.3	1.35		1.68
150	23.2		28.3	1.65		2.00
185	25.8		30.8	2.06		2.43
240	29.0		34.1	2.67		3.09
300	32.1		37.0	3.32		3.77
400	35.8		42.0	4.19		4.83
500	39.6		45.6	5.23		5.92
630	43.8		49.7	6.63		7.42
800	48.3		55.8	8.33		9.50
1000	53.7		61.0	10.44		11.76
2-core cables						
16[a]	18.6	21.9	21.9	0.54	0.98	0.99
25	18.4	22.6	23.0	0.69	1.26	1.28
35	20.1	24.5	24.9	0.95	1.59	1.61
50	22.8	27.4	27.8	1.26	1.99	2.01
70	25.5	30.0	30.4	1.70	2.50	1.52
95	29.3	34.7	35.5	2.31	3.46	3.52
120	31.8	37.2	38.0	2.88	4.12	4.20
150	35.1	40.5	41.3	3.52	4.89	4.96
185	39.1	45.2	46.4	4.39	6.25	6.39
240	43.9	50.0	51.2	5.76	7.86	8.02
300	48.7	54.8	56.4	7.16	9.48	9.71
400	54.2	60.3	61.9	9.04	11.60	11.85

(*cont.*)

PVC INSULATED
DISTRIBUTION CABLES
(BS 6346: 1989)
600/1000 V – copper

Table A13.1 continued

Conductor size (mm^2)	Approximate diameter			Approximate weight		
	Unarmoured (mm)	Armoured (SWA)		Unarmoured (kg/m)	Armoured (SWA)	
		Tape bedding (mm)	Extruded bedding (mm)		Tape bedding (kg/m)	Extruded bedding (kg/m)
3-core cables						
16[a]	19.7	23.1	23.1	0.73	1.20	1.21
25	20.4	24.6	25.0	1.00	1.65	1.67
35	22.4	26.9	27.3	1.30	2.03	2.05
50	25.5	30.1	30.5	1.72	2.56	2.58
70	28.7	34.2	35.0	2.36	3.52	3.59
95	33.3	38.5	39.3	3.33	4.64	4.71
120	36.3	41.4	42.2	4.10	5.51	5.59
150	40.0	46.3	47.5	5.02	6.97	7.11
185	44.6	50.7	51.9	6.26	8.39	8.54
240	50.1	56.2	57.8	8.15	10.55	10.79
300	55.6	61.6	63.2	10.14	12.79	13.04
400	62.2	68.0	69.6	12.86	15.76	16.02
4-core cables						
16[a]	21.6	25.9	26.3	0.92	1.58	1.62
25	22.9	27.4	27.8	1.29	2.03	2.05
35	25.4	30.1	30.5	1.69	2.51	2.53
50	29.2	34.6	35.4	2.25	3.41	3.48
70	33.0	38.4	39.2	3.10	4.40	4.47
95	38.3	43.5	44.3	4.36	5.83	5.90
120	41.8	48.1	49.3	5.38	7.40	7.54
150	46.3	52.4	53.6	6.63	8.81	8.97
185	51.3	57.4	59.0	8.25	10.66	10.89
240	58.0	64.1	65.7	10.73	14.43	13.69
300	64.6	70.4	72.0	13.38	16.33	16.61
400	72.0	79.3	81.3	16.93	21.07	21.48
4-core cables with reduced neutral[b]						
25	22.9	27.4	27.8	1.26	2.00	2.02
35	24.7	29.1	29.5	1.59	2.41	2.43
50	28.3	32.7	33.1	2.12	3.03	3.05
70	32.0	37.2	38.0	2.89	4.15	4.22
95	37.5	42.9	43.7	3.92	5.38	5.46

(cont.)

> **PVC INSULATED**
> **DISTRIBUTION CABLES**
> **(BS 6346: 1989)**
> 600/1000 V – copper

Table A13.1 continued

Conductor size (mm^2)	Approximate diameter			Approximate weight		
	Unarmoured (mm)	Armoured (SWA)		Unarmoured (kg/m)	Armoured (SWA)	
		Tape bedding (mm)	Extruded bedding (mm)		Tape bedding (kg/m)	Extruded bedding (kg/m)
120	41.4	47.8	49.0	4.89	6.85	6.99
150	44.7	50.8	52.0	5.90	8.03	8.18
185	49.9	56.0	57.2	7.40	9.75	9.91
240	56.0	62.1	63.7	9.59	12.23	12.47
300	62.2	68.2	69.8	11.91	14.81	15.08
300	64.2	70.2	71.8	12.19	15.09	15.36
400	69.9	76.6	78.6	15.07	19.07	19.47

[a] Circular conductors
[b] Size of reduced neutral conductor

Phase conductor (mm^2)	25	35	50	70	95	120
Neutral conductor (mm^2)	16	16	25	35	50	70
Phase conductor (mm^2)	150	185	240	300	300	400
Neutral conductor (mm^2)	70	95	120	150	185	185

All cables have stranded conductors.
See appendix A5 for smaller sizes of armoured cables.
The phase conductors are shaped, but for some sizes the neutral conductors are circular.

PVC INSULATED DISTRIBUTION CABLES
(BS 6346: 1989)
600/1000 V – aluminium

Table A13.2 Dimensions and weights

Conductor size (mm²)	Approximate diameter,[a] solid aluminium conductor			Approximate weight			
	Unarmoured (mm)	Armoured (SWA)		Unarmoured (kg/m)	Armoured (SWA)	
		Tape bedding (mm)	Extruded bedding (mm)		Tape bedding (kg/m)	Extruded bedding (kg/m)
Single-core (aluminium wire armoured)						
50	13.8		17.8	0.28		0.46
70	15.4		19.6	0.36		0.57
95	17.6		21.6	0.48		0.70
120	19.0		24.3	0.57		0.89
150	21.0		26.1	0.69		1.03
185	23.3		28.3	0.86		1.22
240	26.1		31.2	1.09		1.50
300	28.9		33.7	1.34		1.77
380[b]	32.4		38.4	1.67		2.25
480[b]	35.7		41.7	2.06		2.70
600[b]	38.7		44.6	2.44		3.13
740[b]	42.2		49.5	2.94		3.89
960[b]	47.4		54.9	3.75		4.79
1200[b]	52.0		59.7	4.58		5.78
2-core cables						
16[c]	17.4	20.7	20.7	0.34	0.34	0.62
25	16.6	20.9	21.3	0.38	0.92	0.94
35	18.0	22.5	22.9	0.49	1.09	1.11
50	20.4	25.1	25.5	0.62	1.31	1.33
70	22.8	27.3	27.7	0.81	1.56	1.58
95	26.2	31.6	32.4	1.08	2.16	2.22
3-core cables						
16[c]	18.4	21.8	21.8	0.42	0.86	0.86
25	19.2	23.5	23.9	0.51	1.14	1.16
35	21.0	25.4	25.8	0.63	1.32	1.34
50	23.8	28.5	28.9	0.81	1.61	1.62
70	26.8	32.2	33.0	1.06	2.16	2.22

(*cont.*)

> **PVC INSULATED**
> **DISTRIBUTION CABLES**
> **(BS 6346: 1989)**
> 600/1000 V – aluminium

Table A13.2 continued

Conductor size (mm^2)	Approximate diameter,[a] solid aluminium conductor			Approximate weight			
	Unarmoured (mm)	Armoured (SWA)		Unarmoured (kg/m)	Armoured (SWA)	
		Tape bedding (mm)	Extruded bedding (mm)		Tape bedding (kg/m)	Extruded bedding (kg/m)
95	31.1	36.3	37.1	1.49	2.73	2.80
120	33.7	38.9	39.7	1.78	3.11	3.19
150	37.2	43.5	44.7	2.16	4.00	4.14
185	41.4	47.5	48.7	2.69	4.69	4.84
240	46.5	52.6	54.2	3.44	5.68	5.90
300	51.6	57.6	59.2	4.25	6.72	6.96
4-core cables						
16[c]	20.4	24.4	24.8	0.50	1.13	1.17
25	21.5	25.9	26.3	0.65	1.36	1.38
35	23.6	28.2	28.6	0.78	1.59	1.61
50	27.1	32.5	33.3	1.05	2.17	2.24
70	30.6	36.0	36.8	1.38	2.62	2.69
95	35.5	40.6	41.4	1.95	3.35	3.42
120	38.6	44.9	46.1	2.32	4.22	4.36
150	42.8	48.9	50.1	2.84	4.90	5.05
185	47.4	53.5	55.1	3.52	5.79	6.01
240	53.5	59.6	61.2	4.53	7.07	7.31
300	59.6	65.4	67.0	5.62	8.39	8.65

[a] The diameter of cables with stranded conductors is the same as for copper (see table A13.1)
[b] Solid sectoral aluminium conductors
[c] Circular conductors

**PVC INSULATED
DISTRIBUTION CABLES
(BS 6346: 1989)
1.9/3.3 kV**

Table A13.3 Dimensions and weights (armoured cables, SWA)

Conductor size (mm^2)	Approximate diameter		Approximate weight	
	Stranded copper or aluminium conductor (mm)	Solid aluminium conductor (mm)	Stranded copper conductor (kg/m)	Solid aluminium conductor (kg/m)
Single-core (aluminium wire armour)				
50	21.0	19.8	0.87	0.55
70	22.8	21.3	1.11	0.65
95	26.0	24.3	1.49	0.85
120	27.6	25.6	1.77	0.97
150	29.4	27.1	2.07	1.10
185	31.3	28.8	2.47	1.25
240	34.1	31.2	3.09	1.50
300	37.0	33.7	3.77	1.77
400	42.0		4.83	
500	45.6		5.92	
630	49.7		7.42	
800	55.8		9.50	
1000	61.0		11.76	
380[a]		38.4		2.25
480[a]		41.7		2.70
600[a]		44.6		3.13
740[a]		49.5		3.89
960[a]		54.9		4.79
1200[a]		59.7		5.78
3-core cables				
16[b]	30.3	29.0	1.80	1.47
25[b]	33.1	31.3	2.25	1.68
35	32.1	30.6	2.57	1.82
50	35.6	33.9	3.33	2.32
70	38.9	36.9	4.14	2.72
95	42.3	40.0	5.11	3.19
120	46.6	44.0	6.45	4.01
150	49.4	46.5	7.43	4.45
185	52.8	49.6	8.73	5.03
240	57.8	54.2	10.83	5.95
300	63.2	59.2	13.09	7.00
400	69.6		16.09	

[a] Solid sectoral aluminium conductors
[b] Circular conductors

The data apply to cables with extruded PVC bedding.

PVC INSULATED DISTRIBUTION CABLES
(BS 6346: 1989)
600/1000 V – armoured

Table A13.4 Sustained ratings

Conductor size (mm^2)	In air Single-core[a] Trefoil (A)	In air Single-core[a] Flat (A)	In air 2-core (A)	In air 3- or 4-core (A)	In ground Single-core[a] Trefoil (A)	In ground Single-core[a] Flat (A)	In ground 2-core (A)	In ground 3- or 4-core (A)
Copper conductors								
16			97	83			119	101
25			128	110			158	132
35			157	135			190	159
50	181	230	190	163	203	211	225	188
70	231	286	241	207	248	257	277	233
95	280	338	291	251	297	305	332	279
120	324	385	336	290	337	341	377	317
150	373	436	386	332	376	377	422	355
185	425	490	439	378	423	417	478	401
240	501	566	516	445	485	469	551	462
300	567	616	592	510	542	515	616	517
400	657	674	683	590	600	549	693	580
500	731	721			660	586		
630	809	771			721	627		
800	886	824			756	648		
1000	945	872			797	679		
Aluminium conductors								
16			71	61			91	77
25			94	80			118	100
35			115	99			142	120
50	131	169	139	119	154	160	168	143
70	168	213	175	151	188	197	209	176
95	205	255	211	186	226	235	250	213
120	238	293		216	257	267		243
150	275	335		250	288	298		272
185	315	379		287	326	332		309
240	372	443		342	377	380		360
300	430	505		399	424	423		407
380[b]	497	551			475	457		
480[b]	568	604			532	501		
600[b]	642	656			586	540		
740[b]	715	707			648	582		
960[b]	808	770			701	608		
1200[b]	880	822			755	644		

[a] Single-core cables with aluminium wire armour, 3-phase circuit
[b] Solid sectoral aluminium conductors

For the relevant conditions see the notes at the beginning of this appendix.

**PVC INSULATED
DISTRIBUTION CABLES
(BS 6346: 1989)**
600/1000 V – unarmoured

Table A13.5 Sustained ratings

Conductor size (mm^2)	In air Single-core[a] Trefoil (A)	In air Single-core[a] Flat (A)	2-core (A)	3- or 4-core (A)
Copper conductors				
16			94	80
25			119	101
35			148	126
50	167	219	180	153
70	216	281	232	196
95	264	341	282	238
120	308	396	328	276
150	356	456	379	319
185	409	521	434	364
240	485	615	514	430
300	561	709	593	497
400	656	852	715	597
500	749	982		
630	855	1138		
800	971	1265		
1000	1079	1420		
Aluminium conductors				
16			73	61
25			89	78
35			111	96
50	128	163	135	117
70	165	210	173	150
95	203	256	210	183
120	237	298		212
150	274	344		245
185	316	394		280
240	375	466		330
300	435	538		381

(*cont.*)

> **PVC INSULATED
> DISTRIBUTION CABLES
> (BS 6346: 1989)**
> 600/1000 V – unarmoured

Table A13.5 continued

Conductor size (mm²)	In air			
	Single-core[a]		2-core (A)	3- or 4-core (A)
	Trefoil (A)	Flat (A)		
380[b]	507	625		
480[b]	590	726		
600[b]	680	837		
740[b]	776	956		
960[b]	907	1125		
1200[b]	1026	1293		

[a] Single-core cables with aluminium wire, 3-phase circuit
[b] Solid sectoral aluminium conductors

For the relevant conditions see the notes at the beginning of this appendix.

**PVC INSULATED
DISTRIBUTION CABLES
(BS 6346: 1989)**
1.9/3.3 V – armoured

Table A13.6 Sustained ratings

Conductor size (mm^2)	In air Single-core[a] Trefoil (A)	In air Single-core[a] Flat (A)	In air 3-core (A)	In ground Single-core[a] Trefoil (A)	In ground Single-core[a] Flat (A)	In ground 3-core (A)
Copper conductors						
16			85			97
25			111			125
35			134			151
50	183	226	163	193	199	178
70	229	281	204	236	242	219
95	284	339	250	282	285	264
120	327	386	290	319	320	299
150	371	433	330	357	354	336
185	426	489	379	401	393	379
240	500	558	446	459	441	436
300	571	620	508	513	483	488
400	649	667	583	566	513	548
500	729	720		621	546	
630	817	780		678	582	
800	881	821		708	599	
1000	949	874		744	626	
Aluminium conductors						
16			64			74
25			84			95
35			102			114
50	134	167	122	147	152	136
70	169	210	154	180	186	168
95	208	256	189	215	222	201
120	240	295	219	244	252	230
150	273	335	249	273	281	257
185	317	380	287	309	313	292
240	375	443	339	357	358	338
300	431	499	388	402	397	381

(*cont.*)

> **PVC INSULATED
> DISTRIBUTION CABLES
> (BS 6346: 1989)**
> 1.9/3.3 V – armoured

Table A13.6 continued

Conductor size (mm^2)	In air Single-core[a] Trefoil (A)	In air Single-core[a] Flat (A)	3-core (A)	In ground Single-core[a] Trefoil (A)	In ground Single-core[a] Flat (A)	3-core (A)
380[b]	497	548		450	429	
480[b]	569	607		503	469	
600[b]	637	661		553	504	
740[b]	719	719		610	541	
960[b]	803	763		658	564	
1200[b]	884	819		707	595	

[a] Single-core cables with aluminium wire armour, 3-phase circuit
[b] Solid sectoral aluminium conductors

For the relevant conditions see the notes at the beginning of this appendix.

PVC INSULATED DISTRIBUTION CABLES (BS 6346: 1989) 600/1000 V

Table A13.7 Voltage drop (50 Hz, per A/m)

Conductor size (mm^2)	Copper Single-core[a] Trefoil (mV)	Copper Single-core[a] Flat (mV)	Copper 2-core (mV)	Copper 3- or 4-core (mV)	Aluminium Single-core[a] Trefoil (mV)	Aluminium Single-core[a] Flat (mV)	Aluminium 2-core (mV)	Aluminium 3- or 4-core (mV)
Armoured cables								
16			2.8	2.4			4.5	3.9
25			1.75	1.5			2.9	2.5
35			1.25	1.1			2.1	1.8
50	0.82	0.86	0.94	0.81	1.35	1.35	1.55	1.35
70	0.58	0.68	0.65	0.57	0.93	1.00	1.05	0.92
95	0.45	0.57	0.50	0.43	0.70	0.80	0.79	0.68
120	0.37	0.50	0.41	0.35	0.57	0.68		0.55
150	0.32	0.45	0.34	0.29	0.47	0.58		0.44
185	0.27	0.41	0.29	0.25	0.39	0.51		0.37
240	0.23	0.37	0.24	0.21	0.32	0.44		0.30
300	0.21	0.34	0.21	0.185	0.27	0.40		0.25
400	0.195	0.32	0.185	0.16				
500	0.18	0.30						
630	0.17	0.28						
800	0.16							
1000	0.155							
380[b]					0.24	0.38		
480[b]					0.22	0.35		
600[b]					0.20	0.32		
740[b]					0.185	0.30		
960[b]					0.175	0.27		
1200[b]					0.17	0.25		
Unarmoured cables								
16			2.8	2.4			4.5	3.9
25			1.75	1.5			2.9	2.5
35			1.25	1.1			2.1	1.8
50	0.82	0.86	0.94	0.81	1.35	1.4	1.55	1.35
70	0.57	0.63	0.65	0.57	0.92	0.96	1.05	0.92

(*cont.*)

PVC INSULATED
DISTRIBUTION CABLES
(BS 6346: 1989)
600/1000 V

Table A13.7 continued

Conductor size (mm^2)	Copper Single-core[a] Trefoil (mV)	Copper Single-core[a] Flat (mV)	2-core (mV)	3- or 4-core (mV)	Aluminium Single-core[a] Trefoil (mV)	Aluminium Single-core[a] Flat (mV)	2-core (mV)	3- or 4-core (mV)
95	0.43	0.51	0.50	0.43	0.69	0.74	0.79	0.68
120	0.36	0.44	0.41	0.35	0.55	0.61		0.55
150	0.30	0.40	0.34	0.29	0.45	0.52		0.44
185	0.26	0.36	0.29	0.25	0.37	0.46		0.37
240	0.22	0.34	0.24	0.21	0.30	0.40		0.30
300	0.19	0.32	0.21	0.185	0.26	0.36		0.25
400	0.175	0.31	0.185	0.16				
500	0.16	0.30						
630	0.15	0.29						
800	0.145	0.29						
1000	0.14	0.28						
380[b]					0.22	0.34		
480[b]					0.195	0.32		
600[b]					0.18	0.31		
740[b]					0.165	0.30		
960[b]					0.155	0.29		
1200[b]					0.15	0.29		

[a] Data for aluminium wire armoured cables, 3-phase circuit
[b] Solid sectoral aluminium conductors

For the relevant conditions see the notes at the beginning of this appendix.

**PVC INSULATED
DISTRIBUTION CABLES
(BS 6346: 1989)
600/1000 V**

Table A13.8 Electrical characteristics

Conductor size (mm^2)	Armoured single-core cables[a] A.C. resistance at 70°C Copper (Ω/km)	Armoured single-core cables[a] A.C. resistance at 70°C Aluminium (Ω/km)	Reactance (50 Hz) Trefoil (Ω/km)	Reactance (50 Hz) Flat[b] (Ω/km)	Armoured or unarmoured multicore cables A.C. resistance at 70°C Copper (Ω/km)	Armoured or unarmoured multicore cables A.C. resistance at 70°C Aluminium (Ω/km)	Reactance (50 Hz) (Ω/km)
16					1.38	2.29	0.087
25					0.870	1.44	0.084
35					0.627	1.04	0.081
50	0.464	0.771	0.112	0.198	0.464	0.770	0.081
70	0.321	0.533	0.107	0.193	0.321	0.533	0.079
95	0.232	0.385	0.103	0.189	0.232	0.385	0.077
120	0.185	0.305	0.103	0.188	0.184	0.305	0.076
150	0.149	0.248	0.101	0.186	0.150	0.248	0.076
185	0.120	0.198	0.099	0.184	0.121	0.198	0.076
240	0.0926	0.152	0.096	0.182	0.0929	0.152	0.075
300	0.0750	0.122	0.094	0.181	0.0752	0.122	0.075
400	0.0600		0.091	0.178	0.0604		0.074
500	0.0484		0.089	0.176			
630	0.0398		0.086	0.173			
800	0.0334		0.086				
1000	0.0290		0.984	0.181			
380[c]		0.0976	0.094	0.179			
480[c]		0.0779	0.092	0.176			
600[c]		0.0643	0.089				
740[c]		0.0522	0.089				
960[c]		0.0415	0.087				
1200[c]		0.0348	0.085				

[a] Aluminium wire armoured
[b] Twice cable diameter spacing between centres
[c] Solid sectoral aluminium conductors

For the relevant conditions see the notes at the beginning of this appendix.

**PVC INSULATED
DISTRIBUTION CABLES
(BS 6346: 1989)
1.9/3.3 kV**

Table A13.9 Electrical characteristics

Conductor size (mm²)	A.C. resistance at 70°C Copper (Ω/km)	A.C. resistance at 70°C Aluminium (Ω/km)	Reactance (50 Hz) Trefoil (Ω/km)	Reactance (50 Hz) Flat[b] (Ω/km)	Capacitance[a] Copper (μF/km)	Capacitance[a] Aluminium (μF/km)
Single-core, aluminium wire armoured						
50	0.464	0.771	0.115	0.173	0.86	0.77
70	0.321	0.533	0.109	0.167	1.00	0.89
95	0.232	0.385	0.107	0.165	1.08	0.96
120	0.185	0.305	0.102	0.160	1.20	1.05
150	0.149	0.248	0.099	0.157	1.30	1.15
185	0.120	0.198	0.096	0.154	1.44	1.26
240	0.926	0.152	0.093	0.151	1.62	1.41
300	0.750	0.122	0.091	0.149	1.69	1.48
400	0.0600		0.091	0.149	1.71	
500	0.0484		0.089	0.147	1.81	
630	0.0398		0.086	0.144	2.03	
800	0.0334		0.086	0.137	2.17	
1000	0.0290		0.084	0.142	2.32	
380[c]		0.0976	0.094	0.152		1.51
480[c]		0.0779	0.092	0.150		1.59
600[c]		0.0643	0.089	0.147		1.74
740[c]		0.0522	0.089	0.147		1.85
960[c]		0.0415	0.087	0.145		2.00
1200[c]		0.0343	0.085	0.143		2.13
3-core, armoured						
16	1.38	2.29	0.107		0.54	0.50
25	0.870	1.44	0.097		0.65	0.57
35	0.627	1.04	0.094		0.71	0.64
50	0.464	0.770	0.090		0.78	0.70
70	0.321	0.533	0.086		0.90	0.80
95	0.232	0.385	0.082		1.01	0.91
120	0.184	0.305	0.080		1.10	0.97
150	0.150	0.248	0.079		1.20	1.06
185	0.121	0.198	0.077		1.32	1.16
240	0.0929	0.152	0.075		1.45	1.27
300	0.0752	0.122	0.075		1.50	1.32
400	0.0604		0.074		1.58	

[a] Not applicable to unarmoured cables
[b] Twice cable diameter spacing between centres
[c] Solid sectoral aluminium conductors

The characteristics apply to cables with extruded PVC bedding.
For the relevant conditions see the notes at the beginning of this appendix.

APPENDIX A14
XLPE Insulated Distribution Cables

CONTENTS

Tables are included for dimensions and weights, sustained ratings and electrical characteristics. Notes on the cable designs and conditions applicable are given before the tables.

The data included cover cables up to 3.3 kV to BS 5467: 1989 and BS 6724: 1990 and cables for 6 kV to 33 kV to BS 6622: 1991. The size range for the 600/1000 V cables is from 16 mm^2 upwards. For smaller cables to BS 5467: 1989 and BS 6724: 1990 see appendix A5.

CABLE DESIGNS

600/1000 V and 1.9/3.3 kV, to BS 5467: 1989 and BS 6724: 1990

Conductors
Stranded copper and solid aluminium circular conductors for single-core and shaped conductors for multicore cables. Stranded aluminium conductors can also be supplied. The dimensions are the same as for cables with copper conductors.

Fig. A14.1 XLPE insulated 4-core 600/1000 V armoured cable

Armour bedding
Extruded PVC for cables to BS 5467: 1989. Extruded special synthetic material for cables to BS 6724: 1990.

A design with PVC tapes is also available for cables to BS 5467: 1989 but dimensions are not included in the tables.

Armour
The tabulated data apply to:

(a) unarmoured and armoured cables to BS 5467: 1989;
(b) armoured only cables to BS 6724: 1990.

Fig. A14.2 XLPE insulated single-core, 6.35/11 kV cable with copper wire screen

Fig. A14.3 XLPE insulated 3-core, 6.35/11 kV wire armoured cable

The armour consists of galvanised steel SWA on multicore cables and aluminium wire SWA on single-core cables.

Oversheath
Extruded PVC for cables to BS 5467: 1989. Special synthetic material for cables to BS 6724: 1990.

3.8/6.6 kV to 19/33 kV inclusive, to BS 6622: 1991 and to IEC 502
Conductors
Stranded copper and stranded aluminium circular conductors for single-core and 3-core cables. Solid aluminium conductors are also available up to and including 11 kV.

The tabulated data only cover circular conductor cables but shaped conductors are also available up to 11 kV.

Screens
Extruded semiconducting layer over conductor and extruded layer over insulation.

Metallic screen component
Copper wires over screen on single-core cables, copper tape on 3-core cables.

Bedding under armour
Extruded PVC on 3-core cables.

Armour
Galvanised steel SWA on 3-core cables.
The tabulated data apply to unarmoured single-core cables and armoured 3-core cables. Aluminium wire armoured single-core cables and unarmoured 3-core cables are also available.

Oversheath
Extruded PVC.

Conductor size range
Ratings are given in the tables for a range of voltages but, particularly at the higher voltages, the smallest conductor sizes may not be manufactured. A footnote indicates when reference should be made to the table of dimensions to check that the size is available.

SUSTAINED RATINGS

The tabulated ratings have been determined in accordance with IEC 287. They are for a single circuit with the standard conditions given below. For other conditions the rating factors included in chapter 8 should be applied.

For cables to BS 5467: 1989 and BS 6724: 1990 (up to 3.3 kV) the ratings given in this appendix are in accordance with ERA Report 69-30, Part V (which is in conformity with IEC 287). For cables in air the ratings are based on an ambient temperature of

30°C to align with the IEE Regulations for Electrical Installations. The ratings actually tabulated in ERA 69-30 are based on 25°C but after application of temperature correction factors the ratings become identical.

Maximum conductor temperature

90°C.

Circuit protection

600/1000 V cables should be selected with a rating of not less than the nominal current of the device providing protection against overload, or not less than 1.38 times this value if the device will not operate within 4 hours at 1.45 times its nominal current.

Installation in air

An ambient temperature of 30°C for cables up to 1.9/3.3 kV and 25°C for higher voltage cables.

The cables are shielded from the direct rays of the sun. Air circulation is not restricted significantly, e.g. if fastened to a wall the cables are spaced at least 20 mm from it; if in a concrete trough they are not covered.

Adjacent circuits are spaced apart (chapter 8) and suitably disposed to prevent mutual heating.

Cables laid direct in ground

Ground temperature
15°C.

Ground thermal resistivity
1.2 K m/W.

Adjacent circuits
At least 1.8 m apart.

Depth of laying
0.5 m for voltages up to 1000 V and 0.8 m for higher voltage cables (measured from ground surface to centre of cable or trefoil group).

Single-core cables

The data apply to three or four single-core cables operating 3-phase.

Bonding
For trefoil installation it is assumed that the armour will be solidly bonded, i.e. bonded at both ends of the run. For very short runs it may be found possible to bond at one end only but consideration must be given to the value of the standing voltage which can occur along the cable length under both normal and fault conditions.

For flat formation the ratings for 600/1000 V and 1.9/3.3 kV cables are based on bonding at both ends and for higher voltage cables on single point bonding.

Trefoil
A close trefoil is assumed with the cables touching.

Flat formation
The ratings and technical data are based on horizontal installation with a spacing between cable centres of twice the overall diameter. Cables installed vertically will have somewhat lower ratings.

ELECTRICAL CHARACTERISTICS

The standard conditions given for sustained ratings also apply to the electrical characteristics. A.C. resistances are tabulated. D.C. resistances are given in table 4.1 of chapter 4.

XLPE INSULATED DISTRIBUTION CABLES
(BS 5467: 1989 or BS 6724: 1990)
600/1000 V – copper

Table A14.1 Dimensions and weights

Conductor size (mm^2)	Approximate diameter Unarmoured (mm)	Armoured (mm)	Approximate weight Unarmoured (kg/m)	Armoured (kg/m)
Single-core (aluminium wire armour)				
50	14.2	17.5	0.54	0.80
70	16.2	20.2	0.75	0.94
95	18.3	22.3	1.01	1.22
120	20.2	24.2	1.25	1.49
150	22.4	27.4	1.53	1.87
185	24.7	30.0	1.90	2.29
240	27.7	32.8	2.47	2.88
300	30.6	35.6	3.08	3.52
400	34.2	40.4	3.89	4.52
500	38.0	44.2	4.97	5.68
630	42.9	48.8	6.37	7.12
800	47.8	55.4	8.07	9.15
1000	53.0	60.6	10.08	11.27
2-core cables				
16[a]	17.2	20.0	0.49	0.90
25[a]	20.8	24.1	0.66	1.05
35[a]	23.2	27.9	0.88	1.48
50	21.0	25.8	1.11	1.80
70	24.0	29.0	1.52	2.32
95	26.9	33.1	2.04	3.16
120	29.9	36.1	2.57	3.79
150	33.4	39.3	3.13	4.50
185	37.1	44.7	3.92	5.82
240	41.7	49.0	5.05	7.22
300	45.8	53.5	6.35	8.71
400	51.6	59.0	8.00	10.65
3-core cables				
16[a]	18.3	21.2	0.64	1.07
25[a]	22.1	26.7	0.92	1.55
35[a]	24.8	29.6	1.19	1.94
50	23.6	28.5	1.58	2.37
70	27.4	32.2	2.22	3.12
95	30.8	37.0	2.98	4.31
120	34.2	40.4	3.73	5.16
150	37.9	45.5	4.58	6.61
185	42.5	49.8	5.74	7.92
240	47.8	55.1	7.45	9.93

(*cont.*)

XLPE INSULATED DISTRIBUTION CABLES
(BS 5467: 1989 or BS 6724: 1990)
600/1000 V – copper

Table A14.1 continued

Conductor size (mm^2)	Approximate diameter Unarmoured (mm)	Approximate diameter Armoured (mm)	Approximate weight Unarmoured (kg/m)	Approximate weight Armoured (kg/m)
300	52.6	60.2	9.23	11.97
400	59.2	66.6	11.72	14.77
4-core cables				
16[a]	20.0	22.9	0.82	1.30
25[a]	24.3	28.9	1.17	1.88
35[a]	27.3	32.1	1.55	2.35
50	26.9	32.0	2.05	2.95
70	31.5	37.7	2.90	4.24
95	35.6	41.7	3.92	5.40
120	39.5	47.1	4.94	7.00
150	44.1	51.4	6.04	8.30
185	49.3	56.6	7.58	10.07
240	55.5	63.0	9.86	12.68
300	61.4	68.8	12.28	15.38
400	68.8	78.1	15.60	19.95
4-core cables with reduced neutral[b]				
25[a]	23.3	28.0	1.11	1.80
35[a]	25.7	30.5	1.39	2.18
50	26.1	31.2	1.85	2.77
70	30.4	36.6	2.55	3.82
95	34.8	41.0	3.46	4.86
120	39.4	45.3	4.41	5.97
150	42.5	50.0	5.28	7.53
185	47.7	55.3	6.67	9.08
240	53.4	61.0	8.70	11.50
300	59.0	66.7	10.75	13.76
300	61.2	68.6		14.10
400	66.4	73.8	13.82	17.27

[a] Circular conductors
[b] Size of reduced neutral conductor

| Phase conductor (mm^2) | 25 | 35 | 50 | 70 | 95 | 120 |
| Neutral conductor (mm^2) | 16 | 16 | 25 | 35 | 50 | 70 |

| Phase conductor (mm^2) | 150 | 185 | 240 | 300 | 300 | 400 |
| Neutral conductor (mm^2) | 70 | 95 | 120 | 150 | 185 | 185 |

All cables have stranded conductors. The phase conductors are shaped, but for some sizes the neutral conductors are circular.

XLPE INSULATED DISTRIBUTION CABLES
(BS 5467: 1989 or BS 6724: 1990)
600/1000 V – aluminium (solid)

Table A14.2 Dimensions and weights

Conductor size (mm^2)	Approximate diameter Unarmoured (mm)	Approximate diameter Armoured (mm)	Approximate weight Unarmoured (kg/m)	Approximate weight Armoured (kg/m)
Single-core (aluminium wire armour)				
50	12.9	16.2	0.24	0.47
70	14.7	18.7	0.32	0.50
95	16.6	20.6	0.41	0.62
120	18.1	22.1	0.50	0.73
150	20.1	25.2	0.61	0.93
185	22.2	27.4	0.75	1.11
240	24.8	29.9	0.95	1.34
300	27.3	32.4	1.17	1.60
380[a]	30.8	37.1	1.45	2.06
480[a]	34.2	40.4	1.81	2.47
600[a]	37.6	43.8	2.19	2.91
740[a]	41.7	49.1	2.71	3.68
960[a]	46.9	54.4	3.47	4.56
1200[a]	52.0	59.7	4.30	5.52
2-core cables				
16[b]	16.0	19.2	0.24	0.72
25[b]	19.1	22.4	0.34	0.75
35[b]	21.1	25.7	0.42	0.95
50	18.7	23.5	0.50	1.12
70	21.3	26.3	0.69	1.42
95	23.8	30.0	0.88	1.92
3-core cables				
16[b]	17.0	20.4	0.32	0.81
25[b]	20.3	24.9	0.42	1.00
35[b]	22.4	27.3	0.57	1.23
50	22.0	26.8	0.70	1.43
70	25.4	30.2	0.94	1.78
95	28.6	34.8	1.21	2.42
120	31.7	37.8	1.48	2.81
150	35.1	42.7	1.81	3.66
185	39.4	46.7	2.30	4.32
240	44.2	51.5	2.73	5.17
300	48.6	56.2	3.56	6.10

(*cont.*)

> **XLPE INSULATED
> DISTRIBUTION CABLES
> (BS 5467: 1989 or BS 6724: 1990)**
> 600/1000 V – aluminium (solid)

Table A14.2 continued

Conductor size (mm^2)	Approximate diameter		Approximate weight	
	Unarmoured (mm)	Armoured (mm)	Unarmoured (kg/m)	Armoured (kg/m)
4-core cables				
16[b]	18.6	21.9	0.40	0.85
25[b]	22.3	26.9	0.53	1.17
35[b]	24.7	29.5	0.67	1.39
50	24.9	30.0	0.90	1.73
70	29.1	35.3	1.22	2.46
95	32.8	39.0	1.58	2.93
120	36.4	44.0	1.94	3.84
150	40.6	47.9	2.38	4.44
185	45.5	52.7	2.96	5.24
240	51.1	58.5	3.85	6.45
300	56.4	63.8	4.71	7.53

[a] Solid sectoral conductors
[b] Circular conductors

**XLPE INSULATED
DISTRIBUTION CABLES
(BS 5467: 1989 or BS 6724: 1990)
1.9/3.3 kV**

Table A14.3 Dimensions and weights

Conductor size (mm^2)	Approximate diameter		Approximate weight	
	Copper (mm)	Solid aluminium (mm)	Copper (kg/m)	Solid aluminium (kg/m)
Single-core (aluminium wire armour)				
50	20.6	19.4	0.81	0.49
70	22.4	20.9	1.04	0.59
95	24.3	22.5	1.33	0.70
120	27.2	25.2	1.68	0.88
150	28.8	26.5	1.97	1.00
185	30.8	28.3	2.37	1.16
240	33.4	30.5	2.96	1.38
300	36.1	32.8	3.61	1.62
400	40.4		4.60	
500	44.2		5.68	
630	48.8		7.16	
800	55.4		9.15	
1000	60.6		11.27	
380[a]		37.1		2.06
480[a]		40.4		2.47
600[a]		43.8		2.91
740[a]		49.1		3.68
960[a]		54.4		4.56
1200[a]		59.7		5.52
3-core cables				
16[b]	28.9	26.0	1.60	1.31
25[b]	32.2	28.4	2.06	1.47
35[b]	35.0	31.5	2.32	1.56
50	34.7	31.0	3.04	2.02
70	38.0	34.0	3.80	2.38
95	41.4	37.9	4.73	2.79
120	45.7	40.9	6.07	3.59
150	48.5	43.4	7.01	4.02
185	51.9	46.5	8.27	4.54
240	56.9	51.8	10.31	5.39
300	61.2	55.8	12.30	6.21
400	66.6		14.77	

[a] Solid sectoral conductors
[b] Circular conductors

**XLPE INSULATED
DISTRIBUTION CABLES
(BS 6622: 1991)
3.8/6.6 kV**

Table A14.4 Dimensions and weights

Conductor size (mm^2)	Approximate diameter — Stranded copper or aluminium (mm)	Approximate weight — Copper (kg/m)	Approximate weight — Stranded aluminium (kg/m)
Single-core (copper wire screened, unarmoured)			
50	23.3	0.9	0.6
70	25.2	1.2	0.7
95	26.9	1.4	0.8
120	28.8	1.7	1.0
150	30.2	2.1	1.2
185	32.2	2.4	1.3
240	34.8	3.0	1.5
300	37.7	3.7	1.8
400	41.6	4.6	2.2
500	45.3	5.7	2.7
630	49.2	7.1	3.2
800	55.4	8.9	3.8
1000	60.1	10.9	4.5
3-core (circular conductors, armoured, SWA)			
25	43.3	3.1	2.7
35	46.1	3.5	3.0
50	50.3	4.5	3.8
70	54.2	5.4	4.3
95	58.3	6.5	5.0
120	62.1	7.6	5.6
150	65.3	8.6	5.9
185	69.6	10.1	6.8
240	75.8	12.4	8.1
300	83.8	15.6	10.0
400	92.2	19.0	12.2

**XLPE INSULATED
DISTRIBUTION CABLES
(BS 6622: 1991)
6.35/11 kV**

Table A14.5 Dimensions and weights

Conductor size (mm^2)	Approximate diameter Stranded copper or aluminium (mm)	Approximate weight Copper (kg/m)	Stranded aluminium (kg/m)
Single-core (copper wire screened, unarmoured)			
50	25.3	1.0	0.7
70	27.0	1.2	0.8
95	29.0	1.5	0.9
120	30.6	1.8	1.0
150	32.2	2.1	1.2
185	34.0	2.5	1.4
240	36.6	3.1	1.6
300	39.1	3.7	1.9
400	42.4	4.7	2.3
500	45.7	5.7	2.7
630	49.6	7.2	3.2
800	55.8	8.9	3.8
1000	60.5	10.9	4.5
3-core cables (circular conductors, armoured, SWA)			
25	48.8	4.0	3.6
35	51.6	4.4	3.9
50	54.6	5.0	4.4
70	58.5	5.9	4.8
95	62.6	7.1	5.6
120	66.6	8.2	6.2
150	69.8	9.2	6.6
185	74.1	10.7	7.4
240	81.2	13.8	9.5
300	86.8	16.2	10.6
400	94.1	19.3	12.5

XLPE INSULATED DISTRIBUTION CABLES (BS 6622: 1991) 8.7/15 kV

Table A14.6 Dimensions and weights

Conductor size (mm²)	Approximate diameter Stranded copper or aluminium (mm)	Approximate weight Copper (kg/m)	Stranded aluminium (kg/m)
Single-core (copper wire screened, unarmoured)			
50	27.5	1.1	0.8
70	29.5	1.3	0.9
95	31.4	1.6	1.0
120	33.0	1.9	1.1
150	34.6	2.3	1.4
185	36.4	2.7	1.5
240	39.0	3.3	1.9
300	41.5	3.9	2.0
400	44.9	4.8	2.4
500	48.1	5.9	2.8
630	52.0	7.4	3.4
800	58.2	9.1	4.0
1000	62.9	11.2	4.8
3-core cables (circular conductors, armoured, SWA)			
25	54.2	4.5	4.1
35	57.0	5.0	4.5
50	60.0	5.7	4.9
70	63.8	6.6	5.5
95	67.9	7.8	6.3
120	71.8	8.9	6.9
150	75.0	10.0	7.3
185	80.0	12.3	9.0
240	86.6	14.7	10.4
300	91.9	17.1	11.4
400	99.4	20.4	13.6

XLPE INSULATED DISTRIBUTION CABLES (BS 6622: 1991) 12.7/22 kV

Table A14.7 Dimensions and weights

Conductor size (mm^2)	Approximate diameter — Stranded copper or aluminium (mm)	Approximate weight — Copper (kg/m)	Approximate weight — Stranded aluminium (kg/m)
Single-core (copper wire screened, unarmoured)			
50	29.8	1.2	0.9
70	31.7	1.4	1.0
95	33.4	1.7	1.1
120	35.2	2.0	1.3
150	36.6	2.4	1.5
185	38.6	2.8	1.6
240	41.2	3.4	1.9
300	43.5	4.0	2.1
400	46.9	5.0	2.6
500	50.1	6.0	3.0
630	54.0	7.5	3.5
800	60.4	9.3	4.2
1000	65.2	11.4	5.0
3-core cables (circular conductors, armoured, SWA)			
35	61.7	5.6	5.1
50	64.5	6.3	5.6
70	68.6	7.2	6.1
95	72.8	8.6	6.9
120	77.8	10.4	8.4
150	81.2	11.6	8.9
185	85.7	13.2	9.8
240	91.3	15.6	11.0
300	96.6	18.0	12.3
400	104.1	21.3	14.3

XLPE INSULATED DISTRIBUTION CABLES (BS 6622: 1991) 19/33 kV

Table A14.8 Dimensions and weights

Conductor size (mm^2)	Approximate diameter Stranded copper or aluminium (mm)	Approximate weight Copper (kg/m)	Approximate weight Stranded aluminium (kg/m)
Single-core (copper wire screened, unarmoured)			
50	35.2	1.5	1.2
70	36.9	1.8	1.3
95	38.8	2.1	1.5
120	40.4	2.4	1.6
150	42.0	2.8	1.9
185	43.8	3.2	2.1
240	46.5	3.8	2.3
300	49.0	4.5	2.6
400	52.3	5.4	3.0
500	55.5	6.6	3.6
630	59.4	8.2	4.2
800	65.7	10.1	4.9
1000	70.6	12.2	5.8
3-core cables (circular conductors, armoured, SWA)			
50	78.2	9.5	8.7
70	82.1	10.3	9.1
95	86.1	11.7	10.1
120	90.0	12.9	10.8
150	93.2	14.1	11.4
185	97.5	16.1	12.7
240	103.3	18.3	13.8
300	108.8	21.5	15.9
400	116.0	24.4	17.4

XLPE INSULATED DISTRIBUTION CABLES
(BS 5467: 1989 or BS 6724: 1990)
600/1000 V – armoured

Table A14.9 Sustained ratings

Conductor size (mm^2)	In air - Single-core[a] Trefoil (A)	In air - Single-core[a] Flat (A)	In air - 2-core (A)	In air - 3- or 4-core (A)	In ground - Single-core Trefoil (A)	In ground - Single-core Flat (A)	In ground - 2-core (A)	In ground - 3- or 4-core (A)
Copper conductors								
16			115	99			141	119
25			152	131			183	152
35			188	162			219	182
50	222	288	228	197	231	242	259	217
70	285	358	291	251	284	295	317	266
95	346	425	354	304	340	350	381	319
120	402	485	410	353	386	395	433	363
150	463	549	472	406	431	434	485	406
185	529	618	539	463	485	482	547	458
240	625	715	636	546	558	545	632	529
300	720	810	732	628	623	597	708	592
400	815	848	847	728	691	637	799	667
500	918	923			765	688		
630	1027	992			841	737		
800	1119	1042			888	760		
1000	1214	1110			942	797		
Aluminium conductors								
16			85	74			108	91
25			112	98			138	116
35			138	120			165	139
50	162	215	166	145	177	185	196	165
70	207	270	211	185	218	227	241	203
95	252	324	254	224	260	270	288	244
120	292	372		264	296	306		278
150	337	424		305	331	339		311
185	391	477		350	374	380		353
240	465	554		418	433	435		409
300	540	626		488	486	483		461

[a] Single-core cables with aluminium wire armour, 3-phase circuit

For relevant conditions see the notes at the beginning of this appendix.

XLPE INSULATED DISTRIBUTION CABLES
(BS 5467: 1989)
600/1000 V – unarmoured

Table A14.10 Sustained ratings

Conductor size (mm^2)	In air Single-core[a] Trefoil (A)	In air Single-core[a] Flat (A)	2-core (A)	3- or 4-core (A)
Copper conductors				
16			115	100
25			149	127
35			185	158
50	209	274	225	192
70	270	351	289	246
95	330	426	352	298
120	385	495	410	346
150	445	570	473	399
185	511	651	542	456
240	606	769	641	538
300	701	886	741	621
400	820	1065	865	741
500	936	1228		
630	1069	1423		
800	1214	1581		
1000	1349	1775		
Aluminium conductors (solid)				
16			91	77
25			108	97
35			135	120
50	159	210	164	146
70	206	271	211	187
95	253	332	257	227
120	296	387		263
150	343	448		304
185	395	515		347
240	471	611		409
300	544	708		471

[a] Single-core cables with aluminium wire armour, 3-phase circuit

For relevant conditions see the notes at the beginning of this appendix.

XLPE INSULATED DISTRIBUTION CABLES
(BS 5467: 1989 or BS 6724: 1990)
1.9/3.3 kV – armoured

Table A14.11 Sustained ratings

Conductor size (mm^2)	In air Single-core[a] Trefoil (A)	In air Single-core[a] Flat (A)	In air 3-core (A)	In ground Single-core[a] Trefoil (A)	In ground Single-core[a] Flat (A)	In ground 3-core (A)
Copper conductors						
16			108			114
25			143			147
35			170			175
50	230	287	204	222	230	207
70	288	357	257	271	279	254
95	353	434	315	324	331	304
120	411	492	365	366	369	345
150	468	553	415	409	409	387
185	534	622	476	460	454	436
240	630	715	560	528	512	502
300	717	793	640	589	560	563
400	817	851	734	651	595	633
500	924	929		720	641	
630	1041	1007		789	684	
800	1131	1054		831	703	
1000	1227	1121		880	735	
Aluminium conductors						
16			82			87
25			108			113
35			128			134
50	173	217	155	170	176	158
70	216	270	194	208	215	194
95	264	328	237	248	256	233
120	308	377	276	282	288	265
150	350	424	313	315	320	297
185	402	483	360	355	359	336
240	475	561	425	410	409	389
300	544	631	489	460	453	439

[a] Single-core cables with aluminium wire armour, 3-phase circuit

For relevant conditions see the notes at the beginning of this appendix.

> **XLPE INSULATED
> DISTRIBUTION CABLES
> (BS 6622: 1991)**
> 3.8/6.6 kV to 8.7/15 kV

Table A14.12 Sustained ratings

Conductor size (mm^2)	In air Single-core[a] Trefoil (A)	In air Single-core[a] Flat (A)	In air 3-core (A)	In ground Single-core[a] Trefoil (A)	In ground Single-core[a] Flat (A)	In ground 3-core (A)
Copper conductors						
25[b]			145			140
35[b]			175			170
50[b]	235	295	220	220	230	210
70	285	370	270	270	280	255
95	360	455	330	320	335	300
120	415	520	375	360	380	340
150	470	600	430	410	430	380
185	540	690	490	460	485	430
240	640	820	570	530	560	490
300	740	940	650	600	640	540
400	840	1100	740	680	730	600
500	990	1280		750	830	
630	1110	1500		830	940	
800	1270	1720		920	1070	
1000	1400	1950		1000	1180	
Aluminium conductors						
25[b]			115			115
35[b]			140			135
50[b]	180	230	170	170	175	160
70	225	290	210	210	215	195
95	280	350	250	250	260	230
120	320	410	295	280	295	265
150	365	465	330	320	330	300
185	425	530	385	360	375	335
240	500	640	450	415	440	380
300	580	730	510	475	495	435
400	670	860	590	540	570	490
500	790	1010		610	650	
630	910	1190		680	750	
800	1060	1330		770	860	
1000	1190	1590		850	970	

[a] Copper wire screened, unarmoured
[b] Not applicable to all voltages. See dimension tables for availability

For cable designs and relevant conditions see the notes at the beginning of this appendix.

XLPE INSULATED DISTRIBUTION CABLES (BS 6622: 1991) 12.7/22 kV and 19/33 kV

Table A14.13 Sustained ratings

Conductor size (mm^2)	In air Single-core[a] Trefoil (A)	In air Single-core[a] Flat (A)	In air 3-core (A)	In ground Single-core[a] Trefoil (A)	In ground Single-core[a] Flat (A)	In ground 3-core (A)
Copper conductors						
35			180			170
50	245	295	225	220	230	210
70	300	365	275	270	280	255
95	360	450	330	320	335	295
120	425	520	380	360	380	335
150	485	590	430	410	430	375
185	550	670	490	460	485	420
240	650	800	570	530	560	480
300	740	920	650	600	640	530
400	850	1070	740	690	730	590
500	980	1250		760	830	
630	1130	1450		850	950	
800	1280	1710		930	1070	
1000	1420	1930		1010	1180	
Aluminium conductors						
35			145			135
50	190	230	175	170	175	160
70	235	285	215	210	215	195
95	280	345	260	250	260	230
120	330	400	300	280	295	260
150	375	455	335	320	330	290
185	430	520	390	360	375	330
240	510	620	460	415	440	380
300	580	710	520	475	495	425
400	680	840	600	550	570	480
500	790	980		610	650	
630	920	1060		690	750	
800	1060	1370		770	860	
1000	1210	1570		860	970	

[a] Copper wire screened, unarmoured

For cable designs and relevant conditions see the notes at the beginning of this appendix.

XLPE INSULATED DISTRIBUTION CABLES
(BS 5467: 1989 or BS 6724: 1990)
600/1000 V

Table A14.14 Voltage drop (50 Hz per A/m)

Conductor size (mm^2)	Copper Single-core[a] Trefoil (mV)	Copper Single-core[a] Flat[b] (mV)	Copper 2-core (mV)	Copper 3- or 4-core (mV)	Aluminium Single-core[a] Trefoil (mV)	Aluminium Single-core[a] Flat[b] (mV)	Aluminium 2-core (mV)	Aluminium 3- or 4-core (mV)
16			2.9	2.5			4.8	4.2
25			1.9	1.65			3.1	2.7
35			1.35	1.15			2.2	1.95
50	0.87	0.90	1.00	0.87	1.4	1.4	1.65	1.45
70	0.62	0.70	0.69	0.60	0.98	1.05	1.15	0.97
95	0.47	0.58	0.52	0.45	0.74	0.83	0.84	0.72
120	0.39	0.51	0.42	0.37	0.60	0.70		0.58
150	0.33	0.45	0.35	0.30	0.49	0.60		0.47
185	0.28	0.41	0.29	0.26	0.41	0.53		0.39
240	0.24	0.37	0.24	0.21	0.34	0.46		0.31
300	0.21	0.34	0.21	0.185	0.29	0.41		0.26
400	0.195	0.33	0.19	0.165				
500	0.18	0.31						
630	0.17	0.29						
800	0.165	0.26						
1000	0.155	0.24						

[a] Data for aluminium wire armoured cables, 3-phase circuit
[b] Twice cable diameter spacing between cores

For the relevant conditions see the notes at the beginning of this appendix.

XLPE INSULATED DISTRIBUTION CABLES
(BS 5467: 1989 or BS 6724: 1990)
600/1000 V – armoured

Table A14.15 Electrical characteristics

Conductor size (mm^2)	A.C. resistance at 90°C Copper (Ω/km)	A.C. resistance at 90°C Aluminium (Ω/km)	Reactance (50 Hz) Armoured single-core[a] Trefoil (Ω/km)	Reactance (50 Hz) Armoured single-core[a] Flat[b] (Ω/km)	Reactance (50 Hz) Armoured or unarmoured multicore (Ω/km)
16	1.47	2.45			0.081
25	0.927	1.54			0.079
35	0.668	1.11			0.077
50	0.494	0.822	0.106	0.164	0.076
70	0.342	0.568	0.103	0.161	0.075
95	0.247	0.411	0.098	0.156	0.073
120	0.197	0.325	0.096	0.154	0.072
150	0.160	0.265	0.097	0.155	0.073
185	0.128	0.211	0.096	0.154	0.073
240	0.0989	0.162	0.092	0.150	0.072
300	0.0802	0.130	0.090	0.148	0.072
400	0.0640		0.090	0.148	
500	0.0515		0.089	0.146	
630	0.0420		0.086	0.144	
800	0.0363		0.086	0.144	
1000	0.0316		0.084	0.142	

[a] Aluminium wire armoured
[b] Twice cable diameter spacing between centres

For the relevant conditions see the notes at the beginning of this appendix.

XLPE INSULATED DISTRIBUTION CABLES
(BS 5467: 1989 or BS 6724: 1990)
1.9/3.3 kV – armoured

Table A14.16 Electrical characteristics

Conductor size (mm^2)	Single-core cables[a]						3-core cables			
	A.C. resistance at 90°C		Reactance (50 Hz)		Capacitance (μF/km)	A.C. resistance at 90°C		Reactance (50 Hz) (Ω/km)	Capacitance (μF/km)
	Copper (Ω/km)	Aluminium (Ω/km)	Trefoil (Ω/km)	Flat[b] (Ω/km)		Copper (Ω/km)	Aluminium (Ω/km)		
16						1.47	2.45	0.104	0.18
25						0.972	1.54	0.094	0.22
35						0.668	1.11	0.091	0.25
50	0.493	0.822	0.116	0.172	0.31	0.494	0.822	0.088	0.27
70	0.342	0.568	0.110	0.165	0.36	0.342	0.568	0.084	0.31
95	0.246	0.410	0.104	0.160	0.42	0.247	0.411	0.081	0.35
120	0.195	0.325	0.104	0.159	0.45	0.197	0.325	0.079	0.38
150	0.160	0.265	0.100	0.156	0.49	0.160	0.265	0.077	0.42
185	0.128	0.211	0.098	0.154	0.54	0.128	0.211	0.076	0.46
240	0.098	0.161	0.094	0.150	0.63	0.099	0.162	0.074	0.51
300	0.080	0.130	0.091	0.147	0.70	0.080	0.130	0.073	0.57
400	0.064		0.090	0.147	0.77				
500	0.051		0.089	0.145	0.80				
630	0.042		0.086	0.143	0.84				

[a] Aluminium wire armoured
[b] Twice cable diameter spacing between centres

For the relevant conditions see the notes at the beginning of this appendix.

XLPE INSULATED DISTRIBUTION CABLES (BS 6622: 1991)
3.8/6.6 kV and 6.35/11 kV

Table A14.17 Electrical characteristics

Conductor size (mm^2)	Single-core cables[a]					3-core cables			
	A.C. resistance at 90°C		Reactance (50 Hz)		Capacitance (μF/km)	A.C. resistance at 90°C		Reactance (50 Hz) (Ω/km)	Capacitance (μF/km)
	Copper (Ω/km)	Aluminium (Ω/km)	Trefoil (Ω/km)	Flat[b] (Ω/km)		Copper (Ω/km)	Aluminium (Ω/km)		
3.8/6.6 kV cables									
16						1.47	2.45	0.126	0.26
25						0.927	1.54	0.117	0.30
35						0.668	1.11	0.109	0.33
50	0.494	0.822	0.121	0.181	0.34	0.493	0.822	0.105	0.36
70	0.342	0.568	0.115	0.174	0.38	0.343	0.568	0.100	0.41
95	0.247	0.411	0.109	0.167	0.43	0.247	0.411	0.095	0.46
120	0.196	0.325	0.105	0.162	0.47	0.196	0.325	0.092	0.50
150	0.159	0.265	0.102	0.159	0.51	0.159	0.265	0.090	0.55
185	0.128	0.211	0.099	0.156	0.56	0.128	0.211	0.087	0.60
240	0.0982	0.162	0.096	0.153	0.61	0.0986	0.162	0.085	0.65
300	0.0791	0.130	0.094	0.151	0.62	0.0798	0.130	0.084	0.67
400	0.0632	0.102	0.092	0.149	0.65	0.0641	0.102	0.082	0.70
500	0.0510	0.804	0.089	0.147	0.69				
630	0.0417	0.0639	0.086	0.144	0.78				

XLPE Insulated Distribution Cables

6.35/11 kV cables

Size										
16						1.47	2.45	0.134	0.21
25						0.927	1.54	0.124	0.24
35						0.668	1.11	0.116	0.26
50	0.494	0.822	0.127	0.185	0.26	0.493	0.822	0.111	0.28
70	0.342	0.568	0.120	0.177	0.30	0.342	0.568	0.106	0.32
95	0.247	0.411	0.114	0.171	0.33	0.247	0.410	0.100	0.36
120	0.196	0.325	0.109	0.166	0.36	0.196	0.325	0.097	0.39
150	0.159	0.265	0.106	0.163	0.39	0.159	0.265	0.094	0.42
185	0.128	0.211	0.103	0.160	0.43	0.128	0.211	0.092	0.46
240	0.0981	0.161	0.099	0.156	0.48	0.0984	0.161	0.089	0.51
300	0.0791	0.130	0.096	0.153	0.52	0.0797	0.130	0.086	0.56
400	0.0632	0.102	0.093	0.150	0.58	0.0639	0.102	0.083	0.62
500	0.0510	0.0804	0.090	0.147	0.66				
630	0.0417	0.0639	0.087	0.145	0.74				

[a] Copper wire screened, unarmoured
[b] Twice cable diameter spacing between centres

For cable designs and relevant conditions see the notes at the beginning of this appendix.

XLPE INSULATED DISTRIBUTION CABLES (BS 6622: 1991) 8.7/15 kV and 12.7/22 kV

Table A14.18 Electrical characteristics

Conductor size (mm²)	Single-core cables[a]					3-core cables			
	A.C. resistance at 90°C		Reactance (50 Hz)		Capacitance (µF/km)	A.C. resistance at 90°C		Reactance (50 Hz) (Ω/km)	Capacitance (µF/km)
	Copper (Ω/km)	Aluminium (Ω/km)	Trefoil (Ω/km)	Flat[b] (Ω/km)		Copper (Ω/km)	Aluminium (Ω/km)		

8.7/15 kV cables

16						1.47	2.45	0.143	0.17
25						0.927	1.54	0.132	0.19
35						0.668	1.11	0.124	0.21
50	0.494	0.822	0.132	0.190	0.21	0.493	0.822	0.118	0.23
70	0.342	0.568	0.125	0.183	0.24	0.342	0.568	0.112	0.26
95	0.247	0.411	0.119	0.176	0.27	0.247	0.410	0.106	0.29
120	0.196	0.325	0.114	0.171	0.29	0.196	0.325	0.102	0.31
150	0.159	0.265	0.111	0.168	0.31	0.159	0.264	0.100	0.34
185	0.128	0.211	0.107	0.164	0.34	0.128	0.211	0.097	0.37
240	0.0979	0.161	0.103	0.160	0.38	0.0982	0.161	0.093	0.41
300	0.0790	0.130	0.100	0.156	0.41	0.0794	0.130	0.090	0.45
400	0.0630	0.102	0.097	0.153	0.46	0.0636	0.102	0.087	0.50
500	0.0507	0.0802	0.093	0.151	0.51				
630	0.0413	0.0636	0.090	0.147	0.57				

12.7/22 kV cables								
16						0.927		0.139	0.17
25						0.668		0.130	0.18
35	0.494	0.822	0.137	0.192	0.18	0.493	1.54	0.124	0.20
50	0.342	0.568	0.130	0.185	0.21	0.342	1.11	0.118	0.22
70							0.822		
95	0.247	0.411	0.123	0.178	0.23	0.247	0.568	0.111	0.24
120	0.196	0.325	0.118	0.173	0.25	0.196	0.410	0.107	0.26
150	0.159	0.265	0.115	0.170	0.27	0.159	0.325	0.104	0.28
185	0.128	0.211	0.111	0.165	0.29	0.127	0.264	0.101	0.31
240	0.098	0.161	0.106	0.161	0.32	0.098	0.211	0.097	0.34
300	0.079	0.130	0.103	0.158	0.35	0.079	0.161	0.094	0.37
400	0.063	0.102	0.0995	0.155	0.39	0.063	0.130	0.090	0.41
500	0.051	0.080	0.0959	0.152	0.43		0.102		
630	0.041	0.064	0.0923	0.149	0.48				

[a] Copper wire screened, unarmoured
[b] Twice cable diameter spacing between centres

For cable designs and relevant conditions see the notes at the beginning of this appendix.

XLPE INSULATED DISTRIBUTION CABLES (BS 6622: 1991) 19/33 kV

Table A14.19 Electrical characteristics

Conductor size (mm²)	A.C. resistance at 90°C Copper (Ω/km)	A.C. resistance at 90°C Aluminium (Ω/km)	Reactance (50 Hz) Trefoil (Ω/km)	Reactance (50 Hz) Flat[a] (Ω/km)	Capacitance (µF/km)
Single-core cables[b]					
70	0.342	0.568	0.143	0.194	0.16
95	0.247	0.411	0.136	0.189	0.18
120	0.196	0.325	0.130	0.184	0.19
150	0.160	0.265	0.127	0.178	0.20
185	0.127	0.211	0.122	0.174	0.22
240	0.0976	0.161	0.117	0.169	0.24
300	0.0785	0.129	0.113	0.166	0.26
400	0.0624	0.101	0.109	0.162	0.29
500	0.0500	0.0797	0.104	0.158	0.32
630	0.0405	0.0630	0.100	0.155	0.35
800	0.0388	0.0509	0.095	0.151	0.40
1000	0.0290	0.0422	0.093		0.44
3-core cables					
70	0.342	0.568	0.129		0.15
95	0.247	0.410	0.122		0.17
120	0.196	0.324	0.117		0.18
150	0.159	0.265	0.114		0.20
185	0.127	0.211	0.110		0.21
240	0.0978	0.161	0.106		0.25
300	0.0789	0.129	0.102		0.27
400	0.0629	0.102	0.098		0.30

[a] Twice cable diameter spacing between centres
[b] Copper wire screened, unarmoured

For cable designs and relevant conditions see the notes at the beginning of this appendix.

APPENDIX A15
PVC Insulated House Service Cables

CABLE DESIGNS

Typical designs are shown in fig. A15.1. The split concentric single-phase copper conductor cables are covered by BS 4553: 1991. Some types of these cables are also available with XLPE insulation – see chapter 18.

Current ratings

The values in the tables are based on:

Maximum sustained conductor temperature	70°C
Ambient air temperature	25°C
Standard ground temperature	15°C
Soil thermal resistivity	1.2 K m/W
Depth of laying	0.5 m

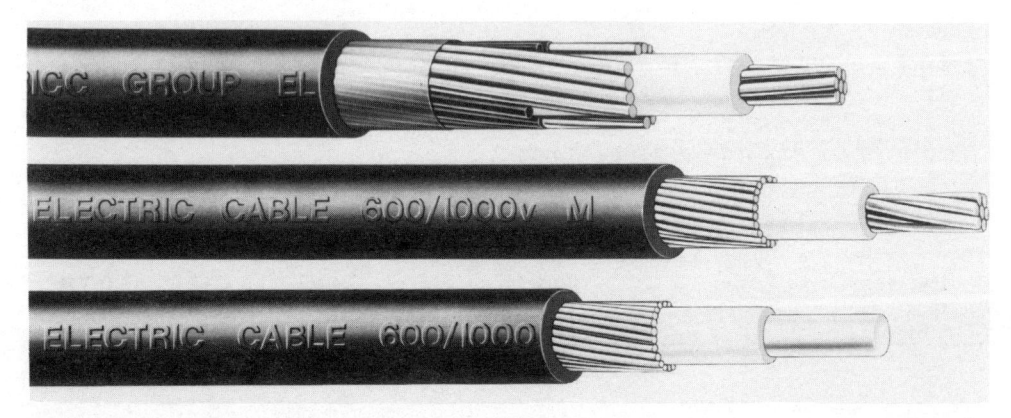

Fig. A15.1 PVC insulated 600/1000 V service cables: split concentric with copper phase conductor (*top*); CNE concentric with copper phase conductor (*middle*); CNE concentric with solid aluminium phase conductor (*bottom*)

> **PVC INSULATED
> SERVICE CABLES
> 600/1000 V**

Table A15.1 Dimensions and weights

Conductor size (mm²)	Number and diameter of copper wires				Approximate diameter (mm)	Approximate weight (kg/m)
	Centre phase conductor (number/mm)	Concentric layer				
		Neutral conductor (number/mm)	Earth conductor (number/mm)	CNE conductor (number/mm)		
Split concentric, copper phase conductor (BS 4553)						
4	7/0.85	7/0.85	3/1.35		10.2	0.206
6	7/1.04	7/1.04	4/1.53		11.7	0.281
10	7/1.35	7/1.35	4/1.78		13.2	0.394
16	7/1.70	7/1.70	4/2.25		15.9	0.583
25	7/2.14	11/1.70	4/2.25		19.4	0.843
35	7/2.52	15/1.70	6/2.25		24.6	1.248
Split concentric, solid aluminium phase conductor						
6		7/0.85	3/1.35		10.1	0.182
10		7/1.04	4/1.53		11.6	0.273
16		7/1.35	4/1.78		13.2	0.365
25		7/1.70	4/2.25		16.0	0.512
35		9/1.70	4/2.25		17.7	0.662
CNE concentric, copper phase conductor						
4	7/0.85			15/0.67	8.7	0.151
6	7/1.04			18/0.67	9.3	0.185
10	7/1.35			18/0.85	11.0	0.280
16	7/1.70			28/0.85	12.1	0.397
25	7/2.14			25/1.13	14.4	0.596
35	7/2.52			24/1.35	16.0	0.789
CNE concentric, solid aluminium phase conductor						
6				16/0.67	8.9	0.137
10				21/0.67	10.2	0.182
16				19/0.85	11.5	0.244
25				27/0.85	13.0	0.334
35				25/1.04	14.4	0.430

**PVC INSULATED
SERVICE CABLES
600/1000 V**

Table A15.2 Ratings and characteristics

Conductor size (mm²)	Current rating In air (A)	Current rating In ground (A)	Impedance (50 Hz) (Ω/km)	Approximate volt drop per A/m (mV)
Split concentric, copper phase conductor				
4	42	53	5.41	11.0
6	54	66	3.61	7.2
10	74	88	2.14	4.3
16	97	115	1.36	2.7
25	130	150	0.86	1.7
35	160	185	0.62	1.2
Split concentric, solid aluminium phase conductor				
6	42	51	5.90	12.0
10	58	69	3.53	7.1
16	73	88	2.22	4.4
25	97	115	1.42	2.8
35	120	140	1.02	2.0
50	140	165	0.76	1.5
CNE concentric, copper phase conductor				
4	42	53	5.41	11.0
6	54	66	3.61	7.2
10	74	88	2.14	4.3
16	97	115	1.36	2.7
25	130	150	0.86	1.7
35	160	185	0.62	1.2
CNE concentric, solid aluminium phase conductor				
6	42	51	5.90	12.0
10	58	69	3.53	7.1
16	73	88	2.22	4.4
25	97	115	1.42	2.8
35	120	140	1.02	2.0
50	140	165	0.76	1.5

APPENDIX A16
Self-contained Fluid-filled Paper Insulated Cables

The information in this appendix covers the following.

(a) Technical data: dimensions, weights, charging current, capacitance, a.c. resistance and reactance for a selected range of conductor sizes (tables A16.2–A16.8).
(b) Power ratings: for single- and 3-core cables, laid direct and in air (figs A16.1–A16.8).
(c) Conditions applicable: table A16.1.

Table A16.1 Conditions applicable

Cables laid direct	
Depth to top of cable[a]	900 mm
Spacing between cable centres (flat)[a]	230 mm
Ground temperature	15°C
Soil thermal resistivity	1.2 K m/W
Conductor operating temperature	90°C
Cables in air	
Spacing between cable centres (flat)[a]	230 mm
Air temperature	25°C
Conductor operating temperature	90°C
(Protection from solar radiation)	

[a] For sketches see figs A16.1–A16.8

(d) Cable design: the data are based on cables described and illustrated in chapter 30 with the following characteristics.
  (i) *Conductors*: circular, except for 33 kV cables with corrugated aluminium sheath, in which case they are oval.
  (ii) *Fluid ducts*: 3-core cables have aluminium ducts but 33 kV cables with aluminium sheath have no ducts or fillers.

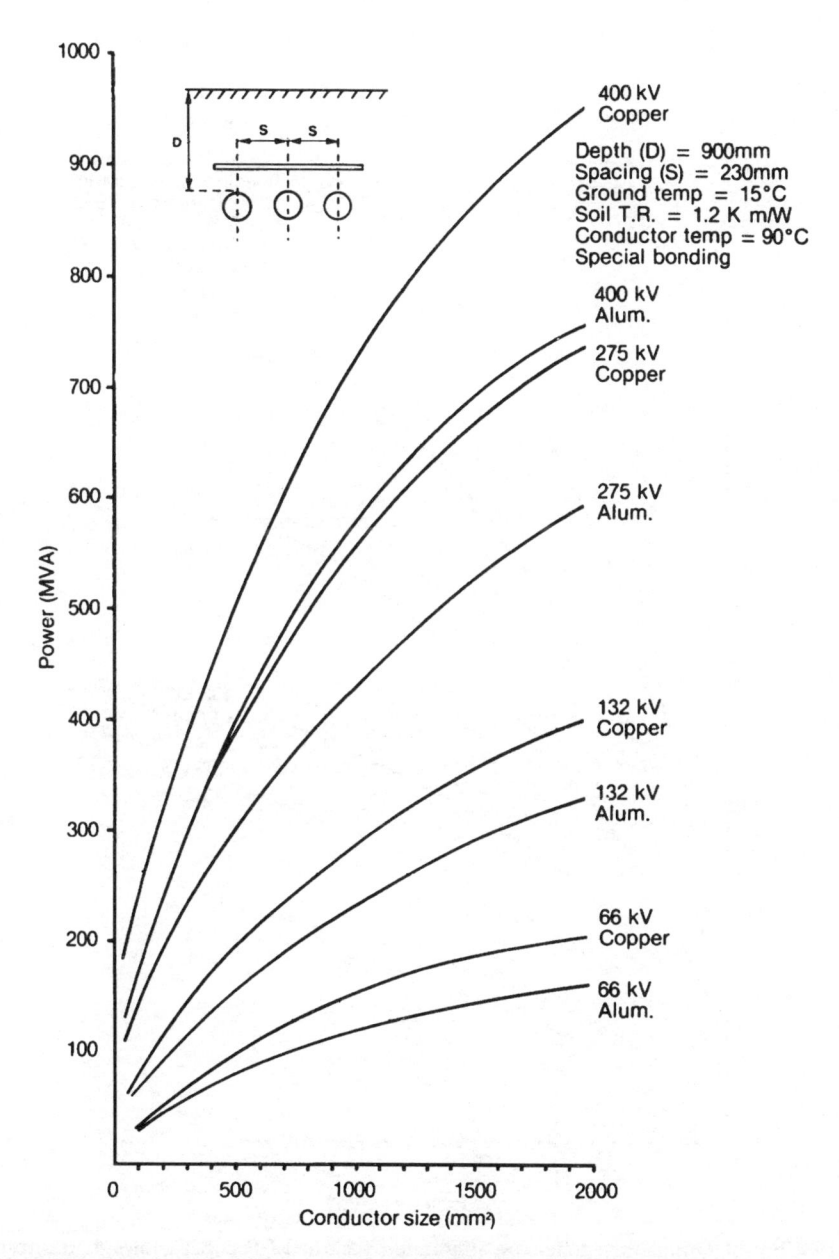

Fig. A16.1 Power ratings for single-core lead sheathed FF paper, laid direct

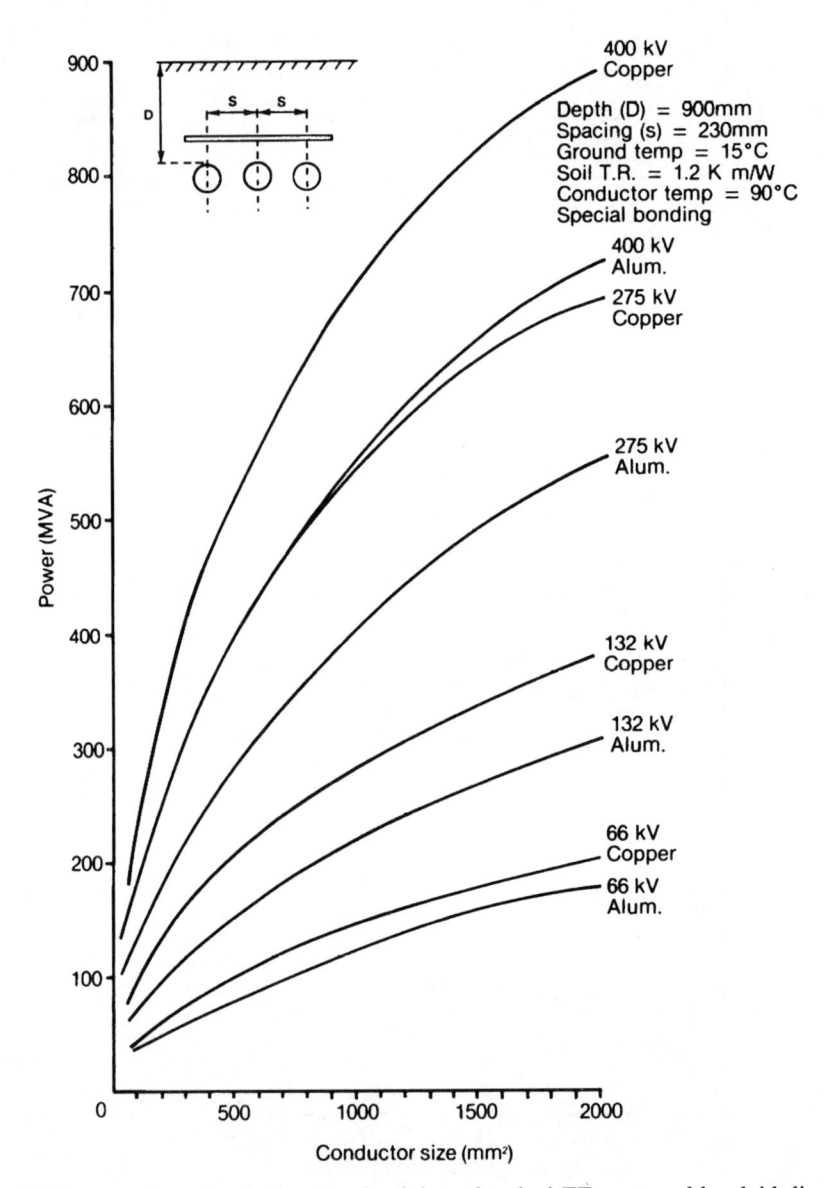

Fig. A16.2 Power ratings for single-core aluminium sheathed FF paper cables, laid direct

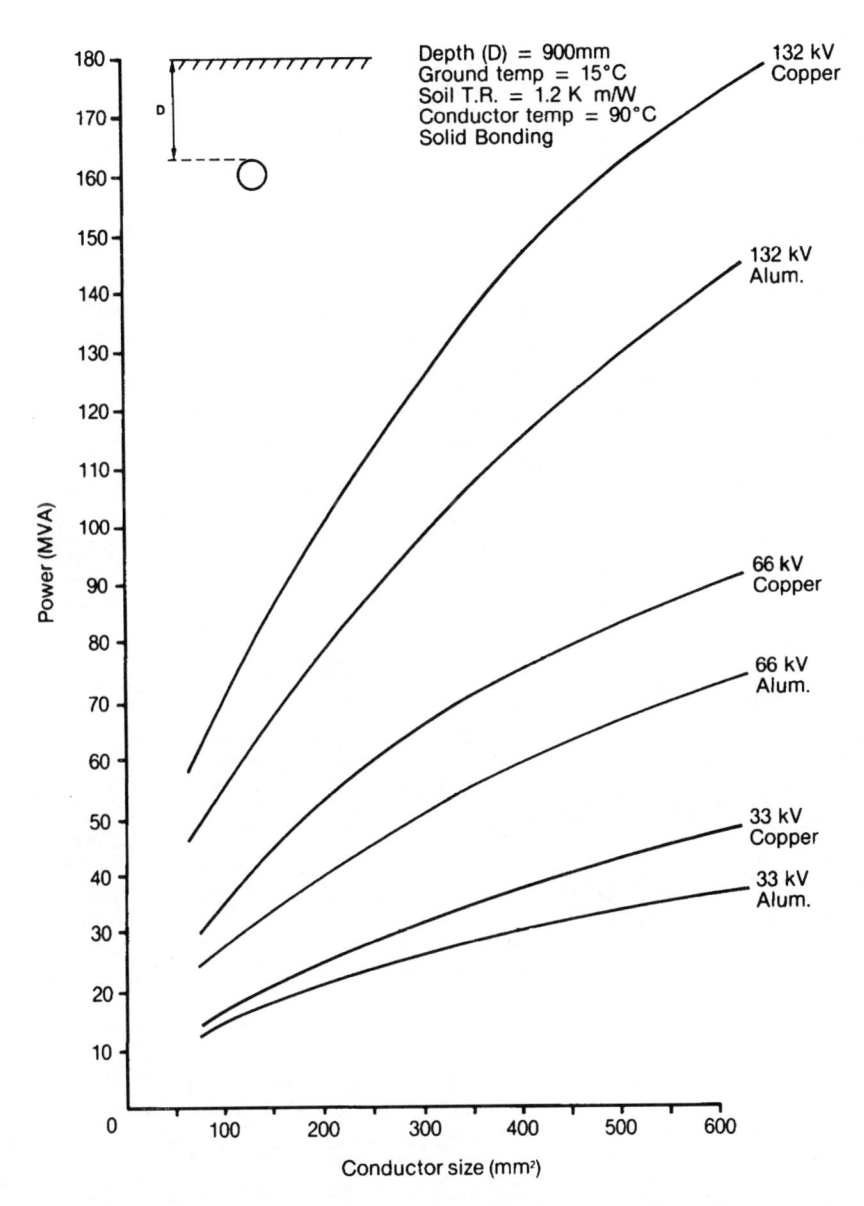

Fig. A16.3 Power ratings for 3-core lead sheathed FF paper cables, laid direct

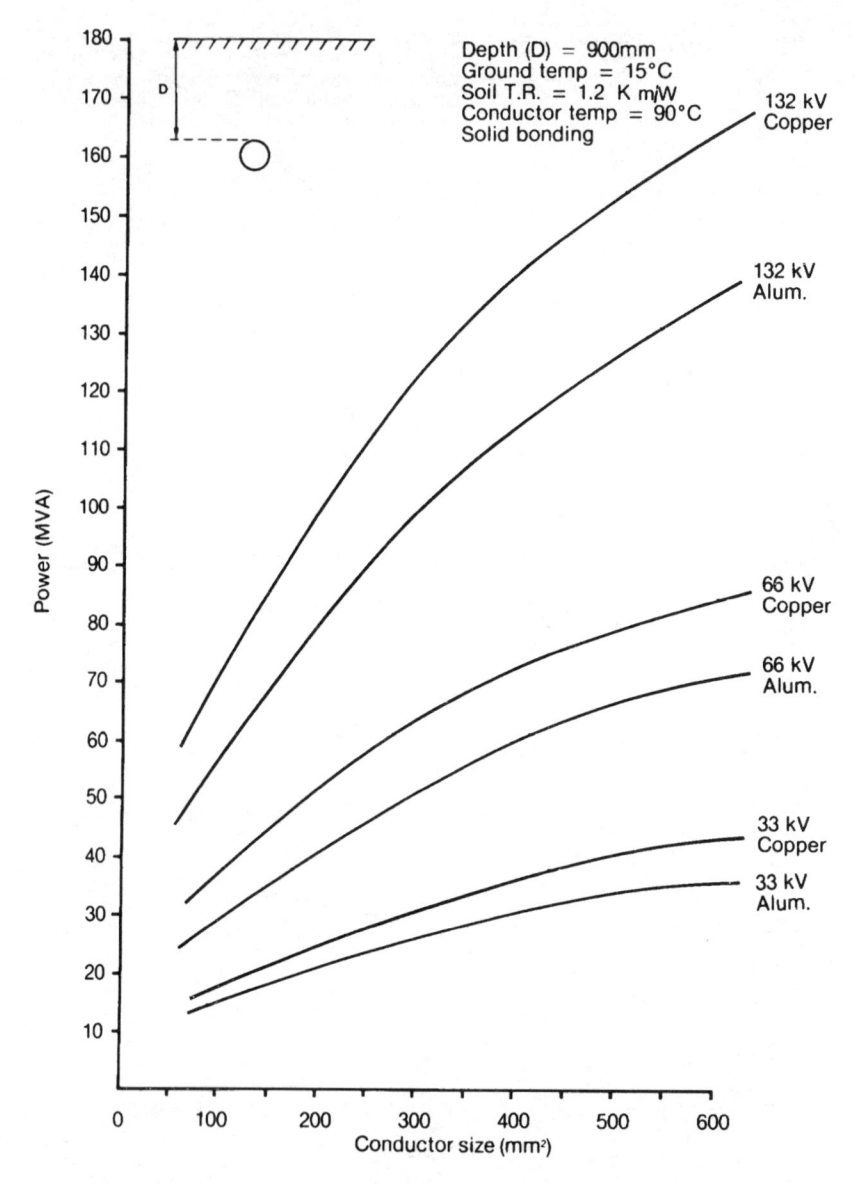

Fig. A16.4 Power ratings for 3-core aluminium sheathed FF paper cables, laid direct

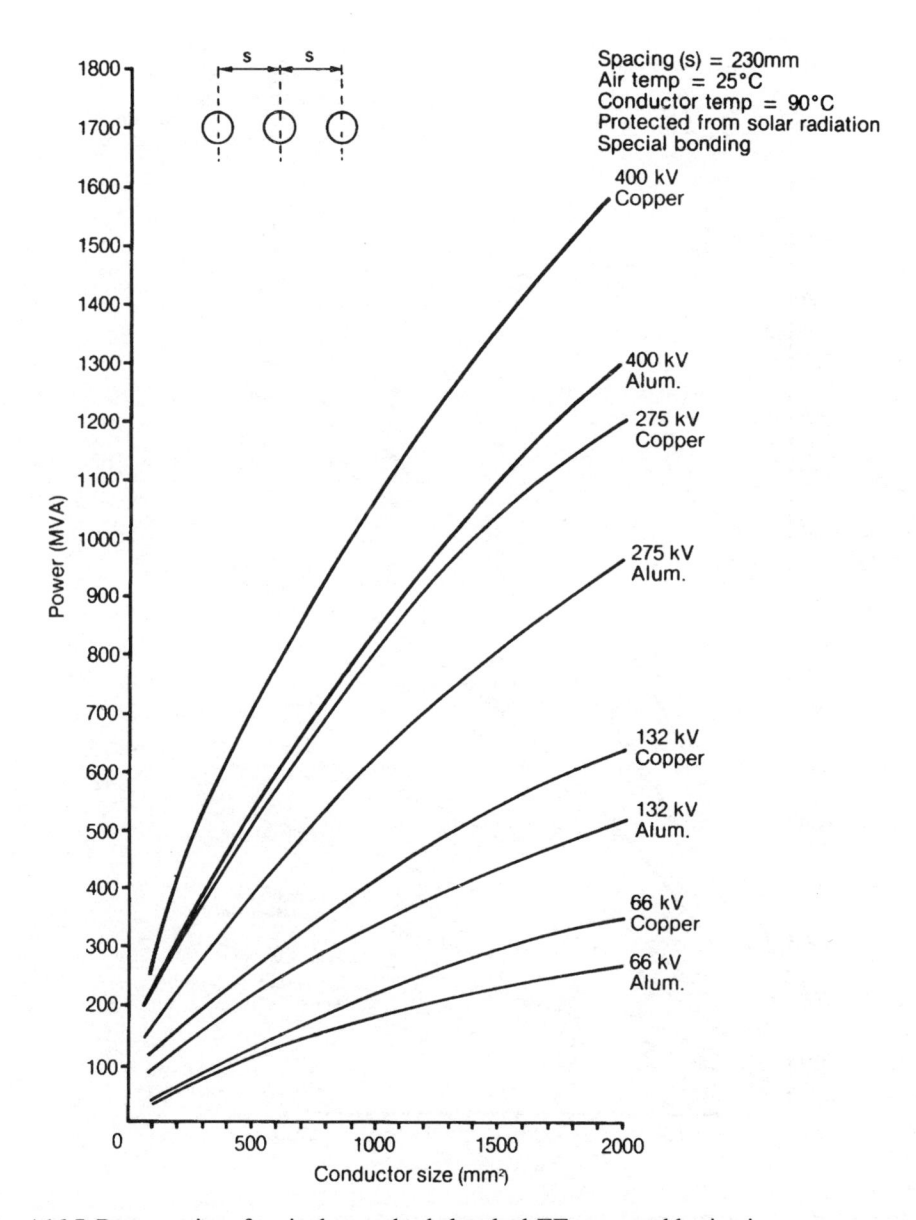

Fig. A16.5 Power ratings for single-core lead sheathed FF paper cables in air

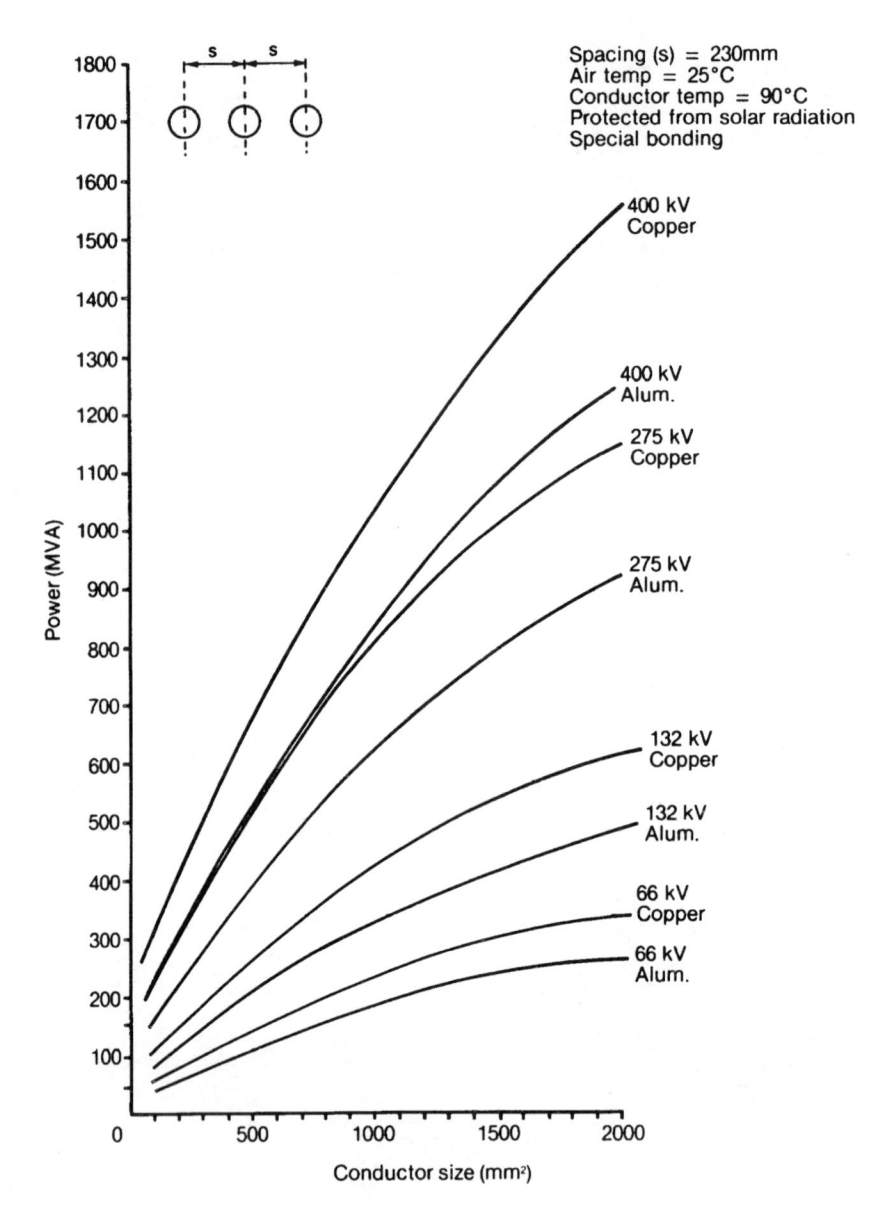

Fig. A16.6 Power ratings for single-core aluminium sheathed FF paper cables in air

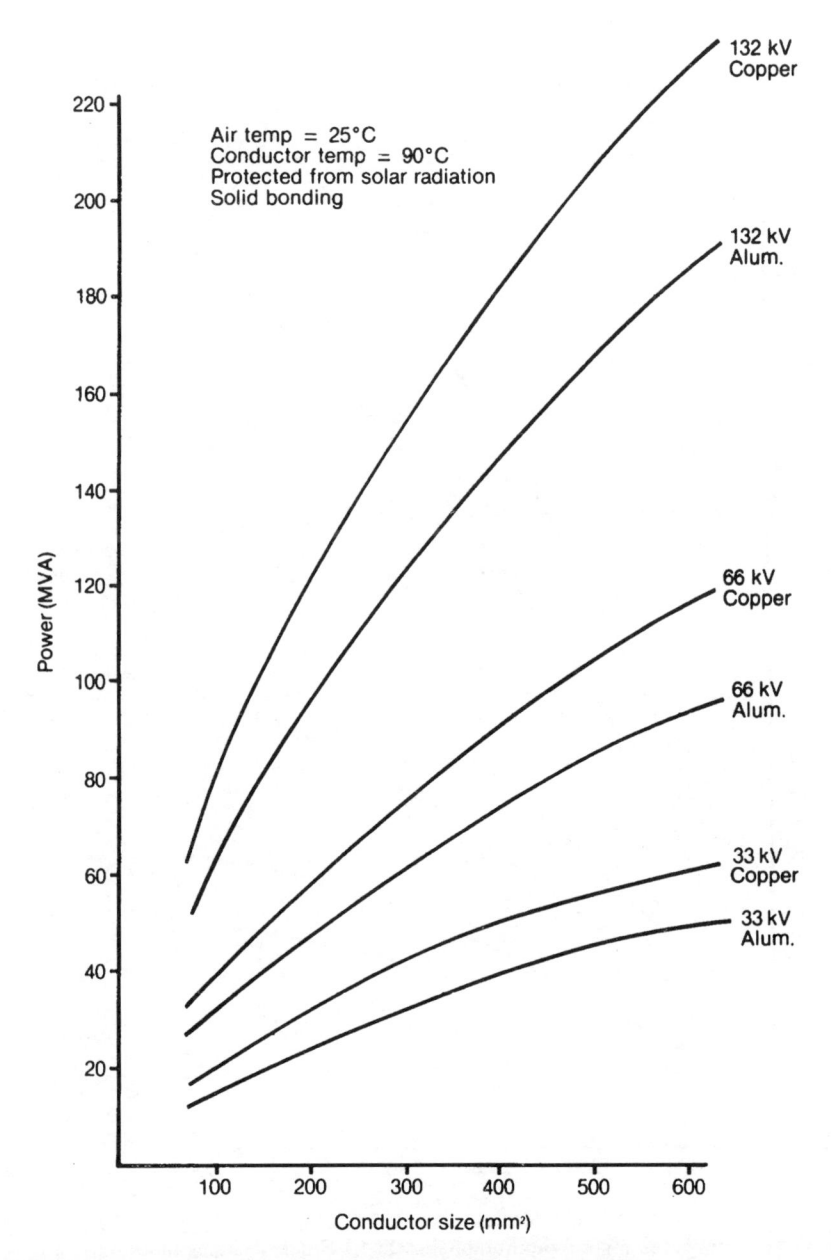

Fig. A16.7 Power ratings for 3-core lead sheathed FF paper cables in air

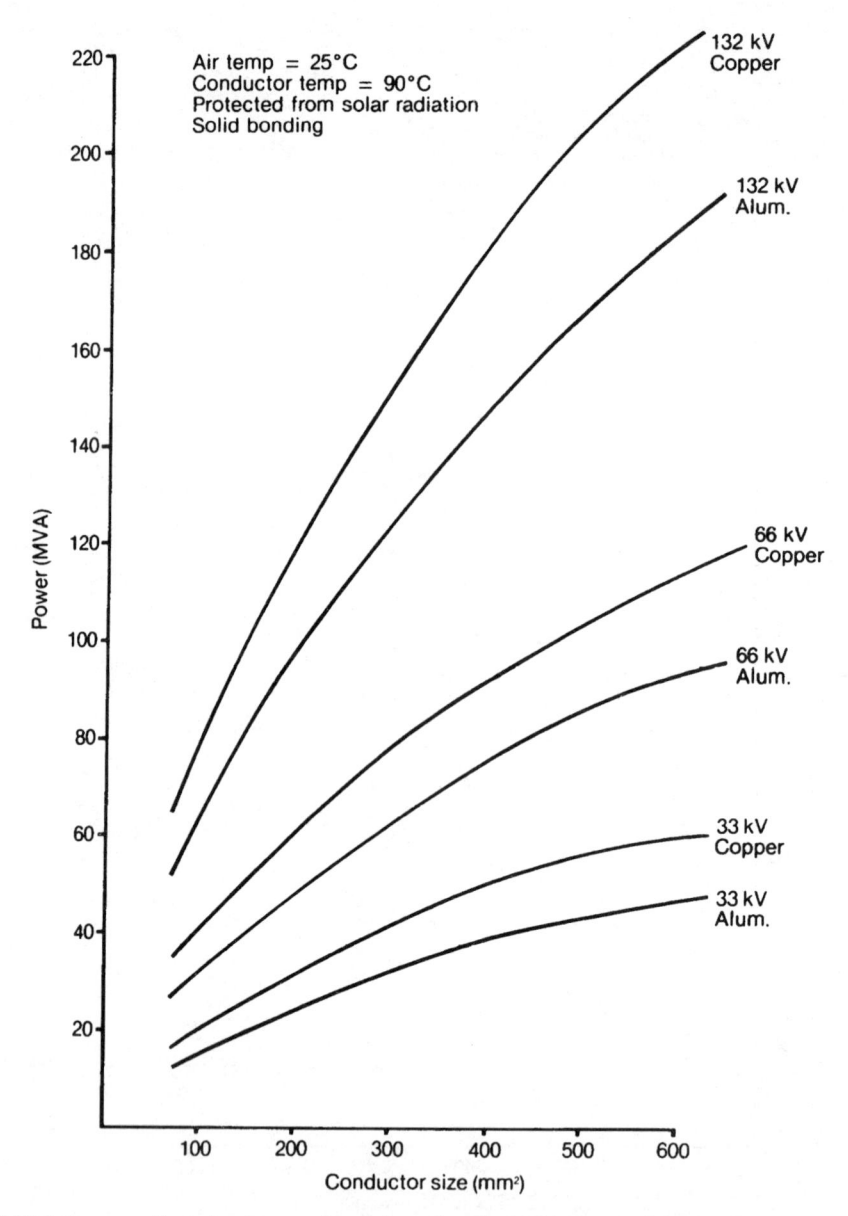

Fig. A16.8 Power ratings for 3-core aluminium sheathed FF paper cables in air

FLUID-FILLED PAPER INSULATED TRANSMISSION CABLES

Table A16.2 Technical data 33 kV 3-core cables

Conductor size (mm²)	Approximate diameter (mm)	Charging current (mA/m)	Capacitance per core (pF/m)	Reactance (μΩ/m)	Copper conductor A.C. (90°C) resistance (μΩ/m)	Copper conductor Approximate weight (kg/m)	Aluminium conductor A.C. (90°C) resistance (μΩ/m)	Aluminium conductor Approximate weight (kg/m)
33 kV 3-core, lead sheath								
70	51	2.4	410	98	342	7.0	555	6.0
185	68	3.5	590	86	128	13.5	212	10.0
240	74	4.0	660	83	98	16.0	162	12.0
400	85	4.9	810	79	64	23.0	102	16.0
630	101	6.0	1010	75	43	34.0	65	22.0
33 kV 3-core, corrugated aluminium sheath								
70	54	2.3	390	95	342	4.5	555	3.0
185	65	3.7	610	86	128	9.0	211	5.5
240	71	4.1	690	83	98	10.5	161	6.0
400	81	5.0	840	79	63	15.5	101	8.0
630	95	6.3	1050	75	42	23.0	64	11.0

For designs and applicable conditions of installation and operation see the introduction to this appendix.

FLUID-FILLED PAPER INSULATED TRANSMISSION CABLES

Table A16.3 Technical data 66 kV single-core cables

Conductor size (mm^2)	Approximate diameter (mm)	Charging current (mA/m)	Capacitance per core (pF/m)	Reactance (μΩ/m)	Copper conductor A.C. (90°C) resistance (μΩ/m)	Copper conductor Approximate weight (kg/m)	Aluminium conductor A.C. (90°C) resistance (μΩ/m)	Aluminium conductor Approximate weight (kg/m)
66 kV single-core, lead sheath								
150	42	6.7	560	217	156	5.2	259	4.2
630	54	10.0	835	194	37	11.0	61	7.0
1000	62	12.3	1025	182	25	15.8	39	9.3
1600	79	15.0	1250	164	15	24.1	24	14.4
2000	85	16.5	1380	158	13	29.0	20	16.4
66 kV single-core corrugated aluminium sheath								
150	45	6.7	560	217	156	3.5	259	2.6
630	58	10.0	835	194	37	8.6	61	4.6
1000	65	12.3	1025	132	25	12.6	39	6.1
1600	82	15.0	1250	164	15	19.3	24	9.6
2000	89	16.5	1380	158	13	23.8	20	11.0

For designs and applicable conditions of installation and operation see the introduction to this appendix.

FLUID-FILLED PAPER INSULATED TRANSMISSION CABLES

Table A16.4 Technical data 66 kV 3-core cables

Conductor size (mm²)	Approximate diameter (mm)	Charging current (mA/m)	Capacitance per core (pF/m)	Reactance (μΩ/m)	Copper conductor A.C. (90°C) resistance (μΩ/m)	Copper conductor Approximate weight (kg/m)	Aluminium conductor A.C. (90°C) resistance (μΩ/m)	Aluminium conductor Approximate weight (kg/m)
66 kV 3-core, lead sheath								
70	62	3.4	280	113	342	9.5	555	8.3
185	74	5.3	445	93	128	14.7	211	11.7
240	79	6.1	510	89	98	17.5	161	13.5
400	90	7.7	645	83	63	24.4	102	17.5
630	105	9.8	815	79	42	35.3	65	23.9
66 kV 3-core, corrugated aluminium sheath								
70	67	3.4	280	113	342	6.2	555	5.0
185	79	5.3	445	93	128	10.4	211	7.2
240	84	6.1	510	89	98	12.6	161	8.3
400	96	7.7	645	83	63	18.0	102	
630	112	9.8	815	79	42	26.8	65	15.3

For designs and applicable conditions of installation and operation see the introduction to this appendix.

FLUID-FILLED PAPER INSULATED TRANSMISSION CABLES

Table A16.5 Technical data 132 kV single-core cables

Conductor size (mm²)	Approximate diameter (mm)	Charging current (mA/m)	Capacitance per core (pF/m)	Reactance (μΩ/m)	Copper conductor A.C. (90°C) resistance (μΩ/m)	Copper conductor Approximate weight (kg/m)	Aluminium conductor A.C. (90°C) resistance (μΩ/m)	Aluminium conductor Approximate weight (kg/m)
132 kV single-core, lead sheath								
150	52	8.1	340	217	156	6.9	259	6.0
630	62	12.1	505	194	37	12.8	61	8.8
1000	70	14.7	615	182	25	17.8	39	11.0
1600	88	16.8	700	164	15	26.9	24	17.2
2000	94	18.4	770	158	13	32.0	20	19.4
132 kV single-core corrugated aluminium sheath								
150	55	8.1	340	217	156	4.6	259	3.7
630	65	12.1	505	194	37	9.6	61	5.6
1000	73	14.7	615	182	25	13.7	39	7.3
1600	92	16.8	700	164	15	21.2	24	11.4
2000	98	18.4	770	158	13	25.5	20	12.9

For designs and applicable conditions of installation and operation see the introduction to this appendix.

FLUID-FILLED PAPER INSULATED TRANSMISSION CABLES

Table A16.6 Technical data 132 kV 3-core cables

Conductor size (mm²)	Approximate diameter (mm)	Charging current (mA/m)	Capacitance per core (pF/m)	Reactance (µΩ/m)	Copper conductor A.C. (90°C) resistance (µΩ/m)	Copper conductor Approximate weight (kg/m)	Aluminium conductor A.C. (90°C) resistance (µΩ/m)	Aluminium conductor Approximate weight (kg/m)
132 kV 3-core, lead sheath								
185	96	6.5	270	112	127	21.8	211	18.6
240	99	7.3	305	106	98	24.4	161	20.1
300	103	8.3	345	101	78	26.9	129	21.8
400	108	9.2	385	97	62	31.0	101	24.1
630	123	11.2	470	91	41	42.9	64	31.5
132 kV 3-core corrugated aluminium sheath								
185	102	6.5	270	112	127	15.1	211	11.8
240	104	7.3	305	106	98	17.1	161	12.7
300	110	8.3	345	101	78	19.4	129	13.9
400	115	9.2	385	97	62	22.5	101	15.5
630	131	11.2	470	91	41	32.1	64	20.4

For designs and applicable conditions of installation and operation see the introduction to this appendix.

FLUID-FILLED PAPER INSULATED TRANSMISSION CABLES

Table A16.7 Technical data 275 kV single-core cables

Conductor size (mm^2)	Approximate diameter (mm)	Charging current (mA/m)	Capacitance per core (pF/m)	Reactance (μΩ/m)	Copper conductor A.C. (90°C) resistance (μΩ/m)	Copper conductor Approximate weight (kg/m)	Aluminium conductor A.C. (90°C) resistance (μΩ/m)	Aluminium conductor Approximate weight (kg/m)
275 kV single-core, lead sheath								
300	74	11.5	230	211	76	12.9	125	11.1
630	78	15.2	305	194	37	16.8	61	12.7
1000	91	16.2	325	182	25	23.6	39	17.1
1600	103	22.0	440	164	15	31.7	24	22.0
2000	108	23.9	480	158	13	36.9	20	24.2
275 kV single-core corrugated aluminium sheath								
300	78	11.5	230	211	76	8.7	125	6.9
630	82	15.2	305	194	37	12.0	61	8.0
1000	95	16.2	325	182	25	17.2	39	10.8
1600	107	22.0	440	164	15	24.1	24	14.3
2000	113	23.9	480	158	13	28.6	20	16.0

For designs and applicable conditions of installation and operation see the introduction to this appendix.

FLUID-FILLED PAPER INSULATED TRANSMISSION CABLES

Table A16.8 Technical data 400 kV single-core cables

Conductor size (mm²)	Approximate diameter (mm)	Charging current (mA/m)	Capacitance per core (pF/m)	Reactance (μΩ/m)	Copper conductor A.C. (90°C) resistance (μΩ/m)	Copper conductor Approximate weight (kg/m)	Aluminium conductor A.C. (90°C) resistance (μΩ/m)	Aluminium conductor Approximate weight (kg/m)
400 kV single-core, lead sheath								
300	105	11.6	160	211	76	22.7	125	20.9
630	102	15.2	210	194	37	24.2	61	20.3
1000	104	18.5	255	182	25	28.3	39	21.9
1600	114	25.4	350	164	15	36.3	24	26.6
2000	119	27.9	385	158	13	41.5	20	28.8
400 kV single-core corrugated aluminium sheath								
300	109	11.6	160	211	76	14.5	125	12.7
630	106	15.2	210	194	37	16.5	61	12.6
1000	109	18.5	255	182	25	20.1	39	13.7
1600	119	25.4	350	164	15	26.6	24	16.9
2000	124	27.9	385	158	13	31.1	20	18.5

For designs and applicable conditions of installation and operation see the introduction to this appendix.

APPENDIX A17
PPL Insulated Self-contained Fluid-filled Transmission Cables

The information in this appendix covers the following.

(a) Technical data: dimensions, weights, charging current, capacitance, a.c. resistance and reactance (tables A17.2–A17.3). Conductor sizes are generally available for the range 1000 mm^2 to 3000 mm^2. Technical data is given for selected sizes.
(b) Power ratings: for single-core cables, laid direct and in air (figs A17.1–A17.4).
(c) Conditions applicable: table A17.1.

Table A17.1 Conditions applicable

Cables laid direct	
Depth to top of cable	900 mm
Spacing between cable centres (flat)	230 mm
Ground temperature	15°C
Soil thermal resistivity	1.2 K m/W
Conductor operating temperature	90°C
Cables in air	
Spacing between cable centres (flat)	230 mm
Air temperature	25°C
Conductor operating temperature	90°C
(Protection from solar radiation)	

(d) Cable design: the data are based on cables described and illustrated in chapter 34.

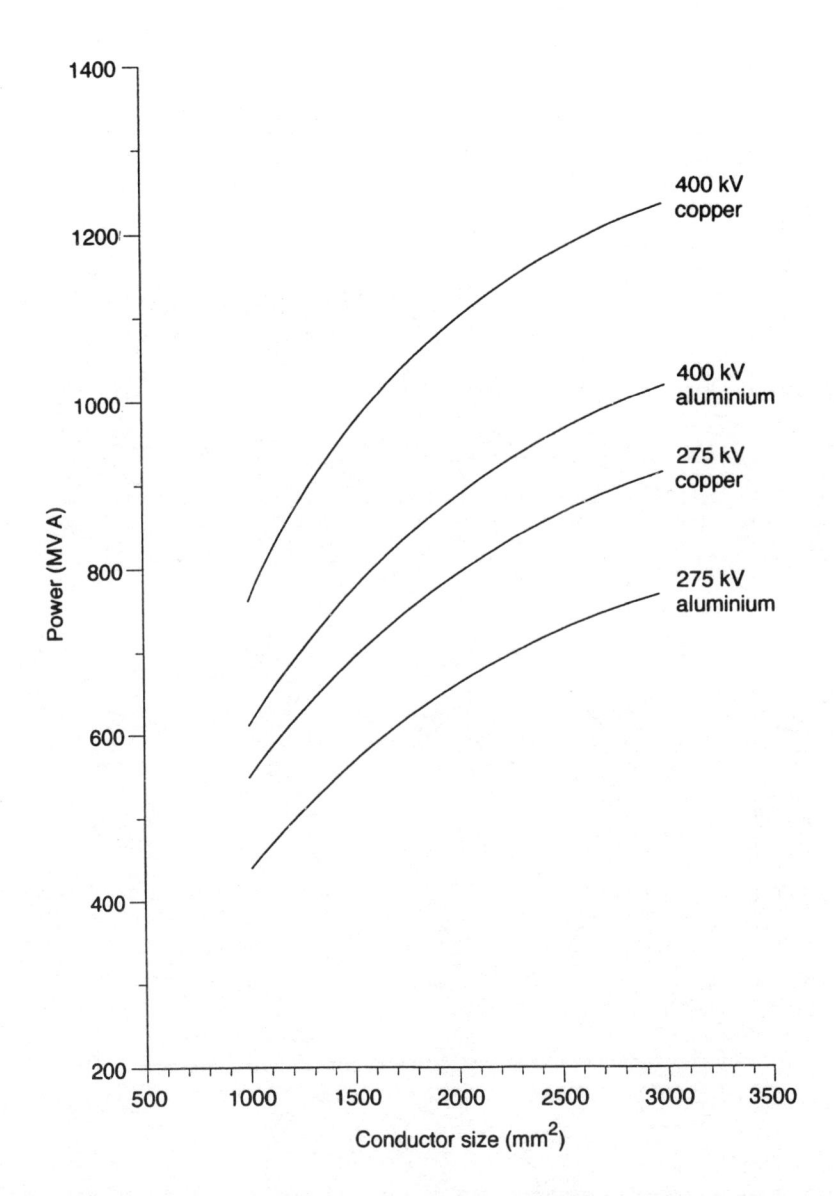

Fig. A17.1 Power ratings for single-core lead sheathed FF PPL cables laid direct

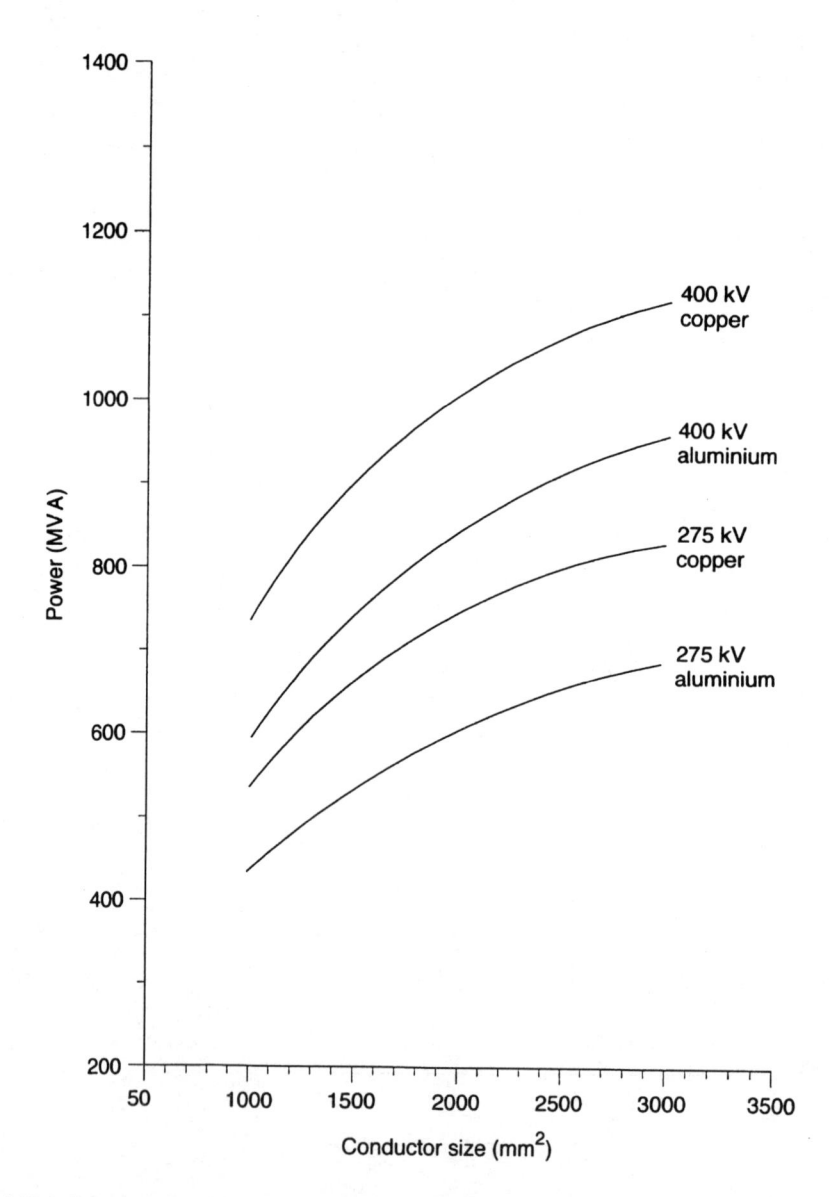

Fig. A17.2 Power ratings for single-core aluminium sheathed FF PPL cables laid direct

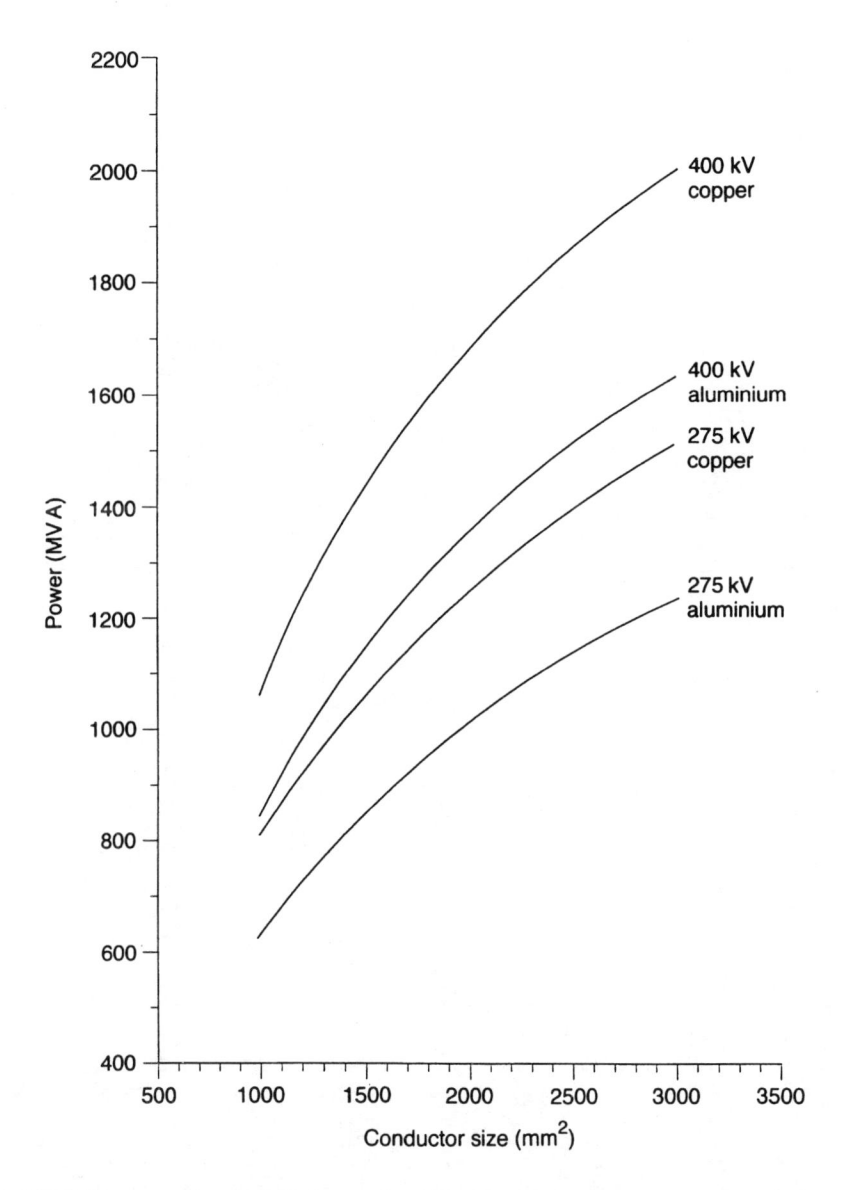

Fig. A17.3 Power ratings for single-core lead sheathed FF PPL cables in air

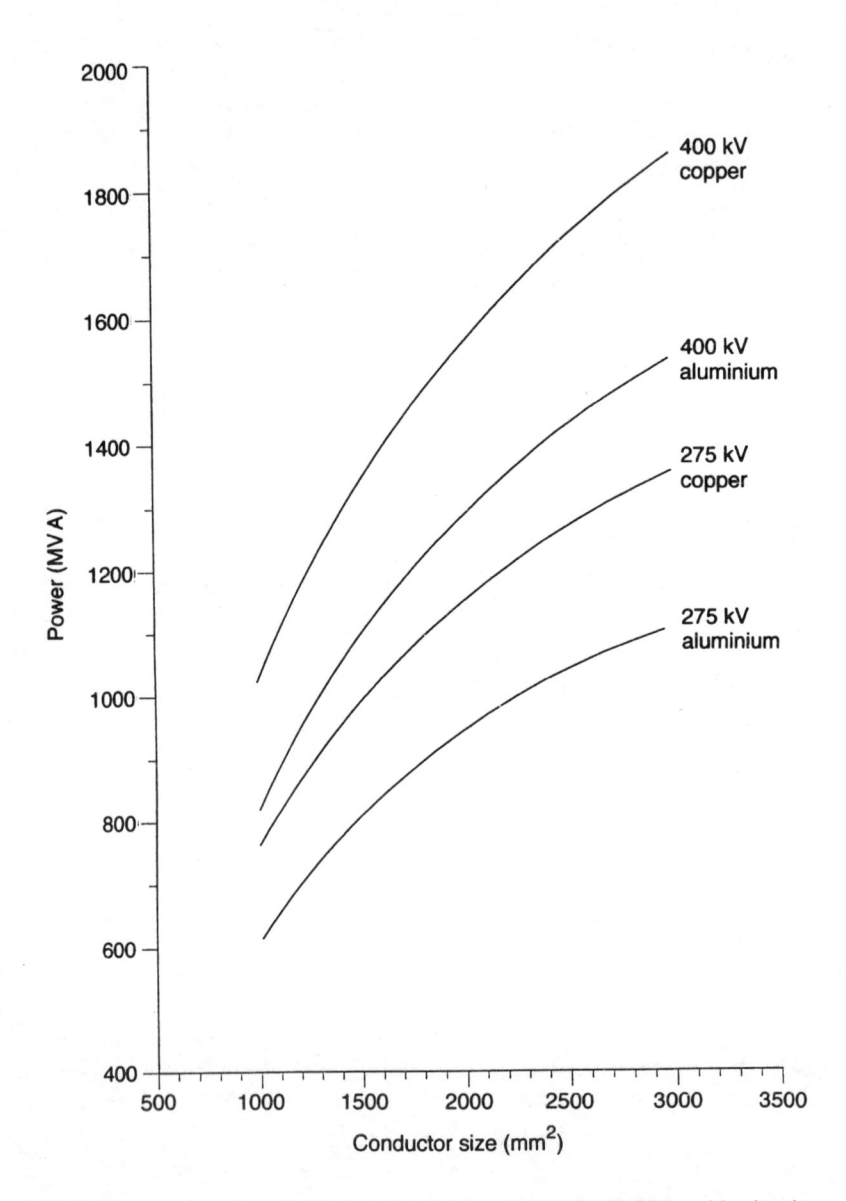

Fig. A17.4 Power ratings for single-core aluminium sheathed FF PPL cables in air

PPL SELF-CONTAINED FLUID-FILLED TRANSMISSION CABLES

Table A17.2 Technical data 275 kV single-core cables

Conductor size (mm^2)	Approximate diameter (mm)	Charging current (mA/m)	Capacitance per core (pF/m)	Reactance (μΩ/m)	Copper conductor A.C. (90°C) resistance (μΩ/m)	Copper conductor Approximate weight (kg/m)	Aluminium conductor A.C. (90°C) resistance (μΩ/m)	Aluminium conductor Approximate weight (kg/m)
275 kV single-core, lead sheath								
1000	85	14.9	292	182	25	24.0	39	15.5
1300	99	15.7	307	170	18	28.0	30	19.6
1600	104	17.4	340	164	15	32.0	24	21.6
2000	109	19.2	375	158	13	37.0	20	24.0
2500	115	21.4	420	150	11	43.3	16	27.9
3000	121	23.1	453	145	9	49.0	12	31.6
275 kV single-core corrugated aluminium sheath								
1000	90	14.9	292	182	25	15.7	39	10.1
1300	104	15.7	307	170	18	20.5	30	12.2
1600	109	17.4	340	164	15	23.9	24	13.6
2000	114	19.2	375	158	13	28.2	20	18.2
2500	121	21.4	420	150	11	31.0	16	20.0
3000	126	23.1	453	145	9	39.1	12	25.2

For designs and applicable conditions of installation and operation see the introduction to this appendix.

PPL SELF-CONTAINED FLUID-FILLED TRANSMISSION CABLES

Table A17.3 Technical data 400 kV single-core cables

Conductor size (mm^2)	Approximate diameter (mm)	Charging current (mA/m)	Capacitance per core (pF/m)	Reactance (μΩ/m)	Copper conductor A.C. (90°C) resistance (μΩ/m)	Copper conductor Approximate weight (kg/m)	Aluminium conductor A.C. (90°C) resistance (μΩ/m)	Aluminium conductor Approximate weight (kg/m)
400 kV single-core, lead sheath								
1000	105	14.9	201	182	25	28.7	39	22.1
1300	111	18.1	245	170	18	32.2	30	23.9
1600	115	20.0	270	164	15	36.2	24	25.9
2000	120	22.1	298	158	13	41.1	20	29.4
2500	126	24.7	334	150	11	47.9	16	32.6
3000	129	26.7	360	145	9	50.0	12	34.0
400 kV single-core corrugated aluminium sheath								
1000	110	14.9	201	182	25	19.5	39	13.9
1300	116	18.1	245	170	18	23.2	30	15.8
1600	120	20.0	270	164	15	26.5	24	17.1
2000	126	22.1	298	158	13	30.8	20	19.0
2500	131	24.7	334	150	11	36.2	16	21.4
3000	137	26.7	360	145	9	41.8	12	22.6

For designs and applicable conditions of installation and operation see the introduction to this appendix.

APPENDIX A18
XLPE Insulated Transmission Cables

The information in this appendix covers the following.

(a) Technical data: dimensions, weights, charging current, capacitance, a.c. resistance and reactance (tables A18.2–A18.5). Conductor sizes are generally available for the range 150 mm² to 3000 mm². Technical data is given for selected sizes.
(b) Power ratings: for single-core cables, laid direct and in air (figs A18.1–A18.4).
(c) Conditions applicable: table A18.1.

Table A18.1 Conditions applicable

Cables laid direct	
Depth to top of cable	900 mm
Spacing between cable centres (flat)	230 mm
Ground temperature	15°C
Soil thermal resistivity	1.2 K m/W
Conductor operating temperature	90°C
Cables in air	
Spacing between cable centres (flat)	230 mm
Air temperature	25°C
Conductor operating temperature	90°C
(Protection from solar radiation)	

(d) Cable design: the data are based on cables described and illustrated in chapter 34.

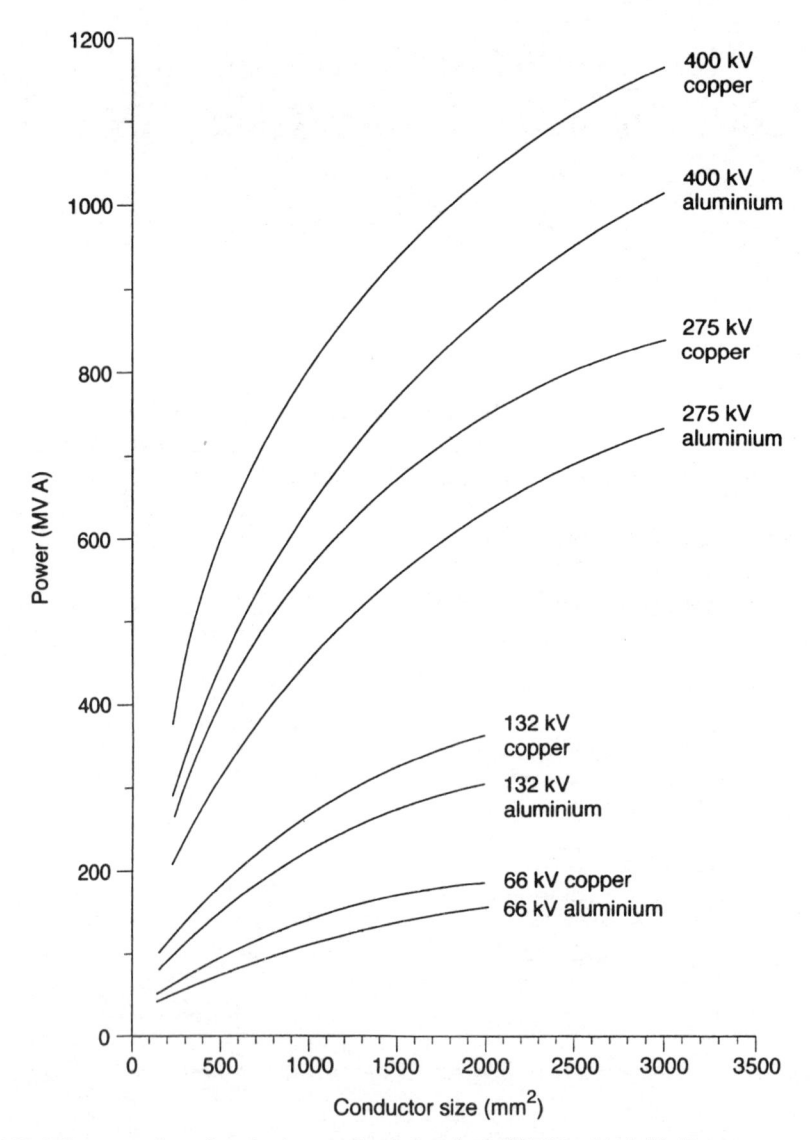

Fig. A18.1 Power ratings for single-core lead sheathed XLPE cable laid direct

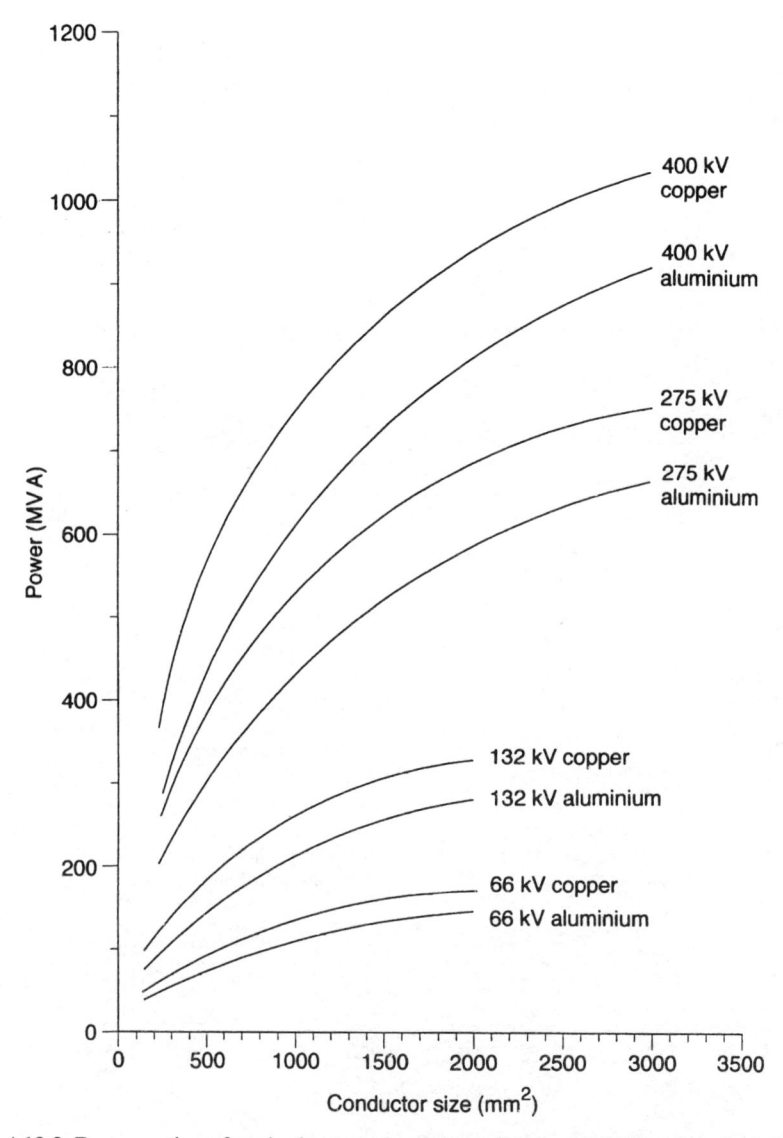

Fig. A18.2 Power ratings for single-core aluminium sheathed XLPE cables laid direct

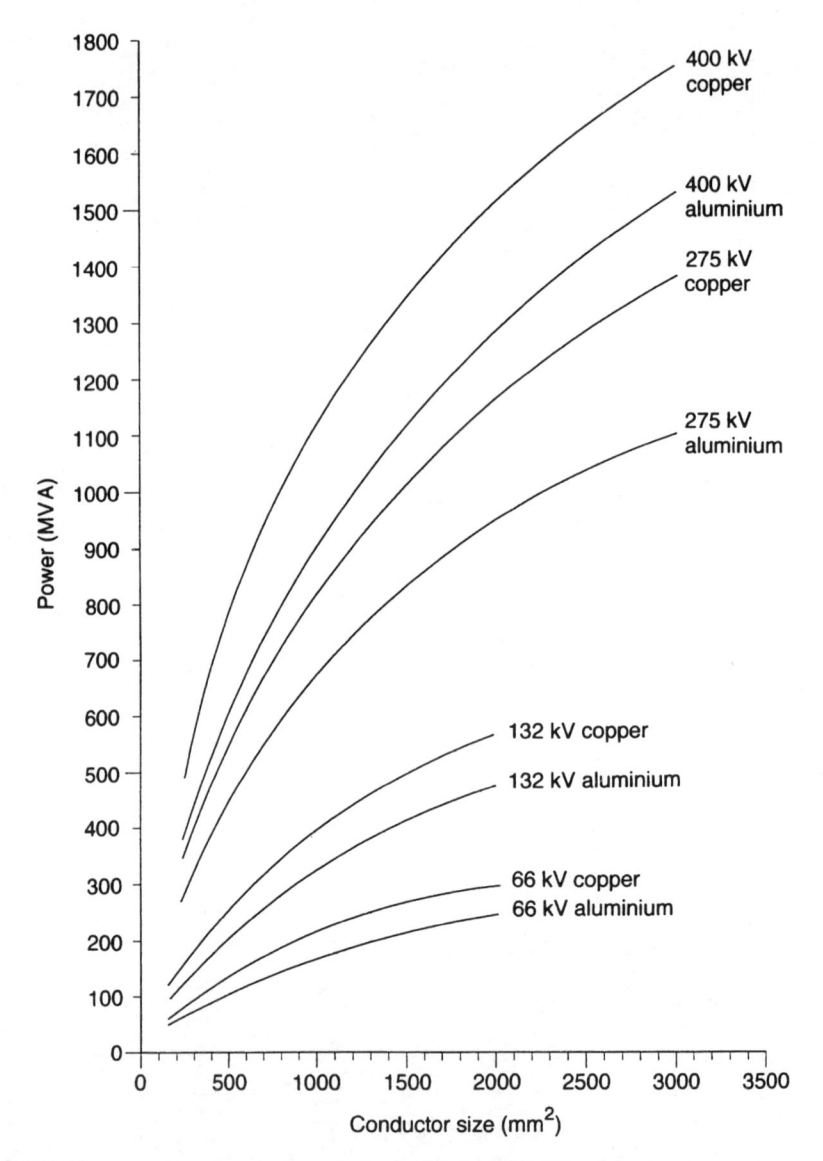

Fig. A18.3 Power ratings for single-core lead sheathed XLPE cables in air

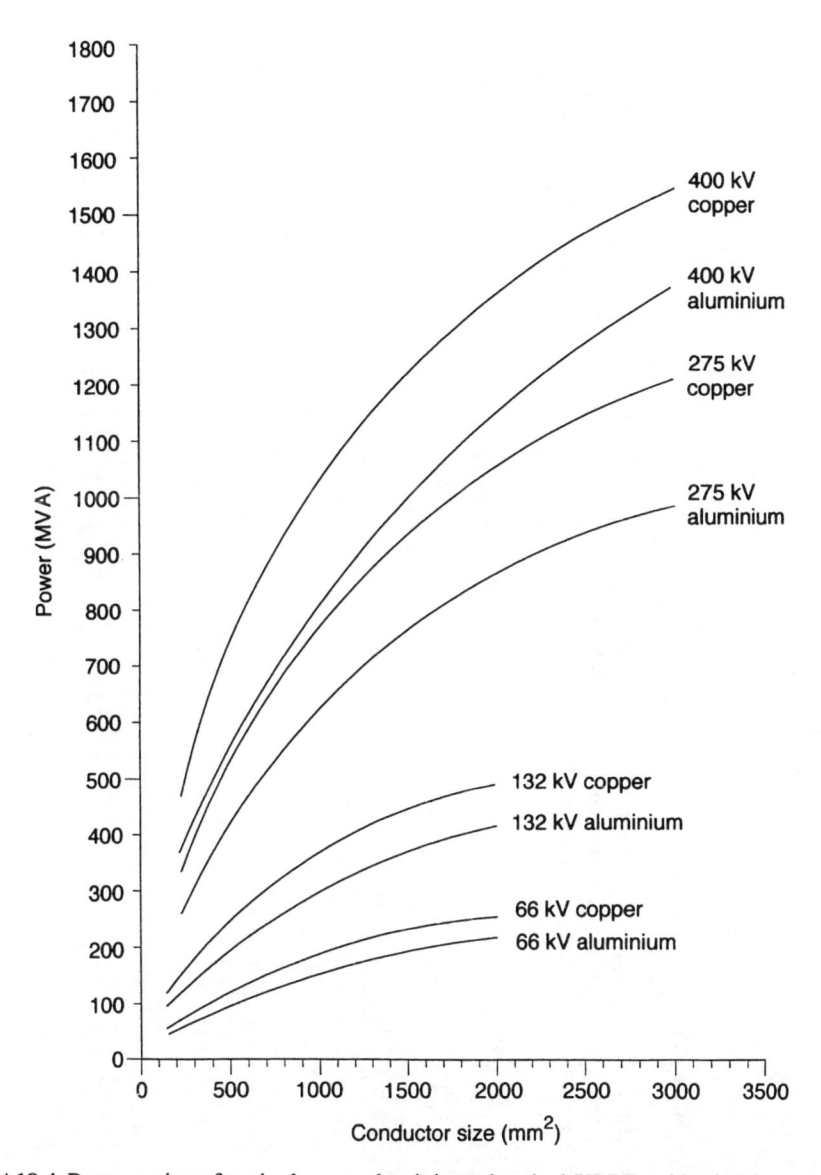

Fig. A18.4 Power ratings for single-core aluminium sheathed XLPE cables in air

XLPE TRANSMISSION CABLES

Table A18.2 Technical data 66 kV single-core cables

Conductor size (mm^2)	Approximate diameter (mm)	Charging current (mA/m)	Capacitance per core (pF/m)	Reactance (μΩ/m)	Copper conductor A.C. (90°C) resistance (μΩ/m)	Copper conductor Approximate weight (kg/m)	Aluminium conductor A.C. (90°C) resistance (μΩ/m)	Aluminium conductor Approximate weight (kg/m)
66 kV single-core lead sheath								
150	55	1.8	142	249	159	6.81	264	5.9
630	74	2.9	230	200	38	14.7	62	10.7
1000	86	3.5	283	185	26	20.7	40	14.2
1600	99	4.3	344	170	18	29.7	26	18.9
2000	104	4.6	366	164	15	34.9	22	21.8
66 kV single-core corrugated aluminium sheath								
150	63	1.8	142	249	159	4.1	264	3.2
630	83	2.9	230	200	38	10.3	62	6.2
1000	94	3.5	283	185	26	14.8	40	8.4
1600	110	4.3	344	170	18	22.8	26	12.2
2000	116	4.6	366	164	15	27.2	22	14.1

For designs and applicable conditions of installation and operation see the introduction to this appendix.

XLPE TRANSMISSION CABLES

Table A18.3 Technical data 132 kV single-core cables

Conductor size (mm^2)	Approximate diameter (mm)	Charging current (mA/m)	Capacitance per core (pF/m)	Reactance (μΩ/m)	Copper conductor A.C. (90°C) resistance (μΩ/m)	Copper conductor Approximate weight (kg/m)	Aluminium conductor A.C. (90°C) resistance (μΩ/m)	Aluminium conductor Approximate weight (kg/m)
132 kV single-core lead sheath								
150	71	2.9	116	249	159	10.2	264	9.3
630	90	4.4	176	200	38	18.9	62	14.9
1000	98	5.1	207	185	26	24.9	40	18.5
1600	113	6.2	248	169	18	34.5	26	23.7
2000	117	6.5	263	164	15	39.9	22	26.9
132 kV single-core corrugated aluminium sheath								
150	80	2.9	116	249	159	6.0	264	5.1
630	99	4.4	176	200	38	12.6	62	8.6
1000	109	5.1	207	185	26	17.2	40	10.8
1600	126	6.2	248	170	18	25.6	26	14.7
2000	131	6.5	263	164	15	30.1	22	17.0

For designs and applicable conditions of installation and operation see the introduction to this appendix.

XLPE TRANSMISSION CABLES

Table A18.4 Technical data 275 kV single-core cables

Conductor size (mm²)	Approximate diameter (mm)	Charging current (mA/m)	Capacitance per core (pF/m)	Reactance (μΩ/m)	Copper conductor A.C. (90°C) resistance (μΩ/m)	Copper conductor Approximate weight (kg/m)	Aluminium conductor A.C. (90°C) resistance (μΩ/m)	Aluminium conductor Approximate weight (kg/m)
275 kV single-core lead sheath								
300	103	5.7	112	226	78	19.8	129	17.8
630	101	8.6	166	200	38	21.8	62	17.6
1000	109	9.5	185	185	23	28.3	38	20.7
2000	127	11.6	225	164	15	43.9	22	27.4
3000	138	13.3	256	152	12	57.8	16	32.9
275 kV single-core corrugated aluminium sheath								
300	109	5.7	112	226	78	11.0	129	9.8
630	108	8.6	166	200	38	13.3	62	10.0
1000	116	9.5	185	185	23	18.2	38	12.0
2000	134	11.6	225	164	15	30.1	22	17.4
3000	145	13.3	256	152	12	41.6	16	21.6

For designs and applicable conditions of installation and operation see the introduction to this appendix.

XLPE TRANSMISSION CABLES

Table A18.5 Technical data 400 kV single-core cables

Conductor size (mm²)	Approximate diameter (mm)	Charging current (mA/m)	Capacitance per core (pF/m)	Reactance (μΩ/m)	Copper conductor A.C. (90°C) resistance (μΩ/m)	Copper conductor Approximate weight (kg/m)	Aluminium conductor A.C. (90°C) resistance (μΩ/m)	Aluminium conductor Approximate weight (kg/m)
400 kV single-core lead sheath								
300	122	7.8	104	226	78	26.4	129	24.6
630	126	9.6	128	200	38	30.3	62	26.3
1000	132	10.7	143	185	23	35.9	38	29.4
2500	148	15.3	204	157	13	55.4	18	39.1
3000	155	16.2	215	152	12	62.5	16	42.9
400 kV single-core corrugated aluminium sheath								
300	130	7.8	104	226	78	14.8	129	13.0
630	133	9.6	128	200	38	18.0	62	14.0
1000	140	10.7	143	185	23	22.7	38	16.2
2500	157	15.3	204	157	13	38.9	18	22.6
3000	164	16.2	215	152	12	44.7	16	25.0

For designs and applicable conditions of installation and operation see the introduction to this appendix.

APPENDIX A19
Minimum Installation Bending Radii

PART 1: BASIS

The various types of cable covered are divided into sections in accordance with the descriptions given in parts 2–4 of the book. The radii quoted are in accordance with British Standards or, for cables not covered by British Standards, represent accepted practice.

Symbols: d_o = cable overall diameter or the major axis for flat cables
d_s = diameter over metal sheath

PART 2: GENERAL WIRING CABLES

Table A19.1 Cables for fixed wiring up to 450/750 V

Insulation	Conductors	Construction	Overall diameter (mm)	Minimum radius
XLPE, PVC or rubber	Aluminium or copper, solid or stranded circular	Unarmoured	Up to 10 10–25 Above 25	$3d_o$ [a] $4d_o$ [b] $6d_o$
Mineral	Copper		Any	$6d_o$

[a] $2d_o$ for single-core cables with circular conductors of stranded construction installed in conduit, ducting or trunking
[b] $3d_o$ for single-core cables with circular conductors of stranded construction installed in conduit, ducting or trunking

Minimum Installation Bending Radii

Table A19.2 EPR insulated cables for installation in ships
Ideally cables should be bent as little as possible and never to radii less than the following:

Type of cable	Minimum bending radius
Instrumentation	$8 \times d_o$
LV power and control	
unarmoured up to 10 mm D	$3 \times d_o$
10–25 mm D	$4 \times d_o$
over 25 mm D	$6 \times d_o$
armoured up to 25 mm D	$4 \times d_o$
over 25 mm D	$6 \times d_o$
HV power	
1.9/3.3 kV	$6 \times d_o$
3.8/6.6 kV and above	
single-core	$20 \times d_o$
3-core	$15 \times d_o$

Table A19.3 PVC and EPR insulated wire armoured mining cables (to BCS 295)

Voltage	Construction	Minimum radius
600/1000 V		$8d_o$
1.9/3.3 kV	Unscreened	$8d_o$
1.9/3.3 kV	Screened	$12d_o$
3.8/6.6 kV		$12d_o$

PART 3: DISTRIBUTION CABLES

Table A19.4 Paper insulated cables

Voltage	Minimum radius			
	Single-core	Multicore	Adjacent to joints and terminations	
			Without former	With former
Up to and including 6.35/11 kV	$15d_o$	$12d_o$		
Above 6.35/11 kV and up to and including 12.7/22 kV	$18d_o$	$15d_o$		
19/33 kV single core	$21d_o$		$20d_o$	$15d_o$
19/33 kV 3-core screened		$18d_o$	$15d_o$	$12d_o$
19/33 kV 3-core SL		$18d_o$	$15d_o$	$12d_o$
19/33 kV cores of SL	$21d_o$		$20d_s$	$15d_s$

Table A19.5 PVC and XLPE insulated cables rated at 600/1000 V and 1.9/3.3 kV

Conductor	Construction	Overall diameter (mm)	Minimum radius
Circular copper	Unarmoured	10–25	$4d_o$
		> 25	$6d_o$
Circular copper	Armoured	Any	$6d_o$
Shaped copper	Armoured and unarmoured	Any	$8d_o$
Solid aluminium	Armoured and unarmoured	Any	$8d_o$

Table A19.6 XLPE insulated cables for 6.6–22 kV

Type of cable	Minimum radius	
	During laying	Adjacent to joints or terminations
Single-core		
(a) unarmoured	$20d_o$	$15d_o$
(b) armoured	$15d_o$	$12d_o$
3-core		
(a) unarmoured	$15d_o$	$12d_o$
(b) armoured	$12d_o$	$10d_o$

PART 4: TRANSMISSION CABLES

Table A19.7 Fluid-filled cables

Voltage	Number of cores	Minimum radius		Adjacent to joints and terminations	
		Laid in ducts	Laid direct or in air	With former	Without former
33–132 kV	1	$35d_o$	$30d_o$	$15d_o$	$20d_o$
	3	$30d_o$	$20d_o$	$12d_o$	$15d_o$
275–525 kV paper insulated	1	$35d_o$	$30d_o$	$20d_o$	$20d_o$
275–525 kV PPL insulated		$35d_o$	$30d_o$	$25d_o$	$25d_o$

Table A19.8 XLPE insulated cables

Type of cable	Minimum radius		Cable placed into position adjacent to joints or terminations	
	Laid in ducts	Laid direct or in air	With former	Without former
33 kV non-metallic sheathed single core	$20d_o$	$20d_o$	$15d_o$	$15d_o$
3-core	$15d_o$	$15d_o$	$12d_o$	$12d_o$
66 kV and 132 kV single-core metal sheathed	$35d_o$	$30d_o$	$15d_o$	$20d_o$
275 kV and 400 kV single-core metal-sheathed	$35d_o$	$30d_o$	$15d_o$	$20d_o$

APPENDIX A20

Weibull Distribution

INTRODUCTION

The electrical breakdown strength, under a.c., d.c., or impulse conditions is often used as a method either to estimate the quality of an electrically stressed device, or to determine 'safe' operating conditions for it. Unfortunately the electrical breakdown strength of a device cannot be defined with a single value as it can be affected by ambient conditions (temperature, humidity), rate of voltage application, volume of material under test, and age of the device. Yet when identical devices are electrically stressed, under identical conditions, the resulting breakdowns do not occur at the same voltage (for ramp tests) or after the same time (for constant stress tests). These results indicate that there is some inherent scatter in the measured breakdown strengths and that the breakdown mechanism is stochastic in nature. Consequently it is not possible to predict the performance of a device precisely, even with knowledge of the exact service conditions.

In many engineering applications there is a need to determine the operating performance and design criteria for electrically stressed devices. The types of devices where this need exists range from metal oxide semiconductor devices (transistors and capacitors) through electric cables to voltage regulators. Therefore to gain a practical understanding of the performance of electrically stressed devices it is essential to perform some form of statistical analysis on the data. The failure of a complex system or a single component tends to follow a few well-defined statistical distributions: exponential, Rayleigh, log normal, normal, and Weibull. The Weibull distribution is often used for the analysis because it can provide good estimates for the other distributions.

THE DISTRIBUTION OF STRESSES UNDER STEP RAMP CONDITIONS

Figure A20.1 shows some results from a step test in which 10 sample plaques of LDPE insulation were subjected to breakdown when using a progressively increasing voltage. The electric fields at failure are represented in a conventional histogram format. The

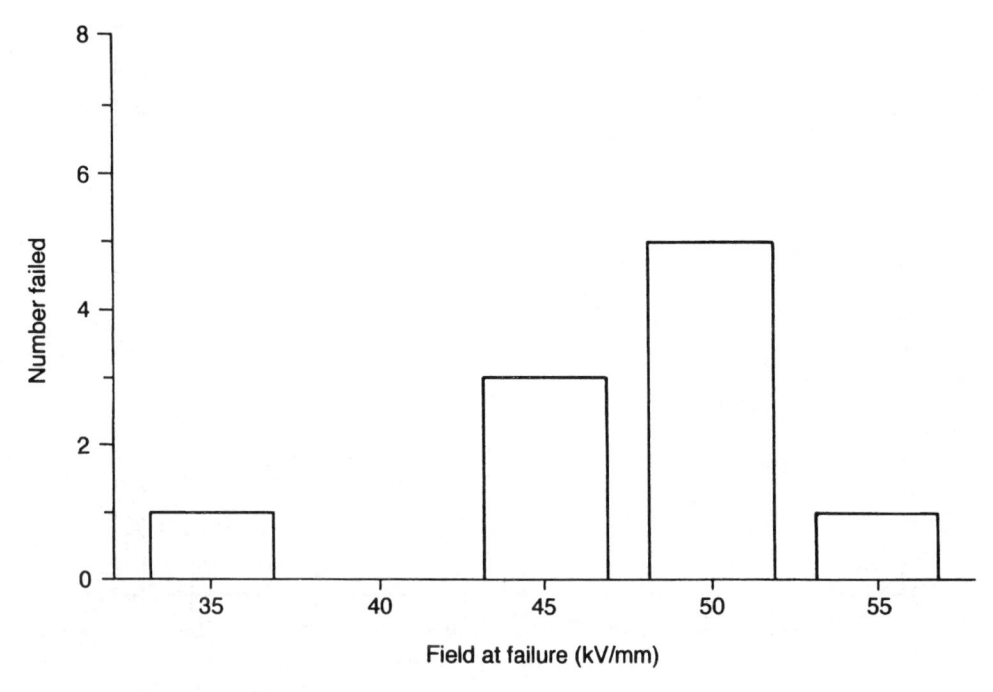

Fig. A20.1 Histogram of fields at failure

left-hand scale shows number of samples broken down within each field range, ±2.5 kV in this case. Given a larger number of samples, it would be reasonable to expect their fields to be distributed in similar proportions.

It is often more useful to consider the proportion of samples which survive to a given field. A manufacturer or utility may well be more interested in establishing the field at which the first 1% of the samples fail rather than the 'scale', i.e. the field at which 63% of the samples fail. In order to do this type of analysis it is useful to re-plot the data in fig. A20.1 as a 'cumulative' plot, i.e. a plot showing the proportion of samples which have broken down at a given field. This is shown in fig. A20.2 for the data from fig. A20.1.

It can be seen that about 10% of samples had broken down by 40 kV/mm and 90% by 55 kV/mm. However it would be difficult from this histogram to estimate the field required for the first 1% to fail, indeed it is not clear how field is related to failure at all. Therefore we need to fit a statistical distribution to the data.

If a very large number of samples had been tested (for example, 1000 instead of just 10) it would have been possible to have drawn a much better histogram – a histogram with narrower field intervals than ±2.5 kV. It may have been possible to draw a smooth line over the tops of the histogram bars to show how the breakdown was related to the applied field. This may have resulted in a curve similar to that shown in fig. A20.2.

In the case of fig. A20.2 we only have 10 data: therefore, to proceed further with the analysis we need to fit a distribution to our data. As we have noted earlier we may have chosen to use either the exponential, Rayleigh, normal, log normal, or three parameter

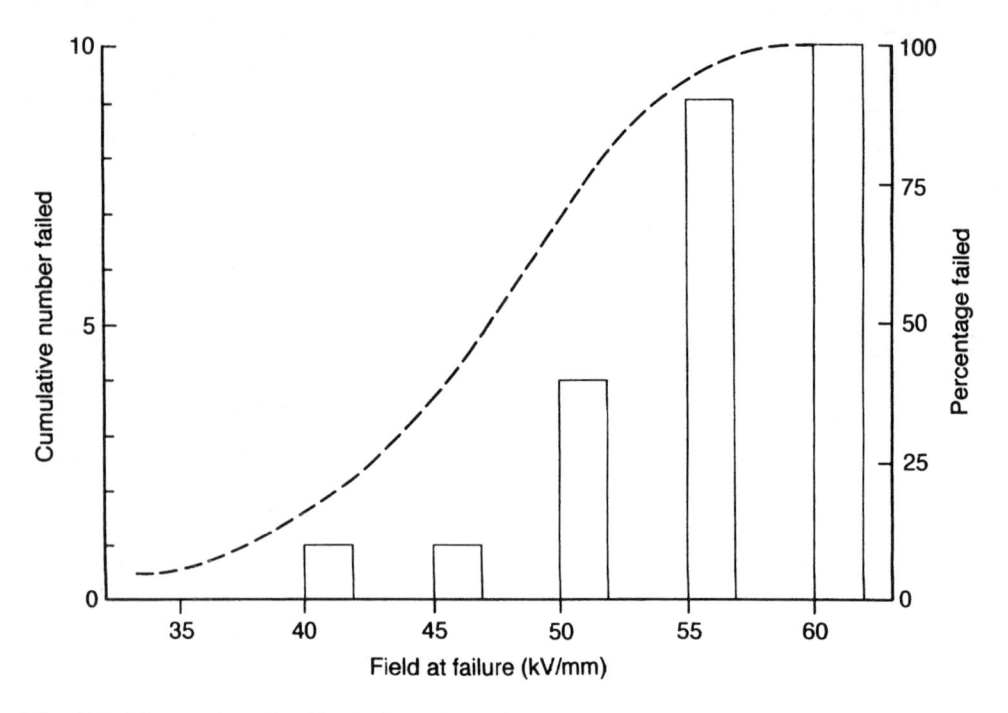

Fig. A20.2 Proportion of cables in fig. A20.1 failed at a given field

Weibull distributions.[1] The critical aspect of the selection of the distribution for analysis is to ensure that the selection is the best that we could have made: to do this we need to calculate the 'goodness of fit'. The goodness of fit for any distribution may be easily checked through the use of standard statistic measures such as the chi-Square or the Kolmogorov-Smirnov tests.[2]

The fit, using the Kolmogorov-Smirnov test, of the Weibull distribution to these data is 0.95: the closer the value to 1 the better the fit of the distribution to the data. However, the fit of alternate distributions are: normal, 0.93 and log normal, 0.89. In the case of the data in fig. A20.1 the fits of the Weibull, log normal, and the normal distributions are all good (each goodness of fit is sufficiently close to 1) for a sample size of 10 and, therefore, we can *choose* which one of these three distributions to use for further analysis. In the majority of cases the electrical test data (ramp and time to failure testing) are best fitted by the Weibull distribution, as indicated here by the goodness of fit and, therefore, we will use this distribution for the analysis of the data.

We have now shown that there is a range of distributions that can be used to analyse breakdown data and that it is possible to assess how well the distribution fits the data. However, there is now the issue of which statistical approach should be adopted if there is no 'good fit' for the distributions. The obvious approach is to collect more data and then to see if an improved fit can be achieved. In many situations it is possible to do this and it is a good solution. However, in many situations it is not practical to collect more data, due either to non-availability of material or the cost of further testing. One method that has proved to be very useful is to avoid making the choice of a distribution and use methods that are termed either non-parametric or distribution free.[3,4]

WEIBULL ANALYSIS

The Weibull distribution was developed by the Swedish statistician Waloddi Weibull. The distribution that bears his name has been used to analyse a very wide range of problems; it has perhaps been most useful in dealing with failure data.[4,5] The applicability of the distribution to failure data has led to the widespread use of the Weibull distribution for the analysis of the times to failure and progressive electrical stress tests on solid insulation systems.[6,7,8] However, care is required at all steps in the analysis as, since Weibull himself noted, the Weibull distribution 'may often render good service'; he did not claim that it worked all of the time. Additionally it should always be remembered that the Weibull approach applies to one failure mechanism at a time; if the data contain contributions from more than one failure mechanism then either an alternative method is required or some pre-processing of the data is needed.

The general form of the Weibull equation, often termed the Two Parameter Weibull equation, is:

$$P_f = 1 - \exp-(G/\alpha)^\beta \qquad (A20.1)$$

Here P_f is the probability of failure and G is the stress that progressively increases and eventually causes the unit to fail. The stress G may either be the time to failure or number of cycles to failure or voltage at failure. Alpha (α) and beta (β) are the characteristic (or scale) and shape values, respectively. These two numbers fully define the Weibull distribution: they are analogous to the mean and standard deviation of the normal distribution. In the case of cables these parameters are estimated from progressive stress or time to failure tests that have been carried out on a number of cables.

In the case of cable testing there are three effects that need to be allowed for:

(1) field induced – as the electric field is progressively increased failure becomes much more likely;
(2) time induced – as the time of application of the electric field is progressively increased failure becomes much more likely; and
(3) volume induced effects – as the volume of insulation under test is increased there is a decrease in both the electric field and time at which failure occur.

These effects require a modification to the simple two-parameter Weibull equation (A20.1) to give a compound equation[9] of the form:

$$P_f = 1 - \exp-[(E/\alpha)^b (t/t_0)^a L r^2 / L_0 r_0^2] \qquad (A20.2)$$

where E = electric field at the cable conductor
 t = time of application of the electric field
 L = length of cable under examination
 r = radius of conductor under examination
 t_0 = characteristic time to failure at constant stress
 a = time exponent of the dielectric (sometimes referred to as the time equivalent)

b = stress exponent (a characteristic constant of the dielectric)
L_0 and r_0 are the dimensions of the reference cable for which α, t_0, a and b were determined.

In practice when conducting tests either the time or stress is held constant during testing such that we can use the simplified equations:

$$\text{Ramp tests} \quad P_f = 1 - \exp-(E/\alpha)^\beta \quad \text{(A20.3)}$$

$$\text{Endurance tests} \quad P_f = 1 - \exp-(t/t_0)^a \quad \text{(A20.4)}$$

These two equations are related by the well known empirical endurance relation (equation (A20.5)), which gives an ageing parameter 'n'. This relation is often used to analyse endurance data collected at different electrical stresses.

$$E^n t = \text{constant} \quad \text{(A20.5)}$$

A PRACTICAL EXAMPLE OF A WEIBULL ANALYSIS

One of the strengths of the Weibull approach is that it provides a graphical display of the probability of failure coupled with a systematic method for determining the distribution parameters. The steps required to produce a Weibull plot, following the general scheme of the IEEE guide, are listed below for the data from fig. A20.1.

(1) The stress at failure is ranked (smallest to largest) and the rank order noted (see table A20.1).

Table A20.1 Data required to produce Weibull plot

Ordered fields (E) at failure (kV/mm)	Rank order	P_f	$\log_{10}(E)$	$\log_{10}(-\log_e(1-P_f))$
35	1	0.067	1.544	−1.157
43	2	0.163	1.633	−0.748
43	3	0.260	1.633	−0.522
46	4	0.356	1.663	−0.357
48	5	0.452	1.681	−0.221
48	6	0.548	1.681	−0.100
51	7	0.644	1.708	0.014
51	8	0.740	1.708	0.130
51	9	0.873	1.708	0.258
57	10	0.933	1.756	0.431

(2) The probability of failure (P_f) is estimated from the rank order. P_f is estimated using the Bernard[9] approximation (equation A20.6). This method of estimating P_f is normally recommended for small sample sizes, $N < 20$, which is often the case for cable test data.

$$P_f = \frac{[i - 0.3]}{[N + 0.4]} \quad \text{(A20.6)}$$

where $i =$ rank order

For the first datum the median rank is given as $(P_f) = (1 - 0.3)/(10 + 0.4) = 0.067$

(3) The Weibull equation can be transformed such that the data form a straight line. The transformation is:

$$\log_{10}[-\log_e(1 - P_f)] \text{ vs } \log_{10} E$$

(4) Plot the data on transformed axis (see fig. A20.3)

(5) The parameters can be extracted from the graph where:
$\beta =$ gradient of the curve
$\alpha =$ field at which $\log_{10} - (\log_e(1 - P_f)) = 0$, and $-\log_e(1 - P_f) = 1$

The straight line on the plot may either be plotted by eye or fitted using numerical methods. The line in fig. A20.3 is the best fit of the Weibull distribution to the data when using the 'maximum likelihood method.' The maximum likelihood method is not the only numerical method[4] but is widely used for parameter estimation (references 1 and 4 give detailed statistical discussions). The results of this analysis are given in table A20.2.

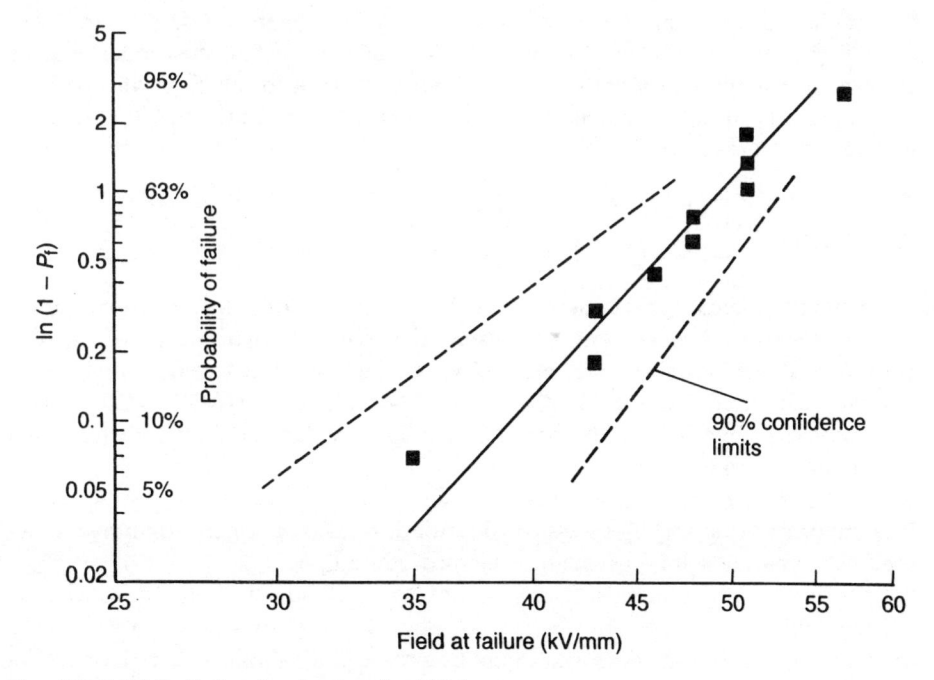

Fig. A20.3 Weibull plot for the data in A20.1

Table A20.2 Weibull parameters estimated for the data of fig. A20.1

	Scale (kV)	Shape	Goodness of fit
Parameter	49.7	9.8	0.947

The Weibull parameters shown in table A20.2 are the 'best fit' to the data. However, there is a range of uncertainty associated with them. The level of the uncertainty can be calculated and is represented by 'confidence limits'. The two dashed lines on fig. A20.3 represent the 90% confidence limits for the fit of the Weibull distribution.[1,5] The way to interpret these limits is that we are 90% confident that the Weibull line lies between these two limits.

Inspection of fig. A20.3 shows that the line associated with Weibull distribution (the solid line) passes through most of the data, thereby indicating that the Weibull distribution is a good fit to the data. This is supported by inspection of the confidence limits. If any of the data had large deviations from the Weibull line or had lain outside the confidence limits, it would have indicated that further investigation was required. The most common reason for deviations or outlying points are errors in the data collection, errors in subsequent calculations, or that the two-parameter Weibull distribution is not an appropriate method of analysis, a more sophisticated distribution being required. In all cases a plot of the data, an estimate of the goodness of fit, and the calculation of confidence limits are invaluable in ensuring high quality analysis of the data.

It is natural to assume that the characteristic value of failure, α, is the most important of these two parameters. The characteristic value represents the voltage by which 63% of the samples have broken down. In fact, the shape parameter, β, is also extremely useful as it provides a great deal of information on the mechanisms underlying the breakdown. The shape is especially important when we wish to calculate when the first 1% (say) are likely to fail: an engineer is more interested in when the first failure might occur than when 63% have failed.

CONCLUSIONS

This appendix presents a basic account of the steps required to perform a simple analysis of electrical breakdown data using the Weibull distribution. A detailed description of the Weibull analysis method is not included but excellent reviews are given in the references, particularly Abernethy, Weibull, and Dissado and Fothergill. The appendix does highlight a number of issues that the analyst should keep in mind when carrying out analysis:

(1) It is important to critically assess the distribution selected for the analysis method used. The analyst will be assisted in this procedure by:
   (a) performing a 'goodness of fit' calculation to check how well the distribution fits the data
   (b) making a plot of the data and examining the relationship between the data and the fitted Weibull line plus associated confidence limits.

(2) When making a Weibull analysis it is essential to make a critical inspection of the curve. The important areas to examine are:
   (a) the gradient of the curve, similar gradients within a group of tests indicate similar mechanisms of failure
   (b) the early failures, i.e. the area generally of most interest to engineers
   (c) if the data includes any outliers (points that lie away from the Weibull line or outside the confidence limits) as these points require detailed examination.

REFERENCES

(1) Dissado, L. A. and Fothergill, J. C. (1992) *Electrical Degradation and Breakdown in Polymers*. Peter Peregrinus Ltd for the IEE.
(2) Miller, J. C. and Miller, J. N. (1989) *Statistics for Analytical Chemistry*. Ellis Horwood Ltd.
(3) Julian, K. and Hampton, R. N. (Nov. 1993) 'Comparison of Weibull statistical methods with other methods of data analysis', IEE Conference on Power Cables and Accessories – 10 kV to 500 kV.
(4) Abernethy, R. B. (1996) *The New Weibull Handbook*. Gulf Publishing Company.
(5) Weibull, W. (1951) *A Statistical Distribution Function of Wide Applicability. Journal of Applied Mechanics*.
(6) Rowlands, S. M., Hill, R. M. and Dissado, L. A. (1986) 'Censored Weibull statistics in the breakdown of thin oxide films', *J. Phys. C.: Solid State Phys.*, **19**, 6263–85.
(7) Jocteur, R., Osty, M., Lemainque, H. and Terramorsi, G. *Research and Development in France in the Field of Extruded Polyethylene Insulated High Voltage Cables*. CIGRE Working group 21-07.
(8) Kubota, T. *et al.* (1994) *Development of 500 kV XLPE Cables and Accessories for Long Distance Underground Transmission Line*. IEE 94 WM 097-6 PWRD.
(9) Bernard, G. (1988) *Application of Weibull Distribution to the Study of Power Cable Insulation*. Report CIGRE WG 21-09 Document 88-22.

APPENDIX A21
Bibliography

PART 1: THEORY AND COMMON ASPECTS

Electrical theory

Starr, A. T. (1957) *Transmission and Distribution of Electric Power*. Pitman.
Buchan, M. F. (1967) *Electricity Supply*. Edward Arnold.
Heinhold, L. (1970) *Power Cables and their Applications*. Siemens Aktiengesellschaft.
Cotton, H. and Barber, H. (1975) *Transmission and Distribution of Electrical Energy*. English Universities Press.
Graneau, P. (1980) *The Science, Technology and Economics of High Voltage Cables*. Wiley.
Weedey, B. M. (1980) *Underground Transmission of Electric Power*. Wiley.
Black, R. M. (1986) *The History of Electric Wires and Cables*. Peter Peregrinus.
Dorf, R. C. (ed.) (1993) *The Electrical Engineering Handbook*. CRC Press.
Daures, M. P. (1995) The major technological issues concerning the electrical systems for the 21st century. Jicable Conference Paper.
Mokry, S. *et al.* (1995) 'Cable fault prevention using dielectric enhancement technology'. Jicable Conference Paper.
Delgardo Lean, M. Z. *et al.* (1996) 'An efficient method for calculation of electric field in three-core power cables'. *IEEE Trans. Power Del.*, **11** (3), 1179–84.
Freundlich, P. and Kolodziej, H. A. (1996) 'The system for complex electric permittivity measurements of solid-state materials in the range of 20 Hz to 1 GHz'. *7th IEE Int. Conf. on Dielectric Materials and Applications*.
Lewis, T. J. *et al.* (1996) 'A new model for electrical ageing and breakdown in dielectrics'. *7th IEE Int. Conf. on Dielectric Materials and Applications*.

Materials

Aluminium
McDonald, F. G. (17 Aug. 1967) 'The hedgehog joint and story of 50 year old aluminium cable'. *Electr. Times*.
(2 July 1970) 'New wiring cable has copper-clad conductors'. *Electr. Times*.

McAllister, D. (10 Sep. 1971) 'Aluminium cables accepted by industry'. *Electr. Times.*
McAllister, D. (Aug. 1976) 'Terminations for aluminium conductor power cable'. *BSI News.*
Chattergee, S. (Aug. 1977) 'Failure of terminations of aluminium conductor cables in house-service meters and domestic wiring installations'. Indian Copper Information Centre and Electrical Contractors' Association of India Paper No. 6.
Roy, M. (1988) Aluminium – a vital element in cable-making. *Middle East Electricity* **12**, (5), 25–6.
United Nations Industrial Development Organization (1989) *Conceptual Design Study on Aluminium Wire Drawing and Stranded Cable Production.* Vienna: United Nations Industrial Development Organization.
Neyens, J., Meurice, D. and Tits, Y. (1997) 'Connectors for aluminium cable: a review of the different technologies and their interaction with cable accessories'. CIRED Paper No. 3.17.

Lead
Hiscock, S. A. (15 Apr., 29 Apr., 13 May 1960) 'American lead alloy cable sheaths'. *Electr. J.*
Hiscock, S. A. (1961) *Lead and Lead Alloys for Cable Sheathing.* Ernest Benn.
Betzer, C. E. (Jun. 1962) 'Determination of the life to fracture by bending of lead sheaths on underground power cable'. *AIEE Summer Meeting.*
Lead Development Association (1964) *Lead Cable Sheathing in N. America.* London.
Lead Development Association (1964) *Lead Alloys for Pressurised Cable Sheaths.* London.
Ball, E. H. and McAllister, D. (Sep. 1968) 'Influence of lead sheath thickness on service performance of cables'. *3rd Int. Conf. on Lead.*
Harvard, D. G. (Jan. 1975) 'Selection of cable sheath lead alloys for fatigue resistance'. *IEEE Power Eng. Soc.*
Hirji, A. (1987) 'Design and performance requirements for transition joint between high voltage paper lead cables and cross linked polyethylene (XLPE) cables'. *CABLE-WIRE – 87. 2nd International Seminar on Cables, Conductors and Winding Wires,* New Delhi. Indian Electr. & Electron. Manuf. Assoc., Bombay, India.
Hey, S. A., Haverkamp, W. L. and Baker, M. (1991) 'A rationalised joint design for MV, polymeric cables and transitions to paper/lead cables up to 36 kV'. *Distribution 2000. Doing it Right for the Future. Insulated Line and Cable Systems. International Conference and Workshop* **2**, p. 293–9. Sydney.
Goodwin, F. E. and Dyba, J. J. (1994) 'Recent technological developments for lead sheathed cables'. *Proc. IEEE/PES Transmission and Distribution Conference,* Chicago.
Smith, J. and Bow, K. (1995) 'Laminated sheathed cable for replacement of lead sheathed cable in medium voltage applications'. Jicable Conference Paper.

Paper and Laminates
Robinson, D. M. (1936) *Dielectric Phenomena in High Voltage Cables.* Chapman and Hall.
Bennett, G. E. (1957) 'Paper cable saturants: European preferences and selection'. *IEEE Trans.* **PAS-81**.
Kelk, E. and Wilson, I. O. (1965) 'Constitution and properties of paper for high-voltage dielectrics'. *Proc. IEE* **112** (3).

Sloat, T. K. (1979) 'Characteristics of insulating oil for electrical application'. EPRI Report No. 577-1, EL-1300.

Chan, J. C. and Hiivala, L. J. (1986) 'Electrical characteristics of synthetic dielectric liquids for oil–paper power cables. *2nd IEE Int. Conf. on Power Cables 10 kV–180 kV*.

Yougong Wang et al. (1988) 'Study of the dielectric properties of oil impregnated PPLP for UHV cable insulation'. *Proc. 2nd International Conference on Properties and Applications of Dielectric Materials* (Cat. No. 88CH2587-4). New York: IEEE.

Robinson, G. (1989) *Interpretation of Discharges in Paper-Insulation Cables*. Capenhurst, UK: Electr. Council Res. Centre.

Broadburst, M. G., De Reggi, A., Dickens, B. and Eichhorn, R. M. (1990) 'Spatial dependence of electric fields due to space charges in films of organic dielectrics used for insulation of power cables.' *Conf. Record of 1990 IEEE International Symposium on Electrical Insulation* (Cat. No. 90-CH2727-6). New York: IEEE.

Robinson, G. (1990) 'Ageing characteristics of paper-insulated power cables'. *Power Engineering Journal* **4** (2), 95–100.

Watson, D. R. and Chan, J. C. (1992) 'Mechanical and electrical performance of polypropylene paper laminates versus cellulose paper for power cables'. *Conf. Record of 1992 IEEE International Symposium on Electrical Insulation* (Cat. No. 92CH3150-0). New York: IEEE.

Gazzana Priaroggia, P., Metra, P. and Miramonti, G. (1992) 'Dielectric phenomena in the breakdown of non pressure assisted, impregnated paper insulated, HVDC cables'. *Proc. of 4th Int. Conf. on Conduction and Breakdown in Solid Dielectrics*. (Cat. No. 92CH3034-6). New York: IEEE.

IEEE (1995) *Guide for the Design, Testing, and Application of Moisture*; impervious, solid dielectric, 5–35 kV power cable using metal plastic laminates. Sponsor: Insulated Conductors Committee of the IEEE Power Engineering Society. New York: Institute of Electrical and Electronics Engineers.

Abdur-Razzaq, M., Auckland, D. W., Chandraker, K. and Varlow, B. R. (1996) 'Frequency and field roles in water absorption in composite dielectrics'. *7th IEE Int. Conf. on Dielectric Materials and Applications*.

Auckland, D. W., Chandraker, K., Shakanti, Z. and Varlow, B. R. (1996) 'Degradation of oil – paper systems due to the electric field enhanced absorption of water'. *7th IEE Int. Conf. on Dielectric Materials and Applications*.

Polymeric materials

Blow, C. M. (1975) *Rubber Technology and Manufacture*. London: Newnes-Butterworths.

Brydson, J. A. (1975) *Plastics Materials*. London: Newnes-Butterworths.

Town, W. L. (1979) 'Using extrudable materials for cable insulation'. *Electr. Times*, 23rd February.

Mason, J. H. (Apr. 1981) 'Assessing the resistance of polymers to electrical treeing'. *Proc. IEE, Part A* **128** (3), 193–201.

Epstein, M. M., Bernstein, B. S. and Shaw, M. T. (1982) 'Ageing and failure in solid dielectric materials'. Paris: CIGRE Paper No. 15-01.

EPRI (1983) 'Examination of distribution cables for chemical and physical changes upon ageing in the field and laboratory'. EPRI Final Report No. EL-3011, Project RP 1357-03.

Occini, E. et al. (1983) 'Thermal, mechanical and electrical properties of EPR insulation in power cable'. *IEEE/PES 1983 Winter Meeting*.

Azaroff, L. V. (1984) 'Materials research on extruded power cables'. Jicable Conference Paper No. B2-3.
Fisher, E. J. (1984) 'Advances in cross-linking polyethylene insulation for medium and high voltage cables'. Jicable Conference Paper A6-6.
Fothergill, J. C. et al. (1984) 'Structure of water trees and their relation to breakdown'. *4th IEE Int. Conf. on Dielectric Materials.*
Swingler, S. G. et al. (1984) 'The dielectric response of polyethylene cable containing water trees'. *4th IEE Int. Conf. on Dielectric Materials.*
Billing, J. (1986) 'Effect of cross-linking residue on the electric strength of XLPE cable insulation'. *5th BEAMA Int. Elect. Ins. Conf., Brighton.*
Simons, M. A. and Gale, P. S. (1986) 'Factors affecting the electric strength of XLPE cable insulation'. *5th BEAMA Int. Elect. Ins. Conf., Brighton.*
Given, M. J. et al. (1987) 'The role of ions in the mechanism of water tree growth'. *IEEE Trans.* **E1-22**, 151–6.
Saure, M. and Kalkner, W. (1987) 'On water tree testing of materials'. *CIGRE 1987.*
Billing, J. W. (1988) 'Thermal history of cable insulation revealed by DSC examination'. *5th IEE Int. Conf. on Dielectric Materials.*
Crine, J.-P., Haridoss, S., Hinrichsen, P., Houdayer, A. and Kajrys, G. (1988) Annual Report. *Conf. on Electrical Insulation and Dielectric Phenomena* (IEEE Cat. No. 88CH2668-2). 'Impurities in electrical trees grown in field-aged cables'.
Das-Gupta, D. K. et al. (1988) 'Measurement of polarisation in XLPE insulated HV cables'. *5th IEE Int. Conf. on Dielectric Materials.*
Head, J. G. et al. (1988) 'Modification of the dielectric properties of polymeric materials'. *5th IEE Int. Conf. on Dielectric Materials.*
Patsch, R. (1988) 'Water treeing in cable insulation'. *5th IEE Int. Conf. on Dielectric Materials.*
Rowland, S. M. and Dissado, L. A. (1988) Measurement of water tree growth in cross-linked polyethylene cables'. *5th IEE Int. Conf. on Dielectric Materials.*
Warren, L. and Paterson, J. R. (1988) 'Studies of irradiation on elastomeric insulation by gamma and neutron flux and thermal radiation degradation'. *5th IEE Int. Conf. on Dielectric Materials.*
Warren, L. and Paterson, J. R. (1988) 'The development of elastomeric insulating materials for use under extreme service conditions'. *5th IEE. Int. Conf. on Dielectric Materials.*
Wasilenko, E. (1988) 'Electrical ageing of polyethylene at impulse and a.c. test voltages'. *5th IEE Int. Conf. on Dielectric Materials.*
Ildstad, E. and Faremo, H. (1991) 'Importance of relative humidity on water treeing in XLPE cable insulation'. *7th International Symposium on High Voltage Engineering*, **2**, p. 207–10. Dresden Univ. Technol, Dresden, Germany.
Schadlich, H., Land, H. G. and Ritschel, C. D. (1991) 'Investigations on dielectric strength of model cables with PE and XLPE insulation'. *7th Int. Symp. on High Voltage Engineering*, **2**, p. 239–41. Dresden Univ. Technol, Dresden, Germany.
Mee, C. and Mackinlay, R. (1992) 'Parameters for tree growth in cable insulation'. *6th International Conference on Dielectric Materials, Measurements and Applications* (Conf. Publ. No. 363). London: IEE.
Nensi, T., Davies, A. E., Vaughan, A. S. and Swingler, S. G. (1992) 'Electrical aging of low density polyethylene cable insulation'. *Proc. of IEEE Conference on Electrical Insulation and Dielectric Phenomena (CEIDP)*, Victoria, Canada. New York: IEEE.

Aladenize, B., Assier, J. C., Galaj, S., Mirebeau, P. and Janah, H. (1995) 'Materials for cables based on polymeric conductors'. Jicable Conference Paper.

Choi, Y. C., Kubota, H., Ohyama, R. I., Kaneko, K. and Lee, J. B. (1995) 'Insulation diagnosis of XLPE cables by expert system using fuzzy interference'. Jicable Conference Paper.

Graham, G. et al. (1995) 'Insulating and semiconductive jackets for medium and high voltage underground power cable applications'. *IEEE Electr. Insul. Mag.* **11** (5), 5–12.

Hemphill, J., Fanichet, L. and Landucci, D. (1995) 'New Polyolefin polymers for low and medium voltage power cables'. Jicable Conference Paper.

Jeffery, A.-M. et al. (1995) 'Dielectric relaxation properties of filled ethylene propylene rubber'. *IEEE Trans. Diel. Electr. Insul.*, **2** (3), 394–408.

Richardson, C. G. (1995) 'The compounding of wire and cable quality polymers'. *Wire Industry*, **62** (739), 393–6.

Auckland, D. W., Su, W. and Varlow, B. R. (1996) 'Non-linear fillers in electrical insulation'. *7th IEE Int. Conf. on Dielectric Materials and Applications.*

Fanggao, C. et al. (1996) 'The effect of hydrostatic pressure and temperature on the permittivity of crosslinked polythylenes'. *7th IEE Int. Conf. on Dielectric Materials and Applications.*

Conductors

Griesser, E. E. (Dec. 1966) 'Sodium as an electrical conductor'. *Wire*.

Humphrey, L. E. et al. (1966) 'Insulated sodium conductors – a future trend'. *IEEE Spectrum*, November.

Steeve, E. J. and Schneider, J. A. (1966) 'Field tests on 15 kV and 600 V sodium cable'. IEEE Paper No. 66-447 (Summer meeting).

Humphrey, L. E. et al. (1967) 'Insulated sodium conductors'. *IEEE Trans.* **PAS-86**.

Watson, P. E. and Ventura, R. M. (Apr. 1967) 'Sodium cables used for all secondaries in URD development'. *Trans. Distr.*

Ball, E. H. and Maschio, G. (1968) 'The a.c. resistance of segmented conductors as used in power cables'. *IEEE Trans.* **PAS-87**.

Kalsi, S. S. and Minnich, S. H. (Mar.–Apr. 1980) 'Calculation of circulating current losses in cable conductors'. *IEEE Trans.* **PAS-99**.

Takaoka, M. et al. (1982) 'Manufacturing method and characteristics of compact segmental conductors with strands insulated with cupric oxide film'. *Proc. 52nd Annual Convention of Wire Associated Int. Inc. 1982.*

Edwards, D. R. (1988) 'Supertension or superconducting cables'. *Proc. IEE, Part C* **135** (1), 9–23.

Boev, M. A. & Zankin, A. I. (1989) 'Procedure for rational selection of cables and conductors for electrical products'. *Elektrotekhnika* **60** (1), 27–30. Translated in *Soviet Electrical Engineering* **60** (1), 40–44.

Kulkarni, A. H. (1991) 'Trends in insulated rectangular copper conductors for electrical machines'. *CABLEWIRE – 91. 3rd Int. Sem. on Cables, Conductors and Winding Wires*. Bombay: Indian Electr. Electron. Manuf. Assoc.

Schoonakker, P. and Willems, H. M. J. (1991) 'Thermomechanical behaviour of HV and EHV cables with copper conductors'. *Jicable Conference Paper.*

Dyos, G. T. & Farrell, T. (ed) (1992) *Electrical Resistivity Handbook*. IEE.

Morgan, V. T. (1992) 'The forced convective heat transfer from a twisted bundle of insulated electrical conductors in air'. *International Journal of Heat and Mass Transfer* **35** (6), 1545–55.

Filipovic-Gledja, V., Morgan, V. T. and Findlay, R. D. (1994) 'A unified model for predicting the electrical, mechanical and thermal characteristics of stranded overhead-line conductors'. *Proc. 1994 Canadian Conf. on Electrical and Computer Engineering*, **1**, 182–5. New York: IEEE.

McClung, L. B. and Ramachandran, S. (1995) 'Improvement in fire safety features of electrical conductors and cables'. *IEEE Transactions on Power Delivery* **10** (1), 43–52.

Ichino, T. *et al.* (1996) 'Measurement of conductor temperature of power cable by optical fibre sensor'. *7th IEE Int. Conf. on Dielectric Materials and Applications*.

Protective finishes

Giblin, J. F. and King, W. T. (May 1954) 'The damage to lead-sheathed cables by rodents and insects'. *Proc. IEE, Part 1* **101**, 129.

Hollingsworth, P. M. (9 Jan., 23 Jan. 1959) 'Protective coverings for lead and aluminium sheathed cables'. *Electr. J.*, 82–87; 221–4.

McAllister, D. (1965) 'Protection of power cables against corrosion'. *Electr. Rev.*, 14th May.

Gay, F. S. and Wetherley, A. H. (1969) 'Laboratory studies of termite resistance'. Part 5, 'The termite resistance of plastics'. Australia: Dept. of Entomology Technical Paper No. 10, Commonwealth Scientific Research Organization.

Siga, T. and Inagaki, Y. (1969) 'Termite damage to cable and its prevention'. *Sumitomo Electr. Techn. Rev.* (12).

Bultman, J. D., Leonard, J. M. and Southwell, C. R. (1972) 'Termite resistance of PVC in southern temperate and tropical environments'. Nat. Res. Lab. (Washington) Report No. 7417.

Beal, R. H., Bultman, J. D. and Southwell, C. R. (1973) 'Resistance of polymeric cable coverings to subterranean termite attack'. *Int. Biodeterior, Bull.* **9** (1–2), 28–34.

Beal, R. H. and Bultman, J. D. (1978) 'Resistance of polymeric cable coverings to subterranean termite attack after 8 years of field testing in the tropics'. *Int. Biodeterior Bull.* **14** (4), 123–127.

Bow, K. E. and Snow, J. H. (1982) 'Chemical/moisture barrier cable for underground systems'. *IEEE Trans.* **PAS-101**, 1942–1949.

Bow, K. E. (1984) 'Moisture barrier power cable with a plastic–metal laminate sheath'. Jicable Conference Paper No. B1.1.

Stoget, H. *et al.* (1985) 'Stresses and behaviour of polyethylene sheaths'. CIRED Paper No. 3.14.

Bungay, E. W. G. (1986) 'Protection of MV underground power cables'. *Electron. Power*, 221–5.

Swingler, S. G. (1986) 'Ensuring satisfactory mechanical performance from power cable sheathing'. *5th BEAMA Int. Elect. Ins. Conf., Brighton*.

Bow, K. E., Bieringa, L., Smith, J. and Zuercher, K. (1991) 'Semiconductive plastic coated metallic shielding and armouring materials for power cables'. Jicable Conference Paper.

Bow, K. E. (1993) 'New development in MV and HV cable with laminate sheaths as moisture barriers'. *IEEE Electr. Insul. Mag.* Sept.–Oct.

Jackson, G. H. and Bartlett, A. D. (1993) 'Damage resistant armoured 3 core polymeric cables for medium and high voltage systems in the UK'. *3rd IEE Int. Conf. on Power Cables and Accessories 10 kV–500 kV.*

CIGRE (1994) 'Prevention of termite attack on HV power cables'. *Electra* (157) December, 69–79.

Bow, K. E. (1995) 'Laminate sheaths as moisture barriers for MV cables'. *Wire Industry*, **62** (741), 481–8.

Cables in fires

Day, A. G. (1975) 'Oxygen index tests: temperature effect and comparisons with other flammability tests'. *Plast. Polym.* **43** (164), 64.

Punderson, J. O. (Sep.–Oct. 1977) 'Toxicity and safety of wire insulation: a state of the art review'. *Wire Tech.*

White, T. M. (Feb. 1977) 'Cable fires in power stations'. *Electron. Power.*

Kourtides, D. A., Gilwee, W. J. and Hilado, C. J. (Jun. 1978) 'Relative toxicity of the pyrolysis products from some thermoset polymers'. *Polym. Eng. Sci.* **18** (8), 674.

National Materials Advisory Board (1978) *Flammability, Smoke, Toxicity and Corrosive Gases of Electric Cable Materials*. Washington: National Academy of Sciences Publication No. NMAB-342.

Sullivan, T. and Willis, A. J. (1978) 'Reducing the hazards from cables in fires'. *Electr. Times*, 17th November.

Einhorn, I. N. and Grunnet, M. L. (1979) 'The physiological and toxicological aspects of the degradation products produced during the combustion of polyvinyl chloride polymers; flammability of solid polymer cable dielectrics'. EPRI Report No. EL-1263, TPS 77–738.

Smith, V. H. (7 Sep. 1979) 'Fire resistant cables for underground railways'. *Electr. Rev. Int.* **205** (9), 54.

Kingsbury, E. R. et al. (Sep. 1980) 'Selection of flame retarded wire and cable for industrial applications'. *Proc. 27th Ann. Pet. Chem. Ind. Conf.*, pp. 55–58. Houston: IEEE.

Barber, M. D., Partridge, P. F. and Gibbons, J. A. M. (1984) 'Reduced fire propagation and low smoke emission requirements for power station cables'. Jicable Conference Paper No. A9-3.

Berry, D. L. and Klamerus, L. J. (1984) 'Cable fire testing to meet realistic design criteria'. Jicable Conference Paper No. A7-1.

Bessei, H. and John, G. (1984) 'Halogen free fire-resistant joints for LV and MV cables'. Jicable Conference Paper No. C-12.

Beyersdorfer, K. W. and Tamplin, P. (1984) 'Accessories for halogen free fire-retardant and fire-resistant cables'. Jicable Conference Paper No. C-13.

Kirchner, F., Guerzio, M. and Villagrasa, F. (1984) 'New cables designed for limiting the risks proceeding from a fire'. Jicable Conference Paper No. A7-3.

Sabiston, A. D. (1984) 'Cables with reduced smoke, toxicity and fire propagation'. Jicable Conference Paper No. A8-2.

Stevens, G. C. and Gibbons, J. A. M. (1984) 'Assessing smoke and gas emission hazards from burning electric cables'. *4th IEE Int. Conf. on Dielectric Materials.*

Beretta, G. (1985) 'Behaviour of cables versus fire risks'. CIRED Paper No. 3.16.
Pays, M. and Simon, C. (1985) 'Improved fire resistant cables'. CIRED Paper No. 3.11.
Dabusti, V. et al. (1986) 'Development and prospects of cables with reduced smoke, toxic and corrosive gas emission'. *5th BEAMA Int. Elec. Ins. Conf., Brighton.*
Day, A. G. (1986) 'Examination of combustion products of electrical insulation using the NES 713 test'. *5th BEAMA Int. Elect. Ins. Conf., Brighton.*
Philbrick, S. E., Bungay, E. W. G., Barber, M. D. and Williamson, A. E. (1986) 'Cables for new power stations'. *2nd IEE Int. Conf. on Power Cables 10 kV–180 kV.*
Stevens, G. C. (1986) 'The appraisal and significance of acidic gas emissions from burning electric cable materials'. *5th BEAMA Int. Elect. Ins. Conf., Brighton.*
Stevens, G. C. et al. (1988) 'Acidic gas emissions from electrical insulation and their influence on electronic components'. *5th IEE Int. Conf. on Dielectric Materials.*
Stevens, G. C. et al. (1988) 'Chemical evaluation of combustion gas toxicity tests applied to electric cable materials'. *5th IEE Int. Conf. on Dielectric Materials.*
Swingler, S. G. et al. (1988) 'Small-scale assessment of flame propagation and heat release rate of electric cable materials'. *5th IEE Int. Conf. on Dielectric Materials.*
Pye, K. (1990) 'Low smoke and fume cables with the performance of PVC equivalents. *6th INSUCON BEAMA International Electrical Insulation Conference.* London: BEAMA.
Daly, J. M. (1991) 'Cable testing for flame propagation and toxicity'. *Wire J. Int.* **24** (9), 221–7.
Bucsi, A. (1992) 'Ignitability and flammability parameters of organic polymers'. *Polymer Degradation and Stability* **38**.
Coaker, A. W. (1992) 'Rate of heat release testing for vinyl wire and cable materials with reduced flammability and smoke – full-scale cable tray tests and small-scale tests'. *Fire Safety Journal* **19**, 19–53.
Fordham, I. L. (1992) 'Low fire hazard polyolefins for the cable industry'. *Conf. Polyethylene the 1990s and Beyond.* London.
Gruner-Nielsen, L., Haslov, P. and Skovgaard, N. H. (1992) 'Short and long term behaviour of halogen free fire retardant cable materials in different environments'. *Proc. of 41st Int. Wire and Cable Symposium*, Eatontown, NJ, USA.
Herbert, M. J. (1992) 'Optimization of ATH filler properties in EVA copolymers for flame retardant cable applications'. *Cables and Fire Protection. Conf. Proc.* (ERA Report 92-0001). Leatherhead: ERA Technol.
Hirschler, M. M. (1992) 'Survey of fire testing of electric cables'. *Fire and Materials* **16**.
Nakagawa, Y. et al. (1992) 'Ignition and flame propagation of electric cables in a laboratory-scale gallery fire test'. *Fire and Materials* **16** (4), 181–6.
Ness, D. E. M. (1992) 'Reduced fire hazard cables'. *Wire Industry*, November.
Nicholls, A. W. and Roberts, D. G. (1992) 'LSF cables safeguard Channel Tunnel'. *Electrotechnology* **3** (1), 9–12.
Weil, E. D. (1992) 'Oxygen index: correlation to other fire tests'. *Fire and Materials* **16**.
Hornsby, P. R. (1994) 'The application of magnesium hydroxide as a fire retardant and smoke-suppressing additive for polymers'. *Fire and Materials,* **18**, 269–76.
Fanichet, L. et al. (1995) 'New developments in halogen-free ignition resistant compounds using Insite technology polymers'. *Jicable Conference*, Versailles, 25–29 June.
Barnes, M. A. et al. (1966) 'A comparative study of the fire performance of halogenated and non-halogenated materials for cable applications'. *Fire and Materials,* **20** (1), 'Part I: Tests on materials and insulated wires', 1–16; Part II: Tests on Cables, 17–37.

Pye, K. (1966) 'The hazards of smoking'. *Cabling World*, December, 18–20.

Karaivanona, M. S. (1997) 'Non-halogen-containing flame-retardant ethylene-propylene copolymer compositions for cable insulation with nitrogen- and sulfur-containing fire retardants'. *J. Appl. Pol. Sci.*, **63** (5), 581–8.

Sustained ratings

Orchard, R. S., Barnes, C. C., Hollingsworth, P. M. and Mochlinski, K. (1960) 'Soil thermal resistivity: a practical approach to its assessment and its influence on the current rating of buried cables'. Paris: CIGRE Paper No. 214.

Arman, A. N., Cherry, D. M., Gosland, L. and Hollingsworth, P. M. (1964) 'Influence of soil moisture migration on power rating of cables in h.v. transmission systems'. *Proc. IEE* **111** (5).

Milne, A. G. and Mochlinski, K. (1964) 'Characteristics of soil affecting cable ratings'. *Proc. IEE* **111** (5).

ERA Technology Ltd (Aug. 1974) 'Heat emission from cables in air'. ERA Report No. 74–27.

Mochlinski, K. (Jan. 1976) 'Assessment of the influence of soil thermal resistivity on the ratings of distribution cables'. *Proc. IEE* **123** (1), 60–72.

Electricity Council (1977) *Engineering Recommendation P17 – Current Rating Guide for Distribution Cables*. London.

Parr, R. G. (Sep. 1980) 'Heat emission from cables on perforated steel trays'. ERA Report 74–28, ERA Technology Ltd.

Deschamps, L. et al. (1983) Thermal and mechanical behaviour of 20 kV cables under overload conditions'. CIRED Paper No. d.08.

Parr, R. G. (1983) 'Circuit protection for cables in groups'. ERA Report No. 83-0078.

Van Hove, C. et al. (1983) 'Overloadability of cable systems'. CIRED Paper No. d.07.

Huchings, E. and Coates, M. (1984) 'Tests to determine data for rating cables in randomly laid stacks on trays'. ERA Report No. 84-0172.

Coates, M. (1985) 'Temperature rise of cables passing through short lengths of thermal insulation'. ERA Report No. 85-0111.

Nelson, R. J. (1988) 'Computer techniques enhance cable ratings'. *Transmission and Distribution* **40** (7), 52–6, 59–60.

Krefter, K.-H. and Niemand, T. (1990) 'Cable ratings can be increased'. *Electrical Review* **223** (10), 16–18.

Machias, A. V. and Kaminaris, S. D. (1990) 'Cables continuous current rating calculation expert system approach'. *Proc. of IASTED International Symposium Artificial Intelligence Application and Neural Networks – AINN '90*. Anaheim: ACTA Press.

(1992) *Current Rating Standards 69-0030 part IX*; sustained current ratings for thermosetting insulated cables up to 70 MM squared to BS 7211 – 1989 in mixed groups in painted steel trunking to BS 4678 part 1 – 1971. Leatherhead: Era Technology.

Tang, L., Ghafurian, R. and Purnhagen, D. W. (1992) 'Dynamic rating of forced cooled transmission cables'. *IEEE Computer Applications in Power* **5** (3), 50–53.

Hiranandani, A. (1993) 'Analysis of cable ampacities in cable systems containing harmonics'. *Wire J. Int.*, September.

Harshe, B. L. et al. (1994) 'Ampacity of cables in single open-top cable trays'. *IEEE Trans. Power Del.* **9** (4), 1733–40.

Anders, G. J. (1996) 'Ratings of cables on riser poles, in trays, in tunnels and shafts – a review'. *IEEE Trans. Power Del.*, **11** (1), 3–11.

Cinquemani, P. L. *et al.* (1996) '105 C/140 C Rated EPR insulated power cables'. *IEEE Trans. Power Del.*, **11** (1), 31–40.

Ferkal, K. *et al.* (1996) 'Proximity effect and eddy current losses in insulated cables'. *IEEE Trans. Power Del.*, **11** (3), 1171–8.

Freitas, D. S. *et al.* (1996) 'Thermal performance of underground power cables with constant and cyclic currents in presence of moisture migration in the surrounding soil'. *IEEE Trans. Power Del.*, **11** (3), 1159–70.

Anders, G. J. (1997) 'Rating of electric cables; ampacity computations for transmission, distribution, and industrial applications'. New York: The Institute of Electrical and Electronics Engineers.

Short-circuit ratings

Gosland, L. and Parr, R. G. (Jan. 1960) 'A basis for short circuit ratings for paper-insulated, lead-sheathed cables up to 11 kV'. ERA Report No. F/T 195.

Buckingham, G. S. (Jun. 1961) 'Short-circuit ratings for mains cables'. *Proc. IEE, Part A* **108**.

Gosland, L. and Parr, R. G. (Jun. 1961) 'A basis for short-circuit ratings for paper-insulated cables up to 11 kV'. *Proc. IEE, Part A* **108**.

Thomas, A. G. (10 Nov. 1961) 'Short-circuit ratings of aluminium cables'. *Electr. Rev.*

Parr, R. G. (1962) 'Bursting currents of 11 kV, 3-core, screened cables (paper-insulated, lead-sheathed)'. ERA Report No. F/T 202.

Parr, R. G. (1964) 'Short-circuit ratings for 11 kV 3-core paper-insulated screened cables'. ERA Report No. 5057.

Parr, R. G. and Yap, J. S. (1965) 'Short-circuit ratings for PVC insulated cables'. ERA Report No. 5056.

Dorison, E. (1991) 'Distribution and influence of the zero-sequence short-circuit currents in the ground'. Jicable Conference Paper.

Klevjer, G. *et al.* (1991) 'Mechanical short-circuit testing for cable installations'. Jicable Conference Paper.

Sens, M. A. *et al.* (1991) 'New technique to detect short circuit current distribution on wire shielding of MV power cables'. Jicable Conference Paper.

Anders, C. J. (1992) 'Transient ratings of buried power cables. Part 1. *IEEE Trans. Power Del.*, **7** (4).

Garros, B., Land, H. G. and Pedersen, J. R. (1995) 'Investigations on breakdown strength of polyethylene at very fast transients'. Jicable Conference Paper.

Gustavsen, B. *et al.* (1996) 'Transient sheath overvoltages in armoured power cables'. *IEEE Trans. Power Del.*, **11** (3), 1594–1600.

PART 2: WIRING AND INDUSTRIAL TYPE CABLES

General wiring cables

Town, W. L. (1972) 'Wiring cables'. *Electr. Times* **161** (25), 25–6; **162** (1), 33–4.

Hollingsworth, P. M. and Town, W. L. (1973) 'Trends in wiring cable design and installation'. *Electr. Rev.* **192** (19), 666–9.

Taylor, F. G. (Mar. 1973) 'Cables for electronics'. *Electrotechnology* **1** (2), 3–10.
Town, W. L. (Apr. 1974) 'A guide to the selection of electrical and electronics wires and cables'. *OEM Design*.
Seccombe, G. H. (1975) 'EEC and cable standards – where are we going?' *Electr. Times* (4312), 6, 13.
Todd, D. (Sep. 1975) 'Electric cable for signalling and track-to-train communications'. *Railw. Eng. J* **4** (5), 71–3.
Bungay, E. W. G. and Hollingsworth, P. M. (2 Apr. 1976) 'Progress with harmonisation of cables standards within CENELEC'. *Electr. Times* (4373), 7–8.
(18 Nov. 1977) 'Selecting the correct cable conductor size based on voltage drop, short-circuit capacity and strength'. *Electr. Times*.
(2 Dec. 1977) 'Selecting cables for the hostile cnditions on construction sites'. *Electr. Times* (4454), 12–14.
(4 Aug. 1978) 'Compatibility of PVC cables'. *Electr. Times* (4485), 3.
Town, W. L. (23 Feb. 1979) 'Using extrudable materials for insulating cables'. *Electr. Times* (4511), 10–12.
Bellini, V., Bragagni, A. and Centurioni, L. N. (1990) 'Low voltage electric cables with high safety and reliability performances. *6th INSUCON 1990 BEAMA International Electrical Insulation Conference*, Brighton.
Carl, A. G. et al. (1994) 'Design, testing and applications of highly flexible control cables'. *Wire*, **44** (5), 414–17.
Aida, F. et al. (1995) 'Development of halogen-free flame-retardant cables for nuclear power plants'. Jicable Conference Paper.
Benard, L., Colombier, S. Galcera, T. and Lepatey, M. (1995) 'Insulation of high temperature cables for car applications'. Jicable Conference Paper.
Journeaux, T. L. (1995) 'Cables for nuclear power stations'. Jicable Conference Paper.
Monjo, J. et al. (1995) 'Technical and economical evaluation of different cable design, for 1 kV distribution cables in Spain'. Jicable Conference Paper.
Musquin, M. and Salzmann, H. (1995) 'Design for an electric cable for coal-cutting machine 5 kV'. Jicable Conference Paper.

Mineral insulated cables

Tomlinson, F. W. and Wright, H. M. (1946) 'Mineral-insulated metal-sheathed conductors'. *IEE J., Part II* **93** (34).
Jordan, C. A. and Eager, G. S. (Jan.–Feb. 1955) 'Mineral-insulated metallic-sheathed cables'. *AIEE Winter Meeting*.
(9 Jun. 1967) 'Protective multiple earthing – economic advantages of sheath return concentric wiring'. *Elect. Rev.* **180** (23).
Lorch, H. R. (1969) 'Earthed concentric wiring – M.I. cable suitable for factory distribution'. *Electr. Times*, 11th September.
Latham, W. B. (1976) 'Mineral insulated cables in hazardous areas'. *Electr. Times*, 2nd April.
Milles, E. (1976) 'Mineral insulated metal sheathed cables'. *Wire Ind.*, May.
Wilson, I. O. (Sep. 1979) 'Magnesium oxide as a high temperaure insulant in insulated cables'. *Proc. 3rd Int. Conf. on Dielectric Materials*, pp. 78–81.
Wilson, I. O. (1981) 'Magnesium oxide as a high temperature insulant'. *Proc. IEE, Part A* **128** (3), 159–64.

Baker, H. and Wilson, I. O. (1987) 'Mineral insulated cable – the inorganic solution'. *Electronics and Power*, J. of the Institution of Electrical Engineers **33** (4), 268–72.

Gill, D. (1987) 'Mineral insulated cables – the ultimate in safety – a continuous production process. *GEC Review* **3** (1), 46–55.

Finlayson, A. J. (1990) 'Low smoke and fume type cables for BNFL Sellafield Works'. *Power Engineering Journal*, **4** (6) 301–7.

Sawada, K., Nishio, M., Inazawa, S. and Yamada, K. (1990) 'Development of new ceramic-insulated wires'. *Sumitomo Electric Technical Review*, **29**, 221–7.

Gadsby, B. (1992) 'Mineral insulated metal sheathed cables'. *Wire Industry*, November.

Gadsby, B. (1992) 'Mineral insulated metal sheathed cables – a new twist'. *Cables and Fire Protection. Conference Proceedings* (ERA Report 92-0001) p. 5.3/1-9.

Baen, P. et al. (1994) 'Safety related considerations for resistance heat tracing cables'. *Proc. of IEEE Petroleum and Chemical Industry Tech. Conf.* (PCIC '94), Vancouver. New York: IEEE.

PART 3: DISTRIBUTION-SYSTEMS AND CABLES

Distribution systems

Taylor, H. G. (1937; 1941) 'The use of protective multiple earthing and ELCBs in rural areas'. *Proc. IEE* **81**; **88** (2).

(9 Jun. 1967) 'Protective multiple earthing – general principles and economic advantages of sheath return concentric wiring'. *Electr. Rev.*

Brown, F. J. and Fisher, J. (1971) *The Development of Interconnected Networks in the City of Liverpool and its Environs.* Liège: CIRED.

Cole, J. E. H. (1972) 'Standardisation of plant and equipment for public electricity supply'. *Proc. IEE* **119** (9), 1319–28.

Ford, D. V. (1972) 'The British Electricity Board's national fault and interruption reporting scheme – objectives, development and operating experience'. *IEEE Power Eng. Soc. Winter Meeting, New York.* Paper No. T72 082-1.

Gosden, J. H. (1973) *Reliability of Overhead Line and Cable Systems in Great Britain.* London: CIRED.

Milne, A. G. (Jan. 1974) 'Distribution of electricity (presidential address)'. *Proc. IEE* **121** (1).

Ross, A. (Nov. 1974) 'Cable practice in electricity board distribution networks: 132 kV and below', *Proc. IEE (IEE Rev).* **121** (11 R).

Cridlin, J. M., Stevens, R. H. and Thue, W. A. (1978) 'Performance of URD primary cable'. *USA Reliability Conference Electric Power Industry*, pp. 80–91.

Dickie, R. A. et al. (Jul. 1978) 'An examination of underground electric power distribution in residential areas'. *IEEE PES Summer Meeting, Los Angeles.*

Freund, A. (1979) 'Distribution systems – grounded or ungrounded?' *Electr. Constr. Maint.*, April, 67–71.

Minsart, G. and Steyaert, R. (1985) 'New trends in the planning of MV supply systems'. CIRED Paper No. 6.11.

Atkinson, W. C. and Ellis, F. E. (1987) 'Electricity distribution – asset replacement considerations'. *Electron Power.*

Kaszowski, D. (1991) 'Machine-aided cable laying in French MV systems. Latest developments'. Jicable Conference Paper.

Koo, J. Y. et al. (1991) 'A study on the continuous permissible current rating in 22.9 kV CN/CV underground distribution power cables'. Jicable Conference Paper.

Mopty, Y., Dupont, A., Tardif, L. and Redolat, R. (1991) 'Compact derivation for LV underground network'. Jicable Conference Paper.

Boniface, P., Brincourt, T. and Pruniere, J.-Y. (1995) 'Development of underground MV network: cable environment and new electrical equipment for rural networks'. Jicable Conference Paper.

Biscaglia, V., Malaguti, C. and Paoletti Gualandi, M. (1997) 'Maintenance planning on MV distribution network'. CIRED Paper No. 3.19.

Grotenhuis, B. J., Kerstens, A. and van Riet, M. J. M. (1997) 'Dutch distribution cable: Improvements in construction reliability, environment and efficiency'. CIRED Paper No. 3.6.

Morgan, A. (1997) 'The technical and economic case for the use of 3 dimensional mapping for the installation of electric power cables'. CIRED Paper No. 3.9.

Power cables (general)

Wanser, G. (1969) 'Experience with plastic-insulated cables in Germany'. *Wire* (99), 10–15.

Lacoste, A., Lagarde, R. and Michel, R. (1970) 'LV cables used in the French distribution system'. *Proc. IEE/ERA Distribution Cable Conf.*, pp. 341–9.

Bax, H. (1971) *Modern LV Cables for the Distribution Networks of Power Supply Companies and Industry (Germany)*. Liège: CIRED.

McAllister, D. and Cox, E. H. (1972) 'Behaviour of MV power distribution cables when subjected to external damage'. *Proc. IEE* 119 (4), 479–86.

Lacoste, A., Lemainque, H. and Schmeltz, J. (1973) *French Distribution Cable techniques – Present Practices and Future Trends*, Part 1, pp. 92–101, London: CIRED.

Ross, A. (Nov. 1974) 'Cable practice in electricity board distribution. 132 kV and below'. *Proc. IEE (IEE Rev.)* 121 (11 R), 1307–44.

Wanser, G. (May–Jun. 1974) 'Power cables – present and future'. *Wire* 24 (E3/74), 135–8.

Blechschmidt, H. H. and Goedecke, H. P. (1975) 'Cables with synthetic insulation in the Federal Republic of Germany'. Liège: CIRED Paper No. 36.

Bungay, E. W. G., Philbrick, S. E., Morgan, A. M. and Sloman, L. M. (1975) 'The development of 11 kV cable systems'. Liège: CIRED Paper No. 37.

Giusseni, A., Maciotta, G., Portineri, G. and Leonardi, E. (1975) 'Present trends and modern design criteria for low and medium voltage distribution cables with extruded insulation in Italy'. Liège: CIRED Paper No. 35.

Gosden, J. H. (1975) 'Reliability of overhead lines and cable systems in Great Britain'. Liège; CIRED Paper No. 40.

Gosden, J. H. and Walker, A. J. (May 1976) 'The reliability of cable circuits for 11 kV and below'. *IEE Conf. on Distribution Cables, and Jointing Techniques, etc.*, pp. 1–4.

Heinhold, L. (1989) 'Power cables and insulated wires'. In *Kabel und Leitungen fur Starkstrom* (3rd rev. ed) Berlin: Siemens Aktiengesellschaft; New York; Wiley.

Bucholz, V. (1993) 'Elevated temperature operation of XLPE distribution cable systems'. *IEEE Trans. Power Del.*, July.

IEE (1993) *3rd Int. Conf. on Power Cables and Accessories 10 kV–500 kV*, 23–25 November 1993. Organised by the Power Division of the Institution of Electrical Engineers in association with CIGRE et al. London: IEE.

Blackwell, A., Davies, A. and Larsen, S. T. (1995) 'Forecasting and probabilistic rating of underground power cables'. Jicable Conference Paper.

Delaunois, G. and Bacnus, R. (1995) 'Development of new 12/20 kV protected aerial lines'. Jicable Conference Paper.

Freitas, D., Prata, A. and De Lima, A. (1995) 'Thermal performance of underground power cables with constant and cyclic currents in presence of moisture migration in the surrounding soil'. Jicable Conference Paper.

Geurts, W. S. M. et al. (1995) 'Water diffusion through sheaths and its effects on cable constructions'. Jicable Conference Paper.

Alison, J. M. and Dissado, L. A . (1996) 'Electric field enhancement within moulded samples of low density polyethylene – a means of failure characterised by voltage ramp tests and space charge measurement'. *7th IEE Int. Conf. on Dielectric Materials and Applications.*

Faremo, H., Benjaminsen, J. T., Larsen, P. B. and Tunheim, A. (1997) 'Service experience for XLPE cables installed in Norway – from Graphite painted insulation screens to axially and radially water tight cable constructions'. CIRED Paper No. 3.2.

Paper insulated cables

Reynolds, E. H. and Rogers, E. C. (Oct. 1961) 'Discharge damage and failure in 11 kV belted cables'. *Trans. S. Afr. Inst. Electr. Eng.* **52** (10).

Terramosi, P. and Couppe, G. L. (1971) 'Developments in non-draining cables'. *Wire Wire Prod.*, May, 95–107.

Swarbrick, P. (1973) 'Developments in 11 kV underground cable systems. Paper-insulated aluminium-sheathed cables and resin filled joints'. *Electr. Rev.*, 14th December.

Bungay, E. W. G. and Philbrick, S. E. (May 1976) 'Paper insulated 11 kV aluminium sheathed cables'. *IEE Conf. on Distribution Cables.*

Bulens, R. et al. (1983) 'Ageing and permissible load of paper insulated medium voltage cables'. CIRED Papr No. 4.06.

Domum, M. (1986) 'Prediction of remaining life of H. V. cables', Part 2, 'Accelerated life tests in the laboratory'. *2nd IEE Int. Conf. on Power Cables, 10 kV–180 kV,*

Harrison, B. J. (1986) 'Some aspects of failure of 33 kV H-type oil–rosin impregnated cables'. *2nd IEE Int. Conf. on Power Cables, 10 kV–180 kV.*

Robinson, G. (1993) 'Low-frequency loss spectra and ageing in paper-insulated cables'. *3rd IEE Int. Conf. on Power Cables and Accessories 10 kV–500 kV.*

Brincourt, T. and Leroy, J.-H. (1995) 'Determination of the internal thermal resistivity of a three-core insulated paper cable'. Jicable Conference Paper.

Cables with combined neutral and earth

Booth, D. H. (1970) 'Some considerations relevant to the design of underground power cable systems for use with PME'. *IEE/ERA Conf. on Distribution, Edinburgh.*

Henderson, J. T. and Swarbrick, P. (1970) 'The Consac cable system'. *IEE/ERA Conf. on Distribution, Edinburgh.*

Hughes, O. I. and Bramley, G. E. A. (1970) 'Development and production of a PME elastomeric insulated MV cable'. *IEE/ERA Conf. on Distribution, Edinburgh*.

Majewski, H. A. (1970) 'Special design features with aluminium-sheathed MV power cables'. *IEE/ERA Conf. on Distribution, Edinburgh*.

Rockliffe, R. H., Hill, E. and Booth, D. H. (1971) 'Protective earthing practices in the UK and their associated underground cable systems'. *IEEE Conf. on Power Distribution*. Paper No. 71C 42-PWR.

McAllister, D. and Cox, E. H. (1972) 'Behaviour of MV power distribution cables when subjected to external damage'. *Proc. IEE* **19** (4), 479–86.

Radcliffe, W. S. and McAllister, D. (1972) 'Cables and joints for PME distribution systems'. *Electr. Rev.*, 24th March.

Garmony, T. H. (15 Nov. 1974) 'Efficiency in electricity distribution. Experience with wave form cables'. *Electr. Rev.*, 15th November.

Baldock, A. T. and Hambrook, L. G. (May 1976) 'Regulations relevant to the design and utilisation of distribution cables for the Electrical Supply Industry and the consumer'. *IEE Conf. on Distribution Cables up to 11 kV*.

Burton, J. M. (May 1976) 'Consac cable system development in the Midlands Electricity Board'. *IEE Conf. on Distribution*.

Geer, P. K. and Sloman, L. M. (May 1976) 'Cables for PME distribution systems'. *IEE Conf. on Distribution*.

Kerney, J. M. (May 1976) 'Experience in Ireland with four-core unscreened elastomeric cables'. *IEE Conf. on Distribution*.

Polymeric insulated cables – 6 kV to 60 kV

Devaux, A., Oudin, J. M., Rerolle, Y., Jocteur, R., Noirclerc, A. and Osty, M. (1968) 'Reliability and development towards high-voltage synthetic insulated cables'. CIGRE Paper No. 21-20 (in two parts).

Tabata, T., Nagai, H., Fukuda, T. and Iwata, Z. (1972) 'Sulphide attack and treeing of polyethylene insulated cables – cause and prevention'. *IEEE Trans.* **PAS-91** (4), 1354–60.

Vahlstrom, W. (1972) 'Investigations of insulation deterioration in 15 kV and 22 kV polyethylene cables removed from service'. *IEEE Trans.* **PAS-91** (3), 1023–1035.

Lawson, J. H. and Vahlstrom, W. (1973) 'Investigation of insulation deterioration in 15 kV and 22 kV cables removed from service' Part 2. *IEEE Trans.* **PAS-92** (2), 824–35.

Bahder, G., Katz, C. and Lawson, J. H. (1974) 'Electrical and electrochemical treeing effect in polyethylene and crosslinked polythylene cables'. *IEEE Trans.* **PAS-93** (3), 977–91.

Tanaka, T., Fukuda, S., Suzuki, Y., Nitta, Y., Goto, H. and Kubato, K. (1974) 'Water trees in crosslinked polyethylene power cables'. *IEEE Trans.* **PAS-93** (2), 693–702.

Hyde, H. B., Philbrick, S. E., Roberts, B. E. and Smith, T. (1976) 'Earth fault spiking tests at system voltage on 11 kV polymeric cables'. *IEE Conf. on Distribution Cables and Jointing Techniques*, pp. 83–6.

McKean, A. L. *et al.* (1976) 'Investigation of mechanism of breakdown in XLPE cables'. EPRI Report No. TD-138 (Final Report).

Eichhorn, R. M. (1977) 'Treeing in solid extruded electrical insulation'. *IEEE Trans.* **EI-12** (1), 2–18.

Bernstein, B. S. (1978) 'Research to determine the acceptable emergency operating temperatures for extruded dielectric cables'. EPRI Report No. EL-938.

Chan, J. C. (1978) 'Electrical performance of oven-dried XLPE cables'. *IEEE Trans.* **EI-13** (6), 444–7.

Densley, R. J. (1978) 'The impulse strength of naturally aged XLPE cables containing water trees'. *IEEE Trans.* **EI-13** (5), 389–91.

Jacobsen, C. T., Attermo, R. and Dellby, B. (1978) 'Experience of dry-cured XLPE-insulated high voltage cables'. Paris: CIGRE Paper No. 21-06.

Srinivias, N. N. and Doepken, H. C. (1978) 'Electrochemical treeing in PE and XLPE insulated cables – frequency effects and impulse degradation'. *IEEE Int. Symp. on Electrical Insulation.* pp. 106–109.

Yoshimitsu, T. and Nakakita, T. (1978) 'New findings on water tree in high polymer insulating materials'. *IEEE Int. Symp. on Electrical Insulation*, pp. 116–21.

Bernadelli, P., Bolognesi, F., Nosca, W. and Zanetti, O. (1979) *'Significance and problems concerned with laboratory tests on power cables and relevant accessories for distribution systems'*. Liège: CIRED Paper No. 30.

Bruggemann, H., Schuppe, W.-D. and Wichmann, H. (1979) *'Polymeric insulated LV cables with XLPE as insulant used in public distribution networks in Germany'*. Liège: CIRED Paper No. 32.

Densley, R. S. *et al.* (1979) 'The surge characteristics of XLPE insulation containing water trees'. *IEEE Symp. on Electrical Insulation*, pp. 204–207.

Doepken, H. C., McKean, A. L. and Singer, M. L. (1979) 'Treeing, insulation material and cable life'. *IEEE Power Engineering Society 7th Transmission and Distribution Conf.*, pp. 299–304.

Ferran, J. and Pinet, A. (1979) 'Development of a new 20 kV cable with synthetic insulation and of its fitting'. Liège: CIRED Paper No. 31.

Horton, W. F. and St. John, A. N. (1979) 'The failure rate of polyethylene-insulated cable'. *IEEE Power Engineering Society 7th Transmission and Distribution Conf.*, pp. 324–8.

Lanctoe, T. P., Lawson, J. H. and McVey, W. L. (1979) 'Investigation of insulation deterioration in 15 kV and 22 kV polyethylene cables removed from service', Part 3. *IEEE Trans.* **PAS-98** (3), 912–25.

Naybour, R. D. (1979) 'The growth of water trees in XLPE at operating stresses and their infuence on cable life'. *IEE 3rd Int. Conf. on Dielectric Materials*, pp. 238–41.

Pinet, A. and Paris, M. (1979) 'New 20 kV cable of the French MV power system'. *IEEE PES Summer Meeting, Vancouver.* Paper No. A79408-6.

Silver, D. A. and Martin, M. A. (1979) 'Progress in overcoming electrochemical treeing'. *Transm. Distr.* **31** (4).

Sletbak, J. (1979) 'A theory of water tree initiation and growth'. *IEEE Trans.* **PAS-98** (4), 1358–66.

Doepken, H. C. and Klinger, Y. (1980) 'Correlation of water tree theories with experimental data'. *IEEE Symp. on Electrical Insulation*, pp. 208, 211.

Fukuda, T. *et al.* (1980) 'Factors governing the voltage breakdown of insulating materials for XLPE insulated cables'. *IEEE Symp. on Electrical Insulation*, pp. 118–21.

Lawson, J. H. and Thue, W. A. (1980) 'Summary of service failure of high voltage extruded dielectric insulated cables in the USA'. *IEEE Symp. on Electrical Insulation*, pp. 100–103.

Martin, M. A. and Hartlein, R. A. (1980) 'Correlation of electrochemical treeing in power cables removed from service and in cables tested in the laboratory'. *IEE Trans.* **PAS-99**, 1597–1605.

Namiki, Y., Shimanuki, H., Aida, F. and Morita, M. (1980) 'A study of microvoids and their filling in crosslinked polyethylene insulated cables'. *IEEE Trans.* **EI-15** (6), 473–80.

Nunes, S. L. and Shaw, M. T. (1980) 'Water treeing in polyethylene – a review of mechanisms'. *IEEE Trans.* **EI-5** (6), 437–50.

Wartusch, J. (1980) 'Increased voltage endurance of polyolefine insulating materials by means of voltage stabilisers'. *IEEE Symp. on Electrical Insulation*, pp. 216–21.

Bahder, G., Katz, C. et al. (1981) 'Life expectancy of XLPE cables rated 15 to 35 kV'. *IEE Trans.* **PAS-100**, 1581–90.

Bulinski, A. and Densly, R. (1981) 'The voltage breakdown characteristics of miniature XLPE cables containing water trees'. *IEE Trans.* **EI-16** (4), 319–26.

Lanfranconi, G. M., Metra, P. and Vecellio, B. (1981) 'MV power cables with extruded insulation. A comparison between XLPE and EPR'. CIRED Paper.

Bahder, G., Garrity, T. et al. (1982) 'Physical model of electric ageing and breakdown of extruded polymeric insulated power cables'. *IEE Trans.* **PAS-100**, 1379–90.

Hayami, T. et al. (1982) 'Relation between water content and bow-tie tree generation in XLPE cables'. *15th Symp. on Electrical Insulating Materials*, pp. 4–7.

Ball, E. H., Bungay, E. W. G. and Sloman, L. M. (1983) 'Polymeric cables for high voltage distribution systems'. CIRED Paper No. d.11.

Bloemer, B. (1983) 'Experiences with 20 kV single-core plastic (XLPE) cables'. CIRED Paper No. d.10.

Brown, M. (1983) 'Performance of EPR in medium and high voltage power cable'. *IEE Trans.* **PAS-102** (2).

Cochini, E. et al. (1983) 'Thermal, mechanical and electrical properties of EPR insulation in power cables'. *IEE Trans.* **PAS-102** (7).

Laar, A. V. D. et al. (1983) 'Experience with XLPE insulated cable with solid aluminium conductors'. CIRED Paper No. d.13.

Occini, E. et al. (1983) 'Thermal, mechanical and electrical properties of EPR insulations in power cables'. *IEE Winter Meeting 1983*. Paper No. WM 00.

Sletbak, J. and Ildstad, E. (1983) 'Effect of service and test conditions on water tree growth in XLPE cables'. *IEE Trans.* **PAS-102** (7), 2069–2074.

Travers, R. and Reidy, P. (1983) 'Medium voltage distribution cables using XLPE insulation – Irish experience'. CIRED Paper No. d.12.

Brown, M. (1984) 'Performance of EPR insulation in medium and high voltage power cables'. Jicable Conference Paper No. B1-2.

Farneti, F. et al. (1984) 'Performance of EPR insulated cables under different laying conditions and unusual thermal stresses'. Jicable Conference Paper No. A3-3.

Franke, H. et al. (1984) 'Testing possibilities and results regarding water ageing of PE/XLPE insulated MV cables'. Jicable Conference Paper No. A6-3.

Katz, C., Eager, G. S., Leber, E. R. and Fischer, F. E. (1984) 'Influence of water on dielectric strength and rejuvenation of in-service aged URD cables'. Jicable Conference Paper No. A6-5.

Nagasaki, S. et al. (1984) 'Life estimation and improvement of water-tree resistivity of XLPE cables'. Jicable Conference Paper No. A6-2.

Ratra, M. C. (1984) 'Some aspects of compatibility and short circuit characteristics of MV polymeric cables'. Jicable Conference Paper No. A3-1.

Ross, A. (1984) 'Developments in medium voltage polymeric cables'. Jicable Conference Paper No. A1-4.

Schuppe, W. F. (1984) 'Progress with XLPE medium voltage cables in the Federal Republic of Germany'. Jicable Conference Paper No. A2-1.

Silver, D. A. and Lukac, R. G. (1984) 'Factors affecting the dielectric strength of extruded dielectric cables in wet environments'. Jicable Conference Paper No. A6-1.

Takenouch, K. *et al.* (1984) 'Experience in service of 66 kV XLPE power cables'. Jicable Conference Paper No. B6-1.

Tanabe, T. *et al.* (1984) 'Water tree deterioration and counter measure of medium voltage XLPE cables'. Jicable Conference Paper No. A6-4.

Ball, E. H., Metra, P. and Ortiz, M. R. (1986) 'Extruded cable insulation for wet locations'. *2nd IEE Int. Conf. on Power Cables 10 kV–180 kV*.

Bergin, T. E. *et al.* (1986) 'Review of 66 kV XLPE cables in the State of Bahrein'. *2nd IEE Int. Conf. on Power Cables 10 kV–180 kV*.

Billing, J. W. (1986) 'Diagnostic investigation into XLPE HV cable insulation'. *2nd IEE Int. Conf. on Power Cables 10 kV–180 kV*.

Field, A. W., Nicholls, A. W. and Marsh, G. C. (1986) 'Effect of water on the life of extruded dielectric cables'. *2nd IEE Int. Conf. on Power Cables 10 kV–180 kV*.

Howard, R. S., Jenkins, T. and Brook, R. T. (1986) 'Operating experience with 11 kV polymeric cable systems in one U.K. Area Board'. *2nd IEE Int. Conf. on Power Cables 10 kV–180 kV*.

Hyde, H. B., Poideven, G. J. and Philbrick, S. E. (1986) 'Development of a single-core polymeric cable for 33 kV distribution systems'. *2nd IEE Int. Conf. on Power Cables 10 kV–180 kV*.

Naybour, R. D. (1986) 'Influence of water on the life of polymeric insulated cables'. *2nd IEE Int. Conf. on Power Cables 10 kV–180 kV*.

Pinet, A. and Ferron, J. (1986) 'Operating experience with the 20 kV cable used in the French network'. *2nd IEE Int. Conf. on Power Cables 10 kV–180 kV*.

Steenis, E. F. and Boone, W. (1986) 'Water treeing in service aged and accelerated aged XLPE cables'. *2nd IEE Int. Conf. on Power Cables 10 kV–180 kV*.

White, T. M., Bungay, E. W. G. *et al.* (1986) '11 kV polymeric insulated triplex cable'. *2nd IEE Int. Conf. on Power Cables 10 kV–180 kV*.

Fernetti, F. *et al.* (1987) 'Characterisation of extruded insulation cables with respect to water'. Jicable Conference Paper.

Harasawa, K. *et al.* (1987) Influence of d.c. voltage application on dielectric performance of XLPE cables'. Jicable Conference Paper.

Jinno, M. *et al.* (1987) 'Present conditions of XLPE cables used in Japan and cable fault analysis'. Jicable Conference Paper.

Ortiz, M. R. *et al.* (1987) 'EPR high voltage cables'. Jicable Conference Paper.

Bulens, R., Beyls, P., Cabaux, J. and Geerts, G. (1989) 'XLPE insulated MV cables; service experience, actual problems and prospects for the future'. *CIRED 10th International Conference on Electricity Distribution* (Conf. Publ. No. 305). London: IEE.

Steennis, E. F., Meijer, G. J., Leufkens, P. P. and Geene, H. T. F. (1991) 'Comparison of thermal effects on XLPE insulated medium voltage cables with solid and stranded aluminium conductors'. *CIRED 11th International Conference on Electricity Distribution*. Liege: AIM.

Patsch, R. (1992) 'Electrical and water treeing'. *IEEE Trans. Elect. Insulation*, **27** (3).
Beament, I. R. and Ford, A. E. W. (1993) 'Effect of system and installation requirements on the introduction of 11 kV polymeric insulated cables in London'. *3rd Int. Conf. on Power Cables and Accessories 10 kV–500 kV* (Conf. Publ. No. 382). London: IEE.
Faremo, H. (1993) 'Water treeing and dielectric loss of WTR-XLPE cable insulation'. *IEE Proc. A* **140** (5).
Fothergill, J. C. (1993) 'Water tree inception and its dependence on electric field, voltage and frequency'. *IEE Proc. A* **140** (5).
Bucci, R. M. et al. (1994) 'Failure prediction of underground distribution feeder cables'. *IEEE Trans. Power Del.* **9** (4), 1943–55.
Kogan, V. I. (1994) 'Explanation for decline in URD cable failure' *IEEE Trans. Power Del.* **9** (1).
Merschel, F. (1994) 'Long-term tests of XLPE insulated medium voltage cables: results, experiences, standardisation'. *Elekrizitatswirtschaft* **93** (19), 1135–43.
Walton, M. et al. (1994) 'Accelerated cable life testing of EPR-insulated medium voltage distribution cables'. *IEEE Trans. Power Del.* **9** (3), 1195–1208.
Calton, P. (1995) 'XLPE insulated MV cables'. *Wire Industry* **62** (739), 389–91.
Katz, C. et al. (1995) 'An assessment of field aged 15 and 35 kV ethylene propylene insulated cables'. *IEEE Trans. Power Del.* **10** (1), 25–33.
Bartnikas, R. et al. (1996) 'Long term accelerated aging tests on distribution cables under wet conditions'. *IEEE Trans Power Del.* **11** (4), 1695–9.
Dang, Chinh et al. (1996) 'Electrical ageing of extruded dielectric cables. Review of existing theories and data'. *IEEE Trans. Diel. Electr. Insul.* **3** (2), 237–47.
David, E. (1996) 'Influence of internal mechanical stress and strain on electrical performance of polyethylene. Electrical treeing resistance'. *IEEE Trans. Diel. Electr. Insul.* **3** (2), 248–56.

Manufacture

Thomas, B. and Bowrey, M. (1977) 'Cross-linked polyethylene insulations using the Sioplas technology'. *Wire J.*, May.
Altonen, M. (1978) 'Completely dry curing and cooling process'. *Wire J.*, June.
Smart, G. (1978) 'PLCV system: pressurised liquid salt continuous vulcanisation'. *Wire J.*
Bickel, H. D., Hellmann, K. and Wiedermann, R. (1980) 'Peroxidal crosslinking procedure for PE insulated cores of 1 kV cables without pressure (salt bath)'. *Wire* **29** (2), 75–80.
Kertscher, E. (1980) 'Continuous methods of polyethylene insulation of conductors'. *Wire World Int.* **22**, Jan–Feb.
Hochstrasser, U. P. (1984) 'A new one-step crosslinking process for MV cables'. Jicable Conference Paper No. A11-3.
Martenson, C. (1993) 'Measurement of wear characteristics of alternative diamond materials in wire draw dies'. *Wire J. Int.*, July.
Zamore, A. M. (1993) 'A comparison between E-beam and CV crosslinking technologies for wire and cable constructions'. *Wire J. Int.*, July.
Boer, P. de (1995) 'Modern SZ processes open new doors'. *Wire*, **45** (4), 219–22.
Mariagrazia, M. (1995) 'Diameter control with hot-cold compensation for CCV lines'. *Wire Industry*, **62** (738), 315–16.

Zamore, A. (1996) 'Moisture curable wire and cable compounds: a comparison with e-beam and CV processing'. *Wire J. Int.*, September, 68–72.

Installation

Holttum, W. (1955) 'The installation of metal sheathed cables on spaced supports'. *Proc. IEE, Part A* **102**, 729–742.

Hazard, M. T. (1980) 'Installation of long lengths (of coal mining cables)'. *Min. Tech.*, January.

Ferron, J., Pinet, A. and Pichon, L. (1985) 'Use of mechanical devices in France for narrow trenches and the laying of underground cables'. CIRED Paper No. 3.08.

Beament, I. R. and Ford, A. E. W. (1991) 'Effect of system and installation requirements on the introduction of 11 kV polymeric insulated cables in London'. *Proc. IEEE.*

Jones, E. W. P. (1993) *'Power Cable Installation Practice'*. Newnes.

Klevjer, G., Lervik, J. K., Tunheim, A. and Voldhaug, L. (1994) 'Guidelines for installation of single core cables in trefoil formation'. *Proc. 6th Int. Symp. on Short-Circuit Currents in Power Systems.*

Journeaux, T. L., Field, C. J. and Corbellini, F. (1995) 'Design and installation of 22 kV power cables for LUL'. Jicable Conference Paper.

Naybour, R. D. and Robinson, G. (1995) 'Examination and diagnostics of service aged 11 kV XLPE insulated cables'. Jicable Conference Paper.

Jointing and accessories

Crossland, J. (1976) 'Joints on 3-core 11 kV paper insulated cables'. *IEE Conf. on Distribution.*

McAllister, D. and Radcliffe, W. S. (1976) 'Joints incorporating mechanical connectors and cast resin filling for 600/1000 V cables'. *IEE Conf. on Distribution.*

Ross, A. (1976) 'Jointing trials and tests on 11 kV aluminium sheathed cables'. *IEE Conf. on Distribution.*

Radcliffe, W. S. and Roberts, B. E. (1978) 'Resin-filled joints for 11 kV paper insulated cables'. *IEE Conf. on Distribution.*

Jorgensen, J. and Nielsen, O. J. (1979) *Straight Joints for Solid Dielectric Insulated Cables, 12–170 kV*. Liège: CIRED.

Naybour, R. D. and Brailsford, J. R. (1979) 'A new type of test for connectors to be used on aluminium conductors'. *Proc. IEE* **126** (10), 991–4.

Ross, A. and Philbrick, S. E. (1981) 'Development of joints for polymeric distribution cables. IEE Conference Publication No. 197, CIRED.

Steckel, R. D. and Eertig, K. (1981) 'Joint boxes for transition from paper to plastic MV cables'. IEE Conference Publication No. 197, CIRED.

Wilk, M., Rupprecht, W. and Bottcher, B. (1981) 'Heat shrinkable terminations and joints for HV power cables'. IEE Conference Publication No. 197, CIRED.

Chatterjee, S. (1984) 'Premoulded cable connector system for today and tomorrow'. Jicable Conference Paper No. C-6.

Sander, D. (1984) 'New solutions in connecting, tap-off, energising and measuring in MV networks'. Jicable Conference Paper No. C-1.

Varner, W. F. (1984) 'Development of factory moulded splices to meet the requirements of advanced synthetic insulated MC cable designs'. Jicable Conference Paper No. C-9.

Bruggemann, H. et al. (1985) 'State of the art of plug-in cable connectors for MV cables in the Federal Republic of Germany'. CIRED Paper No. 3.12.

Philbrick, S. E. et al. (1985) 'Elastic rubber terminations for MV cables'. CIRED Paper No. 3.13.

Bartle, J. and Parr, J. C. (1986) 'Review of installation and service experience of resin joints on 11 kV paper cable'. *2nd IEE Int. Conf. on Power Cables 10 kV–180 kV*.

Franks, R. (1986) 'MV elastic accessories up to 36 kV'. *2nd IEE Int. Conf. on Power Cables 10 kV–180 kV*.

Friday, A. and Banks, V. A. A. (1986) 'Performance of joint designs for 11 kV and 33 kV polymeric cable'. *2nd IEE Int. Conf. on Power Cables 10 kV–180 kV*.

Hey, S. A. and Weatherley, J. W. (1986) 'New separable plant terminations for paper and polymeric cables up to 24 kV'. *2nd IEE Conf. on Power Cables 10 kV–180 kV*.

Ross, A. (1986) 'Practical jointing for medium voltage polymeric cables'. *2nd IEE Int. Conf. on Power Cables 10 kV–180 kV*.

Weatherley, J. W., Parry, M. H. and Hey, S. A. (1986) 'Advances in jointing systems up to 36 kV using heat shrinkable components'. *2nd IEE Int. Conf. on Power Cables 10 kV–180 kV*.

Weatherley, J. W. et al. (1986) 'Heat shrinkable terminations for 66 kV polymeric cables'. *2nd IEE Int. Conf. on Power Cables 10 kV–180 kV*.

Dang, C. and Chaaban, M. (1991) 'Thermal performance modeling of power cable joints'. *Proc. IEE*.

Friday, A. (1991) 'The development of medium voltage cable accessories for ease of installation'. *Proc. IEE*.

Gilbert, P. R. (1991) 'Polymeric insulators for MV cable to overhead line connection'. *Proc. IEE*.

MacDonald, D. S., Weatherley, J. W. and Hollick, D. J. (1991) 'A modern jointing system for medium voltage distribution cables'. *Proc. IEE*.

Muench, F. J., Makal, J. M., DuPont, J. P., Hecker, H. A. and Smith, K. (1991) 'Development and use of a testing regime to predict life of premoulded rubber termination and connection products used on medium voltage power cables'. *Proc. IEE*.

Ratra, M. C. and Nagamani, H. N. (1991) 'Laboratory investigation of failure of medium voltage XLPE cable joints'. *Proc. IEE*.

Crouzet, G. and Charles, F. (1995) 'Developments in MV accessories'. Jicable Conference Paper.

Dzektser, N., Izmailov, V. and Sargsian, L. (1995) 'Welding of aluminium alloy lugs to aluminium cables'. *Electrical Contacts – 1995. Proc. of the 41st IEEE Holm Conference on Electrical Contacts* (Cat. No. 95CB35817) New York: IEEE.

Fara, A., Miola, G. and Vercellotti, U. (1995) 'Fire risk analysis on MV cable networks: the behaviour of joints in case of fault'. Jicable Conference Paper.

Testing

Kreuger, F. H. (1964) *Discharge Detection in High Voltage Equipmnt*. Heywood.

Working Group 21-01 (1968) 'Discharge measurements in long lengths of cable; prevention of errors due to superposition of travelling waves'. Paris: CIGRE Paper No. 21-01, Appendix 4, pp. 23–5.

Mole, G. (Feb. 1970) 'Measurement of the magnitude of internal corona in cables'. *Trans. IEEE* **PAS-89** (2), 204–212.

Working Group 21-03 (1970) 'Elimination of interference in discharge detection'. Paris: CIGRE.

Black, I. A. (1973) 'A pulse discrimination system for discharge measurements on equipment operating in a power system'. *IEE Diagnostic Testing Conf.*, pp. 1–7.

Bowdler, G. W. (1973) *Measurements in High Voltage Test Circuits*. Oxford: Pergamon.

Wilson, A. (1973) 'The application of correlation analysis in partial discharge measurements'. *IEE Diagnostic Testing Conf.* pp. 8–12.

Wilson, A. (1974) 'Discharge detection under noisy conditions'. *Proc. IEE* **121** (9), 993–6.

Mason, J. H. (1975) 'Discharge detection and measurements'. *Proc. IEE* **112** (7), 1407–23.

Smith, A. P. (1984) 'Location of partial discharges in drum length cables'. Jicable Conference Paper No. C-20.

Szaloky, G. and Schwarz, M. (1984) 'Measuring and locating of partial discharges in power cables'. Jicable Conference Paper No. C-21.

Wilson, A. and Swingler, S. G. (1984) 'A.c. test methods used for 11 kV extruded cable in CEGB power stations'. Jicable Conference Paper No. A4-3.

Herstad, K. and Sletbak, J. (1985) 'Effect of increased test volage on the performance of MV XLPE cables'. CIRED Paper No. 3.16.

Hilder, D. A. and Black, I. A. (1988) 'Noise suppression methos for partial discharge measurements on cables'. *5th IEE Int. Conf. on Dielectric Materials*.

Kearley, S. S. and Mackinley, R. R. (1988) 'Discharge measurements in cables using a solid state 30 kV bipolar low frequency generator'. *5th IEE Int. Conf. on Dielectric Materials*.

Mackinley, R. R. and Peters, G. (1988) 'New methods of partial discharge detection and location'. *5th IEE Conf. on Dielectric Materials*.

Simmons, M. A. and White, N. M. (1991) 'Electrical testing of polymeric insulated cables for life prediction'. *IEE Colloquium on Insulating Diagnostics – Methods for Determining Quality, Remnant Life and Proof Testing* (Digest No. 070), London. London: IEE.

Banks, V. A. A. (1992) 'Long term testing of medium voltage polymeric cables: evolution of test regimes and standards'. *IEE Colliquium on A European View of Testing and Assessment of Medium Voltage Polymeric Cables* (Digest No. 155), London. London: IEE.

Papadopulos, M. S. (1992) 'Long term testing of medium voltage polymeric cables'. *IEE Colloquium on A European View of Testing and Assessment of Medium Voltage Polymeric Cables* (Digest No. 155), London. London: IEE.

Clegg, B. (1993) *Underground cable fault location*. London, New York: McGraw-Hill.

Audoux, C., Fondeur, J. H., Leroy, J. H., Pinet, A. and Simeon, E. (1995) 'Long duration immersion tests of cross-linked polyethylene insulated MV cables'. *Jicable Conference Paper*.

Kobayashi, S. et al. (1995) 'Study on detection for the defects of XLPE cable links'. *Jicable Conference Paper*.

Kober, H. (1995) 'Electrical long-term test in Germany. *Jicable Conference Paper*.

Schädlich, H. (1995) 'Comparative long-term testing on 20 kV XLPE cables'. *Jicable Conference Paper*.

Srinivas, N. N. and Bernstein, B. S. (1995) 'Effect of DC testing on XLPE insulated cables'. *Jicable Conference Paper*.

Urtubi, J. M., Reolid, G., Sarda, J. and Valls, F. (1995) 'The behaviour of dry insulation medium voltage cables in a moist environment'. *Jicable Conference Paper*.

Fault incidence and location

Gooding, H. T. and Briant, T. A. (1962) 'Location of serving defects in buried cables'. *Proc. IEE, Part A* **109**, 124–25.

Gooding, H. T. and Briant, T. A. (1962) 'Location of gas leaks in buried pressure cable systems'. *Proc. IEE, Part A* **109**, 126–28.

Gooding, H. T. (1963) 'Cable fault location on power systems'. *Proc. IEE* **113** (1), 11–119.

Gale, P. F. (1975) 'Cable fault location by impulse current method'. *Proc. IEE* **122**, 403–408.

Gosden, J. H. (1975) 'Reliability of overhead lines and cable systems in Great Britain'. Liège: CIRED.

Gosden, J. H. and Walker, A. J. (1976) 'The reliability of cable circuits for 11 kV and below'. *IEE Conf. on Distribution*.

Simmons, M A. and Gale, P. S. (1986) 'Means of investigation of the electric strength of polymeric cable insulation'. *5th BEAMA International Electrical Insulation Conference*, Brighton. London: Federation of British Electrotech. & Allied Manuf. Assoc.

Fiss, H. J. and Schroth, R. G. (1987) 'Service experience and knowledge gained with polymeric-insulated medium-voltage cables over many years'. *CIRED: 9th International Conference on Electricity Distribution*, Liège. Liège: Assoc. Ingenieurs Electr. sortis Inst. Montefiore.

PART 4: TRANSMISSION CABLES AND SYSTEMS

Transmission cables (general)

Burrel, R. W. and Young, F. S. (1971) 'EEI and manufacturers 500/550 kV cable research project, Waltz Mill test facility'. *IEEE trans.* **PAS-90**.

Endacott, J. D. (1973) 'Underground power cables'. *Phil. Trans. R. Soc. A* **275**, 193–203.

Ray, J. J., Arkell, C. A. and Flack, H. W. (1973) '525 kV self-contained oil-filled cable systems for Grand Coulee third powerplant – design and development'. IEEE Paper No. T73, pp. 492–6.

Endacott, J. D., Arkell, C. A., Cox, H. N. and Roulston, R. J. (1974) 'Progress in the use of aluminium in duct and direct buried installations of power transmission cable'. *IEEE Underground Transmission and Distribution Conf.*, pp. 466–74.

Bahder, G., Corry, A. F., Blodgett, R. B., McIlveen, E. E. and McKean, A. L. (1976) '500 kV HPOF pipe cable development in the USA'. Paris: CIGRE Paper No. 21-11.

Miranda, F. J. and Gazzana Prioroggia, P. (1976) 'Self-contained oil-filled cables, a review of progress'. *Proc. IEE* **123** (3), 229–38.

Miranda, F. J. and Gazzana Prioroggia, P. (1977) 'Recent advances in self-contained oil-filled cable systems'. *IEE Electr. Power*, Feb, 136–40.

Occhini, E., Tellarini, M. and Maschio, G. (1978) 'Self-contained oil-filled cable systems for 750 kV and 1100 kV. Design and tests'. Paris: CIGRE Paper No. 21-03.

Shimshock, J. F. (1978) 'Installed cost comparison for self-contained and pipe-type cable'. EPRI Report No. EL-935.

Beale, H. K. (1979) 'Underground cables for HV power transmission'. *CEGB Res.*, June, 24–32.

Bossi, A., Sesto, E., Luoni, G. and Dechini, E. (1979) 'The oil-filled cable for the 1000 kV project – ratings and field tests'. *2nd IEE Conf. Progress Cables for 220 kV and Above.*

Head, J. G., Gale, P. S. and Lawson, W. G. (1979) 'Effects of high temperature and electric stress on the degradation of oil-filled cable insulation'. *3rd Int. Conf. on Dielectric Materials.*

Heumann, H., Oppermann, G., Arkell, C. A. and Mayhew, P. L. (1979) '380 kV oil-filled cable for municipal power supply of Vienna'. *2nd IEE Int. Conf. Progress Cables 220 kV and Above.*

Itoh, H., Nakagawa, M. and Ichino, T. (1979) 'EHV self-contained oil-filled cable insulated with composite paper, DCLP'. *2nd IEE Int. Conf. Progress Cables for 220 kV and Above.*

Lawson, W. G. (1979) 'Fatigue and creep phenomena in oil-filled supertension cables'. *2nd IEE Int. Conf. Progress Cables for 220 kV and Above.*

Rosevear, R. D. and Vecellio, B. (1979) 'Cables for 750/1100 kV transmission'. *2nd IEE Int. Conf. Progress Cables for 220 kV and Above.*

Gibbons, J. A. M., Saunders, S. A. and Stannett, A. W. (1980) 'Role of metal debris in the performance of stop-joints as used in 275 kV and 400 kV self-contained oil-filled cable circuits'. *Proc. IEE, Part C* **127** (6), 406–419.

Skipper, D. J. and Arrighi, R. (1981) 'Progress report of CIGRE Study Committee 21 – HV insulated cables'. *Electra* (54), 9–18.

Hance, G. E. A. (1986) 'The modern self-contained OF cable system for voltages up to 180 kV'. *2nd IEE Int. Conf. on Power Cables 10 kV–180 kV.*

Smee, G. J. and West, R. S. V. (1986) 'Factors influencing the choice between paper and XLPE insulated cables'. *2nd IEE Int. Conf. on Power Cables 10 kV–180 kV.*

Leufkens, P. P. and Wegbrans, B. H. M. (1988) 'Cross-bonding and a special interruption joint for HV XLPE cable'. Paris: CIGRE Paper No. 21-04.

Argaut, P. *et al.* (1991) 'Modelling of power transmission cables'. *Jicable Conference Paper.*

Argaut, P., Protat, F., Bouveret, A., Mathieu, N. and Nguyen, V. (1991) 'Calculation of the heat rate of cables for power transmission'. *Jicable Conference Paper.*

Dorison, E., Lepers, J. and Riot, B. (1991) 'Optimization criteria of the transmission capacity of HV and VHV cables'. *Jicable Conference Paper.*

Bamji, S. S. (1992) 'Threshold voltage for electrical tree inception in underground HV transmission cables'. *IEEE Trans. Electr. Insul.*, **27** (2).

Endersby, T. (1992) 'The application of PPL insulated oil filled cable to EHV and UHV transmission'. *CIGRE* Paper 21-307.

Argaut, P. and Favrie, E. (1995) 'Recent developments in 400 and 500 kV XLPE cables'. *Jicable Conference Paper.*

Aucourt, C., Boisseau, C. and Feldmann, D. (1995) 'Gas insulated cables: from the state of the art to feasibility for 400 kV transmission lines'. *Jicable Conference Paper.*

Bezille, J., Becker, J., Janah, H., Chan, J. and Hartley, M. (1995) 'Electrical breakdown strength evolution of HV XLPE cable after long-term test. Correlation with physical properties'. *Jicable Conference Paper.*

Coelho, R. and Mirebeau, P. (1995) 'On the field distribution in a coaxial DC power cable'. *Jicable Conference Paper*.

Helling, K. *et al.* (1995) 'Prequalification test of 400 kV XLPE cable systems'. *Jicable Conference Paper*.

Janah, H. *et al.* (1995) 'New insulating materials for HV cables'. *Jicable Conference Paper*.

Peschke, E. *et al.* (1995) 'Extension of XLPE cables to 500 kV based on progress in technology'. *Jicable Conference Paper*.

Alquie, C. *et al.* (1996) 'Measurement of space charge distribution in XLPE cables under AC stress'. *7th IEE Int. Conf. on Dielectric Materials and Applications*.

Couderc, D. *et al.* (1996) 'Long-term testing of a low-loss 800 kV PPLP insulated SCFF cable'. *IEEE Trans. Power Del.*, **11** (1), 51–7.

D.C. transmission cables

Maschio, G. and Occhini, E. (1974) 'High voltage d.c. cables – the state of the art'. Paris: CIGRE Paper No. 21.

Bahder, G. *et al.* (1978) 'Development of ±400 kV/±600 kV and medium pressure OF paper insulated d.c. power cable systems'. *IEEE Trans.* **PAS-97**.

Allam, E. M. and McKean, A. L. (1980) 'Design of an optimised ±600 kV d.c. cable system'. *IEEE Trans.* **PAS-99** (5).

Allam, E. M. and McKean, A. L. (1981) 'Laboratory developments of ±600 kV d.c. pipe type cable'. *IEEE Trans.* **PAS-100** (3), 1219–25.

Fukagawa, H. *et al.* (1981) 'Insulation properties of 250 kV d.c. XLPE cables'. *IEE Trans.* **PAS-100** (7), 3175.

Sakamoto, Y. *et al.* (1981) 'Development of 500 kV d.c. self-contained OF cable'. *IEE Trans.* **PAS-100** (4), 1949.

Fukagawa, H. *et al.* (1984) 'Development of a new insulating material for d.c. XLPE cables'. Jicable Conference Paper No. B3-2.

Pays, M. *et al.* (1988) 'Behaviour of extruded HVDC power transmission cables: tests on materials and cables'. Paris: CIGRE Paper No. 21-07.

Alquie, C. and Argaut, P. (1991) 'Space charge under DC voltage. Incidence on the after laying test of the VHV cable links'. *Jicable Conference Paper*.

Benlizidia, F. *et al.* (1991) 'Short time breakdown and life duration under DC voltage and pressure of HDPE'. *Jicable Conference Paper*.

Khalil, M., Procida, I. M. and Lovstrom Sorensen, P. (1991) 'Development of new materials for HVDC polymeric insulated cables'. *Jicable Conference Paper*.

Maekawa, Y., Yamaguchi, A., Ikeda, C. and Hara, M. (1991) 'Research and development of DC XLPE cables'. *Jicable Conference Paper*.

Fleischer, K. *et al.* (1996) 'Power systems analysis for direct current (DC) distribution systems'. *IEEE Trans. Ind. Appl.* **32** (5), 982–9.

Malec, D. *et al.* (1996) 'Space charge and anomalous discharge currents in crosslinked polyethylene'. *IEEE Trans. Diel. Electr. Insul.* **3** (1), 64–9.

Polymeric insulated transmission cables

Mott, R. P. (1975) 'Status of 138 kV solid dielectric cables being evaluated at Waltz Mill'. *IEEE Winter Meeting*. Paper No. C75 007-0.

Corbett, J. T. (1976) 'Experience with 138 kV XLPE insulated cable'. *Transm. Distrib.*, May.

Bahder, E. et al. (1978) 'Development of extruded cables for EHV applications in the range 138–400 kV'. Paris: CIGRE Paper No. 21-11.

Jacobsen, C. T. et al. (1978) 'Experience of dry-cured XLPE insulated high voltage cables'. Paris: CIGRE Paper No. 21-06.

Dellby, B. et al. (1979) 'Design and experience of PEX cables and accessories rated 220 kV'. *2nd IEE Int. Conf. Progress Cables for 220 kV and Above.*

Dorison, E. and Legall, Y. (1979) 'French experience with polyethylene insulated high voltage cables'. *2nd IEE Int. Conf. Progress Cables for 220 kV and Above.*

Favrie, E. and Auclair, H. (1979) '225 kV low density polyethylene insulated cables'. *2nd IEE Int. Conf. Progress Cables for 220 kV and Above.*

Kojima, K. et al. (1980) 'Development and commercial use of 275 kV XLPE power cable'. *IEEE Winter Meeting*, Paper No. F80, pp. 220–224.

Kojima, K. et al. (1981) 'Development and commercial use of 275 kV XLPE power cable'. *IEEE Trans.* **PAS-100** (1), 203–210.

Shinoda, S. et al. (1981) '275 kV XLPE insulated aluminium sheathed power cable for Okynahagi No. 2 power station', *IEEE Trans.* **PAS-100** (3), 1298–1306.

Hosokawa, K. et al. (1982) 'Present situation of XLPE high voltage cables in Japan'. Paris: *CIGRE Paper* No. 21-09.

Nakagawa, H. et al. (1983) 'Installation of 275 kV XLPE cables in the long and steep slope tunnel'. *IEEE Summer Meeting*. Paper No. 83 SM 310.0.

Takoka, M. et al. (1983) 'Development of 275 kV cable system and prospect of 500 kV cable'. *IEEE Trans.* **PAS-102**, 3254–3264.

Ball, E. H. et al. (1984) 'Development of cross-linked polyethylene insulation for high voltage cables'. Paris: CIGRE Paper No. 21-01.

Nakagawa, H. et al. (1984) 'Development and installation of 154 kV XLPE cables'. Jicable Conference Paper No. B6-2.

Takaoka, M. et al. (1984) 'Development of 500 kV bulk power XLPE cables and accessories'. Jicable Conference Paper No. B7-3.

Watanabe, Y. et al. (1984) 'Actual installation of short distance 275 kV XLPE cables and technical development of long distance cables'. Jicable Conference Paper No. B6-3.

Wretemark, S. and Nelin, G. (1984) 'XLPE cable after long time service at EH voltage levels'. Jicable Conference Paper No. B7-1.

Kobayashi, T. et al. (1985) 'Development of 275 kV internally water-cooled XLPE cable'. *IEEE Trans.* **PAS-104**, 775–784.

Nakabasami, T. et al. (1985) 'Investigations of commercial use of 275 kV XLPE cables'. *IEEE Trans.* **PAS-104**, 1938.

Ball, E. H. et al. (1986) 'Development of XLPE insulation for high voltage cables'. Paris: CIGRE Paper No. 21-01.

Benford, D. F. and Ball, E. H. (1986) 'Use of 66 and 132 kV XLPE insulated cables for long circuit connections in London'. *2nd IEE Int. Conf. on Power Cables 10 kV–180 kV.*

Gregory, B. and Vail, J. (1986) 'Accessories for 66 kV and 132 kV XLPE cables'. *2nd IEE Int. Conf. on Power Cables 10 kV–180 kV.*

Harasawa, K. (1986) 'Studies on the application of 275 kV cables to long distance underground transmission in Japan'. Paris: CIGRE Paper No. 21-03.

Jocteur, R. (1986) 'Development of 400 kV links with low density polyethylene insulation'. Paris: CIGRE Paper No. 21-09.

Rosevear, R. D. et al. (1986) 'High voltage XLPE cable and accessories'. *2nd IEE Int. Conf. on Power Cables 10 kV–180 kV*.

Ghindes, H. et al. (1988) '161 kV HDPE insulated cable for Tel Aviv'. Paris: CIGRE Paper No. 21-02.

Head, J. G. et al. (1988) 'Thermo-mechanical characteristics of XLPE HV cable insulation'. *5th IEE Int. Conf. on Dielectric Materials*.

Nagasaki, S. et al. (1988) 'Philosophy of design and experience on high voltage XLPE cables and accessories in Japan'. Paris: CIGRE Paper No. 21-01.

Taralli, C. et al. (1988) 'High voltage EPR insulation cable system – manufacturing and installation characteristics'. Paris: CIGRE Paper No. 21-09.

Ogawa, K., Kosugi, T., Kato, N. and Kawawata, Y. (1990). 'The world's first use of 500 kV XLPE insulated aluminium sheathed power cables at the Shimogo and Imaichi power stations'. *IEEE Transactions on Power Delivery* **5** (1), 26–32.

Densley, J. (1993) 'Multi-stress ageing of extruded insulation systems for transmission cables'. *IEEE Electr. Insul. Mag.* Jan/Feb.

CIGRE (1994) 'Laying and installation of high voltage extruded cable systems: literature evaluation data comparison'. *Electra* (156) 103–113.

Densley, J. (1994) 'Multiple stress ageing of solid dielectric extruded dry-cured insulation'. *IEEE Trans. Power Del.* **9** (1).

Kaneko, T. et al. (1994) 'Investigation on characteristics of 275 kV XLPE cable removed from actual service line'. New York: IEEE.

Kubota, T. et al. (1994) 'Development of 500 kV XLPE cables and accessories for long distance underground transmission line. Part 1: Insulation design of cables'. *IEEE Trans. Power Del.* **9** (4), 1741–9.

Champion, T. C. et al. (1995) 'Long term pre-qualification testing program on a 230 kV XLPE cable system'. *IEEE Trans. Power Del.* **10** (1), 10–17.

Fukawa, N. et al. (1996) 'Development of 500 kV XLPE cables and accessories for long distance underground transmission line. Part III: Electrical properties of 500 kV cables'. *IEEE Trans. Power Del* **11** (2), 627–34.

Kaminaga, K. et al. (1996) 'Development of 500 kV XLPE cables and accessories for long distance underground transmission line. Part V: Long-term performance for 500 kV XLPE cables and joints'. *IEEE Trans. Power Del.* **11** (3), 1185–93.

Ratings and forced cooling

Arman, A. N., Chrry, D. M., Gosland, L. and Hollingsworth, P. M. (1964) 'The influence of soil moisture migration on power rating of cables in HV transmission systems'. *Proc. IEE* **111**, 1000–1016.

Ball, E. H., Occini, E. and Luoni, G. (1965) 'Sheath overvoltages on high voltage cables resulting from special sheath bonding connections'. *IEEE Trans.* **PAS-84**, October.

Cox, H. N. and Coats, R. (1965) 'Thermal analysis of power cables in soils of temperature responsive thermal resistivity'. *Proc. IEE* **112** (12).

Glick, D. et al. (1968) 'Design considerations on the current rating of joints for 275 and 400 oil-filled cables'. *IEE Conf. Progress Cables for 220 kV and Above*.

Gosling, C. T. et al. (1968) 'Practical considerations associated with cable systems cooled by external pipes'. *IEE Conf. Progress Cables for 220 kV and Above*.

Endacott, J. A. et al. (1970) 'Thermal design parameters used for high capacity EHV circuits in Great Britain'. Paris: CIGRE Paper No. 21-03.

Brooks, E. J., Gosling, C. H. and Holdup, W. (1973) 'Moisture control of cable environments with particular reference to suface troughs'. *Proc. IEE* **120** (1).

Cox, H. N. *et al.* (1975) 'Developments in UK cable installation techniques to take account of environmental thermal resistivity'. *Proc. IEE* **122** (11).

Arkell, C. A. *et al.* (1976) 'The development and application of forced cooling techniques to EHV cable systems in the UK'. Paris: CIGRE Paper No. 21-02.

Mochlinski, K. (1976) 'Assessment of the influence of soil thermal resistivity on the ratings of distribution cables'. *Proc. IEE* **123** (1), 60–72.

Williams, R. W. *et al.* (1976) Comprehensive force cooled tests on pipe cables at Waltz Mill'. Paris: CIGRE Paper No. 21-07.

Arkell, C. A., Blake, W. E., Brearley, A. D. R., Hacke, K. J. H. and Hance, G. E. A. (1977) 'The design and construction of the 400 kV cable system for the Severn Tunnel'. *Proc. IEE* **124** (3).

Arkell, C. A., Hutson, R. B. and Nicholson, J. A. (1977) 'Development of internally oil-cooled cable systems'. *Proc. IEE* **124** (3), 317–25.

Arnaud, U. G., Burton, J., Crockett, A. E. and Nicholson, J. A. (1977) 'Development and trials of the integral pipe cooled e.h.v. cable system'. *Proc. IEE* **124** (3), 286–93.

Ball, E. H., Endacott, J. D. and Skipper, D. J. (1977) 'UK requirements and future for force-cooled cable systems'. *Proc. IEE* **124** (3), 334–8.

Brotherton, W., Cox, H. N., Frost, R. F. and Selves, J. (1977) 'Field trials of 400 kV internally oil-cooled cables'. *Proc. IEE* **124** (3), 326–3.

Albrecht, C. V., Mainka, G., Brakelmann, H. and Rasquin, W. (1978) 'High power transmission with conductor cooled cables'. Paris: CIGRE Paper No. 21-10.

Arkell, C. A., Ball, E. H. Barton, A. H., Beale, H. K. and Williams, D. E. (1978) 'The design and installation of cable systems with separate pipe water cooling'. Paris: CIGRE Paper No. 21-01.

Arkell, C. A., Gregory, B. and Smee, G. J. (1978) 'Self-contained oil-filled cables for high power circuits'. *IEEE Trans.* **PAS-97**, March–April.

Mainka, A. G., Brakelmann, H. and Rasquin, W. (1978) 'High power transmission with conductor-cooled cables'. Paris: CIGRE Paper No. 21-10.

Alexander, S. M. and Smee, G. J. (1979) 'Future possibilities for separate pipe cooled 400 kV cable circuits'. *2nd IEE Int. Conf. Progress Cables for 220 kV and Above.* IEE Conference Publication 176.

Alexander, S. M., Smee, G. S., Stevens, D. F. and Williams, D. E. (1979) 'Rating aspects of the 400 kV West Ham – St. Johns Wood cable circits'. *2nd Int. Conf. Progress Cables for 220 kV and Above.* IEE Conference Publication No. 176.

Arkell, C. A. *et al.* (1979) '400 kV self-contained oil-filled cable installation in South London UK (Rowdown Beddington)'. *7th IEEE/PES Transmission Distribution Conf.*

Arkell, C. A., Burton, J., Donelan, J. A. and Nicholson, J. A. (1979) 'Stainless steel sheaths for very heavy duty supertension cables'. *2nd IEE Int. Conf. Progress Cables 220 kV and Above.* IEE Conference Publication No. 176, pp. 208–212.

Bacon, P. E. and Morello, A. (1979) 'Upper limits of power rating of self-contained oil-filled cables'. *2nd IEE Int. Conf. Progress Cables for 200 kV and Above.* IEE Conference Publication No. 176.

Ball, E. H., Reilley, M., Skipper, D. J. and Yates, J. B. (1979) 'Connecting Dinorwic pumped storage power station to the grid system by 400 kV underground cables'. *Proc. IEE* **126** (3), 239–45.

Boone, W., Templeaar, H. G., Voss, C. W. M. and Wiel, G. M. (1979) 'Some results of trials of an externally cooled 400 kV cable system'. *2nd IEE Int. Conf. Progress Cables for 220 kV and Above.* IEE Conference Publication No. 176.

Crockett, A. E. and Yates, G. (1979) 'Cooling of accessories of high voltage cable systems'. *2nd IEE Int. Conf. Progress Cables 220 kV and Above.* IEE Conference Publication No. 176, pp. 256–61.

Donazzi, F., Occhini, R. and Seppi, A. (1979) 'Soil thermal and hydrological characteristics in designing underground cables'. *Proc. IEE* **126** (6), 505–516.

Head, J. G., Gale, P. S. and Lawson, W. G. (1979) 'Effects of high temperatures and electric stresses on the degradation of OF cable insulation'. *IEE 3rd Int. Conf. on Dielectric Materials.*

Preece, R. J. and Hitchcock, J. A (1979) 'Simultaneous diffusion of heat and moisture around a normally buried e.h.v. cable system'. *2nd IEE Int. Conf. Progress Cables 220 kV and Anove.* IEE Conference Publication No. 176, pp. 262–7.

Skipper, D. J. (1979) 'The calculation of continuous ratings for forced cooled cables'. *Electra* (66).

Beale, H. K. (1980) 'Underground cables for high voltage power transmission'. Central Electricity Research Laboratory, Report No. EE 55/T 79-5590.

CIGRE Committee 21 (1981) 'Recommendations for tests on anti-corrosion coverings of self-contained pressure cables and accessories for specially bonded circuits'. *Electra* (75), 41–61.

Kuenisch, H. J. et al. (1982) 'Testing a 110 kV LP OF cable with a water-cooled conductor in Berlin (West). Paris: CIGRE Paper No. 21-01.

CIGRE Study Committee 21, Working Group 7 (1988) 'Guide to the protection of specially bonded cable systems against sheath overvoltages'. Paris: CIGRE.

CIGRE Study Committee 21, Working Group 21-02 (1988) 'Survey of position on the calculation of cyclic ratings with partial drying of the soil'. Paris: CIGRE.

CIGRE Study Committee 21, Working Group 21-08 (1988) 'The steady state thermal behaviour of accessories for cooled cable systems'. Paris: CIGRE.

Argaut, P., Protat, F., Bouveret, A., Mathieu, N. and Nguyen, V. (1991) 'Calculation of the heat rate of cables for power transmission'. *Jicable Conference Paper.*

Dorison, E., Lepers, J. and Riot, B. (1991) 'Optimization criteria of the transmission capacity of HV and VHV cables'. *Jicable Conference Paper.*

Argaut, P. and Lesur, F. (1995) 'Evaluation and optimization of transmission capacities of power link'. *Jicable Conference Paper.*

Namkung, S. (1995) 'Remote monitoring and simulation of 138 kV underground power pipe type cable forced cooling system'. *Jicable Conference Paper.*

Toureille, A., Santana, J. and Berdala, J. (1995) 'Characterization and diagnostic of HV and VHV cables by the thermal step method'. *Jicable Conference Paper.*

Thermomechanical design aspects

Arkell, C. A., Arnaud, U. G. and Skipper, D. J. (1974) 'The thermomechanical design of high power, self-contained cable systems'. CIGRE Paper No. 21-05.

Lawson, W. G., Head, J. G., Lombardi, A. and Anelli, P. (1979) 'Fatigue and creep phenomena in oil-filled supertension cables'. *2nd IEE Int. Conf. Progress Cables for 220 kV and Above.*

Head, J. G., Crockett, A. E., Wilson, A. and Williams, D. E. (1991) 'Thermomechanical behaviour of XLPE cables under normal and short-circuit conditions'. *Jicable Conference Paper.*

Ishii, K., Iwata, Z. and Inoue, T. (1991) 'Design method and analysis for thermomechanical behaviour of 275 kV XLPE cables'. *Jicable Conference Paper.*

Schoonakker, P. and Willems, H. M. J. (1991) 'Thermomechanical behaviour of HV and EHV cables with copper conductors'. *Jicable Conference Paper.*

Installation

Arkell, C. A. and Blake, W. E. (1968) 'Installation of EHV of cables in deep shafts'. *IEE Conf. Progress Overhead Lines and Cables for 220 kV and Above.*

Arkell, C. A. (1976) 'Self-contained oil-filled cable: installation and design techniques'. *IEEE Underground and Transmission Conf.*

Rodenbaugh, T. J. (1979) 'Improvement of civil engineering techniques for buried transmission cables'. EPRI Report No. EL-969.

Ohata, K., Ohno, H. et al. (1980) 'Study on vertical installation methods for high voltage XLPE cables'. *Sumitomo Electrl. Tech. Rev.* (116), 27–41.

Smith, D. G. (1980) 'Calculating pipe cable pulling tensions'. *Transm. Distrib.* **34** (10), 40–44.

Nakagawa, H. et al. (1984) 'Development of various snaking installation methods of cables in Japan'. Jicable Conference Paper No. B7-5.

Foty, S. M., Gupta, B. K., Horrocks, D. J. and Kuffel, J. (1991) 'Installation of a 230 kV XLPE insulated cable at Ontario Hydro'. *Jicable Conference Paper.*

Durand, P. and Le Peurian, S. (1995) 'Mechanized laying of HV cables'. *Jicable Conference Paper.*

Parmar, D., Radhakrishna, H. S. and Steinmanis, J. E. (1995) 'Fluidized thermal backfill for increased ampacity of underground power cables'. *Jicable Conference Paper.*

Toya, A., Katsuta, G., Nakano, T., Adachi, K., Shiba, Y. and Nakanishi, T. (1995) 'Development of short term use repair joint for 275 kV XLPE insulated cable'. *Jicable Conference Paper.*

Jointing and accessories

Ohata, K. et al. (1983) 'Development of XLPE moulded joint for high voltage XLPE cable'. *IEEE Trans.* **PAS-103** (7).

Hedman, L. (1984) 'High voltage accessories for XLPE insulated cables 52–170 kV Stress cones in EPDM rubber'. Jicable Conference Paper No. B5-2.

Gregory, B. and Lindsey, G. P. (1988) 'Improved accessories for supertension cable'. Paris: CIGRE Paper No. 21-03.

Park, J. et al. (1988) 'Application of prefabricated joint on 132 kV XLPE cable for power transmission'. Paris: CIGRE Paper No. 21-05.

Parmigiani, B. (1988) 'Premoulded accessories for high voltage extruded insulation cables'. Paris: CIGRE Paper No. 21-08.

Midoz, J. (1991) 'Temporary cable connections for 90 and 63 kV applications using EPDM insulation'. *Jicable Conference Paper.*

Kubota, T. et al. (1994) 'Development of 500 kV XLPE cables and accessories for long distance underground transmission line. Part 2: Jointing techniques'. *IEEE Trans. Power Del.* **9** (4), 1750–59.

Gahungu, F., Delcoustal, J. M. and Brouet, J. (1995) 'Outdoor and incorporated terminations for extruded synthetic cables up to 400 kV'. *Jicable Conference Paper.*

Karasaki, T. et al. (1995) 'Electrical properties and recommended test methods for HV XLPE and accessory'. *Jicable Conference Paper.*

Mayama, S., Mizutani, Y., Inoue, H. and Fukunaga, S. (1995) 'Development of a jointing machine for XLPE cables'. *Jicable Conference Paper.*

Nakanishi, Y. et al. (1995) 'Development of prefabricated joint for 275 kV XLPE cable'. *IEEE Trans. Power Del.* **10** (3), 1139–47.

Onodi, T. (1995) 'Prefabricated accessories for high voltage cables design, dimensioning and field experience'. *Jicable Conference Paper.*

Schmid, M., Laurent, M. and Gaille, F. (1995) 'Use of elastomeric material for manufacturing of HV and EHV accessories'. *Jicable Conference Paper.*

Takeda, N. (1996) 'Development of 500 kV XLPE cables and accessories for long-distance underground transmission lines. Part IV: Electrical properties of 500 kV extrusion molded joints'. *IEEE Trans. Power Del.* **11** (2), 635–43.

PART 5: SUBMARINE CABLES

Barnes, C. C., Coomber, J. C. E., Rollin, J. and Clavreul, L. (1962) 'The British–French direct current submarine link'. Paris: CIGRE Paper No. 210.

Bjurstrom, B. and Jacobsen, K. (1964) 'D.C. cables for the Knoti-Skan transmission scheme'. *Direct Curr.* **9** (1), 12–17.

Gazzana, Prioroggia, P. and Patlandri, G. L. (1968) '200 kV d.c. submarine cable interconnection between Sardinia and Corsica and between Corsica and Italy'. Paris: CIGRE Paper No. 21-05.

Eyraud, I., Horne, L. R. and Oudin, J. M. (1970) 'The 300 kV d.c. submarine cables transmission between British Columbia mainland and Vancouver Island'. Paris: CIGRE Paper No. 21-07.

Gazzana Prioroggia, P., Piscioheri, J. and Margolin, S. (1971) 'The Long Island Sound submarine cable interconnection'. *IEE Trans.* **PAS-90** (H), July–August.

Oudin, J. M., Eyraud, I. and Constantin, L. (1972) 'Some mechanical problems of submarine cables'. Paris: CIGRE Paper No. 21-08.

Gazzana Prioroggia, P. and Maschio, G. (1973) 'Continuous long length a.c. and d.c. submarine HV power cables'. *IEEE Trans.* **PAS-92** (5), 1744–9.

Eigh, L., Jacobsen, C. T., Bjurstrom, B., Hjalmarsson, G. and Olsson, S. O. (1974) 'The 420 kV a.c. submarine connection between Denmark and Sweden'. Paris: CIGRE Paper No. 21-02.

Hauge, O. and Johnsen, J. N. (1974) 'HV D.C. cable for crossing the Skaggerak Sea between Denmark and Norway'. Paris: CIGRE Paper No. 21-07.

Barnes, C. C. (1977) 'Submarine telecommunication and power cables'. Stevenage; Peter Peregrinus.

Baldwin, D. S. F., Giles, E. G., Hacke, K. J. H., Seamans, J. W. S. and Waterhouse, N. W. (1979) 'Methods for installing buried submarine cables for a 2 GW d.c. cross-channel link'. IEE Conference Publication No. 176, pp. 318–23.

Johnsen, J. N. and Bjorlow-Larsen, K. (1979) 'HVDC submarine cables'. *Progress in Cables and Overhead Lines for 220 kV and Above.* IEE Conference Publication No. 176.

Ledezma, O., Gomez, J. L. and Tsumoto, M. (1979) 'Installation of 115 kV oil-filled submarine cable between Margarita Island and mainland Venezuela'. *IEEE Power Engineering Society Winter Meeting.* Paper No. A79 002-7.

Goddard, S. C. et al. (1980) 'The new 2000 MW interconnection between France and the UK'. Paris: CIGRE Paper No. 14-09.

Minemura, S. et al. (1980) '250 kV d.c. submarine cable for Hokkaido Honshn link'. Paris; CIGRE Paper No. 21-03.

Study Committee 21, Working Group 0.6 (1980) 'Recommendations for mechanical tests on submarine cables'. *Electra* (68), 31–36.

Brown, P. V. (1983) 'Protection of submarine cables'. *British Telecom. Eng.* **2**, October.

Foxall, R. G. et al. (1984) 'Design, manufacture and installation of 525 kV a.c. submarine cable from mainland Canada to Vancouver Island'. Paris: CIGRE Paper No. 21-04.

Giussani, A. and Bracco, G. (1984) 'Review of Italian experience on MV submarine cables insulated with extruded materials'. Jicable Conference Paper No. B4-1.

Rebuffat, L. et al. (1984) 'Installation of submarine power cables in difficult environmental environments. Experience with 400 kV Messina cables'. Paris: CIGRE Paper No. 21-10.

Suden, J. E., Traut, R. T. et al. (1984) 'Testing of a high voltage XLPE cable for dynamic submarine application' Jicable Conference Paper No. B4-3.

Arkell, C. A et al. (1985) 'OF land and sbmarine/land cable transition joints for the cross channel projects'. *IEE A.C. and D.C. Power Transmission Conf., London.*

Yates, J. B. (1985) '2000 MW link England–France submarine cable laying'. *IEE A.C. and D.C. Power Transmission Conf., London.*

Arkell, C. A. and Ball, E. H. (1986) 'Design and installation aspects of 270 kV d.c. cable connections – England to France'. Paris: CIGRE Paper No. 21-02.

Arnold, R. J. et al. (1986) 'Habitat repair facility for the cross channel cable'. Paris: CIGRE.

McConnell, J. et al. (1986) 'Long submarine cables for medium voltage connections'. *2nd IEE Int. Conf. on Power Cables 10 kV–180 kV.*

Voyatzakis, Y. et al. (1986) 'Installation of 150 kV OF submarine cable for Ionion Islands'. *2nd IEE Int. Conf. on Power Cables 10 kV–180 kV.*

Galloway, S. J., Woolmer, D. E. and Woodcock, B. G. (1990) '150 kV Java-Madura submarine cable system interconnection'. *Power Engineering Journal* **4** (1) 7–15.

McConnell, J. (1991) 'Development of 10 kV submarine cable with high mechanical strength'. *Jicable Conference Paper.*

Wan, C. T. et al. (1991) '132 kV cross-linked polyethylene insulated submarine cable'. *Sumitomo Electric Technical Review*, **32**, 86–91.

Bishop, J. and Davies, A. E. (1992) 'The application of electric stress analysis and breakdown statistics to inspection criteria for submarine cable systems'. *Proc. of 41st International Wire and Cable Symposium*, Reno. Eatontown: Int. Wire & Cable Symposium.

Nishimoto, T. et al. (1994) 'XLPE power/optical fiber composite submarine cable'. *Fujikura Technical Review*, **23**, 29–36.

Le Peurian, S., Sin, S., Durand, P. and Auclair, H. (1995) 'Development and qualification of a submarine 90 kV cable system using extruded insulation for the electric supply of Guernsey Island'. *Jicable Conference Paper.*

Nishimoto, T., Miyhara, T., Takenhana, H. and Tateno, F. (1995) 'Development of 66 kV XLPE submarine cable using optical fiber as a mechanical-damage-detection-sensor'. *IEEE Trans. Power Del.* **10** (4) 1711–17.

PART 6: HIGH TEMPERATURE SUPERCONDUCTIVITY

Rao, C. N. R. (ed) (1988) *Chemistry of Oxide Superconductors*. Oxford: Blackwell Science.
Poole, Jr. C. P., Farach, H. A. and Creswick, R. J. (1995) *Superconductivity*. London: Academic Press.
Owens, F. J. and Poole Jr., C. P. (eds) (1996) *The New Superconductors*. New York: Plenum Press.

PART 7: OPTICAL FIBRES IN POWER TRANSMISSION SYSTEM

Optical cables

Murata, H. (1996) *Handbook of Optical Fibres and Cables*. Dekker.
Senior, J. M. (1992) *Optical Fibre Communications: Principles and Practice*. Prentice-Hall.
Miller, S. and Kaminow, I. (eds) (1989) *Optical Fibre Telecommunications*. Academic Press.
Mahlke, G. (1987) *Fiber Optic Cables*. Wiley.
Gilmore, M. (1991) *Fibre Optic Cabling: Theory, Design & Installation Practice*. Butterworth-Heinemann.
Palais, J. C. (1992) *Fiber Optic Communications*. Prentice Hall.

Overhead lines engineering

Gracey, G. C. (1963) *Overhead Electric Power Lines*. Ernest Benn Ltd.
Musgrave, K. C. (ed.) (1967) *Overhead Conductor Design*. BICC Publicity Dept.
CIGRE Study Committee 22, Working Group 22-05 (1981) 'Permanent elongation of conductors. Predictor equation and evaluation methods'. *Electra* (75) March.
Walker, M. (ed.) (1982) *Aluminium Electrical Conductor Handbook*. The Aluminium Association of America.
LaForest, J. J. (ed.) (1982) *Transmission Line Reference Book*. Electric Power Research Institute.
Transmission Line Reference Book, 345 kV and Above 2nd edn (1982) Electric Power Research Institute, Palo Alto, Caifornia.
Price, C. F. and Gibbon, R. R. (1983) 'Statistical approach to thermal rating of overhead lines for power transmission and distribution'. *Proc. IEE*, **130** (5).
IEEE Working Group (1985) 'A simplified method of estimating lightning performance of transmission lines'. *IEEE Trans.* **PAS-104** (4).
Carruthers, R. J. B. (1987) *Planning Overhead Power Line Routes*. Research Studies Press.
Orawski, G. (1993) 'Overhead lines – the state of the art'. *Power Engineering Journal* **7** (5), 221–31.

Tunstall, M. J. (1994) *Operating AAAC at Rated Temperatures above 75°C*. National Grid Company, Technical Report TR(T)128.
Poots, G. (1996) *Ice & Snow Accretion on Structures*. Research Studies Press.
Soares, M. R. *et al.* (1997) 'Optical fiber integrated in the distribution of overhead lines'. CIRED Paper No. 3.33.
Wareing, J. B. (1997) 'The effect of wind, snow and ice on optical fibre systems on overhead line conductors'. CIRED Paper No. 3.35.

PART 8: CABLES FOR COMMUNICATION APPLICATIONS

Royal Signals Handbook of Line Communication (1947) HMSO.
Ware and Reed (1949) *Communication Circuits*. Wiley.
Terman (1951) *Radio Engineering*. McGraw-Hill.
Jackson, W. (1951) *High frequency transmission lines*. Methuen.
Harper, C. A. (1972) *Handbook of Wiring, Cabling, and Interconnecting for Electronics*. McGraw Hill.
Sarch, R. (ed.) (1985) *Basic Guide to Datacommunications*. McGraw Hill.
Schubert, W. (1985) *Communications Cables and Transmission Systems*. Siemens Aktiengesellschaft.
Tsaliovich (1995) *Cable Shielding for Electromagnetic Compatibility*. Van Nostrand Reinhold.

Index

Index

Abbreviations, 795
Abrasion test, 91, 236, 590
Absorption, 695
Accessories, OPGW, 725
Acrylic resins, 410
Aerial, 724, 791
Alkylates, 46, 448
All-dielectric self-supporting (ADSS), 685, 730, 733
Aluminium, 709, 718, 719, 774
   conductors, 34, 78, 366
   corrosion, 37, 40, 84
   extrusion, 376
   sheathed cables, 304, 321
   sheaths, 4, 41, 449, 458, 568, 570
   strip armour, 331
Alumina trihydrate, 97
Analogue, 755
Ancillary equipment (trans. cable), 505, 527, 530
Antimony trioxide, 96
Ants, 89
Aramid yarn, 705, 731, 786
Armaflex cable, 778
Armour
   aluminium strip, 331
   bedding, 306
   braid, 90, 235, 240, 260
   cable manufacture, 378
   conductance, 171, 307, 328
   conductors, 193, 197, 711
   double wire armour, 82, 328, 629
   flexible, 265
   galvanised steel wire, 42, 81, 236, 314, 331
   non-magnetic, 42
   pliable wire armour, 238, 242, 269
   resistance, *see* Armour, conductance
   short-circuit temperature, 200
   steel strip, 42
   steel tape, 42, 81, 314
   submarine cable, 621, 646
Attenuation, 685, 693, 694, 695, 766, 768, 783, 785, 788
Auxiliary cable, 251, 640

Backfills, 137, 392, 480, 553
Balanced conductors, 756, 769, 770, 780, 781
Bandwidth, 685, 757, 785
Baseband, 755
Basic impulse level, 506, 508, 523, 527
Belted construction, 21, 317
Bend losses, 696
Bending radii (installation)
   distribution cable, 324, 392
   general data, 1036
   shipwiring cable, 226
   transmission cable, 551
Bending test, 420, 586
Bentonite, 482, 554
Bitumen, 85
Bituminous finishes, 84

Blow out, 734
Blown fibre, 786
Bond pulling, 557
Bonding of sheaths, 485, 527, 612, 626
Braid, 773, 774
Braided finishes, 90, 233, 235, 260
Breakdown mechanism (insulation)
   crosslinked polyethylene, 338, 353
   impregnated paper, 29, 324
   polymers, 64, 353
   polyvinyl chloride, 51, 333
Breakdown strength
   mineral insulation, 224
   paper, 46, 324, 435
   polymerics, 354
British Approvals Service for Electric Cables, 119
British Coal specifications, 239
British Standards, 108
Broadband, 755
Bundles, 719, 721
Buried cables, 563, 791
Bursting forces, 159
Butt gaps, 46, 316, 431

Cab tyre sheaths, 205
Cable laying, 381, 549, 556–9
Cable types for a.c. transmission, 431
Capacitance, 14, 584, 606, 764, 766, 770, 782, 788
Cast plumbs, 537
Cast resin joints, 410, 416
Category of systems, 301
CCV *see* Crosslinking processes
Ceander cable, 342
Cellular polyethylene, 778
Cenelec, 115, 208, 210, 230, 263
Charging current, 16, 170, 574
Circuit protective conductor, 182, 189, 206
Circulating current *see* Bonding of sheaths
CNE cables, 339, 349
Coal mining cables, 238, 306, 328
Coaxial cables
   conductor, 776
   design, 775
   dielectric, 777
   impedance, 777
   radiating, 777
Coil leads, 248
Cold tails, 227
Cold weather precautions, 267, 392
Combustion gases (from polymers), 95, 101
Communication cable
   applications, 779
   cable screening, 771
   insulation, 771
Communication systems, 755
Compass errors, 624
Composite conductors, 709, 726
Compressed gas insulated cable, 611
Compression connectors, 405, 532
Conductance, 766
Conductivity, 712
Conductor jointing
   connectors, 405
   distribution cable, 37, 405
   transmission cable, 532, 641
Conductors
   a.c. resistance, 11, 126, 574
   aluminium, 35, 78, 462
   areas, 69, 189
   characteristics, 711
   compaction, 73
   compatibility, 697
   compressed gas cable, 611
   cooling, 607
   copper, 34, 78, 462
   copper-clad aluminium, 37, 79
   d.c. resistance, 10, 69, 808
   dimensions, 72
   expansion, 564
   flexible, 75, 232, 256
   fluid-filled cable, 440
   losses, 22, 125
   manufacture, 256, 365
   Milliken, 75, 441, 462, 563
   nickel coated, 232, 256
   protective, 189, 206
   proximity effect, 11, 126, 466, 575
   shapes, 70
   short-circuit stresses, 165
   skin effect, 11, 125, 574

Conductors (*continued*)
  sodium, 79
  solid, 78, 366
  standards, 71
  stranding, 256, 367
  stranding formation, 73
  strength of solder, 155, 165, 405, 532, 571
  superconductors, 614, 659 (*see also* Superconducting cables)
  tinned, 205, 209, 232, 256
Consac cable, 339, 341, 397
Contiguous networks, 288
Conversion factors, 802
Cooling of cable systems, 492, 546, 652
Copper
  clad aluminium, 37, 79
  conductors, 33, 78
  sheaths, 277
Core concentricity, 693
Core identification, 210, 231, 330
Corona, 715
Corrosion
  aluminium, 37, 40, 84
  copper, 273
  corrosive gases from cables, 92, 93, 96, 101, 103, 106
  lead, 40
  protection, 83, 101, 268, 450, 626, 718
  textile degradation, 85
Crane cable, 247
Creep, 717
Critical current (I_c), 673
Critical current density (J_c), 660, 667, 669, 672, 673, 679, 680, 681
Critical field (H_c), 660
  lower critical field (H_{c1}), 660
  upper critical field (H_{c2}), 660
Critical temperature (T_c), 660–63, 666–9, 672, 673
Crosslinked polyethylene, 59, 352, 468, 609
Crosslinked polyethylene cables
  heating cable, 227, 827
  HV distribution cable, 303, 351
  HV submarine cable, 633, 635, 639
  HV transmission cable, 468, 609
  LV distribution cable, 305, 306, 335
  technical data, 973, 1027
Crosslinking processes, 59, 258, 362, 371, 473
Crosstalk, 781
Current ratings
  calculation, 124
  capacity, 713
  conductor diameter effect, 19
  cyclic, 148
  distribution cable, 934, 964, 989, 1003
  fluid-filled cable, 465, 479, 1020
  mineral insulated cable, 274, 903, 906
  offshore oil installations, 147, 223
  published data, 121
  rating factors, 137
  shipwiring cable, 147, 223
  short-circuit, 151, 217, 223
  short-time, 148, 217
  standard conditions, 137
  submarine cable, 624
  sustained (flexible cable), 237
  sustained (general), 123
  sustained (HPOF cable), 466
  sustained (mining cable), 241
  sustained (ship cable), 223
  sustained (transmission cable), 479
  sustained (wiring cable), 217
  trays, 399
  welding cable, 245
  wiring type cable, 140, 217, 811, 819, 879
Cured material, 362
Cutback, 699

Damage to cables
  corrosion *see* Corrosion
  insects and rats, 89, 269
  mechanical (distribution cable), 345
  mechanical (mineral cable), 272
  mechanical (submarine cable), 622, 634, 638, 640
  mechanical (wiring cable), 268
Data
  cables design, 779
  categories, 780
  high performance, 780

Data (*continued*)
 multi-pair, 779
 speeds, 757
 twisted pair, 779
Datacommunication cable, 705, 755
D.C. cables, 574, 610
Degradation, electrical, 737, 741
Degradation of insulants
 paper, 28, 324
 polyethylene, 64, 352
 polyvinyl chloride, 52, 333
Degradation of textiles, 85
Depth of laying, 135, 382
Design basis
 data cables, 779
 distribution cable, 301, 311
 fire resistant cable, 92, 98
 HV polymeric cable, 359
 OPGW, 722
 optical cable, 700
 ribbon cable, 707
 transmission, 435
 wiring cable, 207, 233, 239
Development of transmission cables, 603
Dielectric (communications), 764, 765
Dielectric hysteresis, 23
Dielectric loss angle
 accessories, 545
 definition, 16
 heat generation, 436
 impregnated paper, 28
 test, 420, 582
Dielectric losses
 d.c. cable, 578
 PVC cable, 332
 theory, 16, 22, 125
 transmission cable, 434, 489
Dielectric (pressurised), 434
Digital transmission, 755
Direct current cables, 574, 610
 cost vs. a.c. cable, 579
 rated voltage, 10
 submarine cable, 623, 634, 643
 testing, 583
 transmission cable, 574, 589
Discharge in cables
 detection and location, 599–600
 measurement, 419, 423

mineral insulated, 275
paper insulation, 22, 317, 319, 323, 431
partial, 431, 599–600
polymeric insulation, 353
Dispersion, 694
Distribution networks
 industrial, 294
 public supply, 284
Districable, 343
Disturbances on systems, 294
Drill cable, 244
Driven roller pulling, 557–8
Dropwire, 791
Drum handling, 382
Drum trailers, 382, 550
Dry band arc, 737, 739
Drying of paper cable, 369, 448
Duct filling material, 482, 533, 569
Ducts, 384, 482, 554

Earth conductor, 188, 189, 206
Earth fault isolation, 301
Earthing, 181, 296, 301, 314
Eddy currents (sheaths), 23
Elastomers, 48, 54
Elbow connectors, 407
Electrical Equipment (Safety)
   Regulations, 234
Electrical losses, 22, 123
 conductor, 125
 dielectric (definition), 23, 125
 dielectric (d.c. cable), 578
 dielectric (impregnated paper), 28
 dielectric (PVC cable), 332
 dielectric (transmission cable), 603, 605
 sheaths, 23, 127, 578, 626
Electrical Research Association, 122
Electrical strength
 fillers, 25
 paper, 25, 46
 polyethylene, 354
Elecromagnetic compatibility, 783
Electromagnetic fields, 30
 hysteresis, 31
 magnetic flux density, 31, 32
Electromagnetic forces, 159

Electromagnetic interference, 685
Electronics cables, 765
Emanueli FF cable, 4, 431
Engineering Equipment and Materials Users Association (EEMUA), 219
EPR insulated cable
  distribution, 352, 359, 360
  transmission, 468
  wiring, 209, 226
ERA Technology Ltd, 122
Ethylene acrylic elastomer, 64
Ethylene propylene rubber, 58, 95, 99, 359, 360, 468
Ethyl vinyl acetate, 62, 97
Excavation, 383, 384
Excess current protection, 142, 178
External twisted pair cable
  construction, 790
  insulation, 789
Extrusion of polymers, 257, 370, 473

Fatigue (lead sheath), 38, 449, 569
Fault
  currents (calculation), 175, 178, 713
  currents (distribution systems), 298, 306
  currents (mining cable), 240
  currents (wiring circuits), 178
  incidence, 597
  location, 597
  protective finish, 600
  statistics, 352, 597
Faults, 696, 698, 722, 738
Ferranti 10 kV cable, 4
Ferrules, 405
Fibre loss, 694, 695
Fictitious diameter, 313
Field control, 21
Field theory, 21
Fire alarm, 93, 105
Fire protection, 99
Fire resistance, 94, 102, 105, 225
Flaking of cable, 390
Flame retardance, 93, 96, 97, 239
Flame propagation, 93, 97, 100, 102
Flash ignition, 95
Flexible cables and cords, 230, 852
Flexible conductors 75, 232, 233

Flexible joints, 641
Flexing test, 236
Float-down extrusion, 326, 370
Floor warming cable, 227, 266, 827
Fluctuating load currents, 295
Fluid channel cleanliness, 534
Fluid cooling of cables, 497
Fluid-filled cable
  accessories, 506, 517
  conductor screens, 443
  current carrying capacity, 465, 479, 1004
  design features, 435, 440
  high pressure pipe type, 461
  hydraulic design, 450
  installation, 549
  insulation, 444
  paper laminations, 446
  submarine cable, 621
  system design, 453, 465
  technical data, 1004, 1020
  types, 431
Fluid flow, 506
Fluorine, 691
Fluoropolymers, 55, 258
Force cooling (cable systems), 492, 546, 607
Furnace, 692
Fuses, 178, 181, 192

Galvanised steel, 711, 733
Gas compression cable, 459
Gas filled cable, 457, 638
Gas insulated cable, 611
Gas pressure cable, 431, 456
Gel, 702, 721, 731, 747
Germanium, 691
Glass fibre insulation, 225, 226, 232, 249
Glass tapes and braid, 99, 232, 236
Grading of insulation, 21
Ground, 726
Gutta percha, 4, 205

'HAR' mark, 120, 212
Harmonic currents, 295
Harmonisation documents, 120, 210, 230

Hazard index, 101
Heat generation, 123, 436
Heat treatment of materials, 711
Heating cable, 227, 266, 827
Heat of combustion, 93, 96, 102
Heat resisting insulation, 51, 249, 271, 329
Heat shrinkable materials, 412
Helium, 615
High density polyethylene, 53, 352
High permittivity materials, 402
High power transmission, 603
High pressure FF cable, 461
High temperature superconductors (HTS)
  bismuth-containing compounds, 667–71
  cuprates, 666, 667
  characterisation: a.c. losses, 674, 681; electrical, 673; magnetic, 673, 674; mechanical, 674
  mercury-containing compounds, 666, 667
  thallium-containing compounds, 667, 669
  yttrium barium oxide, 666–9
High voltage
  ADSS systems, 741
  wrap, 751
Historical survey, 4, 205
Hochstadter screen, 4, 21
'H' type cable
  description, 21, 26
  distribution cable, 318
  submarine cable, 623, 635, 637, 638
Hydraulic design (FF cable), 450, 635
Hydrocarbon spillage, 87, 308, 333, 338
Hydrogen chloride, 101, 103
Hydrogen ingress, 697

Ice, 723, 724, 734, 746
IEC, 108, 208, 210, 211, 262
IEE Regulations for ships, 122, 147, 223
IEE Wiring Regulations
  current ratings, 121, 216
  excess current protection 142, 178
  general, 170, 177
  heating cable, 227
  protection time, 180
  protective conductor, 189, 206
  shock, 182
  shock-circuit protection, 179
  voltage drop, 171
  wiring cable installation, 262
Impedance, 12, 171, 183, 185, 192, 199
  communication, 767, 772
  datacommunication, 783
  earth fault currents, 195
  earth fault loop, 183, 186, 193, 199
  positive, negative and zero sequence, 175
  surface transfer, 772, 773
Impregnants (*see also* MIND), 44, 448, 562
Impregnated paper insulation
  alternative dielectrics, 436
  description, 43
  electrical properties, 47, 319, 608
  MIND, 320
  thermal breakdown, 28, 323
  thickness, 308, 319
Impregnated pressure cable, 456
Impregnation, 369, 448, 539
Impulse voltages
  pressure cable design, 435, 446
  testing, 420, 588
Inductance, 11, 764, 783
Industrial type cables
  distribution cables, 305, 327
  flexible cables, 238, 245
  wiring cables, 217, 811
Insect damage, 89
Installation
  ADSS, 733
  datacommunication, 784
  distribution cables, 381
  heating cables, 228
  low temperatures, 86, 267, 392
  OPGW, 725
  pipe type cables, 463
  radii *see* Bending radii
  submarine cables, 621, 626, 646
  transmission cables, 549, 556, 565, 569, 590, 607

Installation (*continued*)
　wiring cables, 225, 262
　wrap cable, 747, 749, 752
　wrapping machine, 747
Insulated sheath cable system, 488
Insulation level, 301
Insulation
　d.c. cables, 575
　co-ordination, 506
　resistance, 13
　retraction, 505
　screen, 471
Insulation thickness
　impregnated paper 308, 319
　polymerics, 311, 359, 468
　submarine cables, 624
Integral sheath cooling, 496
Intemp cable, 232, 236
Internal fluid cooling, 497, 498
Ionisation, 25, 27, 458
Irradiation crosslinking process, 259, 373

Joint bays, 555, 560, 569, 570
Jointing tent, 560
Joints
　cast resin, 410
　cold shrink, 414
　compression, 405
　conductor connection, 405
　design, 401
　fire performance, 408
　force cooling, 546
　heat shrink materials, 412
　low voltage, 408
　mechanical, 407
　medium voltage, 410
　methods (transmission), 532
　paper distribution cable, 408, 410
　polymeric distribution cable, 411, 412
　premoulded, 412
　resin, 410
　sectionalising, 505
　separable, 415
　soldered, 405
　stop joints, 514
　stress control, 401, 508, 511
　transition, 416, 526
　transmission cable, 503, 518, 522
　wiring cable, 269
　wrap cable, 751
Jumper wire, 792
Jute serving, 84, 633

Kapton *see* Polyimide

Laminate insulation, 446, 608
Laying-up process, 260, 375
Lead alloys, 38, 448
Lead burning, 539, 643
Lead sheaths
　alloys, 448
　characteristics, 38, 448
　extrusion, 375
　reinforcement, 450
　supports, 397
Leakage current, 23
Leaks (fluid and gas), 601
Lightning, 714, 722, 728
Linaflex, 875
Local area network (LAN), 758, 780, 785
Loose tube cables, 702, 732, 745
Losses *see* Electrical losses
Low smoke, 786
Low temperature superconductors (LTS)
　ceramics, 662, 666
　inorganic, 662
　organic, 662
　metals, 662, 663, 665
Low Voltage Directive, 120
Lugs, 405

Magnesium hydroxide, 97
Magnesium oxide, 272, 275
Manufacture of cables
　cured material, 362
　distribution, 365
　submarine, 637, 640
　transmission, 440, 472
　wiring, 255, 277
Manufacture of optical fibres
　fibre drawing, 692
　inside vapour deposition, 691

Manufacture of optical fibres
(*continued*)
  measurement, 693
  outside vapour deposition, 691
  preform, 691
  vapour axial deposition, 691
Manufacture of superconductors
  HTS materials, 669, 670
  LTS materials, 663
  powder-in-tube, 669–74
MDCV *see* Crosslinking processes
Mechanical connectors, 407
Mechanical loads, 700, 723
Mechanical stress effects, 268
Meissner effect, 660, 673, 674
Melinex *see* Polyethylene terephthalate
Metal aerial, self-supporting, 686
Metallic coverings, 313
Metallic telephone cable, 755, 770
Metals
  electrical properties, 33, 71, 766, 781
  physical properties, 34
Mica/glass tape insulation, 99
Microbending, 696
Microcrystalline waxes, 44
MIG welding, 441, 533
Milliken conductors, 75, 441, 462
MIND insulation, 44, 320
Mineral insulated cable, 271, 670, 902
Mining cable, 238, 306, 328
Mollerhoj cable, 433
Monosil *see* Sioplas
Motorway lighting, 346
Mylar *see* Polyethylene terephthalate

National Approval Organisation, 120
National Coal Board specifications
  see British Coal specifications
National Grid Company, 719
Neoprene *see* Rubber (chloroprene)
Network voltage adjustment, 295
Networks
  building, 760
  metallic characteristics, 763
  telecommunication, 762

Niobium, 614
  Nb_3Sn, 660–63
  NbTi, 662, 663, 665
Non-draining insulation *see* MIND
Non-linear materials, 403
Nose pulling, 556
Nylon, 55, 89

Offshore oil installations, 221
Oil, 726
Oil refinery cable, 308, 333
Oilostatic cable, 461
Open systems, 759
Operating temperature
  continuous, 132, 323
  short-circuit, 151
  short-time, 148, 216
Optical cables
  bend losses, 696
  construction, 700, 701, 704
  datacommunication, 785
  design, 700, 720, 731, 745, 750
  duplex, 786
  joint, 726
  loose tube, 702, 719, 732, 745
  LSF, 705
  ribbon cable, 706
  slotted core, 706
  strain, 700
  stranded, 702
  strength member, 703, 732, 747
  thermal expansion, 701
  tight buffered, 704, 705
Optical datacommunication cables
  design, 785
  performance, 785
Optical fibres
  fibre count, 685
  graded index, 689
  manufacture, 688, 690
  measurements, 686, 693, 698
  multimode, 689, 695
  propagation, 694
  refractive index, 688
  single mode, 690, 695, 747
  splices, 726
  step index fibres, 688

Optical fibres (*continued*)
  strength, 697
  transmission, 688, 698
  waveguides, 688, 694
Optical power ground wire (OPGW), 685 719, 730
Optical time domain reflectometry, 698
Outer covering, 471
Overhead
  bare conductors, 709
  compacted conductors, 710
  conductors, 709
  power lines, 479, 686, 745
  segmented, 710
  temperatures, 727
Oversheaths
  faults, 583, 597–9, 600
  high density polyethylene, 53, 88, 352
  nylon, 55, 89
  polyethylene, 87
  PVC, 85, 209, 232, 331
  repairs, 394
  rubber, 205, 233
  tests, 90, 582
Oxygen index, 95, 100

Paper, 43
Paper insulated cable
  breakdown mechanism, 25, 28
  design basis, 308, 316
  discharge in voids, 25, 319, 324, 431
  field theory, 21
  industrial applications, 305
  non-draining *see* MIND
  pressurised, 431
  public supply applications, 303
  technical data, 913
  tests, 325, 418
Paper lapping, 368, 444
Parallel transmission, 758
Partial discharge *see* Discharge
Permittivity
  effect on stress, 18
  materials data, 47, 50, 127, 490, 543
  paper, 47
Petrochemical industry installations, 308, 333

Phase
  coefficient, 769
  conductor, 686
Phase conductors (OPPW), 725, 737
Phase wrap cable, 750
Pilot and telephone cable, 251, 640
Pinhole type insulation failure, 324
Pipe cooling, 492
Pipe type cable
  application, 466
  construction, 462
  current carrying capacity, 465
  description, 461
  installation, 463
  insulation, 462
  oilostatic, 461
  rating cp FF cable, 465
  system design, 465
Pipes and ducts, 384
PLCV *see* Crosslinking processes
Plumbed sheath closures, 537
Polarity reversal (d.c. cable), 577
Polar flex cords, 233, 235
Polyester, 703, 731, 772
Polyethylene
  crystallinity, 53
  high density, 53, 87, 352
  insulation (applications), 351, 352, 360, 765, 789
  insulation (high voltage), 468, 609
  material, 53
  screening, 778
  sheathing 87, 703, 720
Polyethylene insulated cable
  auxiliary, 251, 252
  d.c. cable, 579
  distribution, 359
  high voltage, 468
Polyethylene terephthalate, 56, 248, 249
Polyimide, 56
Polymeric cables
  accessories, 522
  design, 469
  transmission, 468, 473, 476, 609
  transmission tests, 591, 595, 596
Polymeric materials
  degradation, 64
  electrical properties, 50, 468

Polymeric materials (*continued*)
  physical properties, 49
  types, 48
Polypropylene, 54, 608
Polypropylene/paper laminate, 446, 558, 582, 606, 608, 635
Polytetrafluoroethylene (PTFE), 56, 95, 96, 97, 258
Polyurethane, 55, 411
Polyurethane casting resin, 410
'Popie' test instrument, 601
Porosity (insulation), 259
Power
  factor, 765 (*see also* Dielectric loss angle)
  optical, 736
  transmission systems, 685
Power system analysis, 300
Pre-connectorised cabling, 787
Preforms, 691, 692
Preparation for laying, 384
Pressure cable types, 431
Pressure in FF cable
  static, 451
  tests (accessories), 584
  transient, 452
Pressure tanks, 453, 530, 585, 638
Production in UK, 4
Propagation
  characteristics, 694
  velocity, 782
Protection
  equipotential bonding, 183
  mining cables, 241
  overload, 178
  shock, 182, 345
  short-circuit, 179
  systems, 298
Protective finishes
  bituminous, 84
  braided, 90, 233, 234, 260
  extruded *see* Oversheaths
  glass fibre, 225, 226, 232, 249
  insect protection, 89
  requirements, 83
  tape, braid and compound, 205
  tests, 90, 584, 590

Protective Multiple Earthing (PME), 182, 297, 399
Proximity effect
  conductors (a.c. resistance), 11, 126, 466
  conductors (d.c. resistance), 574
  cores (pipe cable), 466
Public supply distribution
  cable types, 303
  systems, 283
  voltages, 286, 301
Pulling
  cable laying, 388, 556–9
  eyes, 414
  loads, 395, 396
PVC
  acid-binding, 53
  flame-retardant, 93, 96
  heat-resisting, 250, 329
  insulation, 209, 232, 329
  material, 52, 329
  oversheath, 209, 232, 331
  plasticisers, 52
PVC insulated cable
  auxiliary, 252
  control, 218, 253
  distribution (HV), 307, 336
  distribution (LV), 306, 307
  flexible, 231
  multicore, 252
  wiring type, 213

Quality assurance, 108, 260
Quarry cable, 219, 238
Quarry Regulations, 239

Radial networks, 289
Radiation effects, 274
Railway cables, 220, 454
Rayleigh scattering, 696
Reactance, 12, 177, 224
Reaction to fire, 102, 105
Refractive index, 688, 689
Registration (paper tapes), 46
Reinforced plastic, 731
Reinforcement (pressure cable), 450, 456, 627
Reinstatement, 392

Relative permittivity *see* Permittivity
Repairs to oversheath, 394
Resin, 747
Resin filled joints, 410, 416
Resistance
　communication cables, 764, 769, 788
　conductors (a.c.), 11, 125, 574
　conductors (d.c.), 10, 69, 808
　temperature coefficient, 10
　zero sequence, 177
Return loss, 767
River crossings, 621
Road crossings, 553
Rubber, 231, 233
　acrylonitrile-butadiene, 64
　butyl, 57
　chlorosulphonated (CSP), 63, 239, 243
　ethylene propylene, 58, 232, 243, 352, 355, 356, 359
　fluorocarbon, 64
　heat, oil and flame retardant, 63, 242
　natural, 49, 50, 56, 205
　polychloroprene (PCP), 63, 239, 243
　silicone, 63, 95, 98, 209, 232, 249, 259
　styrene butadiene, 57
Rubber insulated cable
　fixed installation, 214, 216
　flexible, 235, 248
　mining, 239
　welding, 245

Sag, 716, 723, 734
Sand, 480
Scattering, 696
Screening
　conductor, 310, 359, 443, 470
　dielectric, 62, 63, 310, 359, 443
Screening factor (auxiliary cable), 251
Sea-bed conditions, 631, 647
Seals for MI cable, 278
Serial transmission, 757
Service distribution cable, 292, 346, 1001
Severn tunnel, 497
Sheath
　ADDS cables, 730, 738
　bonding, 485, 527

　circulating current, 128
　closure (transmission cable), 536
　communication cables, 702, 705, 786, 791
　composite conductors, 720
　eddy current, 128
　losses *see* Electrical losses
　strain, 569
　unreinforced, 432
　voltage limiters, 487, 489, 527
Shipwiring cable, 221, 835
Short-circuit rating
　asymmetric faults, 156
　cable temperature, 151
　conditions, 151, 178
　derivation, 152
　distribution cable, 154
　electromagnetic forces, 159
　fluid-filled cable, 155
　protection, 179
　shipwiring cable, 223
　thermomechanical force, 165
　wiring type cable, 155
Signal, 695
Silane, 59, 362, 374
Silica, 691, 766
Silicone rubber *see* Rubber
Silicone rubber/glass fibre insulation, 225
Silver, 669–74
Sioplas, 59, 259, 362, 374
Site inspection, 381
Skin effect, 11, 125
Smoke
　density, 105
　evolution, 101, 102, 103, 105
　obscuration, 105
Sodium, 79
Soil
　thermal resistivity *see* Thermal resistivity
　types, 135
Solar radiation, 269
Soot, 691
Specially bonded cable sheaths, 485, 488
Splices, 726
Split-concentric service cable, 346

Index

Stability, 735
Steel tape and wire, 42, 710, 719, 720
Stockings (cable), 384, 395
Stop joints, 514
Strain, 700, 719, 730
Stranding, 73, 256, 367
Stress cones, 403, 508, 511
Stress control (accessories), 402, 508, 511, 543
Stress in insulation
  d.c. cable (transmission), 575
  fluid-filled cable, 435, 444, 508
  gas-filled cable, 459
  polymeric cable (distribution), 359
  polymeric cable (transmission), 470
  theory and derivation, 17
Structured systems, 759, 761, 785
Submarine cable
  a.c. cable, 623, 633, 643
  armour, 629
  cable design, 626, 639
  cable laying, 631, 648
  compass errors, 624
  control cable, 640
  current ratings, 625
  damage, 622, 634, 638, 640
  d.c. cable, 623, 634, 643
  electrical design, 624
  flexible joints, 641
  fluid-filled cable, 635
  fluid pressure, 636, 638
  gas-filled cable, 638
  handling equipment, 648
  hydraulic design, 635
  installation, 621, 626, 646
  lead sheaths, 643
  mechanical design, 626
  paper insulation, 642
  pilot cable, 640
  polymeric ins. cable, 639, 642
  protective finishes, 632
  repair, 655
  river crossing, 621
  screens, 642
  sea bed conditions, 622, 631
  sheath losses, 626
  single-core cables, 637
  site work, 652
  solid type paper cables, 633
  survey, 647
  thermal resistivity of surround, 624
  torque balance, 629, 630
  water pressure, 627, 634
Subsidence, 393
Sulphur hexafluoride
  gas insulated cable, 611, 613
  terminations, 509, 523
Superconducting cables
  environmental benefits, 660
  fault currents in, 674
  high temperature (HTS), 673, 676–82
  low temperature (LTS), 614, 663–5
  market impact, 659
Superconducting machines, 672
Superconductivity
  definition, 660
  high temperature, 661
  low temperature, 661, 662
Supports and spacing, 397, 565, 569
Symbols, 799
System categories, 301
System disturbances, 294
Systems
  blown fibre, 786
  d.c., 579
  communication, 755, 762
  open, 759
  primary parameters, 764
  telecommunication, 788
  wrap, 750

Tapered network, 292
Tapes, 608, 702, 703, 747, 774
Telecommunication
  cabinet, 763
  cable properties, 788
  cables, 755, 788
  exchange, 763
  networks, 762
  primary network, 763
  secondary network, 763
  twisted pair, 788
Temperature
  ambient, 132
  cable operating (continuous), 132

Temperature (*continued*)
  cable operating (flexible), 230
  cable operating (paper), 323
  cable operating (wiring), 265
  index, 95, 100, 104
  lampholders, 267
  low temperature precautions, 267, 392
  optical cables, 700, 713, 722, 727
  overhead lines, 479
  short-circuit, 151, 152, 153
Tensile strength, 715
Teredos, 90
Terminations
  clearances, 404
  cold shrink, 414
  compressed gas, 611
  design, 401, 404
  fluid-filled, 506, 512, 517
  gas-filled, 518
  heat shrink, 412
  paper distribution cable, 408, 410
  polymeric cable, 411, 412, 522
  premoulded, 412
  separable connectors, 415, 416
  SF_6 equipment, 509, 523
  tests, 584, 587, 590
  thermomechanical force, 571
  wiring cable, 269
Termites, 89
Tests
  accessories (transmission cable), 584, 586
  bending, 420, 586
  capacitance, 419
  conductor resistance, 422
  dielectric loss angle, 420, 423, 582, 583, 588, 593
  dielectric security, 424, 589
  discharge, 419, 423
  distribution cables, 418
  drainage, 421
  duration, 422
  flammability, 95, 100
  fire and flame retardance, 421, 426
  high voltage, 419, 582
  impulse, 420, 589
  insulation resistance, 419
  loading cycle, 424, 587, 591
  power factor *see* Dielectric loss angle
  protective finishes, 584, 590, 591
  routine (distribution cable), 422, 432
  routine (transmission cable), 581, 582
  sample tests, 423
  site, 427, 561, 590, 591
  special (distribution cable), 418, 424
  special (transmission cable), 586
  terminations, 584
  thermal resistivity, 589
  thermal stability, 588
  transmission cable systems, 581
  type (paper distribution cable), 418, 424
  type (polymeric distribution cable), 425
  type (transmission cable), 581, 592
Thermal breakdown, 28
Thermal resistivity
  cable materials, 125, 128, 589
  compressed gas cable, 612
  duct, 130
  sea bed, 624
  soil, 130, 135, 481, 482
  special backfill, 481
Thermal stability test, 588
Thermomechanical forces, 322, 504, 532, 562
Thermoplastic materials, 48, 51, 370
Thermosets, 48, 56, 371
Timbering, 383
Touch voltage, 183
Tough rubber sheath, 205
Towers, 733, 734, 748
Toxic fumes, 97, 99
Tracer gas, 601
Trailers, 382
Trailing cables, 239
Transient voltage, 30, 626
Transmission
  balanced cable, 781
  characteristics, 763
  limits in telecommunications, 788
  media, 719
  serial and parallel, 758
  systems, 755
  twisted pair, 782

Transmission accessories, 503
  design, 542, 543
  electrical consideration, 503
  gas-filled cables, 518
  mechanical considerations, 504
  polymeric cables, 522
  pressure equipment, 530
  pressurised cable systems, 505
  single-core cables, 506
  SF_6, 509, 523
  straight joints, 514, 525
  take-off connector, 508, 510
  thermal rating, 544
  three-core cables, 517
  transition joints, 526
Transmission systems, 433, 479, 603
Transportation, 550
Treeing in insulation
  paper, 27
  polymers, 353, 354, 356, 359
Trenching, 384, 387, 551
Triple coaxial apparatus, 773
Triplex cable, 361
Trough installation, 483
Tubing-on extrusion, 326, 370
Twisting of cables, 391, 626, 629

UDF flexible cord, 235
Unarmoured distribution cable, 344
Unbalanced, 756
Underground residential distribution (URD), 293, 350
US practices, 286, 293, 302, 308, 313, 350

VCV *see* Crosslinking processes
Vibration, 716, 736
Voids in insulation, 25, 353, 356, 363, 431
Voltage designation
  cables, 9, 208, 222, 301
  equipment, 303
Voltage drop, 171
Voltage reflection, 767
Vulcanising *see* Crosslinking processes

Water barrier, 702, 789, 790
Water content of XLPE, 362, 363, 473
Water pressure (underwater cables), 594, 627, 634
Water swellable powder, 361
Water trees, 356
Waveconal cable, 305, 342, 397
Waveform cable, 305, 342, 349, 350, 397
Wavelength, 693, 730, 785
Weibull distribution, 355, 1040
Welding cable, 245, 822
Welding of conductors, 408, 533
Welding of sheaths, 538
Winches (pulling), 385
Winch pulling, 395
Wind, 717, 723, 724, 735, 746
Wire drawing, 256, 366
Wrap cable, 685, 686, 745, 749

Zero sequence reactance, 177
Zero sequence resistance, 176